Visual Sensors

Visual Sensors

Special Issue Editors

Oscar Reinoso
Luis Payá

MDPI • Basel • Beijing • Wuhan • Barcelona • Belgrade • Manchester • Tokyo • Cluj • Tianjin

Special Issue Editors
Oscar Reinoso
Miguel Hernandez University
Spain

Luis Payá
Miguel Hernandez University
Spain

Editorial Office
MDPI
St. Alban-Anlage 66
4052 Basel, Switzerland

This is a reprint of articles from the Special Issue published online in the open access journal *Sensors* (ISSN 1424-8220) (available at: https://www.mdpi.com/journal/sensors/special_issues/visualsensors).

For citation purposes, cite each article independently as indicated on the article page online and as indicated below:

LastName, A.A.; LastName, B.B.; LastName, C.C. Article Title. *Journal Name* **Year**, *Article Number*, Page Range.

ISBN 978-3-03928-338-5 (Pbk)
ISBN 978-3-03928-339-2 (PDF)

Contents

About the Special Issue Editors

Oscar Reinoso is full professor at the Miguel Hernández University (Spain). He received his M.S. degree in industrial engineering from the Polytechnic University of Madrid (UPM) in 1991, and his Ph.D. from the Polytechnic University of Madrid in 1996. From 1994 to 1997, he worked in the R&D department of Protos Desarrollo on a visual inspection system. Since 1997, he has been at Miguel Hernandez University. He has been full professor since 2011 in control, robotics, and computer vision. His research interests include robotics, teleoperated robots, computer vision, parallel robots, and visual inspection systems. He has authored over 200 peer-review research articles in international journals, books, and conferences. He has been the associate editor of several journals.

Luis Payá holds a M.Eng. in industrial engineering (Spain, 2002). He obtained his Ph.D. in Industrial Technologies (Spain, 2014) for his work on omnidirectional imaging, global appearance descriptors and topological map building, and localization of mobile robots. He is currently an associate professor of automatic control, electronics, robotics, and computer vision at Miguel Hernández University, Elche, Spain. His current research interests include navigation of mobile robots in social environments, deep learning techniques applied to map building and localization, and deployment of virtual laboratories. He is the author of numerous papers, communications, and books in the cited topics. He has been a visiting researcher at the University of Bristol and at Imperial College London, United Kingdom. He has been an associate editor of *Sensors*, he currently belongs to the Editorial Board of *Mathematical Problems in Engineering*, and he has edited some Special Issues for the journals Sensors and *Applied Sciences*.

Preface to "Visual Sensors"

Visual sensors have characteristics that make them interesting sources of information for any process or system. These small and inexpensive sensors are able to capture precise and high-resolution environment information. These properties have motivated their use for several decades in multiple tasks. This high versatility in their fields of application has increased their use as a source of information to solve a variety of diverse tasks. The main fields of application include robotics, industry, agriculture, quality control, visual inspection, surveillance, autonomous driving, and navigation aid system.

In this book, 36 different proposals of these visual sensors are presented in some fields of application that are of interest today.In the field of visual navigation of mobile robots, simultaneous localization and mapping (SLAM), and visual odometry, different alternatives are presented in the first chapters. So, in the first chapter, an RGB-D SLAM algorithm is presented using the concept of orientation relevance, considering the Manthattan frame estimation. Chapter 2 provides a method for aircraft pose estimation without relying on 3D models, using two widely separated cameras to acquire the pose information. In Chapter 3, a new framework for online visual object tracking is proposed. A motion-aware strategy is employed to predict the possible region and scale of the target in the frame using the previously estimated 3D motion information. The authors in Chapter 4 provide an improved indoor visual SLAM method that considers point and line segment features extracted by stereo cameras, producing robust results. In Chapter 5, an RGB-D sensor is employed with the purpose of constructing a dense 3D semantic mapping of the environment by means of a pixel-voxel network. Chapter 6 proposes a low-overhead real-time ego-motion estimation (visual odometry) system based on either a stereo or RGB-D sensor. With the proposed algorithm, a local map is created, requiring significantly less memory and computational power. The authors in Chapter 7 provide the details of a visual odometry method adapted to the underwater context. They employ the captured stereo image steam to provide real-time navigation and a site coverage map, which was necessary to conduct a complete underwater survey. Chapter 8 presents a visual information fusion approach for robust probability-oriented feature matching. This approach can be used in a more general SLAM procedure. This strategy permits obtaining relevant areas in the image reference system, from which probable matches could be detected.

Image retrieval aims to browse, search, and retrieve images from a large database of digital images. Proposing new descriptors of an image that define the characteristics of the image can be key in this regard. Chapter 9 presents a new texture descriptor booster based on the statistical information of an image. This descriptor is applied to texture-based classification images. In Chapter 10, the authors propose a framework for salient region detection that uses appearance- and regression-based schemes to reduce the computational complexity and focusing on the salient parts of the image. In this sense, Chapter 11 proposes a texture descriptor for image retrieval, designing a local parallel cross pattern in which the local binary pattern map is fused with the color map. Chapter 12 proposes a hybrid histogram descriptor used for image retrieval. The proposed descriptor combines two histograms: a perceptual uniform histogram and a motif co-occurrence histogram including the probability of a pair of motif patterns. Finally, Chapter 13 proposes a method for textile-based image retrieval for indoor environments based on describing the images with different channels (RGB, HSV, etc.) and using the combination of two different descriptors for the image.

Visual sensors can also be an important part of the source of information, providing help and support for other tasks. In Chapter 14, a novel global point cloud descriptor is proposed for reliable object recognition and pose estimation, which can be applied to robot grasping operations. Chapter 15 provides an approach based on depth cameras to robustly evaluate the manipulation success in robot object manipulation. The method proposed allows the robot to accurately detect the presence or absence of contact points between the robot manipulator and a held object. Chapter 16 presents a vision system capable of automatic 3D joint detection. The detection method is applied in a robotic seam tracking system for gas tungsten arc welding.

The calibration of vision systems plays an important role in different applications where these types of sensors are used. Having a well-calibrated system permits more robust results to be obtained in later stages. Chapter 17 presents a simple calibration method for laser range finder systems, needing only a calibration board. In Chapter 18, an alternative approach that uses Gray code patterns displayed on an LCD screen to determine camera parameters is provided. The proposed approach is 1.5 times more precise than using standard calibration with a checkerboard pattern. Finally, Chapter 19 proposes a method that automatically calibrates four cameras of an around-view monitor system in a natural driving situation.

Object recognition is a task in which a vision system is almost always involved. During the past few years, many proposals have been published in this area, including different methods that allow the recognition of the objects present in an image. Chapter 20 presents a method of handshapes recognition based on skeletal data. It encodes the relative differences between vectors associated with the pointing direction of the particular fingers and the palm normal. Chapter 21 presents a new spatio-temporal action localization detector that consists of sequences of per-frame segmentation masks. This proposed detector can pinpoint the starting or ending frame of each action category in untrimmed videos. In Chapter 22, a system for automatically designing the field-of view of a camera, the illumination strength, and the parameters in a recognition algorithm is presented. Chapter 23 proposes a new presentation attack detection method for an iris recognition system using a near-infrared light camera image. This method tries to avoid the effect caused by presentation attack images captured using high-quality printed images in classic iris recognition systems. Chapter 24 presents an approach for pedestrian detection combining different methods previously proposed together with an efficient sliding window classification strategy. The detector achieves fast detection speed combined with state-of-the-art accuracy. Chapter 25 proposes a model to resolve the 3D reconstruction problem for dynamic non-rigid objects with a single RGB-D sensor.

Over the past few years, the field of visual systems has shifted from classical statistical methods to deep learning methods. Video-based person detection and recognition is an important task facing many problems and challenges, such as lighting variation, occlusion, human appearance similarity, etc. In Chapter 26, a video-based person reidentification method with hybrid deep appearance-temporal features is proposed. Another application using deep learning methods is presented in Chapter 27. The authors propose a densely connected fully convolutional network that can determine the true iris boundary even with inferior-quality images using better information gradient flow between the dense blocks. Chapter 28 proposes a method to improve the performance of the star sensor under dynamic conditions based on the ensemble back-propagation neural network.

Scene reconstruction is a key task necessary to handle more complex problems such as mobile robot navigation. Chapter 29 presents a visual inertial odometry as a solution to the robot navigation system. Chapter 30 presents a high-accuracy method for globally consistent surface reconstruction using a single fringe projection profilometry sensor. Lane marking detection and localization are crucial for autonomous driving and lane-based pavement surveys. In Chapter 31, a novel methodology is presented for automated lane marking identification and reconstruction. A case study is provided to validate the proposed methodology. Finally, Chapter 32 proposes an improved method for unmanned aerial vehicle (UAV) image seamline searching. The experimental results show that the proposed method can effectively solve the problems of ghosting and seams in the panoramic UAV images.

One of the most widely discussed topics in vision systems is establishing visual measurements. This theme is the focus of the last chapters of the book. In Chapter 33, the authors present an improved rotation angle measurement method based on geometric moments, suitable for automatic sorting systems. In Chapter 34, a stereo vision system is employed for measuring the ram speed of steam hammers. The systems try to decrease the influence of strong vibration. The accuracy and effectiveness of the method is experimentally verified. Chapter 35 proposes a pose estimation method for sweet pepper detachment. The point cloud acquired is separated in candidate planes that are separately evaluated using a scoring strategy. The last chapter presents a comparative analysis of digital image correlation-based stereo 3D shape measurements.

Oscar Reinoso, Luis Payá
Special Issue Editors

sensors

MDPI

Editorial

Special Issue on Visual Sensors

Oscar Reinoso * and Luis Payá *

Department of Systems Engineering and Automation, Miguel Hernández University, 03202 Elche, Spain
* Correspondence: o.reinoso@umh.es (O.R.); lpaya@umh.es (L.P.)

Received: 4 February 2020; Accepted: 6 February 2020; Published: 8 February 2020

1. Introduction

Visual sensors have characteristics that make them interesting as sources of information for any process or system. On the one hand, they are able to capture a very precise and high-resolution environmental information while occupying a small size and with a reduced price. On the other hand, they are able to capture a large quantity of information from the environment around them. These properties are the reason they have been employed for several decades for the resolution of multiple tasks. This high versatility in their fields of application makes them increasingly used as a source of information to solve a variety of diverse tasks.

Nowadays, a wide variety of visual systems can be found, from the classical monocular systems to omnidirectional, RGB-D, and more sophisticated 3D systems. Every configuration presents some specific characteristics that make them useful to solve different problems. Their range of applications is wide and varied. Among them, we can find robotics, industry, agriculture, quality control, visual inspection, surveillance, autonomous driving, and navigation aid systems.

Visual systems can be used to obtain relevant information from the environment, which can be processed to solve a specific problem. The aim of this Special Issue is to present some of the possibilities that vision systems offer, focusing on the different configurations that can be used and novel applications in any field.

In this Special Issue, 63 contributions were submitted and 36 of them were published (i.e., 57% acceptance rate). The published articles present a very adequate vision of how visual sensors are used in very different fields of application, from mapping for navigation of mobile robots to object recognition or scene reconstruction.

2. Contributions to the Special Issue on Visual Sensors

In the field of visual navigation of mobile robots, SLAM (Simultaneous Localization and Mapping), Visual odometry, etc., we find different alternatives that are presented in some of the papers of the Special Issue. Thus, in [1], an RGB-D SLAM algorithm is presented using the concept of orientation relevance taking into account the Manthattan Frame Estimation. Teng et al. [2] provided a method for aircraft pose estimation without relying on 3D models using two widely separated cameras to acquire the pose information. In [3], a new framework for online visual object tracking is proposed. A motion-aware strategy is employed to predict the possible region and scale of the target in the frame by utilizing the previously estimated 3D motion information. Wang et al. [4] provided an improved indoor visual SLAM method that uses point and line segment features extracted by stereo cameras, achieving robust results. In [5], an RGB-D sensor is employed. In this case, the purpose is to make a dense 3D semantic mapping of the environment by means of Pixel-Voxel network. Aladem et al. [6] proposed a low-overhead real-time ego-motion estimation (visual odometry) system based on either a stereo or RGB-D sensor. By means of the proposed algorithm, a local map is used, requiring significantly less memory and computational power. Nawaf et al. [7] provided the details of a visual odometry method adapted to the underwater context. They employed the captured stereo image to provide real-time navigation and a site coverage map, which is necessary to conduct a complete underwater survey. Valiente et al. [8] presented a visual information fusion approach for robust probability-oriented feature matching. This approach can be used in a more general SLAM procedure. This strategy permits obtaining relevant areas in the image reference system, from which probable matches could be detected.

Image retrieval aims at browsing, searching, and retrieving images from a large database of digital images. Proposing new descriptors of an image that define the characteristics of the image can be key in this regard. García-Olalla et al. [9] presented a new texture descriptor booster based on statistical information of the image. This descriptor is employed in texture-based classification images. Fareed et al. [10] proposed a framework for salient region detection that uses appearance-based and regression-based schemes to reduce the computational complexity and focusing on the salient parts of the image. In this sense, Feng et al. [11] proposed a texture descriptor for image retrieval designing a local parallel cross pattern in which the local binary pattern map is fused with the color map. In addition, Feng et al. [12] proposed a hybrid histogram descriptor used for image retrieval. The proposed descriptor comprises two histograms jointly: a perceptual uniform histogram and a motif co-occurrence histogram including the probability of a pair of motif patterns. Finally, García-Olalla et al. [13] proposed a method for textile based image retrieval for indoor environments based on describing the images with different channels (RGB, HSV, etc.) and using the combination of two different descriptors for the image.

Visual sensors can also be an important source of information to help and support for other tasks. Thus, in [14], a novel global point cloud descriptor is proposed for reliable object recognition and pose estimation, which can be applied to robot grasping operation. Martínez-Martin et al. [15] provided an approach based on depth cameras to robustly evaluate the manipulation success in robot object manipulation. The method proposed allows the robot to accurately detect the presence or absence of contact points between the robot manipulator and a held object. Xue et al. [16] presented a vision system capable of automatic 3D joint detection. The detection method is applied in a robotic seam tracking system for gas tungsten arc welding.

The calibration of vision systems plays a very important role in different applications where these types of sensors are used. Having a well-calibrated system will permit more robust results to be achieved in later stages. Zhang et al. [17] presented a simple calibration method for laser range finder systems needing only a calibration board. In [18], an alternative approach that uses gray-code patterns displayed on an LCD screen to determine camera parameters is provided. The proposed approach is 1.5 times more precise than using standard calibration with a checkerboard pattern. Finally, Choi et al. [19] proposed a method that automatically calibrates four cameras of an around view monitor system in a natural driving situation.

Object recognition is a task in which a vision system is almost always involved. During the past few years, many proposals have been made in this area including different methods that allow the recognition of the objects present in an image. In this way, Kapuscinski et al. [20] presented a method for hand shapes recognition based on skeletal data. It encodes the relative differences between vectors associated with the pointing direction of the particular fingers and the palm normal. Wang et al. [21] presented a new spatiotemporal action localization detector that consists of sequences of per-frame segmentation masks. This proposed detector can pinpoint the starting or ending frame of each action category in untrimmed videos. In [22], a system for automatically designing the field-of view of a camera, the illumination strength, and the parameters in a recognition algorithm is presented. Nguyen et al. [23] proposed a new presentation attack detection method for an iris recognition system using a near infrared light camera image. This method tries to avoid the effect that presentation attack images captured using high-quality printed images can cause in classic iris recognition systems. Fu et al. [24] presented an approach for pedestrian detection combining different methods previously proposed together with an efficient sliding window classification strategy. The detector achieves fast detecting speed at the same time as state-of-the-art accuracy. Wang et al. [25] proposed a model to resolve the 3D reconstruction problem for dynamic non-rigid objects with a single RGB-D sensor.

Over the past few years, the field of visual systems is shifting from classical statistical methods to deep learning methods. Video-based person detection and recognition is an important task with many problems and challenges such as lighting variation, occlusion, human appearance similarity, etc. In [26], a video-based person re-identification method with hybrid deep appearance-temporal features is proposed. Another application using deep learning methods was presented by Arsalan et al. [27]. The authors proposed a densely connected fully convolutional network, which can determine the true iris boundary even with inferior-quality images by using better information gradient flow between the dense blocks. Liu et al. [28] proposed a method to improve the performance of the star sensor under dynamic conditions based on the ensemble back-propagation neural network.

Scene reconstruction is a key task necessary to accomplish more complex problems such as mobile robot navigation. Xia et al. [29] presented a visual inertial odometry as a solution to the robot navigation system. Cheng et al. [30] presented a high-accuracy method for globally consistent surface reconstruction using a single fringe projection profilometry sensor. Lane marking detection and localization are crucial for autonomous driving and lane-based pavement surveys. In [31], a novel methodology is presented for automated lane marking identification and reconstruction. In addition, a case study is given to validate the proposed methodology. Finally, Zhang et al. [32] proposed an improved method for UAV image seamline searching. The experimental results show that the proposed method can effectively solve the problems of ghosting and seams in the panoramic UAV images.

Finally, one of the most widely discussed topics about vision systems is to establish visual measurements. Some of the papers of the Special Issue revolve around this problem. In [33], the authors presented an improved rotation-angle measurement method based on geometric moments that is suitable for automatic sorting systems. In [34], a stereo vision system is employed for measuring the ram speed of steam hammers. The system tries to decrease the influence of strong vibration. The accuracy and effectiveness of the method was experimentally verified. Li et al. [35] proposed a pose estimation method for sweet pepper detachment. The acquired point cloud is separated into candidate planes that are separately evaluated using a scoring strategy. Yang et al. [36] presented a comparative analysis of digital image correlation based stereo 3D shape measurements.

Acknowledgments: This Special Issue would not have been possible without the valuable contributions of the authors, peer reviewers, and editorial team of Sensors. Our most sincere thanks are given to all the authors for their hard work, independently on the final decision about their submitted manuscripts. In addition, all our gratitude is given to the peer reviewers for their help and fruitful feedback to authors. Finally, our warmest thanks go to the editorial team for their untiring support and hard work during all stages of development of this Special Issue and, in general, congratulations are offered on the great success of the journal Sensors.

Conflicts of Interest: The authors declare no conflict of interest.

References

1. Wang, L.; Wu, Z. RGB-D SLAM with Manhattan Frame Estimation Using Orientation Relevance. *Sensors* **2019**, *19*, 1050 [CrossRef] [PubMed]
2. Teng, X.; Yu, Q.; Luo, J.; Zhang, X.; Wang, G. Pose Estimation for Straight Wing Aircraft Based on Consistent Line Clustering and Planes Intersection. *Sensors* **2019**, *19*, 342 [CrossRef] [PubMed]
3. Zhang, Y.; Yang, Y.; Zhou, W.; Shi, L.; Li, D. Motion-Aware Correlation Filters for Online Visual Tracking. *Sensors* **2018**, *18*, 3937 [CrossRef] [PubMed]
4. Wang, R.; Di, K.; Wan, W.; Wang, Y. Improved Point-Line Feature Based Visual SLAM Method for Indoor Scenes. *Sensors* **2018**, *18*, 3559 [CrossRef] [PubMed]
5. Zhao, C.; Sun, L.; Purkait, P.; Duckett, T.; Stolkin, R. Dense RGB-D Semantic Mapping with Pixel-Voxel Neural Network. *Sensors* **2018**, *18*, 3099 [CrossRef] [PubMed]
6. Aladem, M.; Rawashdeh, S. Lightweight Visual Odometry for Autonomous Mobile Robots. *Sensors* **2018**, *18*, 2837 [CrossRef] [PubMed]

7. Nawaf, M.; Merad, D.; Royer, J.; Boï, J.; Saccone, M.; Ben Ellefi, M.; Drap, P. Fast Visual Odometry for a Low-Cost Underwater Embedded Stereo System. *Sensors* **2018**, *18*, 2313 [CrossRef]
8. Valiente, D.; Payá, L.; Jiménez, L.; Sebastián, J.; Reinoso, Ó. Visual Information Fusion through Bayesian Inference for Adaptive Probability-Oriented Feature Matching. *Sensors* **2018**, *18*, 2041 [CrossRef]
9. García-Olalla, Ó.; Fernández-Robles, L.; Alegre, E.; Castejón-Limas, M.; Fidalgo, E. Boosting Texture-Based Classification by Describing Statistical Information of Gray-Levels Differences. *Sensors* **2019**, *19*, 1048 [CrossRef]
10. Fareed, M.; Chun, Q.; Ahmed, G.; Murtaza, A.; Asif, M.; Fareed, M. Appearance-Based Salient Regions Detection Using Side-Specific Dictionaries. *Sensors* **2019**, *19*, 421 [CrossRef]
11. Feng, Q.; Hao, Q.; Sbert, M.; Yi, Y.; Wei, Y.; Dai, J. Local Parallel Cross Pattern: A Color Texture Descriptor for Image Retrieval. *Sensors* **2019**, *19*, 315 [CrossRef] [PubMed]
12. Feng, Q.; Hao, Q.; Chen, Y.; Yi, Y.; Wei, Y.; Dai, J. Hybrid Histogram Descriptor: A Fusion Feature Representation for Image Retrieval. *Sensors* **2018**, *18*, 1943 [CrossRef] [PubMed]
13. García-Olalla, O.; Alegre, E.; Fernández-Robles, L.; Fidalgo, E.; Saikia, S. Textile Retrieval Based on Image Content from CDC and Webcam Cameras in Indoor Environments. *Sensors* **2018**, *18*, 1329 [CrossRef] [PubMed]
14. Wang, F.; Liang, C.; Ru, C.; Cheng, H. An Improved Point Cloud Descriptor for Vision Based Robotic Grasping System. *Sensors* **2019**, *19*, 2225 [CrossRef] [PubMed]
15. Martinez-Martin, E.; del Pobil, A. Vision for Robust Robot Manipulation. *Sensors* **2019**, *19*, 1648 [CrossRef] [PubMed]
16. Xue, B.; Chang, B.; Peng, G.; Gao, Y.; Tian, Z.; Du, D.; Wang, G. A Vision Based Detection Method for Narrow Butt Joints and a Robotic Seam Tracking System. *Sensors* **2019**, *19*, 1144 [CrossRef]
17. Zhang, Z.; Zhao, R.; Liu, E.; Yan, K.; Ma, Y. A Convenient Calibration Method for LRF-Camera Combination Systems Based on a Checkerboard. *Sensors* **2019**, *19*, 1315 [CrossRef]
18. Sels, S.; Ribbens, B.; Vanlanduit, S.; Penne, R. Camera Calibration Using Gray Code. *Sensors* **2019**, *19*, 246 [CrossRef]
19. Choi, K.; Jung, H.; Suhr, J. Automatic Calibration of an Around View Monitor System Exploiting Lane Markings. *Sensors* **2018**, *18*, 2956 [CrossRef]
20. Kapuscinski, T.; Organisciak, P. Handshape Recognition Using Skeletal Data. *Sensors* **2018**, *18*, 2577 [CrossRef]
21. Wang, L.; Duan, X.; Zhang, Q.; Niu, Z.; Hua, G.; Zheng, N. Segment-Tube: Spatio-Temporal Action Localization in Untrimmed Videos with Per-Frame Segmentation. *Sensors* **2018**, *18*, 1657 [CrossRef] [PubMed]
22. Chen, Y.; Ogata, T.; Ueyama, T.; Takada, T.; Ota, J. Automated Field-of-View, Illumination, and Recognition Algorithm Design of a Vision System for Pick-and-Place Considering Colour Information in Illumination and Images. *Sensors* **2018**, *18*, 1656 [CrossRef] [PubMed]
23. Nguyen, D.; Baek, N.; Pham, T.; Park, K. Presentation Attack Detection for Iris Recognition System Using NIR Camera Sensor. *Sensors* **2018**, *18*, 1315 [CrossRef] [PubMed]
24. Fu, X.; Yu, R.; Zhang, W.; Wu, J.; Shao, S. Delving Deep into Multiscale Pedestrian Detection via Single Scale Feature Maps. *Sensors* **2018**, *18*, 1063 [CrossRef] [PubMed]
25. Wang, S.; Zuo, X.; Du, C.; Wang, R.; Zheng, J.; Yang, R. Dynamic Non-Rigid Objects Reconstruction with a Single RGB-D Sensor. *Sensors* **2018**, *18*, 886 [CrossRef] [PubMed]
26. Sun, R.; Huang, Q.; Xia, M.; Zhang, J. Video-Based Person Re-Identification by an End-To-End Learning Architecture with Hybrid Deep Appearance-Temporal Feature. *Sensors* **2018**, *18*, 3669 [CrossRef]
27. Arsalan, M.; Naqvi, R.; Kim, D.; Nguyen, P.; Owais, M.; Park, K. IrisDenseNet: Robust Iris Segmentation Using Densely Connected Fully Convolutional Networks in the Images by Visible Light and Near-Infrared Light Camera Sensors. *Sensors* **2018**, *18*, 1501 [CrossRef]
28. Liu, D.; Chen, X.; Liu, X.; Shi, C. Star Image Prediction and Restoration under Dynamic Conditions. *Sensors* **2019**, *19*, 1890 [CrossRef]
29. Xia, L.; Meng, Q.; Chi, D.; Meng, B.; Yang, H. An Optimized Tightly-Coupled VIO Design on the Basis of the Fused Point and Line Features for Patrol Robot Navigation. *Sensors* **2019**, *19*, 2004 [CrossRef]
30. Cheng, X.; Liu, X.; Li, Z.; Zhong, K.; Han, L.; He, W.; Gan, W.; Xi, G.; Wang, C.; Shi, Y. High-Accuracy Globally Consistent Surface Reconstruction Using Fringe Projection Profilometry. *Sensors* **2019**, *19*, 668 [CrossRef]

31. Li, L.; Luo, W.; Wang, K. Lane Marking Detection and Reconstruction with Line-Scan Imaging Data. *Sensors* **2018**, *18*, 1635 [CrossRef] [PubMed]
32. Zhang, W.; Guo, B.; Li, M.; Liao, X.; Li, W. Improved Seam-Line Searching Algorithm for UAV Image Mosaic with Optical Flow. *Sensors* **2018**, *18*, 1214. [CrossRef] [PubMed]
33. Cao, C.; Ouyang, Q. 2D Rotation-Angle Measurement Utilizing Least Iterative Region Segmentation. *Sensors* **2019**, *19*, 1634 [CrossRef] [PubMed]
34. Chen, R.; Li, Z.; Zhong, K.; Liu, X.; Wu, Y.; Wang, C.; Shi, Y. A Stereo-Vision System for Measuring the Ram Speed of Steam Hammers in an Environment with a Large Field of View and Strong Vibrations. *Sensors* **2019**, *19*, 996 [CrossRef]
35. Li, H.; Zhu, Q.; Huang, M.; Guo, Y.; Qin, J. Pose Estimation of Sweet Pepper through Symmetry Axis Detection. *Sensors* **2018**, *18*, 3083 [CrossRef]
36. Yang, X.; Chen, X.; Xi, J. Comparative Analysis of Warp Function for Digital Image Correlation-Based Accurate Single-Shot 3D Shape Measurement. *Sensors* **2018**, *18*, 1208 [CrossRef]

sensors

MDPI

Article

RGB-D SLAM with Manhattan Frame Estimation Using Orientation Relevance

Liang Wang [1,2,*] and Zhiqiu Wu [1]

1 College of Automation, Faculty of Information Technology, Beijing University of Technology, Beijing 100124, China; s201402158@emails.bjut.edu.cn

2 Beijing Key Laboratory of Computational Intelligence and Intelligent System, Beijing 100124, China

* Correspondence: wangliang@bjut.edu.cn

Received: 31 January 2019; Accepted: 25 February 2019; Published: 1 March 2019

Abstract: Due to image noise, image blur, and inconsistency between depth data and color image, the accuracy and robustness of the pairwise spatial transformation computed by matching extracted features of detected key points in existing sparse Red Green Blue-Depth (RGB-D) Simultaneously Localization And Mapping (SLAM) algorithms are poor. Considering that most indoor environments follow the Manhattan World assumption and the Manhattan Frame can be used as a reference to compute the pairwise spatial transformation, a new RGB-D SLAM algorithm is proposed. It first performs the Manhattan Frame Estimation using the introduced concept of orientation relevance. Then the pairwise spatial transformation between two RGB-D frames is computed with the Manhattan Frame Estimation. Finally, the Manhattan Frame Estimation using orientation relevance is incorporated into the RGB-D SLAM to improve its performance. Experimental results show that the proposed RGB-D SLAM algorithm has definite improvements in accuracy, robustness, and runtime.

Keywords: SLAM; RGB-D; indoor environment; Manhattan frame estimation; orientation relevance; spatial transformation

---

1. Introduction

Simultaneous Localization and Mapping (SLAM), which aims to acquire the structure of an unknown environment and at the same time estimate the sensor pose with respect to this structure, is an essential task for the autonomy of a robot. It can facilitate a wide range of applications from autonomous robots to virtual and augmented reality. In early SLAM algorithms, many types of sensors, such as rotary encoders, inertial sensors, laser range sensors, and cameras, were employed. Recently, the SLAM algorithms based on the compact Red Green Blue-Depth (RGB-D) sensors, such as Kinect or Xtion, became popular [1–6]. This is because RGB-D sensors have the advantages of low price, and appropriate size and weight. More importantly, they can provide direct and dense depth measurements besides the appearance information with the RGB images [7]. Hence, the RGB-D sensors provide opportunities to handle challenges in SLAM systems.

According to the modelling and processing, existing RGB-D SLAM algorithms can be roughly classified into two directories: dense SLAM and sparse SLAM. Newcombe et al. [1,2] firstly introduced dense RGB-D SLAM algorithms in their well-known work, Kinect Fusion. Kinect Fusion can obtain real-time depth measurements and a highly detailed voxel-based map simultaneously. However, their algorithms are only suitable for small workspaces owing to high memory consumption. Moreover, it generally fails when scenes have poor geometric structure. To solve the restricted area problem, Whelan et al. proposed an improved algorithm [3] to densely map large areas in real-time by transforming the voxel grid with sensor pose of each observation. To further improve the efficiency, Keller et al. [4] proposed a point-based fusion representation supporting spatially extended reconstructions with a fused surfel-based model instead of voxel-based representation.

In general, dense SLAM algorithms enable good localization and mapping with high quality scene representation [8,9]. However, they are prone to failure in environments with poor structure and time drift. In addition, their computational costs are very high. To some extent some algorithms utilizing sophisticated equipment such as high-end graphics cards can overcome this deficiency. However, their applications' ranges are constrained.

Instead, sparse RGB-D SLAM algorithms offer a good balance between the computational cost and the quality of pose estimation. Sparse SLAM algorithms are mainly based on the visual odometry, which simply uses visual feature correspondences to compute the motion between the consecutive poses of the RGB-D sensor and then concatenates the pose-to-pose motion. The first RGB-D SLAM algorithm was proposed by Henry et al. [10]. It used features points to estimate sensor poses and then constructed and optimized a graph with nodes representing sensor poses and an edge between two poses being their spatial transformation to refine the localization and mapping. Endres et al. [11] followed the same path and implemented the pose-graph optimization with the G^2o framework [12]. Due to its availability, it is very popular. Indeed, sparse RGB-D SLAM algorithms typically run quickly owing to the sensor's pose estimation based on sparse point features. In addition, such a lightweight implementation ensures a wide range of applications. However, the mapping quality is poor due to limitation of sparse 3D points. More importantly, the mapping result lacks semantic information and there are many repeated and redundant points in the map.

The sparse RGB-D SLAM algorithms have been successful for environments with rich textures. However, they perform poorly and even fail in environments with textureless areas and areas with repetitive textures, which usually exist in indoor scenes with large planar regions [13]. To work well in low-texture environments, researchers begin to show a significant interest in additional high-level geometric information like planar features in recent RGB-D research [14–16], and apply them to RGB-D SLAM algorithms [17–19]. These SLAM approaches show great improvement in robustness. However, the accuracy still needs to be improved.

Three-dimensional planes in indoor environments, which can be easily extracted from point clouds, are extremely common and are generally relevant. Most indoor environments satisfy the Manhattan World (MW) assumption [20], under which the world consists of a set of orthogonal or parallel planes. Then the environment can be represented by three orthogonal directions, i.e., the Manhattan Frame (MF). The early work of MF estimation was mainly taken RGB images as input, which can be called the RGB image-based methods [21,22]. The RGB image-based methods generally utilize perspective property, such as vanishing line, vanishing point, and orientation map, to estimate the MF with a single RGB image. Recently, the RGB-D sensor is applied to estimate the MF. The corresponding RGB-D image-based methods [15,19] take both color image and depth image as input to compute the MF. In general, RGB image-based methods have poor accuracy and robustness since they mainly depend on information of scene structure in two-dimensional RGB image. RGB-D image-based methods generally perform better than RGB image-based methods [19], since not only the RGB image but also the depth information are explored simultaneously. However, the state-of-the-art of RGB-D image-based methods are still unsatisfactory for real applications, especially in accuracy and speed.

Considering the image noise, image blur, the inconsistency between the depth data and the color image, and especially low-texture (i.e., textureless or repeated texture) planar walls dominating the view of observations, some frames could not be matched to any predecessor yet in existing sparse RGB-D SLAM algorithms. Even if the pairwise spatial transformation can be computed, its accuracy and robustness are poor. On the other hand, most indoor environments follow the MW assumption and the MF can be recovered from a single RGB-D image using orientation relevance [15]. Therefore, a new RGB-D SLAM algorithm is proposed by extending Manhattan Frame estimation (MFE) using orientation relevance to RGB-D image sequence. It first performs MFE using the introduced concept of orientation relevance. Then the pairwise spatial transformation in RGB-D SLAM is computed with the estimated MFE. Finally, the sparse RGB-D SLAM is improved by incorporating MFE using orientation relevance. Experiments validate the proposed algorithm. The contributions of this paper are two-fold: I. A novel algorithm for RGB-D SLAM with MFE using orientation relevance is proposed for low-texture indoor environments. II. It improves the performance of sparse RGB-D SLAM in accuracy and robustness.

The remainder of this paper is organized as follows. Section 2 details the proposed algorithm for RGB-D SLAM with MFE using orientation relevance. Experimental results are presented in Section 3. Finally, we summarize and report future works in Section 4.

2. Method

This section presents the proposed RGB-D SLAM method in detail. In the original RGB-D SLAM [11], only point features or all points are used with RANSAC or GICP to estimate the relative spatial transformation between two consecutive observations. Considering the image noise, image blur, and the inconsistency between the depth data and RGB image, some frames could not be matched to any predecessor yet. Even if the pairwise spatial transformation can be computed, its accuracy is not high. It also results in poor robustness or high computational cost. Different from that, the MF of the indoor environment is estimated and used to improve the RGB-D SLAM in the proposed method. In the following, we firstly briefly review the original RGB-D SLAM [11]. Then the algorithm of the Manhattan Frame estimation using orientation relevance is presented. Thirdly, the computation of pairwise spatial transformation with the MFE is presented. Finally, the improved RGB-D SLAM with the Manhattan Frame estimation using orientation relevance is introduced.

2.1. Overview of the Original Method

A schematic overview of sparse RGB-D SLAM is given in Figure 1a [11]. It firstly uses both RGB images and depth data to perform localization and generate the trajectory. Then the mapping is obtained by 3D points registration and voxelization.

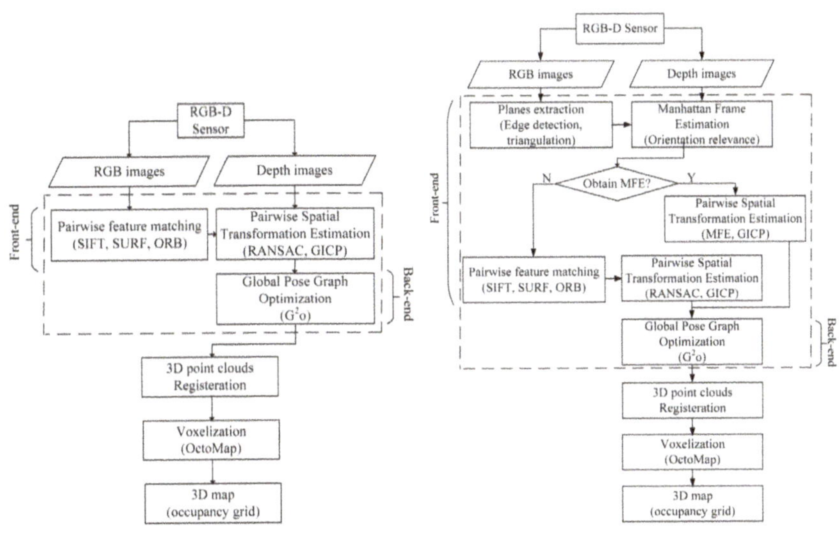

Figure 1. Schematic overview of (**a**) the original Red Green Blue-Depth (RGB-D) Simultaneously Localization And Mapping (SLAM) and (**b**) the proposed RGB-D SLAM.

The trajectory estimation can be further divided into two parts: the front-end and the back-end. The front-end computes spatial transformations between individual observations, and the back-end computes poses of these observations via a graph-based optimization. In the front-end of the sparse RGB-D SLAM, the RGB image of RGB-D sensor is used to detect key points and extract descriptors. Extracted descriptors of detected key points in two consecutive observations are matched to compute the relative pairwise spatial transformation between two observations using RANSAC. In addition, the depth image of RGB-D sensor makes it possible that dense point clouds of two observations are registered in a common coordinate system using RANSAC or GICP. In the back-end, a non-linear cost function defined on a pose graph [12] is optimized to obtain globally optimal poses of all observations, i.e., the trajectory. After obtaining the trajectory, an occupancy voxel grid map is computed.

2.2. Manhattan Frame Estimation Using Orientation Relevance

Due to limitations of RGB-D sensor, the RGB-D SLAM is only applicable for indoor applications. Generally, most man-made indoor environments follow the MW assumption [20], under which the world consists of a set of orthogonal and parallel planes. Three orthogonal directions corresponding to the normal of a set of orthogonal and parallel planes, which are referred to as the MF [15,19], are enough to describe the environment. In RGB-D SLAM, planes in the indoor scene can be detected in each observation. Then candidates of dominant planes can be determined with the constraint of orientation relevance. The MF can be computed by finding the orthogonal dominant planes, which can be described by normal vectors of three orthogonal dominant planes of the scene. It can be further incorporated into RGB-D SLAM to improve the performance of RGB-D SLAM.

Firstly, an edge detection algorithm is run on the input RGB image. Then, end points of detected edges are used to perform 2D Delaunay triangulation to divide the RGB image into several triangles. Next, the triangles are merged according to intensity statistics of pixels in each triangle. Here the intensity statistic, the root mean square error (RMSE) between intensity value of each pixel and the mean intensity of merged area, is taken as measure to merge triangles. Afterwards, the bilateral filter is used to smooth the input depth image. Finally, each plane corresponding to merged triangle in the RGB image, whose area is larger than a threshold, is validated by plane fitting with filtered depth image data. The N ($N = 9$ in our experiments) largest planes are the candidate dominant planes and the normal vector of each candidate plane can be computed with the depth data. These candidate dominant planes are the input of the following MFE using orientation relevance.

An indoor environment satisfying the MW assumption can be denoted by $\mathbf{H} = \{P_1, P_2, \cdots, P_N\}$, where P_n $(1 \leq n \leq N, N \geq 3)$ is one of N detected candidate dominant planes. For each pair of two planes P_i and P_j, their relation can be described by the angle between them θ_{ij}. The closer to 0° or 180° the angle θ_{ij} is, the nearer two planes P_i and P_j are parallel. Otherwise, the closer to 90° the angle θ_{ij} is, the nearer two planes P_i and P_j are perpendicular. Most of planes in $\mathbf{H}$ are mutually perpendicular or parallel and normal vectors of them can be clustered into three directions. These planes are the dominant planes and three directions are the dominant directions corresponding to the MF. Except for dominant planes, lots of little planar regions existing in indoor environment may have parallel or perpendicular relations. This would lead to error result of MFE. So both the normal direction and area of extracted planar regions should be taken into account. We introduce the concept of orientation relevance of extracted dominant planes, which considers both the area of the projection of extracted planes and the angle between them, to evaluate their geometric relations. The orientation relevance consists of parallel relevance and perpendicular relevance.

The parallel relevance of extracted planes is computed by

$$R_{pa}(P_i) = \sum_{n=1}^{N} A(P_n)\sin(\theta_{in}) \tag{1}$$

where $A(P_n)$ is the area of extracted candidate plane P_n, θ_{in} represents the angle between planes P_i and P_n. In fact, $R_{pa}(P_i)$ is the sum of area of all extracted candidate planes' projection on the plane perpendicular to P_i. The larger the quantity and area of extracted candidate planes being parallel to P_i are, the smaller the value of $R_{pa}(P_i)$ is. Otherwise, the larger the value of $R_{pa}(P_i)$ is.

Similarly, the perpendicular relevance is represented by

$$R_{pe}(P_i) = \sum_{n=1}^{N} A(P_n)\cos(\theta_{in}) \tag{2}$$

where $R_{pe}(P_i)$ is the sum of area of all extracted andidate planes' projection on the plane P_i. The larger the quantity and area of extracted candidate planes being perpendicular to P_i are, the smaller the value of $R_{pe}(P_i)$ is. Otherwise, the larger the value of $R_{pe}(P_i)$ is.

In fact, the parallel relevance and the perpendicular relevance are conflict. To make a compromise, we introduce the term orientation relevance,

$$\begin{aligned} R_o(P_i) &= f(R_{pe}(P_i), R_{pa}(P_i)) \\ &= \sum_{n=1}^{N} A(P_n)\cos(\theta_{in})\sin(\theta_{in}) \\ &= \frac{1}{2}\sum_{n=1}^{N} A(P_n)\sin(2\theta_{in}) \end{aligned} \tag{3}$$

where $\theta_{in} \in [0, \frac{\pi}{2}]$ is the angle between the plane P_i and P_n. The orientation relevance can reach the minimum in the domain of definition of θ_{in} when $\theta_{in} = 0$ or $\theta_{in} = \frac{\pi}{2}$. In such cases, the relationship between two planes P_i and P_n is strictly parallel or perpendicular. For indoor environments, one dominant direction may correspond to several parallel dominant planes. Values of the orientation relevance of these parallel dominant planes should be equal in theory. However, they are slightly different from each other in practice due to inevitable noise. Here the dominant direction corresponding to the MF is computed using the dominant plane with the minimal orientation relevance.

$$\tilde{R}_o = min\{R_o(P_i)\} \tag{4}$$

In some cases, it is a planar surface of clutter object rather than a wall that reaches the minimum of orientation relevance. To avoid this case, the area of planar surface is also taken into account,

$$\hat{R}_o = min\{R_o(P_i) - \lambda A(P_i)\} \tag{5}$$

where λ is a coefficient to balance two terms, which usually takes an empirical value of 5000. Then, when the orientation relevance shown in Equation (5) reaches the minimum, the corresponding plane, P_D , is one of the MW's dominant planes. The normal of the plane P_D corresponds to one axis of the MF.

Then, we determine the other two axes of the MF. Since each detected candidate plane usually differs in position and area, their corresponding values of orientation relevance computed by Equation (5) are different from each other. However, for each of three dominant directions, the corresponding dominant plane should have the minimal orientation relevance among all detected planes sharing this dominant direction. So planes corresponding to the N smallest orientation relevance are initially taken as candidates, where N takes 9 in our implementation. Furthermore, the N smallest orientation relevance are sorted in ascending order. Here, the minimal corresponds to the dominant plane P_D. Additionally, check whether the normal of other $N-1$ planes is perpendicular to the normal of P_D in turn. And take the normal of the first plane whose satisfies the aforementioned condition, P'_D, as the second dominant direction, i.e., the second axis of the MF. Finally, the third dominant direction, i.e., the third axis of the MF can be computed by taking cross product of the first dominant direction and the second dominant direction. By now, three orthogonal directions, i.e., the MF of the indoor environment, are recovered.

2.3. Computation of Pairwise Spatial Transformation with the MFE

Once the MF of one observation is computed, it can be used to compute the pairwise spatial transformation of current pose relative to its previous one, and then be incorporated into the RGB-D SLAM to improve its performance.

The MF can be described by unit normal vectors of dominant orthogonal planes. Generally, two unit normal vectors of two orthogonal dominant planes are enough. For example, the unit normal vector of two orthogonal dominant planes is denoted by $\boldsymbol{m}_1$ and $\boldsymbol{m}_2$ respectively. They correspond to two orthogonal directions of the MF. The third direction of the MF can be computed by

$$\boldsymbol{m}_3 = \boldsymbol{m}_1 \times \boldsymbol{m}_2 \tag{6}$$

Then the MF of current observation can be described by unit normal vectors of three orthogonal dominant planes

$$\boldsymbol{M}_1 = [\boldsymbol{m}_1 \quad \boldsymbol{m}_2 \quad \boldsymbol{m}_3] \tag{7}$$

Similarly, the MF of the previous observation can be described as

$$\boldsymbol{N}_1 = [\boldsymbol{n}_1 \quad \boldsymbol{n}_2 \quad \boldsymbol{n}_3] \tag{8}$$

For an RGB-D SLAM application, the MF of the indoor scene is fixed. However, there are relative translation and rotation between two consecutive observations for RGB-D sensor, which make the computed MFs $\boldsymbol{M}_1$ and $\boldsymbol{N}_1$ are different in two local coordinate systems of two observations. The spatial transformation between two consecutive observations in RGB-D SLAM, $\boldsymbol{T}$, consists of $\boldsymbol{R}$ and $\boldsymbol{t}$.

$$\boldsymbol{T} = \begin{bmatrix} \boldsymbol{R} & \boldsymbol{t} \\ \boldsymbol{0}^T & 1 \end{bmatrix} \tag{9}$$

where $\boldsymbol{R}$ and $\boldsymbol{t}$ is the relative rotation matrix and translation vector between two observations respectively. The relative rotation $\boldsymbol{R}$ between two observations can be computed with the MFs estimated in local coordinate system of two observations.

$$\boldsymbol{R} \cdot \boldsymbol{m}_i = \boldsymbol{n}_i \quad (s.t. \boldsymbol{R}^T\boldsymbol{R} = \boldsymbol{I} \quad and \quad det(\boldsymbol{R}) = 1) \quad (i = 1,2,3) \tag{10}$$

As Equation (10) shows, the corresponding MFs of two observations can provide 9 equations to compute unknowns in $\boldsymbol{R}$. However $\boldsymbol{R}$ is a unit orthogonal matrix, some constraints, such as $\boldsymbol{R}^T\boldsymbol{R} = \boldsymbol{I}$ and $det(\boldsymbol{R}) = 1$ (where $\boldsymbol{I}$ is an identity matrix, $det(\cdot)$ denotes the determinant of a matrix), should be satisfied, which results in a complex constrained optimization problem. For each pair of consecutive observations, $\boldsymbol{R}$ can be firstly computed by linearly solving equation system $\boldsymbol{R} \cdot \boldsymbol{m}_i = \boldsymbol{n}_i \quad (i = 1, 2, 3)$, and then enforced the constraints $\boldsymbol{R}^T\boldsymbol{R} = \boldsymbol{I}$ and $det(\boldsymbol{R}) = 1$. Once the rotation matrix $\boldsymbol{R}$ is obtained, the point cloud corresponding to the current observation can be transformed to the local coordinate system of the previous observation using the obtained $\boldsymbol{R}$. Then the translation vector $\boldsymbol{t}$ can be computed by GICP with the transformed point cloud of current observation and the point cloud of previous observation.

The spatial transformation between each pair of consecutive observations, **T**, can be further optimized by bundle adjustment by solving the following unconstrained optimization problem

$$e = \min_{\xi} \frac{1}{2} \sum_{i=1}^{N} \| \boldsymbol{p}_i - exp(\xi^{\wedge}) \boldsymbol{q}_i \|_2^2 \tag{11}$$

where $\boldsymbol{p}_i$ and $\boldsymbol{q}_i$ is the 3D point in the point cloud of previous observation and that of current observation respectively, $\xi = \begin{bmatrix} \rho \\ \boldsymbol{\phi} \end{bmatrix} \in \mathbb{R}^6$ is the Lie algebraic representation of transformation and the relation between the spatial translation and the its Lie algebraic representation follows

$$\boldsymbol{T} = exp(\xi^{\wedge}) = \begin{bmatrix} exp(\boldsymbol{\phi}^{\wedge}) & \boldsymbol{J}\rho \\ \boldsymbol{0} & 1 \end{bmatrix} \tag{12}$$

where

$$exp(\boldsymbol{\phi}^{\wedge}) = exp(\theta \boldsymbol{a}^{\wedge}) = cos\theta \boldsymbol{I} + (1 - cos\theta)\boldsymbol{a}\boldsymbol{a}^T + sin\theta \boldsymbol{a}^{\wedge} \tag{13}$$

$$\boldsymbol{J} = \frac{sin\theta}{\theta}\boldsymbol{I} + (1 - \frac{sin\theta}{\theta})\boldsymbol{a}\boldsymbol{a}^T + \frac{1 - cos\theta}{\theta}\boldsymbol{a}^{\wedge} \tag{14}$$

$$\theta = arccos\frac{tr(\boldsymbol{R}) - 1}{2} \tag{15}$$

$$\boldsymbol{R}\boldsymbol{a} = \boldsymbol{a} \tag{16}$$

$$\boldsymbol{t} = \boldsymbol{J}\rho \tag{17}$$

The Lie algebra $\mathfrak{se}(3) = \{\xi = \begin{bmatrix} \rho \\ \boldsymbol{\phi} \end{bmatrix} \in \mathbb{R}^6, \rho \in \mathbb{R}^3, \boldsymbol{\phi} \in \mathbb{R}^3, \xi^{\wedge} = \begin{bmatrix} \boldsymbol{\phi}^{\wedge} & \rho \\ \boldsymbol{0}^T & 0 \end{bmatrix} \in \mathbb{R}^{4\times4}\}$, which corresponds to the tangent space of the Lie group $SE(3) = \{\mathbf{T} = \begin{bmatrix} \boldsymbol{R} & \boldsymbol{t} \\ 0 & 1 \end{bmatrix} \in \mathbb{R}^{4\times4} | \boldsymbol{R} \in \mathbb{R}^{3\times3}, \boldsymbol{R}^T\boldsymbol{R} = \boldsymbol{I}, det(\boldsymbol{R}) = 1, \boldsymbol{t} \in \mathbb{R}^3\}$, describes the local derivatives. Here we use the Lie algebraic representation to optimize the spatial transformation. On one hand, with the Lie algebra, the obtained unconstrained optimization problem is relatively easier to solve than the corresponding constrained one. On the other hand, the Lie algebra representation makes the computation of derivatives easier during the optimization process. The unconstrained optimization problem Equation (11) can be solved by the Gaussian-Newton method or Levenberg-Marquardt algorithm. Then the pairwise spatial transformation $\boldsymbol{T}$ is obtained.

2.4. Improved RGB-D SLAM

Considering the RGB-D SLAM is only applicable for indoor applications and the MF of the indoor scene is fixed, the MF can be used as a reference to compute the pairwise spatial transformation. So a new algorithm of RGB-D SLAM shown in Algorithm 1 is proposed, in which the aforementioned pairwise spatial transformation computation with MFE using orientation relevance is incorporated into the original RGB-D SLAM [11] to improve its performance as shown in Figure 1b.

Algorithm 1 RGB-D SLAM with MFE Using Orientation Relevance

Input: RGB-D sequences
Output: Trajectory of RGB-D sensor and reconstructed environment.

Step 1. Extract planes from the RGB image using edge detection and triangulation of end points of detected edges.

Step 2. Estimate Manhattan Frame using orientation relevance with dominant planes determined by cross validation on depth information and planes extracted from RGB image.

Step 3. Determine whether the MFE is available. If it's available, compute the pairwise spatial transformation with MFE and GICP, and then jump to Step 5. Otherwise, go to Step4.

Step 4. Compute the pairwise spatial transformation following the routine of the original RGB-D SLAM.

Step 5. Optimize the trajectory.

Step 6. Registrate 3D point clouds.

Step 7. Voxelize the registrated 3D point clouds.

Step 8. Reconstruct the 3D map.

return Trajectory and 3D map.

Different from conventional RGB-D SLAM, which uses correspondences of feature points to compute the pairwise spatial transformation between two consecutive observations, the proposed RGB-D SLAM exploits the information of dominant planes. This makes the computation of pairwise spatial transformation more robust and accurate. In addition, in conventional RGB-D SLAM, the estimated trajectory is usually divided into several fragments due to the failure of feature matching of detected key points in pairwise spatial transformation computation caused by image noise, image blur and the inconsistency between the depth data and RGB image, which increases the complexity of the optimization problem of the back-end of RGB-D SLAM. Whereas, the proposed improved RGB-D SLAM is more robust and can reduce the number of trajectory fragments which makes the corresponding optimization problem more easily and rapidly converge to the global optimum.

3. Experiments

To validate the proposed RGB-D SLAM algorithm, some experiments are performed on a computer with an AMD Phenom II X6 1055T 3.36GHZ CPU and 8GB RAM with the RGB-D dataset and benchmark [23], which provides a dataset of RGB-D sequences from the Kinect and synchronized ground truth pose estimates from the motion capture system. These sequences are captured in a typical indoor environment. Furthermore, the benchmark provides an evaluation tool to compute the RSME. For the convenience of comparison, we use the benchmark tool to evaluate the proposed algorithm. To make a comparison, experiments using the original RGB-D SLAM [11] without the MF estimation are also performed. To show the comparison results in different scenes and different complexity of motion, experiments of 3 sequences are reported here. Critical details of 3 sequences are shown in Table 1. The structure and appearance of each scene can be seen in the following mapping results in the form of volumetric 3D model shown in Figures 2–4a, respectively.

Table 1. Details of sequences from the Red Green Blue-Depth (RGB-D) Simultaneously Localization And Mapping (SLAM) dataset [23].

Sequence	Frames	Duration (s)	Length (m)	Avg. Trans. Velocity (m/s)	Avg. Rot. Velocity (°/s)	Range (m^3)
fr1/360	745	28.69	5.82	0.21	41.60	$0.54 \times 0.46 \times 0.47$
fr3/long_office _household	2585	87.09	21.45	0.25	10.19	$5.12 \times 4.89 \times 0.54$
fr1/floor	1214	49.87	12.57	0.258	15.07	$2.30 \times 1.31 \times 1.58$

The fr1/360 scene is a typical indoor office which includes walls, floor, table and clutters. Table 2 shows the trajectory results of original RGB-D SLAM [11] and the proposed improved RGB-D SLAM. To make a comparison, results of RGB-D SLAM with RMFE algorithm are also reported in Table 2, which are directly cited from [19]. As can be seen from this table, the proposed improved RGB-D SLAM outperforms the original RGB-D SLAM and RGB-D SLAM with RMFE in RMSE of translation, RMSE of rotation and runtime. The most obvious improvement is in runtime, which dramatically drops from 145 s for the original algorithm to 100 s for the improved algorithm. It has about 31% relative improvement (RI) with respect to the corresponding parameter of the original RGB-D SLAM. The RMSE of translation drops from 0.103 m to 0.082 m, which has about 20% RI. The RMSE of rotation drops from 3.41 degrees to 3.10 degrees, which has about 9% RI. Results of estimated trajectory for fr1/360 are shown in Figure 3a. It can be seen that the trajectory estimated by the proposed algorithm is much closer to the ground truth than that of the original RGB-D SLAM. We could not find the source code and detailed parameters of RGB-D SLAM with RMFE. In fairness, we do not show the estimated trajectory of the RGB-D SLAM with RMFE implemented by us to make comparisons since results of RMFE [19] implemented by us are inferior to MFE using orientation relevance as shown in Ref. [15].

Table 2. Trajectory results of RGB-D SLAM with fr1/360 sequence.

Method	Translation		Rotation		Runtime	
	RMSE (m)	RI	RMSE (°)	RI	(s)	RI
original method [11]	0.103	–	3.41	–	145	–
method with RMFE [19]	0.107	−3.9%	3.37	1.2%	112	23%
proposed method	0.082	20%	3.10	9%	100	31%

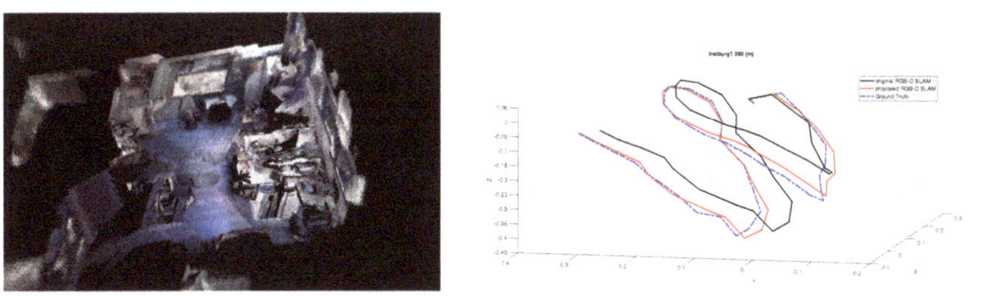

(a) (b)

Figure 2. Experimental results of the proposed RGB-D SLAM with sequence fr1/360. (**a**) Mapping results in the form of volumetric 3D model. (**b**) Estimated trajectories.

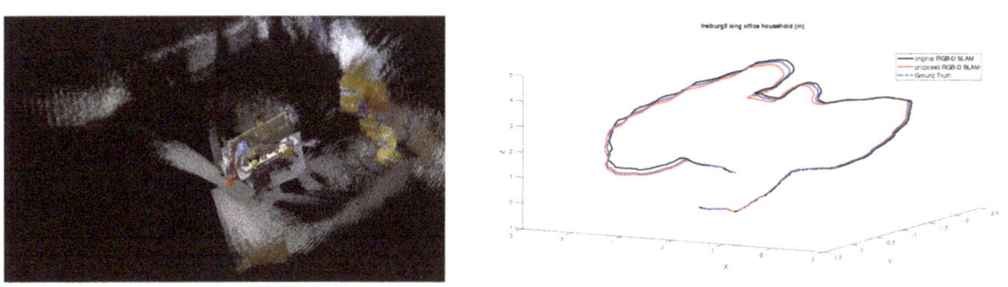

(a) (b)

Figure 3. Experimental results of the proposed RGB-D SLAM with sequence fr3/long_office_household. (**a**) Mapping results in the form of volumetric 3D model. (**b**) Estimated trajectories.

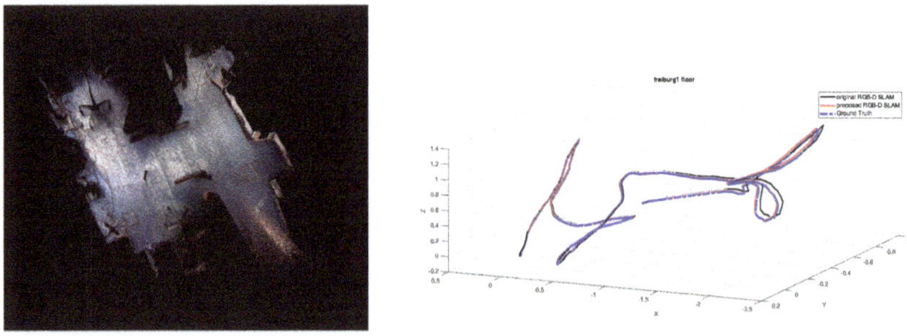

(**a**) (**b**)

Figure 4. Experimental results of the proposed RGB-D SLAM with sequence fr1/floor. (**a**) Mapping results in the form of volumetric 3D model. (**b**) Estimated trajectories.

To further validate the proposed method, experiments are also performed on sequence of fr3/long_office_household and fr1/floor. Considering reasons mentioned above and results shown in Table 1 that the proposed method outperforms the RGB-D SLAM with RMFE, results of the RGB-D SLAM with RMFE implemented by us are not reported here. The sequence of fr3/long_office_household mainly focuses on an office table and its indoor environment. The office table is in the center of this scene, which is surrounded by white walls. Since the range of the scene is so large that the wall and floor far from the table are out of the measurement range of RGB-D sensor, there are some areas with lots of missing data. Results of estimated trajectory of fr3/long_office_household are shown in Figure 3b. As can be seen, the trajectory estimated by the proposed method is much closer to the ground truth than that of the original RGB-D SLAM. From Table 3 we can see that the runtime drops 211 s which results in about 29% RI, the RMSE of translation drops 0.03 m which brings in about 37% RI, and the RMSE of rotation drops 0.11 degrees which brings in about 7% RI. The sequence of fr1/floor mainly focuses on the indoor floor which is marked with blue color, and there is some clutter on the floor. The results of the estimated trajectory for fr1/floor are shown in Figure 4b, where the trajectory estimated by the proposed method is much closer to the ground truth than that of the original RGB-D SLAM. As can be seen from Table 4, the runtime drops 86 s which brings in about 18% RI, the RMSE of translation drops 0.006 m which results in about 10% RI, and the RMSE of rotation drops 0.03 degrees which results in about 1% RI. It is noted that since the scene range becomes larger, and the visual difference between trajectories becomes slighter in comparison with Figure 2b. However, improvements brought by the proposed method are obvious.

Table 3. Trajectory results of RGB-D SLAM with fr3/long_office_householdsequence.

Method	Translation		Rotation		Runtime	
	RMSE (m)	RI	RMSE (°)	RI	(s)	RI
original method [11]	0.082	–	1.63	–	722	–
proposed method	0.052	37%	1.52	7%	511	29%

Table 4. Trajectory results of RGB-D SLAM with fr1/floor sequence.

Method	Translation		Rotation		Runtime	
	RMSE (m)	RI	RMSE (°)	RI	(s)	RI
original method [11]	0.061	–	2.72	–	488	–
proposed method	0.054	11%	2.69	1%	402	18%

From experimental results, we can see that the proposed method consistently outperforms the original RGB-D SLAM. The improvement brought by the proposed RGB-D SLAM on sequence of fr3/long_office_household and fr1/360 are larger than that on sequence of fr1/floor. The reason is mainly because that the focus of sequence of fr1/floor is floor and images containing two or more orthogonal dominant planes are relatively less. Furthermore, it is hard to find enough orthogonal dominant planes to perform MFE in these sequences. As shown in Figure 1b, pairwise spatial transformation estimation with MFE using orientation relevance will fail and conventional routine of the original RGB-D SLAM, which performs pairwise spatial transformation estimation with detection and matching of feature points and registration of 3D point clouds with RANSAC scheme, will function in this case. So in the worst case where the the MW assumption does not hold, the proposed method degrades to the original RGB-D SLAM. Fortunately, the conventional routine of the original RGB-D SLAM is fully functioning in most of these cases since clutter in a small measurement range provide rich texture. So although the trajectory segments of the degraded proposed method coincide with those of the original method in the above experiments, rich textures ensure that the trajectory segments of the original RGB-D SLAM are very close to the ground truth as seen in Figures 3b and 4b. When there are a few low-texture walls corresponding to two or more orthogonal dominant planes in observations of RGB-D SLAM, the performance of the original RGB-D SLAM will degrade. While the proposed method fulfils its function and performs well. In summary, the proposed RGB-D SLAM can bring in obvious improvements in runtime and accuracy of trajectory in comparison with the original RGB-D SLAM and RGB-D SLAM with RMFE. The reasons may be as follows: (1) Using MF estimation with orientation relevance instead of conventional detection and matching of feature points with RANSAC scheme to compute the pairwise spatial transformation in the front-end of RGB-D SLAM can bring in performance improvement. (2) The optimization problem of the back-end of RGB-D SLAM becomes easier since the aforementioned reason leads to a good initialization and less trajectory fragments, which also improves the performance and reduces runtime. Experiments also show that the proposed method is suitable for sequences with different duration, range, and motion velocity. Hence, the proposed method is valid and reliable.

4. Conclusions

A new method of RGB-D SLAM is proposed, which computes the pairwise spatial transformation with the MFE using orientation relevance instead of the conventional routine of the original RGB-D SLAM, which uses detection and matching of point correspondences and registration of 3D point clouds with the RANSAC scheme. It can overcome the deficiency of the original RGB-D SLAM that some observations of RGB-D sensor could not be matched to any predecessor due to image noise, image blur, inconsistency between the depth data and the RGB image, and especially low-texture (i.e., textureless or repeated texture) planar walls dominating the view of observations. Experiments on an open dataset benchmark validate the proposed method. It can bring in obvious improvements in runtime and accuracy of trajectory in comparison with the original RGB-D SLAM and RGB-D SLAM with RMFE. In the future, we will further improve the proposed method to be suitable for real-time applications and extend it to more complex indoor environments such as the Atlanta world [24]. We will also further improve the RGB-D SLAM to be applicable to dynamic environments.

Author Contributions: Conceptualization, L.W. and Z.W.; Methodology, L.W. and Z.W.; Software, Z.W.; Validation, Z.W.; Formal analysis, L.W. and Z.W.; Investigation, L.W. and Z.W.; Resources, L.W.; Data curation, Z.W.; Writing—original draft preparation, Z.W.; Writing—review and editing, L.W. and Z.W.; Visualization, L.W. and Z.W.; Supervision, L.W.; Project administration, L.W.; Funding acquisition, L.W.

Funding: This research was funded by the National Natural Sciences Foundation of China (NSFC) under Grant No. 61772050 and the China Scholarship Council (CSC) under Grant No. 201706545026.

Acknowledgments: We would like to thank computer vision group, faculty of informatics, Technical University of Munich, for their open RGB-D SLAM Dataset and Benchmark.

Conflicts of Interest: The authors declare no conflict of interest.

Abbreviations

The following abbreviations are used in this manuscript:

SLAM	Simultaneously Localization And Mapping
RGB-D	Red Green Blue-Depth
3D	three Dimensional
MW	Manhattan World
MF	Manhattan Frame
MFE	Manhattan Frame Estimation
RANSAC	RANdom SAmple Consensus
GICP	Generalized Iterative Closest Point
RMFE	Robust Manhattan Frame Estimation
RMSE	Root Mean Square Error
RI	Relative Improvement

References

1. Newcombe, R.A.; Izadi, S.; Hilliges, O.; Molyneaux, D.; Kim, D.; Davison, A.J.; Kohi, P.; Shotton, J.; Hodges, S.; Fitzgibbon, A. Kinectfusion: Real-time dense surface mapping and tracking. In Proceedings of the International Symposium on Mixed and Augmented Reality, Basel, Switzerland, 26–29 October 2011; pp. 127–136.
2. Izadi, S.; Kim, D.; Hilliges, O.; Molyneaux, D.; Newcombe, R.; Kohli, P.; Shotton, J.; Hodges, S.; Freeman, D.; Davison, A.; et al. Kinectfusion: Real-time 3D reconstruction and interaction using a moving depth camera. In Proceedings of the ACM Symposium on User Interface Software and Technology, Santa Barbara, CA, USA, 16–19 October 2011; pp. 559–568.
3. Whelan, T.; Kaess, M.; Johannsson, H.; Fallon, M.; Leonard, J.J.; McDonald, J. Real-time large-scale dense rgb-d slam with volumetric fusion. *Int. J. Robot. Res.* **2015**, *34*, 598–626. [CrossRef]
4. Keller, M.; Lefloch, D.; Lambers, M.; Izadi, S.; Weyrich, T.; Kolb, A. Real-time 3D reconstruction in dynamic scenes using point-based fusion. In Proceedings of the Conference on 3D Vision, Seattle, WA, USA, 29 June–1 July 2013; pp. 1–8.
5. Fan, Y.; Feng, Z.; Mannan, A.; Khan, T.U.; Shen, C.; Saeed, S. Estimating tree position, diameter at breast height, and tree height in real-time using a mobile phone with RGB-D SLAM. *Remote Sens.* **2018**, *10*, 1845. [CrossRef]
6. Guo, R.; Peng, K.; Zhou, D.; Liu, Y. Robust visual compass using hybrid features for indoor environments. *Electronics* **2019**, *8*, 220. [CrossRef]
7. Cai, Z.; Han, J.; Liu, L.; Shao, L. RGB-D datasets using Microsoft Kinect or similar sensors: A survey. *Multimedia Tools Appl.* **2017**, *76*, 4313–4355. [CrossRef]
8. Meng, X.R.; Gao, W.; Hu, Z.Y. Dense RGB-D SLAM with multiple cameras. *Sensors* **2018**, *18*, 2118. [CrossRef] [PubMed]
9. Fu, X.; Zhu, F.; Wu, Q.; Sun, Y.; Lu, R.; Yang, R. Real-time large-scale dense mapping with surfels. *Sensors* **2018**, *18*, 1493. [CrossRef] [PubMed]
10. Henry, P.; Krainin, M.; Herbst, E.; Ren, X.; Fox, D. RGB-D mapping: Using kinect-style depth cameras for dense 3D modeling of indoor environments. *Int. J. Robot. Res.* **2012**, *31*, 647–663. [CrossRef]
11. Endres, F.; Hess, J.; Sturm, J.; Cremers, D.; Burgard, W. 3-D mappingwith an RGB-D camera. *IEEE Trans. Robot.* **2014**, *30*, 177–187. [CrossRef]
12. Kümmerle, R.; Grisetti, G.; Strasdat, H.; Konolige, K.; Burgard, W. G^2o: A general framework for graph optimization. In Proceedings of the IEEE International Conference on Robotics and Automation, Shanghai, China, 9–13 May 2011; pp. 3607–3613.
13. Yang, S.; Song, Y.; Kaess, M.; Scherer, S. Pop-up SLAM: Semantic monocular plane SLAM for low-texture environments. In Proceedings of the IEEE/RSJ International Conference on Intelligent Robots and Systems, Daejeon, Korea, 9–14 October 2016; pp. 1222–1229.
14. Wang, L.; Shen, C.; Duan, F.Q.; Lu, K. Energy-based automatic recognition of multiple spheres in three-dimensional point cloud. *Pattern Recognit. Lett.* **2016**, *83*, 287–293. [CrossRef]

15. Wu, Z.; Wang, L. Recovering the Manhattan Frame from a single RGB-D image by using orientation relevance. In Proceedings of the Chinese Control and Decision Conference, Chongqing, China, 28–30 May 2017; pp. 4574–4579.
16. Wang, L.; Shen, C.; Duan, F.Q.; Guo, P. Energy-based multi-plane detection from 3D point clouds. In *Neural Information Processing. ICONIP 2016. LNCS, vol 9948*; Hirose, A., Ozawa, S., Doya, K., Ikeda, K., Lee, M., Liu, D., Eds.; Springer: Cham, Switzerland, 2016; pp. 715–722.
17. Hsiao, M.; Westman, E.; Zhang, G.; Kaess, M. Keyframe-based dense planar SLAM. In Proceedings of the IEEE International Conference on Robotics and Automation, Singapore, 29 May–3 June 2017; pp. 5110–5117.
18. Le, P.H.; Košecka, J. Dense piecewise planar RGB-D SLAM for indoor environments. In Proceedings of the IEEE/RSJ International Conference on Intelligent Robots and Systems, Vancouver, BC, Canada, 24–28 September 2017; pp. 4944–4949.
19. Ghanem, B.; Thabet, A.; Niebles, J.C. Robust Manhattan frame estimation from a single RGB-D image. In Proceedings of the IEEE Conference on Computer Vision and Pattern Recognition, Boston, MA, USA, 7–12 June 2015.
20. Coughlan, J.M.; Yuille, A.L. Manhattan world: Compass direction from a single image by Bayesian inference. In Proceedings of the International Conference on Computer Vision, Kerkyra, Greece, 20–27 September 1999.
21. Lee, D.C.; Hebert M.; Kanade, T. Geometric reasoning for single image structure recovery. In Proceedings of the IEEE Conference on Computer Vision and Pattern Recognition, Miami, FL, USA, 20–25 June 2009; pp. 2136–2143.
22. Lee, D.C.; Gupta, A.; Hebert, M.; Kanade, T. Estimating spatial layout of rooms using volumetric reasoning about objects and surfaces. In Proceedings of the Conference on Neural Information Processing Systems, Vancouver, BC, Canada, 6–9 December 2010.
23. Sturm, J.; Engelhard, N.; Endres, F.; Burgard, W.; Cremers, D. A benchmark for the evaluation of RGB-D SLAM systems. In Proceedings of the IEEE/RSJ International Conference on Intelligent Robots and Systems, Vilamoura, Portugal, 7–12 October 2012; pp. 573–580.
24. Joo, K.; Oh, T.H.; Kweon, I.S.; Bazin, J.C. Globally optimal inlier set maximization for Atlanta frame estimation. In Proceedings of the IEEE Conference on Computer Vision and Pattern Recognition, Salt Lake City, UT, USA, 18–22 June 2018; pp. 5726–5734.

sensors

MDPI

Article

Pose Estimation for Straight Wing Aircraft Based on Consistent Line Clustering and Planes Intersection

Xichao Teng [1,*], Qifeng Yu [1], Jing Luo [2], Xiaohu Zhang [1,3] and Gang Wang [1]

1 College of Aerospace Science and Engineering, National University of Defense Technology, Changsha 410073, China; yuqifeng@vip.sina.com (Q.Y.); zhangxiaohu@nudt.edu.cn (X.Z.); wanggang_nudt@163.com (G.W.)
2 High-Tech Institute, Qing Zhou 262500, China; luoj11@tsinghua.org.cn
3 School of Aeronautics and Astronautics, Sun Yat-Sen University, Guangzhou 510000, China
* Correspondence: tengari@buaa.edu.cn; Tel.: +86-159-7317-5049

Received: 27 December 2018; Accepted: 11 January 2019; Published: 16 January 2019

Abstract: Aircraft pose estimation is a necessary technology in aerospace applications, and accurate pose parameters are the foundation for many aerospace tasks. In this paper, we propose a novel pose estimation method for straight wing aircraft without relying on 3D models or other datasets, and two widely separated cameras are used to acquire the pose information. Because of the large baseline and long-distance imaging, feature point matching is difficult and inaccurate in this configuration. In our method, line features are extracted to describe the structure of straight wing aircraft in images, and pose estimation is performed based on the common geometry constraints of straight wing aircraft. The spatial and length consistency of the line features is used to exclude irrelevant line segments belonging to the background or other parts of the aircraft, and density-based parallel line clustering is utilized to extract the aircraft's main structure. After identifying the orientation of the fuselage and wings in images, planes intersection is used to estimate the 3D localization and attitude of the aircraft. Experimental results show that our method estimates the aircraft pose accurately and robustly.

Keywords: pose estimation; straight wing aircraft; structure extraction; consistent line clustering; parallel line; planes intersection

---

1. Introduction

Since the 3D pose parameters of aircraft could provide a lot of valuable information about the aircraft's flight status, effective and accurate pose estimation is a key technique in many aerospace applications, such as autonomous navigation [1], auxiliary landing [2], collision avoidance [3], accident analysis, and testing of a flight control system [4,5]. In recent years, with the development of imaging technology and computer vision, vision-based pose estimation has become a research hotspot, and a lot of methods have been proposed in the literature to estimate the pose of an aircraft using visual sensors. Visual sensors could be successfully applied in aircraft pose estimation since vision-based methods have the advantages of strong anti-interference ability, low cost, and high precision [6].

Vision-based pose estimation methods can be divided into two categories—on-board vision and external vision—depending on the mounting position of the visual sensors. On-board monocular, depth, or stereo cameras can be used in on-board vision methods to estimate the relative pose between the aircraft and a particular target or marker, while external vision methods utilize external cameras to acquire the pose of an aircraft from its 2D projected images.

Among the on-board vision methods, a binocular stereovision model established by Chen et al. [7] used stereo vision and the RANSAC (RANdom SAmple Consensus) algorithm to measure the pose of a non-cooperative target. Li et al. [8] used parallel binocular cameras to estimate the pose of a non-cooperative target based on stereo matching and 3D restructuring. Zhang et al. [9,10] proposed optimization-based methods to estimate the relative pose using stereo cameras, and the geometric structure of the non-cooperative target was exploited to improve the accuracy. Deng et al. [11] implemented an on-board binocular vision-based system to estimate the pose of Unmanned Aerial Vehicles (UAVs) for autonomous aerial refueling. Zhuang et al. [12] used the line features of the airport and the monocular camera on board to provide pose information for UAV autonomous landing. Benini et al. [13] estimated the pose of a UAV by detecting a marker composed of known circles for autonomous takeoff and landing. For scenes without known landmarks, the structure from motion (SFM) [14–16] method or the simultaneous localization and mapping (SLAM) [17,18] method can be leveraged to estimate the relative pose for aircraft navigation. A sequence of images is processed in these techniques, and a Kalman filter [19,20] is often used to reduce the pose error.

For external vision methods, the aircraft's pose is often estimated using its 2D projected images captured by external imaging devices. Monocular cameras are widely used in external vison systems because the distance between aircraft and cameras is usually large. It is hard to estimate a 3D pose from a single 2D image without prior information such as 3D models of aircraft, synthetic aircraft image datasets, or acquired image sequences. Considering that pose estimation using complete aircraft models viewed from all aspects is storage- and time-consuming, feature extraction and pattern matching methods are proposed to reduce the dimension of pose estimation.

Hmam et al. [21] recognized aircraft based on a geometry-based reasoning system, and a generic model description of the aircraft was used for pose estimation. Wang et al. [22] combined a mathematical morphological algorithm and the Radon transform to extract the aircraft's structure and used the average value of ordinary aircraft as a reference to calculate 3D pose parameters from 2D images. The use of a generic model of aircraft makes these algorithms more efficient and flexible, but this also leads to a reduction in the accuracy and robustness of pose estimation.

Breuers and Reus [23] used a Fourier-descriptor-based algorithm to estimate aircraft pose information. The method computes a Fourier descriptor to characterize the aircraft contour, and the pose information is estimated by comparing this Fourier descriptor to a reference database. Fu et al. [24] estimated the relative pose parameters of aircraft based on a contour model. The method first acquires 2D projections of a 3D model from different views and establishes a database; then, contour matching is employed to derive relative pose parameters. Wang et al. [25] estimated the pose of commercial aircraft in a runway end safety area using geometry structure features. This image-based method obtains aircraft pose information using the central moments of extracted geometry structure features and identifies an aircraft's particular pose by a two-step feature matching strategy. Yuan et al. [26] proposed an aircraft pose recognition method based on locally linear embedding (LLE). In this method, LLE is applied for feature extraction and dimension reduction, and aircraft pose is recognized by propagation neural networks and nearest-neighbor algorithms. Although these methods reduce the complexity of the problem by feature extraction and pattern matching, 3D models of different aircraft are still needed, and a large amount of high-quality training data is necessary for pattern recognition to achieve accurate pose estimation, which reduces the flexibility and efficiency.

While there are a lot of features related to geometric structure to describe the pose information of aircraft, many of them were proposed for swept wing aircraft. Methods for straight wing aircraft structure extraction [27] and pose estimation are seldom addressed, despite the fact that the straight wing and its variants are the most common wing planform for low-speed aircraft [28]. With the rapid development of high-altitude long-endurance (HALE) UAVs, which often adopt a large-aspect-ratio straight wing design in order to increase lift [29,30], pose estimation of straight wing aircraft is of great importance.

In this article, we use a vision system located on the ground to estimate the pose of model-unknown straight wing aircraft. The vision system needs at least two monocular cameras to estimate the 3D pose. There are usually multiple cameras distributed in the flight test site or airport area, which allows our method's requirements to be easily met. Compared to methods which rely on the use of 3D models and/or classifiers, only two 2D images obtained at the same time and some prior assumptions are explored in our approach to achieve accurate pose estimation for straight wing aircraft. We first identify the orientation of the fuselage and wings in an image pair using consistent line clustering; then, the planes intersection method is used to calculate the 3D pose information of the aircraft.

In the application scenario of this article, the dual-station photoelectric theodolite at a flight test site was used to estimate the absolute pose of an aircraft. The photoelectric theodolite tracks the aircraft, captures a sequence of images, and records the camera pose for every image frame. The image pair captured by the dual-station photoelectric theodolite was used to estimate the pose of the aircraft. To improve the measurement range and accuracy, two photoelectric theodolites with large baseline were selected and distributed on both sides of aircraft trajectory. Because of the large baseline and long-distance measurement, it is very difficult to obtain corresponding invariant features, and self-occlusion at certain angles would make feature matching more unreliable. In order to identify the main structure (fuselage and wings) of the aircraft in image pairs efficiently and robustly, the general geometry features of straight wing aircraft were analyzed.

In our method, line features extracted by the line segment detector (LSD) algorithm are used to describe the structure of the aircraft. The spatial and length consistency of line features is exploited to eliminate the disturbance of the background and unrelated parts of the aircraft, and parallel line segments are grouped into orientation-consistent clusters which represent the structure of the straight wing aircraft. To extract the aircraft's structure accurately and robustly, a density-based clustering method is adopted according to the characteristics of the data. Mean shift and image moment methods are also used to improve the localization accuracy of the aircraft's center in images. After recognizing the main structure of the straight wing aircraft in 2D images, the planes intersection method is used to determine the 3D pose. Our algorithm provides a universal framework to estimate the 3D pose of straight wing aircraft without relying on 3D models, cooperative markers, or other datasets.

The remainder of the paper is organized as follows: Section 2 introduces the coordinate system definition. Our pose estimation algorithm is explained in detail in Section 3. In Section 4, the experimental results of structure extraction and pose estimation are presented to validate our algorithm. Finally, Section 5 concludes this article.

2. Coordinate System Definition

In this section, we define several coordinate systems related to pose estimation. There are three major coordinate systems which are shown in Figure 1.

The world coordinate system (see Figure 1a) helps us track the aircraft and determine its position and attitude. We used the East-North-Up (ENU) coordinate system as the world frame.

The camera coordinate system, shown in Figure 1b, is attached to a camera which tracks the aircraft in the image plane. The origin of the camera frame is located at the optical center of the camera; the x axis is parallel to the horizontal axis of the image plane in the right direction, and the z axis is the optical axis of the camera in the right-handed coordinate system. Two cameras are used in our algorithm and are calibrated with respect to the world coordinate system; their poses are known for each image frame they record.

For the body coordinate system of straight wing aircraft shown in Figure 1c, the origin is located at the center of the aircraft. The x axis points along the fuselage reference line; the y axis is perpendicular to the fuselage plane of symmetry, directed to the right; and the z axis is perpendicular to the plane where the fuselage and wings are located in the right-handed coordinate system. For a straight wing or its variants, the wing edge lines are approximately parallel to each other, while line segments along the fuselage are approximately parallel to the fuselage reference line. Based on these geometry structure features, our pose estimation algorithm extracts the orientation of the fuselage reference line and the wing axis from which we can determine the orientation of the body coordinate axes of straight wing aircraft.

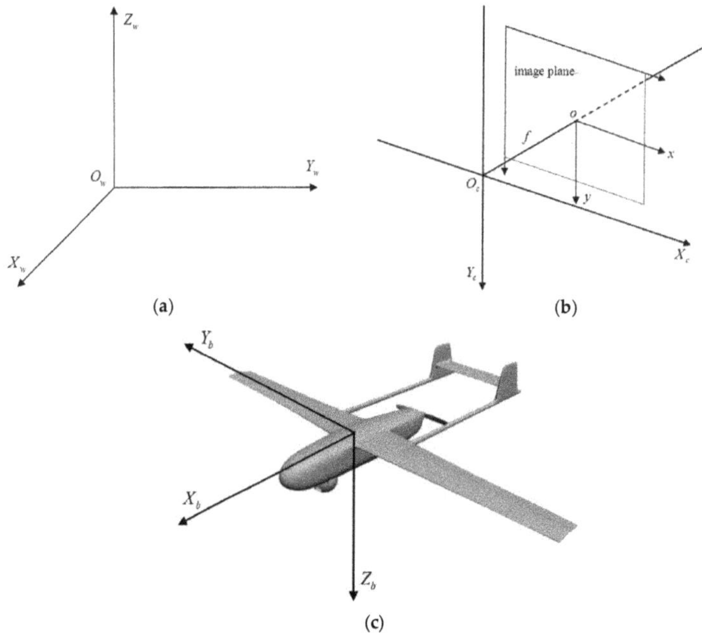

Figure 1. Coordinate systems: (**a**) World frame; (**b**) Camera frame; (**c**) Body frame.

3. Pose Estimation Algorithm

The pose of an aircraft is represented by the transformation using a rotation matrix and translation vector which transform points in the body coordinate frame into points in the world coordinate frame. Our algorithm acquires the 3D pose information of an aircraft by determining the orientation of the body coordinate axes and the position of the body coordinate frame origin with respect to the world coordinate frame.

Our pose estimation algorithm first extracts the orientation of the fuselage and the wings in 2D image pairs, then uses plane–plane intersection to determine the 3D pose of the straight wing aircraft. The 2D pose information acquired by the structure extraction method is used as input to the planes intersection method to acquire the 3D pose of the aircraft. In the process of pose estimation, the initial pose information of the aircraft is needed to avoid ambiguity. In the following sections, the structure extraction and planes intersection methods will be explained in detail.

3.1. Structure Extraction Method

We propose a novel structure extraction method to identify the orientation of the fuselage and wings of straight wing aircraft in a 2D image without needing 3D models or other datasets. Due to the long-range imaging of the aircraft, reliable feature point correspondence is difficult to obtain, especially with ambiguities, extreme poses, or self-occlusions. To obtain the 2D pose information accurately and robustly, we use line features to describe the structure of the aircraft; line features are usually more accurate and robust than feature points in our application scenarios and also adapt to self-occlusions to some extent.

The geometric relations between line features are exploited to recognize the main structure of the aircraft. The most important geometric constraints used in our algorithm are the parallel constraints:

1. Line features distributed along the wing axis are approximately parallel to each other;
2. Line features along the fuselage reference line are approximately parallel to each other.

In addition, line features are concentrated in the area of the aircraft, and the lengths of line features on the main structure (fuselage and wings) of the aircraft are often larger than those on other parts of the aircraft. The main idea of our structure extraction method is to cluster line features based on these geometry constraints to acquire an accurate and robust estimation of the orientation of the aircraft's main structure.

3.1.1. Line Feature Extraction

The state-of-the-art line segment detector (LSD) algorithm is utilized here to extract line features. The LSD algorithm, introduced by Gioi et al. [31], is a linear-time line segment detector giving results to subpixel accuracy, and a comparative study of line extraction methods by Zhang et al. [32] revealed that LSD is an optimal algorithm at different scales, blur degrees, and illumination. We detected 2D line segments using the LSD algorithm in an image to describe the structure of a straight wing aircraft. The result of the line feature extraction is shown in Figure 2a, where red line segments represent the detected line features. We denote the set of detected line segments as S_L.

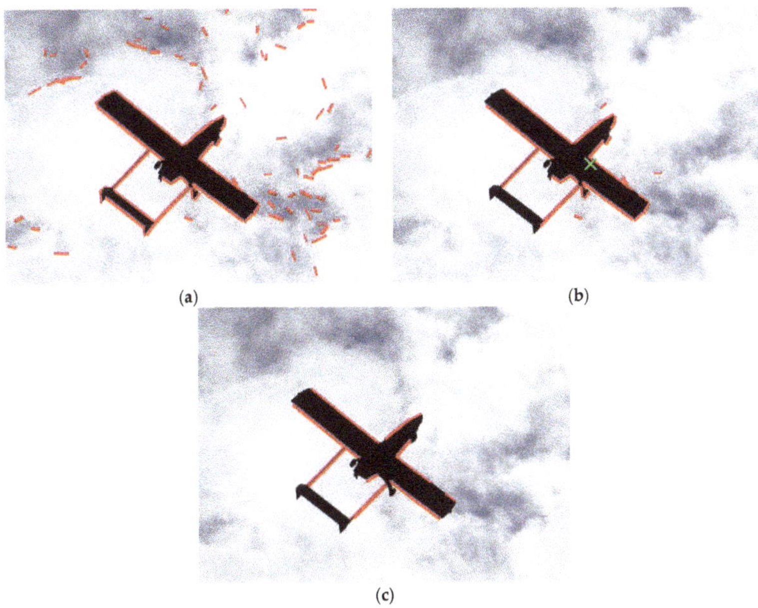

Figure 2. The results of line feature extraction and line clustering based on spatial and length consistency: (**a**) Line feature extraction; (**b**) Spatially consistent line clustering; (**c**) Length-consistent line clustering.

3.1.2. Spatially Consistent Line Clustering

Our algorithm is performed under the condition that the photographic image only contain one aircraft, which is common in actual application scenarios such as flight test, landing, or taking off. As the aircraft is a salient object in the image, detected line segments will be concentrated in the region of the aircraft and close to each other compared to irrelevant line segments, i.e., the density of line segments in the aircraft's region is very high. Based on the location constraint of line segments, we performed spatially consistent line clustering to identify the center of the aircraft and rule out irrelevant line segments caused by the background.

We used the mean shift [33] algorithm to identify the center of the aircraft. Mean shift is a procedure for locating the maxima of a density function given discrete data sampled from that function.

It is useful for detecting the modes of this density, which indicate the spatial consistency of line segments. The set of detected line segments S_L was used as the input of the mean shift algorithm, and a Gaussian kernel **K** on the distance was used to determine the weight for re-estimation of the center. An image pixel **x** on a line segment which belongs to S_L is represented by (x, y) where x and y are the horizontal and vertical coordinates of the pixel, respectively. The clustering center obtained by the mean shift algorithm is considered the aircraft's center. The kernel function **K** and the weighted mean $\mathbf{m}(\mathbf{x})$ of the density can be represented as follows:

$$\begin{aligned} &\mathbf{K}(\mathbf{x}_i - \mathbf{x}) = e^{-c\|\mathbf{x}_i - \mathbf{x}\|^2} \\ &\mathbf{m}(\mathbf{x}) = (\Sigma_{x_i \in n(x)} K(x_i - x) \cdot x_i) \cdot (\Sigma_{x_i \in n(x)} K(x_i - x))^{-1} \end{aligned} \tag{1}$$

where c is the weight of the kernel function and $n(\mathbf{x})$ represents the neighborhood of point **x**. After determining the center of the aircraft, line segments within a certain distance of the clustering center are considered to belong to the aircraft, and other line segments are removed. Figure 2b shows the result of spatially consistent line clustering. The green cross in Figure 2b represents the cluster centroid of the mean shift, and red line segments indicate the reserved line features which are close to the estimated aircraft's center. As we can see, many line segments which do not belong to the aircraft are rejected by spatially consistent line clustering.

The centroid of the aircraft can also be calculated via image moments, which is given by

$$\widetilde{x}_m = \frac{\sum_{i=1}^{N} x_i}{N} \widetilde{y}_m = \frac{\sum_{i=1}^{N} y_i}{N}. \tag{2}$$

Here, $(\widetilde{x}_m, \widetilde{y}_m)$ is the image coordinates of the aircraft's center obtained by the image moment method, and N represents the number of pixels on the line segments. Although the image moment method can identify the centroid of S_L without iteration, it is difficult to obtain the actual center of the aircraft robustly against a cluttered background. Figure 3 shows a comparison of the results of the image moment method and the mean shift algorithm, in which the estimated centroids of the aircraft are indicated by green crosses. The extracted 2D line features (red line segments in Figure 3) were used as the input of both methods. As we can see from Figure 3a, the result of the image moment method deviates from the actual center of the aircraft because of disturbance from the background, while the mean shift algorithm obtained a more accurate aircraft center and is partly resistant to a cluttered background. In the case of a cluttered background, the mean shift algorithm will be used to identify the center of the aircraft, and spatially consistent clustering provides an initial position estimation of the aircraft's center which is then updated in the parallel line clustering.

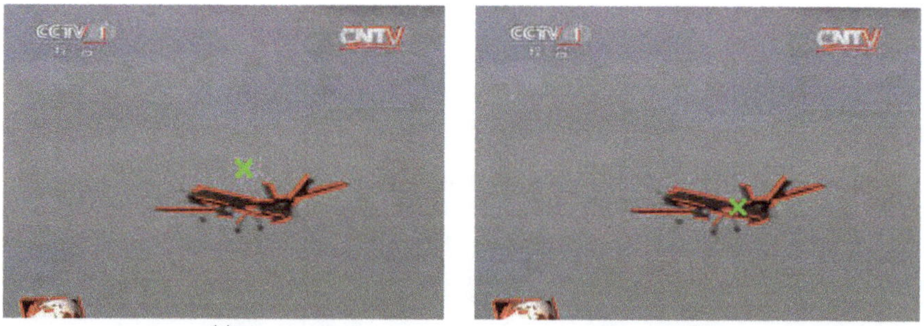

(a) (b)

Figure 3. A comparison of the results of the methods: (**a**) Image moment method; (**b**) Mean shift algorithm.

3.1.3. Length-Consistent Line Clustering

Although many noisy line segments are removed by spatially consistent line clustering, there are still some irrelevant line segments, as shown in Figure 2b. In this section, we use a length consistency criterion to further rule out irrelevant line segments. As the aircraft's main structure (fuselage and wings) is usually larger than other parts such as tail or nose, line segments shorter than a certain threshold can be removed from the set of line segments. As the length of a line segment decreases, the uncertainty of its direction increases, i.e., a small position error of the endpoint causes greater direction error for shorter line segments. Excluding shorter line segments would improve the accuracy of 2D pose estimation in the following parallel line clustering.

The result of length-consistent line clustering is shown in Figure 2c. Compared to Figure 2b, the irrelevant line segments are excluded further, which is of benefit for the following clustering.

3.1.4. Parallel Line Clustering

Parallel line clustering is the key step in the structure extraction algorithm and can acquire the directions of the fuselage and wings without relying on 3D models, other datasets, or cooperative markers. Line segments with similar directions are divided into one orientation-consistent line cluster. In the parallel line clustering process, the direction of a line segment is represented by the angle θ_i between the straight line that it belongs to and the horizontal axis of the image plane. The set of directions of line segments $\Theta = \{\theta_1, \theta_2, \ldots, \theta_N\}$ is used as input to the parallel line clustering. The directions of the fuselage and the wings are denoted θ_f and θ_w respectively.

Weak perspective projection (scaled orthographic projection) is employed in parallel line clustering. As the size of the aircraft is small compared with its distance to the optical center along the optical axis, weak perspective approximation is valid. If line segments on the aircraft are parallel to each other in 3D space (the world frame), then this geometry feature of the corresponding line segments projected into the image plane remains unchanged under weak perspective projection.

For straight wing aircraft, line segments distributed along the wings are roughly parallel to each other, and angle values of these line segments are tightly concentrated around θ_w (small standard deviation), while angle values of line segments along the fuselage are concentrated around θ_f. The directions of the fuselage and the wings can be extracted by clustering the high-density regions of Θ. According to the orientation feature of the line segments, a density-based clustering algorithm, density-based spatial clustering of application with noise (DBSCAN) [34], was used to group the parallel line segments into one orientation-consistent cluster containing the orientation information of the fuselage or the wings.

The data points used in DBSCAN clustering are the directions of line segments θ_i. There are two parameters required to be specified in the DBSCAN algorithm, both of which are used to measure the density of data points. The first parameter is a distance threshold ε within which two data points close to each other will be grouped into one cluster. The distance threshold is the absolute difference between angle values in our algorithm. The second parameter is the minimum number of data points $minPts$ needed to form an orientation-consistent cluster. Based on these two parameters, the data points are classified into three types, as shown in Figure 4:

- Core points: If a data point's ε neighborhood contains at least $minPts$ points, it is a core point (red points in Figure 4);
- Border points: If a data point's ε neighborhood contains fewer than $minPts$ points, but it is reachable from a certain core point (as indicated by one-way arrows in Figure 4), it is a border point (yellow points in Figure 4, the edge of a cluster);
- Noise points: If a data point is neither a core point nor a border point, it is a noise point (blue points in Figure 4).

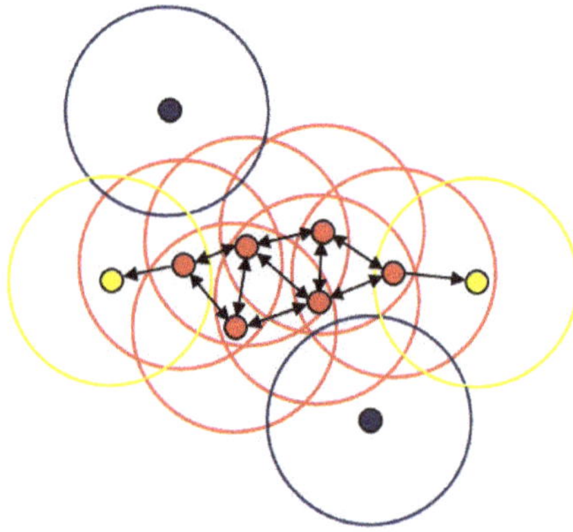

Figure 4. The explanation of the density-based spatial clustering of application with noise (DBSCAN) algorithm: the red points represent the core points, the yellow points represent the border points, and the blue points represent the noise points. The radius of the circle represents the distance threshold.

The steps of the DBSCAN algorithm used for parallel line clustering are briefly described as follows:

1. For every data point θ_i, search points in its ε neighborhood and use $minPts$ to determine the core points in the set Θ.
2. Ignore all non-core points and group core points into parallel line clusters based on the connected components on the neighborhood graph (as indicated by two-way arrows in Figure 4).
3. For every non-core point, if it is in the ε neighborhood of a cluster, it is the border point of the cluster; otherwise, it is a noise point.

In contrast to traditional clustering methods such as *k*-means++ [35], the DBSCAN algorithm does not need to specify the number of clusters in advance and is robust to outliers. It forms clusters based solely on the spatial density of the data. The two orientation-consistent clusters with minimum interclass variance represent the structure of the fuselage and the wings which contain the orientation information of the aircraft in the image. The directions of the fuselage and the wings are obtained by extracting the centers of the parallel line clusters.

In our method, the directions of the fuselage and the wings are distinguished based on an initial pose constraint. The approximate orientation of the aircraft needs to be specified in the initial frame of the image sequence to avoid ambiguity and to help identify the actual pose of the aircraft. With this condition, the directions of the fuselage and the wings can be distinguished in the initial frame, and the orientation information of the current frame will be used in the next frame. In the application scenarios of our algorithm, such as take-off, landing, or flight testing, this condition is easily met. In practice, the pitch angle (or the yaw angle) and the roll angle of the aircraft are provided, or the approximate positions of the nose and one wing tip are marked in one image of the initial image pair.

After obtaining the orientation-consistent clusters, irrelevant line segments are removed from the set S_L, and the position of the aircraft's center is then re-estimated from the set S_L based on the image moment method. Since only line segments on the main structure of the aircraft are left, it is possible to identify the center of the aircraft with higher precision.

The results of parallel line clustering are shown in Figure 5. As shown in Figure 5a, the red straight lines indicate the directions of the fuselage and the wings, and the green cross indicates the estimated centroid of the aircraft. The directions of the fuselage and the wings were correctly extracted by parallel line clustering. However, in Figure 5b, there is only one cluster with enough parallel line segments for this extreme pose. In this case, the direction of only the fuselage or the wings can be acquired by parallel line clustering, and the unknown direction needed for pose estimation is replaced by the corresponding orientation information of the previous frame.

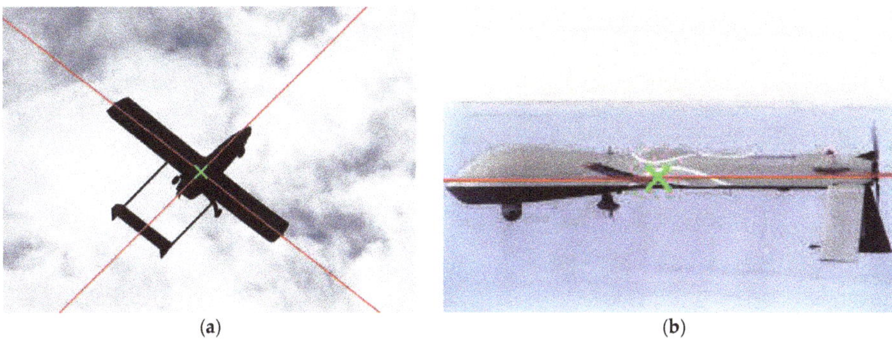

Figure 5. The results of parallel line clustering. (**a**) The directions of the fuselage and the wings are extracted correctly; (**b**) The direction of the wings cannot be extracted for this extreme pose.

3.2. Planes Intersection Method

After the structure extraction method determines the pixel coordinates of the aircraft's center and the directions of the fuselage and the wings in an image pair, the planes intersection method is used to estimate the 3D pose of the aircraft.

Two cameras were used in the intersection measurement and are indicated by their projection matrices $\mathbf{P}_1$ and $\mathbf{P}_2$. The camera projection matrices are of the form

$$\mathbf{P}_1 = \mathbf{K}_1[\mathbf{R}_1|\mathbf{t}_1] \qquad \mathbf{P}_2 = \mathbf{K}_2[\mathbf{R}_2|\mathbf{t}_2] \tag{3}$$

where $\mathbf{K}_i$ $(i = 1,2)$ is the camera intrinsic matrix of the camera, and $\mathbf{R}_i$ and $\mathbf{t}_i$ represent the rotation and translation, respectively, of the corresponding camera with respect to the world frame. We assume that the cameras are calibrated with respect to the world frame and that the $\mathbf{P}_i$ are known.

The camera model is represented as

$$z\mathbf{x} = \mathbf{P}_i\mathbf{X} \tag{4}$$

where $\mathbf{X} = (X, Y, Z, 1)^{\mathrm{T}}$ is the world coordinates and $\mathbf{x} = (u, v, 1)^{\mathrm{T}}$ is the image coordinates of $\mathbf{X}$. As weak perspective projection is employed, z is a positive constant.

The image pair captured by the two cameras at the same time is denoted $\langle I_1, I_2\rangle$. The center of the aircraft obtained by the structure extraction method in the image I_i $(i = 1,2)$ is represented as $AC_i = (x_i, y_i)$ where x_i and y_i are the horizontal and vertical coordinates of the image, and the directions of the fuselage and the wings are represented as θ_i^f and θ_i^w, respectively.

Figure 6 explains the geometric constraint of the planes intersection method. As shown in Figure 6, the two cameras are indicated by their optical centers $\mathbf{C}_1$ and $\mathbf{C}_2$ and by image planes. The 3D line in the world coordinate system is represented as $\mathbf{L}$, which is the line of intersection of the two planes $\boldsymbol{\pi}_1$ and $\boldsymbol{\pi}_2$; l_i $(i = 1,2)$ is the projected line of $\mathbf{L}$ in the image plane; and the plane $\boldsymbol{\pi}_i$ is determined by the line $\mathbf{L}$ and the optical center $\mathbf{C}_i$. Let the normalized vector $\mathbf{V}$ of $\mathbf{L}$ represent the direction of the fuselage or the wings; the planes intersection method estimates the 3D attitude of the aircraft by obtaining the solution of $\mathbf{V}$.

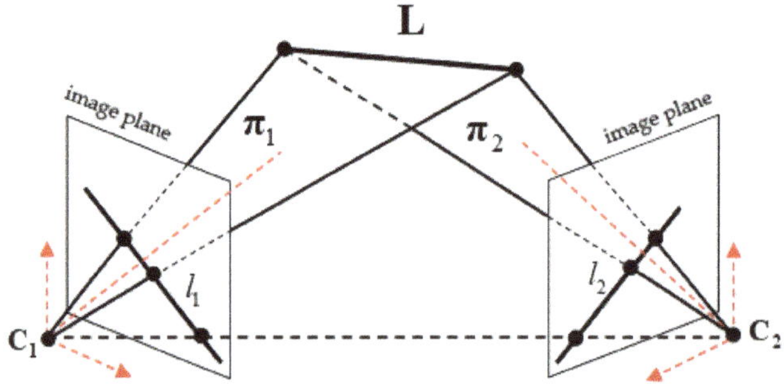

Figure 6. Geometric constraint of plane–plane intersection.

The projected line l_i in the image plane is identified by AC_i and θ_i^f (or θ_i^w); an analytical expression of l_i is

$$a_i u + b_i v + c_i = 0. \tag{5}$$

Equation (5) can be represented in vector form as the following:

$$\begin{bmatrix} a_i & b_i & c_i \end{bmatrix} \mathbf{x} = 0. \tag{6}$$

By substituting Equation (6) into Equation (4) for each camera, we obtain

$$z \begin{bmatrix} a_i & b_i & c_i \end{bmatrix} \mathbf{x} = \begin{bmatrix} a_i & b_i & c_i \end{bmatrix} \mathbf{P}_i \mathbf{X} = 0, \tag{7}$$

and the plane π_i can be expressed as

$$\begin{gathered} [\, A_i \quad B_i \quad C_i \quad D_i \,]\mathbf{X} = 0 \\ \vec{\mathbf{n}}_i = \begin{bmatrix} A_i & B_i & C_i \end{bmatrix} \end{gathered} \tag{8}$$

where $\vec{\mathbf{n}}_i$ is the normalized vector of the plane π_i. Note that Equations (7) and (8) have the same form, and $\begin{bmatrix} a_i & b_i & c_i \end{bmatrix} \mathbf{P}_i$ is already known; the normalized vector $\vec{\mathbf{n}}_i$ is derived from Equations (7) and (8). After we obtain the normalized vectors $\vec{\mathbf{n}}_1$ and $\vec{\mathbf{n}}_2$, the normalized vector $\mathbf{V}$ which contains the orientation information of the aircraft is solved as follows:

$$\mathbf{V} = \vec{\mathbf{n}}_1 \times \vec{\mathbf{n}}_2. \tag{9}$$

As the 3D line $\mathbf{L}$ can be parametrized in the world coordinate frame by the two planes π_1 and π_2 as a 2×4 matrix, let $\mathbf{L}_f$ be the 3D line parallel to the fuselage reference line and $\mathbf{L}_w$ be the 3D line parallel to the wing edge lines. The point of intersection of $\mathbf{L}_f$ and $\mathbf{L}_w$ is the center of the aircraft. By calculating the respective normalized vectors of $\mathbf{L}_f$ and $\mathbf{L}_w$ using Equation (9), we can obtain the 3D attitude of the straight wing aircraft. The rotation matrix is calculated by singular value decomposition, and the initial pose constraint is used to avoid reflective ambiguity. In order to obtain the world coordinates of the point of intersection of $\mathbf{L}_f$ and $\mathbf{L}_w$, which determine the 3D position of the aircraft, overdetermined equations are established as follows.

$$\begin{gathered} \mathbf{AX} = 0 \\ \mathbf{A} = \begin{bmatrix} \mathbf{L}_f \\ \mathbf{L}_w \end{bmatrix} \end{gathered} \tag{10}$$

Here, $\mathbf{A}$ is a 4×4 matrix and $\mathbf{X}$ represents the point of intersection of the two lines (the translation vector). The overdetermined equations $\mathbf{AX} = 0$ can be solved by singular value decomposition, and the solution is the singular vector corresponding to the smallest singular value of $\mathbf{A}$. Before solving the overdetermined equations, an optimal estimator for the center point based on the epipolar constraint can be used to reduce the geometric error [36].

The 3D attitude of the aircraft is determined by the normalized vectors of $\mathbf{L}_f$ and $\mathbf{L}_w$, and the 3D position of the aircraft is determined by the point of intersection $\mathbf{X}$ of the two lines. As we assume that $\mathbf{L}_f$ and $\mathbf{L}_w$ are coplanar in our pose estimation algorithm, the ambiguity will occur during the process of pose estimation, and the initial pose constraint will be used to determine the unique solution. Based on the results of the structure extraction method, the planes intersection method can acquire the 3D pose of the straight wing aircraft. Moreover, our pose estimation algorithm can easily be extended to multiple camera views.

3.3. Algorithm Summary

In this section, we summarize the whole pose estimation algorithm as is shown in Algorithm 1.

Algorithm 1: Pose estimation based on consistent line clustering and planes intersection	
Input:	The image pair $\langle I_1, I_2 \rangle$, the two camera matrices $\mathbf{P}_1$, $\mathbf{P}_2$, and the initial pose constraint.
Output:	The 3D position and 3D attitude of the straight wing aircraft.
Step 1	Extract line features in image pairs using the LSD algorithm;
Step 2	Locate the center of the aircraft in the 2D images and cluster spatially consistent line segments;
Step 3	Rule out line segments shorter than a certain threshold;
Step 4	Classify line segments into orientation-consistent clusters, extract the directions of the fuselage and the wings in the image pair, and re-estimate the center of the aircraft;
Step 5	Calculate the 3D attitude and 3D location using the plane–plane intersection method.

The flowchart of the algorithm is shown in Figure 7.

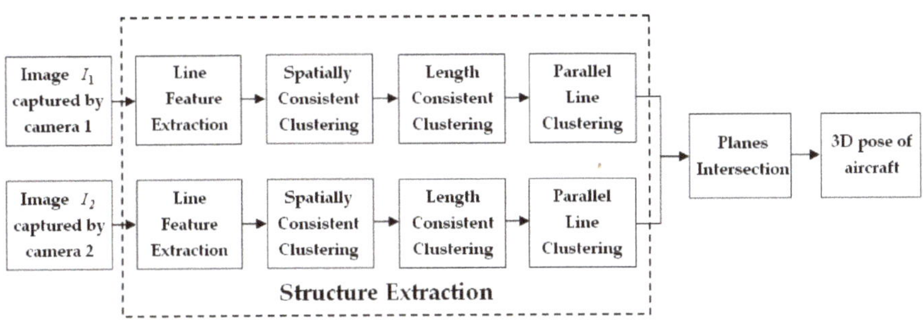

Figure 7. The flowchart of our pose estimation algorithm.

4. Experiments and Results

Experiments were performed to validate the effectiveness and accuracy of the proposed structure extraction and pose estimation methods. Real images of different straight wing aircraft downloaded from the Internet were used to demonstrate the effectiveness and universality of the structure extraction method, and simulated images of straight wing aircraft were exploited to evaluate the accuracy of our pose estimation algorithm. Our method was implemented using MATLAB on a laptop equipped with an Intel Core i7 CPU with a 2.80 GHz processor and 8.00 GB of RAM.

4.1. Experimental Results of Structure Extraction

The qualitative evaluation of our structure extraction method was performed using real images downloaded from the Internet. A total of 60 images of different sizes were downloaded and used in the experiment. Each image contains one aircraft whose planform is the straight wing or its variant, and the structure extraction method was used to identify the orientation of the aircraft's main structure in a single 2D image. Among these images, some are challenging for structure extraction since they contain a cluttered background, other objects, random noise, or perspective effects.

In the experiment, the directions of the fuselage and the wings in 51 of the 60 images were correctly identified. Figure 8 shows some of the results of structure extraction. As we can see, our structure extraction algorithm can be applied flexibly to different types of straight wing aircraft without needing 3D models of aircraft or other datasets, and it can also deal with different aircraft poses effectively and robustly extract the main structure under self-occlusion or a cluttered background. Moreover, the parallel assumption does not need to hold strictly. Even if perspective effects exist or the line segments are not strictly parallel to each other, our algorithm can still recognize the main structure of the aircraft.

Figure 8. Results of the structure extraction method.

While the algorithm achieved good results in most downloaded images, Figure 9 shows some cases in which our structure extraction method obtained incorrect results. There are two main reasons for these incorrect results:

1. The structure of the aircraft (fuselage or wings) does not satisfy the assumption of parallel line clustering, i.e., the line segments distributed along this structure are not parallel to each other in the image (see row 1, Figure 9).
2. Some parts of the aircraft (tail or external mounts) or the background affect the consistent line clustering (see row 2, Figure 9).

Changing weather or light conditions may also affect the success rate of our algorithm. When the weather condition or brightness/darkness level changes, the edges of the aircraft's main structure may be blurred during image acquisition, and unreliable line features will be detected. Changing weather or light conditions may affect the accuracy and robustness of the line feature detection, which in turn disturbs the consistent line clustering results and reduces the accuracy and success rate of our algorithm. In our method, the LSD algorithm used to detect line features can adapt to optical blur and illumination changes to some extent.

Figure 9. Some incorrect results from our structure extraction algorithm.

The situations shown in Figure 9 are uncommon in our application scenarios, and despite the fact that our algorithm is mainly for estimating the pose of a straight wing aircraft at long distance, the experimental results show that the proposed algorithm is able to recognize the aircraft's main structure robustly even at close range.

4.2. Experimental Results of Pose Estimation

Simulated image pairs were used to test our pose estimation algorithm. Two models were used in our experiment to simulate straight wing aircraft, as shown in Figure 10. These two models were created using Autodesk 3ds Max [37], which is a professional 3D computer graphics program for making 3D animations, models, and images. Model 1 (see Figure 10a) represents a general commercial UAV with standard straight wings while Model 2 (see Figure 10b) is a full-size simulation of the MQ-9 unmanned aircraft which has straight tapered wings (a variant of the standard straight wing). The size of Model 1 is 3.4 m × 5.0 m × 0.7 m (length, width, height), and the size of Model 2 is 10.4 m × 24.8 m × 3.1 m (length, width, height).

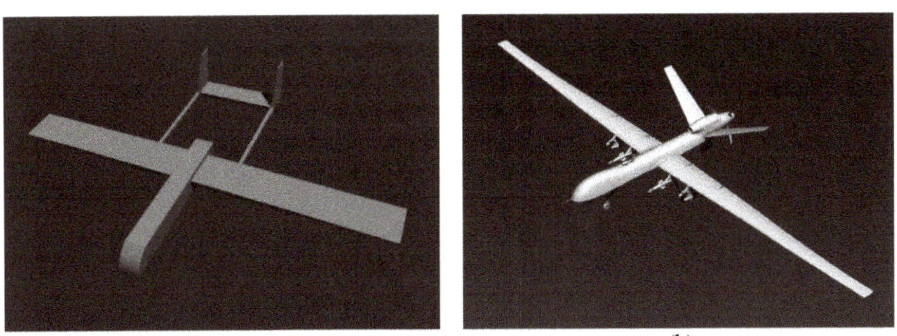

(a) (b)

Figure 10. Two aircraft models: (**a**) Model 1; (**b**) Model 2.

Two cameras in 3ds Max were used to simulate the dual-station photoelectric theodolite at the flight test site. The internal parameters and spatial layouts of the cameras for aircraft pose estimation are shown in Table 1. As we can see from Table 1, flight simulation scenarios were established for Model 1 (Scene 1, see Table 1) and Model 2 (Scene 1, see Table 1).

Table 1. The internal parameters and spatial layouts of the cameras in Scene 1 and Scene 2.

	Camera	Focal Length	Field of View	Image Resolution	Location (x,y,z)
Scene 1	1	70 mm	28.842° × 21.832°	1280 × 960	$(-15\text{ m}, -25\text{ m}, 0)$
	2	75 mm	26.991° × 20.408°	1280 × 960	$(-20\text{ m}, 30\text{ m}, 0)$
Scene 2	1	300 mm	6.867° × 5.153°	1280 × 960	$(350\text{ m}, 550\text{ m}, 30\text{ m})$
	2	275 mm	7.49° × 5.621°	1280 × 960	$(170\text{ m}, -390\text{ m}, 0)$

In our simulation experiments, cameras with different internal parameters and spatial layouts were used to test the performance of our algorithm, and the two cameras in the scene were located on both sides of the aircraft trajectory. The location coordinates of the cameras were in the East-North-Up (ENU) coordinate system. In Scene 1, the baseline between the two cameras was 55.23 m, while the two cameras in Scene 2 had a baseline of 957.55 m. The image pairs were generated by the two cameras in the scenes, and it is very difficult to obtain reliable feature correspondences in these wide-baseline images.

In order to test the performance of our pose estimation algorithm on different poses in simulation image pairs, we rotated Model 1 around the x, y, and z axes to simulate changes in the roll angle γ, pitch angle ψ, and yaw angle φ, respectively. Table 2 shows the selected rotation angles $(\theta_x, \theta_y, \theta_z)$ of Model 1, where θ_x represents the roll angle, θ_y represents the pitch angle, and θ_z represents the yaw angle. As detailed in Table 2, 13 image pairs were generated for Model 1, and the selected angle range was reasonable considering actual flight situations. The translation vector of Model 1 in Scene 1 was $\mathbf{T}_{true} = (0, 0, 20\text{ m})$.

Table 2. The selected rotation angles of Model 1 in Scene 1.

	1	2	3	4	5	6	7	8	9	10	11	12	13
θ_x	0°	0°	0°	0°	0°	0°	0°	0°	0°	0°	0°	−15°	15°
θ_y	0°	0°	0°	0°	0°	0°	0°	−30°	30°	−15°	15°	0°	0°
θ_z	0°	−30°	30°	−60°	60°	−90°	90°	0°	0°	0°	0°	0°	0°

For Model 2 in Scene 2, an aircraft trajectory was designed to simulate the flight. During the flight simulation, the pitch angle of Model 2 varied from −15° to 15°, the roll angle varied from −10° to 10°, and the translation vector was $\mathbf{T}_{true} = (x, 0, 200\text{ m})$, where x ranged from 0 to 600 m. The simulated flight path was rendered into 13 image pairs in steps of 50 m in Scene 2, and the rotation angles of each step are shown in Table 3.

Table 3. The rotation angles of Model 2 in Scene 2.

	1	2	3	4	5	6	7	8	9	10	11	12	13
θ_x	0°	0°	0°	−10°	−5°	−5°	10°	0°	5°	5°	0°	0°	0°
θ_y	0°	0°	0°	0°	15°	10°	−5°	5°	−5°	0°	−15°	−10°	0°
θ_z	0°	0°	0°	0°	0°	0°	0°	0°	0°	0°	0°	0°	0°

We used the 3ds Max rendering engine to generate the simulated image pairs of these two models; these are shown in Figure 11a,b. In Figure 11a, the top row and the bottom row represent the simulated images of Model 1 captured by Camera 1 and Camera 2, respectively, in Scene 1, and every column represents an image pair captured at the same time. In Figure 11b, the top row and the bottom row represent the simulated images of Model 2 captured by Camera 1 and Camera 2, respectively, in Scene 2, and every column represents an image pair captured at the same time. The rotation angles $(\theta_x, \theta_y, \theta_z)$ of the aircraft in each shot are also displayed in Figure 11.

In order to make the simulation scenes more realistic, a sky background with clouds and different types of natural light was also simulated (see Figure 11). As shown in Figure 11, the wide-baseline image pairs contain aircraft with different scales, poses, and self-occlusion, and optical blur exists due to long-range imaging. Under these challenging circumstances, a robust algorithm is needed to obtain accurate pose information from a single image pair.

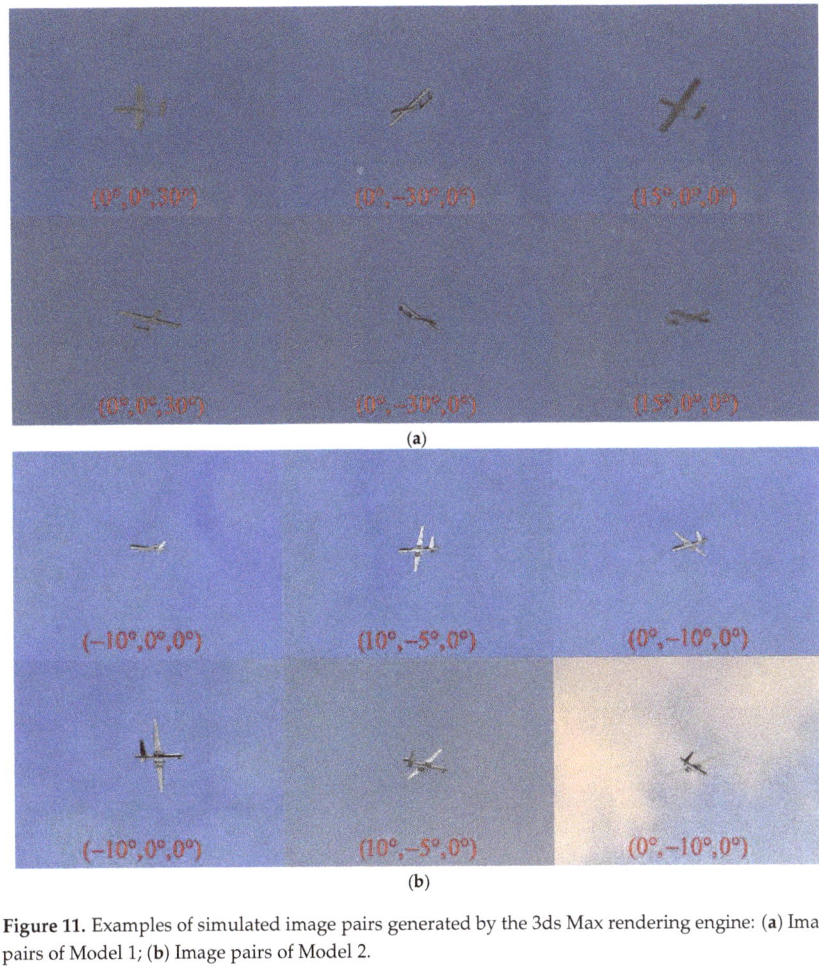

Figure 11. Examples of simulated image pairs generated by the 3ds Max rendering engine: (**a**) Image pairs of Model 1; (**b**) Image pairs of Model 2.

Figure 12 presents the results of our structure extraction method on the simulated image pairs shown in Figure 11. The directions of the fuselage and the wings are indicated by the red lines in Figure 12. As we can see, the main structure of the aircraft was correctly extracted by our structure extraction method, and the results further validate the performance of our method. The 3D pose of the aircraft can be obtained effectively only when the 2D pose information in the image pair is extracted robustly and accurately.

We compared our pose estimation algorithm with Li's method [8] and pose estimation errors were used to evaluate the algorithms. In Li's method, the 3D pose of a non-cooperative target is estimated by a stereo camera based on a triangulation method, and the feature points obtained by the line feature extraction are used for stereo matching and 3D reconstruction. The triangulation method is typically applied to estimate 3D position in computer vision, and the pose estimation pipeline of Li's method is also widely used, so it was selected for comparison to validate our proposed method.

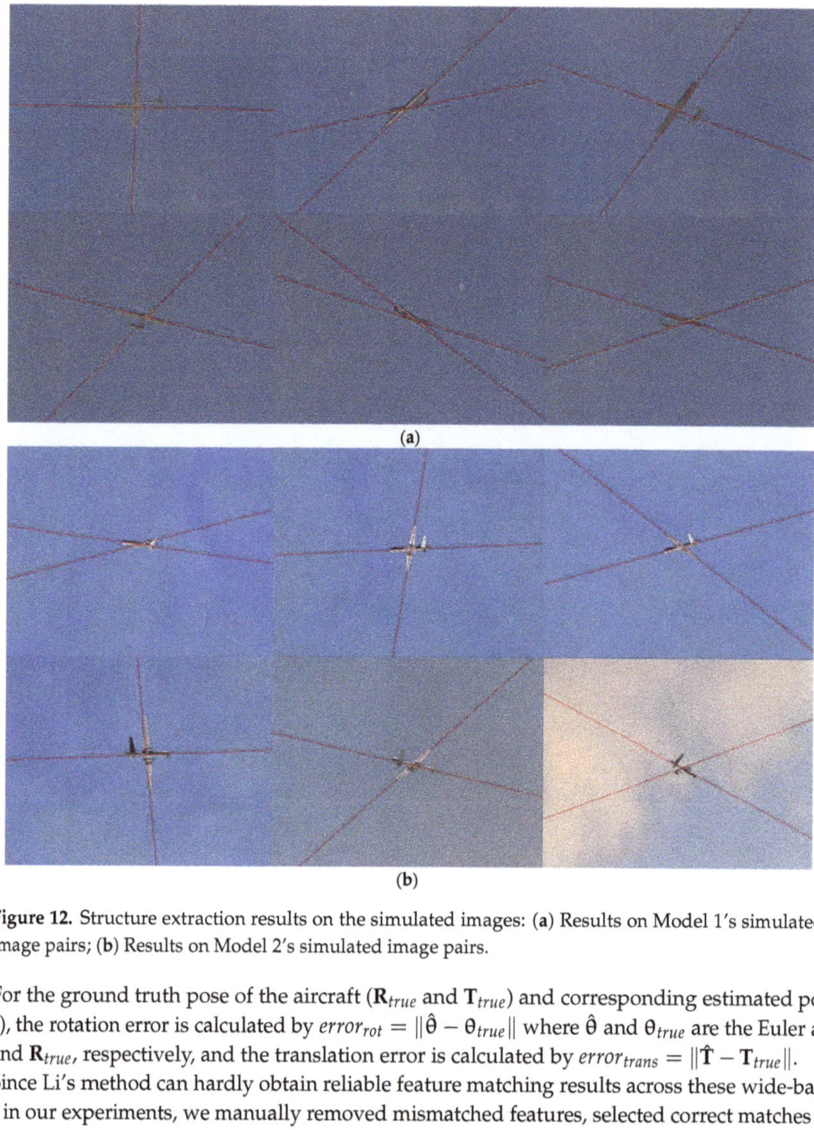

Figure 12. Structure extraction results on the simulated images: (**a**) Results on Model 1's simulated image pairs; (**b**) Results on Model 2's simulated image pairs.

For the ground truth pose of the aircraft ($\mathbf{R}_{true}$ and $\mathbf{T}_{true}$) and corresponding estimated pose ($\hat{\mathbf{R}}$ and $\hat{\mathbf{T}}$), the rotation error is calculated by $error_{rot} = \|\hat{\theta} - \theta_{true}\|$ where $\hat{\theta}$ and θ_{true} are the Euler angles of $\hat{\mathbf{R}}$ and $\mathbf{R}_{true}$, respectively, and the translation error is calculated by $error_{trans} = \|\hat{\mathbf{T}} - \mathbf{T}_{true}\|$.

Since Li's method can hardly obtain reliable feature matching results across these wide-baseline views in our experiments, we manually removed mismatched features, selected correct matches in the image pairs, and confirmed that there were enough corresponding feature points for pose estimation. The 3D models are also used in Li's method to obtain the absolute pose of the aircraft, while our algorithm is model free and acquires the 3D pose information automatically.

Figure 13 shows the pose estimation errors of our algorithm and Li's method for the simulated images of Model 1. In Figure 13a, the rotation errors are presented, and the translation errors are shown in Figure 13b. As we can see from Figure 13, the translation accuracy of our method is similar to that of Li's method, and our pose estimation method performs consistently better than the compared method in the estimation of the rotation angle. The rotation angle errors of our method are within 1°, the average rotation error is 0.47°, and the average translation error is 177.91 mm.

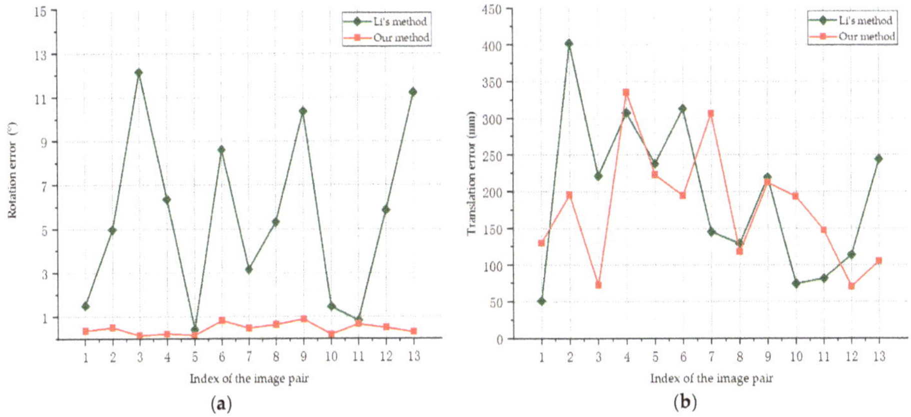

Figure 13. Pose estimation errors for the simulated images of Model 1: (**a**) Rotation errors; (**b**) Translation errors.

Figure 14 shows the pose estimation errors of our algorithm and the compared method for the simulated images of Model 2. In Figure 14a, the rotation errors are presented, and the translation errors are shown in Figure 14b. Model 2 is more complex than Model 1 and there is a greater imaging distance in Scene 2, while our algorithm still achieves accurate and stable pose estimation results compared to the results for Model 1. In Figure 14, the proposed method outperforms the compared method in the estimation of the rotation angle and translation vector, which is due to the accuracy and robustness of our structure extraction and planes intersection methods. The large fluctuations in the result curves indicate that the triangulation process used in Li's method is sensitive to various errors. The triangulation method uses the intersection of two lines to estimate the 3D position; with the measure distance increasing, the uncertainty increases, making the results more sensitive to noise. The average rotation error of our method is 1.21°, and the average translation error is 336.49 mm. The experimental results indicate that our method can extract the structure and estimate the pose accurately.

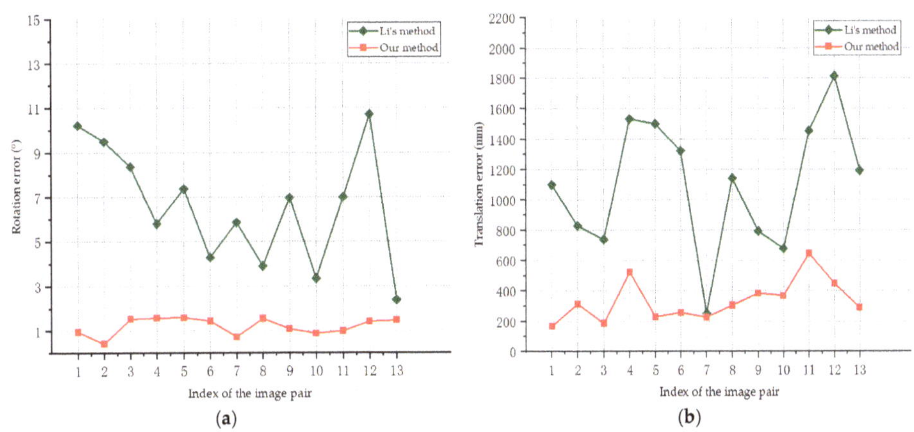

Figure 14. Pose estimation errors for the simulated images of Model 2: (**a**) Rotation errors; (**b**) Translation errors.

In addition, our method is also efficient. We ran our method 1000 times and recorded the execution time. The average execution time was 30.74 ms (including the structure extraction and planes intersection methods), which means that our algorithm can estimate the 3D pose efficiently.

The simulation experiment results show that our algorithm estimates the pose of the straight wing aircraft more accurately and robustly than does the compared method. Meanwhile, our method is efficient and flexible and can be applied to different types of straight wing aircraft.

5. Conclusions

An accurate and robust pose estimation method for straight wing aircraft was proposed in this paper. The geometry structure features of straight wing aircraft were utilized for structure extraction and the pose information was acquired by the planes intersection method. Our method establishes a universal framework for pose estimation of straight wing aircraft without relying on 3D models or other datasets, unlike other existing methods, and can be extended to other targets with similar geometric constraints. For an aircraft without similar geometric constraints to straight wing aircraft, our proposed method is unable to extract its main structure robustly and accurately. In the case of a swept wing aircraft, only the fuselage contains enough parallel lines can be detected effectively, while the wings cannot be extracted accurately. Extending our algorithm to aircraft with different wing planforms will be the focus of our future research.

Our method can also provide initial pose information for algorithms with higher precision efficiently. For an image sequence captured during flight, our future work will also focus on using an extended Kalman filter or particle filter to improve the accuracy of our algorithm.

Author Contributions: Conceptualization, X.T.; Methodology, X.T., Q.Y. and J.L.; Software, X.T.; Validation, X.Z.; Investigation, G.W.; Resources, X.Z.; Writing—Original Draft Preparation, X.T.; Writing—Review and Editing, X.T., Q.Y. and J.L.; Visualization, X.T.

Funding: This research was funded by Scientific Research Program of National University of Defense Technology (number: ZK16-03-27), Development and application of vision based micron-sized high speed quality tester (Grant number: 2013YQ140517), and National Natural Science Foundation of China (Grant number: 11727804).

Conflicts of Interest: The authors declare no conflict of interest. The funders had no role in the design of the study; in the collection, analyses, or interpretation of data; in the writing of the manuscript; or in the decision to publish the results.

References

1. Kok, M.; Hol, J.D.; Schön, T.B. Using inertial sensors for position and orientation estimation. *Found. Trends Signal Process.* **2018**, *11*, 1–153. [CrossRef]
2. Santos, N.P.; Melício, F.; Lobo, V.; Bernardino, A. A ground-based vision system for UAV pose estimation. *Int. J. Robot. Mechatron.* **2015**, *1*, 138–144. [CrossRef]
3. Langelaan, J.W. State Estimation for Autonomous Flight in Cluttered Environments. *J. Guid. Control Dyn.* **2007**, *30*, 1414–1426. [CrossRef]
4. Ward, D.G.; Monaco, J.F.; Bodson, M. Development and flight testing of a parameter identification algorithm for reconfigurable control. *J. Guid. Control Dyn.* **1998**, *21*, 948–956. [CrossRef]
5. Proud, A.W. Close formation flight control. *J. Guid. Control Dyn.* **1999**, *24*, 246–254. [CrossRef]
6. Yang, Z.; Li, C. Review on vision-based pose estimation of UAV based on landmark. In Proceedings of the IEEE International Conference on Frontiers of Sensors Technologies (ICFST), Shenzhen, China, 14–16 April 2017.
7. Chen, L.; Guo, B.; Sun, W. Relative pose measurement algorithm of non-cooperative target based on stereo vision and RANSAC. *Int. J. Soft Comput. Softw. Eng.* **2012**, *2*, 26–35. [CrossRef]
8. Li, R.; Zhou, Y.; Chen, F.; Chen, Y. Parallel vision-based pose estimation for non-cooperative spacecraft. *Adv. Mech. Eng.* **2015**, *7*. [CrossRef]
9. Zhang, L.; Zhu, F.; Hao, Y.; Pan, W. Optimization-based non-cooperative spacecraft pose estimation using stereo cameras during proximity operations. *Appl. Opt.* **2017**, *56*, 4522–4531. [CrossRef] [PubMed]

10. Zhang, L.; Zhu, F.; Hao, Y.; Pan, W. Rectangular-structure-based pose estimation method for non-cooperative rendezvous. *Appl. Opt.* **2018**, *57*, 6164–6173. [CrossRef] [PubMed]
11. Deng, Y.; Xian, N.; Duan, H. A Binocular Vision-based Measuring System for UAVs Autonomous Aerial Refueling. In Proceedings of the IEEE International Conference on Control and Automation (ICCA), Kathmandu, Nepal, 1–3 June 2016.
12. Zhuang, L.; Han, Y.; Fan, Y.; Cao, Y.; Wang, B.; Zhang, Q. Method of pose estimation for UAV landing. *Chin. Opt. Lett.* **2012**, *10*, S20401. [CrossRef]
13. Benini, A.; Rutherford, M.J.; Valavanis, K.P. Real-time GPU-based Pose Estimation of a UAV for Autonomous Takeoff and Landing. In Proceedings of the IEEE International Conference on Robotics and Automation (ICRA), Stockholm, Sweden, 16–21 May 2016.
14. Tai, J.M.; Fieguth, P.W. Incremental Shape Reconstruction using Stereo Image Sequences. In Proceedings of the International Conference on Image Processing, Vancouver, BC, Canada, 10–13 September 2000.
15. Alix, D.; Walli, K.; Raquet, J. Error characterization of flight trajectories reconstructed using Structure from Motion. In Proceedings of the IEEE Applied Imagery Pattern Recognition Workshop, Washington, DC, USA, 14–16 October 2014.
16. Huang, Y.P.; Sithole, L.; Lee, T.T. Structure from motion technique for scene detection using autonomous drone navigation. *IEEE Trans. Syst. Man Cybern. Syst.* **2017**, 1–12. [CrossRef]
17. Schneider, J.; Eling, C.; Klingbeil, L.; Kuhlmann, H.; Förstner, W.; Stachniss, C. Fast and effective online pose estimation and mapping for UAVs. In Proceedings of the IEEE International Conference on Robotics and Automation (ICRA), Stockholm, Sweden, 16–21 May 2016.
18. Andert, F.; Mejias, L. Improving monocular SLAM with altimeter hints for fixed-wing aircraft navigation and emergency landing. In Proceedings of the IEEE International Conference on Unmanned Aircraft Systems (ICUAS), Denver, CO, USA, 9–12 June 2015.
19. Mary, A.; Gerhard, H. Pose estimation of a mobile robot based on fusion of IMU data and vision data using an extended Kalman filter. *Sensors* **2017**, *17*, 2164. [CrossRef]
20. Konovalenko, I.; Kuznetsova, E.; Miller, A.; Miller, B.; Popov, A.; Shepelev, D.; Stepanyan, K. New Approaches to the Integration of Navigation Systems for Autonomous Unmanned Vehicles (UAV). *Sensors* **2018**, *18*, 3010. [CrossRef] [PubMed]
21. Hmam, H.; Kim, J. Aircraft recognition and pose estimation. In Proceedings of the Visual Communications and Image Processing, Perth, Australia, 20–23 June 2000.
22. Wang, L.; Xing, C.; Yan, J. Aircraft pose estimation based on mathematical morphological algorithm and Radon transform. In Proceedings of the International Conference on Fuzzy Systems and Knowledge Discovery, Shanghai, China, 26–28 July 2011.
23. Breuers, M.; Reus, N. Image-based aircraft pose estimation: A comparison of simulations and real-world data. In Proceedings of the Automatic target recognition XI, Orlando, FL, USA, 17–20 April 2001.
24. Fu, T.; Sun, X. The relative pose estimation of aircraft based on contour model. In Proceedings of the International Conference on Optical and Photonics Engineering, Xi'an, China, 14–17 October 2016.
25. Wang, X.; Yu, H.; Feng, D. Pose estimation in runway end safety area using geometry structure features. *Aeronaut. J.* **2016**, *120*, 675–691. [CrossRef]
26. Yuan, W.; Peng, J.; Wang, L.; Lin, S. Aircraft Pose Recognition Using Locally Linear Embedding. In Proceedings of the International Conference on Measuring Technology and Mechatronics Automation, Zhangjiajie, China, 11–12 April 2009.
27. Luo, J.; Teng, X.; Zhang, X.; Zhong, L. Structure extraction of straight wing aircraft using consistent line clustering. In Proceedings of the International Conference on Image, Vision and Computing (ICIVC), Chengdu, China, 2–4 June 2017.
28. Sadraey, M.H. *Aircraft Design: A Systems Engineering Approach*, 1st ed.; John Wiley & Sons: Hoboken, NJ, USA, 2013; pp. 160–249, ISBN 9781119953401.
29. Kholish, R.K.; Aditya, P.; Mochammad, A.M. Design of high altitude long endurance UAV: Structural analysis of composite wing using finite element method. *J. Phys. Conf. Ser.* **2018**, *1005*, 012025. [CrossRef]
30. Park, K.; Han, J.W.; Lim, H.J.; Kim, B.S.; Lee, J. Optimal design of airfoil with high aspect ratio in unmanned aerial vehicles. *Int. J. Aerosp. Mech. Eng.* **2008**, *2*, 381–387. [CrossRef]
31. Gioi, R.G.; Jakubowicz, J.; Morel, J.; Randall, G. LSD: A Fast Line Segment Detector with a False Detection Control. *IEEE Trans. Pattern Anal. Mach. Intell.* **2010**, *32*, 722–732. [CrossRef] [PubMed]

32. Zhang, Y.; Liu, Y.; Zou, Z. Comparative study of line extraction method based on repeatability. *J. Comput. Inf. Syst.* **2012**, *8*, 10097–10104.
33. Comaniciu, D.; Meer, P. Mean shift: A robust approach toward feature space analysis. *IEEE Trans. Pattern Anal. Mach. Intell.* **2002**, *24*, 603–619. [CrossRef]
34. Ester, M.; Kriegel, H.; Sander, J.; Xu, X. A Density-Based Algorithm for Discovering Clusters in Large Spatial Databases with Noise. In Proceedings of the International Conference on Knowledge Discovery and Data Mining, Portland, OR, USA, 2–4 August 1996.
35. Arthur, D.; Vassilvitskii, S. K-Means++: The Advantages of Careful Seeding. In Proceedings of the Eighteenth Annual ACM-SIAM Symposium on Discrete Algorithms, New Orleans, LA, USA, 7–9 January 2007.
36. Hartley, R.; Zisserman, A. *Multiple View Geometry in Computer Vision*, 2nd ed.; Cambridge University Press: Cambridge, UK, 2004; pp. 310–324, ISBN 9780521540513.
37. AUTODESK. Available online: https://www.autodesk.com/ (accessed on 6 December 2017).

sensors

MDPI

Article

Motion-Aware Correlation Filters for Online Visual Tracking

Yihong Zhang *, Yijin Yang *, Wuneng Zhou, Lifeng Shi and Demin Li

College of Information Science and Technology, Engineering Research Center of Digitized Textile & Fashion Technology, Ministry of Education, DongHua University, Shanghai 201620, China; wnzhou@dhu.edu.cn (W.Z.); 2161238@mail.dhu.edu.cn (L.S.); deminli@dhu.edu.cn (D.L.)

* Correspondence: zhangyh@dhu.edu.cn (Y.Z.); 2171318@mail.dhu.edu.cn (Y.Y.); Tel.: +86-138-1782-6259 (Y.Z.); +86-150-0073-3231 (Y.Y.)

Received: 25 September 2018; Accepted: 7 November 2018; Published: 14 November 2018

Abstract: The discriminative correlation filters-based methods struggle deal with the problem of fast motion and heavy occlusion, the problem can severely degrade the performance of trackers, ultimately leading to tracking failures. In this paper, a novel Motion-Aware Correlation Filters (MACF) framework is proposed for online visual object tracking, where a motion-aware strategy based on joint instantaneous motion estimation Kalman filters is integrated into the Discriminative Correlation Filters (DCFs). The proposed motion-aware strategy is used to predict the possible region and scale of the target in the current frame by utilizing the previous estimated 3D motion information. Obviously, this strategy can prevent model drift caused by fast motion. On the base of the predicted region and scale, the MACF detects the position and scale of the target by using the DCFs-based method in the current frame. Furthermore, an adaptive model updating strategy is proposed to address the problem of corrupted models caused by occlusions, where the learning rate is determined by the confidence of the response map. The extensive experiments on popular Object Tracking Benchmark OTB-100, OTB-50 and unmanned aerial vehicles (UAV) video have demonstrated that the proposed MACF tracker performs better than most of the state-of-the-art trackers and achieves a high real-time performance. In addition, the proposed approach can be integrated easily and flexibly into other visual tracking algorithms.

Keywords: visual tracking; correlation filters; motion-aware; adaptive update strategy; confidence response map

1. Introduction

Visual object tracking is one of the most popular fields in computer vision for its wide applications including unmanned vehicles, video surveillance, UAV, and human-computer interaction, where the goal is to estimate the locus of the object given only by an initial bounding box from the first frame in the video stream [1]. Although significant progress has been achieved in recent decades, accurate and robust online visual object tracking is still a challenging problem due to the parameters of fast motion, scale variations, partial occlusions, illumination changes and background clutters [2].

In recent decades, visual object tracking has been widely studied by researchers resulting in a large body of work. The most relevant works, which had been tested on the benchmark datasets of OTB-50 [3], OTB-100 [4], and the Visual Object Tracking benchmarks of VOT-2014 [5], and VOT-2016 [6], are discussed below.

In general, visual object tracking approaches can be broadly classified into two categories, generative methods [7–13] and discriminative methods [14–25]. The generative methods use the features extracted from the previous frame to establish the appearance model of the target, and then search for the most similar region and locate the position of the target in the current frame.

Robust Scale-Adaptive Mean-Shift for Tracking (ASMS) [8] and Distractor-Aware Tracker (DAT) [7] are the two most representative trackers in generative methods. ASMS is a real-time algorithm using the color histogram features for visual tracking where a scale estimation strategy is added to the classical mean-shift framework. However, it is easily distracted by similar objects in the surroundings. The improved method DAT is a distractor-aware tracking algorithm based on the color probabilistic model of the foreground and the background. It uses the Bayesian method to determine the probability of each pixel belonging to the foreground or background to suppress similar objects in the vicinity. However, these methods make the trend of scale shrink for the use of color features where the edge pixels are always overlooked. Meanwhile, the discriminative approaches which are also called as 'track-by-detection methods' are popular for their high accuracy, robustness, and real-time performance. These methods employ machine-learning techniques to train classifiers by numbers of positive and negative samples extracted from the previous frame, and then use the trained classifiers to find the optimal area of the target and locate the position of the target. Among the discriminative approaches, the Discriminative Correlation Filter-based (DCF-based) approach is one of the most popular approach.

1.1. DCF-Based Trackers

Lately, Discriminative Correlation Filters (DCFs) have been extensively applied to visual object tracking in computer vision. It was introduced into the visual tracking fields by Bolme and colleagues in the article visual object tracking using adaptive correlation filters [1]. It named by Minimum Output Sum of Squared Error (MOSSE) which produced astonishing results with tracking speed reaching about 700 Frames Per Second (FPS). Thereafter, numerous improved algorithms [14–17,26–28] based on DCFs have been published with accurate and robust tracking results by sacrificing the tracking speed. The DCF technique is a computationally efficient process in the frequency domain transformed by fast Fourier transform (FFT) [1,29,30]. It is a supervised method for learning a linear classifier or a linear regressor, which trains and updates DCFs online with only one real sample given by the bounding box and various synthetic samples generated by cyclic shift windows. Then the trained DCFs are used to detect the position and scale of the target in the subsequent frame.

Currently, DCF-based methods such as Discriminative Scale Space Tracking (DSST) [16], Fast Discriminative Scale Space Tracking (FDSST) [16], and Spatially Regularized Discriminative Correlation Filters (SRDCF) [26] have demonstrated excellent performance on the popular benchmarks OTB-100 [4], VOT-2014 [5], and VOT-2016 [6]. The DSST trains separate translation and scale correlation filters by the Histogram of Oriented Gradient (HOG) features. And the trained correlation filters are used to respectively detect the position and scale of the target. Then the improved FDSST use the principal component analysis (PCA) method to reduce the dimension of the features to speed up the DSST. However, all these methods detect the target by exploiting a limited search region usually smaller than the whole figure. Although it can reduce computational costs, it can result in tracking failures when the target moves out of the search region due to fast motion or heavy occlusion.

Generally, to reduce the computation costs, the standard DCF-based method tracks the target using a padding region which is several times larger than the target but with size limited. In addition, it multiplies a cosine window with the same size of padding region to emphasize on the target [1,14,16,17,26,31]. Despite its excellent properties, the DCF approach cannot detect the position of the target correctly when the target moves to the boundaries of the padding region. Additionally, it fails to track the target when the target moves out of the padding region due to fast motion or heavy occlusion. The dilemma between a larger padding region which is more computationally expensive and a smaller padding region which lacks the ability to track the target, significantly influences the capabilities of the DCF methods. Furthermore, most of the state-of-the-art DCF-based methods [7,16,17,26,27,32] estimate the scale of the target by using a limited number of scales of various sizes. It results in scale tracking failures when the scale changes significantly due to the fast motion. The dilemma between the exhaustive scale search strategies resulting in higher computational costs and the finite number of scale estimation method leading to failures of scale estimation, severely reduced the robustness of the DCF algorithm. Resolving these two dilemmas are the main aims of the present paper.

1.2. Solutions to the Problem of Fast Motion

To solve the dilemmas, a concise and efficient instantaneous motion estimation method (which is implemented by the differential velocity and acceleration between frames) is applied to predict the possible position and scale of the detected target. Nevertheless, the noises existing in the detected results can dramatically affect the performance of this method. For eliminating the noises of the detected results, we prefer to choose the optimal Kalman filter [33–37] which is a highly efficient autoregressive filter. It can estimate the state of a dynamic system in a combination of many uncertainties. In addition, it is a powerful and versatile tool which is appropriate for changing constantly systems. In recent decades, Kalman filters have been widely used in the field of visual object tracking due to the advantage of a small memory footprint (just retaining the previous state) and computational efficiency. It is ideal for real-time problems and embedded systems [33,34,36,38–41], which can improve the performance of trackers without sacrificing the real-time property.

1.3. Our Contributions

This paper, inspired by the works [16,42–45], proposes a novel Motion-Aware Correlation Filters (MACF) visual tracker which aims to solve the two dilemmas described in Section 1.1. The proposed approach initializes the joint instantaneous motion estimation Kalman filters by using the parameters of the bounding box given by the first frame. Then the improved Kalman filters are used to predict the probable position and scale of the target in the subsequent frame. This makes the target near the center of the padding region which improves the robustness and accuracy of the tracker. The DCFs-based tracker [16] is chosen as the fundamental framework to train correlation filters to detect the location and scale of the target based on the predicted results. For the convenience of computation and integration, the Kalman Filters are decomposed into two parts including a two-dimensional in-plane motion estimation filter and a one-dimensional depth motion estimation filter [46]. In addition, a novel function is proposed to compute the confidence of the response map to determine whether to update the correlation filters. The lower the confidence score is, the higher probability the model is corrupted. Hence, the score below the set threshold means that the target has been occluded or has changed greatly. Then, the learning rate is reduced according to the confidence of the response map to overcome the problem. In this paper, all the implementation and testing codes are all open source in the following Github web: https://github.com/YijYang/MACF.git.

In summary, the main contributions of this paper include:

1. A novel tracking framework named MACF which corrects the padding region using motion cues predicted by separated joint instantaneous motion estimation Kalman filters, one for in-plane position prediction and the other for scale prediction;
2. An attractive confidence function of the response map to identify the situation where the target is occluded or corrupted and an adaptive learning rate to prevent the model from being corrupted.
3. Qualitative and quantitative experiments on OTB-50, OTB-100 and UAV video have demonstrated that our approach outperforms most of the state-of-the-art trackers.

2. The Reference Tracker

In this section, the reference framework of the FDSST tracker is introduced in detail. In contrast to the FDSST, the proposed MACF tracker has been improved on this baseline tracker and achieved a significant progress on the benchmarks as shown in Figure 1.

Figure 1. The comparison of tracking results between our MACF tracker (in red) and the standard FDSST tracker (in green) in three sequences on OTB-100 benchmark. Our tracker performs better than FDSST in the example frames which are shown from the "Board" of fast motion (**top** row), "Gym1" of scale change (**middle** row) and "Human4.2" of heavy occlusion (**bottom** row) videos.

The FDSST tracker is chosen as the baseline of the proposed MACF framework due to its superior performance on VOT-2014. Unlike the other DCFs-based methods, the FDSST tracker learns 1-dimensional scale estimation correlation filters and 2-dimensional translation estimation correlation filters separately, which is implemented by adjusting the feature extraction procedure only for each case [16]. The objective function of correlation filter f can be denoted as follows including a response score function (1) and an L^2 error function (2) with t samples:

$$S_f(x) = \sum_{l=1}^{d} x^l * f^l \tag{1}$$

$$\varepsilon(f) = \sum_{k=1}^{t} \left\| S_f(x_k) - g_k \right\|^2 + \lambda \sum_{i=1}^{d} \left\| f^l \right\|^2 \tag{2}$$

where * denotes circular convolution operation and x denotes the HOG features extracted from the target samples. In function (1), l indicates the l-dimensional HOG features and d represents the total dimension of the HOG features. In function (2), the desired output g_k presents a 2-dimentional Gaussian function with the same size of f and x, and k denotes the k represents the kth sample of the input. The second term in Equation (2) is a regularization term with a parameter λ $(\lambda \geq 0)$.

The function (2) is a linear least square problem which can be solved efficiently in frequency domain transformed by FFT. Therefore, through minimizing the function (2), the final solution can be computed by Equation (5), which is equivalent to solving a system of linear equations as follows:

$$A_t^l = \overline{G} F^l \tag{3}$$

$$B_t = \sum_{k=1}^{d} \overline{X}_t^k X_t^k + \lambda \tag{4}$$

$$F_t^l = \frac{A_t^l}{B_t}, l = 1, 2, \ldots, d \tag{5}$$

where the capital letters denote the FFT and F_t denotes the correlation filter in the Fourier domain. In Equations (3) and (4), A_t denotes the numerator of the filter, and B_t denotes the denominator of the filter. The overbar of $\overline{X}$ denotes the complex conjugation of X.

For computational efficiency, the size of the filter F_t is the same as the padding region which is twice the size of the bounding box. An optimal update strategy is utilized to the numerator A_t in Equation (6) and the denominator B_t in Equation (7) of the filter F_t with a new sample feature X_t as follows:

$$A_t^l = (1 - \eta_0) A_{t-1}^l + \eta_0 \overline{G} F^l \tag{6}$$

$$B_t = (1 - \eta_0) B_{t-1} + \eta_0 \sum_{k=1}^{d} \overline{X}_t^k X_t^k \tag{7}$$

where the scalar η_0 is a parameter of the learning rate.

To detect the variations of position P_t and scale S_t of the target, the FDSST firstly learns a 2-dimensional DCF for position estimation and then learns a 1-dimensional DCF for scale estimation. The responding scores y_t for a new frame can be formulated by function (8).

$$y_t = F^{-1} \left\{ \frac{\sum_{l=1}^{d} \overline{A}_{t-1}^l Z^l}{B_{t-1} + \lambda} \right\} \tag{8}$$

where Z^l denotes the l-dimensional HOG features extracted from the frame of pending detection. F^{-1} represents the Inverse Fast Fourier Transform (IFFT). In Algorithm 1, the capital letter $Y_{t,trans}$ denotes the response scores of translation model and $Y_{t,scale}$ denotes the response scores of scale model. By computing the IFFT, the obtained spatial distribution of the response map is used to determine the spatial location and scale of the target.

Consequently, the position or the scale of the target is determined by the maximal value of the scores y of the corresponding DCFs. In addition, to ultimately reduce the computational costs, the principal component analysis (PCA) method is utilized to decrease the dimension of Histogram of Oriented Gradient (HOG) features. For further details see references [5,6].

3. Our Approach

In this section, two different approaches for motion estimation of the target is introduced, including the instantaneous motion estimation method and Kalman Filters-based motion estimation method. Then the proposed MACF framework is introduced in detail. Firstly, the Joint instantaneous motion estimation Kalman filters for motion prediction are investigated. Secondly, an update scheme with an adaptive learning rate to prevent the model corrupted by heavy occlusion or fast motion is presented. Finally, the algorithm framework of MACF is described in Algorithm 1.

3.1. Instantaneous Motion Estimation between Three Adjacent Frames

A single scheme for incorporating motion estimation is to estimate instantaneous velocity and acceleration between three contiguous frames as shown in Figure 2. Firstly, this method initializes the parameters of position and scale to (x_1, y_1, s_1), and sets the velocity and acceleration of the x-axis, y-axis, and z-axis $(v_{x_1}, v_{y_1}, v_{s_1})$, $(a_{x_1}, a_{y_1}, a_{s_1})$ to (0, 0, 0) in the first frame. Secondly, these parameters are utilized to predict the possible region of the target by Equation (11) in the second frame. Then the FDSST is used to detect the position (x_2, y_2) and the scale s_2 of the target to update $(v_{x_2}, v_{y_2}, v_{s_2})$ by function (9). In the third frame, the accelerations $(a_{x_2}, a_{y_2}, a_{s_2})$ are updated by function (10). Finally, it continuously predicts and detects the location and scale of the target until the last frame of the video stream.

$$\begin{cases} v_{x_t} = x_t - x_{t-1} \\ v_{y_t} = y_t - y_{t-1} \\ v_{s_t} = s_t - s_{t-1} \end{cases} \tag{9}$$

$$\begin{cases} a_{x_t} = v_{x_t} - v_{x_{t-1}} \\ a_{y_t} = v_{y_t} - v_{y_{t-1}} \\ a_{s_t} = v_{s_t} - v_{s_{t-1}} \end{cases} \tag{10}$$

$$\begin{cases} Px_{t+1} = x_t + v_{x_t} \cdot \Delta t + 0.5 \cdot a_{x_t} \cdot \Delta t^2 \\ Py_{t+1} = y_t + v_{y_t} \cdot \Delta t + 0.5 \cdot a_{y_t} \cdot \Delta t^2 \\ Ps_{t+1} = s_t + v_{s_t} \cdot \Delta t + 0.5 \cdot a_{s_t} \cdot \Delta t^2 \end{cases} \tag{11}$$

where Δt denotes time step, $\Delta t = 1$ is used to facilitate the calculation, (x, y, s) denote the results of detection, and (Px, Py, Ps) denote the results of the prediction.

However, this approach can be affected easily by the noise of the detected results. In addition, the basic tracker FDSST has quite a fine scale detection. Hence, the error scale estimation, which is caused by measurement noise, probably leads to tracking failures.

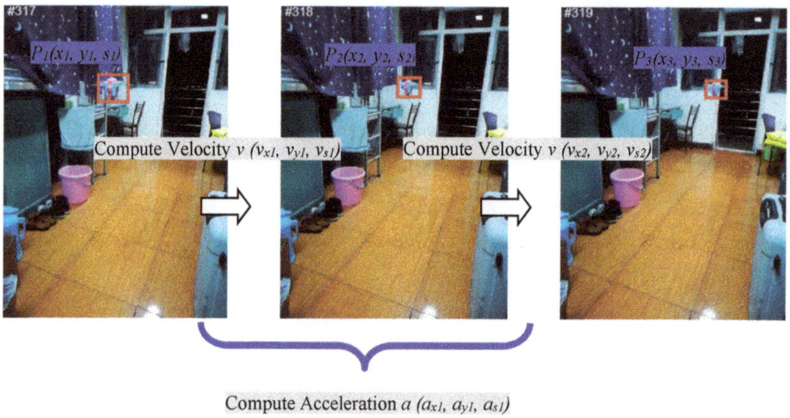

Figure 2. Illustrating of instantaneous motion estimation on the test sequence of UAV.

3.2. Kalman Filters-Based Motion Estimation

For high accuracy of the motion prediction, Kalman Filters serve as a strategy of motion estimation [38,39]. Assuming that the motion model of the target is a constant acceleration model, the motion model can be described by the linear stochastic differential functions as follows:

$$P(t) = AP(t-1) + BM(t) + W(t) \tag{12}$$

$$Z(t) = HP(t) + V(t) \tag{13}$$

In the above two equations, $P(t)$ is the target state of the t-th frame of the video sequence, and $M(t)$ is the motion model of the target in the t-th frame. In function (12), A and B are the parameters of the motion model. In Formula (13), $Z(t)$ is the measured value of the target state of the t-th frame and H is the parameter of the measurement system. In the two equations, $W(t)$ and $V(t)$ represent the process and measured noise respectively and they are assumed to be White Gaussian Noise. Their covariances are Q and R which are assumed not to change with the system state. Q and R respectively represent the confidence of the predicted value and the measured value. It can affect the weight of the predicted value and the measured value through affecting the value of the Kalman gain in the Equation (16). When the value of R is larger, the confidence of the measured value is smaller.

3.2.1. Prediction

For a system which satisfies the above conditions, the Kalman Filter is the optimal information processor. Firstly, the motion model of the target is used to separately predict the position and scale of the target in the next state. Secondly, the current system state is t, the function (14) can be used to predict the position or scale in the current state based on the previous state $P(t-1|t-1)$ of the target. Finally, the current covariance of $C(t-1|t-1)$ can be updated by Equation (15).

$$P(t|t-1) = AP(t-1|t-1) + BM(t) \tag{14}$$

$$C(t|t-1) = AC(t-1|t-1)A' + Q \tag{15}$$

where, $P(t|t-1)$ is the current predicted position or scale of the target, and $P(t-1|t-1)$ is the result of the previous state optimization. In Equation (15) $C(t|t-1)$ is the covariance corresponding to $P(t|t-1)$ and $C(t-1|t-1)$ is covariance corresponding to $P(t-1|t-1)$. In formula (15), A' denotes the transpose matrix of A and Q is the covariance of the motion model which has been set in the first frame.

3.2.2. Measurement and Correction

The position and scale of the target detected by FDSST mentioned in Section 3.1 is used as the measurement value $Z(t)$. Combined with the prediction result $P(t|t-1)$, the measurement value $Z(t)$, and the Kalman gain calculated by Equation (16), the optimal estimate of the current position $P(t|t)$ is achieved using Equation (17).

$$Kg(t) = \frac{C(t|t-1)H'}{HC(t|t-1)H' + R} \tag{16}$$

$$P(t|t) = P(t|t-1) + Kg(t)[Z(t) - HP(t|t-1)] \tag{17}$$

where $Kg(t)$ is the Kalman gain in current frame and H' denotes the transpose matrix of H, and R denotes the measuring error. In short, Q and R respectively represent the confidence of the predicted value and the measured value and can affect the weight of the predicted value and the measured value by affecting the value of the Kalman gain $Kg(t)$. The larger the R, the less the confidence is the measured value.

To keep the Kalman filter running until the last frame of the video streaming [47], the new covariance of $C(t|t)$ is updated by function (18).

$$C(t|t) = [I - Kg(t)H]C(t|t-1) \tag{18}$$

where, I is a unit matrix.

3.3. *Motion-Aware in Our Framework*

Assuming that White Gaussian Noises exist in the measured velocity and acceleration in Equations (9) and (10), the measured results are utilized to predict the position of the target by a linear Equation (11). Obviously, the predictions include the White Gaussian Noises which potentially result in tracking failures. Therefore, the joint instantaneous motion estimation Kalman Filters are utilized to filter out the noise of the predicting results. It means that the predicted values by Equation (11) are taken as the observed input value of the Kalman filter and then output an optimal prediction by Equation (17).

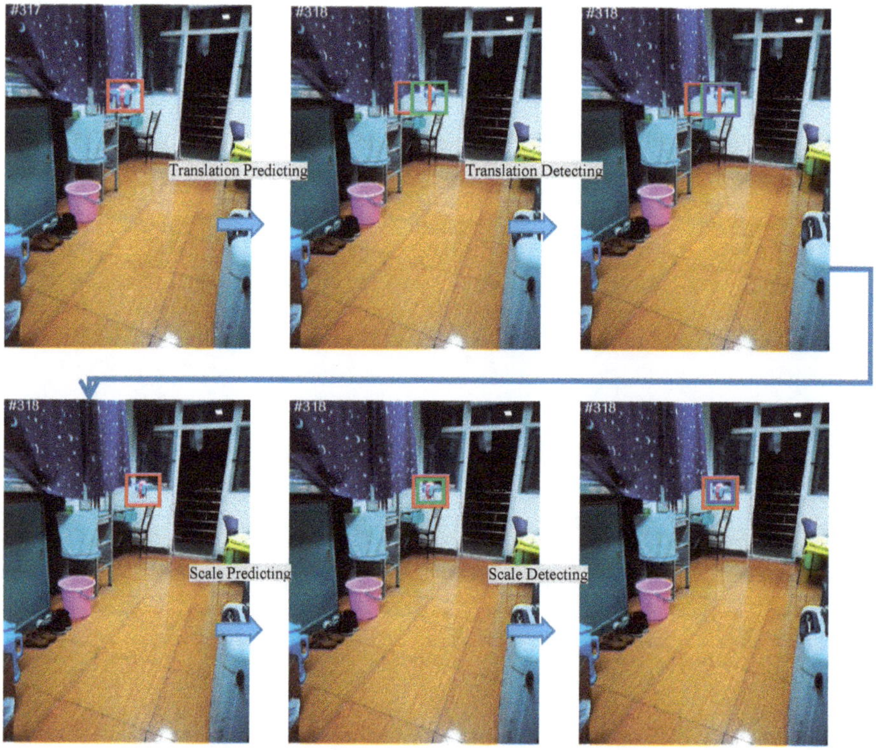

Figure 3. Visualization of the separate translation and scale prediction and detection on the video sequence of UAV. The previous position and scale are indicated by red bounding box, and the predicted position and scale are denoted in green, and the detected position and scale are shown with blue bounding box.

As mentioned in Section 3.2, the instantaneous motion estimation method is affected greatly by the noise, but it can deal with the nonlinear motion model. However, the Kalman Filter filters out the noises, but cannot solve the nonlinear motion model. Hence, for achieving the advantages of both methods, the two methods are combined for Motion estimation of the target. Additionally, for convenient and efficient computation, the optimal Kalman Filters are set up separately for position and scale prediction as shown in Figure 3.

(I) The position prediction filter is responsible for the prediction of the target location and noise filtering. First, motion parameters ($v_{x_{t-1}}, v_{y_{t-1}}, a_{x_{t-1}}, a_{y_{t-1}}$) are employed in the previous frame to predict the translation PP_t (Px_t, Py_t) of the target in the next frame through Equation (11). After that, the two-dimensional Kalman position filter is utilized to eliminate the noises of the prediction by function (17).

(II) The scale prediction filter is employed to predict accurately and reliably the scale of the target by filtering noises. The prediction parameters ($v_{x_{s-1}}, a_{s_{t-1}}$) are first utilized in the front frame to predict the scale Ps_t of the target in the following frame by Equation (11). Afterwards, the one-dimensional Kalman scale filter is employed to remove the noises of the prediction by function (17).

3.4. Position and Scale Detection

The two-dimensional translation correlation filter $F_{t,trans}$ of the FDSST (described in Section 3.1) is used to detect the position of the target in a small padding region based on the filtered predictions. Then, the results of detection (x_t, y_t) is utilized to update the in-plane motion model parameters

$(v_{x_t}, v_{y_t}, a_{x_t}, a_{y_t})$ via Equations (9) and (10). Similarly, for estimating the scale of the target, the scale correlation filter $F_{t,scale}$ is utilized to correct the scale of the target on the foundation of the predicted scale. Then, the estimated scale s_t is utilized to update the deep motion model parameters (v_{s_t}, a_{s_t}) by Equations (9) and (10).

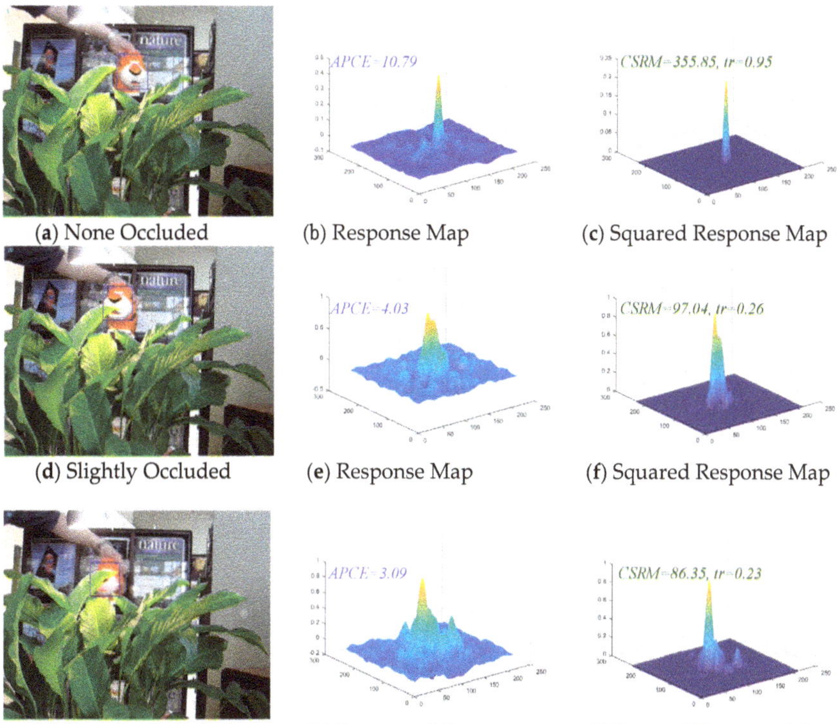

(**a**) None Occluded (**b**) Response Map (**c**) Squared Response Map
(**d**) Slightly Occluded (**e**) Response Map (**f**) Squared Response Map
(**g**) Heavily Occluded (**h**) Response Map (**i**) Squared Response Map

Figure 4. The Confidence of the Squared Response Map (CSRM) in the proposed MACF comparing with the Average Peak-to-Correlation Energy (APCE) of the response map. The example frames are from the sequence "Tiger1" on OTB-100 benchmark. The higher value of the CSMR, the more confident the response map is. The value of parameter *tr* determine the adaptive learning rate which compute by Equation (20). From the figure, the gap of CSMR is larger than APCE between the slightly occluded, heavily occluded and none occluded target.

3.5. A Novel Model Update Strategy

After the study of Average Peak-to-Correlation Energy (APEC) in [42], a novel confidence function (19) of the responding map is proposed in the MACF algorithm in this paper. In [42], APEC is defined as APEC $= R_{\max}/E(R)$, here, $R_{\max}$ denotes the max value of the response scores, and $E(R)$ denotes the expected value of the response scores. APCE indicates the fluctuated degree of response maps and the confidence level of the detected targets. Figure 4b,e,h illustrate that if the target apparently appears in the detection scope, there is a sharper peak in the response map and the value of APEC becomes smaller. On the contrary, if the object is occluded, the peak in response map appears smoother, and the relative value of APEC becomes larger.

Unlike the APCE, the proposed method in this article squared the value of response map (the proof is given in Appendix A) and then calculated the value of Confidence of Squared Response Map (CSRM). CSRM stands for the fluctuated degree of the response maps and the confidence level of the detected targets. The numerator of the CSRM represents the peak of the response map, and the denominator of CSRM represents the mean square value of the response map. Figure 4c,f,i illustrate that if the target is not occluded or contaminated, the corresponding response map presents a sharp peak. It is concluded that when the peak value is larger and the mean square value is smaller, and the result is that the corresponding CSRM value is larger. On the contrary, if the target is occluded or contaminated, the corresponding response map will present a smoother peak and even multiple peaks. It could be concluded that when the peak value is smaller and the mean square value is larger, and the result is that the corresponding CSRM value is smaller. This increases the gap between the confidence response and the diffident response as shown in Figure 4, making it easier to find the threshold between them. Consequently, a threshold is set to distinguish whether the target is occluded or contaminated and an adaptive learning rate η is set by Equation (20) to prevent the model from being corrupted. In addition, Equation (20) is effective and accurate for model learning which can be readily and neatly integrated into DCF-based trackers to improve the tracking performance.

$$\mathrm{CSRM}_t = \frac{\left|R_{\max}^2 - R_{\min}^2\right|^2}{\frac{1}{MN}\sum_{i=1}^{M}\sum_{j=1}^{N}\left|R_{ij}^2 - R_{\min}^2\right|^2} \tag{19}$$

$$\begin{cases} tr_t = \frac{\mathrm{CSRM}_t}{\mathrm{CSRM}_0} \\ \eta_t = \eta_0,\ tr_t > tr_0 \\ \\ \eta_t = \eta_0 \cdot tr_t, others \end{cases} \tag{20}$$

where, CSRM_0 is the Confidence of the Squared Response Map in the initial frame where the response is identified as the most confidence response, CSRM_t is the confidence of the squared response map in the t-th frame, and tr_0 is the threshold to decide the learning rate. In Equation (19), the response map R is a two-dimensional $M * N$ matrix.

Algorithm 1. MACF tracking algorithm

Input:
1: Image I_t.
2: Predicted target position PP_t and scale Ps_t in previous frame.
Output:
1: Detected target position P_t and scale S_t in current frame.
2: Predicted target position PP_{t+1} and scale Ps_{t+1} subsequent frame.
Loop:
1: Initialize the Translation model $A_{1,trans}$, $B_{1,trans}$ and Scale model $A_{1,scale}$, $B_{1,scale}$ in the first frame by Equations (3) and (4), and initialize the Confidence of the Squared Response Map CSRM_0 in the initial frame by Equation (19).
2: for $t \in \left[2, t_f\right]$ **do**.
3: **Position detection and prediction:**
4: Extract pending sample feature $Z_{t,trans}$ from I_t at PP_t and Ps_t.
5: Compute correlation scores $Y_{t,trans}$ by Equation (8).
6: Set P_t to the target position that maximizes $Y_{t,trans}$.
7: Predict the position PP_{t+1} of the target of subsequent frame by joint Equations (11) and (17).
8: **Scale detection and prediction:**
9: Extract pending sample feature $Z_{t,scale}$ from I_t at P_t and Ps_t.
10: Compute correlation scores $Y_{t,scale}$ by Equation (8).
11: Set S_t to the target scale that maximizes $Y_{t,scale}$.
12: Predict the position Ps_{t+1} of the target of subsequent frame by joint Equations (11) and (17).

Model update:
Compute the Confidence of the Squared Response Map $CSRM_t$ in current frame by Equation (17).
    Compute the adaptive learning rate η_t by Equation (18).
    Extract sample features $X_{t,trans}$ and $X_{t,scale}$ from I_t at P_t and S_t.
    Update motion parameters $(v_{x_t}, v_{y_t}, v_{s_t})$, $(a_{x_t}, a_{y_t}, a_{s_t})$ by Equations (9) and (10).
    Update Kalman filters by Equation (18).
    Update the translation model $A_{t,trans}$, $B_{t,trans}$ by adaptive learning rate η_t.
    Update the scale model $A_{t,scale}$, $B_{t,scale}$ by adaptive learning rate η_t.
Return P_t, and PP_{t+1}, Ps_{t+1}.
end for.

---

4. Experiments and Results

In this section, firstly, the implement details and parameter settings are introduced clearly. Then the comprehensive experiments have been tested on the popular benchmark OTB-50, OTB-100 and UAV video, and the results have demonstrated that our MACF approach surpasses most of the state-of-the-art methods.

4.1. Implement Details

All the methods compared in this paper are implemented in MATLAB R2016a, and all experiments run on an INTEL i3-3110 CPU with 6 GB memory.

State-of-the-art trackers: for other trackers compared to our MACF tracker in this paper, we follow the parameter settings in their papers.

Trackers proposed in this paper: Introduced in Section 3.1, the FDSST is employed as the basic tracker. Thus, all parameters of FDSST remain the same as in the paper [16] except for the regularization term λ, learning rate η, search region *padding*, and scale factor α. In our proposed trackers, the regularization term parameter is set to $\lambda = 0.02$, the padding region is set to $padding = 1.8$, the scale factor is set to $a = 1.03$ and the adaptive learning rate is calculated from Equation (20) with a threshold $tr_0 = 0.6$. For two-dimensional translation Kalman Filter, the covariances of motion and measured noise in Equations (12) and (13) are set to $Q = [25, 10, 1]$, $R = 25$. In the one-dimensional scale Kalman Filter, the covariances are set to $Q = [2.5, 1, 0.1]$, $R = 2.5$. However, there are some different parameter settings about the adaptive learning rate enable parameter, the Kalman position filter enable parameter, the Kalman scale filter enable parameter and the instantaneous motion estimation enable parameter. As described in subsequent Section 4.2, in the proposed MACF tracker, these parameters are respectively set to (1, 1, 1, 1). In the IME_CF tracker, these parameters are respectively set to (0, 0, 0, 1). In the KE_CF tracker, these parameters are respectively set to (0, 1, 1, 0). In the ALR_CF tracker, these parameters are respectively set to (1, 0, 0, 0).

4.2. Ablation Experiments

To validate the effectiveness of the strategy proposed in this paper, an ablation experiment is performed on OTB-50, and the MACF is compared with the standard FDSST introduced in Section 2, based on instantaneous motion estimation CFs (IME_CF) discussed in Section 3.1, based on Kalman filters CFs (KF_CF) described in Section 3.2 and based adaptive learning rate CFs (ALR_CF) proposed in Section 3.5. Obviously, Table 1 indicates that the proposed schemes all achieved varying degrees of the tracking performance improvement compared to the standard FDSST. Overall, the proposed MACF achieves a gain of 2.3%, 4.8% and 4.1% in OPE, TRE and SRE, respectively, of LET at 20 pixels and a gain of 1.7%, 1.4% and 2.9% in OPE, TRE and SRE, respectively, of OT at 0.5 compared to the standard FDSST. Furthermore, the proposed MACF run at a real-time speed of 51 FPS in my i3-3110 CPU. However, the strategy of adaptive learning rate achieves the best results instead of our fused MACF. That's because motion-aware strategy is more suitable to track the target of fast motion in a gradient background. Nevertheless, most video sequences on OTB-50 dataset are with the background of dramatic changes.

Table 1. The comparison of ablation results on OTB-50 dataset. Clearly, the success plots (SP) of one pass evaluation (OPE), temporal robustness evaluation (TRE), and spatial robustness evaluation (SRE) utilizing the location error threshold (LET) and the precision plots (PP) of OPE, TRE and SRE using overlap threshold (OT) and the tracking speed are shown in the table below. And the best results are in red and the second results are in blue.

Trackers	Precision Plots (AUC%)			Success Plots (AUC%)			Speed (FPS)
	OPE	SRE	TRE	OPE	SRE	TRE	
FDSST	62.8	55.6	60.8	50.6	44.2	53.0	49
IME_CF	63.7	58.6	63.3	52.6	46.7	53.7	48
KF_CF	64.5	59.3	65.5	54.4	46.9	54.0	46
ALR_CF	65.0	61.1	66.6	54.6	48.6	54.6	55
MACF	65.1	59.7	65.6	52.3	47.1	54.4	51

4.3. Experiment on OTB-50

OTB-50 is an influential benchmark with 50 sequences which are all labeled manually. Th proposed MACF is evaluated on this dataset and compared to 11 state-of-the-art trackers from the works: Tracking-Learning-Detection (TLD) [2], DSST [17], FDSST [16], Compressive Tracking (CT) [20], exploiting the Circulant Structure of tracking-by-detection with Kernels (CSK) [21], high-speed tracking with Kernelized Correlation Filters (KCF) [22], Long-term Correlation Tracking (LCT) [45], Locally Orderless Tracking (LOT) [48], Least Soft-threshold Squares tracking (LSS) [49], robust visual tracking via Multi-Task sparse learning (MIT) [50], Distribution Fields for Tracking (DFT) [19]. Only the ranks for the top eight trackers are reported.

As is shown in Figure 5, the proposed MACF obtains the top ranks 51.5%, 61.9% and 65.2% among the top eight trackers in 3 different attributes of occlusion, motion blur and fast motion and significantly outperforms the standard FDSST. In other words, the proposed adaptive learning rate scheme is accurate and robust for tracking when the target is occluded or blurred. Furthermore, the proposed motion-aware strategy can effectively track the target of fast motion.

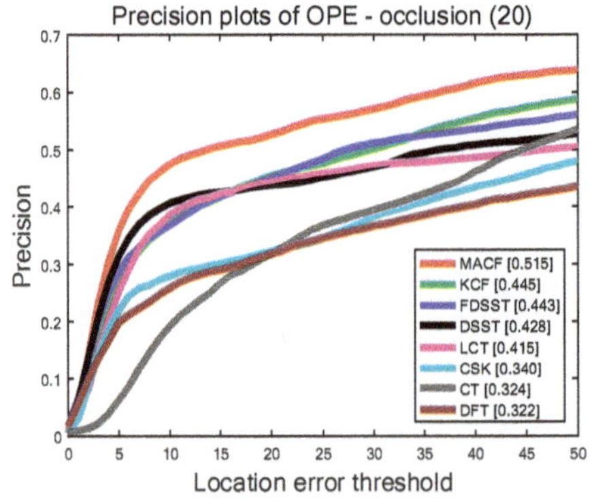

(**a**) The PP of OPE on attribute Occlusion.

Figure 5. *Cont.*

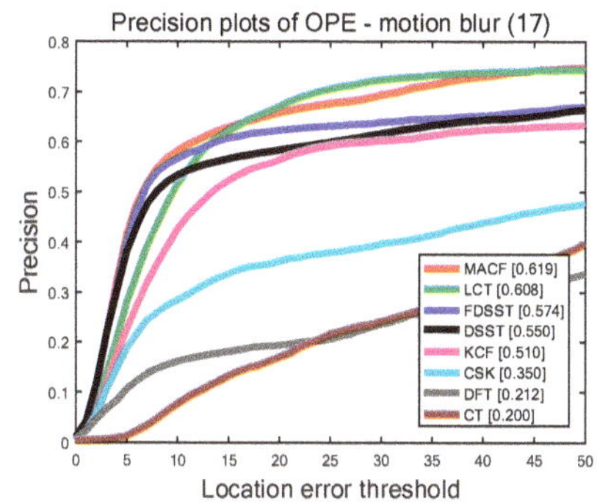

(**b**) The PP of OPE on attribute Motion Blur.

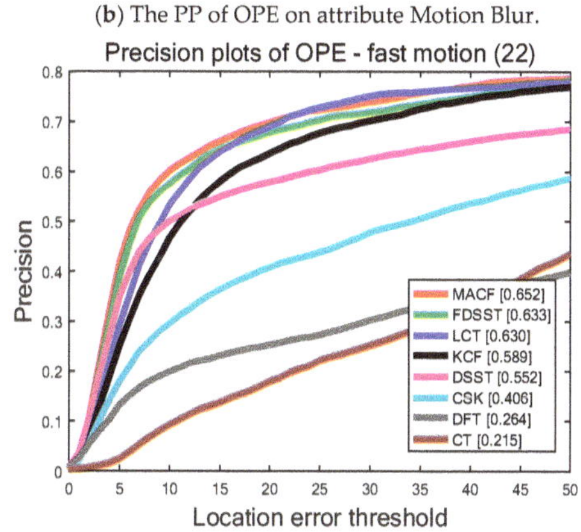

(**c**) The PP of OPE on attribute Fast Motion

Figure 5. The Precision Plots (PP) of One Pass Evaluation (OPE) on OTB-50 benchmark for the top eight trackers determined by 3 different attributes: occlusion, motion blur and fast motion. Among the top eight trackers our MACF obtains the best results on all 3 attributes.

Figure 6 and Table 2 show the SP of OPE, TRE, and SRE utilizing the LET. The PP of OPE, TRE and SRE using OT with the total 50 sequences on OTB-50 are also shown in Figure 6. Generally, the proposed MACF acquires the best results of the top eight trackers including 65.1%, 59.7% and 65.1% in OPE, TRE and SRE, respectively, of LET at 20 pixels and 52.3%, 47.1% and 54.4% in OPE, SRE and TRE, respectively, of OT at 0.5. Furthermore, the proposed MACF achieves a visibly gain of 4.3%, 4.1% and 2.3% in OPE, SRE and TRE, respectively, of LET at 20 pixels and a gain of 1.7%, 2.9% and 1.4% in OPE, SRE and TRE, respectively, of OT at 0.5 compared to the standard FDSST.

Table 2. The Success Polts (SP) and Precision Plots (PP) of One Pass Evaluation (OPE) for the proposed MACF and the other 7 top trackers on the OTB-50 dataset. The best results are highlighted in red and the second results are highlighted in blue.

Trackers	OPE		SRE		TRE	
	SP (%)	PP (%)	SP (%)	PP (%)	SP (%)	PP (%)
MACF	52.3	65.1	47.1	59.7	54.4	65.1
FDSST	50.6	60.8	44.2	55.6	53.0	62.8
LCT	49.3	61.3	44.0	57.4	52.9	63.0
DSST	47.0	58.5	44.0	56.1	53.8	64.6
KCF	43.9	59.7	40.1	54.7	49.0	61.8
CSK	36.5	47.0	33.2	43.2	43.4	53.7

Table 2. *Cont.*

Trackers	OPE		SRE		TRE	
	SP (%)	PP (%)	SP (%)	PP (%)	SP (%)	PP (%)
CT	25.4	32.1	26.1	33.6	29.0	36.7
DEF	26.4	32.3	25.4	32.3	33.1	39.4

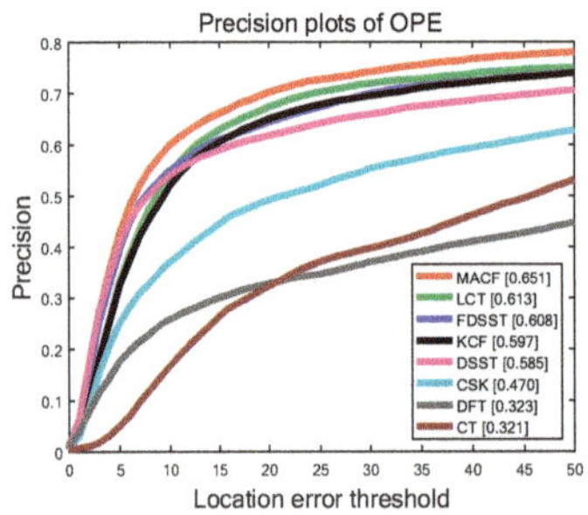

(**a**) The Precision Plots of OPE

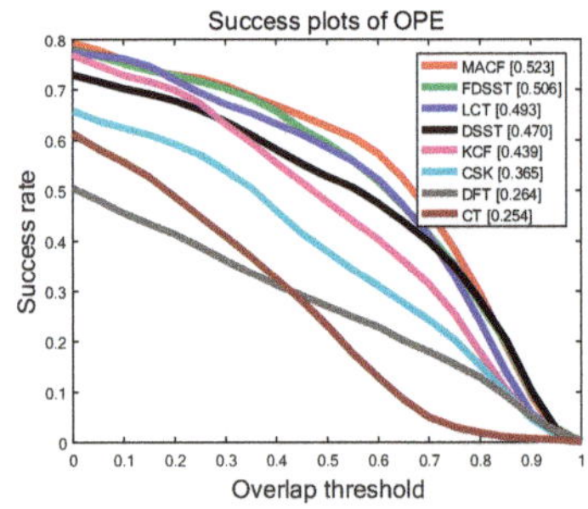

(**b**) The Success Plots of OPE

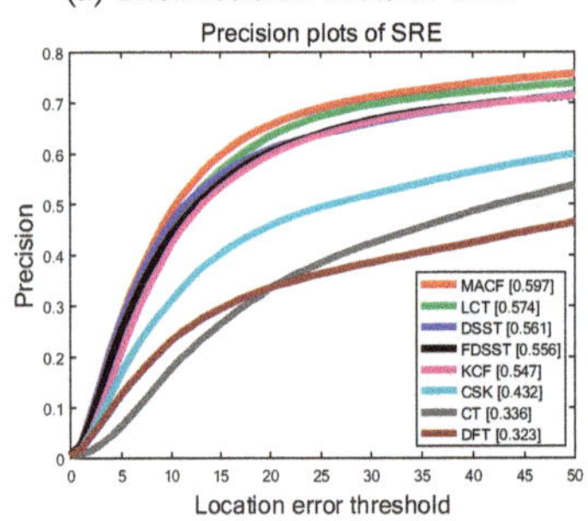

(**c**) The Precision Plots of SRE

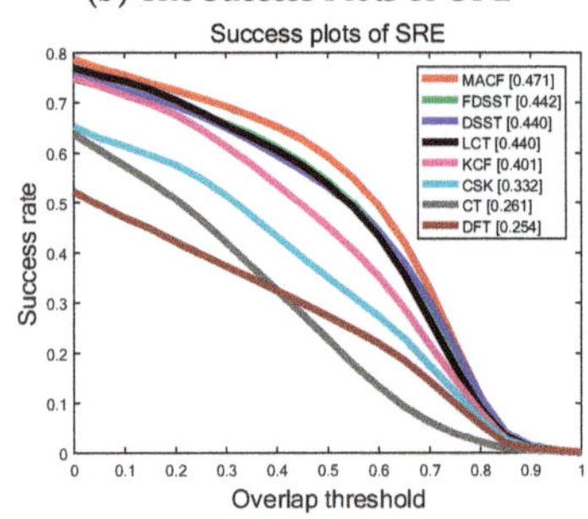

(**d**) The Success Plots of SRE

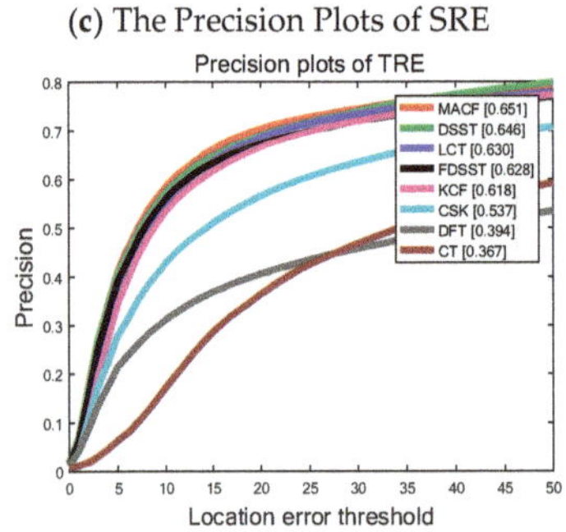

(**e**) The Precision Plots of TRE

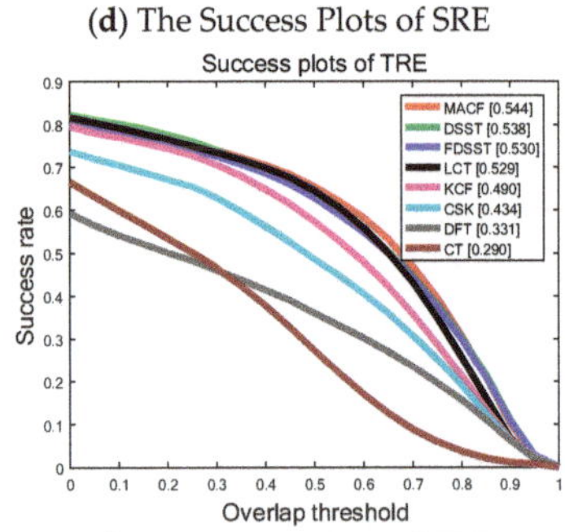

(**f**) The Success Plots of TRE

Figure 6. The Success Plots (SP) and Precision Plots (PP) of One Pass Evaluation (OPE), Temporal Robustness Evaluation (TRE) and Spatial Robustness Evaluation (SRE) using by Overlap Threshold (OT) and Location Error Threshold (LET) comparing MACF with the state-of-the-art trackers on OTB-50 benchmark. The ranks for the top 8 trackers are reported with the Area Under the Curve (AUC) marked in brackets.

4.4. Experiment on OTB-100

OTB-100 is a more challenging benchmark with 100 sequences which are extended by OTB-50. The proposed MACF is evaluated on this dataset and compared to 11 state-of-the-art trackers from the works: TLD [2], DSST [17], FDSST [16], CT [20], CSK [21], KCF [22], LCT [45], LOT [48], LSS [49], MIT [50], DFT [19]. Only the ranks for the top eight trackers are reported.

Figure 7 shows SP of OPE, TRE, and SRE utilizing the LET. The PP of OPE, TRE and SRE using OT with the whole 100 sequences on OTB-100 are shown in Figure 5 as well. Overall, the proposed MACF obtain the top ranks of the top eight trackers including 69.6%, 69.5% and 64.1% in OPE, TRE and SRE, respectively, of LET at 20 pixels and 56.6%, 58.1% and 50.4% in OPE, TRE and SRE, respectively, of OT at 0.5. In addition, the proposed MACF achieves a gain of 1.9%, 0.7% and 1.8% in OPE, TRE and SRE, respectively, of LET at 20 pixels and a gain of 0.5%, 0.5% and 1.7% in OPE, TRE and SRE, respectively, of OT at 0.5 compared to the standard FDSST. However, compared to the experiment on OTB-50, the gains go down due to the extent of 50 video sequences are more challenging with dynamic background. Hence, the additional experiments are conducted on the UAV video in Section 4.6 to validate the accurate and robust gains of the MACF on the video streams with static background.

Table 3 shows the PP of TRE for the top eight trackers determined by 11 different attributes. Among the top eight trackers, the proposed MACF obtains the best results on 8 out of 11 attributes of TRE. Table 4 shows the PP of OPE for the top eight trackers determined by 11 different attributes. Of the top eight trackers the proposed MACF acquires the best ranks on 9 of the 11 attributes of OPE. Table 5 demonstrates the PP of SRE for the top eight trackers determined by 11 different attributes. Of the top eight trackers the proposed MACF achieves the best results on 7 out of 11 attributes of SRE.

Figure 8 qualitatively evaluates the representative frames from four videos successfully tracked by the MACF compared to the top five trackers. From the example frames of Skater1 (the situation of fast motion), it is obvious that the proposed MACF approach performs better than the other four trackers during fast motion and it can be seen from the frames of "Human2" (the situation of occlusion), "Human6" (the situation of occlusion and scale changing greatly), and "Tiger1" (the situation of fast motion and occlusion), the proposed MACF approach is more accurate and robust of the five state-of-the-art trackers when the target is occluded.

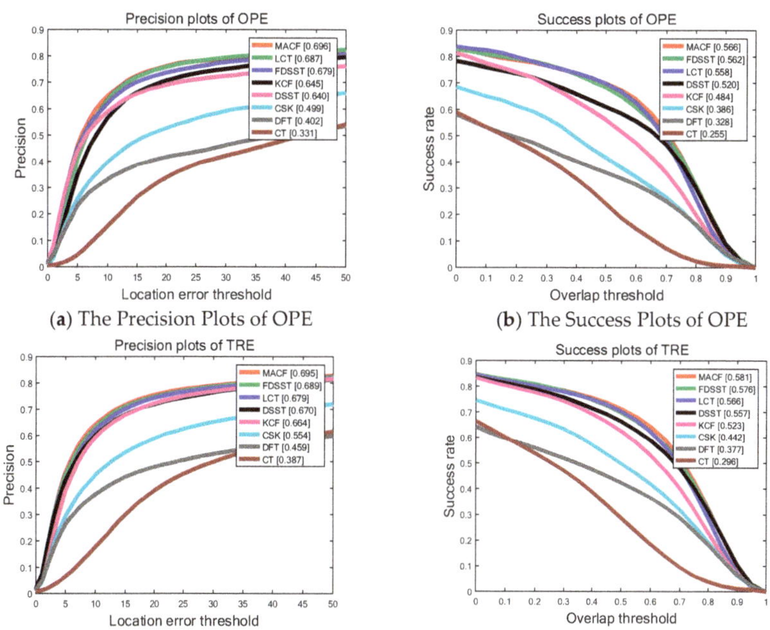

(**a**) The Precision Plots of OPE (**b**) The Success Plots of OPE

(**c**) The Precision Plots of TRE (**d**) The Success Plots of TRE

Figure 7. *Cont.*

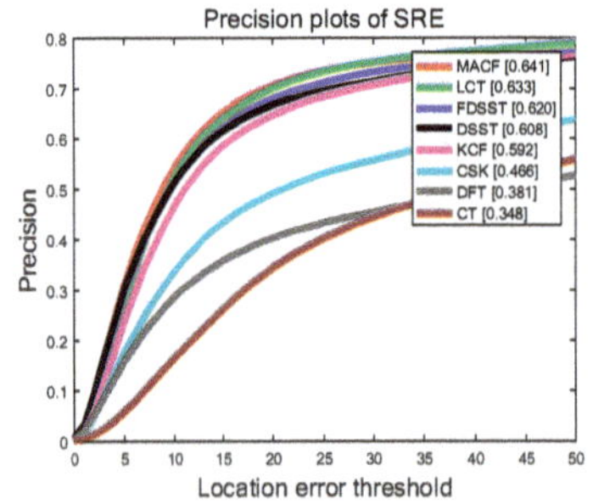

(**e**) The Precision Plots of SRE

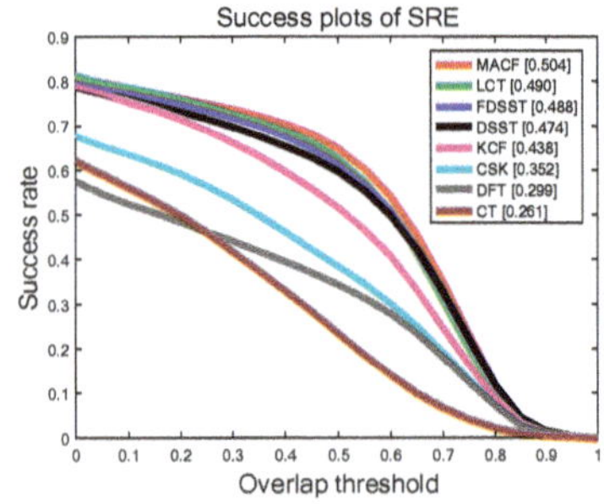

(**f**) The Success Plots of SRE

Figure 7. The Success Plots (SP) and Precision Plots (PP) of One Pass Evaluation (OPE), Temporal Robustness Evaluation (TRE) and Spatial Robustness Evaluation (SRE) using by Overlap Threshold (OT) and Location Error Threshold (LET) comparing the MACF with the state-of-the-art trackers on OTB-100 benchmark. In addition, the ranks for the top 8 trackers are reported with the Area Under the Curve (AUC) marked in brackets.

Table 3. Success plots of Temporal Robustness Evaluation (TRE) for the MACF and the other 7 top trackers on different attributes: scale variation (SV), illumination variation (IV), out-of-plane rotation (OPR), occlusion (OCC), background cluttered (BC), deformation (DEF), motion blur (MB), fast motion (FM), in-plane rotation (IPR), out-of-view (OV), and low resolution (LR). The last column is the Area Under the Curve (AUC). The best results are in red and the second results are in blue.

Trackers	SV	IV	OPR	OCC	BC	DEF	MB	FM	IPR	OV	LR	AUC
MACF	64.6	68.4	64.9	65.6	70.8	64.3	62.1	61.0	65.8	55.6	72.1	69.5
FDSST	62.6	68.0	63.4	63.5	71.4	61.6	63.7	62.5	65.2	49.4	69.9	67.7
LCT	61.3	67.4	64.1	62.1	68.9	63.5	60.8	57.9	65.3	45.8	66.8	68.6
DSST	62.0	67.5	61.9	60.2	67.4	60.8	56.8	54.3	63.6	46.4	68.3	64.7
KCF	60.5	65.2	63.1	60.4	71.6	60.8	56.3	57.0	63.4	46.5	65.1	63.7
CSK	50.3	54.6	52.0	48.7	57.0	51.4	42.7	41.7	52.9	33.5	54.7	55.7
CT	38.4	35.8	40.9	38.2	38.3	39.9	22.9	25.9	39.9	31.8	49.3	38.7
DEF	38.9	43.5	46.1	43.5	47.6	45.7	35.2	34.9	45.3	29.4	41.6	46.0

Table 4. Success plots of One Pass Evaluation (OPE) for the MACF and the other 7 top trackers on different attributes: SV, IV, OPR, OCC, BC, DEF, MB, FM, IPR, OV, and LR. The last column is the AUC. The best results are in red and the second results are in blue.

Trackers	SV	IV	OPR	OCC	BC	DEF	MB	FM	IPR	OV	LR	AUC
MACF	66.2	71.8	65.5	63.5	72.6	62.5	63.6	63.9	67.0	57.3	65.2	69.6
FDSST	61.8	68.4	62.8	59.5	70.5	58.5	61.6	63.2	66.9	50.4	64.7	68.8
LCT	61.9	67.8	66.6	60.2	66.1	61.6	60.2	62.0	69.9	52.0	64.3	68.1
DSST	59.3	67.5	60.6	56.6	63.8	52.8	52.0	51.0	62.8	43.1	63.6	66.9
KCF	58.2	64.2	62.9	60.0	65.2	58.6	55.3	58.1	63.8	48.0	62.3	66.5
CSK	44.2	47.3	46.7	42.0	52.7	42.5	34.9	38.7	49.5	27.7	43.8	49.3
CT	32.8	29.7	35.6	32.4	35.8	31.5	20.7	21.1	34.9	30.8	40.3	33.0
DEF	34.5	39.7	43.0	41.6	43.1	40.4	27.6	30.5	41.4	34.4	41.9	40.6

Table 5. Success plots of Spatial Robustness Evaluation (SRE) for the MACF and the other 7 top trackers on different attributes: SV, IV, OPR, OCC, BC, DEF, MB, FM, IPR, OV, and LR. The last column is the AUC. The best results are in red and the second results are in blue.

Trackers	SV	IV	OPR	OCC	BC	DEF	MB	FM	IPR	OV	LR	AUC
MACF	60.6	64.5	59.8	58.7	63.4	56.1	56.9	57.9	62.3	51.7	67.5	64.1
FDSST	56.9	60.0	57.8	55.7	62.5	51.2	56.1	58.4	61.4	44.2	64.1	61.8
LCT	57.4	61.8	62.0	56.5	59.7	58.8	52.7	49.9	64.8	44.7	62.7	63.3
DSST	56.8	62.7	56.8	53.7	60.9	50.1	49.0	55.1	59.6	42.5	64.3	60.6
KCF	53.7	58.9	57.0	52.9	60.1	53.8	48.8	53.1	58.3	39.7	56.9	59.4
CSK	41.2	44.7	45.0	41.5	45.7	38.9	33.4	36.4	46.8	28.9	45.9	46.2
CT	35.0	30.9	35.8	33.7	31.6	32.8	22.2	24.6	36.4	30.2	42.6	34.4
DEF	31.9	34.8	38.7	36.0	40.3	35.3	28.9	30.4	40.3	28.3	34.5	37.8

Figure 8. The representative frames from four videos successfully tracked by the MACF (in red) compared to the top 5 trackers including FDSST (in green), LCT (in blue), DSST (in black) and KCF (in purple). From top to bottom, the sequences are "Human6", "Human2", "Skater1" and "tiger1" on the OTB-100 benchmark.

4.5. Comparation on Raw Benchmark Results

The proposed MACF algorithm is compared to Efficient Convolution Operators for tracking (ECO) [51], Multi-Domain convolutional neural Networks for visual tracking (MDNet) [52], Structure-Aware Network for visual tracking (SANet) [53], Continuous Convolution Operators for visual Tracking (C-COT) [54], Fully-Convolutional Siamese networks for object tracking (SiamFC_3s) [55], Multi-task Correlation Particle Filter for robust object tracking (MCPF) [56], Deep learning features based SRDCF (DeepSRDCF) [26], ECO based on Hand-Crafted features (ECO-HC) [51], Discriminative Correlation Filter Tracker with Channel and Spatial Reliability (CSR-DCF) [25] and FDSST [16] on the raw benchmark results. In addition, all the raw benchmark results are open source on the web. Furthermore, the proposed MACF framework is integrated into the ECO-HC tracker (ECO-HC+MACF) and have been tested on the datasets of OTB-50 and OTB-100. The implementation codes are also open source in our Github https://github.com/YijYang/MACF-ECO_HC.

As shown in Table 6, the fused ECO-HC + MACF tracker achieves a gain of 1.5% and 3.2% in SP and PP of OPE on OTB-50 and a gain of 1.3% and 1.9% in SP and PP of OPE compared to the ECO-HC standard FDSST. In addition, it runs at a real-time speed of 19 FPS compared to the ECO-HC tracker with a speed of 21 FPS. Hence, it indicates that the proposed MACF can be integrated easily and flexibly into other visual tracking algorithms, and with little loss of real-time performance while improving the accuracy. Most trackers based on deep learning features are more accurate than the proposed MACF method. However, these trackers usually have a lower running speed than MACF except SiamFC_3s method which runs at 86 FPS on a GPU. The proposed MACF achieves a trade-off between the tracking speed and the accuracy. Hence, it is suitable for the embedded real-time systems (for instance, UAV surveillance or unmanned vehicles) which have strict memory and speed limitation.

Table 6. SP and PP of OPE for the proposed MACF, ECO-HC + MACF and the other 10 top trackers on the raw benchmark results of OTB-50 and OTB-100. The last column is the performance of Real-Time and the results are from the original paper, not tested on the same platform. The column of Deep Learning indicates whether the tracker is based on deep learning features. The best results are in red and the second results are in blue.

Trackers	OTB-50		OTB-100		Deep Learning	Real Time (FPS)
	SP of OPE (%)	PP of OPE (%)	SP of OPE (%)	PP of OPE (%)		
ECO	64.3	87.4	69.4	91.0	Y	N (6)
MDNet	64.5	89.0	67.8	90.9	Y	N (1)
SANet	-	-	69.2	92.8	Y	N (1)
C-COT	61.4	84.3	67.1	89.8	Y	N (0.3)
SiamFC_3s	51.6	69.2	58.2	77.1	Y	Y (86)
MCPF	58.3	84.3	62.8	87.3	Y	N (0.5)
DeepSRDCF	56.0	77.2	63.5	85.1	Y	N (<1)
CSR-DCF	59.7	66.7	59.8	73.3	N	Y (13)
ECO-HC + MACF	60.7	84.6	65.6	87.5	N	Y (19)
ECO-HC	59.2	81.4	64.3	85.6	N	Y (21)
MACF	52.3	65.1	56.6	69.6	N	Y (51)
FDSST	50.6	60.8	56.2	67.9	N	Y (49)

4.6. Experiment on UAV Video

4.6.1. Materials and Conditions

The UAV video is taken by a high-definition camera without calibration in the mobile phone. The tested UAV is a high-effective drone from Attop company. The specific parameters of the camera and UAV are illustrated in the Table 7. The UAV video is converted to multi-frame images which have the format of JPG file with three channels, and its resolution is 480 × 640 pixels. In the further research, if the camera for experiment is calibrated, the relative experiment results will be improved [57,58].

Table 7. The parameters of the tested camera and UAV.

Camera Parameters		**UAV Parameters**	
Aperture size	F2.2	Product number	W5
Number of Pixel	1200 W	Expand Size	15.5 × 15.5 × 10 cm
Size of Pixel	1.25 μm	Color	Red
Focusing speed	0.23 s	Type of Control Signal	Wireless Fidelity (Wi-Fi)
Image dimensions	3	Others	No Antivibration used and No gimbal [59] used

4.6.2. Results and Analysis

As mentioned above, our adaptive learning rate compute by CSRM scheme is greatly suitable for the scenes of occlusion, motion blur, defocus blur and so on when the appearance model of the target is corrupted. Therefore, it can obtain significant gains on OTB-50 and OTB-100. Nevertheless, the motion-aware scheme proposed in this paper is more propitious to the video sequences with static background and target of fast motion. Hence, in order to validate this point, the MACF is compared with the state-of-the-art trackers including Efficient Convolution Operators with HOG feature and Color name feature (ECO-HC) [51], Background-Aware Correlation Filters (BACF) [14], fast tracking via Spatio-Temporal Context learning (STC) [28], Sum of Template And Pixel-wise LEarners (Staple) [27], learning Spatially Regularized Discriminative Correlation Filters (SRDCF) [26], Distractor-Aware Tracking (DAT) [7] and FDSST [16] on the test video which include the target of UAV of fast motion with static background. The results have been shown in Figure 9, which demonstrate that the proposed MACF is more accurate and robust in scale and translation detection when tracking a fast-moving target. It runs at a high speed of 56 FPS.

Figure 9 and Table 8 indicate that the proposed MACF tracker outperforms most of state-of-the-art trackers when undergoes the situation of fast motion. Figure 10 shows the predicted trajectory by the MACF approach is almost coincides with the actual trajectory. It illustrates our motion-aware strategy is accurate for predicting the position and scale of fast-moving target with a static background. As shown in Figure 10a,b, there are still small burrs in the predicted trajectory. However, after correcting by Kalman filters, the trajectory becomes smoother and more accurate as shown in Figure 10e,f.

Table 8. The Success Plots (SP) and Precision Plots (PP) of One Pass Evaluation (OPE) for the proposed MACF and the other 7 top trackers on the UAV video. The best results are in red and the second results are in blue.

Trackers	MACF	ECO_HC	BACF	STC	STAPLE	SRDCF	DAT	FDSST
SP of OPE (%)	100.0	94.8	20.6	31.4	21.8	98.7	23.1	46.8
PP of OPE (%)	100.0	98.6	24.1	32.2	29.1	99.5	32.5	52.6

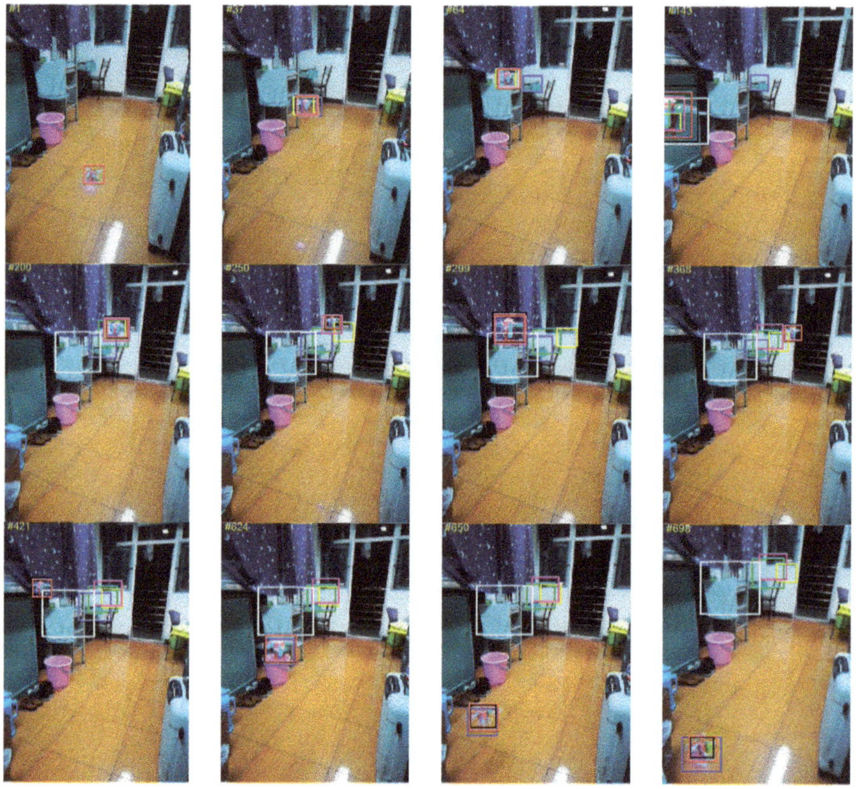

Figure 9. The qualitative experiment comparing the MACF (in red) with state-of-the-art trackers ECO-HC (in blue), BACF (in cyan), STC (in white), Staple (in green), SRDCF (in black), DAT (in yellow) and FDSST (in pink) on UAV video sequence with static background.

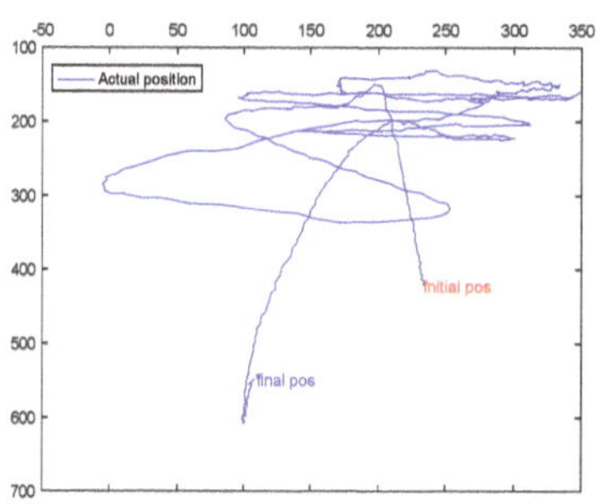

(**a**) The actual position of the UAV in the plane

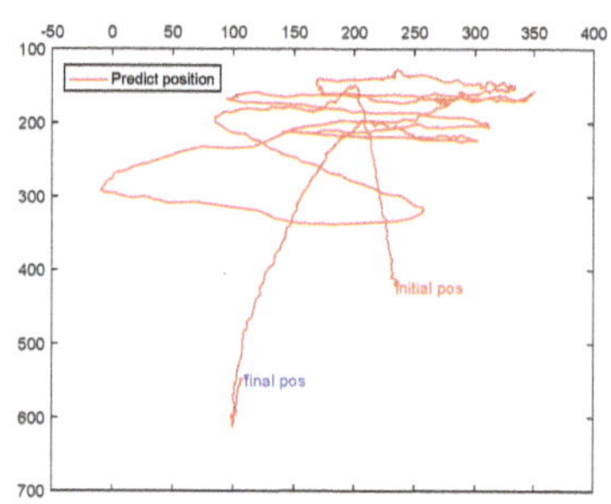

(**b**) The predicted position of the UAV in the plane

Figure 10. *Cont.*

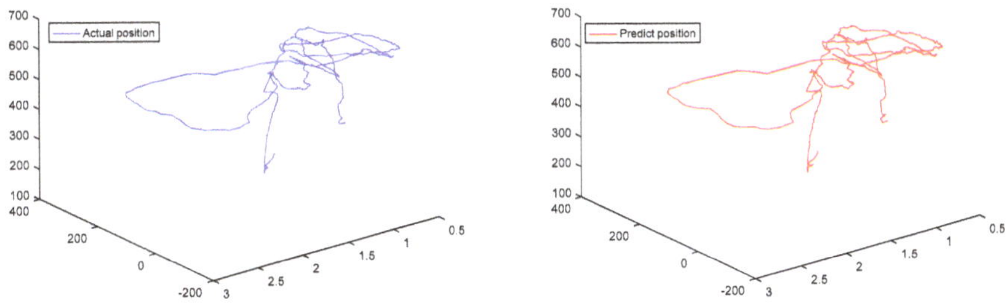

(**c**) The actual position of the UAV in the space (**d**) The predicted position of the UAV in the space

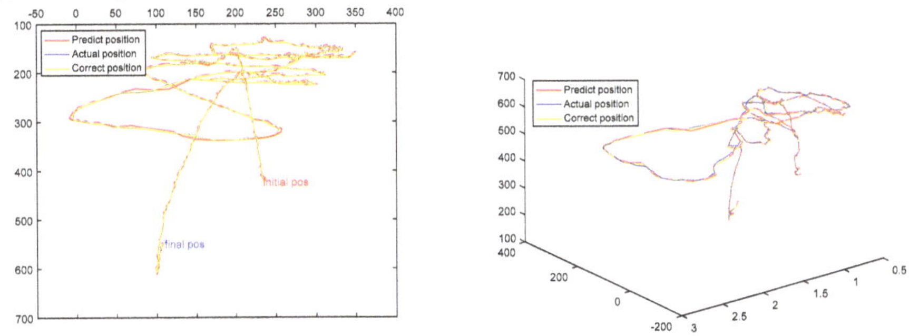

(**e**) The compared results with the predicted, actual and corrected positions in the plane (**f**) The compared results with the predicted, actual and corrected positions in the space

Figure 10. Illustration of the accuracy of the predicted position and scale of the UAV. Here, the Actual position (in blue) is the actual UAV position which is calibrated by manual in the video, the Predict position (in red) is predicted by the instantaneous motion estimation method and the Correct position (in green) is filtered position by Kalman filters. (**a**,**b**) indicate respectively in-plane predicted and actual positions. (**c**,**d**) show apart 3D predicted and actual positions where the scale represents the dept motion. (**e**,**f**) display the overall results.

5. Conclusions

In this paper, a novel tracking framework called MACF is proposed in detail, which fuses the motion cues with the FDSST algorithm for accurately estimating the position and scale of the target. The proposed approach utilizes the instantaneous motion estimation method to predict the position and scale of the target in the next frame. The optimal Kalman Filters are employed to filter noises, and then the FDSST tracker is used to detect the position and scale based on the predictions. Moreover, an improved confidence function of response map is further proposed to determine whether the results of detection are accurate enough to update. Then an adaptive learning rate is set according to the confidence function to prevent model corrupted by occlusions. Furthermore, the proposed MACF framework is flexible and can be readily incorporated into other visual tracking algorithms. Numerous experiments on popular benchmark OTB-50, OTB-100 and UAV video indicate that the proposed MACF achieve a significant improvement among the compared trackers. In this work, the situation where the target is occluded is detected by utilizing the confidence function. Then it prevents model drifting by reducing the learning rate. It is suitable for handling the situations of incomplete occlusions. When the target is severely occluded or completely occluded, the proposed MACF sets the learning rate to 0, hence, the model of the target is not be degraded by occlusions. However, if the target comes out of the other side of the occlusion object and moves out of the current search area, the tracking will fail. Therefore, in future work, a re-detect method is expected to track the target when the target is severely occluded or completely occluded to ensure robust tracking. For instance, when the object is completely occluded, the search area should be extended, and the position and scale of the target can be predicted by the previous velocity and acceleration until the target is re-detected judging by the confidence function.

Author Contributions: Y.Z. and Y.Y. conceived the main idea, designed the main algorithm and wrote the manuscript. Y.Y. designed the main experiments under the supervision of Y.Z., W.Z. and D.L., and the experimental results were analyzed by Y.Z. and L.S. W.Z. and D.L. provided suggestions for the proposed algorithm.

Funding: This research was funded by [the Fundamental Research Funding for the Central Universities of Ministry of Education of China] grant number [18D110408] and [the Special Project Funding for the Shanghai Municipal Commission of Economy and Information Civil-Military Inosculation Project "Big Data Management System of UAVs"] grant number [JMRH-2018-1042]. And the APC was funded by [the Fundamental Research Funding for the Central Universities of Ministry of Education of China].

Conflicts of Interest: The authors declare no conflict of interest.

Appendix

In this section, the expressions bellow are used to prove that the squared response map has a significant effect on confidence calculation. As described in Section 3.5, the Confidence of the Squared Response Map function (CSRM) is defined as follows:

$$\text{CSRM} = \frac{\left|R_{\max}^2 - R_{\min}^2\right|^2}{\frac{1}{MN}\sum_{i=1}^{M}\sum_{j=1}^{N}\left|R_{ij}^2 - R_{\min}^2\right|^2}$$

The Confidence of Response Map function (CRM) is defined as follows:

$$\text{CRM} = \frac{\left|R_{\max} - R_{\min}\right|^2}{\frac{1}{MN}\sum_{i=1}^{M}\sum_{j=1}^{N}\left|R_{ij} - R_{\min}\right|^2}$$

Hence, the difference between the CSRM and CRM compute by follows:

$$\begin{aligned}
\text{CSRM-CRM} &= \frac{\left|R_{\max}^2-R_{\min}^2\right|^2}{\frac{1}{MN}\sum_{i=1}^{M}\sum_{j=1}^{N}\left|R_{ij}^2-R_{\min}^2\right|^2} - \frac{\left|R_{\max}-R_{\min}\right|^2}{\frac{1}{MN}\sum_{i=1}^{M}\sum_{j=1}^{N}\left|R_{ij}-R_{\min}\right|^2} \\
&\geq \frac{\left|R_{\max}^2-R_{\min}^2\right|}{\frac{1}{MN}\sum_{i=1}^{M}\sum_{j=1}^{N}\left|R_{ij}^2-R_{\min}^2\right|} - \frac{\left|R_{\max}-R_{\min}\right|}{\frac{1}{MN}\sum_{i=1}^{M}\sum_{j=1}^{N}\left|R_{ij}-R_{\min}\right|} \\
&= \frac{\left|R_{\max}-R_{\min}\right|*\sum_{i=1}^{M}\sum_{j=1}^{N}\left|(R_{ij}-R_{\min})*(R_{\max}-R_{\min})\right| - \left|R_{\max}-R_{\min}\right|*\sum_{i=1}^{M}\sum_{j=1}^{N}\left|R_{ij}^2-R_{\min}^2\right|}{\frac{1}{M^2N^2}\sum_{i=1}^{M}\sum_{j=1}^{N}\left|R_{ij}^2-R_{\min}^2\right|*\sum_{i=1}^{M}\sum_{j=1}^{N}\left|R_{ij}-R_{\min}\right|} \\
&\geq \frac{\left|R_{\max}-R_{\min}\right|*\sum_{i=1}^{M}\sum_{j=1}^{N}\left|(R_{ij}-R_{\min})*(R_{\max}-R_{ij})\right|}{\frac{1}{M^2N^2}\sum_{i=1}^{M}\sum_{j=1}^{N}\left|R_{ij}^2-R_{\min}^2\right|*\sum_{i=1}^{M}\sum_{j=1}^{N}\left|R_{ij}-R_{\min}\right|} \\
&\geq 0
\end{aligned}$$

Therefore, CSRM $\geq$ CRM and the difference between them increases as the value of $R_{\max}$ increases. Furthermore, the larger value of $R_{\max}$ means the higher confidence score. Hence, this increases the gap between the confidence response and the diffident response.

References

1. Bolme, D.S.; Beveridge, J.R.; Draper, B.A.; Lui, Y.M. Visual object tracking using adaptive correlation filters. In Proceedings of the 2010 IEEE Computer Society Conference on Computer Vision and Pattern Recognition (CVPR), San Francisco, CA, USA, 13–18 June 2010; Volume 119, pp. 2544–2550.
2. Kalal, Z.; Mikolajczyk, K.; Matas, J. Tracking-Learning-Detection. *IEEE Trans. Pattern Anal. Mach. Intell.* **2012**, *34*, 1409–1422. [CrossRef] [PubMed]

3. Wu, Y.; Lim, J.; Yang, M.H. Online Object Tracking: A Benchmark. In Proceedings of the IEEE Conference on Computer Vision and Pattern Recognition CVPR, Portland, OR, USA, 23–28 June 2013; Volume 9, pp. 2411–2418.
4. Wu, Y.; Lim, J.; Yang, M.H. Online Object Tracking: A Benchmark. *IEEE Trans. Pattern Anal. Mach. Intell.* **2015**, *37*, 1834–1848. [CrossRef] [PubMed]
5. Kristan, M.; Pflugfelder, R.; Leonardis, A.; Matas, J.; Porikli, F.; Cehovin, L.; Nebehay, G.; Fernandez, G.; Vojir, T.; Gatt, A.; et al. The Visual Object Tracking VOT2014 Challenge Results. In Proceedings of the IEEE European Conference on Computer Vision Workshops (ECCVW), Zurich, Switzerland, 6–12 September 2014; Volume 8926, pp. 191–217.
6. Kristan, M.; Pflugfelder, R.; Leonardis, A.; Matas, J.; Porikli, F.; Cehovin, L.; Nebehay, G.; Fernandez, G.; Vojir, T.; Gatt, A.; et al. The Visual Object Tracking VOT2016 Challenge Results. In Proceedings of the IEEE European Conference on Computer Vision Workshops (ECCVW), Amsterdam, The Netherlands, 8–10 October 2016; pp. 777–823.
7. Possegger, H.; Mauthner, T.; Bischof, H. In defense of color-based model-free tracking. In Proceedings of the IEEE Conference on Computer Vision and Pattern Recognition (CVPR), Boston, MA, USA, 7–12 June 2015; Volume 2015, pp. 2113–2120.
8. Vojir, T.; Noskova, J.; Matas, J. Robust Scale-Adaptive Mean-Shift for Tracking. *Pattern Recognit. Lett.* **2014**, *49*, 250–258. [CrossRef]
9. Danelljan, M.; Khan, F.S.; Felsberg, M.; Weijer, J.V.D. Adaptive Color Attributes for Real-Time Visual Tracking. In Proceedings of the IEEE Conference on Computer Vision and Pattern Recognition (CVPR), Columbus, OH, USA, 23–28 June 2014; Volume 2014, pp. 1090–1097.
10. Hare, S.; Saffari, A.; Torr, P.H.S. Struck: Structured output tracking with kernels. *IEEE Trans. Pattern Anal. Mach. Intell.* **2015**, *38*, 2096–2109. [CrossRef] [PubMed]
11. He, S.; Yang, Q.; Lau, R.W.H.; Wang, J.; Yang, M.H. Visual Tracking via Locality Sensitive Histograms. In Proceedings of the IEEE Conference on Computer Vision and Pattern Recognition, Portland, OR, USA, 23–28 June 2013; Volume 2013, pp. 2427–2434.
12. Lu, H.; Jia, X.; Yang, M.H. Visual tracking via adaptive structural local sparse appearance model. In Proceedings of the IEEE Computer Vision and Pattern Recognition CVPR, Providence, RI, USA, 16–21 June 2012; Volume 2012, pp. 1822–1829.
13. Learnedmiller, E.; Sevillalara, L. Distribution fields for tracking. In Proceedings of the IEEE Computer Vision and Pattern Recognition CVPR, Providence, RI, USA, 16–21 June 2012; Volume 2012, pp. 1910–1917.
14. Galoogahi, H.K.; Fagg, A.; Lucey, S. Learning Background-Aware Correlation Filters for Visual Tracking. In Proceedings of the 2017 IEEE International Conference on Computer Vision Wrokshop (ICCVW), Venice, Italy, 22–29 October 2017; Volume 2017, pp. 1144–1152.
15. Galoogahi, H.K.; Sim, T.; Lucey, S. Correlation filters with limited boundaries. In Proceedings of the IEEE IEEE Conference on Computer Vision and Pattern Recognition (CVPR), Boston, MA, USA, 7–12 June 2015; Volume 2015, pp. 4630–4638.
16. Danelljan, M.; Häger, G.; Khan, F.S.; Felsberg, M. Discriminative Scale Space Tracking. *IEEE Trans. Pattern Anal. Mach. Intell.* **2017**, *39*, 1561–1575. [CrossRef] [PubMed]
17. Danelljan, M.; Häger, G.; Khan, F.S.; Felsberg, M. Accurate Scale Estimation for Robust Visual Tracking. In Proceedings of the British Machine Vision Conference, Nottingham, UK, 1–5 September 2014; Volume 2014, pp. 65.1–65.11.
18. Li, Y.; Zhu, J. A Scale Adaptive Kernel Correlation Filter Tracker with Feature Integration. In Proceedings of the IEEE European Conference on Computer Vision (ECCV), Zurich, Switzerland, 6–12 September 2014; Volume 8926, pp. 254–265.
19. Li, F.; Yao, Y.; Li, P.; Zhang, D.; Zuo, W.; Yang, M.H. Integrating Boundary and Center Correlation Filters for Visual Tracking with Aspect Ratio Variation. In Proceedings of the 2017 IEEE International Conference on Computer Vision Wrokshop (ICCVW), Venice, Italy, 22–29 October 2017; Volume 2017, pp. 2001–2009.
20. Zhang, K.; Zhang, L.; Yang, M.H. Real-Time Compressive Tracking. In Proceedings of the IEEE European Conference on Computer Vision (ECCV), Florence, Italy, 7–13 October 2012; Volume 2012, pp. 864–877.
21. Rui, C.; Martins, P.; Batista, J. Exploiting the circulant structure of tracking-by-detection with kernels. In Proceedings of the IEEE European Conference on Computer Vision (ECCV), Florence, Italy, 7–13 October 2012; Volume 2012, pp. 702–715.

22. Joao, F.H.; Rui, C.; Pedro, M.; Jorge, B. High-Speed Tracking with Kernelized Correlation Filters. *IEEE Trans. Pattern Anal. Mach. Intell.* **2015**, *37*, 583–596.
23. Xu, L.; Luo, H.; Hui, B.; Zheng, C. Real-Time Robust Tracking for Motion Blur and Fast Motion via Correlation Filters. *Sensors* **2016**, *16*, 1443. [CrossRef] [PubMed]
24. Li, F.; Zhang, S.; Qiao, X. Scene-Aware Adaptive Updating for Visual Tracking via Correlation Filters. *Sensors* **2017**, *17*, 2626. [CrossRef] [PubMed]
25. Lukezic, A.; Vojir, T.; Zajc, L.C.; Matas, J.; Kristan, M. Discriminative Correlation Filter Tracker with Channel and Spatial Reliability. *Int. J. Comput. Vis.* **2018**, *126*, 671–688. [CrossRef]
26. Danelljan, M.; Hager, G.; Khan, F.S.; Felsberg, M. Learning Spatially Regularized Correlation Filters for Visual Tracking. In Proceedings of the IEEE International Conference on Computer Vision (ICCV), Santiago, Chile, 7–13 December 2015; Volume 2015, pp. 4310–4318.
27. Bertinetto, L.; Valmadre, J.; Golodetz, S.; Miksik, O.; Torr, P.H. Staple: Complementary Learners for Real-Time Tracking. In Proceedings of the IEEE Conference on Computer Vision and Pattern Recognition (CVPR), Boston, MA, USA, 7–12 June 2015; Volume 38, pp. 1401–1409.
28. Zhang, K.; Zhang, L.; Yang, M.-H.; Zhang, D. Fast Tracking via Spatio-Temporal Context Learning. Computer. *Science* **2013**, *1*, 25–32.
29. Yang, R.; Wei, Z. Real-Time Visual Tracking through Fusion Features. *Sensors* **2016**, *16*, 949.
30. Shi, G.; Xu, T.; Guo, J.; Luo, J.; Li, Y. Consistently Sampled Correlation Filters with Space Anisotropic Regularization for Visual Tracking. *Sensors* **2017**, *17*, 2889. [CrossRef] [PubMed]
31. Li, Y.; Zhu, J.; Hoi, S.C.H. Reliable Patch Trackers: Robust visual tracking by exploiting reliable patches. In Proceedings of the IEEE Conference on Computer Vision and Pattern Recognition (CVPR), Boston, MA, USA, 7–12 June 2015; Volume 2015, pp. 353–361.
32. Fan, H.; Ling, H. Parallel Tracking and Verifying: Parallel Tracking and Verifying: A Framework for Real-Time and High Accuracy Visual Tracking. In Proceedings of the 2017 IEEE International Conference on Computer Vision Wrokshop (ICCVW), Venice, Italy, 22–29 October 2017; Volume 2017, pp. 5487–5495.
33. Hao, W.; Chen, S.X.; Yang, B.F.; Chen, K. Robust cubature Kalman filter target tracking algorithm based on genernalized M-estiamtion. *Acta Phys. Sin.* **2015**, *64*, 1–7.
34. Gao, S.; Liu, Y.; Wang, J.; Deng, W.; Heekuck, O. The Joint Adaptive Kalman Filter (JAKF) for Vehicle Motion State Estimation. *Sensors* **2016**, *16*, 1103. [CrossRef] [PubMed]
35. Su, L.M.; Hojin, J.; Woo, S.J.; Gook, P.C. Kinematic Model-Based Pedestrian Dead Reckoning for Heading Correction and Lower Body Motion Tracking. *Sensors* **2015**, *15*, 28129–28153. [CrossRef] [PubMed]
36. Pajares Redondo, J.; Prieto González, L.; García Guzman, J.; López Boada, B.; Díaz López, V. VEHIOT: Design and Evaluation of an IoT Architecture Based on Low-Cost Devices to Be Embedded in Production Vehicles. *Sensors* **2018**, *18*, 486. [CrossRef] [PubMed]
37. Ettlinger, A.; Neuner, H.; Burgess, T. Development of a Kalman Filter in the Gauss-Helmert Model for Reliability Analysis in Orientation Determination with Smartphone Sensors. *Sensors* **2018**, *18*, 414. [CrossRef] [PubMed]
38. Li, P.; Zhang, T.; Ma, B. Unscented Kalman filter for visual curve tracking. *Image Vis. Comput.* **2004**, *22*, 157–164. [CrossRef]
39. Funk, N. *A Study of the Kalman Filter Applied to Visual Tracking*; University of Alberta: Edmonton, AB, Canada, 2003.
40. Yoon, Y.; Kosaka, A.; Kak, A.C. A New Kalman-Filter-Based Framework for Fast and Accurate Visual Tracking of Rigid Objects. *IEEE Trans. Robot.* **2008**, *24*, 1238–1251. [CrossRef]
41. Bewley, A.; Ge, Z.; Ott, L.; Ramos, F.; Upcroft, B. Simple online and realtime tracking. In Proceedings of the IEEE International Conference on Image Processing, Phoenix, AZ, USA, 25–28 September 2016; Volume 2016, pp. 3464–3468.
42. Wang, M.; Liu, Y.; Huang, Z. Large Margin Object Tracking with Circulant Feature Maps. In Proceedings of the IEEE Conference on Computer Vision and Pattern Recognition (CVPR), Honolulu, HI, USA, 21–16 July 2017; Volume 2017, pp. 4800–4808.
43. Gladh, S.; Danelljan, M.; Khan, F.S.; Felsberg, M. Deep Motion Features for Visual Tracking. In Proceedings of the 2016 23rd International Conference on Pattern Recognition (ICPR), Cancun, Mexico, 4–8 December 2016.
44. Danelljan, M.; Bhat, G.; Gladh, S.; Khan, F.S.; Felsberg, M. Deep Motion and Appearance Cues for Visual Tracking. *Pattern Recognit. Lett.* **2018**. [CrossRef]

45. Ma, C.; Yang, X.; Zhang, C.; Yang, M.H. Long-term correlation tracking. In Proceedings of the IEEE Conference on Computer Vision and Pattern Recognition (CVPR), Boston, MA, USA, 7–12 June 2015; Volume 2015, pp. 5388–5396.
46. Wojke, N.; Bewley, A.; Paulus, D. Simple Online and Realtime Tracking with a Deep Association Metric. In Proceedings of the 2017 IEEE International Conference on Image Processing (ICIP), Beijing, China, 17–20 September 2017; Volume 2017, pp. 3645–3649.
47. Kronhamn, T. Geometric illustration of the kalman filter gain and covariance update algorithms. *IEEE Control Syst. Mag.* **2003**, *5*, 41–43. [CrossRef]
48. Oron, S.; Bar-Hillel, A.; Levi, D.; Avidan, S. Locally Orderless Tracking. In Proceedings of the 2012 IEEE Conference on Computer Vision and Pattern Recognition, Providence, RI, USA, 16–21 June 2012; Volume 2012, pp. 1940–1947.
49. Wang, D.; Lu, H.; Yang, M.H. Least Soft-Threshold Squares Tracking. In Proceedings of the IEEE Computer Vision and Pattern Recognition CVPR, Portland, OR, USA, 23–28 June 2013; Volume 2013, pp. 2371–2378.
50. Ahuja, N. Robust visual tracking via multi-task sparse learning. In Proceedings of the 2012 IEEE Conference on Computer Vision and Pattern Recognition, Providence, RI, USA, 16–21 June 2012; Volume 2012, pp. 2042–2049.
51. Danelljan, M.; Bhat, G.; Khan, F.S.; Felsberg, M. ECO: Efficient Convolution Operators for Tracking. In Proceedings of the IEEE Conference on Computer Vision and Pattern Recognition, Honolulu, HI, USA, 21–26 July 2017; Volume 2017, pp. 6931–6939.
52. Nam, H.; Han, B. Learning Multi-Domain Convolutional Neural Networks for Visual Tracking. In Proceedings of the IEEE Conference on Computer Vision and Pattern Recognition CVPR, Boston, MA, USA, 7–12 June 2015; Volume 2015, pp. 4293–4302.
53. Fan, H.; Ling, H. SANet: Structure-Aware Network for Visual Tracking. In Proceedings of the IEEE Conference on Computer Vision and Pattern Recognition Workshops CVPRW, Honolulu, HI, USA, 21–26 July 2017; Volume 2017, pp. 2217–2224.
54. Danelljan, M.; Robinson, A.; Khan, F.S.; Felsberg, M. Beyond Correlation Filters: Learning Continuous Convolution Operators for Visual Tracking. In Proceedings of the IEEE European Conference on Computer Vision (ECCV), Amsterdam, The Netherlands, 8–10 October 2016; Volume 2016, pp. 472–488.
55. Bertinetto, L.; Valmadre, J.; Henriques, J.F.; Vedaldi, A.; Torr, P.H.S. Fully-Convolutional Siamese Networks for Object Tracking. In Proceedings of the IEEE European Conference on Computer Vision (ECCV), Amsterdam, The Netherlands, 8–10 October 2016; Volume 2016, pp. 850–865.
56. Zhang, T.; Xu, C.; Yang, M.H. Multi-task Correlation Particle Filter for Robust Object Tracking. In Proceedings of the IEEE Conference on Computer Vision and Pattern Recognition CVPR, Honolulu, HI, USA, 21–26 July 2017; Volume 2017, pp. 4819–4827.
57. Gašparović, M.; Gajski, D. Two-step camera calibration method developed for micro UAV's. In Proceedings of the XXIII ISPRS Congress, Prague, Czech Republic, 12–19 July 2016; Volume 2016, pp. 829–833.
58. Pérez, M.; Agüera, F.; Carvajal, F. Low Cost Surveying Using AN Unmanned Aerial Vehicle. *ISPRS Int. Arch. Photogramm. Remote Sens. Spat. Inf. Sci.* **2013**, *40*, 311–315. [CrossRef]
59. Gašparović, M.; Jurjević, L. Gimbal Influence on the Stability of Exterior Orientation Parameters of UAV Acquired Images. *Sensors* **2017**, *17*, 401. [CrossRef] [PubMed]

sensors

MDPI

Article

Improved Point-Line Feature Based Visual SLAM Method for Indoor Scenes

Runzhi Wang [1,2], Kaichang Di [1], Wenhui Wan [1,*] and Yongkang Wang [3]

1 State Key Laboratory of Remote Sensing Science, Institute of Remote Sensing and Digital Earth, Chinese Academy of Sciences, No. 20A, Datun Road, Chaoyang District, Beijing 100101, China; wangrz@radi.ac.cn (R.W.); dikc@radi.ac.cn (K.D.)
2 College of Resources and Environment, University of Chinese Academy of Sciences, Beijing 100049, China
3 School of Environment Science and Spatial Informatics, China University of Mining and Technology, Xuzhou 221116, China; ts16160030a3@cumt.edu.cn
* Correspondence: wanwh@radi.ac.cn; Tel.: +86-10-6480-7987

Received: 2 September 2018; Accepted: 16 October 2018; Published: 20 October 2018

Abstract: In the study of indoor simultaneous localization and mapping (SLAM) problems using a stereo camera, two types of primary features—point and line segments—have been widely used to calculate the pose of the camera. However, many feature-based SLAM systems are not robust when the camera moves sharply or turns too quickly. In this paper, an improved indoor visual SLAM method to better utilize the advantages of point and line segment features and achieve robust results in difficult environments is proposed. First, point and line segment features are automatically extracted and matched to build two kinds of projection models. Subsequently, for the optimization problem of line segment features, we add minimization of angle observation in addition to the traditional re-projection error of endpoints. Finally, our model of motion estimation, which is adaptive to the motion state of the camera, is applied to build a new combinational Hessian matrix and gradient vector for iterated pose estimation. Furthermore, our proposal has been tested on EuRoC MAV datasets and sequence images captured with our stereo camera. The experimental results demonstrate the effectiveness of our improved point-line feature based visual SLAM method in improving localization accuracy when the camera moves with rapid rotation or violent fluctuation.

Keywords: indoor visual SLAM; adaptive model; motion estimation; stereo camera

1. Introduction

Simultaneous localization and mapping (SLAM) is used to incrementally estimate the pose of a moving platform and simultaneously build a map of the surrounding environment [1–3]. Owing to its ability of autonomous localization and environmental perception, SLAM has become a key prerequisite for robots to operate autonomously in an unknown environment [4]. Visual SLAM, a system that uses a camera as its data input sensor, is widely used in platforms moving in indoor environments. Compared with radar and other range-finding instruments, a visual sensor has the advantages of low power consumption and small volume, and it can provide more abundant environmental texture information for a moving platform. Consequently, visual SLAM has drawn increasing attention in the research community [5]. As a unique example, integration of visual odometry (VO) with these strategies has been applied successfully to planet rover localization of many planetary exploration missions [6–9], and has assisted the rovers to travel through challenging planetary surfaces by providing high-precision visual positioning results. Subsequently, many researchers attempted to improve the efficiency and robustness of SLAM methods. In terms of improving efficiency, some feature extraction algorithms such as Speeded-Up Robust Features (SURF) [10], Binary Robust Invariant Scalable Keypoints (BRISK) [11], and oriented FAST and rotated BRIEF (ORB) [12] were proposed. Further, some systems introduced parallel computing to improve efficiency, such as Parallel Tracking and Mapping (PTAM) for small Augmented Reality (AR) workspaces [13]. This is the first SLAM system to separate feature tacking and mapping as two threads, realizing real-time SLAM. As for improving accuracy and robustness, some SLAM systems have introduced the bag-of-words model [14] for the detection of loop closure. Once a loop closure is detected, the closure error is greatly reduced. In recent years, the ability of autonomous localization and environmental perception has rendered visual SLAM an important method, especially in global navigation satellite system (GNSS) denied environments such as indoor scenes [15,16].

Visual SLAM can be implemented using a monocular camera [17–20], multi-camera [21–23], and RGB-D camera [24–26] setups. The iterative closet point (ICP) algorithm is used in motion estimation from consecutive frames containing dense point clouds and has been applied effectively in RGB-D-based SLAM [27,28]. However, dense point clouds, produced by dense matching and triangulation of stereo or multi-camera, have uncertainties and invalid regions in environments of low texture and illumination change [29], so in most of the visual SLAM methods, sparse feature extraction and matching are employed to calculate pose of the moving platform. Point and line segments are the two types of primary features used in visual SLAM. Point features have been predominantly used because of their convenient parameterization and implementation in feature tracking between consecutive frames. The visual SLAM systems based on point features estimate camera pose and build an environmental map by minimizing the reprojection error of the observed and corresponding reprojected point features. Furthermore, this optimization process is often solved using the general graph optimization algorithm [30]. ORB-SLAM2 is a representative state-of-the-art visual SLAM method based on point feature [31]; it supports monocular cameras, stereo cameras, and RGB-D cameras, and can produce high-precision results in real time.

In addition to point feature-based visual SLAM systems, line feature-based SLAM systems have been developed recently. Although a line feature is not as easily parameterized as a point feature, as a higher-dimensional feature than a point feature, it can express more environmental information in indoor scenes. Zhang et al. built a graph-based visual SLAM system using 3D straight lines instead of a point feature for localization and mapping [32]. StructSLAM used structure lines of buildings and demonstrated the advantage of a line feature in an indoor scene with many artificial objects [33]. Although the line features can provide more structural information, their endpoints are instable. This problem has been tackled in [34] by utilizing relaxed constraints on their positions.

The above systems use point and line features separately. Some visual SLAM methods combine point and line features. For example, a semi-direct monocular VO, named PL-SVO [35], can obtain more robust results in low-textured scenes by combining points and line segments. The PL-SVO uses the photometric difference between pixels of the same 3D line segment point to estimate the pose increment. The authors of PL-SVO also proposed a robust point-line feature-based stereo VO [36]. In this stereo system, the camera motion is recovered through non-linear minimization of the projection errors of two kinds of features. Based on the work of [35,36], the authors extended [36] with loop closure detection algorithm, and developed a stereo SLAM system named PL-SLAM [37]. Note that there is also a real-time monocular visual SLAM [38], which combines point and line features for localization and mapping, and the nonlinear least square optimization model of point and line features is similar to [37]. The major difference between them is that the former uses a monocular camera and the latter uses a stereo camera. In literature [39], the authors proposed a tightly-coupled monocular visual-inertial odometry (VIO) system exploiting both point-line features and inertial measurement units (IMUs) to estimate the state of camera. In those point-line feature based VO or SLAM systems, the distances from the two re-projected endpoints to the observed line segments are often used as the values to be optimized. However, the structural information of line segments, such as the angle between the re-projected and observed line segments, is not considered in the process of optimization. Furthermore, only the VO system in [36] weighted the errors of different features according to their covariance matrices, and other reported systems do not consider the distribution of weight among different features. Di et al. obtained the inverse of the error as the weights of different data sources in RGB-D SLAM and achieved good results [40], but the motion information of the camera was not considered.

In this paper, an improved point-line feature based visual SLAM method in indoor scenes is proposed. First, unlike the traditional nonlinear least square optimization model of line segment features, an improvement of our method is the addition of the minimization of angle observation, which should be close to zero between the line segments of observation and re-projection. Compared with the traditional nonlinear least square optimization model, which includes distances between the re-projected endpoints and the observed line segment, our method combines angle observation and distance observation and shows a better performance at large turns. Second, our visual SLAM method builds an adaptive model in motion estimation so that the pose estimation model is adaptive to the motion state of a camera. With these two improvements, our visual SLAM can fully utilize point and line segment features irrespective of whether the camera is moving or turning sharply. Experimental results on EuRoC MAV datasets and sequence images captured with our stereo camera are presented to verify the accuracy and effectiveness of this improved point-line feature-based visual SLAM method in indoor scenes.

2. Methodology

Figure 1 illustrates our method in a simplified sequence flowchart, which consists of the following main parts: (1) extraction and matching of point and line segment features; (2) building nonlinear least square optimization models of the two kinds of features; (3) motion estimation with an adaptive model. Technical details of the algorithms and models are given in the following sub-sections.

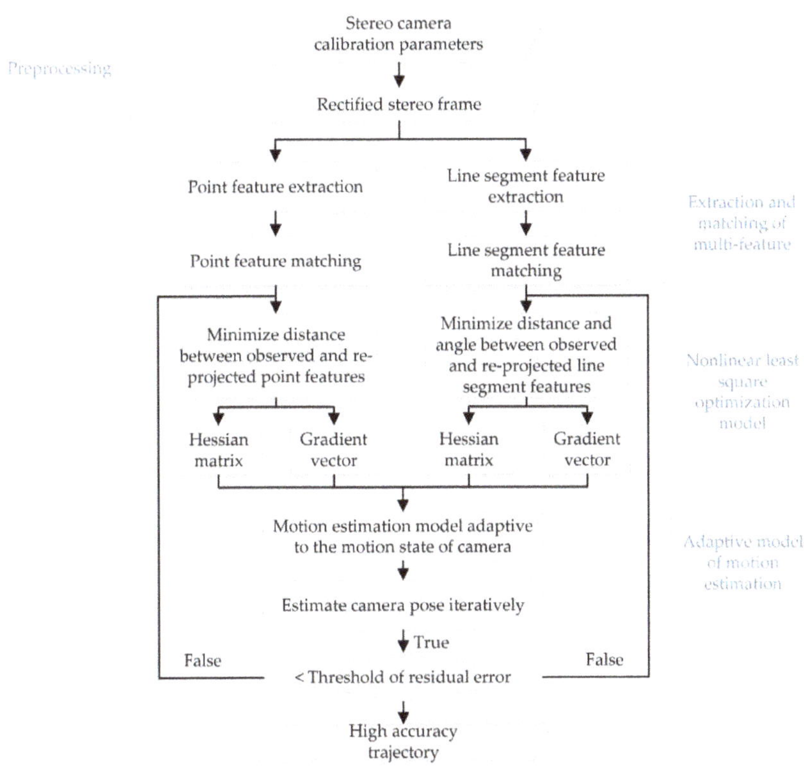

Figure 1. Flowchart of our proposed visual SLAM method.

2.1. Extraction and Matching of Point and Line Segment Features

In point feature tracking, the ORB algorithm [12] is adopted in our method to extract 2D point features and create binary descriptors for initial matching. The matching of the extracted point features in consecutive frames is followed by random sample consensus (RANSAC) algorithm and a fundamental matrix constraint, which is used to eliminate some erroneous corresponding keypoints from the matched results. The fundamental matrix constraint is also called epipolar constraint. That is, if the point m of the left image is obtained, its corresponding point on the right image will be constrained on the epipolar line l' like Figure 2 shows. As a stereo camera has a baseline, we can calculate the depths and 3D coordinates of all the keypoints with respect to the optical center of the left camera.

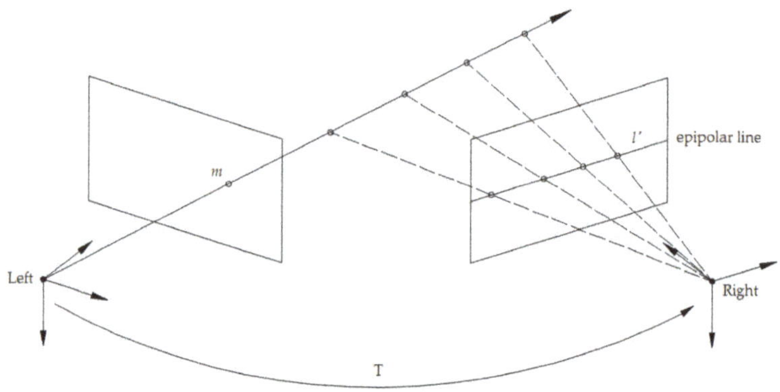

Figure 2. Illustration of the fundamental matrix constraint.

We use the line segment detector (LSD) algorithm [41] for the extraction of line segment features. It can extract line segment features from indoor scenes in linear time, satisfying the real-time requirement of SLAM. Although there are many low-texture environments such as white walls in indoor scenes, the LSD algorithm can stably extract line features, as shown in Figure 3. Furthermore, the line band descriptor method [42] is employed to match line segment features in stereo and consecutive frames with binary descriptors. Similar to the matched point features, we can obtain the 3D coordinates of two endpoints of line segment features and their 2D coordinates.

Figure 3. Common low-texture environment in an indoor scene with white walls, and point features (green x marks) extracted using ORB algorithm and line segment features (red lines) extracted using LSD algorithm.

2.2. Nonlinear Least Square Optimization Models for Motion Estimation

Once the 3D and 2D coordinates of the point and line segment features are obtained, the relationships between two consecutive frames can be established using the homologous features. Subsequently, the point and line segment features are back-projected from the previous frame to the current frame. Subsequently, back-projection error models are built for both the points and the line segments. As these error models are nonlinear, the camera motion should be iteratively estimated using the nonlinear least square optimization method. In this study, we use the Gauss–Newton algorithm for the minimization of the back-projection errors. This section presents the nonlinear least square optimization models of point and line segment features.

2.2.1. Optimization Model of Point Features

For the point features, we use the perspective-n-point method to optimize the camera pose. The error model is the re-projection error, which is the difference between the re-projected 2D position and the observed (matched) 2D position. The optimization process can be divided into three steps. First, 3D map points at the current frame are obtained from stereo image matching and space intersection computation. The world 3D map points are transformed into the coordinate system of the current frame using the iteratively estimated pose of the camera. Subsequently, these 3D map points are re-projected into the image coordinate system of the current frame. Finally, by minimizing the distance error between the re-projected points and their corresponding observed points on the current frame, the error model of point features can be established. This process is illustrated in Figure 4.

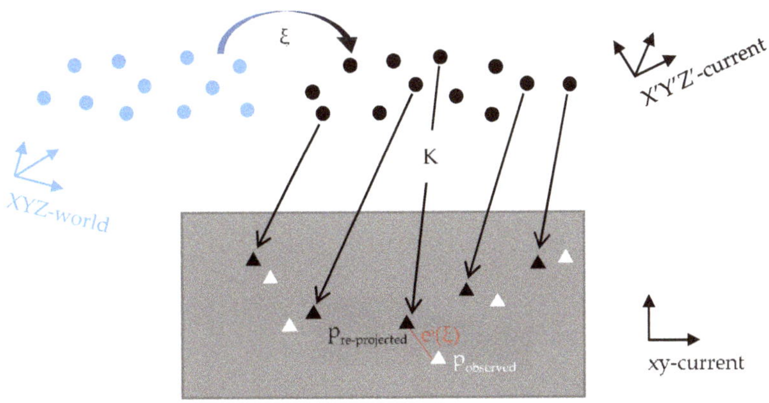

Figure 4. Process of building an error model of point features. The blue dots and black dots represent the world 3D map points and current 3D map points, respectively. The black triangles represent the re-projected 2D points and the white triangles represent the observed 2D points.

In the re-projection error model, the error of the i-th point feature can be described as follows:

$$\mathbf{e}_p^i(\zeta) = \mathbf{K} \cdot \mathbf{T}(\zeta) \cdot P_{XYZ-world} - p. \quad (1)$$

Here, ζ is a six-dimensional vector of Lie algebras that represents the motion of the camera, and $\mathbf{T}(\zeta)$ represents the transformation matrix from the world coordinate system $P_{XYZ-world}(X,Y,Z)$ to the current coordinate system $P'(X',Y',Z')$ based on the pose of the camera. $\mathbf{K}$ represents the internal parameters of the camera and $p(x,y)$ is the corresponding observed point of the re-projected point $p'(x',y')$. $\mathbf{e}_p^i(\zeta)$ is the resultant error vector.

To use the Gauss–Newton method, the partial derivative of the error function with respect to the variables is required, which is the Jacobian matrix $\frac{\partial \mathbf{e}_p^i(\zeta)}{\partial \zeta}$ and can be obtained via the chain rule:

$$\frac{\partial \mathbf{e}_p^i(\zeta)}{\partial \zeta} = \frac{\partial \mathbf{e}_p^i(\zeta)}{\partial p'}\frac{\partial p'}{\partial P'}\frac{\partial P'}{\partial \zeta} = \frac{\partial \mathbf{e}_p^i(\zeta)}{\partial p'}\frac{\partial p'}{\partial \zeta}. \tag{2}$$

By calculating $\frac{\partial p'}{\partial P'}$ and $\frac{\partial P'}{\partial \zeta}$, we can obtain $\frac{\partial p'}{\partial \zeta}$ as follows:

$$\frac{\partial p'}{\partial \zeta} = \begin{bmatrix} f_x \frac{1}{Z'} & 0 & -f_x \frac{X'}{Z'^2} & -f_x \frac{X'Y'}{Z'^2} & f_x(1+\frac{X'^2}{Z'^2}) & -f_x \frac{Y'}{Z'} \\ 0 & f_y \frac{1}{Z'} & -f_y \frac{Y'}{Z'^2} & -f_y(1+\frac{Y'^2}{Z'^2}) & f_y \frac{X'Y'}{Z'^2} & f_y \frac{X'}{Z'} \end{bmatrix}. \tag{3}$$

As for $\frac{\partial \mathbf{e}_p^i(\zeta)}{\partial p'}$, it is a function matrix whose independent variables are the pixel coordinates x' and y'. Thus, the following equation is obtained:

$$\frac{\partial \mathbf{e}_p^i(\zeta)}{\partial p'} = \begin{bmatrix} \frac{\partial(x'-x)}{\partial x'} & \frac{\partial(x'-x)}{\partial y'} \\ \frac{\partial(y'-y)}{\partial x'} & \frac{\partial(y'-y)}{\partial y'} \end{bmatrix} = \begin{bmatrix} 1 & 0 \\ 0 & 1 \end{bmatrix} = \mathbf{I}_{2*2}. \tag{4}$$

After $\frac{\partial \mathbf{e}_p^i(\zeta)}{\partial p'}$ and $\frac{\partial p'}{\partial \zeta}$ are calculated, we can obtain the Jacobian matrix of point features $\frac{\partial \mathbf{e}_p^i(\zeta)}{\partial \zeta}$. In this study, the Gauss–Newton algorithm is used for the iterative estimation of the camera motion. Therefore, we must calculate the Hessian matrix $\mathbf{H}_p^i$ and gradient vector $\mathbf{g}_p^i$ required by the Gauss–Newton algorithm. The Jacobian matrix is represented by $\mathbf{J}_p$ and the Hessian matrix and gradient vector can be obtained as follows:

$$\begin{cases} \mathbf{H}_p^i = \mathbf{J}_p{}^{\mathrm{T}} \cdot \mathbf{P} \cdot \mathbf{J}_p \\ \mathbf{g}_p^i = -\mathbf{J}_p{}^{\mathrm{T}} \cdot \mathbf{P} \cdot \mathbf{e}_p^i(\zeta) \end{cases}, \tag{5}$$

where $\mathbf{P}$ is the weight matrix of a point feature and can be defined as:

$$\mathbf{P} = \begin{bmatrix} \frac{1}{1+\|\mathbf{e}_p^i(\zeta)\|} & 0 \\ 0 & \frac{1}{1+\|\mathbf{e}_p^i(\zeta)\|} \end{bmatrix}. \tag{6}$$

Thus, we can add $\mathbf{H}_p^i$ and $\mathbf{g}_p^i$ of each point and obtain the Hessian matrix $\mathbf{H}_p$ and gradient vector $\mathbf{g}_p$ of all point features in the current frame:

$$\mathbf{H}_p = \sum_{i=1}^{n} \mathbf{H}_p^i,\ \mathbf{g}_p = \sum_{i=1}^{n} \mathbf{g}_p^i. \tag{7}$$

Through the above steps, the optimization model of point features is established.

2.2.2. Optimization Model of Line Segment Features

As for the error model of line segment features, we use two kinds of error functions. One is the traditional minimization of the distances from the re-projected endpoints to the observed line segment. The other is our proposal for use in this study: the error of angle observation, which should be close to zero between the line segments of observation and re-projection. Each line segment feature has two distance errors and two angle observation errors. The process of establishing an error model of line segment features is shown in Figure 5. Consequently, by minimizing both the distance errors and the angle observation errors, the optimization model of line segment features can be established.

The error function of the j-th line segment feature is an 4×1 error vector $\mathbf{e}_l^j(\zeta)$, which can be expressed as follows:

$$\mathbf{e}_l^j(\zeta) = \begin{bmatrix} e^1(\zeta) \\ e^2(\zeta) \\ e^3(\zeta) \\ e^4(\zeta) \end{bmatrix} = \begin{bmatrix} a \times x_{p'} + b \times y_{p'} + c \\ a \times x_{q'} + b \times y_{q'} + c \\ (\overrightarrow{\mathbf{qp}'} \cdot \overrightarrow{\mathbf{qp}}) / (\|\overrightarrow{\mathbf{qp}'}\| \times \|\overrightarrow{\mathbf{qp}}\|) - 1 \\ (\overrightarrow{\mathbf{pq}'} \cdot \overrightarrow{\mathbf{qp}}) / (\|\overrightarrow{\mathbf{pq}'}\| \times \|\overrightarrow{\mathbf{qp}}\|) - (-1) \end{bmatrix}, \tag{8}$$

where:

$$\begin{cases} p'(x_{p'}, y_{p'}) = \mathbf{K} \cdot \exp(\zeta^\wedge) \cdot P_{XYZ-world} \\ q'(x_{q'}, y_{q'}) = \mathbf{K} \cdot \exp(\zeta^\wedge) \cdot Q_{XYZ-world} \end{cases}.$$

In Equation (8), a, b, and c are the three coefficients of the general equation of the observed line. Point $p(x_p, y_p)$ and point $q(x_q, y_q)$ are the starting and ending endpoints of the observed line, respectively. Similarly, point $p'(x_{p'}, y_{p'})$ and point $q'(x_{q'}, y_{q'})$ represent the endpoints of the re-projected line segment. The error $e^1(\zeta)$ can be considered the distance between point p' and the observed line pq, and error $e^2(\zeta)$ is the distance between point q' and the observed line. In addition to the two error functions of distance, we add the error functions of angle. Here, $e^3(\zeta)$ is the cosine of the angle between vector $\overrightarrow{\mathbf{qp}'}$ and vector $\overrightarrow{\mathbf{qp}}$, and $e^4(\zeta)$ is the cosine of the angle between vector $\overrightarrow{\mathbf{pq}'}$ and vector $\overrightarrow{\mathbf{qp}}$.

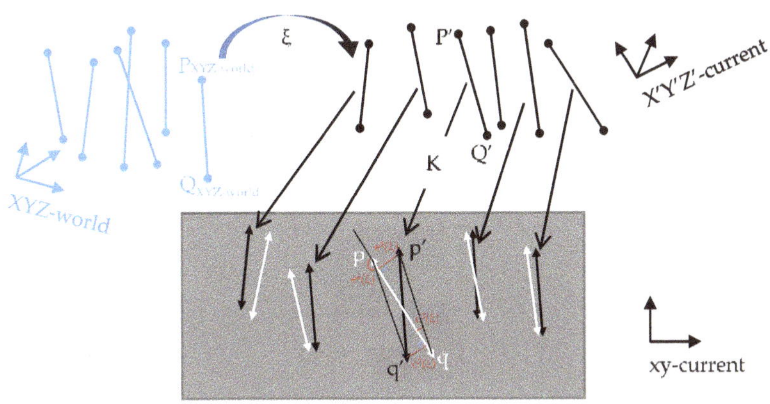

Figure 5. Process of building an error model of line segment features. The blue lines and black lines represent the world 3D map line segments and current 3D map line segments, respectively. The black and white lines with triangular endpoints represent the re-projected and observed line segments, respectively.

Similar to the point features, we use the chain rule to calculate the Jacobian matrix $\frac{\partial \mathbf{e}_l^j(\zeta)}{\partial \zeta}$ of line segment features as follows:

$$\frac{\partial \mathbf{e}_l^j(\zeta)}{\partial \zeta} = \begin{bmatrix} \frac{\partial e^1(\zeta)}{\partial p'} \frac{\partial p'}{\partial P'} \frac{\partial P'}{\partial \zeta} \\ \frac{\partial e^2(\zeta)}{\partial q'} \frac{\partial q'}{\partial Q'} \frac{\partial Q'}{\partial \zeta} \\ \frac{\partial e^3(\zeta)}{\partial p'} \frac{\partial p'}{\partial P'} \frac{\partial P'}{\partial \zeta} \\ \frac{\partial e^4(\zeta)}{\partial q'} \frac{\partial q'}{\partial Q'} \frac{\partial Q'}{\partial \zeta} \end{bmatrix} = \begin{bmatrix} \frac{\partial e^1(\zeta)}{\partial p'} \frac{\partial p'}{\partial \zeta} \\ \frac{\partial e^2(\zeta)}{\partial q'} \frac{\partial q'}{\partial \zeta} \\ \frac{\partial e^3(\zeta)}{\partial p'} \frac{\partial p'}{\partial \zeta} \\ \frac{\partial e^4(\zeta)}{\partial q'} \frac{\partial q'}{\partial \zeta} \end{bmatrix}. \tag{9}$$

The 3D coordinates of the endpoints $P'(X_{P'}, Y_{P'}, Z_{P'})$ and $Q'(X_{Q'}, Y_{Q'}, Z_{Q'})$ are obtained using the pose transformation $\exp(\zeta^\wedge)$. Using Equation (3), we can calculate $\frac{\partial p'}{\partial \zeta}$ and $\frac{\partial q'}{\partial \zeta}$. The subsequent step is to calculate $\frac{\partial e^1(\zeta)}{\partial p'}$, $\frac{\partial e^2(\zeta)}{\partial q'}$, $\frac{\partial e^3(\zeta)}{\partial p'}$, and $\frac{\partial e^4(\zeta)}{\partial q'}$. They are functions of points $p'(x_{p'}, y_{p'})$ or $q'(x_{q'}, y_{q'})$ and can be calculated as follows:

$$\begin{cases} \frac{\partial e^1(\zeta)}{\partial p'} = [a, b] \\ \frac{\partial e^2(\zeta)}{\partial q'} = [a, b] \end{cases}, \begin{cases} \frac{\partial e^3(\zeta)}{\partial p'} = [f_{p'x}, f_{p'y}] \\ \frac{\partial e^4(\zeta)}{\partial q'} = [f_{q'x}, f_{q'y}] \end{cases}, \tag{10}$$

where a and b are the coefficients of the general equation of the observed line; $f_{p'x}$, $f_{p'x}$, $f_{p'x}$, and $f_{p'x}$ are partial derivatives of the coordinates of the re-projected points $p'(x_{p'}, y_{p'})$ and $q'(x_{q'}, y_{q'})$.

$$\begin{cases} f_{p'x} = \frac{(x_p - x_q)\times\|\vec{\mathbf{qp}}\|\times\|\vec{\mathbf{qp}'}\| - ((x_p - x_q)\times(x_{p'} - x_q) + (y_p - y_q)\times(y_{p'} - y_q))\times\|\vec{\mathbf{qp}}\|\times(x_{p'} - x_q)/\|\vec{\mathbf{qp}'}\|}{\|\vec{\mathbf{qp}}\|\times\|\vec{\mathbf{qp}}\|\times\|\vec{\mathbf{qp}'}\|\times\|\vec{\mathbf{qp}'}\|} \\ f_{p'y} = \frac{(y_p - y_q)\times\|\vec{\mathbf{qp}}\|\times\|\vec{\mathbf{qp}'}\| - ((x_p - x_q)\times(x_{p'} - x_q) + (y_p - y_q)\times(y_{p'} - y_q))\times\|\vec{\mathbf{qp}}\|\times(y_{p'} - y_q)/\|\vec{\mathbf{qp}'}\|}{\|\vec{\mathbf{qp}}\|\times\|\vec{\mathbf{qp}}\|\times\|\vec{\mathbf{qp}'}\|\times\|\vec{\mathbf{qp}'}\|} \\ f_{q'x} = \frac{(x_p - x_q)\times\|\vec{\mathbf{qp}}\|\times\|\vec{\mathbf{pq}'}\| - ((x_p - x_q)\times(x_{q'} - x_p) + (y_p - y_q)\times(y_{q'} - y_p))\times\|\vec{\mathbf{qp}}\|\times(x_{q'} - x_p)/\|\vec{\mathbf{pq}'}\|}{\|\vec{\mathbf{qp}}\|\times\|\vec{\mathbf{qp}}\|\times\|\vec{\mathbf{qp}'}\|\times\|\vec{\mathbf{qp}'}\|} \\ f_{q'y} = \frac{(y_p - y_q)\times\|\vec{\mathbf{qp}}\|\times\|\vec{\mathbf{pq}'}\| - ((x_p - x_q)\times(x_{q'} - x_p) + (y_p - y_q)\times(y_{q'} - y_p))\times\|\vec{\mathbf{qp}}\|\times(y_{q'} - y_p)/\|\vec{\mathbf{pq}'}\|}{\|\vec{\mathbf{qp}}\|\times\|\vec{\mathbf{qp}}\|\times\|\vec{\mathbf{qp}'}\|\times\|\vec{\mathbf{qp}'}\|} \end{cases} \tag{11}$$

Thus, using Equations (3), (10), and (11), we can obtain the Jacobian matrix $\frac{\partial \mathbf{e}_l^j(\zeta)}{\partial \zeta}$ of the j-th line segment features.

Similar to the point features, the Hessian matrix $\mathbf{H}_l^j$ and gradient vector $\mathbf{g}_l^j$ of line segment features are also required by the Gauss–Newton algorithm. The Jacobian matrix is represented by $\mathbf{J}_l$ and the Hessian matrix and gradient vector can be obtained as follows:

$$\begin{cases} \mathbf{H}_l^j = \mathbf{J}_l^{\mathrm{T}} \cdot \mathbf{P} \cdot \mathbf{J}_l \\ \mathbf{g}_l^j = -\mathbf{J}_l^{\mathrm{T}} \cdot \mathbf{P} \cdot \mathbf{e}_l^j(\zeta) \end{cases}, \tag{12}$$

where $\mathbf{P}$ is the weight matrix of the j-th line segment feature. As the dimensions of the two distance error functions and the two angle error functions are different, they are weighted in two ways. Thus, $\mathbf{P}$ can be defined as:

$$\mathbf{P} = \begin{bmatrix} \frac{1}{1+\|e^1(\zeta)\|} & 0 & 0 & 0 \\ 0 & \frac{1}{1+\|e^2(\zeta)\|} & 0 & 0 \\ 0 & 0 & \frac{1}{\|e^3(\zeta)\|} & 0 \\ 0 & 0 & 0 & \frac{1}{\|e^4(\zeta)\|} \end{bmatrix}. \tag{13}$$

Subsequently, we add $\mathbf{H}_l^j$ and $\mathbf{g}_l^j$ of each line segment and obtain the Hessian matrix $\mathbf{H}_l$ and gradient vector $\mathbf{g}_p$ of all the line segment features in the current frame as follows:

$$\mathbf{H}_l = \sum_{j=1}^{m} \mathbf{H}_l^j,\ \mathbf{g}_l = \sum_{j=1}^{m} \mathbf{g}_l^j. \tag{14}$$

Through these steps, the optimization model of line segment features is established. Compared with the traditional error model of line segment features used in literature [38,39], we add the angular error functions. We have tested our error model on the EuRoC datasets [43] and compare the results with those obtained from the traditional error model. Figure 6 shows the resulting trajectories of the two models tested on dataset Vicon room 1 "medium". As shown in Figure 6, A and B are two big turns. From the accuracy heat map of the positioning results of these two places, our extended error model with added angular error functions is observed to be superior to the traditional error model.

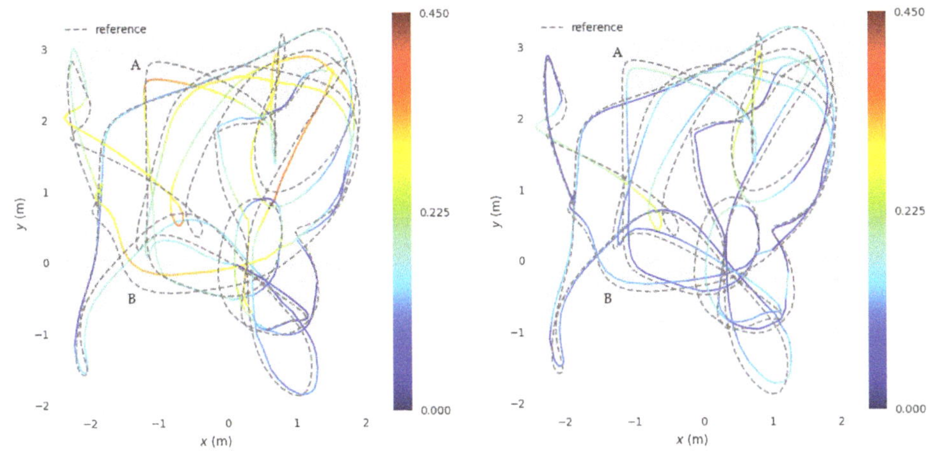

(**a**) (**b**)

Figure 6. Positioning accuracy heat maps of the two error models. (**a**) The resulting trajectory of the traditional error model used in reference SLAM system; (**b**) The resulting trajectory of the extended error model used in our proposed SLAM system. The gray dotted line represents the ground-truth. The color solid lines represent the accuracy of the trajectories. A change in color from blue to red indicates a gradual increase in the positioning error.

We use the relative pose error (RPE) as the evaluation metric, which describes the error between pairs of timestamps in the estimated trajectory file. Then we calculate the average RPE at these two big turns A and B to represent the average drift rate between the estimated trajectory and ground-truth. As shown in Table 1, the average RPE of our extended error model is less at A and B than that of traditional error model, meaning that our proposed error model has less drift rate at A and B. This also shows that proposed error model has good robustness at large turns. More detailed results will be given in the experimental results section.

Table 1. Average RPE results at A and B turns of traditional error model and our proposed error model. The numbers in bold indicate that these terms are better than those of another model. The unit of RPE is meter.

Turns in Figure 6	Traditional Error Model	Proposed Error Model
A	0.149500	**0.138527**
B	0.142878	**0.131200**

2.3. *Adaptive Weighting Model of Motion Estimation*

After the Hessian matrices and gradient vectors of both point features and line segment features are established, our motion estimation model, which is adaptive to the motion state of a camera, is applied to build a new recombined Hessian matrix and gradient vector for iterated pose estimation.

As shown in Figure 7, we have collected the residual errors of nearly 1000 positions and their corresponding motion states of a camera with respect to the previous frame. We use the displacement of the current frame relative to the previous frame as a measure of the current motion state. The blue line in Figure 7 is the linear fitting curve according to these data. It can be observed from Figure 7 that there is a certain degree of correlation between the positioning residual errors and the motion state of the camera. Thus, we calculated the correlation coefficient between them and obtained the result of 0.57. The correlation coefficient is greater than 0.5, indicating that the positioning residual errors and the motion state of the camera are strongly correlated. In other words, the motion state of the camera will affect the positioning result to some extent. However, the reference method does not consider this. Therefore, it can be observed from the accuracy heat maps in the following experiment section that reference method has a large absolute trajectory error (ATE) when the camera moves with large rotation or rapid fluctuation. Hence, we build an adaptive model in the iterative motion estimation. The model is adaptive to the motion state of the camera.

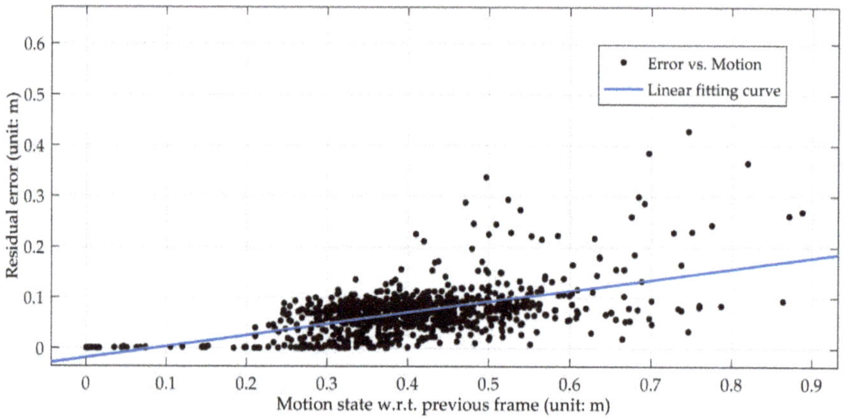

Figure 7. Linear fitting curve between the residual error of the positioning result and the current motion state of a camera. There are 994 motion states and their corresponding residual errors to fit the blue linear curve.

With each iteration, we can obtain the motion of the current frame relative to the previous one. As the frame rate of the camera is fixed, the motion state on the three axes can be represented by the change of camera position $\Delta P(\Delta X, \Delta Y, \Delta Z)$. If the motion of the camera relative to the previous frame is greater in the image plane direction than in the direction perpendicular to the image plane, i.e., ΔX and ΔY are larger than ΔZ, this indicates that the camera is shaking, which may result in blurred or weakened image texture. According to our experience, line segment features can provide significant structural information of the environment, and hence, the detection of the line segment is more robust than the detection of a point feature in such poor texture scenes. It can also be observed from the experimental results in Figure 6 that the line segment features play an important role when the camera makes a big turn. Thus, in such situations, the weight of the line segment features should be larger than the weight of the point features according to the experimental results and experience. If the motion of the camera relative to the previous frame is greater in the direction perpendicular to the image plane than in the image plane direction, ΔZ will be larger. According to the experiments, the point features in this situation are relatively rich and stable. Hence, the weight of the point features should be larger than the weight of the line segment features. Moreover, we use the inverse of the average re-projection error as a factor in weighting the point feature and line segment feature. Based on the comparative experiments and the above analysis, we propose the following adaptive weighting model of motion estimation:

$$\begin{cases} W_p = \dfrac{\exp(sqrt((\Delta X \times \text{fps})^2 + (\Delta Y \times \text{fps})^2))}{(\sum_{i=1}^{n} \left\| \mathbf{e}_p^i(\zeta) \right\|)/n} \\ W_l = \dfrac{\exp(sqrt((\Delta Z \times \text{fps})^2))}{(\sum_{j=1}^{m} \left\| \mathbf{e}_l^j(\zeta) \right\|)/m} \end{cases} . \quad (15)$$

With Equation (15), a new recombined Hessian matrix **H** and gradient vector **g** can be obtained as follows:

$$\begin{cases} \mathbf{H} = \mathbf{H}_p \times W_p + \mathbf{H}_l \times W_l \\ \mathbf{g} = \mathbf{g}_p \times W_p + \mathbf{g}_l \times W_l \end{cases} . \quad (16)$$

Thus, we can use the Gauss–Newton algorithm to estimate the motion of the camera iteratively. With new frames acquired sequentially, our point-line based visual SLAM system calculates the new positions according to our adaptive weighting model of motion estimation.

3. Experimental Results

In this section, to verify the actual performance of the proposed method, we have performed a series of experiments using two types of datasets: public datasets with ground-truth, and sequence images captured using our stereo camera. We also compared our method with the reference method adopted in literature [38,39], which uses the traditional error model of line feature and its weighting model is based on residual errors. All the experiments were performed on a desktop computer with an Intel Core i7-6820HQ CPU with 2.7 GHz and 16 GB RAM without GPU parallelization. The results of the experiments are described in detail below.

3.1. EuRoC MAV Datasets

The EuRoC MAV datasets were collected by an on-board micro aerial vehicle (MAV) [43]. They contain two batches of datasets. The first batch was recorded in the Swiss Federal Institute of Technology Zurich (ETH) machine hall and the second batch was recorded in two indoor rooms. They were both captured with a global shutter camera at 20 FPS. Each dataset contains stereo images and accurate ground-truth. Furthermore, calibration parameters such as the intrinsic and extrinsic parameters of the stereo camera are provided in the datasets. We compared our proposed method with the reference method adopted in recent paper and changed the optimization parameters of the reference method to better adapt to different scenarios for fair comparison. We use the absolute trajectory error (ATE) as the evaluation metric, which directly calculates the error between the estimated trajectory and the ground truth [44]. And we calculate both translation and rotation part of ATE as an evaluation of six degree-of-freedom (DoFs).

Figure 8 shows the accuracy of the three coordinate axes on several different datasets. The dotted line represents the ground truth of the dataset. The solid lines in blue and red represent the results of reference method and our proposed method, respectively. As shown in Figure 8a,b, when the Z-axis values have a large fluctuation while the X-axis and Y-axis are stably changing, our proposed method is superior to reference method in the Z-axis. Further, as shown in Figure 8c, when the values of all the three axes fluctuate greatly, our proposed method is more stable and accurate than reference method in these quivering parts. For example, in the X-axis section of Figure 8c from 55–70 s and in the 40–60 s part of the Z-axis, our estimated trajectory is much closer to the ground truth than that of reference method. And then we calculate the average RPE at these places in Figure 8 where camera has rapid fluctuation. The average RPE can represent the average drift rate between the estimated trajectory and ground-truth. As can be seen in Table 2, the average RPE of our proposed method is less than that of reference method, which means our proposed method has less drift rate at these quivering parts.

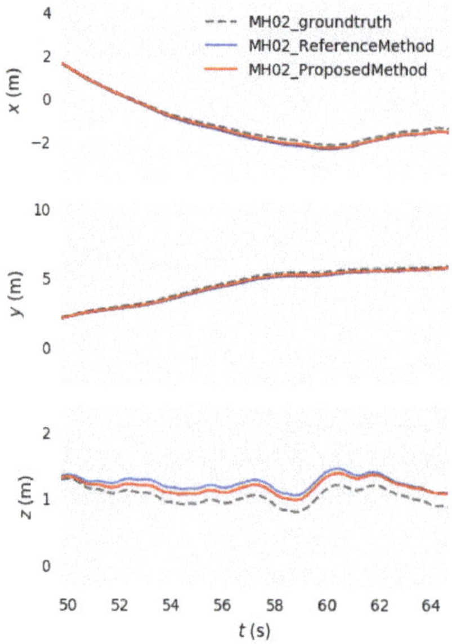

(**a**)

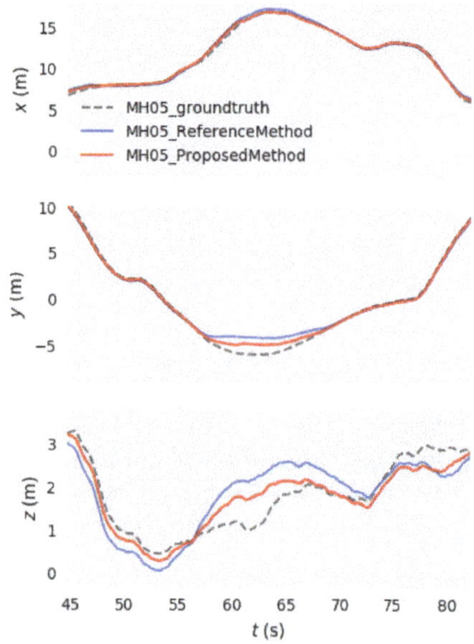

(**b**)

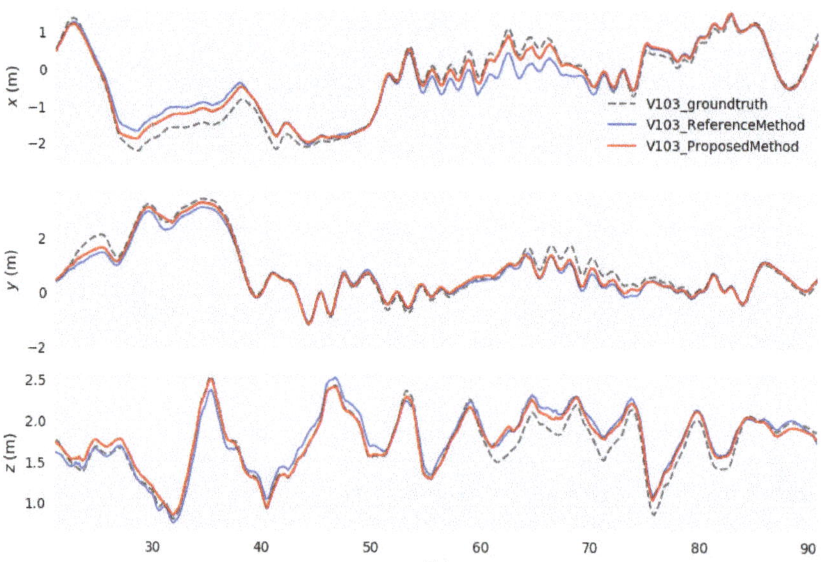

(**c**)

Figure 8. Accuracy of reference method (blue) and our proposed method (red) on the three coordinate axes. (**a**) 50–64 s part of MH_02_easy dataset; (**b**) 45–80 s part of MH_05_difficult dataset. (**c**) 25–90 s part of V1_03_difficult dataset. The trajectories of our proposed method are closer to the ground truth than those of reference method when the camera has large fluctuation.

Table 2. Average RPE results at three quivering parts of reference method and our proposed method. The numbers in bold indicate that these terms are better than those of another method. The unit of RPE is meter.

Part of Datasets	Reference Method	Proposed Method
MH02 50–64 s	0.086233	**0.084906**
MH05 45–80 s	0.224745	**0.152861**
V103 25–90 s	0.182875	**0.113256**

These good performances in the case of rapid fluctuation of camera are mainly attributed to our extended error model and adaptive weighting model of motion estimation. The adaptive weighting model considers both the average re-projection error and the motion state between frames. Therefore, it can better utilize the advantages of different features in different motion states, so as to obtain better positioning results. Figure 8 and Table 2 also confirm that our proposed method performs better when the camera shakes quickly. For quantitative evaluation, we employed the open-source package evo, an easy-to-use evaluation tool (github.com/MichaelGrupp/evo), to evaluate reference method and our proposed method. Table 3 shows the root mean square error (RMSE) of the translation part and rotation part of ATE. Histograms of RMSE and the range of the translation part of ATE are also provided in Figure 9.

Table 3. Translation parts and rotation parts of ATE of the two methods on several EuRoC MAV datasets. The numbers in bold indicate that these terms are better than those of another method. The unit of translation part is meter and the unit of rotation part is degree.

EuRoC MAV Datasets	Reference Method		Proposed Method	
	ATE (Tran.)	ATE (Rot.)	ATE (Tran.)	ATE (Rot.)
MH_01_easy	**0.103648**	2.642111	0.127967	**1.545874**
MH_02_easy	0.211978	1.367287	**0.187721**	**1.180007**
MH_03_medium	0.173194	1.406898	**0.131740**	**0.861475**
MH_04_difficult	1.088483	5.847468	**0.538066**	**3.039995**
MH_05_difficult	0.742775	6.197293	**0.428191**	**3.260860**
V1_01_easy	0.124277	1.522630	**0.106866**	**1.057048**
V1_02_medium	0.208179	2.397838	**0.139928**	**1.747601**
V1_03_difficult	0.434853	2.777067	**0.304662**	**2.514817**
V2_01_easy	0.179175	2.817506	**0.145582**	**2.487347**
V2_02_medium	0.193462	4.059261	**0.183889**	**3.992289**

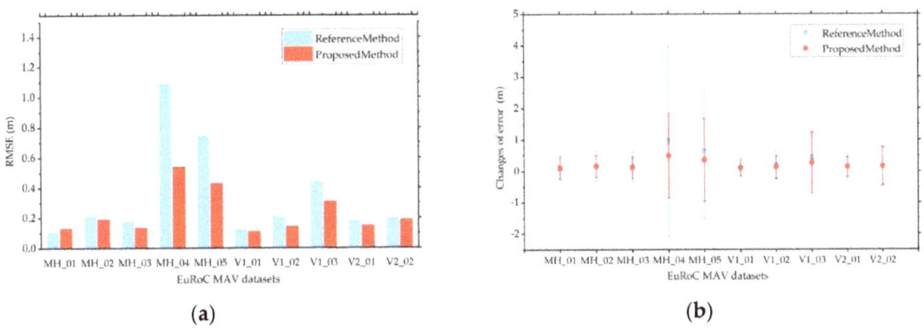

Figure 9. Comparison of RMSEs and the range of translation parts of ATE for reference method and our proposed method using the EuRoC MAV datasets. (**a**) RMSEs of reference method (blue) and our proposed method (red); (**b**) The range of translation parts of ATE of the two methods. The three points on each error bar from top to bottom are the maximum ATE error, average ATE error, and minimum ATE error respectively.

Table 3 shows that our proposed method performs better in almost all scenes of EuRoC MAV datasets for the RMSE in terms of the translation parts and rotation parts of ATE. From Figure 9a, in easy and medium scenes, such as MH_02_easy, V1_01_easy, and V2_02_medium, our proposal shows slightly improved accuracy of the results. However, our method can greatly improve the accuracy in difficult scenes, such as MH_04_difficult, MH_05_difficult, and V1_03_difficult. The main reason for such situations is that the camera shakes rapidly in these difficult scenes. Our method considers this situation and better utilizes the respective advantages of point and line segment features through the adaptive weighting model of motion estimation. Furthermore, as shown in Figure 9b, our proposed method has a smaller range of translation parts of ATE, indicating that the motion estimation is relatively stable.

To demonstrate the results intuitively, several accuracy heat maps of trajectories estimated using reference method and our proposed method are shown in Figure 10. The gray dotted line represents the ground-truth. The color solid lines represent the estimated trajectories. The color bar represents the size of the translation part of ATE. A change in color from blue to red indicates a gradual increase in translation part of ATE. Each row shows the results of the two methods with the same dataset, and the two color bars of each row have the same maximum error and minimum error. Comparing the three trajectories, we can observe that our method shows better accuracy in some areas with large rotations of camera. This also shows that the angular error function added in our model shows a good performance at large turns. Thus, we can conclude that our proposed method with an adaptive motion model and angular error functions can yield smaller errors than reference method when the camera moves with large rotation or rapid fluctuation.

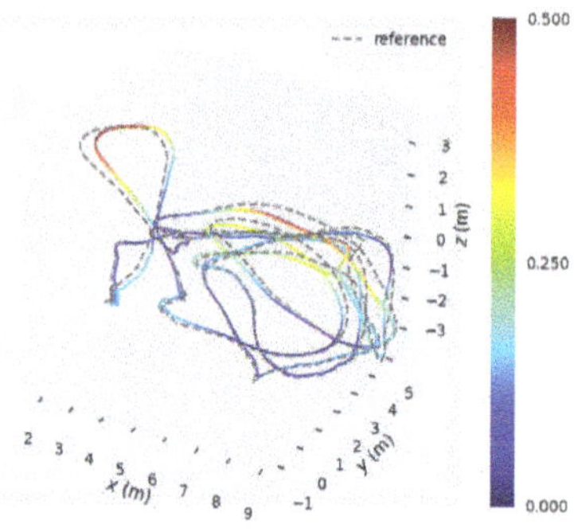

(a)

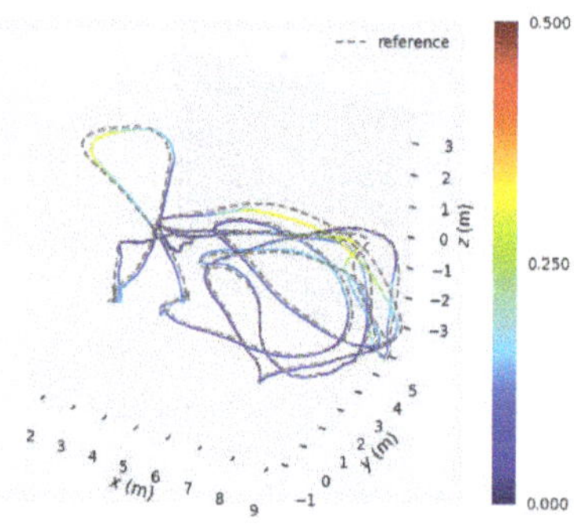

(b)

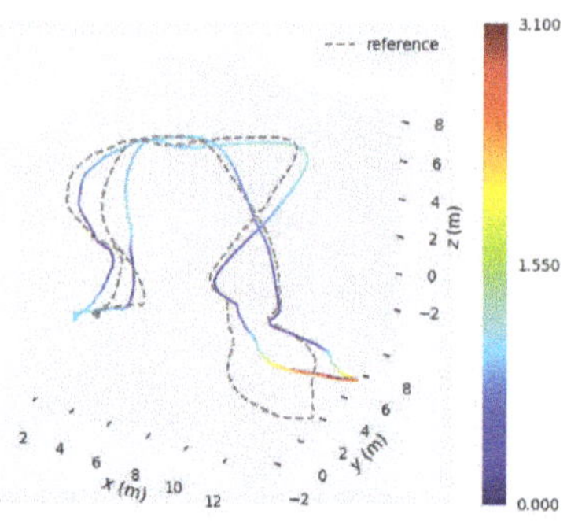

(c)

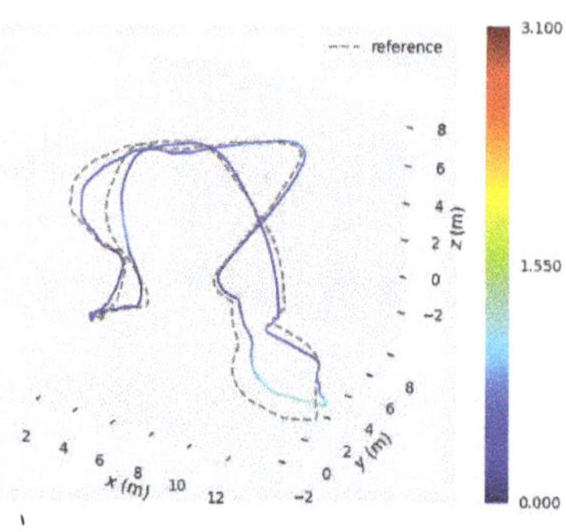

(d)

Figure 10. *Cont.*

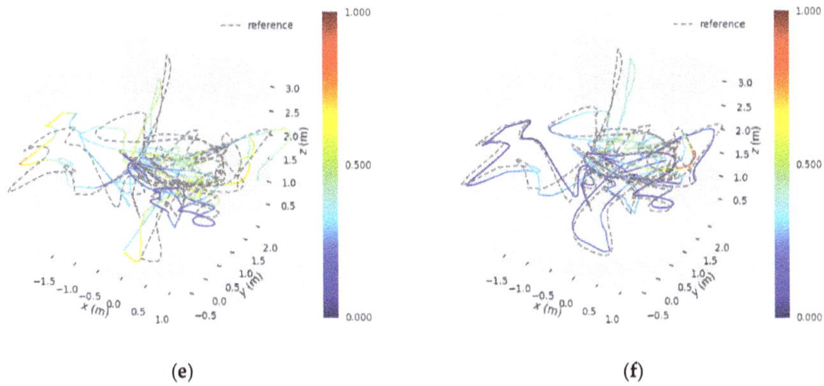

Figure 10. Comparison of several trajectories estimated using reference method and our proposed method. The three accuracy heat maps of the left column are estimated using reference method on the (**a**) MH_03_medium, (**c**) MH_04_difficult, and (**e**) V1_03_difficult sequences. The three accuracy heat maps of the right column are estimated using our proposed method on the (**b**) MH_03_medium, (**d**) MH_04_difficult, and (**f**) V1_03_difficult sequences. The two color bars of each row have the same maximum error and minimum error. The redder the color is, the larger the translation part of ATE is.

3.2. Sequence Images Captured by Our Stereo Camera

In addition to testing the performance and accuracy of our proposed visual SLAM method on public datasets with ground-truth, we also test the universality with sequence images captured with our stereo camera. The sequence images acquired using the stereo camera should be rectified first in order to use them in high-accuracy SLAM processing. In this experiment, a ZED stereo camera is adopted as our data input sensor. It can capture images with a resolution of 720p at up to 60 fps. Although the ZED camera has been adjusted in production, it does not satisfy the requirements of the experiment. We used Stereo Camera Calibrator, a MATLAB-based software package, to complete the camera calibration process, through which the calibration parameters of the stereo camera including lens distortion coefficients and internal and external parameters were calculated. The calibration results are shown in Tables 4 and 5. Using these parameters, we can obtain the rectified stereo sequence images.

Table 4. Internal parameters of the left and right camera. The units of f_x, f_y, c_x, and c_y are pixels.

Camera	f_x	f_y	c_x	c_y	k_1	k_2	k_3	p_1	p_2
Left Camera	659.38	659.51	605.17	375.78	0.00093	−0.00041	0.0	0.0016	0.00036
Right Camera	659.46	659.52	605.98	375.27	0.00063	0.000037	0.0	0.0013	0.00084

Table 5. External parameters of the left camera and right camera.

Rotation Angles (°)	0.00015	−0.0012	−0.00077
Translation Vector (mm)	−119.7164	0.0348	−0.22480

Figure 11 shows the ZED stereo camera used in this experiment. Figure 3 shows a typical image acquired in this experiment. In the acquisition of sequence images, an operator (one of the co-authors of this paper) first placed the camera at the start point on the floor. Subsequently, he picked up the camera and went on a quadrilateral path along the indoor corridor. Finally, he returned to the starting point. Thus, the whole sequence images form a loop closure. In this experiment, we also present a simple comparison with the point-to-point ICP method adopted in [45]. As no ground truth of the trajectory is available for the sequence images captured with our stereo camera, we evaluate the performance by comparing the closure errors of ICP method, reference method used in before experiment (hereinafter referred to as reference method) and our proposed method. Furthermore, for a fair comparison of the three methods, we do not use loop closure detection in this experiment.

Figure 11. ZED stereo camera used in this experiment.

The statistical results are shown in Table 6 and the three estimated trajectories are shown in Figure 12. The percentage error of our proposed method is 1.07%, which is better than the error obtained using the ICP method, i.e., 4.64%, and also better than the error obtained using the reference method, i.e., 1.82%. The path length calculated by three methods is 64.393 m, 65.272 m and 65.120 m, respectively. The three path lengths are close to each other. However, our method shows only approximately a quarter of the closure error of ICP method and half the closure error of reference method. As observed from the trajectories depicted in Figure 12a, the operator moved from the start point and went forward to corner A. After passing through corner A, he went straight to corner B. We can observe that the trajectories of the three methods are very close to each other in this part. However, from corner B to corner C, the trajectories of ICP method and reference method have a large deviation, which eventually leads to a larger closure error than our method.

By analyzing the motion states at corner A and corner B, we observe that Corner A starts on frame 295 and ends on frame 330 (35 frames in total), whereas Corner B starts on frame 397 and ends on frame 422 (25 frames in total). As the FPS of the camera was fixed, it took more time at corner A. In other words, the speed of the camera at corner B is higher than that at corner A, and hence, the motion is more intense. Further, as shown in Figure 12b, from the top view of the three trajectories, the trajectories of ICP method and reference method begin to deform after corner B. However, owing to the two improvements of our method, i.e., the adaptive weighting model for motion estimation and the angular error functions, the positioning result of our method is less affected by the rapid rotation at corner B than that of ICP method and reference method. Therefore, our trajectory is closer to the predetermined quadrilateral path. This experiment also shows that our proposed method is applicable to the sequence images captured with our stereo camera.

Table 6. Localization results of ICP method, reference method and our proposed method.

Method	Total Length (m)	Closure Error (m)	Percentage Error
ICP Method	64.393	2.9913	4.64%
Reference Method	65.272	1.1912	1.82%
Proposed Method	65.120	0.6988	1.07%

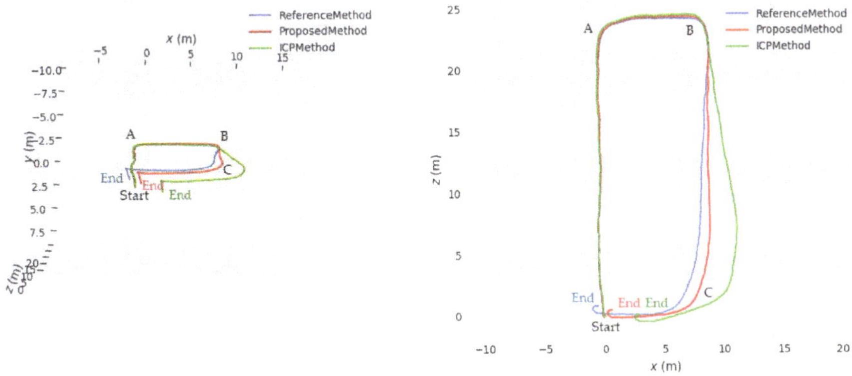

(a) (b)

Figure 12. Estimated trajectories from the three methods. The blue curve represent the trajectory estimated using reference method. The red curve represent the trajectory estimated using our proposed method. The green curve represent the trajectory estimated using ICP method. (**a**) 3D display of the three trajectories; (**b**) Top view of the three trajectories.

4. Conclusions

In this paper, we have presented an improved point-line feature-based visual SLAM method for indoor scenes. The proposed SLAM method has two main innovations: the angular error function added in the optimization process of line segment features, and the adaptive weighting model in iterative pose estimation. Line segment feature is a higher-dimensional feature than point features and has more structural characteristics and geometric constraints. Our optimization model of line segment features with added angular error functions can better utilize this advantage than the traditional optimization model. Furthermore, after the Hessian matrices and gradient vectors of the two kinds of features are established, our model of motion estimation, which is adaptive to the motion state of camera, is applied to build a new recombined Hessian matrix and gradient vector for iterative pose estimation.

We also presented the evaluation results of the proposed SLAM method as compared with the point-line SLAM method developed in [38,39], which uses the traditional error model of line feature and its weighting model is based on residual errors, on both the EuRoC MAV datasets and the sequence images captured with our stereo camera. We also compared the point-to-point ICP method [45] using the sequence images from our stereo camera. According to the experimental results, we arrive at two conclusions. First, the proposed SLAM method has more geometric constraints than the traditional point-line SLAM method and classic ICP method, because the angular error function is added to the optimization model of line segment features. Furthermore, it has good robustness and positioning accuracy at large turns. This is particularly useful for robot navigation in indoor scenes as they include many corners. Second, the adaptive weighting model for motion estimation can better utilize the advantages of point and line segment features in different motion states. Thus, it can improve the system accuracy when the camera moves with rapid rotation or severe fluctuation.

At present, we mainly used the 2D structural constraints of line segment features. In the future, we plan to further improve our SLAM method by introducing the 3D structural constraints of spatial line segment features. Furthermore, topological relations between point features and line segment features will also be considered in our method in the future, so as to better match point and line segment features in indoor environments with repeated textures.

Author Contributions: Runzhi Wang, Kaichang Di and Wenhui Wan conceived the idea and designed the experiments; Runzhi Wang, Kaichang Di and Wenhui Wan designed the methods; Runzhi Wang and Yongkang Wang performed the experiments; Kaichang Di and Wenhui Wan analyzed the data; Runzhi Wang and Wenhui Wan drafted the paper; Kaichang Di revised the manuscript.

Funding: This research was funded by National Key Research and Development Program of China (No.2016YFB0502102). We would like to thank Autonomous Systems Lab from Swiss Federal Institute of Technology Zurich for providing the EuRoC MAV dataset.

Acknowledgments: We also thank the anonymous reviewers for their insightful comments and constructive suggestions on this paper.

Conflicts of Interest: The authors declare no conflict of interest.

References

1. Montemerlo, M.; Thrun, S.; Koller, D.; Wegbreit, B. FastSLAM: A factored solution to the simultaneous localization and mapping problem. In Proceedings of the 18th National Conference on Artificial Intelligence, Edmonton, AB, Canada, 28 July–1 August 2002; pp. 593–598.
2. Durrant-Whyte, H.; Bailey, T. Simultaneous localization and mapping: Part I. *IEEE Robot. Autom. Mag.* **2006**, *13*, 99–110. [CrossRef]
3. Bailey, T.; Durrant-Whyte, H. Simultaneous localization and mapping (SLAM): Part II. *IEEE Robot. Autom. Mag.* **2006**, *13*, 108–117. [CrossRef]
4. Dissanayake, M.W.M.G.; Newman, P.; Clark, S.; Durrant-Whyte, H.F.; Csorba, M.A. Solution to the simultaneous localization and map building (SLAM) problem. *IEEE Trans. Robot. Autom.* **2001**, *17*, 229–241. [CrossRef]
5. Fuentes-Pacheco, J.; Ruiz-Ascencio, J.; Rendón-Mancha, J.M. Visual simultaneous localization and mapping: A survey. *Artif. Intell. Rev.* **2015**, *43*, 55–81. [CrossRef]
6. Cheng, Y.; Maimone, M.W.; Matthies, L. Visual odometry on the Mars exploration rovers—A tool to ensure accurate driving and science imaging. *IEEE Robot. Autom. Mag.* **2006**, *13*, 54–62. [CrossRef]
7. Maimone, M.; Cheng, Y.; Matthies, L. Two years of visual odometry on the mars exploration rovers: Field reports. *J. Field Robot.* **2007**, *24*, 169–186. [CrossRef]
8. Di, K.; Xu, F.; Wang, J.; Agarwal, S.; Brodyagina, E.; Li, R.; Matthies, L. Photogrammetric processing of rover imagery of the 2003 Mars Exploration Rover mission. *ISPRS J. Photogramm. Remote Sens.* **2008**, *63*, 181–201. [CrossRef]
9. Wang, B.F.; Zhou, J.L.; Tang, G.S. Research on visual localization method of lunar rover. *Sci. China Inf. Sci.* **2014**, *44*, 452–460.
10. Bay, H.; Ess, A.; Tuytelaars, T.; Gool, L.V. Speeded-Up Robust Features (SURF). *Comput. Vis. Image Underst.* **2008**, *110*, 346–359. [CrossRef]
11. Leutenegger, S.; Chli, M.; Siegwart, R. BRISK: Binary Robust Invariant Scalable Keypoints. In Proceedings of the 2011 IEEE International Conference on Computer Vision (ICCV), Barcelona, Spain, 6–13 November 2011; pp. 2548–2555.
12. Rublee, E.; Rabaud, V.; Konolige, K. ORB: An efficient alternative to SIFT or SURF. In Proceedings of the 2011 IEEE International Conference on Computer Vision (ICCV), Barcelona, Spain, 6–13 November 2011; pp. 2564–2571.
13. Klein, G.; Murray, D. Parallel tracking and mapping for small AR workspaces (PTAM). In Proceedings of the 2007 6th IEEE and ACM International Symposium on Mixed and Augmented Reality, Washington, DC, USA, 13–16 November 2007; pp. 1–10.
14. Galvez-López, D.; Tardos, J.D. Bags of binary words for fast place recognition in image sequences. *IEEE Trans. Robot.* **2012**, *28*, 1188–1197. [CrossRef]
15. Ido, J.; Shimizu, Y.; Matsumoto, Y.; Ogasawara, T. Indoor Navigation for a Humanoid Robot Using a View Sequence. *Int. J. Robot. Res.* **2009**, *28*, 315–325. [CrossRef]
16. Celik, K.; Chung, S.J.; Clausman, M.; Somani, A.K. Monocular vision SLAM for indoor aerial vehicles. In Proceedings of the 2009 IEEE/RSJ International Conference on Intelligent Robots and Systems (ICRA), St. Louis, MI, USA, 11–15 October 2009; pp. 1566–1573.
17. Davison, A.J.; Reid, I.D.; Molton, N.D.; Stasse, O. MonoSLAM: Real-time single camera SLAM. *IEEE Trans. Pattern Anal. Mach. Intell.* **2007**, *29*, 1052–1067. [CrossRef] [PubMed]

18. Wu, K.; Di, K.; Sun, X.; Wan, W.; Liu, Z. Enhanced monocular visual odometry integrated with laser distance meter for astronaut navigation. *Sensors* **2014**, *14*, 4981–5003. [CrossRef] [PubMed]
19. Lemaire, T.; Lacroix, S. Monocular-vision based SLAM using Line Segments. In Proceedings of the IEEE International Conference on Robotics and Automation (ICRA), Roma, Italy, 10–14 April 2007; pp. 2791–2796.
20. Celik, K.; Chung, S.J.; Somani, A. Mono-vision corner SLAM for indoor navigation. In Proceedings of the 2008 IEEE International Conference on Electro/information Technology, Winsor, ON, Canada, 7–9 June 2008; pp. 343–348.
21. Zou, D.; Tan, P. CoSLAM: Collaborative visual SLAM in dynamic environments. *IEEE Trans. Pattern Anal. Mach. Intell.* **2013**, *35*, 354–366. [CrossRef] [PubMed]
22. Moratuwage, D.; Wang, D.; Rao, A.; Senarathne, N. RFS Collaborative Multivehicle SLAM: SLAM in Dynamic High-Clutter Environments. *IEEE Robot. Autom. Mag.* **2014**, *21*, 53–59. [CrossRef]
23. Kaess, M.; Dellaert, F. Probabilistic structure matching for visual SLAM with a multi-camera rig. *Comput. Vis. Image Understand.* **2010**, *114*, 286–296. [CrossRef]
24. Hu, G.; Huang, S.; Zhao, L.; Alempijevic, A.; Dissanayake, G. A robust RGB-D SLAM algorithm. In Proceedings of the 2012 IEEE/RSJ International Conference on Intelligent Robots and Systems (IROS), Vilamoura, Portugal, 7–12 October 2012; pp. 1714–1719.
25. Kerl, C.; Sturm, J.; Cremers, D. Dense visual SLAM for RGB-D cameras. In Proceedings of the 2013 IEEE/RSJ International Conference on Intelligent Robots and Systems, Tokyo, Japan, 3–7 November 2013; pp. 2100–2106.
26. Ji, Y.; Yamashita, A.; Asama, H. RGB-D SLAM using vanishing point and door plate information in corridor environment. *Intell. Serv. Robot.* **2015**, *8*, 105–114. [CrossRef]
27. Kim, D.H.; Kim, J.H. Image-Based ICP algorithm for visual odometry using a RGB-D sensor in a dynamic environment. *Adv. Intell. Syst. Comput.* **2013**, *208*, 423–430.
28. Steinbrücker, F.; Sturm, J.; Cremers, D. Real-time visual odometry from dense RGB-D images. In Proceedings of the 2011 IEEE International Conference on Computer Vision Workshops (ICCV Workshops), Barcelona, Spain, 6–13 November 2011; pp. 719–722.
29. Jiang, Y.; Chen, H.; Xiong, G.; Scaramuzza, D. ICP Stereo Visual Odometry for Wheeled Vehicles based on a 1DOF Motion Prior. In Proceedings of the 2014 IEEE International Conference on Robotics and Automation (ICRA), Hong Kong, China, 31 May–7 June 2014; pp. 585–592.
30. Kümmerle, R.; Grisetti, G.; Strasdat, H.; Konolige, K.; Burgard, W. G2o: A general framework for graph optimization. In Proceedings of the 2011 IEEE International Conference on Robotics and Automation (ICRA), Shanghai, China, 9–13 May 2011; pp. 3607–3613.
31. Mur-Artal, R.; Tardós, J.D. ORB-SLAM2: An Open-Source SLAM System for Monocular, Stereo, and RGB-D Cameras. *IEEE Trans. Robot.* **2017**, *33*, 1255–1262. [CrossRef]
32. Zhang, G.; Jin, H.L.; Lim, J.; Suh, I.H. Building a 3-d line-based map using stereo SLAM. *IEEE Trans. Robot.* **2015**, *31*, 1364–1377. [CrossRef]
33. Zhou, H.; Zou, D.; Pei, L.; Ying, R.; Liu, P.; Yu, W. StructSLAM: Visual SLAM with building structure lines. *IEEE Trans. Veh. Technol.* **2015**, *64*, 1364–1375. [CrossRef]
34. Micusik, B.; Wildenauer, H. Structure from Motion with Line Segments under Relaxed Endpoint Constraints. In Proceedings of the 2014 International Conference on 3d Vision, Tokyo, Japan, 8–11 December 2014; pp. 13–19.
35. Gomez-Ojeda, R.; Briales, J.; Gonzalez-Jimenez, J. PL-SVO: Semi-direct Monocular Visual Odometry by combining points and line segments. In Proceedings of the 2016 IEEE/RSJ International Conference on Intelligent Robots and Systems (IROS), Daejeon, Korea, 9–14 October 2016; pp. 4211–4216.
36. Gomez-Ojeda, R.; Gonzalez-Jimenez, J. Robust stereo visual odometry through a probabilistic combination of points and line segments. In Proceedings of the 2016 IEEE International Conference on Robotics and Automation (ICRA), Stockholm, Sweden, 16–21 May 2016; pp. 2521–2526.
37. Gomez-Ojeda, R.; Moreno, F.A.; Scaramuzza, D.; Gonzalez-Jimenez, J. PL-SLAM: A Stereo SLAM System through the Combination of Points and Line Segments. *arXiv* **2017**, arXiv:1705.09479.
38. Pumarola, A.; Vakhitov, A.; Agudo, A.; Sanfeliu, A.; Moreno-Noguer, F. PL-SLAM: Real-time monocular visual SLAM with points and lines. In Proceedings of the 2017 IEEE International Conference on Robotics and Automation (ICRA), Singapore, 29 May–3 June 2017; pp. 4503–4508.

39. He, Y.; Zhao, J.; Guo, Y.; He, W.; Yuan, K. PL-VIO: Tightly-Coupled Monocular Visual-Inertial Odometry Using Point and Line Features. *Sensors* **2018**, *18*, 1159. [CrossRef] [PubMed]
40. Di, K.; Zhao, Q.; Wan, W.; Wang, Y.; Gao, Y. RGB-D SLAM based on extended bundle adjustment with 2D and 3D information. *Sensors* **2016**, *16*, 1285. [CrossRef] [PubMed]
41. Grompone, V.G.R.; Jakubowicz, J.; Morel, J.M.; Randall, G. LSD: A fast line segment detector with a false detection control. *IEEE Trans. Pattern Anal. Mach. Intell.* **2010**, *32*, 722–732.
42. Zhang, L.; Koch, R. An efficient and robust line segment matching approach based on LBD descriptor and pairwise geometric consistency. *J. Vis. Commun. Image Represent.* **2013**, *24*, 794–805. [CrossRef]
43. Burri, M.; Nikolic, J.; Gohl, P.; Schneider, T.; Rehder, J.; Omari, S.; Achtelik, M.W.; Siegwart, R. The EuRoC micro aerial vehicle datasets. *Int. J. Robot. Res.* **2016**, *35*, 1157–1163. [CrossRef]
44. Sturm, J.; Engelhard, N.; Endres, F.; Burgard, W.; Cremers, D. A benchmark for the evaluation of RGB-D SLAM systems. In Proceedings of the 2012 IEEE/RSJ International Conference on Intelligent Robots and Systems (IROS), Vilamoura, Portugal, 7–12 October 2012; pp. 573–580.
45. Milella, A.; Siegwart, R. Stereo-based ego-motion estimation using pixel tracking and iterative closest point. In Proceedings of the 2006 IEEE International Conference on Computer Vision Systems, New York, NY, USA, 4–7 January 2006; pp. 21–27.

sensors

MDPI

Article

Dense RGB-D Semantic Mapping with Pixel-Voxel Neural Network

Cheng Zhao [1,*,†], Li Sun [2], Pulak Purkait [3], Tom Duckett [2] and Rustam Stolkin [1]

1 Extreme Robotics Lab, University of Birmingham, Birmingham B15 2TT, UK; R.Stolkin@bham.ac.uk
2 Lincoln Centre for Autonomous Systems (L-CAS), University of Lincoln, Lincoln LN6 7TS, UK; lisunsir@gmail.com (L.S.); tduckett@lincoln.ac.uk (T.D.)
3 Cambridge Research Lab, Toshiba Research Europe, Cambridge CB4 0GZ, UK; pulak.isi@gmail.com
* Correspondence: IRobotCheng@gmail.com; Tel.: +44-742-122-6545
† Current address: School of Computer Science, University of Birmingham, Birmingham B15 2TT, UK

Received: 6 August 2018; Accepted: 11 September 2018; Published: 14 September 2018

Abstract: In this paper, a novel Pixel-Voxel network is proposed for dense 3D semantic mapping, which can perform dense 3D mapping while simultaneously recognizing and labelling the semantic category each point in the 3D map. In our approach, we fully leverage the advantages of different modalities. That is, the PixelNet can learn the high-level contextual information from 2D RGB images, and the VoxelNet can learn 3D geometrical shapes from the 3D point cloud. Unlike the existing architecture that fuses score maps from different modalities with equal weights, we propose a `softmax` weighted fusion stack that adaptively learns the varying contributions of PixelNet and VoxelNet and fuses the score maps according to their respective confidence levels. Our approach achieved competitive results on both the SUN RGB-D and NYU V2 benchmarks, while the runtime of the proposed system is boosted to around 13 Hz, enabling near-real-time performance using an i7 eight-cores PC with a single Titan X GPU.

Keywords: semantic mapping; RGB-D SLAM; visual mapping

---

1. Introduction

Real-time 3D semantic mapping is often desired in a number of robotics applications, such as localization [1,2], semantic navigation [3,4] and human-aware navigation [5]. The semantic information provided with a 3D dense map is more useful than the geometric information [6] itself in robot-human or robot-environment interaction. It enables robots to perform advanced tasks requiring high precision, such as nuclear waste classification [7] and sorting or autonomous package delivery in warehouse environments. For intelligent mobile robotics applications, extending 3D mapping to 3D semantic mapping enables robots not only to localize themselves with respect to the scene's geometrical features, but also to simultaneously understand the higher-level semantic meaning of a complex scene.

A variety of well-known methods such as RGB-D SLAM [8], Kinect Fusion [9] and ElasticFusion [10] can generate a dense or semi-dense 3D map from RGB-D videos. However, these 3D maps contain no semantic-level understanding of the observed scenes. On the contrary, impressive results in semantic segmentation have been achieved with the advancement of convolutional neural networks (CNN). RGB [11–13], RGB-D [14–17] and point cloud [18,19] data have been successfully utilized for semantic segmentation. However, some of those methods are painfully slow due to their high computational demands. Thus, these methods are not yet integrated in real-time systems for robotics applications.

Compared to the well-investigated research on geometric 3D reconstruction and scene understanding, limited literature is available for 3D semantic mapping [20–23]. To date, there are no existing methods that make use of both RGB and point cloud data for semantic mapping. In this paper, we propose a dense RGB-D semantic mapping system with a Pixel-Voxel neural network, which can perform dense 3D mapping, while simultaneously recognizing and semantically labelling each point in the 3D map. The main contributions of this paper can be summarized as follows:

1. A Pixel-Voxel network consuming the RGB image and point cloud is proposed, which can obtain global context information through PixelNet while preserving accurate local shape information through VoxelNet.
2. A `softmax` weighted fusion stack is proposed to adaptively learn the varying contributions of different modalities. It can be inserted into a neural network to perform fusion-style end-to-end learning for arbitrary input modalities.
3. A dense 3D semantic mapping system integrating a Pixel-Voxel network with RGB-D SLAM is developed. Its runtime can be boosted to around 13 Hz using an i7 eight-core PC with Titan X GPU, which is close to the requirements of real-time applications.

The rest of this paper is organized as follows. First, the related work is reviewed in Section 2 followed by the details of the proposed methods in Section 3. The experimental results and analysis are presented in Section 4. Finally, we conclude the paper in Section 5.

2. Related Work

2.1. Dense 3D Semantic Mapping

To the best of our knowledge, the online dense 3D semantic mapping methods can be further grouped into three main sub-categories: semantic mapping based on 3D template matching [20,24], 2D/2.5D semantic segmentation [21,22,25–27] and RGB-D data association from multiple viewpoints [23,28,29].

The first type of methods such as SLAM++ [20] can only recognize known 3D objects in a predefined database. The approach is limited to situations where repeated and identical objects are present for semantic mapping. For the second type of methods, both approaches [21,25] adopt human-designed features with random decision forests to perform per-pixel label predictions of the incoming RGB videos. Then, all of the semantically-labelled images are associated together using visual odometry to generate the semantic map. Because of the state-of-the-art performance provided by the CNN-based scene understanding, SemanticFusion [22] integrates deconvolutional neural networks [30] with ElasticFusion [10] to obtain a real-time-capable (25 Hz) semantic mapping system. All of these three methods require fully connected CRF [31] optimization as an offline post-processing stage, i.e., the best performing semantic mapping methods are not capable of online operation. Zhao et al. [27] proposed the first system to perform simultaneous 3D mapping and pixel-wise material recognition. It integrates CRF-RNN [32] with RGB-D SLAM [8], and a post-processing optimization stage is not required. Keisuke et al. [26] proposed a real-time dense monocular CNN-SLAM method, which can perform depth prediction and semantic segmentation simultaneously from a single image using a deep neural network.

All the above methods mainly focus on semantic segmentation using a single image and perform 3D label refinement through a recursive Bayesian update using a sequence of images. However, they do not take full advantage of the associated information provided by multiple viewpoints of a scene. Yu et al. [23] proposed a data-associated recurrent neural network (DA-RNN) integrated with Kinect Fusion [9] for 3D semantic mapping. DA-RNN employs a recurrent neural network to tightly combine the information contained in multiple viewpoints of an RGB-D video stream to improve the semantic segmentation performance. Ma et al. [28] proposed a multi-view consistency layer, which can use multi-view context information for object-class segmentation from multiple RGB-D views. It utilizes the visual odometry trajectory from RGB-D SLAM [8] to wrap semantic segmentation between two viewpoints. Further, Armin et al. [29] proposed a network architecture for spatially-

and temporally-coherent semantic co-segmentation and mapping of complex dynamic scenes from multiple static or moving cameras.

2.2. Fusion Style Semantic Segmentation

Most of the fusion-style semantic segmentation methods take advantage of both RGB and depth images. FuseNet [14] can fuse RGB and depth cues in a single encoder-decoder CNN architecture for RGB-D semantic segmentation. The long short-term memorized context fusion (LSTM-CF) network [15] fuses contextual information from multiple channels of RGB and depth images through stacking of several convolution layers and a long short-term memory layer. FuseNet normalizes the depth value into the interval of $[0, 255]$ to have the same spatial range as colour images, while the LSTM-CF network encodes depth to a horizontal, height, angle (HHA) image to obtain three channels as the colour image. The HHA representation can improve the depth-based semantic segmentation; however, the HHA representation requires a high computational cost and hence cannot be performed in real time. Spatio-temporal data-driven pooling (STD2P) [33] involves a novel superpixel-based multi-view convolutional neural network for RGB-D semantic segmentation, which uses the spatio-temporal pooling layer to aggregate information over space and time. Locality-sensitive deconvolution networks (LS-DeconvNets) [16] involve a locality-sensitive DeconvNet to refine the boundary segmentation and also a gated fusion layer for combining modalities (RGB and HHA); however the number of input modalities is limited to two. Lin et al. [17] introduced a cascaded feature network (CFN) with a context-aware receptive field (CaRF) with a better control on the relevant contextual information of the learned features for RGB-D semantic segmentation. All of the above RGB-D fusion networks treat the depth image similarly to an RGB image using a CNN with a max-pooling layer. However, this also makes the depth image lose shape information. In contrast, the 3D point cloud should have more 3D geometry information compared to the depth image. We believe there should be the potential to combine RGB and point cloud data for semantic segmentation. The forerunner work PointNet [18] provides a unified architecture for both classification and segmentation, which consumes the raw unordered point clouds as input. PointNet only employs a single max-pooling layer to generate the global feature, which describes the original input clouds; thus, it does not capture the local structures induced by the 3D metric space points live in. The improved version PointNet++ [19] is a hierarchical neural network that applies PointNet recursively on a nested partitioning of the input point set, which can learn local features with increasing contextual scales.

2.3. Discussion

For the task of semantic segmentation, conventional CNN-based methods have struggled with the balance between global and local information. The global context information can alleviate the local ambiguities to improve the recognition performance, while local information is crucial to obtain accurate per-pixel accuracy, i.e., shape information. How to increase the receptive field to get more global context information, while preserving a high resolution feature map, is still an open problem.

Processing the depth image in a similar manner to the RGB image using CNN with max-pooling cannot preserve all the local geometry information. Compared to RGB and RGB-D data, a 3D point cloud can provide rich spatial information. For example, in PointNet [18], a single fully-connected multi-layer network followed by a single global max-pooling layer are used for semantic segmentation of a point cloud. The resolution does not decrease, and it can keep the original spatial information of the data. However, these methods lack the context information because of the usage of a single global max-pooling layer. Intuitively, combining RGB-based and point cloud-based networks together can alleviate each of their drawbacks and leverage each of their advantages. The RGB image can provide global context information as a supplement for point cloud segmentation, while the point cloud can help refine the boundary shape for RGB segmentation.

Moreover, during RGB-D mapping, both the RGB image and point cloud can be obtained directly from an RGB-D camera, which is easily available and enables a potential combination for semantic mapping. This motivated us to utilize a Pixel-Voxel neural network for dense RGB-D semantic mapping.

In addition, the networks in [11,14,15,17] simply fuse the score maps from different modalities using equal weights. The gated fusion in LS-DeconvNets [16] is limited to fusion of the features from (at most) two modalities. However, each modality should have different contributions in different situations for different categories. Therefore, in this paper, a `softmax` weighted fusion stack is proposed for adaptively learning the varying contribution of each modality.

3. Proposed Method

3.1. Overview

The pipeline of the proposed dense RGB-D semantic mapping with a Pixel-Voxel neural network is illustrated in Figure 1. The input RGB image and point cloud pairs of each key-frame are fed into the Pixel-Voxel network. The architecture of the proposed network is displayed in Figure 2. The output of the network—a semantically-labelled point cloud—is combined incrementally according to the visual odometry of RGB-D SLAM. The label probability of each voxel is refined by a recursive Bayesian update. Finally, the dense 3D semantic map is generated. Note that in our current architecture, a voxel consists of just a single 3D point.

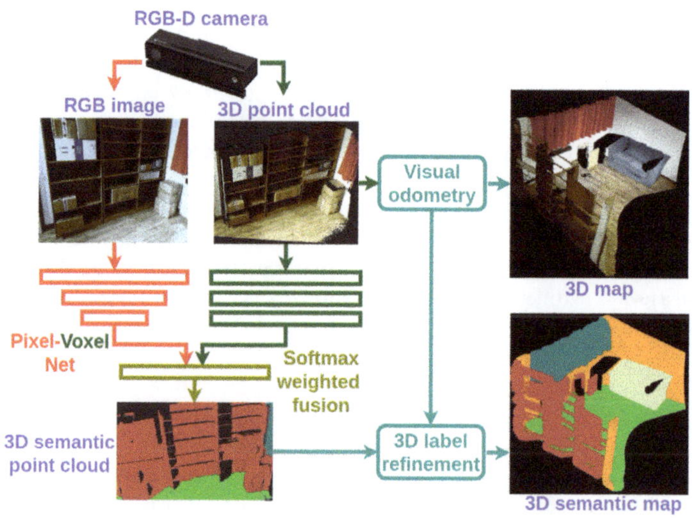

Figure 1. The pipeline of the proposed dense RGB-D semantic mapping with the Pixel-Voxel neural network. The RGB image and 3D point cloud are obtained from an RGB-D camera, Kinect V2. The RGB and point cloud data-pair of each key-frame is fed into the Pixel-Voxel network for semantic segmentation. The semantically-labelled point clouds are then combined incrementally through the visual odometry of RGB-D SLAM. The label probability of each voxel is further refined by a recursive Bayesian update. Finally, the dense 3D semantic map is generated.

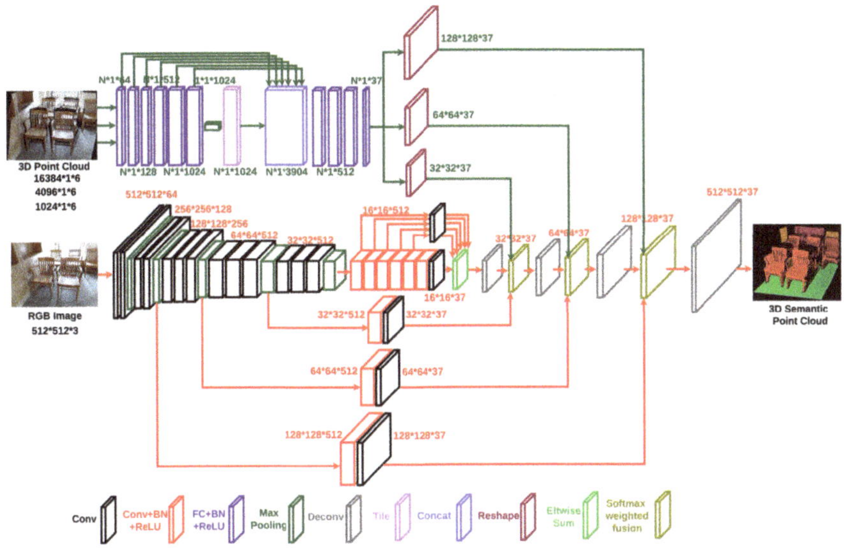

Figure 2. The architecture of the proposed Pixel-Voxel network. The proposed architecture consists of two parallel feed-forward sub-networks: PixelNet and VoxelNet. The PixelNet is comprised of three building blocks: truncated CNN, context stack and skip architecture. The VoxelNet is composed of the following blocks: fully-connected stacks, local and global information combination stack and reshape layer. It obtains global context information through PixelNet while preserving accurate local shape information through VoxelNet. The enlarged architecture of the `softmax` weighted fusion stack can be found in Figure 3. It can fuse the score maps from PixelNet and VoxelNet according to their respective confidence at different resolutions.

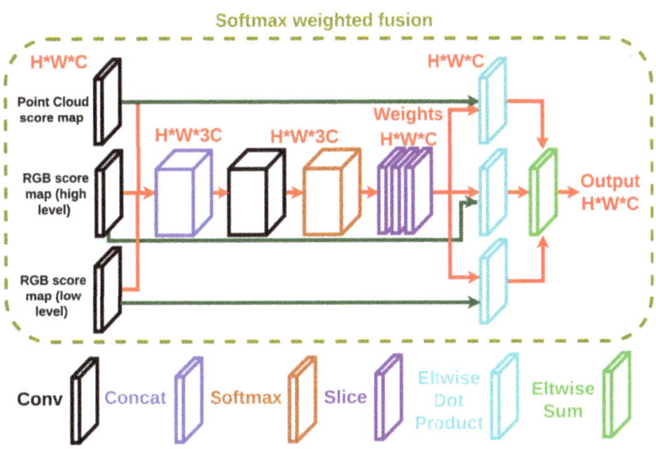

Figure 3. The architecture of the softmax weighted fusion stack. H, W, C are the hight, width and channel number of the feature map. The convolution operation can learn the correlations of the multiple score maps from different modalities to obtain the weight/confidence of each modality.

3.2. Pixel Neural Network

The sub-network PixelNet is comprised of three units: truncated CNN, a context stack similar to [34] and the skip architecture. The input of PixelNet is an RGB image. For the truncated CNN, VGG-16 (http://www.robots.ox.ac.uk/~vgg/research/very_deep/) or ResNe (https://github.com/

KaimingHe/deep-residual-networks) (truncated after *pool*5), pre-trained on ImageNet (http://www.image-net.org/challenges/LSVRC/), can be employed as a baseline. After the truncated CNN, the resolution of the feature maps is decreased 32-times compared with the input image; thus, it drops a significant amount of shape information, which is recovered utilizing the VoxelNet sub-network.

Note that the receptive fields after the *pool*5 layer of VGG-16 are of dimension 212 × 212, which is not large enough to cover the whole 512 × 512 input image. Therefore, a context-stack, composed of a chain of 6 layers of 5 × 5 × 512 convolution stacks $[Conv + BN + ReLU]$, is concatenated on the top of a pre-trained truncated VGG-16 network. The context stack can expand the receptive field progressively, as shown in Figure 4, to cover all the elements in the current feature map (the whole original image). The receptive field of the context stack can be described as:

$$\mathcal{RF}_j = \mathcal{RF}_{j-1} + (k_j - 1) \times \prod_{i=0}^{j-1} S_i, j \in [1, n] \tag{1}$$

where $\mathcal{RF}_j$ and k_j are the receptive field and kernel size of the j-th context stack, S_i refers to the stride of the i-th context stack, $\mathcal{RF}_0$ and S_0 are the receptive field and stride product before the first context stack and $n = 6$ is the number of context stacks. In addition, the score maps of all the context stacks are fused together to aggregate multi-scale context information. Notice that the spatial dimensionality of the feature maps in a context stack is unchanged.

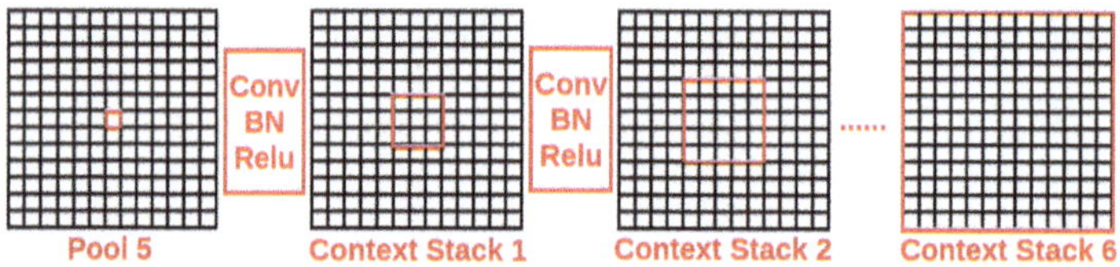

Figure 4. The receptive field (the area of red square) of the context stack is progressively extended to cover all the elements in the feature map.

The skip architecture consists of 3 skip stacks $[Conv + BN + ReLU + Conv\ (score)]$ following *pool*2, *pool*3 and *pool*4 separately. In order to prevent the network training from divergence, conventionally, a smaller learning rate is adopted for the skip architecture during training (similar to [11]). We utilize batch normalization, which stabilizes the back-propagated error signals; thus, a bigger learning rate (0.01) can be employed for training. The skip architecture retains the low-level features of the RGB image.

3.3. Voxel Neural Network

The input of VoxelNet is a point cloud, which is represented as a set of 3D points $\{p_i \mid i = 1, 2 \ldots, n\}$ stored in a vector of length $n \times 6$, where n is the number of points and p_i is a 6-dimensional vector containing position information $(X, Y, Z)^T$ in the world coordinates and pixel colour information $(R, G, B)^T$. Inspired by PointNet [18], we also use max pooling as an invariant function. The max-pooling operation obtains the global feature from all the points, which are concatenated with the pixel features to predict point-wise semantic labels. The higher dimensional feature representation for each point of the subnetwork can be summarized by the following equation.

$$[\mathcal{F}^1_{global} \ldots \mathcal{F}^n_{global}] = \mathcal{T}(\mathcal{M}([f^k_{mlp}(p_1) \ldots f^k_{mlp}(p_n)])) \tag{2}$$

Here, f_{mlp} is the multi-layer perception network, i.e., $FC + BN + ReLU$, and k is the number of multi-layer perception networks before max pooling. Each point shares the same set of fully-connected weights. $\mathcal{M}$ is the max pooling operation with kernel size $n \times 1$, and $\mathcal{T}$ is the tile operation, which restores the shape of the feature map from 1×1 to $n \times 1$.

The output $[\mathcal{F}^1_{global} ... \mathcal{F}^n_{global}]$ is the global feature map of the input set. This is fed to the per-point feature of the multi-layer perception network to concatenate the global and local information.

$$[\mathcal{F}^1_{concat} ... \mathcal{F}^n_{concat}] = Concat([\mathcal{F}^1_{global} ... \mathcal{F}^n_{global}], ... [f^i_{mlp}(p_1) ... f^i_{mlp}(p_n)]), i \in [1,k] \tag{3}$$

Then, the new per-point features are extracted though the multi-layer perception network using the combined global and local point features as:

$$\mathcal{F}^{1...n}_{h \times w} = \mathcal{R}([f^m_{mlp}(\mathcal{F}^1_{concat}) ... f^m_{mlp}(\mathcal{F}^n_{concat})]) \tag{4}$$

where m is the last multi-layer perception network and $\mathcal{R}$ is the reshape operation, which transforms the shape of the score map from $n \times 1$ to $h \times w$ through back-projection (It is worth noting that the distortions are incorporated during the projection to pixel coordinates):

$$d_{u,v} \begin{bmatrix} u \\ v \\ 1 \end{bmatrix} = \begin{bmatrix} f_x \; s \; c_x \\ 0 \; f_y \; c_y \\ 0 \; 0 \; 1 \end{bmatrix} \begin{bmatrix} X \\ Y \\ Z \end{bmatrix} \tag{5}$$

where f_x, f_y are the focal lengths, (c_x, c_y) is the principal point offset, s is the axis skew and (u, v) is the pixel position in the image plane. Here, the radial distortion had been incorporated during the projection to the pixel coordinates. In detail, the feature of the point cloud in (X, Y, Z) can be transformed to the position (u, v) in the image plane, so the score map of VoxelNet can be fused with the score map of PixelNet.

The spatial dimensionality of the features is the same as that of the input data in VoxelNet, so it can preserve all the original shape information. However, if only a single max pooling layer is adopted to generate the global feature, it will drop significant context information from the input point cloud.

3.4. Softmax Weighed Fusion

In contrast to the conventional methods, which simply fuse score maps from different modalities using equal weights, a `softmax` weighted fusion stack, as shown in Figure 3, is designed to learn the varying contribution of each modality in different situations for different categories. To be precise, let us define the score maps by $\mathcal{F}^1, \mathcal{F}^2 ... \mathcal{F}^n \in \mathbb{R}^{c \times h \times w}$, generated from n different modalities, where c is the number of categories and $h \times w$ are the dimensions of the score map. Then, the fusion score map $\mathcal{F}_{fused} \in \mathbb{R}^{n \cdot c \times h \times w}$ can be written as:

$$\mathcal{F}_{fused} = \mathcal{C}([\mathcal{F}^1, \mathcal{F}^2 ... \mathcal{F}^n]) \circledast \mathcal{W}_{conv} \tag{6}$$

where $\circledast$ is the convolution operation, $\mathcal{C}$ is the concatenation operation and $\mathcal{W}_{conv} \in \mathbb{R}^{n \cdot c \times n \cdot c \times 1 \times 1}$ are the weights of the convolution. The convolution operation learns the correlations of the multiple score maps from n different modalities. The channel values of $\mathcal{F}_{fused}$ are further normalized into the interval $[0, 1]$ according to the `softmax` operation. Then, the weights of the score map are obtained through a slice operation as:

$$\mathcal{W}^1, \mathcal{W}^2 ... \mathcal{W}^n = \mathcal{S}[\text{softmax}(\mathcal{F}_{fused})] \tag{7}$$

where $\mathcal{S}$ is the slice operation, $\text{softmax}(\mathbf{x}) = \frac{\exp(\mathbf{x})}{\sum_{i=1}^{n \cdot c} \exp(x_i)}$ and $\mathcal{W}^1, \mathcal{W}^2 ... \mathcal{W}^n \in \mathbb{R}^{c \times h \times w}$ are the corresponding weights of the score maps. The weights signify the confidence of each model. The weighted fusion score map $\mathcal{F}_{score} \in \mathbb{R}^{c \times h \times w}$ can be written as:

$$\mathcal{F}_{score} = \sum_{j=1}^{n} \mathcal{F}^j \odot \mathcal{W}^j, \tag{8}$$

where $\odot$ is the element-wise multiplication operation, $\sum_{j=1}^{n \cdot c} \mathcal{W}^j = \mathbf{1}$ and $\mathbf{1} \in \mathbb{R}^{h \times w}$.

For our problem, the three score maps from PixelNet and VoxelNet are fused together according to their respective confidence levels. Note that the proposed weighted fusion stack can fuse the score maps of an arbitrary number of modalities. Moreover, it can be easily inserted into a neural network that requires fusion of multiple modalities and can be trained end-to-end. Thus, it can potentially be applied to many other similar problems.

3.5. Class-Weighted Loss Function

In most of the datasets for semantic segmentation, we observe highly imbalanced class distributions. Thus, focusing more on the rare classes to boost their recognition accuracy can improve the average recognition performance significantly, while overall recognition performance might decrease a little. We adopt the class-weighted negative log-likelihood as the loss function:

$$\text{loss} = -\sum_{i \in \Theta} (\mathbf{1}_{y_i = j}) 2^{\lceil \log_{10}(\delta / p_j) \rceil} \cdot \log \mathcal{L}\big(\text{softmax}(\mathcal{F}_i), y_i\big) \tag{9}$$

where Θ are the training data, $\mathcal{L}$ is the likelihood function, $\mathcal{F}_i$ is the final score map, y_i refers to the one-hot training label. $\mathbf{1}_{y_i=j}$ is a function that returns 1 if $y_i = j$, or 0 otherwise. p_j is the occurrence frequency of class j, and $2^{\lceil \log_{10}(\delta/p_j) \rceil}$ is the weight of class j. δ is the threshold of frequency criteria for the rare class. $\lceil\rceil$ is the ceiling operation. This will force the network to assign a higher weight to rare classes. The value of δ is set to 2.5% following the 85%–15% rule described in [35], i.e., the frequency sum of all the rare classes is 15%.

3.6. RGB-D Mapping and 3D Label Refinement

RGB-D SLAM [8] is adopted for dense 3D mapping. Its visual odometry can provide the transformation information between two adjacent semantically-labelled point clouds. It is then used for generating a global semantic map and enabling incremental semantic label fusion.

After obtaining the semantically-labelled point clouds from different viewpoints, label hypotheses are fused by a recursive Bayesian update to refine the 3D semantic map. Each voxel in the semantic point cloud stores both the label value and the corresponding discrete probability. The voxels from different viewpoints can be transformed to the same coordinate through the visual odometry of RGB-D SLAM. Then, the voxel's label probability distribution is updated by means of a recursive Bayesian update as:

$$P(x = l_i | I_{1,\ldots,k}) = \frac{1}{Z} P(x = l_i | I_{1,\ldots,k-1}) P(x = l_i | I_k) \tag{10}$$

where l_i is the label prediction, I_k is the k-th frame and Z is the normalizing constant. The label refinement is applied to all label probabilities of each voxel to generate a proper distribution.

4. Experiments

We evaluate the proposed Pixel-Voxel network using two popular indoor scene datasets, i.e., the SUN RGB-D (http://rgbd.cs.princeton.edu/) and NYU V2 (https://cs.nyu.edu/~silberman/datasets/nyu_depth_v2.html) datasets. The former is used to evaluate the semantic segmentation on a single frame, while the latter provides raw RGB-D sequences, which can be used for the semantic segmentation evaluation on multiple frames.

The SUN RGB-D dataset contains 5285 synchronized RGB-D image pairs for training/validation and 5050 synchronized RGB-D image pairs for testing. The RGB-D image pairs with different resolutions are captured by 4 different RGB-D sensors: Kinect V1, Kinect V2, Xtion and RealSense.

The task is to segment 37 indoor scene classes such as table, chair, sofa, window, door, etc. Pixel-wise annotations are available in these datasets. However, the extremely unbalanced distribution of class instances makes the task very challenging. The rareness frequency threshold is set to 2.5% in the class-weighted loss function following the 85–15% rule.

The NYU V2 dataset provides synchronized 1449 pixel-wise annotated RGB-D image pairs captured by Kinect V1, which includes 795 frames for training/validation and 654 frames for testing. The task is to segment 13 classes similar to the SUN RGB-D dataset in an indoor scene. Comparing with the other larger RGB-D datasets, the NYU V2 dataset provides raw RGB-D videos rather than discrete single frames. Therefore, using the odometry of RGB-D SLAM, the semantic segmentation based on multiple frames can be evaluated for the dense semantic mapping.

4.1. Data Augmentation and Preprocessing

For the PixelNet training, all the RGB images are resized to the same resolution 512×512 through bilateral filtering. We randomly flip the RGB image horizontally and rescale the image slightly to augment the RGB training data.

For the VoxelNet training, there is still no available large-scale ready-made 3D point cloud dataset. We generated the point cloud using the RGB-D image pairs and the corresponding intrinsic parameters of the camera through back-projection, e.g., Equation (5) for the SUN RGB-D and NYU V2 datasets. Following [14], 514 training and 558 testing RGB-D image pairs containing invalid values, which might lead to incorrect supervision during training, are excluded from the SUN RGB-D dataset. We also randomly flip the 3D point cloud horizontally to augment the training data. There is huge computational complexity if the original point clouds are used for VoxelNet training. Therefore, we uniformly down-sample the original point cloud to a sparse point cloud in 3 different scales. The numbers of points in these sparse point clouds are 16,384, 4096 and 1024, respectively.

4.2. Network Training

The whole training process can be divided into 3 stages: PixelNet training, VoxelNet training and Pixel-Voxel network training. Firstly, PixelNet and VoxelNet are each trained separately. Then, the pre-trained weights are inherited for the Pixel-Voxel network training.

All the networks are trained using stochastic gradient descent with momentum. The batch size is set to 10, the momentum fixed to 0.9 and the weight decay fixed to 0.0005. The new parameters are randomly initialized from a Gaussian distribution with variance 10^{-2}. The step learning policy is adopted for PixelNet training, and the polynomial learning policy is adopted for PixelNet and Pixel-Voxel Network training. The learning rate is initialized to 10^{-3}, and the learning rate of the newly-initialized parameters is set to 10-times higher than that of the pre-trained parameters. Because there are 3 softmax weighed fusion stacks, 3 rounds of fine-tuning are required during the Pixel-Voxel network training.

4.3. Overall Performance

Following [11], three standard performance metrics for semantic segmentation are used for the evaluation: pixel accuracy, mean accuracy and mean intersection over union (IoU). The three metrics are defined as:

- Pixel accuracy: $\sum_i n_{ii} / \sum_i t_i$
- Mean accuracy: $(1/n_{cl}) \sum_i n_{ii}/t_i$
- Mean IoU: $(1/n_{cl}) \sum_i n_{ii}/(t_i + \sum_j n_{ji} - n_{ii})$

where n_{cl} is the number of classes, n_{ij} is the number of pixels of class i classified as class j and $t_i = \sum_j n_{ij}$ is the total number of pixels belonging to class i.

In the experiment on the SUN RGB-D dataset, the performance of the Pixel-Voxel network and all the baselines are evaluated on a single frame. In the second experiment, the results are obtained by fusing multiple frames (provided by the raw data). To be more specific, visual odometry is employed to associate the pixels in consecutive frames, and then, a Bayesian-update-based 3D refinement is used to fuse all predictions. Similar strategies are used in the baseline methods, i.e., Hermans et al. [21], SemanticFusion [22] and Ma et al. [28].

From Figures 5 and 6, it is clear that after combining VoxelNet with PixelNet, the edge prediction can be improved significantly. Preserving 3D shape information through VoxelNet, the results have accurate boundaries, such as the shape of the bed, toilet and especially the legs of the furniture.

(a) (b) (c) (d)

Figure 5. (**a**,**c**) are the coarse predictions from PixelNet, and (**b**,**d**) are the predictions after combining VoxelNet with PixelNet. It can be seen that the boundary shape is more accurate after the VoxelNet refinement. The colour palette can be found in Figure 6.

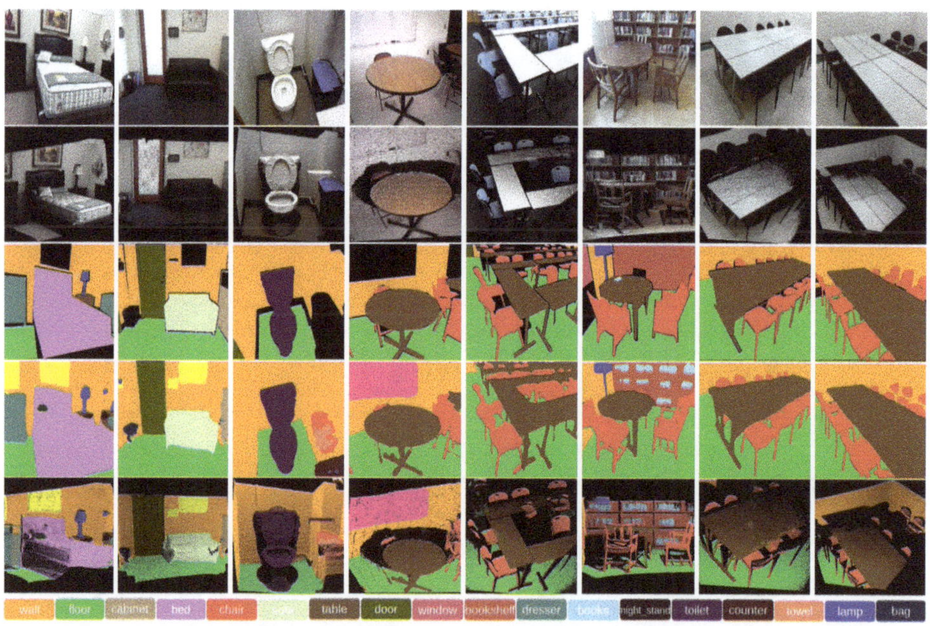

Figure 6. Qualitative results (best viewed in colour) for the Pixel-Voxel network on the SUN RGB-D dataset. For different scenes in each row, the following images are displayed: RGB image (Row 1), 3D point cloud (Row 2), ground truth image (Row 3), 2D semantic image (Row 4) and 3D semantic point cloud (Row 5). The Pixel-Voxel network produces results with accurate boundary shape such as the shape of the bed, toilet and especially the legs of the furniture.

The comparison of overall performance on the SUN RGB-D and NYU V2 datasets are shown in Tables 1 and 2. The class-wise accuracy on the SUN RGB-D and NYU V2 datasets are shown in Tables 3 and 4. The class-wise IoU of the Pixel-Voxel network is also provided. For the SUN RGB-D dataset, we achieved 79.04% for overall pixel accuracy, 57.65% for mean accuracy and 44.24% for mean IoU. After combining VoxelNet edge refinement, the pixel accuracy increased slightly from 77.25%–77.82% for VGG-16 and from 78.30%–78.76% for ResNet101, while the mean accuracy shows a significant increase from 49.33%–53.86% for VGG-16 and from 54.22%–56.81% for ResNet101. For the NYU V2 dataset, we achieved an overall pixel accuracy of 82.53%, a mean accuracy of 74.43% and a mean IoU of 59.30%. After combining VoxelNet edge refinement, the overall accuracy increases slightly from 80.74%–81.50% for VGG-16 and from 81.63%–82.22% for ResNet101, while the mean accuracy shows a significant increase from 70.23%–72.25% for VGG-16 and from 72.18%–73.64% for ResNet101.

Table 1. Comparison of the overall performance on the SUN RGB-D dataset. Some results are copied from [12]. The best performance among the compared methods is marked as bold.

Methods	Pixel Acc.	Mean Acc.	Mean IoU
FCN [11]	68.18%	38.41%	27.39%
DeconvNet [30]	66.13%	33.28%	22.57%
SegNet [12]	72.63%	44.76%	31.84%
DeepLab [13]	71.90%	42.21%	32.08%
Context-CRF [36]	78.4%	53.4%	42.3%
LSTM-CF [15] (RGB-D)	-	48.1%	-
FuseNet [14] (RGB-D)	76.27%	48.30%	37.29%
LS-DeconvNets (RGB-D) [16]	-	58.00%	-
RefineNet-Res101 [37]	80.4%	57.8%	45.7%
RefineNet-Res152 [37]	**80.6%**	**58.5%**	45.9%
CFN (VGG-16, RGB-D) [17]	-	-	42.5%
CFN (RefineNet-152, RGB-D) [17]	-	-	**48.1%**
Pixel Net (VGG-16)	77.25%	49.33%	38.26%
Pixel Net (ResNet101)	78.30%	54.22%	41.73%
Pixel-Voxel Net (VGG-16, without fusion)	77.82%	53.86%	41.33%
Pixel-Voxel Net (ResNet101, without fusion)	78.76%	56.81%	43.59%
Pixel-Voxel Net (VGG-16)	78.14%	54.79%	42.11%
Pixel-Voxel Net (ResNet101)	79.04%	57.65%	44.24%

Table 2. Comparison of overall performance on the NYU V2 dataset. Some results are copied from [28]. The methods with † take advantage of the data from multiple views. The best performance among the compared methods is marked as bold.

Methods	Pixel Acc.	Mean Acc.	Mean IoU
Hermans et al. [21] (RGB-D) †	54.3%	48.0%	-
SemanticFusion [22] †	67.9%	59.2%	-
SceneNet [38]	67.2%	52.5%	-
Eigen et al. [39] (RGB-D)	75.4%	66.9%	52.6%
FuseNet [14] (RGB-D)	75.8%	66.2%	54.2%
Ma et al. [28] (RGB-D) †	79.13%	70.59%	59.07%
Pixel Net (VGG-16) †	80.74%	70.23%	55.92%
Pixel Net (ResNet101) †	81.63%	72.18%	57.78%
Pixel-Voxel Net (VGG-16, without fusion) †	81.50%	72.25%	57.69%
Pixel-Voxel Net (ResNet101, without fusion) †	82.22%	73.64%	58.71%
Pixel-Voxel Net (VGG-16) †	81.85%	73.21%	58.54%
Pixel-Voxel Net (ResNet101) †	**82.53%**	**74.43%**	**59.30%**

Table 3. Comparison of the class-wise accuracy on the SUN RGB-D dataset. Some of the methods in Table 1 do not provide the class-wise accuracy; hence, they are omitted here. The class-wise IoU of the Pixel-Voxel network (PVNet) is also provided. LS, locality-sensitive. The best performance among the compared methods is marked as bold.

Category	Wall	Floor	Cabinet	Bed	Chair	Sofa	Table	Door	Window	Bookshelf	Picture	Counter	Blinds
SegNet [12]	83.42%	93.43%	63.37%	73.18%	75.92%	59.57%	64.18%	52.50%	57.51%	42.05%	56.17%	37.66%	40.29%
LSTM-CF [15]	74.9%	82.3%	47.3%	62.1%	67.7%	55.5%	57.8%	45.6%	52.8%	43.1%	56.7%	39.4%	48.6%
FuseNet [14]	90.20%	94.91%	61.81%	77.10%	78.62%	66.49%	65.44%	46.51%	62.44%	34.94%	67.39%	40.37%	43.48%
LS-DeconvNets [16]	**91.9%**	94.7%	61.6%	**82.2%**	**87.5%**	62.8%	**68.3%**	47.9%	**68.0%**	48.4%	**69.1%**	49.4%	51.3%
PVNet (VGG16)	90.28%	93.21%	66.87%	75.31%	85.45%	**67.37%**	64.81%	58.62%	63.58%	54.54%	64.76%	**51.87%**	**59.23%**
PVNet (ResNet101)	89.19%	**94.94%**	**69.36%**	79.11%	85.70%	66.09%	60.59%	**62.22%**	66.59%	**58.34%**	66.39%	50.56%	53.65%
PVNet (VGG16)$_{IoU}$	76.07%	87.20%	50.66%	68.23%	64.98%	54.17%	**46.07%**	44.83%	46.50%	41.31%	**48.94%**	41.19%	**39.95%**
PVNet (ResNet101)$_{IoU}$	**77.41%**	**87.78%**	**53.44%**	**71.16%**	**66.76%**	**54.61%**	44.46%	**45.19%**	**48.23%**	**41.79%**	46.78%	**41.39%**	35.95%
Category	**Desk**	**Shelves**	**Curtain**	**Dresser**	**Pillow**	**Mirror**	**Floor_Mat**	**Clothes**	**Ceiling**	**Books**	**Fridge**	**TV**	**Paper**
SegNet [12]	11.92%	11.45%	66.56%	52.73%	43.80%	26.30%	0.00%	34.31%	74.11%	53.77%	29.85%	33.76%	22.73%
LSTM-CF [15]	**37.3%**	9.6%	63.4%	35.0%	45.8%	44.5%	0.0%	28.4%	68.0%	47.9%	61.5%	52.1%	36.4%
FuseNet [14]	25.63%	20.28%	65.94%	44.03%	54.28%	52.47%	0.00%	25.89%	84.77%	45.23%	34.52%	34.83%	24.08%
LS-DeconvNets [16]	35.0%	24.0%	**68.7%**	60.5%	**66.5%**	57.6%	0.00%	**44.4%**	**88.8%**	**61.5%**	51.4%	71.7%	37.3%
PVNet (VGG16)	32.05%	23.09%	62.49%	62.13%	54.97%	50.60%	**0.59%**	35.35%	57.78%	41.75%	55.43%	67.60%	35.34%
PVNet (ResNet101)	32.49%	**27.37%**	68.33%	**69.41%**	56.96%	**57.94%**	0.00%	36.45%	68.77%	42.02%	**63.05%**	**72.47%**	**38.11%**
PVNet (VGG16)$_{IoU}$	**26.05%**	12.05%	50.52%	47.43%	36.35%	36.44%	**0.59%**	20.56%	53.61%	28.04%	41.23%	57.36%	24.13%
PVNet (ResNet101)$_{IoU}$	25.30%	**16.86%**	**53.09%**	**50.83%**	**38.16%**	**42.29%**	0.00%	**22.28%**	**63.39%**	**29.21%**	**48.47%**	**60.46%**	**25.20%**
Category	**Towel**	**Shower_Curtain**	**Box**	**Whiteboard**	**Person**	**Night_Stand**	**Toilet**	**Sink**	**Lamp**	**Bathtub**	**Bag**	**Mean**	**-**
SegNet [12]	19.83%	0.03%	23.14%	60.25%	27.27%	29.88%	76.00%	58.10%	35.27%	48.86%	16.76%	31.84%	-
LSTM-CF [15]	36.7%	0.0%	38.1%	48.1%	**72.6%**	36.4%	68.8%	67.9%	58.0%	65.6%	23.6%	48.1%	-
FuseNet [14]	21.05%	**8.82%**	21.94%	57.45%	19.06%	37.15%	76.77%	68.11%	49.31%	73.23%	12.62%	48.30%	-
LS-DeconvNets [16]	**51.4%**	2.9%	**46.0%**	54.2%	49.1%	**44.6%**	**82.2%**	**74.2%**	**64.7%**	**77.0%**	**47.6%**	**58.0%**	-
PVNet (VGG16)	41.12%	4.59%	40.33%	66.56%	60.51%	33.21%	80.62%	69.07%	60.35%	67.78%	28.17%	54.79%	-
PVNet (ResNet101)	48.81%	0.00%	42.15%	**74.22%**	69.40%	38.16%	80.23%	68.20%	61.80%	76.16%	37.63%	57.65%	-
PVNet (VGG16)$_{IoU}$	30.53%	**4.00%**	24.81%	51.10%	48.57%	20.89%	66.31%	48.82%	**43.50%**	55.90%	19.37%	42.11%	-
PVNet (ResNet101)$_{IoU}$	**36.85%**	0.00%	**26.77%**	**54.88%**	**54.77%**	**21.52%**	**66.43%**	**53.15%**	43.00%	**65.00%**	**23.90%**	**44.24%**	-

Table 4. Comparison of the class-wise accuracy on the NYU V2 dataset. Some of the methods in Table 2 do not provide the class-wise accuracy; hence, they are omitted here. The class-wise IoU of the Pixel-Voxel network (PVNet) are also provided. The methods with † take advantage of the data from multiple views. The best performance among the compared methods is marked as bold.

Category	Bed	Books	Ceiling	Chair	Floor	Furniture	Objects	Painting	Sofa	Table	TV	Wall	Window	Mean
Hermans et al. [21] †	68.4%	45.4%	**83.4%**	41.9%	91.5%	37.1%	8.6%	35.8%	28.5%	27.7%	38.4%	71.8%	46.1%	48.0%
SemanticFusion [22] †	62.0%	58.4%	43.3%	59.5%	92.7%	**64.4%**	58.3%	65.8%	48.7%	34.3%	34.3%	86.3%	62.3%	59.2%
PVNet (VGG16) †	**74.85%**	49.93%	82.18%	78.67%	**98.82%**	63.43%	52.57%	63.06%	**70.41%**	74.48%	73.48%	**94.85%**	74.98%	73.21%
PVNet (ResNet101) †	73.85%	**59.60%**	76.14%	**81.99%**	98.33%	58.82%	**59.19%**	**66.27%**	64.07%	**78.41%**	**79.67%**	94.53%	**76.66%**	**74.43%**
PVNet (VGG16)$_{IoU}$ †	**64.17%**	33.34%	**64.05%**	64.25%	**90.39%**	**49.27%**	40.95%	45.17%	**54.78%**	62.83%	**52.31%**	80.62%	58.87%	58.54%
PVNet (ResNet101)$_{IoU}$ †	63.09%	**38.35%**	61.16%	**68.58%**	89.66%	48.07%	**44.34%**	**50.39%**	50.89%	**63.48%**	49.97%	**81.51%**	**61.40%**	**59.30%**

Modelling the global context information and simultaneously preserving the local shape information are the two key problems in CNN-based semantic segmentation. The main idea of Pixel-Voxel net is to leverage the advantages of two complementary modalities, to extract high-level context features from RGB and fuse them with low-level geometric features from the point cloud. The improvement can be attributed to three parts: the hierarchical convolutional stack in PixelNet, the boundary refinement by VoxelNet and the `softmax` weighted fusion stack. First, the hierarchical convolutional stack can learn the high-level contextual information through an incrementally-enlarged receptive field. As shown in Tables 1 and 2, the standalone PixelNet can achieve a very competitive performance. Second, the proposed VoxelNet can refine the 3D object boundaries through learning the low-level geometrical features from the point clouds. As shown in Figure 5, the objects have finer boundaries after combining with VoxelNet. As shown in Tables 1 and 2, the quantitative performance improves significantly through 3D-based shape refinement from VoxelNet. Third, the proposed softmax fusion layer can adaptively learn the confidence of each modality. As a result, the predictions from different modalities can be fused more effectively. As shown in Tables 1 and 2, the quantitative results also increase slightly through the $softmax$ fusion stack. Note that the overall accuracy cannot be improved significantly, as pixels/voxels on the object edge only occupy a very small percentage of the whole pixels/voxels. However, the mean accuracy experiences a substantial improvement due to the increased accuracy on rare classes, for which the edge pixels occupy a relatively large percentage of all pixels.

Most state-of-the-art methods employ multi-scale CRF or a 2D/3D graph to refine the object boundaries. Their main limitation is slowness because of the excessive usage of multi-resolution high computational CRF or graph optimization. Although their performance is slightly better than ours, these methods are unlikely to be applied to real-time robotics applications. Our method can preserve the fine boundary shape through learning the low-level features from 3D geometry data. There is no computational optimization in the Pixel-Voxel network, so it is faster than most state-of-the-art methods.

4.4. Dense RGB-D Semantic Mapping

The dense RGB-D semantic mapping system is implemented under the ROS (http://www.ros.org/) framework and executed on a desktop with i7-6800k (3.4 Hz) 8-core CPU and NVIDIA TITAN X GPU (12G). Kinect V2 is used to obtain the RGB images and point clouds. IAI Kinect2 package2 (https://github.com/code-iai/iaikinect2/) is employed to interface with ROS and calibrate with the Kinect2 cameras. The Pixel-Voxel network is implemented using the Caffe (http://caffe.berkeleyvision.org/) toolbox. The network is trained on a TITAN X GPU, accelerated by CUDA and CUDNN.

The system with a pre-trained network was also tested in a real-world environment, e.g., a living room and bedroom containing a curtain, bed, etc., as shown in Figure 7. It can be seen that most of the point clouds are correctly segmented, and the results have accurate boundaries, but there are still some points on the boundary with wrongly-assigned labels. Some error predictions are caused by upsampling the data through a bilateral filter to the same size as the Kinect V2 data. Furthermore, this network was trained using the SUN RGB-D and NYU V2 datasets, but was tested using the real-world data. Therefore, some errors occur due to illumination variances, category variances, etc. In addition, the noise of the Kinect V2 also causes some errors in predictions.

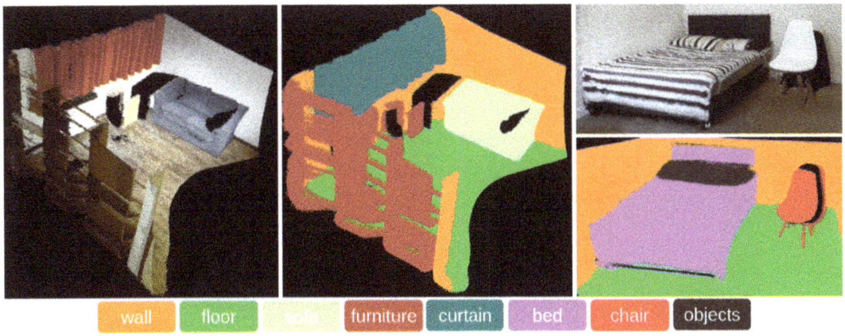

Figure 7. The dense 3D map and dense 3D semantic map (best viewed in colour) of a living room and bedroom.

Using the quad high definition (QHD) data from Kinect2, the runtime performances of our system are 5.68 Hz (VGG16) and 3.23 Hz (ResNet101) when the RGB is resized to 512×512 and the point cloud is down-sampled to three scales, $16{,}384 \times 1$, 4096×1 and 1024×1. During real-time RGB-D mapping, only a few key-frames are used for mapping. Most of the frames are abandoned because of the small variance between two consecutive frames. It is not necessary to segment all the frames in the sequence, but only the key-frames. As mentioned in [21], the 5-Hz runtime performance is nearly sufficient for real-time dense 3D semantic mapping. It is worth noting that the running time can be boosted to 13.33 Hz (VGG16) and 9.01 Hz (ResNet101) using half-sized data with a corresponding decline in segmentation performance. Thus, there is a trade-off between performance requirement and time consumption. The inference running time of Pixel-Voxel Net using different sizes of data can be found in Table 5, and the corresponding decline in performance can be found in Table 6.

Table 5. The average inference runtime of Pixel-Voxel Net (PVNet) using different sizes of data.

Network on the Different Sizes of Data	Inference Runtime	
	Full Size	Half Size
PVNet (VGG-16)	0.176s	0.075s
PVNet (ResNet101)	0.310s	0.111s

Table 6. The declining performance of Poxel-Voxel Net (PVNet) using half-sized data. △ represents the declining performance (in percentage) with half-sized data compared to that with full-sized data.

Network on the Half Size Data	SUN RGB-D			NYU V2		
	△ Pixel acc.	△ Mean acc.	△ Mean IoU	△ Pixel acc.	△ Mean acc.	△ Mean IoU
PVNet (VGG-16)	−1.35%	−1.87%	−1.59%	−1.08%	−0.62%	−1.53%
PVNet (ResNet101)	−1.16%	−2.34%	−1.94%	−1.41%	−0.84%	−1.96%

5. Conclusions

This paper introduced an end-to-end discriminative Pixel-Voxel network for dense 3D semantic mapping. The hierarchical convolutional stack structure in PixelNet can model the high-level contextual information through an incrementally-enlarged receptive field, while the VoxelNet learns geometrical shapes via a non-linear feature transform in order to identify 3D objects with fine object boundaries. More importantly, an adaptive fusion layer, i.e., $softmax$ fusion, can learn the probabilistic confidences in order to fuse features from RGB and depth (3D) modalities in the non-linear fashion. We achieved competitive performance on the SUN RGB-D benchmark (pixel acc.: 79.04%, mean acc.: 57.65% and mean IoU: 44.24%) and NYU V2 benchmark (pixel acc.: 82.53%, mean acc.: 74.43% and mean IoU: 59.30%). Our method is fully parametric without running time optimizations. Consequently, a straightforward inference is used for deployment, which guarantees near-real-time performance.

Our method is faster than most state-of-the-art methods (up to around 13 Hz using an i7 eight-core PC with Titan X GPU) and can be integrated into a SLAM system for near-real-time application in robotics.

For future work, we will investigate the possibility of applying the proposed VoxelNet for semantic segmentation [40] with 3D LiDAR data, where only 3D geometric data are available. Moreover, and we will investigate adopting the proposed semantic mapping method to domestic robot navigation and manipulation tasks. The source code will be published upon acceptance. A real-time demo can be found on the author's Youtube channel (https://youtu.be/UbmfGsAHszc).

Author Contributions: C.Z. propsed the main idea, performed the experiments and implemented the whole system. L.S., P.P., T.D. and R.S. supervised this research and revised the article.

Funding: This work was supported by the DISTINCTIVE scholarship, Toshiba Research Europe and EU Horizon 2020 ILIAD (732737) and RoMaNS (645582) projects.

Acknowledgments: We thank NVIDIA Corporation for generously donating a high-power TITAN X GPU.

Conflicts of Interest: The authors declare no conflicts of interest.

References

1. Purkait, P.; Zhao, C.; Zach, C. SPP-Net: Deep Absolute Pose Regression with Synthetic Views. *arXiv* **2018**, arXiv:1712.03452.
2. Zhao, C.; Sun, L.; Purkait, P.; Duckett, T.; Stolkin, R. Learning monocular visual odometry with dense 3D mapping from dense 3D flow. *arXiv* **2018**, arXiv:1803.02286.
3. Zhao, C.; Mei, W.; Pan, W. Building a grid-semantic map for the navigation of service robots through human-robot interaction. *Digit. Commun. Netw.* **2015**, *1*, 253–266. [CrossRef]
4. Zhao, C.; Hu, H.; Gu, D. Building a grid-point cloud-semantic map based on graph for the navigation of intelligent wheelchair. In Proceedings of the 2015 IEEE International Conference on Automation and Computing (ICAC), Glasgow, UK, 11–12 September 2015; pp. 1–7.
5. Sun, L.; Yan, Z.; Molina, S.; Hanheide, M.; Duckett, T. 3DOF pedestrian trajectory prediction learned from long-term autonomous mobile robot deployment data. In Proceedings of the 2018 IEEE International Conference on Robotics and Automation (ICRA), Brisbane, Australia, 21–25 May 2018; pp. 5942–5948.
6. Valiente, D.; Payá, L.; Jiménez, L.M.; Sebastián, J.M.; Reinoso, Ó. Visual Information Fusion through Bayesian Inference for Adaptive Probability-Oriented Feature Matching. *Sensors* **2018**, *18*, 2041. [CrossRef] [PubMed]
7. Sun, L.; Zhao, C.; Duckett, T.; Stolkin, R. Weakly-supervised DCNN for RGB-D object recognition in real-world applications which lack large-scale annotated training data. *arXiv* **2017**, arXiv:1703.06370.
8. Endres, F.; Hess, J.; Sturm, J.; Cremers, D.; Burgard, W. 3-D mapping with an RGB-D camera. *Trans. Robot.* **2014**, *30*, 177–187. [CrossRef]
9. Newcombe, R.A.; Izadi, S.; Hilliges, O.; Molyneaux, D.; Kim, D.; Davison, A.J.; Kohi, P.; Shotton, J.; Hodges, S.; Fitzgibbon, A. KinectFusion: Real-time dense surface mapping and tracking. In Proceedings of the 2011 10th IEEE International Symposium on Mixed and Augmented Reality, Basel, Switzerland, 26–29 October 2011; pp. 127–136.
10. Whelan, T.; Leutenegger, S.; Salas-Moreno, R.; Glocker, B.; Davison, A. ElasticFusion: Dense SLAM without a pose graph. In Proceedings of the Robotics: Scienceand Systems, Rome, Italy, 13–17 July 2015.
11. Long, J.; Shelhamer, E.; Darrell, T. Fully convolutional networks for semantic segmentation. In Proceedings of the 2015 IEEE Conference on Computer Vision and Pattern Recognition (CVPR), Boston, MA, USA, 7–12 June 2015; pp. 3431–3440.
12. Badrinarayanan, V.; Kendall, A.; Cipolla, R. Segnet: A deep convolutional encoder-decoder architecture for image segmentation. *arXiv* **2015**, arXiv:1511.00561.
13. Chen, L.C.; Papandreou, G.; Kokkinos, I.; Murphy, K.; Yuille, A.L. Deeplab: Semantic image segmentation with deep convolutional nets, atrous convolution, and fully connected crfs. *arXiv* **2016**, arXiv:1606.00915.
14. Hazirbas, C.; Ma, L.; Domokos, C.; Cremers, D. Fusenet: Incorporating depth into semantic segmentation via fusion-based cnn architecture. In *Asian Conference on Computer Vision;* Springer: Cham, Switzerland, 2016; pp. 213–228.

15. Li, Z.; Gan, Y.; Liang, X.; Yu, Y.; Cheng, H.; Lin, L. LSTM-CF: Unifying context modeling and fusion with LSTMS for RGB-D scene labelling. In *European Conference on Computer Vision*; Springer: Cham, Switzerland, 2016; pp. 541–557.
16. Cheng, Y.; Cai, R.; Li, Z.; Zhao, X.; Huang, K. Locality-Sensitive Deconvolution Networks with Gated Fusion for RGB-D Indoor Semantic Segmentation. In Proceedings of the 2017 IEEE Conference on Computer Vision and Pattern Recognition (CVPR), Honolulu, HI, USA, 21–26 July 2017; pp. 3029–3037.
17. Lin, D.; Chen, G.; Cohen-Or, D.; Heng, P.A.; Huang, H. Cascaded Feature Network for Semantic Segmentation of RGB-D Images. In Proceedings of the 2017 IEEE International Conference on Computer Vision (ICCV), Venice, Italy, 22–29 October 2017; pp. 1311–1319.
18. Qi, C.R.; Su, H.; Mo, K.; Guibas, L.J. Pointnet: Deep learning on point sets for 3d classification and segmentation. *arXiv* **2016**, arXiv:1612.00593.
19. Qi, C.R.; Yi, L.; Su, H.; Guibas, L.J. PointNet++: Deep Hierarchical Feature Learning on Point Sets in a Metric Space. *arXiv* **2017**, arXiv:1706.02413.
20. Salas-Moreno, R.F.; Newcombe, R.A.; Strasdat, H.; Kelly, P.H.; Davison, A.J. Slam++: Simultaneous localisation and mapping at the level of objects. In Proceedings of the 2013 IEEE Conference on Computer Vision and Pattern Recognition (CVPR), Sydney, Australia, 1–8 December 2013; pp. 1352–1359.
21. Hermans, A.; Floros, G.; Leibe, B. Dense 3D semantic mapping of indoor scenes from RGB-D images. In Proceedings of the 2014 IEEE International Conference on Robotics and Automation (ICRA), Hong Kong, China, 31 May–7 June 2014; pp. 2631–2638.
22. McCormac, J.; Handa, A.; Davison, A.; Leutenegger, S. Semanticfusion: Dense 3D semantic mapping with convolutional neural networks. In Proceedings of the 2017 IEEE International Conference on Robotics and Automation (ICRA), Singapore, 29 May–3 June 2017; pp. 4628–4635.
23. Xiang, Y.; Fox, D. DA-RNN: Semantic Mapping with Data Associated Recurrent Neural Networks. *arXiv* **2017**, arXiv:1703.03098.
24. Tateno, K.; Tombari, F.; Navab, N. When 2.5 D is not enough: Simultaneous reconstruction, segmentation and recognition on dense SLAM. In Proceedings of the 2016 IEEE International Conference on Robotics and Automation (ICRA), Stockholm, Sweden, 16–21 May 2016; pp. 2295–2302.
25. Vineet, V.; Miksik, O.; Lidegaard, M.; Nießner, M.; Golodetz, S.; Prisacariu, V.A.; Kähler, O.; Murray, D.W.; Izadi, S.; Pérez, P.; et al. Incremental dense semantic stereo fusion for large-scale semantic scene reconstruction. In Proceedings of the 2015 IEEE International Conference on Robotics and Automation (ICRA), Seattle, WA, USA, 26–30 May 2015; pp. 75–82.
26. Tateno, K.; Tombari, F.; Laina, I.; Navab, N. CNN-SLAM: Real-time dense monocular SLAM with learned depth prediction. *arXiv* **2017**, arXiv:1704.03489.
27. Zhao, C.; Sun, L.; Stolkin, R. A fully end-to-end deep learning approach for real-time simultaneous 3D reconstruction and material recognition. In Proceedings of the 2017 IEEE International Conference on Advanced Robotics (ICAR), Hong Kong, China, 10–12 July 2017; pp. 75–82.
28. Ma, L.; Stückler, J.; Kerl, C.; Cremers, D. Multi-view deep learning for consistent semantic mapping with RGB-D cameras. *arXiv* **2017**, arXiv:1703.08866.
29. Mustafa, A.; Hilton, A. Semantically Coherent Co-segmentation and Reconstruction of Dynamic Scenes. In Proceedings of the 2017 Conference on Computer Vision and Pattern Recognition (CVPR), Honolulu, HI, USA, 21–26 July 2017.
30. Noh, H.; Hong, S.; Han, B. Learning deconvolution network for semantic segmentation. In Proceedings of the 2015 IEEE International Conference on Computer Vision (ICCV), Las Condes, Chile, 11–18 December 2015; pp. 1520–1528.
31. Krähenbühl, P.; Koltun, V. Efficient inference in fully connected crfs with gaussian edge potentials. In *Advances in Neural Information Processing Systems*; MIT Press: Cambridge, MA, USA, 2011; pp. 109–117.
32. Zheng, S.; Jayasumana, S.; Romera-Paredes, B.; Vineet, V.; Su, Z.; Du, D.; Huang, C.; Torr, P.H. Conditional random fields as recurrent neural networks. In Proceedings of the 2015 IEEE International Conference on Computer Vision (ICCV), Las Condes, Chile, 11–18 December 2015; pp. 1529–1537.
33. He, Y.; Chiu, W.C.; Keuper, M.; Fritz, M.; Campus, S.I. STD2P: RGBD Semantic Segmentation using Spatio-Temporal Data-Driven Pooling. In Proceedings of the 2017 Conference on Computer Vision and Pattern Recognition (CVPR), Honolulu, HI, USA, 21–26 July 2017.

34. Shuai, B.; Liu, T.; Wang, G. Improving Fully Convolution Network for Semantic Segmentation. *arXiv* **2016**, arXiv:1611.08986.
35. Shuai, B.; Zuo, Z.; Wang, B.; Wang, G. Dag-recurrent neural networks for scene labelling. In Proceedings of the 2016 Conference on Computer Vision and Pattern Recognition (CVPR), Las Vegas, NV, USA, 26 June–1 July 2016; pp. 3620–3629.
36. Lin, G.; Shen, C.; Van Den Hengel, A.; Reid, I. Exploring context with deep structured models for semantic segmentation. *IEEE Trans. Pattern Anal. Mach. Intell.* **2017**, *40*, 1352–1366. [CrossRef] [PubMed]
37. Lin, G.; Milan, A.; Shen, C.; Reid, I. Refinenet: Multi-path refinement networks for high-resolution semantic segmentation. In Proceedings of the 2017 Conference on Computer Vision and Pattern Recognition (CVPR), Honolulu, HI, USA, 21–26 July 2017.
38. Handa, A.; Patraucean, V.; Badrinarayanan, V.; Stent, S.; Cipolla, R. Understanding real world indoor scenes with synthetic data. In Proceedings of the 2016 Conference on Computer Vision and Pattern Recognition (CVPR), Las Vegas, NV, USA, 26 June–1 July 2016; pp. 4077–4085.
39. Eigen, D.; Fergus, R. Predicting depth, surface normals and semantic labels with a common multi-scale convolutional architecture. In Proceedings of the 2015 IEEE International Conference on Computer Vision (ICCV), Las Condes, Chile, 11–18 December 2015; pp. 2650–2658.
40. Sun, L.; Yan, Z.; Zaganidis, A.; Zhao, C.; Duckett, T. Recurrent-OctoMap: Learning State-Based Map Refinement for Long-Term Semantic Mapping With 3-D-Lidar Data. *IEEE Robot. Autom. Lett.* **2018**, *3*, 3749–3756. [CrossRef]

sensors

MDPI

Article

Lightweight Visual Odometry for Autonomous Mobile Robots

Mohamed Aladem and Samir A. Rawashdeh *

College of Engineering and Computer Science, University of Michigan-Dearborn, Dearborn, MI 48128, USA; maladem@umich.edu

* Correspondence: rawashdeh@umich.edu; Tel.: +1-313-593-5466

Received: 19 July 2018; Accepted: 25 August 2018; Published: 28 August 2018

Abstract: Vision-based motion estimation is an effective means for mobile robot localization and is often used in conjunction with other sensors for navigation and path planning. This paper presents a low-overhead real-time ego-motion estimation (visual odometry) system based on either a stereo or RGB-D sensor. The algorithm's accuracy outperforms typical frame-to-frame approaches by maintaining a limited local map, while requiring significantly less memory and computational power in contrast to using global maps common in full visual SLAM methods. The algorithm is evaluated on common publicly available datasets that span different use-cases and performance is compared to other comparable open-source systems in terms of accuracy, frame rate and memory requirements. This paper accompanies the release of the source code as a modular software package for the robotics community compatible with the Robot Operating System (ROS).

Keywords: visual odometry; ego-motion estimation; stereo; RGB-D; mobile robots

1. Introduction

Accurate and real-time motion estimation to enable robot perception and control tasks is a fundamental problem in mobile robotics. Despite being well studied, it remains challenging to provide timely and accurate robot motion estimates for applications where the mobile robot is operating in uncontrolled and previously unknown environments. Examples of such applications include micro-aerial vehicles (MAVs) exploring a new environment, advanced driver-assistance systems (ADAS) in modern automobiles and autonomous self-driving vehicles.

A mobile robot's position estimation can be performed through dead reckoning using a combination of wheel odometers in ground vehicles and inertial measurement units (IMUs). However, continuous integration of measurements from these sensors results in drift in position estimates, rendering them unsuitable for sustained long-term operation. Sonar and ultrasonic sensors can be used for robot localization; however, they are active sensors and can interfere with each other. The global positioning system (GPS) provides absolute position with no error accumulation over time. However, GPS sensors are unavailable in indoor and closed areas and they lose satellite lock in tunnels and urban canyons.

Cameras are attractive sensors for performing the motion estimation task. Cameras are low-cost sensors and can provide a rich stream of information. Furthermore, since cameras are passive sensors, they do not interfere with each other when multiples are deployed. Visual odometry (VO) is the process of estimating the position and orientation of an agent (e.g., a vehicle) using only the input of a single or multiple camera attached to it [1].

In this paper, we present our innovative visual odometry system called lightweight visual tracking (LVT). Unlike typical visual odometry approaches where features are tracked and motion is estimated between consecutive frames only, our system tracks features for as long as possible. This results in a system that is approaching full visual simultaneous localization and mapping (V-SLAM) systems in terms of accuracy while still maintaining low computational overhead. Moreover, the system supports both stereo and RGB-D cameras. For the benefit of the community, the full source code of the system is made publicly available under a permissive license and with support for the Robot Operating System (ROS) at: (https://github.com/SAR-Research-Lab/lvt).

2. Related Work

Visual odometry is an active area of research where many different methods have been developed over the years. A detailed review of the field of visual odometry was published by Scaramuzza and Fraunhofer [1]. The problem of estimating vehicle motion from visual input was first approached by Moravec [2] in the early 1980s. Moravec established the first motion-estimation pipeline, whose main functional blocks are still used today. However, the term visual odometry was first coined by Nister et al. in their landmark paper [3]. Their paper was the first to demonstrate a real-time long-run implementation with a robust outlier rejection scheme.

Kitt et al. presented a visual odometry algorithm based directly on the trifocal geometry between image triples [4]. Howard has presented a visual odometry system in which inliers are detected based on geometric constraints rather than on performing the more commonly used outlier rejection schemes [5]. However, those approaches follow the typical visual odometry scheme of matching or tracking features between consecutive frames and using these features for ego-motion estimation. The first to propose using the whole history of tracked features is the work by Badino et al. [6]. They do so by computing integrated features, which are the sample mean of all previous measured positions of each feature. Integrated features are then used in the motion computation between the two consecutive frames. Our approach differs in that we employ a transient local 3D map for tracking. It consists of a sparse set of 3D points (features) that are used for tracking and, therefore, motion estimation. This local map is internal to the system and is not an attempt to build a global map of the environment. As long as a feature is useful and can be utilized for motion estimation, it is kept alive in this local map.

Related to visual odometry is the simultaneous localization and mapping (SLAM) problem, where the goal is to build an accurate map of the environment while simultaneously localizing within this map. Visual SLAM systems employ sophisticated techniques to improve their accuracy, such as detecting loop-closures, where the system detects a previously visited location and uses this information to correct the map. Early work of bringing vision into SLAM was Davison's MonoSLAM [7], which was a filter-based method utilizing an Extended Kalman Filter (EKF). Filter-based methods were dominant until the development of the Parallel Tracking and Mapping (PTAM) system by Klein and Murray [8]. PTAM separated and parallelized the motion estimation and mapping tasks through a keyframe-based architecture while employing bundle adjustment [9]. An excellent introduction and survey of keyframe-based visual SLAM is by Younes et al. [10]. Unlike typical visual SLAM systems, by employing our local map approach our system is able to achieve high accuracy while maintaining low memory and computation requirements.

Additionally, Visual SLAM and odometry systems can be classified as either direct or feature-based methods. Feature-based methods try to extract distinctive interest points and track them in subsequent frames to estimate camera ego-motion. In contrast, direct methods operate on the whole image directly. That is, the camera is localized by optimizing directly over image pixel intensities. Key representatives of feature-based visual SLAM systems are S-PTAM [11] and ORB-SLAM [12], whereas a representative of the direct visual SLAM systems is LSD-SLAM [13]. Our system is based on point or corner-like features as they enable real-time performance while running completely on a CPU.

With the introduction of commodity depth sensors, RGB-D cameras have become popular in robotics. Huang et al. have introduced a visual odometry system based on RGB-D cameras called Fovis [14]. Their system extracts point features from the image and then each feature's depth is extracted from the depth image. They follow an inlier detection approach similar to Howard's [5] and track features between consecutive frames. On the other hand, DVO by Kerl et al. [15] is a dense visual odometry method that aims to exploit both the intensity and depth information of RGB-D cameras. Yet other approaches use depth information alone for ego-motion estimation. One common approach is by 3D point-cloud registration, which is commonly performed using an iterative closest point (ICP) algorithm [16,17]. KinectFusion by Newcombe et al. [18] is one of earliest and most well-known RGB-D SLAM systems. It fuses depth data into a volumetric dense model that is used for tracking camera motion. Our system supports RGB-D data as well, where features are detected from the RGB image and depth information is extracted from the depth image. The rest of the system remains intact and performs the same operations regardless of the used sensor.

3. System Description

A high-level overview of the visual odometry algorithm is shown in Algorithm 1. In the rest of the paper, camera pose is understood to encompass both position and orientation, that is, the complete 6 degrees-of-freedom transformation. The world reference coordinate frame is set at the pose of the first frame of the sequence.

Algorithm 1. Visual Odometry Algorithm Overview

```
for each frame (stereo or RGB-D)
  Extract features (AGAST, BRIEF)
  if first frame
    Initialize local map
    Set world reference coordinate frame
  else
    Predict new pose
    Track local map
    Estimate pose using tracked measurements
    Update staged map points
    Perform new triangulations if necessary
    Clear no longer trackable map points
  end if
end for
```

The details of different components are illustrated below.

3.1. Image Sequence

The first step is to feed the system a frame that is acquired either from a stereo or RGB-D sensor. In the stereo camera case, the retrieved stereo frame consists of the synchronized images from the left and the right cameras. The stereo frame is assumed to be stereo-rectified. Stereo rectification is the process of virtually transforming the stereo frame so that it appears as if the two cameras of the stereo rig have their image planes aligned to be coplanar. Consequently, epipolar lines are now parallel to the stereo baseline between the cameras. This reduces the stereo correspondence problem to a one-dimensional search, as matching features between the two cameras will lie on the same pixel row, assuming a horizontal stereo rig [19]. In the RGB-D sensor case, the RGB image is converted into a grayscale one and then it is fed along with the depth image to the system.

3.2. Feature Extraction

As a feature-based method, salient point or corner-like features will be extracted and used for subsequent processing. Corner-like features are fast to compute and many good corner detectors are available. In this work, the adaptive and generic accelerated segment test (AGAST) [20] corner detector is used. AGAST builds on the features from accelerated segment test (FAST) [21] corner detector, which enjoys a high degree of repeatability and computational performance. AGAST improves the accelerated segment test that underlies FAST by making it more generic while increasing its performance. No scale pyramid of any image is built and the corners are extracted from the full-size images.

For each detected feature, a feature descriptor is computed. A feature descriptor acts like a signature for that feature. Matching between features can then be performed by comparing their descriptors. In this work, binary robust independent elementary features (BRIEF) [22] descriptors are used. BRIEF descriptors are binary feature descriptors, that is, descriptors that are in the form of a binary string. This means that matching is fast, based on hamming distance. Hamming distance is defined as the number of positions at which the corresponding bits are different. Hamming distance is simple and fast to compute because it is just an exclusive-OR (XOR) operation and a bit count, that is, sum(xor(descriptor1, descriptor2)). However, BRIEF descriptors lack rotational invariance.

An important problem to consider is the distribution of the detected features across the image. Poor distribution where detected features are concentrated in one region of the image can lead to poor results. We follow a two-step process to overcome this problem. First, the image is divided into cells where features will be detected in each cell separately. Then, a technique known as adaptive non-maximal suppression, as described by Brown et al. [23], is performed in each cell. Adaptive non-maximal suppression aims to limit the maximum number of features extracted while at the same time ensuring good distribution across the image. Features are suppressed based on corner strength and only ones that are local maxima in the neighborhood are retained.

3.3. Pose Prediction

Before proceeding to the next step, a prediction of the current camera pose is performed. For ground vehicles, dead reckoning using wheel odometry can be used to perform such prediction since it is usually available. In this work, we will follow a simple motion model where velocity is assumed constant between frames and then computed velocities are averaged over frames. In this simple motion model, a constant frame rate is assumed, that is, time is not considered in the calculations. The motion is calculated as follows: assuming the pose at frame k to be C_k, which consists of orientation represented as a quaternion, $q_k \in SO(3)$ and position $t_k \in \mathbb{R}^3$. C_k is the pose to be predicted. The linear velocity at frame k is computed as:

$$v_k = t_{k-1} - t_{k-2} \tag{1}$$

This linear velocity is then averaged over time:

$$v_k = (v_k + v_{k-1})/2 \tag{2}$$

A similar approach is followed for the rotational component where quaternion multiplication is performed:

$$w_k = q_{k-1} q_{k-2}^{-1} \tag{3}$$

and then the rotational velocity is also averaged over time using the spherical linear interpolation (SLERP) operation:

$$w_k = slerp(w_k, w_{k-1}, 0.5) \tag{4}$$

After computing the new velocities, the predicted pose is easily computed as follows:

$$t_k = t_{k-1} + v_k \tag{5}$$

$$q_k = q_{k-1} w_k \tag{6}$$

The predicted pose is then used as a guide in the next local map-tracking step.

3.4. Tracking Local Map

The goal of this step is to correctly associate visible 3D map points with 2D image features. This 3D-2D data association is used by the following pose estimation step. Note that the associated 2D image features here are, in the stereo camera case, the ones extracted from the left camera image only. The 3D map points are projected onto the image plane of the left camera in the stereo camera case and of the RGB camera in the RGB-D camera case using the pinhole camera model. The neighborhood of each projected 3D map point is then examined for the best matching 2D feature detected in the feature extraction step. In our current implementation, the search neighborhood is set to a 25-pixel radius around each projected feature. If no enough matches are found, the search radius is doubled and the tracking step is performed again. This process is illustrated in Figure 1, where the local 3D map is shown to the left and the detected 2D image features in the current frame are shown to the right.

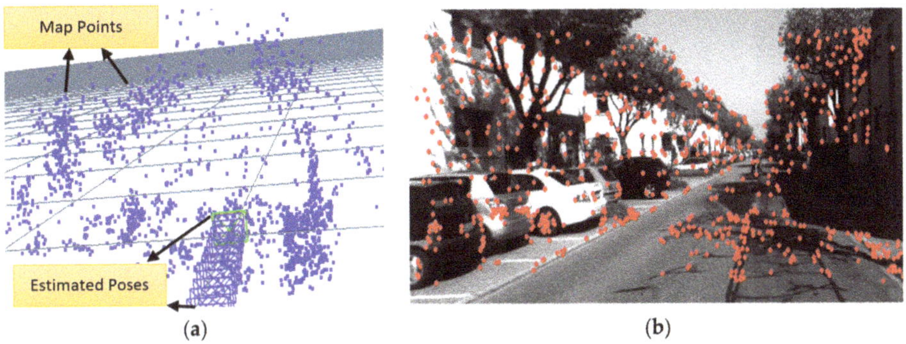

Figure 1. An illustration showing: (**a**) local 3D map and (**b**) detected 2D image features. The goal is to find the correspondence between each 3D map point and detected 2D feature. Sample image from KITTI dataset.

Neighborhood search is accelerated by means of spatial hashing. During the feature extraction step, the image plane is divided into a two-dimensional grid. Each cell of this grid is assigned the list of corners that happen to be within its boundaries. Now, the projected point's prospective grid cell is computed and thus the list of potential matches is readily available. Once candidate neighborhood features are identified, finding the best match of the projected point proceeds by using the widely accepted ratio test proposed by Lowe [24]. The ratio test works by comparing the distance of the closest neighbor to that of the second closest one. The nearest neighbor here is defined as the one with the minimum hamming distance for the BRIEF descriptor. If the nearest feature is much closer than the second nearest one, then it has a higher probability of being the correct match. The ratio test value is set to 0.80 in our implementation.

3.5. Pose Estimation

The found matches are then used for computing the camera pose. Camera pose consists of orientation $\boldsymbol{R} \in SO(3)$ and position $\boldsymbol{t} \in \mathbb{R}^3$. Finding the camera pose is formalized as an optimization problem to find the optimal $\boldsymbol{R}, \boldsymbol{t}$ that minimizes the reprojection error between the matched 3D points and the image 2D features:

$$\{\boldsymbol{R}, \boldsymbol{t}\} = \underset{\boldsymbol{R},\boldsymbol{t}}{\operatorname{argmin}} \sum_{i \in S} \rho\left(||x^i - \pi\left(\boldsymbol{R}X^i + \boldsymbol{t}\right)||^2 \right) \tag{7}$$

where $x^i \in \mathbb{R}^2$ are image features, $X^i \in \mathbb{R}^3$ are world 3D points, for $i \in S$ the set of all matches. ρ is the Cauchy cost function. π is the projection function. This minimization problem is solved iteratively using the Levenberg–Marquardt algorithm. Furthermore, outliers are detected and excluded and the optimization is run for a second time with the inlier set.

3.6. Local Map Maintenance

We maintain a secondary map of points we refer to as staged points, which are also tracked over time but are not used in motion estimation until they are found to be of high quality. After camera pose is computed, staged map points are updated. When a new 3D-points triangulation occurs, these new points are initially stored in this staging area and are not added immediately to the local map. If a staged point is tracked successfully for a specified number of frames, then it is declared of good quality and added to the map. If it fails tracking, then it is removed. However, if the number of map points drops below 1000 point, then staged points will be added immediately to the map in order to always maintain a minimum number of points in the map.

For the purpose of triggering a new points triangulation, we will monitor the number of 2D-3D matches found in the current frame and in the previous two frames. If the number of matches is decreasing, a new triangulation is performed. New points are triangulated from features that were not tracked in the current frame. These newly triangulated points are added to the staging area as described previously. Additionally, map points that fail tracking for a specified number of frames are deleted from the map. Hence, the local map is kept fresh with good immediately useful points from the staging area, while no-longer-trackable points are removed.

When the system initially starts, the local map is empty, so the first frame is used to triangulate the initial set of 3D points, which are added immediately to the local map. This first frame also sets the world reference coordinate frame. All reported poses will be with respect to this world coordinate frame. In the stereo case, triangulations are performed using the Linear-LS method described by Hartley and Sturm [25]. As the stereo frame is rectified, matching corners between the left and right images is greatly simplified, as the search is restricted to the same row in both images. In the RGB-D case, triangulations are performed by extracting depth values directly from the depth image.

3.7. General Implementation Remarks

The algorithm is implemented using the C++ programming language. OpenCV library was used to perform image processing tasks. Corner detector and descriptor extractor implementations available as part of OpenCV library were used. The general graph optimization (g2o) library [26] was used to perform the Levenberg–Marquardt minimization in the pose estimation step.

The algorithm runs purely on the CPU, with no GPU acceleration used. Moreover, the algorithm runs primarily in one thread. However, a parallel thread is spawned to perform the features computation step on the right image in the stereo cameras case while the main thread is busy performing it on the left one.

4. Evaluation

We present results of the evaluation of our visual odometry system on three challenging, publicly available datasets—namely, the KITTI dataset [27], the EuRoC MAV dataset [28] and the TUM RGB-D dataset [29]. Each dataset has its unique characteristics, which will enable comprehensive evaluation of our visual odometry system.

4.1. KITTI Dataset

The KITTI dataset is widely used for evaluating autonomous driving algorithms. The dataset was collected by driving in different traffic scenarios in the city of Karlsruhe, Germany. Some of the challenging aspects of the dataset are the presence of dynamic moving objects (vehicles, cyclists and pedestrians), the different lighting and shadow conditions as the vehicle is moving and the presence of foliage, which results in the detection of many non-stable and challenging-to-track corners.

The presented visual odometry system is also evaluated against two open-source systems, S-PTAM [11] and LIBVISO2 [30]. S-PTAM is a state-of-the-art full V-SLAM system. S-PTAM was compiled without loop-closing capability but all other operations remain intact. LIBVISO2 is a famous stereo visual odometry system that tracks features and estimates motion between consecutive frames only. With S-PTAM being a V-SLAM system and LIBVISO2 a frame-to-frame visual odometry system, they provide a good comparison ground for our presented system, which aims to reach V-SLAM systems accuracy while being lightweight like typical visual odometry systems.

4.1.1. Accuracy

For evaluating the accuracy, two error metrics will be reported. First is the average translation error E_t over all subsequences of length (100, ..., 800) meters as defined in the KITTI dataset paper [27]. We define the other metric ξ as:

$$\xi = RMSE(T_{1:n}) = \left(\frac{1}{n}\sum_{i=1}^{n} T_i^2\right)^{1/2} \tag{8}$$

where T_i is the magnitude of the Euclidean distance along the horizontal plane between the estimated and ground truth pose at frame i.

The computed error metrics for KITTI sequences (00–10) except for sequence 01 are shown in Table 1. Sequence 01 is a challenging highway with unreliable far features. Although our algorithm does not lose tracking, it drifts badly and fails to provide meaningful estimates. Hence, it was excluded. From the presented results, it can be seen how our proposed visual odometry algorithm comes very close to S-PTAM, which is a complete V-SLAM system and surpasses LIBVISO2. Plots of the estimated path against ground truth are shown in Figure 2.

Table 1. Results on KITTI dataset.

Sequence	Length (km)	LVT		S-PTAM		LIBVISO2	
		E_t (%)	ξ (m)	E_t (%)	ξ (m)	E_t (%)	ξ (m)
00	3.7223	1.25	11.01	0.84	8.81	2.74	47.03
02	5.0605	1.33	13.59	0.96	22.20	2.20	69.52
03	0.5590	1.04	2.48	1.14	3.43	2.27	4.66
04	0.3936	0.56	0.78	1.29	2.72	1.08	2.67
05	2.2046	0.89	4.27	0.91	3.01	2.26	19.80
06	1.2326	1.04	2.11	1.28	3.34	1.28	4.20
07	0.6944	0.98	3.17	0.88	2.97	2.34	5.74
08	3.2137	1.25	6.57	1.05	6.69	2.83	44.52
09	1.7025	1.76	9.23	1.21	9.28	2.84	24.32
10	0.9178	1.02	3.81	0.64	3.61	1.39	2.97
Total	19.701	1.23	9.05	0.97	11.73	2.45	43.98

Additionally, we submitted our system for evaluation on the KITTI sequences (11–21) which are test sequences that have no publicly available ground truth. Our system achieved an average translation error of 5.8%. While running the analysis, we suspect the main source of error is sequence 21 which is a highway and suffers from the same issue as sequence 01 before. The problem with highway scenes is that features are far away relative to the stereo baseline causing the stereo cameras to degenerate into a monocular one. This results in the loss of scale information. Moreover, the KITTI evaluation server provides the path plots of a portion of the test sequences which are shown in Figure 2. From these path plots, we can see that our system is able to provide accurate estimates on the test sequences comparable to its estimates on the training ones. This means with the exclusion of the highway sequence 21, we expect our system to attain a comparable average translation error to the one achieved on the training sequences which is 1.23%. Properly handling the problematic far away features is a task for future work.

An important observation is that through our innovative approach of keeping features alive in a local 3D map and tracking them for as long as possible, we were able to greatly improve estimation accuracy compared to LIBVISO2. LIBVISO2 follows the traditional visual odometry approach of tracking features between consecutive frames, that is, from frame to frame only. We will define feature age as the number of frames in which this feature was successfully tracked and used in pose estimation. Average feature age in each frame for KITTI sequence 00 is shown in Figure 3.

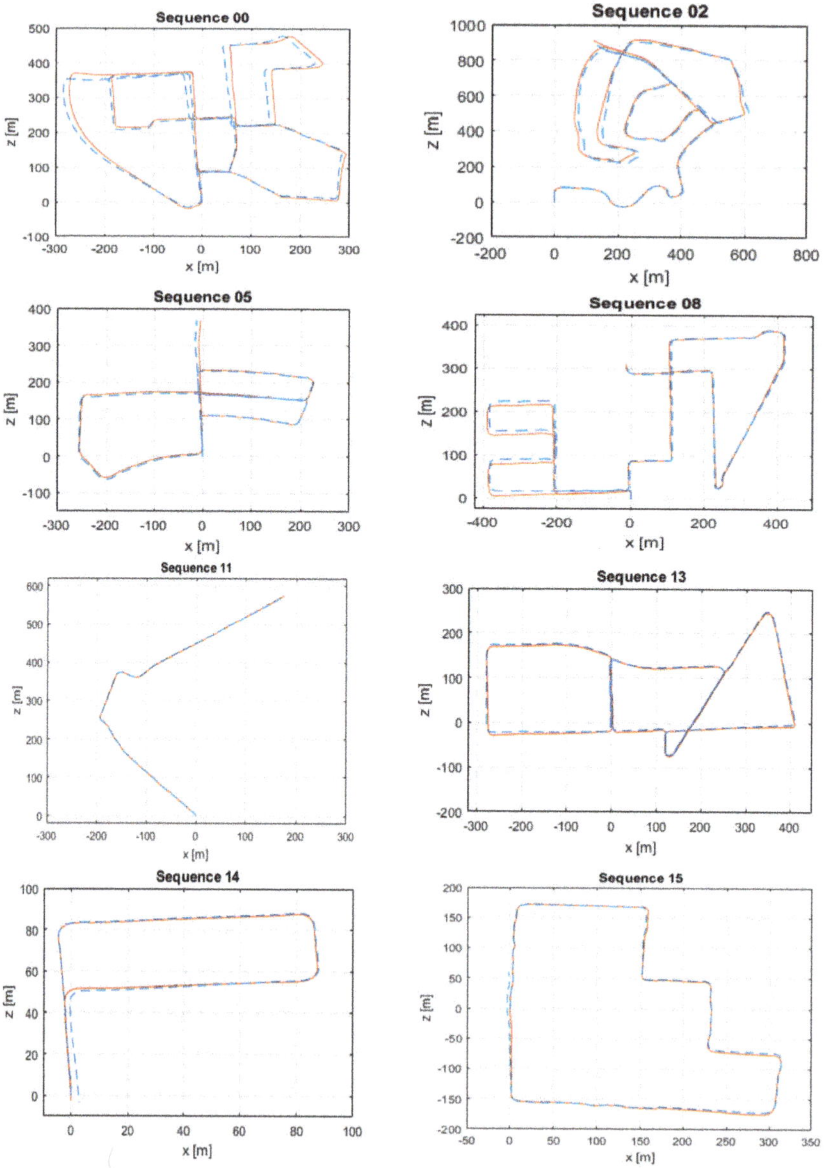

Figure 2. Estimated trajectory (dashed blue) from lightweight visual tracking (LVT) against ground truth (solid red) in KITTI dataset training sequences (00, 02, 05, 08) and evaluation sequences (11, 13, 14, 15).

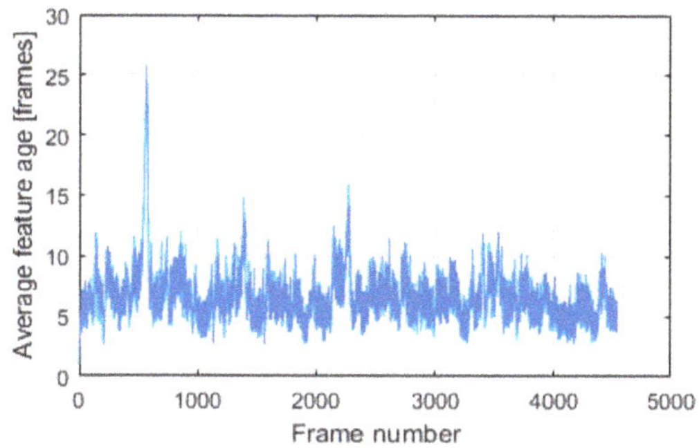

Figure 3. Average feature age in each frame of KITTI sequence 00.

4.1.2. Runtime Performance

Runtime performance evaluation experiments were performed on a laptop computer running the Ubuntu 16.04 operating system, with an Intel i7-7700HQ CPU and 16 GB of memory (RAM). For evaluating the runtime speed, we have timed the processing time of each frame from KITTI sequence 00, excluding the portion where the system is retrieving the stereo image and with no visualization enabled. As for evaluating the memory requirement, memory is read from the System Monitor utility just before the last frame. This process was repeated five times to account for operating system loading variability and the results are listed in Table 2.

Table 2. Runtime performance measurements.

Algorithm	Mean $\pm$ Std (ms)	Memory
LVT	13.82 ± 4.4	22.12 MiB
S-PTAM	36.7 ± 23.8	1.5 GiB
LIBVISO2	23.2 ± 4.5	41.82 MiB

We have timed the four main stages of the visual odometry system and the result is shown in Figure 4.

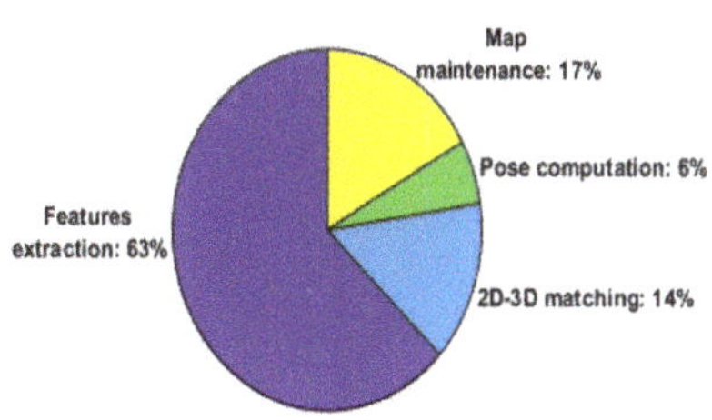

Figure 4. Relative runtime speed of the main stages of the visual odometry system.

The suitability of the presented visual odometry system for constrained real-time applications is evaluated on two embedded Linux single-board computers, namely, the Raspberry PI 3 [31] and the ODROID XU4 [32]. The time to process each frame of KITTI sequence 00 is recorded and computational performance is listed in Table 3.

Table 3. Computational performance on embedded computers.

Single-Board Computer	Mean ± Std (ms)
Raspberry Pi 3	162.37 ± 42.38
ODROID XU4	87.96 ± 20.35

4.2. EuRoC Dataset

The EuRoC datasets [28] were collected on board a micro-aerial vehicle (MAV). Stereo images were collected at a rate of 20 Hz with a stereo camera that provides monochrome WVGA images. The different datasets vary in their level of difficulty based on the flight dynamics and illumination conditions. The fact that AGAST corners and BRIEF descriptors used in LVT are by their design not invariant to in-plane rotations will provide useful insights, given that the MAV is moving in all 6 degrees of freedom. We will use the five sequences collected in an industrial machine hall for evaluation. The VICON room sequences constitute primarily of white plain surfaces and there is not enough texture in the scene for the feature detector to detect corners, thus they were not used in the evaluation. A sample image from the first machine hall dataset showing the detected features is shown in Figure 5.

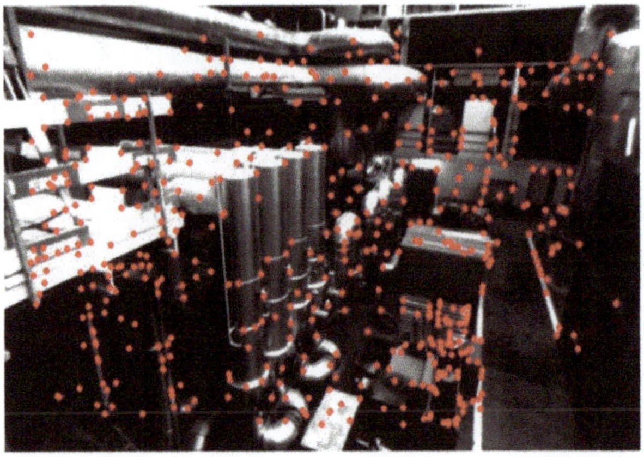

Figure 5. A sample image from Machine Hall 01 of the EuRoC datasets, where detected features are plotted (red dots).

4.2.1. Accuracy

For evaluating the accuracy, two error metrics, absolute trajectory error (ATE) and relative pose error (RPE), as defined Sturm et al. [29], are used. The ATE is well suited for evaluating Visual SLAM systems, as it measures the global consistency of the estimated trajectory. On the other hand, the RPE is more suited for evaluating visual odometry systems, as it measures the local accuracy of a trajectory over a fixed time interval. That is, it measures the drift in a trajectory. In our evaluation, as the stereo images are recorded at a rate of 20 Hz, we will set the fixed time interval for RPE to be 0.05 s. Therefore, RPE will correspond to drift in units of meters per second. We have also run LIBVISO2 [30] on the same dataset. The results are reported in Table 4. Both visual odometry systems are achieving similar drift rates.

Table 4. RMSE values of absolute trajectory error (ATE) and relative pose error (RPE) on EuRoC datasets.

	LVT		LIBVISO2	
Sequence	**ATE (m)**	**RPE (m/s)**	**ATE (m)**	**RPE (m/s)**
MH_01_easy	0.232	0.028	0.234	0.028
MH_02_easy	0.129	0.028	0.284	0.028
MH_03_medium	1.347	0.070	0.86	0.069
MH_04_difficult	1.635	0.069	1.151	0.068
MH_05_difficult	1.768	0.060	0.818	0.060

4.2.2. Effect of Rotational-Invariant Features

AGAST corners and BRIEF descriptors used in our presented visual odometry system do not provide rotational invariance capability. When evaluated previously on the KITTI dataset, this did not pose a problem, as the vehicle is moving on a locally planar ground. However, in the EuRoC datasets, the MAV is flying in all 6 degrees of freedom. In order to evaluate said effect, we have replaced the default AGAST/BRIEF with oriented fast and rotated brief (ORB) [33]. ORB provides rotational-invariant features. With everything else remaining fixed, we have re-run the evaluation and the results are reported in Table 5.

Table 5. RMSE values of ATE and RPE on EuRoC datasets using rotational-invariant oriented fast and rotated brief (ORB) features.

Sequence	ATE (m)	RPE (m/s)
MH_01_easy	0.292	0.029
MH_02_easy	x	x
MH_03_medium	1.328	0.071
MH_04_difficult	2.194	0.070
MH_05_difficult	1.515	0.061

From Table 5, with the same original parameters, LVT now fails to complete the MH_02_easy dataset. The cause is that the system fails to track enough ORB features in a motion-blurred frame. Drift rates are almost the same and no clear improvement in accuracy is attained by the new feature detector alone. We found from this experiment that simply replacing the feature detector with a rotationally-invariant one did not result in any major improvement in the results as was expected.

4.3. TUM RGB-D Dataset

The Technical University of Munich (TUM) RGB-D dataset [29] is a large dataset collected indoors using an RGB-D sensor under different illumination, texture and movement scenarios. The data was collected at a 30 Hz frame rate and a sensor resolution of 640×480. The evaluation results on a subset that is suited for V-SLAM and visual odometry evaluation are reported in Table 6 along with an evaluation against Fovis. Although our visual odometry system was initially designed for stereo cameras, it was, surprisingly, able to achieve low drift rates on RGB-D data. The main difference is that we found it necessary to trigger a new triangulation operation at every frame, unlike the default behavior as described in map maintenance before. We can see from Table 6 that our system is able to achieve similar drift rates to Fovis which was originally designed for RGB-D cameras. However, a fast rotation while the camera faces a low-textured wall during the movement results in the error in fr1_room for our system.

Table 6. RMSE values of ATE and RPE on TUM RGB-D dataset.

	LVT		Fovis	
Sequence	**ATE (m)**	**RPE (m/s)**	**ATE (m)**	**RPE (m/s)**
fr1_desk	0.109	0.010	0.259	0.009
fr1_desk2	0.116	0.012	0.125	0.009
fr1_room	4.138	0.077	0.184	0.007
fr1_xyz	0.035	0.006	0.051	0.006
fr2_desk	0.080	0.003	0.103	0.003
fr2_xyz	0.014	0.002	0.013	0.002
fr3_office	0.132	0.006	0.188	0.005

5. Conclusions and Future Work

In this paper, we have presented our feature-based visual odometry system called LVT, which is compatible with both stereo and RGB-D sensors. Its innovative usage of a transient local map enables it to approach the estimation accuracies common to full V-SLAM systems. The algorithm is designed for real-time operation with low computational overhead and memory requirements. The system was evaluated on KITTI autonomous driving datasets and compared with state-of-the-art V-SLAM and VO systems. Furthermore, it was also evaluated on the EuRoC MAV industrial machine hall and TUM RGB-D datasets as other use cases. This paper accompanies the release of the source code of the system under a permissive license and with support for the Robot Operating System (ROS). The source package contains the example codes used to run our system on the three datasets used in the evaluation along with the used parameters.

Visual SLAM and odometry approaches can be classified as either monocular when one camera is used or stereo when more than one is used. One of the major issues in monocular methods is the scale ambiguity, which means that motion trajectory can be estimated with only an ambiguous scale factor. On the other hand, with a known baseline distance between cameras, stereo methods can estimate the exact motion trajectory, that is, they are able to recover global metric scale. Another issue with monocular methods is that they require careful initialization procedures and usually involve tricky ones, whereas in stereo methods it is easier to achieve a power-and-go system. The main downside of stereo methods is that they degenerate into the monocular case for scenes with very distant objects relative to the stereo camera baseline. This downside was the main failure cause for the KITTI sequence 01, which consists of a highway with far features dominating. Properly handling far away features should be addressed in future work. The loss of scale for highway scenes in our algorithm occurs because 3D point placement is based entirely on stereo disparity. Increasing the stereo camera baseline and/or resolution could provide better depth measurement and address this challenge. Moreover, as far points have large uncertainty in their placement, re-initializing them in subsequent frames is expected to contribute to this problem solution. Another solution would be to use other sensor modalities to estimate scale, such as integrating inertial sensors, or using other odometry techniques based on wheel odometers or LIDAR.

In our experience, building a visual odometry system is tricky as there are many heuristics involved in each part of it. We aimed to build a framework that is adaptable and extensible. For example, switching different feature extractors can be done easily in our system. Furthermore, as the system was initially designed for stereo cameras, the ease of extending it to RGB-D cameras was a pleasant surprise. We expect that extending the system to additional specialized depth sensors in future work, such as infrared (IR) cameras, to be easy as well.

Some challenging frames, such as abrupt movements, lighting changes and low texture, might result in poor pose estimates by visual odometry or even complete loss of tracking. A mechanism to bridge these moments of disruptions is a potential future work. Integrating with an IMU could provide such a mechanism. Additionally, it could reduce drift accumulated by visual odometry.

Author Contributions: Conceptualization, M.A. and S.A.R.; Funding acquisition, S.A.R.; Investigation, M.A.; Methodology, M.A. and S.A.R.; Software, M.A.; Supervision, S.A.R.; Writing—original draft, M.A.; Writing—review & editing, S.A.R.

Funding: This research was carried out using internal resources provided by the College of Engineering and Computer Science (CECS) at the University of Michigan-Dearborn.

Conflicts of Interest: The authors declare no conflict of interest.

References

1. Scaramuzza, D.; Fraundorfer, F. Visual odometry. *IEEE Robot. Autom. Mag.* **2011**, *18*, 80–92. [CrossRef]
2. Moravec, H.P. *Obstacle Avoidance and Navigation in the Real World by a Seeing Robot Rover*; No. STAN-CS-80-813; Stanford University—Department of Computer Science: Stanford, CA, USA, 1980.
3. Nister, D.; Naroditsky, O.; Bergen, J. Visual Odometry. In Proceedings of the 2004 IEEE Computer Society Conference on Computer Vision and Pattern Recognition, Washington, DC, USA, 27 June–2 July 2004; Volume 1.
4. Kitt, B.; Geiger, A.; Lategahn, H. Visual odometry based on stereo image sequences with RANSAC-based outlier rejection scheme. In Proceedings of the 2010 IEEE Intelligent Vehicles Symposium (IV), San Diego, CA, USA, 21–24 June 2010.
5. Howard, A. Real-time stereo visual odometry for autonomous ground vehicles. In Proceedings of the 2008 IEEE/RSJ International Conference on Intelligent Robots and Systems, Nice, France, 22–26 September 2008.
6. Badino, H.; Yamamoto, A.; Kanade, T. Visual Odometry by Multi-frame Feature Integration. In Proceedings of the IEEE International Conference on Computer Vision, Sydney, Australia, 8–12 April 2013; pp. 222–229.
7. Davison, A.J. Real-time simultaneous localisation and mapping with a single camera. In Proceedings of the Ninth IEEE International Conference on Computer Vision, Nice, France, 13–16 October 2003.
8. Klein, G.; Murray, D. Parallel tracking and mapping for small AR workspaces. In Proceedings of the IEEE and ACM International Symposium on Mixed and Augmented Reality (ISMAR), Nara, Japan, 13–16 November 2007; pp. 225–234.
9. Triggs, B.; McLauchlan, P.F.; Hartley, R.I.; Fitzgibbon, A.W. *Bundle Adjustment—A Modern Synthesis*; Springer: Berlin/Heidelberg, Germany, 1999; pp. 298–372.
10. Younes, G.; Asmar, D.; Shammas, E.; Zelek, J. Keyframe-based monocular SLAM: Design, survey, and future directions. *Robot. Auton. Syst.* **2017**, *98*, 67–88. [CrossRef]
11. Pire, T.; Fischer, T.; Castro, G.; de Cristóforis, P.; Civera, J.; Berlles, J.J. S-PTAM: Stereo Parallel Tracking and Mapping. *Robot. Auton. Syst.* **2017**, *93*, 27–42. [CrossRef]
12. Mur-Artal, R.; Montiel, J.M.M.; Tardós, J.D. ORB-SLAM: A Versatile and Accurate Monocular SLAM System. *IEEE Trans. Robot.* **2015**, *31*, 1147–1163. [CrossRef]
13. Engel, J.; Schöps, T.; Cremers, D. LSD-SLAM: Large-Scale Direct Monocular SLAM. In Proceedings of the European Conference on Computer Vision (ECCV), Zurich, Switzerland, 6–12 September 2014; pp. 834–849.
14. Huang, A.S.; Bachrach, A.; Henry, P.; Krainin, M.; Maturana, D.; Fox, D.; Roy, N. *Visual Odometry and Mapping for Autonomous Flight Using an RGB-D Camera*; Robotics Research; Springer: Cham, Switzerland, 2017; Volume 100, pp. 235–252.
15. Kerl, C.; Sturm, J.; Cremers, D. Robust odometry estimation for RGB-D cameras. In Proceedings of the IEEE International Conference on Robotics and Automation (ICRA), Karlsruhe, Germany, 6–10 May 2013.
16. Besl, P.J.; McKay, N.D. A method for registration of 3-D shapes. *IEEE Trans. Pattern Anal. Mach. Intell.* **1992**, *14*, 239–256. [CrossRef]
17. Rusinkiewicz, S.; Levoy, M. Efficient variants of the ICP algorithm. In Proceedings of the Third International Conference on 3-D Digital Imaging and Modeling, Quebec City, QC, Canada, 28 May–1 June 2001.
18. Newcombe, R.A.; Izadi, S.; Hilliges, O. KinectFusion: Real-time dense surface mapping and tracking. In Proceedings of the 10th IEEE International Symposium on Mixed and Augmented Reality (ISMAR), Basel, Switzerland, 26–29 October 2011.
19. Hartley, R.; Zisserman, A. *Multiple View Geometry in Computer Vision*; Cambridge University Press: Cambridge, UK, 2004.

20. Mair, E.; Hager, G.D.; Burschka, D.; Suppa, M.; Hirzinger, G. Adaptive and generic corner detection based on the accelerated segment test. In Proceedings of the European Conference on Computer Vision (ECCV'10), Crete, Greece, 5–11 September 2010.
21. Rosten, E.; Drummond, T. Machine learning for high-speed corner detection. In Proceedings of the 2006 European Conference on Computer Vision, Graz, Austria, 7–13 May 2006; pp. 430–443.
22. Calonder, M.; Lepetit, V.; Strecha, C.; Fua, P. Brief: Binary robust independent elementary features. In Proceedings of the European Conference on Computer Vision, Crete, Greece, 5–11 September 2010; pp. 778–792.
23. Brown, M.; Szeliski, R.; Winder, S. Multi-image matching using multi-scale oriented patches. In Proceedings of the IEEE Computer Society Conference on Computer Vision and Pattern Recognition (CVPR'05), San Diego, CA, USA, 20–25 June 2005; Volume 1, pp. 510–517.
24. Lowe, D. Distinctive image features from scale-invariant keypoints. *Int. J. Comput. Vis.* **2004**, *60*, 91–110. [CrossRef]
25. Hartley, R.; Sturm, P. Triangulation. *Comput. Vis. Image Underst.* **1997**, *68*, 146–157. [CrossRef]
26. Kümmerle, R.; Grisetti, G.; Strasdat, H.; Konolige, K.; Burgard, W. g2o: A general framework for graph optimization. In Proceedings of the IEEE International Conference on Robotics and Automation (ICRA), Shanghai, China, 9–13 May 2011; pp. 3607–3613.
27. Geiger, A.; Lenz, P.; Urtasun, R. Are we ready for autonomous driving? The kitti vision benchmark suite. In Proceedings of the IEEE Conference on Computer Vision and Pattern Recognition (CVPR), Providence, RI, USA, 16–21 June 2012; pp. 3354–3361.
28. Burri, M.; Nikolic, J.; Gohl, P.; Schneider, T.; Rehder, J.; Omari, S.; Achtelik, M.W.; Siegwart, R. The EuRoC micro aerial vehicle datasets. *Int. J. Robot. Res.* **2016**, *35*, 1157–1163. [CrossRef]
29. Sturm, J.; Engelhard, N.; Endres, F.; Burgard, W.; Cremers, D. A Benchmark for the Evaluation of RGB-D SLAM Systems. In Proceedings of the International Conference on Intelligent Robot Systems (IROS), Vilamoura, Portugal, 7–12 October 2012.
30. Geiger, A.; Ziegler, J.; Stiller, C. StereoScan: Dense 3D Reconstruction in Real-time. In Proceedings of the Intelligent Vehicles Symposium (IV), Baden-Baden, Germany, 5–9 June 2011.
31. Raspberry Pi 3 Model B. Available online: https://www.raspberrypi.org/products/raspberry-pi-3-model-b/ (accessed on 27 August 2018).
32. ODROID-XU4. Available online: http://www.hardkernel.com/main/products/prdt_info.php?g_code=G143452239825 (accessed on 27 August 2018).
33. Rublee, E.; Rabaud, V.; Konolige, K.; Bradski, G. ORB: An efficient alternative to SIFT or SURF. In Proceedings of the IEEE International Conference on Computer Vision (ICCV), Barcelona, Spain, 6–13 November 2011.

sensors

MDPI

Article

Fast Visual Odometry for a Low-Cost Underwater Embedded Stereo System †

Mohamad Motasem Nawaf *,‡, Djamal Merad ‡, Jean-Philip Royer ‡, Jean-Marc Boï ‡, Mauro Saccone ‡, Mohamed Ben Ellefi ‡ and Pierre Drap *,‡

Aix-Marseille Université, CNRS, ENSAM , Université De Toulon, LIS UMR 7020, Domaine Universitaire de Saint-Jérôme, Bâtiment Polytech, Avenue Escadrille Normandie-Niemen, 13397 Marseille, France; Djamal.Merad@univ-amu.fr (D.M.); Jean-Philip.Royer@univ-amu.fr (J.-P.R.); Jean-Marc.Boi@univ-amu.fr (J.-M.B.); Mauro.Saccone@univ-amu.fr (M.S.); Mohamed.Ben-Ellefi@univ-amu.fr (M.B.E.)

* Correspondence: mohamad-motasem.NAWAF@univ-amu.fr (M.M.N.); pierre.drap@univ-amu.fr (D.P.); Tel.: +33-4-91-82-85-20 (D.P.)

† This paper is an extended version of our papers published in Nawaf, M.M.; Boï, J.M.; Merad, D.; Royer, J.P.; Drap, P. Low Cost Embedded Stereo System For Underwater Surveys. In Proceedings of the 5th International Workshop LowCost 3D—Sensors, Algorithms, Applications, Hamburg, Germany, 28–29 November 2017, pp. 179–186; Nawaf, M.M.; Drap, P.; Royer, J.P.; Merad, D.; Saccone, M. Towards Guided Underwater Survey Using Light Visual Odometry. In Proceedings of the 7th ISPRS/CIPA 3D Virtual Reconstruction and Visualization of Complex Architectures, Nafplio, Greece, 1–3 March 2017; pp. 527–533.

‡ These authors contributed equally to this work.

Received: 21 June 2018 ; Accepted: 14 July 2018; Published: 17 July 2018

Abstract: This paper provides details of hardware and software conception and realization of a stereo embedded system for underwater imaging. The system provides several functions that facilitate underwater surveys and run smoothly in real-time. A first post-image acquisition module provides direct visual feedback on the quality of the taken images which helps appropriate actions to be taken regarding movement speed and lighting conditions. Our main contribution is a light visual odometry method adapted to the underwater context. The proposed method uses the captured stereo image stream to provide real-time navigation and a site coverage map which is necessary to conduct a complete underwater survey. The visual odometry uses a stochastic pose representation and semi-global optimization approach to handle large sites and provides long-term autonomy, whereas a novel stereo matching approach adapted to underwater imaging and system attached lighting allows fast processing and suitability to low computational resource systems. The system is tested in a real context and shows its robustness and promising future potential.

Keywords: image processing; underwater imaging; embedded systems; stereo vision; visual odometry; 3D reconstruction

1. Introduction

Mobile systems nowadays are undergoing a growing need for self-localization to determine their absolute/relative position over time accurately. Despite the existence of very efficient technologies that can be used on-ground (indoor/outdoor), such as the Global Positioning System (GPS), optical signals, and radio beacons, in the underwater context, most of these signals are jammed so that similar techniques cannot be used. On the other hand, solutions based on active acoustics, such as imaging sonars, water linked GPS or Doppler Velocity Log (DVL) devices remain expensive and require high technical skills for deployment and operation. Moreover, their size specifications prevent their integration within small mobile systems or even the ability to be handheld. The research for an alternative is ongoing; notably, recent advances in embedded systems have led to relatively small, powerful and cheap devices. This opens promising potential to adopt a light visual odometry approach that provides a relative trajectory in real-time using image sensors, and this describes our main research direction. The developed solution is integrated within an underwater archaeological site survey where it plays an important role in facilitating image acquisition.

In underwater survey tasks, mobile underwater vehicles (or divers) navigate over the target site to capture images. The obtained images are treated in a later phase to obtain various information and also to form a realistic 3D model using photogrammetry techniques [1]. In such a situation, the main problem is covering the underwater site totally before ending the mission. Otherwise, we may obtain incomplete 3D models, and the mission cost will rise significantly as further exploitation will bw needed. However, the absence of an overall view of the site, especially under bad lighting conditions, makes the scanning operation blind. In practice, this leads to over-scanning the site which is a waste of time and cost. From another perspective, the quality of the taken images may go below an acceptable limit. This mainly happens in terms of lightness and sharpness, which is often hard to quantify visually on the fly. In this work, we propose solutions for the aforementioned problems. Most importantly, we propose to guide a survey based on a visual odometry approach that runs on a distributed system in real-time. The output ego-motion helps to guide the site scanning task by showing approximate scanned areas. Moreover, overall subjective lightness and sharpness indicators are computed for each image to help the operator control the image quality.

Overall, we provide a complete hardware and software solution for the problem through the conception and realization of a stereo embedded system dedicated to underwater imaging. Two configurations are considered: first, a handheld system to be used by a diver (see Figure 1), and second, a system attached to a customizable Remotely Operated underwater Vehicle (ROV) from BlueRobotics [2] (see Figure 2). Both configurations share similar main architecture (all provided details are for both configurations unless otherwise stated). The system, equipped with two high definition cameras (three cameras in the ROV-attached configuration), can take and store hardware synchronized stereo images while having long-term autonomy. In contrast to other commercially available off-the-shelf products where the system's role ends with image storage, the designed system is based on distributed embedded systems with ARM processors and a Linux operating system and is capable of running most image processing techniques smoothly in real-time. The available optimized open source libraries, such as OpenCV [3] and OpenCL [4], allow straightforward extension of the provided functions and full customization of the system to suit different contexts.

In common approaches of visual odometry, a significant part of the overall runtime is spent on feature point detection, description, and matching, whereas another significant part is dedicated to the optimization process, namely, the bundle adjustment (BA) [5] procedure. In the tested baseline algorithm, feature point matching represents ≈65% of runtime in the local/relative bundle adjustment (BA) approach. Despite their accuracy and successful broad application, modern feature descriptors, such as Scale Invariant Feature Transform (SIFT) [6] and Speeded Up Robust Features (SURF) [7], rely on differences of Gaussians (DoG) and fast Hessian, respectively, for feature detection. These methods are two times slower than the traditional Harris detector [8]. Further, the sophisticated descriptors that are invariant to scale and rotation, which is not necessary for stereo matching, slow down the computation. Moreover, brute force matching is often used which is also time-consuming. In our proposed method, we rely on low-level Harris-based detection and a template matching procedure which significantly speeds up the point matching. Further, whereas in traditional stereo matching the search for correspondence is done along the epipolar line within a specific fixed range, in our method, we proceed first by computing, a priori, a rough depth belief based on image lightness and following the law of light divergence over distance. This is only valid for a configuration in which the only light source is fixed to the system, which is the case here. Hence, our first contribution is that we benefit from rough depth estimation to limit the point correspondence search zone to reduce the processing time. It is worth mentioning that even for the surveys in shallow water where the sunlight provides good visibility, it is preferable to wait for sunset before starting the survey because of the sunlight ripple effect on the scanned site [9] which misleads the photogrammetry process, as it disturbs the photometric consistency.

From another perspective, traditional visual odometry methods based on local BA suffer from rotation and translation drift that grow with time [10]. In contrast, solutions based on using features from the entire image set, such as global BA [5], require more computational resources which are very limited in our case. Similarly, simultaneous localization and mapping (SLAM) approaches [11], which are known to detect loop closure, although being efficient in most robotics applications, suffer from a growing processing time [12], or are not suitable for raster scan trajectories such as hierarchical approaches [13,14]. In our method, we adopt a semi-global approach which proceeds in the same way as local methods for optimizing a subset of image frames. However, it differs in terms of selecting the frame subset, as local methods use the Euclidean distance and deterministic pose representation to select frames, but ours represents the poses in a probabilistic manner and uses a divergence measure to select such subset of frames. The uncertainty of each newly-estimated pose is computed using a novel approach that uses a machine learning technique on the simulated pose estimation system. This is handled by a neural network that is trained to handle a wide range of ego-motion vectors. This will be addressed in detail in Section 5.4.

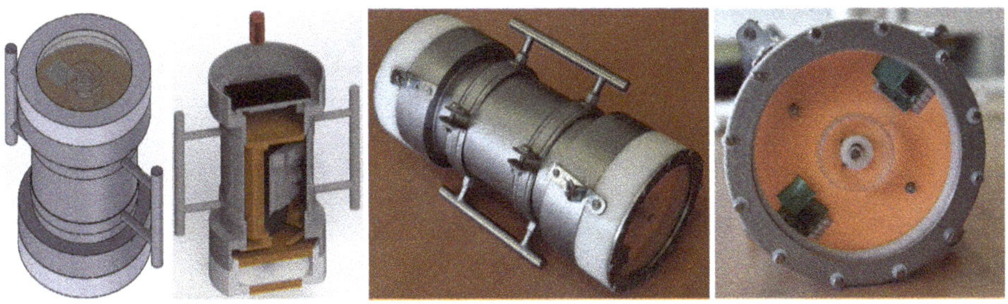

Figure 1. The handheld stereo system design and prototype.

Figure 2. The built trifocal system integrated within a blueROV 2 (the front enclosure).

The rest of the paper is organized as follows: We survey related works in Section 2. In Section 3, we describe the designed hardware platform and the two configurations that we used to implement our solution. The image acquisition and the quality estimation procedure are explained in Section 4. Our proposed visual odometry method is presented in Section 5. The analytical results of the underwater experiments are presented in Section 6. Finally, we present a summary and conclusions. We note that parts of this work have been presented in [15,16].

2. Related Works

In this section, we review related works concerning the two aspects that we mainly aim to improve in our framework: feature point matching and ego-motion estimation.

2.1. Feature Point Matching

Common ego-motion estimation methods rely on feature point matching between several poses [17–24]. Real-time methods tend to use fast feature detectors. The most popular are Features from Accelerated Segment Test (FAST) [25], as in [19,20,23], and Harris-based [26], as in [18,21,22]. These types of detectors are frequently associated with patch descriptors. In general, the choice of approach for matching feature points depends on the context. For instance, feature matching between freely-taken images (six degrees of freedom) with baseline toleration has to be invariant to scale and rotation changes. Scale Invariant Feature Transform (SIFT) [6] and the Speeded Up Robust Features (SURF) [7] are well used in this context [17,24,27,28]. In this case, besides being more computationally expensive, the search for a point's correspondence is generally done using brute force matching.

A new family of feature descriptors that aims to accelerate the extraction process makes use of binary representation computed from image intensity differences tests. The Binary Robust Independent Elementary Features (BRIEF) method [29] is the first in this direction. The method measures the intensity difference on a fixed chosen location pairs around the keypoints which are commonly detected using FAST. An improvement to BRIEF is the Binary Robust Invariant Scalable Keypoints (BRISK) [30], which adds scale and rotation invariance features. This is achieved by introducing multi-scaling and using regular circular pattern around the keypoint. Another difference to BRIEF is that BRISK proposes its own detector, an extension of the AGAST detector [31] (based on FAST) that performs a scale-space search for saliency. Overall, using this over-sampled representation of the keypoint neighborhood makes these methods more sensitive to noise. As this does not has significant inference on terrestrial images, underwater images suffer mostly from turbidity and dust which makes the use of these methods less robust based on our experiments.

In certain situations, some constraints can be imposed to facilitate the matching procedure, in particular, limiting the correspondence search zone. For instance, in the case of pure forward motion, where the focus of expansion (FOE) is a single point in the image, the search for the correspondence of a given point is limited to the epipolar line [32]. Similarly, in the case of sparse stereo matching, the correspondence point lies on the same horizontal line in the case of a rectified stereo or on the epipolar line otherwise. This speeds up the matching procedure, firstly by having fewer comparisons to perform and secondly because low-level features can be used [33,34]. According to our knowledge, no method proposes an adaptive search range following a rough depth estimation from lightness in underwater imaging. We refer to [8] for a comprehensive study of feature point detection and matching.

It is worth mentioning that direct visual odometry methods are well-established when a depth map is available, such as using RGB-D cameras [35]. These featureless methods use geometry transformation between rigid objects in several views to infer ego-motion. Methods that deal with stereo cameras proceed by computing a dense depth estimation that is used to establish a relationship between objects within adjacent views [36], whereas monocular methods [37,38] use a variational approach for estimating pixel-wise depth. The problem is solved under convex assumption using GPU. The main inconvenience of those approaches is the required high computational power and the small baseline between adjacent images which are hard to guarantee in our context.

2.2. Ego-Motion Estimation

Estimating the ego-motion of a mobile system is an old problem in computer vision. Two main categories of methods are developed in parallel, namely, simultaneous localization and mapping (SLAM) [34] and visual odometry [18]. In the following text, we highlight the main characteristics of both approaches.

The SLAM family of methods uses probabilistic models to handle a vehicle's pose. Although this kind of method was developed to handle motion sensors and map landmarks, it works efficiently with solely visual information [24]. In this case, a map of the environment is built, and, at the same time, it is used to deduce the relative pose which is represented using probabilistic models. Several solutions to SLAM involve finding an appropriate representation for the observation model and motion model while preserving an efficient and consistent runtime. Most methods use additive Gaussian noise to handle the uncertainty which is imposed using the extended Kalman filter (EKF) to solve the SLAM problem [34]. In cases where visual features are used, EKF may fail to estimate the trajectory accurately due to the significant uncertainties that appear in large loops [13]. Additionally, runtime and used resources grow constantly for large environments. Later works tried mainly to handle scalability issues.

A remarkable improvement of SLAM is the FastSLAM approach [12] which aims at greater scalability. It uses recursive Monte Carlo sampling to directly represent the non-linear process model. Although the state-space dimensions are reduced when the Rao–Blackwellisation approach is used [39], the method remains not scalable to large autonomy. In the context of long trajectories, several solutions have been proposed to handle relative map representations, such as [22,24,40,41]. In particular, these involve breaking the estimation into smaller mapping regions, called sub-maps, and then computing individual solutions for each sub-map. In the same manner, hierarchical SLAM [13] divides the map into two levels—a lower level that is composed of a set of a sequence of local maps of limited size and an upper level that handles the relative relations between local maps, which are maintained using a stochastic approach. Although these solutions perform well in large environments, sub-mapping is not efficient for raster scanning/motion as this will cause very frequent sub-maps switches. Also, there are some issues in defining the size, overlapping, and the fusion of sub-maps.

In all reviewed SLAM methods, in case of using pure visual information, the measurement noise (such for relative motion estimation) is modeled by a diagonal covariance matrix with equal variances that are set empirically [14]. This modeling leads to the production of uncorrelated measurement error among dimensions. However, the estimated pose should have an associated full degrees of freedom (DOF) uncertainty. Although several works exist in the literature that studied the uncertainty of 3D reconstructed points based on their distance from the camera and the baseline distance between frames, such as [19,42], or the matching error propagation in 3D, such as [9], the effect of the relative motion parameters on the uncertainty of the pose estimation has not been taken into account.

From another perspective, visual odometry methods use structure from motion (SfM) methodology to estimate the relative motion [18]. Based on multiple view geometry fundamentals [43], an approximate relative pose can be estimated. This is followed by a BA procedure to minimize re-projection errors, which yields an improvement in the estimated structure. Fast and efficient BA approaches are proposed to be able to handle a large number of images [44]. However, in the case of longtime navigation, the number of images increases constantly and prevents the application of global BA if real-time performance is needed. Hence, several local BA approaches have been proposed to handle this problem. In local BA, a sliding window copes with motion and select a fixed number of frames to be considered for BA [10]. This approach does not suit the raster scans commonly used in surveys, since the last n frames to the current frame are not necessarily the closest. Another local approach is relative BA, proposed in [45]. Here, the map is represented as a Riemannian manifold-based graph with edges representing the potential connections between frames. The method selects the part of the graph where the BA will be applied by forming two regions—an active region that contains the frames with an average re-projection error changes by more than a threshold, and a static region that contains the frames that have common measurements with frames in the active region. When performing BA, the static region frames are fixed, whereas active region frames are optimized. The main problem with this method is that distances between frames are deterministic, whereas the uncertainty is not considered when computing inter-frame distances.

In the context of underwater robotics, SLAM solutions based on active sensors, such as DVL, the Inertial Navigation Unit (INU) and Side Scan Sonars (SSS) have mostly been proposed [46,47]. An early attempt to use a vision system was proposed in [48], where a fusion is performed between sonar and visual information, and a Lucas Kanade feature tracking is applied to the image stream—it is used to only extract robot's bearing observation which does not generalize to free motion. A more general solution was proposed in [49], in which the ego-motion is estimated by finding the rigid transformation between two point clouds which are generated using a stereo system at two time intervals. The relative motion is then integrated with SLAM which uses an SSS as well. Works relying solely on visual sensors are surprisingly rare; noticeably, they use the same terrestrial SLAM techniques as those reviewed above [50]. The majority of these methods rely on stereo vision to estimate metric trajectory [9,28,51].

3. Hardware Platform

Throughout our hardware design and implementation, we were committed to a low-cost solution. Thanks to the latest developments of single-board computers, power-efficient systems equipped with a high-performance multi-core CPUs, and most modern peripheral networking interfaces are available in the size of a credit card. Being increasingly available and cheap, we chose the popular Raspberry Pi (RPi) [52] (a credit card-sized ARM architecture-based computer with 1.2 GHz 64-bit quad-core CPU and 1 GB of memory, running Rasbain, a Linux-based operating system. We used RPi version 3 in this project) as the main processing unit of our platform. This allowed most image processing and computer vision techniques to be run smoothly. As already mentioned, we designed and implemented two configurations of our system that we present in the following text.

3.1. ROV-Attached Trifocal System

The design here is based on the BlueROV2 from BlueRobotics [2], Figure 2 shows the full system design and implementation. The ROV is equipped with six thrusters (four vectored and two vertical), controlled by a Pixhawk autopilot [53] which allows 4 DoF navigation to be performed—all but pitch and yaw. The ROV is operated from a surface computer (laptop) that also receives the live video feedback. We used 4×1500 lumens diffuse led torches for lighting oriented at a tilt of 135°.

A cylindrical enclosure (34×15 cm) is attached to the front side of the ROV, as shown in Figure 2. It hosts the designed trifocal system, which is composed of three RPi computers; each is connected to one camera module (Sony IMX219 8M Pixel 1/4″ CMOS Image Sensor, 3 mm focal length, f/2 aperture). Using the trifocal system allows three stereo pairs with different baseline distances (set to 5, 10 and 15 cm in our implementation) to be present, which helps to handle image acquisition at different distances. Figure 3 (right) shows the range of each baseline distance. Here, we can deduce that a short baseline is preferred in close-range image acquisition. For instance, with the used configurations, it is difficult to get closer than 80, 53 and 26 cm to the scene using 15, 10 and 5 cm baseline distances, respectively. From another perspective, small baseline distances are less accurate for larger distances. We note here that the visibility limit underwater (≈5 m using our lighting system) is much smaller than the stereo range whatever the used baseline distance. The cameras are synchronized using a hardware trigger connected to the general-purpose input/output (GPIO) interface of the RPi computers. The latter are finally connected to an Ethernet switch that is connected to the surface computer. Figure 4 shows the ROV-attached trifocal system in action.

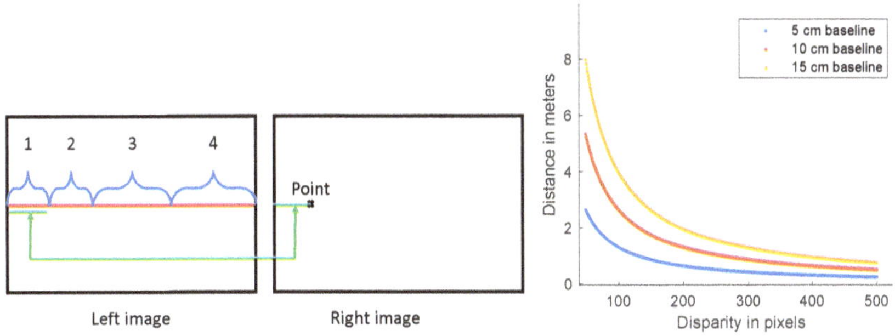

Figure 3. Illustration of stereo disparity ranges (**left**): (1) impossible due to stereo constraint; (2) impossible in deep underwater imaging due to light fading at far distances; (3) possible disparity; (4) the point is very close, so it becomes overexposed, undetectable, or out of focus. At (**right**), the disparity evaluation in pixels as a function of distance (in meters) to the camera for the 3 available baseline distances.

Figure 4. The built trifocal Remotely Operated underwater Vehicle (ROV)-attached system in action.

3.2. *Handheld Stereo System*

An illustration of the built handheld system is shown in Figure 1. It is composed of two RPi computers; each is connected to one camera module to form a stereo pair. The cameras are synchronized using a hardware trigger in the same manner as the previous system. Both RPi computers are connected through Ethernet to the surface. A high contrast monitor is embedded in the same enclosure and is visible from outside (see Figure 5). The monitor is attached to one of the RPi computers and shows real-time preview and diverse information, such as image quality, storage, and connections.

Figure 5. The built handheld stereo system in action.

In both designed systems, the embedded computers are responsible for image acquisition. The captured stereo images are first partially processed on the fly to provide image quality information, as will be detailed in Section 4. Images are then transferred to a central computer which handles the computation of the ego-motion that the system undergoes. This will be detailed in Section 5. We note that our implementation assumed calibrated stereo pairs. Therefore, we employed a traditional but efficient approach that uses an underwater target of chessboard pattern and the camera calibration toolbox in OpenCV [3]. The procedure was performed offline before the mission. After observing stable extrinsic parameters of two trials, we did not perform any further recalibration.

4. Image Acquisition and Quality Estimation

Since underwater images do not tend to be in the best condition, a failing scenario in computing the ego-motion is expected and has to be considered. Here, we could encounter two cases. First, when there is a degenerated configuration that causes a failure to estimate the relative motion, this can be due to poor image quality (blurred, dark or overexposed), lack of textured areas or large camera displacements. This may raise ill-posed problems at several stages. Second, imprecise estimation of the relative motion due to poorly distributed feature points or the dominant presence of outliers in the estimation procedure may occur. While a mathematical analysis can identify the first failure case, the detection of the second case is not trivial. Nevertheless, small errors can be corrected later using the BA procedure.

A real-time image quality estimation provides two benefits: first, it can alert the visual odometry process of having poor image quality. Two reactions can be taken in this case, either pausing the process until the taken image quality goes above a certain threshold or producing position estimation based on previous poses and speed. We went for the first case while leaving the second for further development in the future. Second, the image quality indicator provides direct information to the operator to avoid it going too fast in case of a blur or changing the distance to the captured scene when it is under or over-exposed.

To estimate the image sharpness, we rely on an image gradient measure to detect the high frequencies often associated with sharp images. Thus, we used Sobel kernel-based filtering which computes the gradient with a smoothing effect. This removes the effect of dust commonly present in underwater imaging. Given an image, **I**, we start by computing the image gradient magnitude, **G**, as

$$\mathbf{G} = \sqrt{(SK^\top * \mathbf{I})^2 + (KS^\top * \mathbf{I})^2}, \tag{1}$$

where

$S = [1 \quad 2 \quad 1]^\top$

$K = [-1 \quad 0 \quad 1]^\top$

$*$ is a convolution operator.

We consider our sharpness measure to be the mean value of **G**. A threshold can be easily learned from images by solving a simple linear regression problem. First, we record the number of matched feature points per image versus the sharpness indicator. Then, by fixing the minimum number of matched feature points needed to estimate the ego-motion correctly, we can compute the minimum sharpness indicator threshold (in our experiments, we fixed the number of matches to 100 matches; the obtained threshold was $\approx$20). It is worth noting that several assumptions used in our work, including this measure, do not hold for terrestrial imaging scenarios. In particular, the seabed texture guarantees a minimum sharpness even in object-free scenes.

From another perspective, good scene lighting yields better images, so it influences the accuracy of odometry estimation. Similar to the image sharpness indicator, an image lightness indicator can be integrated into the odometry process as well as helping the operator to take proper actions. To estimate the lightness indicator, we convert the captured images to the CIE-XYZ color space and then to the CIE-LAB color space. We consider the lightness indicator to be the mean value of the lightness channel L (using a percentile-based measure, such as the median, is more representative but it takes around 22 times longer to compute than the mean). The threshold is computed in the same way as for the sharpness.

Both indicators are computed using a sub-sampled image without interpolation. This allows the processing time to be decreased by 80%, with an average time of 60 ms on a single RPi computer, while keeping the accuracy above 95% compared to using the full resolution images.

5. Visual Odometry

After computing and displaying the image quality measures, the images are transferred over the network to the surface computer (average laptop computer). This computer is responsible for hosting the visual odometry process, which will be explained in this section. We begin by introducing the used stereo matching approach, and then we present the ego-motion estimation. Finally, we explain the semi-global BA approach.

5.1. Speeded Up Stereo Matching

Matching feature points between stereo images is essential for the estimation of ego-motion. As the alignment of the two cameras is not perfect, we start by calibrating the camera pair. Hence, for a given point on the right image, we are able to compute the epipolar line containing the corresponding point in the left image. However, based on the known fixed geometry, the corresponding point position is constrained by a positive disparity. Moreover, given that, in deep water, the only light source is the one attached to our system, the farthest distance that feature points can be detected is limited (see Figure 6 for illustration). This means that there is a minimum disparity value that is greater than zero; the red dots in Figure 6 refer to the minimum disparity, for instance. It was at least 130 pixels for the 10 cm baseline stereo pair. Furthermore, when going too close to the scene, parts of the image will become overexposed, undetectable, or out of focus. Similar to the previous case, this imposes a limited maximum disparity. Figure 3 illustrates the constraints mentioned above by dividing the epipolar line into four ranges, in which only one was an acceptable disparity in our context. This range can be directly identified by learning from a set of captured images (oriented at 30° for better coverage).

Figure 6. Examples of underwater images taken with our system and the computed rough depth using only the luminance channel. The rough depth was used later to speed up the stereo matching procedure. The red dots show the minimum detectable disparity (≈130 pixels in 10 cm baseline setup), while the blue dots show the maximum disparity (≈450 pixels in 10 cm baseline setup). See Figure 3 for corresponding distances and conversion to other baselines.

In our approach, we propose to constrain the so-defined acceptable disparity range further, which corresponds to the third range in Figure 3(left). Given the used lighting system, we can assume a light diffuse reflection model where the light reflects equally in all directions. Based on the inverse-square law that relates light intensity over distance, image pixels intensities are roughly proportional to their squared disparities. Based on such an assumption, we can use the pixel intensity to constrain the disparity and hence, limit the range of searching for a correspondence. To do so, we used a dataset of rectified stereo images. For each image pair, we performed feature point matching. Moreover, for each matching pair of points (x_i, y_i) and (x'_i, y'_i), x being the coordinate in the horizontal axis, we computed the squared disparity, $d_i^2 = (x_i - x'_i)^2$. Next, we associated each d_i^2 to the mean lightness value, denoted $\bar{l}_{x_i,y_i}$, of a window centered at the given point computed from the lightness channel, L, in the CIE-LAB color space. We assigned a large window size of ($\approx$15) to compensate for using the Harris operator that promotes local minimum intensity pixels as salient feature points. Several examples of the computed rough depth maps are shown in Figure 6. The computed $(\bar{l}_{x_i,y_i}, d_i^2)$ pair shows the linear relationship between the squared disparity and the average lightness. A subset of such pairs is plotted in Figure 7(left).

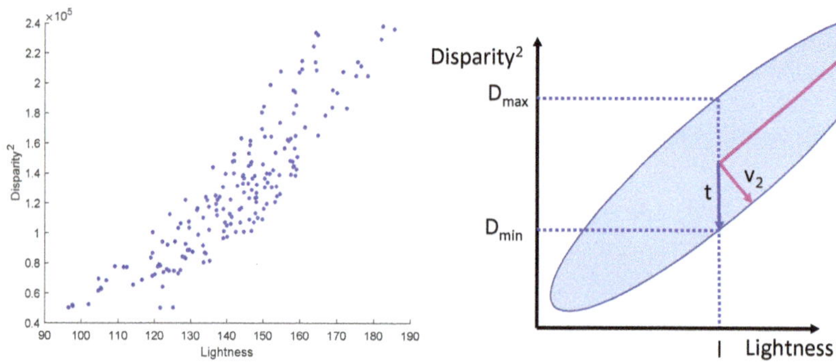

Figure 7. Disparity vs. lightness relationship for a subset of matched points. **Left**: local average pixel lightness vs. squared disparity. **Right**: an illustration of disparity tolerance, t, for a given lightness, l.

In addition to finding the linear relationship between both variables, it was also necessary to capture the covariance that represents how rough our approximation is. More specifically, given the relation shown in Figure 7, we aim to define a tolerance, t, associated with the disparity as a function of the lightness, l. In our method, we rely on the Principal Component Analysis (PCA) technique to obtain this information. In detail, for a given lightness, l, we first compute the corresponding squared disparity, d^2, using a linear regression approach as follows:

$$d^2 = -\alpha l - \beta \tag{2}$$

where

$$\alpha = \frac{Cov(L, D^2)}{Var(L)} \tag{3}$$

$$\beta = \bar{l} - \alpha \bar{d}^2, \tag{4}$$

where D and L , both vectors of $n \times 1$ with n being the data size, are the disparity and the lightness training vectors, respectively, and $\bar{d}$ and $\bar{l}$ are their respective means. Second, let $\mathbf{V}_2 = [v_{2,x}\ v_{2,y}]^\top$ be the computed eigenvector that corresponds to the smallest eigenvalue, λ_2, of the $n \times 2$ matrix $[L\ D^2]$. Based on the illustration shown in Figure 7 (right), the tolerance, t, associated with d^2 can be written as:

$$t = \sqrt{\lambda_2^2 (\frac{v_{2,x}^2}{v_{2,y}^2} + 1)}. \quad (5)$$

By considering a normal error distribution of the estimated rough depth and based on the fact that t is related to the variance of D^2, we define the effective disparity range as

$$d \pm \gamma \sqrt[4]{t}, \quad (6)$$

where γ represents the number of standard deviations. It is trivial that γ is a trade-off between the runtime and the probability of having point correspondences within the chosen tolerance range. We set $\gamma = 2$ which that means there is 95% probability of covering the data. In practice, this translates to less than 100 pixels, which is a significant reduction of the searching range (the used camera has a resolution of 3280×2464, or 1640×1232 in a faster binned-mode).

The proposed methodology deals with errors in the rough depth estimation. For example, the rock, which appears in the first image of the second row in Figure 6, is farther away from where it is estimated in the rough depth map. This is due, generally, to variable surface reflectance among underwater objects or the angle of light incidence. We note that a general indication of the rough depth estimation quality is the eigenvalue (a smaller value means better depth estimation) that corresponds to the eigenvector, V_2, as it represents a deviation in the lightness value from the linear relationship given in Equation (2). An illustration of a true depth (computed from disparity) vs. the depth estimated from the lightness is shown in Figure 8 (left). The residuals of this estimation are illustrated in Figure 8 (right). We reiterate that the range defined in Equation (6) leaves a sufficient margin to account for the deviation from the true value.

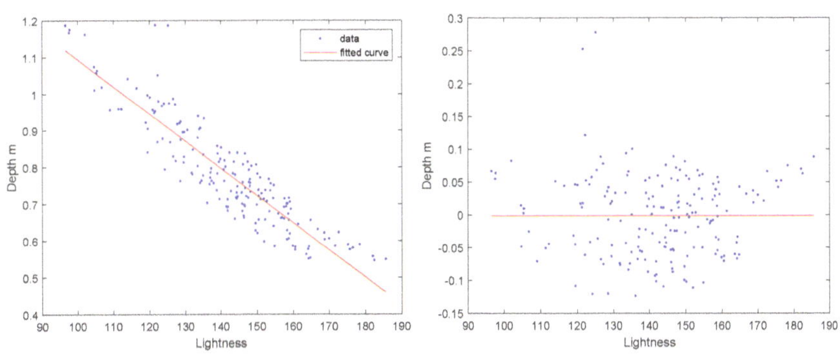

Figure 8. Lightness vs. true depth relationship for a subset of matched points (the same as that used in Figure 7). **Left**: local average pixel lightness vs. true depth (10 cm baseline); the red line represents the lightness to depth transformation—it is deduced from Equation (2). **Right**: the residuals of the depth estimation from lightness vs. true depth.

5.2. Initial Ego-Motion Estimation

An initial ego-motion is calculated every time a new image pair is captured. Let (f_1, f_2, f_3, f_4) denote the previous left, previous right, current left, and current right frames, respectively (see Figure 9 for illustration). We consider here that the relative positions of the previous left frame to the current left frame ($f_1 \rightarrow f_3$) represent the system motion. The pipeline of the ego-motion estimation proceeds is as follows:

1. Feature point detection of f_1 using the Harris-based Shi–Tomasi method [26].
2. Perform feature point matching using the patch descriptor (11×11, as advised in [8]), and the normalized sum of squared differences as a distance measure for the frames (f_1, f_2). Given the

camera calibration parameters, the search range across the epipolar lines is reduced using the analysis presented in Section 5.1.

3. The feature points detected in f_1 are tracked in f_3 using the Pyramidal Lucas–Kanade (LK) method [54].
4. The fundamental matrix is computed for the frames (f_1, f_3) using the normalized eight point method with RANSAC as described in [43]. The matrix is used to reject the tracking outliers. This step is optional—although it improves the accuracy slightly, more computation time adds up.
5. Repeat Step 2 for frames (f_3, f_4) using the tracked feature points found in Step 3.
6. Compute two 3D point clouds using triangulation for the matched feature points in frames (f_1, f_2) and (f_3, f_4) respectively. We note that the correspondence between the two point clouds is known.
7. Compute the relative transformation between the two 3D point clouds, which represents the ego-motion that the ROV undergoes (to be explained in the following text).

We note that starting from the second run of the procedure, Step 1 is appended so that the detected feature points are first compared against those of the previous estimation (using a truncated resolution of 0.1 pixel). This yields two groups of points: new detections and the points that have been already processed from the previous run. Their correspondence is already established within the f_2 frame, and their 3D position is computed. Hence, Steps 2–6 will only be computed for new detections.

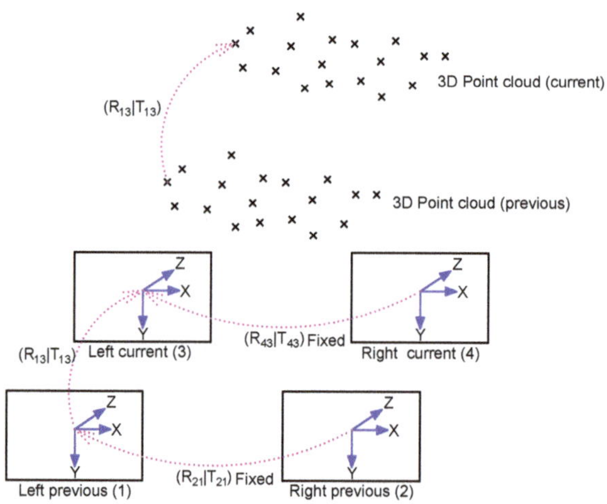

Figure 9. Image quadruplet: the current (left and right) and previous (left and right) frames are used to compute two 3D point clouds. The transformation, $[R_{13}|T_{13}]$, between the two points clouds is equal to the relative motion between the two camera positions.

The choice of using the LK approach is justified by the relatively slow scene change over time, which is reasonably correct due to system mass and smooth motion underwater. Since the LK method employs a closed-form formulation to measure the optical flow, it remains faster than a patch matching scheme. However, it does not suit stereo matching due to the large disparity between corresponding points (up to several hundreds of pixels, as seen earlier).

As there is no scaling problem between the two 3D point clouds, the relative transformation can be expressed as a 3×3 rotation matrix R and a 3×1 translation vector, T, namely $[R_{13}|T_{13}]$ (see Figure 9). The method to compute this transformation is presented in the following text. Let P and P' be the point clouds associated with the image pairs (f_1, f_2) and (f_3, f_4), respectively. Let $p_i \in P$ and $p'_i \in P'$ be two homologous points (correspondence relationship established in Step 3). We have:

$$p'_i = R_{13}\, p_i + T_{13}. \tag{7}$$

We seek a transformation the minimizes the error, r, the sum of squared residuals:

$$r = \sum_{i=1}^{n} \left\| R_{13}\, p_i + T_{13} - p'_i \right\|^2. \tag{8}$$

To solve this problem, we follow the method proposed in [55]. Briefly, a 3×3 matrix C is formed as

$$C = \sum_{i=1}^{n} (p_i - \bar{p})(p'_i - \bar{p}')^\top, \tag{9}$$

where $\bar{p}$ and $\bar{p}'$ are the centers of mass of the 3D point sets, P and P', respectively. Given $C = USV^\top$, the singular value decomposition (SVD) of the matrix, C, the final transformation is computed as

$$R_{13} = VU^\top \tag{10}$$

$$T_{13} = -R_{13}\bar{p} + \bar{p}'. \tag{11}$$

This solution could potentially return reflected rotations, where $det(R_{13}) = -1$. This can be corrected by multiplying the third column of R_{13} by -1.

Once the image pair ,(f_3, f_4), is expressed in the reference system of the image pair, (f_1, f_2), the 3D points can be recalculated using the four observations that we have for each point. A set of verifications are then performed to minimize the pairing errors (verification of the epipolar line, the consistency of the y-parallax, and re-projection residuals). Once validated, this initial ego-motion estimation is used in the BA procedure that will be described later.

5.3. Uncertainty in Visual Odometery

As shown in the literature review, relative visual odometry represents a good solution for long-term autonomy. This kind of approach deals with a selected region of the map at a time. The aim is to reduce the optimization runtime for a new pose. In detail, given a set of frames resulting from camera trajectory. For a new frame at time, t, the pose is estimated w.r.t. frame $t-1$. Here, a relative approach would perform a selection of frames within a certain distance. These frames are assumed to have the largest potential overlap with the current frame. Using these frames, BA is performed to optimize the trajectory. As we have seen earlier, most of the proposed methods assume equal and uncorrelated Gaussian noise for all axes. This is illustrated in Figure 10 (right). In this case, when searching for the nearest frames to be included in the optimization process, the distance, d_2, is larger than d_1, both geometrically and statistically. However, having a full covariance representation of the pose, for instance, as shown in Figure 10 (left), the Euclidean distance measure is no more appropriate. Here, any divergence measure would estimate d_2 to be smaller than d_1, which is more realistic. Since the visual odometry approach suffers from drifting, it is essential to consider an efficient uncertainty measure to represent and determine adjacent frames.

Like any visual odometry estimation, the estimated trajectory using the method mentioned in the previous section is exposed to a computational error, which translates to some uncertainty that grows over time. A global BA may handle this error accumulation; however, it is time-consuming. From another side, a local BA is a tradeoff for accuracy and runtime. The selection of n closest frames is made using the standard Euclidean distance. Loop closure may occur when overlapping with already visited areas which, in return, enhances the accuracy. This approach remains valid as soon as the uncertainty is equal for all estimated variables. However, as the uncertainty varies, the selection of the closest frames based on the Euclidean distance is not suitable. In the following text, we prove that it is the case for any visual odometry method. Also, we provide a formal definition of the uncertainty associated with ego-motion estimation.

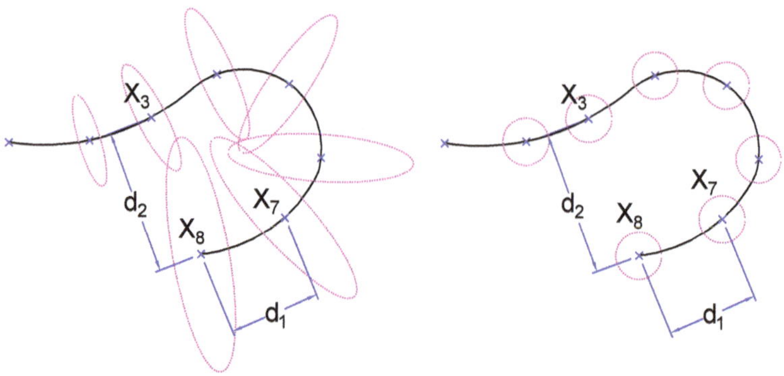

Figure 10. Example of a trajectory with uncertainty modeled by the full covariance matrix (**left**). The distance, d_2, is statistically estimated to be smaller than d_1. In contrast, it is the inverse when the noise is modeled with equal variances (**right**).

Most visual odometry and 3D reconstruction methods rely on matched feature points to estimate the relative motion between two frames. The error in the matched features is resulting from several accumulated errors. These errors are due, non-exclusively, to the following reasons: optical distortion modeling, the discretization of 3D points when projected to image pixels, motion blur, depth of field blur, internal camera noise, salient points detection, and matching. By performing image undistortion, and constraining the point matching with the fundamental matrix, the accumulation of the errors mentioned above can be approximated with a Gaussian distribution. This is implicitly considered in most computer vision fundamentals. Based on this assumption, we can prove that the error distribution of the estimated relative pose is unequal among dimensions. Indeed, it can be fitted to a multivariate Gaussian whose covariance matrix has unequal Eigenvalues, as we will see later.

To better demonstrate this idea, we will take the traditional example of computing the relative pose by means of the fundamental matrix (the results of this analysis also hold for our method, which will be considered in Section 5.4). Formally, a pair of matched points, $\mathbf{m} \leftrightarrow \mathbf{m}'$, between two frames, can be represented by a multivariate Gaussian distribution $\mathcal{N}(\mathbf{m},\Sigma) \leftrightarrow \mathcal{N}(\mathbf{m}',\Sigma)$, where $\Sigma = diag(\sigma^2,\sigma^2)$. The pose estimation procedure relies on the fundamental matrix that satisfies $\mathbf{m}'\ \mathbf{F}\ \mathbf{m} = 0$. Writing $\mathbf{m} = [x\ \ y\ \ 1]^\top$ and $\mathbf{m}' = [x'\ \ y'\ \ 1]^\top$ in homogeneous coordinates, the fundamental matrix constraint for this pair of points can be written as

$$x'xf_{11} + x'yf_{12} + x'f_{13} + y'xf_{21} + y'yf_{22} + y'f_{23} + xf_{31} + yf_{32} + f_{33} = 0, \tag{12}$$

where f_{ij} is the element at row i and column j of $\mathbf{F}$. To show the estimated pose error distribution, we consider one configuration example, the identity camera intrinsic matrix, $K = diag(1\,1\,1)$. Let us now take the case of pure translational motion between the two camera frames, $\mathbf{T} = [T_X\ T_Y\ T_Z]^\top$, and $\boldsymbol{\theta} = [\theta_x\ \theta_y\ \theta_z]^\top = [0\ 0\ 0]^\top$, where $\mathbf{T}$ and $\boldsymbol{\theta}$ are the translation and rotation vectors respectively. The fundamental matrix, in this case, is given as a skew-symmetric matrix of $\mathbf{T}$, denoted $[\mathbf{T}]_\times$. In this case, Equation (12) is simplified to

$$-x'yT_Z + x'T_Y + y'xT_Z - y'T_X - xT_Y + yT_X = 0. \tag{13}$$

By using enough matched points, we can recover the translation vector, $\mathbf{T}$, by solving a linear system. However, the Gaussian error associated with $x, y, x'$and y' will propagate equally to variables T_X and T_Y, with a variance equal to $2\sigma^2$. In contrast to T_Z where the error distribution is different due to the product of two variables, each follows a Gaussian distribution. In addition to not being Gaussian-distributed in general cases, their product's variance is approximate (there is no analytical solution to find the variance of the product of two Gaussian distributed variables). The product of two Gaussian distributed variables follows a normal product distribution; it has been proven that it tends towards a normal distribution when μ/σ is large enough, which is not the case here. Alternatives include numerical integration, Monte Carlo simulation and analytical approximation. The given formula results from the latter case $\sigma^2(x^2 + y'^2 + x'^2 + y^2)$, which largely exceeds the error variance associated with T_X and T_Y.

Moreover, due to the usage of least square approach through an SVD decomposition, as in our method, or two consecutive SVDs (used for fundamental matrix computation and essential matrix decomposition) in traditional visual odometry, the error distributions associated with recovered pose parameters are correlated (even though the observations are uncorrelated), as explained in [56]; this is also demonstrated experimentally in the next sub-section. Overall, this leads to having the estimated pose follow a Gaussian distribution with a full DOF covariance matrix (within the symmetric positive semi-definite constraint).

5.4. Pose Uncertainty Modeling and Learning

Pose uncertainty is difficult to estimate analytically. This is due to the complexity of the pose estimation procedure and the number of variables involved. In particular, the noise propagation through SVD decomposition cannot be analytically modeled. Instead, inspired by the unscented Kalman filter approach [57], we proceed similarly by simulating the noisy input and trying to characterize the output error distribution in this case. This process is illustrated in Figure 11. In our work, we propose to learn the error distribution based on finite but numerous pose samples. This is done using a neural network approach which fits well with our problem as it produces a soft output.

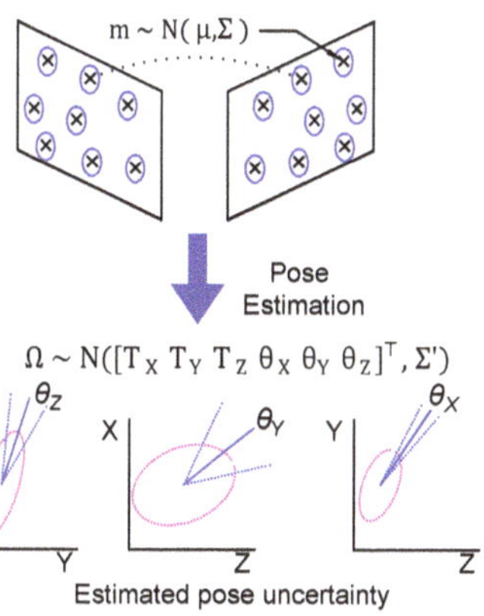

Figure 11. Illustration of error propagation through the pose estimation procedure. The estimated pose uncertainty is shown for each of the six DOF. A full error covariance matrix could result from uncorrelated error distribution of matched 2D feature points.

Two factors play a role in the estimated pose uncertainty. First, the motion, $\Omega = [\mathbf{T}\ \boldsymbol{\theta}]^\top$, between the two frames is expressed by a translation $\mathbf{T}$ and a rotation $\boldsymbol{\theta}$, which is explained in the previous section. Second, is the 3D location of the matched feature points. Although their locations are not computed explicitly in our method, their distances from the camera affect the computation accuracy. In particular, the further the points are from the camera the less accurate the estimated pose is. This is because close points yield larger 2D projection disparity which becomes less sensitive to discretization error. For instance, in a pure translation motion, if all matched points are within the blind zone of the vision system (produce zero-pixel disparity after discretization), the estimated motion will be equal to zero. This problem can be solved with points closer to the camera. Both mentioned factors are correlated to some extent. For instance, given some points in 3D ($n > 7$), the estimated pose accuracy is a function of their depth, but also of the baseline distance and the angle between the two optical centers of the cameras [43] (p. 323). Hence, considering one factor is sufficient. In our work, we consider the motion as a base to predict the uncertainty.

Formally, given a motion vector, $\Omega = [\mathbf{T}\ \boldsymbol{\theta}]^\top$, ideally, we seek to find the covariance matrix that expresses the associated error distribution. Being positive semi-definitive (PSD), an $n \times n$ covariance matrix has $(n^2 + n)/2$ unique entries. Having $n = 6$, in our case, yields 21 DOF, of which six are variances. However, learning this number of parameters freely violates the PSD constraint. Whereas finding the nearest PSD, in this case, distorts the diagonal elements largely because of being mush less. At the same time, we found, experimentally, that the covariance between $\mathbf{T}$ and $\boldsymbol{\theta}$ variables is relatively small compared to that of intra $\mathbf{T}$ and intra $\boldsymbol{\theta}$. Thus, we consider the estimation of two distinct covariance matrices, Σ_T and Σ_θ. So, in total, we have 12 parameters to learn, of which six are the variances.

For the aim of learning Σ_T and Σ_θ, we created a simulation of the pose estimation procedure. For a fixed well-distributed 3D points $X_i \in \mathbb{R}^3 : i = 1..8$, we simulated two cameras (to form a stereo pair) with known intrinsic and extrinsic values. The points were projected according to both cameras' 2D image points. A motion vector, Ω, was applied to the cameras. Then, the 3D points were projected again. All projected points were then disturbed with random Gaussian noise. Next, the ego-motion was estimated by applying the method proposed in Section 5.2 on the disturbed points. Let $\tilde{\Omega} = [\tilde{\mathbf{T}}\ \tilde{\boldsymbol{\theta}}]^\top$ be the estimated motion. Repeating the same procedure (with the same motion Ω) produced a set of motion vectors which represented a point cloud of poses around the real one. Next, we computed the covariance matrices, Σ_T and Σ_θ, of the resulting motion vectors to obtain the uncertainty associated with the given motion, Ω. Furthermore, this procedure was repeated for a large number of motion vectors that covered a wide ranges of its six composing variables (in the performed simulation, we use the range $[0 - 1]$ with a step size of 0.25 for the translation for each of the 3 dimensions. For rotations, we used the range $[0 - \pi/2.5]$ with a step of $\pi/10$. This raised up to 15,625 training samples).

At this stage, having produced the training data by means of motion vectors and the corresponding covariance matrices, we proceeded to build a system to learn the established correspondences (motion $\Leftrightarrow$ uncertainty), so that, in the case of new motion, we would be able to predict the uncertainty. Neural networks offer this soft output by nature, which is the reason why we adopted this learning method. In our experiments, we found that a simple neural network with a single hidden layer [58] was sufficient to fit the data well. The input layer had six nodes that corresponded to the motion vector. The output layer had 12 nodes which corresponded to the unique entries in Σ_T and Σ_θ. Thus, we formed our output vector as

$$O = [\Sigma_T^{11}\ \Sigma_T^{22}\ \Sigma_T^{33}\ \Sigma_T^{12}\ \Sigma_T^{13}\ \Sigma_T^{23}\ \Sigma_\theta^{11}\ \Sigma_\theta^{22}\ \Sigma_\theta^{33}\ \Sigma_\theta^{12}\ \Sigma_\theta^{13}\ \Sigma_\theta^{23}]^\top, \tag{14}$$

where $\Sigma_\cdot^{ij}$ is the element of row i and column j of a covariance matrix. In the learning phase, we used the Levenberg–Marquardt backpropagation which is a gradient-descent based approach, as described in [59]. Further, by using the mean-squared error as a cost function, we were able to achieve around a training error rate of 3%. The obtained parameters were rearranged into two symmetric matrices. In practice, the obtained matrix is not necessarily PSD. We proceeded to find the closest PSD as $Q\Lambda_+ Q^{-1}$, where Q is the eigenvector matrix of the estimated covariance matrix, and Λ_+ is the diagonal matrix of eigenvalues, in which negative values are set to zero.

To validate the training phase, the procedure to generate the training set was repeated but using different values of motion vectors. The validation of this test set using the trained neural network showed an accuracy of 87.6% and a standard deviation of 6.1, which is reasonably acceptable in this context.

5.5. Semi-Global Bundle Adjustment

After initiating the visual odometry, the relative pose estimation at each frame is maintained within a table that contains all poses' related information (18 parameters per pose, in which 6 for the position, and 12 for two covariance matrices). At any time, it is possible to identify the poses in the neighborhood of the current pose being estimated to find potential overlaps to consider while performing BA. Since we were dealing with a statistical representation of the observations, a divergence measure had to be considered. Here, we chose the Bhattacharyya distance as suitable for our problem (modified metric variation could be also used [60]). Formally, the distance between the two poses, $\{\Omega^1, \Sigma^1_T, \Sigma^1_\theta\}$ and $\{\Omega^2, \Sigma^2_T, \Sigma^2_\theta\}$, is given as:

$$D = \frac{1}{8}(\Omega^1 - \Omega^2)^\top \Sigma^{-1}(\Omega^1 - \Omega^2) + \frac{1}{2}\ln(\frac{\det \Sigma}{\sqrt{\det \Sigma^1 + \det \Sigma^2}}), \tag{15}$$

where

$$\Sigma^{\cdot} = \begin{bmatrix} \Sigma^{\cdot}_T & \mathbf{0} \\ \mathbf{0} & \Sigma^{\cdot}_\theta \end{bmatrix}, \ \Sigma = \frac{\Sigma^1 + \Sigma^2}{2}. \tag{16}$$

Having selected the set of frames, $\mathbb{F}$, in the neighborhood of the current pose statistically, we performed BA as follows; First, we divided $\mathbb{F}$ into two subsets similar to [45]. The first subset, $\mathbb{F}_d$, contained the current and previous frames in time, whereas the other subset $\mathbb{F}_s$ contained the remaining frames, mostly resulting from overlap with an already scanned area. Second, BA was performed on both subsets. However, the pose parameters related to $\mathbb{F}_s$ were masked as static, so they were not optimized, in contrast to $\mathbb{F}_d$. This strategy was necessary to reduce the number of variables to optimize.

After determining the error distribution arising with a new pose, it has to be compounded with the error propagated from the previous pose. Similar to SLAM approaches, we propose to use a *Kalman filter* like gain which allows controllable error fusion and propagation. Given an accumulated previous pose estimation, defined as $\{\Omega^p, \Sigma^p_T, \Sigma^p_\theta\}$, and a current one, $\{\Omega^c, \Sigma^c_T, \Sigma^c_\theta\}$, an updated current pose, $\{\Omega^u, \Sigma^u_T, \Sigma^u_\theta\}$, is calculated as:

$$\Omega^u = \Omega^c \tag{17}$$

$$\Sigma^u_\theta = (I - \Sigma^p_\theta(\Sigma^p_\theta + \Sigma^c_\theta)^{-1})\Sigma^p_\theta \tag{18}$$

$$\Sigma^u_T = (I - \Sigma^p_T(\Sigma^p_T + \Sigma^c_T)^{-1})\Sigma^p_T. \tag{19}$$

6. Experimental Results

The first experiments were carried out to test the hardware platform stability, reliability, and autonomy. Snapshots of the operations for both systems are shown in Figures 4 and 5, whereas examples of the taken images are shown in Figure 6. An underwater site was scanned, and the taken images were processed using photogrammetry techniques to validate the quality of the taken images. We use Agisoft Photoscan [61] to perform the 3D reconstruction. Examples of resulting 3D models are shown in Figure 12. Stereo image synchronization was also validated by observing the relative pose estimation between each pair and comparing it with the stereo calibration extrinsic parameters.

It was desired that the proposed visual odometry method would represent a trade-off between accuracy and runtime, the maximum accuracy being the case for global BA, whereas the fastest runtime was an optimization free visual odometry. Moreover, a performance improvement was expected w.r.t the local optimization method due to a better selection of neighboring observations. Therefore, we analyzed the performance of our method from two points of view: runtime and accuracy.

Figure 12. 3D reconstructed models using images captured with the handheld system.

6.1. Runtime Evaluation

We implemented our method using OpenCV [3] bindings in Java. The BA scheme was implemented using the speed optimized BA toolbox proposed in [44]. The image stream processing on the embedded systems including the image quality assessment took around 100 ms per stereo pair to execute. The maximum image acquisition frequency was 3 per second, due to hardware limitations. Therefore, a mid-range laptop computer (with Intel Core i5-7300U CPU @3.50GHz with 16GB RAM) was enough to handle the visual odometry process.

The major improvement that reduced the processing time was the proposed stereo matching method. To demonstrate the time gain, we started by comparing the runtime of our method with an implementation that does not employ any range reduction. The method with range reduction showed an average gain of 72% processing time. Then, we compared this to methods using high-level feature descriptors, in particular, SIFT, SURF, and BRISK. At the same time, we monitored the accuracy for each run. The evaluation was done using the same set of images. In this test, the computational times increased to 342%, 221%, and 142% for SIFT, SURF, and BRISK, respectively. Nevertheless, we noticed a slight gain in accuracy of 1.1% for the average translational error and 0.6% for the average rotational error when using SIFT and SURF, which we do not judge as significant. On the contrary, BRISK performed less accurately, with an increase of 4.1% in average translational error and 0.8% in average rotational error, which is probably due to its sensitivity to water turbidity and dust.

The results above (given in percentages) were more or less consistent across several processing environments, including the running on an RPi computer. Nevertheless, in Table 1, we provide the exact processing times using the same laptop computer mentioned above, applied to full resolution images. The table also shows the percentage of correct matches for each method (our method is not concerned here, as it uses the epipolar geometry to search for matches), which were obtained using the first-order geometric error and a threshold of 0.005. We observe that BRISK features produced less correct matches in underwater images than SIFT and SURF.

Table 1. Performance of used feature matching methods regarding processing time and correct matches. The correct matches are defined as having a first-order geometric error [43] (p. 287) less than 0.005.

Method	Detector	Correct Matches (%)	Processing Time (ms)
Ours	Shi-Tomasi [26]	-	220
Stereo matching (no range reduction)	Shi-Tomasi [26]	-	785
SIFT	DoG	49.5	752
SURF	Fast Hessian	48.7	486
BRISK	AGAST [31]	34.3	313

A break-up of the average time required to run the visual odometry is illustrated in Figure 13. It shows that the stereo matching procedure occupies as little as 16.5% of the total runtime, whereas the BA procedure occupies the majority of time with 42.2%. We note that this result is for using five frames in the optimization phase. More frames can be used to improve the accuracy but with a cost of more time complexity. This will be detailed in the next section.

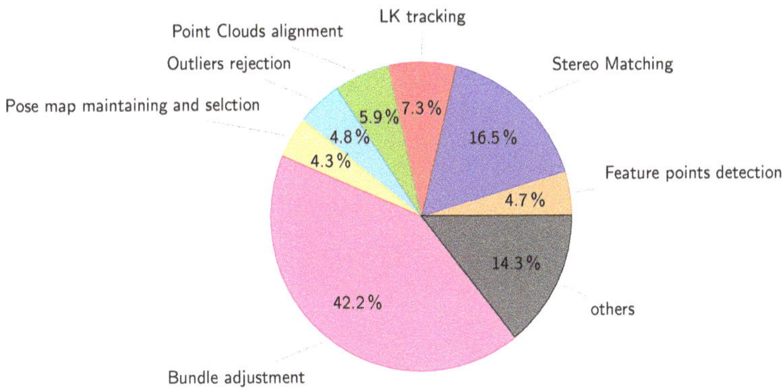

Figure 13. Runtime analysis of the visual odometry system for new pose estimation.

6.2. Visual Odometry Evaluation

Unlike terrestrial odometry datasets that come with ground-truth, there is no such option for underwater odometry. Alternatively, to validate the proposed visual odometry method, we conducted an underwater survey using several scenarios, for instance, long trajectories with loop closures or raster scans. We estimated the overtaken trajectories using Agisoft Photoscan which employs a global optimization approach. We used the best available accuracy settings with large numbers of matching points. We considered the estimated trajectories as a reference for comparison in our experiments. An example of a dense reconstruction of such trajectories is shown in Figure 14. It measures around 5 m $\times$ 12 m and contains several loop closures. We followed a standard evaluation procedure, as described in [62], where, for all test trajectories, we computed translational and rotational errors for all possible sub-trajectories of one-meter length. The errors were measured as percentages for translation and degrees per meter for rotation.

Figure 14. An example of the long trajectory 3D reconstructed model using images captured with the ROV-attached system. Such trajectories were used as ground truths to validate and tune the proposed method.

We first evaluated the effect of varying the number of frames considered in the optimization phase in our method. Table 2 shows the trajectory errors for using 3, 5, and 11 frames. Although the BA running time for the case of using five frames was around half that of using 11 frames, the accuracy gain was not significant for this latter case. Hence, we found that using five frames is the best accuracy vs. runtime trade-off, and also it is the limit to remain within a real-time performance. The caused drift of 3.8% remained acceptable even for large sites. Figure 15 (right) shows the effect of using different numbers of frames on trajectory estimation for the example shown in Figure 14.

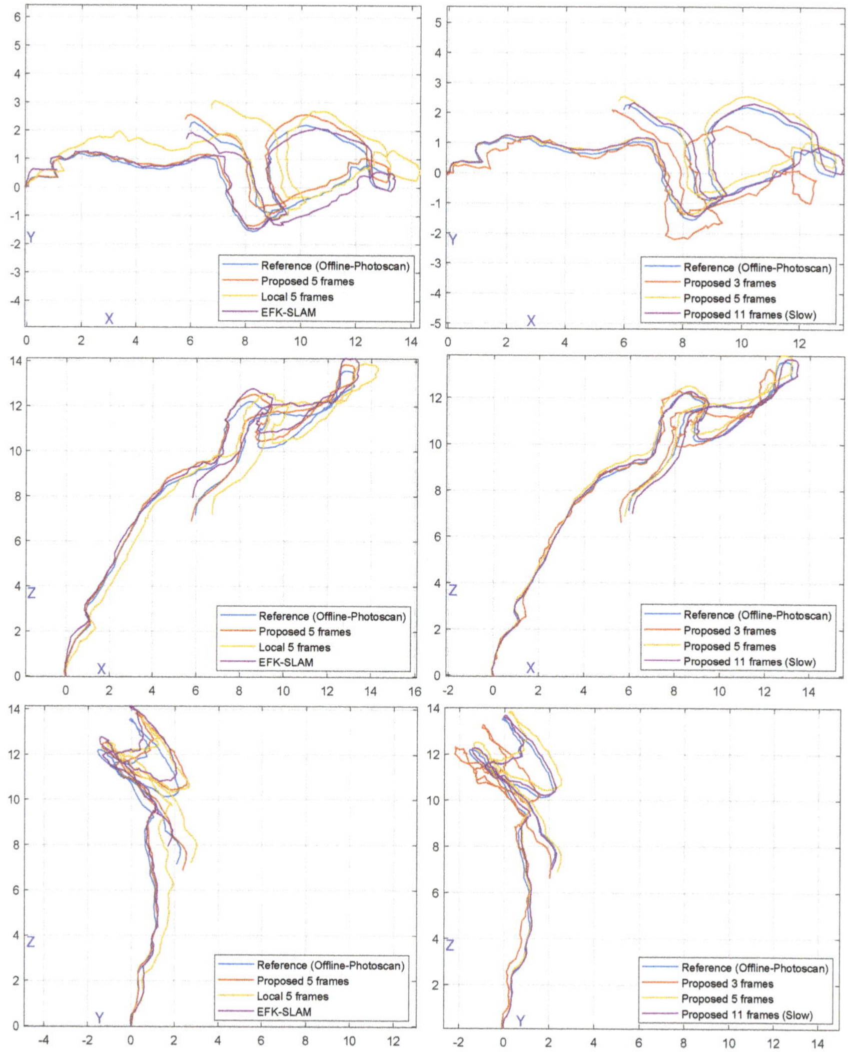

Figure 15. Comparison of trajectory estimation using several methods and parameters. The trajectory obtained using Agisoft Photoscan was considered a reference. It was compared to our method, local optimization using 5 frames, and EFK-SLAM [50] (**left column**). It was also compared to our method in cases where 3, 5, or 11 frames were used for optimization (**right column**). All units are in meters.

Second, we compared our semi-global BA to three cases—using local BA with the same number of frames, using the underwater EFK-SLAM method [50], and without using any BA. Our method (using five frames) remained ahead of these three variations. From another perspective, although we managed to run a BA-free approach entirely on the third RPi computer in the ROV-attached system.

The trajectory drifts were large (with large variance) as shown in Table 2. Trajectory estimation using these methods is shown in Figure 15. From another perspective, although the EFK-SLAM performed better than the local approach, the runtime grew constantly.

Table 2. A comparison of translational and rotational errors for several methods and parameters. The trajectory estimation performed in Agisoft Photoscan was considered to be a reference.

	Translation Error (%)	Rotation Error (deg/m)
Ours (11 frames)—slow	3.8	0.024
Ours (5 frames)	4.3	0.026
Ours (3 frames)	8.2	0.088
EFK-SLAM [50]	5.7	0.032
Local (5 frames)	8.4	0.079
No BA	16.1	0.137

7. Conclusions and Perspectives

In this work, we introduced several improvements to the current traditional visual odometry approach to serve in the context of underwater surveys. The goal was to adapt the approach to low resource systems. The sparse feature points matching, guided with a rough depth estimation using lightness information, is the main factor associated with most of the gain in computation time compared to sophisticated feature descriptors combined with brute-force matching. Also, using stochastic representation and the selection of frames in the semi-global BA improved the accuracy compared to local BA methods while remaining within real-time limits.

The developed hardware platforms represent efficient low-cost solutions for underwater surveys. The live feedback of image quality and navigation helps to achieve better performance and leads to faster reactions on site. Both systems are flexible for upgrades and modifications; new functionalities can be easily added thanks to the compatible optimized image processing libraries.

Our future perspectives are mainly centered on performing the visual odometry within the system. Further code improvements and parallelism are to be considered. Furthermore, at the time of writing this article, new embedded systems that have double the computational power of the used ones have been released, which makes our objective even closer.

On the other hand, dealing with visual odometry failure is an important challenge, especially in the context of underwater imaging, which is mainly due to bad image quality. The ideas of failing scenarios discussed in this paper can be extended to deal with the problem of interruptions in an obtained trajectory.

Author Contributions: Conceptualization, M.M.N. and D.P.; Methodology, M.M.N.; Software, J.-P.R.; Validation, M.M.N., D.P. and J.-M.B.; Formal Analysis, M.M.N; Resources, M.S.; Writing—Original Draft Preparation, M.M.N.; Writing—Review & Editing, D.P.; Visualization, M.B.E.; Project Administration, D.M.; Funding Acquisition, D.P.

Funding: This work was partially supported by both a public grant overseen by the French National Research Agency (ANR) as part of the program *Contenus numériques et interactions (CONTINT) 2013* (reference: ANR-13-CORD-0014), GROPLAN project (Ontology and Photogrammetry; Generalizing Surveys in Underwater and Nautical Archaeology) (http://www.groplan.eu), and the French Armaments Procurement Agency (DGA), DGA RAPID LORI project (LOcalisation et Reconnaissance d'objets Immergés).

Acknowledgments: We wish to thank David Scaradozzi and his group for building the handheld system, and Olivier Bianchimani for carrying out the tests underwater.

Conflicts of Interest: The authors declare no conflict of interest. The founding sponsors had no role in the design of the study; in the collection, analyses, or interpretation of data; in the writing of the manuscript, and in the decision to publish the results.

Abbreviations

The following abbreviations are used in this manuscript:

ROV	Remotely Operated underwater Vehicle
ARM	Advanced RISC Machine
GPS	Global Positioning System
DVL	Doppler Velocity Logs
3D	Three Dimensional
GPU	Graphics Processing Unit
DOF	Degree Of Freedom
BA	Bundle Adjustment
SLAM	Simultaneous Localization And Mapping
RANSAC	RANdom SAmple Consensus
RGB-D	Red Green Blue Depth
RPi	Raspberry Pi
INU	Inertial Navigation Unit
SONAR	SOund Navigation And Ranging
SSS	Side Scan Sonar
SVD	Singular Value Decomposition
CPU	Central Processing Unit
PSD	Positive Semi-Definitive

References

1. Drap, P. *Underwater Photogrammetry for Archaeology*; INTECH Open Access Publisher: Vienna, Austria, 2012.
2. BlueRobotics. 2018. Available online: https://www.bluerobotics.com (accessed on 1 July 2018).
3. Itseez. Open Source Computer Vision Library. 2018. Available online: https://github.com/itseez/opencv (accessed on 1 July 2018).
4. Stone, J.E.; Gohara, D.; Shi, G. OpenCL: A Parallel Programming Standard for Heterogeneous Computing Systems. *Comput. Sci. Eng.* **2010**, *12*, 66–73. [CrossRef] [PubMed]
5. Triggs, B.; McLauchlan, P.; Hartley, R.; Fitzgibbon, A. Bundle adjustment: A modern synthesis. In *Vision Algorithms: Theory and Practice*; Springer: Berlin/Heidelberg, Germany, 2000; pp. 153–177.
6. Lowe, D. Distinctive image features from scale-invariant keypoints. *In. J. Comput. Vis.* **2004**, *60*, 91–110. [CrossRef]
7. Bay, H.; Tuytelaars, T.; Van Gool, L. Surf: Speeded up robust features. In Proceedings of the European Conference on Computer Vision, Graz, Austria, 7–13 May 2006; Springer: Berlin/Heidelberg, Germany, 2006; pp. 404–417.
8. Gauglitz, S.; Höllerer, T.; Turk, M. Evaluation of interest point detectors and feature descriptors for visual tracking. *Int. J. Comput. Vis.* **2011**, *94*, 335–360. [CrossRef]
9. Bellavia, F.; Fanfani, M.; Colombo, C. Selective visual odometry for accurate AUV localization. *Autono. Robot.* **2017**, *41*, 133–143. [CrossRef]
10. Mouragnon, E.; Lhuillier, M.; Dhome, M.; Dekeyser, F.; Sayd, P. Generic and real-time structure from motion using local bundle adjustment. *Image Vis. Comput.* **2009**, *27*, 1178–1193. [CrossRef]
11. Thrun, S.; Burgard, W.; Fox, D. *Probabilistic Robotics*; MIT Press: Cambridge, MA, USA, 2005.
12. Montemerlo, M.; Thrun, S. *FastSLAM: A Scalable Method for the Simultaneous Localization and Mapping Problem in Robotics*; Springer: Berlin, Germany, 2007.
13. Estrada, C.; Neira, J.; Tardós, J.D. Hierarchical SLAM: Real-time accurate mapping of large environments. *IEEE Trans. Robot.* **2005**, *21*, 588–596. [CrossRef]
14. Clemente, L.A.; Davison, A.J.; Reid, I.D.; Neira, J.; Tardós, J.D. Mapping Large Loops with a Single Hand-Held Camera. In Proceedings of the Robotics: Science and Systems III, Atlanta, GA, USA, 27–30 June 2007.
15. Nawaf, M.M.; Boï, J.M.; Merad, D.; Royer, J.P.; Drap, P. Low Cost Embedded Stereo System for Underwater Surveys. In Proceedings of the 5th International Workshop LowCost 3D—Sensors, Algorithms, Applications, Hamburg, Germany, 28–29 November 2017; pp. 179–186.

16. Nawaf, M.M.; Drap, P.; Royer, J.P.; Merad, D.; Saccone, M. Towards Guided Underwater Survey Using Light Visual Odometry. In Proceedings of the 7th ISPRS/CIPA 3D Virtual Reconstruction and Visualization of Complex Architectures, Nafplio, Greece, 1–3 March 2017; pp. 527–533.
17. Se, S.; Lowe, D.; Little, J. Mobile robot localization and mapping with uncertainty using scale-invariant visual landmarks. *Int. J. Robot. Res.* **2002**, *21*, 735–758. [CrossRef]
18. Nistér, D.; Naroditsky, O.; Bergen, J. Visual odometry. In Proceedings of the 2004 IEEE Conference on Computer Vision and Pattern Recognition (CVPR 2004), Washington, DC, USA, 27 June–2 July 2004; p. I-652.
19. Eade, E.; Drummond, T. Scalable monocular SLAM. In Proceedings of the 2006 IEEE Conference on Computer Vision and Pattern Recognition, New York, NY, USA, 17–22 June 2006; Volume 1, pp. 469–476.
20. Williams, B.; Klein, G.; Reid, I. Real-time SLAM relocalisation. In Proceedings of the 2007 IEEE 11th International Conference on Computer Vision, Rio de Janeiro, Brazil, 14–21 October 2007; pp. 1–8.
21. Chekhlov, D.; Pupilli, M.; Mayol, W.; Calway, A. Robust real-time visual SLAM using scale prediction and exemplar based feature description. In Proceedings of the 2007 IEEE Conference on Computer Vision and Pattern Recognition (CVPR'07), Minneapolis, MN, USA, 18–23 June 2007; pp. 1–7.
22. Davison, A.J.; Reid, I.D.; Molton, N.D.; Stasse, O. MonoSLAM: Real-time single camera SLAM. *IEEE Trans. Pattern Anal. Mach. Intell.* **2007**, *29*, 1052–1067. [CrossRef] [PubMed]
23. Klein, G.; Murray, D. Improving the agility of keyframe-based SLAM. In Proceedings of the European Conference on Computer Vision, Marseille, France, 12–18 October 2008; Springer: Berlin/Heidelberg, Germany, 2008; pp. 802–815.
24. Bourmaud, G.; Megret, R. Robust large scale monocular visual SLAM. In Proceedings of the IEEE Conference on Computer Vision and Pattern Recognition, Boston, MA, USA, 7–12 June 2015; pp. 1638–1647.
25. Rosten, E.; Drummond, T. Fusing points and lines for high performance tracking. In Proceedings of the Tenth IEEE International Conference on Computer Vision (ICCV'05), Beijing, China, 17–21 October 2005; Volume 2, pp. 1508–1515.
26. Shi, J.; Tomasi, C. Good features to track. In Proceedings of the 1994 IEEE Conference on Computer Vision and Pattern Recognition (CVPR'94), Seattle, WA, USA, 21–23 June 1994; pp. 593–600.
27. Nawaf, M.M.; Trémeau, A. Monocular 3D structure estimation for urban scenes. In Proceedings of the 2014 IEEE International Conference on Image Processing (ICIP), Paris, France, 27–30 October 2014; pp. 3763–3767.
28. Negre, P.L.; Bonin-Font, F.; Oliver, G. Cluster-based loop closing detection for underwater slam in feature-poor regions. In Proceedings of the 2016 IEEE International Conference on Robotics and Automation (ICRA), Stockholm, Sweden, 16–21 May 2016; pp. 2589–2595.
29. Calonder, M.; Lepetit, V.; Strecha, C.; Fua, P. Brief: Binary robust independent elementary features. In Proceedings of the European Conference on Computer Vision, Heraklion, Greece, 5–11 September 2010; Springer: Berlin/Heidelberg, Germany, 2010; pp. 778–792.
30. Leutenegger, S.; Chli, M.; Siegwart, R.Y. BRISK: Binary robust invariant scalable keypoints. In Proceedings of the 11th IEEE International Conference on Computer Vision (ICCV), Barcelona, Spain, 6–13 November 2011; pp. 2548–2555.
31. Mair, E.; Hager, G.D.; Burschka, D.; Suppa, M.; Hirzinger, G. Adaptive and Generic Corner Detection Based on the Accelerated Segment Test. In Proceedings of the European Conference on Computer Vision (ECCV'10), Heraklion, Greece, 5–11 September 2010; Springer: Berlin/Heidelberg, Germany, 2010.
32. Yamaguchi, K.; McAllester, D.; Urtasun, R. Robust Monocular Epipolar Flow Estimation. In Proceedings of the 2013 IEEE Conference on Computer Vision and Pattern Recognition, Portland, OR, USA, 23–28 June 2013; IEEE Computer Society: Washington, DC, USA, 2013; pp. 1862–1869.
33. Geiger, A.; Ziegler, J.; Stiller, C. Stereoscan: Dense 3D reconstruction in real-time. In Proceedings of the IEEE Intelligent Vehicles Symposium, Baden-Baden, Germany, 5–9 June 2011; pp. 963–968.
34. Davison, A.J. Real-time simultaneous localisation and mapping with a single camera. In Proceedings of the Ninth IEEE International Conference on Computer Vision Computer Vision, Nice, France, 13–16 October 2003; pp. 1403–1410.
35. Kerl, C.; Sturm, J.; Cremers, D. Dense visual SLAM for RGB-D cameras. In Proceedings of the 2013 IEEE/RSJ International Conference on Intelligent Robots and Systems (IROS), Tokyo, Japan, 3–7 Novermber 2013; pp. 2100–2106.

36. Comport, A.I.; Malis, E.; Rives, P. Accurate quadrifocal tracking for robust 3D visual odometry. In Proceedings of the 2007 IEEE International Conference on Robotics and Automation, Roma, Italy, 10–14 April 2007; pp. 40–45.
37. Stühmer, J.; Gumhold, S.; Cremers, D. Real-time dense geometry from a handheld camera. In *Joint Pattern Recognition Symposium*; Springer: Berlin/Heidelberg, Germany, 2010; pp. 11–20.
38. Pizzoli, M.; Forster, C.; Scaramuzza, D. REMODE: Probabilistic, monocular dense reconstruction in real time. In Proceedings of the 2014 IEEE International Conference on Robotics and Automation (ICRA), Hong Kong, China, 31 May–7 June 2014; pp. 2609–2616.
39. Blanco, J.L.; Fernandez-Madrigal, J.A.; González, J. A novel measure of uncertainty for mobile robot slam with rao-blackwellized particle filters. *Int. J. Robot. Res.* **2008**, *27*, 73–89. [CrossRef]
40. Eade, E.; Drummond, T. Unified Loop Closing and Recovery for Real Time Monocular SLAM. *BMVC* **2008**, *13*, 136.
41. Piniés, P.; Tardós, J.D. Scalable SLAM building conditionally independent local maps. In Proceedings of the 2007 IEEE/RSJ International Conference on Intelligent Robots and Systems (IROS 2007), San Diego, CA, USA, 29 October–2 November 2017; pp. 3466–3471.
42. Montiel, J.; Civera, J.; Davison, A.J. Unified inverse depth parametrization for monocular SLAM. *Analysis* **2006**, *9*, 1.
43. Hartley, R.; Zisserman, A. *Multiple View Geometry in Computer Vision*; Cambridge University Press: Cambridge, UK, 2003.
44. Lourakis, M.I.; Argyros, A.A. SBA: A software package for generic sparse bundle adjustment. *ACM Trans. Mathe. Softw.* **2009**, *36*, 2. [CrossRef]
45. Sibley, D.; Mei, C.; Reid, I.; Newman, P. Adaptive relative bundle adjustment. *Robot. Sci. Syst.* **2009**, *32*, 33.
46. Ribas, D.; Ridao, P.; Tardós, J.D.; Neira, J. Underwater SLAM in a marina environment. In Proceedings of the 2007 IEEE/RSJ International Conference on Intelligent Robots and Systems (IROS 2007), San Diego, CA, USA, 29 October–2 November 2007; pp. 1455–1460.
47. He, X.; Yuille, A. Occlusion boundary detection using pseudo-depth. In *ECCV*; Springer: Berlin/Heidelberg, Germany, 2010; pp. 539–552.
48. Williams, S.; Mahon, I. Simultaneous localisation and mapping on the great barrier reef. In Proceedings of the IEEE International Conference on Robotics and Automation (ICRA '04), New Orleans, LA, USA, 26 April–1 May 2004; Volume 2, pp. 1771–1776.
49. Sáez, J.M.; Hogue, A.; Escolano, F.; Jenkin, M. Underwater 3D SLAM through entropy minimization. In Proceedings of the 2006 IEEE International Conference on Robotics and Automation (ICRA 2006), Orlando, FL, USA, 15–19 May 2006; pp. 3562–3567.
50. Salvi, J.; Petillo, Y.; Thomas, S.; Aulinas, J. Visual slam for underwater vehicles using video velocity log and natural landmarks. In Proceedings of the IEEE OCEANS 2008, Quebec City, QC, Canada, 15–18 September 2008; pp. 1–6.
51. Weidner, N.; Rahman, S.; Li, A.Q.; Rekleitis, I. Underwater cave mapping using stereo vision. In Proceedings of the 2017 IEEE International Conference on Robotics and Automation (ICRA), Singapore, 29 May–3 June 2017; pp. 5709–5715.
52. Raspberry Pi Foundation. 2018. Available online: http://www.raspberrypi.org/ (accessed on 1 July 2018).
53. Meier, L.; Tanskanen, P.; Heng, L.; Lee, G.H.; Fraundorfer, F.; Pollefeys, M. PIXHAWK: A micro aerial vehicle design for autonomous flight using onboard computer vision. *Auton. Robot.* **2012**, *33*, 21–39. [CrossRef]
54. Bouguet, J.Y. Pyramidal implementation of the affine lucas kanade feature tracker description of the algorithm. *Intel Corp.* **2001**, *5*, 4.
55. Eggert, D.W.; Lorusso, A.; Fisher, R.B. Estimating 3-D rigid body transformations: A comparison of four major algorithms. *Mach. Vis. Appl.* **1997**, *9*, 272–290. [CrossRef]
56. Strutz, T. *Data Fitting and Uncertainty: A Practical Introduction to Weighted Least Squares and Beyond*; Vieweg and Teubner: Wiesbaden, Geramny, 2010.
57. Wan, E.A.; Van Der Merwe, R. The unscented Kalman filter for nonlinear estimation. In Proceedings of the IEEE 2000 Adaptive Systems for Signal Processing, Communications, and Control Symposium (AS-SPCC), Lake Louise, AB, Canada, 4 October 2000; pp. 153–158.
58. Bishop, C.M. *Neural Networks for Pattern Recognition*; Oxford University Press: Oxford, UK, 1995.

59. Demuth, H.B.; Beale, M.H.; De Jess, O.; Hagan, M.T. *Neural Network Design*; Martin Hagan: Stillwater, OK, USA, 2014.
60. Comaniciu, D.; Ramesh, V.; Meer, P. Kernel-based object tracking. *IEEE Trans. Pattern Anal. Mach. Intell.* **2003**, *25*, 564–577. [CrossRef]
61. Agisoft PhotoScan. 2018. Available online: http://www.agisoft.com/ (accessed on 1 July 2018).
62. Geiger, A.; Lenz, P.; Urtasun, R. Are we ready for Autonomous Driving? The KITTI Vision Benchmark Suite. In Proceedings of the IEEE Conference on Computer Vision and Pattern Recognition, Providence, RI, USA, 16–21 June 2012; pp. 3354–3361.

sensors

MDPI

Article

Visual Information Fusion through Bayesian Inference for Adaptive Probability-Oriented Feature Matching

David Valiente [1,*,†], Luis Payá [1,†], Luis M. Jiménez [1,†], Jose M. Sebastián [2,†] and Oscar Reinoso [1,†]

1 Department of Systems Engineering and Automation, Miguel Hernández University, Av. de la Universidad s/n. Ed. Innova., 03202 Elche (Alicante), Spain; lpaya@umh.es (L.P.); luis.jimenez@umh.es (L.M.J.); o.reinoso@umh.es (Ó.R.)

2 Centre for Automation and Robotics (CAR), UPM-CSIC, Technical University of Madrid, C/ José Gutiérrez Abascal, 2, 28006 Madrid, Spain; jsebas@etsii.upm.es

* Correspondence: dvaliente@umh.es; Tel.: +34-96-665-9005

† These authors contributed equally to this work.

Received: 5 April 2018; Accepted: 24 June 2018; Published: 26 June 2018

Abstract: This work presents a visual information fusion approach for robust probability-oriented feature matching. It is sustained by omnidirectional imaging, and it is tested in a visual localization framework, in mobile robotics. General visual localization methods have been extensively studied and optimized in terms of performance. However, one of the main threats that jeopardizes the final estimation is the presence of outliers. In this paper, we present several contributions to deal with that issue. First, 3D information data, associated with SURF (Speeded-Up Robust Feature) points detected on the images, is inferred under the Bayesian framework established by Gaussian processes (GPs). Such information represents a probability distribution for the feature points' existence, which is successively fused and updated throughout the robot's poses. Secondly, this distribution can be properly sampled and projected onto the next 2D image frame in $t+1$, by means of a filter-motion prediction. This strategy permits obtaining relevant areas in the image reference system, from which probable matches could be detected, in terms of the accumulated probability of feature existence. This approach entails an adaptive probability-oriented matching search, which accounts for significant areas of the image, but it also considers unseen parts of the scene, thanks to an internal modulation of the probability distribution domain, computed in terms of the current uncertainty of the system. The main outcomes confirm a robust feature matching, which permits producing consistent localization estimates, aided by the odometer's prior to estimate the scale factor. Publicly available datasets have been used to validate the design and operation of the approach. Moreover, the proposal has been compared, firstly with a standard feature matching and secondly with a localization method, based on an inverse depth parametrization. The results confirm the validity of the approach in terms of feature matching, localization accuracy, and time consumption.

Keywords: omnidirectional imaging; visual localization; catadioptric sensor; visual information fusion

1. Introduction

There is a growing tendency for the use of visual sensors, to the detriment of the range sensory data approaches [1,2]. Visual sensors, which are essentially represented by digital cameras, have contributed with valuable advantages to the state of the art [3,4], such as the ability to acquire large amounts of information with only one snapshot. They have become a robust alternative to former sensors, and thus they have been extensively integrated in the framework of localization, in mobile robotics. In particular, they can perform as the main sensor [5–7], where no other sensory data are used, and can assist as a secondary sensor [8,9] where the main sensor is unable to produce measures, for instance under GPS (Global Positioning System)-denied circumstances, in unmanned vehicle applications [10].

We have concentrated on catadioptric systems, such as omnidirectional cameras, due to their ability to capture large scenes and their wider field of view, in comparison to planar cameras. Different omnidirectional visual approaches have been proposed. They can be categorized according to the sort of method that processes the visual content of a scene. First, some approaches make use of specific visual points in an image (local feature methods or visual landmark methods) [7,11]. Additionally, a more recent research line has come up with global appearance or holistic methods, relying on the processing of the image as a whole [12,13]. Despite the fact that these recent advances have evidenced a pronounced growth in the efficiency, we have opted for using local feature methods since they have been vastly accepted and tested in terms of performance [14,15], accuracy [7,16], and robustness [17,18].

Nevertheless, both processing methods are required to associate visual data correctly, regardless of the final application, they are intended for. This is a non-trivial aspect that implies an important challenge, which sometimes results in a general issue in many mobile robotics applications [19–21]. In this sense, visual features matching [22,23] is one of the most extended techniques in order to describe and associate visual features from one image to another, by comparing certain pixel description metric. Unfortunately, the final estimation typically reflects the harmful effect of false positives in the data association, denoted as outliers. A considerable amount of research in this area has been conducted [24–28]. Nevertheless, the rejection methods normally need substantial computational efforts and external requirements [29,30], beyond the specifications of the target system application.

In this work, we propose an adaptive matching approach, which takes the most of the same visual data input used by our former localization approach [31], which is aided by the odometer's prior in order to estimate the scale factor. To that end, the visual data are fused at every motion step of the vehicle by means of a Bayesian technique, namely Gaussian processes (GPs) [32]. Such a scheme permits inferring a model of the environment that accounts for the probability of feature existence in the 3D global reference system. In this manner, obtaining a reliable probability distribution permits identifying relevant areas from which some visual features might be detected, in terms of probability of existence. This idea inherits the foundations from exploration models [33], which are aimed at discovering new areas in the environment, and fusing the new information into the estimated models.

The design of these contributions seeks a more realistic approach, with the objective of obtaining robust results in challenging environments, ensuring computational efficiency. Synthesizing, the main differences and contributions of this paper, in contrast to the previous work [31], are as follows:

- The probability framework considers the 3D global reference system, instead of a 2D image frame representation.
- A 3D probability distribution is computed and projected onto the next image, associated to the next pose of the robot, by means of a filter-motion prediction stage. Such probability projection represents relevant areas on the image, where matching detection is more probable.
- The matching process is performed in a single batch, using the entire set of feature points associated with the probability areas projected on the image, instead of a multi-scaled matching, computed feature by feature.
- The information metric permits modulating the probability values for the probability areas, instead of simply representing a set of less precise coefficients for weighting the former multi-scaled matching.

Finally, since this work pursues the achievement of quantitative benefits, a specific application of visual localization has been considered. In this context, different publicly available real datasets have been used in the experimental section, in order to confirm the validity of the approach, and to evaluate its final performance when producing robust and adaptive probability-oriented visual feature matching. Furthermore, extended comparison results have been generated to reinforce and highlight the improvements of the approach, after embedding the implementation into a visual localization application.

The remainder of this paper is structured as follows. First, a general overview to the omnidirectional vision system is provided in Section 2; Section 3 describes the design of the localization model, which relies on the adaption of the epipolar constraint to the omnidirectional geometry of the vision system [31]; the main contributions, regarding the design of the probability distribution of feature points' existence, are presented in Section 4; Section 5 comprises the experiments conducted with publicly available real datasets, which assess the validity and the reliability of the approach, in contrast to well-recognized methods; Section 6 summarizes the main outcomes extracted from the results. Section 7 outlines the fundamental conclusions derived from this work.

2. Vision System

The equipment used in this work consists of a mobile robot, which is equipped with a laser range finder and a catadioptric vision system [31], as shown in Figure 1. The vision system is constituted by a hyperbolic mirror with a CCD (Charge-Coupled Device) camera jointly assembled. This represents a complete omnidirectional vision system, namely an omnidirectional camera.

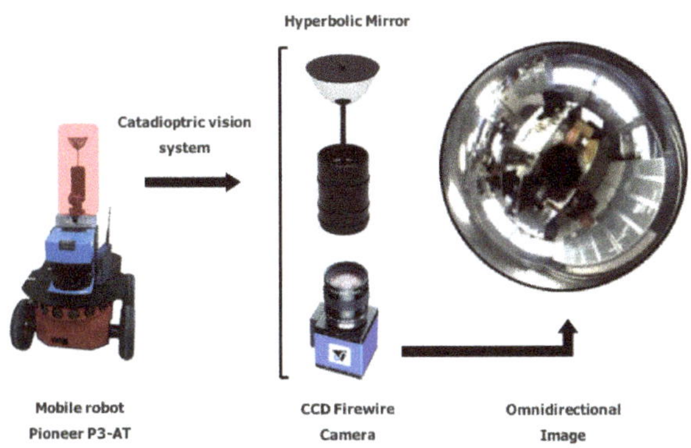

Figure 1. Real equipment constituted by a Pioneer P3-AT robot equipped with an internal odometer, a SICK-LMS200 laser range finder, and a catadioptric vision system, namely an omnidirectional vision system. This vision system is composed of a CCD (Charge-Coupled Device) FireWire DMK21BF04 camera, assembled with a hyperbolic Eizo h Wide 70 Mirror.

According to [34], the projection model of our omnidirectional camera can be posed in terms of a central sphere projection system. Figure 2 reproduces the omnidirectional image generation in terms of such a projection, where a 3D scene point, $Q(x_Q, y_Q, z_Q) \equiv Q$, is projected onto the mirror surface as P, onto the unitary sphere as S, and onto the camera plane as $p(u,v) \equiv p$. The focals of the hyperbolic mirror are F and F', F being coincident with the optical center of the CCD camera, whose optical axis lies in the Z-axis. Notice that the central sphere unifies the notation of the projection vectors for normalization purposes, according to the calibration of the omnidirectional camera [35], regardless of the characteristics of the mirror and its non-linearities. Thus the mapping of a 3D point onto the image plane can be analytically expressed as follows [34]:

$$\rho \begin{bmatrix} p \\ a_0 + a_2||p||^2 + \ldots + a_n||p||^n \end{bmatrix} = CQ \tag{1}$$

where $C \in \mathbb{R}^{3\times4}$ is the projection matrix, denoted as $C = [R|T]$, with R a rotation matrix $\in \mathbb{R}^{3\times3}$ and with $T = [t_x, t_y, t_z]$ a translation $\in \mathbb{R}^3$, between the camera and the global reference system. A Taylor expansion is introduced in order to model the effect of the mirror, whose coefficients $(a_0, a_2, \cdots, a_n)$ are experimentally estimated by a calibration toolbox [35]. Note that the monocular characteristic of this system leads us to include a scale factor, $\rho = |T|$.

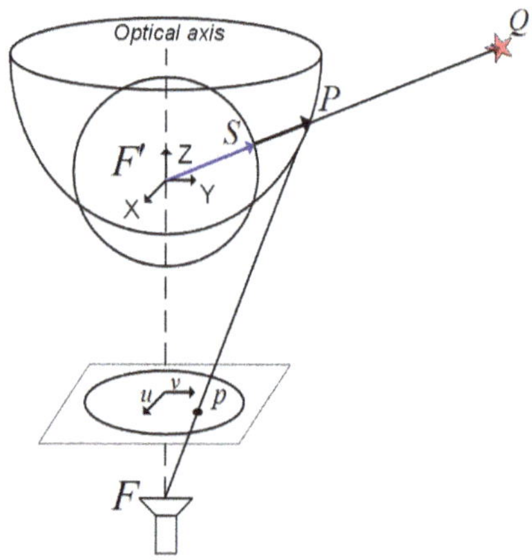

Figure 2. Omnidirectional camera 3D projection model from an XYZ-view.

3. Omnidirectional Visual Localization

The design of the localization model is constrained by the epipolar geometry [34] of two poses of the robot, from which two associated images are acquired. As in our former work [31], a conversion of the standard epipolar constraint is needed in this work in order to adapt it to the geometry of the omnidirectional system.

Solving the epipolar constraint implies estimating the essential matrix, $E_{3\times3}$ [36], in order to extract the motion relation between two poses of the robot. To that aim, a set of matched points between the images acquired from these two poses, has to be introduced into the epipolar constraint:

$$\tilde{x}'^T E \tilde{x} = 0 \tag{2}$$

with $\tilde{x}^T = (x_0, y_0, z_0)$ and $\tilde{x}'^T = (x_1, y_1, z_1)$ being the normalized matched points expressed in the 3D global reference system, using the calibration of the vision system, which has been previously estimated [35].

The essential matrix E can be decomposed into a rotation R and a translation T, as denoted in Section 2. Assuming that the motion of our mobile robot is restricted to a 2D motion plane $\in XY$, E can be expressed by means of the skew symmetric $[T]_x$ and the mentioned rotation R:

$$\begin{aligned} E = [T]_x R &= \begin{bmatrix} 0 & 0 & \sin(\phi) \\ 0 & 0 & -\cos(\phi) \\ -\sin(\phi) & \cos(\phi) & 0 \end{bmatrix} \begin{bmatrix} \cos(\beta) & -\sin(\beta) & 0 \\ \sin(\beta) & \cos(\beta) & 0 \\ 0 & 0 & 1 \end{bmatrix} \\ &= \begin{bmatrix} 0 & 0 & \sin(\phi) \\ 0 & 0 & -\cos(\phi) \\ \sin(\beta-\phi) & \cos(\beta-\phi) & 0 \end{bmatrix} = \begin{bmatrix} 0 & 0 & e_1 \\ 0 & 0 & e_2 \\ e_3 & e_4 & 0 \end{bmatrix} \end{aligned} \tag{3}$$

where $\vec{e}_i$=$[e_1, e_2, e_3, e_4]$ stores the elements in E. Therefore, the motion relation can be recovered as a pair of rotation and translation angles (β, ϕ), between two poses of the robot, up to a scale factor ρ.

3.1. Angular Motion Recovery

More specifically, the retrieval of the rotation and translation angles, is expressed as the following linear system, which results from including Equation (3) into the epipolar constraint, expressed in Equation (2):

$$D_{\mu x4} \cdot \vec{e}_i = \begin{bmatrix} x_0 z_1 & y_0 z_1 & z_0 x_1 & z_0 y_1 \end{bmatrix} \begin{bmatrix} e_1 & e_2 & e_3 & e_4 \end{bmatrix}^T = \vec{0} \;\; \forall i \in [1, \ldots, N] \quad (4)$$

with μ being the total number of pairs of matching points, $\vec{x}^T = (x_0, y_0, z_0)$ and $\vec{x}'^T = (x_1, y_1, z_1)$. Consequently, for each μ-th pair of matching points, the μ-th row of Equation (4) constrains the angular motion by means of the elements in $\vec{e}_i$. It is worth noting that the system may be solved by using a minimum set of $\mu_{min} = 4$ matching pairs. Finally, the application of Singular Value Decomposition (SVD) to Equation (4) allows us to compute the angular pair (β, ϕ). There is a quaternion of possible solutions that is eventually disambiguated by finding the positive ray's intersection in front of the camera.

$$\phi = a\tan\frac{-e_1}{e_2}; \quad \beta = a\tan\frac{e_3}{e_4} + a\tan\frac{-e_1}{e_2}. \quad (5)$$

This angular motion, which finally determines the visual localization of the robot, is graphically depicted in Figure 3, where a univocal image-to-pose equivalence is presented. The same equivalence is expressed in the 3D robot reference system, in Figure 3a, and in the image reference system, in Figure 3b. A 3D point, $Q(x_Q, y_Q, z_Q)$, is represented in the 3D robot reference system and its projections, $p_1(u, v)$ and $p_2(u, v)$, in the corresponding 2D image reference systems, captured from $\vec{x}_1(x_1, y_1, \theta_1)$ and $\vec{x}_2(x_2, y_2, \theta_2)$, which are $\vec{x}_1$ and $\vec{x}_2$, the 2D traversed poses, with orientation θ. The transformation between poses $\vec{x}_1$ and $\vec{x}_2$ is determined by the rotation $R \equiv R(\beta)$, the translation $T \equiv T(\phi)$, and the scale factor ρ.

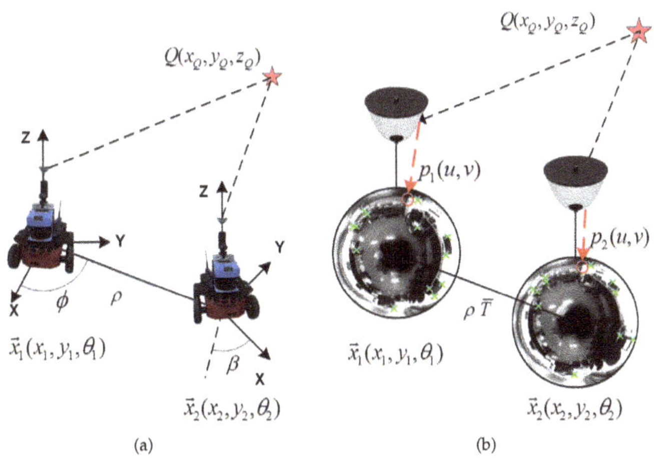

Figure 3. Omnidirectional visual localization between poses $\vec{x}_1$ and $\vec{x}_2$. (**a**) a 3D point $Q(x_Q, y_Q, z_Q)$ is observed from the robot reference system; (**b**) additionally, the projections of Q, $p_1(u, v)$, and $p_2(u, v)$ are presented onto the camera reference system.

3.2. Scale Estimation

The lack of scale can be disambiguated by introducing certain visual patterns or specific objects with well-known 3D dimensions [11]. Since the 2D image projection for such objects or patterns can also be determined over different images, this leads to a triangulation problem [34] sustained by the epipolar constraint, where the 3D real dimensions and the 2D projections are known, and thus the variable to estimate is the scale factor, ρ. However, if such patterns or objects are not seen in the current frame for a long period of time, the estimation might be inaccurate. For this reason, we opted for using the scale estimate provided by the odometer, ρ_{odo}, which we input into the filter-based core system. Thus ρ_{odo} is implicitly present into the prior measure of the filter, represented as the odometer's control input, u_t, as detailed in Section 4.2. That permits obtaining updated estimates of the scale at every t and thus at the baseline between poses.

The odometer provides readings $(\rho_{odo}, \phi_{odo}, \beta_{odo})$, which permit obtaining a relationship between two consecutive 2D poses traversed by the robot, expressed in the odometer's notation as $\vec{x}_{odo_1}(x_1, y_1, \theta_1)$ and $\vec{x}_{odo_2}(x_2, y_2, \theta_2)$, being θ the orientation. As it may be observed in Figure 4, the relation between poses can be stated as follows:

$$\begin{bmatrix} x_2 \\ y_2 \\ \theta_2 \end{bmatrix} = \begin{bmatrix} x_1 \\ y_1 \\ \theta_1 \end{bmatrix} + \begin{bmatrix} \cos(\phi_{odo}) & 0 & 0 \\ \sin(\phi_{odo}) & 0 & 0 \\ 0 & 1 & 1 \end{bmatrix} \begin{bmatrix} \rho_{odo} \\ \beta_{odo} \\ \phi_{odo} \end{bmatrix}. \tag{6}$$

The error model for the odometer is parametrized by a probabilistic motion model [37], which adds zero-mean Gaussian noise, $\mathcal{N}(0, \sigma^2)$:

$$\hat{\rho}_{odo} = \rho_{odo} + \epsilon_\rho \quad \rightarrow \quad \epsilon_\rho \equiv \mathcal{N}(0, \sigma_\rho^2) \tag{7}$$

$$\hat{\phi}_{odo} = \phi_{odo} + \epsilon_\phi \quad \rightarrow \quad \epsilon_\phi \equiv \mathcal{N}(0, \sigma_\phi^2) \tag{8}$$

$$\hat{\beta}_{odo} = \beta_{odo} + \epsilon_\beta \quad \rightarrow \quad \epsilon_\beta \equiv \mathcal{N}(0, \sigma_\beta^2). \tag{9}$$

The standard deviations required to complete the parametrization are computed by using the empiric parameters provided by the manufacturer $(\alpha_1, \alpha_2, \alpha_3, \alpha_4)$, as follows:

$$\sigma_\rho = \alpha_3 \rho_{odo} + \alpha_4(|\phi_{odo}| + |\beta_{odo}|) \tag{10}$$

$$\sigma_\phi = \alpha_1 |\phi_{odo}| + \alpha_2 \rho_{odo} \tag{11}$$

$$\sigma_\beta = \alpha_1 |\beta_{odo}| + \alpha_2 \rho_{odo}. \tag{12}$$

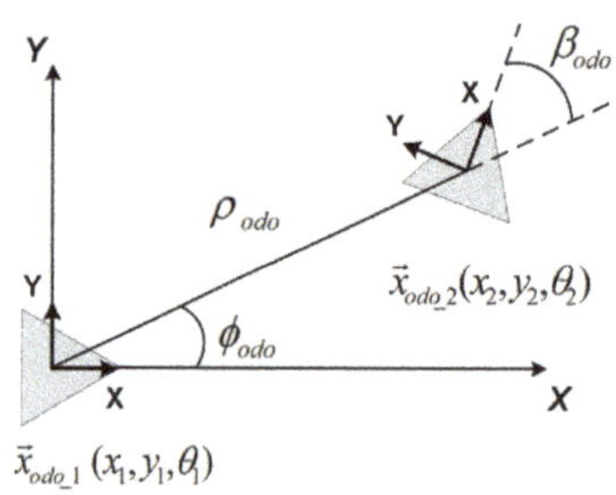

Figure 4. Odometer model.

3.3. *Notation Definitions*

In this subsection, the notation of the localization method is presented. We define a state vector, $s(t)$, that comprises the different variables implied in the estimation. This state vector stores a set of consecutive poses of the robot which are estimated by the localization method. These poses are the result of discretizing the trajectory traversed by the robot. Assuming that an omnidirectional image is captured from a certain 2D pose $\vec{x}_i(x, y, \theta)$, θ being the robot's orientation, such an image can be denoted as a view, encoded as a set of SURF feature points [23] that are extracted from it. The pose of each view is included in the state vector as $\vec{x}_n = (x_n, y_n, \theta_n)^T, \forall n \in [1, \cdots, N]$. The current pose of the robot at time t is expressed as $\vec{x}_t = (x_t, y_t, \theta_t)^T$. Thus the definition of the state vector includes $\vec{x}_t$ and $\vec{x}_n$, with the following 2D structure:

$$s(t) = \begin{bmatrix} \vec{x}_t & \vec{x}_1 & \cdots & \vec{x}_n & \cdots & \vec{x}_N \end{bmatrix}^T. \tag{13}$$

Therefore, the state vector comprises a trajectory with a total number of N views. These variables represent a dual encoding model of the environment. They are expressed in 2D, due to the fact that we work with a robot that is assumed to move in a 2D plane. However, given a specific calibration [35] and the estimation of the scale factor, every 2D point detected inside the views can be back-projected to the 3D global reference system by means of Equation (1). Therefore, re-estimating a view, implies that the entire 3D information of the map is re-estimated at once. This aspect makes the approach more efficient than traditional 3D landmark models [11,38], which need the 3D re-estimation of every single landmark at every t.

Considering this framework within the field of mobile robotics, it is also worth defining a formal observation model, which permits estimating the localization. The procedure has been detailed in the previous subsection. However, it is expressed here in accordance with the state vector's variables, $s(t)$. The motion relation between two poses of the robot can be retrieved in an angular format, as in Equation (5). Transferring the angular localization relation (β, ϕ) into the robot's reference nomenclature, the following observation measurement can be established:

$$z_{t,n} = \begin{bmatrix} \phi \\ \beta \end{bmatrix} = \begin{bmatrix} \arctan\left(\frac{y_n - y_t}{x_n - x_t}\right) - \theta_t \\ \theta_n - \theta_t \end{bmatrix} \tag{14}$$

where the notation corresponds to the one expressed in Equation (13), and thus $z_{t,n}$ represents the angular motion between the current pose of the robot, $\vec{x}_t$, and a view in the state vector, $\vec{x}_n$.

4. Visual Information Fusion

Once the localization model has been described, this section introduces the implementation of the visual information fusion into the system and the rest of the details associated with the main contribution.

The main goal is to obtain a model that accounts for the visual changes in the environment and encodes the probability of existence of visual feature points in the 3D global reference system. Figure 5 illustrates an example of this idea, where the 3D environment is modeled with a specific probability of feature existence. This approach will be extended in order to predict spatial areas from which visual information is more likely to be detected. Such areas can be also projected onto the next image frame so as to map pixel areas where matching is more likely to appear, rather than in other parts of the image, in terms of probability.

A first sketch might consist in recording statistics of feature points that are tracked along the navigation of the robot through the environment. This would lead us to infer a probability distribution for the existence of 3D points along the trajectory of the vehicle, in terms of percentage of occurrence. Nonetheless, a more precise formulation can be introduced as follows, by using Bayesian inference.

A general overview of the entire process can be observed in Figure 6. The main contributions, which try to obtain a probability-oriented feature matching, are present in the following blocks: the 3D back-projection, the GP computation to produce the 3D probability distribution, the probability sampling, and the 2D image projection over the next predicted pose.

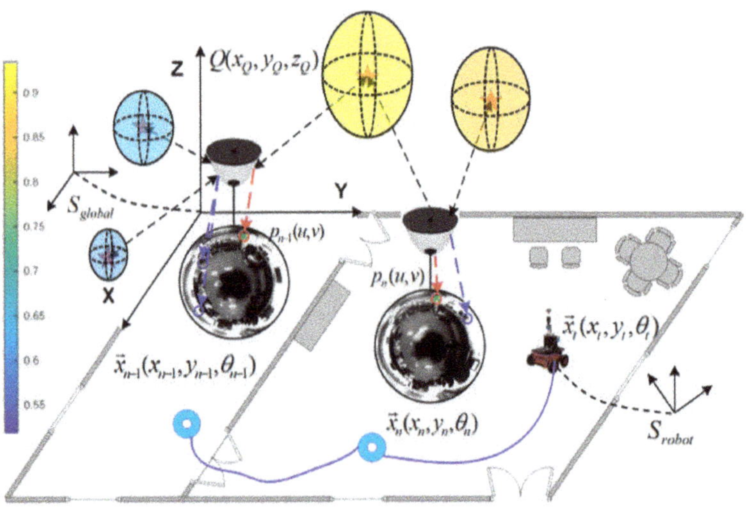

Figure 5. Robot navigation example in an office-like scenario along three poses: $\vec{x}_{n-1}$, $\vec{x}_n$, and $\vec{x}_t$. The 3D probability distribution of feature points' existence permits associating visual feature points with a specific probability, indicated with colored spheres, whose probability values are encoded according to the left-side colorbar. Projections of a 3D point $Q(x_Q, y_Q, z_Q)$, $p_{n-1}(u, v)$ and $p_n(u, v)$, are also indicated. The 3D global reference system is denoted as S_{global}, and the 3D robot reference system as S_{robot}.

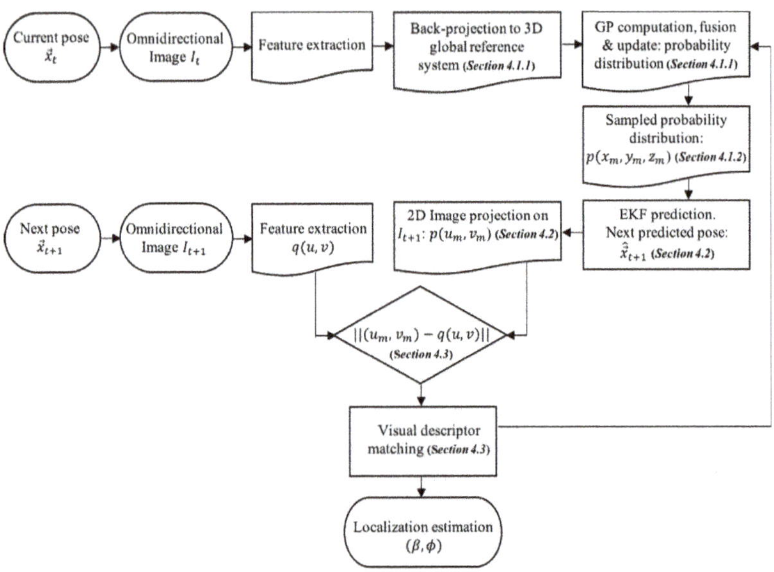

Figure 6. Block diagram of the presented approach.

4.1. 3D Probability Distribution of Feature Existence: GP Computation and 3D Probability Sampling

4.1.1. GP Computation

In this work, we use the same Bayesian regression technique applied in [31], formulated as a Gaussian Process (GP) [32]. However, in this approach, we pursue the probability distribution of feature points' existence in the 3D global reference system rather than in the 2D image frame. GP is able to produce reliable regression results without the need of common associations between inputs and outputs, in comparison to traditional inference techniques [33]. Its general notation is the following:

$$f(x) \sim \mathcal{GP}[m(x), k(x, x')] \tag{15}$$

where the GP function is expressed as $f(x)$, with mean $m(x)$ and covariance $k(x, x')$. The training and test input points, x and x', respectively, represent 3D points at which the value of the function is tested in terms of probability of existence.

Since we intend to obtain a probability distribution in 3D, the nomenclature for the output function has to be adapted to the formulation of our approach, so $f(x) \equiv f[X(x,y,z)]$, $X(x,y,z)$ being a general 3D point in the global reference system, such that $X(x,y,z) \equiv X_{global}$. Thus $f(\cdot)$ evaluates the probability of existence of a feature point over a 3D point in the space. Then $f(\cdot) \in [0,1]$.

The input for the GP is represented by the feature matching between two images associated with two poses traversed by the robot, up to time t. As mentioned above, the Bayesian inference requires data in the 3D global reference system, so the 2D feature matching has to be back-projected to the 3D global reference system. As initially presented in Equation (1), this transformation can be achieved thanks to the scale factor estimation, and the specific camera calibration [35], which establishes the conversion between the 2D image frame and the 3D global reference system.

There is a final step before obtaining the exact 3D global reference system representation. The previous back-projection is expressed in the current 3D robot reference system, therefore we apply the following expression in order to formally convert to the proper 3D global reference system.

$$X_{global} = \rho T + R X_{robot} \tag{16}$$

where a 3D point expressed in the current 3D robot reference system, as X_{robot}, is transformed into the 3D global reference system, expressed as X_{global}, by means of the rotation, R, translation T, and scale factor ρ, presented in Section 3, according to the localization measures.

4.1.2. 3D Probability Sampling

At this point, we have obtained a 3D probability distribution of feature existence up to time t, denoted as $f[X(x,y,z)]$. The next step seeks the reduction of computational resources. To that purpose, a 3D sampling over $f[X(x,y,z)]$ has to be devised in order not to compromise the computational resources of the system. In consequence, a normalization of the 3D probability distribution is carried out. Then, the sampling discretization corresponds to a 3D square grid, as follows:

$$P_{acc} = \iiint_V f[X(x,y,z)]\,dx\,dy\,dz \tag{17}$$

$$f_{norm}[X(x,y,z)] = \frac{f[X(x,y,z)]}{P_{acc}} \tag{18}$$

$$p_{norm} = \sum_{x_m}^{M}\sum_{y_m}^{M}\sum_{z_m}^{M} f_{norm}(x_m, y_m, z_m) = 1 \qquad m \in [1, M] \tag{19}$$

$$p(x_m, y_m, z_m) \equiv f_{norm}(x_m, y_m, z_m) \tag{20}$$

where P_{acc} is the total accumulated probability, which is computed for normalization purposes, so as to obtain $f_{norm}(\cdot)$. Then the definition of the 3D square grid, with M^3-elements, allows us to obtain a sampled normalized probability distribution, $p(x_m, y_m, z_m)$. Additionally, Figure 7 presents a real example of a 3D sampled probability distribution of feature points' existence. Figure 7a shows the complete sampled distribution, $p(x_m, y_m, z_m)$, whereas Figure 7b shows the evaluation of such a distribution at the last feature points observed, as test points, after being back-projected from the 2D image frame to 3D. Notice that, for better illustration, high probability values are represented with a higher radius. These data will be fused into the distribution as the next input data that the GP uses to update the current distribution. The axes represent the 3D global position within the sampling grid (x_m, y_m, z_m) and the 3D probability of feature existence at such positions. $p(x_m, y_m, z_m)$ is expressed by a gradient of color.

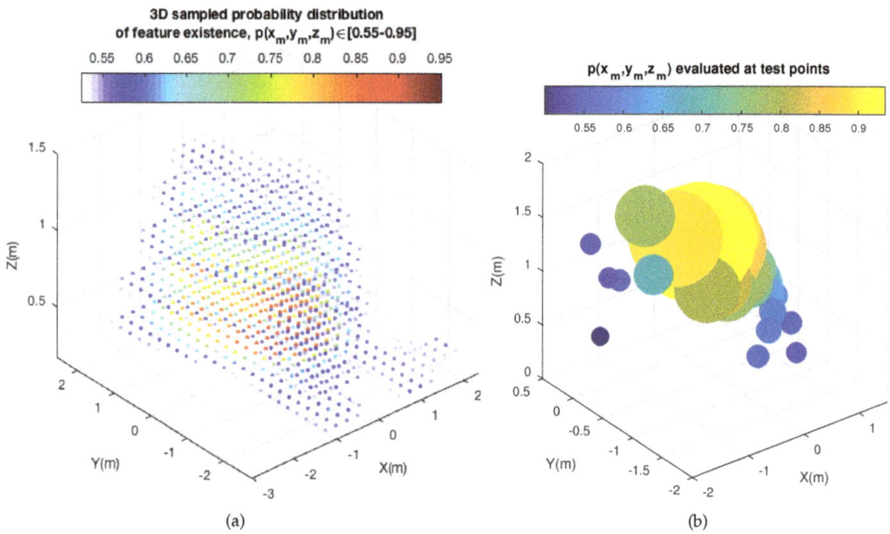

Figure 7. 3D sampled probability distribution of feature existence. (**a**) Complete 3D sampled probability distribution, $p(x_m, y_m, z_m)$; (**b**) $p(x_m, y_m, z_m)$ evaluated at the last feature points observed (test points).

4.2. Motion Prediction and 2D Image Projection

Since the main goal is to predict relevant areas in the 2D image frame, where feature matching is more likely to appear, the resulting 3D probability distribution obtained by GP up to time t, $p(x_m, y_m, z_m)$ after sampling, has to be projected onto the next 2D image frame in time $t+1$, associated with the next pose of the robot. Therefore, the configuration of a prior prediction stage is essential. In this sense, we take the most of an EKF (Extended Kalman Filter)-based filter formulation, similarly to [31,39].

After customizing this configuration properly, we are able to predict the next pose $\hat{\vec{x}}_{t+1}$. As detailed in Section 3.2, the scale factor is disambiguated as the estimate provided by the odometer, ρ_{odo}. It is worth noting that this value is present in the odometer's control input, $u_t \equiv f(\rho_{odo}, \phi_{odo}, \beta_{odo})$, which represents the prior input for the EKF-based system. This can be observed in the notation of the EKF-based prediction stage, listed in Table 1.

Thereby we are able to decompose [34] the predicted motion from pose $\vec{x}_t$ to $\hat{\vec{x}}_{t+1}$, in a rotation $\hat{R}$ and a translation $\hat{T}$.

$$\hat{R} \sim N(\hat{\beta}, \sigma_\beta); \quad \hat{T} \sim N(\hat{\phi}, \sigma_\phi) \tag{21}$$

where σ_β and σ_ϕ are the standard deviations that characterize the proposed localization method described in Section 3. Figure 8 synthesizes the motion prediction process, where $z_{t,n}$ represents the observation measurement of a certain view in the environment, $\vec{x}_n$, computed from the current pose, $\vec{x}_t$, as described in Equations (13) and (14).

Table 1. EKF-based Filter: Prediction stage.

Filter-Based SLAM Stages		
Stage	**Expression**	**Terms**
Prediction	$\hat{\vec{x}}_{t+1\|t} = \mathrm{f}_t(\hat{\vec{x}}_{t\|t}, u_t)$ $\hat{z}_{t+1\|t} = \mathrm{h}_t(\hat{\vec{x}}_{t+1\|t}, \vec{x}_i)$ $P_{t+1\|t} = \frac{\partial f_{t\|t}}{\partial x} P_{t\|t} \frac{\partial f_{t\|t}}{\partial x}^T + W_t$	f_t: relates the odometer's control input u_t and the current state u_t: odometer's control input, initial prior h_t: relates the observation $z_{t,n}$ and the current state P_t: uncertainty covariance W_t: input noise covariance

Finally, the 3D probability distribution $p(x_m, y_m, z_m)$ is projected onto the pixels of the 2D image frame associated with the next pose of the robot, denoted as $p(u_m, v_m)$, by applying Equation (1). Furthermore, a specific probability range with custom values can be defined, $[p_{min}–p_{max}]$, in order to only select points from the distribution with probabilities within that range. The immediate outcome is the generation of probability areas in the image frame, where feature matching is more likely to appear. Figure 9 presents a real example after applying the overall method to the image acquired from the current robot pose. The visualized probability range is $p \in [0.7–1]$. Figure 9a represents the projection of $p(x_m, y_m, z_m)$ onto the image plane, as $p(u_m, v_m)$ in 2D, and Figure 9b in 3D. Figure 9c presents the same 2D projection after transforming the axes in order to generate a histogram representation. This last representation is useful for data processing tasks. Finally, Figure 9d reveals that polar space encoding might produce a better modeling of the distribution rather than Cartesian coordinates. This may be implicitly induced from the elliptical constitution of the epipolar curves in an omnidirectional vision system.

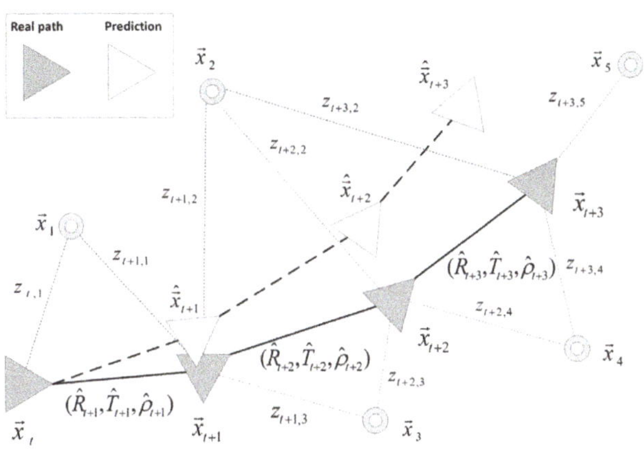

Figure 8. Graph diagram of a robot trajectory. Real path poses, $\vec{x}_t$, and predicted poses, $\hat{\vec{x}}_t$, at each t are indicated, following the notation described in Equations (13) and (14). Observation measurements, $z_{t,n}$, and views in the environment, $\vec{x}_n$, are also depicted.

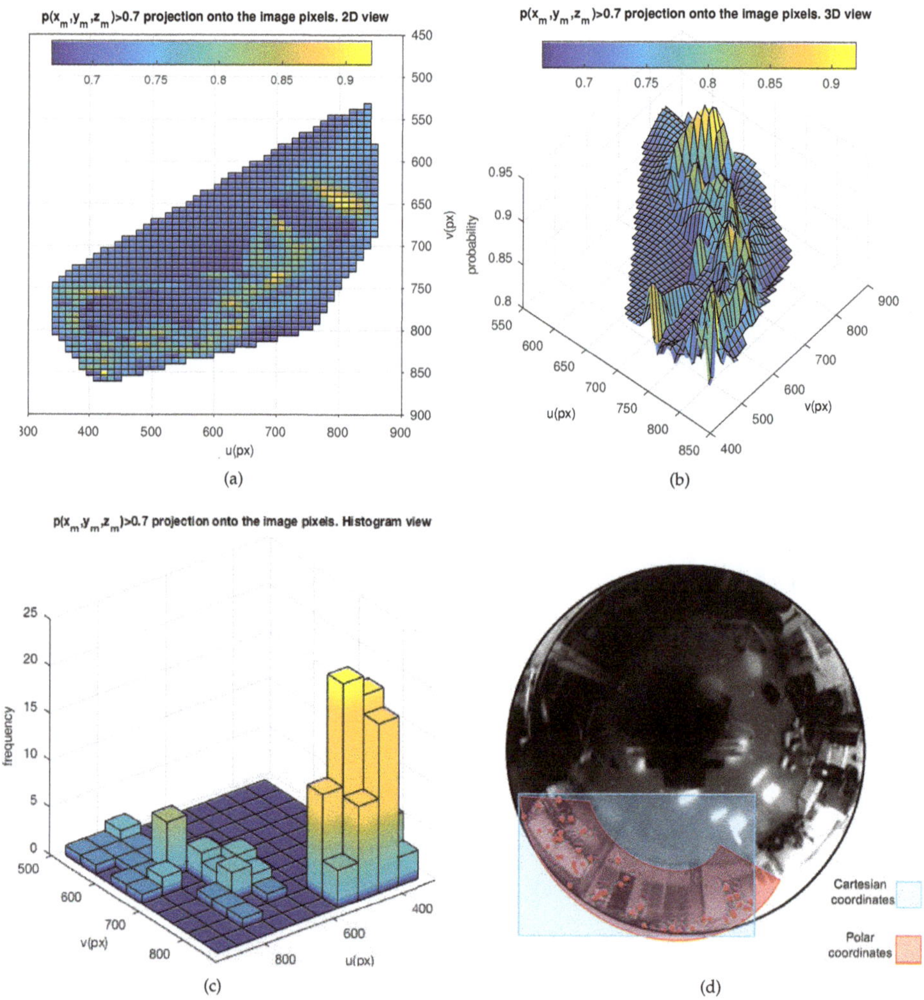

Figure 9. Projection of the 3D sampled probability distribution of feature existence, $p(x_m, y_m, z_m) \in [0.7–1]$, onto the image pixel axes, in t. (**a**) 2D representation, $p(u_m, v_m)$. (**b**) 3D representation with Z-axis expressing probability, $p(x_m, y_m, z_m)$. (**c**) 2D histogram representation. (**d**) Euclidean versus polar coordinates.

4.3. Probability-Oriented Feature Matching

The final stage is intended to perform feature matching. Using the method presented in the previous subsection, probability areas can be detected on the image. Considering this, a straightforward design would entail using a feature detector only on the desired areas, and thus filtering by high probabilities. This would avoid processing the entire image. Nevertheless, it would lead to errors, under certain circumstances, especially when the robot discovers new scenes in the environment. If we assume that there may be substantial changes in the visual appearance as the robot goes through new areas, then it will be necessary to let these new areas be processed in order to detect new features. This is the main reason why we keep detecting features all over the images, so that we allow the GP to update its output when new visual content is discovered. Otherwise the visual content of these new scenes would never be fused into the probability of feature existence, computed by GP.

Taking these last considerations into account, we measure the proximity between the pixels, (u_m, v_m), associated with the sampled projected probability distribution on the 2D image frame, $p(u_m, v_m)$, and all the feature points detected in the next image, $q(u, v)$. Such proximity is computed by means of the Mahalanobis distance [40], $||(u_m, v_m) - q(u, v)||$. Those feature points in $q(u, v)$ are accepted as matching candidates when their pixel distance to (u_m, v_m) meets the confidence threshold established by the chi distribution, $\chi(dof)$, evaluated at the degrees of freedom that represent the dimensionality of the involved variables. Since the image frame is defined at a pixel level, the degrees of freedom are $dof = dim(u, v) = 2$.

$$||(u_m, v_m) - q(u, v)|| \leq \chi[dim(u, v)]. \quad (22)$$

The feature points in $q(u, v)$ that meet Equation (22), namely matching candidates, are then matched through a standard matching process by visual descriptor comparison. In the end, these matching points are the final data which will be used in the localization system in order to obtain an estimation of the current pose of the robot, as previously introduced in Section 3. Finally, the same real example presented in Figure 9 is further detailed in Figure 10, over the corresponding real omnidirectional images, between poses at t and $t + 1$. Here, the proposed approach is compared with a standard matching block [23]. It can be observed how the standard matching (blue circles) produces a significant amount of false positives. Our proposal produces a set of valid matching points (green crosses) under the constraint of the probability area represented by its projection on the image (red dots).

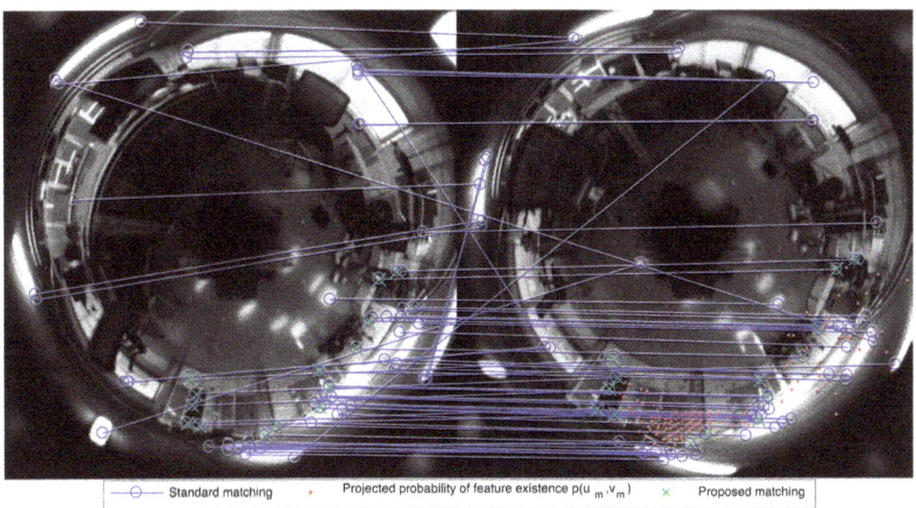

Figure 10. Matching results between images acquired from poses at t and $t + 1$. Standard matching results are indicated with blue circles, and those obtained with the proposed approach are indicated with green crosses. The pixels associated with the projected probability of feature existence $p(u_m, v_m)$ are indicated with red dots.

Regardless of the smaller set of matches obtained, we can rely on the consistency and robustness of these points, since they are highly probable according to the current navigation of the robot. Even under a hypothetical situation where no match is obtained, we can rely on the filter-based estimation until new matches are detected in a subsequent frame. Moreover, as already mentioned, there is no restriction for other new feature points to be matched. A modulation of the selected probability range $p \in [p_{min}–p_{max}]$ is achieved through an adaptive scheme, which is referred to as the current uncertainty of the entire probability distribution $p(x_m, y_m, z_m)$. Similarly to [31], we assess the drifts on uncertainty by evaluating the information gain, using the information-based Kullback–Leibler divergence (KL) [41], with the aid of the standard entropy metric [42]. In this manner, when there are considerable changes in the visual appearance of the scene, the probability distribution produced by GP will change accordingly. The KL measure will encode such change, which will lead the system to modulate the desired range $p \in [p_{min}–p_{max}]$, in order to adapt the final matching to the current uncertainty conditions of the system. Nonetheless, and in contrast to [31], this approach encodes fluctuations in the probability expressed in a 3D global reference system, rather than in the particular 2D image frame of each pose. Furthermore, its application is also different, since we use it to modulate the custom probability ranges, rather than using it as a weighting coefficient for the matching.

5. Results

This section presents a set of experiments conducted with publicly available datasets [43,44]. They assess and compare the performance of the matching and the visual localization, in comparison with well-acknowledged methods [6,23,45]. A public benchmark toolkit [44,46] has also been used to produce such comparisons. Table 2 comprises the characteristics of the datasets.

Table 2. Dataset characteristics.

Real Datasets			
Dataset	**Images**	**Distance**	**Publicly Available**
Dataset 1: *Innova trajectory*	1450	174 m	[43]
Dataset 2: *Bovisa 10-04-2008*	57,733	1310 m	[44]

The computation specifications of the real equipment presented in Figure 1 are: CPU 2 × 1.7 GHz; RAM 2 Gb. The acquisition of omnidirectional images (1280 × 980 px) demands the system to compute estimates at 1.5 Hz. The SICK-LMS200 laser data are utilized to obtain a ground truth estimation for comparison purposes. Finally, the robot is run by the operating platform ROS [47].

5.1. Matching Results

The first set of experiments was conducted with the *Innova trajectory* dataset, in order to evaluate the capability of the approach to produce robust probability-oriented matching results. Each subset within these results' series was configured with a 300-times execution setup, so as to obtain consistent average results. The first set of results evaluates the matching performance between poses of the robot, from which omnidirectional images were captured.

This performance is evaluated through the number of feature matches, the accuracy, and the computation time.

5.1.1. Number of Feature Matches

Figure 11 presents matching results with different configurations. The X-axis represents the minimum range for the probability of feature existence, p_{min}. The distance between consecutive poses has been considered a variable parameter (d_1 to d_4), being $d_i = 0.25i$ meters, and the influence of this distance was tested. The left-side Y-axis represents the number of features obtained by a standard matching technique [23] (grey), the proposed matching candidates (dark blue), and the final number of matches of the proposed approach (light blue). The right-side Y-axis (log) represents the size of $p(x_m, y_m, z_m) < p_{min}$. Additionally, the influence of using either Euclidean coordinates (left column) or polar coordinates (right column) was assessed.

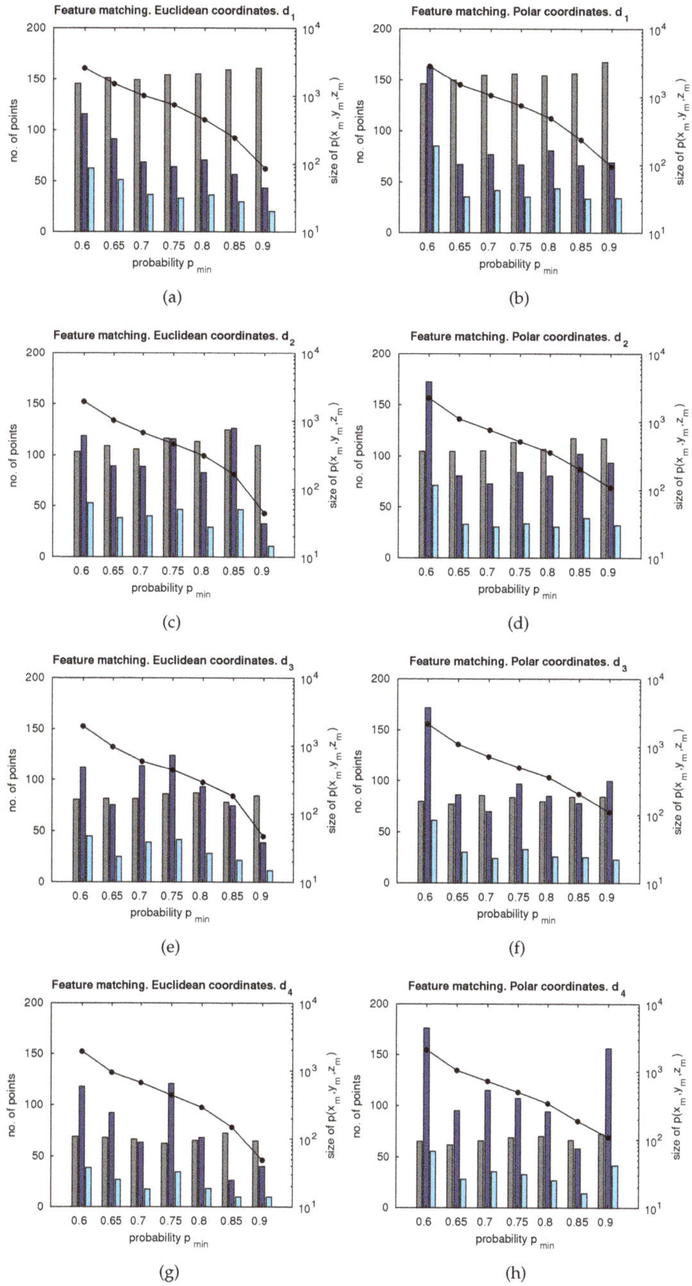

Figure 11. Left axes: number of matches versus p_{min}. Right axes: size of the probability distribution (*log*) versus p_{min}. -•- size[$p(x_m, y_m, z_m)$]. Euclidean coordinates and distance between capture points: (**a**) d_1; (**c**) d_2; (**e**) d_3; (**g**) d_4. Polar coordinates and distance between capture points: (**b**) d_1; (**d**) d_2; (**f**) d_3; (**h**) d_4. Legend: ■ Standard matching; ■ proposed matching candidates; ■ proposed final matching.

Figure 11 evidences that higher distances between images produce a lower number of matching points, considering both the standard matching and the proposed approach. Despite this fact, our proposal provides more matching candidates than the standard approach. Moreover, the drop in the final number of matching points is less accentuated in the proposal, thus ensuring a minimum of matching points, even when images are captured from distant poses. Polar coordinates produce more matches only when p_{min} is high. In other words, when the size of $p(x_m, y_m, z_m)$ is low, the probability areas on the image, where matches may be found, are reduce. In any other case, Euclidean coordinates are more suitable due to their good balance between computational cost and the amount of matching data. According to these results, only Euclidean coordinates and extreme distances, d_1 and d_4, are considered.

5.1.2. Accuracy

The accuracy of the approach is compared with a standard matching technique [23] in Figure 12. Figure 12a,b show the percentage of false positive matches obtained with the standard matching (grey) and the proposed matching (dark blue), respectively. In the same manner, Figure 12c,d compare the resulting localization error (mean of β and ϕ), according to Equation (5) for the standard matching (grey) and the proposed matching (dark blue). All these results show correlated errors due to the fact that the same percentage of false positives is still present in the localization computation.

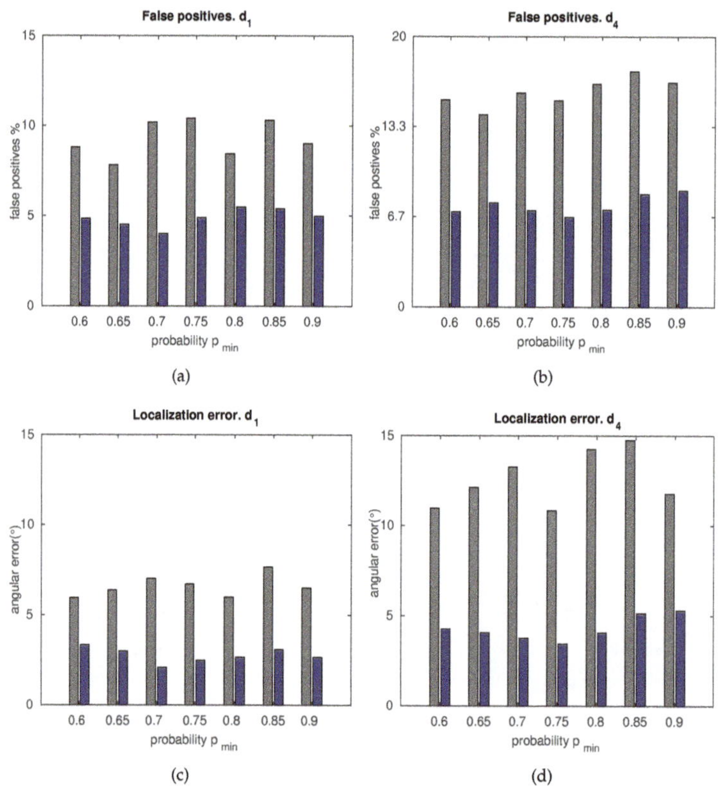

Figure 12. Top row: percentage of false positives. (**a**) Distance d_1; (**b**) distance d_4. Bottom row: localization error (in β and ϕ) versus p_{min}. (**c**) Distance d_1; (**d**) distance d_4. Legend: ■ standard matching; ■ proposed matching.

Besides this, the error decreases with p_{min}, up to an intermediate value, after which it rises slightly. This is due to the fact that high values of p_{min} may restrict the probability of feature existence on the image, to a set of few areas which may be closely arranged. Hence, this may lead the system to focus only on narrow areas of the image, dismissing newer visual information. Although this effect was considered in Section 4, and in the modulation of $p \in [p_{min}–p_{max}]$ by the KL divergence, there is still a subtle influence that can be observed in Figure 12. As a result, it is worth configuring the system with intermediate values such as $p \in [0.65–0.75]$, and then modulating its limits within that range, by means of the evaluation of the current uncertainty through the KL divergence.

5.1.3. Computation Time

Figure 13 compares the computation time for the standard matching and the proposed matching, versus p_{min}, and distances d_1, d_4. The total computation time has been divided into different parts, as indicated in the legend, in order to determine the different contributions:

(a) feature matching;
(b) matching candidates;
(c) final localization estimation.

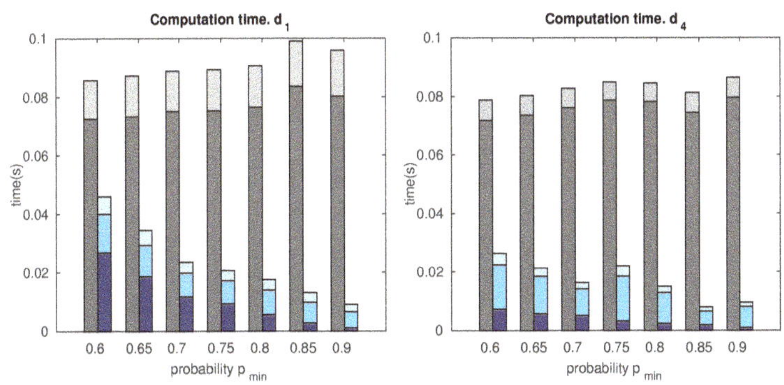

(a) (b)

Figure 13. Computation time versus p_{min}. (**a**) Distance d_1; (**b**) distance d_4. Legend: ■ standard matching: matching computation; ■ standard matching: localization computation; ■ proposed matching: candidates' computation; ■ proposed matching: matching computation; ■ proposed matching: localization computation.

These results are closely related to those presented in Figure 11, since the standard matching and the proposed matching, spend less time when the number of matches is lower. This permits selecting a suitable tradeoff solution between computation and accuracy, as p_{min} = 0.7. In addition to this, the proposed approach is shown to produce more efficient results than the standard matching technique.

5.2. Localization Results

This section deals with the localization estimation, produced by this approach. Figure 14 presents localization results for the *Innova trajectory* dataset. Figure 14a shows a bird's eye view of the estimated poses (plane *XY*, meters). Estimations obtained with the standard matching [22,23] and the proposed matching are compared. Figure 14b carries out the same comparison in terms of the root mean square error (RMSE), for both approaches. It can be observed that our proposal outperforms the localization method with standard matching and is capable of ensuring a bounded error at every t, in contrast to the standard method error, which increases substantially with t. These results validate the design of this approach and its performance, according to the previous subsection.

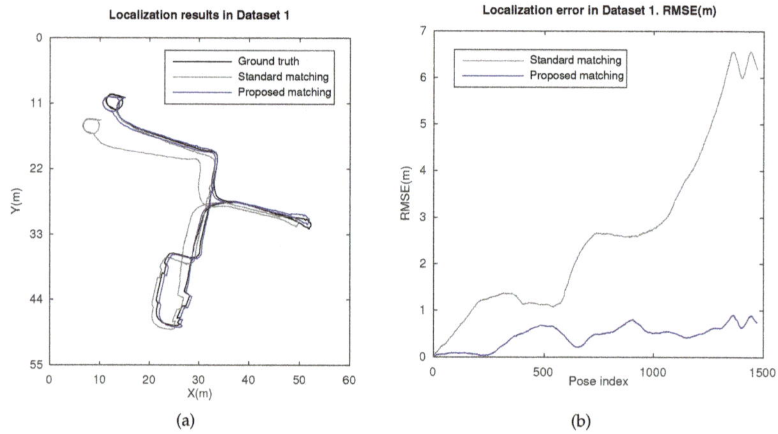

Figure 14. Localization results in Dataset 1, *Innova trajectory*. (**a**) Localization estimation obtained with ground truth (black), standard matching (grey), and the proposed matching (blue); (**b**) RMSE (m) for the localization estimation with standard matching (grey) and the proposed matching (blue).

In addition to the previous localization results, we compared our method with a widely recognized approach, the inverse EKF with depth parametrization [38]. To that end, we used the benchmark toolkit, publicly available in [44], and which also provides Dataset 2, *Bovisa 2008-10-04*. The results generated by the inverse EKF technique can be further consulted in [6,45]. Figure 15 presents localization results in a very challenging outdoor scenario, where dynamic conditions are highly relevant and challenging. Localization estimation results for the inverse EKF (red), the proposal (blue), and the zones where ground data (GPS) are available (black) are depicted. At first inspection, our approach demonstrates an improved reliability.

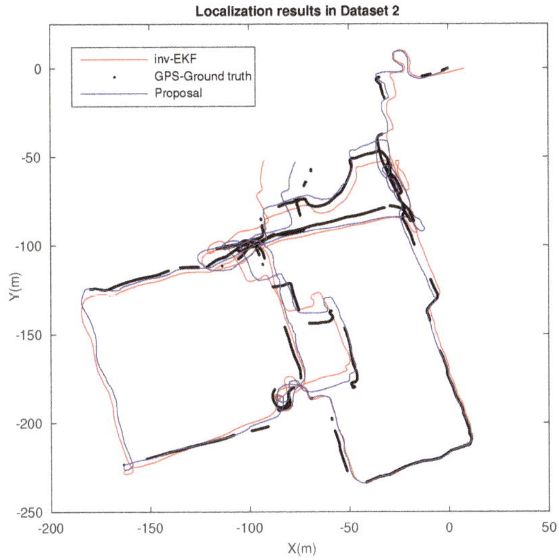

Figure 15. Localization results in Dataset 2, *Bovisa 10-04-2008*. Localization estimation obtained with ground truth (GPS) (black), inverse EKF (red), and the proposed matching (blue).

Moreover, further accuracy results are provided in Figure 16, where histograms of the error at each estimated pose in t, are presented. Two different setups have been considered to obtain these histograms. The inverse EKF does not disambiguate the lack of scale [6,45]. That is the reason why the final estimate only confers reliability on its topological form with respect to the ground truth, but not on its metric form. According to this, the benchmark toolkit provides a Maximum Likelihood Estimator (MLE) that can be applied in order to align the final estimated trajectory, and thus overcoming this issue. Hence, we can enable/disable this alignment method. Therefore, Figure 16a,c, represent the error histograms for the proposed approach, with alignment enabled and disabled, respectively. In the same manner, Figure 16b,d, represent the error histograms for the inverse EKF. It can be noted that the proposed approach improves the accuracy results in contrast to the inverse EKF, regardless of the operation of the alignment method, with average localization errors under 2 m.

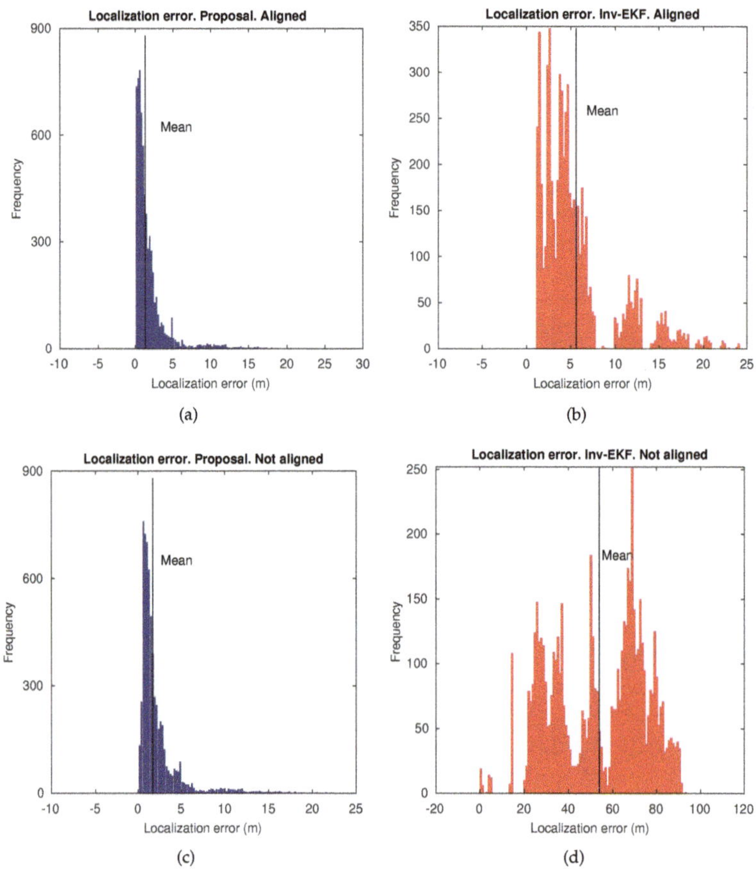

Figure 16. Localization error histograms in Dataset 2, *Bovisa 10-04-2008*. (**a**) Proposed approach with alignment enabled. (**b**) Inverse EKF approach with alignment enabled. (**c**) Proposed approach with alignment disabled. (**d**) Inverse EKF approach with alignment disabled.

6. Discussion

This section analyzes the main aspects regarding the implications extracted from the results. Initially, Figure 11 revealed the capability of the approach to provide probability-oriented matching points, which meet a specific probability distribution of feature existence, according to the Bayesian inference provided by GP. Despite the fact that increasing the distance between capture points implies a substantial decrease on the number of matches found, this proposal proves to keep a stable amount of valid matches, even at long distances, contrarily to a standard matching.

A similar deduction can be made by inspecting Figure 12. In this figure, false positives and localization errors are assessed, and a robust matching procedure is confirmed. Considering that the matching data are then processed into the localization system, it is evident that these results are closely correlated. This approach confirms a good and stable accuracy under the worst situation expected in a matching process, that is, under the presence of false positives. It is worth noting the effect of varying p_{min}. High values of p_{min} may lead the system to narrow on a reduced set of probability areas over the image. This fact may also imply that the visual information contained in new visual spaces discovered by the robot, is dismissed. However, we took this issue into account in order to modulate p_{min}. To that purpose, the localization system is set to work autonomously and computes the information divergence, KL, as a measure of the drifts of the uncertainty of the system. Despite this fact, a subtle influence of this effect is still present, and it can be noticed in the figures. Therefore, an optimal configuration can be selected with values within $p_{min} \in [0.65–0.75]$.

To complete the analysis, the computational costs required by this approach were evaluated. Figure 13 demonstrates that the proposal can be adequately tuned in order to confer valid and robust estimates, which permit working in real time. A relaxed tradeoff can be easily established between accuracy and computation resources. This approach proves to be a more efficient solution than a standard matching technique, at every studied aspect.

Finally, the outcomes of this work have been evaluated in terms of the localization performance. Figure 14 presents suitable results in a large indoor environment. A reliable and robust operation is ensured with stable error, in contrast to the performance offered by a standard matching. Furthermore, the results of a well-acknowledged method are presented in Figure 15 for comparison. Once again, the validity and robustness of our approach in terms of the accuracy of the final estimation, regardless of the challenging conditions in such environment, are reinforced.

Summarizing, the following achievements can be highlighted:

- Adaptive probability-oriented feature matching.
- Stable amount and accurate matches provided, in contrast to standard techniques.
- Efficient approach to work in real time.
- Robust final localization estimate in large and challenging scenarios.

7. Conclusions

This work has presented an information fusion approach for robust probability-oriented feature matching. It uses an omnidirectional vision system for visual localization purposes, and it is an improved extension of [31]. The approach is sustained by visual data fusion through Bayesian inference. The real system is constituted by a mobile robot, equipped with a monocular omnidirectional vision system, which is adequately adapted to work under the constraint of the epipolar geometry between images.

The main goal was to produce a robust approach to obtain relevant and reliable matching points for further localization tasks. To that end, several contributions were designed and implemented. Firstly, the 3D visual information associated to feature points is inferred by a Bayesian technique, represented by GP. Its output, at every t, provides a 3D probability distribution of feature existence in the global reference system. This probability is successively fused and updated while the robot navigates. Secondly, a normalization and sampling is produced in order to alleviate the computation requirements. After that, by taking most of the EKF prediction stage, the sampled probability can be projected in the next 2D image frame, at $t+1$. This is the last step that allows us to map relevant areas in the next image, from which matches with high probability of appearance are expected.

The principal output of the implemented contributions is a dynamic model that adapts the matching according to the visual changes on the scene, by introducing formal probability definitions. The approach has demonstrated to adequately balance the matching between highly relevant areas, in terms of current probability, and new visual spaces discovered by the robot. This has been achieved by modulating the probability areas on the image, by a KL metric over the uncertainty of the system.

The benefits of the contributions presented in this approach have been reinforced by the results obtained with real data, computed with publicly available datasets. The suitability and robustness of the matching proposal have been demonstrated with performance tests, in terms of accuracy and efficiency, in comparison with standard matching techniques. Furthermore, its performance has been further evaluated under a visual localization context, in both large indoor and outdoor scenarios. It has also been shown to outperform a well-acknowledged localization method (the inverse EKF). These results have confirmed the validity and consistency of the proposed approach.

Author Contributions: This work was conceived within a collaboration project established between all the authors. D.V., L.P., and Ó.R. formulated the research lines. D.V., Ó.R., and J.M.S. defined the project objectives. D.V., L.M.J., and L.P. implemented the main software developments. D.V. and L.P. acquired and processed the real data. D.V. performed the experimental tests. D.V., Ó.R., and J.M.S. analyzed the final results. D.V., L.P., and Ó.R. devised the article structure.

Funding: This work was partially supported by the Spanish Government through the project DPI2016-78361-R (AEI/FEDER, UE) and by the Valencian Education Council and the European Social Fund through the post-doctoral grant APOSTD/2017/028, held by D. Valiente, during a research stay at the Technical University of Madrid.

Conflicts of Interest: The authors declare no conflict of interest.

Abbreviations

The following abbreviations are used in this manuscript:

CCD	charge-coupled device
EKF	extended Kalman filter
GP	Gaussian process
GPS	global positioning system
KL	Kullback–Leibler divergence
MLE	maximum likelihood estimator
SURF	speeded-up robust features
SVD	singular value decomposition
RMSE	root mean square error

References

1. Chen, L.C.; Hoang, D.C.; Lin, H.I.; Nguyen, T.H. Innovative Methodology for Multi-View Point Cloud Registration in Robotic 3D Object Scanning and Reconstruction. *Appl. Sci.* **2016**, *6*, 132. [CrossRef]
2. Rodriguez-Cielos, R.; Galan-Garcia, J.L.; Padilla-Dominguez, Y.; Rodriguez-Cielos, P.; Bello-Patricio, A.B.; Lopez-Medina, J.A. LiDARgrammetry: A New Method for Generating Synthetic Stereoscopic Products from Digital Elevation Models. *Appl. Sci.* **2017**, *7*, 906. [CrossRef]
3. Scaramuzza, D.; Fraundorfer, F. Visual Odometry [Tutorial]. *IEEE Robot. Autom. Mag.* **2011**, *18*, 80–92. [CrossRef]
4. Payá, L.; Gil, A.; Reinoso, O. A State-of-the-Art Review on Mapping and Localization of Mobile Robots Using Omnidirectional Vision Sensors. *J. Sens.* **2017**, *2017*, 3497650. [CrossRef]
5. Scaramuzza, D.; Fraundorfer, F.; Siegwart, R. Real-Time Monocular Visual Odometry for On-Road Vehicles with 1-Point RANSAC. In Proceedings of the IEEE International Conference on Robotics & Automation (ICRA), Kobe, Japan, 12–17 May 2009; pp. 4293–4299.
6. Civera, J.; Grasa, O.G.; Davison, A.J.; Montiel, J.M.M. 1-point RANSAC for EKF-based Structure from Motion. In Proceedings of the 2009 IEEE/RSJ International Conference on Intelligent Robots and Systems, St. Louis, MO, USA, 10–15 October 2009; pp. 3498–3504.
7. Mur-Artal, R.; Montiel, J.M.M.; Tardós, J.D. ORB-SLAM: A Versatile and Accurate Monocular SLAM System. *IEEE Trans. Robot.* **2015**, *31*, 1147–1163. [CrossRef]
8. Chow, J.C.; Lichti, D.D.; Hol, J.D.; Bellusci, G.; Luinge, H. IMU and Multiple RGB-D Camera Fusion for Assisting Indoor Stop-and-Go 3D Terrestrial Laser Scanning. *Robotics* **2014**, *3*, 247–280. [CrossRef]
9. Munguia, R.; Urzua, S.; Bolea, Y.; Grau, A. Vision-Based SLAM System for Unmanned Aerial Vehicles. *Sensors* **2016**, *16*, 372. [CrossRef] [PubMed]
10. López, E.; García, S.; Barea, R.; Bergasa, L.M.; Molinos, E.J.; Arroyo, R.; Romera, E.; Pardo, S. A Multi-Sensorial Simultaneous Localization and Mapping (SLAM) System for Low-Cost Micro Aerial Vehicles in GPS-Denied Environments. *Sensors* **2017**, *17*, 802. [CrossRef] [PubMed]
11. Davison, A.J.; Gonzalez Cid, Y.; Kita, N. Real-Time 3D SLAM with Wide-Angle Vision. In *Proceedings of the 5th IFAC/EURON Symposium on Intelligent Autonomous Vehicles*; Elsevier Ltd.: New York, NY, USA, 2004; pp. 117–124.

12. Paya, L.; Reinoso, O.; Jimenez, L.M.; Julia, M. Estimating the position and orientation of a mobile robot with respect to a trajectory using omnidirectional imaging and global appearance. *PLoS ONE* **2017**, *12*, e0175938. [CrossRef] [PubMed]
13. Fleer, D.; Moller, R. Comparing holistic and feature-based visual methods for estimating the relative pose of mobile robots. *Robot. Auton. Syst.* **2017**, *89*, 51–74. [CrossRef]
14. Hu, F.; Zhu, Z.; Mejia, J.; Tang, H.; Zhang, J. Real-time indoor assistive localization with mobile omnidirectional vision and cloud GPU acceleration. *AIMS Electron. Electr. Eng.* **2017**, *1*, 74.
15. Hu, F.; Zhu, Z.; Zhang, J. Mobile Panoramic Vision for Assisting the Blind via Indexing and Localization. In *Computer Vision—ECCV 2014 Workshops*; Agapito, L., Bronstein, M.M., Rother, C., Eds.; Springer International Publishing: Cham, Switzerland, 2015; pp. 600–614.
16. Davison, A.J. Real-Time Simultaneous Localisation and Mapping with a Single Camera. In *ICCV'03 Proceedings of the Ninth IEEE International Conference on Computer Vision*; IEEE Computer Society: Washington, DC, USA, 2003; Volume 2, pp. 1403–1410.
17. Karlsson, N.; di Bernardo, E.; Ostrowski, J.; Goncalves, L.; Pirjanian, P.; Munich, M.E. The vSLAM Algorithm for Robust Localization and Mapping. In Proceedings of the 2005 IEEE International Conference on Robotics and Automation, Barcelona, Spain, 18–22 April 2005; pp. 24–29.
18. Chli, M.; Davison, A.J. Active matching for visual tracking. *Robot. Auton. Syst.* **2009**, *57*, 1173–1187. [CrossRef]
19. Neira, J.; Tardós, J.D. Data association in stochastic mapping using the joint compatibility test. *IEEE Trans. Robot. Autom.* **2001**, *17*, 890–897. [CrossRef]
20. Rasmussen, C.; Hager, G.D. Probabilistic data association methods for tracking complex visual objects. *IEEE Trans. Pattern Anal. Mach. Intell.* **2001**, *23*, 560–576. [CrossRef]
21. Li, Y.; Li, S.; Song, Q.; Liu, H.; Meng, M.H. Fast and Robust Data Association Using Posterior Based Approximate Joint Compatibility Test. *IEEE Trans. Ind. Inform.* **2014**, *10*, 331–339. [CrossRef]
22. Lowe, D. Distinctive Image Features from Scale-Invariant Keypoints. *Int. J. Comput. Vis.* **2004**, *60*, 91–110. [CrossRef]
23. Bay, H.; Tuytelaars, T.; Van Gool, L. Speeded Up Robust Features. *Comput. Vis. Image Underst.* **2008**, *110*, 346–359. [CrossRef]
24. Gerrits, M.; Bekaert, P. Local Stereo Matching with Segmentation-based Outlier Rejection. In Proceedings of the 3rd Canadian Conference on Computer and Robot Vision (CRV'06), Quebec City, QC, Canada, 7–9 June 2006; p. 66.
25. Kitt, B.; Geiger, A.; Lategahn, H. Visual odometry based on stereo image sequences with RANSAC-based outlier rejection scheme. In Proceedings of the 2010 IEEE Intelligent Vehicles Symposium, San Diego, CA, USA, 21–24 June 2010; pp. 486–492.
26. Hasler, D.; Sbaiz, L.; Susstrunk, S.; Vetterli, M. Outlier modeling in image matching. *IEEE Trans. Pattern Anal. Mach. Intell.* **2003**, *25*, 301–315. [CrossRef]
27. Liu, M.; Pradalier, C.; Siegwart, R. Visual Homing From Scale With an Uncalibrated Omnidirectional Camera. *IEEE Trans. Robot.* **2013**, *29*, 1353–1365. [CrossRef]
28. Adam, A.; Rivlin, E.; Shimshoni, I. ROR: rejection of outliers by rotations. *IEEE Trans. Pattern Anal. Mach. Intell.* **2001**, *23*, 78–84. [CrossRef]
29. Abduljabbar, Z.A.; Jin, H.; Ibrahim, A.; Hussien, Z.A.; Hussain, M.A.; Abbdal, S.H.; Zou, D. SEPIM: Secure and Efficient Private Image Matching. *Appl. Sci.* **2016**, *6*, 213. [CrossRef]
30. Correal, R.; Pajares, G.; Ruz, J.J. A Matlab-Based Testbed for Integration, Evaluation and Comparison of Heterogeneous Stereo Vision Matching Algorithms. *Robotics* **2016**, *5*, 24. [CrossRef]
31. Valiente, D.; Gil, A.; Payá, L.; Sebastián, J.M.; Reinoso, O. Robust Visual Localization with Dynamic Uncertainty Management in Omnidirectional SLAM. *Appl. Sci.* **2017**, *7*, 1294. [CrossRef]
32. Rasmussen, C.E.; Williams, C.K.I. *Gaussian Processes for Machine Learning*; Adaptive Computation and Machine Learning Series; Massachusetts Institute of Technology: Cambridge, MA, USA, 2006; pp. 1–266.
33. Ghaffari Jadidi, M.; Valls Miro, J.; Dissanayake, G. Gaussian processes autonomous mapping and exploration for range-sensing mobile robots. *Auton. Robots* **2018**, *42*, 273–290. [CrossRef]
34. Hartley, R.; Zisserman, A. *Multiple View Geometry in Computer Vision*; Cambridge University Press: Cambridge, UK, 2004.

35. Scaramuzza, D.; Martinelli, A.; Siegwart, R. A Toolbox for Easily Calibrating Omnidirectional Cameras. In Proceedings of the 2006 IEEE/RSJ International Conference on Intelligent Robots and Systems, Beijing, China, 9–15 October 2006; pp. 5695–5701.
36. Longuet-Higgins, H.C. A computer algorithm for reconstructing a scene from two projections. *Nature* **1985**, *293*, 133–135. [CrossRef]
37. Thrun, S.; Burgard, W.; Fox, D. *Probabilistic Robotics*; The MIT Press: Cambridge, MA, USA, 2005.
38. Civera, J.; Davison, A.J.; Martínez Montiel, J.M. Inverse Depth Parametrization for Monocular SLAM. *IEEE Trans. Robot.* **2008**, *24*, 932–945. [CrossRef]
39. Valiente, D.; Gil, A.; Reinoso, O.; Julia, M.; Holloway, M. Improved Omnidirectional Odometry for a View-Based Mapping Approach. *Sensors* **2017**, *17*, 325. [CrossRef] [PubMed]
40. McLachlan, G. *Discriminant Analysis and Statistical Pattern Recognition*; Wiley Series in Probability an Statistics; Wiley: Hoboken, NJ, USA, 2004.
41. Kulback, S.; Leiber, R.A. On Information and Sufficiency. *Ann. Math. Stat.* **1951**, *22*, 79–86. [CrossRef]
42. Shannon, C.E. A Mathematical Theory of Communication. *SIGMOBILE Mob. Comput. Commun. Rev.* **2001**, *5*, 3–55. [CrossRef]
43. ARVC: Automation, Robotics and Computer Vision Research Group. Miguel Hernandez University. Omnidirectional Image Dataset at Innova Building. Available online: http://arvc.umh.es/db/images/innova_trajectory/ (accessed on 20 March 2018).
44. The Rawseeds Project: Public Multisensor Benchmarking Dataset. Available online: http://www.rawseeds.org (accessed on 20 March 2018).
45. Civera, J.; Grasa, O.G.; Davison, A.J.; Montiel, J.M.M. 1-Point RANSAC for extended Kalman filtering: Application to real-time structure from motion and visual odometry. *J. Field Robot.* **2010**, *27*, 609–631. [CrossRef]
46. Fontana, G.; Matteucci, M.; Sorrenti, D.G. Rawseeds: Building a Benchmarking Toolkit for Autonomous Robotics. In *Methods and Experimental Techniques in Computer Engineering*; Springer International Publishing: Cham, Switzerland, 2014; pp. 55–68.
47. Quigley, M.; Gerkey, B.; Conley, K.; Faust, J.; Foote, T.; Leibs, J.; Berger, E.; Wheeler, R.; Ng, A. ROS: An open-source Robot Operating System. In Proceedings of the IEEE International Conference on Robotics and Automation (ICRA), Workshop on Open Source Robotics, Kobe, Japan, 12–17 May 2009.

sensors

MDPI

Article

Boosting Texture-Based Classification by Describing Statistical Information of Gray-Levels Differences

Óscar García-Olalla [1], Laura Fernández-Robles [2,3], Enrique Alegre [1,3], Manuel Castejón-Limas [2] and Eduardo Fidalgo [1,3,*]

1 Department of Electrical, Systems and Automation, Universidad de León, 24007 León, Spain; ogaro@unileon.es (O.G.-O.); ealeg@unileon.es (E.A.)
2 Department of Mechanical, Computer Science and Aerospace Engineering, Universidad de León, 24007 León, Spain; l.fernandez@unileon.es (L.F.-R.); mcasl@unileon.es (M.C.-L.)
3 Spanish National Cybersecurity Institute (INCIBE), 24005 León, Spain
* Correspondence: efidf@unileon.es

Received: 21 January 2019; Accepted: 25 February 2019; Published: 1 March 2019

Abstract: This paper presents a new texture descriptor booster, Complete Local Oriented Statistical Information Booster (CLOSIB), based on statistical information of the image. Our proposal uses the statistical information of the texture provided by the image gray-levels differences to increase the discriminative capability of Local Binary Patterns (LBP)-based and other texture descriptors. We demonstrated that Half-CLOSIB and M-CLOSIB versions are more efficient and precise than the general one. H-CLOSIB may eliminate redundant statistical information and the multi-scale version, M-CLOSIB, is more robust. We evaluated our method using four datasets: KTH TIPS (2-a) for material recognition, UIUC and USPTex for general texture recognition and JAFFE for face recognition. The results show that when we combine CLOSIB with well-known LBP-based descriptors, the hit rate increases in all the cases, introducing in this way the idea that CLOSIB can be used to enhance the description of texture in a significant number of situations. Additionally, a comparison with recent algorithms demonstrates that a combination of LBP methods with CLOSIB variants obtains comparable results to those of the state-of-the-art.

Keywords: CLOSIB; statistical information of gray-levels differences; Local Binary Patterns; texture classification; texture description; Visual Sensors

1. Introduction

Texture description is one of the main and active fields of research in computer vision [1] and it has a high impact in several research areas connected to image processing and pattern recognition. Texture description is a challenging task that deals with several open problems, e.g., highly discriminate inter-class textures while achieving robustness to intra-class variations. Moreover, the same texture can be displayed in different images under very different appearance due to modifications in the luminance, quality of the camera, snapshot angle, occlusions, and so on. It is for all of these reasons that texture description is still an open problem. Many experimental datasets [2,3] have been created to analyze and fairly compare new methods about texture description. Two of such datasets are UIUC and USPTex, and recent proposals on computer vision are tested on them. Among the most relevant ones, in 2017, Backes et al. [4] obtained discriminative texture signatures by using the LBP approach and fractal dimension to calculate features from the LBP sources of information resulting in an accuracy of 72.50% and 86.52% for the UIUC and USPTex respectively. Florindo et al. [5] in 2016 computed a connectivity index within a local image neighborhood that corresponded to the number of pixels more closely related to the central pixel yielding a success rate of 88.6% on UIUC dataset. In 2017, Cernadas et al. [6] tested different normalization algorithms for color texture classification on USPTex dataset achieving an accuracy of 95.6%. Casanova et al. [7] in 2016 expressed the complexity of the relations among color channels and obtained a hit rate of 97.04% on USPTex dataset. A wide range of applications needs an appropriate texture description of the regions of interest, such as quality control in factories [8], pedestrian detection in crowded streets [9], medical image diagnoses [10] or geographical analyses of optical remote sensing (RS) images [11].

Material recognition is an important field of visual recognition. Even though it differs from texture recognition since one pattern can be made of different materials, texture features are commonly used for material description. Plenty of industries need quality control of their manufactured products, and the use of cameras multiply the speed of this process, avoiding the possibility of subjective interpretations by an operator. In this line of work, González et al. [12] proposed an adaptative method based on pattern spectrum texture descriptor, in which the structural element shape depends on a distance criterion using euclidean and geodesic metrics. Alegre et al. [13] used texture features based on the Laws filters information to evaluate the surface roughness of inserts in milling head tools. In this paper, we evaluate our proposed method for material recognition purposes using KTH Tips2-a [14] dataset. This dataset, created by Caputo et al. and presented in [15], is very popular for material recognition. Chen et al. [16] proposed a method called Weber Local Descriptor (WLD) based on the Weber's Law that achieved a 64.7% of hit rate on this dataset. Hussain et al. in 2012 presented a method called Local Quantified Patterns (LQP) [17] which yielded an accuracy of 64.2% on the same dataset whereas Hafiane et al. [18] in 2015 achieved a 70.3% of accuracy using a method, Adaptive Median Binary Pattern (AMBP), based on LBP. Due to the high intra-class variation of the classes in KTH TIPS2-a, the accuracy obtained in different works of the literature for this dataset is still moderate. Face recognition is another interesting field for many commercial applications in which texture description has demonstrated to be very useful [19]. Faces are highly variable even though the geometry and appearance are not too complicated. Due to the difficulty of the face recognition task, the number of techniques proposed is large and diverse [20,21]. In this paper, we used the JApanese Female Facial Expression (JAFFE) dataset [22] which was developed by Lyons et al. in 1998 and is still extensively used not only for facial expression but also for facial recognition tasks [23]. Rangaswamy et al. proposed a new technique for face recognition based on a fusion of Wavelet and Fourier features [24]. In the same line of work, Wan et al. [25] achieved a 79% of hit rate using a new method called Quasi-Singular Value Decomposition Random Weight Network (Q-SVD + RWN). Zang et al. [26] yielded an accuracy equal to 86.42%, employing Elastic Preserving Projections (EPP) algorithm.

In recent years, local descriptors have been widely used for multiple problems with very promising results. LBP is one of the most popular methods since Ojala et al. introduced it in Ref. [27]. It presents a low computational cost and complexity and a high capability to describe the texture. Nowadays, there are plenty of research groups studying and proposing new methods based on LBP. The original research group at Oulu University proposed several modifications such as Local Binary Patterns Histogram Fourier Features [28], the spatio-temporal LBP –Volume LBP (VLBP) and LBP on Three Orthogonal Planes (LBP-TOP)—[29] or Lineal Configuration Pattern model (LCP) [30]. Guo and his research group at Honk Kong University developed several variants aiming to add extra information to LBP descriptors. Some of their methods can be found in [31–33] and are briefly explained in the related works section. Specifically, Completed LBP (CLBP) has proven to be one of the best performing non-parametric texture description operators by independent authors [34]. Many more variants of LBP exist; we refer the reader to a recent review on this topic for further details [35]. However, none of the previous works deals with the study of the variations of the gray-level differences at several orientations of the image. García-Olalla et al. studied methods that make use of the statistical information of the image and combine them with LBP [36–38], developing a new booster algorithm which improved previous results [39]. Although this booster outperformed LBP and other state-of-the-art methods, it is very specific and is only able to evaluate one statistical order, the mean of the gray level differences along several orientations, for a unique neighborhood configuration.

In this paper, we present a novel method following the work carried out in Ref. [39] that we name CLOSIB, that stands for Complete Local Oriented Statistical Information Booster [40]. CLOSIB is a new texture booster which extracts statistical information of the gray-scale differences in several orientations of the image. Therefore, it can be fused with other descriptors in order to comprise statistical information of the image. We compare our method versus LBP and three LBP-based methods (Adaptive LBP (ALBP) [31], LBP Variance (LBPV) [32] and Complete LBP (CLBP) [33]) due to the high confidence and performance of these methods in a wide application range. Furthermore, we propose and discuss three new variants of CLOSIB based on multi-scale and feature selection: Half CLOSIB (H-CLOSIB), Multi-scale CLOSIB (M-CLOSIB) and Half Multi-scale CLOSIB (HM-CLOSIB). In order to evaluate the performance of CLOSIB and its efficiency in combination with LBP-based descriptors, we tested our method with four texture datasets. Specifically, KTH Tips2-a [15] for material recognition, UIUC [41] and USPTex [42] for general texture classification, and JAFFE [22] for face recognition. At the moment of the submission of this paper, we have already published results using CLOSIB combined with HOG features [43], where we worked on textile retrieval from images obtained on indoor environments. Regardless of this publication [43], where we applied CLOSIB to the mentioned specific problem, in this paper, we present and explain, for the first time, the whole method and context. We evaluate it on four publicly available datasets, comparing it against 18 handcrafted and three deep learning-based state-of-the-art approaches.

2. Related Works

In this section we review the four LBP variants studied. Due to the number of parameters introduced in our study, we include a table of notations, Table 1, to improve the equation readability.

Table 1. Local Binary Patterns (LBP) variants notation.

Parameter	Meaning
g_c	Gray value of the central pixel
g_p	Gray value of neighbor p
P	Number of neighbors
R	Radius of the neighborhood
w_p	Weight element used to minimize the directional difference
w	Weight, it is a constant between 0 to the maximum gray level value difference
N	Number of rows in the image
M	Number of columns in the image
k	A bin of a histogram
K	Maximum value of LBP
u	Mean over the neighbors
c	Threshold, mean value of the differences between the central pixel and neighbors

2.1. Descriptors Based on LBP

2.1.1. Local Binary Patterns (LBP)

LBP [44] describes the texture of gray-scale images extracting their local spatial structure and using a very simple computation. For each pixel, a pattern code is obtained by comparing its value with the value of its neighbors:

$$LBP_{P,R} = \sum_{p=0}^{P-1} s(g_p - g_c)2^p \ , \ s(x) = \begin{cases} 1 & \text{if } x \geq 0 \\ 0 & \text{if } x < 0 \end{cases} \tag{1}$$

where g_c is the gray value of the central pixel, g_p is the value of its neighbor p, P is the number of neighbors and R is the radius of the neighborhood. An image is described by means of a histogram of the LBP values at each pixel of the image. Ojala et al. [44] introduced the rotation invariant uniform operator, $LBP_{P,R}^{riu2}$, which is invariant to monotonic transformations of the gray scale and to rotation, and it is defined as:

$$LBP_{P,R}^{riu2} = \begin{cases} \sum_{p=0}^{P-1} s(g_p - g_c) & \text{if} \quad U(LBP_{P,R}) \leq 2 \\ P+1 & \text{otherwise} \end{cases} \tag{2}$$

where

$$\begin{aligned} U(LBP_{P,R}) &= |s(g_{P-1} - g_c) - s(g_0 - g_c)| \\ &+ \sum_{p=1}^{P-1} |s(g_p - g_c) - s(g_{p-1} - g_c)| \end{aligned} \tag{3}$$

There are only $P+1$ uniform patterns U ("pattern"), which are defined as the ones presenting a number of bit-wise transitions less than or equal to 2, in a neighbor of P pixels. On the other hand, all non-uniform patterns are labelled under the same category. Finally, a histogram of $P+2$ bins is built by computing $LBP_{P,R}^{riu2}$ for each pixel of the image, yielding the feature set of the image. In this work, we use $LBP_{P,R}^{riu2}$ but, for simplicity, we call it LBP henceforth.

2.1.2. Adaptive Local Binary Patterns (ALBP)

ALBP [31] was motivated by the lack of information about the orientation in LBP. It takes into account the mean and the standard deviation along different orientations over all the pixels in order to improve the robustness against changes in the local spatial structure at the matching step. Guo et al. proposed a scheme to minimize the directional differences between the gray levels of the concerned pixels. This scheme allows softening the variations of the mean and standard deviation of the directional differences. The objective function is defined as follows:

$$w_p = arg_w \min \left\{ \sum_{i=1}^{N} \sum_{j=1}^{M} |g_c(i,j) - w \cdot g_p(i,j)|^2 \right\} \tag{4}$$

where w_p is the weight element used to minimize the directional difference, w is in the range from 0 to the maximum gray level value difference, and N and M are the number of rows and columns in the image respectively. Each weight w_p is estimated along one orientation $2p\pi/P$ for the whole image.

The ALBP output is defined as:

$$ALBP_{P,R} = \sum_{p=0}^{P-1} s(g_p - w_p \cdot g_c)2^p \ , \ s(x) = \begin{cases} 1 & \text{if } x \geq 0 \\ 0 & \text{if } x < 0 \end{cases} \tag{5}$$

In this paper, we compute ALBP using the uniform rotation invariant approach explained in Section 2.1.1, $ALBP_{P,R}^{riu2}$.

2.1.3. Local Binary Patterns Variance (LBPV)

LBPV [32] combines LBP and a contrast distribution method. First, the uniform LBP is calculated in the whole image. Then, the variance of the image is used as an adaptive weight to adjust the contribution of the LBP code in the histogram calculation. The LBPV histogram is computed as:

$$LBPV_{P,R}(k) = \sum_{i=1}^{N} \sum_{j=1}^{M} w(LBP_{P,R}(i,j),k), k \in [0,K] \tag{6}$$

where k represents a bin of the histogram, K the maximum value of LBP and w is defined as:

$$w(LBP_{P,R}(i,j),k) = \begin{cases} VAR_{P,R}(i,j), & LBP_{P,R}(i,j) = k \\ 0 & \text{otherwise} \end{cases} \tag{7}$$

$VAR_{P,R}$ is the variance of the neighborhood.

$$VAR_{P,R} = \frac{1}{P}\sum_{p=0}^{P-1}(g_p - u)^2 \tag{8}$$

where u represents the mean over the different neighbors:

$$u = 1/P\sum_{p=0}^{P-1} g_p \tag{9}$$

In this work, we calculate the uniform rotation invariant LBPV, $LBPV_{P,R}^{riu2}$.

2.1.4. Completed Local Binary Patterns (CLBP)

A local region is represented by its center pixel and a Local Difference Sign—Magnitude Transform (LDSMT). LDSMT decomposes the local structure of an image into two complementary components: the difference signs and the difference magnitudes. In order to code both components, Guo et al. [33] introduced two operators, CLBP-Sign (CLBP_S) and CLBP-Magnitude (CLBP_M). We concatenate both operators to form the final CLBP histogram. CLBP_S is identically defined as the original LBP in Equation (1), and CLBP_M is defined in Equation (10).

$$CLBP\_M_{P,R} = \sum_{p=0}^{P-1} t(m_p, c)2^p \ , \ t(x,c) = \begin{cases} 1 & \text{if } x \geq c \\ 0 & \text{if } x < c \end{cases} \tag{10}$$

where c is a threshold that we set to the mean value of the differences between the central pixel and its neighbors, following [33].

In this paper, we use the uniform rotation invariant CLBP, $CLBP_{P,R}^{riu2}$. Note that Guo et al. also presented a third operator CLBP-Center (CLBP_C) that extracts the image local gray level but, for simplicity, we have not used it in this work.

3. Method

In this section, we describe in detail the booster that we propose, CLOSIB. Then, we present different CLOSIB variants which are very interesting in terms of accuracy (M-CLOSIB and HM-CLOSIB) and in terms of computational cost (H-CLOSIB). We include a summary of the notation used in Section 3 in Table 2.

Table 2. CLOSIB variants notation.

Parameter	Meaning
I	Image
c	Central pixel
p	Neighbor pixel
g_c	Gray value of pixel c
g_p	Gray value of neighbor pixel p
R	Radius of the neighborhood
Δ_p	Absolute difference image at bearing p
$\mu_{i,p}$	$i-th$ moment of image Δ_p
$\|\|$	Concatenation function
θ	Order of the statistical moment considered
η	Variable that allows choosing between CLOSIB and H-CLOSIB

3.1. Overview

In this subsection we present a brief description of CLOSIB with the support of Figure 1. Let us consider a given relative position of a pixel in the image with respect to the central pixel, for example the pixel that is placed to the right ($P = 1$) next to ($R = 1$) the central pixel. The absolute differences of the gray-scale values of pixels placed at a given position with respect to a central pixel $|g_1 - g_c|$ are computed and stored at the position of the central pixel. This operation is done for every pixel of the image being considered as the central pixel of the image, which outputs the image Δ_1. The values of Δ_1 are represented in a histogram of absolute differences for a given relative position. Then, some statistical measure is computed on the histogram (mean, standard deviation). CLOSIB descriptor is made up of the statistical measures obtained when considering a set of relative positions around the central pixels.

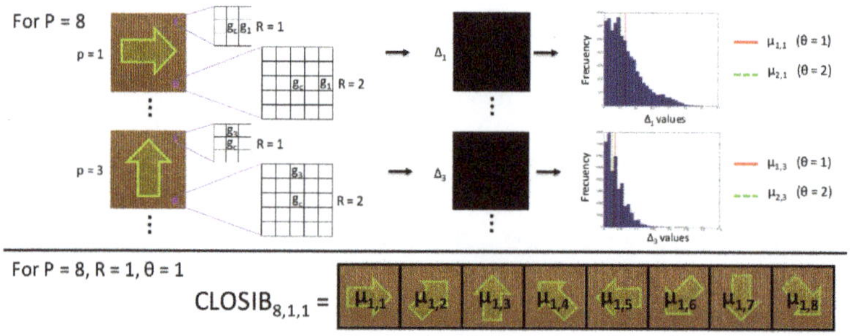

Figure 1. Overview of Complete Local Oriented Statistical Information Booster (CLOSIB) method. Example about the calculation of $\text{CLOSIB}_{8,1,1}$ on an image.

LBP-based descriptors describe the texture of gray-scale images extracting their local spatial structure, whereas CLOSIB extracts statistical information of the gray-scale differences of an image. Thus, the nature of the information provided by LBP-based descriptors is completely different to the one provided by CLOSIB. This difference can be clearly noticed since LBP-based descriptors are local descriptors of the image, but CLOSIB is a global descriptor. Up to our knowledge, this is the first time the statistical information of the image is exploited on the basis of LBP approach.

3.2. Complete Local Oriented Statistical Information Booster (CLOSIB)

We propose a new enhancer that we name Complete Local Oriented Statistical Information Booster (CLOSIB). CLOSIB aims at improving the description performance of image feature descriptors.

CLOSIB is computed from the statistical information of the gray-scale differences of each pixel of the image. The gradient information of an image has been used in several texture descriptors in state-of-the-art. However, the statistical information of the gray-levels differences is infrequently taken into account for the description of an image. CLOSIB is conceptually simple and straightforward to implement.

Let us consider an image I, a particular pixel $c \in I$ and a circularly symmetric set $N = \{p \mid p \in [1, \ldots, P]\}$ where each p represents an equally spaced bearing around c. Let g_c and g_p be the gray values of pixel c located at (x_c, y_c) and its neighbor pixel (x_p, y_p) at bearing p on a circle of radius R respectively. Equation (11) states this relationship between g_c and g_p explicitly.

$$(x_p, y_p) = (x_c + R\cos(2\pi p/P), y_c - R\sin(2\pi p/P)) \tag{11}$$

The gray value of neighbors that are not located in the centers of pixels is estimated by interpolation of their connected pixels.

We define Δ_p as the absolute difference image at bearing p:

$$\Delta_p = \left[\mid g_c - g_p \mid \right], \forall g_c \in I \tag{12}$$

Figure 2 shows an example of the Δ_p images representing the absolute differences of the gray values for $P = 8$ orientations in a neighborhood of radii $R = 1$ and $R = 2$.

Figure 2. Δ_p images showing the absolute differences of the gray values for $P = 8$ orientations in a neighborhood of radii $R = 1$ and $R = 2$. The original image I is shown in the center. The main change in the intensity of the original image occurs in the horizontal direction $p = 1$ and $p = 5$.

Let $\mu_{i,p}$ represent the $i-$th moment of image Δ_p:

$$\mu_{i,p} = \frac{1}{N} \sum_{\forall g_c \in I} \mid g_c - g_p \mid^i \tag{13}$$

where N represents the number of pixels of image I.

We define the CLOSIB vector of image I for P bearings on a circle of radius R and θ-th moment:

$$CLOSIB_{P,R,\theta} = \overset{P/\eta}{\underset{p=1}{\Big|\Big|}} \left((\theta - 1)\mu_{2,p} - (-1)^\theta (\mu_{1,p})^\theta \right)^{1/\theta} \tag{14}$$

where $||$ represents the concatenation function, $\theta \in \{1,2\}$ is the order of the statistical moment considered, and η is a factor that controls the portion of the considered orientations in the quantized angular space. If not specified, we set $\eta = 1$. Therefore, CLOSIB is a feature set of dimensionality P/η.

CLOSIB allows to adjust three parameters: the order of the statistical moment θ, the radius of the neighborhood R and the quantization of the angular space P.

The order of the statistical moment, θ, determines the statistical measure that is used to compute CLOSIB. For $\theta = 1$, CLOSIB is a feature set whose elements are the means of the Δ_p images representing the absolute differences of the gray values for each orientation and every pixel in the image. In the case of $\theta = 2$, the elements of CLOSIB are the standard deviations of the Δ_p images.

Parameter R determines the spatial resolution of the booster. Small radii are quite useful in images with a high level of heterogeneity. As the size of the neighborhood increases, noise is reduced but at the expense of a possible loss of valuable information, especially in images with high variability of the pixel values.

P controls the quantization of the angular space. A higher value of P means that a greater number of orientations are considered in the computation of CLOSIB. As the texture becomes more heterogeneous, the number of orientations should increase in order to capture all the variety of the image. However, using an excessive number of orientations on homogeneous textures may be counter-productive due to the loss of weight of the important ones.

3.3. *CLOSIB Variants*

In the literature, LBP is typically computed for (P, R) pairs of values equals $(8, 1)$, $(16, 2)$ or a concatenation of both. Likewise CLOSIB can be computed for (P, R, θ) triples of values equals $(8, 1, 1)$, $(8, 1, 2)$, $(16, 2, 1)$, $(16, 2, 2)$ or a concatenation of several of them. CLOSIB can also be computed for any other triple of values. We indicate the concatenation of several CLOSIBs with the symbol $||$. For example, the concatenation of $\text{CLOSIB}_{8,1,1}$ and $\text{CLOSIB}_{16,2,1}$ is represented as $\text{CLOSIB}_{8,1,1\,||\,16,2,1}$.

In this section, we propose and describe three specific ways of obtaining CLOSIB.

3.3.1. Multi-Scale CLOSIB (M-CLOSIB)

Chang et al. in [45] proposed a multi-scale LBP (MSLBP) method for face detection that benefits from the multi-resolution information captured from the regional histogram. MSLBP has been extended and applied to other fields in the literature [46,47].

Similarly, we introduce a multi-scale CLOSIB which we name M-CLOSIB. M-CLOSIB is a concatenation of the CLOSIBs obtained for a fixed number of orientations P and several radii of the neighborhood R. Figure 3 shows an schema of the computation of $\text{M-CLOSIB}_{8,1,\theta\,||\,8,2,\theta\,||\,8,3,\theta}$, which results from the concatenation of $\text{CLOSIB}_{8,1,\theta}$, $\text{CLOSIB}_{8,2,\theta}$ and $\text{CLOSIB}_{8,3,\theta}$.

91	7	25	88	13	134	120	200	121
12	**14**	25	7	**26**	44	124	**124**	121
12	14	**0**	1	**2**	56	**125**	200	11
12	7	88	**7**	**0**	**22**	100	123	55
91	182	23	134	28	**173**	**22**	**220**	12
172	87	123	54	86	66	66	12	67
0	0	123	134	77	65	193	211	67
0	12	4	88	19	7	5	41	79
124	22	243	21	13	134	1	211	122

$\text{CLOSIB}_{(8,1,\Theta)}$	$\text{CLOSIB}_{(8,2,\Theta)}$	$\text{CLOSIB}_{(8,3,\Theta)}$
$\text{H-CLOSIB}_{(8,1,\Theta)}$	**$\text{H-CLOSIB}_{(8,2,\Theta)}$**	**$\text{H-CLOSIB}_{(8,3,\Theta)}$**

Figure 3. (Better viewed in color) Neighborhood around a center pixel, (28), considered for the computation of $\text{M-CLOSIB}_{8,1,\theta\,||\,8,2,\theta\,||\,8,3,\theta}$ and $\text{HM-CLOSIB}_{8,1,\theta\,||\,8,2,\theta\,||\,8,3,\theta}$. CLOSIB considers $P = 8$ orientations, while H-CLOSIB only $P = 4$ orientations. In the figure, neighbour pixels considered for H-CLOSIB are shown in bold.

3.3.2. Half CLOSIB (H-CLOSIB)

For even values of P, CLOSIB encompasses statistical information of the absolute differences of the gray values $d_p(x_c, y_c)$ along directions that differ in π radians. Figure 4a shows this fact for $P = 8$. The statistical information along directions that differ in π radians is usually very similar. Figure 4b,c illustrates two examples.

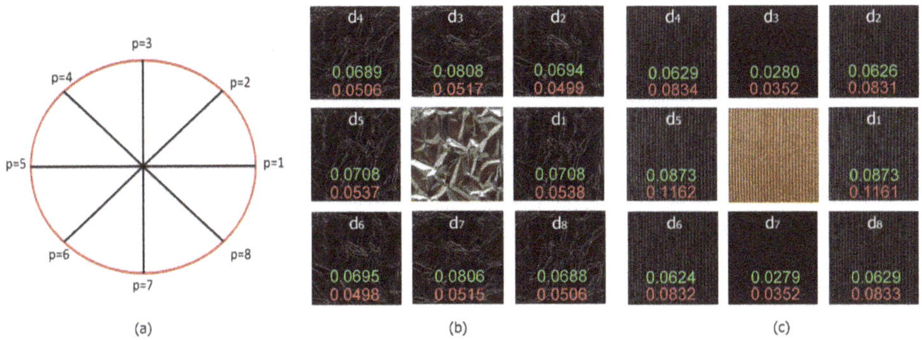

Figure 4. (**a**) Circumference that represents the neighborhood considered for the computation of CLOSIB with $P = 8$. Four pairs of neighbors differ in π radians, such as the neighbors for values $p = 2$ and $p = 6$. (**b**,**c**) Schemas that represent the computation of $\text{CLOSIB}_{8,1,\theta}$ for two different images. We show the original image in the centre and the eight images of the absolute differences of the gray values $d_p(x_c, y_c)$ in the outer layer. The red and green numbers indicate the values of each element of $\text{CLOSIB}_{8,1,1}$ and $\text{CLOSIB}_{8,1,2}$ feature set, respectively, obtained for the corresponding p values of $d_p(x_c, y_c)$. Note that the values of the elements of CLOSIB computed for neighbors that differ in π radians diverge in only a maximum of 0.0002 units whereas the ones that differ in a different angle diverge in at least 0.0006 units.

We define a Half CLOSIB (H-CLOSIB) following Equation (14) with $\eta = 2$. The angular space is yet quantized in P equal parts but only the first P/η orientations are taken into account for the computation of H-CLOSIB. Figure 5 shows an example of the orientations considered when computing $\text{CLOSIB}_{8,1,\theta}$ and $\text{H-CLOSIB}_{8,1,\theta}$.

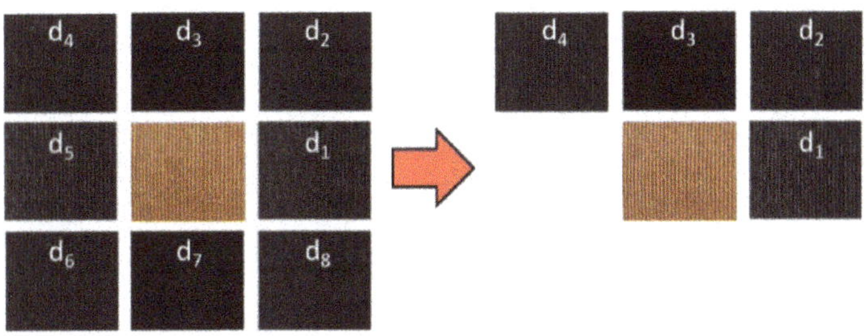

Figure 5. Schemas of the computation of $\text{CLOSIB}_{8,1,\theta}$ (**left**) and $\text{H-CLOSIB}_{8,1,\theta}$ (**right**) using the example of Figure 4c.

H-CLOSIB presents two main characteristics. First, it may eliminate redundant statistical information. As the algorithm computes the magnitude of the first derivative, without sign, the absolute value of the differences between any pair of pixels is the same, without matter the direction of the gray-levels. In Figure 6, can be seen an example that allows to understand this fact better. In Figure 6a, the gray level values of the original image can be found. In Figure 6b,d are the values obtained applying $p = 1$ and $R = 1$ in the first case and $p = 5$ and $R = 1$ in the second. As can be seen, the value of the first order moment, denoted in this Figure as *mean* is the same in both cases, 23.9. Figure 6c presents the same calculation as in (b) but keeping the sign of the derivative. In this case, the mean has a different value of −14.1. The second characteristic is that the size of H-CLOSIB is half of the equivalent CLOSIB. The dimensionality might be decisive in some cases when the amount of memory or computational time are critical, such as in embedded systems with little RAM.

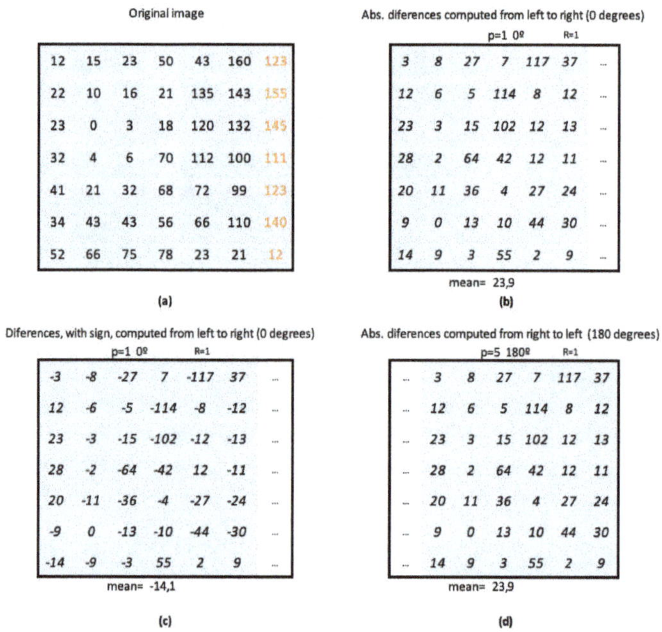

Figure 6. (**a**) Example of gray values of an Original Image. (**b**) Matrix obtained from the first difference of the gray values at 0 degrees. It corresponds with the first element of $CLOSIB_{8,1,1}$. (**d**) Matrix obtained from the first difference of the gray values at 180 degrees. It corresponds with the fifth element of $CLOSIB_{8,1,1}$. (**c**) Differences, with sign, at 0 degrees. CLOSIB uses the absolute value of the differences, therefore these values with sign are never computed.

3.3.3. Half Multi-Scale CLOSIB (HM-CLOSIB)

We propose a Half Multi-scale CLOSIB (HM-CLOSIB) which is obtained as a M-CLOSIB when $\eta = 2$. This variant combines the advantages and disadvantages of both M-CLOSIB and H-CLOSIB. Figure 3 shows a schema of the computation of $HM\text{-}CLOSIB_{8,1,\theta||8,2,\theta||8,3,\theta}$, which results of the combination of $H\text{-}CLOSIB_{8,1,\theta}$, $H\text{-}CLOSIB_{8,2,\theta}$ and $H\text{-}CLOSIB_{8,3,\theta}$.

4. Experiments and Results

4.1. Datasets

4.1.1. KTH TIPS2-a

KTH TIPS2-a dataset (http://www.nada.kth.se/cvap/databases/kth-tips/download.html) aims at evaluating algorithms for classifying materials [15]. It includes a total of 4608 images grouped into 11 classes. The dataset contains four physical samples of each of the 11 materials. The dataset presents a high intra-class variation regarding texture and colour. All samples were taken at nine scales and three poses under four different illumination conditions, which makes the dataset very challenging.

4.1.2. UIUC

The University of Illinois Urbana-Champaign (UIUC) texture dataset (http://www-cvr.ai.uiuc.edu/ponce_grp/data/index.html) [41] contains 25 different texture classes within 40 images per class, giving a total 1000 un-calibrated, unregistered gray-scale images of resolution 640 × 480 pixels.

The database contains materials, fabrics and other textures such as water. Within each class, significant viewpoints variations, scale changes and non-rigid deformations are strongly present [2]. This dataset contains a few numbers of images per class but a high intra-class variability, being a challenging dataset regarding scale and other viewpoint variations.

4.1.3. USPTex

USPTex dataset (http://fractal.ifsc.usp.br/dataset/USPtex.php) [42] contains 191 different classes of 24-bit color png images of general scenes like roads, vegetation, walls, clouds and materials such as seeds, rice or tissues. The most challenging feature of this dataset is the low number of images per class (12), their low resolution (128 × 128 pixels) and the high number of classes included [3].

4.1.4. JAFFE

JAFFE dataset (http://www.kasrl.org/jaffe.html) [22] comprises 213 images of 7 facial expressions (6 basic facial expressions and 1 neutral) posed by 10 Japanese female models. Each subject appears in 20 to 23 images. The images were taken from a frontal pose, and tungsten lights were used to create even illumination on the face. All images are 256 × 256 pixels in size. In this paper, we use JAFFE dataset for face recognition instead of expression recognition. Therefore, we are dealing with a multiclass classification that comprehends 10 classes.

4.2. Experimental Setup

For KTH-TIPS 2a dataset, we used the experimental protocol developed by Caputo et al. [15,16], which is 4-fold cross-validation along the samples of each material. For each fold, we used all images of one sample of each material for testing and the rest for training. This experimental setup is more challenging than a random division of the images into training and test sets due to the high inter-sample variation. We report the results as the average hit rate over the four runs. We define the hit rate as the number of correctly classified images divided by the total number of images in the test set. We used a Support Vector Machine (SVM) to classify the images with the Least Squares training algorithm and a polynomial kernel of order 2. We used the one-vs-one paradigm [48] in which $n(n-1)/2$ binary classifiers are trained for a n-way multi-class problem; each receives the samples of a pair of classes. For testing, all binary classifiers are applied to an unseen sample, and the class that gets the highest number of predictions for all binary classifiers gets predicted.

Concerning UIUC and USPTex datasets, we carried out random sub-sampling cross-validation with 10 repetitions to avoid overfitting. In each iteration, the model is fit to a training set of 75% of the images, and predictive accuracy is assessed using the rest of the images. The results were averaged over the splits. We used an SVM trained using Least Square algorithm and a linear kernel.

Regarding JAFFE dataset, we used the same evaluation setup proposed by Sharma et al. [49]. Specifically, one random image of each facial expression and person forms the test set, and the rest define the training set. We repeat the classification 10 times to avoid biased results due to the random process. We used the multi-block approach introduced by Zang et al. [50] for describing a face using LBP-based descriptors and CLOSIB. We split the image into 8 × 8 blocks and compute a descriptor for each block. We define the descriptor of the image as the concatenation of the descriptors of the blocks. We performed two sets of experiments with JAFFE dataset. On the one hand, we used the images provided in the dataset. On the other hand, we automatically cropped the face of the images using Viola-Jones method [51] and used the cropped images to carry out the experiments. Tables 3 and 4 show the CLOSIBs used in the experiments for CLOSIB and H-CLOSIB, and M-CLOSIB and HM-CLOSIB, respectively.

Table 3. Each row describes the parameters used to compute different CLOSIBs and H-CLOSIBs in the experiments. In column "order", values 1 and 2 indicate that we obtained CLOSIB as a concatenation of the CLOSIBs for each statistical moment, $CLOSIB_{P,R,1\,||\,P,R,2}$.

Radius (*R*)	Neighbors (Orientations) (*P*)	Order (θ)
1	8	1
1	8	2
2	16	1
2	16	2
1	8	1,2
2	16	1,2

Table 4. Each row describes the parameters used to compute different M-CLOSIBs and HM-CLOSIBs in the experiments. Several values for a parameter indicate that we obtained CLOSIB as a concatenation of the CLOSIBs for each single value.

Radius (*R*)	Neighbors (Orientations) (*P*)	Order (θ)
1,2,3	8	1
1,2,3,4,5	8	1
1,2,3	8	2
1,2,3,4,5	8	2
2,3,4	16	1
2,3,4,5,6	16	1
2,3,4	16	2
2,3,4,5,6	16	2
1,2,3	8	1,2
1,2,3,4,5	8	1,2
2,3,4	16	1,2
2,3,4,5,6	16	1,2

In the following subsections, we present and discuss the results obtained using this experimentation. We aim to check if CLOSIB enhances the performance of LBP-based descriptors on several public texture datasets for different fields. For KTH Tips2-a, a more thorough review of the performance of CLOSIB is introduced in order to better understand its behavior.

4.3. Results for KTH Tips2-a

4.3.1. CLOSIB versus LBP-Based Descriptors

We developed CLOSIB as an enhancer of texture descriptors. However, in this section, we show the performance of CLOSIB as a descriptor itself. Figure 7 presents the results that we obtained when describing the images with CLOSIB and with descriptors based on LBP. For all CLOSIB variants, we achieved the best performance using a concatenation of the CLOSIBs for the first and second statistical moments. For all LBP-based descriptors, we obtained the best results for $R = 2$ pixels and $P = 16$ neighbors. It is remarkable that we achieved the highest performance using HM-CLOSIB$_{16,2,1\,||\,16,2,2\,||\,16,3,1\,||\,16,3,2\,||\,16,4,1\,||\,16,4,2}$ which yielded a hit rate of 67.93%. Therefore, the proposed enhancer by itself outperforms some of the state-of-the-art LBP-based descriptors.

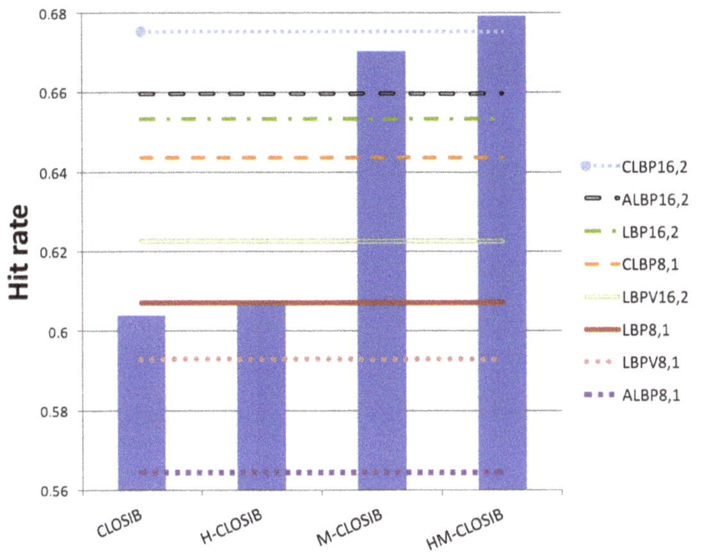

Figure 7. Hit rates when we describe KTH Tips2-a images with different CLOSIBs and LBP-based descriptors. For each CLOSIB variant –CLOSIB (standard), M-CLOSIB, H-CLOSIB and HM-CLOSIB–, we only represent the best result obtained among the results with different combinations of parameters.

4.3.2. CLOSIB and LBP-Based Descriptors

The following experiment consists of combining LBP-based descriptors with CLOSIB. The combination is done by means of a concatenation. Figure 8 and Table 5 graphically and numerically show the results.

Table 5. Hit rates (in %) obtained with a given LBP-based descriptor (LBP, ALBP, LBPV and CLBP) and the concatenations of the descriptor with CLOSIB variants. The best results for each LBP-based descriptor are highlighted in bold. The best overall results are underlined. D stands for Descriptor and C for CLOSIB.

Descriptor (D)	D	D||C	D||H-C	D||M-C	D||HM-C
$LBP_{8,1}$	60.71	68.20	67.80	**71.95**	71.86
$LBP_{16,2}$	65.53	70.52	70.33	71.78	**72.50**
$ALBP_{8,1}$	56.46	64.86	64.84	68.97	**69.15**
$ALBP_{16,2}$	65.97	68.96	68.88	69.84	**70.16**
$LBPV_{8,1}$	59.30	67.05	67.51	69.89	**71.15**
$LBPV_{16,2}$	62.27	69.00	69.24	70.14	**71.17**
$CLBP_{8,1}$	64.37	69.63	69.95	**71.97**	71.76
$CLBP_{16,2}$	67.53	71.95	**<u>72.54</u>**	72.01	**<u>72.54</u>**

In all experiments, we achieved the best results using CLOSIB as an enhancer of LBP-based descriptors in opposition to only using LBP-based descriptors. We obtained the highest hit rates equal to 72.54% with $CLBP_{16,2}$ ||$HM\text{-}CLOSIB_{16,2,1||16,2,2||16,3,1||16,3,2||16,4,1||16,4,2}$ and $CLBP_{16,2}$ ||$H\text{-}CLOSIB_{16,2,1||16,2,2}$ closely followed by $LBP_{16,2}$ ||$HM\text{-}CLOSIB_{8,1,1||8,1,2||8,2,1||8,2,2||8,3,1||8,3,2||8,4,1||8,4,2||8,5,1||8,5,2}$ with a hit rate of 72.50%. For 6 out of the 8 LBP-based descriptors, we achieved the best results with the concatenation of HM-CLOSIB.

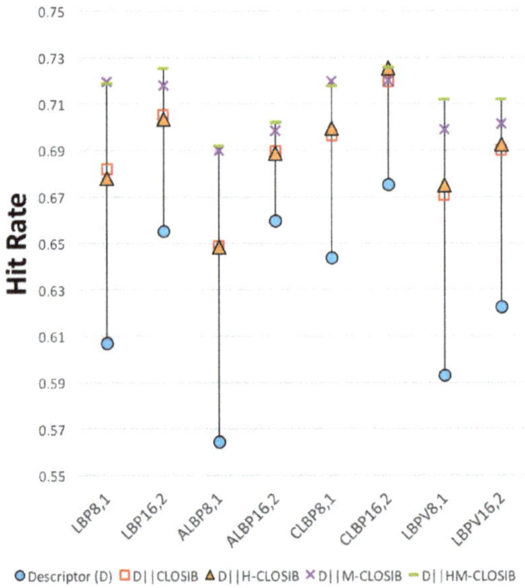

Figure 8. Hit rates obtained with a given LBP-based descriptor (LBP, ALBP, LBPV and CLBP) and the concatenations of the descriptor with CLOSIB variants.

4.3.3. No Multi-Scale versus Multi-Scale LBP-Based Descriptors

The good performance of multi-scale CLOSIB leads us to reproduce the experiments for multi-scale LBP-based descriptors. Figure 9 shows the comparison between the results obtained with LBP-based descriptors and their multi-scale versions.

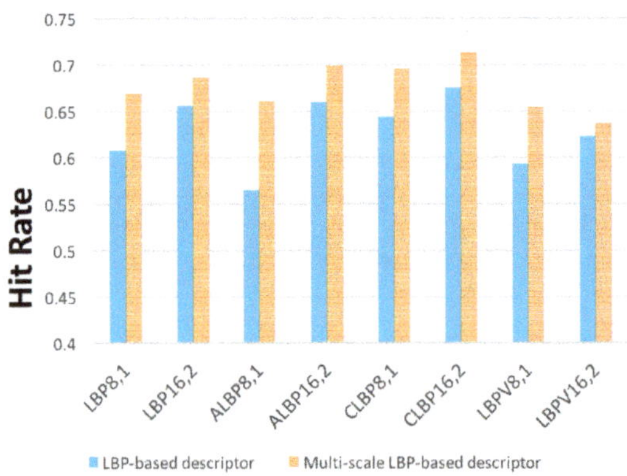

Figure 9. Hit rates for LBP-based descriptors LBP, ALBP, CLBP and LBPV and their multi-scale versions.

We defined a multi-scale LBP as a concatenation of the LBP descriptors obtained with different neighborhood radii (R) and the same number of neighbors (P). We used $R = \{1,2,3\}$ for $P = 8$ and $R = \{2,3,4\}$ for $P = 16$. The hit rate fairly improves with multi-scale LBP-based descriptors in all cases. Therefore, multi-scale descriptors are very interesting for texture retrieval.

Best result was obtained with $\text{CLOSIB}_{16,2||16,3||16,4}$ with a 71.28%. This result means a 5.55% of improvement compared with the standard $\text{CLOSIB}_{16,2}$. However, our proposed descriptor HM-CLOSIB combined with $\text{CLBP}_{16,2}$ gets the best performance so far.

4.3.4. HM-CLOSIB + Multi-Scale LBP-Based Descriptors

Finally, we evaluated the combination of multi-scale LBP-based descriptors with HM-CLOSIB. We selected HM-CLOSIB due to the high performance achieved in terms of accuracy and computational time in previous experiments.

Figure 10 shows the hit rate of the concatenation of multi-scale LBP-based descriptors with HM-CLOSIB. Furthermore, we also present the hit rate of the (non multi-scale) LBP-based descriptors combined with HM-CLOSIB to represent the improvement in accuracy.

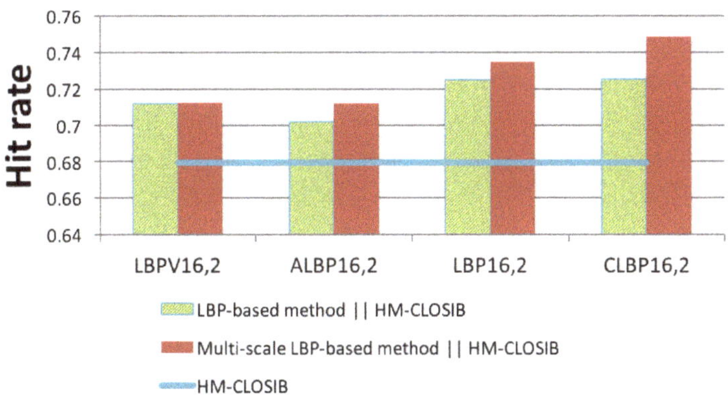

Figure 10. Hit rates obtained with the concatenation of multi-scale LBP-based descriptors and HM-CLOSIB. The horizontal line represents the hit rate of HM-CLOSIB descriptor.

$CLBP_{16,2}||HM\text{-}CLOSIB_{16,2,1||16,2,2||16,3,1||16,3,2||16,4,1||16,4,2}$ outperformed the rest of the methods with a hit rate of 74.83%, which represents an improvement of at least 3.16% in hit rate with respect to the rest of descriptors.

4.3.5. Comparative with the State-of-the-Art

Several authors tested their algorithms using KTH TIPS2-a dataset. In Table 6, we can see the results achieved by 23 state-of-the-art methods, including three that are based on deep learning approaches. To the best of our knowledge, the best result has been yielded by by LFV+FC-CNN [52], an approach where deep features are extracted. The second and third positions are obtained by another deep features approach NmzNet [53], followed by a handcrafted one, IFV [54].

The classification performance of the proposed descriptor $CLBP_{16,2}||HM\text{-}CLOSIB_{16,2,1||16,2,2||16,3,1||16,3,2||16,4,1||16,4,2}$ is the fourth over the 20 methods based on handcrafted approaches. Furthermore, using just the straightforward HM-CLOSIB as a descriptor, we yielded a higher hit rate than most of these studies.

Note that a direct comparison among the results reported by these methods cannot be made due to the different approaches that were taken for preprocessing the images—here no preprocessing has been done—and for carrying out the experiments. It can be found that our booster achieves comparable results to those of the state-of-the-art and that it can be successfully used in combination with LBP-based methods to enhance their performance. As we mentioned in Section 1, we published results using CLOSIB booster together with HOG features [43], demonstrating that CLOSIB could be successfully combined with several handcrafted features, not only LBP-based methods.

Table 6. Hit rates obtained by the proposed booster, the combination of the booster with CLBP and the reported classification scores for 18 state-of-the-art methods on the KTH-TIPS2-a. Scores are as originally reported. Our proposal is highlighted in bold, together with the best proposal of the Deep Features.

Descriptor—Handcrafted	Hit Rate (%)	Reference
WLD	56.4	[16]
MWLD	64.7	[16]
SIFT	52.7	[16]
LTP	60.7	[17]
LQP	64.2	[17]
WLBP	64.4	[55]
LHS	73.0	[49]
CMLBP	73.1	[56]
CMR	69.4	[57]
PC	71.5	[57]
DRLTP	62.6	[58]
DRLBP	59.0	[58]
HoPS	75.0	[59]
IFV	82.2	[54]
AMBP	70.3	[18]
MS4C	70.5	[60]
$CRDP_{3D} - 2$ (NNC)	73.8	[61]
$CRDP_{3D} - 2$ (SVM)	78.0	[61]
HM-CLOSIB	67.9	Ours
$CLBP_{16,2}$ || HM-CLOSIB	**74.8**	Ours
Descriptor—Deep Features	**Hit Rate (%)**	**Reference**
DeCAF	78.4	[54]
LFV + FC-CNN	**82.6**	[52]
NmzNet	82.4	[53]

4.4. Results for UIUC and USPTex

Figure 11 shows the hit rates that we obtained on UIUC and USPTex datasets, respectively. In both cases, every combination of LBP-based descriptors with any CLOSIB variant yielded higher hit rates than the LBP-based descriptors alone. M-CLOSIB outperformed the rest of CLOSIB variants. For UIUC, we obtained the highest hit rate (85.51%), whereas for USPTex, we achieved the highest hit rate (72.91%), in both cases using $CLBP_{16,2}$ $||$M-$CLOSIB_{16,2,1||16,2,2||16,3,1||16,3,2||16,4,1||16,4,2}$.

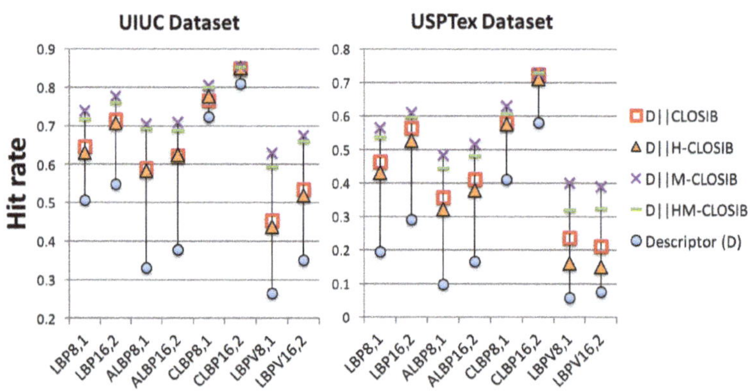

Figure 11. Results using the concatenation of LBP-based descriptors with CLOSIB variants (CLOSIB, H-CLOSIB, M-CLOSIB and HM-CLOSIB) on UIUC (**left**) and USPTex (**right**) dataset.

4.5. Results for JAFFE

We carried out two sets of experiments with JAFFE dataset: with the original images and with automatically cropped images.

Figure 12 shows the hit rates that we obtained in the first experiment, using the original images. In all cases, the LBP-based descriptors achieved worse results than the combination of the LBP-based descriptors with any CLOSIB variant. The combination with M-CLOSIB yielded the highest hit rates in most of the cases, except for $LBP_{16,2}$ and $LBPV_{8,1}$ in which the combination with CLOSIB outperformed the others. We achieved the best results using $LBPV_{16,2} || \text{M-CLOSIB}_{8,1,1||8,1,2||8,2,1||8,2,2||8,3,1||8,3,2||8,4,1||8,4,2||8,5,1||8,5,2}$ with a hit rate of 82.71%.

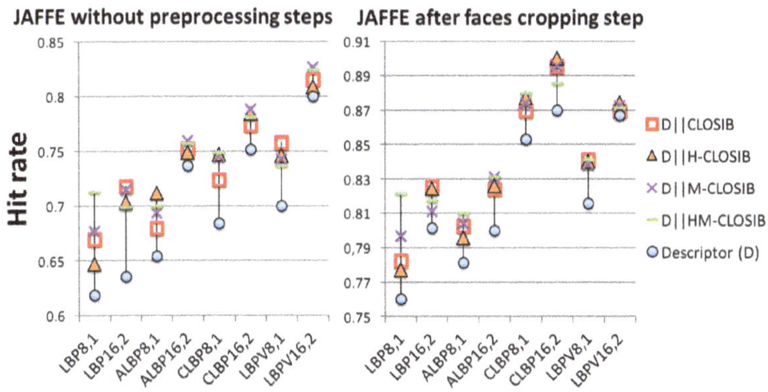

Figure 12. Results using the concatenation of LBP-based descriptors with CLOSIB variants (CLOSIB, H-CLOSIB, M-CLOSIB and HM-CLOSIB) on the original images (**left**) and the cropped ones (**right**) of JAFFE dataset.

Regarding the second experiment, Figure 12 shows the hit rates achieved using the cropped images. Again, tests using LBP-based descriptors yielded worse results than when combined with any CLOSIB variant. In this case, we obtained the highest hit rate, 90.00%, using $CLBP_{16,2} || \text{H-CLOSIB}_{16,2,1||16,2,2}$. It is important to notice that carrying out the preprocessing step, the performance improves up to 8.81%.

4.6. Computational Cost of CLOSIB and LBP Variants

Finally, we present in Tables 7 and 8 the average computational time per image employed for the extraction of CLOSIB and LBP variants descriptors, respectively, on the four datasets studied. The fastest descriptors per dataset are shown in bold.

Table 7. Computational times, in seconds, for the extraction of LBP variants on the four datasets evaluated. LBP variants with underscored parameters *neighborhood*, *radius*.

Dataset	$LBP_{8,1}$	$LBP_{16,2}$	$ALBP_{8,1}$	$ALBP_{16,2}$	$LBPV_{8,1}$	$LBPV_{16,2}$	$CLBP_{8,1}$	$CLBP_{16,2}$
UIUC	**0.09183**	0.17636	0.20309	0,44195	0.15119	0.34011	0.1157	0.23574
USPTex	**0.00351**	0.00488	0.00605	0.01051	0.00428	0.00918	0.00361	0.00594
KTH-TIPS2-a	0.00914	0.01084	0.01297	0.02555	0.0114	0.02283	**0.00754**	0.01326
JAFFE	0.01116	0.01695	0.02179	0.04263	0.01628	0.03312	**0.01013**	0.0183

Table 8. Computational times, in seconds, for the extraction of CLOSIB variants on the four datasets evaluated. CLOSIB (**C**) and H-CLOSIB (**H-C**) with underscored parameters $(radius, neighbors, order)$. M-CLOSIB (**M-C**) and HM-CLOSIB (**HM-C**) with underscored parameters $(minRadius, maxRadius, neighbors, order)$.

Dataset	$C_{1,8,1}$	$C_{2,16,2}$	H-$C_{1,8,1}$	H-$C_{2,16,2}$	M-$C_{1,3,8,1}$	M-$C_{2,4,16,2}$	HM-$C_{1,3,8,1}$	HM-$C_{2,4,16,2}$
UIUC	0.08655	0.18975	**0.08630**	0.18975	0.25392	0.59675	0.25142	0.56396
USPTex	0.00393	0.00594	**0.00377**	0.00564	0.00938	0.01584	0.00912	0.01471
KTH TIPS2-a	0.00921	0.01782	**0.00838**	0.01682	0.02284	0.05158	0.02266	0.04846
JAFFE	0.02640	0.02790	**0.01442**	0.02594	0.03833	0.07177	0.03582	0.06767

In Table 7 we can observe how $LBP_{8,1}$ is the fastest choice for UIUC and USPTex datasets, with 0.09183 and 0.00351 seconds per image respectively, while $CLBP_{8,1}$ is for KTH TIPS2-a and JAFFE with 0.00754 and 0.01013 seconds, respectively. In Table 8, it can be noticed that CLOSIB variants require similar or even less computational time than the LBP variants for equal values of neighbors and order. Regarding CLOSIB variants, the shortest times are achieved by H-$CLOSIB_{1,8,1}$ proposal, obtaining an average of 0.0863, 0.00377, 0.00838 and 0.01442 seconds per image on UIUC, USPTex, KTH TIPS2-a and JAFFE datasets, respectively.

5. Conclusions

We proposed a new texture descriptor booster, called CLOSIB, which is based on the statistical information provided by the gray-level differences of the image. Furthermore, we presented three variants of CLOSIB: H-CLOSIB, useful for embedded systems or machines with a low RAM; M-CLOSIB, a multi-scale descriptor which extracts information for consecutive neighborhoods; and HM-CLOSIB, which is a multi-scale H-CLOSIB. The experiments demonstrated that H-CLOSIB is a little bit more efficient than the general version in terms of precision and computational cost. We also saw that a description of the image at several scales, using the M-CLOSIB, always produces comparable or better results than the general version of CLOSIB. Those differences are very significative in some of the used datasets. We evaluated CLOSIB in three applications: material recognition using KTH TIPS2-a dataset, general texture recognition using UIUC and USPTex datasets and face recognition using JAFFE dataset.

Regarding material recognition, HM-CLOSIB outperformed some of the state-of-the-art LBP-based descriptors. To check the performance of CLOSIB as an enhancer of other texture descriptors, we used a concatenation of LBP-based descriptors with CLOSIB variants. All tested combinations of LBP-based descriptors with CLOSIB yielded better results than the individual descriptors. Moreover, we proved that the classification results for material recognition improves when using multi-scale LBP-based descriptors. We obtained the best result using a concatenation of a multi-scale CLBP and HM-CLOSIB yielding a hit rate of 74.83%. Finally, this method outperformed some relevant state-of-the-art methods tested on KTH TIPS2-a images. Concerning general texture recognition (UIUC and USPTex), every concatenation of LBP-based descriptors with CLOSIB variants yielded higher hit rates than the individual LBP-based descriptors. In relation to face recognition, the combination of LBP-based descriptors with CLOSIB variants outperformed the individual descriptors as well. We obtained the highest hit rate of 90% using a combination of $CLBP_{16,2}$ and H-CLOSIB when automatically cropping the images of the dataset by means of the Viola-Jones method.

All in all, in this paper we introduced a new, efficient and powerful texture descriptor enhancer that adds statistical information about the gray-level differences of the pixels of the image employing a straightforward implementation. Based on the results obtained, we consider that CLOSIB can be regarded as a descriptor enhancer of broad purpose that, when fused with other descriptors, provides new and relevant information that improves the classification results.

In the future, we will evaluate the performance obtained when combining CLOSIB with other different texture descriptors to determine with which ones it works better and its limitations, if any.

We also will propose a HM-CLOSIB for color images and we will evaluate how a rotational invariant codification performs. Among the methods used for pornography detection, skin detection approach uses texture descriptors [62]. In the context of the 4NSEEK European Project, we will explore how the combination of CLOSIB booster and texture descriptors affects the accuracy of a system used for porn detection, and more specifically, for the fight against Child Sexual Abuse (CSA).

Author Contributions: Conceptualization, Ó.G.-O., L.F.-R., E.A. and M.C.-L.; Data curation, Ó.G.-O.; Investigation, Ó.G.-O., L.F.-R., E.A., M.C.-L. and E.F.; Methodology, Ó.G.-O. and L.F.-R.; Software, Ó.G.-O., E.F.; Supervision, L.F.-R., E.A. and E.F.; Validation, Ó.G.-O., L.F.-R., E.A. and M.C.-L.; Visualization, E.F.; Writing—original draft, Ó.G.-O.; Writing— review & editing, Ó.G.-O., L.F.-R., E.A., M.C.-L. and E.F.

Funding: This research was funded by Spanish Government grant number DPI2012-36166 and AP2010-0947; INCIBE grant INCIBEI-2015-27359, Addendum 22 and 01. This research has been partially funded with support from the European Commission under the 4NSEEK project with Grant Agreement 821966. This publication reflects the views only of the author, and the European Commission cannot be held responsible for any use which may be made of the information contained therein.

Acknowledgments: This work was supported by the Spanish Government [grant number DPI2012-36166]; the Spanish Government via the pre-doctoral FPU fellowship program [AP2010-0947]; the INCIBE grant INCIBEI-2015-27359 corresponding to the Ayudas para la Excelencia de los Equipos de Investigación avanzada en ciberseguridad and also by the framework agreement between the University of León and INCIBE (Spanish National Cybersecurity Institute) under Addendum 22 and 01. This research has been partially funded with support from the European Commission under the 4NSEEK project with Grant Agreement 821966. This publication reflects the views only of the author, and the European Commission cannot be held responsible for any use which may be made of the information contained therein.

Conflicts of Interest: The authors declare no conflict of interest.

Abbreviations

The following abbreviations are used in this manuscript:

LBP	Local Binary Pattern
ALBP	Adaptive Local Binary Pattern
ALBPV	Adaptive Local Binary Pattern Variance
CLOSIB	Complete Local Oriented Statistical Information Booster
H-CLOSIB	Half Complete Local Oriented Statistical Information Booster
M-CLOSIB	Multi-scale Complete Local Oriented Statistical Information Booster
ASASEC	Advisory System Against Sexual Exploitation of Children
CDC	Compact Digital Cameras
CNN	Convolutional Neural Networks
CSA	Child Sexual Abuse

References

1. Liu, J.; He, J.; Zhang, W.; Xu, P.; Tang, Z. TCvBsISM: Texture Classification via B-Splines-Based Image Statistical Modeling. *IEEE Access* **2018**, *6*, 44876–44893. [CrossRef]
2. Hossain, S.; Serikawa, S. Texture databases—A comprehensive survey. *Pattern Recognit. Lett.* **2013**, *34*, 2007–2022. [CrossRef]
3. Bianconi, F.; Fernández, A. An appendix to "Texture databases—A comprehensive survey". *Pattern Recognit. Lett.* **2014**, *45*, 33–38. [CrossRef]
4. Backes, A.R.; de Mesquita Sá Junior, J.J. LBP maps for improving fractal based texture classification. *Neurocomputing* **2017**, *266*, 1–7. [CrossRef]
5. Florindo, J.B.; Landini, G.; Bruno, O.M. Three-dimensional connectivity index for texture recognition. *Pattern Recognit. Lett.* **2016**, *84*, 239–244. [CrossRef]
6. Cernadas, E.; Fernández-Delgado, M.; González-Rufino, E.; Carrión, P. Influence of normalization and color space to color texture classification. *Pattern Recognit.* **2017**, *61*, 120–138. [CrossRef]
7. Casanova, D.; Florindo, J.; Falvo, M.; Bruno, O. Texture analysis using fractal descriptors estimated by the mutual interference of color channels. *Inf. Sci.* **2016**, *346*, 58–72. [CrossRef]

8. Dutta, S.; Pal, S.K.; Sen, R. On-machine tool prediction of flank wear from machined surface images using texture analyses and support vector regression. *Precis. Eng.* **2016**, *43*, 34–42. [CrossRef]
9. Castrillón-Santana, M.; Lorenzo-Navarro, J.; Ramón-Balmaseda, E. Fusion of Holistic and Part Based Features for Gender Classification in the Wild. In *New Trends in Image Analysis and Processing–ICIAP 2015 Workshops*; Murino, V., Puppo, E., Sona, D., Cristani, M., Sansone, C., Eds.; Springer International Publishing: New York, NY, USA, 2015; Volume 9281, pp. 43–50.
10. Kavitha, M.S.; An, S.Y.; An, C.H.; Huh, K.H.; Yi, W.J.; Heo, M.S.; Lee, S.S.; Choi, S.C. Texture analysis of mandibular cortical bone on digital dental panoramic radiographs for the diagnosis of osteoporosis in Korean women. *Oral Surg. Oral Med. Oral Pathol. Oral Radiol.* **2015**, *119*, 346–356. [CrossRef] [PubMed]
11. Lan, Z.; Liu, Y. Study on Multi-Scale Window Determination for GLCM Texture Description in High-Resolution Remote Sensing Image Geo-Analysis Supported by GIS and Domain Knowledge. *Int. J. Geo-Inf.* **2018**, *7*, 175. [CrossRef]
12. González-Castro, V.; Alegre, E.; García-Olalla, O.; Fernández-Robles, L.; García-Ordás, M.T. Adaptive pattern spectrum image description using Euclidean and Geodesic distance without training for texture classification. *IET Comput. Vis.* **2012**, *6*, 581–589. [CrossRef]
13. Alegre, E.; Barreiro, J.; Suárez-Castrillón, S. A new improved Laws-based descriptor for surface roughness evaluation. *Int. J. Adv. Manuf. Technol.* **2012**, *59*, 605–615. [CrossRef]
14. Mallikarjuna, P.; Targhi, A.T.; Fritz, M.; Hayman, E.; Caputo, B.; Eklundh, J.O. THE KTH-TIPS2 Database. 2006. Available online:http://www.nada.kth.se/cvap/databases/kth-tips/kth-tips2.pdf (accessed on 28 February 2019).
15. Caputo, B.; Hayman, E.; Mallikarjuna, P. Class-specific material categorisation. In Proceedings of the 10th IEEE International Conference on Computer Vision (ICCV'05), Beijing, China, 17–20 October 2005; pp. 1597–1604.
16. Chen, J.; Shan, S.; He, C.; Zhao, G.; Pietikainen, M.; Member, S.; Chen, X.; Member, S.; Gao, W. WLD: A robust local image descriptor. *IEEE Trans. Pattern Anal. Mach. Intell.* **2010**, *32*, 1705–1720. [CrossRef] [PubMed]
17. Hussain, S.; Triggs, B. Visual Recognition Using Local Quantized Patterns. In *Computer Vision ECCV 2012*; Lecture Notes in Computer Science; Fitzgibbon, A., Lazebnik, S., Perona, P., Sato, Y., Schmid, C., Eds.; Springer: Berlin/Heidelberg, Germany, 2012; pp. 716–729.
18. Hafiane, A.; Palaniappan, K.; Seetharaman, G. Joint Adaptive Median Binary Patterns for texture classification. *Pattern Recognit.* **2015**, *48*, 2609–2620. [CrossRef]
19. Xiang, Z.; Tan, H.; Ye, W. The Excellent Properties of a Dense Grid-Based HOG Feature on Face Recognition Compared to Gabor and LBP. *IEEE Access* **2018**, *6*, 29306–29318. [CrossRef]
20. Elaiwat, S.; Bennamoun, M.; Boussaid, F.; El-Sallam, A. A Curvelet-based approach for textured 3D face recognition. *Pattern Recognit.* **2015**, *48*, 1235–1246. [CrossRef]
21. De Marsico, M.; Nappi, M.; Riccio, D.; Wechsler, H. Robust face recognition after plastic surgery using region-based approaches. *Pattern Recognit.* **2015**, *48*, 1261–1276. [CrossRef]
22. Lyons, M.; Akamatsu, S.; Kamachi, M.; Gyoba, J. Coding facial expressions with Gabor wavelets. In Proceedings of the Third IEEE International Conference on Automatic Face and Gesture Recognition, Nara, Japan, 14–16 April 1998, pp. 200–205.
23. Yang, B.; Cao, J.; Ni, R.; Zhang, Y. Facial Expression Recognition using Weighted Mixture Deep Neural Network Based on Double-channel Facial Images. *IEEE Access* **2017**. [CrossRef]
24. Shenoy, P.D.; Iyengar, S.S.; Raja, K.B.; Buyya, R.; Patnaik, L.M.; Rangaswamy, Y.; Raja, K.; Venugopal, K. FRDF: Face Recognition Using Fusion of DTCWT and FFT Features. *Procedia Comput. Sci.* **2015**, *54*, 809–817.
25. Wan, W.; Zhou, Z.; Zhao, J.; Cao, F. A novel face recognition method: Using random weight networks and quasi-singular value decomposition. *Neurocomputing* **2015**, *151 Pt 3*, 1180–1186. [CrossRef]
26. Zang, F.; Zhang, J.; Pan, J. Face recognition using Elasticfaces. *Pattern Recognit.* **2012**, *45*, 3866–3876. [CrossRef]
27. Ojala, T.; Pietikainen, M.; Harwood, D. Performance evaluation of texture measures with classification based on Kullback discrimination of distributions. In Proceedings of the 12th International Conference on Pattern Recognition, Jerusalem, Israel, 9–13 October 1994; Volume 1, pp. 582–585.

28. Ahonen, T.; Matas, J.; He, C.; Pietikainen, M. Rotation Invariant Image Description with Local Binary Pattern Histogram Fourier Features. In *Image Analysis*; Salberg, A.B., Hardeberg, J.Y., Jenssen, R., Eds.; Springer: Berlin/Heidelberg, Germany, 2009; pp. 61–70.
29. Zhao, G.; Pietikainen, M. Dynamic texture recognition using local binary patterns with an application to facial expressions. *IEEE Trans. Pattern Anal. Mach. Intell.* **2007**, *29*, 915–928. [CrossRef] [PubMed]
30. Guo, Y.; Zhao, G.; Pietikainen, M. Texture Classification using a Linear Configuration Model based Descriptor. In Proceedings of the British Machine Vision Conference, Dundee, UK, 29 August–2 September 2011.
31. Guo, Z.; Zhang, L.; Zhang, D.; Zhang, S. Rotation invariant texture classification using adaptive LBP with directional statistical features. In Proceedings of the 2010 17th IEEE International Conference on Image Processing (ICIP), Hong Kong, China, 26–29 September 2010, pp. 285–288.
32. Guo, Z.; Zhang, L.; Zhang, D. Rotation invariant texture classification using LBP variance (LBPV) with global matching. *Pattern Recognit.* **2010**, *43*, 706–719. [CrossRef]
33. Guo, Z.; Zhang, L.; Zang, D. A Completed Modeling of Local Binary Pattern Operator for Texture Classification. *IEEE Trans. Image Process.* **2010**, *19*, 1657–1663. [PubMed]
34. Fernández, A.; Álvarez, M.X.; Bianconi, F. Texture Description Through Histograms of Equivalent Patterns. *J. Math. Imaging Vis.* **2013**, *45*, 76–102. [CrossRef]
35. Liu, L.; Fieguth, P.; Guo, Y.; Wang, X.; Pietikäinen, M. Local binary features for texture classification: Taxonomy and experimental study. *Pattern Recognit.* **2017**, *62*, 135–160. [CrossRef]
36. García-Olalla, O.; Alegre, E.; Fernández-Robles, L.; García-Ordás, M.T. Vitality assessment of boar sperm using an adaptive LBP based on oriented deviation. In *Computer Vision—ACCV 2012 Workshops*; Park, J.I., Kim, J., Eds.; Springer: Berlin/Heidelberg, Germany, 2013; pp. 61–72.
37. García-Olalla, O.; Alegre, E.; Fernández-Robles, L.; García-Ordás, M.T.; García-Ordás, D. Adaptive Local Binary Pattern with oriented standard deviation (ALBPS) for texture classification. *EURASIP J. Image Video Process.* **2013**, *2013*, 31. [CrossRef]
38. García-Olalla, O.; Alegre, E.; García-Ordás, M.T.; Fernández-Robles, L. Evaluation of LBP Variants using several Metrics and kNN Classifiers. In *Similarity Search and Applications*; Brisaboa, N., Pedreira, O., Zezula, P., Eds.; Springer: Berlin/Heidelberg, Germany, 2013; pp. 151–162.
39. García-Olalla, O.; Alegre, E.; Fernández-Robles, L.; González-Castro, V. Local Oriented Statistics Information Booster (LOSIB) for Texture Classification. In Proceedings of the 2014 22nd International Conference on Pattern Recognition (ICPR), Stockholm, Sweden, 24–28 August 2014; pp. 1114–1119.
40. García-Olalla Olivera, O. Methods for Improving Texture Description by Using Statistical Information Extracted from the Image Gradient. Ph.D. Thesis, Universidad de León, León, Spain, 2017.
41. Lazebnik, S.; Schmid, C.; Ponce, J. A sparse texture representation using local affine regions. *IEEE Trans. Pattern Anal. Mach. Intell.* **2005**, *27*, 1265–1278. [CrossRef] [PubMed]
42. Backes, A.R.; Casanova, D.; Bruno, O.M. Color texture analysis based on fractal descriptors. *Pattern Recognit.* **2012**, *45*, 1984–1992. [CrossRef]
43. García-Olalla, O.; Alegre, E.; Fernández-Robles, L.; Fidalgo, E.; Saikia, S. Textile retrieval based on image content from CDC and webcam cameras in indoor environments. *Sensors* **2018**, *18*, 1329. [CrossRef] [PubMed]
44. Ojala, T.; Pietikainen, M.; Maenpaa, T. Multiresolution gray-scale and rotation invariant texture classification with Local Binary Patterns. *IEEE Trans. Pattern Anal. Mach. Intell.* **2002**, *24*, 971–987. [CrossRef]
45. Chan, C.H.; Kittler, J.; Messer, K. Multi-scale Local Binary Pattern Histograms for Face Recognition. In *Advances in Biometrics*; Lee, S.W., Li, S.Z., Eds.; Springer: Berlin/Heidelberg, Germany, 2007; pp. 809–818.
46. Jia, X.; Yang, X.; Cao, K.; Zang, Y.; Zhang, N.; Dai, R.; Zhu, X.; Tian, J. Multi-scale local binary pattern with filters for spoof fingerprint detection. *Inf. Sci.* **2014**, *268*, 91–102. [CrossRef]
47. Wen, Z.; Li, Z.; Peng, Y.; Ying, S. Virus image classification using multi-scale completed local binary pattern features extracted from filtered images by multi-scale principal component analysis. *Pattern Recognit. Lett.* **2016**, *79*, 25–30. [CrossRef]
48. Hsu, C.W.; Lin, C.J. A Comparison of Methods for Multiclass Support Vector Machines. *IEEE Trans. Neural Netw.* **2002**, *13*, 415–425. [PubMed]
49. Sharma, G.; ul Hussain, S.; Jurie, F. Local Higher-Order Statistics (LHS) for Texture Categorization and Facial Analysis. In *Computer Vision ECCV 2012*; Lecture Notes in Computer Science; Fitzgibbon, A., Lazebnik, S., Perona, P., Sato, Y., Schmid, C., Eds.; Springer: Berlin/Heidelberg, Germany, 2012; Volume 7578, pp. 1–12.

50. Zhang, L.; Chu, R.; Xiang, S.; Liao, S.; Li, S.Z. Face Detection Based on Multi-Block LBP Representation. In *Advances in Biometrics*; Lee, S.W., Li, S.Z., Eds.; Springer: Berlin/Heidelberg, Germany, 2007; pp. 11–18.
51. Viola, P.; Jones, M. Rapid object detection using a boosted cascade of simple features. In Proceedings of the 2001 IEEE Computer Society Conference on Computer Vision and Pattern Recognition, Kauai, HI, USA, 8–14 December 2001; Volume 1.
52. Song, Y.; Zhang, F.; Li, Q.; Huang, H.; Odonnell, L.J.; Cai, W. Locally-Transferred Fisher Vectors for Texture Classification. In Proceedings of the IEEE International Conference on Computer Vision, Venice, Italy, 22–29 October 2017; pp. 4922–4930.
53. Vu, N.S.; Nguyen, V.L.; Gosselin, P.H. A Handcrafted Normalized-Convolution Network for Texture Classification. In Proceedings of the 2017 IEEE International Conference on Computer Vision Workshops, ICCVW 2017, Venice, Italy, 22–29 October 2017; pp. 1238–1245.
54. Cimpoi, M.; Maji, S.; Kokkinos, I.; Mohamed, S.; Vedaldi, A. Describing Textures in the Wild. In Proceedings of the 2014 IEEE Conference on Computer Vision and Pattern Recognition, Columbus, OH, USA, 23–28 June 2014; pp. 3606–3613.
55. Liu, F.; Tang, Z.; Tang, J. WLBP: Weber local binary pattern for local image description. *Neurocomputing* **2013**, *120*, 325–335. [CrossRef]
56. Li, W.; Fritz, M. Recognizing Materials from Virtual Examples. In *Proceedings of the 12th European Conference on Computer Vision—Volume Part IV*; Springer: Berlin/Heidelberg, Germany, 2012; pp. 345–358.
57. Zhang, J.; Zhao, H.; Liang, J. Continuous rotation invariant local descriptors for texton dictionary-based texture classification. *Comput. Vis. Image Underst.* **2013**, *117*, 56–75. [CrossRef]
58. Satpathy, A.; Jiang, X.; Eng, H.L. LBP-Based Edge-Texture Features for Object Recognition. *IEEE Trans. Image Process.* **2014**, *23*, 1953–1964. [CrossRef] [PubMed]
59. Voravuthikunchai, W.; Crémilleux, B.; Jurie, F. Histograms of pattern sets for image classification and object recognition. In Proceedings of the IEEE Computer Society Conference on Computer Vision and Pattern Recognition, Columbus, OH, USA, 23–28 June 2014, pp. 224–231.
60. Li, W. Learning Multi-scale Representations for Material Classification. In *Pattern Recognition*; Jiang, X., Hornegger, J., Koch, R., Eds.; Springer International Publishing: Cham, Switzerland, 2014; pp. 757–764.
61. Wang, K.; Bichot, C.E.; Li, Y.; Li, B. Local Binary Circumferential and radial derivative pattern for texture classification. *Pattern Recognit.* **2017**, *67*, 213–229. [CrossRef]
62. Gangwar, A.; Fidalgo, E.; Alegre, E.; González-Castro, V. Pornography and child sexual abuse detection in image and video: A comparative evaluation. In Proceedings of the 8th International Conference on Imaging for Crime Detection and Prevention (ICDP 2017), Madrid, Spain, 13–15 December 2017; pp. 37–42.

sensors

MDPI

Article

Appearance-Based Salient Regions Detection Using Side-Specific Dictionaries

Mian Muhammad Sadiq Fareed [1], Qi Chun [1,*], Gulnaz Ahmed [2,*], Adil Murtaza [3,*], Muhammad Rizwan Asif [1] and Muhammad Zeeshan Fareed [2]

1 School of Electronic and Information Engineering, Xi'an Jiaotong University, Xi'an 710049, China; sadiqfareed@mail.xjtu.edu.cn (M.M.S.F.); rizwanasif@ciitlahore.edu.pk (M.R.A.)
2 School of Management, Xi'an Jiaotong University, Xi'an 710049, China; zeeshan.fareed@ist.edu.pk
3 School of Science, MOE Key Laboratory for Non-equilibrium Synthesis and Modulation of Condensed Matter, State Key Laboratory for Mechanical Behaviour of Materials, Xi'an Jiaotong University, Xi'an 710049, China
* Correspondence: qichun@mail.xjtu.edu.cn (Q.C.); gulnaz@mail.xjtu.edu (G.A.); adilmurtaza91@mail.xjtu.edu.cn (A.M.)

Received: 29 November 2018; Accepted: 5 January 2019; Published: 21 January 2019

Abstract: Image saliency detection is a very helpful step in many computer vision-based smart systems to reduce the computational complexity by only focusing on the salient parts of the image. Currently, the image saliency is detected through representation-based generative schemes, as these schemes are helpful for extracting the concise representations of the stimuli and to capture the high-level semantics in visual information with a small number of active coefficients. In this paper, we propose a novel framework for salient region detection that uses appearance-based and regression-based schemes. The framework segments the image and forms reconstructive dictionaries from four sides of the image. These side-specific dictionaries are further utilized to obtain the saliency maps of the sides. A unified version of these maps is subsequently employed by a representation-based model to obtain a contrast-based salient region map. The map is used to obtain two regression-based maps with LAB and RGB color features that are unified through the optimization-based method to achieve the final saliency map. Furthermore, the side-specific reconstructive dictionaries are extracted from the boundary and the background pixels, which are enriched with geometrical and visual information. The approach has been thoroughly evaluated on five datasets and compared with the seven most recent approaches. The simulation results reveal that our model performs favorably in comparison with the current saliency detection schemes.

Keywords: salient region detection; appearance based model; regression based model; human visual attention; background dictionary

---

1. Introduction

Salient Region Detection (SRD) is a procedure to confine the image according to human visual attention and discovers the most useful and informative portion of an image. This procedure tries to approximate the possibility that the image region that is taking more attention comes out as a salient object. It is also a very helpful step because it is applied in many computer vision applications to reduce the computational complexity by only focusing on the salient parts of the image. The conventional saliency methods are separated into two groups as the bottom-up [1] and top-down [2]. The first category is a bottom-up method, which is a stimuli-driven approach and it only depends on the prior knowledge of the object and the background. Whereas, the second category is a top-down approach, which is data-driven and does not need prior information to detect the saliency.

The major portion of SRD literature [3–5] is comprised of the bottom-up approaches [1], as these methods only consider low-level features and demonstrate a remarkable performance. The dense and sparse appearance-based models are separately applied in [6,7] for the salient region computation. The dense reconstruction error-based methods [8] have persuasive results when the image border is large and contains the sparsely connected regions. However, these methods lose their efficiency when the background contains a latent pattern or the background is complicated with small-scale high-contrast patterns. The dense appearance-based models [7] provide a more expressive and generic description of the background. These methods are more sensitive towards the background noise. So, the dense representation error-based models are very less useful in detecting the salient objects with a cluttered background. The methods based on a background template set [9–11], and co-similarity matrix [7] have convincing results whenever the salient objects pop out closer to the center part of the scene. However, when the salient objects significantly touch the image boundary, parts of them are wrongly considered as background. Consequently, the extracted saliency is less accurate when the salient object part is popping out or touching the boundary. In this case, the foreground parts of the image are mistakenly considered as the reconstructive dictionary and obtain zero weights, and the salient objects in the remaining parts of the image are found to be less accurate.

In this paper, we introduce a novel SRD method which fuses the compact appearance and discrimination of the individual scenes into a combined framework. Firstly, the input images are segmented into superpixels. Secondly, we employ the appearance-based model to measure the rareness of the features. Thirdly, we apply the regression-based model to rank the previously computed results on the basis of the foreground and the background multi-feature cues, respectively. Finally, we utilize an optimization method to produce an even and accurate salient region map. Our appearance-based model is very simple and easily detects the objects closer to the boundary of the scene. Our regression-based model makes the initial saliency map smoother and it is very helpful in highlighting the salient object part. The proposed method utilizes the visual, geometrical and location information for SRD and shows improved results as compared to the previous contrast-based methods. To fuse the previously obtained results, we applied an enhancement procedure to compute more even and precise salient region maps. We compare our method visually as well as graphically against the seven current SRD methods on the five benchmark databases. From the qualitative and quantitative evaluation, we found that our method performance remains very consistent on all the selected databases. The main contributions of our method are summarized as follows:

- The designed model is robust and easily handles the cluttered and noisy background which was a problem for dense appearance-based models. Also, the side-specific dictionaries of the proposed model are helpful in detecting the salient objects adjacent to the boundary.
- Sometimes the small segments from the background are extremely highlighted and affect the computed saliency. The averaging process of the proposed model is very helpful to overcome this issue by measuring the saliency of a superpixel as an average residual in this segment.
- To enhance the discrimination between the foreground and the background, we engage a multi-feature graph-learning procedure which incorporates the intrinsic weight of regions to implement the uniformity among the similar image patches by utilizing the prior information.
- Furthermore, we optimize the salient regions map by applying the guided filter, which removes the artifacts and further improves the qualitative as well as the quantitative results.

The remaining part of the paper is organized as follows. The current literature about the SRD is discussed in Section 2. In Section 3, different stages of our method like dictionary construction, saliency detection, and refinement processes are discussed in detail. The comparison of our model with the seven most recent methods is given in Section 4. The conclusion of our method is summarized in Section 5.

2. Related Work

Several computational methods are proposed for SRD. The majority of the preceding schemes are appearance-based models, these models mainly depend upon the global or local contrast for their saliency map computation.

2.1. Dictionary Learning-Based SRD

The dictionary-based approaches [2,12–15] facilitate learning multifaceted labeling procedures and represent the image in a space where it can be easily processed. In [12], the basis vector is computed on the belief that the repeatedly activated bases contain less energy as compared to the rare bases. This model works selectively because the unpredicted bases are selected as salient clues. A dictionary for an image patch is constructed from a depository of natural images in [6]. Then, the sparse representation is utilized to find the contrast between each image patch. Shen et al. [13] optimize the objective of feature transformation and low-rank decomposition for training the dictionary. However, these methods manually trained their dictionaries using the top-down way. In [1,14], the authors constructed the dictionary by only utilizing the center-surrounded patches without any training. However, the saliency results are not satisfactory because the inner-region of the salient object is not detected properly. In recent dictionary-based method [8], the author utilized the boundary information to extract the background dictionary. The saliency computed through this background dictionary is not clear because only the boundary information for background dictionary construction is insufficient. Currently, some methods engaged the center-remaining strategy [16], while other used the more background regions [17] to construct their background dictionary. However, most of the time, the background templates contain limited information that leads to incorrect SRD.

2.2. Sparse Representation-Based SRD

The image boundary is always standing out as a part of the background. So, it can be very helpful in constructing the background template set [8–10]. The authors computed the sparse representation error through this background template set. However, the computed results are not significant when the salient object is touching the image boundary. The center-surrounded strategy is helpful in detecting, so the authors in [16] engaged the center-remaining procedure to extract the dictionary. Then, the sparse reconstruction error is calculated through this dictionary. The computed saliency results averaged and improved through a multi-label inference process. To enhance the difference between the salient object and the background, a sparse coding-based generative model is discussed in [17]. To capture all information related to the image a superpixel sparse reconstruction-based model is defined in [9]. However, the results generated by these models are not very clear because these methods only utilizing the local image information for SRD. Consequently, all these methods improved their results through an enhancement process, which recovers the lost information.

2.3. Global or Local Measures-Based SRD

The previously designed SRD techniques are broadly divided into two categories, local and global methods. The local methods compute the saliency by the rarity of neighbors or surrounded regions. While the global methods extract saliency using the uniqueness of features over the entire scene. In [14], the authors computed the saliency as the center-remaining difference of many features. Graph-based SRD method [18] exploits the rarity of different local features to compute the saliency map. A fuzzy growing approach is utilized to compute the saliency with the contrast of neighboring superpixels [19]. Ming Lin et al. [20] proposed the saliency of superpixels by incorporating the global features, namely spatial distribution and uniqueness. They used the PCA method to incorporate color and pattern distinctness to find the SRD. In [7], the authors computed the saliency by the global contrast between the image patches and their spatial position. They performed sampling based on the conventional three-color cues maps and PCA to extract the main features of the image patches.

To extract a saliency map with high resolution that is dependent on color contrast, a Histogram Contrast (HC) method is defined in [21]. In [22], a non-local histogram approach is engaged to improve the efficiency of the method, and a smoothing procedure is applied to get rid of quantization artifacts. However, these proposed techniques are only suitable for simple natural images and lose their accuracy for highly patterned and textured images.

2.4. Multiple Feature-Based SRD

The existing approaches for SRD are mainly focusing on the color features while ignoring the other features like texture, structure, and the orientation. Therefore, these types of methods are not successful when dealing with an image that contains rich textural features. Many approaches for SRD use the RGB color model and few of them depending upon LAB or YC_bC_r color space for their result calculation. The authors consider the near-infrared region with the RGB color model for SRD [23], as tthe near-infrared region provides more clues for recognition and categorization than the RGB color model. SRD using sparsity-based and graph-based models is proposed in [24]; the authors combine the multi-features of colors with sparse representation model to compute the saliency. A method for SRD by combining multiple features of color distribution and contrast is proposed in [25], the authors exploited a multi-features color difference measure, a multi-features color distribution measure, and a multi-features salient object measure to compute the saliency. To exploit the multi-features constructing through image manifold of the different feature, a multi-feature enhancement procedure is discussed in [16]. However, these methods add some high contrast pixels with the salient object that lead to insignificant detection.

2.5. Foreground or Background-Based SRD

The discriminative schemes are also very important because these schemes help in enhancing the contrast between the background and foreground regions for SRD [25]. A number of discriminative strategies based models have appeared in current years. Shuang Li et al., [26] suggested that the saliency of a region is computed by the distance from the most assured background and foreground seeds. Hongyang Li et al., [27] proposed that the saliency of an object is estimated through propagating the cues extracted mainly from the certain object regions and background. The graph-based methods can capture more grouping features in the scene with the graph likeness. Graph similarity typically controls the performance of a graph-based method [11]. Some of them used the semi-supervised learning to approximate the similarities by incorporating local-grouping features deduced from the whole image. The foreground represents appearance consistency and uniformity, while the background many times reveals global or local connectivity with each of the four image boundaries [28]. In [17], a two-stage saliency scheme is defined which is based on relevance to the given query. After that, they used the graph-based manifold ranking procedure to rank the foreground and background cues. However, if the contrast is far from being between the foreground and the background, the computed saliency results are not accurate. Furthermore, it is very difficult to choose the position and the number of salient queries because these cues are generated through the random walks on the graphs, especially for the images that contain, unlike salient objects.

2.6. Deep Convolutional Neural Networks-Based SRD

Since Deep Convolutional Neural Networks (CNN)-based methods [29–31] are engaged for SRD, tremendous progress has been achieved because of the availability of large visual datasets and GPU computing resources. The development of deeper and larger DCNNs [29–31] that could automatically learn more and more powerful feature representations with multiple levels of abstraction from big data. Significant progress has been made in the past few years to boost the accuracy levels of SRD [29–31], but existing solutions often rely on computationally expensive feature representation and learning approaches, which are too slow for numerous applications. In addition to the opportunities they offer, the large visual datasets also lead to the challenge of scaling up while retaining the efficiency of learning approaches and representations for both handcrafted and deeply learned features. In addition, given sufficient amount of annotated visual data, some existing features, especially DCNN features [29–31], have been shown to yield high accuracy for visual recognition. However, there are many applications where only limited amounts of annotated training data can be available or collecting labeled training data is too expensive. Such applications impose great challenges to many existing features.

3. The Proposed Salient Regions Detection Approach

In this section, we present the particulars of our proposed approach in detail. In the first stage, we employ the Appearance-Based Model (ABM) to compute the coarse dense salient region map. In the second stage, we engage the Regression-Based Model (RBM) to enhance the discrimination between the foreground and background cues, respectively. Each of the individual stages of the proposed salient region detection method is illustrated in Figure 1.

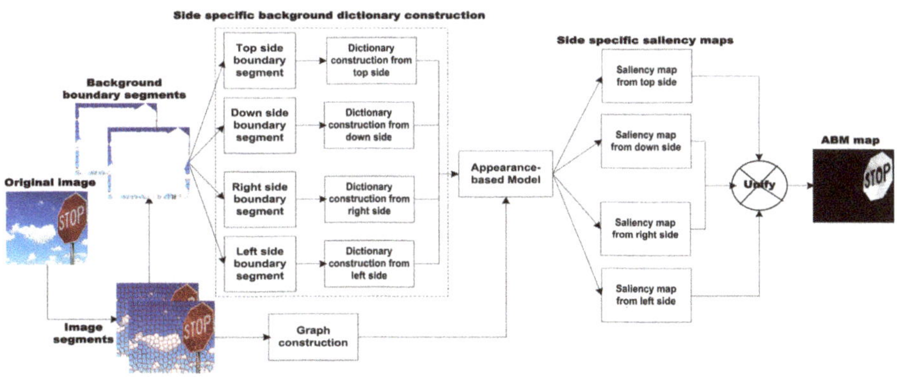

Figure 1. The pipeline of proposed salient region detection model.

3.1. The Visual Feature Extraction

To encode and accomplish better structural information regarding the image, we first segment the input image into superpixels by utilizing the Simple Linear Iterative Clustering (SLIC) mechanism [32]. SLIC adapts a k-means clustering approach to efficiently generate superpixels. Despite its simplicity, SLIC adheres to boundaries as well as or better than previous methods. At the same time, it is faster and more memory efficient, improves segmentation performance, and is straightforward to extend to super voxel generation. SLIC algorithm group pixels into perceptually meaningful atomic regions which can be used to replace the rigid structure of the pixel grid. SLIC captures image redundancy, provide a convenient primitive from which to compute image features, and greatly reduce the complexity of subsequent image processing tasks. Superpixels present a better method for obtaining the features of an image. As discussed in [6], the conventional color model is supportive for SRD because the colors surround the major part of the image. To capture more information relating to the image, we used the mean of the RGB and CIE Lab color space to represent a superpixel as $Z = [R\ G\ B\ L\ a\ b\ x\ y\ g_i\ u_i]$, where R, G, B, and L, a, b express the values of RGB color model and CIE Lab color space, respectively while the x and y express the coordinates of the pixels. Whereas u_i is used to indicate the density of edges. Where g_i is used to highlight the salient object part through the following Gaussian function:

$$g_i = \exp[-(\frac{x_i - x_c}{2\sigma_x^2} - \frac{y_i - y_c}{2\sigma_y^2})] \tag{1}$$

where, $\sigma_x = x_c$ and $\sigma_y = y_c$ are the image center co-ordinates, x_i and y_i indicate the superpixel co-ordinates, s_i and s_j are the ith and jth superpixels of the image. Sometimes due to less contrast or same color of the foreground and the background part is mistakenly considered as foreground. To overcome this issue, our focus is salient object instead of image center. To achieve this objective, we calculate salient object center using the following equation:

$$s_c = \begin{cases} x_c = \sum_{i,j=1}^{n} \frac{s_i}{\sum_{j=1}^{n} s_j} x_i \\ y_c = \sum_{i,j=1}^{n} \frac{s_i}{\sum_{j=1}^{n} s_j} y_i \end{cases} \tag{2}$$

Subsequently, the image is presented as $Z = [z_1, z_2, z_3, ..., z_n] \epsilon R^{D \times N}$, where N and D are the number of segments and features dimensions of the image, respectively. As a result, the calculated saliency maps with textural information have more effective representation as shown in Figure 2b,c, respectively.

(a) OI **(b)** SRD map **(c)** Updated SRD map **(d)** GT

Figure 2. The need for visual features for extracting a good saliency result is obvious from the depicted results. It is worth noting that the results in the second column are comparably less significant and missing a lot of real image information.

3.2. Heuristic Background Dictionary

In current SRD schemes, the background contrast, background prior, and boundary information is used to compute their SRD map. Following the previous assumptions, we also assembled a part of the background and boundary clues and named it as a Heuristic Background Dictionary (HBD). Since constructing this HBD, we also used the idea of center-remaining difference to capture high contrast around the salient objects near the center of the image. The HBD has persuasive results for simple natural images, however, for complex natural images, the resultant map contains a large amount of background noise. When the foreground region and background regions are implicated, and the contrast is much smaller, the HBD is less helpful for finding the foreground region. Consequently, when the background is complex it is difficult for ABM to train the HBD which is not capable of extracting complete information from the background, as a result, the salient region map contains background noises. To achieve improved SRD results, we accumulate the accurate background and boundary clues as for the dictionary bases. We computed the value of a segment i through the following expression:

$$U_{seg(i)} = \frac{\sum_{L=\{right,left,top,down\}} S_{i,L} \cdot \varphi(seg(i) \notin seg_L)}{\sum_{L=\{right,left,top,down\}} \varphi(seg(i) \notin seg_L)} \tag{3}$$

where, $\varphi(.)$ and seg_L represent the indicator function boundary segment set, respectively. According to [33,34], the different dataset contains the different size of the salient part and the largest salient object contains the 35% of the image. In a 15-pixel wide narrow border region, 98% belongs to the background [35]. Using this information, we selected the 30% of background pixels for constructing the dictionary. We used the dictionary-learning procedure to avoid the redundant sampling and computational problem in which the background samples are directly utilized as dictionary bases. This training procedure computes more compact heuristic background dictionary $T = [t_1, t_2, t_3, ..., t_n] \in Q^{p \times n}$. We use the following function to compute HBD as:

$$J_{T,E} = \arg\min_{T,E} \left\{ \|Y - TE\|_F^2 + \nu \|E\|_1 \right\} \quad s.\ t. \quad t_j^\top t_j = 1, \forall j \tag{4}$$

where, $Y \in R^{p \times n}$ used to signify the background segments sets, E is Representation-Coefficient Matrix (RCM) of Y based on T, while ν is used to balance the $\ell_F - norm$ and $\ell_1 - norm$ terms. The Equation (4) represents a joint-optimization function of T and E. Firstly, the T is initialized and fixed after that E is solved using [36] as it becomes a standard optimization problem. Then, we update T by fixing E through the Lagrange multiplier. This procedure is iterated till the values of $J_{T,E}$ are close enough and at that time, we are able to obtain a more reconstructive dictionary.

The compact appearance frameworks construct their background coefficient matrix which detains all of the fundamental characteristics of the background part, however, it is very sensitive to background noises. The dense appearance models provide more meaningful and basic descriptions of the background region as compared to the foreground region. For messy and complicated scenes, the ABM is less useful in computing the salient objects. So, we use the background contrast from four sides of the image boundary and designed four HBDs. Suppose, if the HBD cannot capture all of the information from one side of the image it will definitely collect some background information from the other sides. The salient objects are more accurately captured if we apply the clues and seed extracted from the four sides of the image. The proposed model HBD is designed to handle these issues. In view of the fact that the distinctive border of the image may possibly enclose a component of the salient object parts, the HBD is very effective and capable of appreciably eradicating these regions of the image that are considered as background noises as revealed in Figure 3. Subsequently, the left behind a set of superpixels is preferred as HBDs, which contain additional stable and consistent background information.

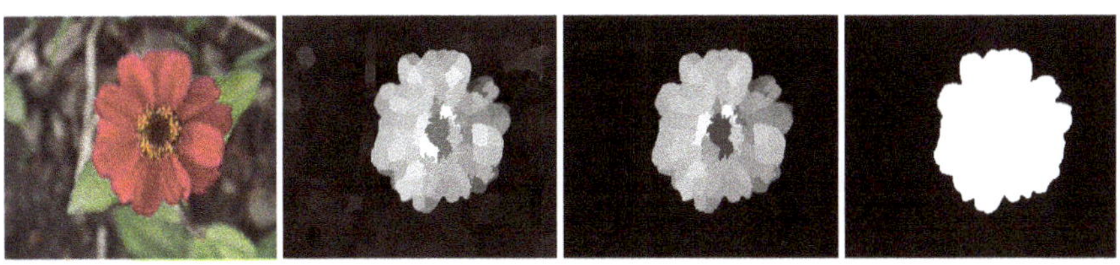

(a) OI **(b)** Result without HBD. **(c)** SRD map with HBD. **(d)** Ground truth

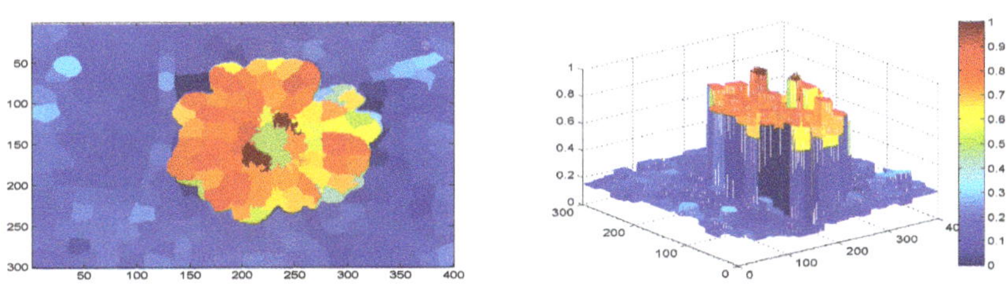

(e) SRD map with background noise. **(f)** SRD map with background noise.

Figure 3. *Cont.*

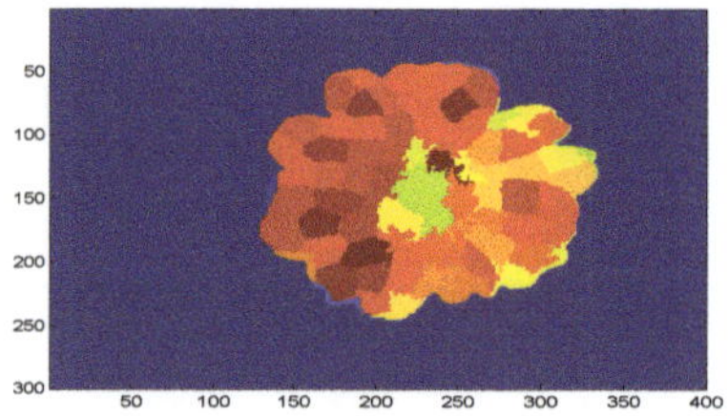

(g) SRD map without background noise.

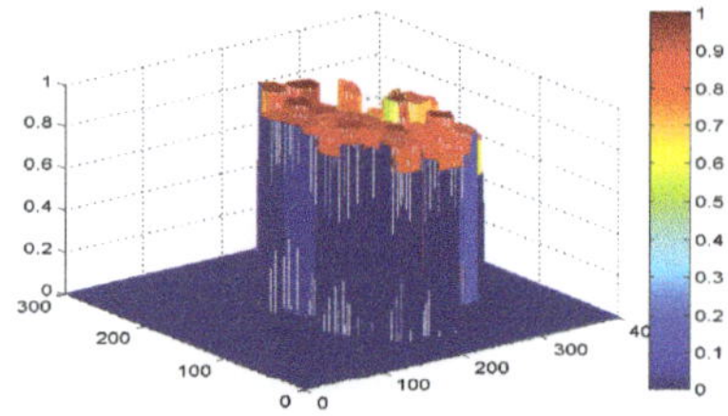

(h) SRD map without background noise.

Figure 3. The effectiveness of the heuristic background dictionary for highly precise and exact salient object maps extraction.

3.3. Appearance-Based Salient Region Detection

Superpixels appearance based saliency computation is the most important step of our model. The image boundary superpixel contains very important information which can be engaged to obtain the saliency maps. The methods based on a background dictionary [9–11] have convincing results whenever the salient objects pop out closer to the center part of the scene. However, when the salient objects significantly touch the image boundary and parts of them are wrongly considered as background. However, our designed HBD $T = [t_1, t_2, ..., t_m]$ has D-dimensional cues of boundary and 35% of background segments. We apply this reconstructive background dictionary to remove the background noise and to compute ABM saliency map. The classical SRD method [7,8] computes the dissimilarity between the coefficient of segment i as follows:

$$\alpha_i = V_T^{\top}(z_i - \bar{z}) \tag{5}$$

where, $\bar{z} = \sum_{i=1}^{n} z_i$ is the mean feature of Z and the eigenvalue and eigenvector is calculated via the normalized covariance matrix of T, $V_T = [v_1, v_2, v_3 ....v_{\pounds}]$. Then, the largest eigenvalues are selected to form the PCA bases for the reconstructive background dictionary. The corresponding saliency of segment i can be calculated using the following expression:

$$e_i = \|z_i - (V_T \alpha_i + \bar{z})\|_2^2 \tag{6}$$

We believe that the dense representation is more expressive to the background features, and it is more sensitive towards the noise. In general, the background part of the image is comparably uniform, sparse, on the contrary, the foreground part is comparably lesser and dense. The key motive for selecting the PCA framework is this when the salient objects are located at the image boundaries. In these typical cases, the background is the main ingredient. So, PCA can easily detect the foreground and filter outs the background. The PCA only deals with simple natural images, however, for complex natural images the resultant map contains a large amount of foreground noise. For cluttered images, the ABM is less effective in measuring salient regions. Dense appearance models, data points through a multivariate Gaussian distribution in feature space, and therefore, it is very difficult to detain multiple scattered patterns particularly when the number of examples is limited. To accomplish better performance of salient region detection, we need to accumulate more correct background information as reconstructive background dictionary bases. We use the background contrast from four sides of the image boundary and designed four HBDs. By utilizing the reconstructive background coefficient set from the top side, we compute the dense representation co-efficient of segment i as follows:

$$\alpha_{i,right} = V_{S_{i,right}}^{\top}(z_i - \bar{z}) \tag{7}$$

The saliency value of each segment is proportional to the dense representation. The dense representation of segment i using the topside dictionary can be calculated using the following expression:

$$e_{i,right} = \|z_i - (V_{S_{i,right}} \alpha_{i,right} + \bar{z})\|_2^2 \tag{8}$$

Particularly, the coarse salient region map of each superpixel z in a region r is extracted as follows:

$$S_{i,top}^{ABM} = \frac{1}{|r|} \sum_{z_i \in r} (1 - d_i) \times e_{i,top} \tag{9}$$

where d_i is the Euclidean distance of the superpixel x_i from the center part of the image, and $|r|$ express the numbers of superpixels in r. At the end, we normalize $S_{i,right}^{ABM}$, $i = 1, ..., n$ in the range $[0, 1]$ to generate the coarse salient region map from topside. Then the saliency maps are generated from remaining sides likewise and combined to generate S^{ABM} salient region map as depicted in Figure 4. Commonly, the salient part of the image is compact and restricted in a small part which is similar in appearance and consistency, whilst the background part is spread over the whole scene with the same pattern and uniformity. Thus, the superpixels in their correspondences sharing their geometrical appearance information and also their saliency scores. This thing specifies that the average remaining in a superpixel is equal to the saliency values in each region. Additionally, this averaging framework is designed to get rid of the most basic issue in saliency like: a number of small segments having higher contrast values are described through high saliency values sometimes, so the overall saliency of the entire salient object is comparably decreased.

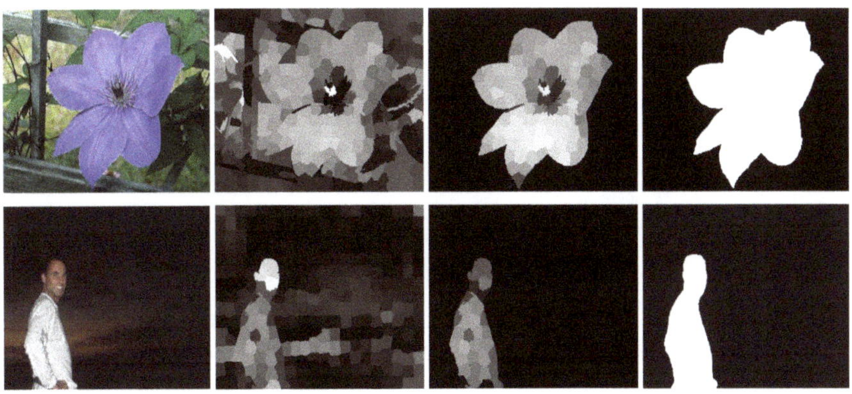

(a) OI (b) DRE map (c) ABM map (d) GT

Figure 4. The validity of obtaining a background coefficient matrix is noticeable from the demonstrated results. The results are arranged as OI, the dense representation error map, ABM map, and the GT.

3.4. Saliency Enhancement through a Regression-Based Model

We compute a graph $G = (V, E)$, where V is set of superpixels and E represents the boundary edges of the image. In [16,24,25], the following function is used to determine the saliency of all the superpixels as:

$$F = \arg\min_F \left(\underbrace{\sum_{i,j=1}^{n} w_{ij}(p_i - q_j)^2}_{Smoothness} + \beta \underbrace{\sum_{i=1}^{n} (F_i - r_i)^2}_{Fitting} \right) \tag{10}$$

where r_i is the ranking value for ith superpixel, $p_i = \frac{F_i}{\sqrt{g_{ii}}}$ is saliency of ith superpixel, and $q_j = \frac{F_j}{\sqrt{g_{jj}}}$ is the saliency of jth superpixel. $W = (w_{ij})_{n \times n}$ is the weight among two superpixels in the CIE LAB color space and is defined as follows:

$$w_{ij} = \exp - \frac{\| c_i - c_j \|}{2\sigma_w^2} \tag{11}$$

while c_i and c_j represent mean of superpixels i and j in a color model, respectively. Here σ_w is engage to balance the color weight. Equation (10) illustrates the energy function, the first expression in the Equation (10) is smoothness constraint while the second part is fitting constraint. Therefore, the ranking values of unranked data are computed by solving the above function as:

$$C = (D - \varphi W)^{-1} \tag{12}$$

where, $D = diag\{d_{11}, .., d_{nn}\}$, and $d_{ii} = \sum_j W_{ij}$ are degree matrix and weight matrix, respectively. While the parameter φ keeps a balance between the smoothness constraint and the fitting constraint. Basically, the optimized graph affinities are described through the inverse matrix C, these graph-affinities are extracted from the prearranged data signified as a graph through semi-supervised learning without integrating. It also specifies the overall weight between two connected superpixels and extracts their grouping information for SRD. We suppose that an image contains k types of features, so weight matrix and degree matrix are computed for k features as: $W_k = (w_{ij}^k)_{n \times n}$, and $D_k = (d_{ij}^k)_{n \times n}$. In our designed cost function, we take two $n \times 1$ vectors U and V, which are attained from the previous saliency results by normalizing in the interval of $0 \sim 1$. After that, we introduce two diagonal matrices $v = [v_{ii}] = diag(V)$ and $u = [u_{ii}] = diag(U)$. To combine numerous features in a single salient region map containing the smoother foreground and suppress background, we define our novel pairwise potential model as:

$$F_l = \arg \min_{F_l, l=1..,k} \sum_{l=1}^{k} \left(\underbrace{\lambda \sum_{i=1}^{n} \sum_{j=1}^{m} w_{ij}^l (F_i^l - F_j^l)^2}_{Smoothness} + \underbrace{\sum_{i=1}^{n} u_{ii}^l (F_i^l - 1)^2}_{Foreground} + \underbrace{\sum_{i=1}^{n} (1 - v_{ii}^l)(F_i^l)^2}_{Background} \right) \tag{13}$$

where, F_i and F_j are saliency values of segment i and segment j, respectively. While the λ is a balancing parameter. The first term on the right-hand side in energy function is the smoothness constraint. For a good saliency map the salient object should be even and smooth. The second term is used here to assign higher values to the foreground region. We employ this term for multi-features foreground computation and highlighting the foreground part. The last defined constraint is background constraint which assigns less weight to the background regions and also helps in creating well-defined boundaries of the salient objects. Previously designed methods are dependent on the color information for computing their saliency. However, the computed images lose their accuracy when the salient objects are pattern objects. To fully capture the salient objects, we combine the boundary, texture, geometry and spatial information to obtain our saliency results. The mean of color features are obtained from the superpixels and utilized after normalizing it. While the textural features like HOG and LBP feature are also extracted from the superpixels but after normalizing their histogram. The sum of texture and color discontinuities is computed through gradient G and utilized it as the boundary information. All of the above features are utilized to compute the weights of superpixels as:

$$w_{ij}^{l} = \exp - \Big(\sum_{l=1}^{k} (\frac{||c_i^l - c_j^l||}{\sigma_w^2}) + \beta \sum_{l=1}^{k} d_L(L_i^l, L_j^l) + \gamma \sum_{l=1}^{k} d_H(H_i^l, H_j^l) \Big) \tag{14}$$

where, the β and γ are used to control the weights between the superpixels. Here, we assign highest weight to the color parameter because it is more reliable than other features. We take the value of $k = 2$, because in this framework we are only dealing with two features. After putting the value of k this optimization function can be written as:

$$\begin{aligned} F_1, F_2 = \arg\min_{F_1,F_2} \frac{1}{2} (\lambda F_1^\top (D_1 - W_1) F_1 + \frac{1}{2} u (F_1 - D_1^{-1})^\top D_1 (F_1 - D_1^{-1}) + \lambda F_2^\top (D_2 - W_2) F_2 + \\ \frac{1}{2} u (F_2 - D_2^{-1})^\top D_2 (F_2 - D_2^{-1}) + \frac{1}{2} F_1 (1 - v) F_1^\top) \end{aligned} \tag{15}$$

We took the value of $k = 2$ to compute the optimal solution of this energy function. We take the derivative of this function with respect to F_1 and F_2 and putting it equal zero. Then we obtained the following expression as:

$$E_1 = 2\lambda(D_1 - W_1) + uD_1 + (I - v) \tag{16}$$

$$E_2 = 2\lambda(D_2 - W_2) + uD_2 + (I - v) \tag{17}$$

Motivated from [6], which observed the paired advantages of Lab and RGB color models for salient region detection, we engaged two types of visual information like E_1 and E_2 to extract our results. After that, we take the average of the salient region maps and normalize the computed result between the range $[0, 1]$ to obtain the final saliency region map. Figure 5 demonstrates the computed results through the proposed model with single and multi-featured.

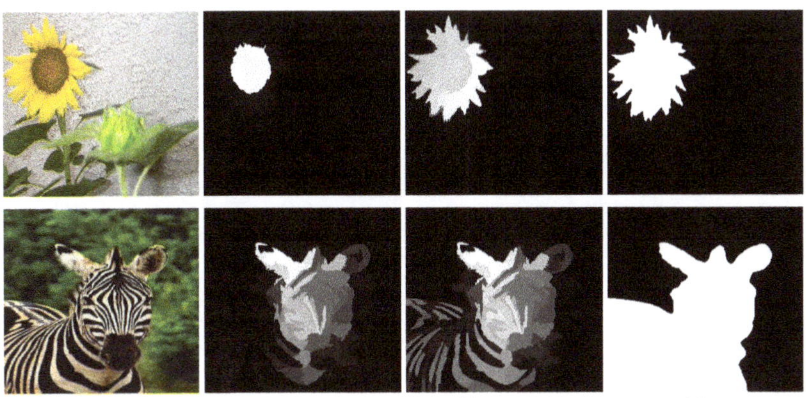

(**a**) OI (**b**) ABM map (**c**) RBM map (**d**) GT

Figure 5. Some examples demonstrating the difference between single and multi-level cues integration step. The results are arranged as OI, salient region map with single feature integration, and the saliency map extracted through multi-label features incorporation.

Instinctively, a region with higher contrast in representation to the neighboring elements always receives high saliency scores. However, the proposed multi-feature inference mechanism not only processes the salient regions of the image depending upon their degree of relevance but also assigns higher saliency scores computed from multi-features spaces. This property effects in highlighting the salient object parts more uniformly and suppressing the background regions. We can note that the ABM is more robust in dealing with the salient object at the image boundary. However, for complex natural images, the resultant map contains a large amount of foreground noise. The RBM is more efficient in dealing with the complex background but loses its efficiency when the objects are at the boundary of the image. Consequently, both the RBM and ABM are essential for computing a good salient region map as shown in Figure 6. In very complex background images, sometimes, background pixels included in the results, we can see artifacts in the computed maps due to the pre-processing. So, in order to remove these artifacts and background pixels, we engage the guided filter [37]. The guided filter produces the background and artifacts free smooth result as revealed in Figure 7.

(a) OI (b) Segmented map (c) ABM map (d) RBM map (e) Optimized map

Figure 6. We individually compare the salient region map of each stage of the proposed method by using ASD database [38]. The results are organized as OI, the segmented image, ABM salient region map, enhanced salient region map through RBM, and the final salient region map .

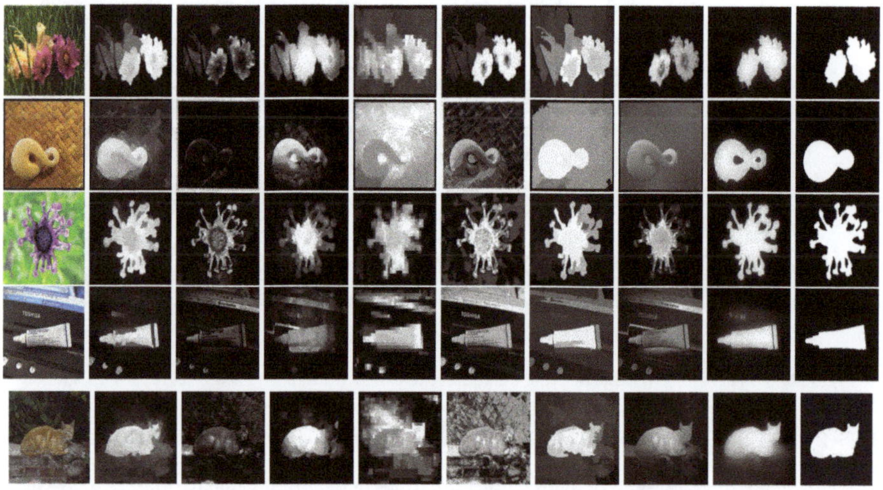

(a) OI (b) MI (c) RS (d) AM (e) BD (f) MC (g) HS (h) UC (i) Our (j) GT

Figure 7. Visual comparison of our scheme with some recent approaches using the ASD database. The SRD results are arranged as OI, MI, RS, AM, BD, MC, HS, UC, our scheme, and the GT. We can note that the SRD maps of our proposed scheme are very close to the GT.

4. Experimental Results

We analyzed and investigate our model on the five largest benchmark datasets against the seven state-of-the-art methods. For performance assessment, four evaluation measures are selected to completely analyze the proposed algorithm against seven preceding schemes. In the next section, we discuss the details of the selected benchmark datasets for performance evaluations.

4.1. Benchmark Datasets

To analyze the computed saliency results, there are many databases available that differ from one and another in size, number objects, and background. We employ a different database to assess and analyze the performance of our proposed algorithm. We assess our salient region detection model on five different standard databases that are: (1) ASD [38], (2) ECSSD [39], (3) DUT-OMRON [28], (4) SED2 [40], and (5) MSRA [41]. We prefer these databases for the following reasons: (1) background nature, (2) complexity level, (3) a large number of images, (4) the different number of objects in a scene, and (5) potential benchmark databases. Firstly, we test the performance of the proposed model in the ASD database. The images in this database have a large variety in the background structure like a simple, smooth, complex, and multifaceted nature. The ASD database contains 1000 images with pixel-wise annotated ground truths. The purpose to include SED2 databases is to assess the performance of our model with an image contains multiple objects. Lastly, we analyze our model over Extended Complex Scene Saliency Data-set (ECSSD), which contains 1000 images that are semantically meaningful, however, having complex and natural images.

4.2. Preceding Methods Selected for Comparison

Our SRD model is compared against seven state-of-the-art models. We first visually and then graphically compare to check and validate our framework. The schemes we compare with our method are chosen due to the following four reasons: (1) recency, (2) citations, (3) computation complexity, and (4) variety. These models are: AM [29], BD [42], RS [43], MC [44], MI [30], HS [39], and UC [31]. The source codes of some of the above-defined approaches are easily accessible for public. While other we obtained from the saliency results generated by Cheng et al. [34]. Only a few of the source codes are directly downloaded from the author's web, therefore, we utilized their source codes to extract the saliency results for comparison purpose.

4.3. Evaluation Metrics

Numerous techniques are applied to evaluate the concurrence between the obtained results and the GT. Before computing the evaluation metrics, the produced salient region maps should be changed in binary form to estimate the generated map. We also apply the adaptive threshold as discussed in [34], the thresholding is used to get the binary mask of salient region map S, that is calculated as:

$$T_h = \frac{1}{w \times h} \sum_{a=1}^{h} \sum_{b=1}^{w} S(a,b) \tag{18}$$

whereas, w and h represent the height and width of saliency map, respectively.

4.3.1. Precision-Recall

The saliency map S is converted to the binary-mask M using the given ground truth T. The PR-curve is computed using this expression:

$$Precision = \frac{|M \bigcap T|}{|M|}, Recall = \frac{|T \bigcap M|}{|T|} \tag{19}$$

4.3.2. F-Score

F-score is calculated using the Precision-Recall, the evaluation of the SRD is not complete without F-score. The F-score is computed using the following expression:

$$F_v = \frac{(1 + v^2) \times Precision \times Recall}{v^2 \times (Precision + Recall)} \quad (20)$$

All of the compared method take the value of $v = 0.3$. So, we have take the value of $v = 0.3$ for a fair comparison.

4.3.3. Receiver Operating Characteristics

The ROC-curve is obtained using the binary mask M with a fixed threshold as:

$$TPR = \frac{|\bar{M} \bigcap T|}{|\bar{M}|}, FPR = \frac{|M \bigcap \bar{T}|}{|\bar{T}|.} \quad (21)$$

where, $\bar{T}$ is opposite of T and $\bar{M}$ is opposite of M. The ROC-curve is obtained through TPR and FPR with changing the value of the fixed threshold.

4.3.4. Mean Absolute Error

To check the worth of SRD maps might have high significance as compared to binary mask. We also applied the MAE between the continuous SRD map S and the ground truth T, both are normalized in the range $[0, 1]$. The MAE value is defined as:

$$MAE = \frac{1}{w \times h} \sum_{a=1}^{h} \sum_{b=1}^{w} |\bar{S}(a,b) - \bar{T}(a,b)| \quad (22)$$

4.4. Implementation and Analysis

We visually and graphically analyze the designed algorithm against preceding algorithms. We also assess the performance of the proposed model with different parameters using PR-curves. In the next section, we describe the comparison of our model with existing schemes.

4.4.1. Parameter Settings

The performance of our model is affected by different parameters. When we are comparing the performance of our model, we used the following parameter settings: $\beta = 0.10$, $\gamma = 10$, $\lambda = 0.35$, $\sigma_w = 0.05$, and $N = 200$, where N represents the number of superpixels. Figure 8 demonstrates the effect of these balancing parameters on the performance of our model. We execute simulations 5 times repetitively to avoid any uncertainty due to the arbitrary initialization.

4.4.2. Evaluation of Our Algorithm

In this section, we evaluate different elements of the designed framework and their impact on the performance in detail. The PR-curves with and with the single and multi-features are also demonstrated in Figure 9. We can also see that the final map with the multi-features is little higher than the final map with a single feature. The final map with a single-feature loses some information during pre-processing. We evaluate the proposed method against two most recent SRD schemes: NS [45], and MSC [46] in Table 1. We used the F-measure, AUC, and MAE to check the performance of our model against these two schemes. We notice that our model outperforms than the opponent schemes in selected metrics with higher F-score, AUC and lesser MAE.

Table 1. The performance comparison of our model with recent schemes.

Models	ECSSD			SED2			DUT-OMRON			ASD		
	NS	MSC	Our	NS	MSC	Our	NS	MSC	Our	NS	MSC	Our
F-score	0.710	0.713	0.73	0.775	0.791	0.802	0.616	0.60	0.699	0.870	0.92	0.93
AUC	0.90	0.89	0.907	0.85	0.859	0.861	0.887	0.883	0.895	0.935	0.952	0.953
MAE	0.245	0.229	0.222	0.182	0.155	0.145	0.149	0.126	0.125	0.095	0.080	0.070

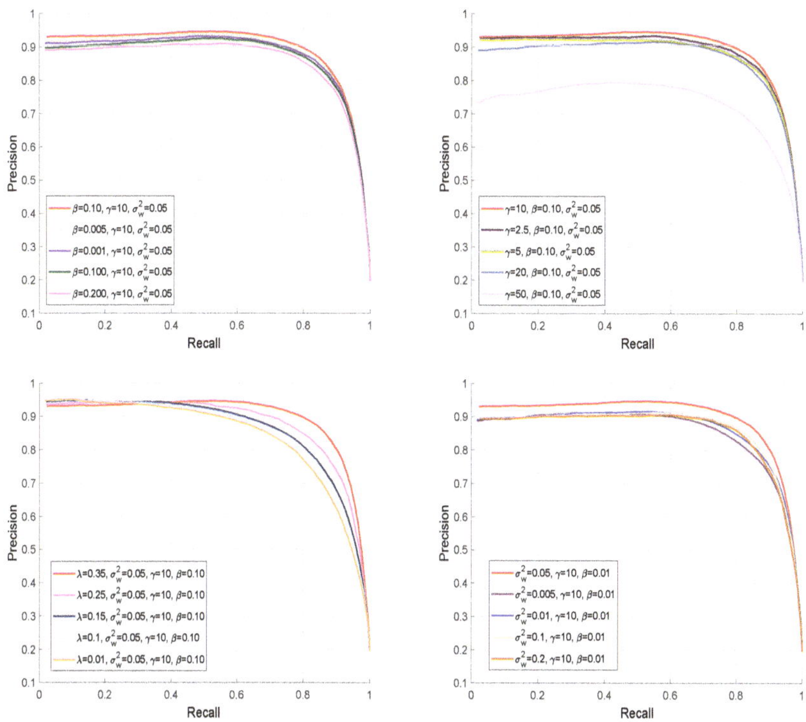

Figure 8. PR-curves to validate our proposed method with different parameters values for the MSRA database. The balancing parameter is tuned at different values to verify the refinement function and their effect on the final SRD map.

4.4.3. ASD Database

We assess the performance of our scheme with previous methods using the ASD dataset as revealed in Figure 10. The reason for selecting the ASD database is to investigate the behavior of our scheme with images having different complexity levels and diversified pattern. We examine and evaluate the proposed method against seven most well-known SRD schemes such as: AM [29], BD [42], RS [43], MC [44], MI [30], HS [39], and UC [31]. We used the ROC-curve, F-measure, PR-curve, and MAE to check the performance of our model. We notice that our model outperforms than the opponent schemes in selected metrics with a higher precision, recall, F-measure, and lesser mean absolute error. The RS [43], HS [39], and MC [44] also achieved good. We considers three latest deep learning-based models for evaluation like [29–31]. We can note from the Figure 10 that proposed model obtains similar precisions with most deep-learning methods and suppresses the recalls, so the proposed method yields relatively lower F-measure scores. However, the proposed model is without preparing expensive ground truth annotations for training the model and overall performs comparable with these deep-learning methods. The proposed method is free of computing power and ground truth annotations and can provide simplicity and easy-to-use generality in many practical inexpensive applications. From the results, we observe that our SRD approach is more efficient in highlighting the salient objects as compared to the other recent models.

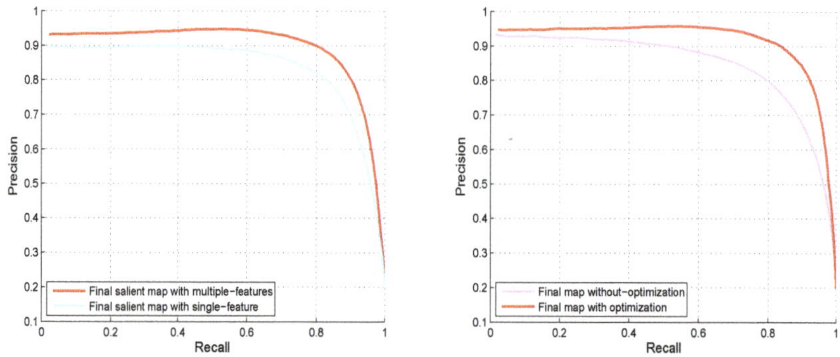

Figure 9. Graphical performance comparison of different stages of our method using PR-curves to validate the single feature, multi-featured, and enhanced results using the MSRA dataset.

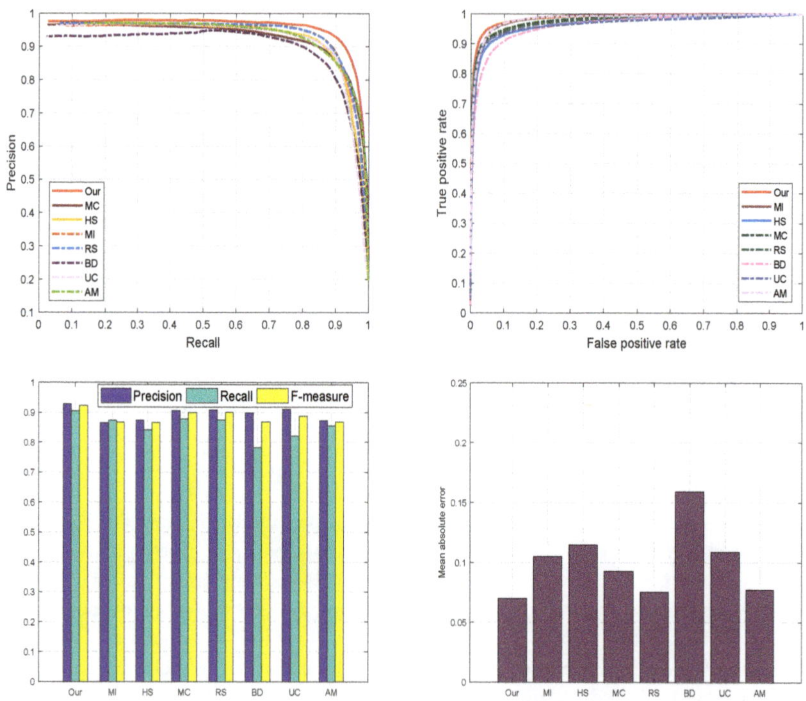

Figure 10. The graphical assessment of our model against seven current approaches AM [29], BD [42], RS [43], MC [44], MI [30], HS [39], UC [31] and our proposed model using the ASD dataset.

4.4.4. DUT-OMRON Database

We also evaluate the performance of the proposed model on a DUT-OMRON database. The motive for electing DUT-OMRON database is this, it contains a large number of images with different complexity levels of the background. Most probably all SRD approaches utilize this database to analyze their methods, therefore, this database is our first priority to evaluate our proposed approach as shown in Figure 11. We verify the performance of our proposed model graphically using the preprocessing and post-processing results. We choose PR and ROC-curve to assess the performance of our proposed method. The resulting graphs are illustrated in Figure 11. Nevertheless, MC [44], RS [43], and BD [42] also demonstrate persuasive results. We notice from our analysis that our approach is more effective and more efficient in highlighting the salient objects than the other discussed methods.

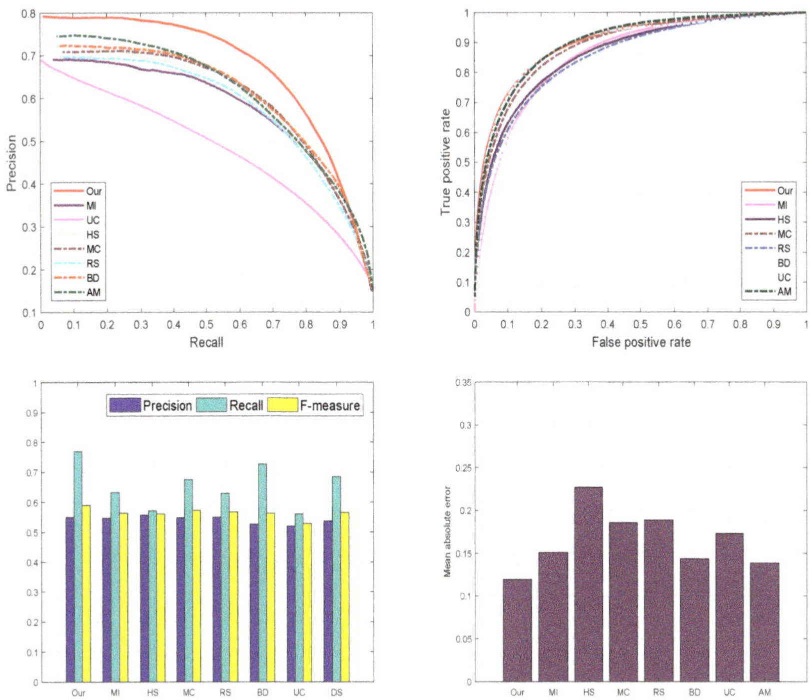

Figure 11. The graphical evaluation of our method with seven current approaches such as AM [29], BD [42], RS [43], MC [44], MI [30], HS [39], UC [31] and our proposed model on the DUT-OMRON database.

4.4.5. ECSSD Database

Moreover, we as well engaged ECSSD database [39] to assess our mechanism graphically. ECSSD database contains more natural images with a diversified pattern for both foreground and background. The reason for selecting ECSSD database is to investigate the behavior of our scheme with images having different complexity levels and diversified pattern. We examine and evaluate the proposed method against seven most well-known SRD schemes such as: AM [29], BD [42], RS [43], MC [44], MI [30], HS [39], and UC [31] on the ECSSD database to declare the strength of our algorithm. We pick four different criteria which are mainly used in the literature to assess the performance of SRD methods. These criteria are PR-curve, ROC curve, F-score, and MAE to check the performance of our proposed approach. From the series of experiments, we found that our proposed method achieves very good results as compared to above-defined approaches. On the other hand, RS [43], BD [42], and UC [31] as well accomplished fine results on all four SRD metrics. Our approach remains very unswerving in all defined evaluation measures and demonstrates significant performance as shown in Figure 12.

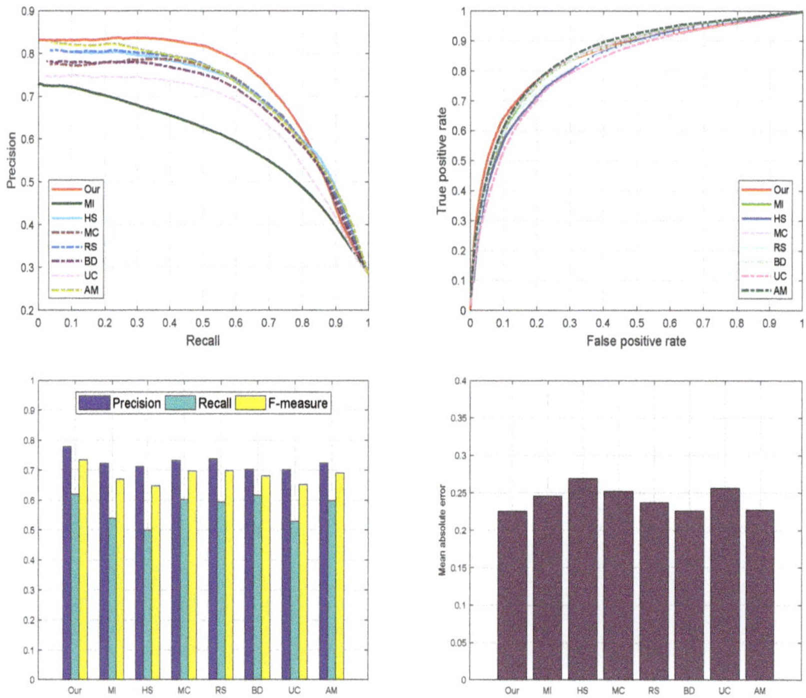

Figure 12. Graphical evaluation of our model using the PR-curve, F-measure, ROC-curve, and MAE with seven most recent models.

4.4.6. SED2 Data-Set

Additionally, we employed SED2 dataset [40] to evaluate and validate the proposed method graphically. The motive for electing SED2 database is to assess the performance of our scheme through an image with two objects. We analyze and compare the proposed method against fourteen most famous state-of-the-art approaches such as: AM [29], BD [42], RS [43], MC [44], MI [30], HS [39], and UC [31] on SED2 database to assure the validity of our algorithm. We choose four different criteria like PR-curve, ROC curve, F-measure, and MAE to estimate the strengths and bounds of our SRD approach. Our SRD model remains very consistent in all the define evaluation measures and shows a remarkable performance as illustrated in Figure 13.

4.4.7. Limitations

The designed method outperforms against above-discussed state-of-the-art SRD methods with the higher PR values. However, the performance of our scheme is not very acceptable in some cases. These typical cases are shown in Figure 14. The proposed method has not achieved very persuasive results when the color of the foreground is similar to the background; in this situation, the salient object is not salient accurately, some of the background pixels are combined with the obtained results and size of the results do not remain significant.

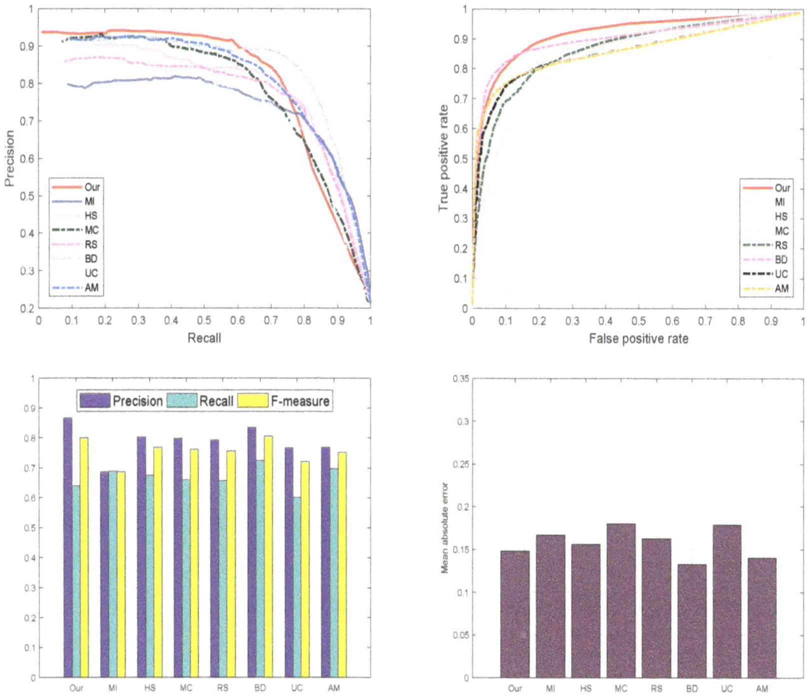

Figure 13. The graphical analysis of our SRD using four different saliency measures with other techniques.

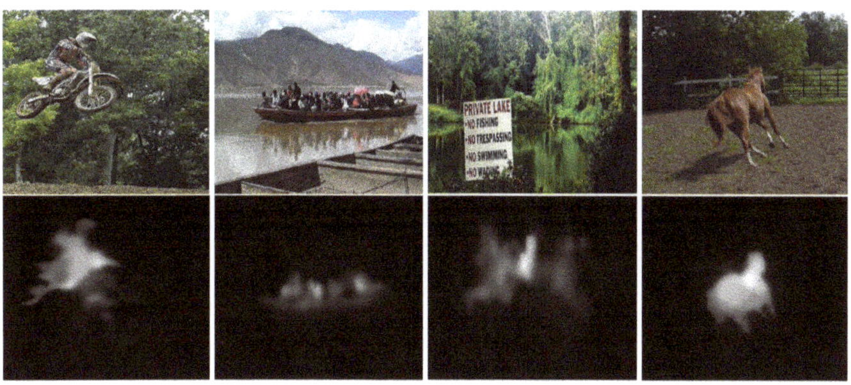

Figure 14. A few cases where our model performance is not very persuasive.

4.4.8. Execution Time

The execution time/image of the proposed model with some previous methods by using MATLAB implementation using the ECSSD data set is elaborated in Table 2. The running time of all the schemes described in the table is achieved through the machine having the Intel Dual Core $i3-2310M$, 2.10 GHz CPU, and 4 GB RAM. Our designed framework is robust than the other state-of-the-art SRD methods. Specially, the SLIC [32] consumes 0.16 s almost 50% of the original time.

Table 2. The comparison of our model with seven state-of-the-art techniques for average running time (seconds per image).

Method	Time(s)	Code
AM [29]	0.185	Matlab
BD [42]	0.453	Matlab
MC [44]	0.547	Matlab
MI [30]	0.025	Matlab
UC [31]	0.495	Matlab
RS [43]	0.108	Matlab
HS [39]	25.3	Matlab
Our	0.32	Matlab

5. Conclusions

In this work, we have introduced a new density-based and regression-based salient regions detection model. To capture the useful structural information, we segmented the image into multiple uniform segments. To obtain more background information and to evenly suppress the background, we constructed side-specific dictionaries. Then, we calculated the more effective contrast-based salient region map using our ABM. To strengthen the generated results, we use RBM to generate the multi-label cues rarity for each segment. To incorporate pre-computed results followed by an optimization method that construct more even, accurate and precise salient regions map. Some previous approaches exploit the single-feature of the background or foreground to produce their saliency results. However, the proposed model infers multi-label color features and demonstrates better performance as compared to the preceding appearance-based learning schemes.

Author Contributions: M.M.S.F. developed the main idea of the proposed scheme, performed simulation and wrote the manuscript. G.A. performed mathematical modeling and helped in simulation. All the editing is done by A.M., M.R.A. and M.Z.F. The refinement of the article is completed under the supervision of Q.C.

Funding: This research is supported by the National Natural Science Foundation of China (Grant No. 61572395 and 6167516).

Acknowledgments: This research is supported by the National Natural Science Foundation of China (Grant No. 61572395 and 61675161) and partly supported by the National Natural Science Funds for International Young Scientists (Grants No. 51850410517).

Conflicts of Interest: The authors declare no conflict of interest.

References

1. Han, B.; Zhu, H.; Ding, Y. Bottom-up saliency based on weighted sparse coding residual. In Proceedings of the ACM International Conference on Multimedia, Scottsdale, AZ, USA, 28 November–1 December 2011; pp. 1117–1120.
2. Yang, J.; Yang, M.-H. Top-down visual saliency via joint CRF and dictionary learning. In Proceedings of the IEEE Conference on Computer Vision and Pattern Recognition, Providence, RI, USA, 16–21 June 2012; pp. 2296–2303.
3. Mehmood, I.; Sajjad, M.; Ejaz, W.; Baik, S.W. Saliency-directed prioritization of visual data in wireless surveillance networks. *Inf. Fusion* **2015**, *24*, 16–30. [CrossRef]
4. Sajjad, M.; Ullah, A.; Ahmad, J.; Abbas, N.; Rho, S.; WookBaik, S. Integrating salient colors with rotational invariant texture features for image representation in retrieval system. *Multimed. Tools Appl.* **2018**, *77*, 4769–4789. [CrossRef]
5. Sajjad, M.; Ullah, A.; Ahmad, J.; Abbas, N.; Rho, S.; WookBaik, S. Saliency-weighted graphs for efficient visual content description and their applications in real-time image retrieval systems. *J. Real-Time Image Process.* **2017**, *13*, 431–447.
6. Borji, A.; Itti, L. Exploiting local and global patch rarities for saliency detection. In Proceedings of the 2012 IEEE Conference on Computer Vision and Pattern Recognition (CVPR), Providence, RI, USA, 16–21 June 2012; pp. 478–485.

7. Duan, L.; Wu, C.; Miao, J.; Qing, L.; Fu, Y. Visual saliency detection by spatially weighted dissimilarity. In Proceedings of the 2011 IEEE Conference on Computer Vision and Pattern Recognition (CVPR), Colorado Springs, CO, USA, 20–25 June 2011; pp. 473–480.
8. Lu, H.; Li, X.; Zhang, L.; Ruan, X.; Yang, M.H. Dense and Sparse Reconstruction Error Based Saliency Descriptor. *IEEE Trans. Image Process.* **2016**, *25*, 1592–1603. [CrossRef] [PubMed]
9. Huo, L.; Yang, S.; Jiao, L.; Wang, S.; Shi, J. Local graph regularized coding for salient object detection. *Infrared Phys. Technol.* **2016**, *77*, 124–131. [CrossRef]
10. Huo, L.; Yang, S.; Jiao, L.; Wang, S.; Wang, S. Local graph regularized sparse reconstruction for salient object detection. *Neurocomputing* **2016**, *194*, 348–359. [CrossRef]
11. Yang, C.; Zhang, L.; Lu, H. Graph Regularized Saliency Detection With Convex-Hull-Based Center Prior. *IEEE Signal Process. Lett.* **2013**, *20*, 637–640. [CrossRef]
12. Hou, X.; Zhang, L. Dynamic visual attention: Searching for coding length increments. Advances in Neural Information Processing Systems 21. In Proceedings of the 22nd Annual Conference on Neural Information Processing Systems, Vancouver, BC, Canada, 8–11 December 2008; pp. 681–688.
13. Shen, X.; Wu, Y. A unified approach to salient object detection via low rank matrix recovery. In Proceedings of the IEEE Conference on Computer Vision and Pattern Recognition, Providence, RI, USA, 16–21 June 2012; pp. 853–860.
14. Li, Y.; Zhou, Y.; Xu, L.; Yang, X.; Yang, J. Incremental sparse SRD. In Proceedings of the IEEE International Conference on Image Processing, Cairo, Egypt, 7–10 November 2009; pp. 3093–3096.
15. Sajjad, M.; Mehmood, I.; Baik, S.W. Image super-resolution using sparse coding over redundant dictionary based on effective image representations. *J. Vis. Commun. Image Represent.* **2015**, *26*, 50–65. [CrossRef]
16. Zhang, L.; Zhao, S.; Liu, W.; Lu, H. SRD via sparse reconstruction and joint label inference in multiple features. *Neurocomputing* **2015**, *155*, 1–11. [CrossRef]
17. Jia, C.; Qi, J.; Li, X.; Lu, H. Saliency detection via a unified generative and discriminative model. *Neurocomputing* **2015**, *173*, 406–417. [CrossRef]
18. Harel, J.J.; Koch, C.; Perona, P. Graph-based visual saliency. Advances in Neural Information Processing Systems 19. In Proceedings of the Twentieth Annual Conference on Neural Information Processing Systems, Vancouver, BC, Canada, 4–7 December 2006; pp. 545–552.
19. Ma, Y.-F.; Zhang, H.-G. Contrast-based image attention analysis by using fuzzy growing. In Proceedings of the Eleventh ACM International Conference on Multimedia (MULTIMEDIA '03), Berkeley, CA, USA, 2–8 November 2003; ACM: New York, NY, USA, 2003; pp. 374–381.
20. Lin, M.; Zhang, C.; Chen, Z. Global feature integration based salient region detection. *Neurocomputing* **2015**, *159*, 1–8. [CrossRef]
21. Cheng, M.-M.; Zhang, G.-X.; Mitra, N.J.; Huang, X.; Hu, S.-M. Global contrast based salient region detection. In Proceedings of the IEEE Conference on Computer Vision and Pattern Recognition, Colorado Springs, CO, USA, 20–25 June 2011; pp. 409–416.
22. Cheng, M.; Mitra, N.J.; Huang, X.; Torr, P.H.S.; Hu, S. Global Contrast Based Salient Region Detection. *IEEE Transa. Pattern Anal. Mach. Intell.* **2015**, *37*, 569–582. [CrossRef] [PubMed]
23. Wang, Q.; Zhu, G.; Yuan, Y. Multi-spectral dataset and its application in saliency detection. *Comput. Vis. Image Understand.* **2013**, *117*, 1748–1754. [CrossRef]
24. Lin, M.; Zhang, C.; Chen, Z. Predicting salient object via multi-level features. *Neurocomputing* **2016**, *205*, 301–310. [CrossRef]
25. Wang, H.; Dai, L.; Cai, Y.; Sun, X.; Chen, L. Salient object detection based on multi-scale contrast. *Neural Netw.* **2018**, *101*, 47–56. [CrossRef] [PubMed]
26. Li, S.; Lu, H.; Lin, Z.; Shen, X.; Price, B. Adaptive Metric Learning for SRD. *IEEE Trans. Image Process.* **2015**, *24*, 3321–3331. [CrossRef] [PubMed]
27. Li, H.; Lu, H.; Lin, Z.; Shen, X.; Price, B. Inner and Inter Label Propagation: Salient Object Detection in the Wild. *IEEE Trans. Image Process.* **2015**, *24*, 3176–3186. [CrossRef] [PubMed]
28. Yang, C.; Zhang, L.; Lu, H.; Ruan, X.; Yang, M.-H. SRD via Graph-Based Manifold Ranking. In Proceedings of the 2013 IEEE Conference on Computer Vision and Pattern Recognition (CVPR), Portland, OR, USA, 23–28 June 2013; pp. 3166–3173.

29. Zhang, P.; Wang, D.; Lu, H.; Wang, H.; Ruan, X. Amulet: Aggregating Multi-level Convolutional Features for Salient Object Detection. In Proceedings of the 2017 IEEE International Conference on Computer Vision (ICCV), Venice, Italy, 22–29 October 2017; pp. 202–211.
30. Huang, F.; Qi, J.; Lu, H.; Zhang, L.; Ruan, X. Salient Object Detection via Multiple Instance Learning. *IEEE Trans. Image Process.* **2017**, *26*, 1911–1922. [CrossRef] [PubMed]
31. Zhang, P.; Wang, D.; Lu, H.; Wang, H.; Yin, B. Learning Uncertain Convolutional Features for Accurate Saliency Detection. In Proceedings of the 2017 IEEE International Conference on Computer Vision (ICCV), Venice, Italy, 22–29 October 2017; pp. 212–221.
32. Achanta, R.; Shaji, A.; Smith, K.; Lucchi, A.; Fua, P.; Susstrunk, S. SLIC superpixels compared to state-of-the-art superpixel methods. *IEEE Trans. Pattern Anal. Mach. Intell.* **2012**, *34*, 2274–2282. [CrossRef] [PubMed]
33. Borji, A.; Cheng, Mi.; Jiang, H.; Li, J. Salient Object Detection: A Survey. *arXiv* **2014**, arXiv:1411.5878.
34. Borji, A.; Cheng, M.-M.; Jiang, H.; Li, J. Salient Object Detection: A Benchmark. *IEEE Trans. Image Process.* **2015**, *24*, 5706–5722. [CrossRef] [PubMed]
35. Jiang, H.; Wang, J.; Yuan, Z.; Wu, Y.; Zheng, N.; Li, S. Salient object detection: A discriminative regional feature integration approach. In Proceedings of the IEEE Conference on Computer Vision and Pattern Recognition, Portland, OR, USA, 23–28 June 2013; pp. 2083–2090.
36. Yang, M.; Zhang, L.; Yang, J.; Zhang, D. Metaface learning for sparse representation based face recognition. In Proceedings of the 2010 IEEE International Conference on Image Processing, Hong Kong, China, 26–29 September 2010; pp. 1601–1604.
37. He, K.; Sun, J.; Tang, X. Guided Image Filtering. *IEEE Trans. Pattern Anal. Mach. Intell.* **2013**, *35*, 1397–1409. [CrossRef]
38. Achanta, R.; Hemami, S.; Estrada, F.; Susstrunk, S. Frequency-tuned salient region detection. In Proceedings of the IEEE Conference on Computer Vision and Pattern Recognition, Miami, FL, USA, 20–25 June 2009; pp. 1597–1604.
39. Yan, Q.; Xu, L.; Shi, J.; Jia, J. Hierarchical saliency detection. In Proceedings of the 2013 IEEE Conference on Computer Vision and Pattern Recognition (CVPR), Portland, OR, USA, 23–28 June 2013; pp. 1155–1162.
40. Alpert, S.; Galun, M.; Basri, R.; Brandt, A. Image Segmentation by Probabilistic Bottom-Up Aggregation and Cue Integration. In Proceedings of the 2007 IEEE Conference on Computer Vision and Pattern Recognition, Minneapolis, MN, USA, 17–22 June 2007.
41. Liu, T.; Sun, J.; Zheng, N.-N.; Tang, X.; Shum, H.-Y. Learning to detect a salient object. In Proceedings of the 2007 IEEE Conference on Computer Vision and Pattern Recognition, Minneapolis, MN, USA, 17–22 June 2007; pp. 1–8.
42. Wang, Z.; Xiang, D.; Hou, S.; Wu, F. Background-Driven Salient Object Detection. *IEEE Trans. Multimedia* **2017**, *19*, 750–762. [CrossRef]
43. Zhang, L.; Yang, C.; Lu, H.; Ruan, X.; Yang, M. Ranking Saliency. *IEEE Trans. Pattern Anal. Mach. Intell.* **2017**, *39*, 1892–1904. [CrossRef] [PubMed]
44. Zhang, L.; Ai, J.; Jiang, B.; Lu, H.; Li, X. Saliency Detection via Absorbing Markov Chain with Learnt Transition Probability. *IEEE Trans. Image Process.* **2018**, *27*, 987–998. [CrossRef]
45. Zhang, Y.Y.; Zhang, S.; Zhang, P.; Zhang, X. Saliency detection via background and foreground null space learning. *Signal Process. Image Commun.* **2019**, *70*, 271–281. [CrossRef]
46. Ji, Y.; Zhang, H.; Tseng, K.-K.; Chow, T.W.S.; Wu, Q.M.J. Graph model-based salient object detection using objectness and multiple saliency cues. *Neurocomputing* **2019**, *323*, 188–202. [CrossRef]

sensors

Article

Local Parallel Cross Pattern: A Color Texture Descriptor for Image Retrieval

Qinghe Feng [1], Qiaohong Hao [2], Mateu Sbert [2,3], Yugen Yi [4], Ying Wei [1,*] and Jiangyan Dai [5,*]

[1] College of Information Science and Engineering, Northeastern University, Shenyang 110004, China; 1510377@stu.neu.edu.cn
[2] College of Intelligence and Computing, Tianjin University, Tianjin 300350, China; qiaohonghao@gmail.com
[3] Institute of Informatics and Applications, University of Girona, 17017 Girona, Spain; mateusbert@mac.com
[4] School of Software, Jiangxi Normal University, Nanchang 330022, China; yiyg510@jxnu.edu.cn
[5] School of Computer Engineering, Weifang University, Weifang 261061, China
* Correspondence: weiying@ise.neu.edu.cn (Y.W.); daijy@wfu.edu.cn (J.D.); Tel.: +86-024-83688326 (Y.W.)

Received: 2 December 2018; Accepted: 11 January 2019; Published: 14 January 2019

Abstract: Riding the wave of visual sensor equipment (e.g., personal smartphones, home security cameras, vehicle cameras, and camcorders), image retrieval (IR) technology has received increasing attention due to its potential applications in e-commerce, visual surveillance, and intelligent traffic. However, determining how to design an effective feature descriptor has been proven to be the main bottleneck for retrieving a set of images of interest. In this paper, we first construct a six-layer color quantizer to extract a color map. Then, motivated by the human visual system, we design a local parallel cross pattern (LPCP) in which the local binary pattern (LBP) map is amalgamated with the color map in "parallel" and "cross" manners. Finally, to reduce the computational complexity and improve the robustness to image rotation, the LPCP is extended to the uniform local parallel cross pattern (ULPCP) and the rotation-invariant local parallel cross pattern (RILPCP), respectively. Extensive experiments are performed on eight benchmark datasets. The experimental results validate the effectiveness, efficiency, robustness, and computational complexity of the proposed descriptors against eight state-of-the-art color texture descriptors to produce an in-depth comparison. Additionally, compared with a series of Convolutional Neural Network (CNN)-based models, the proposed descriptors still achieve competitive results.

Keywords: visual sensor; image retrieval; human visual system; local parallel cross pattern

1. Introduction

Since a huge number of image corpora have been produced by visual sensor equipment, an increasing demand for efficient encoding and indexing of these image corpora has attracted the attention of a considerable number of researchers [1–7]. Thanks to the investigators' breakthroughs, a myriad of methods [8–28] have continuously been developed for encoding and indexing.

In early work, the local binary pattern (LBP) [8], a grayscale texture descriptor, was first proposed for encoding the center pixel and its neighborhood pixels. Afterwards, owing to the disadvantage of losing global information, the LBP was extended to the LBP variance (LBPV) [9] which was amalgamated with global rotation-invariant matching for texture classification. Further, in order to completely detail the local differences among the central pixel and its neighborhood pixels, the completed local binary pattern (CLBP) operator [10] was designed for rotation-invariant feature representation. The local derivative pattern (LDP) [11] was then produced by refining the magnitude difference in local neighborhoods. Along another line, taking into account the situation of non-uniform lighting conditions, the local ternary pattern (LTP) [12] was introduced, and it was combined with kernel principal component analysis (KPCA) to improve its robustness to illumination. Later, the LTP was further modified into the local tetra pattern (LTrP) [13] by using the first-order derivatives in the vertical and horizontal directions. After that, the LTP was again extended to the local maximum edge binary patterns (LMEBP) [14], and the LMEBP were combined with the Gabor transform used for image retrieval and object tracking. Besides these, to achieve robustness to uniform and Gaussian noises, the noise-resistant LBP (NRLBP) [15] was constructed to preserve local structure information. Inspired by the fusion strategy, the local neighborhood difference pattern (LNDP) [16] was concatenated with the LBP map to integrate the local intensity difference information and the local binary information in a parallel manner. However, all the above methods are confined within grayscale image processing, so the major drawback of these construction processes is the inevitable loss of color information.

In recent years, a series of color texture descriptors [17–29] have been sequentially developed for color image processing. Among them, the local oppugnant color texture pattern (LOCTP) [17], a variant of the multi component LTrP, was combined with the colored pattern appearance model (CPAM) in the YCbCr, HSV, and RGB color spaces. In reference [18], according to the adder and decoder concepts, the multi-channel adder local binary pattern (maLBP) and the multi-channel decoder local binary pattern (mdLBP) were designed to combine the LBP maps in the R, G, and B components. After that, a class of pairwise-based local binary patterns [19,20] and a series of color-edge approaches [21–23] were introduced to classify and retrieve natural color images. Recently, on the basis of intra- and inter-channel encoding concepts, the opponent color local binary patterns (OCLBP) were proposed by Mäenpää et al. [24]. Further, in reference [25], Bianconi et al. extended the OCLBP to the improved opponent color local binary patterns (IOCLBP), in which the point-to-point thresholding was replaced by point-to-average thresholding. Considering the graph-based fusion framework, the bag-of-words of local features and the color local Haar binary pattern were integrated by Li et al. [26]. Quite recently, in order to systematically analyze the robustness to illumination changes, a bag of color texture descriptors [27] was studied under 46 lighting conditions. At the same time, with the help of a non-linear support vector machine, the orthogonal combination of local binary patterns (OC-LBP) was concatenated with the color histogram (CH) [28].

In this paper, we present the main following contributions:

1. We design a six-layer color quantizer that is applied to quantize the a* and b* components for color map extraction.
2. We construct a local parallel cross pattern (LPCP) in which the LBP map and the color map are integrated into a whole framework.
3. We further extend the LPCP to the uniform local parallel cross pattern (ULPCP) and the rotation-invariant local parallel cross pattern (RILPCP) to reduce the computational complexity and achieve robustness to image rotation.
4. We benchmark the comparative experiments with eight state-of-the-art color texture descriptors on eight benchmark datasets to illustrate the effectiveness, efficiency, robustness, and computational complexity of the proposed descriptors.
5. We additionally develop a weight-based optimization scheme that shows better improvement.

The rest of this paper is organized as follows. Section 2 briefly introduces the local binary pattern and the color distribution prior in the L*a*b* color space. Section 3 details the feature representation. The experiments and discussion are presented in Section 4. Section 5 concludes this paper and indicates future directions.

2. Related Work

2.1. Local Binary Pattern

In reference [8], Ojala et al. first designed the local binary pattern (LBP) for texture feature representation. Given a gray pixel $G(i, j)$, the computational results among $G(i, j)$ and its neighbors $G_x(i, j)$ are encoded as the LBP value. The formula for the LBP value in pixel $G(i, j)$ is as follows:

$$LBP_{n,r}(i,j) = \sum_{x=0}^{n-1} \vartheta(G(i,j) - G_x(i,j)) \times 2^x, \tag{1}$$

$$\vartheta(t) = \begin{cases} 1, & t \geq 0 \\ 0, & t < 0 \end{cases}, \tag{2}$$

where $G_x(i, j)$ is a pixel $G(i, j)$'s x-th neighbor, and n and r represent the number of neighbors and the radius of the neighborhood, respectively.

To reduce the computational complexity, Ojala et al. [8] defined the uniform LBP, in which each LBP pattern has, at most, two bitwise changes among its neighbors. The measure operator "U" of an LBP pattern is defined as the number of bitwise changes. Mathematically, the uniform LBP is defined as follows:

$$U(LBP_{n,r}(i,j)) = |\vartheta(G(i,j) - G_0(i,j)) - \vartheta(G(i,j) - G_{n-1}(i,j))| + \sum_{x=1}^{n-1} |\vartheta(G(i,j) - G_x(i,j)) - \vartheta(G(i,j) - G_{x-1}(i,j))|. \tag{3}$$

If U($LBP_{n,r}(i,j)$) ≤ 2, then $LBP_{n,r}(i,j)$ is classified as the uniform LBP pattern; otherwise, $LBP_{n,r}(i,j)$ belongs to the non-uniform LBP pattern.

To achieve robustness to image rotation, Pietikäinen et al. [29] designed the rotation-invariant LBP, in which all types of the same transition were considered as one pattern. The measure operator ROR ($LBP_{n,r}(i, j)$, x) is defined as a circular bitwise right shift for x times on the n-bit number $LBP_{n,r}(i, j)$. Mathematically, the rotation-invariant LBP is expressed as follows:

$$LBP_{n,r}^{ri}(i,j) = \min\left\{ \mathrm{ROR}(LBP_{n,r}(i,\ j),\ x) \ \middle|\ x \in 0, 1, \ldots, n-1 \right\}. \tag{4}$$

For details, please refer to references [8,29,30]. For simplicity, referring to reference [19], n and r were set to 8 and 1, respectively. In the rest of this paper, we refer to the LBP map as $LBP(i, j)$, the uniform LBP map as $ULBP(i, j)$, and the rotation-invariant LBP map as $RILBP(i, j)$.

2.2. The Selection of the Color Space

The selection of the color space is acknowledged as an important preprocessing stage [1]. Currently, RGB, HSV, HIS, CMYK, YUV, and L*a*b* are widely adopted in feature representation. Among these, the most commonly used color space is RGB. However, the inferiority of the RGB color space can be summarized as follows: (1) the yellow is lost; (2) there is a plethora from green to blue; and (3) it is not suited for the visual perception mechanism. Different from RGB, the superiority of the L*a*b* color space can be summarized as follows: (1) the L*a*b* remedies the missing yellow in RGB; (2) there is no plethora from green to blue; and (3) it is suited for the visual perception mechanism. Besides this, L*a*b* provides excellent decoupling between color (represented by the a* and b* components) and intensity (represented by the L* component) [31]. Therefore, all images are transformed from RGB to the L*a*b* color space in the preprocessing stage. Referring to reference [32], the standard RGB to L*a*b* transformation is carried out as follows:

$$\begin{cases} L* = 116(\frac{Y}{Y_n})^{1/3} - 16 & \text{for } \frac{Y}{Y_n} > 0.08856 \\ L* = 903.3(\frac{Y}{Y_n})^{1/3} & \text{for } \frac{Y}{Y_n} \leq 0.08856 \end{cases}, \tag{5}$$

$$a* = 500(f(\frac{X}{X_n}) - f(\frac{Y}{Y_n})), \tag{6}$$

$$b* = 500(f(\frac{X}{X_n}) - f(\frac{Y}{Z_n})), \tag{7}$$

with

$$\begin{cases} f(u) = u^{1/3} & \text{for } u > 0.08856 \\ f(u) = 7.78u + \frac{Y}{Y_n} & \text{for } u \leq 0.08856 \end{cases}, \tag{8}$$

where

$$\begin{bmatrix} X \\ Y \\ Z \end{bmatrix} = \begin{bmatrix} 0.412453 & 0.357580 & 0.180423 \\ 0.212671 & 0.715160 & 0.072169 \\ 0.019334 & 0.119193 & 0.950227 \end{bmatrix} \begin{bmatrix} R \\ G \\ B \end{bmatrix}, \tag{9}$$

where X_n, Y_n, and Z_n are set to 0.950450, 1.000000, and 1.088754.

*2.3. Color Distribution Prior Knowledge in the L*a*b* Color Space*

In reference [1], the color distribution prior knowledge in the L*a*b* color space was analyzed and summarized for different color image sets. In Figure 1a,b, an example of the Stex database [33] is illustrated. It can be seen that the frequency of pixels is mainly concentrated in the middle range of the a* and b* components. To validate the consistency of this prior knowledge, extensive experiments were performed on different color image sets, and these results show that the prior knowledge is consistent.

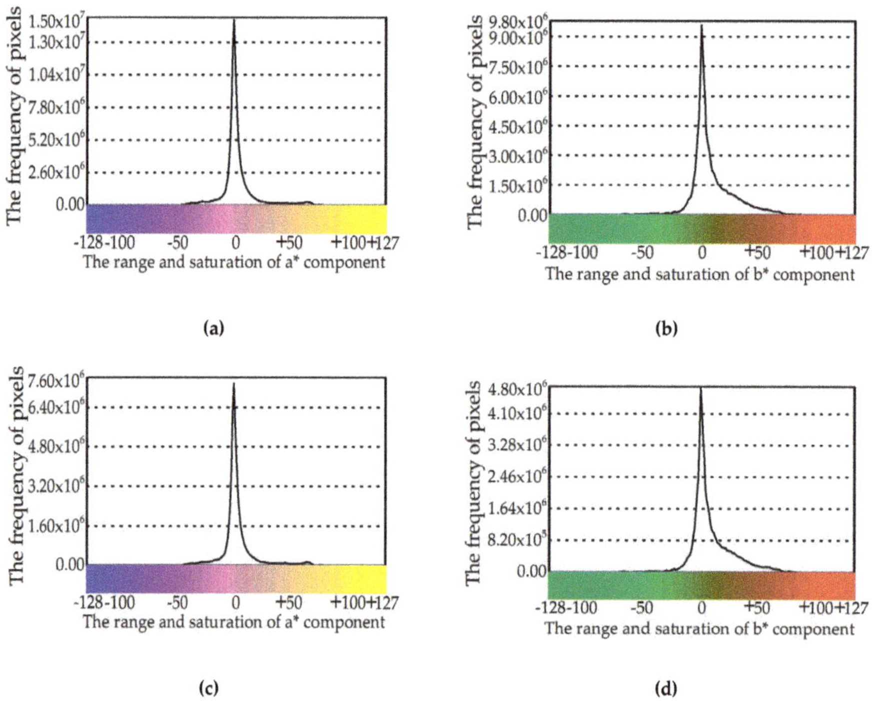

Figure 1. *Cont.*

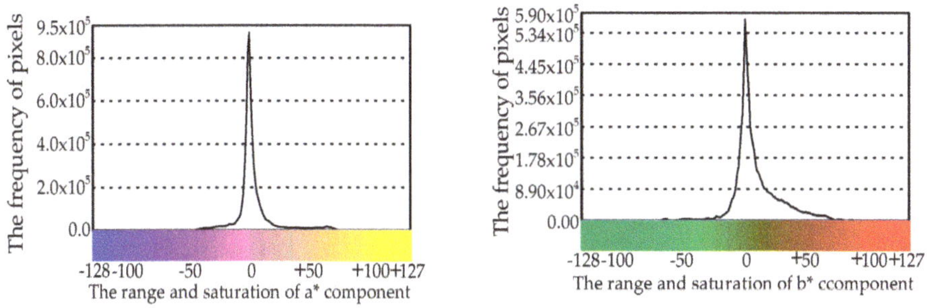

(e) (f)

Figure 1. The frequency of pixels on Stex and its subsets: (**a**,**b**) Stex, (**c**,**d**) 50% of Stex, and (**e**,**f**) 10% of Stex.

Further, the stability of this prior knowledge was also studied when the image database was changed. Examples of 50% and 10% of the Stex image datasets are presented in Figure 1c–f. From those figures, except for the frequency of pixels, we can easily see that the pixels are still distributed in the middle range of the a* and b* components. These phenomena illustrate that the prior knowledge is stable.

The reason for this was studied more deeply in reference [20]. As depicted in Figure 1a–f, the frequency of pixels in the a* and b* components gradually declines from the middle to both sides, but the saturation in the a* and b* components inversely goes up from the middle to both sides. Through extensive experiments, we propose that the color probability distribution has a negative correlation with the saturation of the a* and b* components because higher saturation occurs with a lower frequency.

3. Feature Representation

3.1. Six-Layer Color Quantizer

Inspired by the color distribution prior knowledge in the L*a*b* color space, a novel six-layer color quantizer was designed, as shown in Figure 2, in which each layer includes a set of bins and its corresponding indices. In the proposed quantizer, the original range $[-128, +127]$ is first divided into two equal bins, $2^8/3$, on both sides and two refined bins, $2^7/3$, in the middle. Sequentially, the indices are named 0, 1, 2, and 3 at layer 1. Second, to further refine the two middle bins, $2^7/3$, they are uniformly divided into four equal bins, $2^6/3$, from layers 1 to 2. Meanwhile, the remaining bins are copied from the layer 1 to 2. Third, the operators "Copy" and "Divide" are continuously repeated until the two middle bins at the layer 6. Finally, combining the layer 1 to 6, we construct a six-layer color quantizer. The quantization layers in the a* and b* components are denoted W_{a^*} and W_{b^*}, where W_{a^*}, $W_{b^*} \in \{1, 2, \ldots, 6\}$, and the indices are denoted $\hat{W}_{a^*}$ and $\hat{W}_{b^*}$, $\hat{W}_{a^*} \in \{0, 1, \ldots, \ddot{W}_{a^*}\}$ and $\hat{W}_{b^*} \in \{0, 1, \ldots, \ddot{W}_{b^*}\}$, where $\ddot{W}_{a^*} = 2(W_{a^*} + 1) - 1$ and $\ddot{W}_{b^*} = 2(W_{b^*} + 1) - 1$ respectively.

Referring to reference [34], we discuss the quantization error under different quantization layers, W_{a^*} and W_{b^*}. Figure 3 shows the quantization errors on Stex, 50% of Stex, and 10% of Stex. From the figures, it can be seen that along with the refinement of layers W_{a^*} and W_{b^*}, where W_{a^*}, $W_{b^*} \in \{1, 2, \ldots, 6\}$, the quantization error decreases obviously. This phenomenon illustrates the effectiveness of the proposed quantizer. Moreover, we also note that the values of the quantization errors on Stex, 50% of Stex, and 10% of Stex are extremely close to each other. This phenomenon confirms the stability and consistency of the six-layer color quantizer. In particular, the quantization error from layers 5 to 6 decreases only slightly on the three datasets. These results demonstrate that stopping the quantization layers W_{a^*} and W_{b^*} at the 6th layer is appropriate.

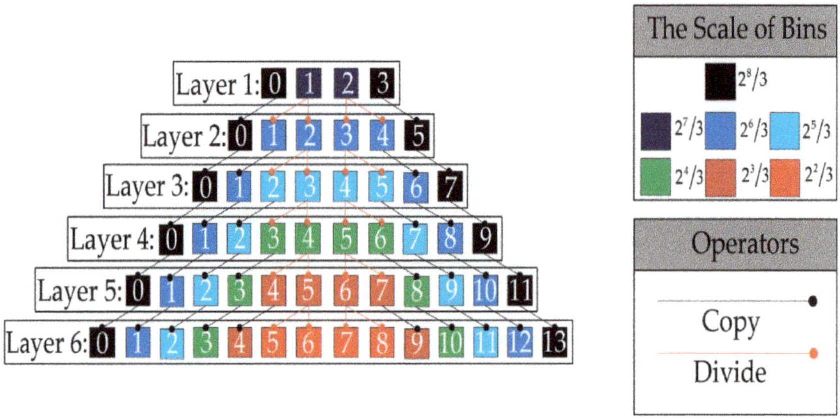

Figure 2. The details of the six-layer color quantizer.

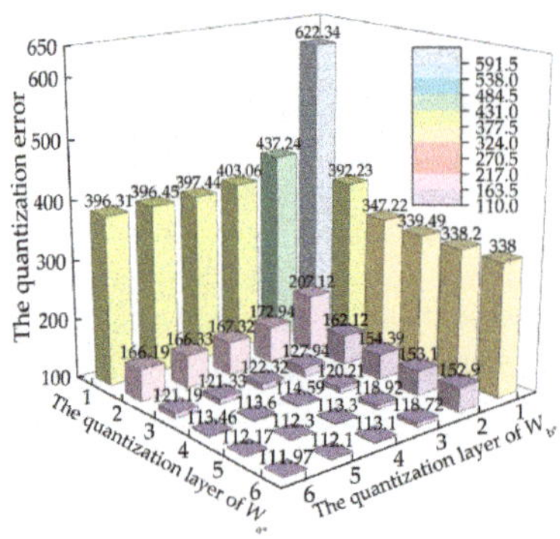

(**a**) Stex

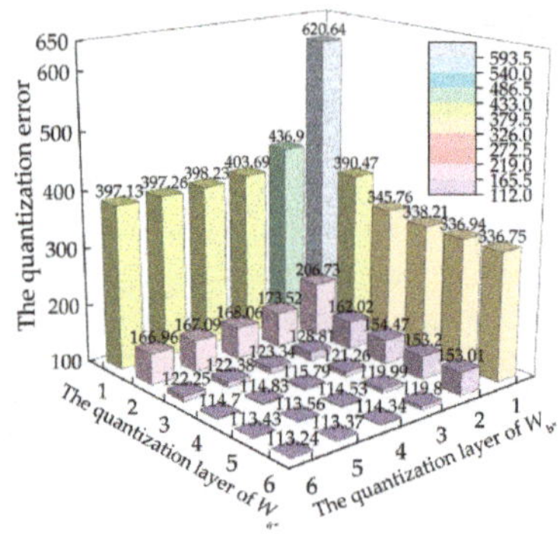

(**b**) 50% of Stex

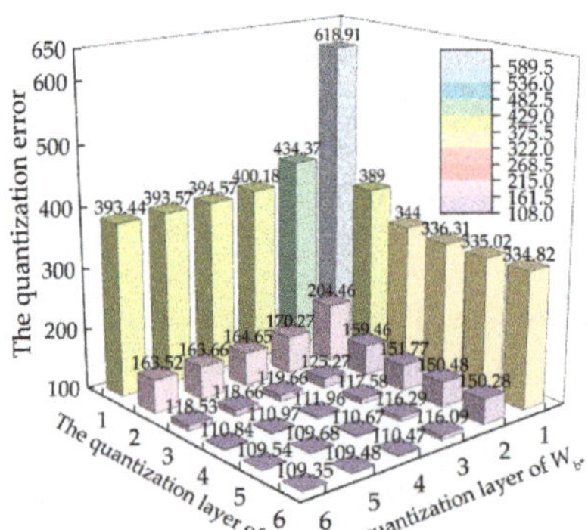

(**c**) 10% of Stex

Figure 3. The quantization errors under different quantization layers, W_{a*} and W_{b*}, on Stex and its subsets: (**a**) Stex, (**b**) 50% of Stex, and (**c**) 10% of Stex.

Furthermore, referring to the visual perception mechanism in reference [35], the original range [0, +100] in the L* component is divided into three bins, namely, [0, +25], [+26, +75], and [+76, +100]. Herein, the quantization layer of the L* component is denoted $\boldsymbol{W}_{\mathbf{L}^*}$, where $W_{L^*} = 1$, and the indices of the three bins are denoted $\hat{W}_{L^*}$, $\hat{W}_{L^*} \in \{0, 1, \ldots, \ddot{W}_{L^*}\}$, where $\ddot{W}_{L^*} = 2W_{L^*}$. For a pixel (i, j) in image I, we combine the indices of W_{L^*}, W_{a^*}, and W_{b^*} to construct the color map $C(i, j)$, and the index of $C(i, j)$ is denoted $\hat{C}$, $\hat{C} \in \{0, 1, \ldots, \check{C}\}$, where $\check{C} = 3 \times 2(W_{a^*} + 1) \times 2(W_{b^*} + 1) - 1$.

3.2. Local Parallel Cross Pattern

As elucidated in Gray's Anatomy [36], the human visual system is an important pathway that codes low-layer visual cues to construct the high-layer semantics perception in parallel and cross manners. On the basis of the human visual system, we propose a novel local parallel cross pattern (LPCP) to integrate the color map and the LBP map as a unified framework in "parallel" and "cross" manners.

Given an original map $I(i, j)$, the central point and its eight neighbors are denoted $I_0(i, j)$ and $I_k(i, j)$, where $k \in \{1, 2, \ldots, 8\}$. Firstly, we extract the LBP map $LBP(i, j)$ (see Section 2.1) and the color map $C(i, j)$ (see Section 3.1). Secondly, all eight neighbors of the LBP map and the color map are mutually crossed to construct the LBP and color cross maps. Thirdly, we calculate the frequency of each neighborhood in the LBP and color cross maps to construct the LBP and color frequency maps, respectively. Finally, the values of the maximum frequency in the LBP and color frequency maps are considered the feature vectors, and the central values in the LBP and color frequency maps are flagged as the indices. For clarity, Figure 4 presents a detailed schematic diagram of the local parallel cross pattern, in which LPCP is encoded as $LPCP_{LBP}(3) = 4$ and $LPCP_{color}(7) = 5$. Mathematically, LPCP is defined as follows:

$$LPCP_{LBP}(LBP_0(i,j)) = \arg\max \mathrm{Fr}\{C_k(i,j)|k = 1,\ 2,\ \ldots,\ 8\}, \tag{10}$$

$$LPCP_{color}(C_0(i,j)) = \arg\max \mathrm{Fr}\{LBP_k(i,j)|k = 1,\ 2,\ \ldots,\ 8\}, \tag{11}$$

where $\mathrm{Fr}\{\cdot\}$ represents the frequency of each neighbor. In particular, when the value of $LBP_0(i, j)$ is equal to $C_0(i, j)$, $LBP_0(i, j)$ and $C_0(i, j)$ can still be separated and encoded as $LPCP_{LBP}(LBP_0(i, j))$ and $LPCP_{color}(C_0(i, j))$, respectively. Herein, the feature dimensions of $LPCP_{LBP}$ and $LPCP_{color}$ are 256 and $3 \times 2(W_{a^*} + 1) \times 2(W_{b^*} + 1)$. Given a color image dataset T, the optimal quantization layers W_{a^*} and W_{b^*} are calculated based upon the retrieval accuracy score. This process is defined as the maximization problem as follows:

$$\max_{W_{a*},W_{b*}} \mathrm{Acc}(T|W_{a*}, W_{b*}), W_{a*},\ W_{b*} \in \{1,2,\ldots,6\}, \tag{12}$$

where $\mathrm{Acc}(T \mid W_{a^*},\ W_{b^*})$ represents the retrieval accuracy score. We provide the optimal color quantization layers W_{a^*} and W_{b^*} in Section 4.4.

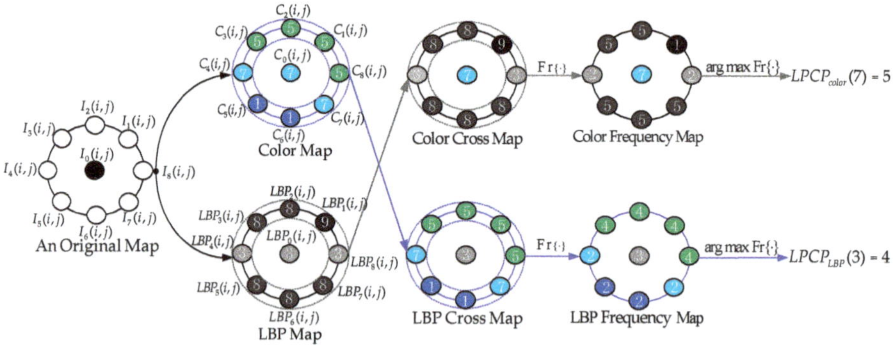

Figure 4. Schematic diagram of the local parallel cross pattern (LPCP).

To reduce the computational complexity, we define the uniform local parallel cross pattern (ULPCP) in which $ULBP(i, j)$ replaces $LBP(i, j)$. To achieve robustness to image rotation, we design the rotation-invariant local parallel cross pattern (RILPCP) in which $RILBP(i, j)$ replaces $LBP(i, j)$.

As presented in Figure 5, RILPCP achieves exactly the same feature encoding in spite of the image rotation. For Figure 5a, we first extract the RILBP map and the color map. Then, according to Equations (10) and (11), RILPCP is encoded as [$\text{RILPCP}_{RILBP}(3) = 4$; $\text{RILPCP}_{color}(7) = 4$]. When Figure 5a is rotated 90° to Figure 5b, we can also extract the same RILBP and color map because both color and RILBP are rotation invariant. Further, RILPCP is still calculated as [$\text{RILPCP}_{RILBP}(3) = 4$; $\text{RILPCP}_{color}(7) = 4$]. In fact, an image can be rotated by an arbitrary degree.

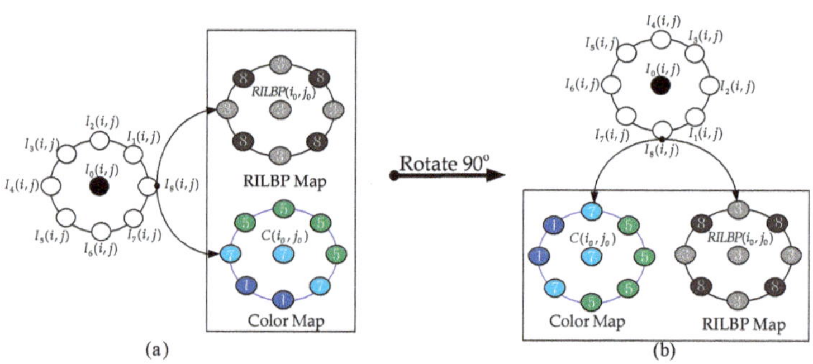

Figure 5. Illustration of the rotation invariance of the rotation-invariant local parallel cross pattern (RILPCP).

4. Experiments and Discussion

4.1. Distance Metric

The distance metric is considered as the measure of similarity between a query image and a database image. In our retrieval framework, we first convert the query image and database images into feature vectors, and we then calculate the distance measure between the query image and database images. Referring to references [1,18,20,21,37], the Extended Canberra Distance is exploited, and it is given as

$$R(F_q, F_d) = \sum_{v=1}^{k} \frac{|F_q(v) - F_d(v)|}{|F_q(v) + \mu_q| + |F_d(v) + \mu_d|}, \tag{13}$$

where F_d, F_q, and k respectively denote the image in the database, the query image by the user, and the feature vector length, and R is the measurement result between the query image F_q and the image F_d in the database. Further, $\mu_q = \sum_{v=1}^{k} F_q(v)/k$ and $\mu_d = \sum_{v=1}^{k} F_d(v)/k$.

4.2. Evaluation Criteria

In this section, we introduce the most popular evaluation criteria, such as the precision rate, the recall rate, the average precision rate (APR) value, the average recall rate (ARR) value, and the precision-recall (PR) curve to validate the proposed descriptors.

First, the precision and recall rates are formulated as follows:

$$Precision = \frac{1}{N_\alpha}\sum_{d=1}^{N_\alpha} \xi(\eta(F_q), \eta(F_d)), \tag{14}$$

$$Recall = \frac{1}{N_\beta}\sum_{d=1}^{N_\beta} \xi(\eta(F_q), \eta(F_d)), \tag{15}$$

$$\xi(\eta(F_q), \eta(F_d)) = \begin{cases} 1, & \eta(F_q) = \eta(F_d) \\ 0, & \eta(F_q) \neq \eta(F_d) \end{cases}, \tag{16}$$

where N_β, N_α, and $\eta(\cdot)$ denote the total number of images in the same category, the total number of returned images, and the category information. $\xi(\cdot)$ is defined as the binarized function. If $\xi(\eta(F_q)$, $\eta(F_d))) = 1$, then $\eta(F_q)$ and $\eta(F_d)$ are determined to be in the same category; if $\xi(\eta(F_q), \eta(F_d))) = 0$, then $\eta(F_q)$ and $\eta(F_d)$ do not belong to the same category.

Then, the average precision rate (APR) and average recall rate (ARR) values are given by the following equations:

$$APR = \frac{\sum_{\hat{n}=1}^{\hat{N}} Precision(\hat{n})}{\hat{N}}, \tag{17}$$

$$ARR = \frac{\sum_{\hat{n}=1}^{\hat{N}} Recall(\hat{n})}{\hat{N}}, \tag{18}$$

where $\hat{n}$ and $\hat{N}$ are the total number of query images and the $\hat{n}$th query image, respectively.

Finally, the precision-recall (PR) curve can be considered an ancillary criterion that evaluates the dynamic precision by computing the threshold recall. Mathematically, the PR curve is defined as follows:

$$PR(\tau) = \frac{N_\beta}{N_\tau} \cdot \tau \times 100\%, \tag{19}$$

where N_β and N_τ are the total number of images in the same category and the number of retrieval images at the recall of τ, where $\tau \in \{1, 2, \ldots, N_\alpha - 1\}$.

4.3. Image Databases

In our experiments, the eight benchmark databases reported in Table 1, comprising one natural object image database (Coil-100 [38]), two facial image databases (Face95 [39] and Face96 [40]), and five color textural image databases (Outex-00031 [41], Outex-00032 [41], Outex-00033 [41], Outex-00034 [41], and MIT-VisTex [42]), were used to provide a comprehensive evaluation.

Table 1. Summary of image databases.

No.	Name	Image Size	Class	Images in Each Class	Images Total	Format	Website
1	Coil-100 (Rotation)	128 × 128	100	72	7200	JPG	http://www.cs.columbia.edu/CAVE/software/softlib/coil-100.php
2	Face95	180 × 200	72	20	1440	JPG	https://cswww.essex.ac.uk/mv/allfaces/faces95.html
3	Face96	196 × 196	91	19 or 20	1814	JPG	https://cswww.essex.ac.uk/mv/allfaces/faces96.html
4	Outex-00031 (Scale)	128 × 128	68	40	2720	BMP	http://lagis-vi.univ-lille1.fr/datasets/outex.html
5	Outex-00032 (Noise)	128 × 128	68	40	2720	BMP	http://lagis-vi.univ-lille1.fr/datasets/outex.html
6	Outex-00033 (Blur)	128 × 128	68	40	2720	BMP	http://lagis-vi.univ-lille1.fr/datasets/outex.html
7	Outex-00034 (illumination)	128 × 128	204	20	4080	BMP	http://lagis-vi.univ-lille1.fr/datasets/outex.html
8	MIT-VisTex	128 × 128	40	16	640	PPM	http://vismod.media.mit.edu/pub/VisTex/

The Coil-100 (No. 1) database was produced by a charge-coupled device (CCD)-camera (Sony XC-77P). It has 7200 images in 100 objects. Each object contains 72 images, each with a size of 128 × 128 in JPG format. Because each image was collected by rotating an object at 5 degrees, the Coil-100 not only evaluates the effectiveness, but also investigates the robustness to rotation. Some samples are depicted in Figure 6a in which each row represents the same semantic category.

(a)

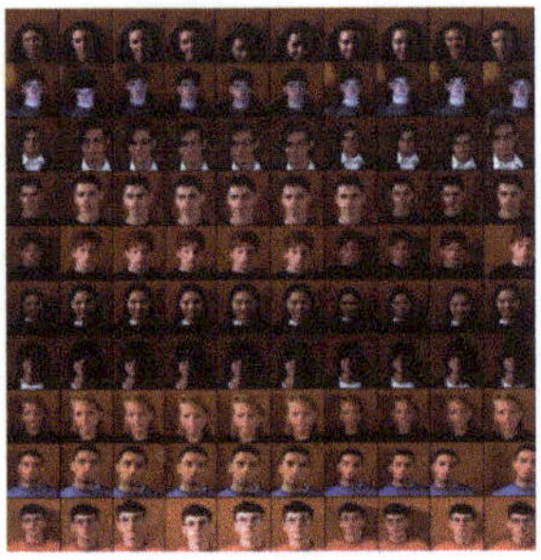
(b)

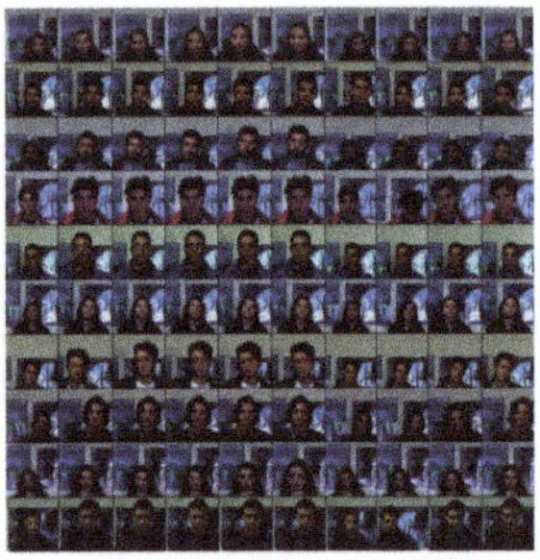
(c)

(d)

(e)

(f)

Figure 6. *Cont.*

(g)

(h)

Figure 6. Some sample images from the eight databases: (**a**) Coil-100; (**b**) Face95; (**c**) Face96; (**d**) Outex-00031; (**e**) Outex-00032; (**f**) Outex-00033; (**g**) Outex-00034; and (**h**) MIT-VisTex.

The Face 95 (No. 2) and Face 96 (No. 3) databases were collected by an S-VHS camcorder. Among them, the Face 95 consists of 1440 images in 72 male and female subjects. For each subject, there are 20 images, each with a size of 180 × 200 in JPG format. The Face 96 has 91 male and female subjects. For each subject, there are 1814 images with a size of 196 × 196 in JPG format. Specifically, all images are variations of head turns, head scales, face expressions, and illumination changes. Some samples are presented in Figure 6b,c.

The Outex-00031 (No. 4), Outex-00032 (No. 5), Outex-00033 (No. 6), and Outex-00034 (No. 7) databases were produced by a charge-coupled device (CCD)-camera (Sony DXC-775P), and the MIT-VisTex (No. 8) was collected from real-world photographs and videos. Some samples of these databases are presented in Figure 6d–h. As documented in Table 1, the Outex-00031, Outex-00032, and Outex-00033 consist of 2720 images in 68 categories. Each category includes 40 images, each with a size of 128 × 128 in BMP format. Differently, the Outex-00034 has 4048 images in 204 categories. Each category includes 20 images, each with a size of 128 × 128 in BMP format. Next, the MIT-VisTex consists of 640 images in 40 categories. Each category includes 16 images, each with a size of 128 × 128 in PPM format. There are resolution differences on Outex-00031, noise differences on Outex-00032, blur differences on Outex-00033, and illumination differences on Outex-00034. Thus, these four databases were used to evaluate the robustness to resolution, noise, blur, and illumination.

In addition, all above databases can be freely downloaded from the corresponding websites. To guarantee accuracy and reproducibility, we chose all images as the query images in the dataset. Referring to references [1,18,20,37], if not specified, the total number of returned images was set to 10 in all of the following experiments.

4.4. Evaluation of Color Quantization Layers

Table 2 shows the highest APR values of LPCP, RILPCP, and ULPCP with the optimal color quantization layers (W_{a^*}, W_{b^*}) on the eight databases. The best values are shown in bold. On Coil-100, three facts are noted: (1) LPCP achieves the highest APR value of 99.47% when ($W_{a^*} = 6$, $W_{b^*} = 5$); (2) RILPCP achieves the highest APR value of 99.44% when ($W_{a^*} = 5$, $W_{b^*} = 5$); and (3) ULPCP yields the highest APR value of 99.52% when ($W_{a^*} = 4$, $W_{b^*} = 6$). Next, on Face 95 and Face 96, LPCP, RILPCP, and ULPCP all achieve their top APR values when ($W_{a^*} = 6$, $W_{b^*} = 6$). Similarly, on the remaining color textural databases, the corresponding color quantization layers W_{a^*} and W_{b^*} result in the best accuracy scores. As noted above, it can be easily summarized that when LPCP, RILPCP, and ULPCP are used on the same image database, the coefficients W_{a^*} and W_{b^*} are extremely close to each other. These phenomena again confirm the stability and consistency of the color distribution prior knowledge. Moreover, it is noteworthy that a single color quantization layer cannot be suitable for all image databases. In the following experiments, the optimal color quantization layers, W_{a^*} and W_{b^*}, were adaptively selected according to the different databases.

Table 2. The highest average precision rate (APR) values of LPCP, RILPCP, and uniform local parallel cross pattern (ULPCP) with the optimal color quantization layers (W_{a^*}, W_{b^*}) on the eight databases.

Method	Performance	Dataset							
		Coil-100	Face95	Face96	Outex-00031	Outex-00032	Outex-00033	Outex-00034	MIT-VisTex
LPCP	(W_{a^*}, W_{b^*})	(6, 5)	(6, 6)	(6, 6)	**(6, 1)**	**(5, 1)**	**(6, 4)**	**(6, 4)**	**(5, 4)**
	APR (%)	99.47	92.33	97.03	**89.62**	**84.86**	**87.80**	**84.36**	**98.33**
RILPCP	(W_{a^*}, W_{b^*})	(5, 5)	**(6, 6)**	**(6, 6)**	(5, 1)	(5, 1)	(6, 3)	(4, 4)	(5, 2)
	APR (%)	99.44	**97.12**	**97.77**	89.53	84.40	87.55	84.06	97.22
ULPCP	(W_{a^*}, W_{b^*})	**(4, 6)**	(6, 6)	(6, 6)	(5, 1)	(5, 1)	(6, 3)	(5, 4)	(3, 2)
	APR (%)	**99.52**	96.65	97.45	89.44	84.07	87.59	84.20	97.39

4.5. Comparison with LBP-Based Descriptors

Table 3 reports comparisons among the proposed descriptors and the LBP-based descriptors in terms of the APR and ARR. The best {APR, ARR} values are shown in bold. As documented in this table, the {APR, ARR} values of LPCP are far better than those of LBP by {9.64%, 1.34%} on Coil-100, {28.88%, 14.43%} on Face95, {32.10%, 16.12%} on Face96, {11.63%, 2.90%} on Outex-00031, {15.44%, 3.86%} on Outex-00032, {12.26%, 3.10%} on Outex-00033, {38.07%, 19.04%} on Outex-00034, and {4.96%, 3.10%} on MIT-VisTex. Analogous to the proposed LPCP, the {APR, ARR} values of ULPCP and RILPCP are also definitely higher than those of ULBP and RILBP on all eight databases. Meanwhile, three points can be observed below: (1) ULPCP has the maximum {APR, ARR} value of {99.52%, 13.82%} on Coil-100; (2) RILPCP produces the best results of {97.12%, 48.56%} on Face95, and {97.77%, 49.04%} on Face96; and (3) LPCP has the highest performance of {89.62%, 22.40%} on Outex-00031, {84.86%, 21.21%} on Outex-00032, {87.80%, 21.99%} on Outex-00033, {84.36%, 42.18%} on Outex-00034, and {98.33%, 61.46%} on MIT-VisTex. According to the above results, it can be asserted that the improvements given by the proposed descriptors are very considerable. The main reason for this is that the LBP information is amalgamated with the color information.

Table 3. The performance comparisons among the proposed descriptors and the local binary pattern (LBP)-based descriptors on the eight databases.

Method	Performance	Dataset							
		Coil-100	Face95	Face96	Outex-00031	Outex-00032	Outex-00033	Outex-00034	MIT-VisTex
LBP	APR (%)	89.83	63.45	64.93	77.99	69.42	75.54	46.29	93.37
	ARR (%)	12.48	31.73	32.55	19.50	17.35	18.89	23.14	58.36
RILBP	APR (%)	84.53	59.78	61.19	76.57	67.00	74.13	45.93	89.75
	ARR (%)	11.74	29.89	30.68	19.14	16.75	18.53	22.96	56.09
ULBP	APR (%)	85.40	58.25	59.42	76.03	65.68	73.04	45.51	90.83
	ARR (%)	11.86	29.12	29.79	19.01	16.42	18.26	22.75	56.77
LPCP	APR (%)	99.47	92.33	97.03	**89.62**	**84.86**	**87.80**	**84.36**	**98.33**
	ARR (%)	13.82	46.16	48.67	**22.40**	**21.21**	**21.99**	**42.18**	**61.46**
RILPCP	APR (%)	99.44	**97.12**	**97.77**	89.53	84.40	87.55	84.06	97.22
	ARR (%)	13.81	**48.56**	**49.04**	22.38	21.10	21.89	42.03	60.76
ULPCP	APR (%)	**99.52**	96.65	97.45	89.44	84.07	87.59	84.20	97.39
	ARR (%)	**13.82**	48.33	48.88	22.36	21.02	21.90	42.10	60.87

4.6. Comparison with Other Color Texture Descriptors

To evaluate the effectiveness, efficiency, robustness, and computational complexity, the proposed descriptors were compared with eight state-of-the-art color texture descriptors in terms of the average precision rate (APR) value, the average recall rate (ARR) value, the precision-recall (PR) curve, the feature vector length, and the memory consumption. All experiments were performed under the leave-one-out cross-validation principle. For clarity, all comparative color texture methods are summarized as follows:

- Multi-channel adder local binary patterns (mdLBP) [18]: The 2048-dimensional color texture descriptor in the RGB color space.
- Multi-channel decoded local binary patterns (maLBP) [18]: The 1024-dimensional color texture descriptor in the RGB color space.
- Color difference histogram (CDH) [21]: The 90-dimensional color histogram and the 18-dimensional edge orientation histogram in the L*a*b* color space.
- Multi-texton histogram (MTH) [22]: The 64-dimensional color histogram and the 18-dimensional edge orientation histogram in the HSV color space.
- Micro-structure descriptor (MSD) [23]: The 72-dimensional color histogram and the 6-dimensional edge orientation histogram in the HSV color space.
- Opponent color local binary patterns (OCLBP) [24]: The 1536-dimensional color texture descriptor in the RGB color space.
- Improved opponent color local binary patterns (IOCLBP) [25]: The 3072-dimensional color-texture descriptor in the RGB color space.
- Orthogonal combination of local binary patterns and color histogram (OC-LBP + CH): The 12-dimensional color histogram in the L*a*b* color space and the 96-dimensional LBP variation in the gray-scale space [28].
- Local parallel cross pattern (LPCP).
- Rotation-invariant local parallel cross pattern (RILPCP).
- Uniform local parallel cross pattern (ULPCP).

Table 4 lists the APR and ARR values for the proposed descriptors and the existing descriptors on the eight databases. The best values are shown in bold. On Coil-100, the {APR, ARR} value of RILPCP is significantly superior to those of mdLBP, maLBP, CDH, MTH, MSD, OCLBP, IOCLBP, and OC-LBP+CH by {7.10%, 0.98%}, {11.11%, 1.54%}, {1.26%, 0.17%}, {0.99%, 0.14%}, {1.43%, 0.20%}, {14.04%, 1.95%}, {10.32%, 1.43%}, and {3.81%, 0.53%}, respectively. Meanwhile, the {APR, ARR} value of RILPCP is slightly improved (by {+0.03%, +0.01%}) by using LPCP. In this scenario, ULPCP acquires a higher {APR, ARR} value than RILPCP: {99.44%, 13.81} compared with {99.52%, 13.82}. Similar to Coil-100, there are more competitive results on Face95, Face96, Outex-00031, Outex-00032, Outex-00033, Outex-00034, and MIT-VisTex. As a consequence of the above results, it can be summarized that the effectiveness of the proposed descriptors is demonstrated in terms of APR and ARR. Remarkably, there are rotation differences on Coil-100, resolution differences on Outex-00031, noise differences on Outex-00032, blur differences on Outex-00033, and illumination differences on Outex-00034. Thus, the robustness is also illustrated to a certain extent.

Table 4. The performance comparisons between the proposed descriptors and the existing descriptors in terms of APR and ARR on the eight databases.

Method	Performance	Dataset							
		Coil-100	Face95	Face96	Outex-00031	Outex-00032	Outex-00033	Outex-00034	MIT-VisTex
mdLBP	APR (%)	92.34	72.97	79.09	69.51	59.60	65.63	48.66	97.05
	ARR (%)	12.83	36.49	39.66	17.38	14.90	16.41	24.33	60.65
maLBP	APR (%)	88.33	67.94	73.99	70.42	60.24	65.42	44.53	95.80
	ARR (%)	12.27	33.97	37.10	17.60	15.06	16.36	22.27	59.87
CDH	APR (%)	98.18	94.69	94.97	85.49	79.65	82.90	74.03	94.03
	ARR (%)	13.64	47.35	47.63	21.37	19.91	20.73	37.02	58.77
MTH	APR (%)	98.45	89.15	92.28	80.74	74.63	79.98	65.10	91.97
	ARR (%)	13.67	44.57	46.28	20.18	18.66	20.00	32.55	57.48
MSD	APR (%)	98.01	92.97	89.94	77.16	72.46	76.16	66.32	92.20
	ARR (%)	13.61	46.48	45.11	19.29	18.12	19.04	33.16	57.63
OCLBP	APR (%)	85.40	64.40	64.55	80.76	69.25	77.99	56.13	92.42
	ARR (%)	11.86	32.20	32.37	20.19	17.31	19.50	28.07	57.76
IOCLBP	APR (%)	89.12	66.47	73.24	83.84	74.54	80.75	73.58	95.59
	ARR (%)	12.38	33.24	36.73	20.96	18.63	20.19	36.79	59.75
OC-LBP+CH	APR (%)	95.63	80.50	88.67	85.07	75.93	81.33	72.18	92.20
	ARR (%)	13.28	40.25	44.46	21.27	18.98	20.33	36.09	57.63
LPCP	APR (%)	99.47	92.33	97.03	**89.62**	**84.86**	**87.80**	**84.36**	**98.33**
	ARR (%)	13.82	46.16	48.67	**22.40**	**21.21**	**21.99**	**42.18**	**61.46**
RILPCP	APR (%)	99.44	**97.12**	**97.77**	89.53	84.40	87.55	84.06	97.22
	ARR (%)	13.81	**48.56**	**49.04**	22.38	21.10	21.89	42.03	60.76
ULPCP	APR (%)	**99.52**	96.65	97.45	89.44	84.07	87.59	84.20	97.39
	ARR (%)	**13.82**	48.33	48.88	22.36	21.02	21.90	42.10	60.87

Figure 7a–h depict the precision-recall (PR) curves of all the comparative descriptors on the eight databases. As we can see from Figure 7a, the curves of LPCP, RILPCP, and ULPCP are higher than those of former color texture descriptors. As shown in Figure 7b, although the curves of ULPCP, MTH, and MSD are interleaved with one another, both LPCP and RILPCP are superior to all other descriptors. One possible reason for this is that the Face95 database emphasizes the importance of the color information. As depicted in Figure 7c, considering the complex background, the curves of the proposed descriptors are better than all remaining descriptors on the Face96 database. Analogous to the results for the Face96 database, LPCP, RILPCP, and ULPCP are much better than other descriptors on the Outex-00031 (see the Figure 7d), Outex-00032 (see the Figure 7e), Outex-00033 (see the Figure 7f), and Outex-00034 (see the Figure 7g) databases, respectively. As depicted in Figure 7h, although mdLBP provides a competitive performance, LPCP achieves the highest curve. The main reason for this is that the MIT-VisTex database contains more textural structure information. Based on all the above observations and analyses, it can be easily noticed that the proposed descriptors are effective in terms of the precision-recall (PR) curve in most of the cases.

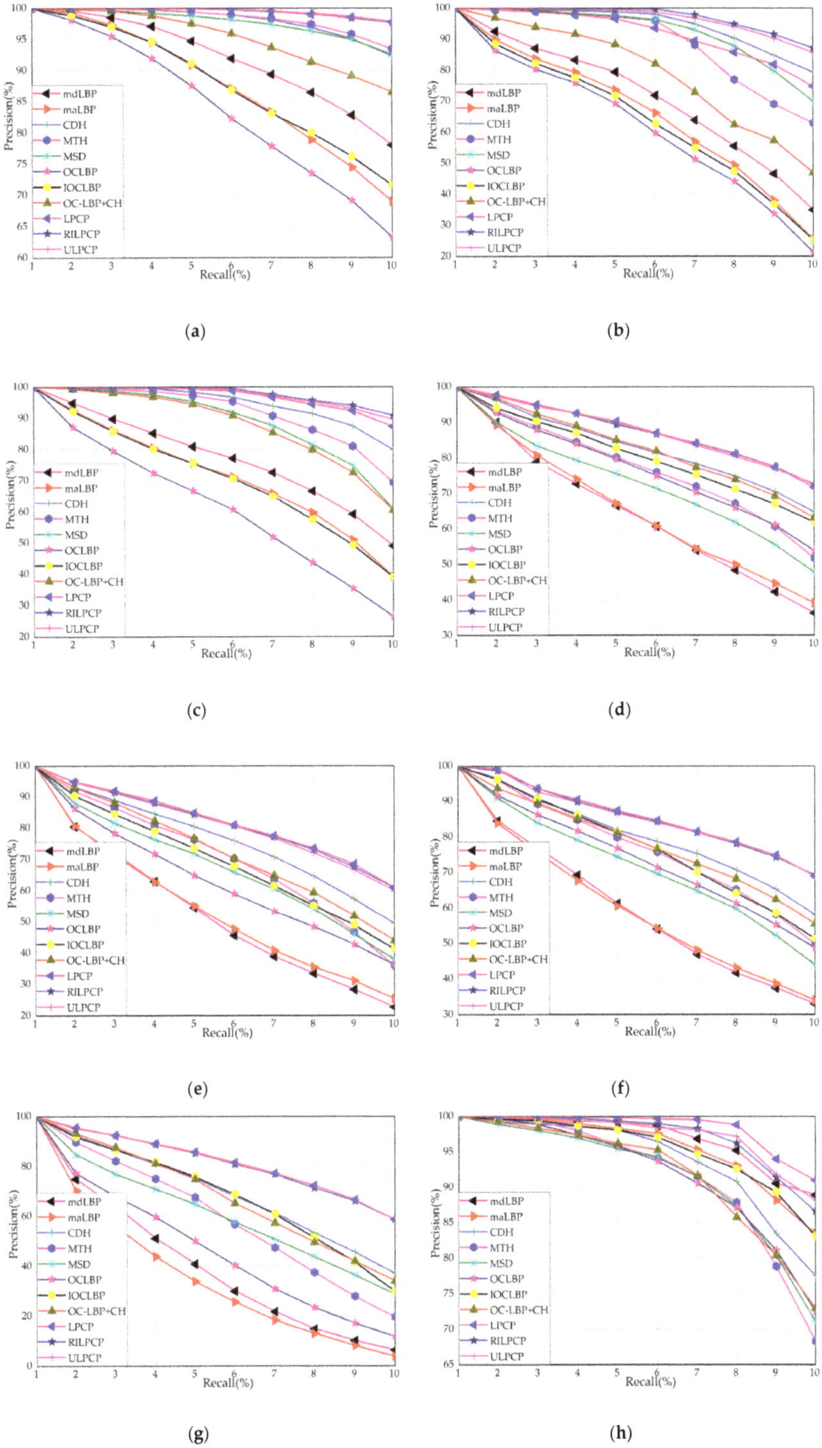

Figure 7. The precision-recall curves of eleven descriptors on the eight databases: (**a**) Coil-100; (**b**) Face95; (**c**) Face96; (**d**) Outex-00031; (**e**) Outex-00032; (**f**) Outex-00033; (**g**) Outex-00034; and (**h**) MIT-VisTex.

Table 5 compares the computational complexity and memory cost in terms of the feature vector length by dimensionality (D) and memory consumption in kilobytes (Kb). All the experiments were performed on a 4.20 GHz four-core CPU with 16 GB of memory. Herein, analogous to RILPCP and ULPCP, 760/844/844/424/400/676/676/616 (D) and 5.94/6.59/6.59/3.31/3.13/5.28/5.28/4.81 (Kb) show that LPCP performs retrieval requiring 760 D and 5.94 Kb on Coil-100, 844 D and 6.59 Kb on Face95, 844 D and 6.59 Kb on Face96, 424 D and 3.31 Kb on Outex-00031, 400 D and 3.13 Kb on Outex-00032, 676 D and 5.28 Kb on Outex-00033, 676 D and 5.28 Kb on Outex-00034, and 616 D and 4.81 Kb on MIT-VisTex. As documented in Table 5, the feature vector length and memory consumption of the proposed descriptors are inferior to those of CDH, MTH, MSD, and OC-LBP+CH, but they are superior to those of mdLBP, mdLBP, OCLBP, and IOCLBP. From the results, although the computational complexity is larger than those of some existing methods, there are several superiorities of LPCP, RILPCP, and ULPCP as follows:

1. The added computational complexity is effective because the retrieval accuracy is enhanced by a large margin.
2. The proposed descriptors can adaptively code the color and texture information from different image databases.
3. The practicability and feasibility of the proposed descriptors, with acceptable feature vector length and competitive memory consumption, are well shown for a realistic system configuration.

Table 5. Feature vector length (D) and memory consumption (Kb) among the proposed descriptors and other previous descriptors.

Method	Feature Vector Length (D)	Memory Consumption (Kb)
mdLBP	2048	16.00
maLBP	1024	8.00
CDH	108	0.84
MTH	82	0.64
MSD	78	0.61
OCLBP	1535	11.99
IOCLBP	3072	24.00
OC-LBP+CH	108	0.84
LPCP	760/844/844/424/400/676/676/616	5.94/6.59/6.59/3.31/3.13/5.28/5.28/4.81
RILPCP	468/624/624/180/180/372/336/252	3.66/4.88/4.88/1.41/1.41/2.91/2.63/1.97
ULPCP	479/647/647/203/203/395/419/203	3.74/5.05/5.05/1.59/1.59/3.09/3.27/1.59

4.7. Comparison with CNN-Based Descriptors

On a different note, we also compared the proposed descriptors with emerging CNN-based descriptors including VGGm, VGGm128, VGGm1024, VGGm2048, ALEX, GoogleNet, and Inception-v3 [43]. Referring to references [44,45], the last fully connected layer was first extracted from the pretrained models. Then, L2 normalization was performed on the extracted fully connected layer. Finally, the distance measure was calculated on the normalized feature vector. For fairness, all images in the database were chosen as the query images, and the number of returned images was set to 10.

Figure 8 compares the proposed descriptors and the CNN-based descriptors. On the Coil-100 database, VGGm and ALEX perform slightly better than the proposed LPCP, RILPCP, and ULPCP methods. On the Face95, Face96, Outex-00031, Outex-00032, Outex-00033, and Outex-00034 databases, more significant APR values are obtained by using the proposed LPCP, RILPCP, and ULPCP methods. On the MIT-VisTex database, VGGm achieves a performance that is competitive with those of RILPCP and ULPCP, but LPCP achieves the highest APR value. Although VGGm and ALEX yield relatively competitive APR values, the superior abilities of LPCP, RILPCP, and ULPCP are revealed as follows:

1. The CNN-based descriptors must be pretrained on a large-scale and annotated dataset (e.g., ImageNet), while the proposed LPCP, RILPCP, and ULPCP methods do not need any pretraining process.
2. The pretrained CNN-based descriptors are computationally expensive (e.g., cloud servers and mainframe computers), but LPCP, RILPCP, and ULPCP can be performed on almost all realistic systems (e.g., personal smartphones and home security cameras).
3. LPCP, RILPCP, and ULPCP are more effective than the CNN-based descriptors on six datasets out of the eight examined.

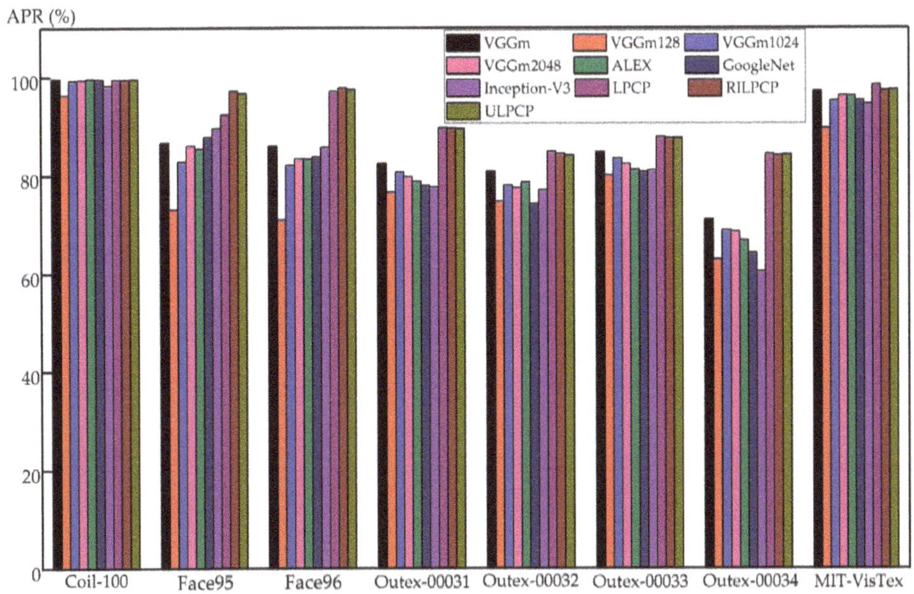

Figure 8. The average precision rate (APR) comparisons among the proposed descriptors and the CNN-based descriptors on the eight databases.

4.8. Additional Experiments on a Weight-Based Optimization Scheme

To further investigate the optimized coefficient scheme, we added weighting-based experiments on the eight databases. Referring to references [35,37], we defined the weighting parameter v, where $v \in \{0, 0.01, \ldots, 1.00\}$, and the weighting local parallel cross pattern ($LPCP_{weight}$) was formulated as follows:

$$LPCP_{weight} = [(1-v)\cdot LPCP_{LBP}, v\cdot LPCP_{color}], \tag{20}$$

Observe that $LPCP_{weight}$ degenerates into $LPCP_{LBP}$ when $v = 0.00$ and into $LPCP_{color}$ when $v = 1.00$. Further, LPCP is derived from $LPCP_{weight}$ when v is set to 0.50. We also extended $LPCP_{weight}$ to the weighting uniform local parallel cross pattern ($ULPCP_{weight}$) and the weighting rotation-invariant local parallel cross pattern ($URILPCP_{weight}$).

Figure 9 shows the curves of the average precision rate (APR) under the weighting parameter v by using $LPCP_{weight}$, $ULPCP_{weight}$, and $RILPCP_{weight}$ on the eight databases. On the Coil-100, Face95, and Face96 databases, with an increasing value v, the APR values first rapidly go up and then gradually become stable. These results illustrate that a higher proportion of color information is crucial for object and facial images. On the five color textural databases, with the addition of the color information, the APR values firstly show fast growth for v near 0.00, gradually become stable, and then abruptly decrease for v near 1.00. These phenomena demonstrate that an appropriate weighting optimization scheme can achieve greater enhancements.

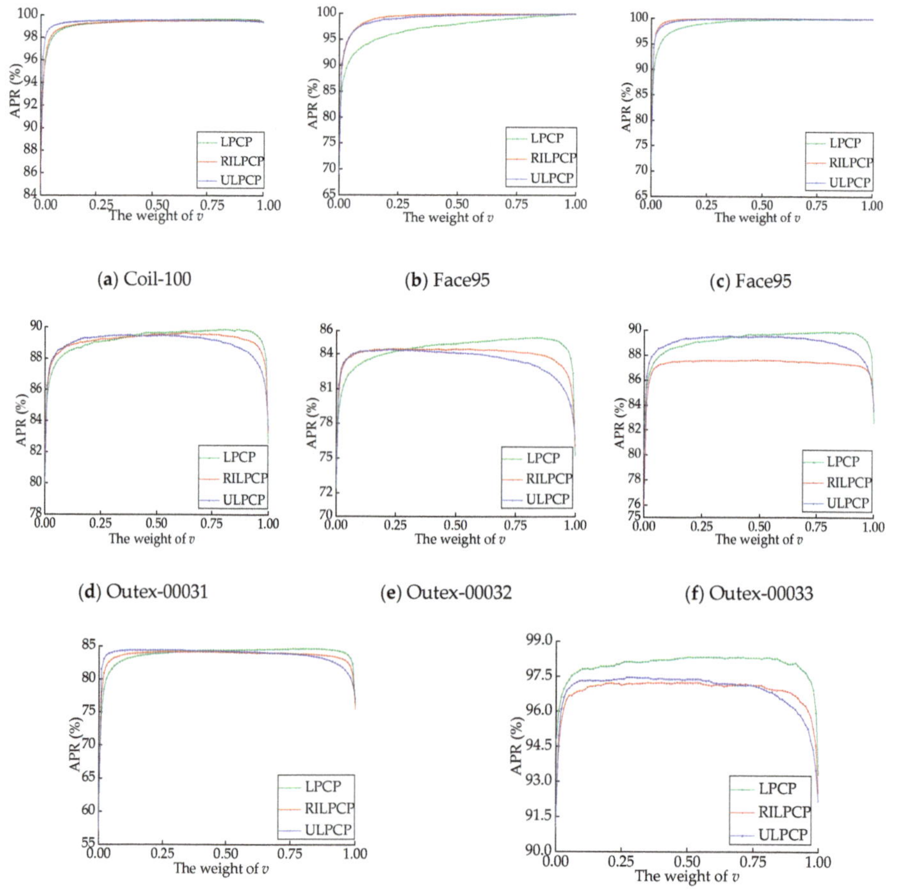

(**a**) Coil-100 (**b**) Face95 (**c**) Face95

(**d**) Outex-00031 (**e**) Outex-00032 (**f**) Outex-00033

(**g**) Outex-00034 (**h**) MIT-VisTex

Figure 9. The average precision rate (APR) under the weight value v by using LPCP, RILPCP, and ULPCP on the eight databases.

In summary, an optimized parameter v is not only beneficial to integrating the merits of $LPCP_{color}$ and $LPCP_{LBP}$, but also yields more notable improvements.

5. Conclusions

In this paper, a color texture method called local parallel cross pattern (LPCP) was proposed for encoding the LBP and color information into a unified framework. Three major contributions were summarized in this paper. First, based on the color prior knowledge in the L*a*b* color space, we designed a six-layer color quantizer that is to be used for color map extraction. Second, inspired by the human visual system, we proposed the local parallel cross pattern (LPCP) for combining the color map and the LBP map in "parallel" and "cross" manners. Third, to improve the computational complexity and provide rotation invariant, LPCP was further extended to the uniform local parallel cross pattern (ULPCP) and the rotation-invariant local parallel cross pattern (RILPCP), respectively.

We performed a series of experiments on the eight databases to evaluate the proposed descriptors. Depending on the average precision rate (APR) results, the optimal quantization layers were chosen from the six-layer color quantizer. Compared with the LBP-based descriptors, LPCP, ULPCP, and RILPCP achieved promising APR and ARR results. To evaluate the effectiveness, efficiency, robustness, and computational complexity, we performed comparative experiments among the proposed methods and eight state-of-the-art color texture descriptors in terms of the average precision rate (APR) value, the average recall rate (ARR) value, the precision-recall (PR) curve, the feature vector length, and the memory consumption. Moreover, the proposed approaches were also compared with a series of CNN-based models and achieved competitive results. Additionally, the weight-based optimization scheme yielded more notable improvements.

In the future, Locality Sensitive Hashing [46] and feature selection [47–50] will be considered to cut down the computation complexity and memory consumption. Meanwhile, Query Expansion (QE) [51] and Graph Fusion (GF) [52] will be integrated into the image retrieval system to retrieve more target images. Moreover, normalization methods [53] also will be considered to achieve illumination invariance.

Author Contributions: Q.F. conceived the research idea. Q.F. performed the experiments. Q.F. wrote the paper. Q.F., Q.H., M.S., Y.Y., Y.W., and J.D. gave many suggestions and helped revise this manuscript.

Funding: This research was funded by the National Nature Science Foundation of China, grant number [61871106], the Fundamental Research Grant Scheme for the Central Universities, grant number [130204003], the Project of Shandong Province Higher Educational Science and Technology Program, grant number [J16LN68], the Shandong Province Natural Science Foundation, grant number [ZR2017QF011] and the National Key Technology Research and Development Programme of the Ministry of Science and Technology of China, grant number [2014BAI17B02].

Conflicts of Interest: The authors declare no conflict of interest.

References

1. Feng, Q.; Hao, Q.; Chen, Y.; Yi, Y.; Wei, Y.; Dai, J. Hybrid histogram descriptor: A fusion feature representation for image retrieval. *Sensors* **2018**, *187*, 1943. [CrossRef] [PubMed]
2. Yang, M.; Song, W.; Mei, H. Efficient retrieval of massive ocean remote sensing images via a cloud-based mean-shift algorithm. *Sensors* **2017**, *17*, 1693. [CrossRef] [PubMed]
3. Liu, S.; Wu, J.; Feng, L.; Qiao, H.; Liu, Y.; Lou, W.; Wang, W. Perceptual uniform descriptor and ranking on manifold for image retrieval. *Inf. Sci.* **2017**, *424*, 235–249. [CrossRef]
4. Smeulders, A.W.M.; Worring, M.; Santini, S.; Gupta, A.; Jain, R. Content-based image retrieval at the end of the early years. *IEEE Trans. Pattern Anal. Mach. Intell.* **2002**, *22*, 1349–1380. [CrossRef]
5. Piras, L.; Giacinto, G. Information fusion in content based image retrieval: A comprehensive overview. *Inf. Fusion* **2017**, *37*, 50–60. [CrossRef]
6. Liu, L.; Chen, J.; Fieguth, P.; Zhao, G.; Chellappa, R.; Pietikäinen, M. From BoW to CNN: Two decades of texture representation for texture classification. *Int. J. Comput. Vis.* **2018**, 1–36. [CrossRef]
7. Fernández, A.; Álvarez, M.X.; Bianconi, F. Texture description through histograms of equivalent patterns. *J. Math. Imaging Vis.* **2013**, *45*, 76–102. [CrossRef]
8. Ojala, T.; Pietikäinen, M.; Maenpaa, T. Multi resolution gray-scale and rotation invariant texture classification with local binary patterns. *IEEE Trans. Pattern Anal. Mach. Intell.* **2002**, *24*, 971–987. [CrossRef]
9. Guo, Z.; Zhang, L.; Zhang, D. Rotation invariant texture classification using LBP variance (LBPV) with global matching. *Pattern Recognit.* **2010**, *43*, 706–719. [CrossRef]
10. Guo, Z.; Zhang, L.; Zhang, D. A completed modeling of local binary pattern operator for texture classification. *IEEE Trans. Image Process.* **2010**, *19*, 1657–1663.
11. Zhang, B.; Gao, Y.; Zhao, S.; Liu, J. Local derivative pattern versus local binary pattern: Face recognition with high-order local pattern descriptor. *IEEE Trans. Image Process.* **2010**, *19*, 533–544. [CrossRef] [PubMed]
12. Tan, X.; Triggs, B. Enhanced local texture feature sets for face recognition under difficult lighting conditions. *IEEE Trans. Image Process.* **2010**, *19*, 1635–1650. [PubMed]
13. Murala, S.; Maheshwari, R.P.; Balasubramanian, R. Local tetra patterns: A new feature descriptor for content-based image retrieval. *IEEE Trans. Image Process.* **2012**, *21*, 2874–2886. [CrossRef] [PubMed]

14. Subrahmanyam, M.; Maheshwari, R.P.; Balasubramanian, R. Local maximum edge binary patterns: A new descriptor for image retrieval and object tracking. *Signal Process.* **2012**, *92*, 1467–1479. [CrossRef]
15. Ren, J.; Jiang, X.; Yuan, J. Noise-resistant local binary pattern with an embedded error-correction mechanism. *IEEE Trans. Image Process.* **2013**, *22*, 4049–4060. [CrossRef] [PubMed]
16. Verma, M.; Raman, B. Local neighborhood difference pattern: A new feature descriptor for natural and texture image retrieval. *Multimed. Tools Appl.* **2018**, *77*, 11843–11866. [CrossRef]
17. Jeena Jacob, I.; Srinivasagan, K.G.; Jayapriya, K. Local oppugnant color texture pattern for image retrieval system. *Pattern Recognit. Lett.* **2014**, *42*, 72–78. [CrossRef]
18. Dubey, S.R.; Singh, S.K.; Singh, R.K. Multichannel decoded local binary patterns for content-based image retrieval. *IEEE Trans. Image Process.* **2016**, *25*, 4018–4032. [CrossRef]
19. Qi, X.; Xiao, R.; Li, C.; Qiao, Y.; Guo, J.; Tang, X. Pairwise rotation invariant co-occurrence local binary pattern. *IEEE Trans. Pattern Anal. Mach. Intell.* **2014**, *36*, 2199–2213. [CrossRef]
20. Hao, Q.; Feng, Q.; Wei, Y.; Sbert, M.; Lu, W.; Xu, Q. Pairwise cross pattern: A color-LBP descriptor for content-based image retrieval. In Proceedings of the 19th Pacific Rim Conference on Multimedia, Hefei, China, 21–22 September 2018.
21. Liu, G.; Yang, J. Content-based image retrieval using color difference histogram. *Pattern Recognit.* **2013**, *46*, 188–198. [CrossRef]
22. Liu, G.; Li, Z.; Zhang, L.; Xu, Y. Image retrieval based on micro-structure descriptor. *Pattern Recognit.* **2011**, *44*, 2123–2133. [CrossRef]
23. Liu, G.; Zhang, L.; Hou, Y.; Li, Z.; Yang, J. Image retrieval based on multi-texton histogram. *Pattern Recognit.* **2010**, *43*, 2380–2389. [CrossRef]
24. Mäenpää, T.; Pietikäinen, M. Texture analysis with local binary patterns. In *Handbook of Pattern Recognition and Computer Vision*; World Scientific: Singapore, 2005; pp. 197–216.
25. Bianconi, F.; Bello-Cerezo, R.; Napoletano, P. Improved opponent color local binary patterns: An effective local image descriptor for color texture classification. *J. Electron. Imag.* **2017**, *27*. [CrossRef]
26. Li, L.; Feng, L.; Yu, L.; Wu, J.; Liu, S. Fusion framework for color image retrieval based on bag-of-words model and color local Haar binary patterns. *J. Electron. Imag.* **2016**, *25*. [CrossRef]
27. Cusano, C.; Napoletano, P.; Schettini, R. Evaluating color texture descriptors under large variations of controlled lighting conditions. *J. Opt. Soc. Am. A* **2016**, *33*, 17–30. [CrossRef]
28. Singh, C.; Walia, E.; Kaur, K.P. Enhancing color image retrieval performance with feature fusion and non-linear support vector machine classifier. *Optik* **2018**, *158*, 127–141. [CrossRef]
29. Pietikäinen, M.; Ojala, T.; Xu, Z. Rotation-invariant texture classification using feature distributions. *Pattern Recognit.* **2000**, *33*, 43–52. [CrossRef]
30. Bianconi, F.; González, E. Counting local n-ary patterns. *Pattern Recognit. Lett.* **2018**, *177*, 24–29. [CrossRef]
31. Sarrafzadeh, O.; Dehnavi, A.M. Nucleus and cytoplasm segmentation in microscopic images using k-means clustering and region growing. *Adv. Biomed. Res.* **2015**, *4*, 174.
32. Gonzalez, R.C.; Woods, R.E. *Digital Image Processing*, 3rd ed.; Publishing House of Electronics Industry: Beijing, China, 2010; pp. 455–456. ISBN 9787121102073.
33. Salzburg Texture Image Database. Available online: http://www.wavelab.at/sources/STex/ (accessed on 22 August 2014).
34. Kolesnikov, A.; Trichina, E.; Kauranne, T. Estimating the number of clusters in a numerical data set via quantization error modeling. *Pattern Recognit.* **2015**, *48*, 941–952. [CrossRef]
35. Zhang, M.; Zhang, K.; Feng, Q.; Wang, J.; Kong, J. A novel image retrieval method based on hybrid information descriptors. *J. Vis. Commun. Image Represent.* **2014**, *25*, 1574–1587. [CrossRef]
36. Standring, S. *Gray's Anatomy: The Anatomical Basis of Clinical Practice*, 41st ed.; Elsevier Limited: New York, NY, USA, 2016; pp. 686–708. ISBN 9780702068515.
37. Guo, J.; Prasetyo, H.; Lee, H.; Yao, C. Image retrieval using indexed histogram of void-and-cluster block truncation coding. *Signal Process.* **2016**, *123*, 143–156. [CrossRef]
38. Nene, S.A.; Nayar, S.K.; Murase, H. *Columbia Object Image Library (COIL-100)*; Technical Report CUCS; Department of Computer Science, Columbia University: New York, NY, USA, 1996.
39. Libor Spacek's Facial Image Databases "Face 95 Image Database". Available online: https://cswww.essex.ac.uk/mv/allfaces/faces95.html (accessed on 8 August 2014).

40. Libor Spacek's Facial Image Databases "Face 96 Image Database". Available online: https://cswww.essex.ac.uk/mv/allfaces/faces96.html (accessed on 8 August 2014).
41. Outex Texture Image Database. Available online: http://lagis-vi.univ-lille1.fr/datasets/outex.html (accessed on 5 October 2017).
42. MIT Vision and Modeling Group. Available online: http://vismod.media.mit.edu/pub/ (accessed on 12 August 2014).
43. Szegedy, C.; Vanhoucke, V.; Loffe, S.; Shlens, J.; Wojna, Z. Rethinking the inception architecture for computer vision. In Proceedings of the IEEE Conference on Computer Vision and Pattern Recognition, Las Vegas, NV, USA, 27–30 June 2016.
44. Napoletano, P. Hand-crafted vs. learned descriptors for color texture classification. In Proceedings of the International Workshop on Computational Color Imaging, Milan, Italy, 29–31 March 2017.
45. Zou, Q.; Ni, L.; Zhang, T.; Wang, Q. Deep learning based feature selection for remote sensing scene classification. *IEEE Trans. Geosci. Remote Sens. Lett.* **2015**, *12*, 2321–2325. [CrossRef]
46. Indyk, P.; Motwani, R. Approximate nearest neighbors: Towards removing the curse of dimensionality. In Proceedings of the IEEE Conference on Multimedia Information Analysis and Retrieval, Dallas, TX, USA, 24–26 May 1998; pp. 604–613.
47. Yi, Y.; Zhou, W.; Liu, Q.; Luo, G.; Wang, J.; Fang, Y.; Zheng, C. Ordinal preserving matrix factorization for unsupervised feature selection. *Signal Process. Image Commun.* **2018**, *67*, 118–131. [CrossRef]
48. Yi, Y.; Chen, Y.; Dai, J.; Gui, X.; Chen, X.; Lei, G.; Wang, W. Semi-supervised ridge regression with adaptive graph-based label propagation. *Appl. Sci.* **2018**, *12*, 2636. [CrossRef]
49. Yi, Y.; Qiao, S.; Zhou, W.; Zheng, C.; Liu, Q.; Wang, J. Adaptive multiple graph regularized semi-supervised extreme learning machine. *Soft Comput.* **2018**, *22*, 3545–3562. [CrossRef]
50. Qi, M.; Wang, T.; Liu, F.; Zhang, B.; Wang, J.; Yi, Y. Unsupervised feature selection by regularized matrix factorization. *Neurocomputing* **2018**, *273*, 593–610. [CrossRef]
51. Chum, O.; Mikulik, M.; Perdoch, M.; Matas, J. Total recall II: Query expansion revisited. In Proceedings of the IEEE Conference on Computer Vision and Pattern Recognition, Colorado Springs, CO, USA, 20–25 June 2011; pp. 889–896.
52. Zhang, S.; Yang, M.; Cour, T.; Yu, K.; Metaxas, D. Query specific rank fusion for image retrieval. *IEEE Trans. Pattern Anal. Mach. Intell.* **2015**, *37*, 803–815. [CrossRef]
53. Cernadas, E.; Fernández-Delgado, M.; González-Rufino, E. Influence of normalization and color space to color texture classification. *Pattern Recognit.* **2017**, *61*, 120–138. [CrossRef]

sensors

MDPI

Article

Hybrid Histogram Descriptor: A Fusion Feature Representation for Image Retrieval

Qinghe Feng [1], Qiaohong Hao [2], Yuqi Chen [3], Yugen Yi [3], Ying Wei [1,*] and Jiangyan Dai [4,*]

[1] College of Information Science and Engineering, Northeastern University, Shenyang 110004, China; 1510377@stu.neu.edu.cn

[2] School of Computer Science and Technology, Tianjin University, Tianjin 300350, China; qiaohonghao@gmail.com

[3] School of Software, Jiangxi Normal University, Nanchang 330022, China; 005090@jxnu.edu.cn (Y.C.); yiyg510@jxnu.edu.cn (Y.Y.)

[4] School of Computer Engineering, Weifang University, Weifang 261061, China

* Correspondence: weiying@ise.neu.edu.cn (Y.W.); daijy@wfu.edu.cn (J.D.); Tel.: +86-024-83688326 (Y.W.)

Received: 18 May 2018; Accepted: 12 June 2018; Published: 15 June 2018

Abstract: Currently, visual sensors are becoming increasingly affordable and fashionable, acceleratingly the increasing number of image data. Image retrieval has attracted increasing interest due to space exploration, industrial, and biomedical applications. Nevertheless, designing effective feature representation is acknowledged as a hard yet fundamental issue. This paper presents a fusion feature representation called a hybrid histogram descriptor (HHD) for image retrieval. The proposed descriptor comprises two histograms jointly: a perceptually uniform histogram which is extracted by exploiting the color and edge orientation information in perceptually uniform regions; and a motif co-occurrence histogram which is acquired by calculating the probability of a pair of motif patterns. To evaluate the performance, we benchmarked the proposed descriptor on RSSCN7, AID, Outex-00013, Outex-00014 and ETHZ-53 datasets. Experimental results suggest that the proposed descriptor is more effective and robust than ten recent fusion-based descriptors under the content-based image retrieval framework. The computational complexity was also analyzed to give an in-depth evaluation. Furthermore, compared with the state-of-the-art convolutional neural network (CNN)-based descriptors, the proposed descriptor also achieves comparable performance, but does not require any training process.

Keywords: visual sensors; image retrieval; hybrid histogram descriptor; perceptually uniform histogram; motif co-occurrence histogram

1. Introduction

In the past decades, affordable visual sensor equipment (e.g., surveillance cameras, smart phones, digital cameras and camcorders) has become widespread in our daily lives. Due to the growing number of images collected from these visual sensors, how to accurately and quickly retrieve the image-of-interest has become a hot topic [1–6]. Compared with text-based image retrieval (TBIR), content-based image retrieval (CBIR) is widely considered as an effective and efficient technology that not only extracts low-level visual cues (e.g., color, shape and texture) automatically, but also bridges high-level semantic comprehension. Until now, the feature representation descriptors, such as independent feature descriptor and fusion-based feature descriptor, have been increasing and developing in the CBIR community.

Color information plays an important role in the feature representation. Currently, color moment [7], color set [8], color coherence vector [9], color correlogram [10] and color histogram [11–18] have been developed for color feature representation continuously. In [11], the color layout descriptor (CLD), scalable color descriptor (SCD), color structure descriptor (CSD) and dominant color descriptor (DCD) are constructed as the color feature descriptors. Subsequently, in [12–15], a series of equal-interval color quantization models are used for the extraction of color histograms. Recently, in [16], Bayesian Information Criterion (BIC), Expectation Maximization (EM) and Gaussian Mixture Models (GMM) are integrated into a universal color quantization framework. More recently, in [17,18], the combined color histogram is proposed for color feature representation. However, the above methods are confined to quantizing the range of different color channels, and a few consider the color probability distribution of different color channels. In addition, several methods (e.g., Fourier transforms [19], moment invariant [20] and edge orientation detection [13–15,21–25]) have been developed for shape-based representation. In [21], edge orientation detection is equipped with different gradient operators for the orientation information computation on grey-scale images. With the appearance of color images, in [13–15], a series of edge detection and quantization strategies is applied to capture the geometry and orientation information from color images in different color spaces. In [22–25], a class of local edge orientation detection descriptors is developed for edge orientation histogram extraction. In short, edge orientation detection and quantization are widely considered as the effective and correct approaches that not only achieve stable performances but also exploit the geometry and orientation information with less computational complexity.

Along other research lines, many strategies [17,18,26–29] have been designed to represent textural features. For example, the local binary pattern (LBP) [26] is first proposed to code the center pixel and its neighborhood pixels as a binary label in eight directions. Later, the LBP is extended to the local extrema pattern (LEP) [17], which computes the index values between the center pixel and its eight neighbors in four directions. Afterwards, the LEP is modified to the local extrema co-occurrence pattern (LEcP) [18], which reveals the relationship of mutual occurrence patterns in the V channel of the HSV color space. Furthermore, the concept of texton or motif [27] is first defined to analysis the elements of texture perception and their interactions. Recently, a grey-level co-occurrence matrix (GLCM) [28] is treated as a co-occurrence-based relation descriptor that computed the occurrence frequencies of a pair of grey-pixels. More recently, the motif co-occurrence matrix (MCM) [29] is defined as a 3D matrix, in which six motif patterns are designed to calculate the probability of a pair of motif patterns in a pre-defined direction. However, using six motif patterns is incomplete, because the perceptually uniform motif patterns are not further discussed and analyzed.

Although the above-mentioned methods have proven to be effective, independent feature descriptors are inadequate to meet the demands of feature representation. Many studies have proven that fusion-based descriptors are more powerful than independent feature descriptors. In [13–15], the color histogram and the edge orientation histogram are treated as a pair of mutual information descriptors, calculated by a color difference operator. In [17], the color histogram is combined with the local extrema pattern histogram used for object tracking in the RGB color space. In [18], the local extrema co-occurrence pattern (LEcP) is transformed into an independent feature vector; then, LEcP is combined with the joint color histogram for feature representation. Again, in [30], a multi-channel decoded local binary pattern (mdLBP) and a multi-channel adder local binary pattern (maLBP) are simultaneously constructed by combining three LBP maps, which are calculated in the RGB color space. Recently, in [31], the local neighborhood difference pattern (LNDP) and the LBP is explored to capture local intensity difference information for the natural and texture image retrieval. In [32], Bianconi et al. provided a general framework and taxonomy of color texture descriptors. In [33], Cusano et al. suggested an evaluation of color texture descriptors under large variations of controlled lighting conditions, whereas Qazi et al. investigated pertinent color spaces for color texture characterization [34]. At the same time, in [35], user relevance feedback, feature re-weight and weight optimization are used to further improve the accuracy of image retrieval.

In this study, the main contributions are summarized as follows:

1. We designed the pyramid color quantization model, which is based on the powerful color probability distribution prior in the L*a*b* color space.

2. We constructed the perceptually uniform histogram, which integrates color and edge orientation as a whole by exploiting a color difference operator.
3. We developed the motif co-occurrence histogram in which the perceptually uniform motif patterns are further discussed and analyzed.
4. We proposed the hybrid histogram descriptor that is comprised of the perceptually uniform histogram and the motif co-occurrence histogram.

The remainder of this paper is organized as follows. Preliminaries are introduced in Section 2, and the feature representation is described in Section 3. Experiments and evaluations are presented in Section 4. Section 5 provides conclusions.

2. Preliminaries

2.1. The Color Space Selection

The selection of the color space is a crucial step before feature representation. In the past decades, several types of color spaces (e.g., RGB, L*a*b*, HSV, CMYK, YUV and HSI) have been widely used for CBIR. Among them, the RGB is recognized as one of the most popular color spaces. It is derived from three colors of light, namely, red (R), green (G) and blue (B) [36]. Nevertheless, its disadvantages are often ignored: (1) the redundancy between blue and green; (2) the missing yellow between red and green; and (3) the non-uniform perception of human eye. Consequently, Hering defined the L*a*b* color space, which includes three pairs of color channels consisting of the white–black pair of the L* channel (ranging from 0 to 100), the yellow–blue pair of the a* channel (ranging from −128 to +127), and the red–green pair of the b* channel (ranging from −128 to +127) [37]. Compared with the RGB, the advantages of the L*a*b* color space are summarized as follows: (1) the L*a*b* remedies the redundant and missing information of the RGB; (2) it conforms to human eye's perception mechanism; and (3) it provides excellent decoupling between intensity (represented by the L* channel) and color (represented by the a* and b* channels) [38]. Therefore, our scheme transforms all images from RGB to L*a*b* color space before the feature representation stage. The details of this transformation are defined using standard RGB to L*a*b* transformations as follows [15,39]:

$$\begin{cases} L* = 116(\frac{Y}{Y_n})^{1/3} - 16 & \text{for } \frac{Y}{Y_n} > 0.08856 \\ L* = 903.3(\frac{Y}{Y_n})^{1/3} & \text{for } \frac{Y}{Y_n} \leq 0.08856 \end{cases}, \tag{1}$$

$$a* = 500(f(\frac{X}{X_n}) - f(\frac{Y}{Y_n})), \tag{2}$$

$$b* = 500(f(\frac{X}{X_n}) - f(\frac{Y}{Z_n})), \tag{3}$$

with

$$\begin{cases} f(u) = u^{1/3} & \text{for } u > 0.08856 \\ f(u) = 7.78u + \frac{Y}{Y_n} & \text{for } u \leq 0.08856 \end{cases}, \tag{4}$$

where

$$\begin{bmatrix} X \\ Y \\ Z \end{bmatrix} = \begin{bmatrix} 0.412453 & 0.357580 & 0.180423 \\ 0.212671 & 0.715160 & 0.072169 \\ 0.019334 & 0.119193 & 0.950227 \end{bmatrix} \begin{bmatrix} R \\ G \\ B \end{bmatrix}, \tag{5}$$

where X_n, Y_n and Z_n are the values of X, Y and Z for the illuminant and $[X_n, Y_n, Z_n]$ = [0.950450, 1.000000, 1.088754] in accordance with illuminant D65 [15].

2.2. Probability Distribution Prior in L*a*b* Color Space

In the previous color quantization models [12–15,17,18], three color channels are uniformly mapped into the fixed intervals. However, during the process of quantization, these models lose some useful color information. Hence, reducing the loss of the useful color information is a serious concern. Inspired by this motivation, we have explored and summarized the color probability distribution of the a* and b* channels in different image databases. The example of the AID image database [40] is shown in Figure 1a,b. The frequency of pixels mainly focuses on the center region of the a* and b* channels.

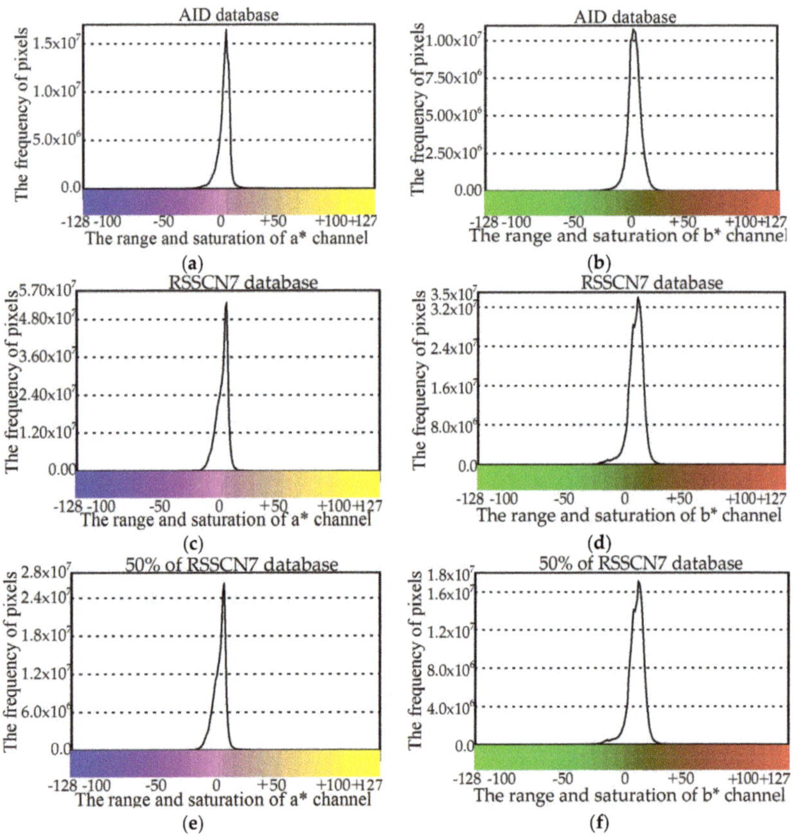

Figure 1. The frequency of pixels over different databases: (**a**,**b**) AID; (**c**,**d**) RSSCN7; and (**e**,**f**) 50% of RSSCN7.

To verify the validity of this prior knowledge, we calculated the color probability distribution statistics of the a* and b* channels on hundreds of image databases. The results show that the proposed prior is stable and consistent. Even if an image database has been changed, the property of the color probability distribution prior is still fairly consistent. For example, the color probability distribution of the a* and b* channels in the RSSCN7 [41] dataset and its subset (50% of the RSSCN7 dataset) is shown in Figure 1c–f. Obviously, there is almost no change between RSSCN7 and its subset, except for the pixel frequency.

3. Feature Representation

3.1. Perceptually Uniform Histogram

3.1.1. Pyramid Color Quantization Model

Inspired by the above prior knowledge, we designed a novel pyramid color quantization model (as shown in Figure 2), in which every layer represents a quantized scheme (including a group of intervals and indexes). The original range (−128, +127) of a* or b* is first projected into two equal intervals in Layer 1, and the indexes of two intervals are flagged as 0 and 1 from left to right, correspondingly. Then, considering the pixels focus on the middle, two middle intervals from Layers 2–7 are split into four equal intervals from the up-layer to down-layer until two middle intervals cannot be split in Layer 7. Finally, the remaining intervals are copied from the up-layer to down-layer, sequentially. In this manner, we refine and retain the color information in the middle of the a* or b* channels effectively. We define the quantization layer of the a* and b* channels as Y_{a^*} and Y_{b^*}, where $Y_{a^*}, Y_{b^*} \in \{1, 2, \ldots, 7\}$, and the indexes are denoted as $\tilde{Y}_{a^*}$ and $\tilde{Y}_{b^*}$, $\tilde{Y}_{a^*} \in \{0, 1, \ldots, \ddot{Y}_{a^*}\}$ and $\tilde{Y}_{b^*} \in \{0, 1, \ldots, \ddot{Y}_{b^*}\}$, where $\ddot{Y}_{a^*} = 2Y_{a^*} - 1$ and $\ddot{Y}_{b^*} = 2Y_{b^*} - 1$, respectively.

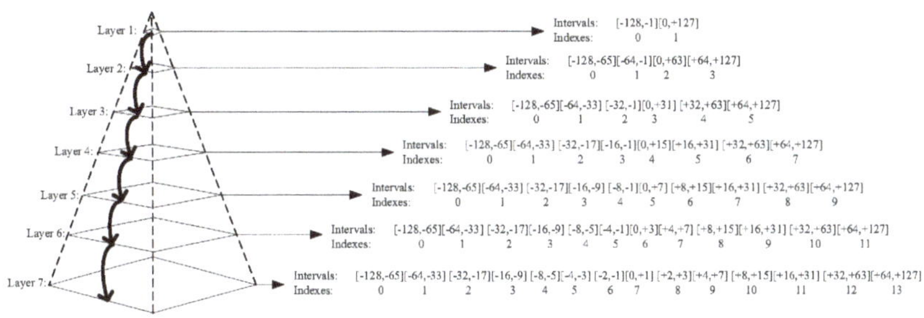

Figure 2. Pyramid color quantization model.

In addition, considering the human visual intensity perception mechanism in [5], the L* channel is quantized into three intervals (0, +25), (+26, +75) and (+76, +100). We define the quantization layer of the L* channel as Y_{L^*}, where $Y_{L^*} = 1$, and the index is flagged as $\tilde{Y}_{L^*}$, $\tilde{Y}_{L^*} \in \{0, 1, \ldots, \ddot{Y}_{L^*}\}$, where $\ddot{Y}_{L^*} = 2Y_{L^*}$. In summary, combining the indexes of the L*, a* and b* channels, the color map of an image $f(x, y)$ is defined as $C(x, y)$, and the index is flagged as $\widetilde{C}$, $\widetilde{C} \in \{0, 1, \ldots, \hat{C}\}$, where $\hat{C} = 2Y_{a^*} \times 2Y_{b^*} \times 3 - 1$.

3.1.2. Perceptually Uniform Histogram Definition

The Gestalt Psychology Theory elucidates that the human visual perception mechanism tends to group elements into a local region where the elements share a homologous or approximate property [42]. Based on this theoretical foundation, perceptually uniform regions can be described as a certain visual feature space in which visual elements have the same rule (e.g., color and edge orientation). For the visual feature space $\tilde{I}$, an element ξ and its neighborhoods ξ_g within $\tilde{I}$ are flagged as $\tilde{I}(\xi)$ and $\tilde{I}(\xi_g)$. Mathematically, the discrimination function $\varphi(\cdot)$ is formulated as follows:

$$\varphi(\widetilde{I}(\xi), \widetilde{I}(\xi_g)) = \begin{cases} 1, & \widetilde{I}(\xi) = \widetilde{I}(\xi_g) \\ 0, & \widetilde{I}(\xi) \neq \widetilde{I}(\xi_g) \end{cases}, g \in \left\{1, 2, \ldots, \ddot{N}\right\}, \tag{6}$$

where $\ddot{N}$ represents the number of neighborhoods. If $\varphi(\tilde{I}(\xi), \tilde{I}(\xi_g)) = 1$, $\tilde{I}\,\tilde{I}(\xi_g)$ belongs to the perceptually uniform region; if $\varphi(\tilde{I}(\xi), \tilde{I}(\xi_g)) = 0$, $\tilde{I}(\xi_g$ does not belong to the perceptually uniform region.

With subject to the perceptually uniform region, we construct the perceptually uniform histogram by exploiting the color difference operator [15,43,44] between the color and edge orientation. Herein, given an image $f(x, y)$, the edge orientation map $O(x, y)$ is first extracted by using the Prewitt operator, due to its advantages of extracting the geometry and boundary information from the observed content. Then, experimentally, the edge orientation value is quantized uniformly into four bins to construct the edge orientation map $O(x, y)$ because it is time consuming and unnecessary to consider all edge orientation values. Finally, the edge orientation map $O(x, y)$ and the color map $C(x, y)$ are divided into the overlapping 3×3 windows in which the central pixel is flagged as (x, y) and its eight neighbors are flagged as (x_g, y_g), $g \in \{1, 2, \ldots, 8\}$. The perceptually uniform histogram (PUH) is defined as follows:

$$PUH^{colour}(O(x,y)) = \sum_{g=1}^{8} \sqrt{\sum_{\psi \in L*,a*,b*} (\Delta f_{\psi})^2} \; sub.t. \; \varphi(C(x,y), C(x_g, y_g)) = 1, \tag{7}$$

$$PUH^{ori}(C(x,y)) = \sum_{g=1}^{8} \sqrt{\sum_{\psi \in L*,a*,b*} (\Delta f_{\psi})^2} \; sub.t. \; \varphi(O(x,y), O(x_g, y_g)) = 1, \tag{8}$$

where Δf represents the color differences among the central pixel (x, y) and its eight neighbors (x_g, y_g) in ψ channels, $\psi \in$ L$*$, a$*$, b$*$. The feature vector length of $PUH^{color}(O(x, y))$ and $PUH^{ori}(C(x, y))$ are 4 and $2Y_{a*} \times 2Y_{b*} \times 3$, respectively. For an image dataset D, the fitness quantization layers of Y_{a*} and Y_{b*} are computed depending upon the retrieval accuracy score Acc(D | Y_{a*}, Y_{b*}). This procedure is expressed as the maximization problem as follows:

$$\max_{Y_{a*}, Y_{b*}} \mathrm{Acc}(D | Y_{a*}, Y_{b*}), Y_{a*}, Y_{b*} \in \{1, 2, \ldots, 7\}, \tag{9}$$

We present the detailed evaluation of different color quantization layers of Y_{a*} and Y_{b*} in Section 4.4.

3.2. Motif Co-Occurrence Histogram

The perceptually uniform histogram only extracts the color and edge orientation information, but the texture information is ignored to some extent. Fortunately, the motif pattern, which depicts the texture information by the pre-defined spatial structure model, can remedy this shortcoming.

3.2.1. Motif Patterns

The motif co-occurrence matrix (MCM) is investigated in [29] where the first six types of motif patterns shown in Figure 3, starting from the top-left point P1, are generated because they represent a completed set of space filling curves. However, using merely six motif patterns is insufficient because the perceptually uniform motif patterns (PUMP) are ignored.

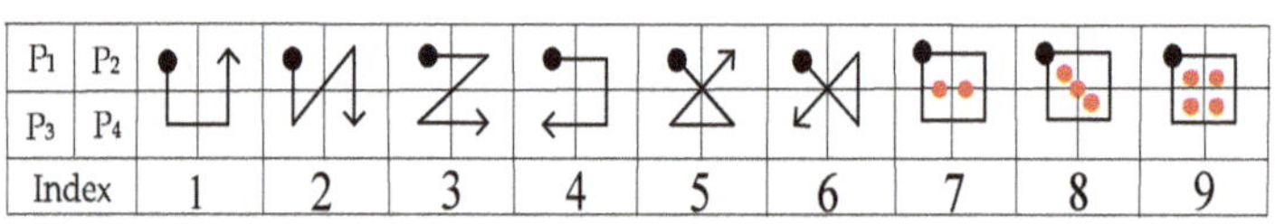

Figure 3. Nine types of motif patterns.

To depict the consistency of spatial structure information, we propose three perceptually uniform motif patterns into which all types of perceptually uniform motif patterns are separated based on the number of equal pixels. Combining the previous six motif patterns, nine motif patterns are obtained, as shown in Figure 3, in which the red dots represent the number of equal pixels in the motif patterns 7, 8 and 9.

3.2.2. Motif Co-Occurrence Histogram Definition

Since the L*a*b* color space provides excellent decoupling between intensity (represented by the L* channel) and color (represented by the a* and b* channels) [38], the L* channel is applied to extract the motif co-occurrence histogram. For simplicity, a 5 × 5 mini-numerical map in Figure 4a is adopted to illustrate the proposed method. In our scheme, each pixel (apart from the lower and right boundary pixels) in the map is divided into the overlapping 2 × 2 grids in Figure 4b. Then, each grid is transformed into a motif pattern with the minimized local gradient to obtain the motif map shown in Figure 4c, which is used to calculate the motif co-occurrence histogram shown in Figure 4d. For example, the red circle in Figure 4c is a pair of motif patterns, indexed as (3, 2), in the 0° direction, corresponding to the red bar "*MCH*(3, 2) = 1" in the motif co-occurrence histogram in Figure 4d. Mathematically, the probability of co-occurrence of a pair of motif patterns is expressed as follows:

$$MCH(MP^{e1}, MP^{e2}) = \Pr\left\{M(i,j) = MP^{e1}, M(i,j+1) = MP^{e2}\right\}, \tag{10}$$

where Pr is the probability of co-occurrence of a pair of motif patterns corresponding to (i,j) and its neighbor $(i,j+1)$ within the motif map $M(x,y)$. MP^{e1} and MP^{e2} represent the indexes of a pair of motif patterns, where $MP^{e1}, MP^{e2} \in \{1,2,\ldots,9\}$. The feature vector length of the motif co-occurrence histogram is 81. We will perform the detailed evaluation of different motif co-occurrence schemes between the motif co-occurrence matrix [29] and the proposed motif co-occurrence histogram in Section 4.5.

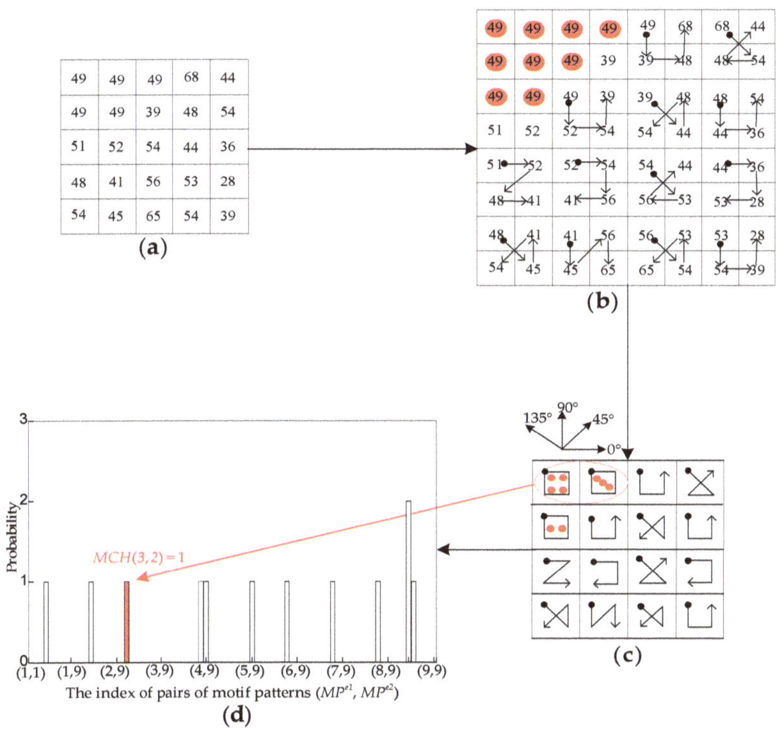

Figure 4. Schematic diagram of the motif co-occurrence histogram: (**a**) a 5 × 5 mini-numerical map; (**b**) the overlapping 2 × 2 grids of (**a**); (**c**) the motif map; and (**d**) the motif co-occurrence histogram.

3.3. Hybrid Histogram Descriptor Definition

It is widely recognized that an image possesses a rich semantic content that goes beyond the description by its metadata [2]. Hence, it is necessary to take a fusion-based feature descriptor into account because it can integrate the merits of the subjective aspects of image semantics. From this point of view, the hybrid histogram descriptor (HHD) is proposed by concatenating the perceptually uniform histogram and the motif co-occurrence histogram, and it is expressed as follows:

$$HHD = [PUH, MCH], \tag{11}$$

We present the detailed evaluation of the proposed descriptors among the perceptually uniform histogram, the motif co-occurrence histogram and the hybrid histogram descriptor in Section 4.6.

4. Experiments and Discussion

4.1. Distance Metric

The distance metric serves as an important step to measure the feature vector dissimilarity. In the CBIR framework, the query image and database images are converted into feature vectors in the form of histogram descriptors, and they are sent to the distance measure for measuring the dissimilarity. In this paper, the Extended Canberra Distance [15,32] is used, and it is defined as follows:

$$T(D,Q) = \sum_{\mu=1}^{K} \frac{|D_\mu - Q_\mu|}{|D_\mu + l_D| + |Q_\mu + l_Q|}, \tag{12}$$

where Q, D, K, and T represent the query image, the database image, the feature vector dimension, and the distance metric result, respectively, where $l_D = \sum_{\mu=1}^{K} D_\mu / K$ and $l_Q = \sum_{\mu=1}^{K} Q_\mu / K$.

4.2. Evaluation Criteria

The final goal of image retrieval is to search a set of target images from the image database [35]. For a query image I_Q and a database image I_D, the precision (*Pre*) and recall (*Rec*) values are given as follows:

$$Pre = \frac{1}{N_\sigma} \sum_{D=1}^{N_\sigma} \varsigma(\vartheta(I_Q), \vartheta(I_D)) \times 100\%, \tag{13}$$

$$Rec = \frac{1}{N_\tau} \sum_{D=1}^{N_\tau} \varsigma(\vartheta(I_Q), \vartheta(I_D)) \times 100\%, \tag{14}$$

$$\varsigma(\vartheta(I_Q), \vartheta(I_D)) = \begin{cases} 1, & \text{if } \vartheta(I_Q) = \vartheta(I_D) \\ 0, & \text{otherwise} \end{cases}, \tag{15}$$

where $\vartheta(\cdot)$, N_σ, and N_τ represent the image category information, the number of retrieved images, and the number of images in each category, respectively. The discrimination function $\varsigma(\cdot)$ is used to determine the category information between the query image and the database images. In the experiments, to guarantee accuracy and reproducibility, all images were chosen as the query image. Referring to the parameter setting in [30,32], the number of retrieved images was set to 10. For ETHZ-53 [45], the number of retrieved images was set to 5.

Further, for N query images, the average precision rate (APR) and average recall rate (ARR) values are defined as follows:

$$APR = \frac{\sum_{n=1}^{N} Pre(n)}{N} \times 100\%, \tag{16}$$

$$ARR = \frac{\sum_{n=1}^{N} Rec(n)}{N} \times 100\%, \tag{17}$$

where n is the nth query image.

Furthermore, considering the order of the retrieved images, the precision–recall curve denotes an auxiliary evaluation criterion that measures the dynamic precision with the threshold recall. Mathematically, the precision–recall curve is formulated as follows:

$$PR(\chi) = \frac{N_\tau}{N_\chi} \cdot \chi \times 100\%, \tag{18}$$

where N_τ and N_χ represent the number of images in each category, and the total number of the shown images at the recall of χ, $\chi \in \{1, 2, \dots, N_\sigma - 1\}$. A higher precision–recall curve indicates a more accurate retrieval performance.

4.3. Image Databases

Extensive experiments were conducted on five benchmark databases, including two remote sensing image databases (RSSCN7 and AID), two textural image datasets (Outex-00013 and Outex-00014), and one object image database (ETHZ-53). The details of these datasets are summarized as follows:

1. RSSCN7 database

The RSSCN7 [41] is a publicly available remote sensing dataset produced by different remote imaging sensors. It consists of seven land-use categories, such as industrial region, farm land, residential region, parking lot, river lake, forest and grass land. For each category, there are 400 images with size of 400 × 400 in JPG format. Some sample images are shown in Figure 5a, in which each row represents one category. Note that there are images with rotation and resolution differences in the same category. Thus, the RSSCN7 dataset can not only verify the effective of the proposed descriptor but also inspect the robustness of different rotations and resolutions. The RSSCN7 dataset can be downloaded from https://www.dropbox.com/s/j80iv1a0mvhonsa/RSSCN7.zip?dl=0.

2. AID database

The aerial image dataset (AID) [40] is also a publicly available large-scale remote sensing dataset produced by different remote imaging sensors. It contains 10,000 images in 30 categories, for example, airport, bare land, meadow, beach, park, bridge, forest, railway station, and baseball field. Each category includes different numbers of images varying from 220 to 420 with size of 600 × 600 in JPG format. Some sample images are shown in Figure 5b, in which each row is one category. Similar to RSSCN7, there are images with rotation and resolution differences in the same category. The AID dataset can be downloaded from http://www.lmars.whu.edu.cn/xia/AID-project.html.

3. Outex-00013

The Outex-00013 [46] is a publicly available color texture dataset produced by an Olympus Camedia C-2500 L digital camera. It contains 1360 images in 68 categories, for example, wool, fabric, cardboard, sandpaper, natural stone and paper. Each category includes 20 images, each with size of 128 × 128 in BMP format. Some sample images from Outex-00013 are shown in Figure 5c, in which each row represents one category. There is no difference in the same category. The Outex-00013 dataset can be downloaded from http://www.outex.oulu.fi/index.php?page=classification.

4. Outex-00014

The Outex-00014 [46] is also a publicly available color texture dataset produced by an Olympus Camedia C-2500 L digital camera. It contains 4080 images in 68 categories, for example, wool, fabric, cardboard, sandpaper, natural stone, and paper. Each category includes 20, each with size of 128 × 128 images in BMP format. Some sample images from Outex-00014 are shown in Figure 5d, in which each row represents one category. All images are produced under three different illuminants: the 4000 K fluorescent TL84 lamp, the 2856 K incandescent CIE A and the 2300 K horizon sunlight. The Outex-00014 dataset can also be downloaded from http://www.outex.oulu.fi/index.php?page=classification.

5. ETHZ-53

The ETHZ-53 [45] is a publicly available object dataset collected by a color camera. It contains 265 images in 53 objects, such as cup, shampoo, vegetable, fruit, and car model. Each object includes 5 images, each with size of 320 × 240 in BNG format. Some sample images are shown in Figure 5e, in which each row represents one category. Note that each object is with 5 different angles. The ETHZ-53 dataset can be downloaded from http://www.vision.ee.ethz.ch/en/datasets/.

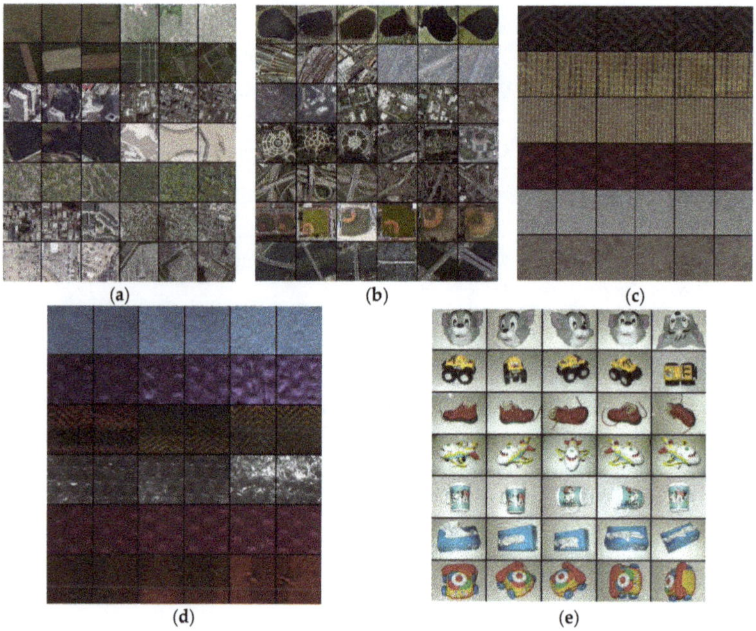

Figure 5. Some sample images from different databases: (**a**) RSSCN7; (**b**) AID; (**c**) Outex-00013; (**d**) Outex-00014; and (**e**) ETHZ-53.

4.4. Evaluation of Different Color Quantization Layers

Tables 1–5 show the average precision rate (APR) of the proposed descriptor on the RSSCN7, AID, Outex-00013, Outex-00014 and ETHZ-53 datasets under different color quantization layers of Y_{a^*} and Y_{b^*}, where $Y_{a^*}, Y_{b^*} \in \{1, 2, \ldots, 7\}$. Bold values highlight the best values. As reported in Tables 1 and 2, i when $Y_{a^*} = 6$ and $Y_{b^*} = 5$, the HHD achieves the best APR = 79.57% on RSSCN7 and the best APR = 58.13% on AID, respectively. As documented in Tables 3 and 4, when $Y_{a^*} = 6$ and $Y_{b^*} = 2$, the HHD achieves the best APR = 84.21% on Outex-00013 and the best APR = 82.82% on Outex-00014, respectively. As listed in Table 5, when $Y_{a^*} = 5$ and $Y_{b^*} = 6$, the HHD achieves the best APR = 97.89% on ETHZ-53. In addition, we can also see that the simplest color quantization scheme (e.g., $Y_{a^*} = 1$ and $Y_{b^*} = 1$) does not lead to the lowest APR on RSSCN7 and Outex-00013, and the most refined color quantization scheme (e.g., $Y_{a^*} = 7$ and $Y_{b^*} = 7$) does not guarantee the highest APR. This phenomenon demonstrates that it is necessary to adaptively select the fitness quantization layers of Y_{a^*} and Y_{b^*}. Depending upon the retrieval accuracy score, the fitness quantization layers of Y_{a^*} and Y_{b^*} will be used in the following experiments.

Table 1. Average precision rate (APR) of different color quantization layers on RSSCN7.

The Color Quantization Layer for Y_{a^*}	The Color Quantization Layer for Y_{b^*}						
	$Y_{b^*} = 1$	$Y_{b^*} = 2$	$Y_{b^*} = 3$	$Y_{b^*} = 4$	$Y_{b^*} = 5$	$Y_{b^*} = 6$	$Y_{b^*} = 7$
$Y_{a^*} = 1$	76.65	76.62	76.65	77.18	77.59	77.69	77.56
$Y_{a^*} = 2$	76.61	76.59	76.56	77.21	77.59	77.81	77.61
$Y_{a^*} = 3$	76.64	76.56	76.45	77.12	77.51	77.81	77.64
$Y_{a^*} = 4$	77.32	77.24	77.09	77.42	77.88	78.18	78.11
$Y_{a^*} = 5$	78.20	78.21	78.18	78.43	79.05	79.34	79.12
$Y_{a^*} = 6$	79.00	79.08	79.10	79.20	**79.57**	79.54	79.26
$Y_{a^*} = 7$	78.75	78.90	78.91	78.94	79.26	79.24	78.68

Table 2. Average precision rate (APR) of different color quantization layers on AID.

The Color Quantization Layer for Y_{a^*}	The Color Quantization Layer for Y_{b^*}						
	$Y_{b^*} = 1$	$Y_{b^*} = 2$	$Y_{b^*} = 3$	$Y_{b^*} = 4$	$Y_{b^*} = 5$	$Y_{b^*} = 6$	$Y_{b^*} = 7$
$Y_{a^*} = 1$	53.07	53.19	53.40	55.01	55.75	55.72	55.52
$Y_{a^*} = 2$	53.19	53.31	53.54	55.15	55.96	55.90	55.74
$Y_{a^*} = 3$	53.30	53.47	53.70	55.19	56.02	55.96	55.81
$Y_{a^*} = 4$	54.52	54.65	54.85	56.05	56.74	56.79	56.71
$Y_{a^*} = 5$	56.18	56.27	56.35	57.17	57.68	57.79	57.83
$Y_{a^*} = 6$	56.83	57.02	57.06	57.81	**58.13**	57.99	57.76
$Y_{a^*} = 7$	56.68	56.83	56.91	57.75	57.99	57.71	57.50

Table 3. Average precision rate (APR) of different color quantization layers on Outex-00013.

The Color Quantization Layer for Y_{a^*}	The Color Quantization Layer for Y_{b^*}						
	$Y_{b^*} = 1$	$Y_{b^*} = 2$	$Y_{b^*} = 3$	$Y_{b^*} = 4$	$Y_{b^*} = 5$	$Y_{b^*} = 6$	$Y_{b^*} = 7$
$Y_{a^*} = 1$	83.41	83.52	83.21	82.60	82.28	81.55	81.21
$Y_{a^*} = 2$	83.52	83.61	83.23	82.79	82.38	81.73	81.39
$Y_{a^*} = 3$	83.54	83.72	83.20	82.99	82.54	81.79	81.31
$Y_{a^*} = 4$	83.43	83.55	83.13	82.87	82.44	81.62	81.28
$Y_{a^*} = 5$	83.38	83.59	83.10	82.76	82.36	81.54	81.14
$Y_{a^*} = 6$	84.11	**84.21**	83.78	83.32	82.84	82.00	81.76
$Y_{a^*} = 7$	83.87	83.92	83.65	83.26	82.82	81.84	81.35

Table 4. Average precision rate (APR) of different color quantization layers on Outex-00014.

The Color Quantization Layer for Y_{a^*}	The Color Quantization Layer for Y_{b^*}						
	$Y_{b^*} = 1$	$Y_{b^*} = 2$	$Y_{b^*} = 3$	$Y_{b^*} = 4$	$Y_{b^*} = 5$	$Y_{b^*} = 6$	$Y_{b^*} = 7$
$Y_{a^*} = 1$	79.22	79.33	80.11	81.44	81.60	81.34	81.02
$Y_{a^*} = 2$	79.33	79.43	80.20	81.59	81.71	81.49	81.19
$Y_{a^*} = 3$	79.36	79.45	80.17	81.62	81.85	81.57	81.21
$Y_{a^*} = 4$	80.71	80.76	80.84	81.84	81.92	81.69	81.35
$Y_{a^*} = 5$	82.00	82.22	81.99	82.43	82.35	82.06	81.80
$Y_{a^*} = 6$	82.71	**82.82**	82.59	82.69	82.56	82.31	82.13
$Y_{a^*} = 7$	82.54	82.68	82.59	82.72	82.58	82.35	82.09

Table 5. Average precision rate (APR) of different color quantization layers on ETHZ-53.

The Color Quantization Layer for Y_{a^*}	The Color Quantization Layer for Y_{b^*}						
	$Y_{b^*} = 1$	$Y_{b^*} = 2$	$Y_{b^*} = 3$	$Y_{b^*} = 4$	$Y_{b^*} = 5$	$Y_{b^*} = 6$	$Y_{b^*} = 7$
$Y_{a^*} = 1$	81.21	81.96	86.87	91.47	92.83	93.06	93.36
$Y_{a^*} = 2$	80.98	81.58	87.32	91.40	92.68	93.43	93.13
$Y_{a^*} = 3$	84.68	85.36	90.19	93.21	94.49	94.87	94.49
$Y_{a^*} = 4$	89.81	89.43	92.68	95.62	96.53	96.91	96.75
$Y_{a^*} = 5$	92.98	93.21	95.55	97.21	97.74	**97.89**	97.66
$Y_{a^*} = 6$	93.36	93.13	95.77	97.13	97.58	97.58	97.43
$Y_{a^*} = 7$	81.21	81.96	86.87	91.47	92.83	93.06	93.36

4.5. Evaluation of Different Motif Co-Occurrence Schemes

Table 6 shows the average precision rate (APR) and average recall rate (ARR) values on the RSSCN7, AID, Outex-00013, Outex-00014 and ETHZ-53 datasets by using the motif co-occurrence matrix (MCM) and the motif co-occurrence histogram (MCH), respectively. Bold values highlight the best values. In Table 6, the {APR, ARR} of MCH greatly outperforms MCM by {18.14%, 0.45%} on RSSCN7, {15.21%, 0.47%} on AID, {41.75%, 20.87%} on Outex-00013 and {24.63%, 12.32%} on Outex-00014. One possible reason is that MCH takes three perceptually uniform motif patterns. Based on the above results, it can be concluded that MCH is more effective than MCM.

Table 6. Average precision rate (APR) and average recall rate (ARR) of different motif co-occurrence histograms.

Descriptor	Performance (%)	Data Set				
		RSSCN7	AID	Outex-13	Outex-14	ETHZ-53
MCM	APR	45.96	22.83	26.85	16.28	29.13
	ARR	1.15	0.68	13.43	8.14	29.13
MCH	APR	**64.10**	**38.04**	**68.60**	**40.91**	**48.38**
	ARR	**1.60**	**1.15**	**34.30**	**20.46**	**48.38**

4.6. Evaluation of the Proposed Descriptors

Table 7 shows the average precision rate (APR) and average recall rate (ARR) values on the RSSCN7, AID, Outex-00013, Outex-00014 and ETHZ-53 datasets by using the motif co-occurrence histogram (MCH), the perceptually uniform histogram (PUH) and the hybrid histogram descriptor (HHD). Bold values highlight the best values. As listed in Table 7, the {APR, ARR} of HHD outperforms MCH by {15.47%, 0.39%} on RSSCN7, by {20.09%, 0.61%} on AID, by {15.61%, 7.80%} on Outex-00013, by {41.91%, 20.95%} on Outex-00014 and by {49.51%, 49.51%} on ETHZ-53. Meanwhile, it can also be observed that the {APR, ARR} of HHD outperforms PUH by {7.35%, 0.18%} on RSSCN7, by {7.09%, 0.21%} on AID, by {4.81%, 2.40%} on Outex-00013, by {6.68%, 3.34%} on Outex-00014, and by {0.08%, 0.08%} on ETHZ-53, respectively. The main reason is that HHD integrates the merits of PUH and MCH effectively. Based on the above results, it can be asserted that HHD performs better than MCH and PUH significantly.

Table 7. Average precision rate (APR) and average recall rate (ARR) of the proposed descriptors.

Descriptor	Performance (%)	Data Set				
		RSSCN7	AID	Outex-13	Outex-14	ETHZ-53
MCH	APR	64.10	38.04	68.60	40.91	48.38
	ARR	1.60	1.15	34.30	20.46	48.38
PUH	APR	72.22	51.04	79.40	76.14	97.81
	ARR	1.81	1.55	39.70	38.07	97.81
HHD	APR	**79.57**	**58.13**	**84.21**	**82.82**	**97.89**
	ARR	**1.99**	**1.76**	**42.10**	**41.41**	**97.89**

4.7. Comparison with Other Fusion-Based Descriptors

To illustrate the effectiveness and robustness of hybrid histogram descriptor (HHD), it is compared with nine fusion-based feature descriptors and the fusion of the perceptually uniform histogram and motif co-occurrence matrix (flagged as "PUH + MCM") on the RSSCN7, AID, Outex-00013, Outex-00014 and ETHZ-53 datasets. All comparative methods are detailed as follows:

(1) mdLBP [30]: The 2048-dimensional multichannel adder local binary patterns by combining three LBP maps extracted from the R, G and B channels.
(2) maLBP [30]: The 1024-dimensional multichannel decoded local binary patterns by combining three LBP maps extracted from the R, G and B channels.
(3) CDH [15]: The 90-dimensional color histogram obtained by quantizing the L*a*b* color space and the 18-dimensional edge orientation histogram extracted from the L*a*b* color space.
(4) MSD [14]: The 72-dimensional color histogram obtained by quantizing the HSV color space and the 6-dimensional edge orientation histogram extracted from the HSV color space.
(5) LNDP + LBP [31]: The 512-dimensional local neighborhood difference pattern extracted from the grey-scale space and the 256-dimensional LBP extracted from the grey-scale space.
(6) MPEG-CED [25]: The 256-dimensional color histogram descriptor (CHD) extracted from the RGB color space, and the 5-dimensional edge histogram extracted from the HSV color space.
(7) Joint colorhist [12]: The 512-dimensional color histogram obtained by combining the quantized R, G and B channels.
(8) OCLBP [47]: The fusion of the 1536-dimensional opponent color local binary patterns extracted from the RGB color space.
(9) IOCLBP [46]: The fusion of the 3072-dimensional improved opponent color local binary patterns extracted from the RGB color space.
(10) PUH + MCM: The fusion of the 148/364-dimensional perceptually uniform histogram (PUH) extracted from the L*a*b* color space and the 36-dimensional motif co-occurrence matrix (MCM) extracted from the grey-scale space.
(11) HHD: The fusion of the 148/364-dimensional perceptually uniform histogram (PUH) and the 81-dimensional motif co-occurrence histogram (MCH) extracted from the L* channel.

Quantitative and Qualitative performance valuations are performed from the following seven perspectives: the average precision rate (APR) value, the average recall rate (ARR) value, the average precision rate versus number of top matches (APR vs. NTM), the average recall rate versus number of top matches (ARR vs. NTM), the top-10 retrieved images, the precision–recall curve and the computational complexity. Meanwhile, the robustness of rotation, illumination and resolution is also illustrated in our comparative experiments. To guarantee the accuracy of the experiments, all experiments are performed under the principle of leave-one-out cross-validation.

Table 8 reports the comparisons between the proposed descriptors and the former schemes in terms of average precision rate (APR) and average recall rate (ARR). Bold values highlight the best values. In Table 8, it can be seen that HHD yields the highest APR and ARR compared to all former existing schemes on five datasets. For example, the {APR, ARR} of HHD on RSSCN7 outperforms mdLBP, maLBP, CDH, MSD, LNDP + LBP, MPEG-CED, OCLBP, IOCLBP and PUH + MCM by {6.47%, 0.16%}, {8.69%, 0.22%}, {5.97%, 0.15%} and {11.13%, 0.28%}, {10.11%, 0.25%}, {4.18%, 0.11%}, {6.75%, 0.17%}, {8.87%, 0.24%}, {9.61%, 0.24%} and {5.63%, 0.14%}, respectively. Similarly, more significant values are reported over AID, Outex-13, Outex-14 and ETHZ-53. From these results, the effectiveness of the proposed descriptor is demonstrated by comparing with other fusion-based feature descriptors in terms of APR and ARR. In addition, since there are various rotation and resolution differences on RSSCN7 and AID datasets (see Figure 5a,b), and various illumination differences on Outex-00014 dataset (see Figure 5d), the robustness of the rotation, resolution and illumination is also well illustrated to some extent.

Table 8. Average precision rate (APR) and average recall rate (ARR) of different methods over RSSCN7, AID, Outex-00013, Outex-00014 and ETHZ-53.

Descriptor	Performance (%)	Data Set				
		RSSCN7	AID	Outex-13	Outex-14	ETHZ-53
mdLBP	APR	73.10	50.81	61.00	48.66	61.43
	ARR	1.83	1.54	30.50	24.33	61.43
maLBP	APR	70.88	47.40	62.54	44.53	55.17
	ARR	1.77	1.43	31.27	22.27	55.17
CDH	APR	73.60	49.50	79.27	74.03	88.53
	ARR	1.84	1.51	39.64	37.02	88.53
MSD	APR	68.44	47.76	70.46	66.32	91.09
	ARR	1.71	1.45	35.23	33.16	91.09
LBP + LNDP	APR	69.46	44.12	70.24	43.86	52.45
	ARR	1.74	1.33	35.12	21.93	52.45
MPEG-CEH	APR	75.39	53.86	78.48	74.41	94.79
	ARR	1.88	1.63	39.24	37.21	94.79
Joint Colorhist	APR	72.82	50.97	77.46	72.97	93.74
	ARR	1.82	1.55	38.73	36.48	93.74
OCLBP	APR	70.70	41.60	77.82	56.13	42.57
	ARR	1.75	1.26	38.91	28.06	42.57
IOCLBP	APR	69.96	44.78	79.58	73.58	45.51
	ARR	1.75	1.35	39.79	36.79	45.51
PUM + MCM	APR	73.94	52.45	81.03	78.13	97.74
	ARR	1.85	1.59	40.51	39.06	97.74
HHD	APR	**79.57**	**58.13**	**84.21**	**82.82**	**97.89**
	ARR	**1.99**	**1.76**	**42.10**	**41.41**	**97.89**

Figure 6a–j shows the performance comparison between HHD and existing approaches in terms of average precision rate versus number of top matches (APR vs. NTM) and average recall rate versus number of top matches (ARR vs. NTM). To guarantee the accuracy and reproducibility, the number of top matches is set to 100, 200, 20, 20 and 5 on RSSCN7, AID, Outex-00013, Outex-00014 and ETHZ-53, respectively. In Figure 6a,b, HHD achieves an obviously higher performance than all other fusion-based feature descriptors on RSSCN7. Meanwhile, we also note that the APR vs. NTM and ARR vs. NTM curves of mdLBP, maLBP, CDH, MSD, LNDP + LBP, MPEG-CED, Joint Colorhist, OCLBP, IOCLBP and PUH + MCM are close to one another extremely. The reason is that only seven land-use categories are very challenging to retrieve the targeted images from RSSCN7. As shown in Figure 6c,d, the APR vs. NTM and ARR vs. NTM curves of HHD achieve an obviously higher curvature than all other descriptors on AID. This phenomenon illustrates that the proposed descriptor can acquire better performance on the large-scale dataset. As expected, as shown in Figure 6e–j, HHD still outperforms all other existing descriptors over Outex-00013, Outex-00014 and ETHZ-53, respectively. Specifically, PUM + MCM and HHD are superior to other descriptors on ETHZ-53 obviously. The main reason is that they not only combine the color and edge information, but also integrate the texture information. Based on the above results, the effectiveness of the proposed descriptor is demonstrated by comparing with other fusion-based methods in terms of APR vs. NTM and ARR vs. NTM.

Figure 7a–e shows the performance comparison of the top-10 retrieved images using different methods. The leftmost image in each row of Figure 7a–e is the query image, and the remaining images are a set of retrieved images ordered in ascending order from left to right. For clarity, if a retrieved image owns the same group label as the query, it is flagged as a green frame; otherwise, it is flagged as a red frame. In Figure 7a, there are 7 related images to the query image "River Lake" from RSSCN7 using mdLBP, 8 using maLBP, 8 using CDH, 4 using MSD, 9 using LNDP + LBP, 3 using MPEG-CED, 3 using Joint Colorhist, 8 using OCLBP, 7 using IOCLBP, 4 using PUH + MCM and 10 using HHD. Note that, although the images from "Forest" have a similar color to "River Lake", leading to the error results by most of the existing schemes, HHD can retrieve the targeted images accurately. In Figure 7b, for the query image "Baseball Field" from AID, the number of targeted images using mdLBP, maLBP, CDH, MSD, LNDP + LBP, MPEG-CED, Joint Colorhist, OCLBP, IOCLBP, PUH + MCM, and HHD descriptors are 7, 7, 9, 6, 5, 9, 5, 8, 9, 9 and 10, respectively. It can be seen that HHD not only displays a better retrieval result than all other descriptors, but also shows the robustness of rotation and resolution differences. In Figure 7c, for the query image "Rice" from Outex-00013, the precision achieved by using mdLBP, maLBP, CDH, MSD, LNDP + LBP, MPEG-CED, Joint Colorhist, PUH + MCM, and HHD descriptors are 40%, 40%, 80%, 70%, 30%, 80%, 80%, 90% and 100%, respectively. In comparison, we can see that although all retrieved images show a similar content appearance, yet HHD still outperforms all other descriptors. In Figure 7d, for the query image "Carpet" from Outex-00014, the precision obtained by using mdLBP, maLBP, CDH, MSD, LNDP + LBP, MPEG-CED, Joint Colorhist, OCLBP, IOCLBP, PUH + MCM, and HHD descriptors are 40%, 30%, 70%, 10%, 30%, 40%, 30%, 70%, 50%, 50% and 100%, respectively. As shown in Figure 7e, for the query image "Paper Bag" from ETHZ-53, HHD still outperforms all other existing descriptors. From the above results, we can conclude that HHD not only depicts the image semantic information with similar textural structure appearance but also discriminates the color and texture differences, effectively. In summary, the effectiveness of the proposed descriptor is demonstrated by comparing with existing approaches in terms of the top-10 retrieved images.

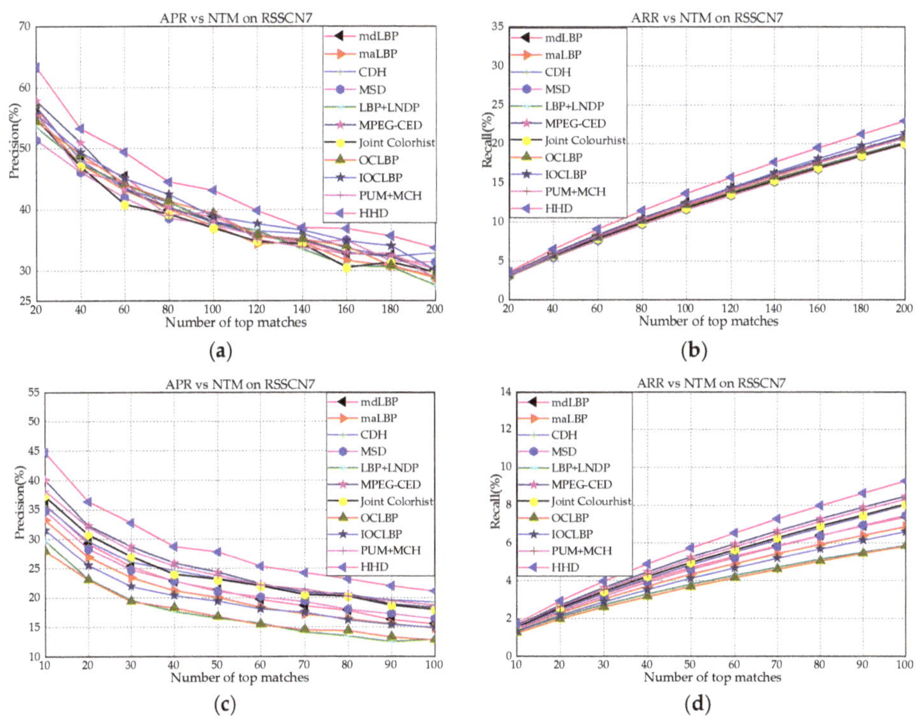

Figure 6. *Cont.*

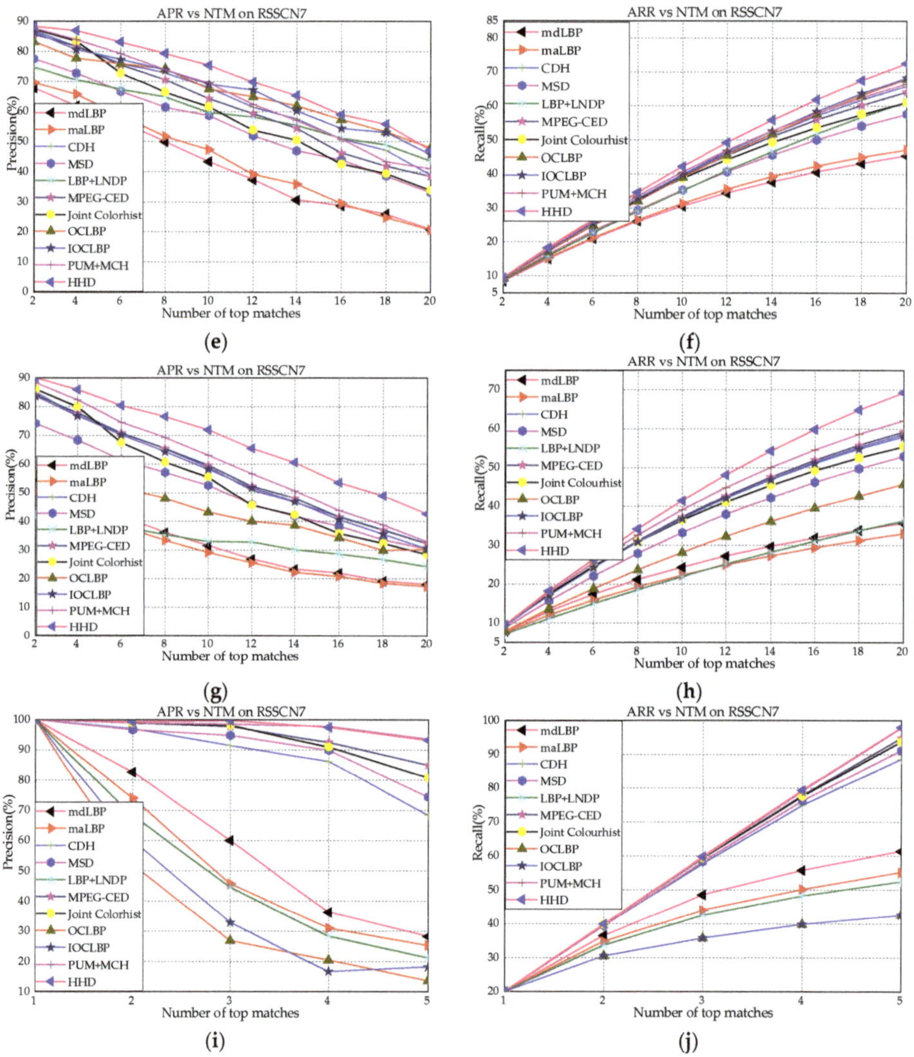

Figure 6. Precision vs. number of top matches (APR vs. NTM) and Recall vs. number of top matches (APR vs. NTM) using different methods over: (**a**,**b**) RSSCN7; (**c**,**d**) AID; (**e**,**f**) Outex-00013; (**g**,**h**) Outex-00014; and (**i**,**j**) ETHZ-53.

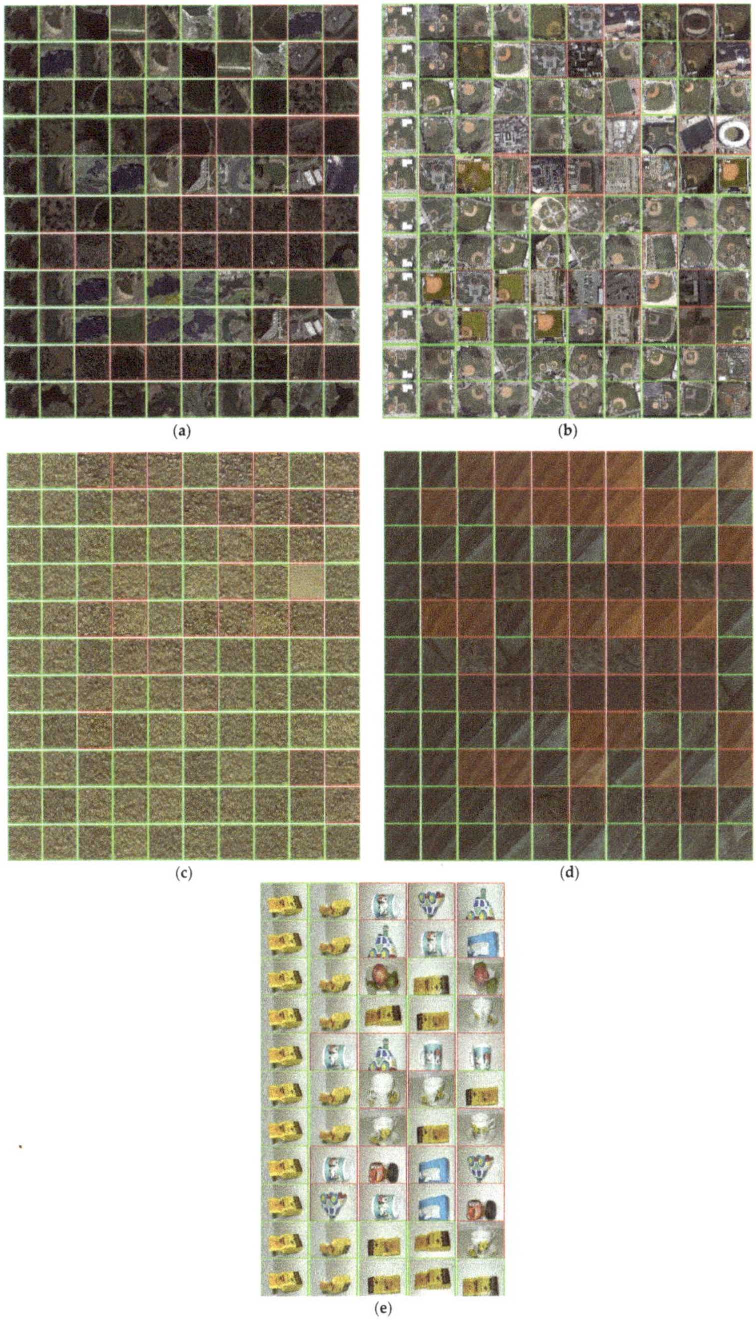

Figure 7. Results of the top-10 retrieved images by considering different query images: (**a**) "River Lake"; (**b**) "Baseball Field"; (**c**) "Rice"; (**d**) "Carpet"; and (**e**) "Paper Bag" using different descriptors (Row 1 using mdLBP, Row 2 using maLBP, Row 3 using CDH, Row 4 using MSD, Row 5 using LNDP + LBP, Row 6 using MPEG-CED, Row 7 using Joint Colorhist, Row 8 using OCLBP , and Row 9 using IOCLBP, Row 10 using PUH + MCM and Row 11 using HHD).

Figure 8a–e shows the performance comparison of the proposed HHD with existing approaches over RSSCN7, AID, Outex-00013 and Outex-00014 in terms of the precision–recall curve. According to Figure 8a,b, it can be observed that the precision–recall curve of HHD is obviously superior to all other fusion-based approaches. According to Figure 8c,d, it can be seen that the precision–recall curve of other fusion-based approaches is inferior to HHD over Outex-00013 and Outex-00014 obviously. Moreover, as shown in Figure 8e, both HHD and PUH + MCM are higher than mdLBP, maLBP, CDH, MSD, LNDP + LBP, OCLBP, IOCLBP, and Joint Colorhist on ETHZ-53. The reasons can be summarized as follows:

(1) Joint Colorhist, mdLBP, maLBP and LNDP + LBP only extract an independent color or texture information.
(2) CDH, MSD and MPEG-CED consider the color and edge orientation information from different channels, while the texture information is ignored.
(3) OCLBP and IOCLBP combine the color and texture information, but the edge orientation information is lost.
(4) Although PUH + MCM integrates the color, edge orientation and texture information as a whole, the perceptually uniform motif patterns are lost.
(5) HHD not only integrates the merits of the color, edge orientation and texture information, but also considers the perceptually uniform motif patterns.

Depending upon the above results and analyses, the effectiveness of the proposed descriptor is demonstrated by comparing with other fusion-based methods in terms of the precision–recall curve.

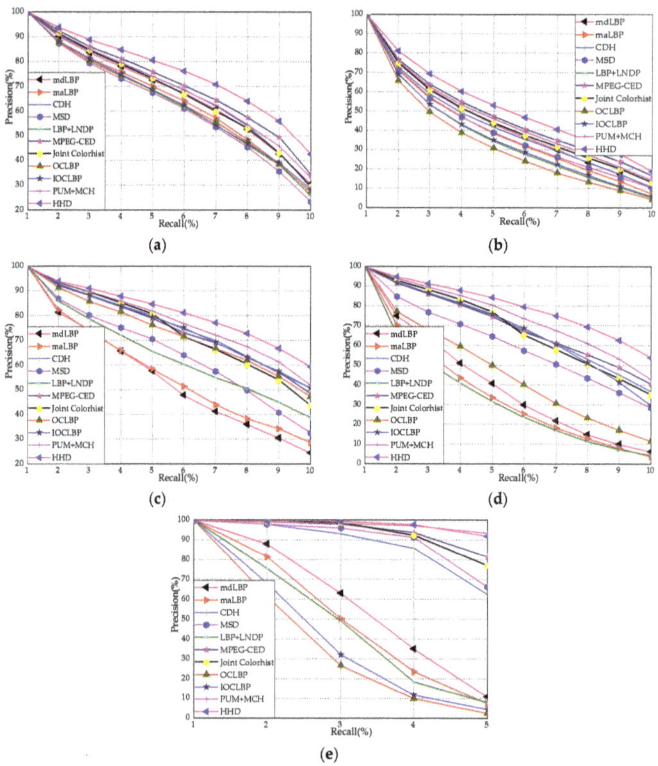

Figure 8. Precision–recall curve of different descriptors over five databases: (**a**) Outex-00013; (**b**) Outex-00014; (**c**) RSSCN7; (**d**) AID; and (**e**) ETHZ-53.

Table 9 shows the feature vector length, average retrieval time, and memory cost per image of different descriptors to provide an in-depth evaluation of the computational complexity. All experiments are carried out on a computer with Intel Core i7-7700K@4.20 GHz CPU processor, 4 cores active and 16 GB RAM. The feature vector length is compared by dimension (D). The average retrieval time is analyzed by seconds (S). The memory cost per image is measured in kilobytes (KB). Similar to PUM + MCM, the items of 445/229 (D) and 3.48/1.79 (KB) represent HHD with 445 dimensions and 3.48 kilobytes performing retrieval over RSSCN7, AID and ETHZ-53 databases, as well as HHD with 229 dimensions and 1.79 kilobytes performing retrieval over Outex-00013 and Outex-00014 databases. For RSSCN7, AID and ETHZ-53, the feature vector length and the memory cost per image of HHD are inferior to those of MSD, CDH, MPEG-CED and PUM + MCM, while HHD are superior to Joint Colorhist, maLBP, mdLBP, OCLBP, IOCLBP and LNDP + LBP For Outex-00013 and Outex-00014, the feature vector length and the memory cost per image of HHD are worse than MSD, CDH and PUM + MCM, but it is better than MPEG-CED, Joint Colorhist, maLBP, mdLBP, OCLBP, IOCLBP and LNDP + LBP. For the average retrieval time, HHD is more than MSD, CDH, MPEG-CED and PUM + MCM, yet HHD is less than Joint Colorhist, maLBP, mdLBP, OCLBP, IOCLBP and LNDP + LBP. The main reason is that the RSSCN7, AID and ETHZ-53 databases have more complex contents as compared with the Outex-00013 and Outex-00014 image databases. Although HHD does not outperform all other fusion-based descriptors, the usability and practicability of HHD is indicated under the content-based image retrieval framework configuration: adaptive feature vector length, competitive average retrieval time, and acceptable memory cost per image.

Table 9. Feature vector length (D), average retrieval time (s) and memory cost per image (KB) of different descriptors.

Method	Feature Vector Length (D)	Average Retrieval Time (s)	Memory Cost per Image (KB)
mdLBP	2048	3.45	16.00
maLBP	1024	1.74	8.00
CDH	108	0.17	0.84
MSD	78	0.15	0.61
LBP + LNDP	768	1.28	6.00
MPEG-CEH	261	0.45	2.04
Joint Colorhist	512	0.88	4.00
OCLBP	1535	2.55	11.99
IOCLBP	3072	5.24	24.00
PUM + MCM	400/184	0.65	3.13/1.44
HHD	445/229	0.72	3.48/1.79

4.8. Comparison with CNN-Based Descriptors

Apart from the fusion-based descriptors, HHD is also compared with emerging deep neural networks techniques. Referring to the experimental setting in [48], we first extracted the last full-connected layer from the pre-trained CNN model (e.g., VGGM1024 and VGGM128). Then, the extracted feature vectors were L2 normalized. Finally, the normalized feature vectors were sent to perform the distance measure. To guarantee a fair comparison, the number of query images were identically set as all images, and the number of retrieved images were set to 10 on RSSCN7, AID, Outex-00013 and Outex-00014, and 5 on ETHZ-35.

Figure 9 shows the comparisons between the proposed descriptors and the CNN-based schemes. In the case of the RSSCN7, Outex-00013, Outex-00014 and ETHZ-35 datasets, HHD performs better than the VGGM1024 and VGGM128 descriptors, and it achieves the highest performance. Particularly, PUM + MCM also outperforms the VGGM1024 and VGGM128 descriptors on the four datasets. Regarding the AID dataset, HHD is worse than VGGM1024. This makes sense because the pre-trained CNN models which are trained on the large-scale imageset, are suitable for the large-scale AID dataset. In contrast to the CNN-based descriptors, the advantages of HHD can be summarized as follows:

(1) HHD does not require any training process in the feature representation.

(2) The pre-trained CNN-based models have a high memory cost which limits its application.
(3) HHD performs better than the CNN-based descriptors in four datasets out of five.

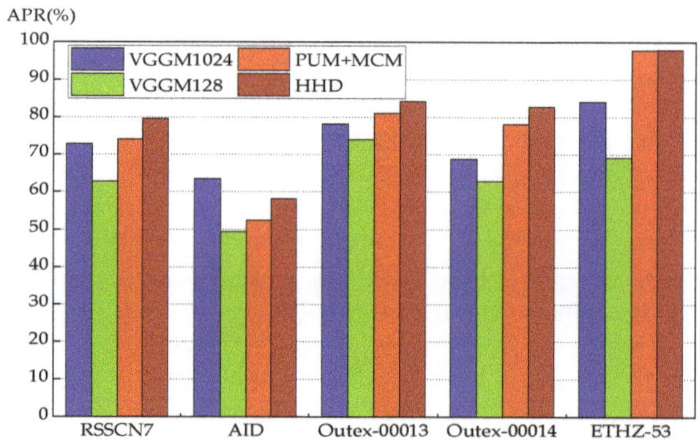

Figure 9. Comparison of the proposed descriptors with the CNN-based schemes over Outex-00013, Outex-00014, RSSCN7, AID and ETHZ-53.

5. Conclusions

In this paper, we propose a fusion method called hybrid histogram descriptor (HHD), which integrates the perceptually uniform histogram and the motif co-occurrence histogram as a whole. The proposed descriptor was evaluated under the content-based image retrieval framework on the RSSCN7, AID, Outex-00013, Outex-00014 and ETHZ-53 datasets. From the experimental results, it can be concluded that the fitness quantization layers of Y_{a^*} and Y_{b^*} are computed depending upon the retrieval accuracy score. It is also deduced that the motif co-occurrence histogram (MCH) exhibits significantly higher performance than the motif co-occurrence matrix (MCM). The performance of the proposed descriptor is much improved by confusing the perceptually uniform histogram (PUH) and the motif co-occurrence histogram (MCH). The performance of the proposed descriptor is superior to ten fusion-based feature descriptors in terms of the average precision rate (APR), the average recall rate (ARR), the average precision rate versus number of top matches (APR vs. NTM), the average recall rate versus number of top matches (ARR vs. NTM), and the top-10 retrieved images. Meanwhile, the feature vector length, the average retrieval time, and the memory cost per image were also analyzed to give an in-depth evaluation of the computational complexity. Moreover, compared with the CNN-based descriptors, the proposed descriptor also achieves comparable performance, but does not require any training process.

The increased dimension of the proposed descriptor slows down the retrieval time, which will be addressed in future research, especially using Locality-Sensitive Hashing [49]. Meanwhile, user relevance feedback, feature re-weight and weight optimization will be considered to further improve the accuracy of image retrieval. In addition, we will further investigate the generalization of the proposed method, especially using RawFooT [50] that includes changes in the illumination conditions.

Author Contributions: Q.F. conceived the research idea. Q.F. and Q.H. performed the experiments. Q.F. wrote the paper. Y.C., Y.Y., Y.W. and J.D. gave many suggestions and helped revise the manuscript.

Funding: This research was funded by Fundamental Research Grant Scheme for the Central Universities, grant number [130204003], the Shandong Provincial Natural Science Foundation of China, grant number [BS2015DX001] and the National Key Technology Research and Development Programme of the Ministry of Science and Technology of China, grant number [2014BAI17B02].

Conflicts of Interest: The authors declare no conflict of interest.

References

1. Yang, M.; Song, W.; Mei, H. Efficient retrieval of massive ocean remote sensing images via a cloud-based mean-shift algorithm. *Sensors* **2017**, *17*, 1693. [CrossRef] [PubMed]
2. Piras, L.; Giacinto, G. Information fusion in content based image retrieval: A comprehensive overview. *Inf. Fusion* **2017**, *37*, 50–60. [CrossRef]
3. Smeulders, A.W.M.; Worring, M.; Santini, S.; Gupta, A.; Jain, R. Content-based image retrieval at the end of the early years. *IEEE Trans. Pattern Anal. Mach. Intell.* **2002**, *22*, 1349–1380. [CrossRef]
4. Bala, A.; Kaur, T. Local texton xor patterns: A new feature descriptor for content-based image retrieval. *Eng. Sci. Technol. Int. J.* **2016**, *19*, 101–112. [CrossRef]
5. Zhang, M.; Zhang, K.; Feng, Q.; Wang, J.; Kong, J. A novel image retrieval method based on hybrid information descriptors. *J. Vis. Commun. Image Represent.* **2014**, *25*, 1574–1587. [CrossRef]
6. Swain, M.J.; Ballard, D.H. Color indexing. *Int. J. Comput. Vis.* **1991**, *7*, 11–23. [CrossRef]
7. Stricker, M.A.; Orengo, M. Similarity of color images. *Proc. SPIE* **1995**, *2420*, 381–392.
8. Bimbo, A.D.; Mugnaini, M.; Pala, P.; Turco, F. Visual querying by color perceptive regions. *Pattern Recognit.* **1998**, *31*, 1241–1253. [CrossRef]
9. Pass, G.; Zabih, R.; Miller, J. Comparing images using color coherence vectors. In Proceedings of the Forth ACM International Conference on Multimedia, Boston, MA, USA, 18–22 November 1996.
10. Huang, J.; Kumar, S.R.; Mitra, M.; Zhu, W.J. Image indexing using color correlograms. In Proceedings of the IEEE Conference on Computer Vision and Pattern Recognition, San Juan, Spain, 17–19 June 1997.
11. Manjunath, B.S.; Ohm, J.R.; Vasudevan, V.V.; Yamada, A. Color and texture descriptor. *IEEE Trans. Circuits Syst. Video Technol.* **2001**, *11*, 703–715. [CrossRef]
12. Ning, J.; Zhang, L.; Zhang, D.; Wu, C. Robust object tracking using joint color-texture histogram. *Int. J. Pattern Recog. Artif. Intell.* **2009**, *23*, 1245–1263. [CrossRef]
13. Liu, G.H.; Zhang, L.; Hou, Y.K.; Li, Z.Y.; Yang, J.Y. Image retrieval based on multi-texton histogram. *Pattern Recognit.* **2010**, *43*, 2380–2389. [CrossRef]
14. Liu, G.H.; Li, Z.Y.; Zhang, L.; Xu, Y. Image retrieval based on micro-structure descriptor. *Pattern Recognit.* **2011**, *44*, 2123–2133. [CrossRef]
15. Liu, G.H.; Yang, J.Y. Content-based image retrieval using color difference histogram. *Pattern Recognit.* **2013**, *46*, 188–198. [CrossRef]
16. Zeng, S.; Huang, R.; Wang, H.B.; Kang, Z. Image retrieval using spatiograms of colors quantized by gaussian mixture models. *Neurocomputing* **2016**, *171*, 673–684. [CrossRef]
17. Murala, S.; Wu, Q.M.J.; Balasubramanian, R.; Maheshwari, R.P. Joint histogram between color and local extrema patterns for object tracking. In Proceedings of the IS&T/SPIE Electronic Imaging, Burlingame, CA, USA, 3–7 February 2013.
18. Verma, M.; Raman, B.; Murala, S. Local extrema co-occurrence pattern for color and texture image retrieval. *Neurocomputing* **2015**, *165*, 255–269. [CrossRef]
19. Kuhl, F.P.; Giardina, C.R. Elliptic Fourier features of a closed contour. *Comput. Graph. Image Process.* **1982**, *18*, 236–258. [CrossRef]
20. Yap, P.T.; Paramesran, R.; Ong, S.H. Image analysis using Hahn moments. *IEEE Trans. Pattern Anal. Mach. Intell.* **2007**, *29*, 2057–2062. [CrossRef] [PubMed]
21. Torre, V.; Poggio, T.A. On edge detection. *IEEE Trans. Pattern Anal. Mach. Intell.* **1986**, *PAMI-8*, 147–163. [CrossRef]
22. Wang, Y.; Zhao, Y.; Cai, Q.; Li, H.; Yan, H. A varied local edge pattern descriptor and its application to texture classification. *J. Vis. Commun. Image Represent.* **2016**, *34*, 108–117. [CrossRef]
23. Song, Q.; Wang, Y.; Bai, K. High dynamic range infrared images detail enhancement based on local edge preserving filter. *Infrared Phys. Technol.* **2016**, *77*, 464–473. [CrossRef]
24. Li, J.; Sang, N.; Gao, C. LEDTD: Local edge direction and texture descriptor for face recognition. *Signal Process. Image Commun.* **2016**, *41*, 40–45. [CrossRef]
25. Won, C.S.; Park, D.K.; Park, S.J. Efficient use of MPEG-7 edge histogram descriptor. *ETRI J.* **2002**, *24*, 23–30. [CrossRef]
26. Ojala, T.; Pietikainen, M.; Maenpaa, T. Multiresolution gray-scale and rotation invariant texture classification with local binary patterns. *IEEE Trans. Pattern Anal. Mach. Intell.* **2002**, *24*, 971–987. [CrossRef]

27. Julesz, B. Textons, the elements of texture perception, and their interactions. *Nature* **1981**, *290*, 91–97. [CrossRef] [PubMed]
28. Haralick, R.M.; Shanmugam, K. Texture features for image classification. *IEEE Trans. Syst. Man Cybern.* **1973**, *SMC-3*, 610–621. [CrossRef]
29. Jhanwar, N.; Chaudhuri, S.; Seetharaman, G. Content based image retrieval using motif co-occurrence matrix. *Image Vis. Comput.* **2004**, *22*, 1211–1220. [CrossRef]
30. Dubey, S.R.; Singh, S.K.; Singh, R.K. Multichannel decoded local binary patterns for content-based image retrieval. *IEEE Trans. Image Process.* **2016**, *25*, 4018–4032. [CrossRef] [PubMed]
31. Verma, M.; Raman, B. Local neighborhood difference pattern: A new feature descriptor for natural and texture image retrieval. *Multimed. Tools Appl.* **2018**, *77*, 11843–11866. [CrossRef]
32. Bianconi, F.; Harvey, R.; Southam, P.; Fernández, A. Theoretical and experimental comparison of different approaches for color texture classification. *J. Electron. Imag.* **2010**, *20*, 043006. [CrossRef]
33. Cusano, C.; Napoletano, P.; Schettini, R. Evaluating color texture descriptors under large variations of controlled lighting conditions. *J. Opt. Soc. Am. A* **2016**, *33*, 17–30. [CrossRef] [PubMed]
34. Qazi, I.U.H.; Alata, O.; Burie, J.C.; Moussa, A.; Fernandez-Maloigne, C. Choice of a pertinent color space for color texture characterization using parametric spectral analysis. *Pattern Recognit.* **2011**, *44*, 16–31. [CrossRef]
35. Guo, J.M.; Prasetyo, H.; Lee, H.; Yao, C.C. Image retrieval using indexed histogram of void-and-cluster block truncation coding. *Signal Process.* **2016**, *123*, 143–156. [CrossRef]
36. Young, T. The bakerian lecture: On the theory of light and colors. *Philos. Trans. R. Soc. Lond. B* **1802**, *92*, 12–48. [CrossRef]
37. Hurvich, L.M.; Jameson, D. An opponent-process theory of color vision. *Psychol. Rev.* **1957**, *64*, 384–404. [CrossRef] [PubMed]
38. Sarrafzadeh, O.; Dehnavi, A.M. Nucleus and cytoplasm segmentation in microscopic images using k-means clustering and region growing. *Adv. Biomed. Res.* **2015**, *4*, 174. [PubMed]
39. Gonzalez, R.C.; Woods, R.E. *Digital Image Processing*, 3rd ed.; Publishing House of Electronics Industry: Beijing, China, 2010; pp. 455–456. ISBN 9787121102073.
40. Xia, G.S.; Hu, J.; Hu, F.; Shi, B.; Bai, X.; Zhong, Y.; Zhang, L. AID: A benchmark data set for performance evaluation of aerial scene classification. *IEEE Trans. Geosci. Remote Sens.* **2017**, *55*, 3965–3985. [CrossRef]
41. Zou, Q.; Ni, L.; Zhang, T.; Wang, Q. Deep learning based feature selection for remote sensing scene classification. *IEEE Trans. Geosci. Remote Sens. Lett.* **2015**, *12*, 2321–2325. [CrossRef]
42. Köhler, W. Gestalt psychology. *Psychol. Res.* **1976**, *31*, XVIII–XXX.
43. Lam, C.F.; Lee, M.C. Video segmentation using color different histogram. In *Multimedia Information Analysis and Retrieval. MINAR 1998*; Ip, H.H.S., Smeulders, A.W.M., Eds.; Lecture Notes in Computer Science; Springer: Berlin/Heidelberg, Germany, 1998; Volume 1464, pp. 159–174.
44. Panda, D.K.; Meher, S. Detection of moving objects using fuzzy color difference histogram based background subtraction. *IEEE Trans. Signal Proc. Lett.* **2016**, *23*, 45–49. [CrossRef]
45. Kang, Y.; Li, X. A novel tiny object recognition algorithm based on unit statistical curvature feature. In Proceedings of the European Conference on Computer Vision, Amsterdam, The Netherlands, 8–16 October 2016.
46. Bianconi, F.; Bello-Cerezo, R.; Napoletano, P. Improved opponent color local binary patterns: An effective local image descriptor for color texture classification. *J. Electron. Imag.* **2017**, *27*, 011002. [CrossRef]
47. Mäenpää, T.; Pietikäinen, M. Texture analysis with local binary patterns. In *Handbook of Pattern Recognition and Computer Vision*; Word Scientific: Singapore, 2005; pp. 197–216.
48. Napoletano, P. Hand-crafted vs. learned descriptors for color texture classification. In Proceedings of the International Workshop on Computational Color Imaging, Milan, Italy, 29–31 March 2017.
49. Indyk, P.; Motwani, R. Approximate nearest neighbors: Towards removing the curse of dimensionality. In Proceedings of the IEEE Conference on Multimedia Information Analysis and Retrieval, Dallas, TX, USA, 24–26 May 1998; pp. 604–613.
50. RawFooT DB: Raw Food Texture Database. Available online: http://projects.ivl.disco.unimib.it/minisites/rawfoot/ (accessed on 28 December 2015).

sensors

MDPI

Article

Textile Retrieval Based on Image Content from CDC and Webcam Cameras in Indoor Environments

Oscar García-Olalla [1], Enrique Alegre [1,3], Laura Fernández-Robles [2,3], Eduardo Fidalgo [1,3,*] and Surajit Saikia [1,3]

1 Department of Electrical, Systems and Automation, Universidad de León, 24071 León, Spain; ogaro@unileon.es (O.G.-O.); ealeg@unileon.es (E.A.); ssai@unileon.es (S.S.)
2 Department of Mechanical, Computer Science and Aerospace Engineering, Universidad de León, 24071 León, Spain; l.fernandez@unileon.es
3 Researcher at INCIBE (Spanish National Cybersecurity Institute), 24005 León, Spain
* Correspondence: efidf@unileon.es; Tel.: +34-987-293-521

Received: 21 March 2018; Accepted: 21 April 2018; Published: 25 April 2018

Abstract: Textile based image retrieval for indoor environments can be used to retrieve images that contain the same textile, which may indicate that scenes are related. This makes up a useful approach for law enforcement agencies who want to find evidence based on matching between textiles. In this paper, we propose a novel pipeline that allows searching and retrieving textiles that appear in pictures of real scenes. Our approach is based on first obtaining regions containing textiles by using MSER on high pass filtered images of the RGB, HSV and Hue channels of the original photo. To describe the textile regions, we demonstrated that the combination of HOG and HCLOSIB is the best option for our proposal when using the correlation distance to match the query textile patch with the candidate regions. Furthermore, we introduce a new dataset, TextilTube, which comprises a total of 1913 textile regions labelled within 67 classes. We yielded 84.94% of success in the 40 nearest coincidences and 37.44% of precision taking into account just the first coincidence, which outperforms the current deep learning methods evaluated. Experimental results show that this pipeline can be used to set up an effective textile based image retrieval system in indoor environments.

Keywords: content-based image retrieval; textile retrieval; textile localization; texture retrieval; texture description; visual sensors

1. Introduction

The process of automatically finding objects, textiles, faces, or other patterns in images and videos is one of the most studied topics in computer vision. Nowadays, with the huge amount of digital images and videos, it becomes even more critical. Visual sensors are able to acquire a large quantity of visual information from the surroundings around them. Content Based Image Retrieval (CBIR) consists of retrieving images using their content properties from a collection that match a user's query [1] based on a similarity measure [2]. Many research fields, e.g., medical image [3–5], human retrieval [6], biological analysis [7,8], agricultural retrieval [9] and biometric security [10], achieved interesting results using CBIR techniques.

Most works related to CBIR aim at finding objects in datasets of images. Research groups face this problem using different approaches such as invariant local features (SIFT [11,12], SURF [13]), color description [14,15], template matching [16,17] or, more recently, deep learning techniques [18–20]. Nonetheless, these techniques may fail when the object does not present a rigid shape or it has a plain shape, as it is the case of textiles. The same textile can appear in images with very skewed shapes. Moreover, textile retrieval shares the difficulties of object retrieval such as the variety in illumination conditions, occlusions, lack of texture information, etc.

The need of retrieving textiles from image collections captured under a variety of visual sensors can be motivated by many applications [21]—for example, for marketing studies in textile stores that suggest the products that fit a decorated room to users. The recently published book "Applications of computer vision in fashion and textiles" [22] deals with three aspects related to computer vision techniques applied to textile industry: (i) textile defect detection and quality control, (ii) fashion recognition and 3D modeling, and (iii) 2D and 3D human body modeling for improving clothing fit. One of its chapters [23] reviews the computer vision state-of-the-art techniques for fashion textile modeling, recognition, and retrieval. A completely different approach in which textiles are needed to be retrieved is to connect evidence of different crime scenes.

In our case, this work is framed in the Advisory System Against Sexual Exploitation of Children (ASASEC) project, a European project that fights child pornography using forensic analysis, data mining and computer vision techniques. It was demonstrated that perverts usually use the same bedrooms to take their pictures or videos [24]. A way to link two images, and consequently provide relationships among the many cases of child pornography, is finding the exact same textiles such as carpets, blankets or any other repeated texture. In our specific case, we aim at evaluating textiles in order to retrieve images in huge datasets (thousands of images) of past proven cases of child abuse connected with a query textile of interest.

The rest of the paper is organized as follows. The related work is presented in Section 2. Section 3 describes the pipeline of our proposed method for textile based image retrieval. Section 4 introduces the TextilTube dataset, the evaluation metrics, the evaluation set-up and the decision evaluation of a distance measure. We show the results of all the experiments carried out in Section 5. Finally, Section 6 draws the conclusions of the paper.

2. Related Research

Material retrieval is related to textile retrieval in some aspects and it is more broadly studied in the literature. Zhu and Brilakis [25] presented a system for detecting concrete based on the account of the colour of the regions in the image. After that, they described the regions using color features and trained a machine learning classifier to determine if the region contains concrete or not. This method cannot deal with very heterogeneous regions due to the way of creating the image partitions. Son et al. [26] proposed a method based on ensemble classifiers in order to distinguish between concrete, steel and wood. One of the main disadvantages of this work is the necessity of uniform areas of the same material for the segmentation step, which consists of dividing the original image into sub-regions of a fixed size. If the material region is smaller than the grid division, a lot of information of the background is processed as a material resulting in a non-accurate description. In [27,28], the authors proposed two methods able to identify multiple materials in object surfaces without the need of segmentation. They first recognized the object class and then used correlations of material labels for such object. In this approach, the correct definition of detailed semantic cues of objects and materials is needed. In 2017, Xue et al. [29] focused on material recognition of real-world outdoor surfaces for which they presented a new very useful dataset for autonomous agents. They exploited the idea of extracting characteristics of materials encoded in the angular and spatial gradients of their appearance from images taken with small angular variations. We refer the reader to [30] for an overview of methods and applications for the automatic characterization of the visual appearance of materials. Material retrieval systems are effective when construction materials are involved, but they may fail with other kind of textiles. The main three differences with general textiles are: the classes of construction materials are well defined, the texture of the construction materials is more homogeneous and the image patches of construction materials are usually big and present regular shapes.

Besides material retrieval, there are also few textile retrieval works in the bulk of the literature. Bashar et al. [31] proposed a system based on three wavelet-domain based features called symmetry, regularity and directionality. In this paper, the authors demonstrated outperformance of the combination of the three features versus just the isolated descriptors using two datasets formed by 150 and 300 images of curtain patterns. Similarly, in 2009, Carbunaru et al. [32] proposed a method that applies independent component analysis over wavelet-domain images. In that case, the researchers chose a dataset composed of images of 30 different fabrics, obtaining an average recognition rate of 94.86%. Recently, in Chun et al. [33], a new method which uses composite feature vectors of color from spatial domain and texture from wavelet-transformed domain is proposed. In contrast with the other papers described before, Chun et al. carried out a retrieval system using a large dataset composed of 1343 textile images. In 2014, Huang and Lin [34] proposed a system based on the combination of color, texture and shape features in order to retrieve textiles over more than 4000 images downloaded from Globle-Tex Co., (http://www.globle-tex.com/). The retrieval system was based on a signature process extracted by different k-means clusters achieving an 83% of success rate. Nevertheless, in all cases, the material or textile datasets are already segmented, usually as a plain piece of fabric, and the system is only focused in the retrieval process. On the contrary, in our proposed work, the textiles are located in real environments presenting diverse shapes, under a wide range of capturing conditions and exposed to occlusions. Recently, a bunch of papers deal with query targets such as cloth worn on human bodies [35–40]. It is quite challenging to extract robust features from different images presenting different poses. Our paper encompasses a wider definition of the word textiles, and it is used to retrieve not only cloth on human bodies but any other textile that may appear in indoor environments.

The segmentation of regions of interest is thus critical for an efficient method. In 2016, Zheng and Sarem proposed a method called NAMES, which stands for Non-symmetry and Anti-packing Model and Extended Shading and is based on the idea of packing pixels with a very high performance in terms of time [41]. In 2015, Yang et al. [42] presented a method based on color histogram segmentation using HSV color space. In 2004, Matas et al. [43] proposed a method based on the extraction of Maximally Stable Extremal Regions (MSER) taking into account a binary threshold that varies along all the gray scale spectrum. In our work, we segmented our images using the latter method due to its tolerance against regions with little changes of intensity and the possibility of adjusting it using the binary threshold correctly. After the segmentation step, the description of the regions is another key step of our CBIR system. The detected textile regions can be described using texture descriptors, which are widely used for texture analysis. Texture analysis is a challenging and still open problem in computer vision that consists of detecting and describing the gray level spatial variations of the image pixels. Nowadays, there are multiple fields that profit from automatic texture retrieval, as it makes processes faster with no need for many qualified staff. For this purpose, local descriptors are yet extensively employed for texture description due to their high performance in terms of time and accuracy. Histogram of Oriented Gradients (HOG) is a very popular texture descriptor since Dalal and Triggs presented it in 2005 [44]. This method has demonstrated a great performance in multiple fields, such as pedestrian detection [44] or face recognition [45]. Another very popular descriptor is Local Binary Pattern (LBP) proposed by Ojala et al. [46] due to their simplicity and high capability to extract the intrinsic features from the textures. Guo et al. developed several modifications to LBP such as LBP variance (LBPV) [47], complete LBP (CLBP) [48] or adaptive LBP (ALBP) [49]. García-Olalla et al. introduced algorithms to enhance LBP description [50–52], developing a new booster method that can be fused with LBP in order to improve accuracy results [53]. We refer the reader to [54,55] for a general framework and a taxonomy of local binary patterns variants. Recent approaches are focusing on deep Convolutional Neural Networks (CNN) such as AlexNet [56], GoogleNet [57] or VGG-Net [58]. The activations generated at the fully connected layers are used as feature descriptors for image understanding [59], scene recognition [60], semantic segmentation [61], among others.

In this work, we propose a new method for textile based image retrieval in indoor scenes under diverse capturing conditions and subjected to different shapes and occlusions. In accordance with that goal, we present in this paper a new labeled dataset that we created and made publicly available (http://pitia.unileon.es/varp/node/483). It is composed of 684 images extracted from videos with 67 different classes of textiles and it is called TexilTube. We used videos recorded with different visual sensors, such as compact digital cameras (CDC) and webcams. This dataset reproduces, at a small scale, a typical scenario of image evidence related to child pornography. We used MSER on high pass filetered images of the RGB, HSV and Hue channels of the original images to extract the regions of textiles. To describe the textile regions, we used local texture features, i.e., LBP, ALBP, HOG and Faster R-CNN (Region based Convolutional Neural Network) [62]. To enhance the description of the texture descriptors, we used Complete Local Oriented Statistical Information Booster (CLOSIB) booster. We evaluated several distance measures, i.e., Spearman rank, Cityblock, Euclidean and Correlations distances and two evaluation metrics, i.e., precision at n and success at n. We consider the following contributions of this work: (i) we propose a method for extracting the regions of interest based on computing MSER on high pass filtered images of the RGB, HSV and Hue channels of the original images; (ii) we evaluated the performance several descriptors; (iii) we assessed several distance measures by means of a voting schema; and (iv) we present a new dataset for textile retrieval.

3. Method

3.1. Overview

In Figure 1, we illustrate the pipeline of our new method for textile based image retrieval. We can divide the method in two main stages: feature extraction and matching. By means of the experimentation carried out, we were able to determine that the regions in a scene containing textiles have to be extracted from three different transformations of the picture: a high pass filtered image of the RGB, HSV and Hue channels of the original image. We took this into consideration for building up the pipeline of the method.

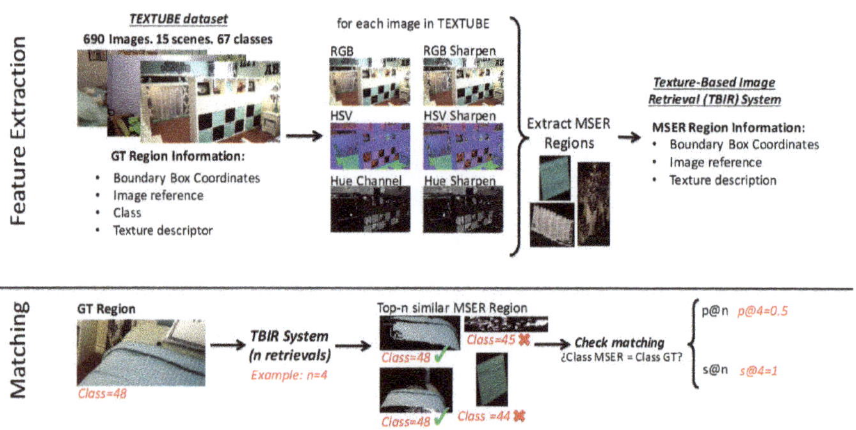

Figure 1. Scheme of the textile based image retrieval system.

The feature extraction is comprised of four steps. First, we convert the images to RGB and HSV colour spaces and we also extract the Hue channel. Then, we sharpen the image representations to increase the contrast along the edges where different colors meet. We adopt the unsharp masking method in which an image is sharpened by subtracting a blurred (unsharp) version of the image from itself. We use a Gaussian lowpass filter of standard deviation 1.5 for blurring the image. We use these three image representations for the extraction and description of the regions of interest in the images. The MSER [43] of the sharpened image representations define the regions of interest of the images. Finally, we describe the regions of interest by computing texture descriptors on the gray scale patch. We create a database in which we store for each detected region: the image coordinates of the bounding box of the region of interest in the image of reference, the images of reference themselves and the descriptors.

The matching stage allows for retrieving a given number of images that present the most similar regions to a query region (textile) of interest. It is made up of three steps. First, we describe the gray scale query region by means of the same texture descriptors. Second, we compute some distance measures among the descriptors of the query region and the descriptors of the database. Finally, the hit list is ranked by sorting the regions of the database in ascending order in relation with the distance measure.

Below, we briefly describe the methods used to build the novel pipeline.

3.2. Region Extraction: MSER

We use the MSER method [43] to automatically extract the regions (textiles) of interest due to the good results achieved in preliminary tests. Other methods apart from MSER could be evaluated for finding distinguished regions that possess some distinguishing or singular properties and allow for repeatedly detecting them over a range of image conditions, such as "sieve" [63]. However, an intensive evaluation of such methods is out of the scope of this paper.

MSER is a method for blob detection that extracts from an image a number of co-variant regions called MSERs. These high contrast regions are connected areas characterized by almost uniform intensity, surrounded by a contrasting background. MSERs are constructed by binarizing the image at multiple threshold levels and selecting the connected components that maintain their sizes over a large set of thresholds.

Experimentally, we chose to extract a great diversity of sizes for the areas of the regions of interest, specifically, comprehended between 3000 and 540,000 pixels, step size between intensity threshold levels equal to 3, and a maximum area variation between extremal regions of 0.7.

3.3. Region Description

We use the following methods to describe the textiles: LBP [64], ALBP [49] and HOG [44], and early fusion concatenations of the previous descriptors with CLOSIB and Half Complete Local Oriented Statistical Information Booster (HCLOSIB) enhancers [65].

3.3.1. Local Binary Pattern (LBP)

LBP describes the texture of gray scale images by means of the local spatial structure on the image. For every pixel, a pattern code is computed by comparing its gray level value with the value of its neighbors.

In this work, we used uniform rotational invariant LBP [64] with 16 neighbors and a radius of two pixels, $\text{LBP}_{16,2}^{riu2}$. The dimension of the descriptor is $P + 2$, in this case, $16 + 2 = 18$ elements. However, for simplicity, we call it LBP henceforth.

3.3.2. Adaptive Local Binary Pattern (ALBP)

Guo et al. [49] presented a variation of LBP that considers the mean and the standard deviation along given orientation of the pixels in the image. This information is used in the matching step and makes it more robust against changes in the local spatial structure of the images.

We consider also the uniform rotation invariant version, $\text{ALBP}_{16,2}^{riu2}$, and we call it ALBP for simplicity.

3.3.3. Histogram of Oriented Gradients (HOG)

Histograms of Oriented Gradients [44] evaluates local histograms of image gradient orientations over a grid. HOG characterizes the local appearance of objects taking into account the local edge direction distributions. The method is implemented by dividing the image into small uniform regions called cells, often overlapped. Then, for each cell, a histogram of the gradient orientations over the pixels is extracted. The final descriptor is yielded by concatenation of the gradients along all the cells. In this work, we use overlapped cells of size 64 × 64 pixels on images resized to 256 × 256 pixels.

3.3.4. Complete Local Oriented Statistical Information Booster (CLOSIB) Variants

CLOSIB [65] is obtained from the statistical information of the gray scale gradient magnitude of each pixel of the image. The statistical information of the gradient magnitudes is rarely taken into account to describe the image and provides useful information for texture classification. Equation (1) shows how to compute the CLOSIB enhancer of an image:

$$\text{CLOSIB}_{P,R,\theta} = \Big\|_{p=1}^{P/\eta} \left((\theta-1)\mu_2^p - (-1)^\theta (\mu_1^p)^\theta\right)^{1/\theta}, \tag{1}$$

where $||$ symbolizes the concatenation function, $\theta \in \{1,2\}$ is the order of the statistical moment considered, μ_1^p and μ_2^p are the first and second statistical raw moments, respectively, defined in Equation (2) and η is a factor that controls the portion of the considered orientations in the quantized angular space. We set $\eta = 1$ for CLOSIB and $\eta = 2$ for Half CLOSIB (HCLOSIB):

$$\mu_i^p = \frac{1}{N}\sum_{c=1}^{N} (|g_c - g_p|)^i, \tag{2}$$

where N is the number of pixels in the image, g_c the gray value of the center pixel and g_p the gray value of the neighbor, located at a distance R with orientation $2\pi p/P$ from the center pixel. In this work, we use the boosters $CLOSIB_{16,2,1}||CLOSIB_{16,2,2}$ and $HCLOSIB_{16,2,1}||HCLOSIB_{16,2,2}$, which we name CLOSIB and HCLOSIB, respectively, henceforth. Furthermore, we also use the early fusion (concatenation) of LBP, ALBP and HOG descriptors with CLOSIB and HCLOSIB enhancers to describe the texture of the textile regions. We denote the concatenation with the symbol +. For instance, LBP + CLOSIB stands for the early fusion of LBP and CLOSIB.

3.3.5. Faster R-CNN

Faster R-CNN [62] is a Region based Convolutional Neural Network (R-CNN) that generates region of interest proposals by a Region Proposal Network (RPN). Faster R-CNN is basically composed of two parts: a RPN for creating a list of region proposals and a Fast R-CNN network [66] for classifying the regions into objects.

For the RPN, we applied a sliding window of size 3 × 3 on the features obtained at the last convolution layer, which yields an intermediate layer of dimension 512. We fed the intermediate layer into a box classification layer and a box regression layer. We fed the region proposals into a Fully Connected (FC) layer and we extracted the neural codes. Similarly to the MSER approach, we saved a database with the coordinates of the region proposals, the image reference and the neural codes. The matching step remains the same as for MSER approach. We used Faster R-CNN algorithm with VGG-16 [58] architecture pre-trained with an MS-COCO [67] dataset.

3.4. Distance Measures

We use five distance measures to compute the distances among the descriptors of the query region and the descriptors of the automatically detected regions of interest of the database. These are: Spearman, Cosine, Cityblock, Euclidean and Correlation distances. Spearman rank correlation coefficient is a nonparametric measure of rank correlation and it measures the strength and direction of association between two ranked variables. This measure uses a variable's rank which is the average of their positions in the ascending order of the values. Spearman rank correlation coefficient of two vectors A and B is mathematically defined in Equation (3):

$$d_{spe}(A,B) = 1 - \frac{((r(A) - \overline{r(A)})(r(B) - \overline{r(B)})^T}{\sqrt{((r(A) - \overline{r(A)})(r(A) - \overline{r(A)})^T}\sqrt{((r(B) - \overline{r(B)})(r(B) - \overline{r(B)})^T}}, \quad (3)$$

where $\overline{r(A)}$ and $\overline{r(B)}$ are the mean value of the ranked vector A, $r(A)$, and ranked vector B, $r(B)$, respectively. The superscript T indicates the transpose of the matrix. Hereafter, we keep the same notation. Cosine distance calculates the angular cosine between two vectors following Equation (4):

$$d_{cos}(A,B) = 1 - \frac{AB^T}{\sqrt{(AA^T)(BB^T)}}. \quad (4)$$

Cityblock distance is calculated using Equation (5) and is defined by the sum of the absolute distances of every coordinate between two vectors. n is the dimension of the vectors. This measure distance depends on the rotation of the coordinate system but is invariant to reflection and translation:

$$d_{cit}(A,B) = \sum_{j=1}^{n} |A_j - B_j|. \quad (5)$$

Euclidean distance (see Equation (6)) is the most commonly used distance measure and calculates the length of the straight segment that connects two vectors:

$$d_{euc}(A,B) = \sqrt{((A-B)(A-B)^T}. \quad (6)$$

The Correlation distance is obtaining by dividing the distance covariance of two vectors by the product of their distance standard deviations. See Equation (7):

$$d_{cor}(A,B) = 1 - \frac{(A-\overline{A})(B-\overline{B})^T}{\sqrt{(A-\overline{A})(A-\overline{A})^T}\sqrt{(B-\overline{B})(B-\overline{B})^T}}. \quad (7)$$

4. Experiments

4.1. TextilTube Dataset

Textile retrieval in real environments is a poorly investigated research field besides fashion cloth retrieval. Up to our knowledge, there is no publicly available dataset that focuses on the recognition of rigid and non-rigid textiles presented in different sizes, shapes and capturing conditions. For this reason, we created a new dataset for the retrieval of textiles in bedrooms (http://pitia.unileon.es/varp/node/483).

The dataset is composed of 684 images of sizes that range between 480×360 and 1280×720 pixels obtained from 15 videos of YouTube. The videos were recorded in bedrooms with different visual sensors, such as CDCs and webcams. The videos contain plenty of textiles, different camera poses, illumination conditions, occlusions, etc., which makes the textile retrieval task very challenging. The dataset contains 67 classes of textiles such as curtains, carpets, sofas, shirts or dresses, among others. In one image, several classes of textiles may appear. Figure 2 shows a mosaic encompassing one region sample of each class and indicates the number of regions in each class. The number of elements of each class varies from 4 to 116. There is a total of 1913 regions. Therefore, the dataset is highly skewed, simulating a real scenario.

We labelled the dataset in order to provide a ground truth that allows the user to automatically evaluate the performance of a method on the dataset. The ground truth includes the bounding box coordinates and the class labels of each textile region in the images of the dataset. We provide the ground truth in the form of an XML file. We show the diversity in terms of type, size, pose, etc. of some textile classes of the dataset in Figure 3.

TextilTube dataset can be very interesting in fields like child sexual abuse or robbery to connect evidence of different investigations and also for marketing studies in textile stores to suggest the products that best fit the decoration of users' rooms.

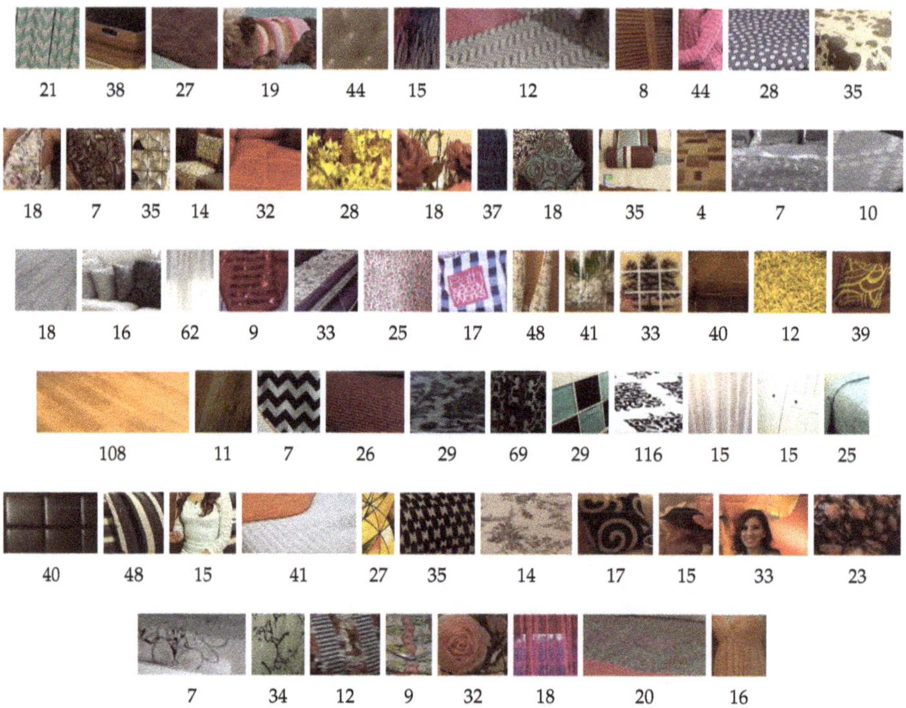

Figure 2. A region sample of each of the 67 classes in TexilTube dataset. The number underneath indicates the amount of regions that belong to that class.

Figure 3. In rows, images that contain the same textile class in TextilTube dataset. The yellow rectangles overlaid in the images indicate the bounding boxes of the textile regions of the ground truth.

4.2. Performance Evaluation Metrics

In retrieval systems, it is important that the retrieved images are ranked according to their relevance to the query region forming a hit list, rather than being returned as a set. The most relevant hits must be within the top images of the hit list returned for a query region. To account for the quality of ranking the hits in the hit list, we used relevance ranking measures, i.e., precision at n and success at n.

4.2.1. Precision at n

Precision at n, $p@n$, is the rate of the top-n images of the hit list correctly classified in relation to the class of the query region. Likewise, the precision at a cut-off of n elements of the hit list. We define $HitList_n$ as the set that contains the n images with smallest distance to the query region, q. Equation (8) presents the mathematical definition of precision at n:

$$p@n = \frac{\#H(q)}{n}, \tag{8}$$

where $\#H(q)$ is the cardinal of $HitList_n$ in which the query class is actually present in the image and the detected region overlaps the bounding box of the ground truth. It is formally defined in Equation (9):

$$H(q) = \{h_i / (h_i \in HitList_n) \wedge (class(h_i) = class(q)) \mid i = 1...n\}, \tag{9}$$

where h_i is i-th retrieved image in the hit list.

4.2.2. Success at n

There are occasions in which the user does not need to see many relevant images but is disappointed by a completely irrelevant top-n [68]. This is the case of the ASASEC project, in which finding at least one hit in all the hit list would be a satisfactory result. Success at n, $s@n$, measures if a relevant image was retrieved within the top-n hits of the hit list. Success at n is equal to 1 if the top-n images contain a relevant document and 0 otherwise (see Equation (10)):

$$s@n = \begin{cases} 1, & \text{if } H(q) \neq \emptyset, \\ 0, & \text{otherwise,} \end{cases} \tag{10}$$

where $H(q)$ is the set of images defined in Equation (9).

4.3. Experimental Setup

We applied the method described in Section 3 to the 684 images of TextilTube dataset, extracting a total of 58,031 regions. In order to evaluate the performance of our method, we used the ground truth textile regions as query regions of interest. For each query region, we calculated $p@n$ and $s@n$ metrics for the retrieved hit list when computing a given distance measure among the texture descriptors of the query region and the analogous texture descriptors of the database. Experiments using Faster R-CNN were developed using the Caffe [69] deep learning framework in Nvidia Titan X GPU https://www.nvidia.com/en-us/geforce/products/10series/titan-x-pascal/.

4.4. Distance Measure Evaluation

In order to determine the best distance measure and present uniform results, we carried out the following voting system. For each texture descriptor in Section 3.2, we computed $s@n$ for $n \in \mathbb{N} \mid n = \{1, 2, \ldots, 40\}$ with all distance measures described in Section 3.4. We assigned three, two and one points to the distance measures that achieved the highest, second highest and third highest $s@n$, respectively, for each experiment. Finally, we summed up the points along all combinations. Figure 4 shows a scheme of the procedure. We disregarded a voting system that only relies on the best distance measure of each experiment because the results for the different distance measures were not enough distinctive.

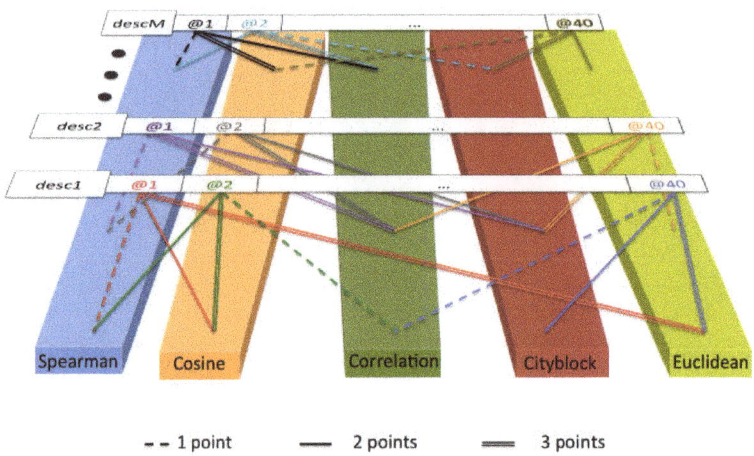

Figure 4. Scheme of the voting procedure to determine the best distance measure.

Figure 5 presents the results in parts per unity achieved with each distance measure. Correlation distance achieved the best results with a 32% of votes, followed by Cosine distance (27%) and Spearman rank correlation coefficient (20%). The commonly used Euclidean distance only yielded a 7% of the votes. Therefore, we carried out our experiments using the Correlation distance.

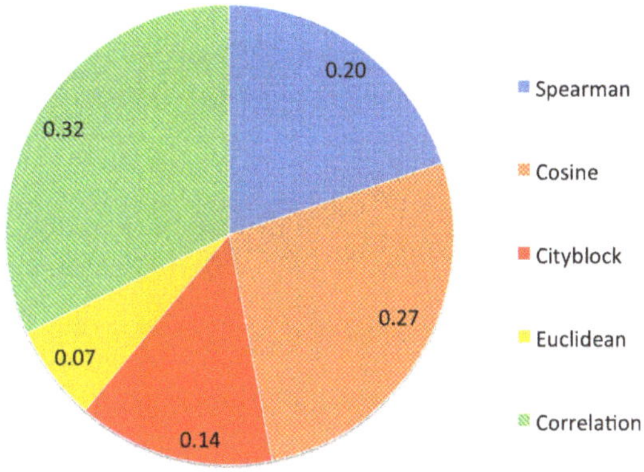

Figure 5. Results of the voting process in parts per unity for the different distance measures.

5. Results

In this section, we present the results obtained following the proposed method and experimentation for each evaluated texture descriptor and the neural codes extracted by Faster R-CNN.

Figure 6 shows the precision at n ($p@n$) achieved for all texture descriptors. We used values of $n \in \mathbb{N} \mid n = \{1, 2, \ldots, 40\}$. For $n \leq 11$, HOG + HCLOSIB descriptor outperformed the rest with a precision of 37.17% for $n = 1$. The early fusion of CLOSIB and HCLOSIB with HOG outperforms HOG alone. However, the early fusion of CLOSIB and HCLOSIB with LBP obtained the worst results. In the case of ALBP, the descriptor alone outperforms the early fusion for small values of n, whereas the opposite is true for high values of n. It is worth noting the better performance of ALBP (28.83% for $n = 1$) versus LBP (16.60% for $n = 1$). At a cut of 20, the precision at n values starts to stabilize. We present the numerical results for precision at cuts $\{1, 2, \ldots, 20\}$ in Table 1. For high cuts of the hit list, Faster-RCNN slightly outperformed the rest. The best performance was not achieved by some LBP variant as we expected, but by HOG combined with HCLOSIB. HOG is oriented to gather the external and internal shape and HCLOSIB represents the statistical distributions of the texture. The combination of both represents both the shape of the textile's texture (HOG) and how this texture is organized along the evaluated patch.

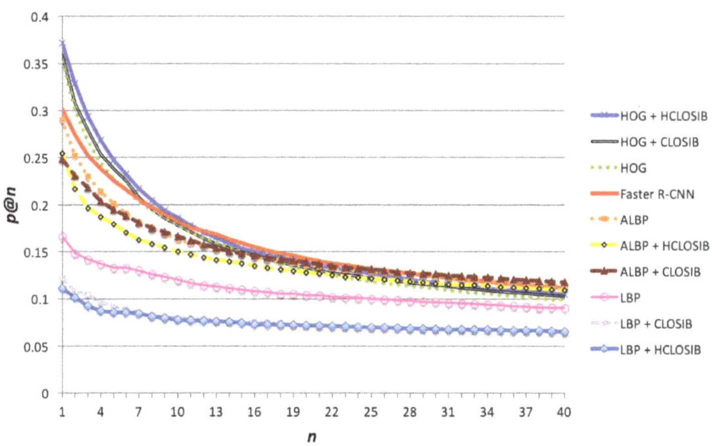

Figure 6. Precision at n ($p@n$) for all texture descriptors using Correlation distance and $n \in \mathbb{N} \mid n = \{1, 2, \ldots, 40\}$.

Table 1. Precision at n ($p@n$) for all texture descriptors using Correlation distance and $n \in \mathbb{N} \mid n = \{1, 2, \ldots, 20\}$. Results highlighted in bold mark the best results per cut of the hit list.

Descriptor	n									
	1	2	3	4	5	6	7	8	9	10
HOG + HCLOSIB	**37.2**	**32.8**	**29.3**	**26.8**	**24.7**	**23.2**	**21.7**	**20.5**	**19.4**	**18.6**
HOG + CLOSIB	35.9	30.8	27.9	25.4	23.8	22.5	20.8	19.5	18.5	17.7
HOG	35.2	30.0	26.7	24.4	22.8	21.5	20.2	19.4	18.7	17.8
Faster R-CNN	30.1	27.4	25.2	23.8	22.5	21.6	20.6	19.7	19.0	18.2
ALBP	28.9	25.2	23.0	21.4	20.1	19.0	18.1	17.5	16.9	16.3
ALBP + HCLOSIB	25.5	21.7	19.6	18.7	17.9	17.0	16.2	15.8	15.4	15.0
ALBP + CLOSIB	24.8	23.1	21.8	20.4	19.5	18.8	18.1	17.5	17.1	16.6
LBP	16.6	14.8	14.2	13.7	13.3	13.3	13.0	12.6	12.3	12.0
LBP + CLOSIB	12.2	10.8	10.4	9.6	9.2	8.9	8.5	8.3	8.1	7.9
LBP + HCLOSIB	11.1	10.1	9.3	8.7	8.6	8.5	8.4	8.1	8.0	7.7

Descriptor	n									
	11	12	13	14	15	16	17	18	19	20
HOG + HCLOSIB	**17.7**	17.0	16.4	15.8	15.3	15.0	14.6	14.4	14.0	13.7
HOG + CLOSIB	17.1	16.4	15.8	15.4	14.9	14.6	14.2	13.9	13.6	13.3
HOG	17.0	16.4	15.9	15.5	14.9	14.5	14.1	13.7	13.4	13.0
Faster R-CNN	17.6	**17.1**	**16.8**	**16.3**	**15.9**	**15.5**	**15.1**	**14.8**	**14.6**	**14.3**
ALBP	15.9	15.6	15.2	15.0	14.7	14.4	14.2	14.0	13.8	13.5
ALBP + HCLOSIB	14.7	14.4	14.1	14.0	13.7	13.5	13.3	13.1	13.0	12.8
ALBP + CLOSIB	16.2	15.7	15.4	15.1	14.8	14.6	14.4	14.2	14.0	13.9
LBP	11.7	11.4	11.3	11.1	11.0	10.8	10.7	10.5	10.5	10.4
LBP + CLOSIB	7.9	7.8	7.7	7.6	7.6	7.5	7.4	7.4	7.3	7.2
LBP + HCLOSIB	7.7	7.7	7.6	7.5	7.5	7.3	7.3	7.2	7.2	7.2

Figure 7 illustrates the success at n ($s@n$). As expected, $s@1$ is the same as $p@1$ and for higher cuts of the hit list the success metric increases. In Table 2, we show the numerical results for success at cuts $\{1, 2, \ldots, 20\}$. For values of $n \leq 5$, HOG + HCLOSIB yielded the best results, whereas for higher values of n, Faster R-CNN outperformed the others with a 84.94% of s@40 (74.86% with HOG + HCLOSIB). ALBP is the second best descriptor for $n = 40$ reaching 82.00% of success. CLOSIB enhancer improves the performance of HOG and decreases the performance of LBP and ALBP.

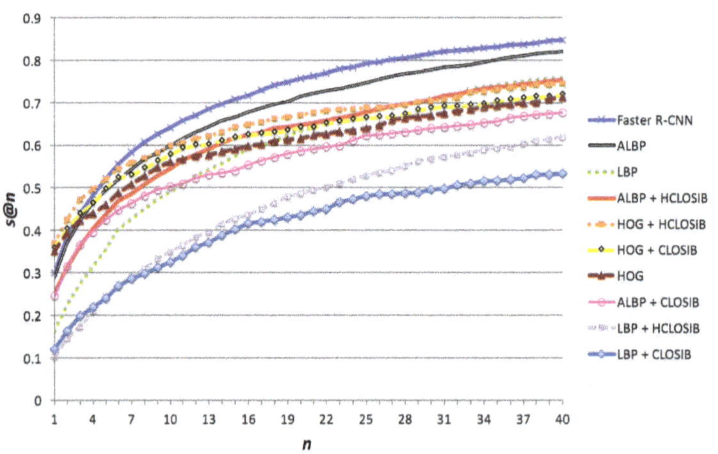

Figure 7. Success at n ($s@n$) for all texture descriptors using Correlation distance and $n \in \mathbb{N} \mid n = \{1, 2, \ldots, 40\}$.

Table 2. Success at n ($p@n$) for all texture descriptors using Correlation distance and $n \in \mathbb{N} \mid n = \{1, 2, \ldots, 20\}$. Results highlighted in bold mark the best results per cut of the hit list.

Descriptor	n									
	1	**2**	**3**	**4**	**5**	**6**	**7**	**8**	**9**	**10**
Faster R-CNN	30.1	38.7	44.3	48.4	**52.2**	**55.7**	**58.6**	**60.9**	**62.8**	**64.3**
ALBP	28.9	37.0	42.3	46.7	49.5	51.9	54.4	56.8	58.2	60.0
LBP	16.6	22.6	27.6	31.7	35.5	40.5	42.8	45.2	47.5	49.2
ALBP + HCLOSIB	25.5	31.4	36.5	40.6	44.0	47.1	48.8	51.0	53.0	54.7
HOG + HCLOSIB	**37.2**	**42.8**	**47.5**	**50.0**	**52.2**	54.6	56.1	57.1	58.4	60.1
HOG + CLOSIB	35.9	40.6	44.0	46.6	49.8	52.5	53.2	54.8	56.4	57.9
HOG	35.2	39.5	42.4	44.1	46.1	48.9	50.9	52.9	54.9	56.1
ALBP + CLOSIB	24.8	31.7	36.7	39.8	42.7	49.0	46.6	48.3	49.8	50.5
LBP + HCLOSIB	10.4	14.7	17.4	21.1	24.4	27.1	29.0	31.1	33.4	34.9
LBP + CLOSIB	12.2	16.5	20.0	22.0	24.1	26.9	28.7	29.9	31.3	32.6
Descriptor	n									
	11	**12**	**13**	**14**	**15**	**16**	**17**	**18**	**19**	**20**
Faster R-CNN	**65.9**	**67.2**	**68.7**	**69.9**	**71.1**	**71.9**	**73.1**	**74.3**	**75.0**	**75.9**
ALBP	61.5	63.0	64.3	65.8	66.5	67.8	68.7	69.6	70.2	71.5
LBP	50.9	53.0	54.5	55.8	57.6	59.4	60.2	61.0	62.4	63.5
ALBP + HCLOSIB	56.6	57.9	59.2	60.8	61.9	62.6	63.1	64.2	64.5	64.9
HOG + HCLOSIB	60.8	61.7	62.3	63.3	64.5	65.0	65.6	66.5	66.8	67.3
HOG + CLOSIB	59.5	60.2	60.4	61.2	62.0	62.6	62.9	63.2	63.7	64.3
HOG	56.8	57.5	57.7	58.6	59.1	59.9	60.3	61.0	61.4	61.6
ALBP + CLOSIB	51.2	52.5	53.3	53.7	54.2	55.6	56.6	57.4	58.1	58.8
LBP + HCLOSIB	36.5	38.5	39.7	41.7	42.9	43.9	45.1	46.6	48.1	48.7
LBP + CLOSIB	34.2	36.1	37.2	38.9	40.4	41.8	42.1	42.7	43.2	43.7

In order to get a unique value to evaluate the performance of each descriptor, we computed the arithmetic mean of the success and precision for three intervals of $n \in \mathbb{N} \mid n = \{1, 2, \ldots, j\}$, where $j = \{10, 20, 40\}$. Figure 8 and Table 3 show the arithmetic mean of the success and the precision in these intervals of values of n. Regarding $p@n$, HOG + HCLOSIB outperformed the rest of descriptors for the intervals with $j = 10$ and $j = 20$, whereas Faster-RCNN obtained the best results for $j = 40$. With respect to $s@n$, HOG + HCLOSIB yielded the best results for the intervals with $j = 10$, whereas Faster R-CNN did for $j = 20$ and $j = 40$. In such a difficult dataset, the outlined method using HOG + HCLOSIB descriptor and Correlation distance measure yielded an arithmetic mean of the precision at 10 of 24.80% and an arithmetic mean of the success at 10 of 51.08%.

Table 3. Arithmetic mean of precision and success at n for intervals of n from 1 to 10, from 1 to 20 and from 1 to 40. Results highlighted in bold mark the best results per performance metric.

Descriptor	Precision			Success		
	Mean (1–10)	**Mean (1–20)**	**Mean (1–40)**	**Mean (1–10)**	**Mean (1–20)**	**Mean (1–40)**
HOG + HCLOSIB	**24.8**	**19.5**	15.1	**51.1**	57.3	63.9
HOG + CLOSIB	23.7	18.8	14.7	48.7	54.9	61.4
HOG	23.1	18.5	14.3	46.6	52.6	59.4
Faster R-CNN	22.5	18.8	**15.3**	50.3	**59.9**	**69.8**
ALBP	20.3	17.2	14.5	47.5	56.4	66.1
ALBP + HCLOSIB	18.1	15.7	13.5	42.2	50.9	60.1
ALBP + CLOSIB	19.6	17.0	14.6	40.7	47.3	55.1
LBP	13.5	12.2	10.8	34.1	44.4	56.1
LBP + CLOSIB	9.3	8.4	7.5	23.4	30.6	39.0
LBP + HCLOSIB	8.8	8.1	7.4	22.8	31.3	42.0

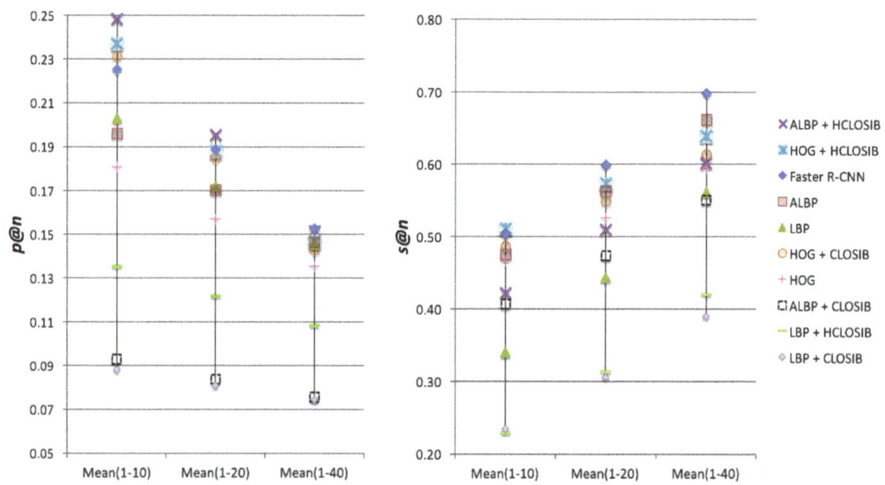

Figure 8. Arithmetic mean of precision and success at n for intervals of n from 1 to 10, from 1 to 20 and from 1 to 40.

Figures 9 and 10 show the visual results for the first five retrieved images in the hit list using HOG + HCLOSIB and Faster R-CNN, respectively. The third textile, artificial flower textile, is a difficult case. HOG + HCLOSIB manages to get one correctly retrieved images for $n = 5$. The first two retrieved images are similar textiles of a rug. For the same query image, Faster R-CNN does not retrieve any correct images at a cut of hit list of $n = 5$. Pre-trained deep neural networks are trained to classify objects instead of textiles. When a textile does not present a patterned texture, a pre-trained Faster R-CNN is not appropriate to retrieve such queries. However, Faster R-CNN manages to retrieve textiles correctly that present distinctive patterns.

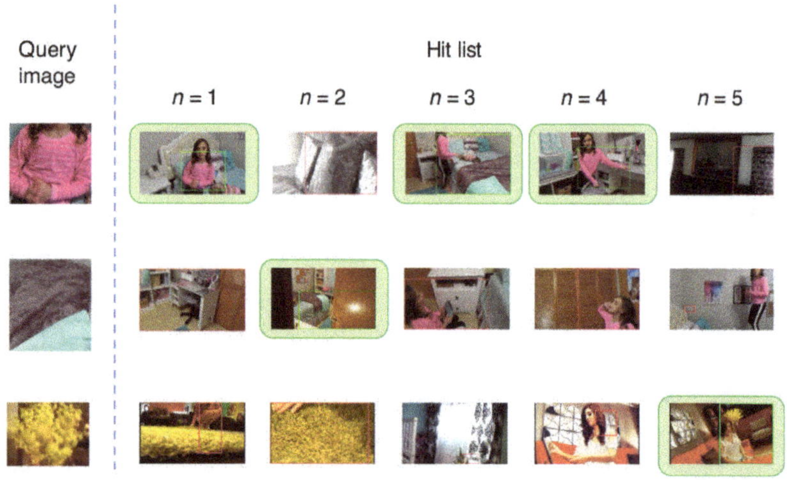

Figure 9. First five retrieved images in the hit list for three query samples using HOG + HCLOSIB.

Figure 10. First five retrieved images in the hit list for three query samples using Faster R-CNN.

6. Conclusions

In this paper, we presented a new application for textile based image retrieval in indoor environments. Textile based image retrieval is barely studied and when doing so, it is usually applied to fashion cloth retrieval. We introduced a new framework of study, the fight of child sexual abuse. Law enforcement agencies are interested in relating evidence by using textile queries in order to retrieve images or videos that contain the same textile in proven cases of child pornography, usually taken from rooms of houses. We proposed a new effective method for textile based image retrieval in rooms based on texture description of the MSER regions of the images. We assessed LBP, ALBP, HOG and their combination with CLOSIB for describing the image patches and several distance metrics for sorting the hit list. We also evaluated the Faster R-CNN algorithm with VGG-16 architecture pre-trained with MS-COCO dataset. Furthermore, we created and introduced a new public dataset, TextilTube, which consists of 684 frames from 15 Youtube videos of rooms recorded with different visual sensors. The dataset contains 1913 regions of interest that highly vary in terms of capturing conditions, occlusions, illuminations, etc. Moreover, textiles appearing in the images are not rigid and present different shapes. Correlation distance proved to be the most discriminant distance measure based on a voting system analysis. Correlation distance achieved 32% of the votes followed by cosine distance with 27%. HOG + HCLOSIB yielded the best results for low cuts of the hit list, whereas Faster R-CNN performed better for high cuts, closely followed by ALBPS. Taking into account just the most similar image retrieved, HOG + HCLOSIB achieved a precision of 37.17%, which is remarkable due to the number of classes in the dataset (67 classes) and their high intra-class variability. Taking into account the success at n metric, Faster R-CNN achieved a 84.94% retrieving 40 images (ALBP obtained a 82%), which means that about 85 out of 100 images have at least one correspondence in the top 40 retrieved images. This is a very interesting result that can be presented as an application for the criminal police in order to let them evaluate a grid of 40 images at a glance to check if there is a real match to the query image within the hit list. For the application at hand, it is interesting to achieve a high precision at low cuts of the hit list in order to reduce the number of images to visually inspect. To measure this fact, we computed the arithmetic mean of precision at n from 1 to 10. HOG + HCLOSIB outperformed the rest yielding a 24.8% hit rate. The main problem in this application is to find regions containing textiles, rather than objects. To the best of our knowledge, all the deep learning region proposal models are oriented and trained to detect objects. Objects are usually non-homogeneous regions as opposed to textiles. The reason is that CBIR systems are oriented to retrieve objects but not textiles or similar surfaces. Similarly, the deep learning models, are trained with datasets such as MS-COCO or ImageNet, among others, that contain objects and different classes of objects and they are oriented for instance retrieval. In future works, we will train a model for proposing regions with a textile dataset to strengthen the use of deep learning for textile retrieval. Besides evaluating other Region Proposal Networks in future works, different alternatives to MSER for finding distinguished regions, such as Sieve, will be also tested.

Author Contributions: Oscar García-Olalla and Enrique Alegre conceived the proposed method and designed the experiments; Surajit Saikia performed the experiments for CNNs and Oscar García-Olalla performed the rest of experiments; Oscar García-Olalla, Enrique Alegre, Laura Fernández-Robles and Eduardo Fidalgo analyzed the data; Laura Fernández-Robles and Oscar García-Olalla wrote the paper; Eduardo Fidalgo, Enrique Alegre and Laura Fernández-Robles substantively revised the paper, Eduardo Fidalgo provided the visual results.

Funding: This research was funded by [the Spanish Government] grant number [DPI2012-36166] and by [INCIBE (Spanish National Cybersecurity Institute)] grant number [INCIBEC-2015-02493, Addendum 22].

Acknowledgments: This work has been supported by grant DPI2012-36166, the pre-doctoral FPU fellowship program from the Spanish Government (AP2010-0947), grant INCIBEC-2015-02493 corresponding to "Ayudas para la Excelencia de los Equipos de Investigación avanzada en ciberseguridad", the framework agreement between the University of León and INCIBE (Spanish National Cybersecurity Institute) under Addendum 22, and the PIRTU program of the Regional Government of Castilla y León. We gratefully acknowledge the support of Nvidia Corporation for their kind donation of GPUs (GeForce GTX Titan X and K-40) that were used in this work.

Conflicts of Interest: The authors declare no conflict of interest.

Abbreviations

The following abbreviations are used in this manuscript:

LBP	Local Binary Pattern
ALBP	Adaptive Local Binary Pattern
HOG	Histogram of Oriented Gradients
CLOSIB	Complete Local Oriented Statistical Information Booster
HCLOSIB	Half Complete Local Oriented Statistical Information Booster
MSER	Maximally Stable Extremal Regions
CBIR	Content Based Image Retrieval
ASASEC	Advisory System Against Sexual Exploitation of Children
HSV	Hue, Saturation, Value
CDC	Compact Digital Cameras
p@n	precision at *n*
s@n	success at *n*
CNN	Convolutional Neural Networks
R-CNN	Region based Convolutional Neural Network
RPN	Region Proposal Network
FC	Fully Connected
RGB	Red, Green, Blue
SIFT	Scale-Invariant Feature Transform
SURF	Speeded Up Robust Features
VGG	Visual Geometry Group
MS-COCO	MicroSoft Common Objects in COntext
XML	eXtensible Markup Language

References

1. Czúni, L.; Rashad, M. Lightweight Active Object Retrieval with Weak Classifiers. *Sensors* **2018**, *18*, 801. [CrossRef] [PubMed]
2. Domínguez, S. Saliency-based similarity measure. *Rev. Iberoam. Autom. Inform. Ind.* **2012**, *9*, 359–370. [CrossRef]

3. Faria, A.V.; Oishi, K.; Yoshida, S.; Hillis, A.; Miller, M.I.; Mori, S. Content-based image retrieval for brain MRI: An image-searching engine and population-based analysis to utilize past clinical data for future diagnosis. *NeuroImage Clin.* **2015**, *7*, 367–376. [CrossRef] [PubMed]
4. Srinivas, M.; Naidu, R.R.; Sastry, C.; Mohan, C.K. Content based medical image retrieval using dictionary learning. *Neurocomputing* **2015**, *168*, 880–895. [CrossRef]
5. Bugatti, P.H.; Kaster, D.S.; Ponciano-Silva, M.; Traina, C., Jr.; Azevedo-Marques, P.M.; Traina, A.J. PRoSPer: Perceptual similarity queries in medical CBIR systems through user profiles. *Comput. Biol. Med.* **2014**, *45*, 8–19. [CrossRef] [PubMed]
6. Jung, J.; Yoon, I.; Lee, S.; Paik, J. Normalized Metadata Generation for Human Retrieval Using Multiple Video Surveillance Cameras. *Sensors* **2016**, *16*, 963. [CrossRef] [PubMed]
7. Feng, L.; Bhanu, B.; Heraty, J. A software system for automated identification and retrieval of moth images based on wing attributes. *Pattern Recognit.* **2016**, *51*, 225–241. [CrossRef]
8. Mallik, J.; Samal, A.; Gardner, S.L. A content based image retrieval system for a biological specimen collection. *Comput. Vis. Image Underst.* **2010**, *114*, 745–757. [CrossRef]
9. Liu, B.; Yue, Y.M.; Li, R.; Shen, W.J.; Wang, K.L. Plant Leaf Chlorophyll Content Retrieval Based on a Field Imaging Spectroscopy System. *Sensors* **2014**, *14*, 19910–19925. [CrossRef] [PubMed]
10. Iqbal, K.; Odetayo, M.O.; James, A. Content-based image retrieval approach for biometric security using colour, texture and shape features controlled by fuzzy heuristics. *J. Comput. Syst. Sci.* **2012**, *78*, 1258–1277. [CrossRef]
11. Chang, L.; Duarte, M.M.; Sucar, L.; Morales, E.F. A Bayesian approach for object classification based on clusters of SIFT local features. *Expert Syst. Appl.* **2012**, *39*, 1679–1686. [CrossRef]
12. Fidalgo, E.; Alegre, E.; González-Castro, V.; Fernández-Robles, L. Compass radius estimation for improved image classification using Edge-SIFT. *Neurocomputing* **2016**, *197*, 119–135. [CrossRef]
13. Chen, L.C.; Hsieh, J.W.; Yan, Y.; Chen, D.Y. Vehicle make and model recognition using sparse representation and symmetrical SURFs. *Pattern Recognit.* **2015**, *48*, 1979–1998. [CrossRef]
14. Li, H.; Liu, Z.; Huang, Y.; Shi, Y. Quaternion generic Fourier descriptor for color object recognition. *Pattern Recognit.* **2015**, *48*, 3895–3903. [CrossRef]
15. Zhu, J.; Yu, J.; Wang, C.; Li, F.Z. Object recognition via contextual color attention. *J. Vis. Commun. Image Represent.* **2015**, *27*, 44–56. [CrossRef]
16. Shih, H.C.; Yu, K.C. SPiraL Aggregation Map (SPLAM): A new descriptor for robust template matching with fast algorithm. *Pattern Recognit.* **2015**, *48*, 1707–1723. [CrossRef]
17. Tan, M.; Hu, Z.; Wang, B.; Zhao, J.; Wang, Y. Robust object recognition via weakly supervised metric and template learning. *Neurocomputing* **2016**, *181*, 96–107. [CrossRef]
18. Salvador, A.; Giró i Nieto, X.; Marqués, F.; Satoh, S. Faster R-CNN Features for Instance Search. In Proceedings of the 2016 IEEE Conference on Computer Vision and Pattern Recognition Workshops (CVPRW), Las Vegas, NV, USA, 26 June–1 July 2016; pp. 394–401.
19. Han, J.; Zhang, D.; Cheng, G.; Liu, N.; Xu, D. Advanced Deep-Learning Techniques for Salient and Category-Specific Object Detection: A Survey. *IEEE Signal Process. Mag.* **2018**, *35*, 84–100. [CrossRef]
20. Saikia, S.; Fidalgo, E.; Alegre, E.; Fernández-Robles, L. Query Based Object Retrieval Using Neural Codes. In *Advances in Intelligent Systems and Computing*; Springer: Berlin/Heidelberg, Germany, 2018; Volume 649, pp. 513–523.
21. D'Amato, J.P.; Mercado, M.; Heiling, A.; Cifuentes, V. A proximal optimization method to the problem of nesting irregular pieces using parallel architectures. *Rev. Iberoam. Autom. Inform. Ind.* **2016**, *13*, 220–227. [CrossRef]
22. Wong, C. *Applications of Computer Vision in Fashion and Textiles*, 1st ed.; The Textile Institute Book Series; Woodhead Publishing: Cambreidge, UK; Elsevier Science: New York, NY, USA, 2017.
23. Wen, J.; Wong, W. Chapter 2—Fundamentals of common computer vision techniques for fashion textile modeling, recognition, and retrieval. In *Applications of Computer Vision in Fashion and Textiles*; Wong, W., Ed.; The Textile Institute Book Series; Woodhead Publishing: Cambreidge, UK, 2018; pp. 17–44.
24. Gangwar, A.; Fidalgo, E.; Alegre, E.; González-Castro, V. Pornography and Child Sexual Abuse Detection in Image and Video: A Comparative Evaluation. In Proceedings of the 8th International Conference on Imaging for Crime Imaging for Crime Detection and Prevention, Madrid, Spain, 13–15 December 2017.

25. Zhu, Z.; Brilakis, I. Parameter optimization for automated concrete detection in image data. *Autom. Constr.* **2010**, *19*, 944–953. [CrossRef]
26. Son, H.; Kim, C.; Hwang, N.; Kim, C.; Kang, Y. Classification of major construction materials in construction environments using ensemble classifiers. *Adv. Eng. Inform.* **2014**, *28*, 1–10. [CrossRef]
27. Xie, X.; Yang, L.; Zheng, W.S. Learning object-specific DAGs for multi-label material recognition. *Comput. Vis. Image Underst.* **2016**, *143*, 183–190. [CrossRef]
28. Yang, L.; Xie, X. Exploiting object semantic cues for Multi-label Material Recognition. *Neurocomputing* **2016**, *173*, 1646–1654. [CrossRef]
29. Xue, J.; Zhang, H.; Dana, K.; Nishino, K. Differential Angular Imaging for Material Recognition. In Proceedings of the 2017 IEEE Conference on Computer Vision and Pattern Recognition (CVPR), Honolulu, HI, USA, 21–26 July 2017; pp. 6940–6949.
30. González, E.; Bianconi, F.; Álvarez, M.X.; Saetta, S.A. Automatic Characterization of the Visual Appearance of Industrial Materials through Colour and Texture Analysis: An Overview of Methods and Applications. *Adv. Opt. Technol.* **2013**, *2013*, 1–11. [CrossRef]
31. Bashar, M.; Ohnishi, N.; Matsumoto, T.; Takeuchi, Y.; Kudo, H.; Agusa, K. Image retrieval by pattern categorization using wavelet domain perceptual features with LVQ neural network. *Pattern Recognit. Lett.* **2005**, *26*, 2315–2335. [CrossRef]
32. Carbunaru, A.E.; Coltuc, D.; Jourlin, M.; Frangu, L. A texture descriptor for textile image retrieval. In Proceedings of the 2009 International Symposium on Signals, Circuits and Systems, Iasi, Romania, 9–10 July 2009; pp. 1–4.
33. Chun, J.C.; Kim, W.G. Textile Image Retrieval Using Composite Feature Vectors of Color and Wavelet Transformed Textural Property. *Appl. Mech. Mater.* **2013**, *333*, 822–827. [CrossRef]
34. Huang, Y.F.; Lin, S.M. *Searching Images in a Textile Image Database*; Springer International Publishing: Cham, Switzerland, 2014; pp. 267–274.
35. Yamaguchi, K.; Kiapour, M.H.; Ortiz, L.E.; Berg, T.L. Retrieving Similar Styles to Parse Clothing. *IEEE Trans. Pattern Anal. Mach. Intell.* **2015**, *37*, 1028–1040. [CrossRef] [PubMed]
36. Liang, X.; Lin, L.; Yang, W.; Luo, P.; Huang, J.; Yan, S. Clothes Co-Parsing Via Joint Image Segmentation and Labeling with Application to Clothing Retrieval. *IEEE Trans. Multimed.* **2016**, *18*, 1175–1186. [CrossRef]
37. Sun, G.L.; Wu, X.; Peng, Q. Part-based clothing image annotation by visual neighbor retrieval. *Neurocomputing* **2016**, *213*, 115–124. [CrossRef]
38. Chen, Q.; Huang, J.; Feris, R.; Brown, L.M.; Dong, J.; Yan, S. Deep domain adaptation for describing people based on fine-grained clothing attributes. In Proceedings of the 2015 IEEE Conference on Computer Vision and Pattern Recognition (CVPR), Boston, MA, USA, 7–12 June 2015; pp. 5315–5324.
39. Huang, J.; Feris, R.; Chen, Q.; Yan, S. Cross-Domain Image Retrieval with a Dual Attribute-Aware Ranking Network. In Proceedings of the 2015 IEEE International Conference on Computer Vision (ICCV), Santiago, Chile, 7–13 December 2015; pp. 1062–1070.
40. Kiapour, M.H.; Han, X.; Lazebnik, S.; Berg, A.C.; Berg, T.L. Where to Buy It: Matching Street Clothing Photos in Online Shops. In Proceedings of the 2015 IEEE International Conference on Computer Vision (ICCV), Santiago, Chile, 7–13 December 2015; pp. 3343–3351.
41. Zheng, Y.; Sarem, M. A fast region segmentation algorithm on compressed gray images using Non-symmetry and Anti-packing Model and Extended Shading representation. *J. Vis. Commun. Image Represent.* **2016**, *34*, 153–166. [CrossRef]
42. Yang, B.; Yu, H.; Hu, R. Unsupervised regions based segmentation using object discovery. *J. Vis. Commun. Image Represent.* **2015**, *31*, 125–137. [CrossRef]
43. Matas, J.; Chum, O.; Urban, M.; Pajdla, T. Robust wide-baseline stereo from maximally stable extremal regions. *Image Vis. Comput.* **2004**, *22*, 761–767. [CrossRef]
44. Dalal, N.; Triggs, B. Histograms of oriented gradients for human detection. In Proceedings of the 2005 IEEE Computer Society Conference onComputer Vision and Pattern Recognition (CVPR 2005), San Diego, CA, USA, 20–25 June 2005; Volume 1, pp. 886–893.
45. Li, B.; Huo, G. Face recognition using locality sensitive histograms of oriented gradients. *Opt. Int. J. Light Electron Opt.* **2016**, *127*, 3489–3494. [CrossRef]

46. Ojala, T.; Pietikainen, M.; Harwood, D. Performance evaluation of texture measures with classification based on Kullback discrimination of distributions. In Proceedings of the 12th International Conference on Pattern Recognition, Jerusalem, Israel, 9–13 October 1994; Volume 1, pp. 582–585.
47. Guo, Z.; Zhang, L.; Zhang, D. Rotation invariant texture classification using LBP variance (LBPV) with global matching. *Pattern Recognit.* **2010**, *43*, 706–719. [CrossRef]
48. Guo, Z.; Zhang, L.; Zhang, D. A completed modeling of local binary pattern operator for texture classification. *IEEE Trans. Image Process.* **2010**, *19*, 1657–1663. [PubMed]
49. Guo, Z.; Zhang, L.; Zhang, D.; Zhang, S. Rotation invariant texture classification using adaptive LBP with directional statistical features. In Proceedings of the 2010 17th IEEE International Conference on Image Processing (ICIP), Paris, France, 27–30 October 2010; pp. 285–288.
50. García-Olalla, O.; Alegre, E.; Fernández-Robles, L.; García-Ordás, M.T. Vitality assessment of boar sperm using an adaptive LBP based on oriented deviation. In Proceedings of the Computer Vision—ACCV 2012 Workshops, Daejeon, Korea, 5–9 November 2012; Park, J.I., Kim, J., Eds.; Springer: Berlin/Heidelberg, Germany, 2013; pp. 61–72.
51. Garcia-Olalla, O.; Alegre, E.; Fernandez-Robles, L.; Garcia-Ordas, M.T.; Garcia-Ordas, D. Adaptive local binary pattern with oriented standard deviation (ALBPS) for texture classification. *EURASIP J. Image Video Process.* **2013**, *2013*, 31. [CrossRef]
52. García-Olalla, O.; Alegre, E.; García-Ordás, M.T.; Fernández-Robles, L. Evaluation of LBP Variants using several Metrics and kNN Classifiers. In *Similarity Search and Applications*; Brisaboa, N., Pedreira, O., Zezula, P., Eds.; Springer: Berlin/Heidelberg, Germany, 2013; pp. 151–162.
53. Garcia-Olalla, O.; Alegre, E.; Fernandez-Robles, L.; Gonzalez-Castro, V. Local Oriented Statistics Information Booster (LOSIB) for Texture Classification. In Proceedings of the 2014 22nd International Conference on Pattern Recognition (ICPR), Stockholm, Sweden, 24–28 August 2014; pp. 1114–1119.
54. Fernández, A.; Álvarez, M.X.; Bianconi, F. Texture Description Through Histograms of Equivalent Patterns. *J. Math. Imaging Vis.* **2013**, *45*, 76–102. [CrossRef]
55. Liu, L.; Zhao, L.; Long, Y.; Kuang, G.; Fieguth, P. Extended local binary patterns for texture classification. *Image Vis. Comput.* **2012**, *30*, 86–99. [CrossRef]
56. Krizhevsky, A.; Sutskever, I.; Hinton, G.E. ImageNet Classification with Deep Convolutional Neural Networks. In *Advances in Neural Information Processing Systems 25*; Pereira, F., Burges, C.J.C., Bottou, L., Weinberger, K.Q., Eds.; Curran Associates, Inc.: Red Hook, NY, USA, 2012; pp. 1097–1105.
57. Szegedy, C.; Liu, W.; Jia, Y.; Sermanet, P.; Reed, S.; Anguelov, D.; Erhan, D.; Vanhoucke, V.; Rabinovich, A. Going deeper with convolutions. In Proceedings of the 2015 IEEE Conference on Computer Vision and Pattern Recognition (CVPR), Boston, MA, USA, 7–12 June 2015; pp. 1–9.
58. Simonyan, K.; Zisserman, A. Very Deep Convolutional Networks for Large-Scale Image Recognition. *arXiv* **2014**, arXiv:1409.1556. [CrossRef]
59. Vallet, A.; Sakamoto, H. Convolutional Recurrent Neural Networks for Better Image Understanding. In Proceedings of the 2016 International Conference on Digital Image Computing: Techniques and Applications (DICTA), Gold Coast, Australia, 30 November–2 December 2016; pp. 1–7.
60. Donahue, J.; Jia, Y.; Vinyals, O.; Hoffman, J.; Zhang, N.; Tzeng, E.; Darrell, T. DeCAF: A Deep Convolutional Activation Feature for Generic Visual Recognition. In Proceedings of the 31st International Conference on Machine Learning, Bejing, China, 21–26 June 2014; Xing, E.P., Jebara, T., Eds.; PMLR: Bejing, China, 2014; Volume 32, pp. 647–655.
61. Girshick, R.; Donahue, J.; Darrell, T.; Malik, J. Rich Feature Hierarchies for Accurate Object Detection and Semantic Segmentation. In Proceedings of the 2014 IEEE Conference on Computer Vision and Pattern Recognition (CVPR '14), Columbus, OH, USA, 23–28 June 2014; IEEE Computer Society: Washington, DC, USA, 2014; pp. 580–587.
62. Ren, S.; He, K.; Girshick, R.; Sun, J. Faster R-CNN: Towards Real-Time Object Detection with Region Proposal Networks. *IEEE Trans. Pattern Anal. Mach. Intell.* **2017**, *39*, 1137–1149. [CrossRef] [PubMed]
63. Bangham, J.A.; Harvey, R.W.; Ling, P.D.; Aldridge, R.V. Morphological scale-space preserving transforms in many dimensions. *J. Electron. Imaging* **1996**, *5*, 5–17. [CrossRef]
64. Ojala, T.; Pietikainen, M.; Maenpaa, T. Multiresolution gray-scale and rotation invariant texture classification with Local Binary Patterns. *IEEE Trans. Pattern Anal. Mach. Intell.* **2002**, *24*, 971–987. [CrossRef]

65. García-Olalla Olivera, O. Methods for Improving Texture Description by Using Statistical Information Extracted from the Image Gradient. Ph.D. Thesis, Universidad de León, León, Spain, 2017.
66. Girshick, R. Fast R-CNN. In Proceedings of the 2015 IEEE International Conference on Computer Vision (ICCV), Santiago, Chile, 7–13 December 2015; pp. 1440–1448.
67. Lin, T.Y.; Maire, M.; Belongie, S.; Hays, J.; Perona, P.; Ramanan, D.; Dollár, P.; Zitnick, C.L. Microsoft COCO: Common Objects in Context. In Proceedings of the Computer Vision—ECCV 2014, Zurich, Switzerland, 6–12 September 2014; Fleet, D., Pajdla, T., Schiele, B., Tuytelaars, T., Eds.; Springer International Publishing: Cham, Switzerland, 2014; pp. 740–755.
68. Liu, L.; Zsu, M.T. *Encyclopedia of Database Systems*, 1st ed.; Springer: Berlin/Heidelberg, Germany, 2009.
69. Jia, Y.; Shelhamer, E.; Donahue, J.; Karayev, S.; Long, J.; Girshick, R.; Guadarrama, S.; Darrell, T. Caffe: Convolutional Architecture for Fast Feature Embedding. In Proceedings of the 22nd ACM International Conference on Multimedia (MM '14), Orlando, FL, USA, 3–7 November 2014; ACM: New York, NY, USA, 2014; pp. 675–678.

sensors

MDPI

Article

An Improved Point Cloud Descriptor for Vision Based Robotic Grasping System

Fei Wang [1,*], Chen Liang [1], Changlei Ru [2] and Hongtai Cheng [3]

[1] Faculty of Robot Science and Engineering, Northeastern University, Shenyang 110169, China; 1700951@stu.neu.edu.cn
[2] College of Information Science and Engineering, Northeastern University, Shenyang 110819, China; 1870652@stu.neu.edu.cn
[3] School of Mechanical Engineering and Automation, Northeastern University, Shenyang 110819, China; chenght@me.neu.edu.cn
* Correspondence: wangfei@mail.neu.edu.cn; Tel.: +86-1394-005-8702

Received: 1 March 2019; Accepted: 9 May 2019; Published: 14 May 2019

Abstract: In this paper, a novel global point cloud descriptor is proposed for reliable object recognition and pose estimation, which can be effectively applied to robot grasping operation. The viewpoint feature histogram (VFH) is widely used in three-dimensional (3D) object recognition and pose estimation in real scene obtained by depth sensor because of its recognition performance and computational efficiency. However, when the object has a mirrored structure, it is often difficult to distinguish the mirrored poses relative to the viewpoint using VFH. In order to solve this difficulty, this study presents an improved feature descriptor named orthogonal viewpoint feature histogram (OVFH), which contains two components: a surface shape component and an improved viewpoint direction component. The improved viewpoint component is calculated by the orthogonal vector of the viewpoint direction, which is obtained based on the reference frame estimated for the entire point cloud. The evaluation of OVFH using a publicly available data set indicates that it enhances the ability to distinguish between mirrored poses while ensuring object recognition performance. The proposed method uses OVFH to recognize and register objects in the database and obtains precise poses by using the iterative closest point (ICP) algorithm. The experimental results show that the proposed approach can be effectively applied to guide the robot to grasp objects with mirrored poses.

Keywords: vision-guided robotic grasping; object recognition; pose estimation; global feature descriptor; iterative closest point

1. Introduction

Three-dimensional (3D) machine vision is a key technology in the field of robotics. Although the rise of 3D vision technology [1,2] is later than two-dimensional (2D) vision technology [3,4], it presents some advantages that 2D vision does not have when performing some complex visual tasks in 3D space. For example, 3D point cloud can provide a wealth of geometric (3D coordinates, curvature variations, surface normals, depth boundaries) and luminosity (color, color variations, transparency, reflectance intensity) information, helping to achieve better results in recognizing objects with less appearance information (e.g., textureless objects) and directly estimating the full 6 degrees of freedom (DOF) object pose. In addition, under unfavorable lighting conditions, 3D data provided by infrared laser technology can achieve better results than 2D images. In recent years, low-cost real-time 3D sensors such as Microsoft Kinect and Asus Xtion have become low-cost consumer devices accessible to ordinary users. These sensors can be used to generate color 3D point clouds on the surface of a given scene in real time, which also promotes the research of 3D object recognition and registration.

For 3D point cloud, there are two kinds of object recognition methods: local feature descriptors [5–7] and global feature descriptors [8–12]. Global feature descriptors describe the geometry, appearance or both of the object point cloud, which is more advantageous in object recognition and pose estimation [13]. Moreover, compared with local methods, global methods compute less descriptors and have a simpler recognition pipeline to speeding up the recognition speed [14], which is very important for applications designed to run in nearly real time [15]. The viewpoint feature histogram (VFH) [8] is a global feature descriptor which can be used for object recognition and pose estimation in a 6DOF robot grasping operation. Previous work has proved the computational efficiency and high recognition ability of this feature [16,17]. However, it is difficult for VFH to distinguish the mirrored poses of the object having the mirrored structure with respect to the viewpoint. Paper [18] proposes a modified viewpoint feature histogram (MVFH), which applies the calculation method of extended FPFH to the calculation of viewpoint component. Although this work is beneficial, it does not theoretically explain the reasons for the effect. In order to solve this difficulty, a novel global feature descriptor named orthogonal viewpoint feature histogram (OVFH) is proposed in this paper. The OVFH uses an extended fast point feature histogram (FPFH) to describe the surface shape of the object. Its viewpoint component is calculated by using the orthogonal vector of the viewpoint direction, which is calculated based on the reference frame of the object point cloud, to obtain the distinguishing ability for the mirrored poses.

The main contributions of this paper are (1) a novel and efficient global feature descriptor for object recognition and pose estimation; (2) an evaluation of the object recognition rate and the ability to distinguish the mirrored poses; (3) a visual guidance method for a robotic grasping system.

The rest of the paper is organized as follows: Section 2 introduces the proposed improved global feature descriptor OVFH. Section 3 describes the object recognition and pose estimation algorithms for robotic grasping operation. Section 4 details the experimental device and the effectiveness of the proposed algorithm in publicly available datasets and robotic grasping experiments. Finally, conclusions and future work are discussed in Section 5.

2. Improved Global Feature Descriptor

In this section, VFH is reviewed and used to derive the proposed OVFH, which is consistent with our goal of performing well in object recognition and distinguishing mirrored poses. The specific calculation procedures are detailed in the following subsections.

2.1. Global Feature Descriptor VFH

The object's global feature descriptor is a high dimensional representation of the object's 3D shape and is designed for object recognition and pose retrieval. VFH [8] is an effective point cloud feature which is used for the applications about object recognition and 6DOF pose estimation. VFH is a combined histogram containing viewpoint direction features and an extended FPFH [5], which represents the distributions of the four angles representing the geometric characteristics of the point cloud. In point cloud P, let $\boldsymbol{p}_v$ denote the position of the viewpoint, $\boldsymbol{p}_c$ denote the center of gravity of of all points $\boldsymbol{p}_i$ in the point cloud, and $\boldsymbol{n}_c$ donate the average normal vector of all normal $\boldsymbol{n}_i$ at point $\boldsymbol{p}_c$. $\boldsymbol{p}_c$ and $\boldsymbol{n}_c$ are calculated as follows:

$$\boldsymbol{p}_c = \frac{1}{n}\sum_{i=1}^{n} \boldsymbol{p}_i, \tag{1}$$

$$\boldsymbol{n}_c = \frac{1}{n}\sum_{i=1}^{n} \boldsymbol{n}_i, \tag{2}$$

For each $\boldsymbol{p}_i$ in the cloud P, a local coordinate system is defined at point $\boldsymbol{p}_c$, as in Equation (3).

$$\begin{cases} \boldsymbol{u} = \boldsymbol{n}_c \\ \boldsymbol{v} = \frac{\boldsymbol{p}_i - \boldsymbol{p}_c}{\|\boldsymbol{p}_i - \boldsymbol{p}_c\|_2} \times \boldsymbol{u} \\ \boldsymbol{w} = \boldsymbol{u} \times \boldsymbol{v} \end{cases}, \tag{3}$$

Using the local coordinate system uwv defined above, the relative deviation between the centroid vector n_c and the unit normal vector n_i of the point p_i can be represented by a set of angles, as shown in Figure 1.

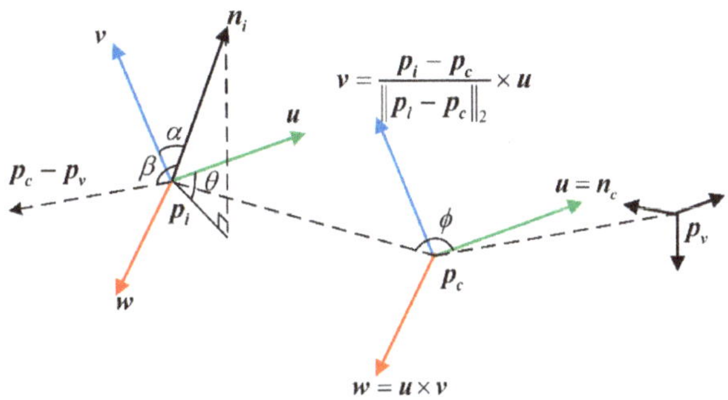

Figure 1. Relative relationship between points described by a set of angles.

The feature descriptor for each point in the point cloud can be represented by a quintuple (α, ϕ, θ, d, β), which is calculated as follows:

$$\begin{cases} \cos\alpha = v \cdot n_i \\ \cos\beta = n_i \cdot \frac{p_c - p_v}{\|p_c - p_v\|_2} \\ \cos\phi = u \cdot \frac{p_i - p_c}{d} \\ \theta = \arctan\frac{w \cdot n_i}{u \cdot n_i} \\ d = \|p_i - p_c\|_2 \end{cases}, \quad (4)$$

The percentages of the values of $\cos\alpha$, $\cos\phi$, θ, d, and $\cos\beta$ of each point in the point cloud P falling in different bins are counted, respectively corresponding to the curves on the abscissa ranges [1, 45], [46, 90], [91, 135], [136, 180], [181, 308] of the feature histogram. Since the distance d between points gradually increases along the viewpoint direction and the density of the local points will affect the feature result, d is often omitted in the 2.5-dimensional data acquired by the robot for better robustness. Finally, the VFH descriptor describes the point cloud using a total of 263 bins. Figure 2 shows an example of the VFH.

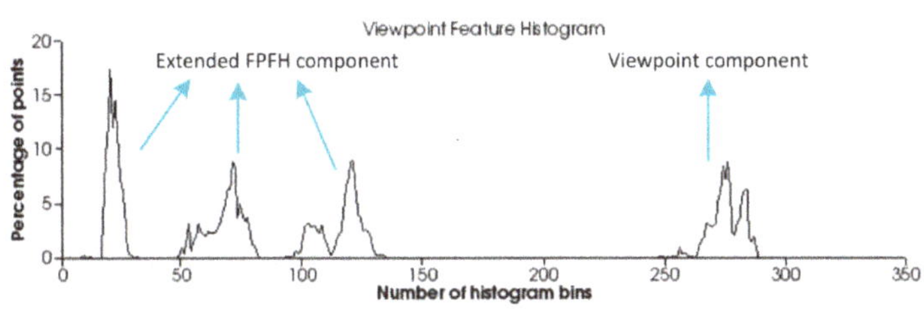

Figure 2. Viewpoint feature histogram.

Although Rusu et al. [8] have obtained promising results in using VFH as a 3D feature for object recognition and 6DOF pose estimation, they have encountered limitations of accurate 3D pose estimation. For example, if the surface of the object has mirror symmetry, it will get similar VFHs in symmetrical poses, as shown in Figure 3. The vector n_i is the normal vector at point p_i. Although the two poses are different poses mirrored with respect to the viewpoint direction, the two VFHs are highly similar because their surface normals of each point have similar or identical angular deviations from the viewpoint direction. In the case shown in the figure, the kind of object can be correctly identified using VFH, but the mirrored poses are confused.

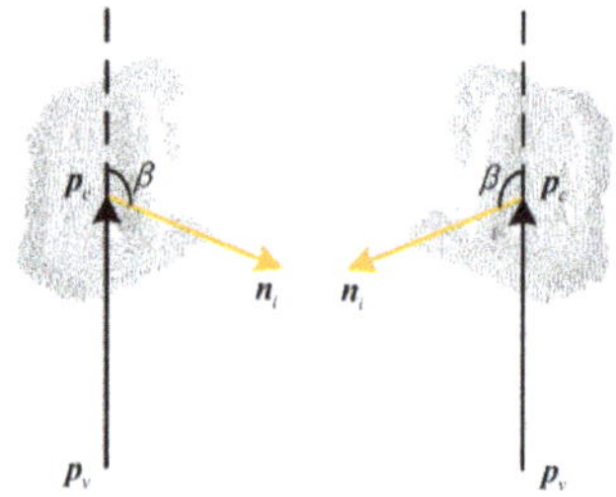

Figure 3. The same viewpoint features of viewpoint feature histograms (VFHs) in mirrored poses.

2.2. *Improved Global Feature Descriptor OVFH*

In order to overcome this drawback, it is necessary to modify the existing VFH descriptor. Chen et al. [18] have did some groundbreaking work in this area and proposed the MVFH descriptor to improve the viewpoint direction component of VFH. The MVFH gives three components to the viewpoint direction component using the method similar to estimating the extended FPFH. However, just as the geometric features characterized by the extended FPFH are very similar in the mirrored poses, the viewpoint direction component counted by this method still cannot explain theoretically how to distinguish the mirrored poses.

The core idea of solving this problem is to change the way to calculate the viewpoint direction component so that statistical angle values are different in the mirrored poses. According to this idea, an OVFH descriptor is proposed, which is described in detail as follows.

First, the reference frame of the point cloud needs to be estimated using the principal component analysis (PCA) method, which will help to compute the orthogonal vector of the viewpoint direction. All the points $\boldsymbol{p}_i$ belonging to the point cloud are given to represent the view of the object, where $i \in \{1,\ldots, n\}$. Their centroid $\boldsymbol{p}_c$ are calculated according to Equation (1) and used as the origin of object reference frame. After that, the covariance matrix C of all points is calculated by $\boldsymbol{p}_i$ and $\boldsymbol{p}_c$ as the following equation:

$$C = \frac{1}{n}\sum_{i=1}^{n} (p_i - p_c)(p_i - p_c)^T, \tag{5}$$

Then, the eigenvalue λ_j of C and its corresponding eigenvector $\boldsymbol{v}_j$ that satisfy $C\boldsymbol{v}_j = \lambda_j \boldsymbol{v}_j$, where $j \in \{1, 2, 3\}$, are computed. The eigenvector $\boldsymbol{v}_{min}$ which is corresponding to the minimum eigenvalue λ_{min} is taken as the z-axis of the reference frame. In order to eliminate the ambiguity in the z-axis direction, if the angle between $\boldsymbol{v}_{min}$ and the observation direction is in the range of [−90°, 90°], the opposite vector of $\boldsymbol{v}_{min}$ is taken. This ensures it points to the observer all the time. The eigenvector $\boldsymbol{v}_{max}$ which is corresponding to the maximum eigenvalue λ_{max} is taken as the x-axis of the reference frame. After that, the y-axis is computed by $\boldsymbol{y} = \boldsymbol{v}_{min} \times \boldsymbol{v}_{max}$. The reference frame estimated for a given partial view is shown in Figure 4.

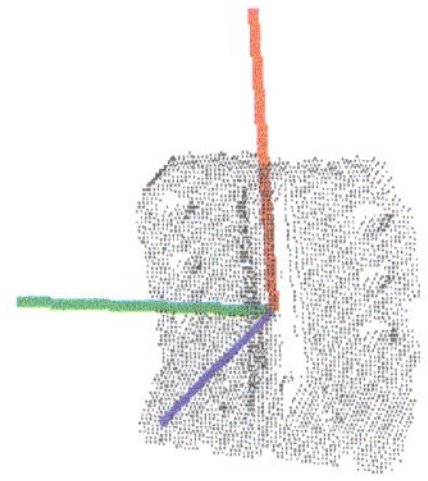

Figure 4. Reference frame estimated from three-dimensional (3D) point cloud: x-axis (red), y-axis (green), and z-axis (blue).

The z-axis pointing to the observer is obtained after determining the reference frame representing the overall point cloud of the partial view. The cross product of the z-axis and the viewpoint direction is calculated by Equation (6) to obtain an orthogonal vector of the viewpoint direction:

$$v_O = z \times (p_c - p_v), \tag{6}$$

Different from VFH [8], OVFH calculates the viewpoint component by counting a histogram, which is a statistic of the angles between the orthogonal vector of the viewpoint direction and each normal, so that the viewpoint direction component of the mirrored poses are different from each other, as shown in Figure 5. The angular deviation $\cos\beta_O$ between the orthogonal viewpoint vector and each normal n_i is calculated by Equation (7):

$$\cos\beta_O = n_i \cdot \frac{v_O}{\|v_O\|_2}, \tag{7}$$

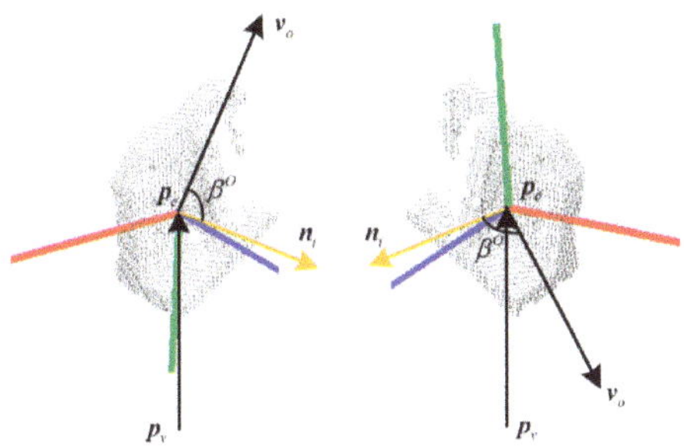

Figure 5. The different viewpoint features of orthogonal viewpoint feature histograms (OVFHs) in mirrored poses.

Using the speed and discriminative power of the FPFH to ensure the strong recognition result of the OVFH, the FPFH is extended to estimate the entire object point cloud. The difference between the normal n_i of each point p_i and the central normal n_c can be represented by $\cos\alpha_O$, $\cos\phi_O$, and θ_O, which represent relative pan, tilt, and yaw angles, respectively. They are given by the following equations:

$$\begin{cases} \cos\alpha_O = v \cdot n_i \\ \cos\phi_O = u \cdot \frac{p_i - p_c}{\|p_i - p_c\|_2} \\ \theta_O = \arctan\frac{w \cdot n_i}{u \cdot n_i} \end{cases}, \tag{8}$$

In summary, the proposed OVFH descriptor contains two components: one is the surface shape component constituted of the extended FPFH, and the other is the viewpoint component improved by the orthogonal vector of the viewpoint direction. The OVFH uses 45 bins for each value of the extended FPFH by default, and another 128 bins for the improved view component, thus, the OVFH descriptor has 263 dimensions. Figure 6 shows the principle and result of OVFH. Figure 7 shows the calculation results of VFH and OVFH in the cases that the object faces the viewpoint and deviates from the viewpoint direction by +60° and –60° yaw angle, respectively. As shown in the figure, although both VFH and OVFH assign 128 bins to encode the viewpoint direction component, the viewpoint direction information of VFH is only distributed in 64 bins. This is because the normals of the point cloud always point to the sensor, and their dot product with the central viewpoint direction must be in the range [0, 1]. The dot product of the point cloud normals and the orthogonal vector of the central viewpoint direction is in the range of [−1, 1]; thus, the viewpoint component of OVFH is distributed in all 128 bins and reserves more viewpoint information than VFH. The VFHs of mirrored poses are very similar, which may cause their corresponding poses are misidentified because of the similar match scores in the template matching phase. The OVFH descriptors of mirrored poses have distinctly different viewpoint direction components in the mirrored poses, so that the false mirrored pose can be avoided in the template matching phase.

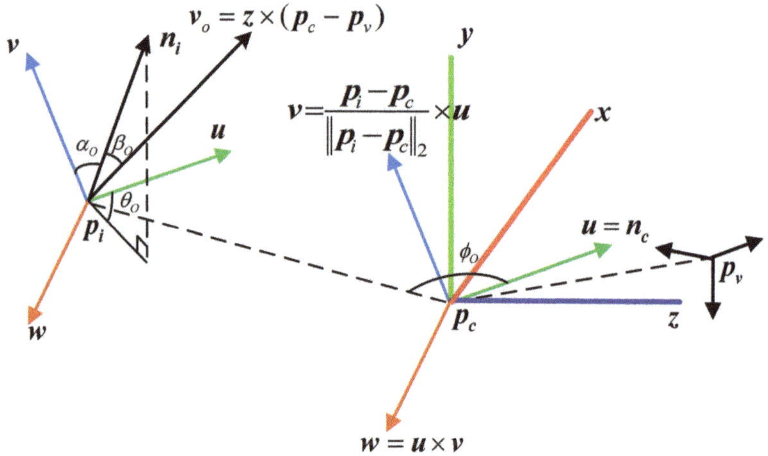

(a)

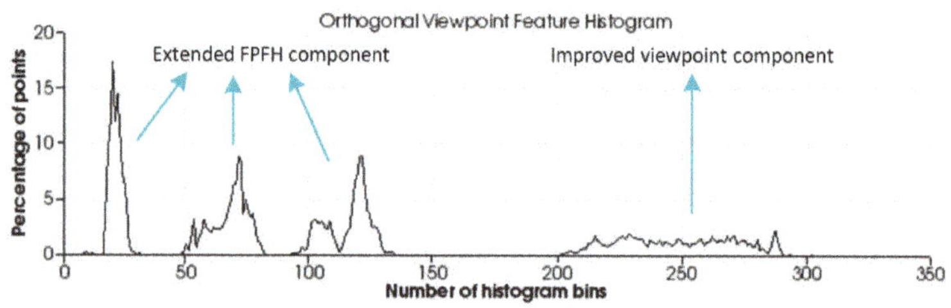

(b)

Figure 6. OVFH concept map: (**a**) improved feature description; (**b**) orthogonal viewpoint feature histogram.

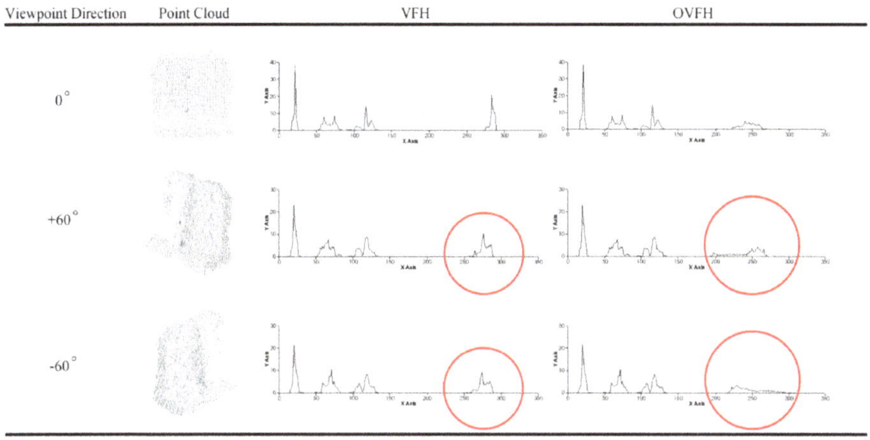

Figure 7. Comparison of two feature descriptors.

3. Visual Guidance Algorithm for the Robotic Grasping System

The robotic visual grasping algorithm includes two phases, offline and online, as shown in Figure 8. In the offline phase, a database that has complete poses of experimental objects is created by changing the sensor viewpoint. The object poses under each viewpoint are combined with the available grasping poses to teach the robot how to grasp the object in different poses. After that, all relevant information is stored in the database, including point clouds, feature descriptors, classification information, and grasping poses for each viewpoint. In the online phase, the scene point cloud is captured using a depth camera. After filtering and segmenting the scene point cloud, the global feature descriptor of the object is calculated to match the database. And the recognition result is the sample that has the most similar feature histogram with the object. Finally, the iterative closest point (ICP) algorithm [19] is used to calculate the precise pose to generate the trajectory and pose for robotic grasping operation.

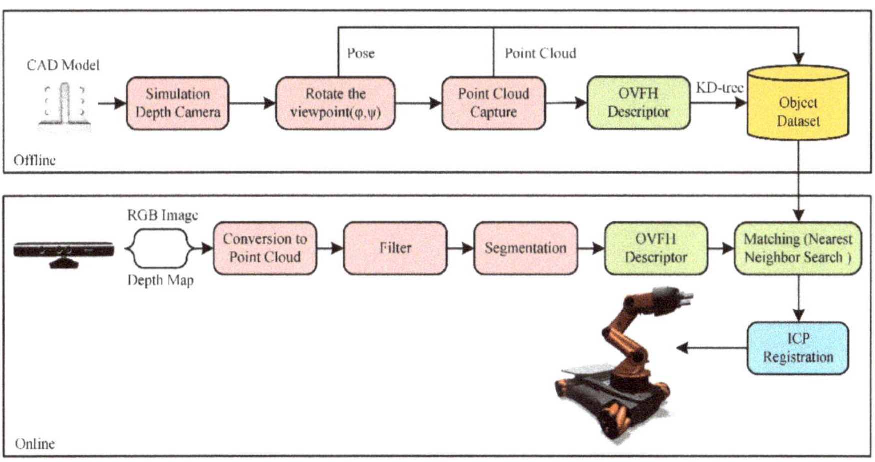

Figure 8. Algorithm for the robotic grasping system.

3.1. Creation of the Database

The database mainly consists of two parts: a multi-view point cloud database and a grasping pose database.

A multi-view point cloud database is created to contain all the poses of the object to be grabbed. Point cloud data from different perspectives can be captured by building a rotating platform such as [8], but creating a training database for a reasonable number of objects using a real device can be a cumbersome task, and even difficult if one wants to have all the different views and poses of an object. Alternatively, as described in [9], if the object has an available 3D computer aided design (CAD) model, a virtual camera can be placed directly around the object in the rendering system and all desired viewpoints can be obtained without calibrating the system and a time-consuming capture process. This is a database creation method which is low-cost and easy to extend object set; therefore, this paper uses this method to create a database.

Taking CAD model as the center of the sphere, the bounding sphere with radius r is established. A virtual depth camera [20] is set up on the viewpoints uniformly selected on the spherical surface to capture the object point clouds. As shown in Figure 9a, first, to ensure the coverage and uniformity of viewpoint selection, viewpoints are selected on the bounding sphere at every 15° yaw angle and pitch angle, denoted as (φ,ψ). Then, the virtual depth camera is set up at each viewpoint to capture the corresponding object point cloud, and record the pose data of the object model in the camera coordinate system. Finally, the point cloud data is processed offline. The OVFH descriptors of the point clouds under each viewpoint are calculated and the feature files are saved in the database.

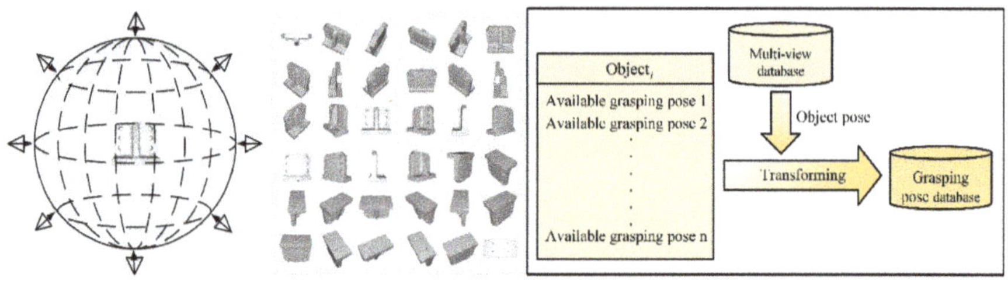

(**a**) (**b**)

Figure 9. Database creation: (**a**) multi-view point cloud database; (**b**) grasping pose database.

Objects usually have multiple stable poses. It is impossible for the robot to grip the object in the same grasping pose in any cases because of the limitation of the environment and robot's working space. Therefore, it is necessary to teach the robot how to grab objects in different poses. A plurality of stable robot grasping poses relative to the object are recorded in the database, and the robot grasping poses in different object poses can be obtained by rotating and transforming the object poses under different viewpoints in the multi-view point cloud database. Figure 9b shows the database that records the grasping poses.

3.2. Object Recognition and Pose Estimation

A 640 × 480 pixels RGB-D image captured by the Kinect v1 sensor is processed with the PCL library and converted into a point cloud containing 307,200 points (see Figure 10a). It requires initial filtering before the segmentation process. First, invalid points (NaN) that are useless for 3D processing without depth information due to factors such as specular surface, occlusion or transparency are removed. Then, a passthrough filter is used to remove all the points located outside the defined range. Experiments have shown that it is impossible to identify small objects reliably outside the range of 0.4–1.5 m away from the sensor. Therefore, the cut-off distance of the passthrough filter along the Z-axis is set as this range. The appropriate X and Y axis ranges are set to confine the isolated region of interest (ROI) to the graspable workspace of the robotic arm (see Figure 10b). Then, the random sample consensus (RANSAC) algorithm is used to detect the principal plane of the remaining point cloud and remove the inlier points of the plane, that is, to remove the ground points. Finally, the cluster of the target object is obtained by Euclidean cluster segmentation, and the noise clusters are eliminated by setting an appropriate threshold of the number of the cluster points (see Figure 10c).

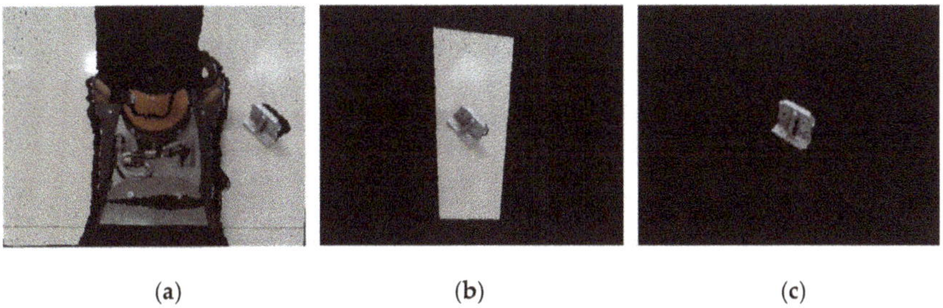

Figure 10. Point cloud pre-processing: (**a**) scene point cloud; (**b**) passthrough filter; (**c**) object segmentation.

The OVFH feature of the target point cloud is calculated after the target point cloud in the scene is obtained through the above preprocessing. The obtained OVFH descriptor is compared with the multi-view point cloud database by the k-nearest neighbor search based on K dimension (K-D) tree, and the winning result is selected as the object recognition result with rough pose estimation. Then, the point cloud of winning pose is translated to the centroid of the target object and iteratively optimized by ICP algorithm. This algorithm iteratively modifies the rigid transformation matrix between two point clouds to minimize their distance until the iteration error is less than the threshold or the current iteration number is greater than the maximum number. Figure 11 shows the registration effect of the template point cloud and the target point cloud. Finally, the robotic manipulator grasps the object based on the object refined pose after using ICP.

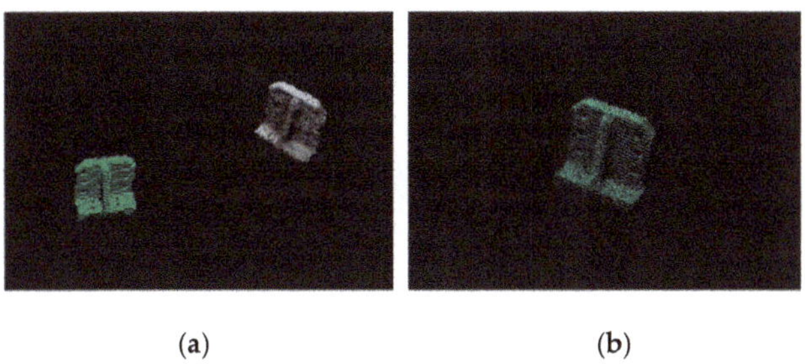

Figure 11. Registration results: (**a**) Initial relative pose; (**b**) Refined pose after the iterative closest point (ICP).

4. Experimental Results

4.1. Experimental Results on the Data Set

The hardware used in the evaluation experiment was a computer with Intel Core i5-7500 CPU @ 3.40 GHz processor and 16 GB RAM. In order to prove that the proposed descriptor OVFH is improved in pose retrieval compared with VFH, a publicly available dataset [21] was used for test experiments. Each object in the data set has 600 point clouds which are captured from five polar angles and 120 turntable positions, with an azimuth equidistance of 3°. The 12 objects shown in Figure 12 were selected from the BIGBIRD data set for a comparative experiment of pose retrieval using VFH and OVFH. A complete pose database ($12 * 5 * 24 = 1440$ in total) was created by selecting poses at 24 equally spaced 15° azimuths from each polar angle of each object. In order to verify the pose identification ability of OVFH and VFH, 24 point clouds with mirrored poses of each object were selected as the test set. Figure 13 shows the object recognition accuracy using two kinds of feature descriptors. As far as the test was concerned, OVFH had similar accuracy to VFH in object recognition. The average recognition rates of OVFH and VFH were both 94.44% for all 12 objects. Figure 14 shows the mirrored pose distinction accuracy using two kinds of feature descriptors. For mirrored poses, the average distinction rate of OVFH (95.49%) was significantly higher than that of VFH (82.6%). Table 1 presents the computation time for each procedure when using two kinds of descriptors to test. Although OVFH descriptor required an additional step of reference frame estimation, OVFH could avoid false pose recognition and provide a better initial pose for ICP, which greatly reduced the time consuming of refining pose. Table 1 shows that the average computation time was 856.866 ms when object recognition and rough pose estimation were performed using VFH and the pose was further refined using ICP. OVFH reduced the average computation time to 546.212 ms because reducing the number of ICP iterations. The distance root mean squared error (RMSE) of corresponding point pairs in the point cloud registration process is used to describe the error of the pose estimation. It can be seen from the Table 1 that OVFH can obtain higher average precision of pose estimation by providing better initial values for ICP. From Figures 13 and 14 and Table 1, OVFH enhances the distinction capability of the mirrored pose while preserving the identification capability of VFH, and is more computationally efficient and precise in the accurate pose estimation combined with the ICP. Experimental results show that the proposed feature (OVFH) can be used to improve the performance of pose retrieval.

Figure 12. 12 tested objects.

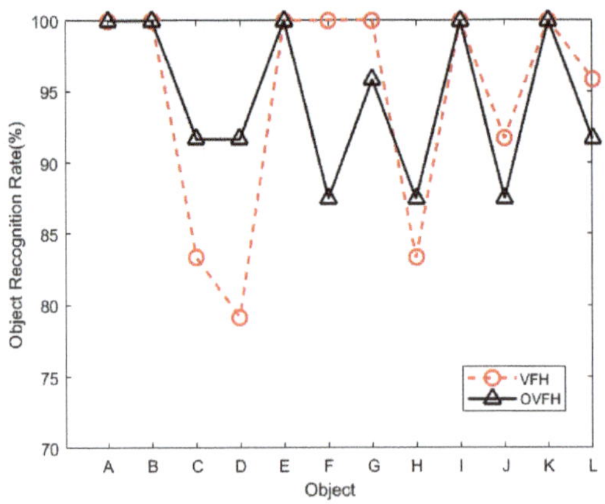

Figure 13. Object recognition rate.

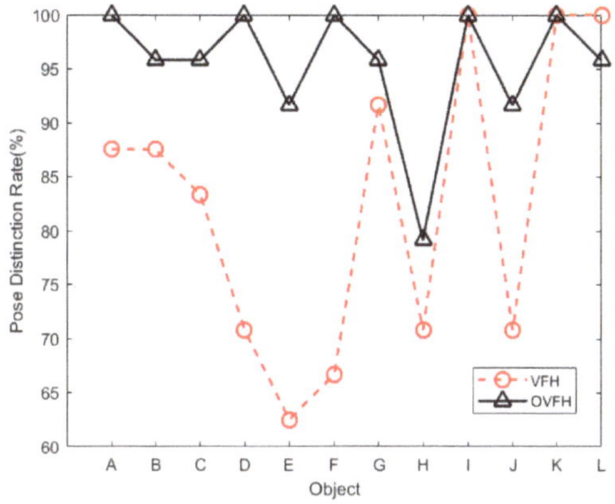

Figure 14. Mirrored pose distinction rate.

Table 1. Computation time for processing a single object instance.

Method	Procedure	Average Time (ms)	Average RMSE (m)
VFH + ICP	Description	1.694	3.219×10^{-5}
	Pose refinement	855.172	
OVFH + ICP	Reference frame estimation	10.767	1.391×10^{-5}
	Description	1.832	
	Pose refinement	533.613	

4.2. Robotic Grasping Experiment

Figure 15 shows the hardware setup of the robotic grasping experiment. A KUKA Youbot robot was used to grasp objects, and Microsoft Kinect v1 was mounted on the robot's stand to capture point clouds for robotic vision. The computer used for object recognition and pose estimation was equipped with an Intel i5 CPU and 16 GB RAM. Eight objects used for the grasping experiment are shown in Figure 16, where objects A-E have mirrored structure. A multi-view point cloud database was built by the CAD models of all experimental objects, and the robot was taught how to capture objects of different poses.

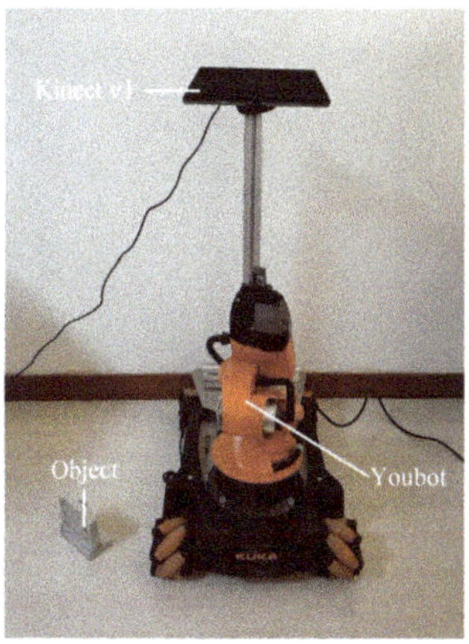

Figure 15. The hardware setup of the robotic grasping experiment.

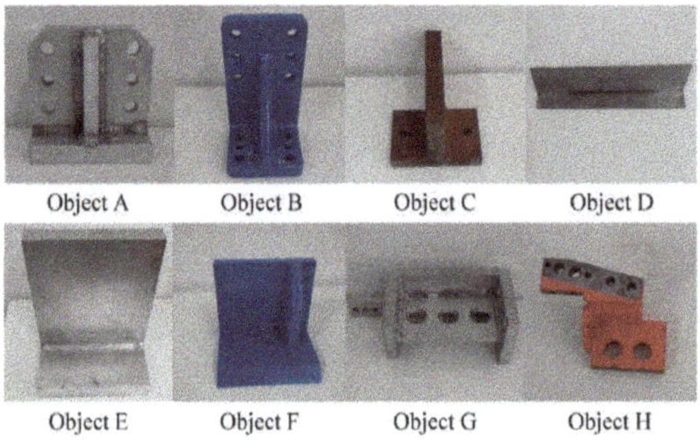

Figure 16. The objects used in robotic grasping experiment.

Experiments were performed on eight objects, each of which was set to 10 initial poses in the robot's workspace. Figure 17 shows the results of object recognition and pose estimation. As shown in the figure, the point cloud collected online was precisely registered with the recognition result in database. To test the effectiveness of the proposed OVFH descriptor, the first nine matching scores were used to determine the distinction capability of the mirrored poses, as shown in Figure 18. Figure 18a,b, respectively, show the results of object recognition using the VFH and OVFH descriptor, when the object A was placed in a pose of −60° relative to the viewpoint. The result in the lower left corner was the best match. When using VFH, the true pose and its mirrored poses were alternately arranged, and a false positive was generated for the mirrored pose. When using OVFH, the true pose was correctly identified, and the matching scores of the mirror poses and the true pose were quite different, which would avoid the false positive effectively. The experimental results of eight objects are recorded in Table 2. For objects F-H without mirrored structure, VFH and OVFH exhibited similar performance. For objects A-E with mirrored structure, VFH could not distinguish its mirror poses. Identifying the wrong initial pose would increase the registration time of ICP and the pose estimation error, and even lead to registration failure. Since OVFH avoided false pose Identification, the convergence was faster when using ICP to refine the pose, and the average computation time was reduced to 0.523 s. At the same time, correct pose recognition made the model point cloud and the target point cloud better fit, thus obtaining higher pose estimation accuracy. After object recognition and registration, the refined pose was used to guide the robot to grasp the object. Figure 19 shows the grasping experiment results of an object with mirrored poses.

Figure 17. The result for object recognition and registration.

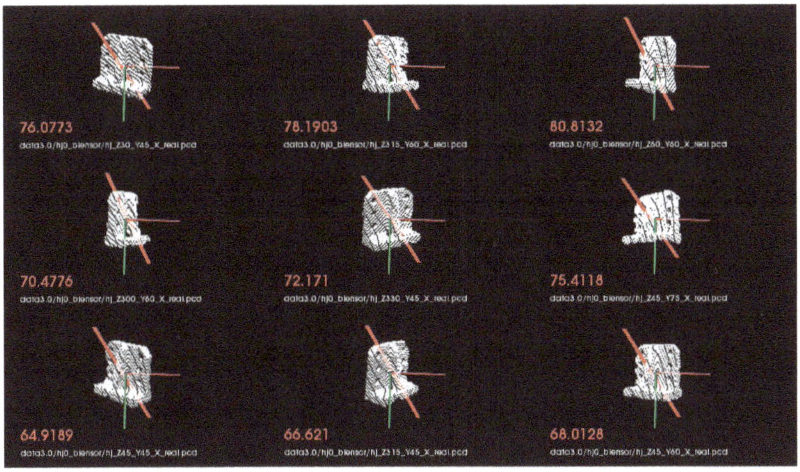

(a)

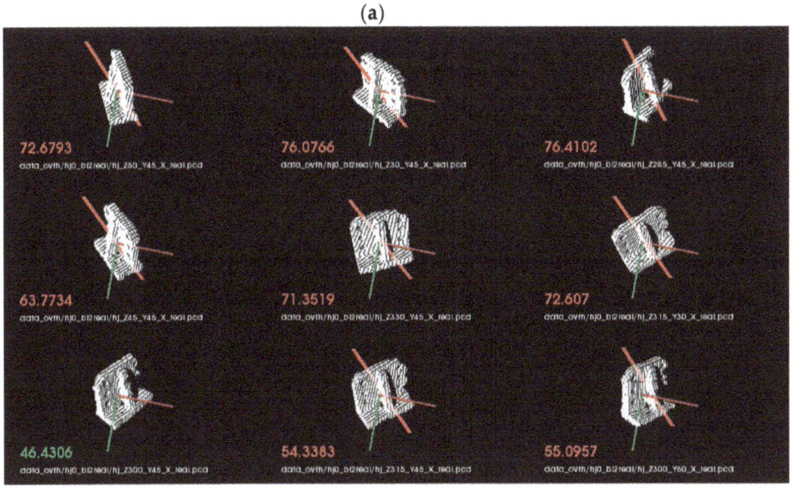

(b)

Figure 18. Matching results for (**a**) VFH descriptor and (**b**) OVFH descriptor.

Table 2. Experimental results of eight objects.

Object	VFH + ICP			OVFH + ICP		
	Pose Distinction Rate (%)	Average Time (s)	Average RMSE (m)	Pose Distinction Rate (%)	Average Time (s)	Average RMSE (m)
A	60	0.878	3.388×10^{-5}	100	0.428	1.543×10^{-5}
B	70	0.973	6.895×10^{-5}	100	0.574	2.258×10^{-6}
C	90	0.676	1.674×10^{-5}	100	0.471	2.852×10^{-6}
D	70	0.747	1.044×10^{-5}	90	0.338	2.994×10^{-6}
E	80	1.079	6.611×10^{-5}	100	0.775	2.415×10^{-6}
F	100	0.501	4.317×10^{-6}	100	0.494	4.062×10^{-6}
G	100	0.672	6.213×10^{-6}	100	0.647	6.429×10^{-6}
H	100	0.446	5.379×10^{-6}	100	0.457	5.240×10^{-6}
Average Value	83.75	0.747	2.650×10^{-5}	98.75	0.523	5.210×10^{-6}

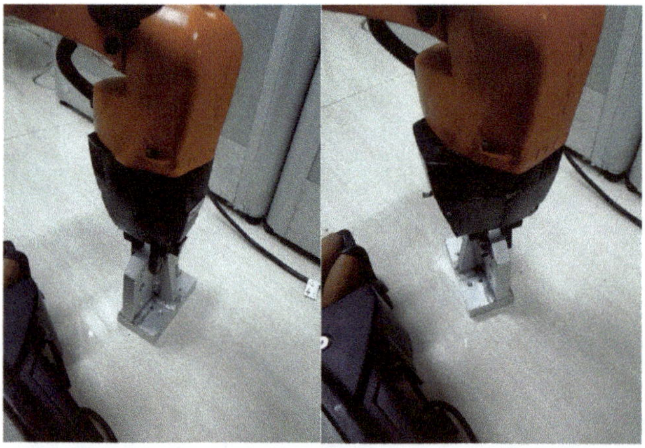

Figure 19. The grasping results for object with mirrored poses.

5. Conclusions and Future Work

In order to correctly distinguish the mirrored poses relative to the viewpoint, an effective global feature descriptor OVFH is proposed in this paper, which was successfully applied to object recognition and pose estimation. The proposed method computes an orthogonal vector of the viewpoint direction by using a reference frame estimated for the entire point cloud. This orthogonal vector is used to improve the viewpoint component of the feature descriptor. Experimental results in public data set show that OVFH descriptor can characterize object poses well and enhance the ability to distinguish mirrored poses. Based on OVFH descriptor, an object recognition and pose estimation method for vision-guided robotic grasping system is designed. The experimental results show that the proposed vision-guided robotic grasping method can effectively distinguish the mirrored poses and guide the robot to grasp different objects.

For future work, the proposed feature descriptor can be extended by color description to obtain the ability to recognize objects with the same geometries and different patterns. It will also be studied to combine the proposed idea of calculating orthogonal viewpoint direction with other feature descriptors to obtain better results.

Author Contributions: F.W. contributed the scientific issues and the research ideas; C.L. designed and performed the experiments; H.C. provided hardware support and supervised experiments; C.L. and C.R. wrote and revised the paper.

Funding: This work was supported in part by the Fundamental Research Funds for the Central Universities of China under Grant N172608005, N182608003 and N182612002, Liaoning Provincial Natural Science Foundation of China under Grant 20180520007.

Conflicts of Interest: The authors declare no conflict of interest.

References

1. Luo, R.C.; Kuo, C.W. A scalable modular architecture of 3D object acquisition for manufacturing automati on. In Proceedings of the 2015 IEEE 13th International Conference on the Industrial Informatics (INDIN), Cambridge, UK, 22–24 July 2015; pp. 269–274.
2. Rusu, R.B.; Marton, Z.C.; Blodow, N.; Beetz, M. Learning informative point classes for the acquisition of object model maps. In Proceedings of the 10th International Conference on Control, Automation, Robotics and Vision (ICARCV), Hanoi, Vietnam, 17–20 December 2008.
3. Canny, J.F. Finding Edges and Lines in Images. Master's Thesis, Massachusetts Institute of Technology, Artificial Intelligence Laboratory, Cambridge, MA, USA, 1983.

4. Lowe, D.G. Distinctive image features from scale-invariant keypoints. *Int. J. Comput. Vis.* **2004**, *60*, 91–110. [CrossRef]
5. Rusu, R.B.; Blodow, N.; Beetz, M. Fast point feature histograms (FPFH) for 3D registration. In Proceedings of the IEEE International Conference on Robotics and Automation, Kobe, Japan, 12–17 May 2009; pp. 3212–3217.
6. Tombari, F.; Salti, S.; Stefano, L.D. A combined texture-shape descriptor for enhanced 3D feature matching. In Proceedings of the 2011 18th IEEE International Conference on Image Processing, Brussels, Belgium, 11–14 September 2011; pp. 809–812.
7. Nascimento, E.R.; Oliveira, G.L.; Campos, M.F.M.; Vieira, A.W.; Schwartz, W.R. Brand: A robust appearance and depth descriptor for rgb-d images. In Proceedings of the 2012 IEEE/RSJ International Conference on Intelligent Robots and Systems, Vilamoura, Portugal, 7–12 October 2012; pp. 1720–1726.
8. Rusu, R.B.; Bradski, G.; Thibaux, R.; Hsu, J. Fast 3d recognition and pose using the viewpoint feature histogram. In Proceedings of the 2010 IEEE/RSJ International Conference on the Intelligent Robots and Systems (IROS), Taipei, Taiwan, 18–22 October 2010; pp. 2155–2162.
9. Aldoma, A.; Vincze, M.; Blodow, N.; Gossow, D.; Gedikli, S.; Rusu, R.B.; Bradski, G. Cad-model recognition and 6dof pose estimation using 3D cues. In Proceedings of the 2011 IEEE International Conference on Computer Vision Workshops (ICCV Workshops), Barcelona, Spain, 6–13 November 2011; pp. 585–592.
10. Aldoma, A.; Tombari, F.; Rusu, R.B.; Vincze, M. Our-cvfh–oriented, unique and repeatable clustered viewpoint feature histogram for object recognition and 6dof pose estimation. Proceedings of Pattern Recognition: Joint 34th DAGM and 36th OAGM Symposium, Graz, Austria, 28–31 August 2012; pp. 113–122.
11. Aldoma, A.; Tombari, F.; Prankl, J.; Richtsfeld, A.; Stefano, L.D.; Vincze, M. Multimodal cue integration through hypotheses verification for rgb-d object recognition and 6dof pose estimation. In Proceedings of the 2013 IEEE International Conference on Robotics and Automation, Karlsruhe, Germany, 6–10 May 2013; pp. 2104–2111.
12. Lima, J.P.S.D.M.; Teichrieb, V. An efficient global point cloud descriptor for object recognition and pose estimation. In Proceedings of the 2016 29th SIBGRAPI Conference on Graphics, Patterns and Images (SIBGRAPI), Sao Paulo, Brazil, 4–7 October 2016.
13. Zhang, K.; Zhang, L. 3D Object recognition and 6DoF pose estimation in scenes with occlusions and clutter based on C-SHOT 3D descriptor. *J. Comput. Aided Des. Comput. Graph.* **2017**, *29*, 846–853. (In Chinese)
14. Shan, S.A.A.; Bennamoun, M.; Boussaid, F. Iterative deep learning for image set based face and object recognition. *Neurocomputing* **2016**, *174*, 866–874.
15. Nafouki, K. *Object Recognition and Pose Estimation from An Rgb-D Image*; Technical Report; Technical University of Munich: Munich, Germany, 2016.
16. Alodma, A.; Marton, Z.; Tombari, F.; Wohlkinger, W.; Potthast, C.; Zeisl, B.; Rusu, R.B.; Gedikli, S.; Vincze, M. Tutorial: Point cloud library: Three-dimensional object recognition and 6 dof pose estimation. *IEEE Robot. Autom. Mag.* **2012**, *19*, 80–91. [CrossRef]
17. Luo, J.; Jiang, M. Object recognition method based on RGB-D image kernel descriptors. *J. Comput. Appl.* **2017**, *37*, 255–261. (In Chinese)
18. Chen, C.-S.; Chen, P.-C.; Hsu, C.-M. Three-dimensional object recognition and registration for robotic grasping systems using a modified viewpoint feature histogram. *Sensors* **2016**, *16*, 1969. [CrossRef] [PubMed]
19. Besl, P.J.; Mckay, N.D. A method for registration of 3-D shapes. *IEEE Trans. Pattern Anal. Mach. Intell.* **1992**, *14*, 239–256. [CrossRef]
20. Gschwandtner, M.; Kwitt, R.; Uhl, A.; Pree, W. BlenSor: Blender sensor simulation toolbox. In Proceedings of the International Conference on Advances in Visual Computing, Las Vegas, NV, USA, 26–28 September 2011; Springer: Berlin, Germany, 2011; pp. 199–208.
21. BigBIRD: (Big) Berkeley Instance Recognition Dataset. Available online: http://rll.berkeley.edu/bigbird/ (accessed on 12 January 2019).

sensors

MDPI

Article

Vision for Robust Robot Manipulation

Ester Martinez-Martin [1,*] and Angel P. del Pobil [2,3,*]

1 RoViT, University of Alicante, 03690 San Vicente del Raspeig (Alicante), Spain
2 RobInLab, Jaume I University, 12071 Castello de la Plana, Spain
3 Interaction Science Dept., Sungkyunkwan University, Jongno-Gu, Seoul 110-745, Korea
* Correspondence: ester@ua.es (E.M.-M.); pobil@uji.es (A.P.d.P.)

Received: 24 December 2018; Accepted: 3 April 2019; Published: 6 April 2019

Abstract: Advances in Robotics are leading to a new generation of assistant robots working in ordinary, domestic settings. This evolution raises new challenges in the tasks to be accomplished by the robots. This is the case for object manipulation where the detect-approach-grasp loop requires a robust recovery stage, especially when the held object slides. Several proprioceptive sensors have been developed in the last decades, such as tactile sensors or contact switches, that can be used for that purpose; nevertheless, their implementation may considerably restrict the gripper's flexibility and functionality, increasing their cost and complexity. Alternatively, vision can be used since it is an undoubtedly rich source of information, and in particular, depth vision sensors. We present an approach based on depth cameras to robustly evaluate the manipulation success, continuously reporting about any object loss and, consequently, allowing it to robustly recover from this situation. For that, a Lab-colour segmentation allows the robot to identify potential robot manipulators in the image. Then, the depth information is used to detect any edge resulting from two-object contact. The combination of those techniques allows the robot to accurately detect the presence or absence of contact points between the robot manipulator and a held object. An experimental evaluation in realistic indoor environments supports our approach.

Keywords: robotics; robot manipulation; depth vision

---

1. Introduction

Advances in Robotics are leading to a new generation of assistant robots working in ordinary domestic settings, such as healthcare and rehabilitation [1,2], agriculture [3], emergency situations [4,5], or guidance assistance [6]. In this context, the ability to autonomously manipulate objects is of critical importance. Though there exist a wide research on robot grasping (e.g., Refs. [7–11]), it is mainly focused on object location, along with motion and grasp planning. Only a few efforts have been devoted to monitoring the grasp action for error recovery, an issue that is, however, crucial to achieve the required level of autonomy in the robotic system.

Along this line, a state-of-the-art solution is to equip the robot gripper with tactile sensors. In this way, the presence or absence of a grasped object can be easily perceived through pressure distribution measure or contact detection [12,13]. For that reason, a wide variety of tactile sensors for robot hands have been developed [14]. However, the existing tactile technologies have multiple limitations. First, most of the existing sensors are too bulky to be used without sacrificing the system dexterity. Another reason is that they are too expensive, slow, fragile, sensitive to temperature, or complex to manufacture. They may also lack elasticity, mechanical flexibility or robustness. Therefore, it is necessary to have an alternative or complementary sensing approach to robustly detect errors in object grasping.

Alternatively, information about joint position, joint velocity or joint torque (*proprioception*), has been often used for robot grasping [15,16]. Nevertheless, the grasp stability may be affected by several parameters such as the configuration of the robotic gripper, the (mis)alignment of the joint axes, or inaccurate data (e.g., open/close instead of the exact grip aperture). These drawbacks limit the suitability of this approach for service robots.

As a solution, we propose to use computer vision since it can provide more accurate information than other robot sensors. Thus, the evaluation of a manipulation action may be mediated by a proper recognition of both the gripper and the held object. To the best of our knowledge, no other approach exists in which vision is used for error detection after an attempt to pick up an object. For instance, taking the Amazon Picking Challenge as a test case, none of the over 60 teams that participated in its three editions (2015–2017) reported the use of vision for detecting grasping errors [17,18]. Often grasping errors were not detected at all or error detection was based on a vacuum sensor when a suction cup was used [19], as well as weight checking [20].

A wide range of approaches for gripper and/or object recognition varying in complexity and functionality can be found in the literature. Currently, the most popular approach is *deep learning* [21–26]. This approach could be described as computational models composed of multiple processing layers that allows it to learn representations of data with multiple levels of abstraction. Nevertheless, as a training stage is required, all the manipulated objects (including the robot gripper) must be known in advance. In addition, the use of elastically deformable objects or grippers can lead to a failure of this approach since a sufficiently large number of visual appearances may not be available for system training. Furthermore, the high requirements of current deep learning solutions in terms of memory and computational resources make it infeasible for robot tasks.

With the purpose of real-time operation, visual local features could be used. One of the most implemented technique is SIFT [27,28]. This approach shares many features with neuron responses in primate vision. Basically, SIFT transforms visual input into linear scale-invariant coordinates that are relative to local features. In this way, an object can be located in an image that contains many other objects. The main drawback of this approach (and its alternatives [29–31]) is that a certain amount of texture in the objects to be detected is required, a requirement that cannot be always guaranteed in ordinary, domestic settings. Moreover, the grasping action may result in a great object occlusion making the object visually undetectable.

In this context, traditional Computer Vision techniques could fit since they allow us to extract simple image features like colour or shape that can be used for a proper robot gripper monitoring. In particular, similarly to the human vision system, this paper proposes a technique to combine simple visual features (e.g., motion, orientation, colour, etc.) for gripper monitoring. More specifically, edge, depth and colour are properly combined to detect a contact between a robot gripper and any grasped object.

This paper is organised as follows: Section 2 overviews the robot grasp task, while Section 3 introduces our approach for grasping monitoring. Experimental results are presented and discussed in Section 4. Finally, conclusions and future work can be found in Section 5.

2. The Grasping Task

Any grasping task involves a device to hold and manipulate objects that can be in the form of simple grippers or highly dexterous robotic hands (see Figure 1 for some examples). So, these devices have evolved according to the Robotics demands. Firstly, the two-finger grippers were designed to satisfy the industrial assembly needs. From that starting point, different designs have been proposed in the literature to properly fulfill robot service tasks. In addition, the wide variety of objects to deal with has also led to the use of different materials allowing the robot manipulator to flexibly adapt itself to the most varied shapes (see Figure 2). This flexibility results in a deformation (sometimes permanent) and, as a consequence, recognizing the gripper turns into a much more difficult task. In addition, techniques based on a model of the gripper or its shape become impractical due to the complexity in modelling the many different ways a gripper or its fingers can deform.

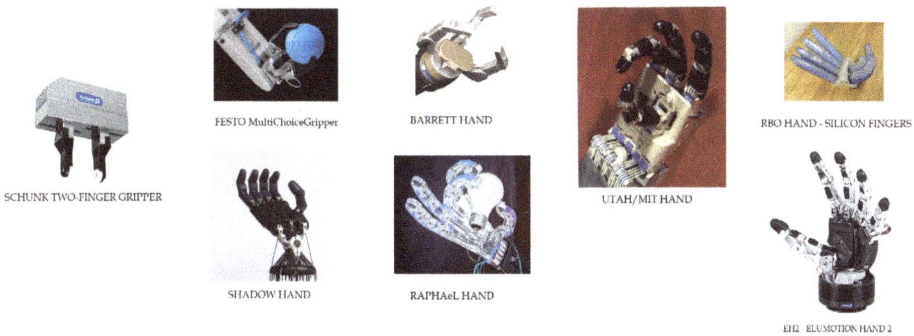

Figure 1. A sample of the evolution of robotic manipulators.

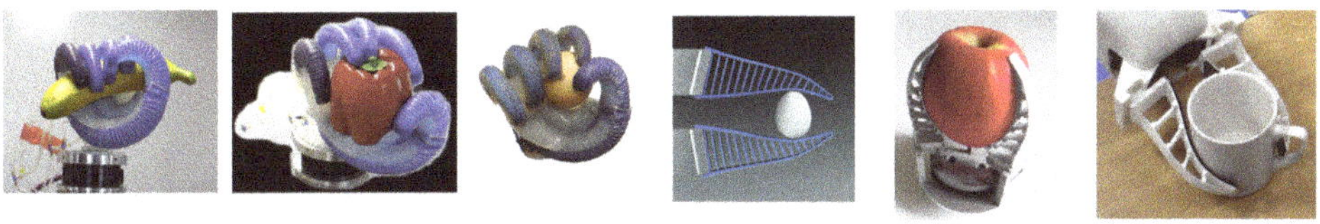

Figure 2. A sample of deformations when flexible robotic grippers grasp an object.

For that reason, an abstraction is required. Generally speaking, a *grasp* can be defined as a set of contacts between a robot manipulator and the surface of any held object (see Figure 3). From this definition, the grasping action could be detected as the contact between the object and the manipulator. Therefore, a solution could be to properly detect both the object and the manipulator and find their contact points. However, there are several issues to be overcome such as detecting them in different environments, the wide variety of objects (some of them could be quite similar to the others), and a great manipulator diversity. In addition, using only an RGB input can lead to *tricky* situations where the manipulator and the object are not in contact, but the visual system may wrongly identify contact points. As illustrated in Figure 4, given the visual alignment between the robotic manipulator and the object, the robot may be unable to distinguish if they are in contact or not. What is more, colour-based object recognition highly depends on illumination conditions; so, with the purpose of reducing its influence, different colour models have been investigated in terms of sensitivity to image parameters [32]. From this study, the *Lab* colour space is the best alternative due to its invariance under different conditions.

Figure 3. Contact points resulting from robotic grippers grasping an object.

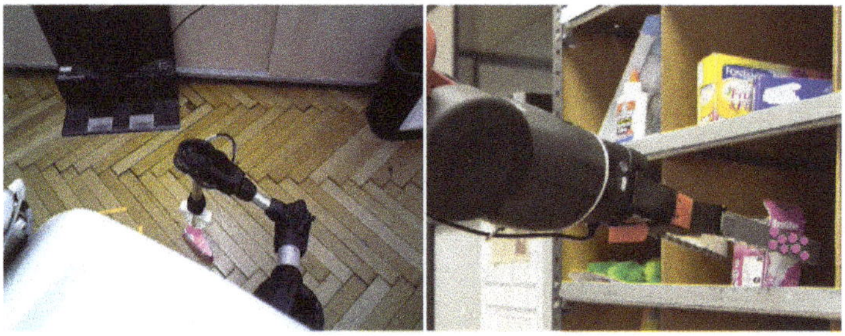

Figure 4. *Tricky* RGB situations of non-grasping contact points where a visual aligment between an object and the robotic manipulator can be confusing.

Note that the colour coordinates are experimentally set for each robotic manipulator. For that, several images under different environmental conditions (five images in our experiments) are required to properly adjust the *Lab* range. However, a colour-based segmentation extracts all the elements within the scene satisfying those colour coordinates, as illustrated in Figure 5. Thus, more information is required to properly identify the robot gripper so that a robust detection of grasp contact points is achieved and, as a consequence, the grasping action itself is more dependable. In this paper, we propose to fuse *Lab* data with depth information to achieve this goal. This data could be obtained from an RGB-D camera, a popular device in the last years due to its low price and the enriched information it provides. As explained in the following section, this sensory fusion also solves the detection of the gripper and the object.

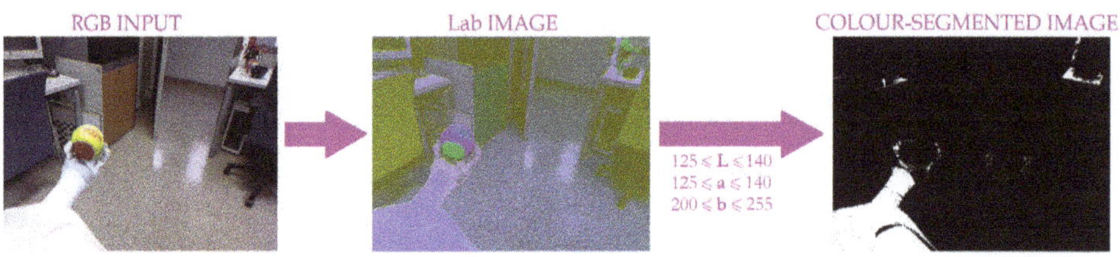

Figure 5. Image segmentation based on *Lab* colour model.

3. Grasp Monitoring

As mentioned above, the proposed approach is based on the fusion of two visual inputs: *RGB* and depth. So, *RGB* information provides an early, coarse image segmentation. As previously shown in Figure 5, the *RGB* input is first converted into its corresponding *Lab* image. Then, a segmentation based on *Lab* gripper coordinates is applied. Given that real environments are considered, several elements could present the same colour distribution and, as a consequence, they also appear in the segmentation result. This is the case of Pepper's robot that is homogeneously coloured and consequently, all the robot parts are present in the colour-based segmentation result as depicted in Figure 5. For that reason, an additional cue is required to properly identify the robot gripper and, therefore, the grasping task.

In this sense, depth data has been used to overcome the colour segmentation issues. Thus, on the one hand, the depth cue provides information about an object's position with respect to its neighbours. This allows the robot to robustly detect the contact points (or their absence) between the scene objects. In this way, the real contact points can be properly identified based on the depth difference between two touching objects. Nonetheless, this approach detects any contact point between two objects. So, for instance, apart from the grasping contact points, it obtains the contact points between a table and any object on it, those between two objects in touch or overlapping, or even the contact points between different parts of the same object, as shown in Figure 6. Due to the sensor limitation, there is a noteworthy amount of pixels without depth information. For example, too close pixels like the robot's body, are missing in the depth map. In addition, other visual objects are *vanished* as it is the case of the door. As a result, the number of contact points is reduced although more information is necessary to accurately isolate gripper-object contact points.

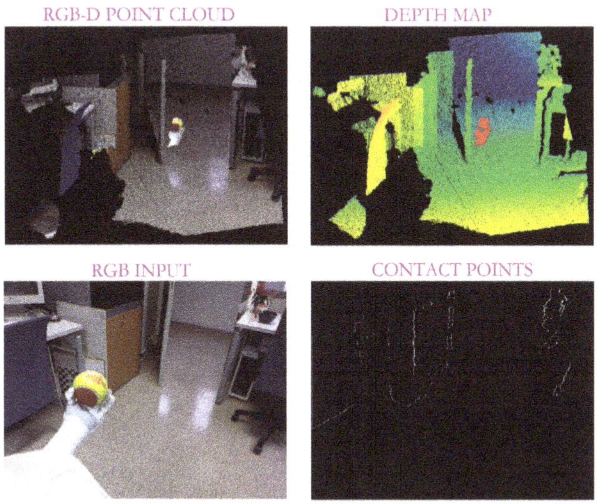

Figure 6. Contact points detected by the depth-based map.

To overcome this difficulty, the gripper recognition is applied to properly identify the grasping contact points and, consequently, evaluate the robot grasping task and detect its possible errors. With that aim, a contour extraction is performed, that is, the contours are obtained from depth changes. A pixel is classified as a contour when there is a leap between the depth information for that pixel and one of its neighbours. In our case, that jump was limited to 0.01 depth units (approximately 1 cm). Note that to achieve this, a critical issue is the missing depth points mainly resulting from the distance with respect to the sensor and the object's thinness. As a solution, the border pixels in terms of presence/absence of information have been also considered as contours (see Figure 7).

Figure 7. Results of our edge detection from depth information such that the bottom centre column represents the first contour segmentation, while the last one shows the contour segmentation after refinement.

This fact leads to all the object's contours in the scene. Consequently, an edge refinement is necessary to adequately isolate the robot gripper. Given that the vision system is always located at the top of the robot and looking ahead, the robot actuator contour emerges from the bottom part of the image. Therefore, all the contours out of the image bottom are discarded as shown in Figure 7.

Once the contours are obtained, they are combined with the colour segmented image. In this way, the gripper is properly identified within the visual scene. The last step is to check the presence or absence of contact points with a held object. For that, only the objects contained between the robot *fingers* are considered.

Therefore, the whole approach combines all the abovementioned methods to properly check the grasping status at any time. So, as illustrated in Figure 8 and sketched in Algorithm 1, our approach concurrently performs three raw segmentations: the first is based on the *Lab* gripper components; the second obtains all the contact points between two objects separated by less than 5 cm, while the last one outputs an image with all the object contours. As all the object contours are obtained, the last segmentation is refined such that only the ones that start at the bottom of the image are considered. This information, together with the colour segmentation, allows the system to properly isolate the robot gripper. Finally, the overlap between this last image and the raw contact points segmentation provides the robot with the information about the presence or absence of contact points and, consequently, the status of the grasping task. Note that the proposed approach only depends on two parameters: the *Lab* components, corresponding to the robot gripper; and the depth threshold. So, on the one hand, the *Lab* components are defined by an interval of values for each component obtained from a Lab-component analysis of the robot gripper under different illumination conditions. On the other hand, the depth threshold must be set from camera information such that it approximately corresponds to 1 cm.

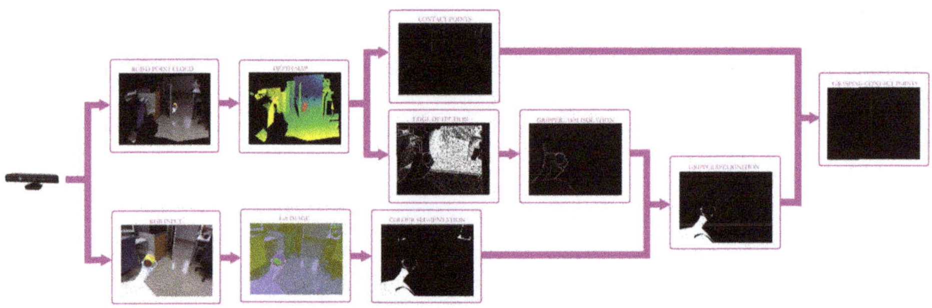

Figure 8. Our approach flowchart for robust grasping monitoring.

Algorithm 1: Our grasping monitoring approach.

```
Data: RGB-D image
Result: a boolean indicating an object is grasped
nContactPoints ← 0;
for each pixel do
    LAB components are obtained from RGB coordinates;
    if L_min ≤ L_pixel ≤ L_max && a_min ≤ a_pixel ≤ a_max && b_min ≤ b_pixel ≤ b_max then
        LAB_segmented ← 255;
    else
        LAB_segmented ← 0;
    end
    if Depth_pixel is NaN then
        Edge_detection ← neighbourhood_classification;
    else
        if distance(Depth_pixel, Depth_neighbourhood) ≤ Depth_threshold then
            Edge_detection ← 255;
        else
            Edge_detection ← 0;
        end
    end
    if Edge_detection && Bottom_Edge then
        Arm_detection ← 255;
    else
        Arm_detection ← 0;
    end
    if Arm_detection && Depth_pixel ≤ Contact_threshold then
        Contact_points ← 255;
    else
        Contact_points ← 0;
    end
    if LAB_segmented && Contact_points then
        nContactPoints ← nContactPoints + 1;
    end
end
```

4. Experimental Results

With the purpose of validating our approach, three different robot platforms have been used: the *Baxter* robot [33], the *Pepper* robot [34] and the *Hobbit* robot [35] (see Figure 9). The *Baxter* robot is a two-armed robot designed for industrial automation. On the contrary, the *Pepper* and *Hobbit* robots are social platforms designed to interact with people. So, *Pepper* is a commercial semi-humanoid robot being adapted to several applications like a guide assistant, while the *Hobbit* is a socially assistive robot aimed at helping seniors and elderly people at home. All these robot platforms are endowed with multiple sensors, providing the robot with perceptual data, and actuators, allowing the system to perform its tasks.

Figure 9. The three robot platforms used to evaluate the performance of our approach: the *Baxter* robot [33] (left), the *Pepper* robot [34] (center) and the *Hobbit* robot [35] (right).

There are several differences between them to be taken into account for grasping tasks. On the one hand, the robot gripper is quite different in each robot. In particular, *Baxter* is provided with a parallel jaw gripper intended to perform industrial tasks such as packaging, material handling or machine tending. On its behalf, *Pepper* emulates a human hand with a five-finger gripper, whereas *Hobbit* is endowed with a gripper based on FESTO *Fin Ray Effect*; in this design, the two soft, triangular fingers with hard crossbeams can buckle and deform to conform around grasped objects. This allows us to evaluate the performance of our approach not only with rigid grippers but also with continuously shape-changing grippers.

On the other hand, the camera location varies between the platforms. Indeed, the visual input is provided by a pan-tilt RGB-D camera (i.e., Microsoft Kinect) mounted on the *head* of each robot and, therefore, it is approximately located at a height of 160 cm (*Baxter*), 110 cm (*Pepper*), and 130 cm (*Hobbit*).

With the aim to accurately evaluate the approach performance, the three robots were located at different unstructured scenarios (seven in total) carrying out different tasks. So, *Baxter* is performing a pick-and-place task (see Figures 10 and 11), while *Pepper* and *Hobbit* execute assistive tasks as depicted in Figures 12 and 13. A total of twenty objects were used in our experiments including challenging ones such as keys, a bottle of water, a pack of gum, or a headphone's bag.

As shown in Figure 10, several contact points are detected within a scene. So, all the objects on the bin present contact points. However, thanks to the gripper recognition module, only the *grasping* contact points are considered for evaluating the status of the grasping task. Another critical issue is the missing depth data, clearly present in Figure 10. The combination of colour and depth cues and the inclusion of non-data points allows our approach to successfully detect the presence or absence of contact points between the robotic manipulator and the object, as shown in Figure 10 and its partial version in Figure 11.

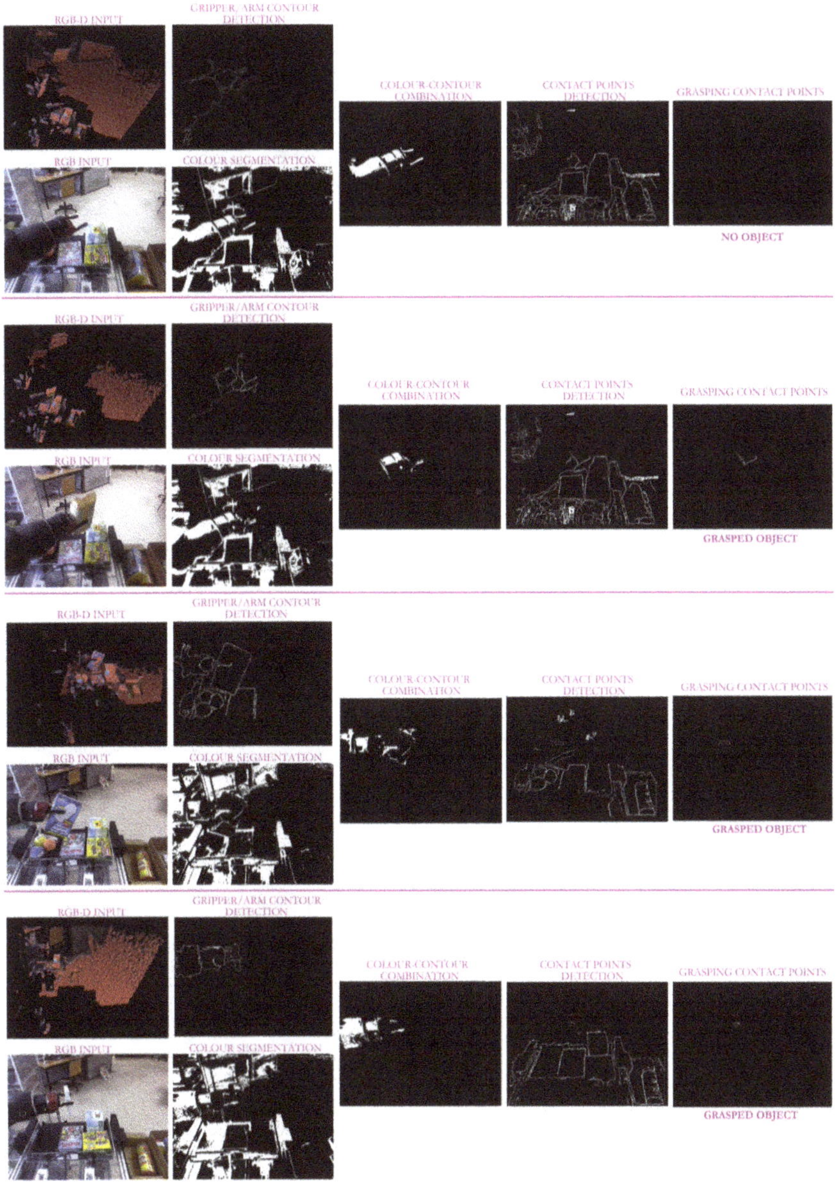

Figure 10. Some experimental results of our approach with the *Baxter* robot in pick-and-place tasks. The first column corresponds to the taken RGB-D image given as an RGB image and a depth map. The second column illustrates the *Lab* segmentation with the robot arm contour obtained from the contour segmentation refinement. The third column illustrates the combination of the images in the two columns. The next column depicts all the two-object contact points based on depth proximity. The last image shows the contact points obtained from the overlap between the results in the third and fourth columns.

Figure 11. Some experimental results of our approach with the *Baxter* robot in pick-and-place tasks: the left column corresponds to the input RGB image; the middle column illustrates the detected contact points; and the last column shows the overlapping between the original image and the detected contact points (in pink).

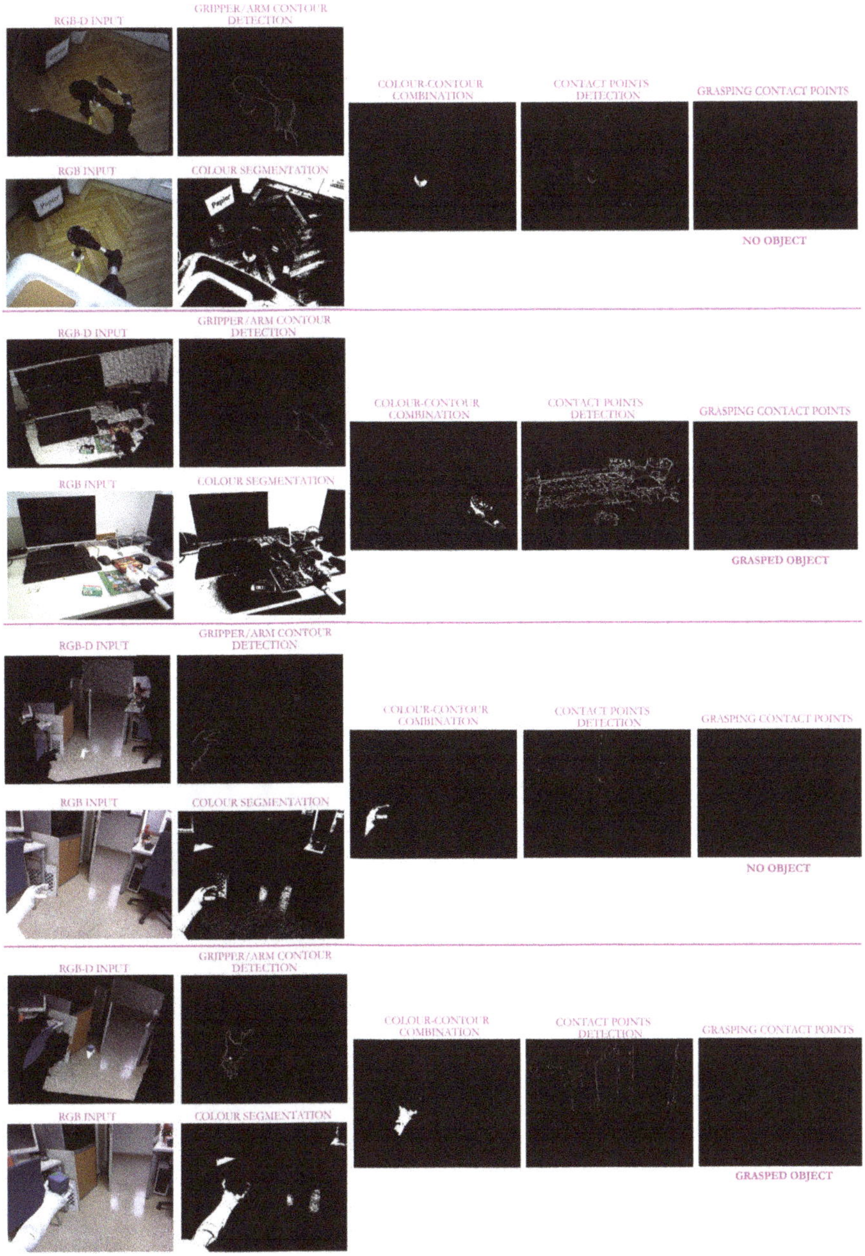

Figure 12. Some experimental results of our approach with the *Hobbit* and *Pepper* robots in assistive tasks. The first column corresponds to the taken RGB-D image given as an RGB image and a depth map. The second column illustrates the *Lab* segmentation with the robot arm contour obtained from the contour segmentation refinement. The third column illustrates the combination of the images in the two columns. The next column depicts all the two-object contact points based on depth proximity. The last image shows the contact points obtained from the overlap between the results in the third and fourth columns.

On its behalf, Figure 12, and the partial version in Figure 13, highlight the resolution of the visual ambiguities since no false positive *grasping* contact points are obtained, even when the robot gripper is close to the ground and its visibility is poor. Thin objects can be also properly detected when they are grasped as in the case of the chewing gum pack. In addition, it can be observed that neither the changing shape of *Hobbit*'s gripper nor the use of different robot grippers affect the approach results.

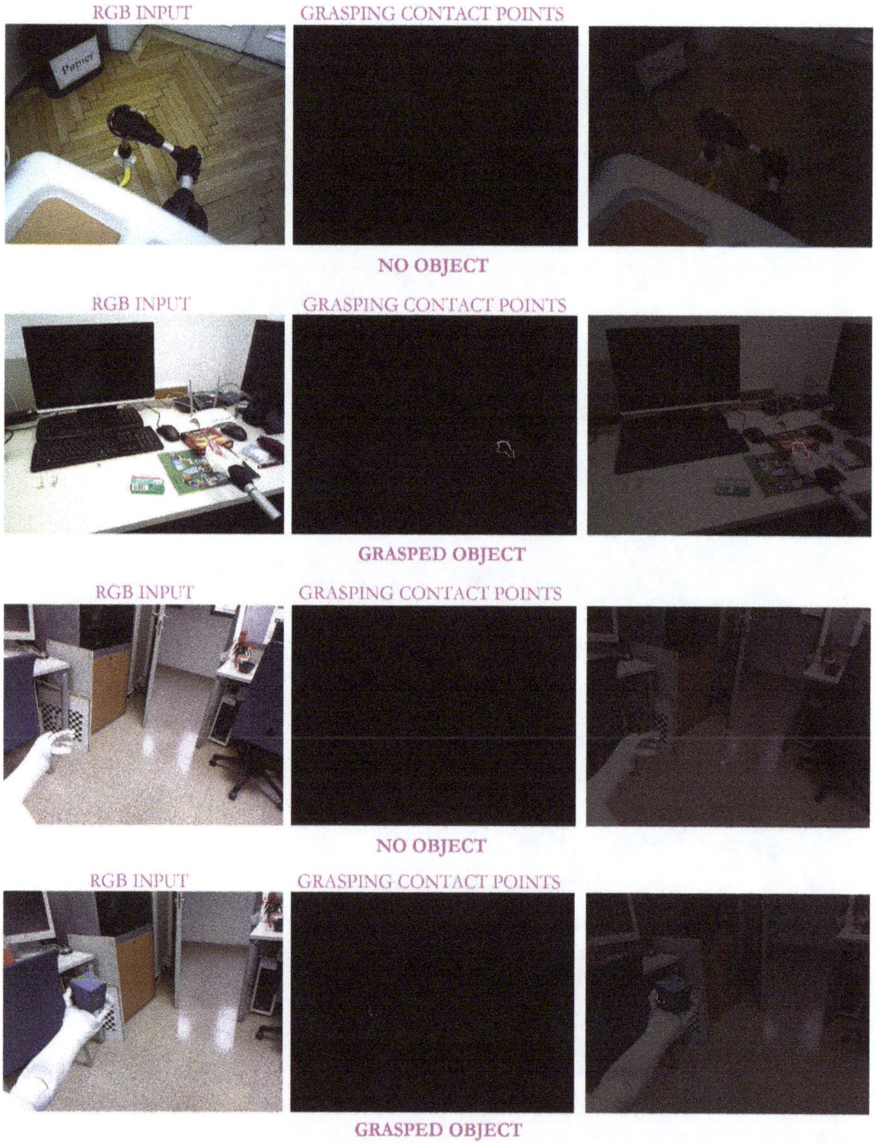

Figure 13. Some experimental results of our approach with the *Hobbit* and *Pepper* robots in assistive tasks: the left column corresponds to the input RGB image; the middle column illustrates the detected contact points; and the last column shows the overlapping between the original image and the detected contact points (in pink).

The approach's performance has been analysed by means of a comparison between its output in terms of presence or absence of a grasped object and the images manually labelled considering seven scenarios, three robot platforms, and twenty objects with different visual features. With a total of one thousand 640×480 images, the algorithm was able to successfully evaluate the grasping status with an accuracy of 97.5% at a speed of 160 ms per image. Note that this speed allows the robot to work in real-time, what is crucial for service robots. The main errors were a consequence of handling small and/or thin objects in specific configurations.

5. Conclusions

Reliable grasping is a decisive task for any robotic application from industrial pick-and-place to service assistance. For that reason, it is critical to successfully perform any grasp and properly recover for any error. This is, however, not straightforward due to the great variety of robot manipulators and, especially, those with a design that prevents the use of other devices like touch sensors.

In this paper, we propose a novel vision approach for monitoring the grasping tasks and verifying any lost of the held object. The underlying idea is the recognition of the contact points between the robot manipulator and the grasped object. For that, all the contact points between two objects within the scene are obtained from depth data. Then, it is checked whether any contact point corresponds to the inner part of the gripper. With that aim, a griper recognition method based on the fusion of depth and colour cues is presented.

So, on the one hand, the input RGB image is segmented according to the *Lab*-colour manipulator coordinates. At the same time, edge information is extracted from depth data. An edge refinement under the assumption of the manipulator boundary comes from the bottom of the image, allows our approach to extract the robot arm contour. Finally, the colour-contour combination together with the contact point map determines the grasping status at any time.

With the aim of properly evaluating the performance of our approach, three different robot platforms have been used: Baxter, Pepper and Hobbit. So, its performance was evaluated in different scenarios, with different objects and with several head poses. The experiment results highlight the good performance, obtaining an accuracy of 97.5%. It is noteworthy that the erroneous cases are present when thin or small objects are manipulated and only in some manipulator configurations. For that reason, the approach should improve to cover these cases. In addition, the proposed approach runs in real-time, which is an issue particularly problematic for robot applications.

As future work, other visual features will be analysed with the aim of overcoming the problems detected with small or thin objects without constraining the robot's autonomy.

Author Contributions: Conceptualization, E.M.-M. and A.P.d.P.; Methodology, E.M.-M. and A.P.d.P.; Validation, E.M.-M. and A.P.d.P.; Resources, E.M.-M. and A.P.d.P.; Writing, E.M.-M. and A.P.d.P.

Funding: This research was partially funded by Ministerio de Economía y Competitividad grant number DPI2015-69041-R.

Acknowledgments: This paper describes research done at UJI Robotic Intelligence Laboratory. Support for this laboratory is provided in part by Ministerio de Economía y Competitividad and by Universitat Jaume I (UJI-B2018-74).

Conflicts of Interest: The authors declare no conflict of interest.

References

1. Costa, A.; Martinez-Martin, E.; Cazorla, M.; Julian, V. PHAROS-physical assistant robot system. *Sensors* **2018**, *18*, 2633. [CrossRef] [PubMed]
2. Gomez-Donoso, F.; Orts-Escolano, S.; Garcia-Garcia, A.; Garcia-Rodriguez, J.; Castro-Vargas, J.A.; Ovidiu-Oprea, S.; Cazorla, M. A robotic platform for customized and interactive rehabilitation of persons with disabilities. *Pattern Recogn. Lett.* **2017**, *99*, 105–113. [CrossRef]

3. Duckett, T.; Pearson, S.; Blackmore, S.; Grieve, B.; Chen, W.H.; Cielniak, G.; Cleaversmith, J.; Dai, J.; Davis, S.; Fox, C.; et al. Agricultural robotics: The future of robotic agriculture. *arXiv* **2018**, arXiv:1806.06762.
4. Robinette, P.; Li, W.; Allen, R.; Howard, A.M.; Wagner, A.R. Overtrust of robots in emergency evacuation scenarios. In Proceedings of the 2016 11th ACM/IEEE International Conference on Human-Robot Interaction (HRI), Christchurch, New Zealand, 7–10 March 2016. [CrossRef]
5. Tang, B.; Jiang, C.; He, H.; Guo, Y. Human mobility modeling for robot-assisted evacuation in complex indoor environments. *IEEE Trans. Hum. Mach. Syst.* **2016**, *46*, 694–707. [CrossRef]
6. Azenkot, S.; Feng, C.; Cakmak, M. Enabling building service robots to guide blind people a participatory design approach. In Proceedings of the 2016 11th ACM/IEEE International Conference on Human-Robot Interaction (HRI), Christchurch, New Zealand, 7–10 March 2016. [CrossRef]
7. Graña, M.; Alonso, M.; Izaguirre, A. A panoramic survey on grasping research trends and topics. *Cybern. Syst.* **2019**, *50*, 40–57. [CrossRef]
8. Mahler, J.; Matl, M.; Satish, V.; Danielczuk, M.; DeRose, B.; McKinley, S.; Goldberg, K. Learning ambidextrous robot grasping policies. *Sci. Robot.* **2019**, *4*, eaau4984. [CrossRef]
9. Morrison, D.; Corke, P.; Leitner, J. Closing the loop for robotic grasping: A real-time, generative grasp synthesis approach. *arXiv* **2018**, arXiv:1804.05172.
10. Laskey, M.; Lee, J.; Chuck, C.; Gealy, D.; Hsieh, W.; Pokorny, F.T.; Dragan, A.D.; Goldberg, K. Robot grasping in clutter: Using a hierarchy of supervisors for learning from demonstrations. In Proceedings of the 2016 IEEE International Conference on Automation Science and Engineering (CASE), Fort Worth, TX, USA, 21–24 August 2016. [CrossRef]
11. Nogueira, J.; Martinez-Cantin, R.; Bernardino, A.; Jamone, L. Unscented Bayesian optimization for safe robot grasping. In Proceedings of the 2016 IEEE/RSJ International Conference on Intelligent Robots and Systems (IROS), Daejeon, Korea, 9–14 October 2016. [CrossRef]
12. Howe, R.D. Tactile sensing and control of robotic manipulation. *Adv. Robot.* **1993**, *8*, 245–261. [CrossRef]
13. Prats, M.; del Pobil, A.P.; Sanz, P.J. *Robot Physical Interaction through the Combination of Vision, Tactile and Force Feedback*; Springer: Berlin, Germany, 2013; Volume 84.
14. Kappassov, Z.; Corrales, J.A.; Perdereau, V. Tactile sensing in dexterous robot hands—Review. *Robot. Auton. Syst.* **2015**, *74*, 195–220. [CrossRef]
15. Chen, T.; Ciocarlie, M. Proprioception-based grasping for unknown objects using a series-elastic-actuated gripper. *arXiv* **2018**, arxiv:1803.09674.
16. Homberg, B.S.; Katzschmann, R.K.; Dogar, M.R.; Rus, D. Robust proprioceptive grasping with a soft robot hand. In *Autonomous Robots*; Springer: Berlin, Germany, 2018. [CrossRef]
17. Eppner, C.; Höfer, S.; Jonschkowski, R.; Martín-Martín, R.; Sieverling, A.; Wall, V.; Brock, O. Lessons from the Amazon Picking Challenge: Four Aspects of Building Robotic Systems. In Proceedings of the Twenty-Sixth International Joint Conference on Artificial Intelligence, IJCAI-17 Melbourne, Australia, 19–25 August 2017; pp. 4831–4835. [CrossRef]
18. Correll, N.; Bekris, K.E.; Berenson, D.; Brock, O.; Causo, A.; Hauser, K.; Okada, K.; Rodriguez, A.; Romano, J.M.; Wurman, P.R. Analysis and observations from the first amazon picking challenge. *IEEE Trans. Autom. Sci. Eng.* **2018**, *15*, 172–188. [CrossRef]
19. Hernandez, C.; Bharatheesha, M.; Ko, W.; Gaiser, H.; Tan, J.; van Deurzen, K.; de Vries, M.; Mil, B.V.; van Egmond, J.; Burger, R.; et al. Team Delft's robot winner of the amazon picking challenge 2016. In *RoboCup 2016: Robot World Cup XX*; Springer International Publishing: Berlin, Germany, 2017; pp. 613–624. [CrossRef]
20. Del Pobil, A.P.; Kassawat, M.; Duran, A.J.; Arias, M.; Nechyporenko, N.; Mallick, A.; Cervera, E.; Subedi, D.; Vasilev, I.; Cardin, D.; et al. UJI RobInLab's approach to the amazon robotics challenge 2017. In Proceedings of the 2017 IEEE International Conference on Multisensor Fusion and Integration for Intelligent Systems (MFI), Daegu, Korea, 16–18 November 2017. [CrossRef]
21. Nicodemou, V.C.; Oikonomidis, I.; Argyros, A. Single-shot 3D hand pose estimation using radial basis function networks trained on synthetic data. In *Pattern Analysis and Applications*; Springer: Berlin, Germany, 2019. [CrossRef]
22. Pham, T.H.; Kyriazis, N.; Argyros, A.A.; Kheddar, A. Hand-object contact force estimation from markerless visual tracking. *IEEE Trans. Pattern Anal. Mach. Intell.* **2018**, *40*, 2883–2896. [CrossRef] [PubMed]

23. Yuan, S.; Garcia-Hernando, G.; Stenger, B.; Moon, G.; Chang, J.Y.; Lee, K.M.; Molchanov, P.; Kautz, J.; Honari, S.; Ge, L.; et al. Depth-based 3D hand pose estimation: From current achievements to future goals. In Proceedings of the 2018 IEEE/CVF Conference on Computer Vision and Pattern Recognition, Salt Lake City, UT, USA, 18–22 June 2018. [CrossRef]
24. Voulodimos, A.; Doulamis, N.; Doulamis, A.; Protopapadakis, E. Deep learning for computer vision: A brief review. *Comput. Intell. Neurosci.* **2018**, *2018*, 1–13. [CrossRef] [PubMed]
25. Guo, Y.; Liu, Y.; Oerlemans, A.; Lao, S.; Wu, S.; Lew, M.S. Deep learning for visual understanding: A review. *Neurocomputing* **2016**, *187*, 27–48. [CrossRef]
26. Bengio, Y.; Courville, A.; Vincent, P. Unsupervised feature learning and deep learning: A review and new perspectives. *arXiv* **2012**, arXiv:1206.5538v1.
27. Lowe, D.G. Distinctive image features from scale-invariant keypoints. *Int. J. Comput. Vis.* **2004**, *60*, 91–110. [CrossRef]
28. Lowe, D. Object recognition from local scale-invariant features. In Proceedings of the Seventh IEEE International Conference on Computer Vision, Kerkyra, Greece, 20–27 September 1999.
29. Alahi, A.; Ortiz, R.; Vandergheynst, P. FREAK: Fast retina keypoint. In Proceedings of the 2012 IEEE Conference on Computer Vision and Pattern Recognition, Providence, RI, USA, 16–21 June 2012. [CrossRef]
30. Rublee, E.; Rabaud, V.; Konolige, K.; Bradski, G. ORB: An efficient alternative to SIFT or SURF. In Proceedings of the 2011 International Conference on Computer Vision, Barcelona, Spain, 6–13 November 2011. [CrossRef]
31. Bay, H.; Tuytelaars, T.; Gool, L.V. SURF: Speeded up robust features. In *Computer Vision—ECCV 2006*; Springer: Berlin/Heidelberg, Germany, 2006; pp. 404–417. [CrossRef]
32. Martinez-Martin, E.; del Pobil, A.P. Visual object recognition for robot tasks in real-life scenarios. In Proceedings of the 10th International Conference on Ubiquitous Robots and Ambient Intelligence (URAI), Jeju, Korea, 30 October–2 November 2013; pp. 644–651. [CrossRef]
33. Rethink Robotics—Baxter Robot. Available online: https://www.rethinkrobotics.com/baxter/ (accessed on 22 October 2018).
34. Softbank Robotics—Pepper. Available online: https://www.softbankrobotics.com/emea/en/pepper (accessed on 22 October 2018).
35. HOBBIT—The mutual care robot. Available online: http://hobbit.acin.tuwien.ac.at/ (accessed on 22 October 2018).

sensors

MDPI

Article

A Vision Based Detection Method for Narrow Butt Joints and a Robotic Seam Tracking System

Boce Xue [1,2], Baohua Chang [1,2], Guodong Peng [1,2], Yanjun Gao [3], Zhijie Tian [3], Dong Du [1,2,*] and Guoqing Wang [3,*]

1 Department of Mechanical Engineering, Tsinghua University, Beijing 100084, China; xbc17@mails.tsinghua.edu.cn (B.X.); bhchang@tsinghua.edu.cn (B.C.); pgd16@mails.tsinghua.edu.cn (G.P.)
2 Key Laboratory for Advanced Materials Processing Technology, Ministry of Education, Beijing 100084, China
3 Capital Aerospace Machinery Ltd., Beijing 100076, China; gaoyanjun274507@163.com (Y.G.); tzhj_2004@126.com (Z.T.)
* Correspondence: dudong@tsinghua.edu.cn (D.D.); guoqingwang211@soho.com (G.W.)

Received: 18 January 2019; Accepted: 1 March 2019; Published: 6 March 2019

Abstract: Automatic joint detection is of vital importance for the teaching of robots before welding and the seam tracking during welding. For narrow butt joints, the traditional structured light method may be ineffective, and many existing detection methods designed for narrow butt joints can only detect their 2D position. However, for butt joints with narrow gaps and 3D trajectories, their 3D position and orientation of the workpiece surface are required. In this paper, a vision based detection method for narrow butt joints is proposed. A crosshair laser is projected onto the workpiece surface and an auxiliary light source is used to illuminate the workpiece surface continuously. Then, images with an appropriate grayscale distribution are grabbed with the auto exposure function of the camera. The 3D position of the joint and the normal vector of the workpiece surface are calculated by the combination of the 2D and 3D information in the images. In addition, the detection method is applied in a robotic seam tracking system for GTAW (gas tungsten arc welding). Different filtering methods are used to smooth the detection results, and compared with the moving average method, the Kalman filter can reduce the dithering of the robot and improve the tracking accuracy significantly.

Keywords: robotic welding; seam tracking; visual detection; narrow butt joint; GTAW

1. Introduction

In automatic welding, it is necessary to align the welding torch with the center of the joint to ensure the welding quality. Nowadays, the motion path of a welding robot is usually set by offline programming or manual teaching. However, during the welding process, the actual joint trajectory may deviate from the path set before welding due to factors, such as machining error, assembly error, and thermal deformation. In view of the abovementioned reason, it is necessary to perform automatic joint detection.

In the welding field, visual detection is widely used for the monitoring of weld defects [1], recognition of the weld joint [2–4], etc. For automatic joint detection, the structured light method based on optical triangulation is commonly used to detect the 3D position of joints with large grooves. Zou et al. projected a structured laser on the workpiece surface and extracted laser stripes from images strongly disturbed by an arc to calculate the 3D position of the joint in the world frame and control the motion of the welding torch in real time [5]. Li et al. proposed a robust automatic welding seam identification and tracking method by utilizing structured light vision, which can identify deformed laser stripes in the complex welding environment and find the position of the welding joint in the pixel coordinate [6]. Some companies have released commercial joint detection sensors based on the structured light method [7,8].

However, for the butt joint with a narrow gap (with a width less than 0.2 mm), the deformation of structured light stripes almost disappears, so it is difficult to detect the position of the narrow butt joint with the structured light method [9]. To solve this problem, researchers have proposed a variety of methods. Xu et al. developed a passive visual sensing method to capture the image of a molten pool and extracted the edge of the molten pool with an improved Canny operator to calculate the deviation of the joint relative to the torch [10]. Gao et al. tried to capture the image of a molten pool with an infrared camera and calculated the deviation of the joint relative to the torch by the shape of the weld pool. Then, an adaptive Kalman filter and Elman neural network were used to improve the accuracy [11]. Nilsen et al. estimated the offset of the joint relative to the torch in laser welding by the image of the keyhole and the spectrum of the plasma sprayed from the keyhole, respectively, and combined these two methods to construct a composite sensing system [12]. Shah et al. used an auxiliary light source to illuminate the workpiece. Considering the uneven brightness on the surface of the workpiece, the local thresholding method was used to extract the position of the joint in pixel coordinates [13]. Nele et al. constructed an image acquisition system, which was combined with the pattern learning algorithm to detect the position of the butt joint relative to the torch and corrected the torch position in real time [14]. Kramer et al. distinguished the boundaries between the two surfaces of the workpiece to be welded by their texture information, thereby finding the nearly invisible narrow line imaged by the joint gap [15]. Gao et al. introduced a novel method, in which the deviation of the weld joint relative to the torch was detected according to the magneto-optical effect [16].

The above methods for the detection of a narrow butt joint can only detect its 2D position. However, in the welding of a butt joint with a width less than 0.2 mm and with a 3D trajectory, the 3D position of the joint is required. Furthermore, the welding torch should also maintain a proper orientation relative to the workpiece surface to ensure the welding quality, so the normal vector of the workpiece surface also needs be obtained. Fang et al. presented a visual seam tracking system in which the deviation of the joint relative to the torch in the horizontal direction was detected according to the position of the joint in the image under natural light illumination, and the deviation in the vertical direction was detected using the structured light method, but the method was incapable of detecting the orientation of the workpiece surface [17]. Shao et al. projected three laser stripes onto the workpiece surface, blended the 2D information of the joint in the image with the 3D information of the structured light, calculated the 3D position of the joint and the normal vector of the workpiece surface, and adjusted the position and orientation of the torch in real time [18]. However, this method still relied on the deformation of laser stripes. Because of the machining error and assembly error, the width of the joint can be uneven and the gap of some points on the joint will disappear. Under this circumstance, this method will miss some joint points.

Zeng et al. designed a narrow butt detection sensor, which projected uniform light and crosshair structured light onto the surface of a workpiece and captured images alternately. Then, 2D and 3D information were combined to calculate the 3D position of the joint and the normal vector of the workpiece surface in a world frame, and the position of the torch was corrected in real time [9,19]. Based on this method, Peng et al. tried to fit the workpiece surface with the moving least squares (MLS) method in the calculation process to improve the fitting accuracy [20]. However, Zeng's method [9,19] has certain limitations. To conveniently extract the 2D information of the joint from the image, the method requires strict lighting conditions for the auxiliary light source. When the auxiliary light source is on, the grayscale of the workpiece surface in the image needs to be almost saturated, but this requirement can be achieved only with a specular reflection workpiece surface and the workpiece surface needs to be close to the auxiliary light source. The working distance of those commercial joint detection sensors based on the structured light method can reach more than 100 mm [7,8], while the working distance in [9] does not exceed 40 mm generally. In the case of a diffuse reflection workpiece surface or remote detection, the illumination intensity of the LED light source is insufficient as the auxiliary light source. To achieve the desired high grayscale of the workpiece surface in the image, the exposure time of the camera needs to be extended, which will deteriorate the detection speed. If a laser light source is used as the auxiliary light source, speckle in the image will affect the image quality and make it difficult to extract the 2D information of the joint from the image. In this paper, a vision based detection method for a narrow butt joint is proposed, which reduces the requirements for the lighting conditions of the auxiliary light source, and the proposed method is used in building a robotic seam tracking system for GTAW (gas tungsten arc welding).

The rest of the paper is organized as follows. In Section 2, the processes and details of the proposed detection method are presented. To apply this detection method in the robotic seam tracking system, Section 3 introduces the necessary coordinate transformation. In Section 4, the configurations of the joint detection sensor and the robotic seam tracking system for GTAW are detailed. In Section 5, the detection results of the proposed method are presented, and different filtering methods are used to smooth the detection results to reduce the dithering of the robot and improve the seam tracking accuracy. Finally, Section 6 gives the conclusions of this paper.

2. Detection Method for the Narrow Butt Joint

In this section, the principle of the detection method is introduced first, then details of the method are discussed, including the grabbing of images with an appropriate grayscale distribution, image processing, and the calculation of the position and orientation of the joint. Finally, applications of the proposed method are discussed.

2.1. Principle of the Method

The basic experimental setup of the proposed detection method for a narrow butt joint is shown in Figure 1. A crosshair laser is projected onto the workpiece surface and the LED (light-emitting diode) auxiliary light source is used to illuminate the workpiece surface continuously. Images with an appropriate grayscale distribution are grabbed by using the auto exposure function of the camera to adjust the exposure time. Then, the joint region and laser stripe region can be extracted by different gray thresholds. The laser stripe region provides the 3D information of the workpiece surface (normal vector included) and the joint region provides the 2D information of the joint. By combining the 2D and 3D information together, the 3D position of the joint can be obtained. To improve the processing speed, different processing flows are used for the first frame and the subsequent frames. The flowchart of the proposed joint detection method is shown in Figure 2, and its details will be described in the following subsections.

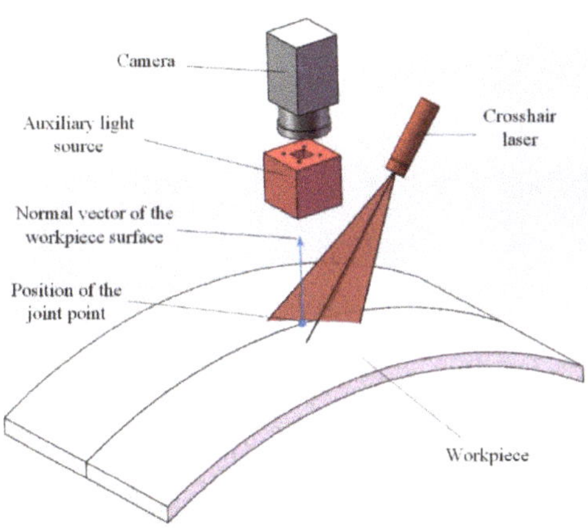

Figure 1. Basic experimental setup of the proposed method.

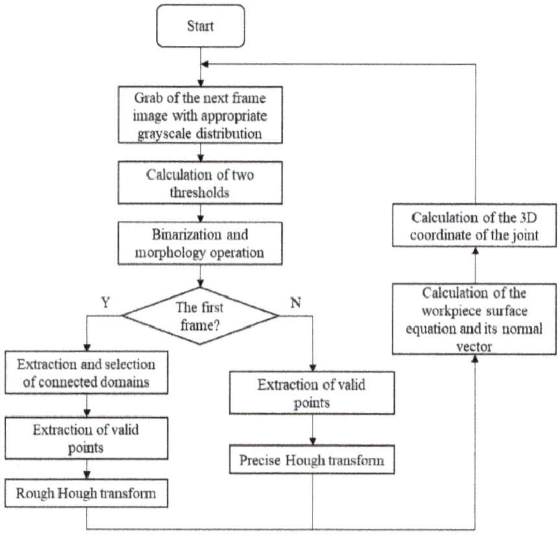

Figure 2. Flowchart of the proposed joint detection method.

2.2. Grabbing of Images with an Appropriate Grayscale Distribution

The ROI (region of interest) of the grabbed image is shown in Figure 3, and the field-of-view corresponding to it is 16.25 × 10 mm. The grayscale of the background region is affected by the illumination intensity of the auxiliary light source. The laser stripe region is created by the projection of the crosshair laser. Ideally, the grayscale for the joint region should be very low (close to 0) and that for the laser stripe region should be very high (close to 255) in the image. So, in the auto exposing of the camera, the expected average grayscale of the ROI is set to 128 to ensure that the joint region and laser stripe region can be differentiated clearly from the background region according to different gray thresholds. However, if the illumination intensity of the auxiliary light source is too strong, the exposure time will be reduced greatly, so the grayscale for the laser stripe region will be obviously smaller than 255 and be close to that of the background region when the expected average grayscale of the ROI is set to 128, which will make it difficult to differentiate the laser stripe region from the background region. So, to differentiate the laser stripe region from the background region, the illumination intensity of the auxiliary light source is controlled to be obviously weaker than that of the crosshair laser by adjusting its supply voltage to make sure that the grayscale for the laser stripe region is close to 255. In fact, this can easily be achieved, especially under the circumstance of remote detection, because the orientation of the LED light is much worse than that of the laser. The average grayscale of the ROI shown in Figure 3 is 128.4. The pixel coordinate system {P} is established on the ROI.

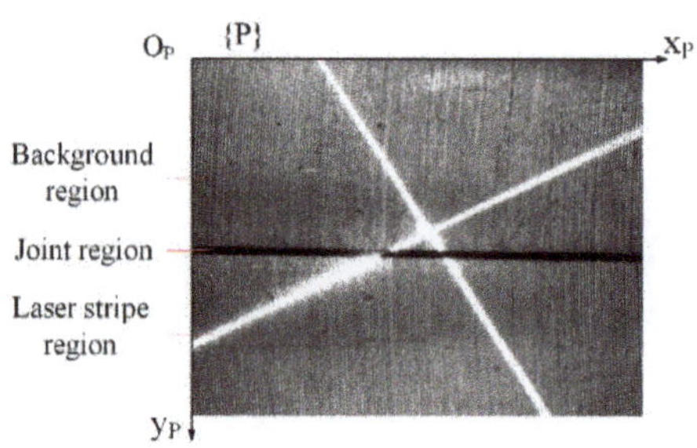

Figure 3. The grabbed ROI (region of interest).

2.3. *Image Processing*

2.3.1. Determination of Thresholds for Binarization

The thresholds for the extraction of the laser stripe region and the joint region are determined according to the histogram, $h(r)$, of the ROI, where r represents the grayscale and h represents the number of pixels with a grayscale of r. To eliminate false valleys caused by accidental factors, a Gaussian filter with a length of 5 is used to smooth the original histogram first, and the smoothed histogram is shown in Figure 4. The background region with a medium grayscale results in peak 2 in the histogram. For the laser stripe region, its grayscale is very high and its area is not very small, which causes peak 3 in the high grayscale region of the histogram. So, the grayscale corresponding to valley 2, which is between peak 2 and peak 3, can be regarded as the threshold, t_{high}, and it can be used for extracting the laser stripe region from the background region. The grayscale corresponding to valley 2 is found to be 234 in the smoothed histogram, so $t_{\text{high}} = 234$.

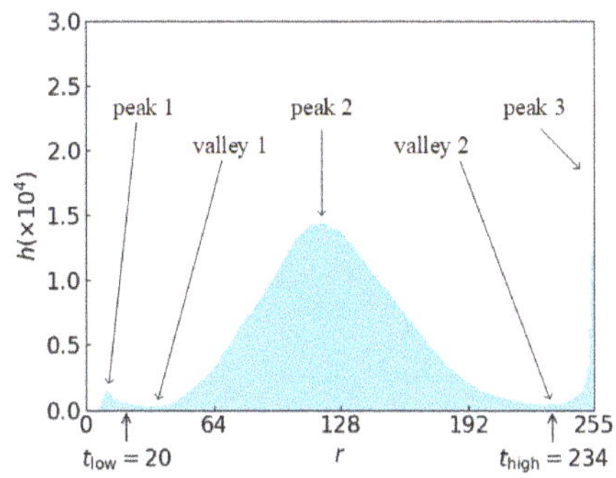

Figure 4. Smoothed histogram of the grabbed ROI.

For the joint region, the binarization threshold cannot be determined in the same way as the laser stripe region. This is because when the joint is extremely narrow, the area of the joint region in the ROI can be very small. In this case, peak 1 in the histogram does not exist at all, and neither does valley 1. To ensure the robustness of the threshold determination method, Equation (1) is used to determine the threshold, t_{low}, for extracting the joint region from the background region:

$$\begin{aligned} t_{\text{low}} &= \max r \\ \text{s,t.} \ \sum_{i=0}^{r} h(i) &\leq Sa \end{aligned} \tag{1}$$

where S is the area of the ROI and a is a ratio, which is $a = 0.01$. We get $t_{\text{low}} = 20$. The value of a should be near the percentage of the joint region's area in the ROI. If the value is set too high, the background region is not eliminated effectively. On the contrary, if the value is set too low, the main part of the joint region is not kept completely.

2.3.2. Binarization and Morphology Operation

The binarization of the images is processed with the thresholds, t_{high} and t_{low}, respectively, and the binary ROI are shown in Figure 5. In the two images of Figure 5, there are some disconnected small regions, so the morphology of the close operation is used to connect them in the two images, and the images after close operation are shown in Figure 6.

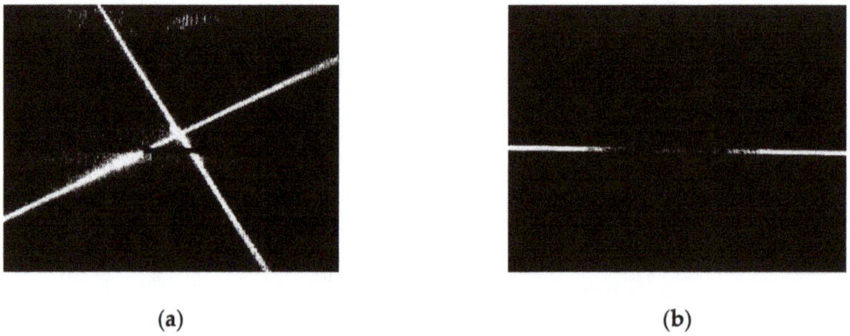

Figure 5. ROI after binarization. (a) Laser stripe region. (b) Joint region.

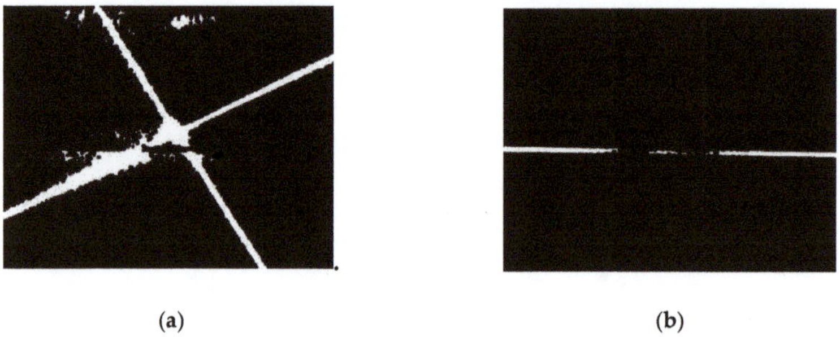

Figure 6. ROI after close operation. (a) Laser stripe region. (b) Joint region.

2.3.3. Extraction and Selection of Connected Domains

When the image is the first frame, the connected domains in the two images of Figure 6 are extracted. Because the laser stripe region and the joint region, ideally, do not have holes, only the outermost contours are kept if there are nested contours. For the laser stripe region and joint region, the connected domains have a relatively large area and a slender shape, while those falsely kept regions always have a relatively small area or a less slender shape, so we can retain the laser stripe region and joint region according to their area and circularity ratio [21] (pp. 844–845):

$$\begin{cases} A > A_{\min} \\ R_c < R_{c\max} \end{cases} \tag{2}$$

where A is the area of the connected domains and $A_{\min}$ is the area threshold. The values of $A_{\min}$ are selected by several attempts to keep the main part of the laser stripe region or the joint region. For the laser stripe region, $A_{\min}$ is set to 2000 pixel^2 and for the joint region, $A_{\min}$ is set to 1000 pixel^2. This is because the area of the laser stripe region is obviously larger than that of the joint region. R_c represents the circularity ratio of the connected domains and $R_{c\max}$ is the circularity ratio threshold. R_c is defined as:

$$R_c = \frac{4\pi A}{P^2} \tag{3}$$

where P is the perimeter of the connected ratio. The circularity ratio can represent the slenderness degree of a region. It is 1 for a circular region and 0 for a line, so the circularity ratio threshold, $R_{c\max}$, is set to 0.5 for both the laser stripe region and the joint region and we find this value effective.

The connected domains kept are shown in Figure 7. It can be found that there is a false connected domain kept in Figure 7a, which results from the area threshold being not large enough. In fact, if the area threshold is set to 3000 pixel^2, this false connected domain will be eliminated. However, even if it is kept, the laser stripe can still be extracted successfully in the following steps.

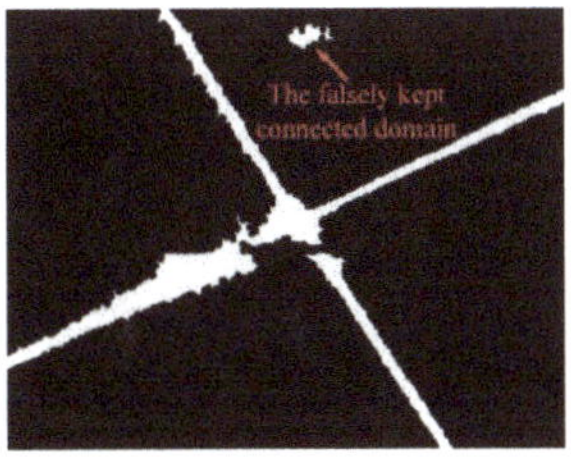

(**a**)

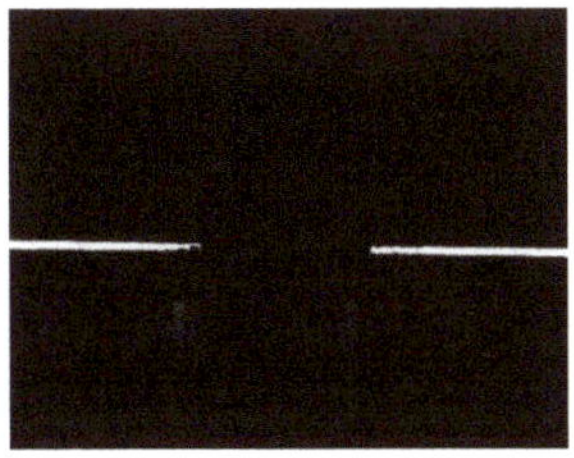

(**b**)

Figure 7. Kept connected domains. (**a**) Laser stripe region. (**b**) Joint region.

Because the extraction and selection of the connected domains are time-consuming tasks, to improve the speed of image processing, they are only performed for the first frame. For the rest of the frames, those falsely kept regions do not influence the detection result because of the robustness of the image processing method, which is introduced in the content below.

2.3.4. Extraction of Valid Points

For the images in Figure 7, each column is scanned from top to bottom to find every line segment whose length is greater than 10, and then the midpoint of each found line segment is marked as a valid point, as shown in Figure 8. The valid points of the laser stripe region are marked in red and the valid points of the joint region are marked in blue. The use of a length threshold of the line segments is to eliminate those line segments located at the boundary of the connected domains.

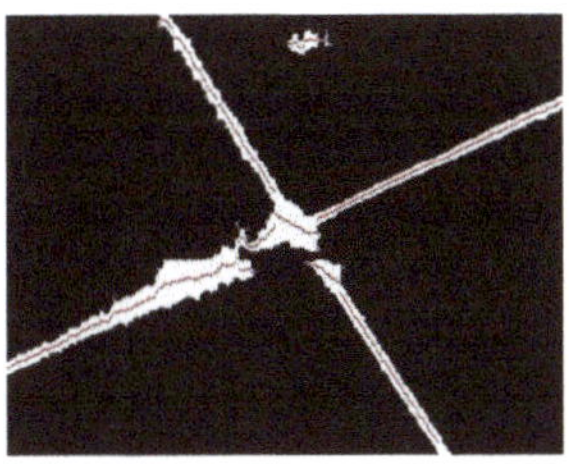

(**a**)

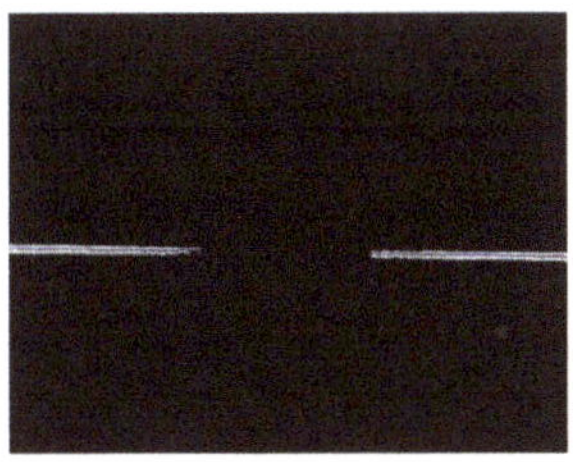

(**b**)

Figure 8. Extracted valid points. (**a**) Laser stripe region. (**b**) Joint region.

2.3.5. Line Extraction

Because the field-of-view corresponding to the ROI is small (with a size of 16.25 × 10 mm), the joint in the ROI can be approximated as a straight line when its trajectory does not change sharply. When the curvature of the joint trajectory increases, the size of the field-of-view corresponding to the ROI should be reduced to make sure that the joint region can be regarded as a straight line. Similarly, the workpiece surface in the ROI can be approximated as a plane when its curvature is small, so the laser stripes can be approximated as two straight lines.

The Hough transform [22] is a common method for the detection of straight lines. A line can be represented as $\rho = x\cos\theta + y\sin\theta$, where ρ is the perpendicular distance from the origin to the line and θ is the angle formed by this perpendicular line and the horizontal axis, so any line can be represented with (ρ, θ). A 2D array or accumulator is created with a resolution of $\Delta\rho \times \Delta\theta$. For every point in the image, θ is changed within its domain of definition with a step size of $\Delta\theta$ and a different ρ is obtained. For every (ρ, θ) pair, the value of the bin corresponding to it in the accumulator increases. Finally, the parameters of the bin corresponding to the maximum value in the accumulator are regarded as the extracted line.

However, the computational cost of the Hough transform increases with the improvement of its detection accuracy. To ensure the detection accuracy of the Hough transform and meanwhile ensure the speed of detection to meet the requirements of real-time performance, different parameter settings are adopted for the first frame and the subsequent frames. If the speed of detection is not fast enough, the detected point would not be dense enough so the detection accuracy of the joint trajectory will deteriorate.

For the first frame, a rough Hough transform is used. The Hough transform is applied to extract two laser stripe lines and the joint line, respectively, and only valid points in Figure 8 are considered. To increase the detection speed, the resolution, $\Delta\rho$ and $\Delta\theta$, are set with relatively large values, which means that the accuracy of the line extraction is relatively low. $\Delta\rho = \Delta\rho_{\mathrm{f}} = 10$ pixel and $\Delta\theta = \Delta\theta_{\mathrm{f}} = 0.1$ rad are set here. Because we currently do not know any information about the line's position, the value range of ρ is set to $[-diag,\ diag]$, where $diag = 1640$ pixel is the diagonal length of the ROI and the value range of θ is set to $[-\frac{\pi}{2},\ -\frac{\pi}{2}]$. For the valid points of the laser stripe region, the accumulator's bins with maximum values in the range of $\theta > 0$ and $\theta < 0$ are searched, respectively, and the parameters of these two found bins represent the two laser stripe lines. For the valid points of the joint region, the accumulator's bin with the maximum value is searched for in the whole parameter space of ρ and θ, and the parameters of the found bin represent the joint line. The found laser stripe lines and the joint line are drawn in Figure 9a,b, respectively, from which it can be seen that the accuracy of the line extraction is inadequate.

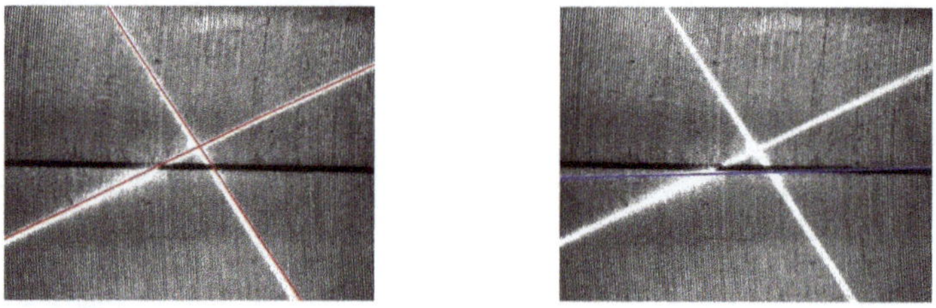

(a) (b)

Figure 9. Extracted lines with the rough Hough transform. (**a**) Laser stripe lines. (**b**) Joint line.

For the subsequent frames, a precise Hough transform is used. During two successive frames, there is no significant relative motion between the camera and the workpiece, so the position of the laser stripe region and the joint region only changes a little. Therefore, from the second frame, the value ranges of the parameters of the Hough transform only need to be near the extracted lines of the last frame. For example, suppose that the parameters of the extracted joint line in one frame are $(\rho_{\mathrm{l}}, \theta_{\mathrm{l}})$. Then, the value of ρ is restricted to $[\rho_{\mathrm{l}} - \rho_{\mathrm{n}}, \rho_{\mathrm{l}} + \rho_{\mathrm{n}}]$ and θ is restricted to $[\theta_{\mathrm{l}} - \theta_{\mathrm{n}}\ ,\ \theta_{\mathrm{l}} + \theta_{\mathrm{n}}]$ when extracting the joint line in the subsequent frame. ρ_{n} and θ_{n} represent half of the value range of ρ and θ in the Hough transform of the subsequent frame, which are set to be 100 pixels and 0.2 rad, respectively. In the same way, the value ranges for the extraction of two laser stripe lines are determined. Since the value ranges of the parameters become smaller, the value of $\Delta\rho$ and $\Delta\theta$ can be reduced to increase the extraction accuracy, so it is set that $\Delta\rho = \Delta\rho_{\mathrm{s}} = 1$ pixel and $\Delta\theta = \Delta\theta_{\mathrm{s}} = 0.005$ rad. The laser stripe lines and the joint line extracted with this method are shown in Figure 10, from which we can see that the accuracy of the line extraction obviously increases.

It is worth noting that the extraction and selection of the connected domains described in Section 2.3.3 are not performed for the subsequent frames, so there may be some valid points falsely extracted. The restriction for the value ranges of the parameters of the Hough transform in the subsequent frames can eliminate the effect of these falsely extracted valid points on line extraction, which therefore increases the robustness of the line extraction method.

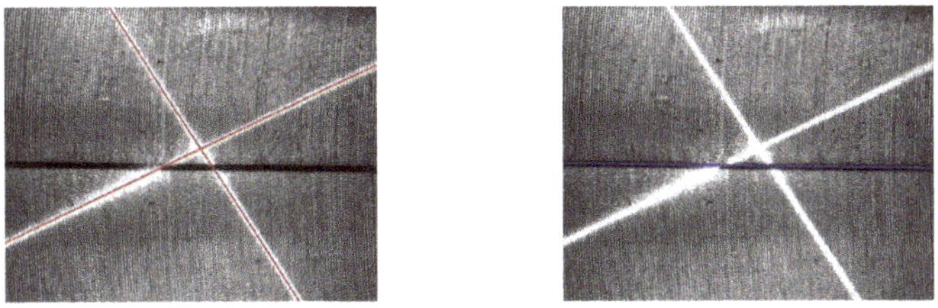

(**a**) (**b**)

Figure 10. Extracted lines with a precise Hough transform. (**a**) Laser stripe lines. (**b**) Joint line.

Below is a comparison of computational complexity between the rough Hough transform and the precise Hough transform. When the number of valid points is fixed, the computational complexity of the Hough transform is $O(M_1)$ and the computational complexity of searching for the accumulator's bin with the maximum value is $O(M_1M_2)$, where M_1 and M_2 are the number of possible values for θ and ρ, respectively.

For the first frame:

$$M_1 = \frac{\pi}{\Delta\theta_\mathrm{f}} \approx 63, \qquad M_2 = \frac{2diag}{\Delta\rho_\mathrm{f}} \approx 328 \tag{4}$$

For the subsequent frames:

$$M_1 = \frac{2\theta_\mathrm{n}}{\Delta\theta_\mathrm{s}} = 80, \qquad M_2 = \frac{2\rho_\mathrm{n}}{\Delta\rho_\mathrm{s}} \approx 200 \tag{5}$$

It can be seen that compared with the first frame, the computational complexity of the precise Hough transform for the subsequent frames does not change much though its accuracy increases significantly.

2.4. Calculation of the 3D Coordinates of the Joint and the Normal Vectors of the Workpiece Surface

By performing calibration in advance [23], the relationship between a point $(x^\mathrm{P}, y^\mathrm{P})$ in the pixel coordinate system {P} and its corresponding point $(x^\mathrm{C}, y^\mathrm{C}, z^\mathrm{C})$ in the camera coordinate system {C} is obtained as:

$$\begin{cases} x^\mathrm{C} = z^\mathrm{C} S_\mathrm{x}(x^\mathrm{P}, y^\mathrm{P}) \\ y^\mathrm{C} = z^\mathrm{C} S_\mathrm{y}(x^\mathrm{P}, y^\mathrm{P}) \end{cases} \tag{6}$$

where $S_\mathrm{x}(x^\mathrm{P}, y^\mathrm{P})$ and $S_\mathrm{y}(x^\mathrm{P}, y^\mathrm{P})$ represent the transformation function between {P} and {C}, which are determined by the camera itself. The equations of the two light planes of the crosshair laser source in {C} can also be obtained through calibration:

$$A_i x^\mathrm{C} + B_i y^\mathrm{C} + C_i z^\mathrm{C} + D_i = 0,\ i = 1, 2 \tag{7}$$

Two laser stripe lines are denoted as l_1 and l_2. N points centered on the intersection of l_1 and l_2 are selected with an equal distance on l_1 and l_2 in {P}, respectively, and these selected points are denoted as $\left(x_{ij}^P, y_{ij}^P\right)$, $j = 1, 2, \ldots, N$. N is set to 50 and the distance between two adjacent points is set to 10 pixels. For each selected point $\left(x_{ij}^P, y_{ij}^P\right)$, its corresponding coordinate, $\left(x_{ij}^C, y_{ij}^C, z_{ij}^C\right)$, in {C} can be solved by combining Equations (6) and (7):

$$\begin{cases} z_{ij}^\mathrm{C} = \dfrac{-D_i}{A_i S_\mathrm{x}\left(x_{ij}^\mathrm{P}, y_{ij}^\mathrm{P}\right) + B_i S_\mathrm{y}\left(x_{ij}^\mathrm{P}, y_{ij}^\mathrm{P}\right) + C_i} \\ x_{ij}^\mathrm{C} = z_{ij}^\mathrm{C} S_\mathrm{x}\left(x_{ij}^\mathrm{P}, y_{ij}^\mathrm{P}\right) \\ y_{ij}^\mathrm{C} = z_{ij}^\mathrm{C} S_\mathrm{y}\left(x_{ij}^\mathrm{P}, y_{ij}^\mathrm{P}\right) \end{cases},\quad i = 1, 2,\ j = 1, 2, \ldots, 50 \tag{8}$$

Because the field-of-view corresponding to the ROI is small, the workpiece surface in the ROI can be approximated as a plane in {C} when its curvature is small, and its least-squares plane can be estimated with $\left(x_{ij}^{C}, y_{ij}^{C}, y_{ij}^{C}\right)$, which can be represented with Equation (9):

$$A_{w}^{C}x^{C} + B_{w}^{C}y^{C} + C_{w}^{C}z^{C} + D_{w}^{C} = 0 \tag{9}$$

Naturally, the normal vector of the workpiece surface in {C} can be represented as $n_{w}^{C} = \left[A_{w}^{C}, B_{w}^{C}, C_{w}^{C}\right]^{T}$, which can also be regarded as the normal vector (or orientation) of the joint point.

Among the points on the joint line, only the point located at the middle along the y direction of the ROI is selected and calculated as the joint point here, which is denoted as $\left(x_{s}^{P}, y_{s}^{P}\right)$ in {P}. Then, its corresponding 3D coordinate, $(x_{s}^{C}, y_{s}^{C}, z_{s}^{C})$, in {C} can be solved by combining Equations (6) and (9):

$$\begin{cases} z_{s}^{C} = \dfrac{-D_{w}^{C}}{A_{w}^{C}S_{x}\left(x_{s}^{P}, y_{s}^{P}\right) + B_{w}^{C}S_{y}\left(x_{s}^{P}, y_{s}^{P}\right) + C_{w}^{C}} \\ x_{s}^{C} = z_{s}^{C}S_{x}\left(x_{s}^{P}, y_{s}^{P}\right) \\ y_{s}^{C} = z_{s}^{C}S_{y}\left(x_{s}^{P}, y_{s}^{P}\right) \end{cases}. \tag{10}$$

2.5. Applications of the Proposed Detection Method

The proposed detection method can be used both before welding and during welding. On the one hand, it can be used before welding to correct the path of the robot when the trajectory of the joint changes after teaching. Under this circumstance, it can be applied to welding methods, like laser welding, GMAW (gas metal arc welding), and GTAW. On the other hand, it can be used during welding to guide the motion of the torch in real time. Since this method requires images of the joint with little disturbance, it can be applied to welding methods that include almost no spatter, like GTAW.

3. Coordinate Transformation

In Section 2, the position and orientation of the joint point in the camera coordinate system {C} was calculated with the proposed joint detection method. When applying this method in the robotic seam tracking system, the position and orientation in the camera coordinate system {C} need to be transformed into the base coordinate system {B} of the robot to guide the motion of the robot.

Coordinate systems involved in coordinate transformation include the camera coordinate system {C} fixed to the camera, the base coordinate system {B} attached to the robot base, and the tool coordinate system {T} fixed to the welding torch, as shown in Figure 11. A coordinate transformation can be described with a homogenous transformation matrix [24]. To describe the transformation relationship of {C} with respect to {B}, a homogenous transformation matrix, ${}_{C}^{B}\boldsymbol{T}$, is needed.

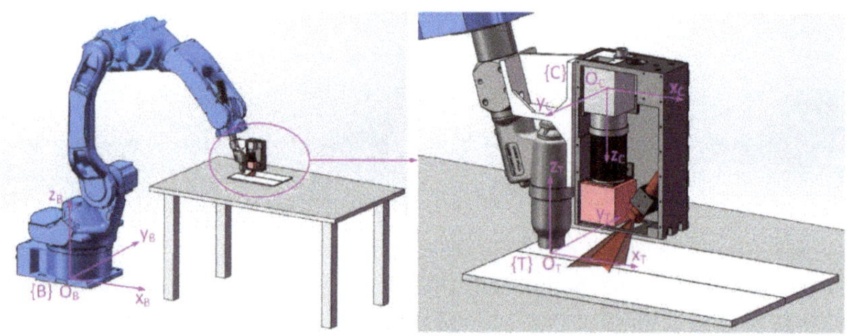

Figure 11. Coordinate systems involved in the coordinate transformation.

To obtain a homogenous transformation matrix, ${}^{\mathrm{B}}_{\mathrm{C}}\boldsymbol{T}$, a homogenous transformation matrix, ${}^{\mathrm{T}}_{\mathrm{C}}\boldsymbol{T}$, is required, which describes the transformation relationship of {C} with respect to {T}, and ${}^{\mathrm{B}}_{\mathrm{T}}\boldsymbol{T}$, which describes the transformation relationship of {T} with respect to {B}. Then, ${}^{\mathrm{B}}_{\mathrm{C}}\boldsymbol{T}$ can be derived from:

$$ {}^{\mathrm{B}}_{\mathrm{C}}\boldsymbol{T} = {}^{\mathrm{B}}_{\mathrm{T}}\boldsymbol{T}\,{}^{\mathrm{T}}_{\mathrm{C}}\boldsymbol{T} \tag{11} $$

Since in the robotic seam tracking system the camera is fixed on the welding torch, ${}^{\mathrm{T}}_{\mathrm{C}}\boldsymbol{T}$ can be predetermined by calibration. The origin of {T} is denoted as TCP (tool center point). ${}^{\mathrm{B}}_{\mathrm{T}}\boldsymbol{T}$ is related to the position and orientation of the robot, which can be represented as $\left(x_{\mathrm{T}}^{\mathrm{B}}, y_{\mathrm{T}}^{\mathrm{B}}, z_{\mathrm{T}}^{\mathrm{B}}, \alpha_{\mathrm{T}}^{\mathrm{B}}, \beta_{\mathrm{T}}^{\mathrm{B}}, \gamma_{\mathrm{T}}^{\mathrm{B}}\right)$, where $\left(x_{\mathrm{T}}^{\mathrm{B}}, y_{\mathrm{T}}^{\mathrm{B}}, z_{\mathrm{T}}^{\mathrm{B}}\right)$ represents the position of the TCP in {B} and $\left(\alpha_{\mathrm{T}}^{\mathrm{B}}, \beta_{\mathrm{T}}^{\mathrm{B}}, \gamma_{\mathrm{T}}^{\mathrm{B}}\right)$ represents the orientation of the welding torch in Euler angles. $\alpha_{\mathrm{T}}^{\mathrm{B}}$, $\beta_{\mathrm{T}}^{\mathrm{B}}$, and $\gamma_{\mathrm{T}}^{\mathrm{B}}$ are the roll, pitch, and yaw angles of {T} relative to {B}. With these six parameters known, ${}^{\mathrm{B}}_{\mathrm{T}}\boldsymbol{T}$ can be derived, which is not detailed here. Additionally, ${}^{\mathrm{B}}_{\mathrm{C}}\boldsymbol{T}$ can be derived from Equation (11). Then, the coordinates of the joint point, $\left(x_{\mathrm{s}}^{\mathrm{B}}, y_{\mathrm{s}}^{\mathrm{B}}, z_{\mathrm{s}}^{\mathrm{B}}\right)$, in {B} can be derived from:

$$ \begin{bmatrix} x_{\mathrm{s}}^{\mathrm{B}} \\ y_{\mathrm{s}}^{\mathrm{B}} \\ z_{\mathrm{s}}^{\mathrm{B}} \\ 1 \end{bmatrix} = {}^{\mathrm{B}}_{\mathrm{C}}\boldsymbol{T} \begin{bmatrix} x_{\mathrm{s}}^{\mathrm{C}} \\ y_{\mathrm{s}}^{\mathrm{C}} \\ z_{\mathrm{s}}^{\mathrm{C}} \\ 1 \end{bmatrix}. \tag{12} $$

With ${}^{\mathrm{B}}_{\mathrm{C}}\boldsymbol{T}$ derived, the normal vector of the joint point in {B} $\boldsymbol{n}_{\mathrm{w}}^{\mathrm{B}}$ can be derived since $\boldsymbol{n}_{\mathrm{w}}^{\mathrm{C}}$ is known. Suppose that $\boldsymbol{n}_{\mathrm{w}}^{\mathrm{B}}$ can be represented as:

$$ \boldsymbol{n}_{\mathrm{w}}^{\mathrm{B}} = \left[A_{\mathrm{w}}^{\mathrm{B}},\ B_{\mathrm{w}}^{\mathrm{B}},\ C_{\mathrm{w}}^{\mathrm{B}}\right]^{\mathrm{T}} \tag{13} $$

Therefore, the orientation of the joint point in {B} can be described with the Euler angles, $\gamma_{\mathrm{s}}^{\mathrm{B}}$ and $\beta_{\mathrm{s}}^{\mathrm{B}}$, as shown in Figure 12, which are given as follows:

$$ \begin{cases} \gamma_{\mathrm{s}}^{\mathrm{B}} = \arctan\left(B_{\mathrm{w}}^{\mathrm{B}} / C_{\mathrm{w}}^{\mathrm{B}}\right) \\ \beta_{\mathrm{s}}^{\mathrm{B}} = \arctan\left(A_{\mathrm{w}}^{\mathrm{B}} / C_{\mathrm{w}}^{\mathrm{B}}\right) \end{cases} \tag{14} $$

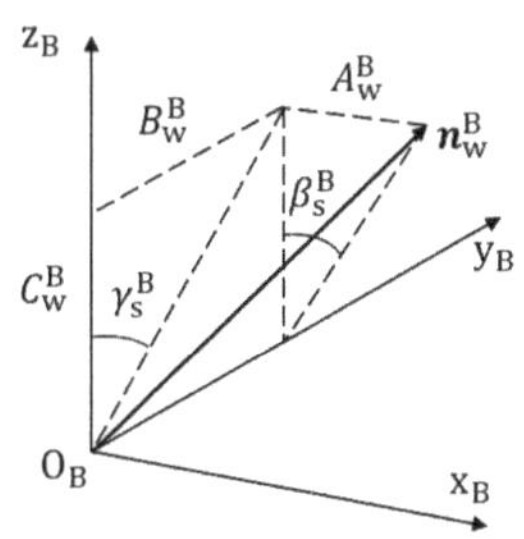

Figure 12. Illustration of $\gamma_{\mathrm{s}}^{\mathrm{B}}$ and $\beta_{\mathrm{s}}^{\mathrm{B}}$.

So, the position and orientation of the joint point in {B} can be represented with $\left(x_{\mathrm{s}}^{\mathrm{B}}, y_{\mathrm{s}}^{\mathrm{B}}, z_{\mathrm{s}}^{\mathrm{B}}, \beta_{\mathrm{s}}^{\mathrm{B}}, \gamma_{\mathrm{s}}^{\mathrm{B}}\right)$.

In the welding process, besides a relative position with the joint, the welding torch also needs to maintain a desired relative orientation with the workpiece surface to ensure the welding quality. Suppose that the welding torch should be perpendicular to the workpiece surface. Then, the axis of the welding torch, z_{T}, needs to be parallel with $\boldsymbol{n}_{\mathrm{w}}^{\mathrm{B}}$. In the experiments described below, the joints are mainly along the x_{B} direction and there is no sharp change of their trajectories, so the welding torch does not need to rotate around its axis, z_{T}. Thus, $\alpha_{\mathrm{T}}^{\mathrm{B}}$ is constant at 0. γ_{s}^{B} and β_{s}^{B} are regarded as the target values of $\gamma_{\mathrm{T}}^{\mathrm{B}}$ and $\beta_{\mathrm{T}}^{\mathrm{B}}$ in the seam tracking process, respectively. Therefore, in seam tracking, the target values of the position and orientation, $(x_T^B, y_T^B, z_T^B, \beta_T^B, \gamma_T^B)$, of the robot are that of the joint point, $(x_S^B, y_S^B, z_S^B, \beta_S^B, \gamma_S^B)$, when α_T^B is fixed to 0.

4. Experiment Setup

The configuration of the joint detection sensor designed according to the abovementioned detection method for a narrow butt joint of GTAW is shown in Figure 13. The Gig (Gigabit Ethernet) camera has a resolution of 1600 × 1200 pixel and offers the auto exposure function. With a working distance (the distance from the bottom of the sensor to the detected workpiece) of 30 mm, the field-of-view of the camera is 20 × 15 mm. The size of the ROI is set to 1300 × 1000 pixel in the experiments. The square LED diffused light with a central wavelength of 630 nm is used as the auxiliary light source. Since the image captured by the camera is rectangular, and the shell of the sensor is cuboid, a square LED can make better use of the space in the sensor and the area in the image. The crosshair laser has a central wavelength of 635 nm and the narrow bandpass filter has a central wavelength of 635 nm and FWHM (full width at half maximum) of 10 nm. The central wavelength for the square LED diffused light, the crosshair laser, and the narrow bandpass filter can eliminate the effect of the arc light on joint detection in aluminum alloy welding using GTAW, since in the arc light spectrum, the intensity near 635 nm is relatively low [9].

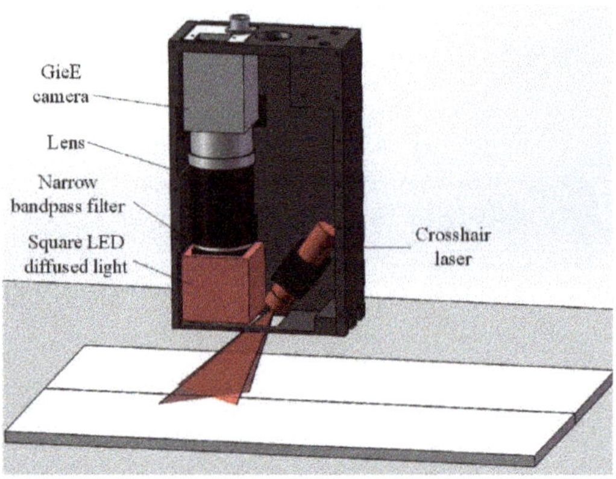

Figure 13. Configuration of the joint detection sensor.

A schematic of the robotic seam tracking system for GTAW is shown in Figure 14. The joint detection sensor is fixed 52.3 mm in front of the welding torch. The welding torch is installed at the end of the robot arm, so its position and orientation can be controlled by changing the position and orientation of the robot arm. The robot is a Yaskawa MA1440 six-axis robot, which can be controlled directly with the DX200 robot cabinet. The industrial computer has 8 G RAM and i7-6700 CPU with a clock frequency of 2.60 GHz. The image of the workpiece surface is grabbed and sent to the industrial computer by the joint detection sensor. Then, the industrial computer performs image processing to obtain the position and orientation of the joint point in the camera coordinate system {C}. By combing them with the current position and orientation of the robot sent by the robot cabinet, the industrial computer performs coordinate transformation and calculates the position and orientation of the joint point in the base coordinate system {B}, namely, the target position and orientation of the robot.

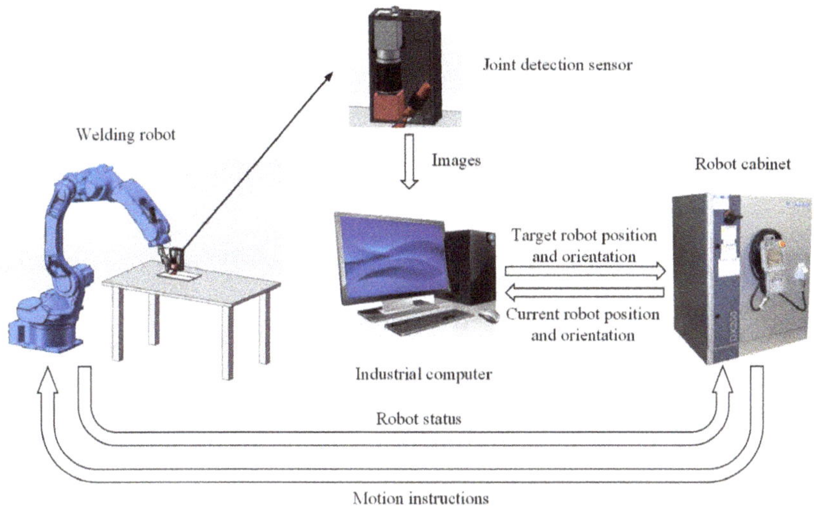

Figure 14. Schematic of the robotic seam tracking system for GTAW (gas tungsten arc welding).

5. Process and Result of the Joint Detection and Seam Tracking Experiment

In this section, a joint detection experiment is carried out with the robotic seam tracking system first. Then, a seam tracking experiment is carried out and in order to smooth the detection results and improve the tracking accuracy, different smoothing methods are used and their effects are compared.

5.1. Process and Results of the Joint Detection Experiment

The joint detection experiment was performed with the robotic seam tracking system described in Section 4, in which the plane workpieces used are shown in Figure 15. The width of the joint between these two workpieces was less than 0.2 mm. The frame rate of the camera was 10 fps, and every image was used to calculate the target position and orientation. It should be noted that in this experiment, the welding torch moved along the x_B axis of the base coordinate system {B} at a constant speed of 5 mm/s and did not change its motion status according to the detected result, so there was only joint detection and no seam tracking. The theoretical and detected results are shown in Figure 16 in which y_s^B, z_s^B, γ_s^B, and β_s^B are plotted against x_s^B, respectively. The theoretical results were calculated according to drawings of the plane workpieces in Figure 15. It can be seen that the position error does not exceed ± 0.15 mm and the angle error does not exceed $\pm 1.5°$, which indicates the effectiveness of the proposed detection method for narrow butt joints.

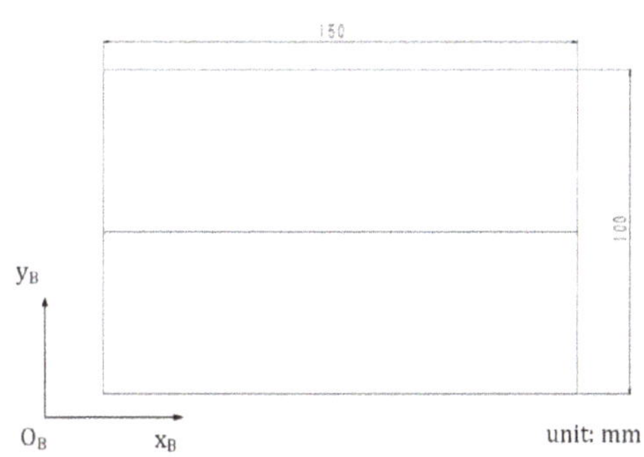

Figure 15. Dimension and orientation of the plane workpieces.

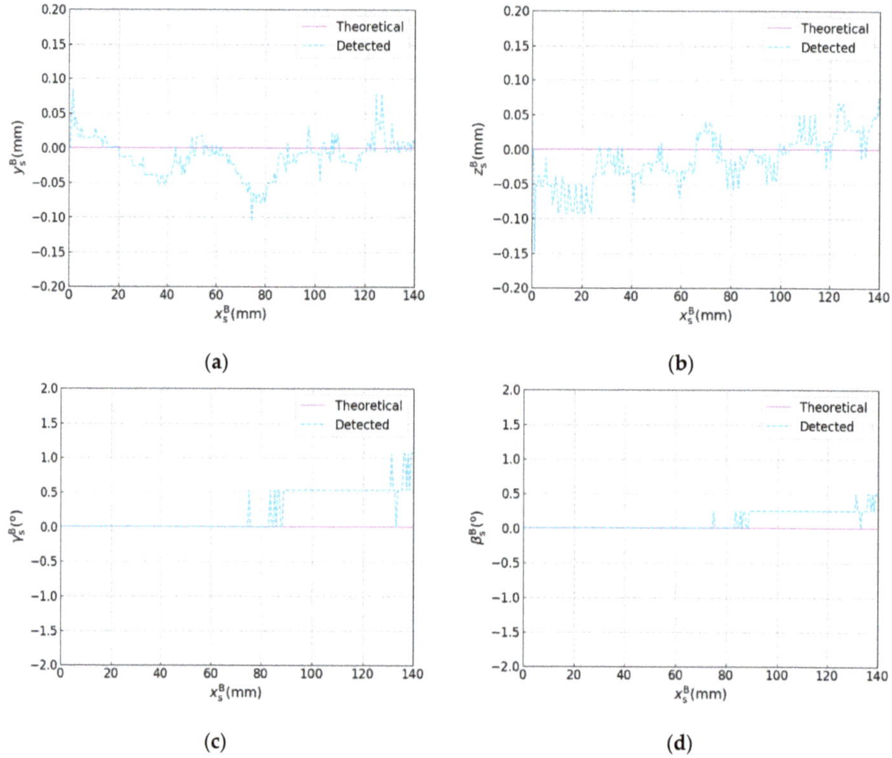

Figure 16. Results of the joint detection experiment with the plane workpieces. (**a**) y_s^B versus x_s^B. (**b**) z_s^B versus x_s^B. (**c**) γ_s^B versus x_s^B. (**d**) β_s^B versus x_s^B.

5.2. Process and Results of the Seam Tracking Experiment

The seam tracking experiment was performed with the robotic seam tracking system, in which the curved workpieces used are shown in Figure 17. The joint to be tracked was a curve with a 3D trajectory and width less than 0.2 mm. The frame rate of the camera was 10 fps and the linear speed and angular speed of the robot were 5 mm/s and 5 °/s, respectively. From the results of the joint detection experiment shown in Figure 16, some fluctuations can be noted. If these detected results $\left(x_s^B, y_s^B, z_s^B, \beta_s^B, \gamma_s^B\right)$ are used to guide the motion of the robot directly, dithering will happen in the robot's motion because the detected results are not smooth enough, which will affect the accuracy of the detection and tracking. So, in the seam tracking experiment, the detected results $\left(x_s^B, y_s^B, z_s^B, \beta_s^B, \gamma_s^B\right)$ need smoothing by a filter. Then, the smoothed results of position and orientation $\left(x_f^B, y_f^B, z_f^B, \beta_f^B, \gamma_f^B\right)$ were sent to a buffer, and the results in the buffer were sent to the robot in sequence to guide its motion. The existence of the buffer is thus necessary. For the robot, we could only send it the next target after it had reached the last target. Because the next target may be detected before the robot had reached its last target, we needed the buffer to store these newly detected targets. The process is shown in Figure 18. In addition, only the filtered results whose positions were at a minimal distance (1.5 mm here) from that of the previous filtered result were sent to the buffer to make sure that positions of the filtered results used to guide the motion of the robot were not too close, otherwise obvious pauses in the motion of the robot would have resulted.

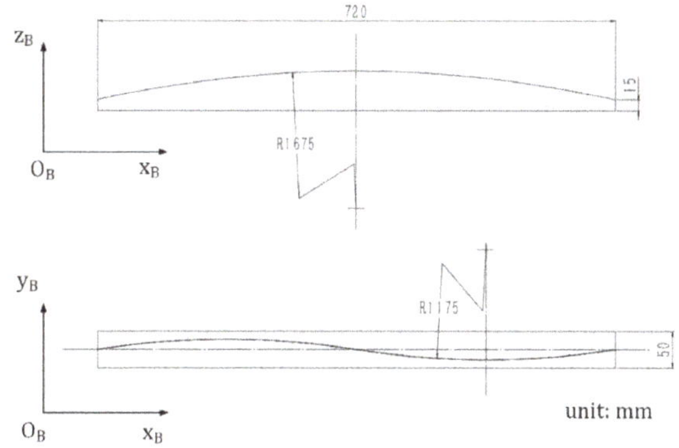

Figure 17. Dimension and orientation of the curved workpieces.

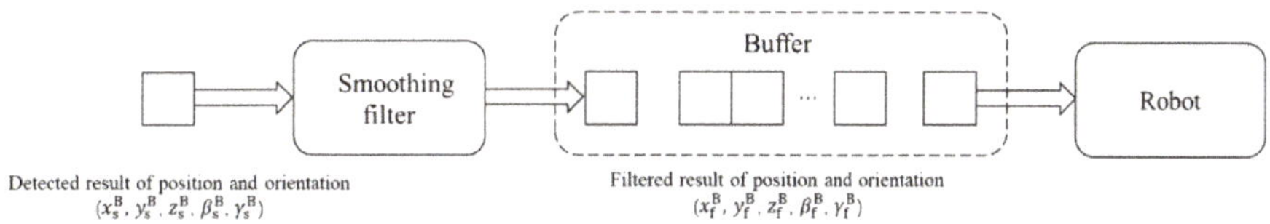

Figure 18. Filtering and sending of the position and orientation results.

Two smoothing methods were used to smooth the detected results, and their effects were compared.

The first smoothing method was the moving average (MA). The recent 10 detected results were taken into account in order to eliminate individual results with large errors. For each dimension in these results, the maximum and minimum were excluded and the average of the rest values were calculated and regarded as the filtered result. Taking x_s^B as an example, the filtered value of x_s^B is denoted as x_f^B, which can be calculated from the following formula:

$$x_f^B(i) = \frac{\sum x_s^B(i+k) - \max\left[x_s^B(i+k)\right] - \min\left[x_s^B(i+k)\right]}{8}, \ k = 0, 1, \ldots, 9 \quad (15)$$

where $x_s^B(i)$ is the ith value of x_s^B and $x_f^B(i)$ is the ith value of x_f^B. For y_s^B, z_s^B, β_s^B, and γ_s^B, the same method is applied and the filtered values are calculated, respectively.

The second smoothing method was the Kalman filter (KF) [25]. The state and measurement equations for a system can be described as:

$$\begin{cases} x_i = Ax_{i-1} + Bu_{i-1} + w_{i-1} \\ z_i = Hx_i + v_i \end{cases} \quad (16)$$

where x_i is the ith value of the variable, w_i and v_i are the process and measurement noise, respectively, and they are assumed to be independent, white, and with normal probability distributions, $p(w) \sim N(0, Q)$ and $p(v) \sim N(0, R)$, respectively, where Q is the process noise covariance and R is the measurement noise covariance. A is the state transformation matrix and B is the control matrix.

The Kalman filter iterated algorithm can be written as:

$$\begin{cases} \hat{x}_i^- = A\hat{x}_{i-1} + Bu_{i-1} \\ P_i^- = AP_{i-1}A^T + Q \\ K_i = P_i^- H^T \left(HP_i^- H^T + R\right)^{-1} \\ \hat{x}_i = \hat{x}_i^- + K_i\left(z_i - H\hat{x}_i^-\right) \\ P_i = (I - K_i H)P_i^- \end{cases} \quad (17)$$

where $\hat{x}_i^-$ is the ith priori state estimate, $\hat{x}_i$ is the ith posteriori state estimate, P_i^- is the ith priori estimate error covariance, P_i is the posteriori estimate error covariance, K_i. is the ith Kalman gain, z_i is the ith measurement, and H is the measurement matrix.

The Kalman filter was applied for each dimension of the detected results $(x_s^B, y_s^B, z_s^B, \beta_s^B, \gamma_s^B)$ to get the filtered results $(x_f^B, y_f^B, z_f^B, \beta_f^B, \gamma_f^B)$. Taking x_s^B as an example, the average of the first 10 values of x_s^B calculated from Equation (19) was regarded as $\hat{x}_0$. $x_s^B(i+10)$ was regarded as the measurement, z_i, so $H = 1$. Because it was unknown and uncontrollable how the position and orientation of the joint point would change, $A = 1$ and $B = 0$ were set. For the other parameters, $P_0 = 0$, $Q = 10^{-5}$, and $R = 0.01$, which were determined from experience. The posteriori estimate, $\hat{x}_i$, was regarded as the filtered value of x_f^B; that is, $x_f^B(i) = \hat{x}_i$.

The theoretical and filtered results are shown in Figure 19. The theoretical results were calculated according to the drawings of the curved workpieces in Figure 17. When MA was used, obvious dithering happens in the robot's motion. This indicates that MA was unable to smooth the detected results effectively, so the fluctuation of the detected results caused dithering of robot's motion as the robot performs seam tracking and its motion follows the filtered position and orientation. Because the calculation and the communication between the robot cabinet and the industrial computer need time, there was some delay (about tens of milliseconds in our experiment) between the grab of the image and the acquisition of the current position and orientation of the robot, which may bring some detection error into the coordinate transformation. When dithering starts to happen, the detection error will increase and in turn aggravate the dithering of the robot's motion. Compared with MA, the KF can smooth the detected results and eliminate the dithering of the robot's motion much more effectively, therefore, increasing the accuracy of joint detection and seam tracking significantly, which indicates that KF is quite applicable for the proposed robotic seam tracking system.

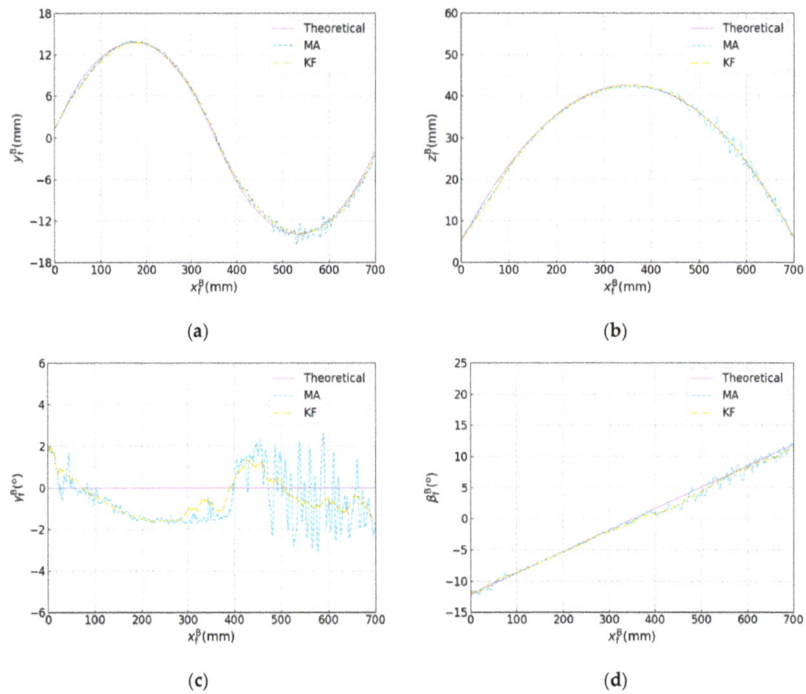

Figure 19. Results of the seam tracking experiment with the curved workpieces. **(a)** y_f^B versus x_f^B. **(b)** z_f^B versus x_f^B. **(c)** γ_f^B versus x_f^B. **(d)** β_f^B versus x_f^B.

Next, the processing time for the joint detection and smoothing (including image processing, coordinate transformation and filtering) were tested and compared. One hundred results of the position and orientation of the joint points were detected and smoothed with MA and KF, respectively. Mean values of the required time were 44.7 ms and 44.5 ms, and the standard deviations were 9.8 ms and 8.0 ms, respectively. It can be found that KF does not lead to an increase in the processing time compared with MA. Suppose the welding speed is 10 mm/s, the distance between two detected points is less than 0.5 mm, so the processing speed of the proposed joint detection method meets the real-time requirements to make the detected trajectory accurate enough.

6. Conclusions

A vision based detection method for a narrow butt joint was proposed in this paper. The proposed method can detect the 3D position of the narrow butt joint with a width of less than 0.2 mm and the normal vector of the workpiece surface simultaneously. The position error does not exceed ±0.15 mm and the angle error does not exceed ±1.5 °. In addition, the proposed detection method was applied in a robotic seam tracking system for GTAW. It was found that the Kalman filter can reduce the dithering of the robot and improve the tracking accuracy significantly compared with the moving average method, which indicates that KF is applicable for the proposed robotic seam tracking system.

Author Contributions: Conceptualization, B.X.; Data curation, Y.G., Z.T. and G.W.; Formal analysis, B.X.; Investigation, G.P.; Supervision, B.C., D.D. and G.W.; Writing—original draft, B.X.; Writing—review & editing, B.C. and D.D.

Funding: This research was funded by the National Natural Science Foundation of China grant number U1537205 and the National Defence Basic Scientific Research Project grant number JCKY2014203A001.

Acknowledgments: The research was financially supported by the National Natural Science Foundation of China (No.U1537205), and the National Defence Basic Scientific Research Project (No.JCKY2014203A001).

Conflicts of Interest: The authors declare no conflict of interest.

References

1. Hong, Y.; Chang, B.; Peng, G.; Yuan, Z.; Hou, X.; Xue, B.; Du, D. In-Process Monitoring of Lack of Fusion in Ultra-Thin Sheets Edge Welding Using Machine Vision. *Sensors* **2018**, *18*, 2411. [CrossRef] [PubMed]
2. Zeng, J.; Chang, B.; Du, D.; Wang, L.; Chang, S.; Peng, G.; Wang, W. A Weld Position Recognition Method Based on Directional and Structured Light Information Fusion in Multi-Layer/Multi-Pass Welding. *Sensors* **2018**, *18*, 129. [CrossRef] [PubMed]
3. Guo, J.; Zhu, Z.; Sun, B.; Yu, Y. Principle of an innovative visual sensor based on combined laser structured lights and its experimental verification. *Opt. Laser Technol.* **2019**, *111*, 35–44. [CrossRef]
4. Guo, J.; Zhu, Z.; Sun, B.; Yu, Y. A novel multifunctional visual sensor based on combined laser structured lights and its anti-jamming detection algorithms. *Weld. World* **2018**. [CrossRef]
5. Zou, Y.; Chen, T. Laser vision seam tracking system based on image processing and continuous convolution operator tracker. *Opt. Laser Eng.* **2018**, *105*, 141–149. [CrossRef]
6. Li, X.; Li, X.; Ge, S.S.; Khyam, M.O.; Luo, C. Automatic Welding Seam Tracking and Identification. *IEEE Trans. Ind. Electron.* **2017**, *64*, 7261–7271. [CrossRef]
7. Non-Contact Seam Tracking TH6D System. Available online: https://www.scansonic.de/en/products/th6d-optical-sensor (accessed on 29 December 2018).
8. Smart Laser Probe. Available online: http://meta-vs.com/slpr.html (accessed on 29 December 2018).
9. Zeng, J.; Chang, B.; Du, D.; Hong, Y.; Chang, S.; Zou, Y. A Precise Visual Method for Narrow Butt Detection in Specular Reflection Workpiece Welding. *Sensors* **2016**, *16*, 1480. [CrossRef] [PubMed]
10. Xu, Y.; Fang, G.; Chen, S.; Zou, J.J.; Ye, Z. Real-time image processing for vision-based weld seam tracking in robotic GMAW. *Int. J. Adv. Manuf. Technol.* **2014**, *73*, 1413–1425. [CrossRef]
11. Gao, X.; You, D.; Katayama, S. Seam Tracking Monitoring Based on Adaptive Kalman Filter Embedded Elman Neural Network During High-Power Fiber Laser Welding. *IEEE Trans. Ind. Electron.* **2012**, *59*, 4315–4325. [CrossRef]

12. Nilsen, M.; Sikström, F.; Christiansson, A.; Ancona, A. Vision and spectroscopic sensing for joint tracing in narrow gap laser butt welding. *Opt. Laser Technol.* **2017**, *96*, 107–116. [CrossRef]
13. Shah, H.N.M.; Sulaiman, M.; Shukor, A.Z.; Kamis, Z.; Rahman, A.A. Butt welding joints recognition and location identification by using local thresholding. *Robot. CIM-Int. Manuf.* **2018**, *51*, 181–188. [CrossRef]
14. Nele, L.; Sarno, E.; Keshari, A. An image acquisition system for real-time seam tracking. *Int. J. Adv. Manuf. Technol.* **2013**, *69*, 2099–2110. [CrossRef]
15. Kramer, S.; Fiedler, W.; Drenker, A.; Abels, P. Seam tracking with texture based image processing for laser materials processing. *Proc. SPIE* **2014**, *8963*, 89630P.
16. Gao, X.; Liu, Y.; You, D. Detection of micro-weld joint by magneto-optical imaging. *Opt. Laser Technol.* **2014**, *62*, 141–151. [CrossRef]
17. Fang, Z.; Xu, D.; Tan, M. Visual seam tracking system for butt weld of thin plate. *Int. J. Adv. Manuf. Technol.* **2010**, *49*, 519–526. [CrossRef]
18. Shao, W.J.; Huang, Y.; Zhang, Y. A novel weld seam detection method for space weld seam of narrow butt joint in laser welding. *Opt. Laser Technol.* **2018**, *99*, 39–51. [CrossRef]
19. Zeng, J.; Chang, B.; Du, D.; Peng, G.; Chang, S.; Hong, Y.; Wang, L.; Shan, J. A Vision-Aided 3D Path Teaching Method before Narrow Butt Joint Welding. *Sensors* **2017**, *17*, 1099. [CrossRef] [PubMed]
20. Peng, G.; Xue, B.; Gao, Y.; Tian, Z.; Hong, Y.; Chang, B.; Du, D. Vision sensing and surface fitting for real-time detection of tight butt joints. *J. Phys. Conf. Ser.* **2018**, *1074*, 12001. [CrossRef]
21. Gonzalez, R.C.; Woods, R.E. *Digital Image Processing*; Publishing House of Electronics Industry: Beijing, China, 2007; pp. 844–845.
22. Hough, P.V.C. Method and Means for Recognizing Complex Patterns. U.S. Patent 3069654, 18 December 1962.
23. Zhang, Z. A flexible new technique for camera calibration. *IEEE Trans. Pattern Anal.* **2000**, *22*, 1330–1334. [CrossRef]
24. de Graaf, M.; Aarts, R.; Jonker, B.; Meijer, J. Real-time seam tracking for robotic laser welding using trajectory-based control. *Control Eng. Pract.* **2010**, *18*, 944–953. [CrossRef]
25. Welch, G.; Bishop, G. *An Introduction to the Kalman Filter*; University of North Carolina: Chapel Hill, NC, USA, 1995; pp. 1–16.

sensors

MDPI

Article

Presentation Attack Detection for Iris Recognition System Using NIR Camera Sensor

Dat Tien Nguyen, Na Rae Baek, Tuyen Danh Pham and Kang Ryoung Park *

Division of Electronics and Electrical Engineering, Dongguk University, 30 Pildong-ro 1-gil, Jung-gu, Seoul 100-715, Korea; nguyentiendat@dongguk.edu (D.T.N.); naris27@dongguk.edu (N.R.B.); phamdanhtuyen@gmail.com (T.D.P.)

* Correspondence: parkgr@dongguk.edu; Tel.: +82-10-3111-7022; Fax: +82-2-2277-8735

Received: 28 March 2018; Accepted: 20 April 2018; Published: 24 April 2018

Abstract: Among biometric recognition systems such as fingerprint, finger-vein, or face, the iris recognition system has proven to be effective for achieving a high recognition accuracy and security level. However, several recent studies have indicated that an iris recognition system can be fooled by using presentation attack images that are recaptured using high-quality printed images or by contact lenses with printed iris patterns. As a result, this potential threat can reduce the security level of an iris recognition system. In this study, we propose a new presentation attack detection (PAD) method for an iris recognition system (iPAD) using a near infrared light (NIR) camera image. To detect presentation attack images, we first localized the iris region of the input iris image using circular edge detection (CED). Based on the result of iris localization, we extracted the image features using deep learning-based and handcrafted-based methods. The input iris images were then classified into real and presentation attack categories using support vector machines (SVM). Through extensive experiments with two public datasets, we show that our proposed method effectively solves the iris recognition presentation attack detection problem and produces detection accuracy superior to previous studies.

Keywords: iris recognition; presentation attack detection; convolutional neural network; support vector machines

1. Introduction

Over recent decades, biometric technology has gained much attention and is widely used in various applications to enhance user convenience and the security level of recognition systems compared to traditional recognition methods [1–9]. However, researchers have recently indicated that biometric recognition systems are vulnerable to attack by attackers presenting fake samples to data collecting systems [2,10–16]. Using appropriate artificial biometric features, an unauthorized person can be recognized as authorized by a biometric recognition system using either direct or indirect attack methods [16]. As a result, presentation attack detection methods are required to protect a biometric recognition system from attackers and enhance its security level.

Among the many biometric features, the iris pattern has been recently used for recognition because of its reliability and high security [3,9]. However, several studies have indicated that a fake iris pattern can be made by recapturing a real iris pattern or by printing an iris pattern on a contact lens to fool iris recognition systems. To address this problem, we propose a new presentation attack detection method for an iris recognition system by using hybrid image features and offer a classification method to overcome the limitations of previous research. Our proposed method is novel in five ways compared to previous research.

- First, this is the first approach to use a deep CNN model for iPAD to overcome the limitation of previous studies which adopted only shallow CNN networks. The trained CNN model can extract discriminative features for classifying real and presentation attack images because it is trained using a large amount of augmented training images.

- Second, since presentation attack images have special characteristics such as noise or discrete patterns of textures, we applied a multi-level local binary pattern (MLBP) method to extract these images features. The handcrafted image features can be seen as a complement to the deep features to enhance the classification result.
- Third, we combined the detection results based on MLBP and deep features to enhance the accuracy of the iPAD method. The combination was performed using feature level fusion and score level fusion. This is the first approach to combine handcrafted and deep features for iPAD.
- All previous research showed the performances of iPAD according to the individual iPAD dataset such as printed or contact lenses. However, we present the robustness of our method irrespective of the kinds of iPAD datasets through the evaluation with the fused datasets of printed and contact lenses.
- Finally, we made our trained models and algorithms for iPAD available to other researchers for comparison purposes [17].

2. Related Works

Previously, several methods have been proposed for detecting presentation attack images for iris recognition systems [18–24]. Generally, these studies can be classified into two groups, including iPAD methods based on expert-knowledge (handcrafted) image features and iPAD methods based on learning-based image features.

In the first group, authors mainly designed several feature extraction methods based on their expert knowledge of the problem. With the extracted image features, they performed classification methods such as support vector machines to detect real and presentation attack images [18–20]. One example of the first group for the iPAD method is the work by Gragnaniello et al. [18]. In this work, several local descriptors were used to detect iris images. Local descriptors such as the local binary pattern (LBP) and its variants, local phase quantization (LPQ), binarized statistical image features (BSIF), and shift-invariant descriptors (SID) were proven to be effective for detecting presentation attack images. However, as shown in their experimental results, the detection accuracy varied according to the kind of feature extraction methods and working datasets and reduced the reliability of the detection system. The BSIF feature extraction method was successfully used in a study by Doyle et al. [19] for detecting the textured contact lenses in an iris recognition system. One important result obtained from this study was that the accurate segmentation of the iris region is not required to obtain accurate detection results. In a study by Komogortsev et al. [21], the eye movement information was used for iris liveness detection. However, eye movements can be simulated by imposters who have expert-knowledge of the problem. Instead of using a gray-textured image, Raja et al. [22] used the information from different color channels to detect a presentation attack ocular image. As indicated from these studies, the handcrafted image features were effective for detecting presentation attack iris images.

In the second group, authors leave the details of feature extraction and classification behind the scenes by applying a learning-based method on a large amount of training data to train a detection model. For example, Silva et al. [23] used a convolutional neural network (CNN) called spoofnet to detect textured cosmetic contact lenses. Experimental results using the Notre Dame Contact Lens (NDCL-2013) dataset showed that the CNN method produced state-of-the-art detection results. However, using the IIIT-Delhi dataset, the CNN method produced less than state-of-the-art results. In addition, the spoofnet used in this research was relatively shallow (two convolution layers and one fully connected layer). This problem can affect the detection accuracy. Similar to this research, Menotti et al. [24] used a CNN network by applying two optimization schemes including structure optimization and filter optimization. They validated the detection performance for various biometric features such as face, fingerprint, and iris. Their proposed method combining the architecture and filter optimizations worked well for the fingerprint benchmark. However, their face and iris benchmarks produced detection results just comparable with state-of-the-art results. Again, the CNN networks used in this research were relatively shallow with two convolution layers and one fully connected dense layer. The results of these studies demonstrate that a deep convolutional neural network is effective for detecting presentation attack images for biometric recognition systems. However, in addition to the scarceness of training data, the use of a shallow network architecture can be a limitation of these studies. In Table 1, we summarize previous studies by considering the detection methods with their strengths and weaknesses.

Table 1. Summary of previous studies on iPAD systems.

Category	Method	Strength	Weakness
Expert-knowledge-based feature extraction methods	- Uses local descriptors such as LBP, LPQ, and BSIF for detecting presentation attack image [18–20]; Eye movement information [21]; and color information [22].	- Easy to implement. - Do not require a large amount of training data.	- Detection accuracy varies according to dataset. - Cross-sensor problems.
Learning-based feature extraction methods	- Uses convolutional network to extract image features and neural network with SoftMax regression for classification [23]. - Uses CNN with structure and filter optimization [24].	- Good detection accuracy. - Image features are learned using a large amount of training data similar to that of human brain.	- More complex than use of handcrafted image features. - Over-fitting problem. - Requires large amount of real and presentation attack images to successfully train CNN network.

The rest of our paper is organized as follows. In Section 3, we present the main structure of our proposed iPAD method and a detailed description of the technique. In Section 4, we perform various experiments using two public datasets to evaluate the detection performance of our proposed iPAD method and compare our experimental results with those of previous research and discuss our results. Finally, we provide concluding remarks in Section 5.

3. Proposed PAD Method for Iris Recognition System

3.1. Overview of Proposed Method

Figure 1 shows the overall flowchart of our proposed iPAD method. Similar to an iris recognition system, we first detected the iris region from the input iris image to localize the iris region. This step was necessary because the iris region can differentiate between a real and presentation attack iris image, while the other regions contain no or less discrimination information according to the attack method. Based on the detection result of this step, we extracted an iris region of interest and used this image to extract features for our proposed method. The detailed explanation of this step is given in Section 3.2.

We then extracted the image features in the localized region of interest produced by the preprocessing step. Our proposed method extracted the handcrafted features and deep features using the MLBP and a CNN method, respectively. The details of these image feature extraction methods are provided in Sections 3.3 and 3.4, respectively. As a result, we obtained a feature vector for the MLBP method and for the CNN method. These two feature vectors were then combined using feature level fusion and score level fusion approaches. A detailed description of each fusion method is provided in Section 3.5. Finally, we used a SVM to classify the input image into real and presentation attack classes using the extracted image features.

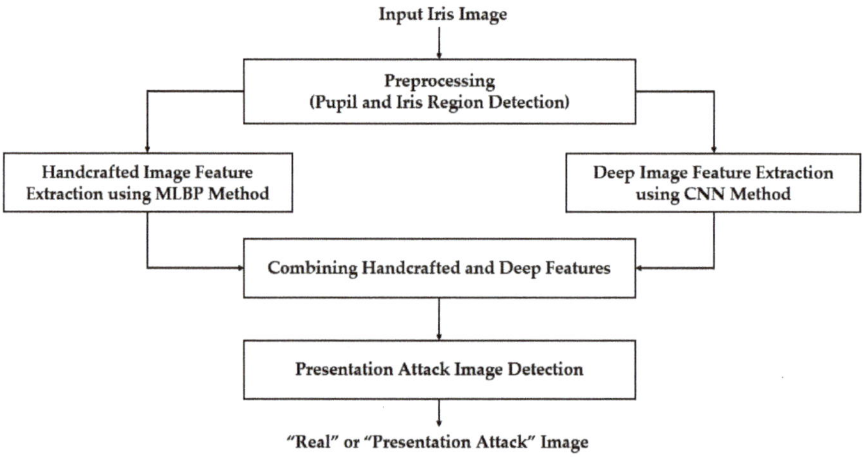

Figure 1. General flowchart of proposed iPAD method.

3.2. *Iris Region Detection Using Circular Edge Detection Method*

Since an iris recognition system uses the iris region to recognize individuals, attackers to this system attempt to create a presentation attack sample that is similar to that of the real image. Therefore, the iris region probably contains more discrimination information between real and presentation attack images than the sclera and skin regions in an iris image. Based on this observation, the first step in our proposed method was designed to detect the iris region in an input iris image. To efficiently detect the iris region, our proposed method used a sub-block-based template matching procedure to roughly detect the pupil region based on the characteristics of the iris image. Based on the result of pupil region detection, we continued to roughly localize the image region in which the iris region exists. Finally, we used the CED method to accurately detect the boundaries of the iris region as shown in Figure 2.

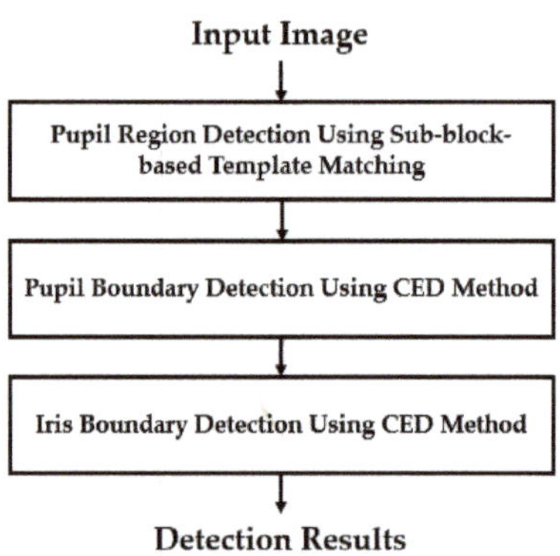

Figure 2. Flowchart of iris segmentation method in our study.

Inspired by the observation that the iris region of the human eye is displayed as a circular shape region in the iris image, the iris boundaries can be effectively detected by the CED method [25]. Although we can detect the iris boundaries using the CED method by searching the entire image, it incurs a long processing time because we must search the boundaries at various center positions and potential radius values. In addition, the effect of noise and abnormal texture can affect the detection result. To overcome this problem, our proposed method used a preprocessing method called the sub-block-based template matching method to detect the pupil region roughly first before detecting the iris boundaries using the CED method. Using NIR light, iris images are normally captured with a pupil region that is darker than other regions such as the iris sclera and skin regions. This characteristic is caused by the different absorption and reflection of NIR light in different regions of the human eye. Based on this characteristic, we used a sub-block-based template matching method to first localize the pupil region in a given iris image. The sub-block-based template matching was performed by measuring the difference in gray-levels of the sub-blocks that surround the pupil region with the center sub-block as shown in Figure 3. In this figure, at a center position (x, y) with block-size (s), we denote $U_{0,x,y,s}$ as the average gray-level of the center sub-block and $U_{i,x,y,s}$ (i = 1, ... , 8) as the average gray-levels of the surrounding sub-blocks. As a result, if the center sub-block contains the pupil region, its average gray level ($U_{0,x,y,s}$) is much smaller than those of the surrounding sub-blocks ($U_{i,x,y,s}$). Based on this observation, we detected the pupil region in a given iris image by using Equation (1) with the condition that $U_{0,x,y,s}$ is smaller than $U_{i,x,y,s}$ (i = 1, ... , 8). Furthermore, to speed up the processing of this step, the integral image was used to quickly calculate the average gray-level of the sub-blocks [26]. An example result of the pupil detection step is shown in Figure 4 with a rectangular bounding box.

$$\underset{x,y,s}{\operatorname{argmax}}\left(\sum_{i=1}^{8}\left(U_{i,x,y,s}-U_{0,x,y,s}\right)\right) \tag{1}$$

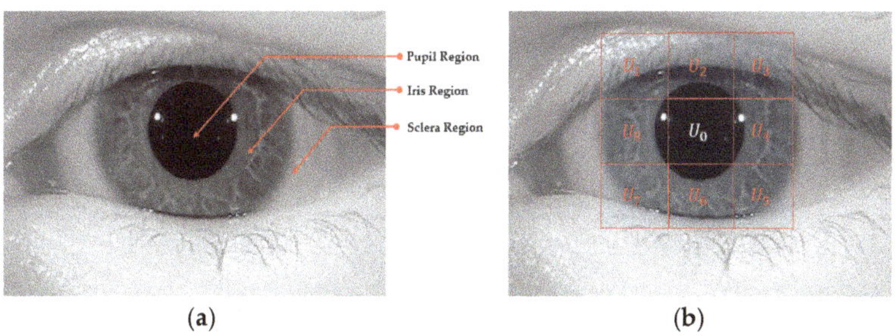

Figure 3. Block-based method for pupil region detection. (**a**) Pupil, iris, and sclera regions of eye image. (**b**) Example of 9 sub-blocks on pupil and iris regions.

We then accurately detected the iris boundaries based on the detection result of the pupil region using the CED method [25–27]. As shown in Figures 3 and 4, the center of the iris and pupil regions are pixels inside the bounding-box of the pupil region. In addition, the radius of the pupil region is smaller than that of the iris region. Based on this observation, we used two circular edge detectors to find the boundaries of the pupil and iris regions. The pupil region normally appears as a complete circle. Therefore, we first used the complete circular edge detector shown in Equation (2) to detect the boundary of the pupil region. In this equation, r and (x_c, y_c) are the radius and center position of the pupil region. However, the iris region can be occluded by some additional regions such as the eyelid, eyelash, or eyebrow. As a result, the boundary of the iris region can be not continuous. To overcome this problem, we used the CED method in a limited circular range. As suggested by previous research [26], we used the circular range of $-45°$ to $+30°$ and $+150°$ to $+225°$ as shown in Equation (3). In this equation, r' and (x'_c, y'_c) are the radius and center position of the iris region. In Figure 4, we show an example of the result of our iris detection method.

$$\underset{x_c,y_c,r}{\operatorname{argmax}}\left[\frac{\partial}{\partial r}\int_{0}^{2\pi}\frac{I(x_c+r\cos\theta,y_c+rsin\theta)}{2\pi r}d\theta\right] \tag{2}$$

$$\underset{x'_c, y'_c, r'}{\operatorname{argmax}} \left[\frac{\partial}{\partial r} \left(\int_{-\frac{\pi}{4}}^{\frac{\pi}{6}} \frac{I(x'_c + r' cos\theta, y'_c + r' sin\theta)}{5\pi r'/12} d\theta + \int_{\frac{5\pi}{6}}^{\frac{5\pi}{4}} \frac{I(x'_c + r' cos\theta, y'_c + r' sin\theta)}{5\pi r'/12} d\theta \right) \right] \quad (3)$$

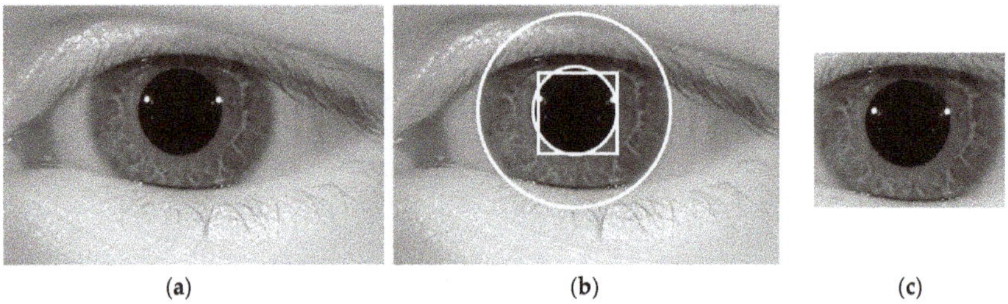

Figure 4. Example of detection results of pupil and iris boundary detection method: (**a**) input iris image; (**b**) detection results of sub-block-based pupil detection (rectangular box) and CED for pupil and iris region detection; and (**c**) final iris image to input iPAD system.

As shown in Figure 1, our proposed iPAD method uses CNN method for extracting deep image features. As we will show in next section, the CNN network requires the 3-channel input images. To make the input images for CNN network, we localized the iris region of interest (ROI) based on the detection results of pupil and iris detection method and made the final iris images for iPAD system by scaling the iris ROIs to the size of 224-by-224-by-3 images using bilinear interpolation method. Because the iris ROI is gray image, we duplicated it into the 3 channels, and obtained the 3-channel image. In Figure 4c, we showed an example of iris image that is used to input into iPAD system in our study.

3.3. *Image Feature Extraction Based on MLBP Method*

In Figure 5, we show an example of one real and two presentation attack iris images according to two different attack methods using a printed image and a contact lens. As shown in this figure, while the real iris image contains very clear iris patterns and fine texture features, the presentation attack images contain dot noise and broken textures (Figure 5b,c) because of the effects of printed iris patterns on paper or on a contact lens. Based on this observation, our proposed method used the LBP method to extract the image features for the iPAD.

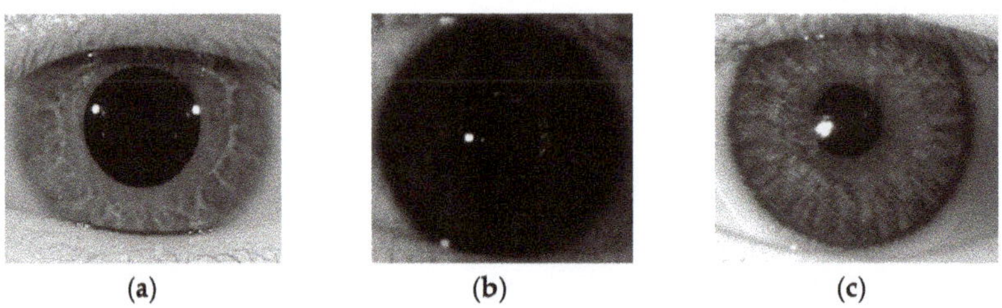

Figure 5. Example of NIR iris images: (**a**) real image; (**b**) presentation attack image obtained by recapturing a printed sample on paper; and (**c**) presentation attack image obtained by recapturing a contact lens.

As indicated by previous studies, the LBP method is a very efficient image feature extraction method in image processing and computer vision research by providing illumination and rotation invariant characteristics to extracted image features [28–30]. Furthermore, the LBP descriptor describes well the micro-texture features such as blob, edge, corner, and flat regions. By definition, the LBP method encodes each center pixel of a given image by a sequence of P (bits) using P surrounding pixels of the center pixel with a radius of R as shown in Equation (4). The LBP operator works as an adaptive thresholding function and offers the illumination invariant to the image features extracted by the LBP method.

$$\text{LBP}_{R,P} = \sum_{i=0}^{P-1} s(g_i - g_c)2^i \quad \text{where} \quad s(x) = \begin{cases} 1\,if\ x \geq 0 \\ 0\,if\ x < 0 \end{cases} \tag{4}$$

To extract the image features for the iPAD, we classified the LBP descriptors of pixels in a given image into two categories of uniform and non-uniform patterns. By definition, the uniform patterns are patterns that have at most two bit-wise transitions from 0 to 1 or 1 to 0, and the non-uniform patterns are those that have more than two bit-wise transitions from 0 to 1 or 1 to 0. The reason for this classification is that the uniform patterns effectively describe various useful micro-texture features such as blob, corner, edge, or flat regions [28–30], while the non-uniform patterns are complex and normally caused by noise and non-uniform texture patterns. In Figure 6, we show an example of the ability of an LBP descriptor to represent several micro-texture features such as blob, corner, and edge. As we explained at the beginning of this section, the definition of the LBP method is suitable for discriminating between real and presentation attack images because the presentation attack iris images can contain dot noise and non-ideal image texture features.

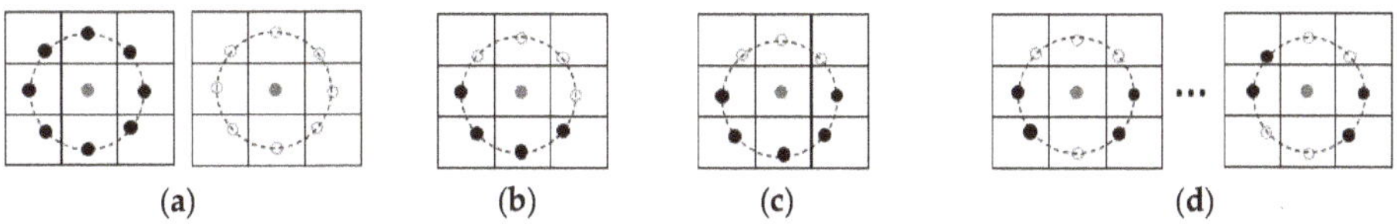

Figure 6. Example of LBP descriptors for representing micro-texture features: (**a**) flat/blob textures; (**b**) edge texture; (**c**) corner texture; and (**d**) complex noise-sensitive texture features.

As the final step, we constructed the image feature vector by accumulating the histogram of uniform and non-uniform patterns over the image. The histogram features effectively describe the characteristics of image texture because the histograms of uniform and non-uniform patterns statistically measure the distribution of micro-texture features over an iris image. Suppose that we used an LBP operator with radius R and number of surrounding pixels P to extract image features, the dimension of the extracted image features is given by Equation (5).

$$DIM_{\text{LBP}} = P \times (P-1) + 3 \tag{5}$$

As suggested from previous studies, our study accumulated the LBP features for an iris image by concatenating histogram features obtained from hyper-parameters such as radius (R) and number of representation pixels (number of surrounding pixels, P). The MLBP method was used to capture richer information from iris images than conventional LBP methods [30]. In our experiment, we used various values for radius (R in range from 1 to 3) and number of surrounding pixels (P of 8, 12, and 16) for MLBP feature extraction method. As a result, we extracted a 933-dimensional image feature vector for iPAD.

3.4. Image Feature Extraction Based on CNN Method

As shown in Figure 1, our proposed method used MLBP and CNN methods to extract image features for iPAD. While the MLBP is a hand-designed feature extraction method, the CNN method is a learning-based feature extraction method based on a learning procedure to learn a model that is applicable for feature extraction and classification. In literature, this method has been successfully used in various computer vision systems such as image classification [31–34], object detection [35,36], face recognition [37], gender recognition [38], and even the PAD problem [2,22,23]. As shown in these studies, the CNN method can produce state-of-the-art results compared to previous hand-designed methods. In the field of iris recognition, the CNN method has also successfully used and provided state-of-the-art recognition accuracy [39,40]. In the study by Gangwar et al. [39], two deep CNN networks named as DeepIrisNet-A (with 8 convolutional layers and 3 fully connected layers) and DeepIrisNet-B (with 5 conventional convolutional layers, 2 inception layers, and 3 fully connected layers) were used for iris recognition. The results of this study show that the CNN method is effective at not only enhancing the recognition accuracy but also robust to cross-sensor recognition. In a recent research conducted by Nguyen et al. [40], they used several pre-trained CNN models including AlexNet, VGGNet, InceptionNet, ResNet, and DenseNet to extract image features for iris recognition. Based on their experimental results, the CNN method outperformed the baseline iris recognition method although the CNN models were trained for a different task. Inspired by these previous studies, we used the CNN method to extract the deep features for iPAD.

In Table 2, we provide a detailed description of the CNN network architecture in our study. The CNN network was based on the very deep network proposed by Simonyan et al. [32] called VGG Net-19. The network architecture is depicted in Figure 7. Generally, a CNN network consists of two main components of convolution layers and fully-connected layers [31,32]. The convolution layers are responsible for image manipulation to extract image features using an image filtering technique, and the fully-connected layers are used to classify the extracted image features into several categories of desired class labels. In addition to these two main components, a CNN model can contain several layers such as activation layers (using sigmoid, tanh, or rectified linear unit (ReLU) functions), pooling layers (max or average pooling), and SoftMax layers. As shown in Table 2 and Figure 7, our CNN network consisted of 19 weight layers (16 convolution layers and three fully-connected layers) followed by several ReLU and max pooling layers. In addition, the last fully-connected layer in our study contained only two neurons which stand for "real" and "presentation attack" classes instead of the 1000 neurons used in the original VGG Net-19 [32]. In this table, we grouped several convolution layers which have same parameters together as denoted as G_1, G_2 ... G_8 in Table 2 and Figure 7. For example, the G_0 group contains two convolutional layers which have same parameters of the number of filters (64 filters), filter size (3 × 3 pixels), stride (1 × 1 pixel) and padding (1 × 1 pixel). The output of convolutional layers is 512 feature maps of the size of 7 × 7 pixels taken at the end of the G_5 group. In total, we obtained 25,088 activation neurons after 16 convolutional layers. These output neurons are connected to 4096 neurons in the next fully connected layer of the G_6 group by fully interconnection based on weighted summation. For example, the value to the 1st one of 4096 neurons is calculated by $w_1 \times o_1 + w_2 \times o_2 + \ldots + w_{25088} \times o_{25088}$ where $o_1, o_2, \ldots o_{25088}$ are the values from 25,088 activation neurons, and $w_1, w_2, \ldots w_{25088}$ are the weights for interconnection.

An optimal CNN model for a given problem can be obtained using a training procedure using a large amount of training data through which the filter's coefficients and weights of fully connected layers are efficiently learned with respect to the ground-truth labels of images. However, the CNN method always faces the problem of over-fitting because the network contains a very large number of parameters (filter coefficients and weights of fully connected layers) and because of the small training dataset and/or poor network parameter initialization. To reduce the over-fitting problem of the CNN network, we applied the dropout method to the first two fully-connected layers with a dropout value of 0.5. In addition, we used a pre-trained model that was successfully trained using ImageNet dataset [32] to initialize the weights of our CNN model. With the initialized network, we re-trained the whole network parameters (training from scratch). This is different procedure form conventional transfer learning [41]. We used the stochastic gradient descent method with momentum to train the CNN models [31]. The detailed parameters of training process are given in Table 3. To extract the image features using the CNN method, we extracted the activations of the second fully-connected layers (G_7 in Figure 7) and used them as the extracted features of the input images. Although it is possible to use the other layers (convolution layers or fully-connected layers) for feature extraction, the use of the deeper layer contains more information than that of the shallower layers. As a result, we extracted a 4096-component feature vector for our proposed iPAD.

Table 2. Description of CNN architecture used for iPAD in our study.

Operation Group	Operation	Layer Name	Number of Filters	Filter Size	Stride Size	Padding Size	Output Size
Group_0 (G_0)	Input image	Input layer	n/a	n/a	n/a	n/a	224 × 224 × 3
Group 1 (G_1)	Convolution (2 times)	Convolution layer	64	3 × 3 × 3	1 × 1	1 × 1	224 × 224 × 64
		ReLU layer	n/a	n/a	n/a	n/a	224 × 224 × 64
	Pooling	Max pooling layer	1	2 × 2	2 × 2	0	112 × 112 × 64
Group_2 (G_2)	Convolution (2 times)	Convolution layer	128	3 × 3 × 64	1 × 1	1 × 1	112 × 112 × 128
		ReLU layer	n/a	n/a	n/a	n/a	112 × 112 × 128
	Pooling	Max pooling layer	1	2 × 2	2 × 2	0	56 × 56 × 128
Group_3 (G_3)	Convolution (4 times)	Convolution layer	256	3 × 3 × 128	1 × 1	1 × 1	56 × 56 × 256
		ReLU layer	n/a	n/a	n/a	n/a	56 × 56 × 256
	Pooling	Max pooling layer	1	2 × 2	2 × 2	0	28 × 28 × 256
Group_4 (G_4)	Convolution (4 times)	Convolution layer	512	3 × 3 × 256	1 × 1	1 × 1	28 × 28 × 512
		ReLU layer	n/a	n/a	n/a	n/a	28 × 28 × 512
	Pooling	Max pooling layer	1	2 × 2	2 × 2	0	14 × 14 × 512
Group_5 (G_5)	Convolution (4 times)	Convolution layer	512	3 × 3 × 512	1 × 1	1 × 1	14 × 14 × 512
		ReLU layer	n/a	n/a	n/a	n/a	14 × 14 × 512
	Pooling	Max pooling layer	1	2 × 2	2 × 2	0	7 × 7 × 512
Group_6 (G_6)	Inner Product	Fully connected layer	n/a	n/a	n/a	n/a	4096
		ReLU layer	n/a	n/a	n/a	n/a	4096
	Dropout	Dropout layer (dropout = 0.5)	n/a	n/a	n/a	n/a	4096
Group_7 (G_7)	Inner Product	Fully connected layer	n/a	n/a	n/a	n/a	4096
		ReLU layer	n/a	n/a	n/a	n/a	4096
	Dropout	Dropout layer (dropout = 0.5)	n/a	n/a	n/a	n/a	4096
Group_8 (G_8)	Inner Product	Output layer	n/a	n/a	n/a	n/a	2
	Softmax	Softmax layer	n/a	n/a	n/a	n/a	2
	Classification	Classification layer	n/a	n/a	n/a	n/a	2

Table 3. Parameters for training CNN models in our experiments.

Momentum	Mini-Batch Size	Initial Learning Rate	Learning Rate Drop Factor	Learning Rate Drop Period (Epochs)	Number of Epochs
0.90	32	0.001	0.1	3	9

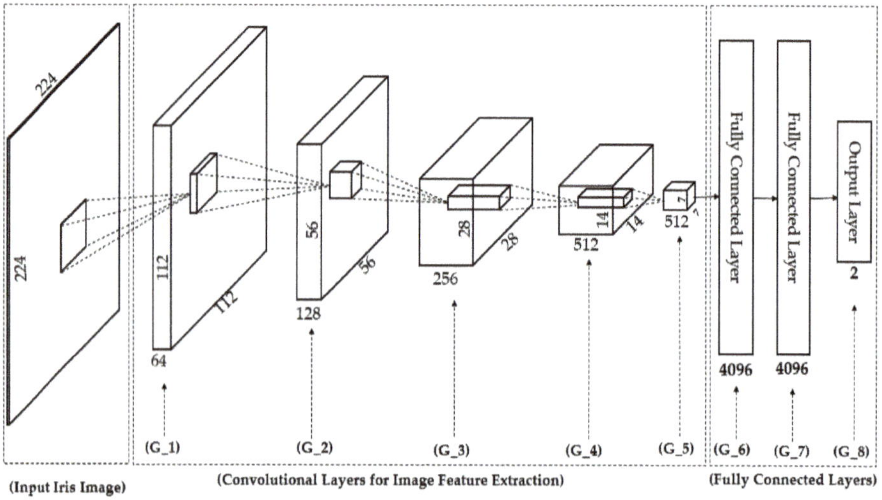

Figure 7. Visualization of convolutional neural network architecture in Table 2.

3.5. Image Feature Extraction and Detection Using SVM Method

Using the two feature extraction methods mentioned in Sections 3.3 and 3.4 (i.e., MLBP and CNN), we extracted two corresponding feature vectors of f_1 and f_2 for the MLBP and CNN features, respectively. These two feature vectors can contain different information for our iPAD because they were extracted using two different methods. As the main contribution of our proposed method, the information from the two feature vectors was combined to enhance the detection accuracy of the iPAD system. As explained in Section 3.1, we used the feature level fusion and score level fusion approaches for this step.

For the first fusion method, we combined the two vectors to form a new feature vector called the hybrid feature vector, to represent the input image. As a result, the flowchart of our proposed method in Figure 1 changed to that of Figure 8. For this purpose, we first normalized each feature vector to a zero-mean and unit standard deviation using the z-score normalization method shown in Equation (6) [28]. In this equation, f_{mean} and σ are the mean and the standard deviation vector obtained by a training dataset, respectively. Using this equation, we normalized the extracted feature vectors f_1 and f_2 and obtained the two corresponding normalized feature vectors, f_1^{norm} and f_2^{norm}. Finally, the hybrid feature f_{hybrid} was formed by simply concatenating the two normalized feature vectors as shown in Equation (7).

$$f^{norm} = \frac{f - f_{mean}}{\sigma} \tag{6}$$

$$f_{hybrid} = [f_1^{norm}, f_2^{norm}] \tag{7}$$

Although we can extract richer information to combat presentation attacks by using the hybrid feature vector rather than using only the MLBP or CNN feature vector, the iPAD system must process data in a higher dimensional space in later steps (classification step) than that of an individual feature vector. This problem increases the processing time for both the training and testing phases and the complexity of the classification model. To overcome this problem, we further reduced the dimension of the hybrid feature vector using a subspace method called principal component analysis (PCA). This well-known method reduces the dimension of data by constructing a low dimensional space in which the original data are well represented [28,30]. Originally, we extracted a 4096-dimensional feature vector using CNN-based method using the second fully-connected layer of CNN network in

Table 2. For the MLBP feature, we extracted image feature using various values of LBP parameters (radius (R) from 1 to 3 and resolution (P) of 8, 12 and 16). Consequently, we extract a feature vector in 933-dimensional space. As a result, the hybrid feature vector is a 5029-dimensional vector. In our experiments, we used the PCA for obtaining the optimal dimension of features before using SVM method for classification. In details, we used the number of principal component of 512 which is much smaller than the dimension of original features. The use of this reduced number of feature dimension helps us to lessen the complexity of classifiers, processing time, and effects of noise. As the final step of this fusion approach, we classified the input image into real and presentation attack classes using extracted image features. For this purpose, we used an up-to-date classification method based on SVMs for the classification problem. Conventionally, the SVM method constructs a classifier using several data points called support vectors and uses it to classify new input features into classes by evaluating the sign of evaluation function in Equation (8). In this equation, x_i and y_i are the support vectors and its corresponding class label, a_i and b are the parameters of the classifier, and $K(x, x_i)$ is the SVM kernel function, a hyper-parameter of the SVM method [42]. These classifier parameters are trained using training data and saved to predict the class label of new input features. In our experiments, we used three different kinds of kernel functions, including the linear, radial basis function (RBF), and polynomial kernel functions as shown in Equations (9)-(11) [42–44].

$$f(x) = sign(\sum_{i=1}^{k} a_i y_i K(x, x_i) + b) \tag{8}$$

$$\text{Linear kernel}: \ K(x_i, x_j) = {x_i}^T x_j \tag{9}$$

$$\text{RBF kernel}: \ K(x_i, x_j) = e^{-\gamma \|x_i - x_j\|^2} \tag{10}$$

$$\text{Polynomial kernel}: \ K(x_i, x_j) = \left(\gamma {x_i}^T x_j + coef\right)^{degree} \tag{11}$$

Moreover, the combination of handcrafted and deep features can be done by another combination method called score level fusion [45]. For this combination method, the overall detection system in Figure 1 changed to that of Figure 9. In this configuration, the handcrafted and deep features are used separately for iPAD. The results of each iPAD system are scored to represent the probability of the input image belonging to either a real or presentation attack class. The two scores are combined by the weighted sum rule to make a final detection result as shown in Equation (12). In this equation, S_1 and S_2 are the decision scores of the PAD system based on only deep or only handcrafted image features, respectively. These scores are combined using two weight values of w_1 and w_2 whose sum is 1 as shown in Equation (13) to produce a final detection score S. In our experiment, we chose the optimal pair of w_1 and w_2 which produced the best classification accuracy of real and presentation attack on training dataset.

$$S = w_1 S_1 + w_2 S_2 \tag{12}$$

$$w_1 + w_2 = 1 \tag{13}$$

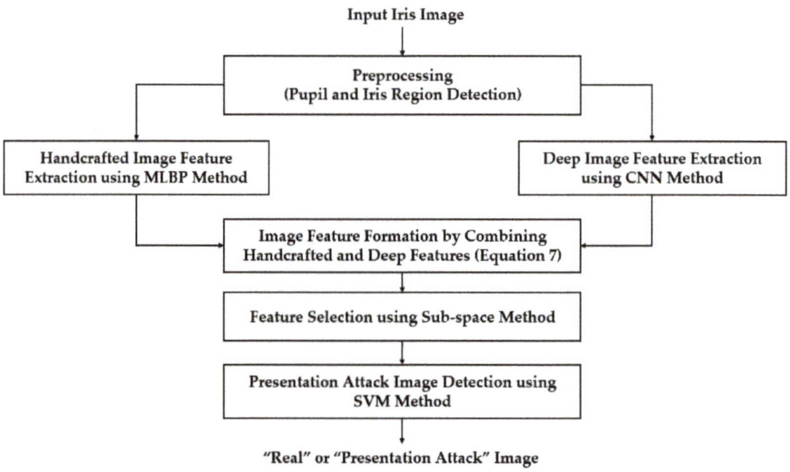

Figure 8. Flow chart of proposed iPAD method based on feature level fusion approach.

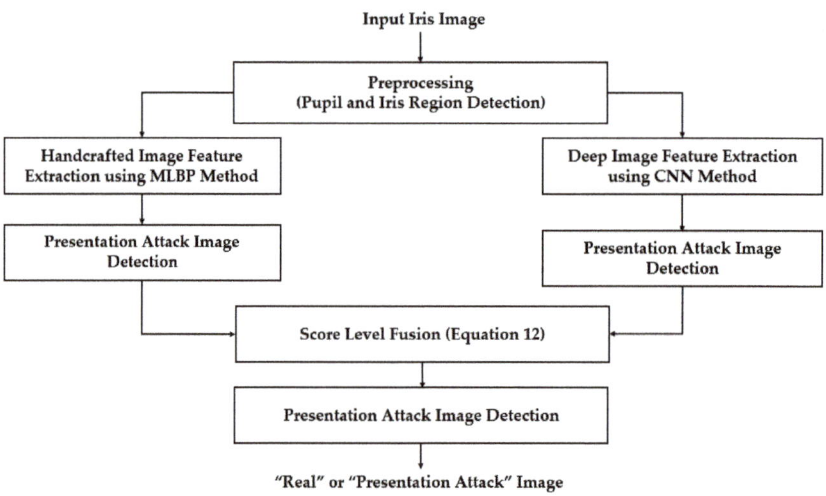

Figure 9. Flow chart of proposed iPAD method based on score level fusion approach.

Using the SVM method, we classified the input images into either the real or presentation attack class. To evaluate the performance of our proposed iPAD method and to compare it with previous studies, we used the standard criteria, called average classification error rate (ACER), to measure the detection performance [2,46–48]. By definition, the ACER is a measurement of the average error rate of the attack presentation classification error rate (APCER) and the bona-fide presentation classification error rate (BPCER). In a PAD system, the APCER indicates the proportion of attack presentation images incorrectly classified as bona-fide presentation attack images, and BPCER indicates the proportion of bona-fide presentation attack images incorrectly classified as attack presentation images. The ACER was measured using Equation (14). Since the ACER indicates the error rate of a detection system, a lower value indicates better detection performance (small error). We used the training data to train the CNN model, PCA coefficients, and the SVM classifier. Consequently, the performance of the detection system (APCER, BPCER, and ACER) was measured using testing data. In experiments, we used the MATLAB environment for constructing and training the CNN model, image feature extraction, PCA, and SVM-based classification [49–51].

$$ACER = \frac{APCER + BPCER}{2} \quad (14)$$

4. Experimental Results

4.1. Datasets

To evaluate the detection performance of our proposed iPAD method, we used two public datasets LivDet-Iris 2017-Warsaw [48] and Notre Dame Contact Lens Detection (NDCLD2015) [48,52]. For convenience, we refer to these datasets as Warsaw2017 and ND2015 in our study. Although there are other presentation attack iris image datasets such as IIITD-WVU, Clarkson [48], and PAVID [53], they were unavailable to us via internet request. In addition, the datasets we chose have been used in previous iPAD studies (LivDet-Iris 2017 competition [48]). The use of these datasets allowed us to compare the detection performance of our proposed method with those of previous studies.

The Warsaw2017 dataset contains 5168 real and 6845 presentation attack iris images obtained from 468 unique iris patterns with an image resolution of 640 × 480 pixels. This dataset was used in the LivDet-Iris 2017 iPAD competition and is the extended version of the two previous datasets of LivDet-Iris 2013 [54] and LivDet-Iris 2015 [52]. The presentation attack iris images in the Warsaw2017 dataset were collected by simulating a simple attack method by which the attackers use a printed sample of an iris pattern on paper to fool an iris recognition system during the image acquisition stage. A general statistical description of the Warsaw2017 dataset is given in the upper part of Table 4. Similar to the Warsaw2017 dataset, the ND2015 dataset was also used in the LivDet-Iris 2017 competition [48]. However, the presentation attack iris images in this dataset were simulated by iris patterns printed on a contact lens. Using this method, the presentation attack iris images look more like real ones than those of the Warsaw2017 dataset. The ND2015 dataset was first collected for the purpose of detecting whether a user used contact lenses [19]. This dataset was further used for detecting the presentation attack iris image in the LivDet-Iris 2017 competition because the fake iris images in this dataset simulate an attack method by which iris patterns are printed on the surface of a contact lens. In the lower part of Table 4, we show the general descriptions of the ND2015 dataset.

Table 4. Description of Warsaw2017 and ND2015 datasets.

Dataset	Number of Real Images	Number of Attack Images	Total	Collection Method
Warsaw2017	5168	6845	12,013	Recaptured printed iris patterns on paper
ND2015	4875	2425	7300	Recaptured printed iris patterns on contact lens

4.2. Detection Performance for Attack Method Based on Printed Samples

As our first experiment, we investigated the detection performance of our proposed iPAD method for the attack method based on printed paper samples. For this purpose, we used the Warsaw2017 dataset. In addition, we also measured the detection performances of iPAD systems that use only CNN method as classifier, CNN or MLBP features for comparison purposes. For evaluating the performance of an iPAD method, the Warsaw2017 dataset was preclassified into the three sub-datasets of training, test-known, and test-unknown. The training sub-dataset was used to construct the classification model, while the two testing sub-datasets were used for evaluating the performance of the trained model. The training and test-known sub-datasets were collected using the same capturing devices (Iris Guard AD 100), while the test-unknown dataset was collected using a different capturing device (a lab mate camera [48]). The use of the test-unknown dataset allowed us to evaluate the performance of the detection system for cross-sensor configuration. A detailed description of these training and testing sub-datasets is provided in Table 5. As shown in this table, we used 4513 images (1844 real and

2669 presentation attack images) for training. To test the detection model, 2990 images (974 real and 2016 presentation attack images) were used for the test-known dataset and 4510 images (2350 real and 2160 presentation attack image) were used for the test-unknown dataset. We generalized the training dataset by artificially making augmented images from original images to reduce the over-fitting tendency of the CNN method. In detail, we artificially made eight additional images from each original presentation attack iris image and an additional 14 images from each real iris image using shifting, cropping, and scaling method. This augmentation method has been also used in previous research [31]. Consequently, we increased the number of training images from 4513 to 51,681 images. The different number of artificial images for real and presentation attack was used because the number of original real iris images was much smaller than that of the presentation attack images. By using a different number of artificial images for each class, we made the number of images of each class similar in order to reduce over-fitting during the training process. A description of these sub-datasets and the corresponding augmented dataset are provided in Table 5. Data augmentation was performed for only the training data, and the testing data remained the same as the original. This approach was used to ensure a fair comparison of detection performance of our study with previous studies. Using the augmented train dataset, we performed the training procedure to train the CNN, PCA, and SVM models for the iPAD system. The experimental results on test datasets are given in Figure 10.

Table 5. Description of training and testing data used with Warsaw2017 dataset.

Dataset	Training Dataset			Testing Dataset					
				Test-Known Dataset			Test-Unknown Dataset		
	Real Image	Attack Image	Total	Real Image	Attack Image	Total	Real Image	Attack Image	Total
Original dataset	1844	2669	4513	974	2016	2990	2350	2160	4510
Augmented dataset	27,660 (1844 × 15)	24,021 (2669 × 9)	51,681	974	2016	2990	2350	2160	4510

As shown in Figure 10, we obtained the best detection error of 0.000% using the test-known dataset for the iPAD systems using only CNN, MLBP, or hybrid features. There are two reasons for this result. First, the presentation attack iris images in the Warsaw2017 dataset were collected by recapturing the printed iris samples on paper. Therefore, the presentation attack iris images inherit many differences from real images such as broken textures and printing noise. In addition, as explained above, the test-known dataset was collected using the same capturing procedure and devices as the training dataset. Consequently, the characteristics of images in the training and testing datasets were very similar. Therefore, we obtained very good detection results using the test-known dataset. However, the situation was little changed using the test-unknown dataset. We obtained an error (ACER) of 0.423% using the iPAD method that used only CNN features with the polynomial kernel of the SVM method. The iPAD method that used only MLBP features produced an error of 0.357% using the polynomial kernel of the SVM method. Our proposed hybrid features iPAD method produced an error of 0.242% using the polynomial kernel of the SVM method. The iPAD system detection errors using the test-unknown dataset were higher than those using the test-known dataset because the test-unknown dataset was collected using different capturing devices than that of the test-known dataset. Consequently, it caused several differences in the characteristics of the images of the two datasets. From these results, we conclude that the hybrid features iPAD method outperformed the conventional CNN and MBLP image features by producing the lowest detection error.

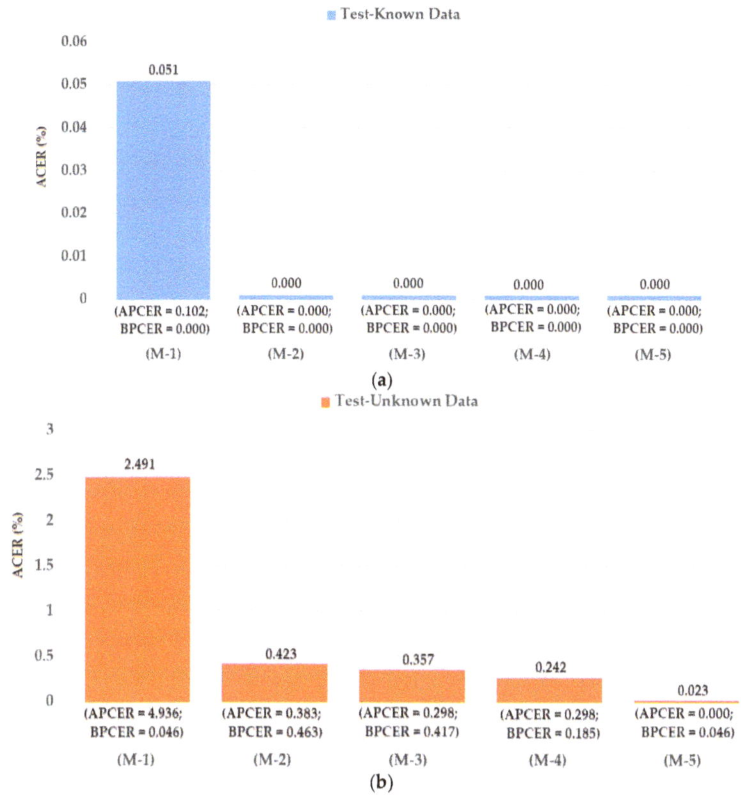

Figure 10. Detection errors of various iPAD methods using Warsaw2017 dataset: (**a**) Using Test-Known dataset; and (**b**) Using Test-Unknown dataset. Note: (M-1) Using CNN as Classifier; (M-2) Using CNN Features with PCA and Polynomial SVM Kernel; (M-3) Using MLBP Features with PCA and Polynomial SVM Kernel; (M-4) Using Feature Level Fusion with PCA and Polynomial SVM Kernel; and (M-5) Using Score Level Fusion with PCA and Polynomial SVM Kernel.

As a next experiment, we measured the detection errors produced by our proposed iPAD method based on score level fusion approach. Using the test-known dataset, we again obtained the same best detection error (ACER) of 0.000% as using the feature level fusion approach. For the test-unknown dataset, we obtained the best detection error of 0.023% using the combination rule of "polynomial-polynomial". This error was much smaller than the error of 0.242% using the feature level fusion approach. Based on the experimental results, we can see that the combination of deep and handcrafted features was effective at enhancing the detection performance of the iPAD system. In addition, the score level fusion approach worked better than the feature level approach on the Warsaw2017 dataset. For demonstration, we show the detection error tradeoff (DET) curves of these experiments in Figure 11. In this figure, we drew the change of APCER according to the change in the bona-fide presentation acceptance rate (BPAR). The BPAR was calculated as 100—BPCER (%). Since the iPAD methods using only CNN, MLBP, or hybrid features perfectly detected presentation attack images for the test-known dataset, DET curves for these cases are meaningless. Therefore, we only show the DET curves of the four detection configurations using the test-unknown data in Figure 11. As shown in Figures 10 and 11, we can see that the iPAD using combined features outperformed the iPAD system using CNN and MLBP features. In addition, the score level fusion outperformed the feature level fusion for the Warsaw2017 dataset. As shown at the beginning bars of

Figure 10, we obtained detection errors of 0.051% and 2.491% using the CNN method as classifier (using the CNN method for directly classifying the real and presentation attack images) on the test-known and test-unknown datasets, respectively. These high detection errors indicate that our approach that uses the PCA for feature selection and SVM for classification is more efficient than the use of CNN method directly for iPAD. The reason is that the CNN network contains a huge number of parameters that make the CNN method usually faces with overfitting problem. As a result, redundant information can exist in extracted deep features, but it can be removed using PCA method.

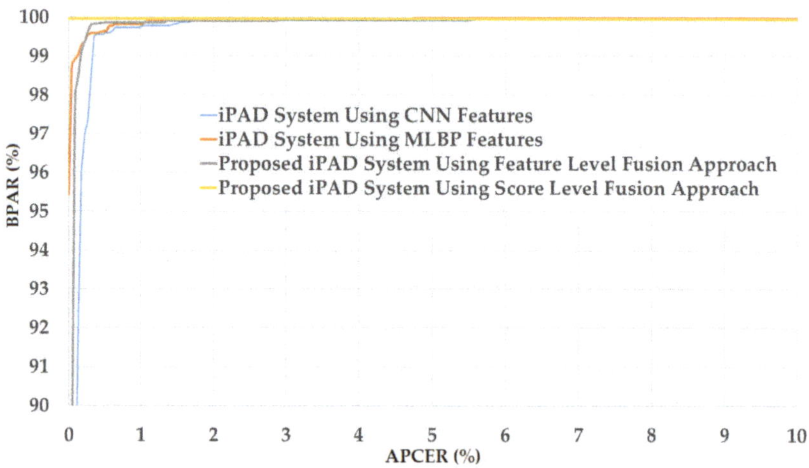

Figure 11. DET curves of iPAD systems based on use of CNN, MLBP, and hybrid image features (feature level fusion and score level fusion approach) using Warsaw2017 test-unknown dataset.

4.3. Detection Performance for Attack Method Based on Contact Lens

As the second experiment in our study, we investigated the detection performance of our proposed iPAD for a presentation attack method based on contact lenses. For this purpose, we used the ND2015 dataset. As explained in Section 4.1, the ND2015 dataset was used in the LivDet-Iris 2017 iPAD competition. In this competition, the images in the ND2015 dataset were classified into training and testing datasets. They used a set of 600 real and 600 presentation attack images for a training dataset and a set of 900 real and 900 presentation attack images for a testing dataset. Similar to the Warsaw2017 dataset, two testing datasets were constructed including a test-known dataset (in which the presentation attack images were collected using the same contact lens manufacturer as that of the training dataset) and a test-unknown dataset (in which the presentation attack images were collected using contact lenses from a different manufacturer than that of the training dataset) [48]. However, the detailed information of how the images were divided into training and testing datasets was not available for us. In addition, the LivDet-Iris 2017 competition did not use the entire ND2015 dataset in its experiments. This approach can bias the detection results because only a small set of the dataset was used. Therefore, in our experiments using the ND2015 dataset, we considered three division methods for dividing the images into training and testing datasets.

For the first division method, we performed the training and testing division approach similar to that of the previous study [48]. For this purpose, we divided images into training and testing datasets by randomly selecting images from the entire ND2015 dataset using the same criteria as the study by Yambay et al. [48]. The training dataset contained 600 real images (with no contacts, either soft or cosmetic) and 600 presentation attack images (with textured contact lenses manufactured by Ciba, UCL, and ClearLab) [48]. The test-known dataset contained 900 real and 900 presentation attack images and used contact lenses made by Ciba, UCL, and ClearLab (same as training data). The test-unknown dataset contained 900 real and 900 presentation attack images and used contact lenses made by Cooper and Johnson & Johnson [48]. The division procedure was performed by ensuring that there were no overlapped images in the three datasets. We iterated the above division procedure two times and performed experiments for measuring the detection performances because the information on dividing images into training and testing datasets in the study by Yambay et al. [48] was not available to us. As a result, the final detection performance was measured by averaging the detection results of the two iterated experiments. By using this division approach, we were able to fairly compare the detection performance of our proposed iPAD method with previous methods. In Table 6 we show the description of datasets used in the experiments, and in Figure 12 we show the experimental results.

Table 6. Description of training and testing data used with ND2015 dataset.

Dataset	Training Dataset			Testing Dataset					
				Test-Known Dataset			Test-Unknown Dataset		
	Real Image	Attack Image	Total	Real Image	Attack Image	Total	Real Image	Attack Image	Total
Original ND2015 dataset	600	600	1200	900	900	1800	900	900	1800
Augmented dataset	29,400 (600 × 49)	29,400 (600 × 49)	58,800	900	900	1800	900	900	1800

In Figure 12, we show the experimental results using our proposed method based on the feature level fusion approach. Using the test-known dataset, we obtained the best detection errors of 0.056%, 0.278% and 0.028% using the iPAD system based on only CNN feature, MLBP features, and hybrid features, respectively. Using the test-unknown dataset, these errors increased to 7.319%, 11.584%, and 4.167%. All these results were obtained using polynomial kernel of SVM method. In addition, we obtained an error of 0.056% for the case of using test-known data and 5.833% for the case of using test-unknown data using the score level fusion approach with 'polynomial-polynomial' combination rule. This detection error was higher than the error produced by the feature level fusion approach. However, this detection error was still lower than the detection errors produced by the iPAD systems using only CNN or MLBP features (ACERs of 7.139% and 11.584%, respectively). These results prove that our proposed iPAD method was effective at enhancing the detection performance of the iPAD system. In addition, the feature level fusion approach worked better than the score level fusion approach in our experiments using the ND2015 dataset. For demonstration purposes, we drew the DET curves of four system configurations using the test-known and test-unknown data in Figure 13. As observed from Figures 12 and 13, we can see that our proposed method outperformed the conventional detection methods based on only CNN or MLBP features.

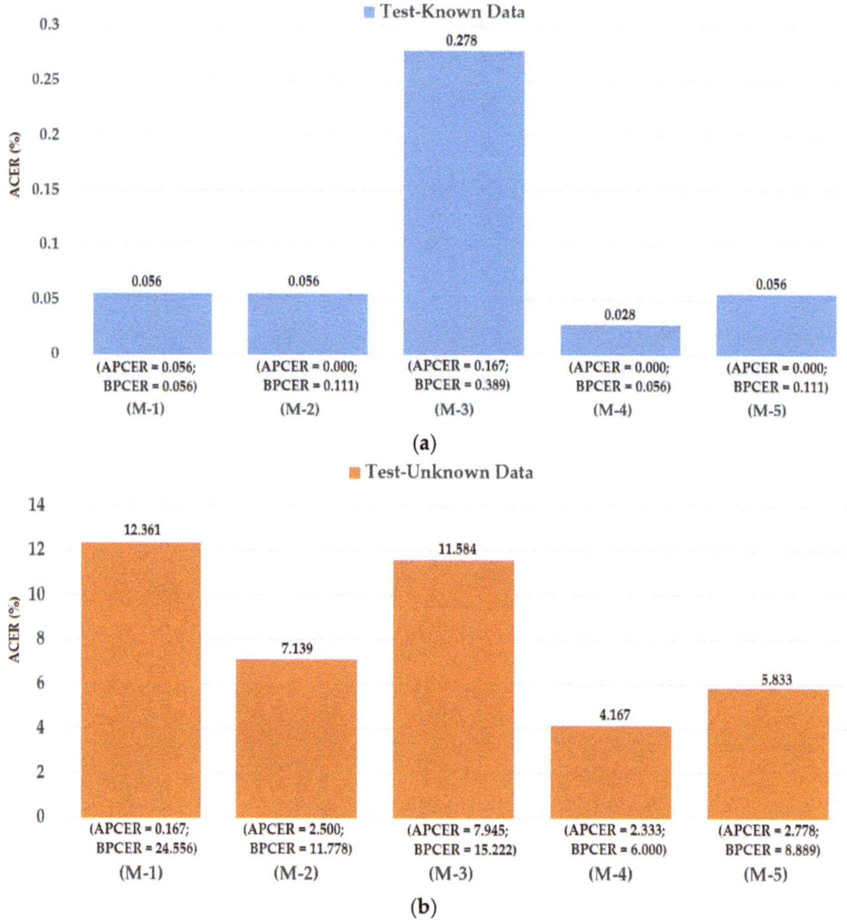

Figure 12. Detection errors of various iPAD methods using the first training-testing division method on ND2015 dataset: (**a**) Using Test-Known dataset; and (**b**) Using Test-Unknown dataset. Note: (M-1) Using CNN as Classifier; (M-2) Using CNN Features with PCA and Polynomial SVM Kernel; (M-3) Using MLBP Features with PCA and Polynomial SVM Kernel; (M-4) Using Feature Level Fusion with PCA and Polynomial SVM Kernel; and (M-5) Using Score Level Fusion with PCA and (Polynomial-Polynomial) SVM Kernels.

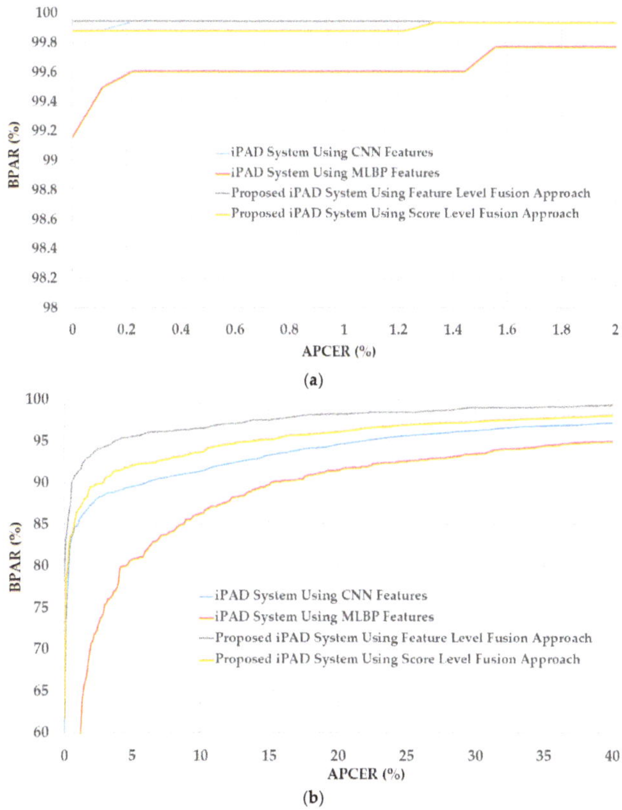

Figure 13. DET curves of iPAD systems based on use of CNN, MLBP, and hybrid image features (feature level fusion and score level fusion approach) using the first division method and ND2015 test-unknown dataset (**a**) DET curves of test-known dataset; and (**b**) DET curves of test-unknown dataset.

The first division method was performed using the same criteria as the division method used in LivDet-iris 2017 competition [48]. As a result, the real images were defined as the iris images without contact lens (with no contacts, either soft or cosmetic). However, there is a case in which users of iris recognition systems wear a soft (transparent) contact lens to protect their eyes or compensate their eye's problem such as myopia or hyperopia. For this case, an iris recognition system should allow users using the system and the consequent iPAD method must consider an iris with soft contact lens as the real image ones. Based on this phenomenon, we re-performed the above experiment by considering the iris images with soft (transparent) contact lens as the real images ones. Similar to the first division method, we randomly selected 600 real images (with no contacts or with soft (transparent) contact) and 600 presentation attack images (with textured contact lenses manufactured by Ciba, UCL, and ClearLab) [48] for training dataset. By similar method, we selected the test-known and test-unknown datasets that contained 900 real and 900 presentation attack images. The number of images in training and testing datasets in this experiment is same as the above experiment and shown in Table 6. The detection results are provided in Figure 14. As shown in this figure, we obtained perfect detection performance (ACER of 0.000%) using the iPAD system based on CNN features or hybrid features on the test-known dataset. Using the MLBP features, the lowest average error of 0.306% was obtained. Similar to our experiments with the Warsaw2017 dataset, the detection error increased when we used the test-unknown dataset. We obtained the lowest detection errors (ACER) of 7.528%

and 11.667% using the iPAD systems that use only CNN or only MLBP features, respectively. Using our proposed method based on the feature level fusion approach, the error was reduced to 5.056% using the polynomial kernel of the SVM method. Using the score level fusion approach, we obtained the lowest detection error of 6.861% using the "linear-polynomial" combination rule. This detection error was higher than the error produced by the feature level fusion approach (ACER of 5.056%). However, this detection error was still lower than the detection errors produced by the iPAD systems using only CNN or MLBP features (ACERs of 7.528% and 11.667%, respectively). These results prove that our proposed iPAD method was effective at enhancing the detection performance of the iPAD system. Furthermore, the feature level fusion approach worked better than the score level fusion approach in our experiments using the ND2015 dataset. For demonstration purposes, we drew the DET curves of four system configurations using the test-unknown data in Figure 15. We do not show the DET curves for the test-known dataset because we obtained perfect detection results using this dataset. As observed from Figures 14 and 15, we can see that our proposed method outperformed the conventional detection methods based on only CNN or MLBP features.

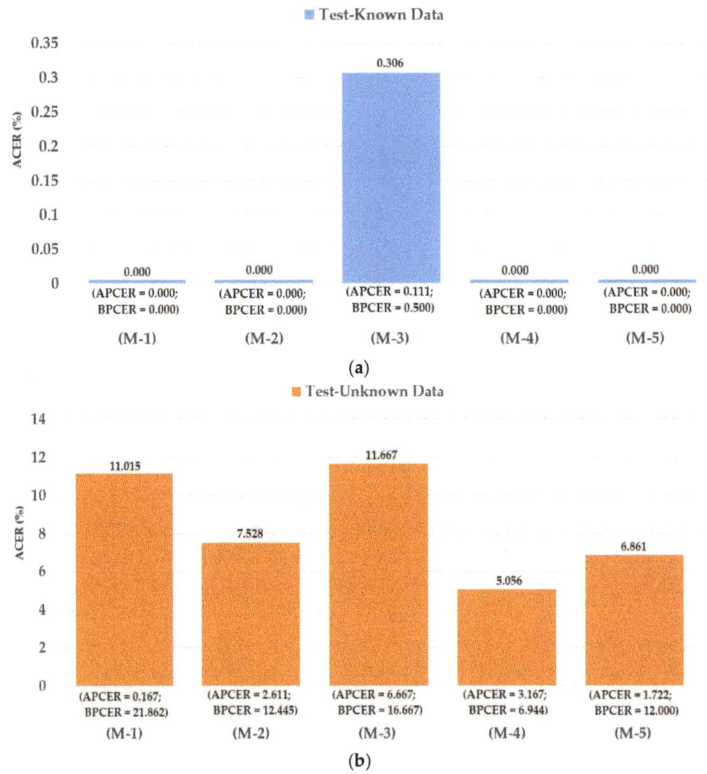

Figure 14. Detection errors of various iPAD methods using the second training-testing division method on ND2015 dataset: (**a**) Using Test-Known dataset; and (**b**) Using Test-Unknown dataset. Note: (M-1) Using CNN as Classifier; (M-2) Using CNN Features with PCA and Polynomial SVM Kernel; (M-3) Using MLBP Features with PCA and Polynomial SVM Kernel; (M-4) Using Feature Level Fusion with PCA and Polynomial SVM Kernel; and (M-5) Using Score Level Fusion with PCA and (Linear–Polynomial) SVM Kernels.

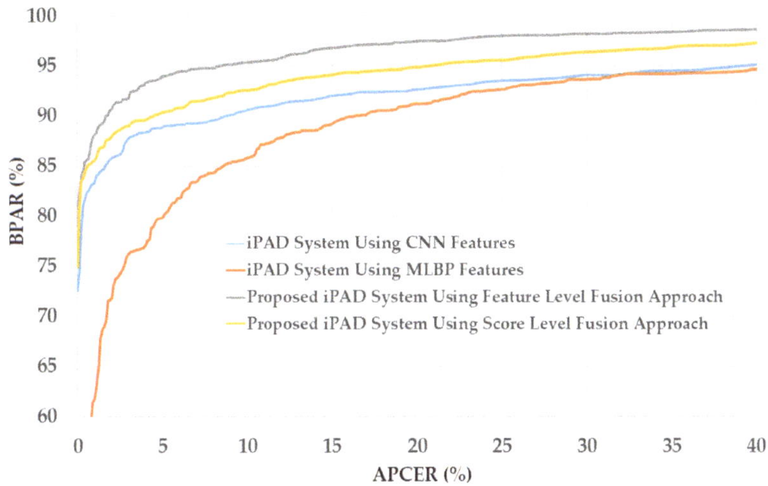

Figure 15. DET curves of iPAD systems based on use of CNN, MLBP, and hybrid image features (feature level fusion and score level fusion approach) using the second division method and ND2015 test-unknown dataset.

For the third division method, we used the entire ND2015 dataset for our experiments. For this purpose, we performed a two-fold cross-validation procedure to measure the detection accuracy of our proposed method. For the first fold, we divided the ND2015 dataset into training and testing datasets of which a half of ND2015 dataset was used for training and the other half for testing. The division was performed by ensuring that the images of the same individual only existed in either the training or the testing dataset. For the second fold, the training and testing datasets in the first fold were exchanged. By dividing the entire ND2015 dataset into training and testing datasets using this criterion, we were able to measure the detection accuracy using the entire dataset. In addition, this division approach divided images into the training and testing datasets without considering the difference in contact lens manufacturers. Therefore, we measured the detection accuracy in general. Based on this division method, we obtained the training and testing datasets as shown in Table 7. Similar to previous experiments, we performed data augmentation to generalize the training data. In Figure 16, we show the experimental results for this experiment. We obtained the best average detection accuracy (ACER) of 1.666% for the iPAD system using only CNN features and 7.539% for the iPAD system using only MLBP features. Both results were obtained using the RBF kernel of the SVM method. By using the feature level fusion approach, the detection error was reduced to 1.559%. The combination of two individual systems based on the score level fusion approach produced the lowest detection errors (ACER) of 1.481% using the RBF kernel in both subsystems. This detection error was lower than those produced by the two individual iPAD systems and the proposed iPAD system based on the feature level fusion approach. As shown in the experimental results in Figures 12, 14 and 16, our approach that uses the PCA for feature selection and SVM for classification on extracted CNN features outperformed the detection method that uses CNN as classifiers. For demonstration purposes, we show the DET curves of these experimental results in Figure 17. As demonstrated in the results, we can see that the proposed method was sufficient for iPAD. In addition, these detection accuracies were much better than those obtained in our previous experiment with the ND2015 dataset. The reason is that, in this experiment, we used a larger dataset for training the detection model, and we trained the detection model by merging all the possible cases of presentation attack images (without considering the known or unknown cases). This result suggests that we can obtain a much better detection accuracy when we collect enough data samples for training and perform testing with an attack method similar to that used in the training phase. However, this requirement is normally difficult to implement in real systems because various possible attack methods can be used in the testing phase that cannot be simulated in the training phase. To enhance the detection accuracy, we should simulate as many attack methods as possible for the training phase of the iPAD system.

Table 7. Description of training and testing data used for entire ND2015 dataset.

Dataset	Training Dataset			Testing Dataset		
	Real Image	Attack Image	Total	Real Image	Attack Image	Total
Original entire ND2015 (1st Fold)	2340	1068	3408	2535	1357	3892
Augmented dataset (1st Fold)	28,080 (2340 × 12)	26,700 (1068 × 25)	54,780	2535	1357	3892
Original entire ND2015 (2nd Fold)	2535	1357	3892	2340	1068	3408
Augmented dataset (2nd Fold)	30,420 (2535 × 12)	33,925 (1357 × 25)	64,345	2340	1068	3408

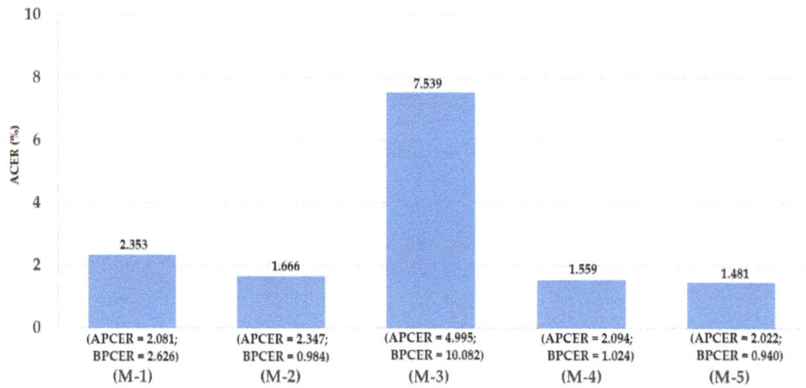

Figure 16. Detection errors of various iPAD methods using the third training-testing division method on ND2015 dataset. Note: (M-1) Using CNN as Classifier; (M-2) Using CNN Features with PCA and RBF SVM Kernel; (M-3) Using MLBP Features with PCA and RBF SVM Kernel; (M-4) Using Feature Level Fusion with PCA and RBF SVM Kernel; and (M-5) Using Score Level Fusion with PCA and (RBF–RBF) SVM Kernels.

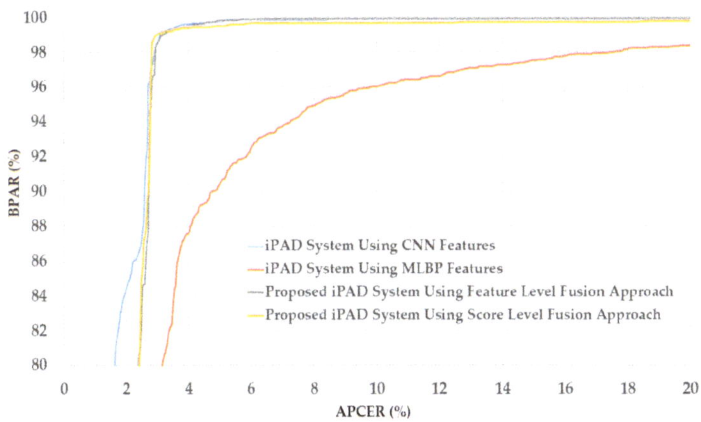

Figure 17. DET curves of iPAD methods based on use of CNN, MLBP, or hybrid image features (feature level fusion and score level fusion approach) using entire ND2015 dataset.

4.4. Detection Performance for Attack Method Based on Both Printed Samples and Contact Lens

As explained in Section 4.1, the presentation attack iris images in the Warsaw2017 and ND2015 datasets were collected by simulating two different attack methods, i.e., using printed samples (in the Warsaw2017 dataset) and contact lens (in the ND2015 dataset). The Warsaw2017 dataset was collected by recapturing the printed samples of real iris images. However, the ND2015 dataset was collected using a more complex attack method—the use of contact lenses. By performing experiments with each attack method, the detection system is only responsible for detecting presentation attack images for that given attack method. To make the detection accuracy robust for several kinds of attack methods, we performed experiments with a new dataset created by merging the Warsaw2017 and ND2015 datasets. By merging the two original datasets, the new dataset, named WARSAW-ND dataset in our study, contained real images captured using various cameras and capturing conditions and presentation attack images captured using two different attack methods as well as various capturing conditions. As a result, the WARSAW-ND dataset was more generalized than the Warsaw2017 and ND2015 datasets for iris presentation attack detection. For our experiment in this section, we combined the Warsaw2017 dataset (Table 5) and the ND2015 dataset (Table 6) to create the WARSAW-ND dataset shown in Table 8. For this experiment, we used the second division approach for dividing ND2015 dataset into training and testing datasets because it is reasonable for real applications. For the training dataset, we used 51,681 images from the Warsaw2017 dataset and 58,800 images from the ND2015 dataset. Using the same method, we created a test-known dataset containing 4790 images and a test-unknown dataset containing 6310 images for the experiment. Similar to the above experiments with the individual Warsaw2017 and ND2015 datasets, we performed experiments with the WARSAW-ND dataset using two system configurations based on feature level fusion and score level fusion. The experimental results are given in Figure 18.

Table 8. Description of training and testing datasets of WARSAW-ND dataset.

Training Dataset			Testing Dataset					
			Test-known dataset			Test-unknown dataset		
Images from Warsaw2017 dataset	Images from ND2015 dataset	Total	Images from Warsaw2017 dataset	Images from ND2015 dataset	Total	Images from Warsaw2017 dataset	Images from ND2015 dataset	Total
51,681	58,800	110,481	2990	1800	4790	4510	1800	6310

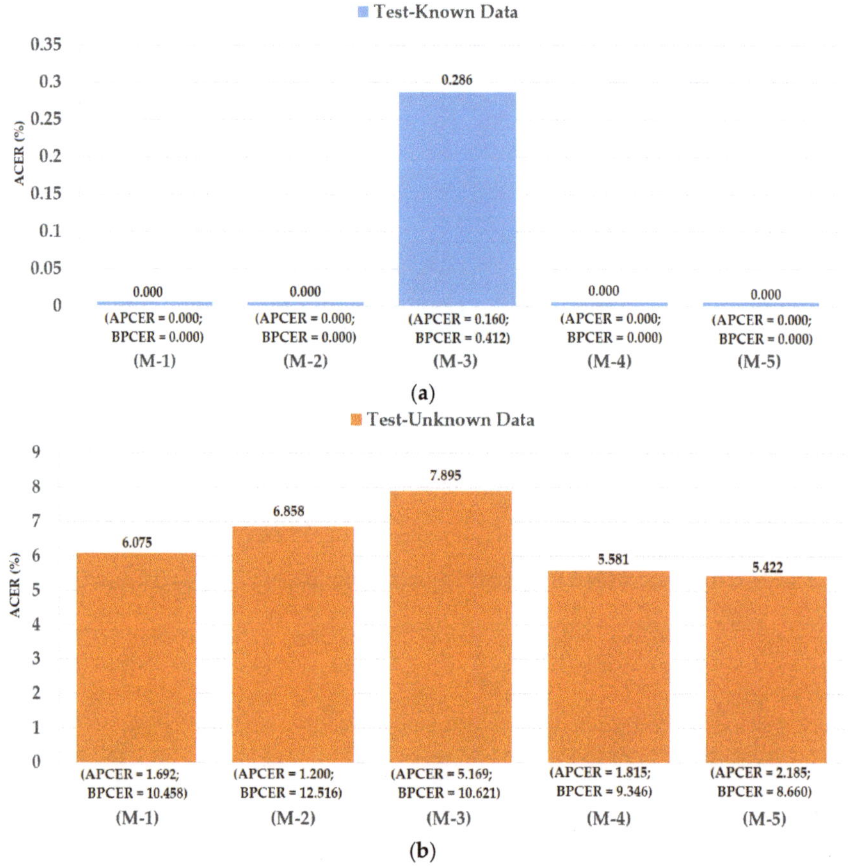

Figure 18. Detection errors of various iPAD methods using the fused dataset of Warsaw2017 and ND2015 datasets: (**a**) Using Test-Known dataset; and (**b**) Using Test-Unknown dataset. Note: (M-1) Using CNN as Classifier; (M-2) Using CNN Features with PCA and Polynomial SVM Kernel; (M-3) Using MLBP Features with PCA and Polynomial SVM Kernel; (M-4) Using Feature Level Fusion with PCA and Polynomial SVM Kernel; and (M-5) Using Score Level Fusion with PCA and (Linear–Polynomial) SVM Kernels.

For the test-known dataset case, we obtained the best detection errors of 0.000%, 0.286%, and 0.000% using iPAD systems that use CNN features, MLBP features, and our proposed hybrid features, respectively. These results show that we obtained perfect detection using the test-known dataset. Similar to the explanations provided in Sections 4.2 and 4.3, this result was caused by the fact that the test-known data were similar to the training data. However, the detection errors increased quickly for the test-unknown data case. We obtained the lowest detection errors of 6.858%, 7.895%, and 5.581% using the iPAD systems that use CNN features, MLBP features, and our proposed hybrid features, respectively. These detection results were much higher than those produced in the test-known data case. Using the score level fusion approach, the combination "linear-polynomial" rule produced the lowest detection errors with an ACER of 0.000% using test-known data and 5.422% using test-unknown data. These detection errors were equal for the test-known data case and lower for the test-unknown data case. However, the difference between the detection errors produced by the feature level fusion and score level fusion approaches was small (5.581% vs. 5.422%). From these results, we conclude that our proposed method is effective for enhancing the detection accuracy of iPAD systems whether they are based on the feature level fusion or the score level fusion approach. In addition, we again confirm that the iPAD system faces a significant problem with the unknown data because of the different capturing devices and contact lens manufacturers. For demonstration purposes, we drew the DET curves of the experimental results in Figure 19. We did not draw the curves for experiments using test-known data because we obtained perfect detection results with this data. This figure again confirms the efficiency of our proposed method over the individual methods based on only CNN or MLBP features.

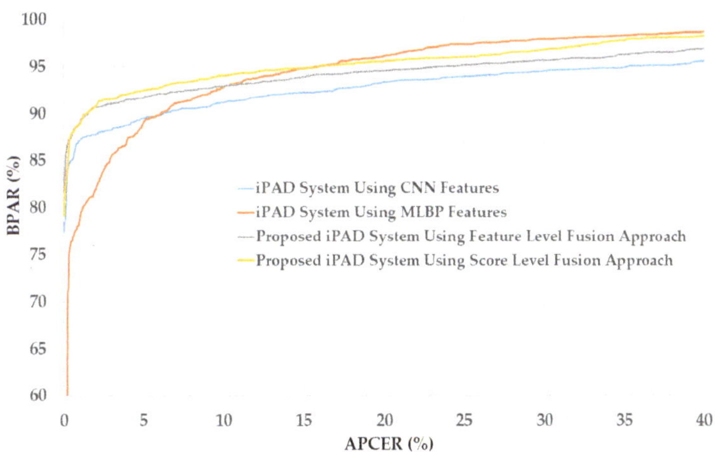

Figure 19. DET curves of iPAD systems based on use of CNN, MLBP, and hybrid image features (feature level fusion and score level fusion approach) using unknown data from a combination of ND2015 and Warsaw2017 datasets.

4.5. Comparisons and Discussion

As explained in Section 4.1, Warsaw2017 and ND2015 datasets were used for the LivDet-Iris 2017 detection competition for iris recognition systems. In this competition, several detection methods were proposed by research groups, including CASIA, Anon1, and UNIA. To validate the detection performance of our proposed method, we performed a comparison of detection performances of our proposed method with those produced by previous methods used in the LivDet-Iris 2017 competition. The detailed comparison is shown in Figure 20. In this figure, the detection performances are given as the weighted average of detection errors of both the test-known and test-unknown datasets.

Using the Warsaw2017 dataset, the study by Yambay et al. [48] showed that the detection errors were about 6.00%, 5.81%, and 7.41% using the CASIA, Anon1, and UNINA methods, respectively. Using our proposed method, we reduced the detection error to 0.142% and 0.016% for the feature level fusion and score level fusion approaches, respectively. These detection errors were also lower than those of 0.263% and 0.224% produced by the iPAD systems using only CNN or MLBP features, respectively.

Using the ND2015 dataset, the work by Yambay et al. [48] obtained the best detection accuracy by using the Anon1 method with a reported detection error of 4.03%. As shown in our experimental results in Figure 20, our study obtained an error of 3.598% using the iPAD system using only CNN features. We obtained an average detection error of 5.931% using only MLBP features, which is still lower than the results obtained by the CASIA and UNINA methods [48]. Although the detection error produced by the iPAD system using only MLBP features was higher than that produced by the Anon1 method, the combination of the MLBP and CNN features using the feature level fusion approach produced an average error of 2.098%, which is much lower than the best detection error of 4.03% produced by a previous study [48]. In addition, although the detection error produced by our proposed method based on score level fusion was higher than that of the feature level fusion approach (ACER of 2.945%), this error was still lower than the best detection error reported by Yambay et al. [48]. From comparison with the very recent study on iPAD using the same datasets, we conclude that our proposed method outperforms previous studies and is an effective method for iPAD.

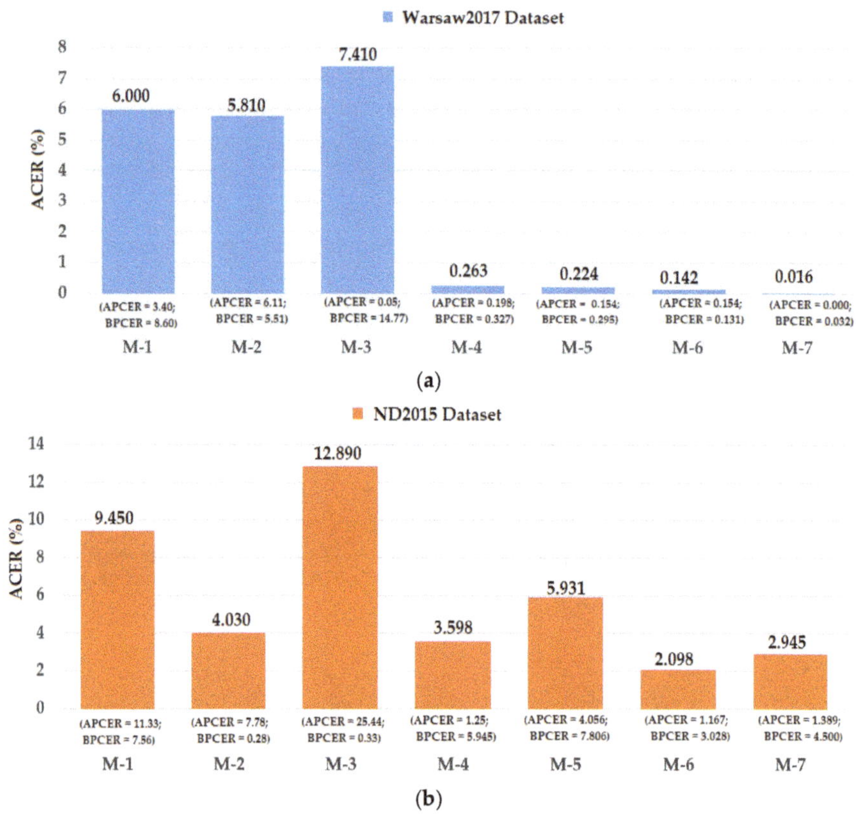

Figure 20. Comparison of detection error (ACER) between proposed method and previous methods using (**a**) Warsaw2017 and (**b**) ND2015 datasets. Note: (M-1) CASIA method [48]; (M-2) Anon1 method [48]; (M-3) UNINA method [48]; (M-4) CNN-based method [32]; (M-5) MLBP-based method [29]; (M-6) Proposed method based on feature level fusion; and (M-7) Proposed method based on score level fusion.

As shown in Figure 20, we obtained a very good detection result with the Warsaw2017 dataset. However, although the detection result for the ND2015 dataset was better than those produced by the previous study [48], it was still high compared to the results of the Warsaw dataset. The reason for this is that the Warsaw2017 dataset uses a very simple attack method and the consequent images in the Warsaw2017 dataset exhibit many noise components such as printing noise and broken texture that are easy to detect as shown in our experimental results in Section 4.2. However, by printing iris patterns on contact lenses for attack purposes, the iris patterns in the captured iris images in the ND2015 dataset display clearly without the additional negative components such as printing noise or broken texture features. In addition, a contact lens does not differentiate between real and presentation attack images on the non-iris regions such as the sclera, eyelid, eyelash, or skin regions. As a result, presentation attack images in the ND2015 dataset are more difficult to detect than those in the Warsaw2017 dataset.

In the CNN method of Yambay et al. [48], called spoofnet, the CNN network architecture with four convolution layers and one inception module was shallower than the CNN architecture of our study. In addition, we used the PCA method to select optimal image features and the SVM method to classify the input images based on extracted image features instead of using fully connected layers directly. As a result, our detection accuracy was higher than that of Yambay's method. As shown in our experimental results, we also see that the cross-sensor or cross contact lens manufacturer is an important factor in an iPAD system. The use of a different capturing device for image acquisition or a different method to create a presentation attack iris image has a strong effect on a detection system by increasing the possibility of a successful attack on an iris recognition system.

5. Conclusions

In this study, we proposed a new PAD method for enhancing the security level of iris recognition systems. The main contribution of our proposed method is that we reduced the limitation of the deep learning-based method by using a combination of handcrafted image features and deep features. Although the deep learning-based method has proven to be effective for solving many computer vision problems, it still has several limitations such as over-fitting caused by the limited number of training data and the huge number of model parameters. As a result, the performance of the deep learning method is limited when applied to a problem which lacks training data. In our work, we used handcrafted image features designed by expert knowledge of PAD for an iris recognition system to extract the image features and extracted image features using the deep learning method. By combining the two kinds of image features, we enhanced the detection accuracy of a PAD system compared to previous studies. Using the popular Warsaw2017 and ND2015 public datasets, we showed that our proposed method outperformed previous methods by producing a much lower detection error rate as shown in Section 4. In addition, the polynomial kernel of SVM method works better than linear and RBF kernels in our experiments with Warsaw2017 and ND2015 datasets. We conclude that our proposed PAD method effectively enhances the security level of iris recognition systems.

Author Contributions: Dat Tien Nguyen and Kang Ryoung Park designed and implemented the overall system, performed experiments, and wrote this paper. Na Rae Baek and Tuyen Danh Pham helped with comparative experiments.

Acknowledgments: This research was supported by the National Research Foundation of Korea (NRF) grant funded by the Korea government (Ministry of Science and ICT) (NRF-2017R1C1B5074062), by the Bio & Medical Technology Development Program of the NRF funded by the Korean government, MSIT (NRF-2016M3A9E1915855), and by the Basic Science Research Program through the National Research Foundation of Korea (NRF) funded by the Ministry of Education (NRF-2017R1D1A1B03028417).

Conflicts of Interest: The authors declare no conflict of interest.

References

1. Jain, A.K.; Ross, A.; Prabhakar, S. An introduction to biometric recognition. *IEEE Trans. Circuits Syst. Video Technol.* **2004**, *14*, 4–20. [CrossRef]
2. Nguyen, D.T.; Yoon, H.S.; Pham, T.D.; Park, K.R. Spoof detection for finger-vein recognition system using NIR camera. *Sensors* **2017**, *17*, 2261. [CrossRef] [PubMed]
3. Nguyen, K.; Fookes, C.; Jillela, R.; Sridharan, S.; Ross, A. Long range iris recognition: A survey. *Pattern Recognit.* **2017**, *72*, 123–143. [CrossRef]
4. Peralta, D.; Galar, M.; Triguero, I.; Paternain, D.; Garcia, S.; Barrenechea, E.; Benitez, J.M.; Bustince, H.; Herrera, F. A survey on fingerprint minutiae-based local matching for verification and identification: Taxonomy and experimental evaluation. *Inf. Sci.* **2015**, *315*, 67–87. [CrossRef]
5. Pham, T.D.; Park, Y.H.; Nguyen, D.T.; Kwon, S.Y.; Park, K.R. Nonintrusive finger-vein recognition system using NIR images sensor and accuracy analyses according to various factors. *Sensors* **2015**, *15*, 16886–16894. [CrossRef] [PubMed]

6. Lin, C.-L.; Wang, S.-H.; Cheng, H.-Y.; Fan, K.-C.; Hsu, W.-L.; Lai, C.-R. Bimodal biometric verification using the fusion of palmprint and infrared palm-dorsum vein images. *Sensors* **2015**, *15*, 31339–31361. [CrossRef] [PubMed]
7. Mirmohamadsadeghi, L.; Drygajlo, A. Palm-vein recognition with local texture patterns. *IET Biom.* **2014**, *3*, 198–206. [CrossRef]
8. Zhou, H.; Milan, A.; Wei, L.; Creighton, D.; Hossny, M.; Nahavandi, S. Recent advances on single modal and multimodal face recognition: A survey. *IEEE Trans. Hum. Mach. Syst.* **2014**, *44*, 701–716. [CrossRef]
9. Shin, K.Y.; Kim, Y.G.; Park, K.R. Enhanced iris recognition method based on multi-unit iris images. *Opt. Eng.* **2013**, *52*, 1–11. [CrossRef]
10. Nguyen, D.T.; Pham, T.D.; Baek, N.R.; Park, K.R. Combining deep and handcrafted image features for presentation attack detection in face recognition using visible light camera sensors. *Sensors* **2018**, *18*, 699. [CrossRef] [PubMed]
11. Sousedik, C.; Busch, C. Presentation attack detection methods for fingerprint recognition system: A survey. *IET Biom.* **2014**, *3*, 219–233. [CrossRef]
12. Galbally, J.; Marcel, S.; Fierrez, J. Biometric antispoofing methods: A survey in face recognition. *IEEE Access* **2014**, *2*, 1530–1552. [CrossRef]
13. Nguyen, D.T.; Park, Y.H.; Shin, K.Y.; Kwon, S.Y.; Lee, H.C.; Park, K.R. Fake finger-vein image detection based on Fourier and wavelet transforms. *Digit. Signal Process.* **2013**, *23*, 1401–1413. [CrossRef]
14. Galbally, J.; Marcel, S.; Fierrez, J. Image quality assessment for fake biometric detection: Application to iris, fingerprint and face recognition. *IEEE Trans. Image Process.* **2014**, *23*, 710–724. [CrossRef] [PubMed]
15. De Souza, G.B.; Da Silva Santos, D.F.; Pires, R.G.; Marana, A.N.; Papa, J.P. Deep texture features for robust face spoofing detection. *IEEE Trans. Circuits Syst. II Express Briefs* **2017**, *64*, 1397–1401. [CrossRef]
16. Akhtar, Z.; Micheloni, C.; Foresti, G.L. Biometric liveness detection: Challenges and research opportunities. *IEEE Secur. Priv.* **2015**, *13*, 63–72. [CrossRef]
17. Dongguk Iris Spoof Detection CNN Model (DFSD-CNN) with Algorithm. Available online: http://dm.dgu.edu/link.html (accessed on 26 March 2018).
18. Gragnaniello, D.; Poggi, G.; Sansone, C.; Verdoliva, L. An investigation of local descriptors for biometric spoofing detection. *IEEE Trans. Inf. Forensic Secur.* **2015**, *10*, 849–863. [CrossRef]
19. Doyle, J.S.; Bowyer, K.W. Robust detection of textured contact lens in iris recognition using BSIF. *IEEE Access* **2015**, *3*, 1672–1683. [CrossRef]
20. Hu, Y.; Sirlantzis, K.; Howells, G. Iris liveness detection using regional features. *Pattern Recognit. Lett.* **2016**, *82*, 242–250. [CrossRef]
21. Komogortsev, O.V.; Karpov, A.; Holland, C.D. Attack of mechanical replicas: Liveness detection with eye movement. *IEEE Trans. Inf. Forensic Secur.* **2015**, *10*, 716–725. [CrossRef]
22. Raja, K.B.; Raghavendra, R.; Busch, C. Color adaptive quantized pattern for presentation attack detection in ocular biometric systems. In Proceedings of the ACM International Conference on Security of Information and Networks, Newark, NJ, USA, 20–22 July 2016; pp. 9–15.
23. Silva, P.; Luz, E.; Baeta, R.; Pedrini, H.; Falcal, A.X.; Menotti, D. An approach to iris contact lens detection based on deep image representation. In Proceedings of the IEEE Conference on Graphics, Patterns and Images, Salvador, Brazil, 26–29 August 2015; pp. 157–164.
24. Menotti, D.; Chiachia, G.; Pinto, A.; Schwartz, W.R.; Pedrini, H.; Falcao, A.X.; Rocha, A. Deep representation for iris, face and fingerprint spoofing detection. *IEEE Trans. Inf. Forensic Secur.* **2015**, *10*, 864–879. [CrossRef]
25. Daugman, J. How iris recognition works. *IEEE Trans. Circuits Syst. Video Technol.* **2004**, *14*, 21–30. [CrossRef]
26. Cho, S.R.; Nam, G.P.; Shin, K.Y.; Nguyen, D.T.; Pham, T.D.; Lee, E.C.; Park, K.R. Periocular-based biometrics robust to eye rotation based on polar coordinates. *Multimed. Tools Appl.* **2017**, *76*, 11177–11197. [CrossRef]
27. Kim, Y.G.; Shin, K.Y.; Park, K.R. Improved iris localization by using wide and narrow field of view cameras for iris recognition. *Opt. Eng.* **2013**, *52*, 103102-1–103102-12. [CrossRef]
28. Choi, S.E.; Lee, Y.J.; Lee, S.J.; Park, K.R.; Kim, J. Age estimation using a hierarchical classifier based on global and local facial features. *Pattern Recognit.* **2011**, *44*, 1262–1281. [CrossRef]
29. Ojala, T.; Pietikainen, M.; Maenpaa, T. Multiresolution gray-scale and rotation invariant texture classification with local binary patterns. *IEEE Trans. Pattern Anal. Mach. Intell.* **2002**, *24*, 971–987. [CrossRef]
30. Nguyen, D.T.; Cho, S.R.; Pham, T.D.; Park, K.R. Human age estimation method robust to camera sensor and/or face movement. *Sensors* **2015**, *15*, 21898–21930. [CrossRef] [PubMed]

31. Krizhevsky, A.; Sutskever, I.; Hinton, G.E. ImageNet classification with deep convolutional neural networks. In Proceedings of the Advances in Neural Information Processing Systems, Lake Tahoe, NV, USA, 3–8 December 2012.
32. Simonyan, K.; Zisserman, A. Very deep convolutional neural networks for large-scale image recognition. In Proceedings of the International Conference on Learning Representations, Kunming, China, 25–27 September 2013.
33. He, K.; Zhang, X.; Ren, S.; Sun, J. Deep residual learning for image recognition. In Proceedings of the IEEE Conference on Computer Vision and Pattern Recognition, Las Vegas, NV, USA, 27–30 June 2016.
34. Huang, G.; Liu, Z.; Weinberger, K.Q.; Van de Maaten, L. Densely connected convolutional networks. In Proceedings of the IEEE Conference on Computer Vision and Pattern Recognition, Honolulu, HI, USA, 22–25 July 2017.
35. Ren, S.; He, K.; Girshick, R.; Sun, J. Faster R-CNN: Towards real-time object detection with region proposal networks. *ArXiv*, 2016.
36. Redmon, J.; Divvala, S.; Girshick, R.; Farhadi, A. You only look one: Unified, real-time object detection. *ArXiv*, 2016.
37. Taigman, Y.; Yang, M.; Ranzato, M.; Wolf, L. Deepface: Closing the gap to human-level performance in face verification. In Proceedings of the IEEE Conference on Computer Vision and Pattern Recognition, Columbus, OH, USA, 23–28 June 2014.
38. Levi, G.; Hassner, T. Age and gender classification using convolutional neural networks. In Proceedings of the IEEE Conference on Computer Vision and Pattern Recognition Workshops, Boston, MA, USA, 7–12 June 2015.
39. Gangwar, A.; Joshi, A. DeepIrisNet: Deep iris representation with applications in iris recognition and cross-sensor iris recognition. In Proceedings of the IEEE International Conference on Image Processing, Phoenix, AZ, USA, 25–28 September 2016.
40. Nguyen, K.; Fookes, C.; Ross, A.; Sridharan, S. Iris recognition with off-the-shelf CNN features: A deep learning perspective. *IEEE Access* **2018**, *6*, 18848–18855. [CrossRef]
41. Pan, S.J.; Yang, Q. A survey on transfer learning. *IEEE Trans. Knowl. Data Eng.* **2010**, *22*, 1345–1359. [CrossRef]
42. Chang, C.-C.; Lin, C.-J. LIBSVM: A library for support vector machines. *ACM Trans. Intell. Syst. Technol.* **2011**, *2*, 27. [CrossRef]
43. LIBSVM Tools for SVM Classification. Available online: https://www.csie.ntu.edu.tw/~cjlin/libsvm/ (accessed on 26 March 2018).
44. Nguyen, D.T.; Kim, K.W.; Hong, H.G.; Koo, J.H.; Kim, M.C.; Park, K.R. Gender recognition from human-body images using visible-light and thermal camera videos based on a convolutional neural network for feature extraction. *Sensors* **2017**, *17*, 637. [CrossRef] [PubMed]
45. Nanni, L.; Ghidoni, S.; Brahnam, S. Handcrafted vs. non-handcrafted features for computer vision classification. *Pattern Recognit.* **2017**, *71*, 158–172. [CrossRef]
46. ISO/IEC JTC1 SC37 Biometrics. *ISO/IEC WD 30107-3: 2014 Information Technology—Presentation Attack Detection—Part 3: Testing and Reporting and Classification of Attacks*; International Organization for Standardization: Geneva, Switzerland, 2014.
47. Raghavendra, R.; Busch, C. Presentation attack detection algorithms for finger vein biometrics: A comprehensive study. In Proceedings of the 11th International Conference on Signal-Image Technology and Internet-Based Systems, Bangkok, Thailand, 23–27 November 2015; pp. 628–632.
48. Yambay, D.; Becker, B.; Kohli, N.; Yadav, D.; Czajka, A.; Bowyer, K.W.; Schuckers, S.; Singh, R.; Vatsa, M.; Noore, A.; et al. LivDet iris 2017—Iris liveness detection competition 2017. In Proceedings of the International Conference on Biometrics, Denver, CO, USA, 1–4 October 2017.
49. Deep Learning Matlab Toolbox. Available online: https://www.mathworks.com/help/nnet/deep-learning-basics.html?s_tid=gn_loc_drop (accessed on 26 March 2018).
50. Principal Component Analysis Matlab Toolbox. Available online: https://www.mathworks.com/help/stats/pca.html (accessed on 26 March 2018).
51. Support Vector Machines (SVM) for Classification. Available online: https://www.mathworks.com/help/stats/support-vector-machine-classification.html (accessed on 26 March 2018).

52. Yambay, D.; Walczak, B.; Schuckers, S.; Czajka, A. LivDet-iris 2015—Iris liveness detection. In Proceedings of the IEEE International Conference on Identity, Security and Behavior Analysis, New Delhi, India, 22–24 February 2017.
53. Presentation Attack Video Iris Dataset (PAVID). Available online: http://nislab.no/biometrics_lab/pavid_db (accessed on 26 March 2018).
54. Yambay, D.; Doyle, J.S.; Bowyer, K.W.; Czajka, A.; Schucker, S. LivDet-iris 2013—Iris liveness detection competition 2013. In Proceedings of the IEEE International Joint Conference on Biometrics, Clearwater, FL, USA, 29 September–2 October 2014.

sensors

MDPI

Article

Camera Calibration Using Gray Code

Seppe Sels *, Bart Ribbens, Steve Vanlanduit and Rudi Penne

Faculty of Applied Engineering Department Electromechanics, Universiteit Antwerpen, Groenenborgerlaan 171, 2020 Antwerpen, Belgium; Bart.Ribbens@uantwerpen.be (B.R.); Steve.Vanlanduit@uantwerpen.be (S.V.); Rudi.Penne@uantwerpen.be (R.P.)

* Correspondence: seppe.sels@uantwerpen.be

Received: 27 November 2018; Accepted: 4 January 2019; Published: 10 January 2019

Abstract: In order to determine camera parameters, a calibration procedure involving the camera recordings of a checkerboard is usually performed. In this paper, we propose an alternative approach that uses Gray-code patterns displayed on an LCD screen. Gray-code patterns allow us to decode 3D location information of points of the LCD screen at every pixel in the camera image. This is in contrast to checkerboard patterns where the number of corresponding locations is limited to the number of checkerboard corners. We show that, for the case of a UEye CMOS camera, the precision of focal-length estimation is 1.5 times more precise than when using a standard calibration with a checkerboard pattern.

Keywords: camera calibration; Gray code; checkerboard

---

1. Problem Statement and Introduction

Commonly, camera calibration is done with checkerboard patterns printed on paper with a standard printer and attached to a flat surface [1,2]. The use of these low-cost checkerboards limits the number of input points of the calibration to the number of checker corners. It is also challenging to obtain input points of pixels located near the edges and corners of the image. This difficulty might lead to high uncertainty on the calculated camera parameters. The primary goal in this work is to obtain more accurate camera calibration by using a new calibration pattern. The proposed pattern needs to give a high number of accurate input points to the calibration algorithm. The input points must be easy to detect and evenly distributed in the image frame to avoid overfitting. In addition, the used calibration pattern must be easy to handle, cheap, and easy to make. Therefore, a standard laptop LCD screen (or any other screen) with a displayed pattern is proposed. A laptop screen is a high-quality flat surface and easy to get by. As a pattern, we propose to use Gray code. In our experiments (Section 2) a Gray-code pattern performs better than a checkerboard pattern. The main advantage of Gray code is that each pixel of the image has a corresponding calibration input point. The use, advantages, and disadvantages of Gray code are further explained in Section 1.3. In Section 1.1, a brief introduction on the standard camera-calibration method is given. Section 2 describes the experiments used to validate the proposed calibration board. Section 2 also compares the proposed Gray-code pattern with a checkerboard pattern. The literature exists where patterns displayed on LCD screens are used [3–5], but they are mostly used in combination with checkerboards and circular patterns, or need additional hardware. To our knowledge, Gray code is not used as a calibration pattern in combination with standard pinhole monocular camera calibration. The closest related work is the paper by Hirooka S. [6]. The work uses Gray code displayed on LCD screens to calibrate a stereo pair of cameras. In contrast to the paper of Hirooka S. [6], this work focuses on monocular calibration. Additionally, we analyze the precision of the calibration and compare it to standard checkerboards displayed on an LCD screen. The standard error measure in camera calibration (reprojection error) of the standard method using checkerboards and the proposed method using Gray code is investigated.

1.1. Camera Calibration

Geometric monocular camera calibration plays an important role in computer vision [7–14]. This work focuses on calculating Intrinsic Parameters I and distortion parameters of the camera. These parameters are needed if images are used for pose estimation of detected objects or 3D reconstruction (scanning) of objects K is the intrinsic matrix according to the pinhole model assuming compensated distortion and zero skew (Equation (1)). The intrinsic matrix contains the focal lengths [pixel] in the x- and y-direction (f_x, f_y) and the principal point (c_x, c_y) [pixel]. The focal length in pixels corresponds to the focal length in meter with $f_{x,y} = s_{x,y} * f$. Where f is the focal length (meters), and $s_{x,y}$ (pixels/meters) is the size in x or y direction of a pixel on the camera sensor. The goal of camera calibration is to calculate this intrinsic matrix and the distortion parameters. As input points the calibration algorithm uses known world co-ordinates and corresponding camera coordinates. Commonly, these co-ordinates are called *imagepoints* (u, v) (Equation (2)) when used as calibration points. *Objectpoints* $(x, y, z)_{world}$ (Equation (3)) are points corresponding with these *imagepoints* defined in a world co-ordinate system.

$$K = \begin{bmatrix} f_x & 0 & c_x \\ 0 & f_y & c_y \\ 0 & 0 & 1 \end{bmatrix} \tag{1}$$

$$\begin{bmatrix} u \\ v \\ 1 \end{bmatrix} = K \begin{bmatrix} x \\ y \\ z \end{bmatrix} \tag{2}$$

H is the homogeneous transformation matrix between the *objectpoints* defined in a world co-ordinate system and the corresponding points in the camera co-ordinate system. For more information about the pinhole model with (radial) distortion parameters, we refer to the work of Zhang Z., OpenCV and Matlab documentation [1,15,16].

$$\begin{bmatrix} x \\ y \\ z \\ 1 \end{bmatrix} = H. \begin{bmatrix} x \\ y \\ z \\ 1 \end{bmatrix}_{world} \tag{3}$$

Currently, *object* and *imagepoints* for camera calibration are commonly obtained using planar checkerboards [11,16,17]. From the images of these boards, the checker corners are calculated and located. This location is used to estimate camera parameters, where the camera parameters contain the intrinsic matrix and distortion parameters. There are various kinds of checkerboards. The most common is a board used in OpenCV tutorials (see Figure 1a). Other types of boards exist (Figure 1b–d) using circles instead of corners. Boards with additional markers to detect partially occluded checkerboards also exist (Figure 1c,d). These types of calibration boards all have a limited number of detectable points per image, e.g., the 9 × 6 OpenCV checkerboard has 54 detectable points [11], where even a low-resolution camera has 76,800 pixels (320 × 240 pixels). Using a limited set of points can cause inaccurate calibrations with a large uncertainty (see Section 2).

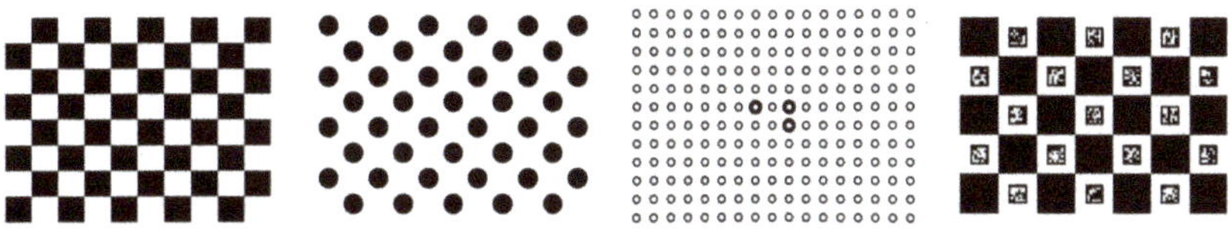

(a) (b) (c) (d)

Figure 1. Calibration Boards. (**a**) Opencv 9 × 6 checkerboard; (**b**) Opencv asymmetric circle calibration board; (**c**) board with additional (bold circles) markers (used by Scan in box Idea software); (**d**) Charuco markers in combination with Checkerboard (OpenCV) [1,18] (edited figure from OpenCV documentation).

1.2. LCD Screen

Commonly planar checkerboards or other patterns are made by printing the pattern on paper using a standard printer and then placing or glueing the paper on a planar surface [15,16]. This manufacturing process can introduce errors caused by the printing or glueing process [2]. To avoid problems with this manufacturing process, the pattern can be displayed on an LCD screen [3,19]. An LCD screen is flat, and the displayed dimensions of the pattern are highly accurate and without distortions. Song Z. [5] states that the flatness of a printed checkerboard easily exceeds 0.1 mm, while the planetary deviations of standard LCD panels are below 0.05 mm. High-quality machine-vision calibration targets exist (using ceramic or glass substrates), but their price range is much higher than calibration boards printed on paper and manually glued on a flat surface. However, in this work, it is not our goal to replace these industrial-grade calibration boards, but only the low-cost printed calibration boards.

1.3. Gray Code

In this work, we replace a standard checker calibration pattern with a pattern that gives more correspondences for calibration. As a new pattern, we propose to use Gray code [20] (reflected binary code). Commonly, Gray code is used for error detection and correction in digital communication. It is also used in encoders for position detection and used in structured light 3D scanners [21–25]. In structured light 3D scanners, Gray code is a type of binary code that uses black and white stripes. Each stripe corresponds to a unique code word. N patterns can code 2^N unique patterns. The sequence of these striped patterns codes 2^n unique locations. In the proposed methodology, each pixel of the camera is mapped to a row/column of the LCD screen. To calculate this mapping, Gray code is displayed on a screen and captured by the camera. Gray code encodes column and row indices in a unique time sequence of black and white patterns. This pattern forms code words (see Figure 2). Each pixel of the camera will have two codewords, one corresponding with the row of the LCD screen it sees, and one with the column. Two neighboring code words have a Hamming distance of one. This property has the advantage that if one frame of the Gray-code sequence is detected wrongly, the corresponding pixel only shifts one row/column [20,26]. In contrast to standard checkerboard detection, multiple frames are recorded to calculate the mapping. This recording has as a disadvantage that the camera needs to remain fixed during recording. The number of displayed frames is equal to $\lceil 2\log_2(w) + 2\log_2(h) \rceil$ where w is the width of the LCD screen and h the height (in pixel). The displayed frames are the bit planes needed for decoding and their inverse. This inverse is used to define a variable threshold to distinguish black (0) and white (1) [26]. Checkerboard detection requires the detection of corners on a subpixel level. Standard techniques use a Harris feature point detector as a rough corner estimation and use a gradient search operation for subpixel corner estimation. As proven by Datta A. [27], this gradient search introduces a bias on the corner location when the calibration boards are not frontoparallel to the camera. Although the work of Datta A. solves this by using an iterative approach, the use of Gray-code patterns eliminates this bias because only a threshold operation is needed for composing the per pixel Gray-code code word. This per-pixel threshold does not use line detection or other spatial information about the calibration pattern that might get distorted in the image. As a downside, Gray code only gives only pixel correspondences in contrast to subpixel correspondences of checkerboard detection. There are, however, a lot more pixel correspondences (all pixels of the image can be used) than the 54 points used by a standard 9×4 checkerboard. The camera can be placed in a position where it only sees the LCD screen because it is not necessary to see the complete pattern. This property gives the advantage that each camera pixel has a correspondence that also forces the correspondences to be evenly distributed.

When using Gray code displayed on an LCD screen, the camera needs to operate in the visible light spectrum, where checkerboards can also be used to calibrate camera and lenses in the infrared and near-infrared spectrum [28]. When recording a screen with a camera, special care needs to be taking to avoid incompatibility with refresh rates of the LCD screen and capturing rate of the camera. In our experiments, aliasing effects like the moiré effect did not occur. The methodology is summarised in Algorithm 1. An example of the first three Gray-code patterns is given in Figure 2.

Algorithm 1: Methodology.

Step 1: Aim camera to LCD screen.
Step 2: Display and Capture Gray code pattern.
Step 3: Decode Gray code.

  Step 3.a: Make mapping between screen pixels and camera pixels.

Step 4: Redo 1–3 for multiple positions.
Step 5: Use OpenCV functions for camera calibration.

  Step 5.a: Sample complete dataset (otherwise large computation time).
  Step 5.b: Use OpenCV function calibrateCamera (inputs) with as input imagepoints (u,v coordinates of image) and objectpoints (corresponding screen pixels).

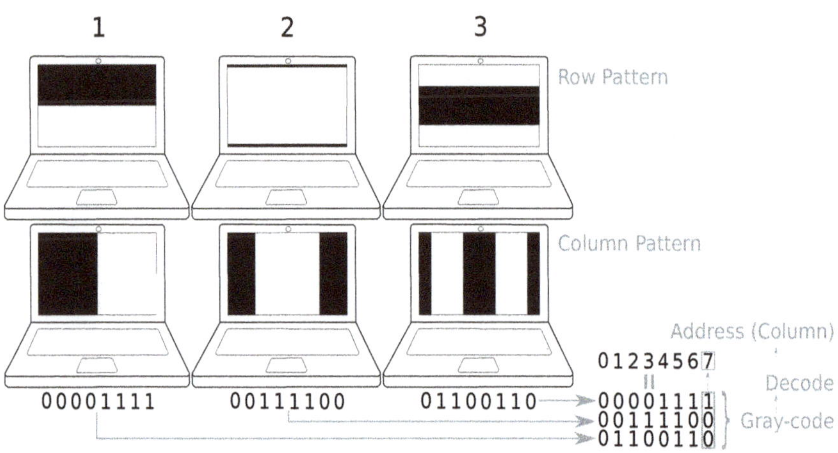

Figure 2. Gray-code pattern (first 3) Top: Pattern for row correspondences (top), column for row correspondences. An example of decodation of the column pattern is given. The example assumes only three patterns are given (eight columns). Figure based on Reference [6].

2. Experiments

2.1. Experimental Setup

In our experimental setup, a UEye CMOS camera with a resolution of 1024 × 1080 pixels was used in combination with a 16 mm lens. The camera sensor has a pixel size of 5.3 μm, which gives a theoretical focal length of 3018 pixels. As LCD screen, the screen of a Dell Lattitude E5550 (1080 × 1920 pixels) was used. In the experiment, the method using Gray code is compared with checkerboards.

Checkerboard detection uses additional circle markers on the board (Figure 3b). These markers are used as a reference to ensure correct labelling (ordering) of the corners when the complete checkerboard is not visible (Figure 3d). In our experiments, we use a 9 × 14 checkerboard. A checkerboard detection is considered correct when minimal 54 points are detected (total points 126). 54 points are used as a threshold to do equally or better than the standard calibration board used in OpenCV tutorials (9 × 6, 54 points) [1].

For each calibration, the camera is repositioned 15 times. We chose this number because OpenCV documentation (version 3.4.1) states that at least 10 positions are needed. Matlab documentation (using the same algorithms) states that 10 to 20 positions are needed. Calibration software like DLR [29] only needs three to 10 images. Halcon (industrial vision software package) also uses 15 positions of a calibration board in its documentation [30]. To obtain proper calibrations, we positioned the camera with angles up to 30° as done by Albarelli A. et al. [2].

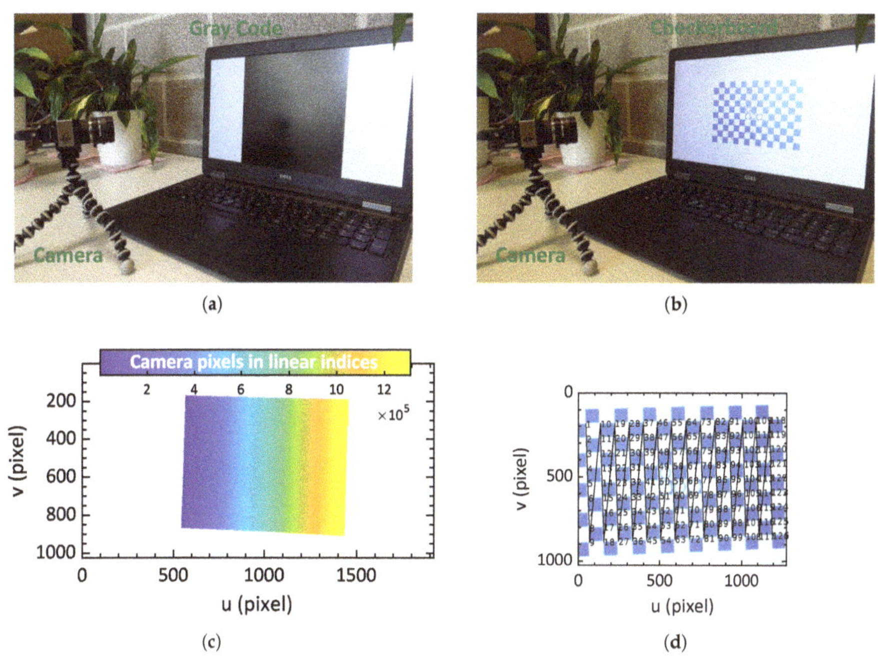

Figure 3. (**a**) Setup with one Gray-code frame displayed (**b**) setup with checkerboard displayed (**c**) recorded frame with Gray code; (**d**) calculated correspondence map between LCD screen and camera. Each pixel on the LCD screen (x and y-axis) has a corresponding pixel in the camera. The corresponding pixels are visualized with colors. The white area indicates the parts of the screen that is not visible in the camera image.

Display and decoding of the Gray code is done using the Matlab Psychtoolbox (Matlab OpenGL wrapper) and code made available by Moreno D. [31] (Figure 3a,c). Calculating the calibration is done using OpenCV (version 3.4.1) functions. The function uses the calibration algorithm of Zhang Z. [15]), and default settings are used. Although not displayed in the text, radial distortion (to the 2nd order) and principal points are calculated. Skew and tangential distortion, and third-order radial distortion are assumed to be negligible.

2.2. Number of Calibration Points

In a first experiment, the effect of a higher number of calibration points was analysed. From a dataset of 15 camera positions, N random number of samples per camera position are taken, and the calibration is executed. In Figure 4 different calibration results (focal length) with a different number of samples are listed. For each sample size, the calibration is repeated 50 times. The samples are randomly selected from a complete dataset containing all samples from 15 different positions of the camera. The experiment shows that, when using more than 100,000 points, standard deviation goes below 0.14 pixels, and the mean focal length does not change. When using 700,000 points or more, standard deviation stays below 0.02 pixels. Therefore, in the following experiments, 700,000 points are used as input sample size. Calculating the calibration with 700,000 points takes approximately two minutes (computation done on 1 core @ 4 GHz). In comparison, a calibration using 54 points per view (checkerboard) takes less than one second.

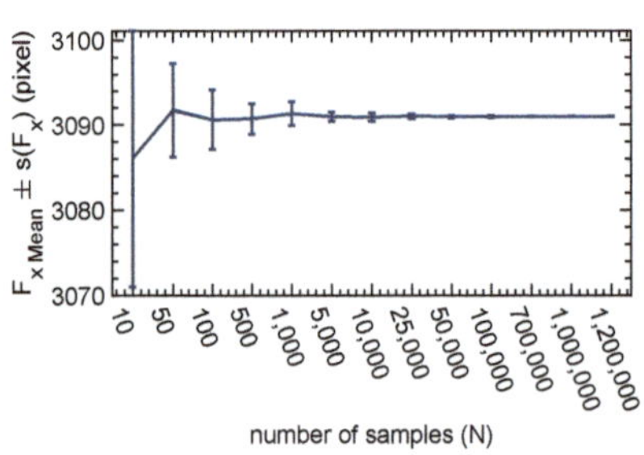

Figure 4. Calibration using Gray code with a different number of inputpoints.

2.3. Comparison between Gray Code and Checkerboard

In this experiment, in 15 different camera positions, Gray code was recorded and subsequently decoded. In each camera position, a checkerboard is also displayed on the screen. Next, the camera was calibrated using both the detected checkerboard corners and a random subset (700,000 points) of the Gray-code pattern. Next, the experiment was repeated seven times to check repeatability.

Table 1 shows the results of the experiment. The reprojection error of the Gray code was higher than the reprojection error calculated with checkerboards. This higher error is expected since the checkerboard detection has subpixel resolution, whereas Gray code only has pixel correspondences. The reprojection error is the root mean square of the distances between the detected *imagepoints* and the projected (using the calculated camera parameters) *objectpoints* (see Section 1.1). The experiments of Albarelli A. et al. [2] and Poulin-Girard A. et al. [32] also prove that low reprojection errors do not necessarily give better calibration. The positioning of the calibration boards proved to be far more important to get good calibrations. For example, large calibration errors with low reprojection errors are obtained using only positions planar to the calibration. Therefore, we used standard deviation on the calibration results to compare results.

Table 1. Experimental results of seven independent calibrations. Each calibration is calculated from 15 (new) positions. f_x is the focal length in the x-direction, f_y is the focal length in the y-direction. $r1, r2$ are the distortion parameters (first and second order radial distortion). c_x and c_y are the principal points. With individual calibrations $s, (parameter_*)$ is the standard deviation estimated from the intrinsic parameters. $s(parameter_*)$ of the mean value is the standard deviation calculated with the respective parameter of each calibration.

Type	f_x [pixel]	$s(f_x)$ [pixel]	f_y [pixel]	$s(f_y)$ [pixel]	Mean Reprojection Error [pixel]	$r1$	$s(r1)$	$r2$	$s(r2)$	c_x [pixel]	$s(c_x)$ [pixel]	c_y [pixel]	$s(c_y)$ [pixel]
Gray-code	3093.1	0.10	3091.7	0.09	1.15	−0.14	0.26	0.21	0.0006	659.80	0.04	511.27	0.07
	3095.0	0.11	3094.7	0.10	1.01	−0.16	0.44	−0.92	0.0006	654.93	0.04	509.43	0.08
	3087.9	0.07	3087.7	0.06	0.88	−0.16	0.46	−1.12	0.0005	655.62	0.03	512.87	0.05
	3091.9	0.08	3091.2	0.07	0.93	−0.13	0.02	2.28	0.0005	658.27	0.03	513.02	0.06
	3090.9	0.10	3089.9	0.09	0.80	−0.14	0.20	0.87	0.0005	657.81	0.04	510.73	0.06
	3085.7	0.07	3091.7	0.09	1.15	−0.14	0.26	0.21	0.0006	659.80	0.04	511.27	0.07
	3091.0	0.07	3090.9	0.06	0.94	−0.15	0.44	−0.76	0.0005	653.75	0.03	514.89	0.05
mean ± s	3090.8 ± 3.1		3091.1 ± 2.1		0.98 ± 0.13	−0.15 ± 0.01		0.11 ± 1.2		657.14 ± 2.4		511.93 ± 1.8	
checkerboard	3080.1	2.13	3079.8	1.87	0.17	−0.16	0.85	−7.73	0.0097	658.80	0.78	510.53	1.74
	3079.8	2.87	3080.0	2.70	0.37	−0.16	0.61	−5.29	0.0188	658.52	1.59	513.45	2.42
	3088.7	1.74	3087.7	1.52	0.20	−0.17	1.02	−7.99	0.0125	659.64	0.85	508.38	1.43
	3078.4	2.07	3078.7	1.87	0.21	−0.15	0.69	−6.76	0.0113	656.52	0.93	519.11	1.38
	3082.0	3.87	3081.9	3.53	0.23	−0.16	1.31	−14.33	0.0165	656.99	1.21	515.91	2.53
	3073.0	7.70	3074.4	7.44	0.39	−0.17	1.28	−12.97	0.0293	656.24	2.83	524.64	4.49
	3077.3	2.22	3078.0	2.06	0.21	−0.16	0.90	−6.73	0.0120	660.55	1.03	519.41	1.46
mean ± s	3079.9 ± 4.8		3080.1 ± 4.1		0.25 ± 0.09	−0.16 ± 0.01		−8.83 ± 3.43		658.18 ±1.64		515.92 ± 5.63	

In Table 1, the comparison between the use of a checkerboard and Gray code is summarised. From the complete set, the mean focal length (in x-direction) calculated with Gray code is 3091 pixels with a standard deviation of 2.3 pixels, while the calculated focal length using a checkerboard pattern is 3080 pixel with a standard deviation of 4.8 pixels. In this dataset, the use of Gray code gives an improvement of 2.5 pixels on standard deviation.

As an extra validation, an LCD screen displaying a checkerboard is mounted on a translation stage. The checkerboard is detected in 20 different positions between 200 mm and 500 mm from the camera. Between each pair of positions, there was a translation of 10 mm. Next, the relative translation of the checkerboard is calculated with the mean calibration parameters of Table 1 of both the method using Gray code and the method using checkerboards. The mean relative error on the translation with the Gray-code parameters was 0.78 mm, with a standard deviation of 0.32 mm. The mean relative error using checkerboard parameters was 1.65 mm, with a standard deviation of 0.93 mm.

The experiment shows that a large number of correspondences has the advantage to give smaller standard deviations on the estimated camera parameters. Where the checkerboard gives standard deviations on the focal length of around 2.0 pixels, the Gray code gives standard deviations of around 0.02 pixels which is an improvement of a factor 100. Therefore, we conclude that using Gray code makes the camera-calibration algorithm more robust than calibration using correspondences from checkerboards. Note that the iterative method of Datta A. [27] (see Section 1.3) has an approximate improvement of a factor 3.

2.4. Repeatability

To further check the repeatability of the calibration, stability analysis was executed. A new calibration set was built with 49 different camerapositions. From this set, seven random camera positions are chosen. In this subset, 700,000 points were randomly selected, and then calibration was calculated. In the case of the checkerboard, the same camera positions are used to calibrate the camera with all detected checkerboard corners. This analysis was repeated 100 times and the focal length (f_x) is saved.

In this analysis (see Figure 5), the standard deviation of the focal length using Gray code was 3.30 pixels, which is lower than the focal length calculated with checkerboard corners (5.87 pixels). Note that in this experiment samples are taken randomly. Consequently, it is possible to build a bad dataset with only low-angled calibration patterns. For that reason, this experiment would most probably give higher deviations than real experiments done by an experienced user. However, since the same sampling is used for both techniques, the experiment can be used as a comparison. Note that in this dataset the lens of the camera was refocused and consequently the lens would have a slightly different focal distance compared to previous experiments.

The used Gray-code pattern uses 44 images (22 for rows, 22 for columns) displayed on the screen per position. As an additional experiment, 44 different checkerboards were also detected per camera position. This experiment did not significantly alter the calibration results with checkerboards and calculating the calibration with this dataset takes between 350 and 60 min. Due to this high calculation time, and the fact that this is not a standard use of checkerboard patterns, this setup of 44 checkerboards per camera position was not further analysed. For all other experiments described in this work, only one checkerboard was detected per camera position.

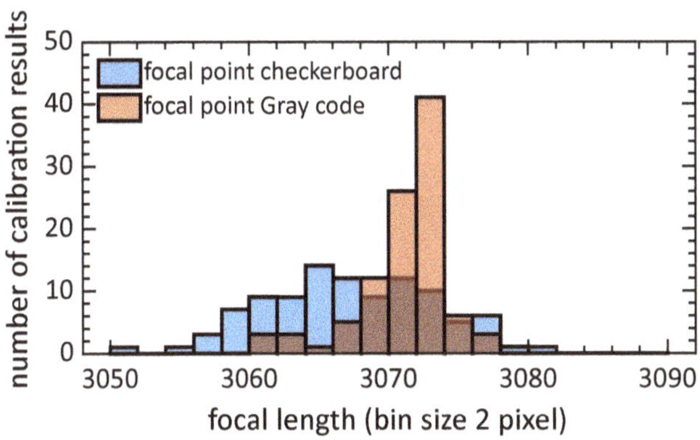

Figure 5. Histogram of focal length ((f_x) x-direction) using a calibration by sampling seven random positions out of 49, executed 100 times. Standard deviation using Gray code is 3.30 and 5.87 pixels using standard checkerboard displayed on a LCD screen

3. Conclusions

A Gray-code pattern displayed on a standard LCD screen can be used for the geometric calibration of a camera. In our experiments, the average reprojection error was around 1 pixel. This error is larger than the error obtained using standard checkerboards (0.25 pixels). This higher error is caused by the pixel-accuracy detection of Gray code, whereas the checkerboard detection uses subpixel detection of the corners. The Gray-code pattern has the advantage of giving a large number of correspondences. This large number made it possible to give each camera pixel a world co-ordinate in our experiments. This is in contrast to the correspondences of a checkerboard that are limited to the checkerboard corners. A large number of correspondences has the advantage to give smaller standard deviations on the estimated camera parameters. Where the checkerboard gives standard deviations on the focal length of around 4.8 pixels, the Gray code gives standard deviations of around 3.1 pixels, which is an improvement of a factor of 1.5. Therefore, we conclude that using Gray code makes the camera-calibration algorithm more robust than calibration using correspondences from checkerboards. Calibration does not need special lab-equipment, only a tripod for mounting the camera and a laptop or LCD monitor.

Author Contributions: S.S.: Conceived and designed the methodology, wrote software, wrote the paper. B.R.: Analysed calibration data + design of validation experiment, editing of original draft. R.P.: Validated calibration routines, editing of original draft. S.V.: Designed methodology, wrote the paper.

Funding: This research was funded by the Industrial Research Fund of the University of Antwerp and the TETRA fund of the Flanders Innovation and Entrepreneurship Agency (VLAIO) (Sensalo HBC.2017.0044).

Conflicts of Interest: The authors declare that there are no conflicts of interest related to this article.

References

1. OpenCV. OpenCV: Detection of ChArUco Corners. Available online: https://docs.opencv.org/3.1.0/df/d4a/tutorial_charuco_detection.html (accessed on 25 July 2018).
2. Albarelli, A.; Rodolà, E.; Torsello, A. Robust Camera Calibration using Inaccurate Targets. In Proceedings of the British Machine Vision Conference 2010, Wales, UK, 30 August–2 September 2010; pp. 16.1–16.10. [CrossRef]
3. Zhan, Z. Camera calibration based on liquid crystal display (lcd). In Proceedings of the International Archives of the Photogrammetry, Remote Sensing and Spatial Information Sciences, ISPRS, Beijing, China, 3–11 July 2008.

4. Tehrani, M.A.; Beeler, T.; Research, D.; Grundhöfer, A. A Practical Method for Fully Automatic Intrinsic Camera Calibration Using Directionally Encoded Light. In Proceedings of the IEEE Conference on Computer Vision and Pattern Recognition, CVPR 2017, Honolulu, HI, USA, 21–26 July 2017; pp. 1106–1114. [CrossRef]
5. Song, Z.; Chung, R. Use of LCD panel for calibrating structured-light-based range sensing system. *IEEE Trans. Instrum. Meas.* **2008**, *57*, 2623–2630. [CrossRef]
6. Hirooka, S.; Nakano, N.; Kazui, M. 3D camera calibration using gray code patterns. In Proceedings of the IEEE International Conference on Consumer Electronics, Digest of Technical Papers, Las Vegas, NV, USA, 9–13 January 2008; pp. 7–8. [CrossRef]
7. Engelke, T.; Keil, J.; Rojtberg, P.; Wientapper, F.; Schmitt, M.; Bockholt, U. Content first a concept for industrial augmented reality maintenance applications using mobile devices. In Proceedings of the 6th Conference on ACM Multimedia Systems—MMSys'15, Portland, OR, USA, 18–20 March 2015; ACM Press: New York, NY, USA, 2015; pp. 105–111. [CrossRef]
8. Hua, G.; Jégou, H. (Eds.) Computer Vision—ECCV 2016 Workshops. In *Lecture Notes in Computer Science*; Springer International Publishing: Cham, Switzerland, 2016; Volume 9915. [CrossRef]
9. Tjaden, H.; Schwanecke, U.; Schömer, E. Real-Time Monocular Pose Estimation of 3D Objects using Temporally Consistent Local Color Histograms. In Proceedings of the IEEE International Conference on Computer Vision ICCV 2017, Venice, Italy, 22–29 October 2017. [CrossRef]
10. Prisacariu, V.A.; Reid, I.D. PWP3D: Real-Time Segmentation and Tracking of 3D Objects. *Int. J. Comput. Vis.* **2012**, *98*, 335–354. [CrossRef]
11. Kaehler, A. *Learning OpenCV 3: Computer Vision in C++ with the OpenCV Library*; O'Reilly Media: Sebastopol, CA, USA, 2016.
12. Wuest, H.; Engekle, T.; Wientapper, F.; Schmitt, F.; Keil, J. From CAD to 3D Tracking—Enhancing & Scaling Model-Based Tracking for Industrial Appliances. In Proceedings of the 2016 IEEE International Symposium on Mixed and Augmented Reality, ISMAR-Adjunct 2016, Yucatan, Mexico, 19–23 September 2016; pp. 346–347. [CrossRef]
13. Fischler, M.A.; Bolles, R.C. Random sample consensus: A paradigm for model fitting with applications to image analysis and automated cartography. *Commun. ACM* **1981**, *24*, 381–395. [CrossRef]
14. Ferrara, P.; Piva, A.; Argenti, F.; Kusuno, J.; Niccolini, M.; Ragaglia, M.; Uccheddu, F. Wide-angle and long-range real time pose estimation: A comparison between monocular and stereo vision systems. *J. Visual Commun. Image Represent.* **2017**, *48*, 159–168. [CrossRef]
15. Zhang, Z. A Flexible New Technique for Camera Calibration (Technical Report). *IEEE Trans. Pattern Anal. Mach. Intell.* **2002**, *22*, 1330–1334. [CrossRef]
16. Bouguet, J.Y. Camera Calibration Toolbox for Matlab. 2015. Available online: http://www.vision.caltech.edu/bouguetj/calib_doc/ (accessed on 20 July 2018).
17. Geiger, A.; Moosmann, F.; Car, O.; Schuster, B. Automatic camera and range sensor calibration using a single shot. In Proceedings of the 2012 IEEE International Conference on Robotics and Automation, St. Paul, MN, USA, 14–18 May 2012; pp. 3936–3943. [CrossRef]
18. Romero-Ramirez, F.J.; Muñoz-Salinas, R.; Medina-Carnicer, R. Speeded up detection of squared fiducial markers. *Image Vis. Comput.* **2018**, *76*, 38–47. [CrossRef]
19. Grossmann, E.; Woodfill, J.; Gordon, G. Display Screen for Camera Calibration. U.S. Patent 8,743,214, 3 June 2014.
20. Gray, F. Pulse Code Communication. U.S. Patent 2,632,058, 17 March 1953.
21. Kimura, M.; Mochimaru, M.; Kanade, T. Projector Calibration using Arbitrary Planes and Calibrated Camera. In Proceedings of the 2007 IEEE Conference on Computer Vision and Pattern Recognition, Minneapolis, MN, USA, 17–22 June 2007; pp. 1–2. [CrossRef]
22. Rodriguez, L.; Quint, F.; Gorecky, D.; Romero, D.; Siller, H.R. Developing a Mixed Reality Assistance System Based on Projection Mapping Technology for Manual Operations at Assembly Workstations. *Proc. Comput. Sci.* **2015**, *75*, 327–333. [CrossRef]
23. Schmalz, C. Robust Single-Shot Structured Light 3D Scanning. Ph.D. Thesis, Technischen Fakultät der Universität Erlangen-Nürnberg, Erlangen, Germany, 2011.
24. Salvi, J.; Pagès, J.; Batlle, J. Pattern codification strategies in structured light systems. *Pattern Recognit.* **2004**, *37*, 827–849. [CrossRef]

25. Garbat, P.; Skarbek, W.; Tomaszewski, M. Structured light camera calibration. *Opto-Electron. Rev.* **2013**, *21*, 23–38. [CrossRef]
26. Lanman, D.; Taubin, G. Build Your Own 3D Scanner: 3D Photography for Beginners. *Siggraph* **2009**, 94. [CrossRef]
27. Datta, A.; Kim, J.S.; Kanade, T. Accurate camera calibration using iterative refinement of control points. In Proceedings of the 2009 IEEE 12th International Conference on Computer Vision Workshops, ICCV Workshops 2009, Kyoto, Japan, 27 September–4 October 2009; pp. 1201–1208. [CrossRef]
28. Usamentiaga, R.; Garcia, D.; Ibarra-Castanedo, C.; Maldague, X. Highly accurate geometric calibration for infrared cameras using inexpensive calibration targets. *Measurement* **2017**, *112*, 105–116. [CrossRef]
29. DLR. *DLR-Institute of Robotics and Mechatronics-ESS-OSS*; DLR: Weßling, Germany, 2015.
30. Halcon. *Camera_calibration [HALCON Operator Reference/Version 12.0.2]*; Halcon: Munich, Germany, 2015.
31. Moreno, D.; Taubin, G. Simple, accurate, and robust projector-camera calibration. In Proceedings of the 2nd Joint 3DIM/3DPVT Conference: 3D Imaging, Modeling, Processing, Visualization and Transmission, 3DIMPVT 2012, Zürich, Switzerland, 13–15 October 2012; pp. 464–471. [CrossRef]
32. Poulin-Girard, A.S.; Thibault, S.; Laurendeau, D. Influence of camera calibration conditions on the accuracy of 3D reconstruction. *Opt. Express* **2016**, *24*, 2678. [CrossRef] [PubMed]

sensors

MDPI

Article

Automatic Calibration of an Around View Monitor System Exploiting Lane Markings

Kyoungtaek Choi [1], Ho Gi Jung [1] and Jae Kyu Suhr [2,*]

[1] Department of Electronic Engineering, Korea National University of Transportation, 50 Daehak-ro, Chungju-si, Chungbuk 27469, Korea; maninquestion75@gmail.com (K.C.); hogijung@ut.ac.kr (H.G.J.)

[2] School of Intelligent Mechatronics Engineering, Sejong University, 209 Neungdong-ro, Gwangjin-gu, Seoul 05006, Korea

* Correspondence: jksuhr@sejong.ac.kr; Tel.: +82-2-3408-3481

Received: 26 July 2018; Accepted: 2 September 2018; Published: 5 September 2018

Abstract: This paper proposes a method that automatically calibrates four cameras of an around view monitor (AVM) system in a natural driving situation. The proposed method estimates orientation angles of four cameras composing the AVM system, and assumes that their locations and intrinsic parameters are known in advance. This method utilizes lane markings because they exist in almost all on-road situations and appear across images of adjacent cameras. It starts by detecting lane markings from images captured by four cameras of the AVM system in a cost-effective manner. False lane markings are rejected by analyzing the statistical properties of the detected lane markings. Once the correct lane markings are sufficiently gathered, this method first calibrates the front and rear cameras, and then calibrates the left and right cameras with the help of the calibration results of the front and rear cameras. This two-step approach is essential because side cameras cannot be fully calibrated by themselves, due to insufficient lane marking information. After this initial calibration, this method collects corresponding lane markings appearing across images of adjacent cameras and simultaneously refines the initial calibration results of four cameras to obtain seamless AVM images. In the case of a long image sequence, this method conducts the camera calibration multiple times, and then selects the medoid as the final result to reduce computational resources and dependency on a specific place. In the experiment, the proposed method was quantitatively and qualitatively evaluated in various real driving situations and showed promising results.

Keywords: around view monitor (AVM) system; automatic calibration; lane marking; parking assist system; advanced driver assistance system (ADAS)

---

1. Introduction

Recently, around view monitor (AVM) systems have become popular as parking aid products because they provide convenience to drivers by showing the surrounding view of the vehicle [1]. The AVM system consists of four cameras located at the centers of the front and rear bumpers and under two side view mirrors as shown in Figure 1 with red triangles [2]. Four images acquired from four cameras are transformed into bird's eye view images, and they are stitched to generate an AVM image. To this end, intrinsic and extrinsic parameters of four cameras should be known. The intrinsic parameters describe the optical properties of the camera, and the extrinsic parameters describe the relationship between the camera and vehicle coordinate systems. Since the intrinsic parameters hardly change, they need to be calibrated only once when the camera is manufactured. On the other hand, the extrinsic parameters can be changed due to various external shocks so that they need to be occasionally calibrated. However, the recalibration is quite inconvenient because, in reality, it should be carried out by skilled workers with a calibration pattern and external equipment as shown in Figure 2. This means that drivers must visit large-scale repair shops equipped with those specialized facilities.

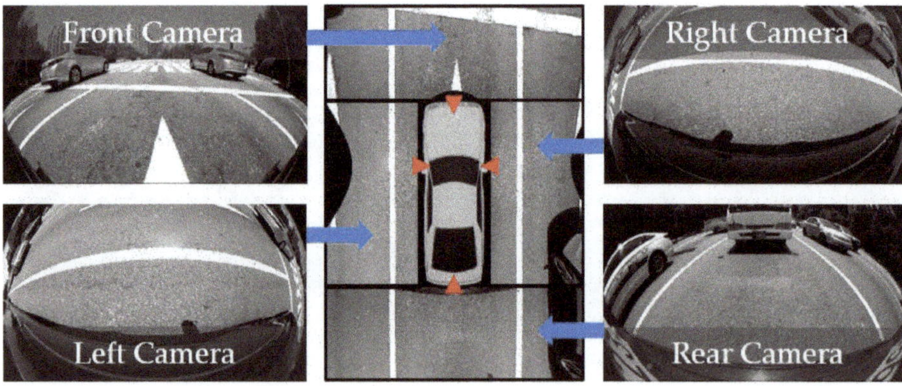

Figure 1. Camera configuration of the around view monitor (AVM) system. Red triangles indicate four cameras of the AVM system.

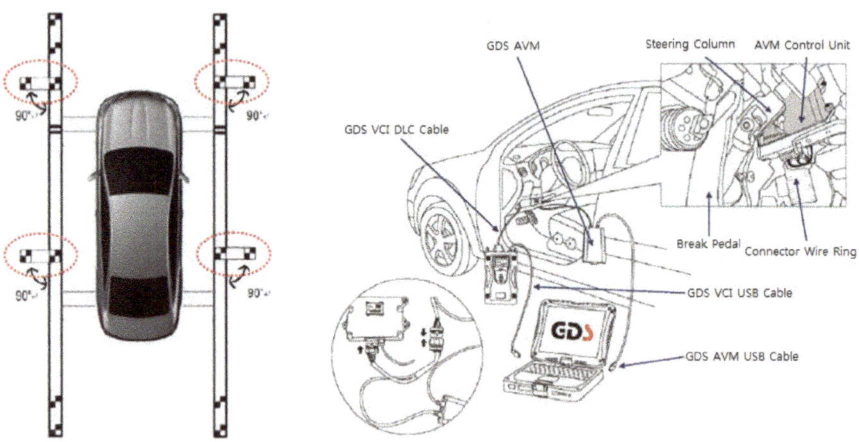

Figure 2. Example of the calibration pattern (**left**) and external equipment (**right**) used for recalibrating the AVM system [3].

To alleviate this inconvenience, an automatic calibration method for AVM systems should be developed. In order for an automatic calibration method to be practical and useful, the following points should be considered. First, it should use visual features that are easily observable in typical road conditions. Second, it should operate at various vehicle speeds that can occur in natural driving situations. Last, it should consider the relationship between adjacent cameras in order to obtain a seamless AVM image. Many methods have been suggested for calibrating vehicle-mounted cameras, and they can be categorized into three approaches: calibration pattern-based, interest point-based, and lane marking-based. The calibration pattern-based approach is inconvenient, because it needs specialized patterns. The interest point-based approach is limited by the vehicle speed. Since four cameras of the AVM system are facing the ground at low heights as shown in Figure 1, point features move too fast to reliably track when the vehicle travels at high speed. In addition, it is difficult to obtain point correspondences between images of adjacent cameras because overlapping areas are severely distorted due to the use of fisheye lenses as shown in Figure 1. On the other hand, the lane marking-based approach is suitable for calibrating the AVM system since lane markings exist in almost all on-road situations and are detectable at both low and high speeds. Furthermore, it is easier to find corresponding lane markings in images of adjacent cameras compared with the point features. However, previous methods in the lane marking-based approach only deal with a single camera and assume that the lane markings are correctly detected.

Based on these analyses, this paper proposes a novel and practical method that can automatically calibrate four cameras of the AVM system using lane markings. The proposed method assumes that four cameras are mounted on the vehicle as shown in Figure 1. Unlike the previous methods, the proposed method calibrates multiple cameras by automatically finding corresponding lane markings across images of adjacent cameras and efficiently handling falsely detected lane markings. Since the camera orientation more severely degrades the quality of the AVM image compared with the camera location [4], this paper focuses on estimating the rotation angles of four cameras. The proposed method starts by detecting lane markings from images captured by four cameras of the AVM system in a cost-effective manner. Since the detection results inevitably include falsely detected lane markings, it identifies and rejects those outliers by utilizing the statistical properties of the detected lane markings. Once the correct lane markings are sufficiently gathered, this method first calibrates the front and rear cameras, and then calibrates the left and right cameras with the help of the calibration results of the front and rear cameras. This two-step approach is essential because side cameras cannot be fully calibrated by themselves, due to insufficient lane marking information. After this initial calibration, this method collects the corresponding lane markings appearing across images of adjacent cameras and simultaneously refines the initial calibration results of four cameras to obtain seamless AVM images. In the case of a long image sequence, this method conducts the above procedure multiple times by dividing the image sequence, and selects the medoid of the multiple calibration results as the final one to reduce both computational resources and dependency on a specific place. Figure 3 shows the flowchart of the proposed method.

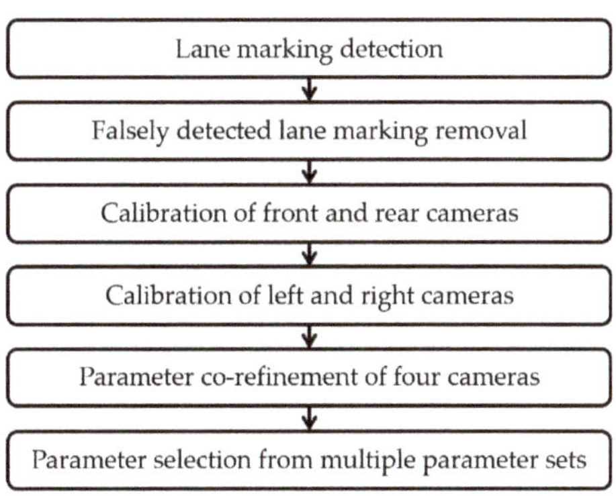

Figure 3. Flowchart of the proposed method.

The rest of this paper is organized as follows. Section 2 briefly explains related research. Sections 3–6 describe the essential procedures of the proposed method: lane marking detection, false detection removal, parameter estimation, and parameter selection. Section 7 presents experimental results and analyses. Finally, this paper is concluded with a summary in Section 8.

2. Related Research

Since camera calibration is a subject that has been extensively researched for a long period of time, it is unreasonable to cover all the aspects in this paper. Thus, this literature review focuses only on the previous methods suggested for calibrating extrinsic parameters of vehicle-mounted cameras. According to which features are used, the previous methods can be categorized into three approaches: calibration pattern-based, interest point-based, and lane marking-based.

2.1. Calibration Pattern-Based Approach

The calibration pattern-based approach estimates camera parameters using special patterns that consist of corners, circles, or lines. Since this approach uses precisely drawn patterns whose configurations are known, it is possible to accurately estimate the camera parameters. However, the methods in this approach are quite inconvenient, because drivers must prepare those patterns themselves, or visit repair shops that can equip those patterns. Chang et al. [5] utilized a pattern composed of multiple rectangles drawn on the ground. This method compares images of rectangles with the reference rectangles to estimate extrinsic parameters of four cameras composing the AVM system. Mazzei et al. [6] used a checkerboard pattern on the ground to calibrate extrinsic parameters of the front view camera by minimizing the reprojection error of the corner locations. Hold et al. [7] used a similar approach using a pattern of circles on the ground. This method finds extrinsic parameters of the front view camera that minimizes the reprojection error of the centers of the circles. Lebraly et al. [8] utilized a pattern composed of multiple circles and dots to estimate the relative pose between rigidly coupled cameras based on the hand–eye calibration scheme and bundle adjustment. Antonelli et al. [9] calibrated a mobile robot mounted camera by utilizing a rectangular parallelepiped pattern, with multiple dots and odometry calculated by wheel encoders. Tan et al. [10] utilized an H-shaped pattern that consisted of three lines (two are parallel and one is perpendicular to the vehicle) to calibrate the front view camera. Li et al. [11] also used an H-shaped pattern to calibrate a rear view camera with a fisheye lens. This method exploits the pattern embedded in parking lot lines. Natroshvili et al. [12] estimated the rotations of four cameras composing the AVM system using point patterns. This paper presents several calibration approaches that use different configurations of the point patterns.

2.2. Interest Point-Based Approach

The interest point-based approach estimates the camera parameters using the interest points extracted and tracked from consecutive images. If the interest points are reliably extracted and tracked, this approach produces accurate calibration results. However, in terms of the AVM system calibration, this approach has several drawbacks. Interest points are hardly tracked through the image sequence when the vehicle moves fast, because the cameras of the AVM system are facing the ground at low heights as shown in Figure 1. It is also difficult to obtain point correspondences between images of adjacent cameras because overlapping areas are severely distorted due to the use of fisheye lenses as shown in Figure 1. This makes it difficult to consider the relationships between adjacent cameras to obtain seamless AVM images. Since this approach is not well suited for AVM systems, this literature review introduces some which are closely related with this paper. Miksch et al. [13] calibrated the orientation of the camera attached to the sideview mirror by using homography and vehicle odometry. They reduce the number of parameters of the homography into one with the help of the vehicle odometry, and estimate them by minimizing the brightness difference around the point correspondences between two consecutive images. They proposed a similar method [14] that uses two point correspondences to estimate the homography. This method assumes that the vehicle moves straight ahead to remove the necessity of the vehicle odometry. Ruland et al. [4] estimated the orientation of the fisheye camera by using homography and vehicle odometry. This method finds the camera's rotation angles that minimize the pixel displacement error between the points transformed by the homography and the points transformed by the vehicle odometry. Tan et al. [15] calibrated a front view camera by combining optical flow, motion stereo, visual odometry, and vehicle odometry. This method estimates the camera's extrinsic parameters by comparing the visual odometry and vehicle odometry. The optical flow-based vanishing point and motion stereo-based 3D points are used as constraints. Chao et al. [16] estimated the relation between odometer and camera using a two-step least square minimization. This method first estimates two orientation parameters and then calculates the remaining parameters based on integrated odometry measurements and point correspondences between consecutive images. Heng et al. [17] calibrated extrinsic parameters between a rig with multiple cameras and vehicle odometry. This method first estimates the camera-odometry transformation for each camera using the tracked interest points, and then optimizes the initial parameters using bundle adjustment. Heng et al. [18] estimated extrinsic parameters of a multicamera system by using a pre-acquired highly accurate map. It finds the camera poses based on the 2D–3D correspondences between each set of synchronized camera images and the map.

2.3. Lane Marking-Based Approach

The lane marking-based approach estimates the camera parameters using lane markings on the ground. Lane markings exist in almost all on-road situations and are detectable at both low and high speeds. However, this approach cannot work on off-road situations where lane markings are not presented, and its accuracy can be degraded when lane markings are worn. Hold et al. [19] calibrated extrinsic parameters of the front view camera. This method detects dashed lane markings at predefined vertical coordinates, and calculates the extrinsic parameters by analyzing the detected lane markings and the measured vehicle velocity. Paula et al. [20] estimated the height, pitch, and roll of the front view camera using a rectangular portion of the road, which is calculated from the lane boundaries and the distance between adjacent dashed lane markings. Wang et al. [21] estimated the extrinsic parameters of the front view camera based on two vanishing points generated from dashed lane markings. One vanishing point is obtained from left and right lane markings and the other one is obtained from lines connecting the corners of the lane segments. These three methods [19–21] require dashed lane markings, and two of them [20,21] need accurate vehicle speed synchronized with the camera. Ribeiro et al. [22] proposed a method similar to the method in [20]. This method finds a trapezoid composed with two lane markings and estimates the parameters which transform the trapezoid to a rectangle. Xu et al. [23] calibrated the extrinsic parameters of the front view camera using parallel lines extracted from lane markings, as well as parallel lines perpendicular to the ground (e.g., edges of buildings). This method manually designates lane markings and requires parallel lines perpendicular to the ground, which are rarely captured by the ground-facing cameras of the AVM system. Catala-Prat et al. [24] calibrated the orientation of the front view camera by finding parameters that minimize the sum of squares of the lane markings' slopes. This method automatically detects lane markings, but manually rejects outliers for reliable parameter estimation. Zhao et al. [25] estimated the pitch and yaw of the vehicle mounted camera. This method uses a vanishing point calculated by applying the weighted least squares approach to detected lane markings. The calculated vanishing point is tracked by Kalman filter to obtain consistency.

As aforementioned in the introduction, the lane marking-based approach is more suitable for calibrating the AVM system in natural driving conditions compared to the other approaches. This is because of the following reasons: lane markings exist in almost all on-road situations, are detectable at both low and high speeds, and appear across images of adjacent cameras. However, previous methods in this approach cannot be directly used to calibrate cameras of the AVM system because they require ideally drawn dashed lanes, both left and right lanes at the same time, or lines perpendicular to the ground. Furthermore, previous methods have not handled falsely detected lane markings that are inevitably included in a real situation, and have not utilized the lane markings appearing across images of adjacent cameras.

In order to overcome the limitations of the previous methods, this paper proposes a novel lane marking-based automatic calibration method for AVM systems. The proposed method can handle both dashed and solid lane markings, calibrate side view cameras where only the left or right lane marking is observed, and does not need rare objects, such as lines perpendicular to the ground. In addition, this method explicitly deals with falsely detected lane markings based on the statistical properties of the detected lane markings, and utilizes the lane markings that appear across images of adjacent cameras to consider the inter-camera relationships.

Besides these three approaches, there is a road geometry-based approach that calibrates vehicle-mounted cameras by estimating the geometry of the road in front of the vehicle [26,27]. Since almost all methods in this approach use stereo cameras to obtain three-dimensional geometry information of the road, it cannot be utilized for calibrating the cameras of the AVM system where there is very little overlap between views of adjacent cameras. Therefore, this literature review does not deal with this approach in depth.

3. Lane Marking Detection

The proposed method detects lane markings by combining simple techniques: edge detection, line estimation, and line pairing. This paper deliberately avoids using sophisticated lane detection methods in order to detect lane markings from images of four cameras in real-time. This method assumes that the angles of the four cameras can be changed within $\pm 5^{\circ}$. This value was empirically set with the help of automotive experts. This method precalculates the ranges of position, orientation, and width for lane markings as geometrical constraints during offline simulation. Figure 4 shows the precalculated position range of the lane markings in the AVM image. Figure 4a–c are the position ranges of the left and right lanes in the images of the front and rear cameras, the left and right lanes in images of the left and right cameras, and the stop lines in the images of the front and rear cameras, respectively. Red, blue, and green regions indicate the position ranges for left lanes, right lanes, and stop lines, respectively. The ranges of orientation and width are not depicted because it is difficult to graphically draw them.

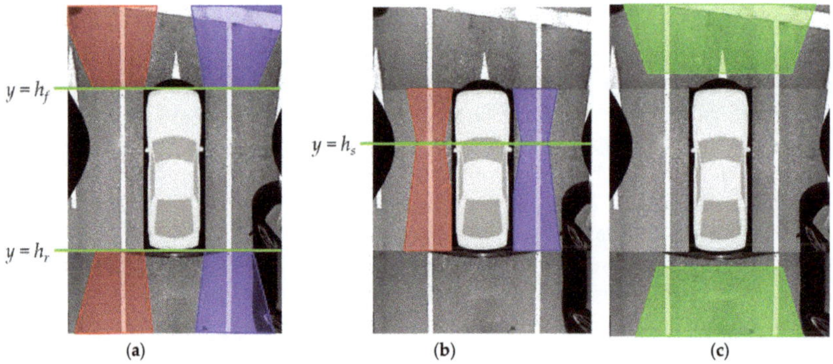

Figure 4. Precalculated position range of lane markings. (**a**) Position ranges of left and right lanes in images of the front and rear cameras; (**b**) Position ranges of left and right lanes in images of the left and right cameras; (**c**) Position ranges of stop lines in images of the front and rear cameras. Red, blue, and green regions indicate the position ranges of left lanes, right lanes, and stop lines, respectively.

The proposed method detects lane markings in bird's-eye view images. Since the angles of the cameras have not yet been calibrated, the bird's-eye view images can erroneously be generated with the initial camera angles. However, it is easier to detect lane markings in bird's-eye view images compared to undistorted images because they are less affected by perspective distortion. Before detecting lane markings, a 3×3 median filter is applied to the bird's-eye view image to remove salt and pepper-like noises that frequently appear on asphalt roads. The image gradient is calculated by the Sobel operator [28] and pixels whose gradient magnitudes are larger than a predetermined value are detected as edge pixels. Each edge pixel is classified according to its gradient orientation. If an edge pixel has a gradient orientation between -90° and 90°, it is classified as a rising edge pixel. Otherwise, it is classified as a falling edge pixel. Figure 5a shows an edge detection and classification result in a bird's-eye view image acquired by the front camera of the AVM system. Red and green dots indicate the rising and falling edge pixels, respectively. Once edge pixels are obtained, straight lines are estimated by random sample consensus (RANSAC) [29]. In this procedure, RANSAC is separately applied to the rising and falling edge pixels. If the detected line satisfies the ranges of position and orientation precalibrated during the offline simulation, it is regarded as valid. Otherwise, it is discarded. Figure 5b shows four straight lines estimated by RANSAC using the edge pixels presented in Figure 5a. Red and green lines are estimated from the rising and falling edge pixels, respectively. Once the lines are estimated, they are paired to compose lane markings based on three conditions. The lines that follow the conditions are used for further calibration procedures, and the lines that do not follow the conditions are discarded. The three conditions are as follows: (1) Two lines should not be crossed within the bird's-eye view image because they are parallel in the real world; (2) The distance between two lines should be within the width range of the lane marking, which is precalculated during the offline simulation; (3) The edge pixels of two lines composing a lane marking should have similar vertical positions. This means that the number of edge pixels whose vertical positions are similar should be larger than a predetermined value. The detection procedure of the stop line is almost the same as that of the lane marking. The only differences are that it uses the precalculated range of the stop line width, and counts the number of edge pixels whose horizontal positions are similar. Figure 6a–d show some examples of the lane marking and stop line detection results. Figure 6a includes two lanes and a stop line detected in a front camera image, Figure 6b includes two lanes detected in a rear camera image, and Figure 6c,d include left and right lanes detected in images of the left and right cameras, respectively.

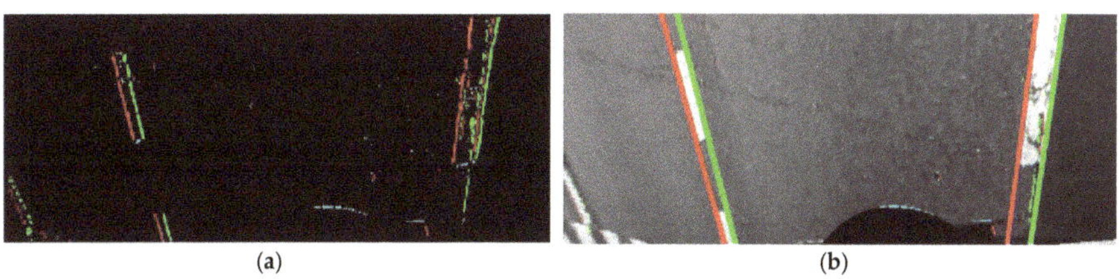

Figure 5. (**a**) Edge detection and classification result (red: rising edge pixels, green: falling edge pixels); (**b**) Lane marking detection result (red: lines from rising edge pixels, green: lines from falling edge pixels).

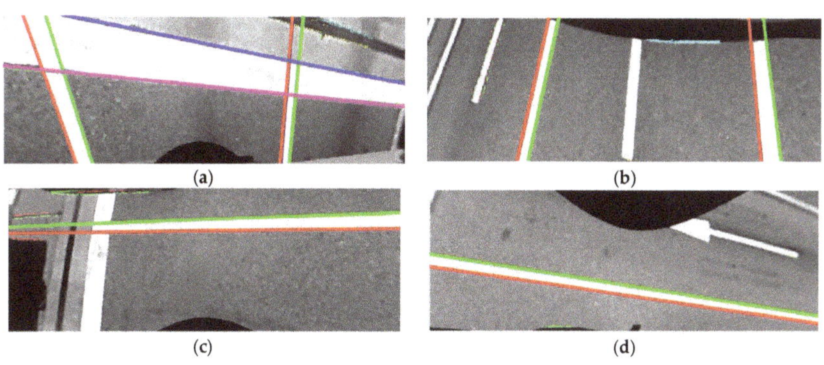

Figure 6. Example detection results of lane marking and stop line. (**a**) Two lanes and a stop line detected in a front camera image; (**b**) Two lanes detected in a rear camera image; (**c**) A left lane detected in a left camera image; (**d**) A right lane detected in a right camera image.

4. Falsely Detected Lane Marking Removal

In real situations, the lane marking detection result inevitably includes false detections, so-called outliers. These outliers should be rejected because they can severely degrade the calibration result. However, the previous methods in the lane making-based approach introduced in Section 2 ignore this important procedure, and simply assume that all lane markings are correctly detected, or manually reject the outliers. Since a fully automated calibration method should include a procedure that handles outliers, this paper suggests a method that identifies and rejects outliers based on the statistical properties of the detected lane markings. The suggested method assumes that the vehicle travels along the lane and that straight driving is more frequent than curved driving. Under these assumptions, this method statistically draws two common properties of the correctly detected lane markings in terms of position and orientation, and then rejects the lane markings that do not follow those properties.

This method first finds the common property of lane marking position, and rejects outliers using it. To draw the positional property, this method calculates a positional histogram based on the intersections between the detected lane markings and the line $y = h$. A total of six positional histograms are generated for left and right lane markings in the front camera images, left and right lane markings in the rear camera images, left lane markings in the left camera images, and right lane markings in the right camera images. The value of h is set to h_f for the lane markings in the front camera images, h_r for the lane markings in the rear camera images, and h_s for the lane markings in left and right camera images. h_f, h_r, and h_s are shown in Figure 4a,b. h_f, h_r, and h_s are selected at the locations where positional variations of the detected lane markings are expected to be small according to the offline simulation. Once six positional histograms are generated, their modes are estimated. If a histogram is multimodal, the mode closest to the vehicle is selected. After obtaining a mode for each histogram, this method rejects the lane markings whose locations at $y = h$ are farther from the mode by the predetermined value. Figure 7a,b show the position-based outlier rejection procedure in the case of the rear camera image. Blue, red, and cyan lines in Figure 7a show the detected lane markings, and two thick magenta curves indicate two positional histograms for left and right lane markings. Using the highest modes of those two histograms (magenta lines), the red lines can be removed because their locations at $y = h_r$ are distant from the corresponding modes. Figure 7b shows the remaining lines after the position-based outlier rejection.

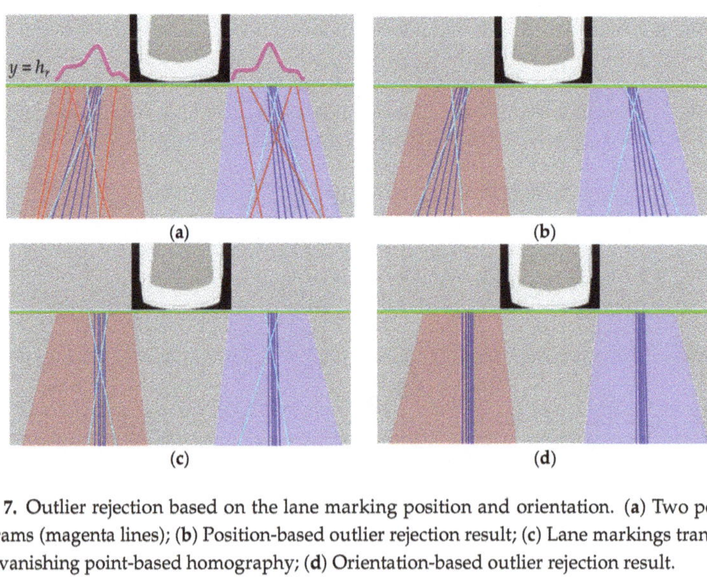

Figure 7. Outlier rejection based on the lane marking position and orientation. (**a**) Two positional histograms (magenta lines); (**b**) Position-based outlier rejection result; (**c**) Lane markings transformed by the vanishing point-based homography; (**d**) Orientation-based outlier rejection result.

The remaining lane markings after the position-based outlier rejection are filtered, once again, based on their orientations. To this end, this method utilizes the vanishing point and homography. In cases of the lane markings detected in images of the left and right cameras, the vanishing point is estimated by RANSAC. This first randomly selects two lines and calculates their intersection point, $\mathbf{v}_l = [u_p\ v_p\ 1]^{\mathrm{T}}$. A rotation matrix, R, that transforms $\mathbf{v}_l$ into a point at infinity, $\mathbf{p}_\infty = [0\ 1\ 0]^{\mathrm{T}}$, is obtained by calculating the rotation axis, $\mathbf{u}$ and angle, θ as

$$\begin{array}{l} \mathbf{u} = \overline{\mathbf{v}}_l \times \mathbf{p}_\infty, \quad \theta = \cos^{-1}\left(\overline{\mathbf{v}}_l \cdot \mathbf{p}_\infty\right) \\ \text{where } \overline{\mathbf{v}}_l = \mathbf{v}_l / \|\mathbf{v}_l\| \end{array}, \tag{1}$$

where $\times$ and $\cdot$ indicate the cross and dot products, respectively. The rotation matrix, R can be derived by $\mathbf{u} = [u_x\ u_y\ u_z]^{\mathrm{T}}$ and θ as

$$R = \begin{bmatrix} \cos\theta + u_x^2(1-\cos\theta) & u_x u_y(1-\cos\theta) - u_z\sin\theta & u_x u_z(1-\cos\theta) + u_y\sin\theta \\ u_y u_x(1-\cos\theta) + u_z\sin\theta & \cos\theta + u_y^2(1-\cos\theta) & u_y u_z(1-\cos\theta) - u_x\sin\theta \\ u_z u_x(1-\cos\theta) - u_y\sin\theta & u_z u_y(1-\cos\theta) + u_x\sin\theta & \cos\theta + u_z^2(1-\cos\theta) \end{bmatrix}. \tag{2}$$

Once R is obtained, the lane markings, $\mathbf{l}$ are transformed by the homography, H as

$$\mathbf{l}' = H^{-\mathrm{T}}\mathbf{l} = \left(KRK^{-1}\right)^{-\mathrm{T}}\mathbf{l}, \tag{3}$$

where K is a 3×3 intrinsic parameters matrix of the virtual camera of the AVM system, that is predetermined when the AVM system is manufactured. After transforming the lane markings via Equation (3), this method counts the number of transformed lane markings, $\mathbf{l}'$ that are parallel to the heading direction of the vehicle (vertical axis of the AVM image) as a consensus set. This procedure is iterated, and H_{max} that maximizes the number of consensus set is selected. Finally, the lane markings excluded from the consensus set of H_{max} are identified as outliers and rejected.

In cases of the lane markings detected in images of the front and rear cameras, the vanishing point is estimated by a voting-based method. If RANSAC is used in these cases, it requires a large amount of computation cost for transforming lane markings and counting consensus sets. This is because the number of lane markings in images of the front and rear cameras are approximately twice that of the side camera. Unlike the side camera, both left and right lane marking can be captured by the front and rear cameras. The locations of the vanishing points calculated from a pair of left and right lane markings are accumulated in 2D voting bins. The final vanishing point is estimated by averaging the locations that have the largest accumulation results. The homography, H_{max}, is calculated from the final vanishing point using Equations (1)–(3). All lane markings are transformed by H_{max} using Equation (3), and the transformed lane markings not parallel to the heading direction of the vehicle (vertical axis of the AVM image) are identified as outliers and rejected. Figure 7c,d show the orientation-based outlier rejection procedure in the case of the rear camera image. Blue and cyan lines in Figure 7c show the lane markings transformed by the vanishing point-based homography in Equation (3). In Figure 7c, the cyan lines are identified as outliers, and rejected, because they are not parallel to the heading direction of the vehicle. Figure 7d shows the remaining lines after the orientation-based outlier rejection. The remaining lines after both the position-based and orientation-based outlier rejection procedures are used for further calibration procedures, and the rejected lines from either of these two procedures are discarded. The outlier rejection procedure for stop lines are also conducted by RANSAC. This procedure will be described in detail in Section 5 since it is conducted during the parameter estimation.

5. Parameter Estimation

The proposed method first calibrates the front and rear cameras, and then calibrates the left and right cameras with the help of the calibration results of the front and rear cameras. This two-step approach is essential for calibrating cameras of the AVM system because side cameras cannot be fully calibrated by themselves due to insufficient lane marking information. As shown in Figure 1, both left and right lane markings are observable in images of the front and rear cameras, so that rotation angles of those cameras can be fully estimated. However, only either the left or right lane marking is observable in images of the left and right cameras, so that rotation angles of those cameras can only be estimated in part. This is the reason why this paper utilizes the two-step approach. Once four cameras are initially calibrated by the two-step approach, this method simultaneously refines the initial calibration results of four cameras using the corresponding lane markings appearing across images of adjacent cameras.

5.1. Calibration of Front and Rear Cameras

The proposed method first calibrates the front and rear cameras. Those two cameras are separately calibrated using the same approach. Figure 8a,b show pitch, yaw, and roll of the front and rear cameras, respectively. This method first estimates the pitch and yaw angles of the front and rear cameras, and then estimates their roll angles. The pitch and yaw angles are estimated by finding the vanishing point. If $\mathbf{l}_i$ is a 3 × 1 parameter vector of the i-th line survived from the outlier rejection procedure, and $\mathbf{v}_l$ is the vanishing point in homogeneous coordinates, then they should be related as $\mathbf{l}_i^{\mathrm{T}}\mathbf{v}_l = 0$. If there are N_L lane markings, it can be expanded as

$$L\mathbf{v}_l = \mathbf{0}, \quad \text{where } L = \begin{bmatrix} \mathbf{l}_1 & \mathbf{l}_2 & \cdots & \mathbf{l}_{2N_L-1} & \mathbf{l}_{2N_L} \end{bmatrix}^{\mathrm{T}}, \tag{4}$$

where L is a $(2N_L) \times 3$ matrix that includes $2N_L$ lines. There are $2N_L$ lines in N_L lane markings because each lane marking consists of two lines as shown in Figure 6. Thus, a least-squares solution of $\mathbf{v}_l$ is calculated by finding the eigenvector of $L^{\mathrm{T}}L$ corresponding to the smallest eigenvalue. The rotation matrix is obtained from $\mathbf{v}_l$ via Equations (1) and (2), and the pitch and yaw angles are estimated by decomposing the obtained rotation matrix. The roll angle obtained from this rotation matrix may not be correct because the vanishing point does not have enough information to estimate the roll angle. The pitch and yaw angles estimated by this closed-form solution are refined using Levenberg–Marquardt (LM) algorithm [30] by minimizing the cost, c_{py}, as

$$c_{py} = \sum_{i=1}^{2N_L} |u_i - d_i| + \sum_{i=1}^{2N_L}\sum_{j=1}^{2N_L} \left| |u_i - u_j| - |d_i - d_j| \right|, \tag{5}$$

where u_i and d_i are shown in Figure 9a. They indicate the horizontal locations of the intersection points between the i-th line and $y = 0$ and $y = h_{max}$, respectively. In Equation (5), the left term is minimized when lines are located upright (parallel to the heading direction of the vehicle), and the right term is minimized when pairs of lines are parallel to each other. Thus, c_{py} is minimized when the lines are both upright and parallel to each other. The vanishing point $\mathbf{v}_l$ is also refined based on the pitch and yaw angles optimized by LM, and the refined vanishing point is denoted by $\mathbf{v}_l'$.

The one remaining angle, roll, is estimated based on either stop line or lane marking width. If enough stop lines are detected, the roll angle is estimated based on the stop lines. The roll angle estimation and false stop line rejection are simultaneously conducted using RANSAC. A stop line consists of two lines, as shown in Figure 9b with red and blue. A pair of two lines composing a stop line is randomly selected, and the vanishing point, $\mathbf{v}_s$, is calculated by finding their intersection point. Since the vanishing point induced by the lane markings, $\mathbf{v}_l'$, is already calculated, the complete rotation matrix that contains all pitch, yaw, and roll angles can be obtained as

$$R = \begin{bmatrix} \mathbf{v}'_l & -\mathbf{v}'_l \times \mathbf{v}_s & \mathbf{v}_s \end{bmatrix}. \tag{6}$$

After calculating R, all pairs of two lines composing stop lines are transformed by homography H in Equation (3). Since the two lines of the correct pair should be parallel to each other after the transformation, the number of line pairs parallel to each other is counted as a consensus set. This procedure is iterated and R_{max} that maximizes the number of consensus set is selected. Finally, the line pairs excluded from the consensus set of R_{max} are identified as outliers and rejected. The roll angle is obtained by decomposing R_{max}. After removing the outliers, the roll angle is refined using LM algorithm by minimizing the cost, c_{rs} as

$$c_{rs} = \sum_{i=1}^{N_S} ||tl_i - bl_i| - |tr_i - br_i||, \tag{7}$$

where N_S is the number of stop lines classified as inliers. tl_i, bl_i, tr_i, and br_i are shown in Figure 9b. tl_i and tr_i are the vertical locations of the intersection points between the top line of the i-th stop line and $x = 0$ and $x = w_{max}$, respectively, and bl_i and br_i are the vertical locations of the intersection points between the bottom line of the i-th stop line and $x = 0$ and $x = w_{max}$, respectively. c_{rs} in Equation (7) is minimized when the top and bottom lines of the stop lines are parallel to each other.

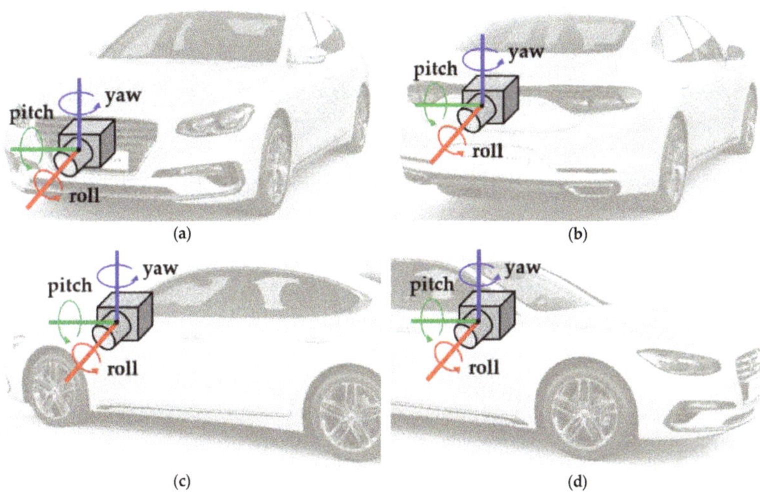

Figure 8. Pitch, yaw, and roll of four cameras composing the AVM system. (**a**) Front camera; (**b**) Rear camera; (**c**) Left camera; (**d**) Right camera.

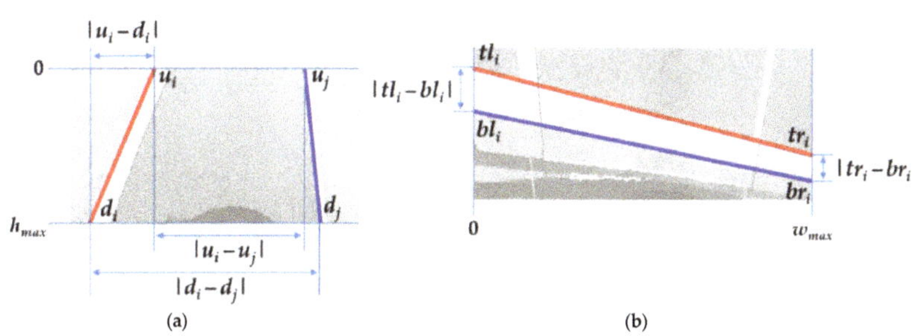

Figure 9. Explanations of cost functions used for the LM algorithm. (**a**) In the case of a lane marking; (**b**) In the case of a stop line.

If the number of detected stop lines are insufficient, the roll angle is estimated by lane markings by assuming that the left and right lane markings detected in the same image have the same width. Since there are some cases where this assumption is not valid, this method first rejects the lane marking pairs whose widths are expected to be different to each other. To this end, the lane markings are first transformed by the homography, which is calculated by the vanishing point of the left and right lane markings via Equations (1)–(3). Figure 10a,b show the lane markings before and after the homography-based transformation, respectively. Since this homography transforms the vanishing point to the point at infinity, $\mathbf{p}_\infty = [0\ 1\ 0]^T$, all lines composing lane markings become parallel to the vertical axis of the image. After the transformation, this method rejects pairs of left and right lane markings based on their width ratio using RANSAC. A pair of left and right lane markings detected in the same image is randomly selected, and the ratio between their widths is calculated. After that, the number of lane marking pairs whose width ratio is similar to this ratio is counted as a consensus set. This procedure is iterated, and the width ratio that maximizes the number of consensus set is selected. The lane marking pairs whose width ratios are different from the selected width ratio are classified as outliers and rejected. After the outlier removal, the roll angle is estimated using LM algorithm by minimizing the cost, c_{rl} as

$$c_{rl} = \sum_{i=1}^{N_P} |wl_i - wr_i|, \tag{8}$$

where N_P is the number of pairs of left and right lane markings detected in the same images. wl_i and wr_i are the widths of the i-th lane marking pair. wl_i is for the left lane marking and wr_i for the right lane marking. Based on this approach, all pitch, yaw, and roll angles of both the front and rear cameras are estimated. Figure 11a shows the AVM image before the camera calibration, and Figure 11b shows the AVM image after calibrating the front and rear cameras. Red, blue, and green lines indicate the lane markings in the front, rear, and side cameras, respectively. It can be noticed that two lane markings in images of the front and rear cameras are upright, and have the same widths as in Figure 11b. This implicitly reveals that the angles of the front and rear cameras are correctly estimated. But, in Figure 11b, the lane markings in the images of the two side cameras are not correctly located because the two side cameras have not been yet calibrated.

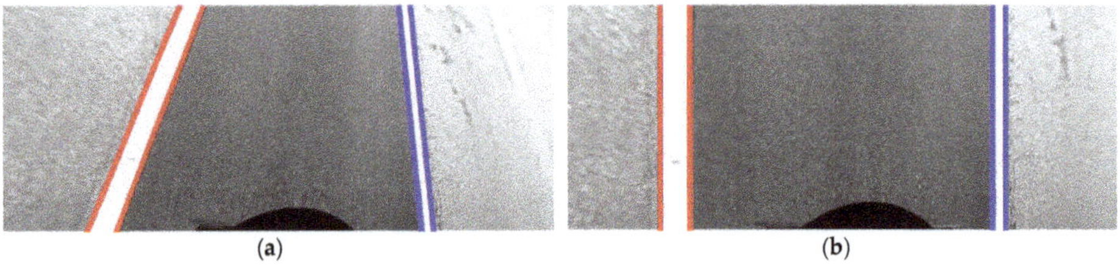

Figure 10. Images before and after the homography-based transformation. (**a**) Before the transformation; (**b**) After the transformation.

5.2. Calibration of Left and Right Cameras

Once the front and rear cameras are calibrated, the two side cameras are then calibrated. Figure 8c,d show the pitch, yaw, and roll of the left and right cameras, respectively. The left and right cameras are separately calibrated using the same method. This method first estimates the yaw and roll angles of the side camera, and then estimates its pitch angle with the help of the calibration results of the front and rear cameras. The yaw and roll angles of the side camera are estimated by the same method used for estimating the pitch and yaw angles of the front and rear cameras in Section 5.1. That is, the yaw and roll angles of the side camera are obtained by finding the vanishing point via

Equation (4) and minimizing the cost in Equation (5). However, in the case of the side camera, the pitch angle cannot be obtained by itself. This is because, unlike the front and rear cameras, only a single lane is observable in an image of the side camera. Figure 11c shows the AVM image after calibrating the yaw and roll angles of the two side cameras. The lane markings in the images of the two side cameras are upright because the yaw and roll angles have been calibrated. However, their locations and widths are not consistent with the lane markings in the images of the front and rear cameras. This is because the pitch angles of two side cameras cannot be calibrated by themselves. To overcome this limitation, this paper proposes an approach that estimates the pitch angles of the two side cameras with the help of the calibration results of the front and rear cameras.

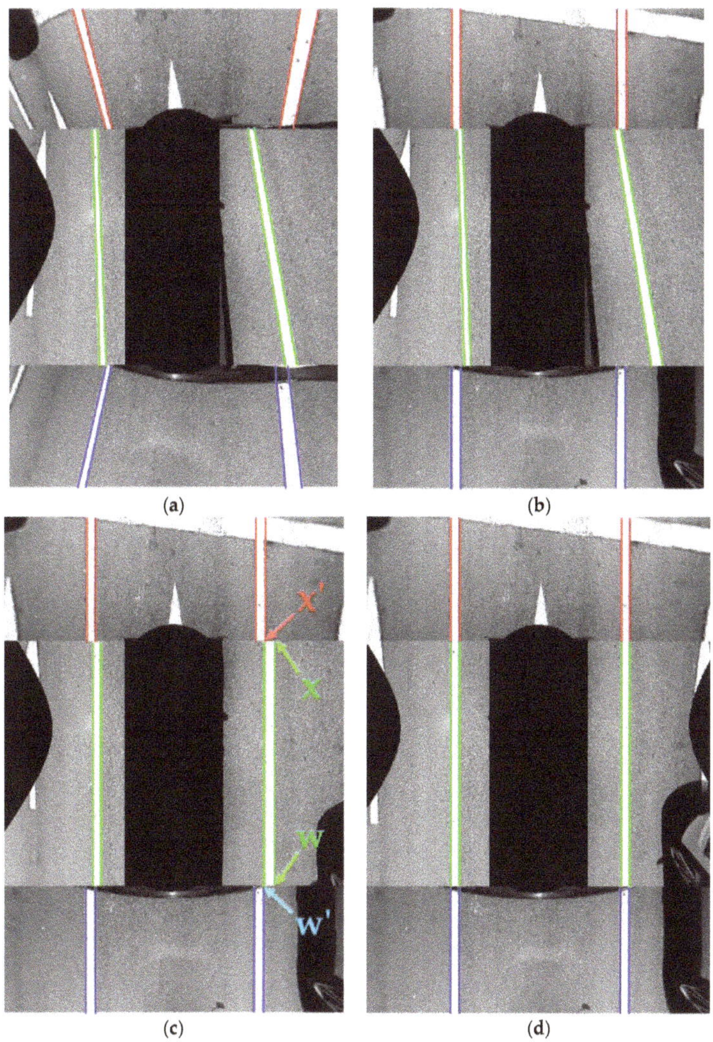

Figure 11. Resulting AVM images at consecutive calibration stages. (**a**) Before the calibration; (**b**) After calibrating the front and rear cameras; (**c**) After estimating the yaw and roll angles of two side cameras; (**d**) After calibrating all pitch, yaw, roll angles of four cameras. Red, blue, and green lines indicate the lane markings in the front, rear, and side cameras, respectively.

The proposed method finds the pitch angle that makes the lane markings appearing across images of adjacent cameras properly connect to each other. This method has two assumptions: one is that some of the lane markings are simultaneously captured by adjacent cameras, and the other is that the detected lane markings are locally straight. Since it cannot be assumed that all lane markings detected in images of adjacent cameras at the same time correspond to each other, this method uses RANSAC to simultaneously find the corresponding lane markings and the pitch angle of the side camera. This method first randomly selects one line detected in an image of the side camera, and finds the lines detected in images of the front or rear camera at the same time. If there are lines detected in images of the side and front cameras at the same time, it is assumed that two lines correspond each other. The upper end point of the line detected in the side camera image is denoted as $\mathbf{x} = [u\ v\ 1]^{\mathrm{T}}$, and the lower end point of the line detected in the front camera image is denoted as $\mathbf{x}' = [u'\ v'\ 1]^{\mathrm{T}}$. The locations of $\mathbf{x}$ and $\mathbf{x}'$ are shown in Figure 11c. Those two points are related with the pitch angle, φ, of the side camera as

$$\mathbf{x}' = KR_{\phi}K^{-1}\mathbf{x}, \tag{9}$$

where K is the intrinsic parameters matrix of the virtual camera of the AVM system, which is predetermined when the AVM system is manufactured. R_{φ} is a 3×3 rotation matrix induced by φ. If K moves from the right side to the left side in Equation (9), it is converted as

$$\begin{gathered} \underbrace{K^{-1}\mathbf{x}'}_{\mathbf{y}'} = R_{\phi}\underbrace{K^{-1}\mathbf{x}}_{\mathbf{y}} \quad . \\ \mathbf{y}' = R_{\phi}\mathbf{y} \end{gathered} \tag{10}$$

Since the vertical locations of $\mathbf{x}$ and $\mathbf{x}'$ are the same as shown in Figure 11c, $\boldsymbol{y}$ and $\boldsymbol{y}'$ can be denoted as $[x\ y\ 1]^{\mathrm{T}}$ and $[x'\ y\ 1]^{\mathrm{T}}$, respectively. Based on these notations, Equation (10) can be rewritten as

$$\begin{bmatrix} x' \\ y \\ 1 \end{bmatrix} = \begin{bmatrix} \cos\phi & 0 & \sin\phi \\ 0 & 1 & 0 \\ -\sin\phi & 0 & \cos\phi \end{bmatrix} \begin{bmatrix} x \\ y \\ 1 \end{bmatrix}. \tag{11}$$

Since the second row of Equation (11) is meaningless, it can be simplified as

$$\begin{bmatrix} x' \\ 1 \end{bmatrix} = \begin{bmatrix} \cos\phi & \sin\phi \\ -\sin\phi & \cos\phi \end{bmatrix} \begin{bmatrix} x \\ 1 \end{bmatrix}. \tag{12}$$

In Equation (12), $[x\ 1]^{\mathrm{T}}$ and $[x'\ 1]^{\mathrm{T}}$ are related with a 2×2 rotation matrix. Thus, the pitch angle, φ of the side camera can be calculated by using Procrustes analysis [31] as

$$\phi = -\tan^{-1}\left(\frac{x - x'}{xx' + 1}\right). \tag{13}$$

The same procedure can be used for estimating the pitch angle, φ using the lower end point of the line detected in the side camera image, $\mathbf{w}$, and the upper end point of the line detected in the rear camera image, $\mathbf{w}'$. The locations of $\mathbf{w}$ and $\mathbf{w}'$ are shown in Figure 11c. Once φ is estimated, all the lines detected in images of the side camera, $\mathbf{l}_s$ are transformed by homography, H_{φ} as

$$\mathbf{l}'_s = H_{\phi}^{-\mathrm{T}}\mathbf{l}_s = \left(KR_{\phi}K^{-1}\right)^{-\mathrm{T}}\mathbf{l}_s. \tag{14}$$

After the transformation, the number of line correspondences between the images of adjacent cameras is counted as a consensus set. This procedure is iterated, and the pitch angle that produces the largest number of consensus set is selected as the pitch angle of the side camera. In addition, the line correspondences included in the largest consensus set are classified as correct corresponding lines between adjacent cameras. The pitch angle estimated by RANSAC, is refined using LM algorithm by minimizing the cost, c_{sp}, as

$$c_{sp} = \sum_{i=1}^{N_C} |u_i - u'_i|, \tag{15}$$

where u_i and u_i' are the horizontal locations of the i-th line correspondence after the transformation using Equation (14). N_C is the number of line correspondences. c_{sp} is minimized when line correspondences detected in images of adjacent cameras are exactly connected to each other. Figure 11d shows the AVM image produced by the calibration results of all four cameras. It can be seen that the lane markings in images of adjacent cameras are properly connected to each other.

5.3. Parameter Co-Refinement of Four Cameras

Once all four cameras of the AVM system are initially calibrated using the two-step approach explained in Sections 5.1 and 5.2, the proposed method simultaneously refines all angles of four cameras using LM algorithm by minimizing the cost, c_{total}, that consists of c_{py} in Equation (5), c_{rs} in Equation (7), c_{rl} in Equation (8), and c_{sp} in Equation (15) as

$$c_{total} = \begin{cases} c_{py,front} + c_{py,rear} + c_{py,left} + c_{py,right} + c_{rs,front} + c_{rs,rear} + c_{sp,left} + c_{sp,right}, & N_S > T \\ c_{py,front} + c_{py,rear} + c_{py,left} + c_{py,right} + c_{rl,front} + c_{rl,rear} + c_{sp,left} + c_{sp,right}, & otherwise \end{cases}, \tag{16}$$

where the second subscription of c indicates the camera used for calculating the cost. N_S and T are the number of stop lines and the predetermined minimum number of stop lines, respectively. The proposed method simultaneously refines the initial calibration results of four cameras, not only using the lane markings detected in images taken from individual cameras, but also using the corresponding lane markings matched between adjacent cameras. This approach can make the produced AVM images more seamless. The corresponding lane markings have been obtained during the RANSAC-based pitch angle estimation of the side cameras in Section 5.2.

6. Parameter Selection

If the proposed calibration procedure is applied to a long image sequence, it requires a huge amount of computational resources (memory and time) because a huge number of lane markings should be stored and processed. To alleviate this limitation, this procedure is conducted multiple times in the case of a long image sequence. That is, the camera parameters are repeatedly estimated whenever a sufficient number of lane markings are gathered. When using this approach, multiple parameter sets are obtained. A parameter set includes 12 camera angles (three angles for each camera), so that it can be considered as a 12-dimensional vector. To find the most appropriate parameter set and prevent it from being overfitted to a specific place, this paper selects the medoid ($\mathbf{a}'$) of multiple parameter sets ($\mathbf{a}_1$, $\mathbf{a}_2$, ... , $\mathbf{a}_{NA}$) as

$$\mathbf{a}' = \underset{\mathbf{b}\in\{\mathbf{a}_1,\mathbf{a}_2,\cdots,\mathbf{a}_{N_A}\}}{\operatorname{argmin}} \sum_{i=1}^{N_A} d(\mathbf{a}_i, \mathbf{b}), \tag{17}$$

where N_A is the number of parameter sets, and d ($\mathbf{a}_i$, $\mathbf{b}$) indicates the Euclidean distance between two parameter sets ($\mathbf{a}_i$ and $\mathbf{b}$).

7. Experiments

7.1. Experimental Setup

The test dataset was acquired by the AVM system mounted on an off-the-shelf vehicle, a Hyundai Genesis G80 [32]. The AVM system consists of four fisheye cameras located at the centers of the front and rear bumpers, and under two side view mirrors, as shown in Figure 1. The resolution, field-of-view, and acquisition frequency of each fisheye camera are 1280 × 720 pixels, 190 degrees, and 30 frames per second, respectively. The test dataset was captured at 10 different sites for 116 min. Figure 12 shows example images included in the test dataset. Only the images taken by the front camera of the AVM system are presented. As shown in this figure, the test dataset was acquired in various real driving situations including congested, uncongested, wide, and narrow roads.

Figure 12. Example images of the test dataset taken at 10 different sites.

In order to quantitatively evaluate the performance of the proposed method, the ground truth angles of four cameras are necessary. To obtain them, the intrinsic parameters of four cameras are first calibrated using the method suggested in [33], and then the extrinsic parameters are estimated using a sophisticatedly designed calibration pattern shown in Figure 2. The camera angles obtained by this procedure are considered as the ground truth camera angles. When applying the proposed method to the test dataset, five degrees of noise with random signs are added to 12 ground truth camera angles (three angles for each camera). Those noisy camera angles are called the initial camera angles, and the proposed method conducts the camera calibration starting from the initial camera angles. Once the proposed method estimates three angles per each camera, those angles are compared with the corresponding ground truth camera angles. Since there are too many camera angles (12 angles), it is difficult to understand the experimental results at a glance if all of their errors are presented. Thus, this paper suggests a single measure called an average camera angle error, which is obtained by taking the average of differences between the estimated camera angles and corresponding ground truth camera angles. Detailed errors on pitch, roll, and yaw will be presented later when summarizing the performance evaluation.

7.2. Performance Evaluation

Table 1 shows the average camera angle error of the proposed two-step approach explained in Sections 5.1 and 5.2 at 10 different sites. The two-step approach consists of the front and rear camera calibration followed by the left and right camera calibration. The results shown in Table 1 are the errors before applying the parameter co-refinement of four cameras explained in Section 5.3. This table reveals that the proposed two-step approach can estimate the angles of the front and rear cameras where both left and right lanes are observable as well as the angles of the left and right cameras where only one of two lanes is observable. The two-step approach without the parameter co-refinement gives 0.48 degrees of the average camera angle error at 10 different sites. It can be seen that the errors of the left and right cameras are relatively larger than those of the front and rear cameras. This is because the front and rear cameras are calibrated using both left and right lanes, but the left and right cameras are calibrated using only one of two lanes. This means that the amount of information used for calibrating the left and right cameras is less than that of the front and rear cameras. It is considered that the difference of lane marking information causes the difference of the average camera angle errors.

Table 1. Average Camera Angle Error without Parameter Co-Refinement (Degree).

Site No.	Front Camera	Rear Camera	Left Camera	Right Camera	Overall
1	0.42	0.40	0.70	0.54	0.51
2	0.47	0.63	0.73	1.16	0.75
3	0.20	0.46	0.22	0.66	0.39
4	0.26	0.39	0.53	0.47	0.41
5	0.38	0.38	0.65	0.52	0.48
6	0.33	0.33	0.51	0.37	0.38
7	0.39	0.43	0.69	0.71	0.56
8	0.36	0.34	0.69	0.46	0.46
9	0.33	0.37	0.52	0.45	0.42
10	0.28	0.35	0.52	0.46	0.40
Overall	**0.34**	**0.41**	**0.58**	**0.58**	**0.48**

Table 2 shows the average camera angle error at 10 different sites after applying the parameter co-refinement of four cameras, which is explained in Section 5.3. The parameter co-refinement procedure simultaneously refines the angles of four cameras estimated by the two-step approach based on the corresponding lane markings appearing across images of adjacent cameras. The proposed method with the parameter co-refinement gives 0.41 degrees of the average camera angle error at 10 different sites. It can be seen that the parameter co-refinement procedure reduces the average camera angle error by 15% (0.08 degrees) compared to the result without this procedure. The co-refinement procedure not only quantitatively reduces the camera angle error, but also qualitatively increases the quality of the AVM image. This is because the co-refinement procedure simultaneously refines the angles of four cameras in the direction that the corresponding lane markings appearing across images of adjacent cameras are smoothly connected. Since drivers cannot directly measure the camera angle error, they are likely to judge the quality of the AVM image by checking if images taken from adjacent cameras are smoothly connected at their boundaries. Figure 13a,b show the resulting AVM images without and with the parameter co-refinement procedure, respectively. It can be seen that the quality of the AVM image with the co-refinement (Figure 13b) is superior to that without it (Figure 13a) based on the fact that the road markings captured from adjacent cameras are smoothly connected to each other at the image boundaries (red lines). Since the AVM image is created for the purpose of showing it to the driver, it is important to find not only the parameters with small average camera angle error, but also the parameters that smoothly stitch the images taken by adjacent cameras.

Table 2. Average Camera Angle Error with Parameter Co-Refinement (Degree).

Site No.	Front Camera	Rear Camera	Left Camera	Right Camera	Overall
1	0.37	0.25	0.53	0.48	0.41
2	0.44	0.38	0.53	0.92	0.57
3	0.20	0.24	0.34	0.31	0.27
4	0.26	0.30	0.73	0.41	0.42
5	0.34	0.27	0.68	0.39	0.42
6	0.35	0.24	0.65	0.43	0.42
7	0.33	0.24	0.65	0.53	0.44
8	0.34	0.24	0.71	0.36	0.41
9	0.34	0.18	0.60	0.37	0.37
10	0.34	0.19	0.55	0.37	0.36
Overall	**0.33**	**0.25**	**0.60**	**0.46**	**0.41**

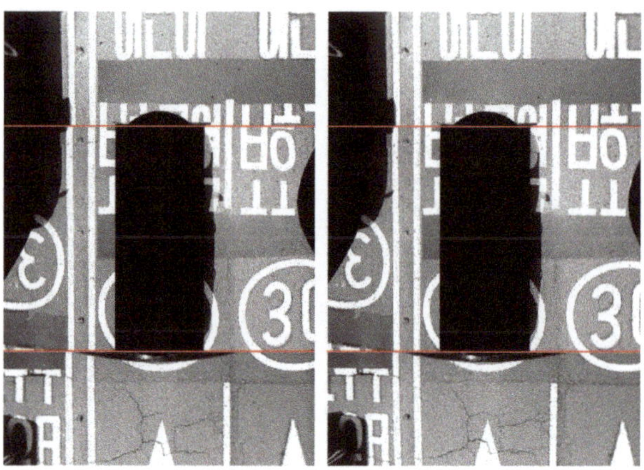

(a) (b)

Figure 13. Resulting AVM images (**a**) without the parameter co-refinement, and (**b**) with the parameter co-refinement. Red lines indicate image boundaries.

Table 3 shows the average camera angle error at 10 different sites after applying both the parameter co-refinement and parameter selection explained in Section 6. The parameter selection procedure selects a set of camera angles by finding the medoid of multiple camera angle sets obtained from a long image sequence. The proposed method with both the parameter co-refinement and parameter selection gives 0.31 degrees of the average camera angle error at 10 different sites. It can be seen that the parameter selection procedure reduces the average camera angle error by 24% (0.10 degrees) compared to the result without this procedure. It reveals that the parameter selection procedure not only reduces computational resources by separating a long image sequence and processing them individually, but also improves the camera angle estimation performance by decreasing the dependency on a specific place. In Table 3, the errors of the left and right cameras are still shown to be larger than those of the front and rear cameras. However, it can be seen that the difference between the errors of the front and rear cameras and the errors of the left and right cameras are remarkably reduced compared to the results in Tables 1 and 2.

Table 3. Average Camera Angle Error with Parameter Co-Refinement and Parameter Selection (Degree).

Site No.	Front Camera	Rear Camera	Left Camera	Right Camera	Overall
1	0.25	0.33	0.25	0.43	0.31
2	0.31	0.34	0.14	0.68	0.37
3	0.20	0.21	0.22	0.27	0.23
4	0.21	0.34	0.57	0.12	0.31
5	0.33	0.28	0.50	0.37	0.37
6	0.39	0.22	0.50	0.24	0.34
7	0.19	0.18	0.50	0.06	0.23
8	0.26	0.25	0.51	0.22	0.31
9	0.34	0.17	0.37	0.43	0.33
10	0.33	0.15	0.37	0.35	0.30
Overall	**0.28**	**0.25**	**0.39**	**0.32**	**0.31**

Table 4 summarizes the average camera angle errors of three different experimental settings shown in Tables 1–3 with detailed pitch, roll, and yaw angle errors. Among three settings, the one with both the co-refinement and parameter selection gives the best performance. In this setting, the left and right cameras give higher errors than the front and rear cameras. In more detail, in terms of the roll and yaw angles, the left and right cameras have a similar amount of errors to the front and rear cameras. This means the error difference mostly comes from the pitch angles. In the last row of Table 4, it can be found that the pitch angle errors of the left and right cameras are higher than those of the front and rear cameras. In case of the front and rear cameras, both left and right lane markings are captured, so that their pitch angles can be estimated by their own lane information as described in Section 5.1. However, in the case of the left and right cameras, only one of two lane markings is captured, so that their pitch angles cannot be estimated by their own lane information, as explained in Section 5.2. Due to this limitation, this paper utilizes the approach that calibrates the pitch angles of the left and right cameras with the help of the calibration results of the front and rear cameras. This means that the pitch angles of the left and right cameras are more severely affected by the errors of the front and rear cameras, compared to the roll and yaw angles that can be estimated without the help of the calibration results of the front and rear cameras. This is considered to be the reason that makes the pitch angle have a higher error than the roll and yaw angles in the case of the left and right cameras.

Table 4. Quantitative Performance Comparison with Different Experimental Settings (Degree).

Experimental Setting	Front Camera	Rear Camera	Left Camera	Right Camera	Overall
	Pitch/Roll/Yaw	Pitch/Roll/Yaw	Pitch/Roll/Yaw	Pitch/Roll/Yaw	Pitch/Roll/Yaw
W/O Co-Refinement	0.34	0.41	0.58	0.58	0.48
W/O Parameter Selection (Table 1)	0.42/0.31/0.30	0.50/0.40/0.32	0.88/0.45/0.40	1.18/0.28/0.28	0.74/0.36/0.33
W/Co-refinement	0.33	0.25	0.60	0.46	0.41
W/O Parameter Selection (Table 2)	0.46/0.22/0.31	0.24/0.24/0.28	0.93/0.42/0.45	0.80/0.31/0.26	0.61/0.30/0.33
W/Co-Refinement	0.28	0.25	0.39	0.32	0.31
W/Parameter Selection (Table 3)	0.42/0.18/0.24	0.28/0.19/0.27	0.68/0.21/0.28	0.51/0.24/0.20	0.47/0.20/0.25

Figure 14 shows the resulting AVM images at 10 different sites. In this figure, the left, middle, and right columns show the AVM images generated by the ground truth camera angles, initial camera angles with five degrees of noise with random signs, and the camera angles estimated by the proposed method, respectively. At this point, the proposed method indicates the two-step approach with both the co-refinement and parameter selection. In the AVM images produced by the proposed method, it can be seen that the road markings including lanes, stop lines, arrows, letters, etc., are well connected across images of adjacent cameras. In addition, it can be noted that the AVM images produced by the proposed method are quite similar to those generated by the ground truth camera angles. Based on these results, it can be said that the proposed method shows a promising performance for estimating the angles of four cameras composing the AVM system.

Site No.	Using the Ground Truth Camera Angles	Using the Initial Camera Angles	Using the Estimated Camera Angles
Site#1			
Site#2	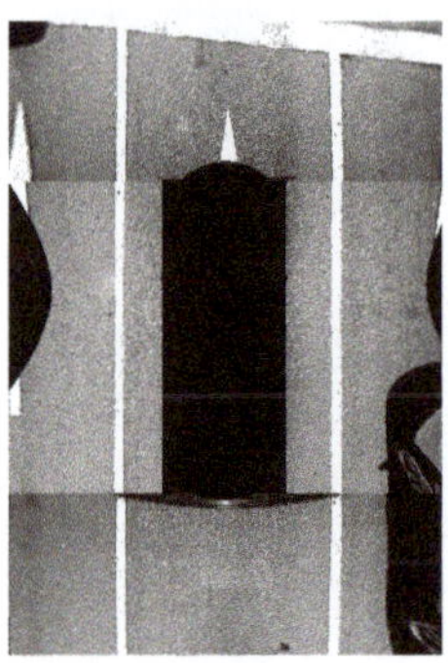		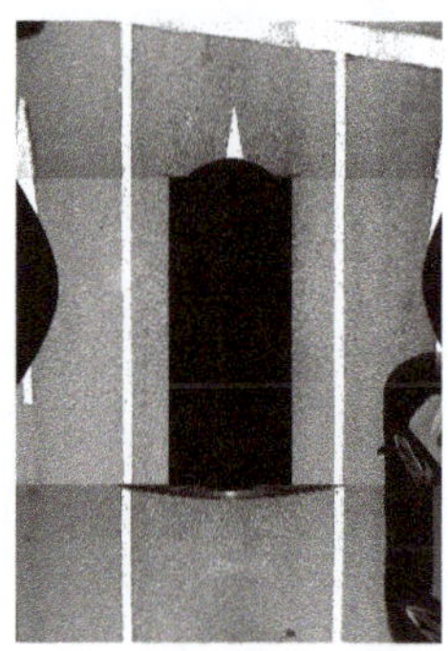
Site#3			

Figure 14. *Cont.*

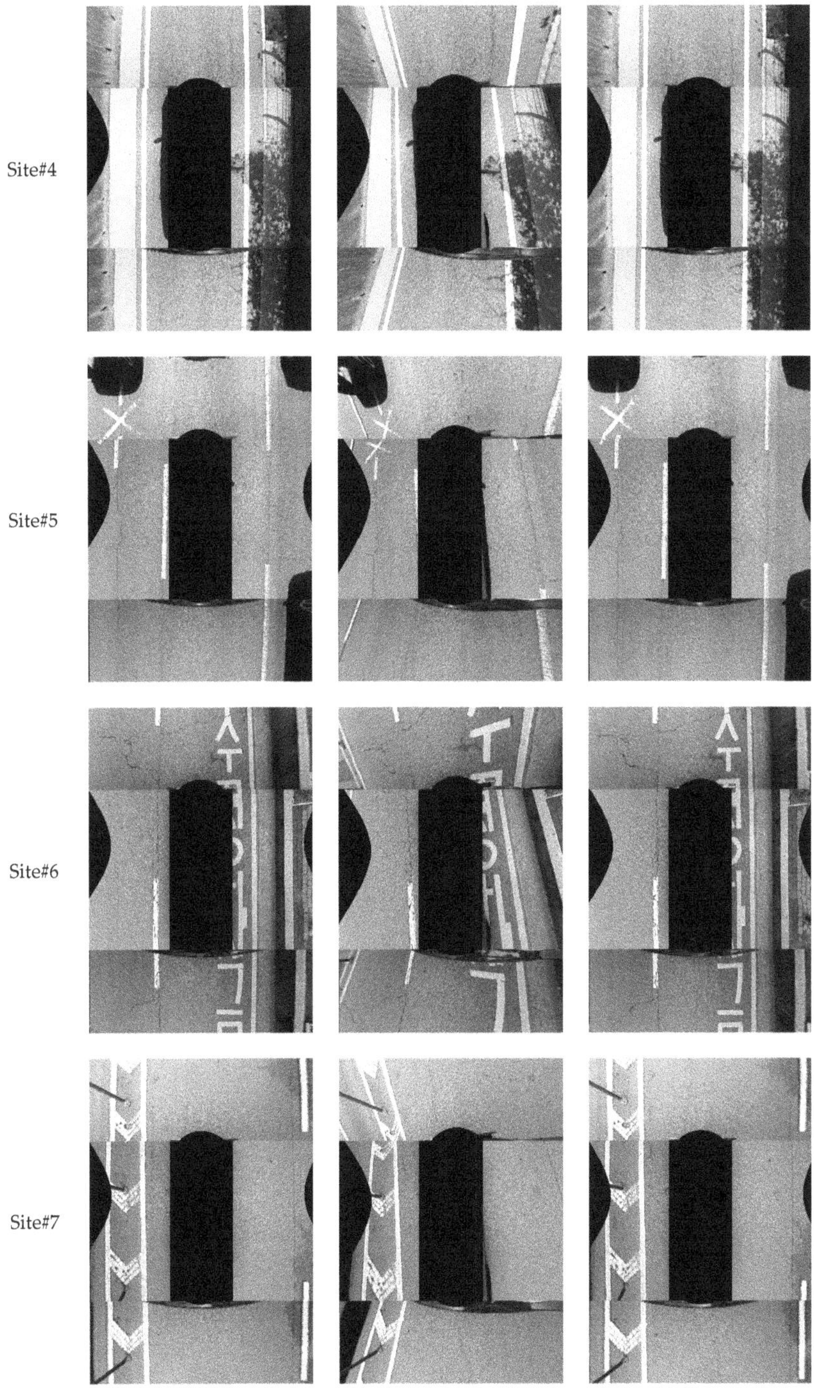

Figure 14. *Cont.*

Site#8

Site#9

Site#10

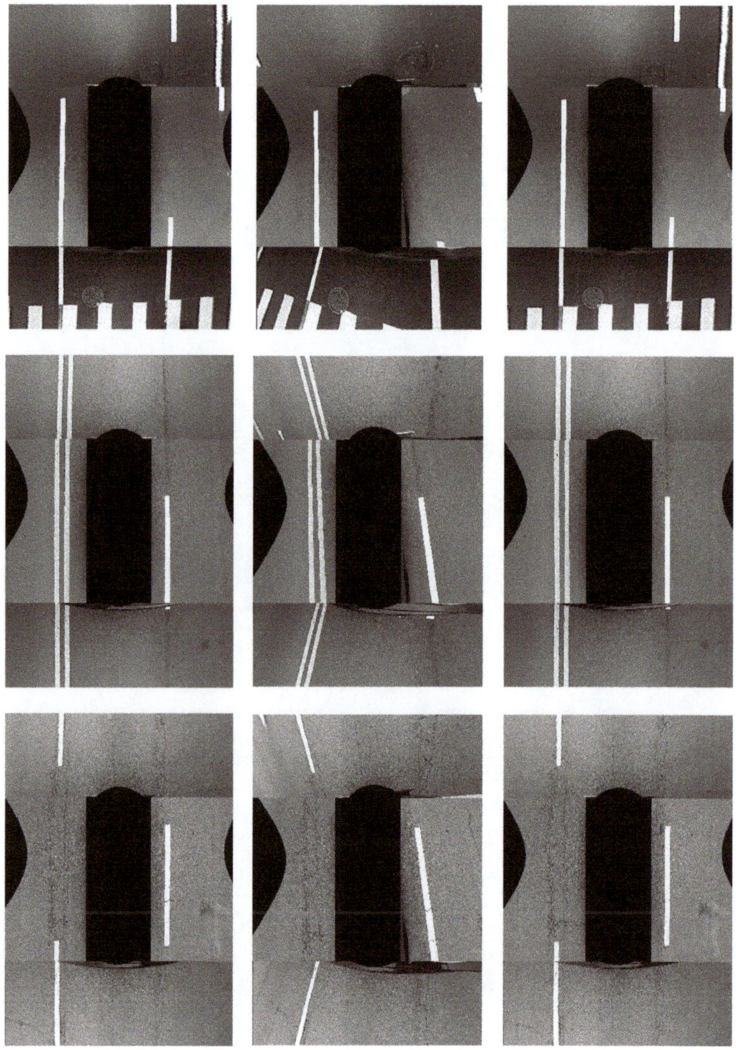

Figure 14. Resulting AVM images at 10 different sites. (**left**) AVM images generated by the ground truth camera angles; (**middle**) AVM images generated by the initial camera angles; (**right**) AVM images generated by the camera angles resulting from the proposed method.

Note that there are mainly two cases where even the AVM images generated by the ground truth camera angles include some discontinuity at image boundaries. The first case is shown in the AVM images at Site#4 and Site#6 of Figure 14. In the left half region of the AVM image at Site#4, and the right half region of the AVM image at Site#6, there are some object regions that include large discontinuities. These discontinuities are not because the camera angles are incorrectly estimated, but because those objects (curbs) have different heights from the ground. In AVM images, only the objects whose heights are the same as the ground are connected across the images of adjacent cameras. The second case is shown in the AVM images at Site#7 and Site#8 of Figure 14. In the left-most lane, marking at Site#7 and the right-most lane marking at Site#8, there are some discontinuities, even though these markings are drawn on the ground. These discontinuities are not because of the incorrect camera calibration, but because of the non-flat ground. One of the most fundamental assumptions of the AVM system is that the ground is flat. This assumption is sometimes invalid in the area distant from the ego-vehicle, and this makes some discontinuities in the AVM image. Note that these two cases of discontinuities are not due to the camera angle estimation error but a fundamental characteristic of the AVM system.

7.3. Execution Time

Table 5 shows the execution time for four main modules of the proposed method. These times were measured on an Intel Core i7-2600 CPU using only a single core. Note that among four modules, only the lane marking detection module is required to be processed in real time. The other three modules need to be processed only once after a sufficient number of lane markings are gathered. In Table 5, the lane marking detection module requires 26.12 ms to process four images acquired from four cameras of the AVM system, which means that this module can process more than 30 frames per second in real time. The other three modules require a total execution time of 101.76 ms. Since those modules need to be processed only once after gathering lane markings, their execution times do not hinder the proposed method from operating in real time.

Table 5. Execution time (ms).

Module	Time
Lane marking detection	26.12
False lane marking removal	22.02
Parameter estimation	79.69
Parameter selection	0.05

7.4. Comparison with Previous Methods

As aforementioned in Section 2, the previous methods suggested for calibrating vehicle-mounted cameras can be categorized into three approaches: calibration pattern-based, interest point-based, and lane marking-based. Table 6 shows the comparison of the previous and proposed methods from the viewpoint of the AVM system calibration. The calibration pattern-based methods can consider the inter-camera relationship and handle side cameras. However, they are inconvenient because the driver must visit a specific place where the pattern is installed, and it cannot be conducted in a natural driving situation. The interest point-based methods do not require the driver to visit a specific place and are applicable to side cameras. However, the vehicle must travel at a very low speed to stably track interest points. In addition, it is hard to obtain point correspondences between images of adjacent cameras because overlapping areas are severely distorted in the case of the AVM system. This makes it difficult to consider the inter-camera relationship. The lane marking-based methods do not require the driver to visit a specific place and can be applied at both low and high vehicle speeds. However, the previous lane marking-based methods cannot consider the inter-camera relationship because they do not utilize the corresponding lane markings that appear across images of adjacent cameras. Furthermore, those methods cannot handle side cameras because they assume that both left and right lane markings are observable in a single camera. In contrast to those methods, the proposed method can consider the inter-camera relationship by using corresponding lane markings across images of multiple cameras, and calibrate all four cameras including the side ones by using the two-step approach.

Table 6. Comparison of the previous and proposed methods.

Method		Driver's Convenience	High Speed Condition	Inter-Camera Relationship	Side Camera Handling
Calibration pattern-based methods		Poor	Poor	Good	Good
Interest point-based methods		Good	Poor	Fair	Good
Lane marking-based methods	Previous methods	Good	Good	Poor	Poor
	Proposed method	**Good**	**Good**	**Good**	**Good**

Unfortunately, since there is no previous method that can calibrate all four cameras of the AVM system in a natural driving situation, including low and high speeds, it is impossible to conduct quantitative performance comparison of the previous and proposed methods with the same dataset. However, the comparison summarized in Table 6 clearly explains that the proposed method is superior to the other previous methods in terms of automatic AVM system calibration.

8. Conclusions

This paper proposes a novel and practical method that calibrates the AVM system in a fully automatic manner using lane markings. The proposed method has the following advantages: First, it is applicable to a natural driving situation where the vehicle travels at both low and high speeds. Second, it calibrates not only the front and rear cameras but also the left and right cameras where only one of two lane markings is captured. Last, it considers the inter-camera relationship using the corresponding lane markings across images of adjacent cameras to produce seamless AVM images. The proposed method was evaluated by the image sequences taken in various real driving conditions and showed a promising performance along with a real-time processing capability. This method is expected to improve the driver's convenience by automatically adjust the AVM system in the case of the posture changes of the cameras.

Author Contributions: K.C. developed the algorithm and performed the experiments. H.G.J. and J.K.S. developed the system architecture and analyzed the experimental results. All three authors wrote the paper together.

Acknowledgments: This research was supported in part by the Hyundai Mobis and the MSIT (Ministry of Science and ICT), Korea, under the ITRC (Information Technology Research Center) support program (IITP-2018-2013-0-00680) supervised by the IITP (Institute for Information & communications Technology Promotion).

Conflicts of Interest: The authors declare no conflict of interest.

References

1. Ehlgen, T.; Pajdla, T.; Ammon, D. Eliminating blind spots for assisted driving. *IEEE Trans. Intell. Transp. Syst. Mag.* **2008**, *9*, 657–665. [CrossRef]
2. Suhr, J.K.; Jung, H.G. A universal vacant parking slot recognition system using sensors mounted on off-the-shelf vehicles. *Sensors* **2018**, *18*, 1234. [CrossRef] [PubMed]
3. AVM Calibration Machine. Available online: http://www.angtec.com/nhome_e/m02_1_07_3.html (accessed on 4 May 2018).
4. Ruland, T.; Loose, H.; Pajdla, T.; Krüger, L. Extrinsic autocalibration of vehicle mounted cameras for maneuvering assistance. In Proceedings of the Computer Vision Winter Workshop, Hove Hrdady, Czech Republic, 3–5 February 2010.
5. Chang, Y.; Hsu, L.; Chen, O. Auto-calibration around-view monitoring system. In Proceedings of the IEEE Vehicular Technology Conference, Las Vegas, NV, USA, 2–5 September 2013.
6. Mazzei, L.; Medici, P.; Panciroli, M. A lasers and cameras calibration procedure for VIAC multi-sensorized vehicles. In Proceedings of the IEEE Intelligent Vehicles Symposium, Alcala de Henares, Spain, 3–7 June 2012.
7. Hold, S.; Nunn, C.; Kummert, A.; Muller-Schneiders, S. Efficient and robust extrinsic camera calibration procedure for lane departure warning. In Proceedings of the IEEE Intelligent Vehicles Symposium, Xi'an, China, 3–5 June 2009.

8. Lebraly, P.; Royer, E.; Ait-Aider, O.; Deymier, C.; Dhome, M. Fast calibration of embedded non-overlapping cameras. In Proceedings of the IEEE International Conference on Robotics and Automation, Shanghai, China, 9–13 May 2011.
9. Antonelli, G.; Caccavale, F.; Grossi, F.; Marino, A. A non-iterative and effective procedure for simultaneous odometry and camera calibration for a differential drive mobile robot based on the singular value decomposition. *Intell. Serv. Robot.* **2010**, *3*, 163–173. [CrossRef]
10. Tan, J.; Li, J.; An, X.; He, H. An interactive method for extrinsic parameter calibration of onboard camera. In Proceedings of the IEEE Intelligent Vehicles Symposium, Baden-Baden, Germany, 5–9 June 2011.
11. Li, S.; Ying, H. Estimating camera pose from H-pattern of parking lot. In Proceedings of the IEEE International Conference on Robotics and Automation, Anchorage, AK, USA, 3–7 May 2010.
12. Natroshvili, K.; Scholl, K. Automatic extrinsic calibration methods for surround view systems. In Proceedings of the IEEE Intelligent Vehicles Symposium, Los Angeles, CA, USA, 11–14 June 2017.
13. Miksch, M.; Yang, B.; Zimmermann, K. Homography-based extrinsic self-calibration for cameras in automotive applications. In Proceedings of the International Workshop on Intelligent Transportation, Hamburg, Germany, 23–24 March 2010.
14. Miksch, M.; Yang, B.; Zimmermann, K. Automatic extrinsic camera self-calibration based on homography and epipolar geometry. In Proceedings of the IEEE Intelligent Vehicles Symposium, San Diego, CA, USA, 21–24 June 2010.
15. Tan, J.; An, X.; Xu, X.; He, H. Automatic extrinsic calibration for an onboard camera. In Proceedings of the Chinese Automation Congress, Changsha, China, 7–8 November 2013.
16. Chao, G.; Faraz, M.; Stergios, R. An analytical least-squares solution to the odometer-camera extrinsic calibration problem. In Proceedings of the IEEE International Conference on Robotics and Automation, Saint Paul, MN, USA, 14–18 May 2012.
17. Heng, L.; Li, B.; Pollefeys, M. Camodocal: Automatic intrinsic and extrinsic calibration of a rig with multiple generic cameras and odometry. In Proceedings of the IEEE/RSJ International Conference on Intelligent Robots and Systems, Tokyo, Japan, 3–7 November 2013.
18. Heng, L.; Burki, M.; Lee, G.; Furgale, P.; Siegwart, R.; Pollefeys, M. Infrastructure-based calibration of a mulit-camera rig. In Proceedings of the IEEE International Conference on Robotics and Automation, Hong Kong, China, 31 May–7 June 2014.
19. Hold, S.; Gormer, S.; Kummert, A.; Meuter, M.; Muller-Schneiders, S. A novel approach for the online initial calibration of extrinsic parameters for a car-mounted camera. In Proceedings of the International IEEE Conference on Intelligent Transportation Systems, St. Louis, MO, USA, 4–7 October 2009.
20. Paula, M.; Jung, C.; Silveira, L. Automatic on-the-fly extrinsic camera calibration of onboard vehicular cameras. *Expert Syst. Appl.* **2014**, *41*, 1997–2007. [CrossRef]
21. Wang, H.; Cai, Y.; Lin, G.; Zhang, W. A novel method for camera external parameters online calibration using dotted road line. *Adv. Robot.* **2014**, *28*, 1033–1042. [CrossRef]
22. Ribeiro, A.; Dihl, L.; Jung, C. Automatic camera calibration for driver assistance systems. In Proceedings of the International Conference on Systems, Signals and Image Processing, Budapest, Hungary, 21–23 September 2006.
23. Xu, H.; Wang, X. Camera calibration based on perspective geometry and its application in LDWS. *Phys. Procedia* **2012**, *33*, 1626–1633. [CrossRef]
24. Catala-Prat, A.; Rataj, J.; Reulke, R. Self-calibration system for the orientation of a vehicle camera. In Proceedings of the ISPRS Image Engineering and Vision Metrology, Dresden, Germany, 25–27 September 2006.
25. Zhao, K.; Iurgel, U.; Meuter, M. An automatic online camera calibration system for vehicular applications. In Proceedings of the International IEEE Conference on Intelligent Transportation Systems, Qingdao, China, 8–11 October 2014.
26. Musleh, B.; Martín, D.; Armingol, J.M.; de la Escalera, A. Pose self-calibration of stereo vision systems for autonomous vehicle applications. *Sensors* **2016**, *16*, 1492. [CrossRef] [PubMed]
27. Wang, Q.; Zhang, Q.; Rovira-Mas, F. Auto-calibration method to determine camera pose for stereovision-based off-road vehicle navigation. *Environ. Control Biol.* **2010**, *48*, 59–72. [CrossRef]
28. Gonzalez, R.C.; Woods, R.E. *Digital Image Processing*, 3rd ed.; Prentice-Hall: Englewood Cliffs, NJ, USA, 2008.
29. Fischler, M.; Bolles, R. Random sample consensus: A paradigm for model fitting with applications to image analysis and automated cartography. *Commun. ACM* **1981**, *24*, 381–395. [CrossRef]

30. Press, W.; Flannery, B.; Teukolsky, S.; Wetterling, W. *Nemerical Recipes in C*; Cambridge University Press: Cambridge, UK, 1988.
31. Gower, J.C.; Dijksterhuis, G.B. *Procrustes Problems*; Oxford University Press: Oxford, UK, 2004.
32. Hyundai, Genesis G80. Available online: https://www.genesis.com/worldwide/en/luxury-sedan-genesis-g80.html (accessed on 4 May 2018).
33. Kannala, J.; Brandt, S.S. A generic camera model and calibration method for conventional, wide-angle, and fish-eye lenses. *IEEE Trans. Pattern Analysis. Mach. Intell.* **2006**, *28*, 1335–1340. [CrossRef] [PubMed]

sensors

MDPI

Article

Handshape Recognition Using Skeletal Data

Tomasz Kapuscinski * and Patryk Organisciak

Department of Computer and Control Engineering, Rzeszow University of Technology, 35-959 Rzeszow, Poland; patrykorganisciak@gmail.com

* Correspondence: tomekkap@prz-rzeszow.pl

Received: 10 July 2018; Accepted: 5 August 2018; Published: 6 August 2018

Abstract: In this paper, a method of handshapes recognition based on skeletal data is described. A new feature vector is proposed. It encodes the relative differences between vectors associated with the pointing directions of the particular fingers and the palm normal. Different classifiers are tested on the demanding dataset, containing 48 handshapes performed 500 times by five users. Two different sensor configurations and significant variation in the hand rotation are considered. The late fusion at the decision level of individual models, as well as a comparative study carried out on a publicly available dataset, are also included.

Keywords: handshape recognition; sign language; finger alphabet; skeletal data

1. Introduction

Handshapes are the basis of so-called finger alphabets that are used by deaf people to express words for which there are no separate signs in sign languages. The same handshapes, shown for various positions and orientations of the hand, are also important components of dynamic signs occurring in sign languages. Moreover, in the case of the so-called minimal pairs, the shape of the hand is the only distinguishing feature. Therefore, building a complete system for automatic recognition of the manual part of sign language is not possible without solving the problem of recognizing static handshapes.

The problem is challenging. Handshapes occurring in finger alphabets are complicated. A projection, that takes place during the image formation in a camera, results in significant loss of information. Individual fingers overlap each other or remain completely covered. In addition, some handshapes are very similar. Moreover, a movement trajectory is not available and therefore a detailed analysis of the shape is required. In the case of typical cameras, including stereo cameras, a big challenge is a dependence on variable backgrounds and lighting conditions. Individual differences in showing particular shapes by different users need to be considered as well. Therefore, systems developed in a controlled and sterile laboratory environment do not always work in demanding real-world conditions.

Currently, there are imaging devices on the market which operate both in the visible and near-infrared and provide accurate and reliable 3D information in the form of point clouds. These clouds can be further processed to extract skeletal information. An example of such a device is the popular Kinect controller, which, along with the included software, provides the skeletal data for the entire body of the observed person. There are similar solutions, with smaller observation area and higher resolution, for obtaining skeletal data for the observed hand. Examples of such devices are some time-of-flight cameras or a Leap Motion controller (LMC). These are in early stages of development but technological progress in this area is fast. For example, the first version of the Leap Motion Software Development Kit (SDK) was able to track only visible parts of the hand, but the version 2 uses some prediction algorithms, and the individual joints of each finger are tracked even when the controller cannot see them. It is expected that sooner or later, newer solutions will emerge. Therefore, it is reasonable to undertake research on the handshape recognition based on skeletal data.

Despite a number of works in this field, the problem remains unresolved. Current works are either dedicated to one device only or deal with a few simple static shapes or dynamic gestures, for which the great support is the distinctive role of the motion trajectory.

In this paper, a method of handshapes recognition, based on skeletal data is described. The proposed feature vector encodes the relative differences between vectors associated with the pointing directions of the fingers and the palm normal. Different classifiers are tested on the demanding dataset, containing 48 handshapes performed by five users. Each shape is repeated 500 times by each user. Two different sensor configurations and significant variation in the hand rotation are considered. The late fusion at the decision level of individual models, as well as comparative study carried out on a publicly available dataset, are also included.

The remainder of this paper is organized as follows. The recent works are characterized in Section 2, Section 3 describes the method, Section 4 discusses the experiment results, and Section 5 summarizes the paper. Appendix A contains the full versions of the tables with the results of leave-one-person-out cross-validation.

2. Recent Works

The suitability of the skeletal data, obtained from the LMC, for Australian Sign Language (Auslan) recognition has been explored in [1]. Testing showed that despite the problems with accurate tracking of fingers, especially when the hand is perpendicular to the controller, there is a potential for the use of the skeletal data, after some further improvement of the provided API.

An extensive evaluation of the quality of skeletal data, obtained from the LMC, was also tested in [2]. Static and dynamic measurements were performed using a high-precision motion tracking system. For static objects, the 3D position estimation with the standard deviation less than 0.5 mm was reported. A spatial dependency of the controller's precision was also tested. In [1,2] the early version of the provided software was used. Recently, the stability of tracking has been significantly improved.

In [3], the skeletal data was used to recognize a subset of 10 letters from American Manual Alphabet. Handshapes were presented 10 times by 14 users. The feature vector was based on the positions and orientations of the fingers measured by the LMC. The multi-class support vector machine (SVM) classifier was used. The recognition accuracy was 80.86%. When the feature vector was augmented by features calculated from the depth map obtained with the Kinect sensor, the recognition accuracy increased to 91.28%.

In [4], the 26 letters of the English alphabet in American Sign Language (ASL) performed by two users were recognized using the features derived from the skeletal data. The recognition rate was 72.78% for the k-nearest neighbor (kNN) classifier and 79.83% for SVM.

Twenty-eight signs corresponding to the Arabic alphabet, performed 100 times by one person were recognized using 12 selected attributes of the hand skeletal data [5]. For the Naive Bayes (NB) classifier, the recognition rate was 98.3% and for the Multilayer Perceptron (MP) 99.1%.

In [6], the 50 dynamic gestures from Arabic Sign Language (ArSL), performed by two persons, were recognized using the feature vector composed of positions of fingers and distances between them and multi-layer perceptron neural network. The recognition accuracy was 88%.

A real-time multi-sensor system for ASL recognition was presented in [7]. The skeletal data, collected from Leap Motion sensors, was fused using multiple sensors data fusion and the classification was performed using hidden Markov models (HMM). The 10 gestures, corresponding to the digits from 0 to 9, were performed by eight subjects. The recognition accuracy was 93.14%.

In [8], the 24 letters from ASL were recognized using the feature vector that consists of the normal vector of the palm, coordinates of fingertips and finger bones, the arm direction vector, and the fingertip direction vector. These features were derived from the skeletal data provided by LMC. The decision tree (DT) and genetic algorithm (GA) were used as the classifier. The recognition accuracy was 82.71%.

Five simple handshapes were used to control a robotic wheelchair in [9]. Skeletal data was acquired by LMC. Feature vector consisted of the palm roll, pitch and yaw angles, and the palm normal direction vector. Block Sparse Representation (BSR) based classification was applied. According to the authors, the method yields accurate results but no detailed information about experiments and obtained recognition accuracy are given.

In [10], 10 handshapes corresponding to the digits in Indian Sign Language were recognized. The feature vector consisted of the distances between the consecutive fingertips and palm center and the distances between the fingertips. The features were derived from skeletal data acquired by LMC. Multi-Layer Perceptron (MP) neural network with back propagation algorithm was used. Each shape was presented by four subjects. The recognition accuracy of 100% is reported in the paper.

In [11], 28 letters of the Arabic Sign Language were recognized using the body and hand skeletal data acquired by Kinect sensor and LMC. One thousand four hundred samples were recorded by 20 subjects. One hundred and three features for each letter were reduced to 36 using the Principal Component Analysis algorithm. For the SVM classifier, the recognition accuracy of 86% is reported.

In [12], 25 dynamic gestures from Indian Sign Language were recognized using a multi-sensor fusion framework. Data was acquired using jointly calibrated Kinect sensor and LMC. Each word was repeated eight times by 10 subjects. Different data fusion schemes were tested and the best recognition accuracy of 90.80% was reported for the Coupled Hidden Markov Models (CHMM).

Twenty-eight handshapes corresponding to the letters of the Arabic alphabet were recognized using skeletal data from LMC and RGB image from Kinect sensor [13]. Gestures were performed at least two times by four users. Twenty-two of 28 letters were recognized with 100% accuracy.

In [14], Rule Based-Backpropagation Genetic Algorithm Neural Network (RB-BGANN) was used to recognize 26 handshapes corresponding to the alphabet in Sign System of Indonesian Language. Thirty-four features, related to the fingertips positions and orientations, taken from the hand skeletal data acquired by LMC, were used. Each gesture was performed five times. The recognition accuracy was 93.8%.

The skeletal data provided by the hand tracking devices LMC and Intel RealSense was used for recognizing 20 of the 26 letters from ASL [15]. The SVM classifier was used. The developed system was evaluated with over 50 individuals, and the recognition accuracy for particular letters was in the range of 60–100%.

In [16], a method to recognize static sign language gestures, corresponding to 26 American alphabet letters and 10 digits, performed by 10 users, was presented. The skeletal data acquired by LMC was used. Two variants of the feature vector were considered: (i) the distances between fingertips and the center of the palm, and (ii) the distances between the adjacent fingertips. The nearest neighbor classifier with four different similarity measures (Euclidean, Cosine, Jaccard, and Dice) was used. The obtained recognition accuracy varied from 70–100% for letters and 95–100% for digits.

Forty-four letters of Thai Sign Language were recognized using the skeletal data acquired by LMC and the decision trees [17]. The recognition accuracy of 72.83% was reported, but the authors do not indicate how many people performed gestures.

In [18], the skeletal data, acquired from two Leap Motion controllers, was used to recognize 28 letters from Arabic Sign Language. Handshapes were presented 10 times by one user. For the data fusion at features level and Linear Discriminant Analysis (LDA) classifier, the average accuracy was about 97.7%, while for classifier level fusion using Dempster-Shafer theory of evidence—97.1%.

Ten static gestures performed 20 times by 13 individuals were recognized using the new feature called Fingertips Tip Distance, derived from LMC skeletal data, and Histogram of Oriented Gradients (HOG), calculated from undistorted, raw sensor images [19]. After dimension reduction, based on Principal Component Analysis (PCA), and feature weighted fusion, the multiclass SVM classifier was used. Several variants of feature fusion were explored. The best recognition accuracy was 99.42%.

In [20], 28 isolated manual signs and 28 finger-spelling words, performed four times by 10 users, were recognized. The proposed feature vector consisted of fingertip positions and orientations derived from the skeletal data obtained with LMC. The SVM classifier was used to differentiate between manual and finger spelling sequences and the Bidirectional Long Short-Term Memory (BLSTM) recurrent neural networks were used for manual sign and fingerspelled letters recognition. The obtained recognition accuracy was 63.57%.

Eight handshapes, that can be used to make orders in a bar, were recognized in [21]. Each shape was presented three times by 20 participants. The feature vector consisted of normalized distances between the tips of the fingers and the center of the palm and was calculated from row skeletal data provided by LMC. Three classification methods: kNN, MP and Multinomial Logistic Regression (MLR) were considered. The best recognition accuracy of 95% was obtained for kNN classifier.

In [22], fingertip distances, fingertip inter-distances, and hand direction, derived from skeletal data acquired by LMC as well as the RGB-D data provided by Kinect sensor were used for sign language recognition in a multimodal system. Ten handshapes, performed 10 times by 14 users were recognized using data-level, feature-level, and decision-level multimodal fusion techniques. The best recognition accuracy of 97.00% was achieved for the proposed decision level fusion scheme.

The current works are summarized in Table 1.

Table 1. Recent works on handshape recognition using skeletal data.

Work	Sign Type	Sign Vocabulary	Users	Device	Method	Accuracy [%]	Data Available
[3]	static	10 letters ASL	14	LMC LMC + Kinect	SVM	80.86 91.28	Yes
[4]	static	26 letters ASL	2	LMC	kNN SVM	72.78 79.83	No
[5]	static	28 letters ArSL	1	LMC	NB MP	98.3 99.1	No
[6]	dynamic	50 sign ArSL	2	LMC	MP	88	No
[7]	static	10 digits ASL	8	2 LMs	HMM	93.14	No
[8]	static	24 letters ASL	1	LMC	DT + GA	82.71	No
[9]	static	5 simple shapes	?	LMC	BSR	?	No
[10]	static	10 digits ISL	4	LMC	MP	100	No
[11]	static	28 letters ArSL	20	LMC + Kinect	SVM	86	No
[12]	dynamic	25 sign ISL	10	LMC + Kinect	CHMM	90.80	Yes
[13]	static	28 letters ArSL	4	LMC + Kinect	kNN	100	No
[14]	static	26 letters SIBI	1	LMC	RB-BGANN + GA	93.8	No
[15]	static	20 letters ASL	50	LMC, RealSense	SVM	60–100	No
[16]	static	26 letters ASL 10 digits ASL	10	LMC	kNN	70–100 95–100	No
[17]	static	44 letters ThSL	?	LMC	DT	72.83	No
[18]	static	28 letters ArSL	1	2 LMs	LDA	97.7	No
[19]	static	10 hand shapes	13	LMC	SVM	99.42	No
[20]	dynamic	28 signs and 28 words ISL	10	LMC	SVM + BLSTM	63.57	No
[21]	static	8 hand shapes	20	LMC	kNN	95	Yes
[22]	static	10 hand shapes	14	LMC + Kinect		97	No

3. Proposed Method

3.1. Hand Skeletal Data

The skeletal hand model considered in this paper is shown in Figure 1.

It consists of bones visualized in the form of straight line sections and connections between them (joints) depicted as numbered balls. There are four kinds of bones in this model: (i) four metacarpals (between joints P_5–P_6, P_{10}–P_{11}, P_{15}–P_{16}, P_{20}–P_{21}), (ii) five proximal phalanges (P_1–P_2, P_6–P_7, P_{11}–P_{12}, P_{16}–P_{17}, P_{21}–P_{22}), (iii) five intermediate phalanges (P_2–P_3, P_7–P_8, P_{12}–P_{13}, P_{17}–P_{18}, P_{22}–P_{23}), and (iv) five distal phalanges (P_3–P_4, P_8–P_9, P_{13}–P_{14}, P_{18}–P_{19}, P_{23}–P_{24}).

In a contactless way, such a model can be acquired directly using LMC released in 2012 [23] or Intel RealSense device released in 2015 [24] and embedded in some laptop models. A simplified version of the model, sufficient to determine the feature vector proposed in this paper, can be also obtained using Softkinetic DepthSense 325 camera along with the Close Interaction Library [25]. LMC has been recently evaluated [26], but there are no many publications about RealSense due to its recent release. It is expected that in the near future these devices and the supplied software will be further improved to allow for reliable skeletal hand tracking.

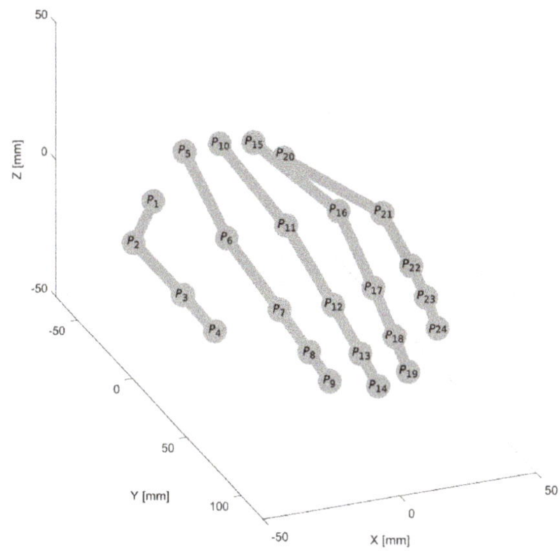

Figure 1. Hand skeletal model.

3.2. Feature Vector

The proposed feature vector encodes the relative differences between vectors associated with the pointing directions of the fingers and the palm normal. Let P_c be the center of the palm, n_c normal to the palm at point P_c, P_i the end of the i-th finger, and n_i the vector pointed by that finger (Figure 2).

The relative position of vectors n_c and n_i can be unambiguously described giving four values determined from the Formulas (1)–(4) [27]:

$$\alpha_i = acos(v_i \cdot n_i) \tag{1}$$

$$\phi_i = acos\left(u \cdot \frac{d_i}{|d_i|}\right) \tag{2}$$

$$\Theta_i = atan\left(\frac{w_i \cdot n_i}{u \cdot n_i}\right) \tag{3}$$

$$d_i = P_i - P_c \tag{4}$$

where the vectors u, v_i, and w_i define the so-called Darboux frame [28]:

$$u = n_c \tag{5}$$

$$v_i = \frac{d_i}{|d_i|} \times u \tag{6}$$

$$w_i = u \times v_i \tag{7}$$

and · indicates the scalar and ×—vector products. Since the d_i vectors depend on the size of the hand, they have been omitted. The feature vector consists of 15 values calculated for individual fingers using the Formulas (1)–(3):

$$V = [\alpha_1, \phi_1, \Theta_1, \alpha_2, \phi_2, \Theta_2, \alpha_3, \phi_3, \Theta_3, \alpha_4, \phi_4, \Theta_4, \alpha_5, \phi_5, \Theta_5] \tag{8}$$

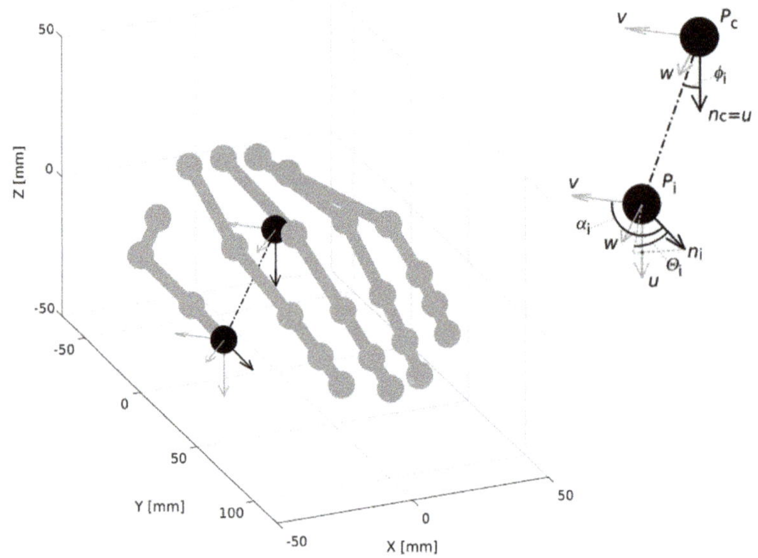

Figure 2. Feature vector construction.

In the case of LMC, the palm center, the palm normal and the pointing directions of the fingers are returned along with the skeletal data. For other devices, they can be derived from the skeletal data using the Formulas (9)–(11) (see Figure 1):

$$P_c = \frac{1}{10}\sum_{j \in J} P_j \tag{9}$$

$$n_c = \overrightarrow{P_{15}P_{16}} \times \overrightarrow{P_5P_6} \tag{10}$$

$$n_i = \overrightarrow{P_{3+5(i-1)}P_{4+5(i-1)}} \tag{11}$$

where $J = \{1, 2, 5, 6, 10, 11, 15, 16, 20, 21\}$.

3.3. Classification

The following classification methods have been tested: decision trees (DT) [29], linear and quadratic discriminants (LD and QD) [30,31], support vector machines with linear, quadratic, cubic and Gaussian kernel function (SVM Lin/Quad/Cub/Gauss) [32–34], different version of k-nearest neighbor classifiers (1 NN, 10 NN, 100 NN, 10 NN Cos, 10 NN W) [35,36], different ensemble classifiers, that meld results from many weak learners into one model (Ens Boost/Bag/RUS/SubD/kNN) [37–40] and fast approximate nearest neighbors with randomized kd-trees (FLANN) [41]. A detailed list of tested classifiers with their initial parameters is provided in Table 2.

Table 2. Tested classifiers and their parameters.

Classifier	Parameter	Value
DT	Maximum number of splits Split criterion	100 Gini's diversity index
LD	Covariance structure	Full
QD	Covariance structure	Full
SVM Lin/Quad/Cub/Gauss	Kernel function Box constraint level Multiclass method	Linear/Quadratic/Cubic/Gaussian 1 One-vs-one
1 NN/10 NN/100 NN	Number of neighbors Distance metric	1 /10/100 Euclidean
10 NN Cos	Number of neighbors Distance metric	10 Cosine
10 NN W	Number of neighbors Distance metric Distance weight	10 Euclidean Squared inverse
Ens Boost/Bag/RUS	Ensemble method Learner type Number of learners	Boosted/bagged/random subspace trees Decision tree 30
Ens Sub D/Sub kNN	Ensemble method Learner type Number of learners Subspace dimension	Subspace Discriminant/1 NN 30 8
FLANN	Number of neighbors Number of trees Number of times the trees should be recursively traversed	1 8 128

4. Experiments

4.1. Datasets

Two datasets were considered.

4.1.1. Dataset 1: Authors' Own Dataset

Forty-eight static handshapes, occurring in Polish Finger Alphabet (PFA) and Polish Sign Language (PSL) were considered (Figure 3) [25].

The gestures were recorded in two configurations: (i) LMC lies horizontally on the table (configuration user-sensor); (ii) the sensor is attached to the monitor and directed towards the signer (configuration user-user). In the configuration (i), two variants were additionally considered: (a) gestures are made with fixed hand orientation (like in PFA); (b) spatial hand orientation changes in a wide range (like in PSL). In the configuration (i) variant (a) five people, designated hereinafter A, B, C, D, and E, participated in the recordings. In other cases, the gestures of person A were recorded. Gestures were shown by each person 500 times. During the data collection, visual feedback was provided, and when an abnormal or incomplete skeleton was observed, the process was repeated to ensure that 500 correct shapes were registered for each class. Incorrect data was observed for approximately 5% of frames. It was also noticed that the device works better when the whole hand with very visible fingers is presented first and then slowly changes to the desired shape.

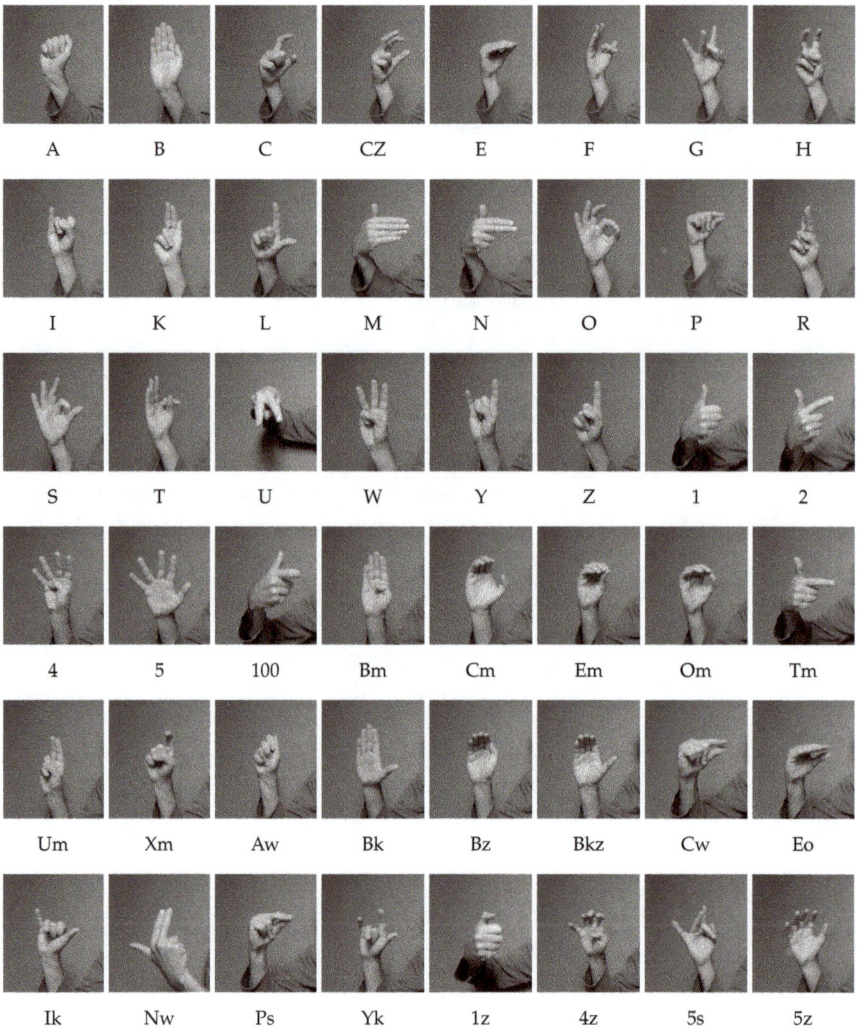

Figure 3. Static handshapes, occurring in Polish Finger Alphabet and Polish Sign Language.

4.1.2. Dataset 2: Microsoft Kinect and Leap Motion Dataset

In order to evaluate the method for more users and to make a comparative analysis, the database provided in the work [3] was used. The database contains the recordings of 10 letters from ASL, performed 10 times by 14 people and acquired by jointly calibrated LMC and depth sensor.

4.2. Results

The results of 10-fold cross-validation for the dataset 1 are shown in Tables 3–5.

For LMC lying on the table (configuration (i)) the best recognition rates (⩾99.5%) were for SVM, kNN, Ens Bag, Ens Sub kNN and FLANN, wherein the results obtained under large variation in hand's rotation (variant (b)) were only slightly worse. For configuration (ii), the results are better. This configuration seems to be more natural for a user accustomed to showing gestures to another person.

Table 3. 10-fold cross-validation results for dataset 1, configuration (i), variant (a).

Classifier	DT	LD	QD	SVM Lin	SVM Quad	SVM Cub	SVM Gauss	1 NN	10 NN
Recognition rate [%]	81.2	72.8	99.7	99.5	100.0	100.0	100.0	100.0	99.9
Classifier	**100 NN**	**10 NN Cos**	**10 NN W**	**Ens Boost**	**Ens Bag**	**Ens Sub D**	**Ens Sub kNN**	**Ens RUS**	**FLANN**
Recognition rate [%]	99.2	99.9	100.0	64.2	100.0	69.9	100.0	39.9	100.0

Table 4. 10-fold cross-validation results for dataset 1, configuration (i), variant (b).

Classifier	DT	LD	QD	SVM Lin	SVM Quad	SVM Cub	SVM Gauss	1 NN	10 NN
Recognition rate [%]	83.1	78.7	97.2	96.7	99.4	99.7	99.1	99.8	98.5
Classifier	**100 NN**	**10 NN Cos**	**10 NN W**	**Ens Boost**	**Ens Bag**	**Ens Sub D**	**Ens Sub kNN**	**Ens RUS**	**FLANN**
Recognition rate [%]	88.9	98.5	99.5	67.1	99.7	77.3	99.8	35.8	99.7

Table 5. 10-fold cross-validation results for dataset 1, configuration (ii).

Classifier	DT	LD	QD	SVM Lin	SVM Quad	SVM Cub	SVM Gauss	1 NN	10 NN
Recognition rate [%]	100.0	100.0	100.0	100.0	100.0	100.0	100.0	100.0	100.0
Classifier	**100 NN**	**10 NN Cos**	**10 NN W**	**Ens Boost**	**Ens Bag**	**Ens Sub D**	**Ens Sub kNN**	**Ens RUS**	**FLANN**
Recognition rate [%]	100.0	100.0	100.0	100.0	100.0	100.0	100.0	43.8	100.0

However, the results of the leave-one-subject-out cross-validation experiment, shown in Tables 6 and A1, are much worse for all considered classification methods. The best recognition rates (⩾50.0%) were for: LD, SVM Lin, Ens Bag.

Table 6. Leave-one-subject-out cross-validation results for dataset 1, configuration (i), variant (1).

Classifier	DT	LD	QD	SVM Lin	SVM Quad	SVM Cub	SVM Gauss	1 NN	10 NN
Recognition rate [%]	41.4	50.9	42.5	52.3	49.1	46.9	14.0	48.6	48.6
Classifier	**100 NN**	**10 NN Cos**	**10 NN W**	**Ens Boost**	**Ens Bag**	**Ens Sub D**	**Ens Sub kNN**	**Ens RUS**	**FLANN**
Recognition rate [%]	48.9	47.9	48.7	43.3	50.8	46.8	49.9	28.7	47.6

The performances of the individual gestures are strongly dependent on the user, and the training set consisting of four people is not sufficiently representative to correctly classify the gestures of the fifth, unknown person.

The results obtained for dataset 2, and shown in Tables 7, 8 and A2, confirm that when the training set consists of more users, the discrepancy between 10-fold cross-validation and leave-one-subject-out cross-validation is significantly lower. However, it should be mentioned that in this case, the number of recognized classes is much smaller.

Table 7. 10-fold cross-validation results for dataset 2.

Classifier	**DT**	**LD**	**QD**	**SVM Lin**	**SVM Quad**	**SVM Cub**	**SVM Gauss**	**1 NN**	**10 NN**
Recognition rate [%]	87.6	84.5	86.4	87.6	86.2	84.1	88.4	88.6	88.0
Classifier	**100 NN**	**10 NN Cos**	**10 NN W**	**Ens Boost**	**Ens Bag**	**Ens Sub D**	**Ens Sub kNN**	**Ens RUS**	**FLANN**
Recognition rate [%]	82.6	87.9	89.1	87.8	88.9	85.1	88.5	86.9	85.9

Table 8. Leave-one-subject-out cross-validation results for the dataset 2.

Classifier	**DT**	**LD**	**QD**	**SVM Lin**	**SVM Quad**	**SVM Cub**	**SVM Gauss**	**1 NN**	**10 NN**
Recognition rate [%]	86.2	84.2	87.5	87.6	86.4	82.3	86.7	89.2	85.5
Classifier	**100 NN**	**10 NN Cos**	**10 NN W**	**Ens Boost**	**Ens Bag**	**Ens Sub D**	**Ens Sub kNN**	**Ens RUS**	**FLANN**
Recognition rate [%]	82.4	85.6	89.6	87.0	87.7	84.1	89.3	86.4	85.4

For the dataset 2 and 10-fold CV, the best results (⩾88.0%) were for SVM Gauss, kNN 1, kNN W, Ens Bag and Ens Sub kNN, whereas for leave-one-subject-out cross-validation the best results (⩾88.0%) were for kNN1, kNN W, Ens Sub kNN.

Because for the most demanding case (Table 6) the best results were obtained for SVM Lin and Ens Bag—the parameters of these two classifiers were further analyzed (see Tables 9 and 10).

The SVM classifier is by nature binary. It classifies instances into one of the two classes. However, it can be turned into a multinomial classifier by two different strategies: one-vs-one and one-vs-all. In one-vs-one, a single classifier for each pair of classes is trained. The decision is made by applying all trained classifiers to an unseen sample and a voting scheme. The class that has been recognized most times is selected. In one-vs-all, a single classifier per class is trained. The samples of that class are positive samples, and all other samples are negatives. The decision is made by applying all trained classifiers to an unseen sample and selecting the one with the highest confidence score.

Table 9. Support Vector Machines classifier with linear kernel function performance when the multiclass method was changed from one-vs-one to one-vs-all.

Training	**Testing**	**SVM Lin**
B, C, D, E	A	36.8
A, C, D, E	B	21.1
A, B, D, E	C	47.8
A, B, C, E	D	52.0
A, B, C, D	E	46.6
Avg		40.8

Table 10. Ens Bag performance for a different number of learners.

Training	**Testing**	**10**	**20**	**30**	**40**	**50**	**100**	**200**	**400**	**800**	**1000**	**2000**
B, C, D, E	A	44.6	47.4	39.6	46.8	45.3	46.8	46.9	46.1	45.4	45.9	46.3
A, C, D, E	B	40.1	40.3	40.5	38.0	38.6	37.0	38.8	39.2	40.0	41.4	38.7
A, B, D, E	C	53.2	56.6	59.0	56.1	55.8	53.9	58.6	58.3	57.4	57.7	58.0
A, B, C, E	D	54.8	54.6	56.1	54.8	58.5	57.9	57.0	57.4	57.9	58.1	57.9
A, B, C, D	E	58.7	60.0	58.7	60.6	58.6	63.4	60.5	61.2	62.5	63.0	61.7
Avg		50.4	51.8	50.8	51.3	51.3	51.8	52.4	52.4	52.6	53.2	52.5

In SVM Lin classifier, the change of the multiclass method from one-vs-one to one-vs-all leads to decrease in the recognition accuracy. For Ens Bag classifier, the recognition accuracy increases with the number of learners, but the response time increases as well (see Figure 4).

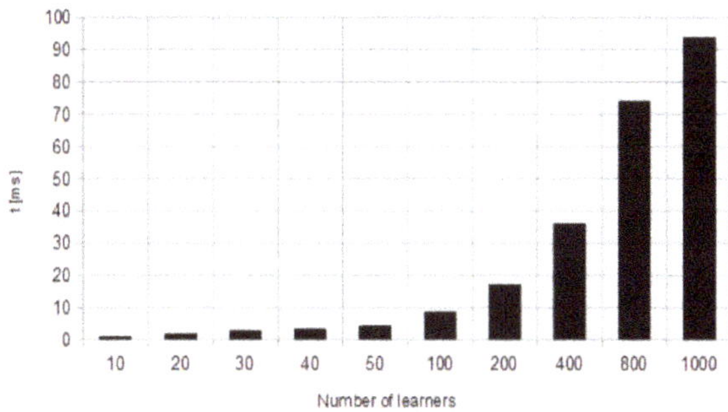

Figure 4. Response time for Ens Bag classifier for different number of learners.

The experiment was stopped when the response time reached 100 ms, i.e., the value at which the typical user will notice the delay [42].

In Table 8 the best results were obtained for 1 NN, 10 NN W, and Ens Sub kNN. The FLANN version of the kNN classifier turned out to be the fastest one. Therefore, a further analysis of kNN classifiers has been carried out.

In Table 11, the nearest neighbor classifier 1 NN with brute-force search in the dataset was compared with the FLANN version with a different number of the randomized trees.

Table 11. 1 NN vs. FLANN with a different number of trees (given in parenthesis).

Training	Testing	1 NN	FLANN (1)	FLANN (2)	FLANN (4)	FLANN (8)	FLANN (16)	FLANN (32)
B, C, D, E	A	43.1	38.0	41.2	39.5	40.4	41.1	40.4
A, C, D, E	B	32.3	31.1	30.9	33.6	33.6	33.0	33.6
A, B, D, E	C	60.6	55.8	56.5	57.4	56.1	57.0	56.4
A, B, C, E	D	52.3	52.1	51.3	51.3	51.1	51.5	51.2
A, B, C, D	E	54.5	55.1	56.6	56.5	55.7	55.5	55.6
Avg		48.6	46.4	47.3	47.6	47.4	47.6	47.4

As should be expected, the results obtained for the exact version are slightly better than for the classifier, which finds the approximate nearest neighbor. However, if we compare the processing times, the FLANN version is over 400 times faster. Therefore, this classifier is a particularly attractive choice in practical applications.

An experiment was also carried out to check whether the late fusion of classifiers, at the decision level of individual models, leads to improved recognition accuracy. A simple method was used, in which every classifier votes for a given class. According to [43], simple unweighted majority voting is usually the best voting scheme. All possible combinations of classifiers were tested. The best result of leave-one-subject-out cross-validation on dataset 1, 56.7%, was obtained when the outputs of the classifiers LD, QD, SVM Lin, Ens Boost, Ens Bag were fused. The result is better than the best result obtained for a single classifier by 4.4%. However, the fusion of classifiers leads to a decrease in the individual classes recognition. The voting deteriorates the prediction in classes F, I, Xm, Yk.

4.3. Computational Efficiency

The average response times of the individual classifiers are shown in Table 12.

Table 12. Average response times of the individual classifiers.

Classifier	DT	LD	QD	SVM Lin	SVM Quad	SVM Cub	SVM Gauss	1 NN	10 NN
Response time [ms]	0.07	0.46	0.42	26.73	31.98	30.12	64.87	24.15	26.64
Classifier	**100 NN**	**10 NN Cos**	**10 NN W**	**Ens Boost**	**Ens Bag**	**Ens Sub D**	**Ens Sub kNN**	**Ens RUS**	**FLANN**
Response time [ms]	29.24	22.50	23.14	3.22	2.95	9.3	47.65	4.14	0.06

Together with the average time needed for data acquisition and feature vectors calculation, which is equal to 6 ms, they do not exceed 100 ms, so the typical user will not notice the time delay between presentation of the given gesture and the predicted response of the system [42]. However, all experiments were carried out on a fairly powerful workstation, equipped with a 2.71 GHz processor, 32 GB of RAM and a fast SSD. For less-efficient systems, e.g., mobile or embedded devices, the preferred choice is FLANN or DT. Moreover, in the case of FLANN classifier, the randomized trees can be searched in parallel.

4.4. Comparative Analysis

According to the authors' knowledge, the only database of static hand skeletal data available on the Internet for which comparative analysis can be carried out is Dataset 2 [3]. Table 13 compares the recognition accuracy obtained for this database.

Table 13. 10-fold cross validation results of different methods obtained for the Dataset 2.

Lp	Reference	Features	Method	Recognition Rate
1	[3]	Fingertips distances, angles and elevations	Multiclass SVM	80.9%
2	[19]	Fingertips Tip distance	Multiclass SVM	81.1%
3	This paper	As described in Section 3.2	SVM Lin	87.6%
4	This paper	As described in Section 3.2	10NN W	89.1%

The first row quotes the results obtained for LMC, without additional data from the Kinect sensor. The proposed feature vector allows obtaining better results even with the same classifier (SVM).

5. Conclusions

Handshape recognition based on its skeleton becomes an important research problem because there are more and more new devices on the market that enable the acquisition of such data. In this paper:

- A feature vector was proposed, which describes the relative differences between the pointing directions of individual fingers and the hand normal vector.
- A demanding dataset containing 48 hand shapes, shown 500 times by five persons in two different sensor placement, has been prepared and made available [44].
- The registered data has been used to perform classification. Seventeen known and popular classification methods have been tested.
- For classifiers SVM Lin and Ens Bag, given the best recognition accuracies, an analysis of the impact of their parameters on the obtained results was carried out.
- It was found that the weaker result for leave-one-person-out validation may be caused by individual character of performances of individual gestures, a difficult dataset, containing as many as 48 classes, among which there are very similar shapes, and imperfections of the LMC, which in the case of individual fingers occlusions tries predict their position and spatial orientation.

It is worth mentioning that other works on static handshape recognition, cited in the literature, concern a smaller number of simpler gestures.

- The proposed feature vector allows obtaining better results.
- It was determined experimentally that although late fusion improves the results, it causes the deterioration of recognition efficiency in some classes, which in some applications may be undesirable.

To recognize complicated handshapes occurring in the sign languages, a feature vector invariant to translation, rotation, and scale, which is sensitive to the subtle differences in shape, is needed. The proposed feature vector is inspired by research on local point cloud descriptors [27]. Angular features, describing the mutual position of two vectors normal to the cloud surface, are used there to form a representation of the local geometry. Such a descriptor is sensitive to subtle differences in shape [45]. In our proposition, the fingertips and the palm center are treated as a point cloud, and the finger directions and the palm's normal are used instead of the surface normals. It is also worth noting that the proposed feature vector is invariant to position, orientation, and scale. This is not always the case in the literature, where the features depending on the hand size or orientation are used. This invariance is particularly important in the case of sign language, where unlike in the finger alphabet, the hand's position and orientation are not fixed. An interesting proposition of an invariant feature vector was proposed in [3] and enhanced in [19]. In Section 4.4, it was compared to our proposal.

Analysis of the confusion matrices obtained for the dataset 1 shows that the most commonly confused shapes are: B-Bm, C-100, N-Nw, S-F, T-O, Z-Xm, Tm-100, Bz-Cm and 4z-Cm. In fact, these are very similar shapes (see Figure 3). In adverse lighting conditions, when they are viewed from some distance or from the side, they can be confused even by a person. When sequences of handshapes, corresponding to fingerspelled words, are recognized, disambiguation can be achieved by using the temporal context. However, this is not always possible because often fingerspelling is used to convey difficult names, foreign words or proper names. If the similar shapes are discarded from the dataset 1, leave-one-subject-out cross-validation gives recognition efficiencies of about 80%.

The proposed system is fast and requires no special background or specific lighting. One of the reasons for the weaker results of leave-one-person-out validation is the imperfection of a sensor, that does not cope well with fingers occlusions. Therefore, as part of further work, the processing of point clouds registered with two calibrated sensors is considered in order to obtain more accurate and reliable skeletal data. Further work will also include recognition of letter sequences and integration of the presented solution with the sign language recognition system.

Author Contributions: T.K. conceived and designed the experiments; P.O. prepared the data; T.K. and P.O. performed the experiments and analyzed the data; T.K. wrote the paper.

Funding: This research received no external funding.

Conflicts of Interest: The authors declare no conflict of interest.

Appendix A. Results of Leave-One-Person-Out Cross-Validation

Table A1. Leave-one-subject-out cross-validation results for dataset 1, configuration (i), variant (1).

Training	Testing	DT	LD	QD	SVM Lin	SVM Quad	SVM Cub	SVM Gauss	1 NN	10 NN
B, C, D, E	A	37.4	51.3	32.1	46.1	41.0	38.8	4.0	43.1	43.3
A, C, D, E	B	31.2	37.6	29.6	33.0	32.9	31.0	5.6	32.3	31.7
A, B, D, E	C	47.9	54.7	41.6	57.8	57.5	55.4	21.8	60.6	61.2
A, B, C, E	D	41.6	53.8	51.1	58.6	54.0	51.7	18.7	52.3	51.3
A, B, C, D	E	48.9	57.1	58.1	66.1	60.1	57.4	19.8	54.5	55.5
Avg		41.4	50.9	42.5	52.3	49.1	46.9	14.0	48.6	48.6

Table A1. *Cont.*

Training	Testing	100 NN	10 NN Cos	10 NN W	Ens Boost	Ens Bag	Ens Sub D	Ens Sub kNN	Ens RUS	FLANN
B, C, D, E	A	45.0	44.3	43.2	38.7	39.6	50.4	42.2	27.6	40.9
A, C, D, E	B	32.6	32.7	31.8	37.6	40.5	31.3	35.9	16.4	33.8
A, B, D, E	C	60.4	55.7	61.6	46.6	59.0	51.0	60.4	35.3	56.5
A, B, C, E	D	48.9	51.7	51.3	45.2	56.1	44.8	52.0	30.1	50.8
A, B, C, D	E	57.7	55.3	55.4	48.5	58.7	56.5	59.0	33.9	56.1
Avg		48.9	47.9	48.7	43.3	50.8	46.8	49.9	28.7	47.6

Table A2. Leave-one-subject-out cross-validation results for the dataset 2.

Training	Testing	DT	LD	QD	SVM Lin	SVM Quad	SVM Cub	SVM Gauss	1 NN	10 NN
2–14	1	85.0	76.0	77.0	77.0	77.0	71.0	79.0	95.0	74.0
1, 3–14	2	89.0	87.0	91.0	90.0	90.0	91.0	91.0	89.0	90.0
1–2, 4–14	3	81.0	88.0	91.0	91.0	84.0	77.0	83.0	82.0	84.0
1–3, 5–14	4	89.0	92.0	97.0	96.0	95.0	87.0	97.0	94.0	94.0
1–4, 6–14	5	90.0	90.0	92.0	91.0	91.0	87.0	92.0	86.0	92.0
1–5, 7–14	6	91.0	92.0	93.0	92.0	92.0	89.0	93.0	93.0	92.0
1–6, 8–14	7	87.0	85.0	88.0	88.0	86.0	80.0	87.0	84.0	89.0
1–7, 9–14	8	86.0	90.0	92.0	92.0	87.0	90.0	93.0	91.0	92.0
1–8, 10–14	9	86.0	84.0	87.0	87.0	84.0	78.0	87.0	88.0	86.0
1–9, 11–14	10	84.0	77.0	89.0	89.0	89.0	86.0	81.0	86.0	85.0
1–10, 12–14	11	79.0	72.0	74.0	80.0	80.0	75.0	73.0	76.0	74.0
1–11, 13–14	12	90.0	94.0	100.0	100.0	100.0	96.0	100.0	95.0	97.0
1–12, 14	13	85.0	76.0	77.0	77.0	77.0	73.0	79.0	95.0	74.0
1–13	14	85.0	76.0	77.0	77.0	77.0	72.0	79.0	95.0	74.0
Avg		86.2	84.2	87.5	87.6	86.4	82.3	86.7	89.2	85.5
Training	**Testing**	**100 NN**	**10 NN Cos**	**10 NN W**	**Ens Boost**	**Ens Bag**	**Ens Sub D**	**Ens Sub kNN**	**Ens RUS**	**FLANN**
2–14	1	73.0	76.0	95.0	77.0	80.0	76.0	95.0	74.0	95.0
1, 3–14	2	89.0	90.0	91.0	91.0	89.0	87.0	90.0	89.0	89.0
1–2, 4–14	3	88.0	84.0	82.0	84.0	83.0	88.0	82.0	90.0	82.0
1–3, 5–14	4	89.0	94.0	95.0	97.0	97.0	92.0	94.0	92.0	94.0
1–4, 6–14	5	90.0	91.0	89.0	92.0	92.0	91.0	86.0	92.0	80.0
1–5, 7–14	6	88.0	92.0	93.0	93.0	93.0	92.0	93.0	93.0	87.0
1–6, 8–14	7	82.0	89.0	87.0	88.0	86.0	88.0	83.0	88.0	78.0
1–7, 9–14	8	88.0	91.0	90.0	92.0	90.0	91.0	91.0	91.0	91.0
1–8, 10–14	9	86.0	85.0	86.0	87.0	88.0	84.0	88.0	87.0	88.0
1–9, 11–14	10	75.0	84.0	84.0	85.0	89.0	76.0	86.0	89.0	86.0
1-10, 12-14	11	74.0	74.0	78.0	78.0	79.0	72.0	75.0	79.0	69.0
1–11, 13–14	12	85.0	96.0	94.0	100.0	100.0	89.0	97.0	98.0	95.0
1–12, 14	13	73.0	76.0	95.0	77.0	82.0	76.0	95.0	74.0	70.0
1–13	14	73.0	76.0	95.0	77.0	80.0	76.0	95.0	74.0	91.0
Avg		82.4	85.6	89.6	87.0	87.7	84.1	89.3	86.4	85.4

References

1. Potter, L.E.; Araullo, J.; Carter, L. The Leap Motion Controller: A View on Sign Language. In Proceedings of the 25th Australian Computer-Human Interaction Conference: Augmentation, Application, Innovation, Collaboration, Adelaide, Australia, 25–29 November 2013; ACM: New York, NY, USA, 2013; pp. 175–178.
2. Guna, J.; Jakus, G.; Pogačnik, M.; Tomažič, S.; Sodnik, J. An Analysis of the Precision and Reliability of the Leap Motion Sensor and Its Suitability for Static and Dynamic Tracking. *Sensors* **2014**, *14*, 3702–3720. [CrossRef] [PubMed]
3. Marin, G.; Dominio, F.; Zanuttigh, P. Hand gesture recognition with leap motion and kinect devices. In Proceedings of the 2014 IEEE International Conference on Image Processing (ICIP), Paris, France, 27–30 October 2014; pp. 1565–1569.
4. Chuan, C.H.; Regina, E.; Guardino, C. American Sign Language Recognition Using Leap Motion Sensor. In Proceedings of the 2014 13th International Conference on Machine Learning and Applications, Detroit, MI, USA, 3–6 December 2014; pp. 541–544.
5. Mohandes, M.; Aliyu, S.; Deriche, M. Arabic sign language recognition using the leap motion controller. In Proceedings of the 2014 IEEE 23rd International Symposium on Industrial Electronics (ISIE), Istanbul, Turkey, 1–4 June 2014; pp. 960–965.
6. Elons, A.S.; Ahmed, M.; Shedid, H.; Tolba, M.F. Arabic sign language recognition using leap motion sensor. In Proceedings of the 2014 9th International Conference on Computer Engineering Systems (ICCES), Cairo, Egypt, 22–23 December 2014; pp. 368–373.
7. Fok, K.Y.; Ganganath, N.; Cheng, C.T.; Tse, C.K. A Real-Time ASL Recognition System Using Leap Motion Sensors. In Proceedings of the 2015 International Conference on Cyber-Enabled Distributed Computing and Knowledge Discovery, Xi'an, China, 17–19 September 2015; pp. 411–414.
8. Funasaka, M.; Ishikawa, Y.; Takata, M.; Joe, K. Sign Language Recognition using Leap Motion Controller. In Proceedings of the 2015 International Conference on Parallel and Distributed Processing Techniques and Applications (PDPTA'15), Las Vegas, NV, USA, 27–30 July 2015; pp. 263–269.
9. Boyali, A.; Hashimoto, N.; Matsumato, O. Hand Posture Control of a Robotic Wheelchair Using a Leap Motion Sensor and Block Sparse Representation based Classification. In Proceedings of the Third International Conference on Smart Systems, Devices and Technologies, Paris, France, 20–24 July 2014; pp. 20–25.
10. Naglot, D.; Kulkarni, M. ANN based Indian Sign Language numerals recognition using the leap motion controller. In Proceedings of the 2016 International Conference on Inventive Computation Technologies (ICICT), Coimbatore, India, 26–27 August 2016; Volume 2, pp. 1–6.
11. Almasre, M.A.; Al-Nuaim, H. Recognizing Arabic Sign Language gestures using depth sensors and a KSVM classifier. In Proceedings of the 2016 8th Computer Science and Electronic Engineering (CEEC), Colchester, UK, 28–30 September 2016; pp. 146–151.
12. Kumar, P.; Gauba, H.; Roy, P.P.; Dogra, D.P. Coupled HMM-based multi-sensor data fusion for sign language recognition. *Pattern Recognit. Lett.* **2017**, *86*, 1 – 8. [CrossRef]
13. Miada A.; Almasre, H.A.N. A Real-Time Letter Recognition Model for Arabic Sign Language Using Kinect and Leap Motion Controller v2. *Int. J. Adv. Eng. Manag. Sci.* **2016**, *2*, 514–523.
14. Nurul Khotimah, W.; Andika Saputra, R.; Suciati, N.; Rahman Hariadi, R. Alphabet Sign Language Recognition Using Leap Motion Technology and Rule Based Backpropagation-Genetic Algorithm Nueral Network (RBBPGANN). *JUTI J. Ilm. Teknol. Inf.* **2017**, *15*, 95–103.
15. Quesada, L.; López, G.; Guerrero, L. Automatic recognition of the American sign language fingerspelling alphabet to assist people living with speech or hearing impairments. *J. Ambient Intell. Humaniz. Comput.* **2017**, *8*, 625–635. [CrossRef]
16. Auti, A.; Amolic, R.; Bharne, S.; Raina, A.; Gaikwad, D.P. Sign-Talk: Hand Gesture Recognition System. *Int. J. Comput. Appl.* **2017**, *160*, 13–16. [CrossRef]
17. Tumsri, J.; Kimpan, W. Thai Sign Language Translation Using Leap Motion Controller. In Proceedings of the International Multi Conference of Engineers and Computer Scientists 2017, Hong Kong, China, 15–17 March 2017; Volume I, pp. 46–51.

18. Mohandes, M.; Aliyu, S.; Deriche, M. Prototype Arabic Sign language recognition using multi-sensor data fusion of two leap motion controllers. In Proceedings of the 2015 IEEE 12th International Multi-Conference on Systems, Signals Devices (SSD15), Mahdia, Tunisia, 16–19 March 2015; pp. 1–6.
19. Du, Y.; Liu, S.; Feng, L.; Chen, M.; Wu, J. Hand Gesture Recognition with Leap Motion. CoRR, 2017. Available online: http://xxx.lanl.gov/abs/1711.04293 (accessed on 9 July 2018)
20. Kumar, P.; Saini, R.; Behera, S.K.; Dogra, D.P.; Roy, P.P. Real-time recognition of sign language gestures and air-writing using leap motion. In Proceedings of the 2017 Fifteenth IAPR International Conference on Machine Vision Applications (MVA), Nagoya, Japan, 8–12 May 2017; pp. 157–160.
21. Toghiani-Rizi, B.; Lind, C.; Svensson, M.; Windmark, M. Static Gesture Recognition using Leap Motion. *arXiv* **2017**, arxiv:1705.05884.
22. Ferreira, P.M.; Cardoso, J.S.; Rebelo, A. Multimodal Learning for Sign Language Recognition. In *Pattern Recognition and Image Analysis*; Alexandre, L.A., Salvador Sánchez, J., Rodrigues, J.M.F., Eds.; Springer: Cham, Switzerland, 2017; pp. 313–321.
23. Leap Motion. Available online: https://www.leapmotion.com/ (accessed on 9 March 2018).
24. Intel RealSense Technology. Available online: https://www.intel.com/content/www/us/en/architecture-and-technology/realsense-overview.html (accessed on 9 March 2018).
25. Close Interaction LIbrary. Available online: https://www.sony-depthsensing.com/products/CIlib (accessed on 9 March 2018).
26. Weichert, F.; Bachmann, D.; Rudak, B.; Fisseler, D. Analysis of the Accuracy and Robustness of the Leap Motion Controller. *Sensors* **2013**, *13*, 6380–6393. [CrossRef] [PubMed]
27. Rusu, R.B.; Marton, Z.C.; Blodow, N.; Beetz, M. Learning informative point classes for the acquisition of object model maps. In Proceedings of the 2008 10th International Conference on Control, Automation, Robotics and Vision, Hanoi, Vietnam, 17–20 December 2008; pp. 643–650.
28. Spivak, M. *A Comprehensive Introduction to Differential Geometry*, 3rd ed.; Publish or Perish: Houston, TX, USA, 1999; Volume III.
29. Quinlan, J.R. Induction of decision trees. *Mach. Learn.* **1986**, *1*, 81–106. [CrossRef]
30. Rayens, W.S. Discriminant Analysis and Statistical Pattern Recognition. *Technometrics* **1993**, *35*, 324–326. [CrossRef]
31. Rencher, A. *Methods of Multivariate Analysis*; Wiley Series in Probability and Statistics; Wiley: Hoboken, NJ, USA, 2003.
32. Press, W.H.; Teukolsky, S.A.; Vetterling, W.T.; Flannery, B.P. *Numerical Recipes 3rd Edition: The Art of Scientific Computing*, 3rd ed.; Cambridge University Press: New York, NY, USA, 2007.
33. Boser, B.E.; Guyon, I.M.; Vapnik, V.N. A Training Algorithm for Optimal Margin Classifiers. In Proceedings of the Fifth Annual Workshop on Computational Learning Theory, Pittsburgh, PA, USA, 27–29 July 1992; ACM: New York, NY, USA, 1992; pp. 144–152.
34. Cortes, C.; Vapnik, V. Support-vector networks. *Mach. Learn.* **1995**, *20*, 273–297. [CrossRef]
35. Cover, T.; Hart, P. Nearest neighbor pattern classification. *IEEE Trans. Inf. Theory* **1967**, *13*, 21–27. [CrossRef]
36. Dudani, S.A. The Distance-Weighted k-Nearest-Neighbor Rule. *IEEE Trans. Syst. Man Cybern.* **1976**, *SMC-6*, 325–327. [CrossRef]
37. Freund, Y.; Schapire, R.E. A Decision-Theoretic Generalization of On-Line Learning and an Application to Boosting. *J. Comput. Syst. Sci.* **1997**, *55*, 119–139. [CrossRef]
38. Breiman, L. Bagging Predictors. *Mach. Learn.* **1996**, *24*, 123–140. [CrossRef]
39. Seiffert, C.; Khoshgoftaar, T.M.; Hulse, J.V.; Napolitano, A. RUSBoost: Improving classification performance when training data is skewed. In Proceedings of the 2008 19th International Conference on Pattern Recognition, Tampa, FL, USA, 8–11 December 2008; pp. 1–4.
40. Ho, T.K. The random subspace method for constructing decision forests. *IEEE Trans. Pattern Anal. Mach. Intell.* **1998**, *20*, 832–844.
41. Muja, M.; Lowe, D.G. Scalable Nearest Neighbor Algorithms for High Dimensional Data. *IEEE Trans. Pattern Anal. Mach. Intell.* **2014**, *36*, 2227–2240. [CrossRef] [PubMed]
42. Card, S.K. The Model Human Processor: A Model for Making Engineering Calculations of Human Performance. *Proc. Hum. Factors Soc. Annu. Meet.* **1981**, *25*, 301–305. [CrossRef]

43. Moreno-Seco, F.; Iñesta, J.M.; de León, P.J.P.; Micó, L. Comparison of Classifier Fusion Methods for Classification in Pattern Recognition Tasks. In *Structural, Syntactic, and Statistical Pattern Recognition*; Yeung, D.Y., Kwok, J.T., Fred, A., Roli, F., de Ridder, D., Eds.; Springer: Berlin/Heidelberg, Germany, 2006; pp. 705–713.
44. Dataset. Available online: vision.kia.prz.edu.pl (accessed on 9 July 2018).
45. Rusu, R.B.; Bradski, G.; Thibaux, R.; Hsu, J. Fast 3D recognition and pose using the Viewpoint Feature Histogram. In Proceedings of the 2010 IEEE/RSJ International Conference on Intelligent Robots and Systems, Taipei, Taiwan, 18–22 October 2010; pp. 2155–2162.

sensors

MDPI

Article

Segment-Tube: Spatio-Temporal Action Localization in Untrimmed Videos with Per-Frame Segmentation

Le Wang [1,*], Xuhuan Duan [1], Qilin Zhang [2], Zhenxing Niu [3], Gang Hua [4] and Nanning Zheng [1]

1 Institute of Artificial Intelligence and Robotics, Xi'an Jiaotong University, Xi'an, Shannxi 710049, China; duanxuhuan0123@stu.xjtu.edu.cn (X.D.); nnzheng@xjtu.edu.cn (N.Z.)
2 HERE Technologies, Chicago, IL 60606, USA; qilin.zhang@here.com
3 Alibaba Group, Hangzhou 311121, China; zhenxing.nzx@alibaba-inc.com
4 Microsoft Research, Redmond, WA 98052, USA; ganghua@microsoft.com
* Correspondence: lewang@xjtu.edu.cn; Tel.: +86-29-8266-8672

Received: 23 April 2018; Accepted: 16 May 2018; Published: 22 May 2018

Abstract: Inspired by the recent spatio-temporal action localization efforts with tubelets (sequences of bounding boxes), we present a new spatio-temporal action localization detector Segment-tube, which consists of sequences of per-frame segmentation masks. The proposed Segment-tube detector can temporally pinpoint the starting/ending frame of each action category in the presence of preceding/subsequent interference actions in untrimmed videos. Simultaneously, the Segment-tube detector produces per-frame segmentation masks instead of bounding boxes, offering superior spatial accuracy to tubelets. This is achieved by alternating iterative optimization between temporal action localization and spatial action segmentation. Experimental results on three datasets validated the efficacy of the proposed method, including (1) temporal action localization on the THUMOS 2014 dataset; (2) spatial action segmentation on the Segtrack dataset; and (3) joint spatio-temporal action localization on the newly proposed ActSeg dataset. It is shown that our method compares favorably with existing state-of-the-art methods.

Keywords: action localization; action segmentation; 3D ConvNets; LSTM

1. Introduction

Joint spatio-temporal action localization has attracted significant attention in recent years [1–18], whose objectives include action classification (determining whether a specific action is present), temporal localization (pinpointing the starting/ending frame of the specific action) and spatio-temporal localization (typically bounding box regression on 2D frames, e.g., [6,12]). Such efforts include local feature based methods [1], convolution neural networks (ConvNets or CNNs) based methods [2,14,15], 3D ConvNets based methods [4,11] and its variants [19–21]. Recently, long short-term memory (LSTM) based recurrent neural networks (RNNs) are added on top of CNNs for action classification [5] and action localization [7].

Despite the successes of the prior methods, there are still several limiting factors impeding practical applications. On the one hand, a large number of methods [2,3,5,13] conduct action recognition only on trimmed videos, where each video contains only one action without interferences from other potentially confusing actions. On the other hand, many methods [1,7–11,15–17] emphasize only on temporal action localization with untrimmed videos, without depicting the spatial locations of the target action in each video frame.

Although there are several tubelet-style (which outputs sequences of bounding boxes) spatio-temporal action localization efforts [6,12,22], they are restricted to trimmed video only. For practical applications, untrimmed videos are much more prevalent, and sequences of bounding boxes might not offer enough spatial accuracy, especially for irregular shapes. This motivated us to propose a practical spatio-temporal action localization method, which is capable of spatially and temporally localizing the target actions with per-frame segmentation in untrimmed videos.

With applications in untrimmed videos with improved spatial accuracy in mind, we propose the spatio-temporal action localization detector Segment-tube, which localizes target actions as sequences of per-frame segmentation masks instead of sequences of bounding boxes.

The proposed Segment-tube detector is illustrated in Figure 1. The sample input is an untrimmed video containing all frames in a pair figure skating video, with only a portion of these frames belonging to a relevant category (e.g., the DeathSpirals). Initialized with saliency [23] based image segmentation on individual frames, our method first performs temporal action localization step with a cascaded 3D ConvNets [4] and LSTM, and pinpoints the starting frame and the ending frame of a target action with a coarse-to-fine strategy. Subsequently, the Segment-tube detector refines per-frame spatial segmentation with graph cut [24] by focusing on relevant frames identified by the temporal action localization step. The optimization alternates between the temporal action localization and spatial action segmentation in an iterative manner. Upon practical convergence, the final spatio-temporal action localization results are obtained in the format of a sequence of per-frame segmentation masks (bottom row in Figure 1) with precise starting/ending frames. Intuitively, the temporal action localization and spatial action segmentation naturally benefit each other.

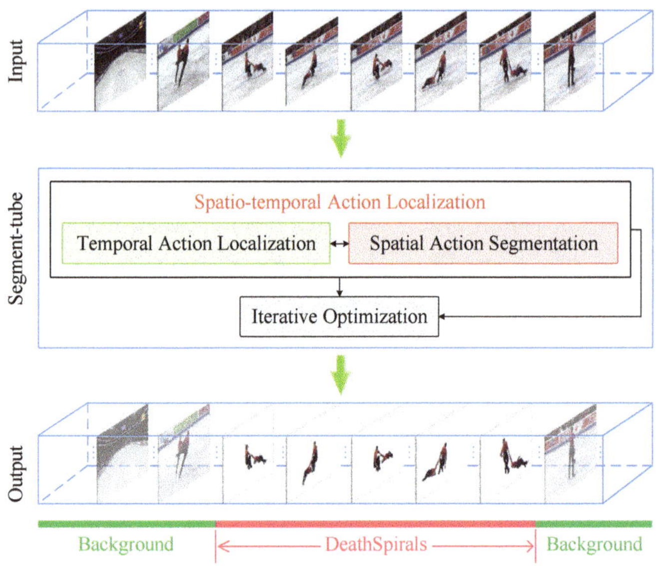

Figure 1. Flowchart of the proposed spatio-temporal action localization detector Segment-tube. As the input, an untrimmed video contains multiple frames of actions (e.g., all actions in a pair figure skating video), with only a portion of these frames belonging to a relevant category (e.g., the DeathSpirals). There are usually irrelevant preceding and subsequent actions (background). The Segment-tube detector alternates the optimization of temporal localization and spatial segmentation iteratively. The final output is a sequence of per-frame segmentation masks with precise starting/ending frames denoted with the red chunk at the bottom, while the background are marked with green chunks at the bottom.

We conduct experimental evaluations (in both qualitative and quantitative measures) of the proposed Segment-tube detector and existing state-of-the-art methods on three benchmark datasets, including (1) temporal action localization on the THUMOS 2014 dataset [25]; (2) spatial action segmentation on the SegTrack dataset [26,27]; and (3) joint spatio-temporal action localization on the newly proposed ActSeg dataset, which is a newly proposed spatio-temporal action localization dataset with per-frame ground truth segmentation masks, and it will be released on our project website. The experimental results show the performance advantage of the proposed Segment-tube detector and validate its efficacy in spatio-temporal action localization with per-frame segmentation.

In summary, the contributions of this paper are as follows:

- The spatio-temporal action localization detector Segment-tube is proposed for untrimmed videos, which produces not only the starting/ending frame of an action, but also per-frame segmentation masks instead of sequences of bounding boxes.
- The proposed Segment-tube detector achieves collaborative optimization of temporal localization and spatial segmentation with a new iterative alternation approach, where the temporal localization is achieved by a coarse-to-fine strategy based on cascaded 3D ConvNets [4] and LSTM.
- To exactly evaluate the proposed Segment-tube and to build a benchmark for future research, a new ActSeg dataset is proposed, which consists 641 videos with temporal annotations and per-frame ground truth segmentation masks.

The remainder of the paper is organized as follows. In Section 2, we review the related work. In Section 3, we present the problem formulation for spatio-temporal action localization with per-frame segmentation. In Section 4, the experimental results are presented with additional discussions. Finally, the paper is concluded in Section 5.

2. Related Works

The joint spatio-temporal action localization problem involves three distinctive tasks simultaneously, i.e., action classification, temporal action localization, and spatio-temporal action localization. Brief reviews of related works on these three topics are first provided. In addition, relevant works in video object segmentation are also introduced in this section.

2.1. Action Classification

The objective of action classification is to determine the presence of a specific action (e.g., jump and pole vault) in a video. A considerable amount of previous efforts are limited to action classification in manually trimmed short videos [2,3,5,13,28,29], where each video clip contains one and only one action, without possible interferences from either proceeding/subsequent actions or complex background.

Many methods [1] rely on handcrafted local invariant features, such as histograms of image gradients (HOG) [30], histograms of flow (HOF) [31] and improved Dense Trajectory (iDT) [28]. Video representations are typically built on top of these features by the Fisher Vector (FV) [32] or Vector of Linearly Aggregated Descriptors (VLAD) [33] to determine action categories. Recently, CNNs based methods [2,14,15] have enabled the replacement of handcrafted features with learned features, and they have achieved impressive classification performance. 3D ConvNets based methods [4,19–21] are also proposed to construct spatio-temporal features. Tran et al. [4] demonstrated that 3D ConvNets are good feature learning machines that model appearance and motion simultaneously. Carreira et al. [19] proposed a new two-stream Inflated 3D ConvNet (I3D) architecture for action classification. Hara et al. [21] discovered that 3D architectures (two-stream I3D/ResNet/ResNeXt) pre-trained on Kinetics dataset outperform complex 2D architectures. Subsequently, long short-term memory (LSTM)-based recurrent neural networks (RNNs) are added on top of CNNs to incorporate longer term temporal information and better classify video sequences [5,7].

2.2. Temporal Action Localization

Temporal action localization aims at pinpointing the starting and ending frames of a specific action in a video. Much progress has been made recently, thanks to plenty of large-scale datasets including the THUMOS dataset [25], the ActivityNet dataset [34], and the MEXaction2 dataset [35]. Most state-of-the-art methods are based on sliding windows [1,11,32,36,37], frame-wise predictions [7,15,38,39], or action proposals [22,40,41].

The sliding window-based methods typically exploit fixed-length temporally sliding windows to sample each video. They can leverage the temporal dependencies among video frames, but they commonly lead to higher computational cost due to redundancies in overlapping windows. Gaidon et al. [36] used sliding window classifiers to locate action parts (actoms) from a sequence of histograms of actom-anchored visual features. Oneata et al. [32] and Yuan et al. [37] both used sliding window classifiers on FV representations of iDT features. Shou et al. [11] proposed a sliding window-style 3D ConvNet for action localization without relying on hand-crafted features or FV/VLAD representations.

The frame-wise predictions-based methods classifies each individual video frame (i.e., predicts whether a specific category of action is present), and aggregate such predictions temporally. Singh et al. [38] used a frame-wise classifier for action location proposal, followed by a temporal aggregation step that promotes piecewise smoothness in such proposals. Yuan et al. [15] proposed to characterize the temporal evolution as a structural maximal sum of frame-wise classification scores. To account for the dynamics among video frames, RNNs with LSTM are typically employed. In [7,39], an LSTM produced detection scores of activities and non-activities based on CNN features at every frame. Although such RNNs can exploit temporal state transitions over frames for frame-wise predictions, their inputs are frame-level CNN features computed independently on each frame. On contrary in this paper, we leverage 3D ConvNets with LSTM to capture the spatio-temporal information from adjacent frames.

The action proposals-based methods leverage temporal action proposals instead of video clips for efficient action localization. Jain et al. [22] produced tubelets (i.e., 2D + t sequences of bounding boxes) by merging a hierarchy of super-voxels. Yu and Yuan [40] proposed the actionness score and a greedy search strategy to generate action proposals. Buch et al. [41] introduced a temporal action proposals generation framework that only needs to process the entire video in a single pass.

2.3. Spatio-Temporal Action Localization

There are many publications about the spatio-temporal action localization problem [6,12,42–45]. Soomro et al. [43] proposed a method based on super-voxel. Several methods [6,44] formulated spatio-temporal action localization as a tracking problem with object proposal detection at each video frame and sequences of bounding boxes as outputs. Kalogeiton et al. [12] proposed an action tubelet detector that takes a sequence of frames as input and produces sequences of bounding boxes with improved action scores as outputs. Singh et al. [45] presented an online learning framework for spatio-temporal action localization and prediction. Despite their successes, all the aforementioned spatio-temporal action localization methods require trimmed videos as inputs, and only output tubelet-style boundaries of an action, i.e., sequences of bounding boxes.

In contrast, we propose the spatio-temporal action localization detector Segment-tube for untrimmed videos, which can provide per-frame segmentation masks instead of sequences of bounding boxes. Moreover, to facilitate the training of the proposed Segment-tube detector and to establish a benchmark for future research, we introduce a new untrimmed video dataset for action localization and segmentation (i.e., ActSeg dataset), with temporal annotations and per-frame ground truth segmentation masks.

2.4. Video Object Segmentation

Video object segmentation aims at separating the object of interest from the background throughout all video frames. Previous video object segmentation methods can be roughly categorized into the unsupervised methods and the supervised counterparts.

Without requiring labels/annotations, unsupervised video object segmentation methods typically exploit features such as long-range point trajectories [46], motion characteristics [47], appearance [48,49], or saliency [50]. Recently, Jain et al. [51] proposed an end-to-end learning framework which combines motion and appearance information to produce a pixel-wise binary segmentation mask for each frame.

Differently, supervised video object segmentation methods do require user annotations of a primary object (i.e., the foreground), and the prevailing methods are based on label propagation [52,53]. For example, Marki et al. [52] utilize the segmentation mask of the first frame to construct appearance models, and the inference for subsequent frames are obtained by optimizing an energy function on a regularly sampled bilateral grid. Caelles et al. [54] adopted the Fully Convolutional Networks (FCNs) to tackle video object segmentation, given the segmentation mask for the first frame.

However, all the above video object segmentation methods assume that the object of interest (or primary object) consistently appears throughout all video frames, which is reasonable for manually trimmed video dataset. On the contrary, for practical applications with user-generated, noisy untrimmed videos, this assumption seldom holds true. Fortunately, the proposed Segment-tube detector eliminates such a strong assumption, and it is robust to irrelevant video frames and can be utilized to process untrimmed videos.

3. Problem Formulation

Given a video $V = \{f_t\}_{t=1}^{T}$ consisting of T frames, our objective is to determine whether a specific action $k \in \{1, \dots, K\}$ appears in V, and if so, temporally pinpoint the starting frame $f_s(k)$ and ending frame $f_e(k)$ for action k. Simultaneously, a sequence of segmentation masks $B = \{b_t\}_{t=f_s(k)}^{f_e(k)}$ within such frame range should be obtained, with b_t being a binary segmentation label for frame f_t. Practically, b_t consists of a series of superpixels $b_t = \{b_{t,i}\}_{i=1}^{N_t}$, with N_t being the total number of superpixels in frame f_t.

3.1. Temporal Action Localization

A coarse-to-fine action localization strategy is implemented to accurately find the temporal boundaries of the target action k from an untrimmed video, as illustrated in Figure 2. This is achieved by a cascaded 3D ConvNets with LSTM. The 3D ConvNets [4] consists of eight 3D convolution layers, five 3D pooling layers, and two fully connected layers. The fully-connected 7th layer activation feature is used to represent the video clip. To exploit the temporal correlations, we incorporate a two-layer LSTM [5] using the Peephole implementation (with 256 hidden states in each layer) with 3D ConvNets.

Coarse Action Localization. The coarse action localization determines the approximate temporal boundaries with a fixed step-size (i.e., video clip length). We first generate a set of U saliency-aware video clips $\{u_j\}_{j=1}^{U}$ with variable-length (e.g., 16 and 32 frames per video clip) sliding window with 75% overlap ratio on the initial segmentation B_o of video V (by using saliency [23]), and proceed to train a cascaded 3D ConvNets with LSTM that couples a proposal network and a classification network.

The proposal network is action class-agnostic, and it determines whether any actions ($\forall k \in \{1, \dots, K\}$) are present in video clip u_j. The classification network determines whether a specific action k is present in video clip u_j. We follow [11] to construct the training data from these video clips. The training details of the proposal network and classification network are presented immediately below in Section 4.2.

Specifically, we train the proposal network (a 3D ConvNets with LSTM) to score each video clip u_j with a proposal score $\mathbf{p}_j^{pro} = \left[\mathbf{p}_j^{pro}(1), \mathbf{p}_j^{pro}(0)\right]^T \in \mathcal{R}^2$. Subsequently, a flag label l_j^{fla} is obtained for each video clip u_j,

$$l_j^{fla} = \begin{cases} 1, & \text{if} \quad \mathbf{p}_j^{pro}(1) > \mathbf{p}_j^{pro}(0), \\ 0, & \text{otherwise}, \end{cases} \tag{1}$$

where $l_j^{fla} = 1$ denotes the video clip u_j contains an action ($\forall k \in \{1, \dots, K\}$), and $l_j^{fla} = 0$ otherwise.

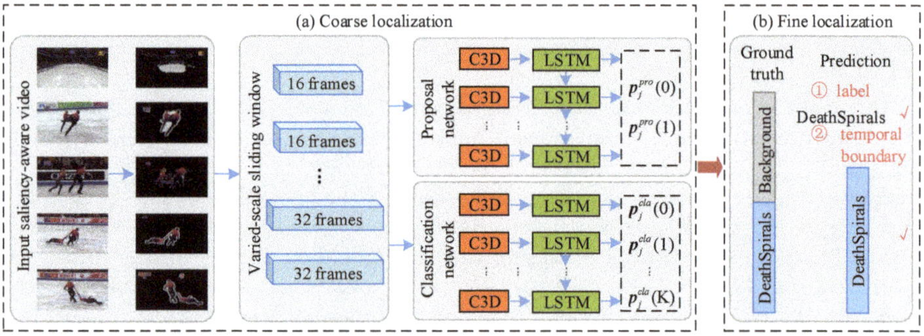

Figure 2. Overview of the proposed coarse-to-fine temporal action localization. (**a**) coarse localization. Given an untrimmed video, we first generate saliency-aware video clips via variable-length sliding windows. The proposal network decides whether a video clip contains any actions (so the clip is added to the candidate set) or pure background (so the clip is directly discarded). The subsequent classification network predicts the specific action class for each candidate clip and outputs the classification scores and action labels. (**b**) fine localization. With the classification scores and action labels from prior coarse localization, further prediction of the video category is carried out and its starting and ending frames are obtained.

A classification network (also a 3D ConvNets with LSTM) is further trained to predict a $(K+1)$-dimensional classification score $\mathbf{p}_j^{cla}$ for each clip that contains an action $\left\{u_j | l_j^{fla} = 1\right\}$, based on which a specific action label $l_j^{spe} \in \{k\}_{k=0}^{K}$ and score $v_j^{spe} \in [0,1]$ for u_j are assigned,

$$l_j^{spe} = \arg\max_{k=0,\ldots,K} \mathbf{p}_j^{cla}(k), \tag{2}$$

$$v_j^{spe} = \max_{k=0,\ldots,K} \mathbf{p}_j^{cla}(k). \tag{3}$$

where category 0 denotes the additional "background" category. Although the proposal network prefilters most "background" clips, a background category is still needed for robustness in the classification network.

Fine Action Localization. With the obtained per-clip specific action labels l_j^{spe} and v_j^{spe}, the fine action localization step predicts the video category k^* ($k^* \in \{1,\ldots,K\}$), and subsequently obtains its starting frame $f_s(k^*)$ and its ending frame $f_e(k^*)$. We calculate the average of specific action scores v_j^{spe} over all video clips for each specific action label l_j^{spe}, and take the label k^* with the maximum average predicted score as the predicted action, as illustrated in Figure 3.

Subsequently, we average specific action scores v_j^{spe} of each frame f_t for the label k^* in different video clips to obtain the action score $\alpha_t(f_t)$ for frame f_t. By selecting an appropriate threshold we can obtain the action label l_t for frame f_t. The action score $\alpha_t(f_t|k^*)$ and the action label l_t for frame f_t specifically are determined by

$$\alpha_t(f_t|k^*) = \frac{\sum\limits_{j \in \{j|f_t \in u_j\}} v_j^{spe}}{|\{j|j \in \{f_t \in u_j\}\}|}, \tag{4}$$

$$l_t = \begin{cases} k^*, & \text{if} \quad \alpha_t > \gamma, \\ 0, & \text{otherwise,} \end{cases} \tag{5}$$

where $|\{\cdot\}|$ denotes the cardinality of set $\{\cdot\}$. We empirically set $\gamma = 0.6$. $f_s(l_t)$ and $f_e(l_t)$ are assigned as the starting and ending frame of a series of consecutive frames sharing the same label l_t, respectively.

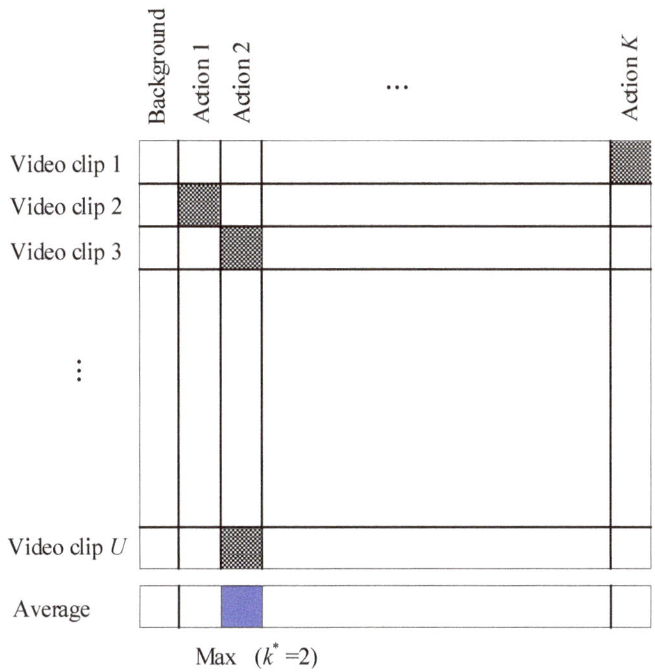

Figure 3. The diagrammatic sketch on the determination of video category k^* from video clips.

3.2. Spatial Action Segmentation

With the obtained temporal localization results, we further conduct spatial action segmentation. This problem is cast into a spatio-temporal energy minimization framework,

$$E(B) = \sum_{s_{t,i} \in V} D_i(b_{t,i}) + \sum_{s_{t,i}, s_{t,n} \in \mathcal{N}_i} S_{in}^{intra}(b_{t,i}, b_{t,n}) + \sum_{s_{t,i}, s_{m,n} \in \bar{\mathcal{N}}_i} S_{in}^{inter}(b_{t,i}, b_{m,n}), \tag{6}$$

where $s_{t,i}$ is the ith superpixel in frame f_t. $D_i(b_{t,i})$ composes the data term, denoting the cost of labeling $s_{t,i}$ with the label $b_{t,i}$ from a color and location based appearance model. $S_{in}^{intra}(b_{t,i}, b_{t,n})$ and $S_{in}^{inter}(b_{t,i}, b_{m,n})$ compose the smoothness term, constraining the segmentation labels to be spatially coherence from a color based intra-frame consistency model, and temporally consistent from a color based inter-frame consistency model, respectively. $\mathcal{N}_i$ is the spatial neighborhood of $s_{t,i}$ in frame f_t. $\bar{\mathcal{N}}_i$ is the temporal neighborhood of $s_{t,i}$ in adjacent frames f_{t-1} and f_{t+1}. We compute the superpixels by using SLIC [55], due to its superiority in terms of adherence to boundaries, as well as computational and memory efficiency. However, the proposed method is not tied to any specific superpixel method, and one can choose others.

Data Term. The data term $D_i(b_{t,i})$ defines the cost of assigning superpixel $s_{t,i}$ with label $b_{t,i}$ from an appearance model, which learns the color and location distributions of the action object and the backgrounds of video V. With a segmentation B for V, we estimate two color Gaussian Mixture

Models (GMMs) and two location GMMs for the foregrounds and the backgrounds of V, respectively. The corresponding data term $D_i(b_{t,i})$ based on color and location GMMs in Equation (6) is defined as

$$D_i(b_{t,i}) = -\log\left(\beta \mathbf{h}^{col}_{b_{t,i}}(s_{t,i}) + (1-\beta)\mathbf{h}^{loc}_{b_{t,i}}(s_{t,i})\right), \tag{7}$$

where $\mathbf{h}^{col}_{b_{t,i}}$ denotes the two color GMMs, i.e., $\mathbf{h}^{col}_1$ for the action object and $\mathbf{h}^{col}_0$ for the background across video V. Similarly, $\mathbf{h}^{loc}_{b_{t,i}}$ denotes the two location GMMs for the action object and the background across V, i.e., $\mathbf{h}^{loc}_1$ and $\mathbf{h}^{loc}_0$. β is a parameter controlling the contributions of color $\mathbf{h}^{col}_{b_{t,i}}$ and location $\mathbf{h}^{loc}_{b_{t,i}}$.

Smoothness Term. The action segmentation labeling B should be spatially consistent in each frame, and meanwhile temporally consistent throughout video V. Thus, we define the smoothness term by assembling an intra-frame consistency model and an inter-frame consistency model.

The intra-frame consistency model enforces the spatially adjacent superpixels in the same action frame to have the same label. Due to the fact that the adjacent superpixels either have similar color or distinct color contrast [56], the well-known standard contrast-dependent function [56,57] is exploited to encourage the spatially adjacent superpixels with similar color to be assigned with the same label. Then, $S^{intra}_{iu}(b_{t,i}, b_{t,n})$ in Equation (6) is defined as

$$S^{intra}_{in}(b_{t,i}, b_{t,n}) = \mathbb{1}_{[b_{t,i} \neq b_{t,n}]}\exp(-||\mathbf{c}_{t,i} - \mathbf{c}_{t,n}||^2_2), \tag{8}$$

where the characteristic function $\mathbb{1}_{[b_{t,i} \neq b_{t,n}]} = 1$ when $b_{t,i} \neq b_{t,n}$, and 0 otherwise. $b_{t,i}$ and $b_{t,n}$ are the segmentation labels of superpixels $s_{t,i}$ and $s_{t,n}$, respectively. $\mathbf{c}$ is the color vector of the superpixel.

The inter-frame consistency model encourages the temporally adjacent superpixels in consecutive action frames to have the same label. As the temporally adjacent superpixels should have similar color and motion, we use the Euclidean distance between the motion distributions of temporally adjacent superpixels along with the above contrast-dependent function in Equation (8) to constrain the labels of them to be consistent. In Equation (6), $S^{inter}_{in}(b_{t,i}, b_{m,n})$ is then defined as

$$S^{inter}_{in}(b_{t,i}, b_{m,n}) = \mathbb{1}_{[b_{t,i} \neq b_{m,n}]}(\exp(-||\mathbf{c}_{t,i} - \mathbf{c}_{m,n}||^2_2) + \exp(-||\mathbf{h}^m_{t,i} - \mathbf{h}^m_{m,n}||_2)), \tag{9}$$

where $\mathbf{h}^m$ is the histogram of oriented optical flow (HOOF) [58] of the superpixel.

Optimization. With $D_i(b_{t,i})$, $S^{intra}_{in}(b_{t,i}, b_{t,n})$ and $S^{inter}_{in}(b_{t,i}, b_{m,n})$, we leverage graph cut [24] to minimize the energy function in Equation (6), and can obtain a new segmentation B for video V.

3.3. Iterative and Alternating Optimization

With an initial spatial segmentation B_o of video V using saliency [23], the temporal action localization first pinpoints the starting frame $f_s(k)$ and the ending frame $f_se(k)$ of a target action k from an untrimmed video V by a coarse-to-fine action localization strategy, and then the spatial action segmentation further produces the spatial per-frame segmentation B by focusing on the action frames identified by the temporal action localization. With the new segmentation B of video V, the overall optimization alternates between the temporal action localization and spatial action segmentation. Upon the practical convergence of this iterative process, the final results B are obtained. Naturally, the temporal action localization and spatial action segmentation benefit each other. In the experiments, we terminate the iterative optimization after practical convergence is observed, i.e., the relative variation between two successive spatio-temporal action localization results are smaller than 0.001.

4. Experiments and Discussion

4.1. Datasets and Evaluation Protocol

We conduct extensive experiments on multiple datasets to evaluate the efficacy of the proposed spatio-temporal action localization detector Segment-tube, including (1) temporal action localization task on the THUMOS 2014 dataset [25]; (2) spatial action segmentation on the SegTrack dataset [26,27]; and (3) spatio-temporal action localization task on the newly proposed ActSeg dataset.

The average precision (AP) and mean average precision (mAP) are employed to evaluate the temporal action localization performance. If an action is assigned the same category label with the ground truth, and, simultaneously, its predicted temporal range overlaps the ground truth at a ratio above a predefined threshold (e.g., 0.5). Such temporal localization of an action is deemed correct.

The intersection-over-union (IoU) value is utilized to evaluate the spatial action segmentation performance, and it is defined as

$$\mathrm{IoU} = \frac{|Seg \cap GT|}{|Seg \cup GT|}, \tag{10}$$

where Seg denotes the binary segmentation result obtained by a detector, GT denotes the binary ground truth segmentation mask, and $|\cdot|$ denotes the cardinality (i.e., pixel count).

4.2. Implementation Details

Training the proposal network. The proposal network is to predict each video clip u_j either contains an action ($l_j^{fla} = 1$) or the background ($l_j^{fla} = 0$), and thus can remove the background video clips, as described in Section 3.1. We build the training data as follows to train the proposal network. For each video clip from trimmed videos, we assign its action label as 1, denoting it contains some action k ($\forall k \in \{1, \ldots, K\}$). For each video clip from untrimmed videos with temporal annotations, we set its label by using the IoU value between it and the ground truth action instances. If the IoU value is higher than 0.75, we assign the label as 1, denoting that it contains an action; if the IoU value is lower than 0.25, we assign the label as 0, denoting that it does not contain an action.

The 3D ConvNets [4] components (as shown in Figure 2) are pre-trained on the training split of the Sports-1M dataset [59], and used as the initializations of our proposal and classification networks. The output of the softmax layer in the proposal network is of two dimensions, which corresponds to either an action or the background. In all the following experiments, the batch size is fixed at 40 during the training phase, and the initial learning rate is set at 10^{-4} with a learning rate decay of factor 10 every 10 K iterations.

For the LSTM component, the activation feature of the fully-connected 7th layer of the 3D ConvNets [4] is used as the input to the LSTM. The learning batch size is set to be 32, where each sample in the minibatch is a sequence of ten 16-frame video clips. We use RMSprop [60] with a learning rate of 10^{-4}, a momentum of 0.9 and a weight decay factor of 5×10^{-4}. The number of iterations depends on the size of the dataset, and will be elaborated in the following temporal action localization experiments.

Training the classification network. The classification network is to further predict whether each video clip u_j contains a specific action ($l_j^{spe} \in \{k\}_{k=0}^{K}$) or not, as described in Section 3.1. The training data for the classification network is built similarly to that of the proposal network. The only difference is that, for the saliency-aware positive video clip, we assign its label as a specific action category $k \in \{1, \ldots, K\}$ (e.g., "LongJump"), instead of 1 for training the above proposal network.

As to the 3D ConvNets [4] (see Figure 2), we train a classification model with K actions plus one additional "background" category. The learning batch size is fixed at 40, the initial learning rate is 10^{-4} and the learning rate is divided by 2 after every 10 K iterations.

To train the LSTM, the activation feature of the fully-connected 7th layer of the 3D ConvNets [4] is fed to the LSTM. We fix the learning batch size at 32, where each sample in the minibatch is a sequence of ten 16-frame video clips. We also use RMSprop [60] with a learning rate of 10^{-4}, a momentum of 0.9 and a weight decay factor of 5×10^{-4}.

4.3. Temporal Action Localization on the THUMOS 2014 Dataset

We first evaluate the temporal action localization performance of the proposed Segment-tube detector on the THUMOS 2014 dataset [25], which is dedicated to localizing actions in long untrimmed videos involving 20 actions. The training set contains 2755 trimmed videos and 1010 untrimmed validation videos. For the 3D ConvNets training, the fine-tuning stops at 30 k for the two networks. For the LSTM training, the number of training iterations is 20 k for two networks. For testing, we use 213 untrimmed videos that contain relevant action instances.

Five existing temporal action localization methods, i.e., AMA [1], FTAP [9], ASLM [10], SCNN [11], and ASMS [15], are included as competing algorithms. AMA [1] combines iDT features and frame-level CNN features to train a SVM classifier. FTAP [9] leverages high recall temporal action proposals. ASLM [10] uses a length and language model based on traditional motion features. SCNN [11] is an end-to-end segment-based 3D ConvNets framework, including proposal, classification and localization network. ASMS [15] localizes actions by searching for the structured maximal sum.

The mAP comparisons are summarized in Table 1, which demonstrate that the proposed Segment-tube detector evidently outperforms the five competing algorithms with IoU being 0.3 and 0.5, and is marginally inferior to SCNN [11] with IoU threshold being 0.4. We also present the qualitative temporal action localization results of the proposed Segment-tube detector for two action instances of the testing split from the THUMOS 2014 dataset in Figure 4, with IoU threshold being 0.5.

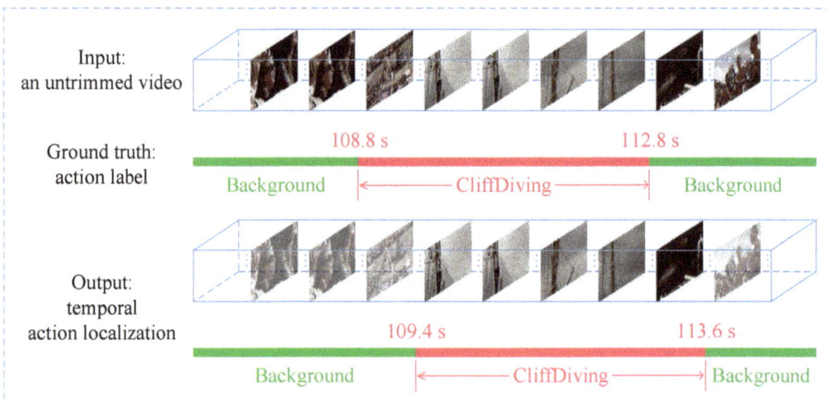

(a) CliffDiving

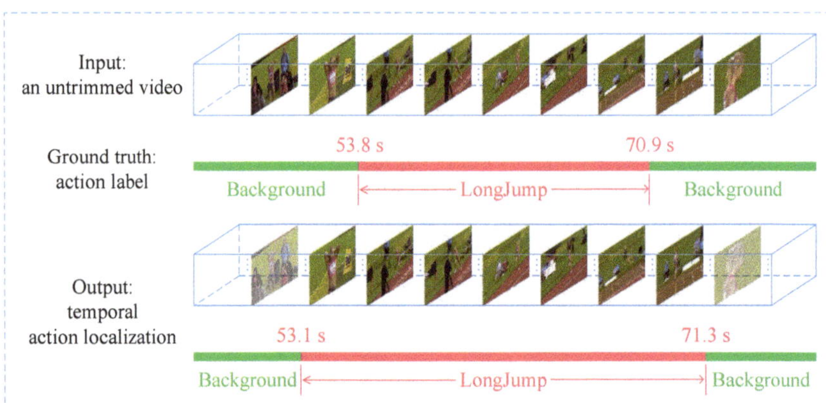

(b) LongJump

Figure 4. Qualitative temporal action localization results of the proposed Segment-tube detector for two action instances, i.e., (**a**) CliffDiving and (**b**) LongJump, in the testing split of the THUMOS 2014 dataset, with intersection-over-union (IoU) threshold being 0.5.

Table 1. Mean average precision (mAP) comparisons of five state-of-the-art temporal action localization methods and our proposed Segment-tube detector on the THUMOS 2014 dataset [25]. mAP values are in percentage. Higher values are better.

IoU Threshold	0.3	0.4	0.5
AMA [1]	14.6	12.1	8.5
FTAP [9]	-	-	13.5
ASLM [10]	20.0	23.2	15.2
SCNN [11]	36.3	28.7	19.0
ASMS [15]	36.5	27.8	17.8
Segment-tube	39.8	27.2	20.7

4.4. Spatial Action Segmentation on the SegTrack Dataset

We then evaluate the performance of spatial action segmentation from trimmed videos on the SegTrack dataset [26,27]. The dataset contains 14 video sequences with lengths varying from 21 to 279 frames. Every frame is annotated with a pixel-wise ground-truth segmentation mask. Due to the limitation of the competing methods [47,48,52], a subset of eight videos are selected, all of which contains only one action object.

We compare our proposed Segment-tube detector with three state-of-the-art video object segmentation methods, i.e., VOS [48], FOS [47] and BVS [52]. VOS [48] automatically discovers and groups key segments to isolate the foreground object. FOS [47] separates the foreground object based on an efficient initial foreground estimation and a foreground-background labeling refinement. BVS [52] obtains the foreground object via bilateral space operations.

The IoU value comparison of VOS [48], FOS [47], BVS [52] and our proposed Segment-tube detector on the SegTrack dataset [26,27] is presented in Table 2. Some example results of them are given in Figure 5, where the predicted segmentation masks are visualized by polygons with red edges. As is shown in Table 2, our method significantly outperforms VOS [47] and FOS [47], and performs better than BVS [52] with a small margin of 2.3. The performance of BVS [52] could possibly due to its exploitation of the first-frame segmentation mask to facilitate the subsequent segmentation procedure.

Table 2. Intersection-over-union (IoU) value comparison of three state-of-the-art video object segmentation methods (VOS [48], FOS [47] and BVS [52]) and our proposed Segment-tube detector on the SegTrack dataset [26,27]. IoU values are in percentage. Higher values are better.

Algorithm	VOS [48]	FOS [47]	BVS [52]	Segment-Tube
bird_of_paradise	92.4	81.8	91.7	93.1
birdfall2	49.4	17.5	63.5	66.7
frog	75.7	54.1	76.4	70.2
girl	64.2	54.9	79.1	81.3
monkey	82.6	65.0	85.9	86.9
parachute	94.6	76.3	93.8	90.4
soldier	60.8	39.8	56.4	64.5
worm	62.2	72.8	65.5	75.2
Average	72.7	57.8	76.5	78.8

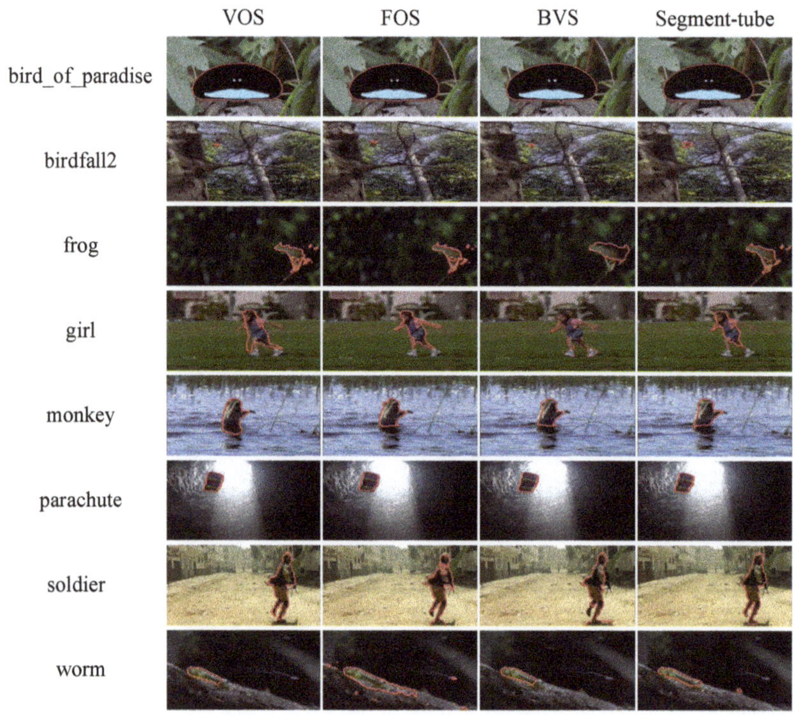

Figure 5. Example results of three state-of-the-art video object segmentation methods (VOS [48], FOS [47] and BVS [52]) and our proposed Segment-tube detector on the SegTrack dataset [26,27].

4.5. Spatio-Temporal Action Localization on the ActSeg Dataset

ActSeg dataset. To fully evaluate the proposed spatio-temporal human action localization detector and to build a benchmark for future research, a new ActSeg dataset is introduced in this paper, including both untrimmed and trimmed videos. The list of action classes are presented in Table 3, including single person actions (e.g., "ArabequeSpin", "PoleVault", "NoHandWindmill") and multi-person actions (e.g., "DeathSpirals").

Table 3. Statistics on total, untrimmed and trimmed videos in each category of the ActSeg dataset.

Number	Total Videos	Untrimmed Videos	Trimmed Videos
ArabequeSpin	68	58	10
CleanAndJerk	73	61	12
UnevenBars	67	57	10
SoccerPenalty	82	67	15
PoleVault	72	59	13
TripleJump	62	50	12
NoHandWindmill	68	57	11
DeathSpirals	78	66	12
Throw	71	56	15
Sum	641	531	110

All raw videos are downloaded from YouTube. Typical untrimmed videos contain approximately 10–120 s of irrelevant frames prior and/or after the specific action. The trimmed videos are pruned so that they only contain relevant action frames. We have recruited 30 undergraduate students to independently decide whether a specific action is present (positive label) in the original video or not (negative label). If four or more positive labels are recorded, the original video is accepted in the ActSeg dataset and the time boundaries of the action are determined as follows. Each accepted video is independently distributed to 3~4 undergraduate students for manual annotation (for both the temporal boundaries and per-frame pixel-wise segmentation labels) and an additional quality comparison is carried out for each accepted video by a graduate student and the best annotation is selected as the ground truth.

The complete ActSeg dataset contains 641 videos in nine human action categories. There are 446 untrimmed videos and 110 trimmed videos in its training split, 85 untrimmed videos and no trimmed video in its testing split. Table 3 presents detailed statistics for the untrimmed/trimmed video distribution in each category. Some typical samples with their corresponding ground truth annotations are illustrated in Figure 6.

Figure 6. Sample frames and their ground truth annotations in the ActSeg dataset. Action frames are marked by green check marks and the corresponding boundaries are marked by polygons with red edges. The background (irrelevant) frames are marked by red cross marks.

Mixed Dataset. To maximize the number of videos in each category (see Table 3), a mixed dataset is constructed by combining videos of identical action categories from multiple datasets. The training split of the mixed dataset consists of all 446 untrimmed videos and 110 trimmed videos in the proposed ActSeg dataset, 791 trimmed videos from the UCF-101 dataset [61], and 90 untrimmed videos from the THUMOS 2014 dataset [25]. The testing split of the mixed dataset consists of all the 85 untrimmed videos from the testing split of the proposed ActSeg dataset.

Temporal Action Localization. SCNN [11] and ARCN [8] are used as competing temporal action localization methods. All three methods are trained on the training split of the mixed dataset. For the 3D ConvNets, the fine-tuning stops at 20 k for the proposal and classification networks.

For LSTM training, the number of training iterations is 10 k for the two networks. Table 4 presents the mAP comparisons of SCNN [11], ARCN [8] and our proposed Segment-tube detector on the testing split of the mixed dataset, with IoU threshold being 0.3, 0.4, and 0.5, respectively. The results show that our proposed Segment-tube method achieves the best mAP with all three IoU thresholds. These manifest the efficacy of the proposed coarse-to-fine action localization strategy and also the Segment-tube detector.

Table 4. Mean average precision (mAP) comparisons of two temporal action localization methods (SCNN [11] and ARCN [8]) and our proposed Segment-tube detector on the testing split of the mixed dataset, with intersection-over-union (IoU) threshold being 0.3, 0.4, and 0.5, respectively. mAP values are in percentage. Higher values are better.

IoU Threshold	0.3	0.4	0.5
ARCN [8]	39.1	33.8	17.2
SCNN [11]	41.0	35.9	18.4
Segment-tube	42.6	37.5	21.2

Spatial Action Segmentation. The spatial action segmentation task is implemented entirely on the ActSeg dataset, with three competing video object segmentation methods, i.e., VOS [48], FOS [47] and BVS [52]. The IoU score comparisons of them are summarized in Table 5. Figure 7 presents some example results of them, where the predicted segmentation masks are visualized by polygons with red edges. Note that the IoU scores are computed only on frames that contain the target action, which are localized by the temporal action localization of the proposed Segment-tube detector.

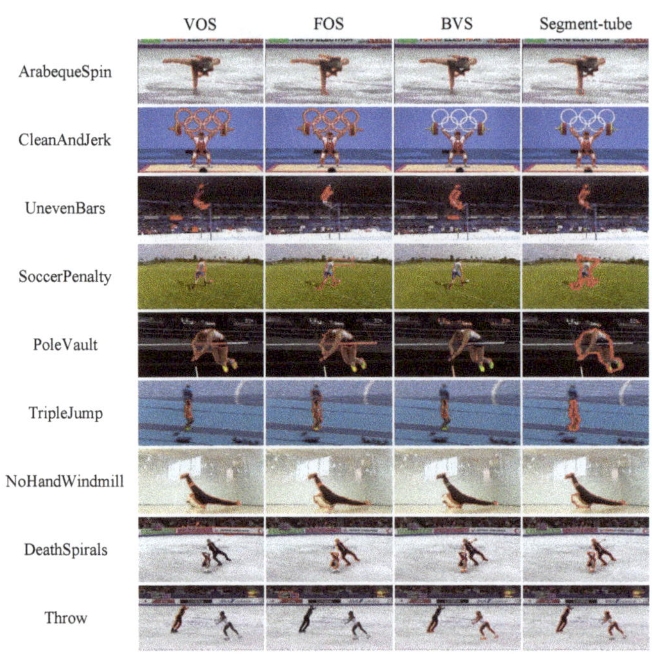

Figure 7. Example results of three video object segmentation methods (VOS [48], FOS [47] and BVS [52]) and our proposed Segment-tube detector on the ActSeg dataset.

The results in Table 5 demonstrate that the Segment-tube detector evidently outperforms VOS [48], FOS [47], and the label propagation based method BVS [52]. On the videos of PoleVault and TripleJump categories, the IoU scores of all the methods are low, which is mainly due to severe occlusions.

Because existing methods either implement temporal action localization or spatial action segmentation, but never achieve both of them simultaneously, we do not include performance comparisons of joint spatio-temporal action localization with per-frame segmentations. To supplement this, we further present the qualitative spatio-temporal action localization results of the proposed Segment-tube for two action instances in the ActSeg dataset (testing split) in Figure 8.

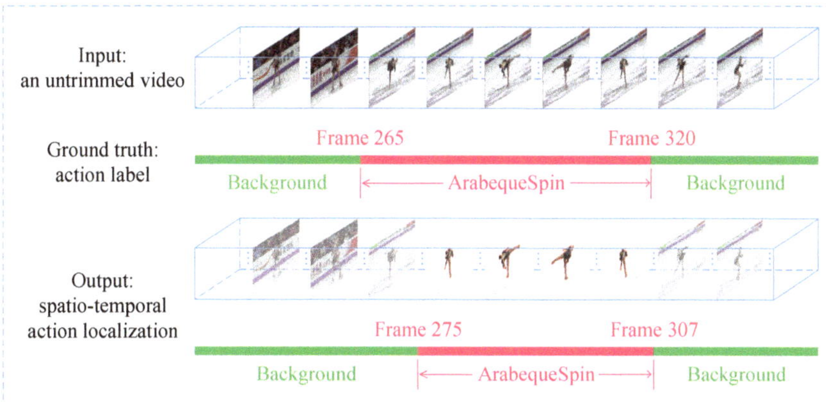

(a) ArabequeSpin

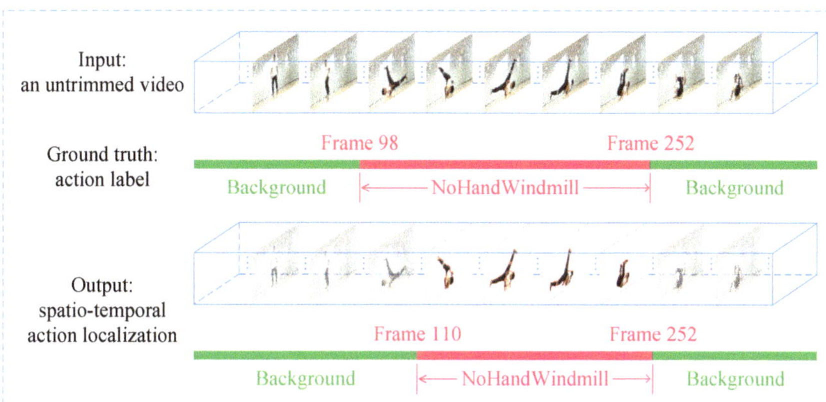

(b) NoHandWindmill

Figure 8. Qualitative spatio-temporal action localization results of the proposed Segment-tube for two action instances, i.e., (**a**) ArabequeSpin and (**b**) NoHandWindmill, in the testing split of the ActSeg dataset, with intersection-over-union (IoU) threshold being 0.5.

To summarize, the experimental results on the above three datasets reveal that the Segment-tube detector produces superior results to existing state-of-the-art methods, which verifies its ability of collaboratively and simultaneously implementing spatial action segmentation and temporal action localization with untrimmed videos.

Table 5. Intersection-over-union (IoU) value comparisons of three video object segmentation methods (VOS [48], FOS [47] and BVS [52]) and our proposed Segment-tube detector on the ActSeg dataset. IoU values are in percentage. Higher values are better.

Video	VOS [48]	FOS [47]	BVS [52]	Segment-Tube
ArabequeSpin	53.9	82.5	64.0	83.4
CleanAndJerk	20.1	50.0	85.9	87.8
UnevenBars	12.0	40.3	59.0	56.5
SoccerPenalty	54.4	38.5	59.8	54.7
PoleVault	38.9	41.2	42.6	49.8
TripleJump	30.6	36.1	33.5	58.4
NoHandWindmill	77.1	73.3	81.8	87.9
DeathSpirals	1	66.7	77.9	66.5
Throw	33.8	2	58.7	56.2
Average	35.8	47.8	62.6	66.8

4.6. *Efficiency Analysis*

The segment-tube detector is highly computational efficient, especially comparing with other approaches that fuse multiple features. Most video clips containing pure background are eliminated by the proposal network, thus the computational cost with the classification network is significantly reduced. On a NVIDIA (NVIDIA Corporation, Santa Clara, CA, USA) Tesla K80 GPU with 12 GB memory, the amortized time of processing one batch (approximately 40 sampled video clips) is approximately one second. Video clips have variable length and 16 frames are uniformly sampled from each video clip. Each input for the 3D ConvNets is a sampled video clip of dimension $3 \times 16 \times 171 \times 128$ (RGB channels $\times$ frames $\times$ width $\times$ height).

5. Conclusions

We propose the spatio-temporal action localization detector Segment-tube, which simultaneously localizes the temporal action boundaries and per-frame spatial segmentation masks in untrimmed videos. It overcomes the common limitation of previous methods that either implement only temporal action localization or just (spatial) video object segmentation. With the proposed alternating iterative optimization scheme, temporal localization and spatial segmentation could be achieved collaboratively and simultaneously. Upon practical convergence, a sequence of per-frame segmentation masks with precise starting/ending frames are obtained. Experiments on three datasets validate the efficacy of the proposed Segment-tube detector and manifest its ability to handle untrimmed videos.

The proposed method is currently dedicated to spatio-temporal localization of a single specific action in untrimmed videos, and we are planning to extend it to simultaneous spatio-temporal localization of multiple actions with per-frame segmentations in our future work. One potential direction is the generation of multiple action category labels in the classification network of the coarse action localization step, followed by independent fine action localization and spatial action segmentation for each action category.

Author Contributions: L.W., X.D. and G.H. conceived the idea and designed the experiments; L.W. and X.D. performed the experiments; L.W., X.D., and Q.Z. analyzed results and wrote the paper; Q.Z., Z.N., N.Z. and G.H. revised and approved the final manuscript.

Acknowledgments: This work was supported partly by National Key Research and Development Program of China Grant 2017YFA0700800, National Natural Science Foundation of China Grants 61629301, 61773312, 61503296, and 91748208, and China Postdoctoral Science Foundation Grants 2017T100752 and 2015M572563.

Conflicts of Interest: The authors declare no conflict of interest.

Abbreviations

The following abbreviations are used in this manuscript:

CNNs	Convolutional Neural Networks
ConvNets	Convolutional Neural Networks
I3D	Inflated 3D ConvNet
ResNet	Deep Residual Convolutional Neural Networks
LSTM	Long-Short Temporal Memory
RNNs	Recurrent Neural Networks
HOG	Histograms of Image Gradients
HOF	Histogram of Flow
iDT	improved Dense Trajectory
FV	Fisher Vector
VLAD	Vector of Linearly Aggregated Descriptors
FCNs	Fully Convolutional Networks
GMMs	Gaussian Mixture Models
HOOF	Histogram of Oriented Optical Flow
AP	Average Precision
mAP	mean Average Precision
IoU	Intersection-over-Union
RGB	Red Green Blue

References

1. Wang, L.; Qiao, Y.; Tang, X. Action recognition and detection by combining motion and appearance features. *THUMOS14 Action Recognit. Chall.* **2014**, *1*, 2.
2. Simonyan, K.; Zisserman, A. Two-stream convolutional networks for action recognition in videos. In Proceedings of the Advances in Neural Information Processing Systems, Montreal, QC, Canada, 8–13 December 2014; pp. 568–576.
3. Wang, L.; Qiao, Y.; Tang, X. Action recognition with trajectory-pooled deep-convolutional descriptors. In Proceedings of the IEEE Conference on Computer Vision and Pattern Recognition, Boston, MA, USA, 7–12 June 2015; pp. 4305–4314.
4. Tran, D.; Bourdev, L.; Fergus, R.; Torresani, L.; Paluri, M. Learning spatiotemporal features with 3D convolutional networks. In Proceedings of the IEEE International Conference on Computer Vision, Santiago, Chile, 11–18 December 2015; pp. 4489–4497.
5. Donahue, J.; Anne Hendricks, L.; Guadarrama, S.; Rohrbach, M.; Venugopalan, S.; Saenko, K.; Darrell, T. Long-term recurrent convolutional networks for visual recognition and description. In Proceedings of the IEEE Conference on Computer Vision and Pattern Recognition, Boston, MA, USA, 7–12 June 2015; pp. 2625–2634.
6. Weinzaepfel, P.; Harchaoui, Z.; Schmid, C. Learning to track for spatio-temporal action localization. In Proceedings of the IEEE International Conference on Computer Vision, Santiago, Chile, 11–18 December 2015; pp. 3164–3172.
7. Ma, S.; Sigal, L.; Sclaroff, S. Learning activity progression in lstms for activity detection and early detection. In Proceedings of the IEEE Conference on Computer Vision and Pattern Recognition, Las Vegas, NV, USA, 26 June–1 July 2016; pp. 1942–1950.
8. Montes, A.; Salvador, A.; Pascual, S.; Giro-i Nieto, X. Temporal activity detection in untrimmed videos with recurrent neural networks. In Proceedings of the 1st NIPS Workshop on Large Scale Computer Vision Systems, Barcelona, Spain, 10 December 2016.
9. Caba Heilbron, F.; Carlos Niebles, J.; Ghanem, B. Fast temporal activity proposals for efficient detection of human actions in untrimmed videos. In Proceedings of the IEEE Conference on Computer Vision and Pattern Recognition, Las Vegas, NV, USA, 26 June–1 July 2016; pp. 1914–1923.
10. Richard, A.; Gall, J. Temporal action detection using a statistical language model. In Proceedings of the IEEE Conference on Computer Vision and Pattern Recognition, Las Vegas, NV, USA, 26 June–1 July 2016; pp. 3131–3140.

11. Shou, Z.; Wang, D.; Chang, S.F. Temporal action localization in untrimmed videos via multi-stage cnns. In Proceedings of the IEEE Conference on Computer Vision and Pattern Recognition, Las Vegas, NV, USA, 26 June–1 July 2016; pp. 1049–1058.
12. Kalogeiton, V.; Weinzaepfel, P.; Ferrari, V.; Schmid, C. Action tubelet detector for spatio-temporal action localization. In Proceedings of the IEEE International Conference on Computer Vision, Venice, Italy, 22–29 October 2017; pp. 4405–4413.
13. Wang, Y.; Long, M.; Wang, J.; Yu, P.S. Spatiotemporal pyramid network for video action recognition. In Proceedings of the IEEE Conference on Computer Vision and Pattern Recognition, Honolulu, HI, USA, 21–26 July 2017; pp. 2097–2106.
14. Girdhar, R.; Ramanan, D.; Gupta, A.; Sivic, J.; Russell, B. ActionVLAD: Learning spatio-temporal aggregation for action classification. In Proceedings of the IEEE Conference on Computer Vision and Pattern Recognition, Honolulu, HI, USA, 21–26 July 2017; pp. 971–980.
15. Yuan, Z.; Stroud, J.C.; Lu, T.; Deng, J. Temporal action localization by structured maximal sums. In Proceedings of the IEEE Conference on Computer Vision and Pattern Recognition, Honolulu, HI, USA, 21–26 July 2017; pp. 3215–3223.
16. Dai, X.; Singh, B.; Zhang, G.; Davis, L.S.; Chen, Y.Q. Temporal context network for activity localization in videos. In Proceedings of the IEEE International Conference on Computer Vision, Venice, Italy, 22–29 October 2017; pp. 5727–5736.
17. Shou, Z.; Chan, J.; Zareian, A.; Miyazawa, K.; Chang, S.F. CDC: Convolutional-de-convolutional networks for precise temporal action localization in untrimmed videos. In Proceedings of the IEEE Conference on Computer Vision and Pattern Recognition, Honolulu, HI, USA, 21–26 July 2017; pp. 1417–1426.
18. Gao, Z.; Hua, G.; Zhang, D.; Jojic, N.; Wang, L.; Xue, J.; Zheng, N. ER3: A unified framework for event retrieval, recognition and recounting. In Proceedings of the IEEE Conference on Computer Vision and Pattern Recognition, Honolulu, HI, USA, 21–26 July 2017; pp. 2253–2262.
19. Carreira, J.; Zisserman, A. Quo vadis, action recognition? A new model and the kinetics dataset. In Proceedings of the IEEE Conference on Computer Vision and Pattern Recognition, Honolulu, HI, USA, 21–26 July 2017; pp. 4724–4733.
20. Hara, K.; Kataoka, H.; Satoh, Y. Learning spatio-temporal features with 3D residual networks for action recognition. In Proceedings of the ICCV Workshop on Action, Gesture, and Emotion Recognition, Venice, Italy, October 2017; Volume 2, p. 4.
21. Hara, K.; Kataoka, H.; Satoh, Y. Can spatiotemporal 3D CNNs retrace the history of 2D CNNs and ImageNet? In Proceedings of the IEEE Conference on Computer Vision and Pattern Recognition, Salt Lake City, UT, USA, 18–22 June 2018.
22. Jain, M.; Van Gemert, J.; Jégou, H.; Bouthemy, P.; Snoek, C. Action localization with tubelets from motion. In Proceedings of the IEEE Conference on Computer Vision and Pattern Recognition, Columbus, OH, USA, 23–28 June 2014.
23. Borji, A.; Cheng, M.M.; Jiang, H.; Li, J. Salient object detection: A benchmark. *IEEE Trans. Image Process.* **2015**, *24*, 5706–5722. [CrossRef] [PubMed]
24. Boykov, Y.; Jolly, M. Interactive graph cuts for optimal boundary & region segmentation of objects in ND images. In Proceedings of the IEEE International Conference on Computer Vision, Vancouver, BC, Canada, 7–14 July 2001; pp. 105–112.
25. Jiang, Y.; Liu, J.; Zamir, A.R.; Toderici, G.; Laptev, I.; Shah, M.; Sukthankar, R. THUMOS Challenge: Action Recognition with A Large Number of Classes. In Proceedings of the European Conference on Computer Vision Workshop, Zurich, Switzerland, 6–12 September 2014.
26. Tsai, D.; Flagg, M.; Nakazawa, A.; Rehg, J.M. Motion coherent tracking using multi-label MRF optimization. *Int. J. Comput. Vis.* **2012**, *100*, 190–202. [CrossRef]
27. Li, F.; Kim, T.; Humayun, A.; Tsai, D.; Rehg, J.M. Video segmentation by tracking many figure-ground segments. In Proceedings of the IEEE International Conference on Computer Vision, Sydney, Australia, 1–8 December 2013; pp. 2192–2199.
28. Wang, H.; Schmid, C. Action recognition with improved trajectories. In Proceedings of the IEEE International Conference on Computer Vision, Sydney, Australia, 1–8 December 2013; pp. 3551–3558.
29. Kataoka, H.; Satoh, Y.; Aoki, Y.; Oikawa, S.; Matsui, Y. Temporal and fine-grained pedestrian action recognition on driving recorder database. *Sensors* **2018**, *18*, 627. [CrossRef] [PubMed]

30. Dalal, N.; Triggs, B. Histograms of oriented gradients for human detection. In Proceedings of the IEEE Conference on Computer Vision and Pattern Recognition, San Diego, CA, USA, 20–26 June 2005; pp. 886–893.
31. Dalal, N.; Triggs, B.; Schmid, C. Human detection using oriented histograms of flow and appearance. In Proceedings of the European Conferemce on Computer Vision, Graz, Austria, 7–13 May 2006; pp. 428–441.
32. Oneata, D.; Verbeek, J.; Schmid, C. Action and event recognition with fisher vectors on a compact feature set. In Proceedings of the IEEE International Conference on Computer Vision, Sydney, Australia, 1–8 December 2013; pp. 1817–1824.
33. Jégou, H.; Douze, M.; Schmid, C.; Pérez, P. Aggregating local descriptors into a compact image representation. In Proceedings of the IEEE Conference on Computer Vision and Pattern Recognition, San Francisco, CA, USA, 13–18 June 2010; pp. 3304–3311.
34. Heilbron, F.C.; Escorcia, V.; Ghanem, B.; Niebles, J.C. ActivityNet: A large-scale video benchmark for human activity understanding. In Proceedings of the IEEE Conference on Computer Vision and Pattern Recognition, Boston, MA, USA, 7–12 June 2015; pp. 961–970.
35. Mexaction2. Available online: http://mexculture.cnam.fr/xwiki/bin/view/Datasets/Mex+action+dataset (accessed on 15 July 2015).
36. Gaidon, A.; Harchaoui, Z.; Schmid, C. Temporal localization of actions with actoms. *IEEE Trans. Pattern Anal. Mach. Intell.* **2013**, *35*, 2782–2795. [CrossRef] [PubMed]
37. Yuan, J.; Pei, Y.; Ni, B.; Moulin, P.; Kassim, A. Adsc submission at thumos challenge 2015. In Proceedings of the IEEE Conference on Computer Vision and Pattern Recognition THUMOS Workshop, Las Vegas, NV, USA, 11–12 June 2015; Volume 1, p. 2.
38. Singh, G.; Cuzzolin, F. Untrimmed video classification for activity detection: Submission to activitynet challenge. *arXiv* **2016**, arXiv:1607.01979.
39. Yeung, S.; Russakovsky, O.; Jin, N.; Andriluka, M.; Mori, G.; Fei-Fei, L. Every moment counts: Dense detailed labeling of actions in complex videos. *Int. J. Comput. Vis.* **2018**, *126*, 375–389. [CrossRef]
40. Yu, G.; Yuan, J. Fast action proposals for human action detection and search. In Proceedings of the IEEE Conference on Computer Vision and Pattern Recognition, Boston, MA, USA, 7–12 June 2015; pp. 1302–1311.
41. Buch, S.; Escorcia, V.; Shen, C.; Ghanem, B.; Niebles, J.C. Sst: Single-stream temporal action proposals. In Proceedings of the IEEE Conference on Computer Vision and Pattern Recognition, Honolulu, HI, USA, 21–26 July 2017; pp. 6373–6382.
42. Gong, J.; Fan, G.; Yu, L.; Havlicek, J.P.; Chen, D.; Fan, N. Joint target tracking, recognition and segmentation for infrared imagery using a shape manifold-based level set. *Sensors* **2014**, *14*, 10124–10145. [CrossRef] [PubMed]
43. Soomro, K.; Idrees, H.; Shah, M. Action localization in videos through context walk. In Proceedings of the IEEE International Conference on Computer Vision, Santiago, Chile, 11–18 December 2015; pp. 3280–3288.
44. Gkioxari, G.; Malik, J. Finding action tubes. In Proceedings of the IEEE Conference on Computer Vision and Pattern Recognition, Boston, MA, USA, 7–12 June 2015; pp. 759–768.
45. Singh, G.; Saha, S.; Sapienza, M.; Torr, P.; Cuzzolin, F. Online real-time multiple spatiotemporal action localisation and prediction. In Proceedings of the IEEE International Conference on Computer Vision, Venice, Italy, 22–29 October 2017; pp. 3637–3646.
46. Palou, G.; Salembier, P. Hierarchical video representation with trajectory binary partition tree. In Proceedings of the IEEE Conference on Computer Vision and Pattern Recognition, Portland, OR, USA, 23–28 June 2013; pp. 2099–2106.
47. Papazoglou, A.; Ferrari, V. Fast object segmentation in unconstrained video. In Proceedings of the IEEE International Conference on Computer Vision, Sydney, Australia, 1–8 December 2013; pp. 1777–1784.
48. Lee, Y.J.; Kim, J.; Grauman, K. Key-segments for video object segmentation. In Proceedings of the IEEE International Conference on Computer Vision, Barcelona, Spain, 6–13 November 2011; pp. 1995–2002.
49. Khoreva, A.; Galasso, F.; Hein, M.; Schiele, B. Classifier based graph construction for video segmentation. In Proceedings of the IEEE Conference on Computer Vision and Pattern Recognition, Boston, MA, USA, 7–12 June 2015; pp. 951–960.
50. Wang, W.; Shen, J.; Porikli, F. Saliency-aware geodesic video object segmentation. In Proceedings of the IEEE Conference on Computer Vision and Pattern Recognition, Boston, MA, USA, 7–12 June 2015; pp. 3395–3402.

51. Jain, S.D.; Xiong, B.; Grauman, K. Fusionseg: Learning to combine motion and appearance for fully automatic segmention of generic objects in videos. In Proceedings of the IEEE Conference on Computer Vision and Pattern Recognition, Honolulu, HI, USA, 21–26 July 2017.
52. Märki, N.; Perazzi, F.; Wang, O.; Sorkine-Hornung, A. Bilateral space video segmentation. In Proceedings of the IEEE Conference on Computer Vision and Pattern Recognition, Las Vegas, NV, USA, 26 June–1 July 2016; pp. 743–751.
53. Tsai, Y.H.; Yang, M.H.; Black, M.J. Video segmentation via object flow. In Proceedings of the IEEE Conference on Computer Vision and Pattern Recognition, Las Vegas, NV, USA, 26 June–1 July 2016; pp. 3899–3908.
54. Caelles, S.; Maninis, K.K.; Pont-Tuset, J.; Leal-Taixé, L.; Cremers, D.; Van Gool, L. One-shot video object segmentation. In Proceedings of the IEEE Conference on Computer Vision and Pattern Recognition, Honolulu, HI, USA, 21–26 July 2017; pp. 221–230.
55. Achanta, R.; Shaji, A.; Smith, K.; Lucchi, A.; Fua, P.; Süsstrunk, S. SLIC superpixels compared to state-of-the-art superpixel methods. *IEEE Trans. Pattern Anal. Mach. Intell.* **2012**, *34*, 2274–2282. [CrossRef] [PubMed]
56. Wang, L.; Hua, G.; Sukthankar, R.; Xue, J.; Niu, Z.; Zheng, N. Video object discovery and co-segmentation with extremely weak supervision. *IEEE Trans. Pattern Anal. Mach. Intell.* **2017**, *39*, 2074–2088. [CrossRef] [PubMed]
57. Zhang, D.; Javed, O.; Shah, M. Video object co-segmentation by regulated maximum weight cliques. In Proceedings of the European Conferemce on Computer Vision, Zurich, Switzerland, 6–12 September 2014; pp. 551–566.
58. Chaudhry, R.; Ravichandran, A.; Hager, G.; Vidal, R. Histograms of oriented optical flow and binet-cauchy kernels on nonlinear dynamical systems for the recognition of human actions. In Proceedings of the IEEE Conference on Computer Vision and Pattern Recognition, Miami, FL, USA, 20–25 June 2009; pp. 1932–1939.
59. Karpathy, A.; Toderici, G.; Shetty, S.; Leung, T.; Sukthankar, R.; Fei-Fei, L. Large-scale video classification with convolutional neural networks. In Proceedings of the IEEE Conference on Computer Vision and Pattern Recognition, Columbus, OH, USA, 23–28 June 2014; pp. 1725–1732.
60. Dauphin, Y.; De Vries, H.; Chung, J.; Bengio, Y. RMSProp and equilibrated adaptive learning rates for non-convex optimization. *arXiv* **2015**, arXiv:1502.04390.
61. Soomro, K.; Zamir, A.R.; Shah, M. UCF101: A Dataset of 101 Human Actions Classes From Videos in the Wild. *Comput. Sci.* **2012**, arXiv:1212.0402.

sensors

MDPI

Article

Automated Field-of-View, Illumination, and Recognition Algorithm Design of a Vision System for Pick-and-Place Considering Colour Information in Illumination and Images

Yibing Chen [1], Taiki Ogata [2,*], Tsuyoshi Ueyama [3], Toshiyuki Takada [3] and Jun Ota [2]

1 Department of Precision Engineering, the University of Tokyo, Tokyo 113-8656, Japan; y.chen@race.u-tokyo.ac.jp
2 Research into Artifacts, Center for Engineering (RACE), The University of Tokyo, Chiba 277-8568, Japan; ota@race.u-tokyo.ac.jp
3 Denso Wave Incorporated, Aichi 470-2298, Japan; tsuyoshi.ueyama@denso-wave.co.jp (T.U.); toshiyuki.takada@denso-wave.co.jp (T.T.)
* Correspondence: ogata@race.u-tokyo.ac.jp; Tel.: +81-4-7136-4252

Received: 28 March 2018; Accepted: 18 May 2018; Published: 22 May 2018

Abstract: Machine vision is playing an increasingly important role in industrial applications, and the automated design of image recognition systems has been a subject of intense research. This study has proposed a system for automatically designing the field-of-view (FOV) of a camera, the illumination strength and the parameters in a recognition algorithm. We formulated the design problem as an optimisation problem and used an experiment based on a hierarchical algorithm to solve it. The evaluation experiments using translucent plastics objects showed that the use of the proposed system resulted in an effective solution with a wide FOV, recognition of all objects and 0.32 mm and 0.4° maximal positional and angular errors when all the RGB (red, green and blue) for illumination and R channel image for recognition were used. Though all the RGB illumination and grey scale images also provided recognition of all the objects, only a narrow FOV was selected. Moreover, full recognition was not achieved by using only G illumination and a grey-scale image. The results showed that the proposed method can automatically design the FOV, illumination and parameters in the recognition algorithm and that tuning all the RGB illumination is desirable even when single-channel or grey-scale images are used for recognition.

Keywords: automated design; vision system; FOV; illumination; recognition algorithm

1. Introduction

Machine vision technologies have been widely applied in the industrial field for automated visual inspection, process control, parts identification, and robotic guidance [1,2]. Designers have been attempting to tune the parameters for a variety of vision systems. A vision system is usually composed of a camera and an illumination and recognition algorithm [3], which are also known as the main design factors of a vision system. In the object recognition system of a pick-and-place robot, for example, the camera position needs to be set to obtain a suitable Field-of-View (hereinafter referred to as FOV), the illumination requires to be changed to strengthen features in targets, and the image recognition process needs to be optimised through parameter tuning. As this creates a number of conflicting variables, the design process must be reiterated until acceptable results are obtained. This is a time-consuming task even when carried out by experts, and even a simple pick-and-place vision system usually takes several days to design.

Previous studies have addressed the automated design of sensor locations [4–12], illumination levels [13–18], and recognition algorithms [19–25]. Some studies proposed a method to automatically determine the place to set a vision sensor for specific features of recognition targets to satisfy the specific constraints of recognition requirements [4–6]. Some other studies focused on the sensing strategies for recognition and localisation of targets with the help of 3D models [7–9]. Moreover, several sensor planning methods were designed respectively based on the vision tasks in [10–12].

Researches on automated planning of illumination parameters have also been carried out. Experiment-based approaches have been proposed to optimise illumination with a set of images of an object captured under different illumination conditions [13,14]. Besides, illumination planning methods based on mathematical models of illumination were proposed [15,16]. More recently, with the help of rendering techniques, illumination planning approaches based on computer simulation were reported [17,18].

Some studies have attempted to automate the image processing procedures. Automated image pre-processing techniques were proposed in [19–21]. Some other studies investigated the automated design for feature extraction [22,23]. Automated generation of discriminators were discussed in [24,25]. Especially, an approach was proposed for automatically designing an image recognition procedure from the aspect of pre-processing, feature extraction, and discriminator [25].

It is clear from the former studies that an overall design approach to vision systems could hardly be found. One reason is probably the interactions among the different design factors. Therefore, in the case of an overall design, the situation becomes more and more complex because design factors influence each other in unpredictable ways. To the best of our knowledge, a design approach to deal with various design factors has been presented only in [26,27]. Experiment-based methods were applied to achieve an automated design of a vision system on the basis of illumination and a recognition algorithm in [26]. By adding FOV, Chen Y., et al. [27] provided a more comprehensive vision system design method. The problem is that in both the studies, the recognition tasks were far from a being practical task because only one or two objects were considered.

Another obstacle in taking out an overall design approach consists of the uncertainties of the real world. Colour is known as one of the uncertainties in image recognition. Because objects' colours change with illumination, colour- and illumination-invariant recognition methods have been postulated [28–31]. The greyscale process transforms colourful multi-channel images into grey and single-channel ones, which could be more easily understood by vision systems. Such multi-channel image encoding approaches were presented in [32–34], while it was also pointed out that greyscale approaches could influence the recognition performance to a great extent [35]. In this study, we have mainly focused on the uncertainties caused by colour information contained in both illumination and grabbed images.

This study transferred a vision system design problem into an optimisation problem and proposed an experiment-based approach to realise an automated vision system design. It was proved in the study that the proposed design could provide vision systems that were effective in pick-and-place tasks with suitable parameters of the FOV, illumination and recognition algorithm. Moreover, we studied one kind of uncertainties from the real world, that is, colour information illuminate from the light source that is absorbed by the camera sensor. Thus, we conducted an experiment of automated designs using our proposed method by changing the colour channels that were utilised for both illumination and recognition. By this experiment, we investigated: (1) whether or not providing colourful illumination improves recognition accuracy when even the vision system reads only the greyscale images and (2) whether or not single-channel images like R-channel images provide better performance in recognition than the greyscale ones.

2. Problem Formulation

2.1. Preconditions

The vision systems applied to a pick-and-place robot are set to the design target in this study. In order to pick objects and place them in the right positions, the vision systems are required to provide the following information:

- *Types*

Before picking up an object, the system must know which kind of object to select. For instance, in some sorting tasks, type refers to information that describes targets' appearance, such as the shape, colour, and which side is facing upwards. By using the type information, the robot is able to distinguish the target objects into several categories. Therefore, a dictionary which contains type information must be available to vision systems for pick-and-place.

- *Position*

For a pick-and-place robot, definitely 'pick' is one of the most important quests. To pick objects up, position information, in other words, the centre of gravity of each target object should be identified. A vision system captures positional information in pixels. In this study, the coordinate origin is set to the left-top of the image, the x-axis forward direction to the right, and the y-axial forward direction is downward.

- *Orientations*

For both 'pick' and 'place' quests, the orientation information is important. That is to say, the vision system must also provide angular information about each object. In this study, we assumed the measurement range to be [0, 360).

To clarify the problem, the working environment for the proposed system is set up based on the following three requirements: choice of camera, assumed scenes of recognition, and image processing software. The details for each of these requirements are given as follows:

- *Camera*

Compared to binocular cameras, monocular cameras are more widely used in pick-and-place tasks. As a result, we used a monocular CMOS (Complementary Metal Oxide Semiconductor) camera in our system. Since the camera is mounted on the end effector of an industrial manipulator, its viewpoint is held perpendicular to the workspace on which the recognition targets are arranged; the FOV is therefore of a 2D type.

- *Scenes*

Based on where to pick objects, pick-and-place tasks could be categorised into two types: tasks in dynamic systems, for example, a moving conveyor and tasks in static systems such as a tray. In this study, we chose the latter one as the option for the proposed system. In this case, all the recognition targets are placed in a limited space. This means that the camera distance at which all objects can be captured in a single image can be specified in advance. Moreover, just like most situations in industrial applications, the recognition targets are placed on the same plane surface, without overlaps. This is also true for industrial applications such as picking objects from a conveyor.

- *Software*

In this study, the proposed system was tested with a commercially available image processing library: MVTec HALCON (MVTec, Seeshaupt, Germany).

2.2. Design Variables

By describing the settings and the working environment, the basic information on the target vision system to be designed was provided. In order to arrive at a proper design of the described vision system, several parameters, or we can call them design variables, are required to be optimised. To further clarify the problem, such design variables are determined in this section.

In general, a vision system could be established by considering three design factors, which are: illumination condition; camera FOV; and the recognition algorithm. The illumination is usually designed for its strength and colour, illuminating the workpieces and repressing the reflections at the same time. Camera FOV determines the resolution with which the targets are recognised and the size of the recognition area. By tuning FOV, accuracy and efficiency of the vision system could be balanced. Besides, in order to maximise the performance of the chosen recognition algorithm, some parameters inside the algorithm also require optimisation.

The design variables of the vision system could also be taken as parameters of the optimisation problem that were defined in the previous section. Categorised by the three design factors, the design variables of this study are addressed as follows (details are given in Table 1):

Table 1. List of design variables.

Design Factor	Name	Description	Range
FOV	Shoot time	Number of images required in one recognition for the entire area	$1, 4, \ldots, n^2$
	Camera distance	Represents FOV size	Determined by shoot time
Illumination	Light strength (Red)	Strength of red component in illumination	[0, 255]
	Light strength (Green)	Strength of green component in illumination	[0, 255]
	Light strength (Blue)	Strength of blue component in illumination	[0, 255]
Recognition algorithm	Discriminator	Thresholds for classifying different kinds of recognition objects	(0, 1)
	Contrast	Contrast value to extract contour model from template	[0, 255]

- *FOV*

FOV is the extent of the observable world that is seen at any given moment. In the case of a camera set up for pick-and-place tasks, the FOV directly determines the number of objects that can be captured in a single image. To maximise the efficiency of recognition, the FOV must be maximised while meeting the required accuracy tolerances. In this study, FOV was balanced from the viewpoint of shoot time and camera distance.

Shoot time means the time taken by the camera to capture figures within the current camera distance. Obviously, with a limited FOV size, the vision system could not comprehend the intricate details of the workspace. Thus, it requires a system based on a moving camera which captures images several times.

On the other hand, camera distance refers to the distance from the camera lens to the plane on which the recognition targets are placed. As mentioned in the preconditions, the camera's viewpoint is held perpendicular to the workspace; the distance therefore reflects the actual FOV size.

- *Illumination*

The illumination variables include the strength of the red, green, and blue components. Increasing the strength may produce reflections, whereas at low strength, some details of the target objects may not be captured. Both will reduce the recognition accuracy. Additionally, some details may be enhanced by selecting the specific colour of illumination. We therefore allowed the strength of each RGB component to be controlled individually. The illumination strength ranges from 0 to 255, and is searched by an increment.

- *Recognition Algorithm*

Not only the recognition algorithm but also the parameters inside the chosen algorithm influence the performance of a vision system. We just focus on the latter to optimise the inner parameters of a given recognition algorithm.

The inner parameters, for example, image pre-processing parameters or parameters for making proper templates, could more or less influence the performance of a recognition algorithm. Since different parameters may have their own properties, the optimisation method should be designed individually.

Moreover, no matter what recognition algorithm is used, a discriminator to classify correct and incorrect detections by the recognition process is required. The discriminators should also be considered as one of the design variables.

2.3. *Inputs and Outputs*

2.3.1. Inputs

Aiming to turn the vision system design process into a fully-automated one, manual operations during the process of design must be minimised. Hence, the inputs to the automated design system should be considered from many aspects which are preparation data for both scenes and templates, ground truth data, and camera calibration data.

- *Preparation Data for Scenes:*

$$S = (S_1, \ldots, S_i, \ldots, S_n),$$

here, S_i denotes the i-th coordinate on the work plane of the position where the corresponding scene was set, and n the total number of scenes. The number of images required to capture one scene depends on the camera distance.

$$S_i = (x_i, \ldots, y_i, \ldots, z_i),$$

Each S_i contains the locations of x, y, and z directions such that the manipulator can hold the camera and capture images of the existing scene. z_i describes the distance from the camera to the plane where the recognition targets are arranged.

- *Preparation Data for Templates:*

$$\boldsymbol{T} = (\boldsymbol{T}_1, \ldots, \boldsymbol{T}_l, \ldots, \boldsymbol{T}_{nT}),$$

where nT represents the total number of recognition target kinds.

$$\boldsymbol{T}_l = (x, y, z, x_l, y_l, w_l, h_l),$$

the l-th template is prepared by automatically cutting the object image from the original image which was obtained by holding the camera at the position (x, y, z). By using the position of the objects in the acquired image, namely x_l and y_l, as well as the predetermined width and height, w_l and h_l, the template could be obtained.

- *Ground Truth Data:*

$$G = (G_{S_1}, \ldots, G_{S_i}, \ldots, G_{S_n}),$$

where G_{S_i} denotes the ground truth data for the i-th scene.

$$G_{S_i} = (G_{S_i,1}, \ldots, G_{S_i,k}, \ldots, G_{S_i,m_i});$$

however, the scene contains many recognition targets; the ground truth data always include information on each recognition target, from the 1st to the m_i-th.

$$G_{S_i,k} = (Type_{i,k}, x_{i,k}, y_{i,k}, \theta_{i,k}),$$

the ground truth data for each object includes the object type, the x and y position in captured images and the orientation angle.

- *Camera Calibration Data:*

$$C = (C_1, \ldots, C_i, \ldots, C_n),$$

where C_i denotes the i-th image for calibration and n the total number of images required for a calibration.

2.3.2. Outputs

The system output is the optimal solution to the set of design variables.

- *Optimal solution:*

$$S_{olution} = (R, G, B, P_{recognition}),$$

where R, G, and B denote the light strength of red, green and blue, and $P_{recognition}$ the set of parameters related to the chosen recognition algorithm. Especially, $P_{recognition}$ consists of:

$$P_{recognition} = (P_1, P_2, \ldots, P_n)$$

the entire number of parameters n is determined by the chosen recognition algorithm.

2.4. Evaluation Function and Constraints

2.4.1. Evaluation Function

The evaluation uses four values which are the FOV size, $F_{measure}$, positional error, and angular error.

The shoot time describes the FOV and largely determines the computing speed, as the time cost increases in line with the number of images and camera movements.

The $F_{measure}$ is used to describe the accuracy of recognition. It considers both the $P_{recision}$ and R_{ecall}, and the definition is given by Equation (1):

$$F_{measure} = \frac{2 \times P_{recision} \times R_{ecall}}{P_{recision} + R_{ecall}}. \tag{1}$$

In this study, the $P_{recision}$ and R_{ecall} values were given by the following equations:

$$P_{recision} = \frac{\sum_{I_i \in I} m_{ci}}{\sum_{I_i \in I} m_i} \tag{2}$$

$$R_{ecall} = \frac{\sum_{I_i \in I} m_{ci}}{\sum_{I_i \in I} m_{di}}. \tag{3}$$

Here, m_i, m_{ci}, and m_{di} refer to the total number of targets, correctly recognised targets, and targets detected by the recognition process for the i-th learning image set, respectively. The $F_{measure}$ value ranges from [0, 1]. A value closer to 1 indicates greater accuracy.

Positional errors (P_{osErr}) were defined as follows:

$$P_{osErr} = \max\{P_{osErr_1}, \ldots, P_{osErr_i}, \ldots, P_{osErr_n}\} \tag{4}$$

$$P_{osErr_i} = \sqrt{(x_i - x_{gti})^2 + (y_i - y_{gti})^2}. \tag{5}$$

The maximum positional error among n targets was used in the evaluation, and each positional error was calculated from the difference between the points detected by the recognition system (x_i, y_i) and the ground truth (x_{gti}, y_{gti}). As the proposed system used a moveable camera, we first transformed the positional results from the camera coordinates to world coordinates and then measured the error in millimetres.

The angular errors (A_{gErr}) were defined as follows:

$$A_{gErr} = \max\{A_{gErr_1}, \ldots, A_{gErr_i}, \ldots, A_{gErr_n}\} \tag{6}$$

$$A_{gErr_i} = \left|(\theta_i - \theta_{gti})(mod\ 360)\right|. \tag{7}$$

The maximum angular error among n targets was used in the evaluation. As the detection range was from [0, 360), the angular error was given by the difference between the angles detected by the recognition system θ_i and the ground truth θ_{gti}. This was given by Equation (7).

We set the following order for evaluation: first, the camera distance; second, the $F_{measure}$; third, the positional error; and finally, the angular error. Accuracy was determined from the minimum $F_{measure}$ values and maximum positional error and angular error values. The system would therefore choose the solution based on the FOV size, $F_{measure}$, positional error, and angular error, successively.

For a vision system in pick-and-place tasks, it is important to do recognition as efficient as possible with the guarantee of accuracy. The positional and angular errors can be often tolerated to some extent by the selection of the manipulator though we did not discuss the type of the manipulator in this paper. If the recognition accuracy is high enough, the higher the image capturing efficiency is, the better. Therefore, FOV size and $F_{measure}$ take the first two priorities for evaluation. If the positional error is too large, the manipulator cannot pick the objects. Therefore, we decided to prioritize positional error over angular error.

2.4.2. Constraints

For the designed vision system to be applied to a pick-and-place task, it is necessary to ensure the minimal performance. In other words, at least the designed vision system could pick up and place the objects without any failure. The constraints are therefore set to guarantee the minimal performance of designed system.

3. Methodology

3.1. Algorithm Overview

Figure 1 shows the algorithm we proposed to solve the problem formulated in Section 2. In general, we prepared respective optimisations for parameters of the three design factors and arranged them hierarchically.

The system first set the FOV size to its maximum, so that all the target objects could be captured into one image. Based on the multi-start nearest neighbour search, which is discussed in more detail in Section 3.3, the illumination search centre was set randomly to (Red_i, $Green_i$, $Blue_i$), and the parameters in recognition algorithm were then designed. After the recognition algorithm design, the system obtained the local solution (Red_i, $Green_i$, $Blue_i$, $Height_i$, $Parameters_{Best}$) and its accuracy evaluation

($F_{measurei}$, P_{osErri}, A_{gErri}) for the corresponding illumination condition. Evaluation was performed repeatedly under neighbour illumination conditions around the selected search centre. After all neighbours were searched, the search centre was moved to its best neighbour until it became the best design. Illumination optimisation was repeated N times, yielding N local optimal solutions. If solutions meeting the design criteria were found, the system chose the optimal solution among N candidates. Otherwise, the system returned to its initial state and narrowed the FOV size by decreasing the camera distance and increasing the shoot time. The methods to apply narrow FOV and estimate FOV size by camera distance are presented in Section 3.2.

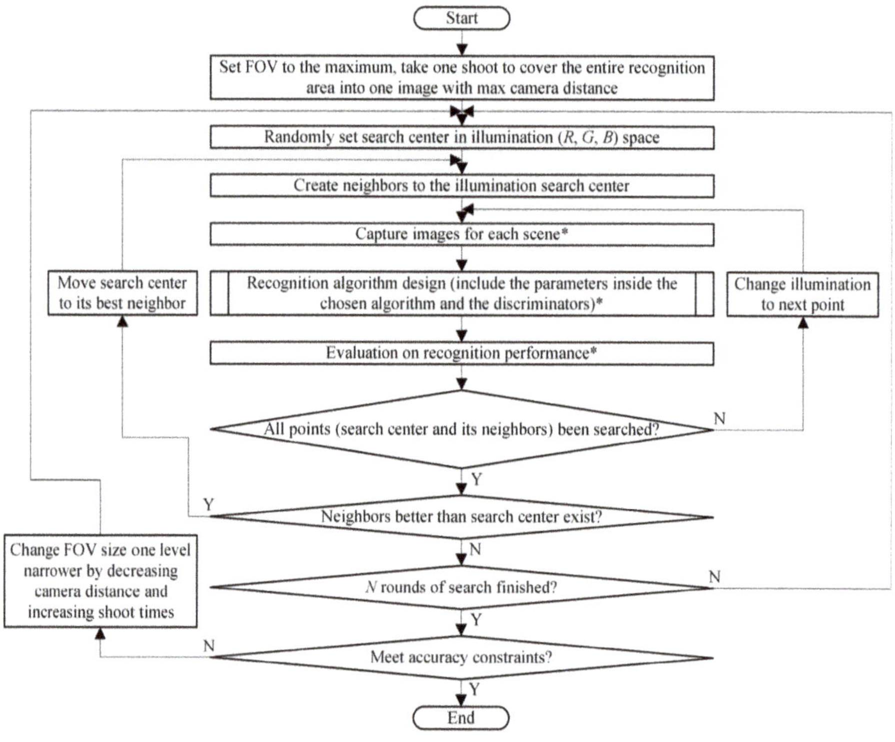

Figure 1. Proposed algorithm to design FOV, illumination and image pre-processing parameters for recognition system. *: The procedures will be skipped if the current selected point has been searched before.

3.2. FOV Design

The FOV is applied to the vision system in the following two ways: first, carry out recognition once with an FOV size and just fit the size of the recognition area and second, carry out recognition by scanning the entire area n^2 times with a FOV of a specific size. Figure 2 shows an example of taking an image of an object placed in the area for recognition. Since the angle between the viewpoint of the camera and the work plane is fixed, which is stated in the preconditions, the FOV size could be easily estimated from the distance between the camera and work plane.

The steps to estimate the FOV size by camera distance are:

(1) *Obtain the mathematical relation between the width of a taken image and camera distance.*

Several images are captured under different camera distances. By adding camera calibrations, the $F_{OVwidth}$, or in other words, the distance of y direction in the taken images, can be measured in millimetres. Repeating this operation several times, the relations between $F_{OVwidth}$ and camera distance could be fitted to a linear one:

$$F_{OV_{width}}(C_{distance}) = a_{width} \times C_{distance} + b_{width}. \tag{8}$$

The $C_{distance}$ denotes the camera distance, a and b are coefficients calculated by experimental data.

(2) *Obtain mathematical relation between the length of a taken image and camera distance.*

Similar to FOV width, relations between $F_{OVlength}$ and camera distance are found using the following expression:

$$F_{OV_{length}}(C_{distance}) = a_{length} \times C_{distance} + b_{length}. \tag{9}$$

(3) *Choose either length or width to represent the F_{OVsize} based on length-width ratios of the recognition area and captured image.*

$$\frac{F_{OV_{length}}(C_{distance})}{F_{OV_{width}}(C_{distance})} > \frac{R_{length}}{R_{width}}. \tag{10}$$

R_{length} and R_{width} denote length and width of the recognition area, respectively. Equation (10) is the criterion to judge whether to use length or width to represent the size of FOV. Based on the condition of inequality applied in this study, if the length-width ratio of FOV is larger than the ratio of the recognition area, then the width should be selected for calculations in later steps. Otherwise, the length should be chosen.

(4) *Calculate desired FOV size.*

Using either length or width to stand for the size, the desired F_{OVsize} could be calculated in addition to the size margin and the scan time.

$$F_{OV_{size}} = \frac{R_{size} + (\sqrt{S_{time}} - 1) \times M_{argin}}{\sqrt{S_{time}}}. \tag{11}$$

R_{size} denotes the length or width of the recognition area, M_{argin} the margin of the FOV size decided by the maximum size of chosen recognition targets and S_{time} the total scan time. Here the square root of S_{time} is used to present the scan time in either the x or y direction.

(5) *Estimate corresponding camera distance.*

By substitution of the calculated desired FOV size into either Equation (8) or Equation (9), the corresponding camera distance for the desired FOV size could be obtained:

$$C_{distance} = \frac{F_{OV_{size}} - b_{size}}{a_{size}}. \tag{12}$$

Here, a_{size} and b_{size} denotes a_{width} and b_{width} in Equation (8) or a_{length} and b_{length} in Equation (9) depend on the truth or false of Equation (10).

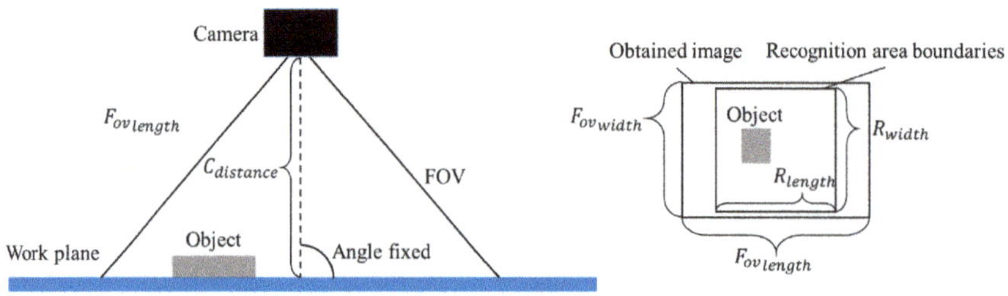

Figure 2. Illustration of the estimation of the FOV size. The positional relation between camera, object(s) and the work plane are shown on left side. The right side shows the length-width ratio of the recognition area and FOV. In the given example, the ratio of FOV is larger than that of the recognition area, which suggests FOV width to represent FOV size.

3.3. Illumination Design

We selected a random multi-start nearest neighbour search, which is one of metaheuristic method, for optimisation of the illumination strength of red, green, and blue. Due to find constraint satisfaction solutions in limited time, we allowed the system choose search centres randomly, even that may result in different optimums in a fixed condition.

The neighbours were generated by changing the value (adding or subtracting the increments shown in Table 1) of one variable, while holding the others constant. The system then created six neighbours for RGB strength in illumination:

$$N_{eighbors} = \{(R + I_{ncrement}, G, B), (R - I_{ncrement}, G, B), (R, G + I_{ncrement}, B), (R, G - I_{ncrement}, B), (R, G, B + I_{ncrement}), (R, G, B - I_{ncrement})\}.$$

3.4. Recognition Algorithm Design

The proposed system is capable of automatically selecting threshold values as discriminators for all kinds of recognition objects.

Suppose that a recognition method has a value E to evaluate its detections, the higher E value indicates the detection is more likely to be a correct one. For a given recognition object, a series of E values are used for n detections after one recognition.

$$\boldsymbol{D} = \{E_{D1}, E_{D2}, \ldots, E_{Dn}\}. \tag{13}$$

Set $\boldsymbol{D}$ was then categorised into two sets with the help of ground truth data; $\boldsymbol{T}$ for correctly detected results, $\boldsymbol{F}$ for incorrectly detected results:

$$\boldsymbol{T} = \{E_{T1}, E_{T2}, \ldots, E_{Tm}\}, \tag{14}$$

$$\boldsymbol{F} = \{E_{F1}, E_{F2}, \ldots, E_{Fl}\}, \tag{15}$$

$$l + m = n. \tag{16}$$

The threshold T_h was then generated as follows:

$$T_h = \max\{E_{Ti} | E_{Ti} \in \boldsymbol{T} \cap E_{Ti} < min\{E_{Fj} | E_{Fj} \in \boldsymbol{F}\}\}. \tag{17}$$

The threshold represents the maximal evaluation in the correctly recognised results, and is smaller than the minimal evaluation of the incorrectly recognised results.

As stated before, the optimisation target and corresponding approach rely on the chosen algorithm for recognition. A contour matching method in HALCON library called shape-based matching was utilised as the recognition algorithm in this study.

The target parameter of the shape-based matching algorithm to be designed is the contrast value to extract contour models from the templates. Figure 3 illustrates two contour models extracted from different contrast values. A too large contrast value decreases the number of contours in the obtained model to a great extent. Matching with less contours therefore yields more possible candidates, and finally results in longer matching time. If the contour model is decreased to just a short line, the matching time can be infinity.

Based on this principle, the design method for the contrast value in shape-based matching is set to traverse all the possible values from the minimum to the maximum and end if the detection number reaches a threshold (Figure 4).

Figure 3. Contour models (marked with red lines) made by different contrast values; the number of contours decrease as the contrast increases. Less contours in a model increase the number of detections that have to be matched and thus result in a longer time taken for matching.

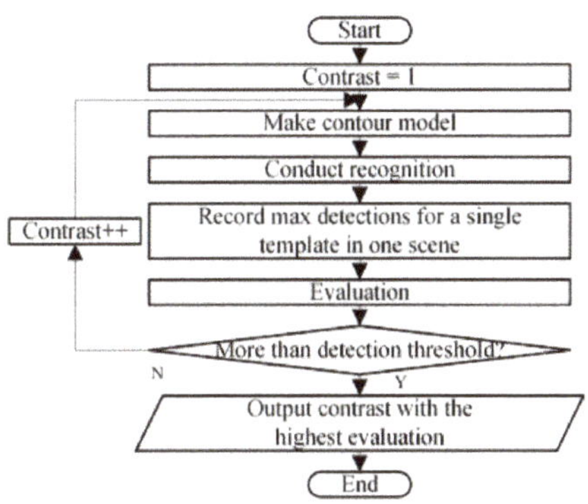

Figure 4. Contour models (marked with red lines) made by different contrast values, the number of contours decrease as the contrast increases. Less contours in a model increase the number of detections that should be matched and thus result in a longer time taken for matching. Algorithm proposed for contrast value design of the chosen recognition algorithm [HALCON (an image processing library of MVTec Company) shape-based matching].

4. Evaluation Experiment

4.1. Experimental Setup

The experimental environment was an industrial manipulator with six DoFs, a ring-shaped illumination device and an industrial monocular camera (Figure 5). The camera and illumination were mounted on the tips of the manipulator using a 3D-printed joint. The processor was an Intel Core i5-5300U@2.30 GHz.

To reflect potential applications, we chose the two sides of a semi-transparent plastic part (Figure 5) as the recognition target. Different from its side at the rear, the face side had a convex structure in the middle. The following constraints were applied: an $F_{measure}$ score no less than 1; a positional error no more than 3 mm; and an angular error no more than 5°.

Three scenes with different functions were arranged on a piece of black cloth below the manipulator (Figure 5). In order to prevent overfitting, two scenes were prepared for recognition. Every time the FOV or illumination changed, the templates were updated; a scene for updating the templates was therefore required. Scene 1 and Scene 2, with two face and two rear side objects in each, were set up for recognition. Scene 3, with a face side and a rear side object, was set up to create templates.

Two different plans for FOV were given to our system: one was to shoot once with a wide FOV at a camera distance of 158 mm, and the other was to shoot four times with narrow FOVs at a camera distance of 105 mm.

We controlled the colour channels utilised in both recognition and illumination, and created three experimental conditions. Recognitions were conducted with greyscale images in Condition I and II, while R-channel images were used in Condition III. The details are listed below. On the other hand, illuminations were changed from only G channel in Condition I, and changed from RGB three channels in Condition II and III. Details of the conditions based on which each experiment was conducted are listed in Table 2.

Table 2. Experimental conditions.

Condition	Illumination Channel(s)	Increment of Illumination Parameter(s)	Recognition Image(s)
I	G only	1	Greyscale
II	RGB	15	Greyscale
III	RGB	15	R-channel

The reason why G illumination was chosen in Condition I is that it is considered to influence the brightness in the obtained images to the greatest extent. Therefore, the dimension of illumination was reduced to a great extent, and the increment of illumination strength was set to 1 in Condition I.

For all aforementioned conditions, we manually measured the ground truth data. For the illumination variables, 16 local optimisation searches were performed. The maximum detection to end contrast value search was set to 4.

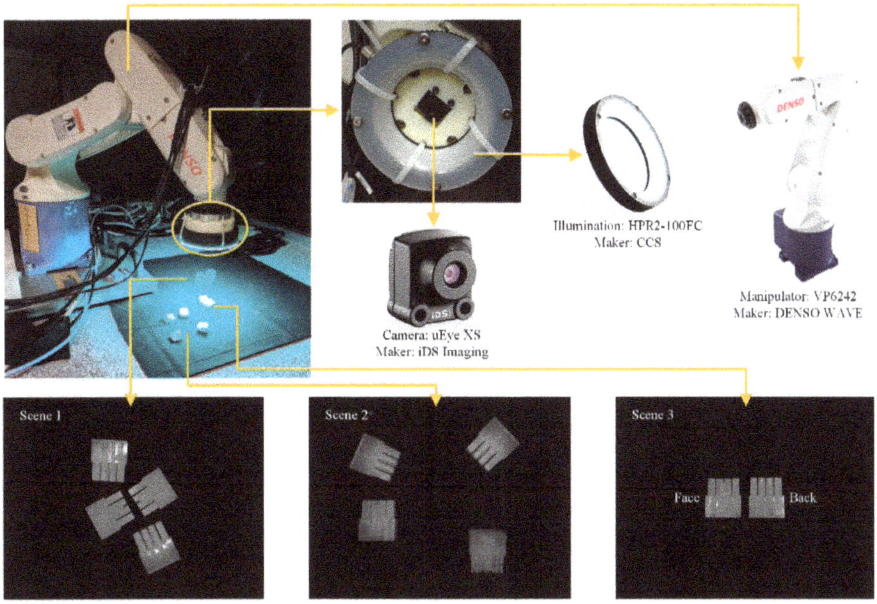

Figure 5. Experimental devices, scenes and recognition targets. Monocular camera and ring-shaped illumination were attached to the end effector of a six-DoF manipulator. Three scenes were prepared on a piece of black cloth; Scene 1 and Scene 2 were for recognition; Scene 3 was for making templates. Two sides of a semi-transparent plastic part (20 mm in length, 20 mm in width, and 8 mm in height) were chosen as the recognition targets.

4.2. Results

The best-three solutions and their evaluations of the three conditions are presented in Tables 3–5 and the images taken under the optimal parameter sets are shown in Figure 6.

Proper design could not be achieved with either with one shot or four shots when tuning illumination from only G component in Condition I. The optimal design was realised with a 79 in green illumination, four shots and a contrast value of 3, which provided a 0.93 $F_{measure}$, about 0.6 mm maximum positional error, and 2.1° maximum angular error.

The optimal design of Condition II corresponded to a (195, 120, 75) illumination RGB strength, four shoots and contour models generated by a contrast value of 4. This set up resulted in an $F_{measure}$ value of 1, a maximum positional error of about 0.6 mm, and a 3.1° maximum angular error.

Replacing the greyscale images with R-channel images, suitable designs were found only with one shoot. The optimal design was (195, 120, and 75) in illumination RGB, 11 in contrast value, and with one shoot. Its evaluation showed an $F_{measure}$ of 1, about 0.3 mm in maximum positional error, and 0.4° in maximum angular error.

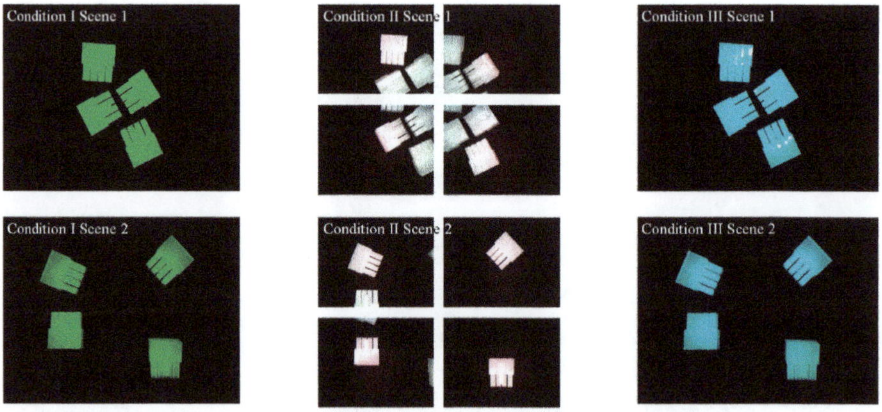

Figure 6. Optimal illumination and FOV conditions designed for the three conditions. Condition I: 1 shoot under strong green illumination. Condition II: four shoots under illumination with red component relatively higher. Condition III: 1 shoot under strong green and blue illumination.

Table 3. Best-three designs of Condition I.

Rank	R	G	B	FOV	Contrast	$F_{measure}$	Positional Error (mm)	Angular Error (°)
1	0	232	0	wide	1	0.93	0.64	2.1
2	0	79	0	narrow	3	0.93	0.58	3.4
3	0	84	0	narrow	3	0.86	0.35	3.6

The results were ranked by their evaluations; higher rank represents better evaluation. A wide FOV denotes one shot at a camera distance of 158 mm, and a narrow FOV denotes four shots at a camera distance of 105 mm.

Table 4. Best-three designs of Condition II.

Rank	R	G	B	FOV	Contrast	$F_{measure}$	Positional Error (mm)	Angular Error (°)
1	195	120	75	narrow	4	1.00	0.60	3.1
2	240	45	75	narrow	3	1.00	0.92	3.0
3	105	30	150	narrow	4	1.00	1.15	2.9

The results were ranked by their evaluations; higher rank represents better evaluation. A wide FOV denotes one shot at a camera distance of 158 mm, and a narrow FOV denotes four shots at a camera distance of 105 mm.

Table 5. Best-three designs of Condition III.

Rank	R	G	B	FOV	Contrast	$F_{measure}$	Positional Error (mm)	Angular Error (°)
1	15	225	240	wide	11	1.00	0.32	0.4
2	0	225	150	wide	11	1.00	0.50	0.4
3	45	225	210	wide	9	1.00	0.62	0.4

The results were ranked by their evaluations; higher rank represents better evaluation. A wide FOV denotes one shot at a camera distance of 158 mm, and a narrow FOV denotes four shots at a camera distance of 105 mm.

5. Discussion

Generally speaking, designs under the accuracy constraints, that is, an $F_{measure}$ of 1 and no more than 3 mm and 5° in positional and angular errors were found in both conditions of illumination tuned from RGB channels. This finding proved that our system is capable of tuning parameters for a vision system used in pick-and-place tasks.

Comparing the results of the first two conditions, designs under accuracy constraints were found when illumination was tuned from RGB, while no proper design was found with an $F_{measure}$ of 1 with illumination tuned only from G. In both conditions, the input images were of the greyscale type, which indicated that although a vision system finally converts colour images into grey, it is still essential to tune the illumination based on the three RGB channels.

On the other hand, from the results of Condition II and Condition III, it was found that using R-channel images could provide better performance in recognition than greyscale ones. Designs with one shot and a wide FOV were found in Condition III, while a narrow FOV was designed with four shoots in Condition II.

In order to further discuss the effects of R-channel images, the R-channel images for the two scenes under the optimal design of Condition III (illumination RGB equals to 15, 225, and 240, 1 shoot) were extracted. We processed the two figures with greyscale; both R-channel and greyscale figures are shown in Figure 7. Moreover, to confirm that the R-channel images perform better than greyscale images under the same situation, an additional design was implemented with greyscale images, as shown in Figure 7. Results showed that the optimal design with the greyscale images could only provide an $F_{measure}$ of 0.93.

To our human eye, it is obvious that the greyscale images are easier to recognise. However, in a vision system, R-channel images are recognised with a higher recognition accuracy. The probable reason might be that in a sufficiently bright image, the noise is also enlarged to a great extent. Vision systems do not detect a picture as humans do; these systems read the limited features in the form of mathematical values in matrixes instead. When the noise is so large that it obscures the useful information indicated in these matrixes, judgements made by the system could be flawed. From this point of view, the key to a 'clear' image for vision systems is that these images must contain little but important information. As an example, though the R-channel images in Figure 7 were really dark, the contours of each object could still be seen clearly. The great contrast between contours and background therefore make the images 'clear'. In some ways, image pre-processing is just a method to serve the vision systems with 'clearer' images.

Moreover, illumination in the designs with high evaluations showed no relations to each other with greyscale images input, while a clear pattern was discovered in the circumstance of R-channel images. Based on Table 4, illuminations of the best-three designs were found with low red illumination (under 50), high green illumination (near 225), and relatively high blue illumination (from 150 to 240). Generally speaking, tuning green and blue illuminations is not effective when the image can only be seen using a red channel. However, Figure 8 shows that even with no red illumination, the objects are visible in the R-channel image. The probable reason may be that the RGB tuned from the illumination side is not the same as the RGB information contained in an image. Because of the wavelength of the illumination device or some reflections, the G and B components could still influence the R-channel image to some extent. Actually, such an influence eventually resulted in 'clearer' R-channel images compared with the greyscale ones. The illumination pattern found in Condition III also confirmed the importance of green and blue illumination. In addition, patterns of illumination indicated that relations might exist between the recognition performance and its illumination conditions, which give rise to possibilities for the application of other optimisation methods.

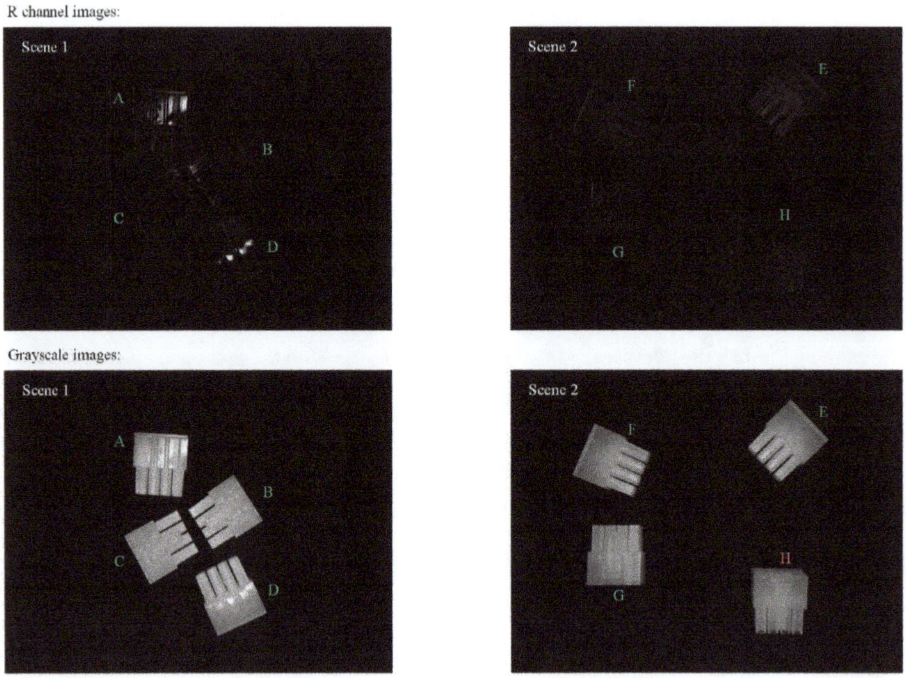

Figure 7. R-channel images, greyscale images for the two scenes under the optimal design of Condition III [illumination RGB (red, green and blue) equals to 15, 225 and 240, with a wide FOV], and their recognition results. Objects labelled in green were those that could be correctly recognised, while the red one could not be recognised.

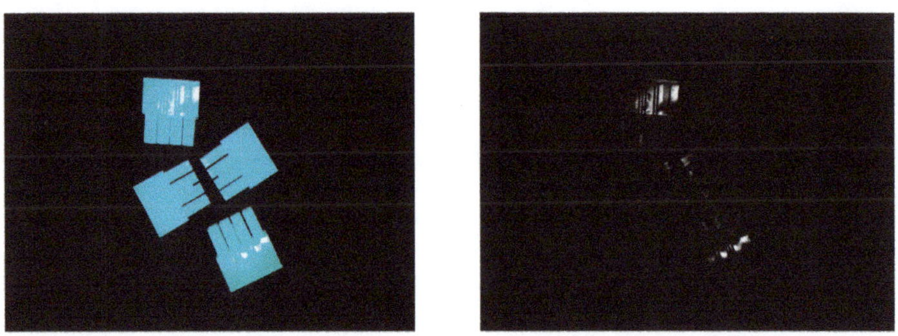

Figure 8. An image taken with no red component in illumination (illumination RGB were set to 0, 225, and 255, respectively), and its R-channel image.

Nevertheless, the experiment was limited under the environment we prepared. We could only state that R-channel image could provide better recognition accuracy under the experimental settings. We cannot affirm that whether this phenomenon could be discovered with other recognition targets, or by recognition with other algorithms. To better explain it, further experiments will be required.

6. Conclusions

In this study, we proposed an automated design approach for vision systems in pick-and-place tasks. The vision system design was first formulated as a parameter optimisation problem and then solved in an experiment-based approach with a hierarchical algorithm. Rather than seeking a suitable parameter set randomly in the solution space, the proposed algorithm separates and sets hierarchies for each optimisation based on the design factors. As one of the uncertainties from the real world, the influence of colour on the recognition performance of the designed vision systems was also investigated through experiments in this research.

It could be seen through the experiments that the proposed system was able to design a vision system with a 100% recognition rate, and a positional and angular error of 0.32 mm and 0.4°, respectively. When using greyscale images for recognition, G illumination resulted in an $F_{measure}$ of only 0.93, which proved the necessity for colourful illumination. Consequently, when RGB illumination was used, designs with R-channel images used only one shot, which indicates that R-channel images provide better recognition accuracy than the greyscale ones.

In future work, from the viewpoint of robustness, it is necessary to improve the prevention against overfitting by increasing the number of scenes for recognition and include the measurement of overfitting in the evaluation of the designed vision system. Aiming to take out better solutions, the selection of recognition algorithm should also be included into the design process. Additionally, further research could be conducted on searching more appropriate optimisation methods, for example, neural networks or genetic algorithms, to provide better solutions that are less time-consuming for the vision system design problem.

Author Contributions: Y.C. and T.O. conceived and designed the study; Y.C. performed the experiments; Y.C. analysed the data; T.U. and T.T. contributed experimental platform; Y.C. wrote the manuscript; J.O. and T.O. reviewed the manuscript.

Conflicts of Interest: The authors declare no conflict of interest.

References

1. Newman, T.S.; Jain, A.K. A survey of automated visual inspection. *Comput. Vis. Image Underst.* **1995**, *61*, 231–262. [CrossRef]
2. Golnabi, H.; Asadpour, A. Design and application of industrial machine vision systems. *Robot. Comput. Integr. Manuf.* **2007**, *23*, 630–637. [CrossRef]
3. Malamas, E.N.; Petrakis, E.G.M.; Zervakis, M.; Petit, L.; Legat, J.D. A survey on industrial vision systems, applications and tools. *Image Vis. Comput.* **2003**, *21*, 171–188. [CrossRef]
4. Cowan, C.K.; Kovesi, P.D. Automatic sensor placement from vision task requirements. *IEEE Trans. Pattern Anal. Mach. Intell.* **1988**, *10*, 407–416. [CrossRef]
5. Tarabanis, K.; Tsai, R.Y.; Allen, P.K. Automated sensor planning for robotic vision tasks. In Proceedings of the IEEE International Conference on Robotics and Automation, Sacramento, CA, USA, 9–11 April 1991; pp. 76–82.
6. Tarabanis, K.; Tsai, R.Y. Computing viewpoints that satisfy optical constraints. In Proceedings of the Computer Vision and Pattern Recognition, Maui, HI, USA, 3–6 June 1991; pp. 152–158.
7. Hutchinson, S.A.; Kak, A.C. Planning sensing strategies in a robot work cell with multi-sensor capabilities. *IEEE Trans. Robot. Autom.* **1989**, *5*, 407–416. [CrossRef]
8. Cameron, A.; Wu, H.L. Identifying and localizing electrical components: A case study of adaptive goal-directed sensing. In Proceedings of the IEEE International Symposium on Intelligent Control, Arlington, VA, USA, 13–15 August 1991; pp. 495–500.
9. Thuilot, B.; Martinet, P.; Cordesses, L.; Gallice, J. Position based visual servoing: Keeping the object in the field of vision. In Proceedings of the IEEE International Conference on Robotics and Automation, Washington, DC, USA, 11–15 May 2002.
10. Rahimian, P.; Kearney, K.J. Optimal camera placement for motion capture systems. *IEEE Trans. Vis. Comput. Graph.* **2017**, *3*, 1209–1221. [CrossRef] [PubMed]

11. Ito, A.; Tsujiuchi, N.; Okada, Y. Multipurpose optimization of camera placement and application to random bin-picking. In Proceedings of the 41st Annual Conference of the Industrial Electronics Society, Yokohama, Japan, 9–12 November 2015; pp. 528–533.
12. Foix, S.; Alenyà, G.; Torras, C. 3D Sensor planning framework for leaf probing. In Proceedings of the Intelligent Robots and Systems (IROS), Hamburg, Germany, 28 September–2 October 2015; pp. 6501–6506.
13. Murase, H.; Nayar, S.K. Illumination planning for object recognition using parametric eigenspaces. *IEEE Trans. Pattern Anal. Mach. Intell.* **1994**, *6*, 1219–1227. [CrossRef]
14. Pfeifer, T.; Wiegers, L. Reliable tool wear monitoring by optimized image and illumination control in machine vision. *Measurement* **2000**, *28*, 209–218. [CrossRef]
15. Yi, S.; Haralick, R.M.; Shapiro, L.G. Automatic sensor and light source positioning for machine vision. In Proceedings of the 10th International Conference on Pattern Recognition, Atlantic City, NJ, USA, 16–21 June 1990; pp. 55–59.
16. Eltoft, T.; deFigueiredo, R.J.P. Illumination control as a means of enhancing image features in active vision systems. *IEEE Trans. Pattern Anal. Mach. Intell.* **1995**, *4*, 1520–1530. [CrossRef] [PubMed]
17. Slusallek, P.; Seidel, H.P. Vision—An architecture for global illumination calculations. *IEEE Trans. Vis. Comput. Graph.* **1995**, *1*, 77–96. [CrossRef]
18. Cang, N.Y.C.; Wu, C.C. Automatic optimal lighting adjustment and control for vision recognition. In Proceedings of the 14th IFToMM World Congress, Taipei, Taiwan, 25–30 October 2015.
19. Aoki, S.; Nagao, T. Automatic construction of tree-structural image transformation using genetic programming. In Proceedings of the IEEE 10th International Conference on Image Analysis and Processing, Venice, Italy, 27–29 September 1999.
20. Shirakawa, S.; Nagao, T. Genetic Image Network (GIN): Automatically construction of image processing algorithm. In Proceedings of the International Workshop on Advanced Image Technology (IWAIT), Bangkok, Thailand, 8–9 January 2007.
21. Bai, H.; Yata, N.; Nagao, T. Automatic finding of optimal image processing for extracting concrete image cracks using features ACTIT. *IEEJ Trans. Electr. Electron.* **2012**, *7*, 308–315. [CrossRef]
22. Lillywhite, K.; Tippetts, B.; Lee, D. Self-tuned evolution-constructed features for general object recognition. *Pattern Recognit.* **2012**, *45*, 241–251. [CrossRef]
23. Lillywhite, K.; Lee, D.; Tippetts, B.; Archibald, J. A feature construction method for general object recognition. *Pattern Recognit.* **2013**, *46*, 3300–3314. [CrossRef]
24. Kumar, R.; Lal, S.; Kumar, S.; Chand, P. Object detection and recognition for a pick and place robot. In Proceedings of the 2014 Asia-Pacific World Congress on Computer Science and Engineering (APWC on CSE), Nadi, Fiji, 4–5 November 2014.
25. Ogata, T.; Tsujimoto, K.; Yukisawa, T.; Huang, Y.J.; Arai, T.; Ueyama, T.; Takada, T.; Ota, J. Automated design of image recognition process for picking system. *Int. J. Autom. Technol.* **2016**, *10*, 737–752. [CrossRef]
26. Ogata, T.; Yukisawa, T.; Arai, T.; Ueyama, T.; Takada, T.; Ota, J. Automated design of image recognition in capturing environment. *IEEJ Trans. Electr. Electron.* **2017**, *12*, S49–S55. [CrossRef]
27. Chen, Y.; Ogata, T.; Ueyama, T.; Takada, T.; Ota, J. Automated design of the field-of-view, illumination, and image pre-processing parameters of an image recognition system. In Proceedings of the 13th IEEE Conference on Automation Science and Engineering (CASE), Xi'an, China, 20–23 August 2017; pp. 1079–1084.
28. Gevers, T.; Smeulders, A.W.M. Color-based object recognition. *Pattern Recognit.* **1999**, *32*, 453–464. [CrossRef]
29. Drew, M.S.; Wei, J.; Li, Z.N. Illumination-invariant color object recognition via compressed chromaticity histograms of color-channel-normalized images. In Proceedings of the 6th International Conference on Computer Vision, Bombay, India, 4–7 January 1998; pp. 533–540.
30. Alferez, R.; Wang, Y.F. Geometric and illumination invariants for object recognition. *IEEE Trans. Pattern Anal. Mach. Intell.* **1999**, *21*, 505–536. [CrossRef]
31. Diplaros, A.; Gevers, T.; Patras, I. Combining color and shape information for illumination-viewpoint invariant object recognition. *IEEE Trans. Image Process.* **2006**, *15*, 1–11. [CrossRef] [PubMed]
32. Bala, R.; Eschbach, R. Spatial color-to-grayscale transform preserving chrominance edge information. In Proceedings of the 12th Color and Imaging Conference, Scottsdale, AZ, USA, 9–12 November 2004; pp. 82–86.
33. Grundland, M.; Dodgson, N. Decolorize: Fast, contrast enhancing, color to grayscale conversion. *Pattern Recognit.* **2007**, *40*, 2891–2896. [CrossRef]

34. Gooch, A.A.; Olsen, S.C.; Tumblin, J.; Gooch, B. Color2Gray: Salience-preserving color removal. *ACM Trans. Graph.* **2005**, *24*, 634–639. [CrossRef]
35. Kanan, C.; Cottrell, G.W. Color-to-grayscale: Does the method matter in image recognition? *PLoS ONE* **2012**, *7*, e29740. [CrossRef] [PubMed]

sensors

MDPI

Article

A Convenient Calibration Method for LRF-Camera Combination Systems Based on a Checkerboard

Zhuang Zhang [1,2,†], Rujin Zhao [1,*,†], Enhai Liu [1], Kun Yan [1,2] and Yuebo Ma [1,2]

[1] Institute of Optics and Electronics of Chinese Academy of Sciences, Chengdu 610209, China; zhangzhuang14@mails.ucas.ac.cn (Z.Z.); leh@ioe.ac.cn (E.L.); yankunioe@163.com (K.Y.); MYB_IOE@163.com (Y.M.)

[2] University of Chinese Academy of Sciences, Beijing 100149, China

* Correspondence: zrj0515@163.com; Tel.: +86-155-3000-3288

† These authors contributed equally to this work.

Received: 6 February 2019; Accepted: 11 March 2019; Published: 15 March 2019

Abstract: In this paper, a simple and easy high-precision calibration method is proposed for the LRF-camera combined measurement system which is widely used at present. This method can be applied not only to mainstream 2D and 3D LRF-cameras, but also to calibrate newly developed 1D LRF-camera combined systems. It only needs a calibration board to record at least three sets of data. First, the camera parameters and distortion coefficients are decoupled by the distortion center. Then, the spatial coordinates of laser spots are solved using line and plane constraints, and the estimation of LRF-camera extrinsic parameters is realized. In addition, we establish a cost function for optimizing the system. Finally, the calibration accuracy and characteristics of the method are analyzed through simulation experiments, and the validity of the method is verified through the calibration of a real system.

Keywords: LRF; camera calibration; extrinsic calibration; sensors combination

1. Introduction

In the field of measurements, a single sensor is seldom able to perform high-precision measurements by itself. Combined multi-sensor measurement schemes can effectively combine the characteristics of each sensor, leveraging the complementary advantages of sensors, and improving the accuracy and robustness of the measurement system. As [1] shows, laser range finders (LRFs) provide high-precision distance information, while camera can provide rich image information. The combination of LRFs and cameras has attracted wide attention, with interesting applications in navigation [2], human detection [3] and 3D texture reconstruction [4].

Compared with the current mainstream schemes combining scanning lasers and vision, the more challenging combination of 1-D laser ranging and vision has attracted the attention of researchers due to its low cost and wide applicability. The Shuttle Radar Topography Mission (SRTM) [5] realizes high-precision measurements of Interferometric Synthetic Aperture Radar (IFSAR) on long-range cooperative targets. Ordez [6] proposed a combination of camera and Laser Distance Meter (LDM) to estimate the length of a line segment in an unknown plane. Wu [7] applied this method to a visual odometry (VO) system and realized the application in a quasi-plane scene. In our previous work, we further extended this method and constructed a complete SLAM method based on laser-vision fusion [1].

Sensor calibration is the premise of data fusion, including the calibration of each sensor's own parameters and the relationship of relative data between each sensor [8]. However, as a necessary prerequisite for high-precision measurements, the calibration technology of 1D laser-camera systems evolves seldom. The existing calibration algorithm based on scanning laser ranging has been unable to apply, but the traditional one-dimensional laser calibration algorithms require a high-precision manipulator laser interferometer and other complex equipment.

In this paper, a simple and feasible high-precision laser and visual calibration algorithm is proposed, which can calibrate the parameters of laser and camera sensors through only simple data processing. Firstly, the camera parameters and distortion coefficients are determined using a non-iterative method. Then, the coordinates of the laser spot in the camera coordinate system are obtained by inversion of the laser image points in the image, and the initial values of external camera and laser ranging parameters are estimated. Finally, the parameters are optimized through the parameterization of the rotation matrix [9] and the Gröbner basis method [10]. Compared with the existing methods, the main contributions of this paper are as follows:

(1) The method proposed in this paper has wider applicability. It can be used for joint calibration of vision sensors and LRF from 1D to 3D.
(2) Compared with existing 1D laser-vision calibration methods, the proposed method can be realized using a simple chessboard lattice, without complicated customized targets and high-precision mechanical structures.
(3) The accuracy and usability of the proposed method are verified by simulation and observation experiments.

This paper is organized as follows: the existing methods related to our work are outlined in the following section. Sections 3 and 4 describe the mathematical model and illustrate the proposed algorithm. In Section 5, we evaluate the solution of the simulation and observation experiments. Finally, conclusions and future are provided in Section 6.

2. Related Work

For the extrinsic parameters between LRF and vision sensors, it is helpful to combine the high-precision distance information of laser ranging with the high lateral resolution of vision to achieve high-precision pose estimation. However, this method is mostly used to calibrate 2D or 3D LRFs and cameras.

Vasconcelos [11] calibrated the camera-laser extrinsic parameters by moving a checkerboard freely. This method assumes that the internal parameters are known and accurate, and converts the external parameter calibration problem into a plane-coplanar alignment problem to reach an exact solution. Similar work includes Scaramuzza [12] and Ha [13]. Ranjith [14] realized the correlation and calibration of 3D LiDAR data and image data through feature point retrieval. Zhang [15] uses mobile LRF and visual camera to achieve self-calibration of their external parameters through motion constraints. Viejo [16] realized the correlation of the two sets of data by arranging control points, and calibrated the external parameters of 3D LiDAR and a monocular camera.

However, the above algorithms are mostly used to calibrate the external parameters of 2D or 3D scanning laser and vision systems, and cannot be used for 1D laser ranging without a scanning mechanism due to the lack of constraints. For the calibration of 1D LRF, the traditional method mostly realizes the correlation between the two by means of complex a manipulator or specific calibration target. For example, Zhu's [17] calibration algorithm is used to calibrate the direction and position parameters of a laser range finder based on spherical fitting. The calibration accuracy is high, but the solution is highly customized and not universal. Lu [18] designed a multi-directional calibration block to calibrate the laser beam direction of a point laser probe on the platform of a coordinate measuring machine. Zhou [19] proposed a new calibration algorithm for serial coordinate measuring machines (CMMs) with cylindrical and conical surfaces as calibration objects. Similar calibration methods are used in the implementation of the LFR-camera slam method [1]. The relative rotation and translation of the two sensors' coordinate systems are estimated through a high-precision laser tracker.

Although this method can achieve high accuracy, it requires the installation of sensors on precision measuring equipment, which has high calibration cost and complex operation, and cannot meet the needs of low-cost and fast landing scenarios such as existing robots. In 2010, Ordez [6] proposed a set method of cameras and LRF to measure short distances in the plane. In another study [20], the author introduces a preliminary calibration method for a digital camera and a laser rangefinder. The experiment involves the artificial adjustment of the projection center of the laser pointer, and only two laser projections are used. The accuracy and robustness of the calibration method are both problematic. After that, Wu et al. [7] proposed a two-part calibration method based on the Ransac scheme, and solved the corresponding linear equation in the image by creating the index table of laser spot. However, this method cannot be well applied to the case where the laser light is close to the optical axis of the camera, and the final accuracy evaluation criteria are not given.

Zhang [21] proposed a simple calibration method for camera intrinsic parameters, where the parameters were determined using a non-linear method, and high accuracy was achieved. Afterwards, based on Zhang's framework, researchers improved accuracy and scene expansion by designing different forms of targets [22–24] and improving the calibration of the algorithm [25–27]. Hartly [28] introduced the distortion division model to correct the imaging distortion. On this basis, Hong [29] further explored the calibration method of large distortion cameras.

Currently, the calibration of omnidirectional cameras has attracted wide attention in order to improve the user's degree of freedom and immersion in the virtual reality and autopilot. Li et al. [30] proposed a multi-phase camera calibration scheme based on random pattern calibration board. Their method supports the calibration of a camera system which comprise normal pinhole cameras. Gwon Hwan [31] proposed a new intrinsic calibration and extrinsic calibration method of omnidirectional cameras based on the Aruco marker and a Charuco board. The calibration structure and method can solve the problem of suing overly complicated procedures to accurately calibrate multiple cameras.

At the same time, the calibration board also plays an important role in the other calibration processs. Liu [32] studied different applications of lasers and cameras. The calibration method of multiple non-common-view cameras by scanning a laser rangefinder is proposed. In the literature, the correlation between laser distance information and camera images is established through a specific calibration plate, so as to realize the relative pose estimation between cameras. Inspired by Liu's work [32], we establish a constraint of 1D laser and monocular vision by combining planar and coplanar constraints, so as to determine related external parameters. Considering that the camera imaging model has a direct impact on the calibration accuracy, we have improved Zhang's method [21] used in camera calibration by replacing the traditional polynomial model with the division distortion model, and solved the linear solution of the iterative optimization using variable least squares on the basis of Hartly [28] and Hong [29]. Thus, the problem of falling into local optimal solutions is avoided, and the calibration speed is greatly improved. Combining the above innovations, a convenient method for calibrating the parameters of the camera-laser measurement system is realized, which can complete the calibration of measurement systems, including camera internal parameters, distortion coefficients and camera-laser external parameters, in one operation.

3. Measurement Model

Previous researchers established relatively mature camera imaging and laser measurement models. We integrate the two mathematical models and construct a complete mathematical description of the coordinate system.

As shown in the Figure 1, O_C is the camera coordinate system, $\overrightarrow{O_C Z_C}$ is the optical axis direction of the camera, $O-uv$ is the image plane of the camera, O_T is the coordinate system of the target itself, point P_l is the spatial position of the laser spot, point P_w is the spatial coordinate of the target control point and O_l represents the coordinate system of 1D laser ranging. We set the camera coordinate system O_c as the measurement coordinate system O_M of the system. In the next part, we introduce the imaging model of the monocular camera and the 1D laser ranging model, and convert and fuse the data through extrinsic parameters $\begin{bmatrix} R_{l2C} & T_{l2C} \end{bmatrix}$.

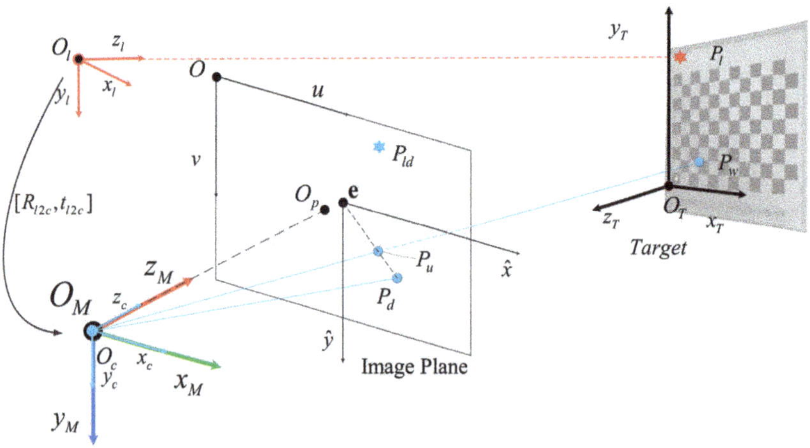

Figure 1. Measurement Model.

3.1. Camera Imaging Model

In order to describe the imaging process of a monocular camera more accurately, we combine the lens distortion model with the aperture imaging model and introduce the shift of the distortion center **e** relative to the image center O_P [28]. In the camera coordinate system, $O_p - xy$ is the physical coordinate system of the phase plane and $O - uv$ represents the image coordinate system. Image center Op denotes the intersection of the optical axis and the image plane.

The ideal imaging process can be described as the process of transforming a point P_i^T ($\mathbf{X}_i^T = \begin{bmatrix} X_i^T & Y_i^T & Z_i^T & 1 \end{bmatrix}^T$) in the world coordinate system to the image plane imaging point P_i^u ($\mathbf{x}_i^u = \begin{bmatrix} u_i^u & v_i^u & 1 \end{bmatrix}^T$) through a projection relationship. The mathematical expression is as follows:

$$\rho_i \mathbf{x}_i^u = \mathbf{A}_C \mathbf{T}_{T2C} \mathbf{X}_i^T = \begin{bmatrix} f_u & s & u_0 \\ 0 & f_v & v_0 \\ 0 & 0 & 1 \end{bmatrix} \begin{bmatrix} r_{11} & r_{12} & r_{13} & t_1 \\ r_{21} & r_{22} & r_{23} & t_2 \\ r_{31} & r_{32} & r_{33} & t_3 \end{bmatrix} \begin{bmatrix} X_i^T \\ Y_i^T \\ Z_i^T \\ 1 \end{bmatrix} \tag{1}$$

where ρ_i is a named depth scale factor, the intrinsic matrix and $\mathbf{A}_C$ is described by a five-parameter model; f_u, f_v are the focal lengths, $\begin{bmatrix} u_0 & v_0 \end{bmatrix}^T$ is the coordinate of the image center O_P, and s is the skew coefficient. $\mathbf{T}_{T2C}$ is the transformation matrix relating O_T to O_C and it can be expressed as a rotation matrix $\mathbf{R}_{T2C}$ combined with the translation vector $\mathbf{t}_{T2C}$.

Due to lens design and processing, the actual imaging process is distorted. We introduce a division distortion model to improve our imaging. The mathematical expressions are as follows:

$$\mathbf{x}_i^u - \mathbf{e} = \frac{\mathbf{x}_i^d - \mathbf{e}}{1 + \lambda_1 \left[r_i^d\right]^2 + \lambda_2 \left[r_i^d\right]^4 + \ldots} \tag{2}$$

$\mathbf{x}_i^d$ represents the actual position of projection point P_i^T, and its coordinates are $\mathbf{x}_i^d = \begin{bmatrix} u_i^d & v_i^d & 1 \end{bmatrix}^T$; λ_1 and λ_2 are the distortion coefficients and r_d represents the distance from point $\mathbf{x}_i^d$ to the distortion center **e**, expressed as $r_i^d = \sqrt{(u_i^d - du_0)^2 + (v_i^d - dv_0)^2}$.

In order to illustrate the method more clearly, the most important parameters used in this paper and their meaning are shown in Table 1.

Table 1. The parameter statement of the system.

	Parameter	Mean
Coordinats	$O_M/O_T/O_C/O_l$	measurement/target/camera/laser-ranging coordinate system
	$O-uv$	image plane coordinate system of camera
	$\mathbf{T}_{T2C}$	transformation matrix relating O_T to O_C
	$\mathbf{R}_{l2M}, \mathbf{t}_{l2M}$	Extrinsic matrix of O_l to O_C
	$\mathbf{R}_{T2C}, \mathbf{t}_{T2C}$	Extrinsic matrix of O_T to O_C
	$\mathbf{T}_{e2\hat{e}}$	transformation matrix relating the distortion center $\mathbf{e}$ to new distortion center $\hat{\mathbf{e}}$
Imaging Geometry	$\mathbf{A}_C$	intrinsic matrix
	λ_1, λ_2	distortion coefficients
	$\mathbf{e}$	distortion center, expressed as $\begin{bmatrix} du_0 & dv_0 \end{bmatrix}^T$
	ρ_i	depth scale factor
	$\mathbf{H}$	homography matrix
	$\hat{\mathbf{H}}$	transformed homography matrix $\mathbf{H}$
	$\mathbf{F_H}$	fundamental matrix of distortion
	$\hat{\mathbf{F}}_\mathbf{H}$	transformed fundamental matrix of distortion
Variable	$\mathbf{X}_i^T$	position of the corner in the target coordinate system,
	$\mathbf{x}_i^u$	ideal position of projection point in image plane
	$\mathbf{x}_i^d$	actual position of projection point in image plane
	$\hat{\mathbf{x}}_i^d$	transformed image coordinates $\mathbf{x}_i^d$
	$\mathbb{Q}(X^C, Y^C, Z^C)/\mathbb{Q}(\ \mathbf{R}_{T2M},\ \ \mathbf{t}_{T2M}\ )$	plane equation of the target in camera coordinate system
	$\mathbb{C}(X^C, Y^C, Z^C)$	linear equation of the laser beam in the camera coordinate system
	$E(\mathbf{A}_C, \lambda_1, \lambda_2, \mathbf{R}_{T2M}, \mathbf{t}_{T2M})$	objective functions to be optimized
	E_{dr}	re-projection error

3.2. LRF Model

The mathematical model of the 1D laser ranging module is relatively simple. The laser ranging module can output single point laser distance information by observing the reflected signal and calculating the optical path using image coherence [33]. The mathematical determination of the origin coordinate and laser direction of the laser ranging module allows the coordinate of the laser in the measurement coordinate system. In order to better represent the measurement results in the system measurement coordinate system O_M, we set up the European three-dimensional coordinate system O_l for the LRF module. The laser emission direction is $\overrightarrow{O_lZ_l}$, the directions of $\overrightarrow{O_lX_l}$ are perpendicular and parallel to the $O_C - xy$ plane, and the directions of $\overrightarrow{O_lY_l}$ are determined by the right-hand rule, as shown in Figure 1. The measured distance information d_i^l represents the distance from the origin O_l to the laser spot P_i^l.

In the process of extrinsic parameter calibration, the coordinate origin O_l of the laser ranging coordinate system and the laser emission direction $\overrightarrow{O_lZ_l}$ need to be calculated. Finally, the conversion relations between camera measurement system O_M and the LRF coordinate system O_l are estimated, the rotation matrix $\mathbf{R}_{l2M}$ and the translation vector $\mathbf{t}_{l2M}$ are determined.

4. Methodology

Calibration of the measurement system is the process of determining the model parameters of the measurement system. For our system, through the measurement and imaging of a specific target, the model parameters of the measurement system are determined using the corresponding relationship between the coordinates of the control points and the image coordinates. The main parameters are the intrinsic parameters of the camera and the extrinsic parameters of between LRF and camera.

The specific calibration process is divided into three main steps: (1) the estimation of the camera distortion center $\mathbf{e}$; (2) the intrinsic parameters $\mathbf{A}_C$ and distortion coefficients $\lambda_1\lambda_2$ are decoupled and determined independently; (3) finding the extrinsic parameters $\begin{bmatrix} \mathbf{R}_{l2M} & \mathbf{t}_{l2M} \end{bmatrix}$ for translating the laser-vision coordinate system to the measurement coordinate system; (4) determining the optimal solution $\left(\mathbf{A}_C, \lambda_1, \lambda_2, \begin{bmatrix} \mathbf{R}_{I2M} & \mathbf{t}_{I2M} \end{bmatrix}\right)$ using the Gröbner basis method. In this section, we elaborate on the above.

4.1. The Center of Distortion

In many studies, it is usually assumed that the distortion center and the main point are in the same position, but Hartley [28] determined experimentally that there is a certain deviation between them. During the calibration process, we use a checkerboard as the calibration object, and extract the corners P_i^T of the checkerboard as the control points for camera calibration. Since the corners are distributed on a plane, we set the $Z_i^T = 0$ in the target coordinate system, in which case the imaging model can be expressed as:

$$\rho_i \mathbf{x}_i^u = \mathbf{P}\mathbf{X}_i^T = \mathbf{A}\begin{bmatrix} \mathbf{r}_1 & \mathbf{r}_2 & \mathbf{r}_3 & \mathbf{t}_1 \end{bmatrix}\begin{bmatrix} X_i^T \\ Y_i^T \\ 0 \\ 1 \end{bmatrix} \tag{3}$$

where $\begin{bmatrix} \mathbf{r}_1 & \mathbf{r}_2 & \mathbf{r}_3 \end{bmatrix}$ is the column vector of rotation matrix $\mathbf{R}_{T2C}$. The above equation can be simplified as:

$$\rho_i \mathbf{x}_i^u = \mathbf{H}\mathbf{X}_i^T = \mathbf{A}_C\begin{bmatrix} \mathbf{r}_1 & \mathbf{r}_2 & \mathbf{t} \end{bmatrix}\begin{bmatrix} X_i^T \\ Y_i^T \\ 1 \end{bmatrix} \tag{4}$$

Matrix $\mathbf{H}$ called the homography matrix, and expresses the mapping relation between the corner of the checkerboard and the image points. The coordinates of P_i^T are abbreviated as $\mathbf{X}_i^T = \begin{bmatrix} X_i^T & Y_i^T & 1 \end{bmatrix}^T$.

From the division model of Equation (2), we obtain:

$$\mathbf{x}_i^d = \mathbf{e} + k_i(\mathbf{x}_i^u - \mathbf{e}), k_i = 1 + \lambda_1\left[r_i^d\right]^2 + \lambda_2\left[r_i^d\right]^4 + \ldots \tag{5}$$

We multiply the left side of the equations by $[\mathbf{e}]_\times$ and combine it with Equation (4). In consideration of $[\mathbf{e}]_\times \mathbf{e} = 0$:

$$[\mathbf{e}]_\times \mathbf{x}_i^d = k_i[\mathbf{e}]_\times \mathbf{H}\mathbf{X}_i^T, [\mathbf{e}]_\times = \begin{bmatrix} 0 & -1 & dv_0 \\ 1 & 0 & -du_0 \\ -dv_0 & du_0 & 0 \end{bmatrix} \tag{6}$$

We then multiply the left sides of the equations by $\left[\mathbf{x}_i^d\right]^T$ and obtain:

$$\left[\mathbf{x}_i^d\right]^T [\mathbf{e}]_\times \mathbf{H}\mathbf{X}_i^T = 0 \tag{7}$$

Let $\mathbf{F_H} = [\mathbf{e}]_\times \mathbf{H}$. $\mathbf{F_H}$ is called the fundamental matrix of distortion and is expressed as follows:

$$\left[\mathbf{x}_i^d\right]^T \mathbf{F_H}\mathbf{X}_i^T = 0,\ \mathbf{F_H} = \begin{bmatrix} F_{11} & F_{12} & F_{13} \\ F_{21} & F_{22} & F_{23} \\ F_{31} & F_{32} & F_{33} \end{bmatrix} \tag{8}$$

We can solve the values of the fundamental matrix $\mathbf{F_H}$ using 8 pairs of corresponding corner points. The equation can be formulated as:

$$\mathbf{A}\mathbf{f_H} = 0 \tag{9}$$

where:

$$\mathbf{A} = \begin{bmatrix} x_1^d X_1^T & x_1^d Y_1^T & x_1^d & y_1^d X_1^T & y_1^d Y_1^T & y_1^d & X_1^T & Y_1^T & 1 \\ \vdots & \vdots & \vdots & \vdots & \vdots & \vdots & \vdots & \vdots & \vdots \\ x_n^d X_n^T & x_n^d Y_n^T & x_n^d & y_n^d X_n^T & y_n^d Y_n^T & y_n^d & X_n^T & Y_n^T & 1 \end{bmatrix}$$
$$\mathbf{f_H} = \begin{bmatrix} F_{11} & F_{12} & F_{13} & F_{21} & F_{22} & F_{23} & F_{31} & F_{32} & F_{33} \end{bmatrix} \tag{10}$$

The corresponding equations are solvable using least square when the number of points is greater than 8 points. The corresponding distortion center $\mathbf{e}$ is the left null vector of $\mathbf{F_H}$:

$$\mathbf{e}^T[\mathbf{e}]_\times = 0 \Leftrightarrow \mathbf{e}^T\mathbf{F_H} = \mathbf{e}^T[\mathbf{e}]_\times\mathbf{H} = 0 \tag{11}$$

So far, we have obtained the image coordinates of the distorted center $\mathbf{e}$. The corresponding homography matrix $\mathbf{H}$ can be obtained using the fundamental matrix $\mathbf{F_H}$.

4.2. Decoupling Camera Parameters

If the image coordinate origin O_P is moved to the distortion center $\mathbf{e}$, the new distortion center after translation is expressed as $\hat{\mathbf{e}} = \begin{bmatrix} 0 & 0 & 1 \end{bmatrix}^T$. In the new coordinate system, Equation (8) is expressed as:

$$\begin{bmatrix} \hat{\mathbf{x}}_i^d \end{bmatrix}^T \hat{\mathbf{F}}_\mathbf{H}\mathbf{X}_i^T = 0 \tag{12}$$

where $\hat{\mathbf{x}}_i^d$ and $\hat{\mathbf{F}}_\mathbf{H}$ represent the transformed image coordinates $\mathbf{x}_i^d$ and the fundamental matrix $\mathbf{F_H}$. The transformation relationship is as follows:

$$\hat{\mathbf{x}}_i^d = \mathbf{T}_{\mathbf{e}2\hat{\mathbf{e}}}\mathbf{x}_i^d, \quad \hat{\mathbf{F}}_\mathbf{H} = \mathbf{T}_{\mathbf{e}2\hat{\mathbf{e}}}\mathbf{F_H}, \quad \text{where } \mathbf{T}_{\mathbf{e}2\hat{\mathbf{e}}} = \begin{bmatrix} 1 & 0 & -du_0 \\ 0 & 1 & -dv_0 \\ 0 & 0 & 1 \end{bmatrix} \tag{13}$$

From the definition of $\hat{\mathbf{F}}_\mathbf{H}$:

$$\hat{\mathbf{F}}_\mathbf{H} = [\mathbf{e}]_\times\hat{\mathbf{H}} = \begin{bmatrix} 0 & -1 & 0 \\ 1 & 0 & 0 \\ 0 & 0 & 0 \end{bmatrix}\hat{\mathbf{H}} \tag{14}$$

Let:

$$\hat{\mathbf{F}}_\mathbf{H} = \begin{bmatrix} \hat{\mathbf{F}}_1 \\ \hat{\mathbf{F}}_2 \\ \hat{\mathbf{F}}_3 \end{bmatrix} \begin{bmatrix} \hat{F}_{11} & \hat{F}_{12} & \hat{F}_{13} \\ \hat{F}_{21} & \hat{F}_{22} & \hat{F}_{23} \\ \hat{F}_{31} & \hat{F}_{32} & \hat{F}_{33} \end{bmatrix} \tag{15}$$

and:

$$\hat{\mathbf{H}} = \begin{bmatrix} \hat{\mathbf{H}}_1 \\ \hat{\mathbf{H}}_2 \\ \hat{\mathbf{H}}_3 \end{bmatrix} \begin{bmatrix} \hat{H}_{11} & \hat{H}_{12} & \hat{H}_{13} \\ \hat{F}_{21} & \hat{F}_{22} & \hat{H}_{23} \\ \hat{H}_{31} & \hat{H}_{32} & \hat{H}_{33} \end{bmatrix} \tag{16}$$

Equations (15) and (16) are then introduced into Equation (14):

$$\hat{\mathbf{H}}_1 = \hat{\mathbf{F}}_2, \hat{\mathbf{H}}_2 = -\hat{\mathbf{F}}_1 \tag{17}$$

So far, the first two rows $\hat{\mathbf{H}}_1\ \hat{\mathbf{H}}_2$ of the homography matrix have been obtained. Referring to Equations (2) and (4), the image distortion after translation can be expressed as follows:

$$\rho_i \frac{\mathbf{x}_i^d}{1+\lambda_1\left[r_i^d\right]^2+\lambda_2\left[r_i^d\right]^4+\dots} = \mathbf{H}\mathbf{X}_i^T \tag{18}$$

An equation set can be obtained after sorting out:

$$\begin{bmatrix} \hat{x}_i^d\left[X_i^T\right]^T\left[-\hat{\mathbf{F}}_2X_i^T\right]\left[\begin{array}{ccc}\left[\hat{r}_i^d\right]^2 & \left[\hat{r}_i^d\right]^4 & \dots\end{array}\right] \\ \hat{y}_i^d\left[X_i^T\right]^T\left[\hat{\mathbf{F}}_1X_i^T\right]\left[\begin{array}{ccc}\left[\hat{r}_i^d\right]^2 & \left[\hat{r}_i^d\right]^4 & \dots\end{array}\right] \end{bmatrix} \begin{bmatrix} \left[\hat{\mathbf{H}}_3\right]^T \\ \lambda_1 \\ \lambda_2 \\ \vdots \end{bmatrix} = \begin{bmatrix} \hat{\mathbf{F}}_2X_i^T \\ -\hat{\mathbf{F}}_1X_i^T \end{bmatrix} \tag{19}$$

For Equation (19) and the combined Equation (17), two equations can be obtained for each pair of corner points. When the number of corresponding points $N >= n+3$ (where n denotes the number of distortion parameters), an overdetermined equation is obtained. This can be achieved by moving the target, as shown in Figure 2. The homography matrix $\hat{\mathbf{H}}$ and the distortion coefficients $\lambda_1\lambda_2$ can be obtained by using the least square method.

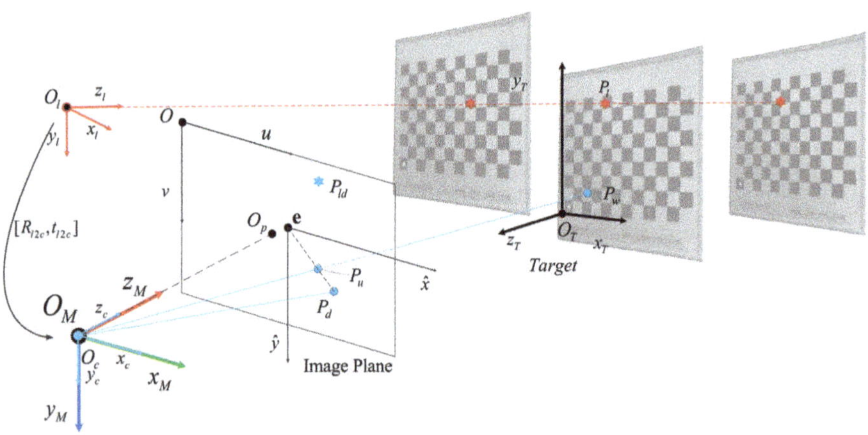

Figure 2. Changing of the azimuth and angle of the calibration plate and performing multiple measurements.

4.3. Parameter Solution

From the perspective projection model, the imaging relationship of the translation sequence can be expressed as follows:

$$\rho_i\mathbf{x}_i^u = \rho_i\left[\mathbf{T}_{\mathrm{e2}\hat{\mathrm{e}}}\right]^{-1}\hat{\mathbf{x}}_i^u = \left[\mathbf{T}_{\mathrm{e2}\hat{\mathrm{e}}}\right]^{-1}\hat{\mathbf{H}}\mathbf{X}_i^T = \mathbf{H}\mathbf{X}_i^T \tag{20}$$

It is known that:

$$\mathbf{H} = \left[\mathbf{T}_{\mathrm{e2}\hat{\mathrm{e}}}\right]^{-1}\hat{\mathbf{H}} = \begin{bmatrix} 1 & 0 & du_0 \\ 0 & 1 & dv_0 \\ 0 & 0 & 1 \end{bmatrix}\hat{\mathbf{H}} \tag{21}$$

Equation (19) can been solved to obtain $\hat{\mathbf{H}}$. The initial homography matrix **H** can be calculated by substituting Equation (21). We set $[\ \mathbf{r}_1 \ \ \mathbf{r}_2 \ \ \mathbf{t}\ ]H = \begin{bmatrix} \mathbf{H}_1 & \mathbf{H}_2 & \mathbf{H}_3 \end{bmatrix} = \mathbf{A}_C[\ \mathbf{r}_1 \ \ \mathbf{r}_2 \ \ \mathbf{t}\ ]$. By using the orthogonality and normality of rotation matrix $[\ \mathbf{r}_1 \ \ \mathbf{r}_2 \ \ \mathbf{r}_3\ ]$, we obtain:

$$\begin{cases} \mathbf{r}_1\mathbf{r}_2 = 0 \\ \mathbf{r}_1\mathbf{r}_1 = \mathbf{r}_2\mathbf{r}_2 \end{cases} \Leftrightarrow \begin{cases} [\mathbf{H}_1]^T[\mathbf{A}_C]^{-T}[\mathbf{A}_C]^{-1}\mathbf{H}_2 = 0 \\ [\mathbf{H}_1]^T[\mathbf{A}_C]^{-T}[\mathbf{A}_C]^{-1}\mathbf{H}_1 = [\mathbf{H}_2]^T[\mathbf{A}_C]^{-T}[\mathbf{A}_C]^{-1}\mathbf{H}_2 \end{cases} \tag{22}$$

Therefore, three images are needed to find five unknowns in the camera intrinsic parameter matrix $\mathbf{A}_C$. If the camera collects n images from different directions for calibration, a set of linear equations containing $2n$ constrained equations can be established, which can be written in matrix form as follows:

$$\mathbf{Vb} = 0 \tag{23}$$

where **V** is the coefficient matrix and **b** is the variable to be solved, with:

$$\mathbf{b} = [\ B_{11} \ \ B_{12} \ \ B_{13} \ \ B_{22} \ \ B_{23} \ \ B_{33}\ ] \tag{24}$$

$$\mathbf{B} = [\mathbf{A}_C]^{-T}[\mathbf{A}_C]^{-1} = \begin{pmatrix} B_{11} & B_{12} & B_{13} \\ B_{21} & B_{22} & B_{23} \\ B_{31} & B_{32} & B_{33} \end{pmatrix} \tag{25}$$

The solvable camera intrinsic parameters are:

$$\begin{aligned} & v_0 = (B_{12}B_{13} - B_{11}B_{23})/(B_{11}B_{22} - B_{12}^2) \\ & \lambda = B_{33} - [B_{13}^2 + v_0(B_{12}B_{13} - B_{11}B_{23})]/B_{11} \\ & f_u = \sqrt{\lambda/B_{11}} \\ & f_v = \sqrt{\lambda B_{11}/(B_{11}B_{22} - B_{12}^2)} \\ & s = -B_{12}f_u^2 f_v/\lambda \\ & u_0 = kv_0/f_v - B_{13}f_u^2/\lambda \end{aligned} \tag{26}$$

Similarly, the camera parameters can be obtained:

$$[\ \mathbf{r}_1 \ \ \mathbf{r}_2 \ \ \mathbf{r}_3 \ \ \mathbf{t}_1\ ] = \left[\ \frac{[\mathbf{A}_C]^{-1}\mathbf{H}_1}{\left|[\mathbf{A}_C]^{-1}\mathbf{H}_1\right|} \ \ \frac{[\mathbf{A}_C]^{-1}\mathbf{H}_2}{\left|[\mathbf{A}_C]^{-1}\mathbf{H}_3\right|} \ \ \mathbf{r}_1 \times \mathbf{r}_2 \ \ \frac{[\mathbf{A}_C]^{-1}\mathbf{H}_3}{\left|[\mathbf{A}_C]^{-1}\mathbf{H}_3\right|} \ \right] \tag{27}$$

In the case of obtaining the parameters outside the target, the spatial coordinate $\mathbf{X}_l^C = \begin{bmatrix} X_l^C & Y_l^C & Z_l^C & 1 \end{bmatrix}^T$ of the laser spot P_l in the camera coordinate system can be found by solving the known plane equation $\mathbb{Q}(X^C, Y^C, Z^C)$ in the direction obtained by connecting the ray and the target from the camera optical center O_C to the ideal image point coordinate $\mathbf{x}_l^u = \begin{bmatrix} u_l^u & v_l^u & 1 \end{bmatrix}^T$. Moving the calibration board along the laser direction, the spatial position of laser spot can be obtained at different distances after multiple acquisitions. By processing the data, the laser beam can be straight in the camera coordinate system.

Through data processing, the linear equation $\mathbb{C}(X^C, Y^C, Z^C)$ of the laser beam in the camera coordinate system can be obtained in the form of Equation (29). By combining the distance information DL obtained through laser ranging, the spatial coordinates of the laser origin in the camera coordinate system can be obtained, and the transformation relationship $\begin{bmatrix} \mathbf{R}_{l2M} & \mathbf{t}_{l2M} \end{bmatrix}$ between the laser system and camera system can be estimated:

$$\mathbb{Q}(X^C, Y^C, Z^C) : \begin{bmatrix} A_Q & B_Q & C_Q & D_Q \end{bmatrix} \begin{bmatrix} X^C & Y^C & Z^C & 1 \end{bmatrix}^T = 0 \tag{28}$$

$$\mathbb{C}(X^C, Y^C, Z^C) : \frac{X_{li}^C - X_{l0}^C}{A_l} = \frac{Y_{li}^C - Y_{l0}^C}{B_l} = \frac{Z_{li}^C - Z_{l0}^C}{C_l} \tag{29}$$

where $X_{l0}^C = \frac{1}{n}\sum X_{li}^C$, $Y_{l0}^C = \frac{1}{n}\sum Y_{li}^C$, $Z_{l0}^C = \frac{1}{n}\sum Z_{li}^C$. Combining with Equation (27), we have:

$$\mathbb{Q}(\ \mathbf{R}_{T2M},\ \ \mathbf{t}_{T2M}\ ) = \begin{bmatrix} A_Q & B_Q & C_Q & D_Q \end{bmatrix}\begin{bmatrix} \mathbf{r}_1 & \mathbf{r}_2 & \mathbf{r}_3 & \mathbf{t}_1 \end{bmatrix}\begin{bmatrix} X^T & Y^T & Z^T & 1 \end{bmatrix}^T \tag{30}$$

By combining with the imaging model, the linear equation between laser spot and camera light center can be expressed in two-point form:

$$\frac{X^C}{u_l^u} = \frac{Y^C}{v_l^u} = \frac{2 \cdot Z^C}{f_u + f_v} \tag{31}$$

Equations (30) and (31) are solved simultaneously, the only solution of which $\mathbf{X}_l^C = \begin{bmatrix} X_l^C & Y_l^C & Z_l^C & 1 \end{bmatrix}^T$ is the coordinate of the laser spot on the target in the camera coordinate system.

After many measurements, the linear equation can be expressed as a series of spatial point sets $\left\{\mathbf{X}_{li}^C \middle| \mathbf{X}_{li}^C = \begin{bmatrix} X_{li}^C & Y_{li}^C & Z_{li}^C & 1 \end{bmatrix}^T, i = 1, 2, 3, \cdots\right\}$, as shown in Figure 2. Constraints can be applied using a point-line relationship to solve the linear equation $\mathbb{C}(X^C, Y^C, Z^C)$ corresponding to laser rays, such as:

$$\begin{bmatrix} Y_{li}^C - Y_{l0}^C & -[X_{li}^C - X_{l0}^C] & 0 \\ 0 & -[Z_{li}^C - Z_{l0}^C] & Y_{li}^C - Y_{l0}^C \end{bmatrix}\begin{bmatrix} A_l & 0 \\ B_l & B_l \\ 0 & C_l \end{bmatrix} = 0 \tag{32}$$

A space point can provide two constraints, and we need at least two space points to solve the equation and estimate the linear equation $\mathbb{C}(X^C, Y^C, Z^C)$. Finally, the laser origin position is determined on the line by calculating the distance information obtained by ranging according to the coordinate system established before and using the relative transformation matrix of laser-camera $\begin{bmatrix} \mathbf{R}_{T2M} & \mathbf{t}_{T2M} \end{bmatrix}$.

4.4. Optimization of Solution

The above process does not involve any iteration. The camera internal parameters and laser-camera external parameters can be found using least squares. The calculation speed is fast and local minima can be effectively avoided effectively. If we want to obtain higher accuracy, we can take the calculated value as the initial value, and further improve the calibration accuracy of the system through the non-linear optimization method.

Given n calibrated images, each image has m corners x_i^d and one laser projection point $\mathbf{x}_l^d$. The following objective functions are then constructed:

$$E(\mathbf{A}_C, \lambda_1, \lambda_2, \mathbf{R}_{T2M}, \mathbf{t}_{T2M}) = \sum_{i=1}^{n}\left(\sum_{j=1}^{m}\left|x_{i,j}^d - Pro(X_j^T)\right| + \gamma\left|x_{l,j}^d - Pro(d_i^l)\right|\right) \tag{33}$$

where $Pro(\mathbf{X}_j^T)$ and $Pro(d_i^l)$ represent the projection functions of corner points $\mathbf{X}_j^T$ and laser spot $\mathbf{X}_l^C$ under the division distortion model, and γ is a named weight coefficient that denotes the contribution of corner and laser points to errors, generally speaking $\gamma = 5$.

Using Cayley-Gibbs-Rodriguez (CGR) [9] to parameterize the rotation matrix $\mathbf{R}$, the latter can be expressed as a function of the CGR parameters $\mathbf{s} = \begin{bmatrix} s_1 & s_2 & s_3 \end{bmatrix}$:

$$\mathbf{R} = \frac{1}{1 + s_1{}^2 + s_2{}^2 + s_3{}^2}\begin{bmatrix} 1 + s_1{}^2 - s_2{}^2 - s_3{}^2 & 2s_1s_2 - 2s_3 & 2s_1s_3 + 2s_2 \\ 2s_1s_2 + 2s_3 & 1 - s_1{}^2 + s_2{}^2 - s_3{}^2 & 2s_2s_3 - 2s_1 \\ 2s_1s_3 - 2s_2 & 2s_2s_3 + 2s_1 & 1 - s_1{}^2 - s_2{}^2 + s_3{}^2 \end{bmatrix} \tag{34}$$

The problem is then transformed into an unconstrained optimization problem. The automatic Gröbner basis method [10] is used to solve Equation (32), and the minimum solution $E_{\min}(\tilde{\mathbf{A}}_C, \tilde{\lambda}_1, \tilde{\lambda}_2, \tilde{\mathbf{R}}_{l2M}, \tilde{\mathbf{t}}_{l2M})$ can be obtained. A nonlinear optimization method is used to further improve the accuracy and stability of the solution.

In this part, we have completed the estimation of the optimal solution of all parameters, including the camera intrinsic parameter matrix $\tilde{\mathbf{A}}_C$, distortion coefficient $\tilde{\lambda}_1\tilde{\lambda}_2$ and laser-camera external parameters $\begin{bmatrix} \tilde{\mathbf{R}}_{l2M} & \tilde{\mathbf{t}}_{l2M} \end{bmatrix}$.

5. Experiment and Analysis

In this part, we evaluate the calibration methods of the camera internal parameters and camera-laser external parameters. The effectiveness and influencing factors of the proposed system calibration algorithm are analyzed through computer simulation experiments, while the measurement system is calibrated through observation experiments. In order to better evaluate the calibration results, we refer to the re-projection error [34] evaluation method in the camera calibration process, and unify the laser spot and target corner to establish the following error evaluation function:

$$E_{dr} = \frac{1}{m \cdot n} \sum_{i=1}^{n} \left(\sum_{j=1}^{m} \left| x_{i,j}^d - Pro(X_j^T) \right| + \gamma \left| x_{l,i}^d - Pro(d_i^l) \right| \right) \tag{35}$$

The re-projection error E_{dr} is an important metric of the calibration results: the smaller E_{dr} is, the better the calibration results are.

5.1. Simulation Result

For the simulation experiment, we used the MATLAB R2016a software for Windows 10. The relevant parameters of the simulation system are shown in Table 2. In the measurement system, the laser direction is parallel to the optical axis of the camera and a 50 mm offset in the $\overrightarrow{O_M X_M}$ direction is arranged.

Table 2. System parameters of simulation.

Parameter	$\mathbf{A}_C$	$\mathbf{e}$	(λ_1, λ_2)	$\mathbf{R}_{l2M}$	$\mathbf{t}_{l2M}$
Set value	$\begin{bmatrix} 850 & s & 512 \\ 0 & 850 & 384 \\ 0 & 0 & 1 \end{bmatrix}$	$\begin{bmatrix} 509 \\ 380 \end{bmatrix}$	$\begin{pmatrix} 6.15 \times 10^{-7} \\ 1.6 \times 10^{-13} \end{pmatrix}$	$\begin{bmatrix} 1 & 0 & 0 \\ 0 & 1 & 0 \\ 0 & 0 & 1 \end{bmatrix}$	$\begin{bmatrix} 50 \\ 0 \\ 0 \end{bmatrix}$
Unit	pixel	pixel	$\begin{pmatrix} \text{pixel}^{-2} \\ \text{pixel}^{-4} \end{pmatrix}$	-	mm

The target is shown in the Figure 3, where the blue dots represent the corners of the checkerboard lattice, evenly distributed in the plane, and the adjacent corners are 15 mm apart. The relative position between the target and the system is randomly generated by the system within a given range.

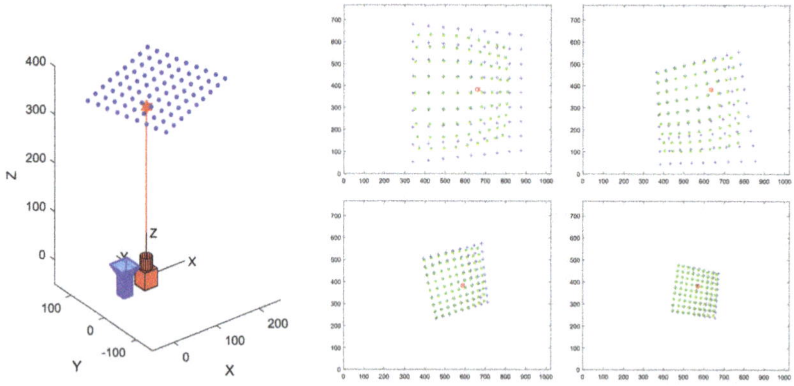

Figure 3. The imaging illustration. (**Left**) Schematic diagram of a simulated scenario. (**Right**) The generated image. The blue dots represent the ideal image points, the green dots represent the added distortion points, the red X represents the ideal image point of laser and the red circle is the distorted laser projection point.

Throughout the experiment, we compare the estimated values from each calculation with the real values set by simulation, and evaluate the accuracy of the algorithm by calculating the deviation between the two. The error is expressed as follows:

$$\begin{cases} E_{fu} = \frac{|\tilde{f}_u - f_u|}{f_u}, E_{fv} = \frac{|\tilde{f}_v - f_v|}{f_v} \\ E_{u0} = \frac{|\tilde{u}_0 - u_0|}{u_0}, E_{u0} = \frac{|\tilde{v}_0 - v_0|}{v_0} \\ E_{\mathbf{R}} = \max_{i=1}^{3} \left| \arccos \left| \tilde{\mathbf{r}}_i \cdot \mathbf{r}_i \right| \right| \\ E_{\mathbf{t}} = \left| \tilde{\mathbf{t}}_{l2M} - \mathbf{t}_{l2M} \right| \end{cases} \quad (36)$$

Kopparapu et al., confirmed [32] that noise has a significant impact on calibration accuracy. We add $\omega_{noise} \sim Gauss(0, \Sigma_{noise})$ Gaussian noise to the simulated projection image, where Σ_{noise} is the standard deviation of the Gaussian distribution. In the simulation, the standard deviation Σ_{noise} of noise increases gradually in the range of 0.1 to 1.5 pixel. For each ω_{noise} X distribution, we performed 100 independent experiments, and obtained the average value of calibration error as the statistical result.

The results are shown in Figure 4. It can be seen that with the increase of noise, the deviation between the calibration parameters and the true value increases linearly. When the corner extraction noise is 0.5 pixels, the system calibration error is about 0.2, the focal length deviation is 0.1%, and the main point deviation is about 0.8%. In terms of extrinsic parameters, the translation error also follows a linear distribution, but the fluctuation is more obvious. It can be seen that the system is sensitive to the internal parameters. At the same time, under the corner extraction error of 0.5 pixels, the translation error is about 1 mm and the rotation error is 0.02 degrees.

In addition, we analyzed the impact of the number of collected data on the calibration accuracy, and set the calibration data to gradually increase from the minimum of three groups of image distance data to 15 groups. The results are shown in the Figure 5. With the increase of calibration data, the re-projection error remains almost stable, but the accuracy of the estimated system variables is significantly improved. When the number of data increases to 8, the decline of the correlation error slows down. Therefore, sufficient calibration data collected in a certain range can help to improve the accuracy of system calibration. However, after reaching a certain number, the effect gradually decreases, and so 8–10 groups of data are appropriate.

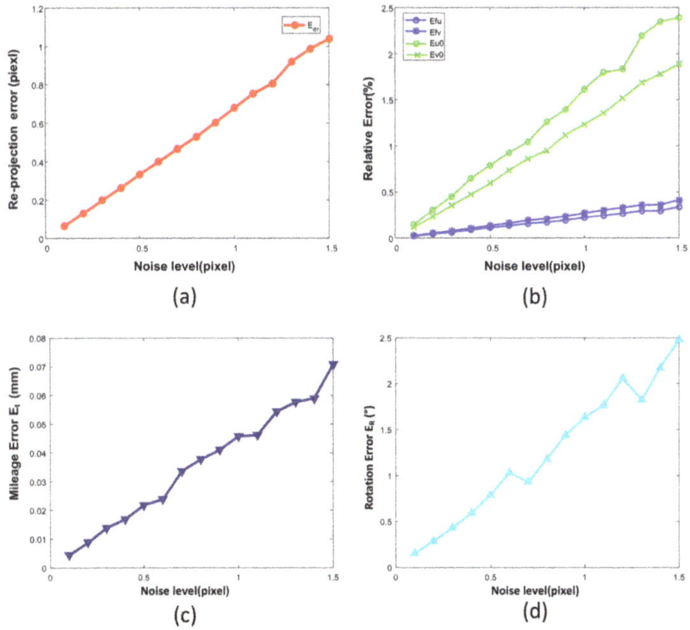

Figure 4. Simulation results for different image noise levels. (**a**) The re-projection error E_{dr}; (**b**) The effects of noise on intrinsic parameters such as E_{fu}, E_{fv}, E_{u0}, E_{u0}; (**c**) mileage error $E_{\mathbf{t}}$ for different noise levels; (**d**) effects of noise on rotation error $E_{\mathbf{R}}$.

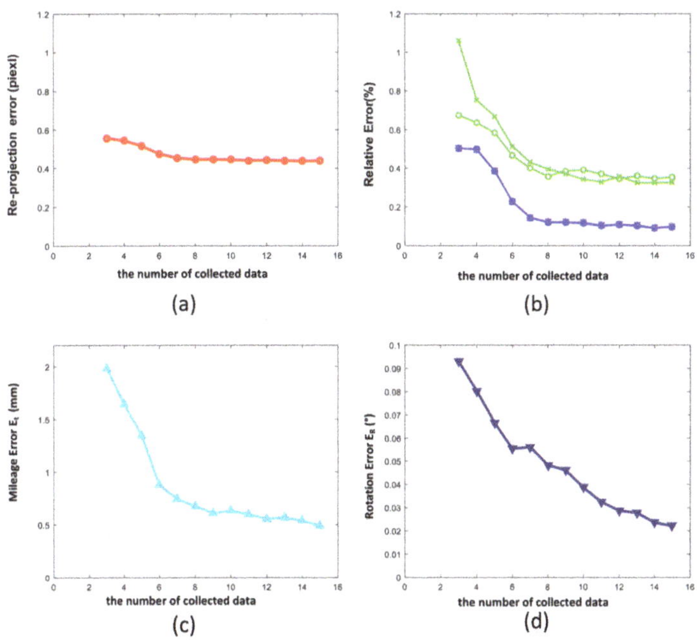

Figure 5. The simulation results for different numbers of collected data. (**a**) Re-projection error E_{dr}; (**b**) Effects of the number of data on intrinsic parameters such as E_{fu}, E_{fv}, E_{u0}, E_{u0}; (**c**) Mileage error $E_{\mathbf{t}}$ for different noise levels; (**d**) Effect of the number of data on rotation error $E_{\mathbf{R}}$.

We also analyzed the influence of the measurement error of the laser ranging system on the calibration accuracy. The Gaussian-distributed noise $\omega_d \sim Gauss(0, \Sigma_d)$ was added to the ranging error, and the standard deviation Σ_d was changed gradually from 1 mm to 15 mm. We calculated the calibration errors of the parameters of the system at each noise level. As shown in Figure 6, except for the linear relationship between translation vector and distance error, the other parameters hardly change with the increase of error.

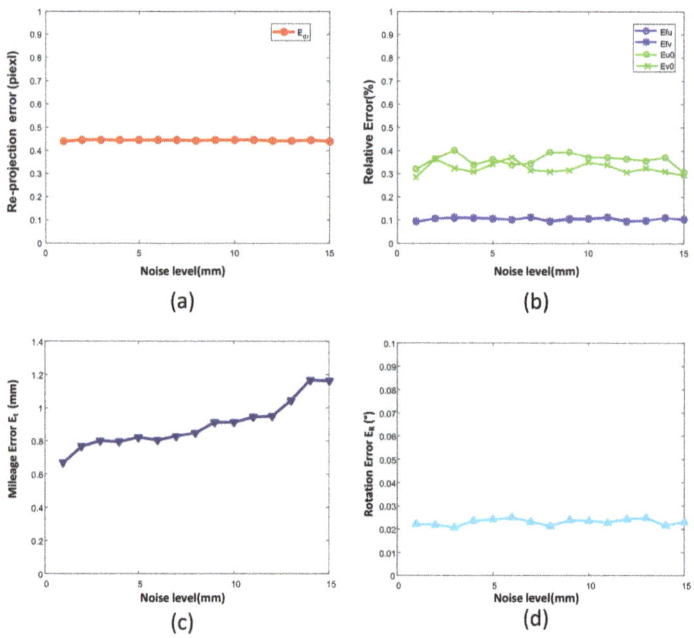

Figure 6. Simulation results of different distance noise levels. (**a**) Re-projection error E_{dr}; (**b**) Effects of noise on intrinsic parameters such as E_{fu}, E_{fv}, E_{u0}, E_{u0}; (**c**) Mileage error $E_{\mathbf{t}}$ with different noise; (**d**) the effects of noise on rotation error $E_{\mathbf{R}}$.

5.2. Real Experiment

In the actual experiment, we built a measurement system with a 1D laser-camera combination, and calibrated the system with the method proposed in this paper. As shown in Figure 7, the system is composed of a MER-131-210U3C camera and a SKD-100 laser ranging system. The related parameters are shown in Table 3.

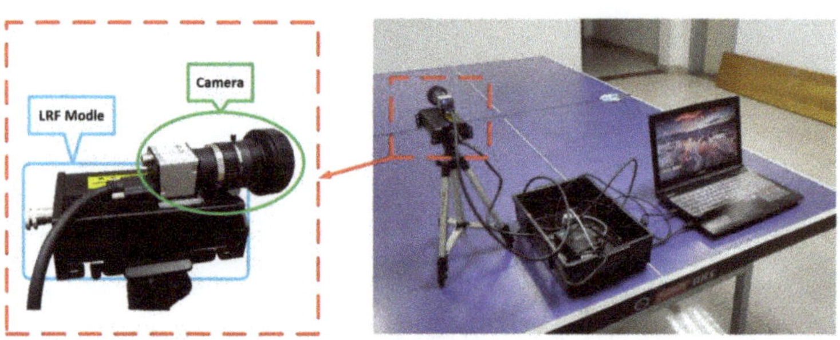

Figure 7. Measurement system combining 1D LRF and camera.

Table 3. The system parameters of 1D LRF and camera.

Sensors	Parameter	Value
MER-131-210U3C (camera)	Sensor Size	1/2″
	Resolution	1280 (H) × 1024 (V)
	Frame Rate	210 *FPS*
	Pixel Size	4.8 µm × 4.8 µm
	Focuses	5 mm
	F (Relative Aperture)	1.4 ∼ 16
SKD-100 (LRF)	Wavelength	635 nm
	Range	1 ∼ 1000 mm
	Accuracy	2 mm

The system was calibrated using a calibration board composed of 11 × 8 square chessboard lattices with a distance of 15 mm between corners. The iterative Harris algorithm was used to extract the checkerboard corner coordinates (red +) from the calibrated image accurately, and the centroid method was used to extract the image coordinates (green ×) of the laser spot. The accuracy can reach sub-pixel level. The results of 12 images collected at different distances from 150 mm to 1500 mm are shown in Figure 8.

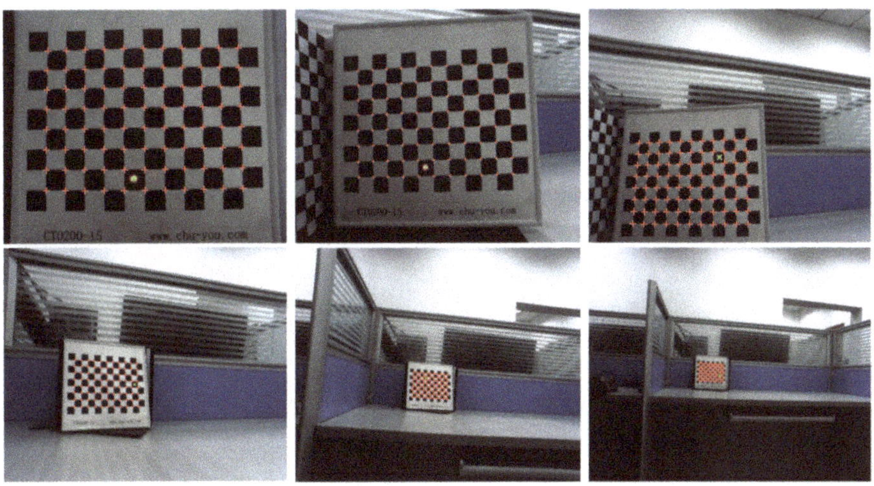

Figure 8. Samples of images used for the real experiment.

In order to verify the accuracy of our calibration method, we compared the internal parameters obtained with the classical Zhang [21] calibration method and the Li's method [30]. In the calibration process of Bo's method, we replaced the original random corner matching process by directly inputting the coordinates of checkerboard lattices into the program, but still retain the complete algorithm for camera parameter determination. The results are shown in Table 4, where the accuracy of the intrinsic parameters obtained by the calibration methods are compared. The calibration accuracy is evaluated using the re-projection error [34] and expressed as:

$$E_{rp} = \frac{1}{m}\sum_{j=1}^{m}\sqrt{\left|x_j^d - Pro(X_j^T)\right|^2} \quad (37)$$

Table 4. System parameters of simulation.

Method	$\mathbf{A}_C$	$\mathbf{e}$	(λ_1, λ_2)	Mean E_{rp}
Zhang [21]	$\begin{bmatrix} 1053.3 & 0 & 643.5 \\ 0 & 1048.0 & 539.7 \\ 0 & 0 & 1 \end{bmatrix}$ pixel	-	$\begin{pmatrix} 0.1348\ \text{mm}^{-2} \\ 0.01661\ \text{mm}^{-4} \end{pmatrix}$	0.09337 pixel
Li [30]	$\begin{bmatrix} 1059.7 & 0.1604 & 649.5 \\ 0 & 1058.9 & 539.2 \\ 0 & 0 & 1 \end{bmatrix}$ pixel	-	$\begin{pmatrix} 0.1372\ \text{mm}^{-2} \\ 0.1993\ \text{mm}^{-4} \end{pmatrix}$	0.07974 pixel
proposed	$\begin{bmatrix} 1055.2 & 0 & 647.2 \\ 0 & 1054.9 & 538.1 \\ 0 & 0 & 1 \end{bmatrix}$ pixel	$\begin{bmatrix} 509 \\ 380 \end{bmatrix}$ pixel	$\begin{pmatrix} 6.15 \times 10^{-7} & \text{pixel}^{-2} \\ 1.6 \times 10^{-13} & \text{pixel}^{-4} \end{pmatrix}$	0.07725 pixel

From the calibration results in Figure 9 and Table 4, we see that our method and Zhang's method [21] have similar calibration results in camera intrinsic parameters. Because the distortion models used by the two methods are different, the physical meanings of the distortion coefficients are different, so it is meaningless to compare them. Judging from the re-projection error, our method is slightly better than Zhang's calibration algorithm. This proves the effectiveness of our calibration algorithm.

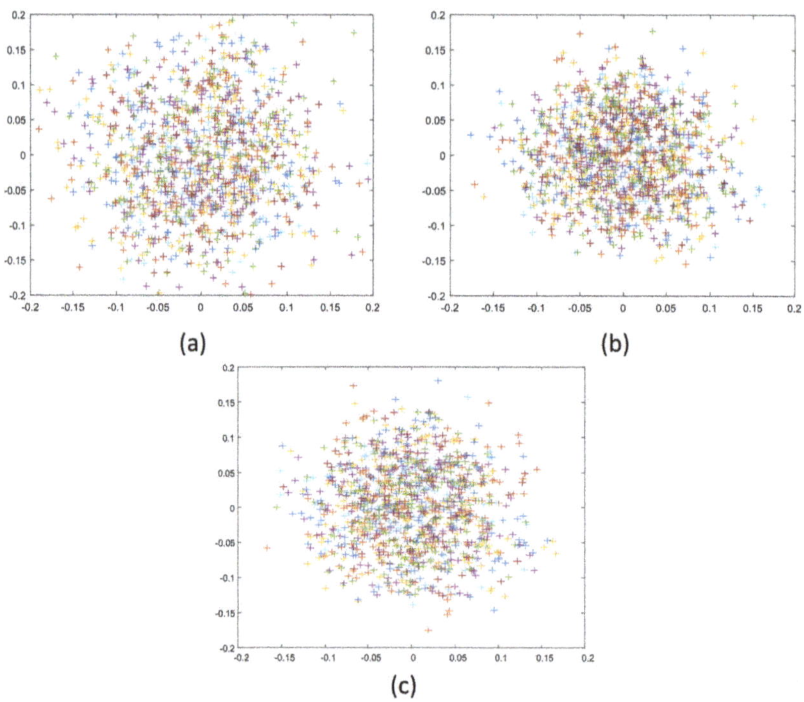

Figure 9. Re-projection error distribution for different images marked as different colors: (**a**) Zhang's method [21]; (**b**) proposed method; (**c**) Li's method [30].

At the same time, the extrinsic parameters $\begin{bmatrix} \tilde{\mathbf{R}}_{l2M} & \tilde{\mathbf{t}}_{l2M} \end{bmatrix}$ of the laser-camera combination of the measurement system are also calculated and the calibration results were evaluated using the evaluation function set Equation (37). The results are shown in Table 5 and Figure 10.

Table 5. System extrinsic parameters and evaluation error.

Parameter	$\mathbf{R}_{I2M}$	$\mathbf{t}_{I2M}$	E_{dr}
Set value	$\begin{bmatrix} 0.9974 & 0.0052 & -0.0715 \\ -0.0052 & 0.9893 & -0.1456 \\ 0.0715 & 0.1456 & 1 \end{bmatrix}$	$\begin{bmatrix} 0.5030 \\ 32.6479 \\ 1.859 \end{bmatrix}$	0.10173
Unit	-	mm	pixel

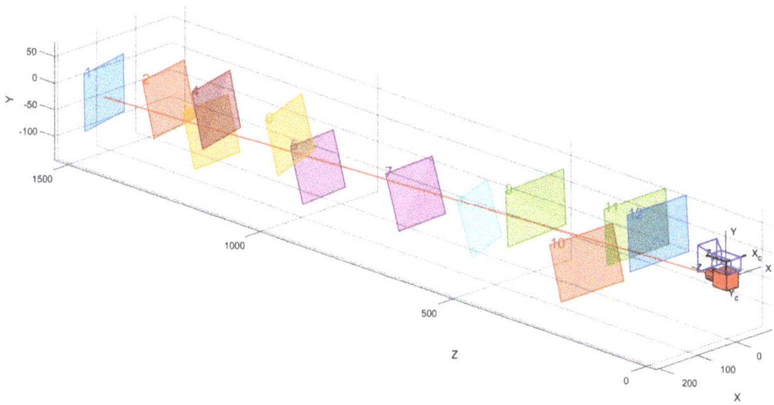

Figure 10. Visualization of extrinsic parameters $\begin{bmatrix} \mathbf{R}_{I2M} & \mathbf{t}_{I2M} \end{bmatrix}$.

Ferrara et al. [35] mentioned that the position of the checkerboard has an effect on the accuracy of calibration. We supplemented a set of calibration data of checkerboard location on the edge of the image to verify the effect of the change of checkerboard location on the accuracy of the proposed method. The data are shown in Figure 11. The calibration results are shown in Table 6. When the image is in the edge position, the calibration results are basically consistent with the internal and external parameters obtained in Tables 3 and 4, and the re-projection errors of the internal and external parameters are slightly increased, but the difference is small. It therefore shown that the method proposed in this paper is also applicable when the collected data lie on the edge of the image.

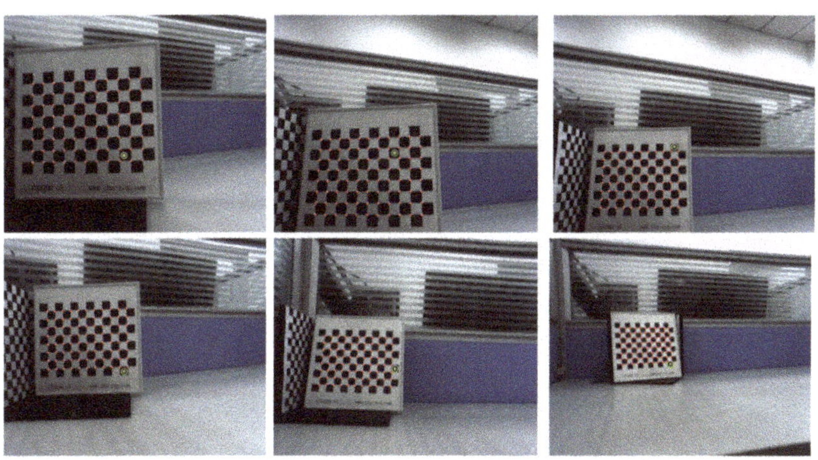

Figure 11. Image sample when the checkerboard is close to the edge.

Table 6. System parameters and evaluation error when checkerboard is close to edge.

Parameter	$\mathbf{A}_C$	$\mathbf{e}$	(λ_1, λ_2)	Mean E_{rp}
Value	$\begin{bmatrix} 1057.9 & 0 & 646.9 \\ 0 & 1057.2 & 536.8 \\ 0 & 0 & 1 \end{bmatrix}$ pixel	$\begin{bmatrix} 507 \\ 379 \end{bmatrix}$ pixel	$\begin{pmatrix} 6.21 \times 10^{-7} & \text{pixel}^{-2} \\ 1.53 \times 10^{-13} & \text{pixel}^{-4} \end{pmatrix}$	0.08371 pixel
Parameter	$\mathbf{R}_{l2M}$	$\mathbf{t}_{l2M}$	E_{dr}	
Set value	$\begin{bmatrix} 0.9962 & 0.0101 & -0.0865 \\ -0.0101 & 0.9902 & -0.1381 \\ 0.0855 & 0.1388 & 0.9999 \end{bmatrix}$	$\begin{bmatrix} 0.8167 \\ 31.5531 \\ 1.6742 \end{bmatrix}$ mm	0.1147 pixel	

6. Conclusions

In this paper, we present a convenient and fast method for calibrating a combined 1D laser ranging and monocular camera measurement system, aiming to realize an accurate measurement system fusing laser and vision. The method is easy to implement and has high calibration accuracy. The fast robust determination of the camera imaging model parameters is achieved by introducing a division distortion model. Then, a linear-plane constraint is formulated to realize robust estimation of the initial value of the laser-vision parameters. Finally, an unconstrained optimization problem is formulated using the rotation matrix parameters, and the high precision calibration of the whole measurement system is realized. The factors affecting the calibration accuracy are analyzed through simulation experiments, and the effectiveness of the proposed method is verified through real scene experiments.

Author Contributions: Z.Z. and R.Z. conceived the methodology and implemented the methodology. Z.Z. designed the simulated experiments and analyzed the data. R.Z. wrote the paper. Z.Z. and R.Z. contributed equally to this work. E.L. designed and guided the experiments. K.Y. and Y.M. performed the experiments.

Funding: This research was funded by the National Natural Science Foundation of China (No. 61501429) and the Youth Innovation Promotion Association CAS (No. 2016335).

Acknowledgments: Thanks to the accompaniers working with us in department of the Institute of Optics and Electronics, Chinese Academy of Sciences.

Conflicts of Interest: The authors declare no conflict of interest.

References

1. Zhang, Z.; Zhao, R.; Liu, E.; Yan, K.; Ma, Y. Scale Estimation and Correction of the Monocular Simultaneous Localization and Mapping (SLAM) Based on Fusion of 1D Laser Range Finder and Vision Data. *Sensors* **2018**, *18*, 1948. [CrossRef] [PubMed]
2. Douillard, B.; Fox, D.; Ramos, F. Laser and Vision Based Outdoor Object Mapping. *Robot. Sci. Syst.* **2008**, 9–16. [CrossRef]
3. Premebida, C.; Ludwig, O.; Nunes, U. *LIDAR and Vision-Based Pedestrian Detection System*; John Wiley and Sons Ltd.: Hoboken, NJ, USA, 2009.
4. Whelan, T.; Kaess, M.; Johannsson, H.; Fallon, M.; Leonard, J.J.; McDonald, J. Real-time large-scale dense RGB-D SLAM with volumetric fusion. *Int. J. Robot. Res.* **2015**, *34*, 598–626. [CrossRef]
5. Duren, R.M.; Wong, E.; Breckenridge, B.; Shaffer, S.J.; Duncan, C.; Tubbs, E.F.; Salomon, P.M. Metrology, attitude, and orbit determination for spaceborne interferometric synthetic aperture radar. In Proceedings of the Acquisition, Tracking, & Pointing XII, Orlando, FL, USA, 30 July 1998.
6. Ordóñez, C.; Arias, P.; Herráez, J.; Rodriguez, J.; Martín, M.T. A combined single range and single image device for low-cost measurement of building façade features. *Photogramm. Rec.* **2010**, *23*, 228–240. [CrossRef]
7. Wu, K.; Di, K.; Sun, X.; Wan, W.; Liu, Z. Enhanced monocular visual odometry integrated with laser distance meter for astronaut navigation. *Sensors* **2014**, *14*, 4981–5003. [CrossRef] [PubMed]
8. Chen, Z.; Yang, X.; Zhang, C.; Jiang, S. Extrinsic calibration of a laser range finder and a camera based on the automatic detection of line feature. In Proceedings of the International Congress on Image and Signal Processing, Biomedical Engineering and Informatics, Datong, China, 15–17 October 2016.

9. Hesch, J.A.; Roumeliotis, S.I. A Direct Least-Squares (DLS) method for PnP. In Proceedings of the IEEE International Conference on Computer Vision, Barcelona, Spain, 6–13 November 2011.
10. Kukelova, Z.; Bujnak, M.; Pajdla, T. Automatic Generator of Minimal Problem Solvers. In Proceedings of the European Conference on Computer Vision, Marseille, France, 12–18 October 2008.
11. Vasconcelos, F.; Barreto, J.P.; Nunes, U. A minimal solution for the extrinsic calibration of a camera and a laser-rangefinder. *IEEE Trans. Pattern Anal. Mach. Intell.* **2012**, *34*, 2097–2107. [CrossRef] [PubMed]
12. Scaramuzza, D.; Harati, A.; Siegwart, R. Extrinsic self calibration of a camera and a 3D laser range finder from natural scenes. In Proceedings of the IEEE/RSJ International Conference on Intelligent Robots and Systems, San Diego, CA, USA, 29 October–2 November 2007.
13. Ha, J.E. Improved algorithm for the extrinsic calibration of a camera and laser range finder using 3D-3D correspondences. *Int. J. Control Autom. Syst.* **2015**, *13*, 1272–1276. [CrossRef]
14. Unnikrishnan, R.; Hebert, M. *Fast Extrinsic Calibration of a Laser Rangefinder to a Camera*; Carnegie Mellon University: Pittsburgh, PA, USA, 2005.
15. Zhang, Q.; Pless, R. Extrinsic calibration of a camera and laser range finder (improves camera calibration). In Proceedings of the IEEE/RSJ International Conference on Intelligent Robots and Systems, Sendai, Japan, 28 September–2 October 2005.
16. Viejo, D.; Navarrete-Sanchez, J.; Cazorla, M. Portable 3D laser-camera calibration system with color fusion for SLAM. *Investigación* **2013**, *3*, 29–45.
17. Zhu, Z.; Tang, B.Q.; Li, J.; Gan, Z. Calibration of laser displacement sensor used by industrial robots. *Opt. Eng.* **2004**, *43*, 12–14.
18. Ke-Qing, L.U.; Wang, W.; Chen, Z.C. Calibration of laser beam-direction for point laser sensors. *Opt. Precis. Eng.* **2010**, *18*, 880–886.
19. Zhou, A.; Guo, J.; Shao, W.; Li, B. A segmental calibration method for a miniature serial-link coordinate measuring machine using a compound calibration artefact. *Meas. Technol.* **2013**, *24*, 065001. [CrossRef]
20. Martínez, J.; Ordóñez, C.; Arias, P.; Armesto, J. Non-contact 3D Measurement of Buildings through Close Range Photogrammetry and a Laser Distance Meter. *Photogramm. Eng. Remote Sens.* **2011**, *77*, 805–811. [CrossRef]
21. Zhang, Z. A Flexible New Technique for Camera Calibration. *IEEE Trans. Pattern Anal. Mach. Intell.* **2000**, *22*, 1330–1334. [CrossRef]
22. Liu, Z.; Wu, Q.; Wu, S.; Pan, X. Flexible and accurate camera calibration using grid spherical images. *Opt. Express* **2017**, *25*, 15269–15285. [CrossRef] [PubMed]
23. Liu, Z.; Wu, Q.; Wu, S.; Pan, X. Camera Calibration from the Quasi-affine Invariance of Two Parallel Circles. *Opt. Express* **2004**, *3021*, 190–202.
24. Wong, K.Y.; Mendonca, P.R.S.; Cipolla, R. Camera Calibration from Surfaces of Revolution. *IEEE Trans. Pattern Anal. Mach. Intell.* **2003**, *25*, 147–161. [CrossRef]
25. Anchini, R.; Beraldin, J.A. Subpixel location of discrete target images in close-range camera calibration: A novel approach. *Proc. SPIE* **2007**, *6491*, 10–18.
26. Wang, Q.S. A Model Based on DLT Improved Three-dimensional Camera Calibration Algorithm Research. Available online: http://www.en.cnki.com.cn/Article_en/CJFDTotal-DBCH201612065.htm (accessed on 14 March 2019).
27. Kukelova, Z.; Pajdla, T. A Minimal Solution to Radial Distortion Autocalibration. *IEEE Trans. Pattern Anal. Mach. Intell.* **2011**, *33*, 2410–2422. [CrossRef] [PubMed]
28. Hartley, R.I.; Kang, S.B. Parameter-free radial distortion correction with centre of distortion estimation. *IEEE Trans. Pattern Anal. Mach. Intell.* **2007**, *29*, 1309–1321. [CrossRef] [PubMed]
29. Hong, Y.; Ren, G.; Liu, E. Non-iterative method for camera calibration. *Opt. Express* **2015**, *23*, 23992–24003. [CrossRef] [PubMed]
30. Li, B.; Heng, L.; Koser, K.; Pollefeys, M. A multiple-camera system calibration toolbox using a feature descriptor-based calibration pattern. In Proceedings of the IEEE/RSJ International Conference on Intelligent Robots & Systems, Tokyo, Japan, 3–7 November 2014.
31. An, G.; Lee, S.; Seo, M.W.; Yun, K.; Cheong, W.S.; Kang, S.J. Charuco Board-Based Omnidirectional Camera Calibration Method. *Electronics* **2018**, *7*, 421. [CrossRef]
32. Liu, Z.; Li, F.; Zhang, G. An external parameter calibration method for multiple cameras based on laser rangefinder. *Measurement* **2014**, *47*, 954–962. [CrossRef]

33. Lichti, D.D. Error modelling, calibration and analysis of an AM–CW terrestrial laser scanner system. *ISPRS J. Photogramm. Remote Sens.* **2007**, *61*, 307–324. [CrossRef]
34. Kopparapu, S.; Corke, P. The Effect of Noise on Camera Calibration Parameters. *Graph. Models* **2001**, *63*, 277–303. [CrossRef]
35. Ferrara, P.; Piva, A.; Argenti, F.; Kusuno, J.; Niccolini, M.; Ragaglia, M.; Uccheddu, F. Wide-angle and long-range real time pose estimation: A comparison between monocular and stereo vision systems. *J. Vis. Commun. Image Represent.* **2017**, *48*, 159–168. [CrossRef]

sensors

MDPI

Article

Delving Deep into Multiscale Pedestrian Detection via Single Scale Feature Maps

Xinchuan Fu [1,*], Rui Yu [2], Weinan Zhang [3], Jie Wu [4] and Shihai Shao [1]

1 National Key Laboratory of Science and Technology on Communications, University of Electronic Science and Technology of China, Chengdu 611731, China; ssh@uestc.edu.cn
2 Department of Computer Science, University College London, London WC1E 6BT, UK; r.yu@cs.ucl.ac.uk
3 Department of Computer Science & Engineering, Shanghai Jiao Tong University, Shanghai 200240, China; wnzhang@sjtu.edu.cn
4 Department of MOE Research Center for Software/Hardware Co-Design Engineering and Application, East China Normal University, Shanghai 200062, China; 52151500020@stu.ecnu.edu.cn
* Correspondence: 201311010310@std.uestc.edu.cn; Tel.: +86-028-6183-0596

Received: 28 February 2018; Accepted: 26 March 2018; Published: 2 April 2018

Abstract: The standard pipeline in pedestrian detection is sliding a pedestrian model on an image feature pyramid to detect pedestrians of different scales. In this pipeline, feature pyramid construction is time consuming and becomes the bottleneck for fast detection. Recently, a method called multiresolution filtered channels (MRFC) was proposed which only used single scale feature maps to achieve fast detection. However, there are two shortcomings in MRFC which limit its accuracy. One is that the receptive field correspondence in different scales is weak. Another is that the features used are not scale invariance. In this paper, two solutions are proposed to tackle with the two shortcomings respectively. Specifically, scale-aware pooling is proposed to make a better receptive field correspondence, and soft decision tree is proposed to relive scale variance problem. When coupled with efficient sliding window classification strategy, our detector achieves fast detecting speed at the same time with state-of-the-art accuracy.

Keywords: pedestrian detection; boosted decision tree; scale invariance; receptive field correspondence; soft decision tree

1. Introduction

Pedestrian detection aims to locate all the pedestrians in an image. It has many real world applications, such as driving assistance and video surveillance. It also serves as a playground for many image processing and machine learning algorithms. There have been well established benchmark datasets [1–4] and a variety of methods have published to address this problem [3,5–10]. In many real applications, detection speed is often as important as accuracy, like in Advanced Driver Assistance Systems (ADAS) [11]. Although recently deep learning methods have achieved the state-of-the-art accuracy in pedestrian detection, the detection speed is often low even with high-end GPU [12,13]. On the other hand, boosted decision tree (BDT) methods remain highly competitive in this area for its efficacy with (light-weight) CPU implementation [14–16]. In this paper, we focus on the BDT methods for pedestrian detection.

Pedestrians in images may exhibit a large range of scale, which constitutes a significant mode of intra-class variability. For example, the heights of the pedestrian samples in the Caltech pedestrian dataset [1] range from 7 to 476 pixels. How to detect pedestrians of different scales becomes a key problem in pedestrian detection.

It is obvious that pedestrians of different scales have different representations in the original image—at least they have different numbers of pixels. A feature will be appropriate for multiscale detection if it is invariant across different scales. Thus for a long time, the community of detection and recognition strives for building the so-called scale-invariant representations. Unfortunately, many features are not scale invariant, including Histograms of Oriented Gradients (HOG) [3] which are widely used in pedestrian detection. This means a feature extracted in a large scale pedestrian is different from a corresponding feature extracted in a small scale pedestrian. Therefore, to achieve scale-invariance, the standard pipeline in pedestrian detection is to construct an image pyramid, compute feature maps at each layer of the pyramid, and finally perform sliding window detection with a trained pedestrian model. As the pedestrian scale exhibits a large scope, the pyramid need to contain many layers and the construction of feature pyramid takes a lot of time, which becomes the bottleneck for fast pedestrian detection.

Could we avoid feature pyramid construction and detecting multiscale pedestrians only using single scale feature maps while still achieve high accuracy? In this paper, we will delve deep into this problem. In fact, some pioneer works [14,17] have shown promising results in this direction. Our work is based on MRFC [14] which is the most recent work on this topic. We analyze some weaknesses of their work and show how to make improvements to it. The main contributions of our paper are as follows:

1. To achieve better receptive field correspondence, we propose to use scale-aware pooling instead of gridwise sampling. Based on it, we use efficient difference integral channels (DICs) to enrich our features.
2. To relieve the scale variance problem, we propose to use a soft decision tree to build a weak classifier in the BDT cascade. Two branches of the soft decision tree specialize in large or small scale instances respectively.
3. Experiments show that when coupled with efficient sliding window classification strategy, our method achieves state of the art accuracy with fast detection speed.

A preliminary version of this work appeared in [18]. The main extension in this paper is as follows. Firstly, we add the content about soft decision tree (Section 4) and sparse grid detection (Section 7). Secondly, motion channels are added to our detector in this paper. Thirdly, more experiment results are given. For example, all the experiments on KITTI dataset [2], the tables listing the performance on different setups of our method and plots which show the performance of our detector under conditions of small scale, atypical aspect ratio and partial occlusion are newly added. Fourthly, on Caltech dataset [1], we achieve a noticeable accuracy improvement compared to that in [18] (average miss rate from 15.89% to 12.96%).

The rest of this paper is organized as follows. We first give a review of the related work of our paper in Section 2. Then based on the shortcomings of MRFC, we describe how to make a better receptive field correspondence using scale-aware pooling in Section 3, how to relive the scale variance problem using soft decision tree in Section 4. To accelerate the speed in the sliding window classification stage, we propose two strategies in Section 5. Experiment results are given in Section 6. Finally, we conclude in Section 7.

2. Related Work

Features play a key role in pedestrian detection [19]. The first popular features for pedestrian detection are HOG features. Many pedestrian and general object detectors adopt these features [20–23]. Based on HOG features, Dollár et al. [24] proposed to use LUV color channels, gradient magnitude channel and 6 orientation channels as feature maps. Since then, these 10 feature maps have become popular and a lot of recent papers are based on these feature maps [5,6,14,25–27]. For its effectiveness and efficiency, many deep learning methods also adopt them to perform region proposal [28–30].

Multiscale detection methods usually aim to find a scale-invariant representation. If the features used are scale-invariant, the detection can be performed using single scale feature maps and multiscale objects are detected by resizing the model [31]. Unfortunately, not all features used in pedestrian detection are scale-invariant. As to the 10 channels we described above, the LUV color channels are scale invariant, but the gradient magnitude channel and the 6 orientation channels are not scale invariant. In such cases, the standard treatment is to use dense image pyramid as shown in Figure 1a, which results in high computational cost. Some papers are published to accelerate this process. FPDW [32] computed a sparse feature pyramid and approximated intermediate feature scales using the exponential scaling law (Figure 1b). In practice, this strategy usually leads to noticeable accuracy degradation. Instead of using feature pyramid, VeryFast [33] trained a sparse classifier pyramid according to different pedestrian scales and approximated intermediate classifier scales using the exponential scaling law (Figure 1c). This strategy actually transforms testing time to training time. FastCF [16] used both feature pyramid and classifier pyramid (Figure 1d).

(a) (b) (c) (d) (e)

Figure 1. Different multiscale detection strategies. (**a**) Dense image pyramid and single classifier; (**b**) Sparse image pyramid and single classifier; (**c**) Single image scale and sparse classifier pyramid; (**d**) Sparse image pyramid and sparse classifier pyramid; (**e**) Single image scale and single classifier.

Recently, another point of view [14,17] was proposed which ignored the requirement for scale invariance. The authors of these papers argued that though there are many sources of intra-class variance for pedestrian detection, like illumination, orientation and occlusion, et al., which lead to significant different representation for the pedestrian class, a BDT is able to handle this intra-class variance and provide a good result. They thought that the feature variance caused by different scales is just another source of intra-class variance and will hopefully be handled well by the BDT. The precondition is that the number of weak classifiers and the training data is large enough. Based on this idea, they just used a single scale feature maps and trained one pedestrian model (Figure 1e). At testing time, they scaned these feature maps with resized pedestrian models.

We adopt this point of view, and our work makes some improvements in comparison with previous works. WordChannels [17] uses 192 feature maps, for each one of which an integral map is computed, which results in a high computational cost. The authors used GPU to implement their algorithm. MRFC [14] is a more recent work from the same authors which achieves fast detection based on CPU. It computed 210 feature maps efficiently without integral map computing. The problem of this method is the features' receptive field correspondence (which will be described in detail in Section 3) between different scales is weak, which limits the accuracy of the method. To make a more accurate receptive field correspondence, we borrow the idea from spatial pyramid pooling (SPP) [34], also known as spatial pyramid matching (SPM) [35]. It partitions the feature map into a predefined number of divisions and performs pooling in each division. Thus feature maps of various sizes are converted into a fixed length vector. This technique is very useful in convolutional neural network (CNN)-based detectors as the fully connected layer needs to be fed in a vector with fixed length. It has become a key component in fast-RCNN [36] and faster-RCNN [37] detectors. While the CNN use max-pooling, we use average-pooling as it can be computed efficiently via integral maps [31].

Both [14,17] ignored the requirement for scale invariance. This is another problem of their strategy. Though their strategy will work, the scale variance problem will undoubtedly degrade performance. In this paper, we relieve the scale variance problem using a divide and conquer strategy for different scales. Some work explicitly models the differences of different scales. Rajaram et al. [38] trained different models for different scales and at test time the detection results were combined. Yan et al. [39] extended the idea to the popular Deformable Part Models (DPM) detector [22], which applied two resolution-aware transformations P_H and P_L for high and low resolution samples respectively.

Park et al. [40] shared the low resolution model for all samples, and for large-scale samples the high resolution model was added. The idea of [28] is close to ours, which uses two built-in subnetworks to detect pedestrians from different scales. While their model is used for CNN, we adapt this idea to BDT by using soft decision tree.

3. Scale-Aware Pooling

Our work is inspired by MRFC. To our knowledge, it is the first CPU-based solution which only uses a single scale feature maps and a single pedestrian model. In this section, we first give a simple description of the basic idea of MRFC, then explain its weakness in receptive field correspondence and show how to improve it by scale-aware pooling.

The base feature maps used in MRFC are the 10 LUV+HOG channels we describe in Section 2. After the 10 channels are computed for the original image, a 3×3 box filter is applied sequentially to the original 10 channels 6 times, obtaining 70 channels. This process can also be taken as applying 6 convolution kernels with different standard deviation σ to the original 10 channels. Then two edge filters (vertical and horizontal) are applied to each of these channels, yielding 210 channels in total, which are the so called multiresolution filtered channels. To detect pedestrians at multiple scales, pedestrian models of different scales are slid on the computed channels. The classification features are extracted by sampling from the channels in a gridwise manner and the space between grids are adapted to the window size, as illustrated in Figure 2a. In this way, the same number of features is obtained for pedestrians of different sizes. At training time, as opposed to resizing the pedestrian to a fixed size like traditional methods, the features are extracted from the original pedestrian size in the image. At testing time, there is no need to compute an image pyramid for multiscale detection. The 210 channels could be computed very efficiently. Moreover, after the multiresolution filtered channels are computed, the feature accessing only needs a single pixel indexing like ACF detector [5]. Thus the speed of the detector is very fast.

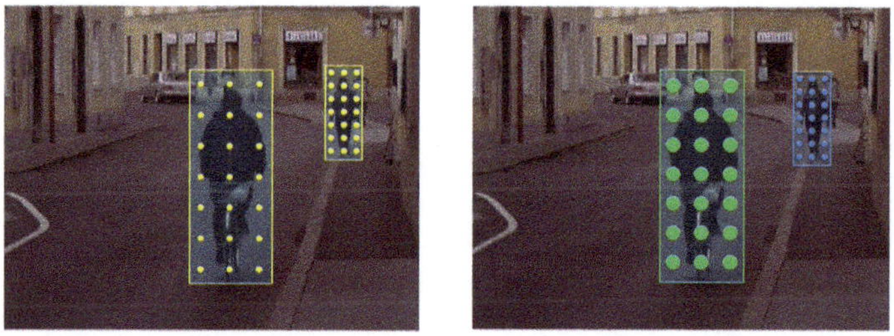

(a) (b)

Figure 2. Illustration of the problem of receptive field correspondence in MRFC method. (**a**) The yellow circles represent the receptive fields of a feature. Note for pedestrians of different scales, the area of the circle do not change, which is unreasonable; (**b**) The ideal circumstance is the receptive field of a feature resizes according to the scale of the pedestrian. Note the areas are different for circles of different colors.

However, there are some shortcomings of the original implementation of the MRFC method. One of them is about the receptive field correspondence. As we stated above, there are totally 7 different receptive field sizes in these channels. The problem is that these 7 sizes are fixed. As shown in Figure 2a, in MRFC's implementation, the receptive field of a feature does not change with the size of pedestrians. Thus a feature's receptive field in a small pedestrian does not correspond to that in a large pedestrian. In this circumstances a feature corresponding to the nose for a large pedestrian may correspond to the whole face for a small pedestrian, which is unreasonable. The ideal circumstance is that the receptive field of a feature resizes along with the scale of the pedestrian, as shown in Figure 2b.

Now we show how to overcome this problem. Similar to SPP, we partition the detection window to $m \times n$ cells, and a feature is computed by average pooling in one or more cells, as illustrated in Figure 3b. In this way, the area of a cell is resized according to the size of the detection window which leads to a better correspondence of features' receptive field. In our work, we use 23×11 cells and the pooling region is constrained to be not larger than a 4×2 cell, which yields 1806 features for each feature map. In MRFC, 7 receptive field sizes are formed by sequentially convolution. In our method, 8 different types of receptive field are formed by combining the basic cells with their areas varying according to the window size.

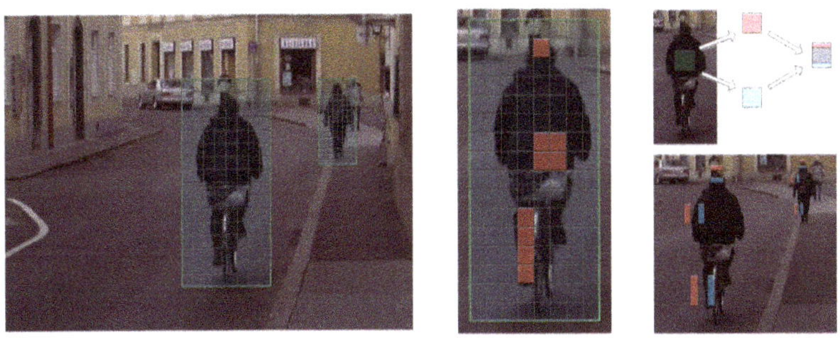

(**a**) (**b**) (**c**)

Figure 3. Illustration of the feature extraction process of our method. (**a**) Feature maps of different scales are divided into the same number of cells whose size vary with the pedestrian size; (**b**) Features are extracted by average pooling in different regions which are composed of one or more cells; (**c**) Top: Pooling in feature gradient maps is equivalent to computing difference of two shifted pooling regions, which has similar effect with Non-Neighboring Features (NNF). Down: Some discriminative features in DICs.

Our baseline implementation also uses the 10 channels and the 20 gradient channels, resulting in 30 feature maps in total. Note unlike other types of convolutional feature maps, the feature gradient maps can be computed very efficiently by using SSE (Streaming SIMD Extensions) instructions. In our experiment, we also test adding motion channels using the method described in [41], which result in another 3 channels (1 base channel and 2 gradient channels). Adding motion channels will increase accuracy, but will significantly slow down detection speed, as shown in Section 6.

To quickly perform average pooling, integral maps [31] are pre-computed before detection, then the sum of feature values in a region will be computed in constant time irrelevant to the region size. The naive implementation computes an integral map for each of the 30 feature maps (10 base feature maps + 20 gradient feature maps), as shown in Figure 4a. However, consider the commutation law for convolution, we have

$$\Omega(x,y) * G(x,y) * u(x,y) = \Omega(x,y) * u(x,y) * G(x,y)\,, \quad (1)$$

where Ω is the feature map, $G(x,y)$ is the gradient filter and $u(x,y)$ is the step function. Note that integration could be taken as convolution with the step function. We only need to compute the integral maps of the original 10 feature maps and then compute 20 gradient maps on the computed integral map, as shown in Figure 4b. We call these channels difference integral channels (DICs). In addition to gradient filters, this strategy could also be used for other linear filters. Features formed by nonlinear transformations, like word channels [17] or CNN based features (after ReLU layer), can not take such an advantage.

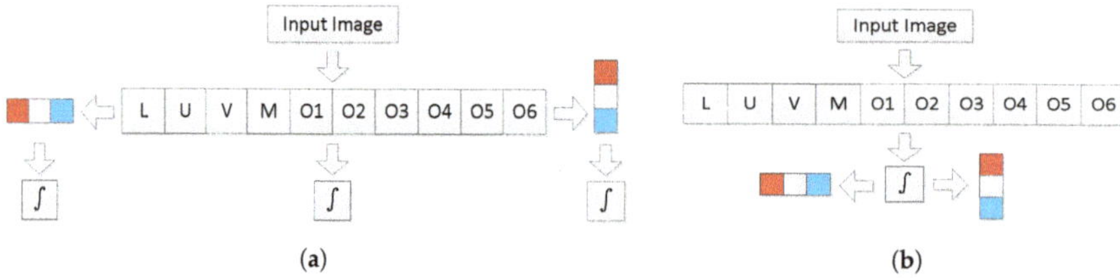

Figure 4. Acceleration strategy for computing integral maps. (**a**) Naive approach. Every feature maps need to be integrated; (**b**) Our method. Only the original 10 feature maps need to be integrated.

In fact, the features extracted in DICs (Figure 3c) resemble the Non-Neighboring Features (NNF) [27] which are demonstrated to be effective for pedestrian detection. NNF are differences of non-neighboring rectangular areas in the same horizontal. Our DICs are superior to NNF in the following aspects. First, NNF use a fixed model size, so an image pyramid is required, whereas we do not construct image pyramid, which saves a lot of computation. Second, accessing a NNF feature needs to compute average pooling for two regions, whereas in our method, we only need to perform average pooling for one region. Third, the NNF implementation only considers two regions in the same horizontal, whereas we also include the two regions in the same vertical which may be useful to represent some pedestrian structure like head, shoulder, and feet. Figure 3c shows some discriminative features in DICs.

By definition, an average pooling feature is computed by first sum all the feature values in the region, then divided by the region size. For efficiency, we do not perform the dividing operation at test time. Instead, we change the threshold of the decision stumps in each decision tree in advance. That is to say, though we only train one detector, we switch this detector to n detectors where n is the number of the template scales. These detectors have the same feature indexes but with different thresholds which are multiplied by their corresponding region size.

4. Soft Decision Tree

The authors of MRFC ignore the requirement for scale invariance and argue that the BDT will handle the intra-class variance caused by using features which are not scale invariant. The experiments in [14] show the feasibility of this idea. However, this strategy undoubtedly leads to a more diverse feature distribution and the decision surface between positive and negative samples become more complex. This leaves a more difficult classification task to the BDT cascade. Hence we believe if we could relive the scale variance problem to some extent, a better result is expected.

A naive solution is to using different models for each scale, which is adopted in [33]. This leads to a more training expense. Furthermore, there is a dilemma in how to use the training data. If we train a specific model using samples of its corresponding scale, we need to split the training data for each scale which leads to insufficient training data. On the other hand, if we train each model using all the samples by resizing all the samples to the corresponding scale, the blurring artifacts caused by the resizing operation [33] leads to unreliable training data.

Here we propose to use a soft decision tree [42] to build each weak classifier in the BDT, where the two branches of the tree specialize in large and small scale pedestrians respectively. In this solution all the samples are involved in training in a single BDT cascade.

While a hard decision node used in hard decision trees deterministically directs a sample x to one of its children, a soft decision node directs a sample x to both its left and right branches with probabilities $P(L|x)$ and $P(R|x)$ and the output of the soft decision node is computed by weighted sum of the output of its two branches, as shown in Figure 5. The soft decision tree in [42] use soft decision node for all the non-leaf nodes of the tree. In testing time, every instance must pass through all the intermediate nodes until it reaches the leaf nodes. This scheme is only suitable for a single decision tree. For BDT with thousands of trees, this scheme leads to high computation cost. Thus we simplify it by only using the soft decision node for the root node. As we are dealing with the multiscale detection problem, we make the left branch responsible for large scale pedestrians' classification and the right branch responsible for small scale pedestrians' classification. The following gate function is used to define the probabilities:

$$P(L|x) = \frac{1}{1 + \exp[\frac{\overline{h} - h_x}{\beta}]} \tag{2}$$

$$P(R|x) = 1 - P(L|x)\,, \tag{3}$$

where $\overline{h}$ is the mean of pedestrian heights in the training set, h_x is the height of the sample x, β is a hyper parameter which control the amount of smoothing for the gate function. As h_x increases, $P(L|x)$ gets larger and sample x direct more weight to the left branch. β reflects the degree of similarity of different scale pedestrians. When β is large, the function $P(L|x)$ is gentle which means we think different scale instances are very similar. The extreme case is $\beta \to \infty$ which means we treat instances of all scales equally and the tree becomes an average ensemble. when β is small, the function $P(L|x)$ is steep which means we think different scale instances have very different representation and need to be tackled differently. The extreme case is $\beta \to 0$ and the node becomes a hard decision node.

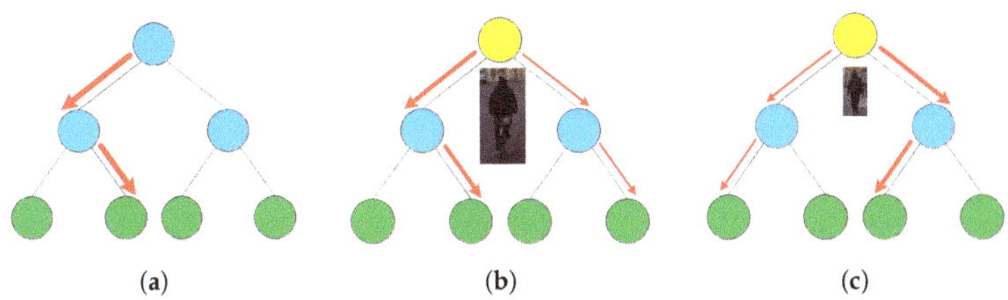

Figure 5. Comparison between hard decision trees and soft decision trees. The blue, yellow and green nodes denote hard decision node, soft decision node and leaf nodes respectively. The red arrows denote the flow of sample weights. (**a**) The hard decision tree is composed of hard decision nodes and leaf nodes. For a given sample, the hard decision node direct all its weight to one of its children; (**b**) The root node of the soft decision tree is a soft decision node which directs the sample weight to both its children according to the sample size. Given a large sample, the soft decision node directs more weight to its left branch. Note the arrow of the left branch is thicker than the arrow of the right branch; (**c**) Another example of the soft decision tree with a small sample. The soft decision node directs more weight to its right branch.

After the sample x is directed to both branches of the root node, it is evaluated by the two branches to get $P(y = 1|x, L)$ and $P(y = 1|x, R)$ where $y \in \{-1, +1\}$ is the sample label. Then, the probability of x to be a positive sample given by the whole soft decision tree is

$$P(y = 1|x) = P(L|x)P(y = 1|x, L) + P(R|x)P(y = 1|x, R)\,. \tag{4}$$

Note that we use RealBoost [43] to train our BDT in which the leaf node do not outputs the probability, but the half log ratio

$$f(x) = \frac{1}{2}\log\frac{p(x)}{1 - p(x)}\,, \tag{5}$$

where $p(x)$ is fraction of the positive sample weight in the leaf node. To get $P(y = 1|x, L)$ and $P(y = 1|x, R)$ in Equation (4), we need to switch the $f(x)$ to its corresponding probability by the inverse function of Equation (5)

$$p(x) = \frac{e^{2f(x)}}{1 + e^{2f(x)}} . \tag{6}$$

Because in Realboost's updating rule each weak classifier should output half log ratio, after we get $P(y = 1|x)$ using Equation (4), we need to switch it to its corresponding half log ratio using Equation (5). At training time, when the half log ratio is acquired for every sample, the sample weights are updated as the commonly used RealBoost.

At testing time, we need to compute Equation (6) two times for each soft decision tree, one for each branch. For efficiency, we switch the half log ratio to its corresponding probability using Equation (6) for all the tree nodes after training. By doing this, we avoid computing Equation (6) at testing time.

5. Accelerating Sliding Window Classifications

The advantage of using a single scale feature maps is that compared with traditional method which construct a dense pyramid, it greatly saves computation cost in the feature maps computation stage. However, apart from feature maps computation, there is another time consuming process in pedestrian detection: sliding windows classification. Because our method is based on regional average pooling, computation cost is higher at this stage compared to some other detectors like ACF [5] and LDCF [25]. For these detectors, after the feature pyramid is constructed, accessing a feature by a tree node only needs one pixel accessing, while for our method, it needs 4 pixel accessing and 3 plus/minus operation. Furthermore, the computation cost is doubled by our soft decision tree. Thus our single scale feature maps strategy must be followed by a efficient sliding window classification strategy, otherwise its advantage is limited because the later one will dominate the computation time. Luckily, based on the characteristics of pedestrian detection problem, there are various ways to accelerate this stage [14,44,45]. To demonstrate the advantage of using single scale feature maps, in this section we introduce two simple strategies to accelerate sliding window classification which will be used in our experiments.

5.1. Ground Plane Constraint

For images captured by a fixed in-vehicle camera, pedestrians of a certain scale will not appear in some positions because of the ground plane constraint (GPC) [19], as shown in Figure 6a.

The key idea of GPC is that under some assumptions [40] which are valid for an in-vehicle camera, the projected height h and the vertical position y of a pedestrian exhibit a linear relationship. GPC has been widely used for pedestrian detection [14,15,23,40,46,47]. There are different ways to use GPC. Some of them adopt a post processing strategy using support vector machine (SVM) [15,40]. This type of methods aims to increase the detection accuracy, but lead to additional computation cost. Since our purpose is to accelerate sliding window classification, this type of methods is not suitable. Here we adopt the method proposed in our previous work [48] in which the possible position (h, y) of a pedestrian is bounded by two straight lines, as shown in Figure 6b. At detection stage, we only scan the possible pedestrian positions hence save computation cost. This method also has positive impact on accuracy for it gets rid of some false positives. Of course, there still exists some risk that in some special cases, some true positives in the testing set are not bounded by the two lines. In practice, we found the positive impact dominates.

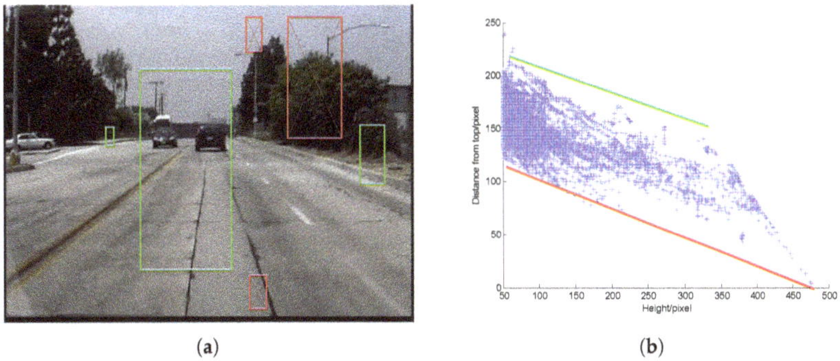

Figure 6. Illustration of GPC. (**a**) A pedestrian may be bounded by the green boxes, but may not be bounded by the red boxes; (**b**) (h, y)s of the pedestrian windows in the Caltech training set. They can be bounded by two straight lines.

5.2. Sparse Grid Detection

For a 480 × 640 image, there are more than 330,000 candidate windows (different scales and different positions) to be classified and most of them belong to background area. Note what we really want is the peak score windows which is the window with the local maximum score. Any windows with lower detection score in its neighbourhood will be suppressed by the peak score window after non-maximum suppression (NMS). To save computation cost, there is no need to evaluate all the candidate windows. We only need to make sure that all the peak score windows are evaluated.

Because the detector responses at nearby locations are correlated, the neighbouring positions of the peak score window usually also have positive responses. In other words, there exists a region of support (ROS) of the peak score window. For BDT cascade, the ROS size decreases with the number of the weak classifiers [44].

Based on this, we begin by evaluate only a sparse grid G_3 with a step size of 3. If a window in G_3 passes k stages of the cascade, every window in its 3 × 3 neighbourhood is triggered to be evaluated. The reason behind this strategy is illustrated in Figure 7. Suppose the window P is a peak score window but not belongs to G_3, there will be a window $x_1 \in G_3$ in its 3 × 3 neighbourhood. Window x_1 tends to have a positive score because of ROS and P will be triggered by x_1. Because G_3 only account for about 1/9 of all the sliding windows and the number of triggered windows tends to be very small, computation cost is greatly reduced.

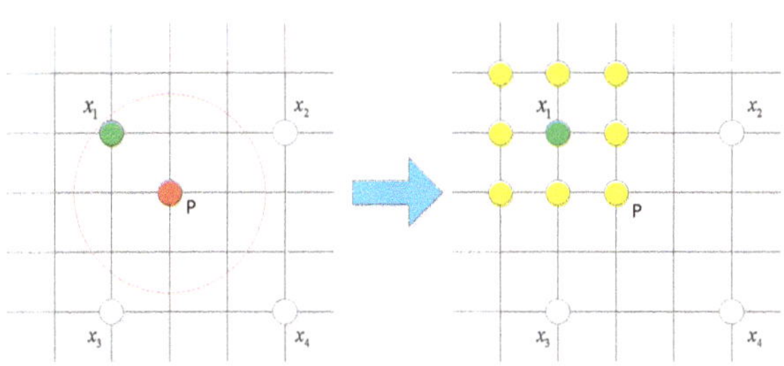

Figure 7. The sparse grid detection strategy. We begin by evaluate only a sparse grid (x_1, x_2, x_3, x_4). Suppose P is a peak score window and its ROS is represented by the red dash line circle. Window x_1 is in the ROS, thus it will passes k stages of the BDT cascade and every window in its 3 × 3 neighbourhood is triggered (yellow circles).

When the k becomes larger, the ROS becomes smaller, the triggered windows become less and the detection speed becomes higher. However, this will increase the probability of missing peak score windows and degrade accuracy. In our experiment, we set $k = 20$ which achieves a good tradeoff between speed and accuracy.

6. Experiments

In this section, we evaluate our proposed method on two standard pedestrian detection datasets: KITTI and Caltech. They are currently the most popular and widely used ones in the literature.

A detected bounding box (bb_d) is taken as a true positive if the Intersection-over-Union (IoU) with a groundtruth bounding box (bb_g) is greater than a threshold. The IoU is defined as

$$\mathrm{IoU}(bb_d, bb_g) = \frac{bb_d \bigcap bb_g}{bb_d \bigcup bb_g}, \tag{7}$$

and for both these two benchmarks, the IoU threshold is set to 0.5. Results on Caltech are compared using miss rate vs. False-Positive-Per-Image (FPPI) curves, which is the well-recognized evaluation metric for pedestrian detection [49]. Methods are ranked by log average miss rate (MR) which is computed by averaging miss rate at 9 FPPI points that are evenly spaced in the log-space ranging from 10^{-2} to 10^{0}. Results on KITTI are compared using precision-recall curves, and methods are ranked by the average precision (AP) at 11 evenly spaced recall points ranging from 0 to 1.

6.1. Experiments on KITTI Dataset

The KITTI object detection benchmark has 7481 training and 7518 test images. It contains three object classes for evaluation: Car, Pedestrian, and Cyclist. Here we only choose pedestrian class for evaluation. KITTI differentiates the difficulty in identifying pedestrians to three levels: easy, moderate and hard, corresponding to different height, occlusion and truncation. Methods are ranked based on the moderate difficult level (the minimum height of bounding box is 25 pixels, the maximum occlusion level is "partly occluded" and the maximum truncation is 0.30).

In order to tune parameter and analysis the impact of different components of our algorithm, we split the training set into training and validation sets. As in [38], we ensure the images of the training and validation sets come from different video sequences and the number of images and pedestrians are comparable. As a result, the training set has 3740 images with 1792 pedestrians and the validation set has 3741 images with 1791 pedestrians. According to the evaluation metric, we need to detect pedestrians taller than 25 pixels. The smallest model size is set as 32×16 (including some background area) and we use 10 scales per octave. The sliding window stride is set to 1/16 of the window height/width in the vertical/horizontal direction. The final classifier is built via three rounds of hard negative mining (starting from a forest with 32 trees, and then 256, 1024, 4096 trees). Realboost [43] are used to train our model and the weak classifiers are level-4 decision trees. In the last round, we switch our RealBoost algorithm to the shrinkage version as is used in [29,50]. The shrinkage parameter is set to 0.5.

As stated in Section 4, the parameter β control the amount of smoothing of the gate function. We first test the effect of different β values. The result is listed in Table 1. From this table we see $\beta = 50$ performs best, hence in the following experiment we set $\beta = 50$.

Table 1. AP on KITTI validation set using different β.

β	40	45	50	55	60
AP	68.02%	68.87%	69.25%	68.10%	67.81%

Next we evaluate the impact of the five different components of our algorithm on accuracy and speed. The speed is tested on a single core of Intel i7 6700K CPU (4 GHz) and is measured as frames per second (FPS). To simplify notation, we use the following abbreviation: scale-aware pooling (SAP), soft decision tree (SDT), ground plane constraint (GPC) and sparse grid detection (SGD). Table 2 shows the impact of different combinations of the five components. We divide the five components into two categories. One category includes SAP, SDT and motion features, which are used for improving accuracy. Another category includes GPC and SGD, which are used for accelerating detection. Any AP/FPS value in the table corresponds to a detector combining some components for accuracy and some components for speed. For example, the AP/FPS value in the last row and the third column (70.53/1.65) denotes the performance of the detector combining SAP, SGD, motion features and GPC. Note SAP serves as our baseline detector, thus all the combinations in the table include SAP.

From this table, we could easily see the impact of different components by comparing the accuracy and speed between rows or columns. For example, By comparing the second and the third rows, we see that using soft decision tree definitely improve accuracy, but decrease speed. This is because for a hard decision tree, a sample is only directed to one branch of the root node, while for a soft decision tree, a sample is directed to both branches of the root node, hence the computation cost is doubled. By comparing the third and the forth rows, we see that adding motion features has the same effect: improving accuracy and decreasing speed. By comparing the second and the third columns, we see that though we use GPC to accelerate detection, it also improves accuracy for it gets rid of some false positives. By comparing the third and the forth columns, we see that using sparse grid evaluation significantly accelerates detection and slightly decreases accuracy. The fast version of our algorithm (67.29%/6.85FPS) is combining SAP, GPC and SGD, while combining SAP, SDT, motion features and GPC achieves the highest AP (70.53%/1.65FPS).

Table 2. AP and FPS on KITTI validation set under different setups of our method.

AP(%)/FPS — Component for Speed / Component for Accuracy	No Acceleration	+GPC	+SGD
SAP	67.17/1.54	67.85/3.85	67.29/6.85
+SDT	69.25/0.68	69.98/1.94	69.73/4.89
+Motion	69.98/0.61	70.53/1.65	70.07/3.40

To test our method on the test set, we train another model using the whole training set and all the components of our algorithm. Because of more training data, level-5 decision tree is used and the resultant detector is a little slower (3.37 fps) than the validation version. We use this model to detect pedestrians in the test images. Because the annotations of the test set are not public, the detection result is submitted to the KITTI evaluation server to get the evaluation result. We compare our method with some state-of-the-art non deep learning methods (which are listed on the KITTI benchmark website http://www.cvlibs.net/datasets/kitti/), as shown in Figure 8. In the precision-recall plot, for each recall point, the precision is the higher the better. From the figure, we see that our method does not achieve the highest precision for the whole recall range. As we stated at the beginning of this section, the KITTI benchmark use AP to rank different methods. The AP values are given at the figure legend. According to the AP values, our method outperforms all the other methods.

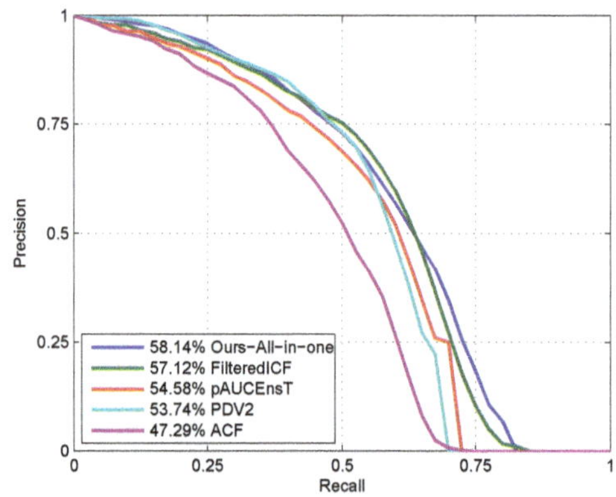

Figure 8. Comparison with non deep learning methods on the KITTI dataset. Our method does not achieve the highest precision for the whole recall range, but based on AP, our method outperforms the other methods.

6.2. Experiments on Caltech Dataset

Caltech pedestrian datasets is currently the largest and the most widely used pedestrian detection dataset. It is more suitable for our method, because the amount of training data is much larger than that of KITTI dataset. It enables a comparison among more than 60 state-of-the-art approaches published during recent years. It consists of 250,000 labeled 640 × 480 frames (in 137 approximately minute long segments) which are divided into 11 sessions. The first 6 sessions are used for training and the last 5 sessions are used for testing. The standard evaluation is performed on every 30th frame of the test set, which yields 4024 images in all. Unless otherwise noted, the results are evaluated using the reasonable difficulty which means the pedestrian is at least 50 pixels in height with a visibility of at least 65%.

Following [25], our training images are collected by sampling one image out of every 4 consecutive frames. Because the evaluation metric only needs to detect pedestrians taller than 50 pixels, the smallest model size is set as 64 × 32. For efficiency, we downsample the channel by 2×, thus the feature map size of the smallest model is 32 × 16. Since there is more training data, we use twice the weak classifier number in each bootstrapping round (that is, starting from a model with 64 trees, and then 512, 2048, 8196 trees).

We compare our method with four representative methods (which are listed on the Caltech benchmark website http://www.vision.caltech.edu/Image_Datasets/CaltechPedestrians/): WordChannels [17], MRFC+Semantic [14], LDCF++ [15] and SDS-RCNN [13]. SDS-RCNN and LDCF++ are currently the best deep learning and non-deep learning methods reported on Caltech benchmark respectively. WordChannels and MRFC+Semantic are another two methods which also based on single scale feature maps like our method. MRFC+Semantic is the MRFC detector enhanced by semantic segmentation channels which need extra data set to train. The original MRFC achieves a MR of 19.09% with a speed of 20FPS. MRFC+Semantic achieves higher accuracy (16.84%) but slow down detection speed (8FPS).

The miss rate vs. FPPI plot is shown in Figure 9. In this plot, for each FPPI point, the miss rate is the lower the better. From the figure we see that among non deep learning methods, our method achieves the lowest miss rate for the whole FPPI range. The MR values are given at the figure legend. According to the MR values, our method outperforms all the other non deep learning methods.

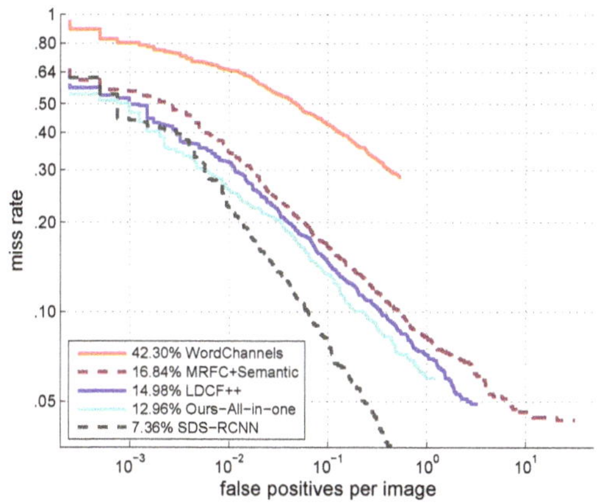

Figure 9. Comparison with the top methods and single scale feature maps-based methods on the Caltech dataset. Our method achieves the lowest miss rate for the whole FPPI range in all the non deep learning methods.

As in the KITTI experiment, we evaluate the impact of the five different components of our algorithm for Caltech dataset. The result is shown in Table 3. The trend is similar with that in the KITTI experiment. The fastest version (15.97%/27.72FPS) of our algorithm is by combining SAP, GPC and SGD, while the most accurate version is by combining SAP, SDT, motion features and GPC (12.62%/3.11FPS). Note that all the detector versions in the table achieve higher accuracy than MRFC + Semantic, including our baseline detector (only use SAP). Some of them are also faster than MRFC + Semantic.

Table 3. MR and FPS on Caltech test set under different setups of our method.

MR(%)/FPS — Component for Speed / Component for Accuracy	No Acceleration	+GPC	+SGD
SAP	16.45/9.62	15.89/18.36	15.97/27.72
+SDT	14.34/3.75	13.60/9.95	13.84/20.15
+Motion	13.08/2.29	12.62/3.11	12.96/3.72

We also evaluate our detector under conditions of small scale, atypical aspect ratio and partial occlusion, as shown in Figure 10. Small scale means the pedestrian height is between 50 px and 80 px. Aspect ratio is computed as w/h, where w and h is the pedestrian width and height respectively. According to [49], the distribution of w/h is centered at 0.41. Atypical aspect ratio is defined as $|w/h - 0.41| \geq 0.1$. Partial occlusion means a pedestrian is occluded, but not more than 35%. From the figure we see that our detector also ranks the first among all the non deep learning methods under these conditions.

Though the accelerating strategies for sliding window classification can also be used in pyramid-based methods, the feature pyramid computing takes too much time and the accelerating effect is limited. For example, the second best non deep learning method LDCF++ [15] needs 0.65 s to construct a feature pyramid for a 480 × 640 image. Thus no matter what strategies are used in the sliding window classification, the detection speed could not be more than 1.54 FPS. While for our method (without motion channels), the feature map computing takes only 0.02 s using the same hardware, thus further speeding up is expected by using a better strategy in sliding window classification stage. For motion channels, though we do not construct pyramid either, even computing a single scale motion channels is very time consuming.

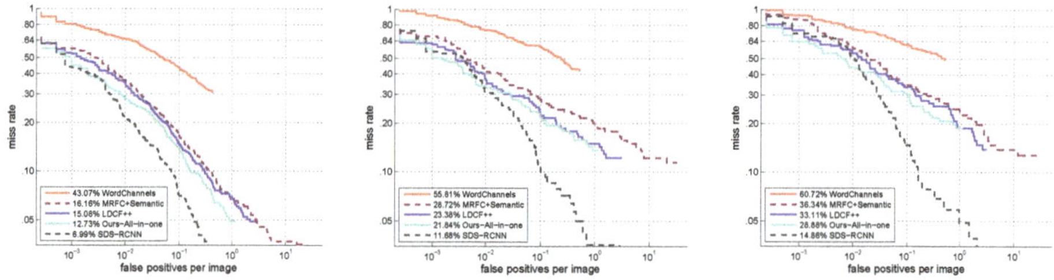

(a) (b) (c)

Figure 10. Evaluation results under conditions of small scale, atypical aspect ratio and partial occlusion. (**a**) Small scale (50 px$\leq h \leq$80 px); (**b**) Atypical aspect ratio ($|w/h - 0.41| \geq 0.1$); (**c**) Patial occlusion (0–35% occluded).

In Figure 11, we show miss rate (the lower the better) versus FPS (the higher the better) of different algorithms which have given their detection time. We include two versions of our algorithm which with and without using motion channels. From the figure, we see both versions of our algorithm achieve high accuracy and competitive speed.

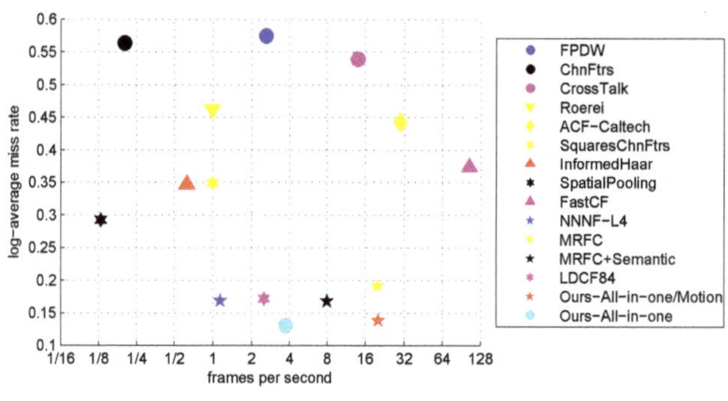

Figure 11. MR versus FPS on the Caltech Dataset.

7. Conclusions

In this paper, we delve deep into the problem of multiscale pedestrian detection via single scale feature maps. Our work is inspired by MRFC, which achieves fast detection using single scale feature maps. We analysis two shortcomings of MRFC and propose scale-aware pooling and soft decision tree to tackle with them respectively. Because our solution leads to more computation cost in sliding window classification stage, we propose to use GPC and SGD to decrease the computation cost in this stage. Experiments on KITTI and Caltech dataset show our solution achieves high accuracy with fast detection speed.

Acknowledgments: This work was supported by National Natural Science Foundation of China (61771107).

Author Contributions: Xinchuan Fu conceived the work, designed the algorithms, performed the experiments and wrote the paper. Rui Yu, Weinan Zhang, Jie Wu contributed through amending the paper. Shihai Shao supervised the work and amend the paper.

Conflicts of Interest: The authors declare no conflict of interest.

References

1. Dollár, P.; Wojek, C.; Schiele, B.; Perona, P. Pedestrian detection: A benchmark. In Proceedings of the IEEE Computer Society Conference on Computer Vision and Pattern Recognition (CVPR 2009), Miami, FL, USA, 20–25 June 2009; pp. 304–311.
2. Geiger, A.; Lenz, P.; Urtasun, R. Are we ready for autonomous driving? The KITTI vision benchmark suite. In Proceedings of the IEEE Conference on Computer Vision and Pattern Recognition, Providence, RI, USA, 16–21 June 2012; pp. 3354–3361.
3. Dalal, N.; Triggs, B. Histograms of Oriented Gradients for Human Detection. In Proceedings of the IEEE Computer Society Conference on Computer Vision and Pattern Recognition (CVPR 2005), San Diego, CA, USA, 20–26 June 2005; pp. 886–893.
4. Ess, A.; Leibe, B.; Schindler, K.; Gool, L.J.V. A mobile vision system for robust multi-person tracking. In Proceedings of the IEEE Computer Society Conference on Computer Vision and Pattern Recognition (CVPR 2008), Anchorage, AK, USA, 24–26 June 2008.
5. Dollár, P.; Appel, R.; Belongie, S.J.; Perona, P. Fast Feature Pyramids for Object Detection. *IEEE Trans. Pattern Anal. Mach. Intell.* **2014**, *36*, 1532–1545.
6. Zhang, S.; Benenson, R.; Schiele, B. Filtered channel features for pedestrian detection. In Proceedings of the IEEE Conference on Computer Vision and Pattern Recognition, CVPR 2015, Boston, MA, USA, 7–12 June 2015; pp. 1751–1760.
7. Baek, J.; Kim, J.; Kim, E. Fast and Efficient Pedestrian Detection via the Cascade Implementation of an Additive Kernel Support Vector Machine. *IEEE Trans. Intell. Transp. Syst.* **2017**, *18*, 902–916.
8. Sun, R.; Zhang, G.; Yan, X.; Gao, J. Robust Pedestrian Classification Based on Hierarchical Kernel Sparse Representation. *Sensors* **2016**, *16*, 1296.
9. Kim, J.H.; Hong, H.G.; Park, K.R. Convolutional Neural Network-Based Human Detection in Nighttime Images Using Visible Light Camera Sensors. *Sensors* **2017**, *17*, 1065.
10. He, M.; Luo, H.; Chang, Z.; Hui, B. Pedestrian Detection with Semantic Regions of Interest. *Sensors* **2017**, *17*, 2699.
11. Gerónimo, D.; López, A.M.; Sappa, A.D.; Graf, T. Survey of Pedestrian Detection for Advanced Driver Assistance Systems. *IEEE Trans. Pattern Anal. Mach. Intell.* **2010**, *32*, 1239–1258.
12. Du, X.; El-Khamy, M.; Lee, J.; Davis, L.S. Fused DNN: A Deep Neural Network Fusion Approach to Fast and Robust Pedestrian Detection. In Proceedings of the IEEE Winter Conference on Applications of Computer Vision, WACV 2017, Santa Rosa, CA, USA, 24–31 March 2017; pp. 953–961.
13. Brazil, G.; Yin, X.; Liu, X. Illuminating Pedestrians via Simultaneous Detection and Segmentation. In Proceedings of the IEEE International Conference on Computer Vision, ICCV 2017, Venice, Italy, 22–29 October 2017; pp. 4960–4969.
14. Costea, A.D.; Nedevschi, S. Semantic Channels for Fast Pedestrian Detection. In Proceedings of the IEEE Conference on Computer Vision and Pattern Recognition, CVPR 2016, Las Vegas, NV, USA, 27–30 June 2016; pp. 2360–2368.
15. Ohn-Bar, E.; Trivedi, M.M. To boost or not to boost? On the limits of boosted trees for object detection. In Proceedings of the 23rd International Conference on Pattern Recognition, ICPR 2016, Cancún, Mexico, 4–8 December 2016; pp. 3350–3355.
16. Costea, A.D.; Vesa, A.V.; Nedevschi, S. Fast Pedestrian Detection for Mobile Devices. In Proceedings of the IEEE 18th International Conference on Intelligent Transportation Systems, ITSC 2015, Gran Canaria, Spain, 15–18 September 2015; pp. 2364–2369.
17. Costea, A.D.; Nedevschi, S. Word Channel Based Multiscale Pedestrian Detection without Image Resizing and Using Only One Classifier. In Proceedings of the IEEE Conference on Computer Vision and Pattern Recognition, CVPR 2014, Columbus, OH, USA, 23–28 June 2014; pp. 2393–2400.
18. Fu, X.; Yu, R.; Shao, S. Fast Pedestrian Detection Using Scale-aware Pooling. In Proceedings of the International Conference on Digital Image Processing, ICDIP 2018, Shanghai, China, 11–14 May 2018; Accepted.
19. Benenson, R.; Omran, M.; Hosang, J.H.; Schiele, B. Ten Years of Pedestrian Detection, What Have We Learned? In Proceedings of European Conference on Computer Vision—ECCV 2014 Workshops, Zurich, Switzerland, 6–7 and 12 September 2014; pp. 613–627.

20. Zhu, Q.; Yeh, M.; Cheng, K.; Avidan, S. Fast Human Detection Using a Cascade of Histograms of Oriented Gradients. In Proceedings of the IEEE Computer Society Conference on Computer Vision and Pattern Recognition (CVPR 2006), New York, NY, USA, 17–22 June 2006; pp. 1491–1498.
21. Maji, S.; Berg, A.C.; Malik, J. Classification using intersection kernel support vector machines is efficient. In Proceedings of the IEEE Computer Society Conference on Computer Vision and Pattern Recognition (CVPR 2008), Anchorage, AK, USA, 24–26 June 2008.
22. Felzenszwalb, P.F.; Girshick, R.B.; McAllester, D.A.; Ramanan, D. Object Detection with Discriminatively Trained Part-Based Models. *IEEE Trans. Pattern Anal. Mach. Intell.* **2010**, *32*, 1627–1645.
23. Marín, J.; Vázquez, D.; López, A.M.; Amores, J.; Leibe, B. Random Forests of Local Experts for Pedestrian Detection. In Proceedings of the IEEE International Conference on Computer Vision, ICCV 2013, Sydney, Australia, 1–8 December 2013; pp. 2592–2599.
24. Dollár, P.; Tu, Z.; Perona, P.; Belongie, S.J. Integral Channel Features. In Proceedings of the British Machine Vision Conference, BMVC 2009, London, UK, 7–10 September 2009; pp. 1–11.
25. Nam, W.; Dollár, P.; Han, J.H. Local Decorrelation For Improved Pedestrian Detection. In Proceedings of the Advances in Neural Information Processing Systems 27: Annual Conference on Neural Information Processing Systems 2014, Montreal, QC, Canada, 8–13 December 2014; pp. 424–432.
26. Zhang, S.; Bauckhage, C.; Cremers, A.B. Informed Haar-Like Features Improve Pedestrian Detection. In Proceedings of the IEEE Conference on Computer Vision and Pattern Recognition, CVPR 2014, Columbus, OH, USA, 23–28 June 2014; pp. 947–954.
27. Cao, J.; Pang, Y.; Li, X. Pedestrian Detection Inspired by Appearance Constancy and Shape Symmetry. In Proceedings of the IEEE Conference on Computer Vision and Pattern Recognition, CVPR 2016, Las Vegas, NV, USA, 27–30 June 2016; pp. 1316–1324.
28. Li, J.; Liang, X.; Shen, S.; Xu, T.; Yan, S. Scale-aware Fast R-CNN for Pedestrian Detection. *IEEE Trans. Multimed.* **2017**, *20*, 985–996.
29. Hu, Q.; Wang, P.; Shen, C.; van den Hengel, A.; Porikli, F.M. Pushing the Limits of Deep CNNs for Pedestrian Detection. *IEEE Trans. Circuits Syst. Video Technol.* **2017**, doi:10.1109/TCSVT.2017.2648850.
30. Cao, J.; Pang, Y.; Li, X. Learning Multilayer Channel Features for Pedestrian Detection. *IEEE Trans. Image Process.* **2017**, *26*, 3210–3220.
31. Viola, P.A.; Jones, M.J. Robust Real-Time Face Detection. *Int. J. Comput. Vis.* **2004**, *57*, 137–154.
32. Dollár, P.; Belongie, S.J.; Perona, P. The Fastest Pedestrian Detector in the West. In Proceedings of the British Machine Vision Conference, BMVC 2010, Aberystwyth, UK, 31 August–3 September 2010; pp. 1–11.
33. Benenson, R.; Mathias, M.; Timofte, R.; Gool, L.J.V. Pedestrian detection at 100 frames per second. In Proceedings of the IEEE Conference on Computer Vision and Pattern Recognition, Providence, RI, USA, 16–21 June 2012; pp. 2903–2910.
34. He, K.; Zhang, X.; Ren, S.; Sun, J. Spatial Pyramid Pooling in Deep Convolutional Networks for Visual Recognition. *IEEE Trans. Pattern Anal. Mach. Intell.* **2015**, *37*, 1904–1916.
35. Lazebnik, S.; Schmid, C.; Ponce, J. Beyond Bags of Features: Spatial Pyramid Matching for Recognizing Natural Scene Categories. In Proceedings of the IEEE Computer Society Conference on Computer Vision and Pattern Recognition (CVPR 2006), New York, NY, USA, 17–22 June 2006; pp. 2169–2178.
36. Girshick, R.B. Fast R-CNN. In Proceedings of the IEEE International Conference on Computer Vision, ICCV 2015, Santiago, Chile, 7–13 December 2015; pp. 1440–1448.
37. Ren, S.; He, K.; Girshick, R.B.; Sun, J. Faster R-CNN: Towards Real-Time Object Detection with Region Proposal Networks. In Proceedings of the Advances in Neural Information Processing Systems 28: Annual Conference on Neural Information Processing Systems 2015, Montreal, QC, Canada, 7–12 December 2015; pp. 91–99.
38. Rajaram, R.N.; Ohn-Bar, E.; Trivedi, M.M. Looking at Pedestrians at Different Scales: A Multiresolution Approach and Evaluations. *IEEE Trans. Intell. Transp. Syst.* **2016**, *17*, 3565–3576.
39. Yan, J.; Zhang, X.; Lei, Z.; Liao, S.; Li, S.Z. Robust Multi-resolution Pedestrian Detection in Traffic Scenes. In Proceedings of the IEEE Conference on Computer Vision and Pattern Recognition, Portland, OR, USA, 23–28 June 2013; pp. 3033–3040.
40. Park, D.; Ramanan, D.; Fowlkes, C.C. Multiresolution Models for Object Detection. In Proceedings of the Computer Vision—ECCV 2010, 11th European Conference on Computer Vision, Heraklion, Crete, Greece, 5–11 September 2010; pp. 241–254.

41. Park, D.; Zitnick, C.L.; Ramanan, D.; Dollár, P. Exploring Weak Stabilization for Motion Feature Extraction. In Proceedings of the IEEE Conference on Computer Vision and Pattern Recognition, Portland, OR, USA, 23–28 June 2013; pp. 2882–2889.
42. Irsoy, O.; Yildiz, O.T.; Alpaydin, E. Soft decision trees. In Proceedings of the 21st International Conference on Pattern Recognition, ICPR 2012, Tsukuba, Japan, 11–15 November 2012; pp. 1819–1822.
43. Friedman, J.H.; Hastie, T.; Tibshirani, R. Additive logistic regression: A statistical view of boosting. *Ann. Stat.* **2000**, *28*, 337–407.
44. Dollár, P.; Appel, R.; Kienzle, W. Crosstalk Cascades for Frame-Rate Pedestrian Detection. In Proceedings of the Computer Vision—ECCV 2012—12th European Conference on Computer Vision, Florence, Italy, 7–13 October 2012; pp. 645–659.
45. Gualdi, G.; Prati, A.; Cucchiara, R. Multi-stage Sampling with Boosting Cascades for Pedestrian Detection in Images and Videos. In Proceedings of the Computer Vision—ECCV 2010—11th European Conference on Computer Vision, Heraklion, Crete, Greece, 5–11 September 2010; pp. 196–209.
46. Hoiem, D.; Efros, A.A.; Hebert, M. Putting Objects in Perspective. *Int. J. Comput. Vis.* **2008**, *80*, 3–15.
47. Kim, H.K.; Kim, D. Robust pedestrian detection under deformation using simple boosted features. *Image Vis. Comput.* **2017**, *61*, 1–11.
48. Fu, X.; Yu, R.; Zhang, W.; Feng, L.; Shao, S. Pedestrian Detection by Feature Selectedf Self-Similarity Features. *IEEE Access* **2018**, *6*, 14223–14237.
49. Dollár, P.; Wojek, C.; Schiele, B.; Perona, P. Pedestrian Detection: An Evaluation of the State of the Art. *IEEE Trans. Pattern Anal. Mach. Intell.* **2012**, *34*, 743–761.
50. Paisitkriangkrai, S.; Shen, C.; van den Hengel, A. Strengthening the Effectiveness of Pedestrian Detection with Spatially Pooled Features. In *European Conference on Computer Vision (ECCV)*; Springer: Cham, Switzerland, 2014; pp. 546–561.

sensors

MDPI

Article

Dynamic Non-Rigid Objects Reconstruction with a Single RGB-D Sensor

Sen Wang [1,†], Xinxin Zuo [1,2,*,†], Chao Du [2], Runxiao Wang [1], Jiangbin Zheng [1] and Ruigang Yang [2,3,*,‡]

1 Northwestern Polytechnical University, Xi'an 710072, China; wangsen1312@gmail.com (S.W.); wangrx@nwpu.edu.cn (R.W.); zhengjb@nwpu.edu.cn (J.Z.)
2 University of Kentucky, Lexington, KY 40506, USA; chao.du@uky.edu
3 Baidu Inc., Beijing 100085, China
* Correspondence: xinxin.zuo@uky.edu (X.Z.); ryang@cs.uky.edu (R.Y.)
† These two authors contribute equally to this paper.
‡ This project is from National Engineering Laboratory of Deep Learning Technology and Application, China.

Received: 21 January 2018; Accepted: 14 March 2018; Published: 16 March 2018

Abstract: This paper deals with the 3D reconstruction problem for dynamic non-rigid objects with a single RGB-D sensor. It is a challenging task as we consider the almost inevitable accumulation error issue in some previous sequential fusion methods and also the possible failure of surface tracking in a long sequence. Therefore, we propose a global non-rigid registration framework and tackle the drifting problem via an explicit loop closure. Our novel scheme starts with a fusion step to get multiple partial scans from the input sequence, followed by a pairwise non-rigid registration and loop detection step to obtain correspondences between neighboring partial pieces and those pieces that form a loop. Then, we perform a global registration procedure to align all those pieces together into a consistent canonical space as guided by those matches that we have established. Finally, our proposed model-update step helps fixing potential misalignments that still exist after the global registration. Both geometric and appearance constraints are enforced during our alignment; therefore, we are able to get the recovered model with accurate geometry as well as high fidelity color maps for the mesh. Experiments on both synthetic and various real datasets have demonstrated the capability of our approach to reconstruct complete and watertight deformable objects.

Keywords: 3D reconstruction; RGB-D sensor; non-rigid reconstruction

1. Introduction

3D scanning or modeling is a challenging task that has been extensively studied for decades due to its vast applications in 3D printing, measurement, gaming, etc. The availability of low cost commodity depth sensors, such as Microsoft Kinect, has made the static scene modeling substantially easier than ever. Many scanning systems has been proposed exploiting rigid alignment algorithms to deal with static objects or scenes, e.g., indoor modeling [1–3]. However, the limitation to static or rigid scenarios prevents broader applications where the scene or the subject might move or deform in a non-rigid way. Considering the much higher dimensionality and complexity of the deformation space than purely rigid motion, non-rigid objects modeling in dynamic scenario is much more difficult than the static case. In this paper, we will tackle the 3D modeling problem of deformable objects with a single RGB-D sensor.

There have been ways to accommodate deformable objects using color images to track the motion and then reconstruct the 3D shapes [4–6]. There are quite a lot of previous works [5,7] that have been done on multi-view stereo that utilize multiple color cameras to reconstruct the 3D model by exploiting the photo consistency constraints together with some smoothness regularizations. Vlasic [6] propose deforming a pre-scanned model under the constraints of multiple images. However, those methods suffer from the ambiguities of appearance matching and also the color variation caused by the illumination effects and view changes. More recently, researchers have taken advantages of the depth sensors while adopting the multi-view setup [8–10]. By now, the most recent state-of-art method using multiple cameras has been proposed in [11], which has exploited the temporal information to generate consistent models in time space. Those systems with multiple depth sensors have demonstrated impressive results on dynamic objects modeling. However, they are not portable and often require very precise calibration between those multiple sensors. This makes the 3D modeling with a single depth sensor more attractive.

Many follow-up systems [12,13] have specifically looked at scanning humans where the user rotates in front of the Kinect while maintaining a roughly rigid pose. There are others that incorporate human template (e.g., SCAPE [14] or skeleton [15]) as the prior information and deform the template to align with the input. In this paper, we will focus on reconstruction of deformable objects without any template and also with no need to keep any certain pose. As compared to some previous dynamic fusion works [16–18] that suffer from the error accumulation problem, the main contribution of this paper is that we address this drifting problem by enforcing the loop closure constraints explicitly. We have also exploited the captured color images to resolve the ambiguity that exists in non-rigid surface alignment with purely geometric information.

We propose a global non-rigid registration and fusion optimization framework to deal with the error accumulation problem utilizing both geometric and appearance information. In more detail, first, we decompose the input sequence into continuous segments and fuse the frames in each segment to get a partial model (fragment). Those neighboring fragments can be aligned pairwise under our non-rigid registration approach. Next, we detect the loop between those fragments and establish correspondences between fragments that form a loop. Correspondences from the loop closure constraints together with those achieved from pairwise registration procedure are fed into a global non-rigid registration framework. We will get a fused model after the global registration. Then, this fused model will be taken as a proxy model, which is used to facilitate better alignment between those fragments so that the proxy model gets updated to be confronted with all of those fragments. Finally, we are able to generate a watertight 3D model with consistent and clear color maps.

We have evaluated the proposed approach on both a synthetic dataset and several real datasets of deformable objects captured with an RGB-D sensor. As shown in Figure 1, the experimental results demonstrate that our approach is capable of generating high quality and complete 3D models with high fidelity of recovered color maps.

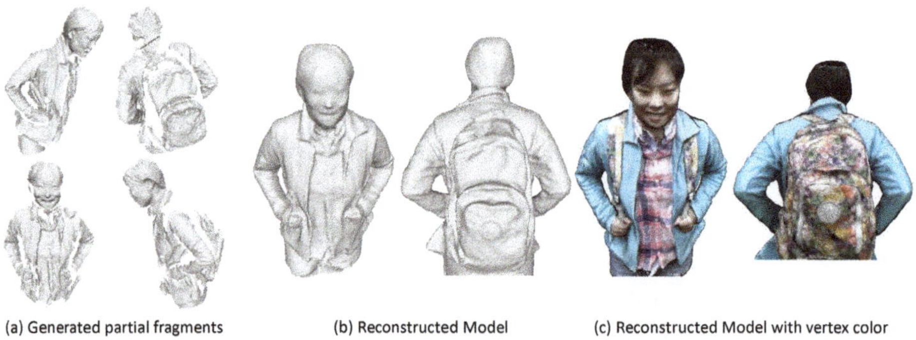

Figure 1. The reconstructed 3D model with our approach. (**a**) some sampled partial scans or fragments; (**b**) the reconstructed model using our approach; (**c**) is our recovered mesh model shown with color maps.

2. Related Works

In this paper, we focus on the 3D modeling of non-rigid deformable objects. It is an even harder problem as compared with the rigid object reconstruction problem considering the more complex non-rigid motion. Researchers have proposed various ways to address this problem.

We review some related approaches that use only a single depth sensor for the non-rigid object reconstruction. There are some papers that specify their modeling targets as some pre-scanned models or human body. The pre-scanned model makes the occlusion problem easier to handle as the overall shape is already available. For example, in Ref. [19], the template is pre-scanned and built up first and then got deformed to fit the input acquired from a depth sensor. Later on, Guo [20] improves the surface tracking performance by incorporating both L0 and L2 regularizations. Refs. [21,22] focus on 3D modeling of human body and exploits the prior knowledge in the form of SCAPE model. Therefore, instead of tracking the deformation of all those vertices on the surface, they solve the coefficients of a SCAPE model. Those prior information or human template are enforced to reduce the search space of the overall solution. Another way of reducing the complexity of the non-rigid reconstruction problem and making it more tractable is to set some restrictions on the movement of the target. For example, Li [12] and Zhu [23] have presented the system that asks the user to rotate in front of the sensor while keeping a certain static pose. In addition, the user is also assumed to perform a loop closure explicitly at the end of the sequence. This is restrictive and it may not be easy to hold the same pose during rotation.

Recently, as an extension to the KinectFusion system, Newcombe et al. [16] has proposed the dynamic fusion approach that takes non-rigid motion into account with a non-rigid warp field updated with respect to every frame. The current input gets fused to the canonical model under the current warp field. Later on, Ref. [17] incorporates sparse feature matches into the framework to resolve the ambiguities in alignment. Guo [24] takes advantage of the dense color information to improve the robustness of surface tracking. They have also decomposed the lighting effect from the image to eliminate the color variation affected by the environment lighting. Yu et al. [25] enforce the skeleton constraints in the typical fusion pipeline to get better performance on both surface fusion and skeleton tracking. Those methods allow the user to move more freely. However, they haven't dealt with the loop closure problem, which makes them not suitable for complete model recovery considering the almost inevitable drifting problem as the sequence proceeds. This issue has been addressed in paper [26] with the proposed non-rigid bundle adjustment method. They have obtained some pleasant results, but the bundle adjustment could be quite expensive and time-consuming due to the number of unknowns and also the search space being quite large. In addition, the recovered color maps of the 3D model is blurry as they haven't incorporated any color information. In this paper, we will deal with the loop closure problem in a more efficient way. Finally, a complete 3D model together with clear color maps will get reconstructed.

3. Pipeline

We illustrate the overall pipeline of our method in Figure 2.

First, given an RGB-D sequence as input, instead of trying to fuse them continuously altogether, we partition it into several segments. We are able to reconstruct a locally precise surface fragment or partial scan from each such segment [17]. Then, those partial scans will be aligned with their neighboring pieces under our pairwise non-rigid registration procedure. Next, we apply a globally non-rigid registration procedure to align those pieces altogether. This is accomplished first by our loop detection process. When the loop is detected, we try to align these pieces that form a loop. Correspondences established between these pieces are enforced in the global registration process. After that, we will get a fused proxy model by merging all the pieces. Finally, in our model update step, we use the proxy model as a starting point to refine the correspondences so as to achieve better alignments afterwards. During the registration process, we have exploited both geometric and color information to register partial pieces and align them altogether. Therefore, we arrive at a complete high quality 3D model together with consistent color maps eventually.

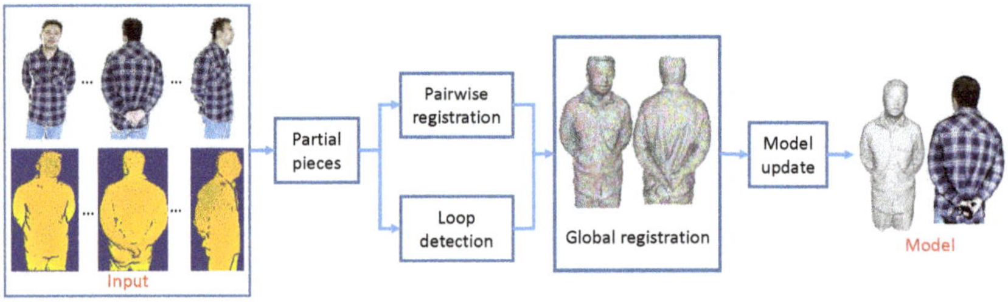

Figure 2. The pipeline of our method.

4. Our Approach

In this section, we will describe our framework step by step with partial pieces generation, pairwise non-rigid registration, loop closure detection, global registration and finally the model-update.

4.1. Partial Pieces Generation and Pairwise Non-Rigid Registration

4.1.1. Partial Pieces Generation

We begin our approach by dividing the input RGB-D sequence into N continuous segments and extract high quality but only partial scans of the model from each segment exploiting the free form dynamic fusion method [17]. In this method, the working space is defined by a volume with each voxel containing the signed distance value with respect to the surface. A rotation and translation vector is also associated with each voxel to describe its motion or deformation from the canonical space. The surface is represented by these signed distance functions. Typically, the first frame of each segment is taken as the canonical frame. For every input frame, the motion field will be calculated and optimized to get the deformed surface to be confronted with the input depth map. Afterwards, the input depth data can be fused into the canonical model under the guidance of the motion field. The signed distance value in each voxel gets updated and the voxel color is also fused. As the sequence proceeds, the canonical model will get enhanced with some geometric details and occlusion parts revealed. Some examples of reconstructed partial scans are demonstrated in Figure 3. More details can be referred in [17]. We denote those reconstructed canonical meshes as $M_1 \sim M_N$. In the meantime, as we keep tracking the motion of each voxel, we will also get the deformed models corresponding to the last frame of every segment. We denote those deformed surfaces as S_1–S_N. Those deformed models will be used to guide the pairwise registration in the next section.

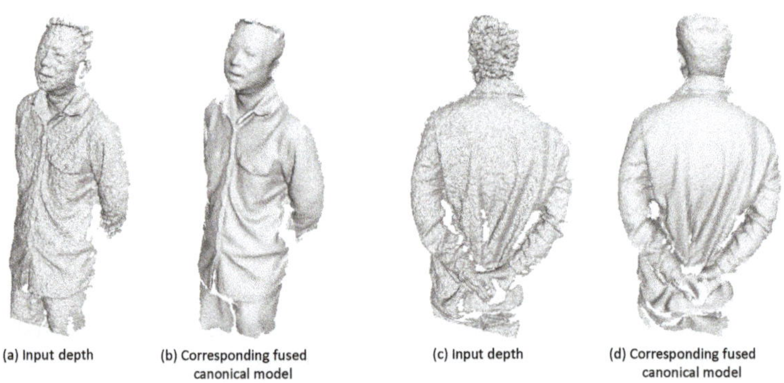

Figure 3. Some sampled partial pieces generated from the fusion procedure. (**a**,**c**) are some input frames; (**b**,**d**) are corresponding partial scans.

Thereafter, our goal is to fuse all those partial pieces M_1–M_N to generate a complete 3D model. We will achieve this in three steps: pairwise registration, global non-rigid registration with loop closure and finally model update/refinement process. Next, we will illustrate each of these steps in detail in the following sections.

4.1.2. Pairwise Non-Rigid Registration

In this section, we describe our approach to register those canonical models non-rigidly and pairwise with their neighboring frames. That is, we try to compute the dense deformation field from M_{k-1} to M_k so that M_{k-1} is aligned with its following neighoring piece M_k. The reason for this pairwise registration is that we want to find the reliable matches between neighboring pieces that can be enforced during the global non-rigid registration process. We accomplish this by exploiting the Embedded Deformation Model [27] to parametrize the deformation of mesh M_{k-1}. The key point is that we do not need to specify and calculate the motion parameters for each vertex. Instead, a set of graph nodes $\mathbf{g}_1$–$\mathbf{g}_l$ are uniformly sampled throughout the mesh and, for each node $\mathbf{g}_i$, it has an affine transformation specified by a 3×3 matrix $\mathbf{A}_i$ and a 3×1 translation vector $\mathbf{t}_i$. For each vertex $\mathbf{v}$, it gets deformed as driven by its K nearest graph nodes with a set of weights $\omega_j(v)$:

$$\phi(\mathbf{v}) = \sum_{j=1}^{K} \omega_j(v)[\mathbf{A}_j(\mathbf{v} - \mathbf{g}_j) + \mathbf{g}_j + \mathbf{t}_j]. \tag{1}$$

In our case, we take M_{k-1} as the source mesh and M_k as the target mesh. We randomly sample a set of graph nodes ($\mathbf{g}_1$–$\mathbf{g}_l$) on the mesh M_{k-1} to build up the embedded graph. In order to find the optimal alignment from mesh M_{k-1} to M_k, deformation parameters $\mathbf{A_1}$–$\mathbf{A_l}$ (denoted as $\mathcal{A}$) and $\mathbf{t_1}$–$\mathbf{t_l}$ (denoted as $\mathcal{T}$) are optimized by minimizing the following objective function:

$$E(\mathcal{A}, \mathcal{T}) = \alpha_r E_r(\mathcal{A}) + \alpha_s E_s(\mathcal{A}, \mathcal{T}) + \alpha_g E_g(\mathcal{A}, \mathcal{T}) + \alpha_c E_c(\mathcal{A}, \mathcal{T}), \tag{2}$$

where α_r, α_s, α_g, α_c are the weights for each term. Next, we explain each of those constraints in detail.

First, the term $E_r(\mathcal{A})$ serves as the as-rigid-as-possible term that specifies that the affine transformations ($\mathbf{A}_1 \sim \mathbf{A}_l$) should try to keep properties of a rotation matrix so as to prevent arbitrary surface distortion:

$$E_r(\mathcal{A}) = \sum_{i=1}^{l} ||\mathbf{A}_i^T \mathbf{A}_i - \mathbf{I}||_F^2. \tag{3}$$

Next, the smoothness constraints $E_s(\mathcal{A}, \mathcal{T})$ assure the similarity of the local transformations between connected graph nodes. This ensures the smooth deformation of neighboring nodes:

$$E_s(\mathcal{A}, \mathcal{T}) = \sum_{(i,j) \in \mu} ||\mathbf{A}_i(\mathbf{g}_j - \mathbf{g}_i) + \mathbf{g}_i + \mathbf{t}_i - (\mathbf{g}_j + \mathbf{t}_j)||_2^2. \tag{4}$$

Finally, the critical part will be how to collect correspondences between the source and target mesh. In this paper, we have incorporated correspondences extracted from both geometric cues and color cues.

For the geometric term E_g, the correspondences between M_{k-1} and M_k are established via the deformed mesh S_{k-1} that we have recorded during the partial piece fusion procedure in Section 4.1.1. The deformed mesh S_{k-1} is supposed to have roughly good initial alignment with M_k, since the mesh S_{k-1} has actually been optimized to confront with the last frame of sequence $k-1$ from the canonical mesh M_{k-1} and this last frame of segment $k-1$ is just the first frame for kth segment. Therefore, we can establish the correspondences between S_{k-1} and M_k using nearest search. After that, those correspondences will be transferred from S_{k-1} to M_{k-1}, given that the corresponding vertices of M_{k-1} and S_{k-1} share the same vertices indexes. This will make the alignment between neighboring segments much easier to achieve. The correspondences are updated after several iterations during the optimization in an ICP manner. We define this term in Equation (5) with C_g denotes the correspondences set:

$$E_g(\mathcal{A}, \mathcal{T}) = \sum_{(\mathbf{p}_i, \mathbf{q}_i) \in C_g} ||\phi(\mathbf{p}_i) - \mathbf{q}_i||_2^2. \tag{5}$$

However, as the nearest searching strategy is not guaranteed to provide the correct matches (as shown in Figure 4b), we also exploit the color information to resolve the ambiguity. Specifically, we need to compute the dense 3D flow from the colored mesh S_{k-1} to M_k. For the classical scene flow computation from two input RGB-D images, we use the two-dimensional image plane as the parameterization domain to optimize the flow field. However, in our case, the unstructured point clouds do not provide a natural parametrization domain.

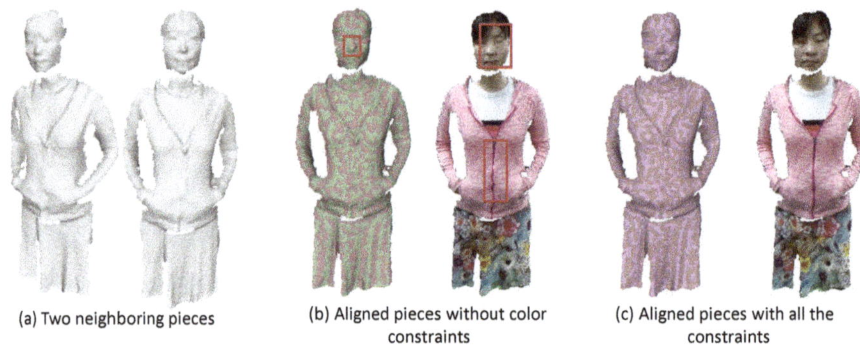

Figure 4. Illustration of pairwise alignment results. (**a**) two neighboring pieces achieved from the above partial piece generation step; (**b**) the alignment results without the color information. These two pieces are shown with different colors in the left of (**b**) so that we can see the alignment result more clearly. The misalignment in the appearance is visualized in the right figure of (**b**) where the two meshes colored by the captured color images are overlaid; (**c**) demonstrates the alignment result with all those constraints in Equation (2) where we can see that the two meshes are well-aligned.

We address this problem by defining a virtual image on the tangent plane of every point and projecting the colored vertices around that point onto the virtual plane. In more detail, for every vertice $\mathbf{p}$ in mesh S_{k-1}, we gather its K nearest connected faces around $\mathbf{p}$ and render this neighboring colored mesh piece orthogonally onto the plane $Pl_{\mathbf{p}}$ defined by the vertice $\mathbf{p}$ and its normal $\mathbf{n}_{\mathbf{p}}$. The rendered image patch I_p can be seen as a local approximation of the colored mesh around vertice $\mathbf{p}$. We parametrize the colored mesh locally by the virtual plane. For the vertice $\mathbf{p}$, the corresponding nearest vertice in mesh M_k has been computed from the geometric term, and we denote it as $\mathbf{q}$. Similarly, we gather the faces around $\mathbf{q}$ and, by rendering this mesh fragment onto the plane $Pl_{\mathbf{p}}$, we get the rendered image patch as I_q. The procedure is illustrated in Figure 5. The dense matches between I_p and I_q are found by the calculation of the flow field between these two image patches followed by a cross check validation step to filter out outliers. Those matches provide us correspondences between mesh S_{k-1} and M_k as we keep track of the mapping from 3D vertices to the rendered 2D image patches.

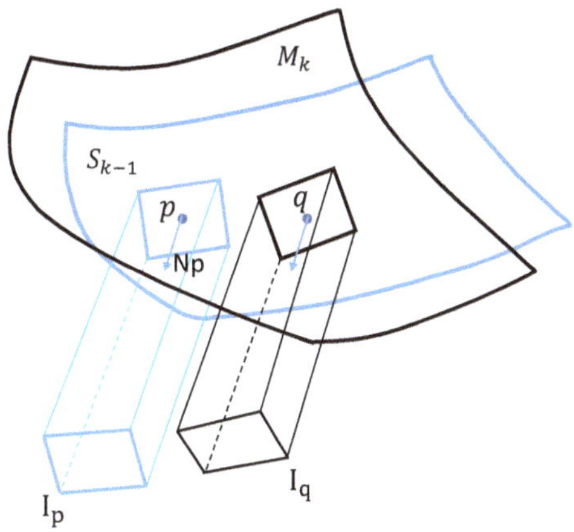

Figure 5. Illustration of the virtual plane to define color matches. $\mathbf{p}$ is a vertice on mesh S_{k-1} with its normal $\mathbf{n_p}$. $\mathbf{q}$ is the nearest vertice to $\mathbf{p}$ on mesh M_k. The neighboring verices around $\mathbf{p}$ and $\mathbf{q}$ are projected under the direction of $\mathbf{n_p}$ to get the rendered image I_p and I_q, respectively. The neighboring might not form a rectangular, and we just use a rectangle box for simplification of illustration.

Another issue that will arise in the above procedure is that for some vertice $\mathbf{p}$ in mesh S_{k-1}, more than one correspondence may be found in mesh M_k since it might be collected by multiple vertices as neighbors. Therefore, we set the correspondence as the median of the multiple corresponding vertices to reduce the affect of outliers. Finally, we get the correspondence set C_c between these two colored meshes after the above process and arrive at the color energy term defined as follows:

$$E_c(\mathcal{A}, \mathcal{T}) = \sum_{(\mathbf{p}_i, \mathbf{q}_i) \in C_c} ||\phi(\mathbf{p}_i) - \mathbf{q}_i||_2^2. \quad (6)$$

We demonstrate the effectiveness of the color correspondences matching term in Figure 4.

By now, with all the constraints defined, we can minimize the objective function of Equation (2) to get the unknown deformation parameters $\mathcal{A}$ and $\mathcal{T}$. This optimization problem can be solved with the Levenberg–Marquardt algorithm. Afterwards, we can apply the deformation field to all the vertices in M_{k-1} via Equation (1) and we will get the deformed mesh T_{k-1} that is aligned with M_k.

4.2. Loop Detection and Global Non-Rigid Registration

After the above pairwise alignments of meshes from neighboring segments, we are ready to find reliable correspondences between neighboring pieces. We can certainly align those canonical models incrementally into the first piece using techniques as described in Section 4.1.2. However, the drifting problem is almost inevitable during the sequential alignment. As shown in Figure 6, the large gaps between the first and last pieces stops the sequential alignment strategy from getting complete and visually plausible models. We argue that the two key aspects of assembling those pieces are loop detection and global non-rigid registration. We describe these two procedures in the following sections.

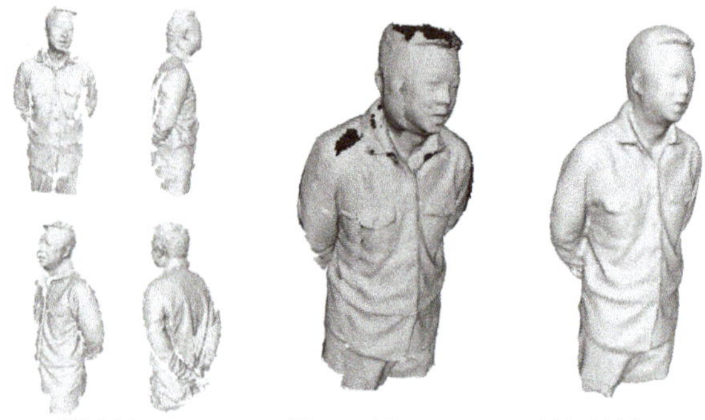

Figure 6. Illustration of drifting effect of incremental alignment and comparison with the result after our global registration. (**a**) some sampled partial pieces; (**b**) the result from incremental alignment where large gaps exist; (**c**) the result after global registration.

4.2.1. Loop Detection

In our case, loop detection is to find the partial pieces that have sufficiently large overlap with the first piece, that is, while the subject rotates in front of the sensor, we want to find the piece where he/she has rotated all around and arrived back to the first frame. In this section, we develop the loop detection strategy exploiting those SIFT features as similar to the loop detection in the SLAM system [3], where the Bag of Words has often been used. During the partial pieces generation procedure (Section 4.1.1), the SIFT features have been extracted and matched to assist the tracking [17]. For each frame, the matched features are lifted and stored in the 3D canonical space. Therefore, for each mesh M_k, we have some sparse features associated with it.

Our goal will be to find some pieces among $M_2 \sim M_N$ that have great overlap with M_1 given those canonical models $M_1 \sim M_N$ with sparse SIFT feature descriptors attached.

First, for each model M_k ($k = 2$ to N), we find the matches of SIFT features within certain matching threshold between mesh M_1 and M_k. Then, we evaluate the degree of coverage of those matches with respect to the surface. It is assumed that, if the matches reside only on a small part of the model, it implies that these two models do not have sufficient overlap. Otherwise, these matches would spread over the surface. However, it is still not sufficient to simply use this to define the extent of overlap, as the SIFT vertices might scatter over the surface unevenly, which will also cause the uneven distribution of overlap over the surface.

Thus, taking both factors into consideration, we evaluate the coverage adaptively on different regions depending on the distribution of the SIFT vertices. First, we sample a set of vertices ($\mathcal{P}$) over the surface and we compute the coverage degree of SIFT features around each sampled vertice. The coverage degree f_s of SIFT features for each vertice $\mathbf{v}$ is measured via $f_s(\mathbf{v}) = exp(-d_s(\mathbf{v})^2/\sigma_{ds}^2)$, where $d_s(\mathbf{v})$ is the distance to the nearest feature on the mesh. Similarly, the coverage degree of matches $f_m(\mathbf{v})$ is computed with $f_m(\mathbf{v}) = exp(-d_m(\mathbf{v})^2/\sigma_{dm}^2)$, where $d_m(\mathbf{v})$ is the distance to the nearest match. The coverage score is defined as $S = 1/|\mathcal{P}| \sum_{\mathbf{v}\in\mathcal{P}}[f_m(\mathbf{v})/f_s(\mathbf{v})]$. The larger the score, the larger the coverage of matches. Ideally, if the two meshes are identical, the score should be equal to 1. Therefore, we select L ($L = 2$) pieces from $M_2 \sim M_k$ that have the largest coverage score with respect to the mesh M_1.

4.2.2. Global Non-Rigid Registration

In this section, we present how to enforce those loop constraints to achieve a global registration. Similar to the pairwise registration part, the Embedded Deformation Model is also employed here to extrapolate the deformation field. First, we build up and embed a deformation graph for every piece of mesh (M_1 to M_N). Our goal will be to optimize the deformation parameters ($\mathbb{A} = \mathcal{A}_1 \sim \mathcal{A}_N$, $\mathbb{T} = \mathcal{T}_1 \sim \mathcal{T}_N$) of all those graphs altogether. For each $\mathcal{A}_i$ and $\mathcal{T}_i$, they are a set of affine matrices and translation vectors, respectively, which are associated with the deformation graph embedded in mesh M_i. Following the technique in [28], the objective function is formulated as

$$E(\mathbb{A},\mathbb{T}) = \sum_{i=1}^{N}[\alpha_{rigid}E_r(\mathcal{A}_i,\mathcal{T}_i) + \alpha_{smooth}E_s(\mathcal{A}_i,\mathcal{T}_i)] + \alpha_{corr}E_{corr}. \quad (7)$$

The first two terms in the above equation are the as-rigid-as-possible term and smooth term, respectively, as defined in Equations (3) and (4). We have the third term E_{corr} enforcing the correspondences' constraints, which is the most critical part. In our case, we have two sets of correspondences including the those established between neighboring pieces (E_{corr_nei}) and those from pieces that form a loop (E_{corr_loop}):

$$\begin{aligned} E_{corr} = E_{corr\_nei} + E_{corr\_loop} &= \sum_{k=1}^{N-1} \sum_{(p_i,q_i)\in C_k} ||\phi(M_k^{p_i},\mathcal{A}_k,\mathcal{T}_k) - M_{k+1}^{q_i}||_2^2 \\ &+ \sum_{k\in Lp(1)} \sum_{(p_i,q_i)\in C_k} ||\phi(M_k^{p_i},\mathcal{A}_k,\mathcal{T}_k) - M_1^{q_i}||_2^2. \end{aligned} \quad (8)$$

For the first part, it incorporates the constraints between neighboring pieces M_k and M_{k+1}. After the pairwise registration from Section 4.1.2, we are ready to find correspondences of neighboring pieces by the nearest search since those pieces have already been aligned. In practice, we do not need to enforce all those matches; instead, we randomly sample about 300–400 correspondences for every two pieces. We denote the correspondence set between M_k and M_{k+1} as C_k.

Second, for those pairs of pieces that have been marked as a loop, the correspondences between them play an important role in global registration by enforcing the loop closure constraints (E_{corr_loop}). The key problem now is how to register those pairs of pieces to get the correspondences.

Finding reliable matches between those pairs of pieces is not a trivial problem due to the more complicated non-rigid deformation and also the drifting issue. The real correspondences might have quite a large distance, which makes the nearest searching strategy not proper in this case. Therefore, we cannot simply apply the pairwise registration algorithm proposed in Section 4.1.2. Another possible solution would be to adopt the sparse SIFT features to match correspondences. However, the problem is that there might not be features extracted in textureless regions, and to make things even worse, we cannot guarantee those matches to be reliable. To deal with this issue, we propose here to exploit the dense flow information of those two colored meshes.

Now, suppose we want to align the colored mesh M_c to the first piece M_1. First, we apply rigid registration between those two meshes to make them roughly aligned. Afterwards, we generate two color images I_c and I_1 by rendering M_c and M_1, respectively, under the camera projection of mesh M_1. Instead of searching correspondences locally as in the pairwise registration approach, we compute the dense optical flow globally from the rendered image I_c to I_1. Considering that the flow displacement might be quite large under the non-rigid deformation, we exploit the method from paper [29] to adopt HOG features into the flow computation framework to handle large displacement flow.

Next, to validate those matches and remove outliers, we exploit the 3D geometry information of the two meshes. The intuitive way is to reject those candidate matches for which the distance is quite large in 3D space. However, given that the subject is experiencing non-rigid deformation, we cannot be sure how large the deformation would be. We might actually remove some potential true correspondences if we set the threshold of the distance to be small; on the other hand, outliers might not be filtered out if we set it to be large. To handle this issue, we propose a more intelligent filtering strategy under an as-rigid-as-possible principle.

Now, suppose we have a pixel p in I_c that has its corresponding pixel q in I_1, which has been acquired from the computed flow field. For pixel p, we have its corresponding vertice on mesh M_c denoted as $\mathbf{v_p}$. Its neighboring vertices $\mathbf{N_v}$ within some certain distance on the mesh can be extracted. In addition, we make use of the geodesic distance here to keep the extracted neighboring vertices to be connected. The corresponding pixels for those vertices $\mathbf{N_v}$ on the rendered image I_c are denoted as N_p. With the computed flow field, we can obtain the correspondences of $\mathbf{N_v}$ on the mesh M_1. Those corresponding vertices are denoted as $\mathbf{N_v}^{'}$.

From the corresponding vertices set $\mathbf{N_v}$ and $\mathbf{N_v'}$, we approximate the rigid transformation $\mathbf{R_v}$, $\mathbf{T_v}$ ($\mathbf{R_v}$ is a 3×3 rotation matrix and $\mathbf{T_v}$ is a 3×1 translation vector) by minimizing the following energy function:

$$E(\mathbf{R_v}, \mathbf{T_v}) = \sum_{i=1}^{|N_v|} ||(\mathbf{R_v}\mathbf{N_v}^i + \mathbf{T_v}) - \mathbf{N_v}'^i||_2^2. \tag{9}$$

To eliminate the affect of outliers, we adopt a RANSAC procedure to find the best rigid transformation that will align those two vertice sets. Afterwards, under the assumption of locally as-rigid-as-possible deformation, if the deformation of vertice $\mathbf{v_p}$ confronts the estimated transformation $\mathbf{R_v}$, $\mathbf{T_v}$, we would say that this is potentially a good match. Otherwise, the match we get from the flow field for pixel p will be regarded as an outlier. We measure the deformation consistency using the following equation:

$$M_d = exp(-\frac{||(\mathbf{R_v}\mathbf{v_p} + \mathbf{T_v}) - \mathbf{v_p}^{'}||_2^2}{2\sigma_M^2}). \tag{10}$$

We remove matches with M_d smaller than a threshold.

To this point, all the constraints in Equation (7) have been built up and we are ready to solve the optimization problem to get the optimal deformation parameters that will align all those pieces together and form a complete 3D model. The results after this registration are shown in Figure 6c. Deformed meshes after the global non-rigid registration are represented as $M_1^g \sim M_N^g$.

We demonstrate the evolution process for the global registration optimization in Figure 7 showing the curve of the energy cost with respect to number of iterations. The optimization gets converged after a few iterations.

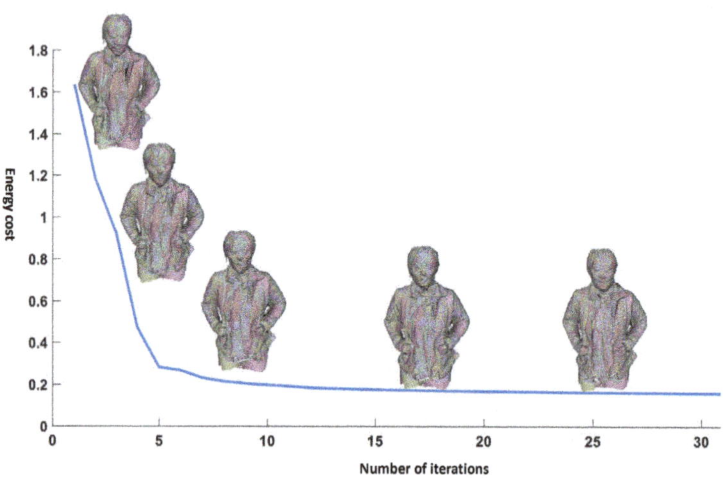

Figure 7. Illustration of the global registration evolution process.

4.3. Model Update

At this point, we have got a fairly good 3D model of the deformable object, whereas there are still some artifacts caused by misalignment as shown in Figure 8a,c. We found out that the major reason for the misalignment is surface occlusion. Specifically, if some part of the subject has been captured and modeled in piece M_k, which is then being occluded in the next piece M_{k+1}, the part reappears in piece M_{k+2}. Then, misalignment might show up between M_{k+1} and M_{k+2} in the overlapping region since we haven't enforced any constraints explicitly between these two pieces during the previous alignment procedure. One simple and naïve way to handle this would be to apply non-rigid pairwise alignment between M_k and M_{k+2} to establish reliable correspondences. However, first, we wouldn't know when this kind of misalignment will occur and, second, the overlap between M_k and M_{k+2} might be fairly small, which makes it even harder to find reliable correspondences.

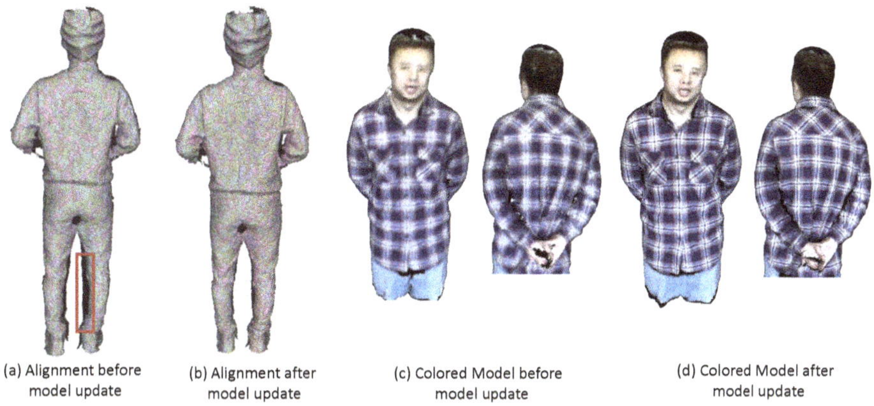

Figure 8. Illustration of results before and after model update. (**a**,**b**) show all the aligned pieces before and after the model update step, respectively. The misalignment still exists around the legs as marked red after the global registration. The alignment has gotten better after our model update as shown in (**b**). Some colored models are shown in (**c**,**d**). The appearance before model update (**c**) is quite blurry, while more clear and consistent color maps have been achieved after the model update (**d**).

Therefore, to deal with this kind of misalignment, we take advantage of the current model (denoted as $\mathcal{V}_0$) that we have reconstructed after the global non-rigid alignment step and take it as a starting point to update and refine the model. Essentially, we want to find the optimal model that will confront all those pieces both in its geometry and appearance. Instead of exploiting the expensive bundle adjustment strategy, we intend to update the model iteratively by deforming those pieces onto this proxy model. Algorithm 1 shows our procedure for updating the model.

First of all, at the initialization step, we deform the first piece M_1^g to the current proxy model $\mathcal{V}_0$, which is a trivial problem since $\mathcal{V}_0$ is recovered with the first piece as the canonical frame. Afterwards, we attach the vertices color to the proxy model from the region covered by M_1^g via nearest search. That is, for each vertice v in $\mathcal{V}_0$, we find its nearest vertice v_m in M_1^g and set the color of v to be same as M_1 if $|v - v_m| < Thres$. After this initialization step, we get the proxy model that is partially colored.

For step 2, we update the proxy model $\mathcal{V}_0$ with respect to each of those pieces. $\mathcal{V}_0$ covers the whole model while each piece M_k^g only covers part of the model. Therefore, instead of deforming $\mathcal{V}_0$ to align with the mesh M_k^g (k starts from 2 to N), we align those meshes towards the current model $\mathcal{V}_0$ exploiting the method proposed in Section 4.1.2 that utilizes both geometric and color information to achieve better alignment. We denote the deformed mesh of M_k^g as $M_k^{g'}$.

Then, correspondences between the mesh M_k^g and the proxy model are established via nearest search between $M_k^{g'}$ and $\mathcal{V}_0$. We deform the geometry of the proxy model under the guidance of those correspondences with Laplacian constraints. In the meantime, the vertices color in M_k^g can be transferred to the proxy model as described in the initialization step. In addition, we update the appearance (the vertex color) of the proxy model as the weighted average of the current vertices color and the vertices color acquired from M_k^g.

For step 3, after finishing the iteration for each piece of the segment in step 2, we re-apply the global non-rigid registration for the pieces M_k^g (k from 1 to N). The correspondence term in Equation (7) is built up by nearest search between every two pieces of the deformed meshes $M_1^{g'} - M_N^{g'}$.

We iteratively go over the above steps for better alignment and updating of the model. In addition, we will finally arrive at our reconstructed colored model that has good quality in both geometry and appearance as shown in Figure 8d.

Algorithm 1: Model-update algorithm.

Input: $M_1^g \sim M_N^g$: deformed mesh after the global registration;
  $\mathcal{V}_0$: the proxy model;
Output: Updated model;
while *not converged* **do**
  Initialization step;
  for $k = 2; k \leq N; k++$ **do**
    align mesh M_k^g to $\mathcal{V}_0$ and get the deformed mesh $M_k^{g'}$;
    build up the correspondence set C_k between M_k^g and $\mathcal{V}_0$;
    deform mesh $\mathcal{V}_0$ under those correspondences C_k
  end
  Align those meshes M_k^g globally and set $\mathcal{V}_0$ to be the new proxy model
end

4.4. *Implementation Details*

The overall pipeline is performed offline while the partial pieces generation part can be done in real time [17]. It takes about 40 min overall to run in Matlab 2016a on a desktop with 8-core 3.6 G Hz Intel CPU and 16 GB memory. In more detail, the pairwise registration part takes about 5 min and around 20 min are taken in global registration part. The final model update part takes about 15 min. The loop detection part takes little time as compared to those registration procedures.

The parameters used in the paper are set with $\alpha_r = 100, \alpha_s = 1000, \alpha_g = 1.0, \alpha_c = 1.0$, $\alpha_{rigid} = 50, \alpha_{smooth} = 500, \alpha_{corr} = 1.0$.

5. Experiments

We demonstrate the effectiveness of our approach in the experimental part with both quantitative and qualitative results. Furthermore, we present an application of model completion using our recovered 3D model.

5.1. *Quantitative Evaluation on Rigid Objects*

Even though we target on the non-rigidly deformable objects, it does not stop us from implementing our approach on the rigid objects. It is more convenient to take advantage of the rigid objects for quantitative evaluation. Here, we use a textured mesh model scanned by a multi-view scanner system as the groundtruth and synthesize a sequence of depth and color images by moving a virtual camera around the 3D mesh model. We run both the VolumeDeform [17] and our method on this synthetic data with the results shown in Figure 9. We plot the error map to show the geometric error of our reconstructed model as compared with the groundtruth model. The error for each vertice is computed via a nearest search from this vertice to the groundtruth mesh model. As we can see from the 3D error map in Figure 9c, the most largest error (about 0.0041 m) comes from the part of arms and hands, which have relative thin structure and are more difficult to track and align. From the overall model, we get the mean error as 0.0023 m. The result demonstrates that we can get a recovered model that is fairly accurate.

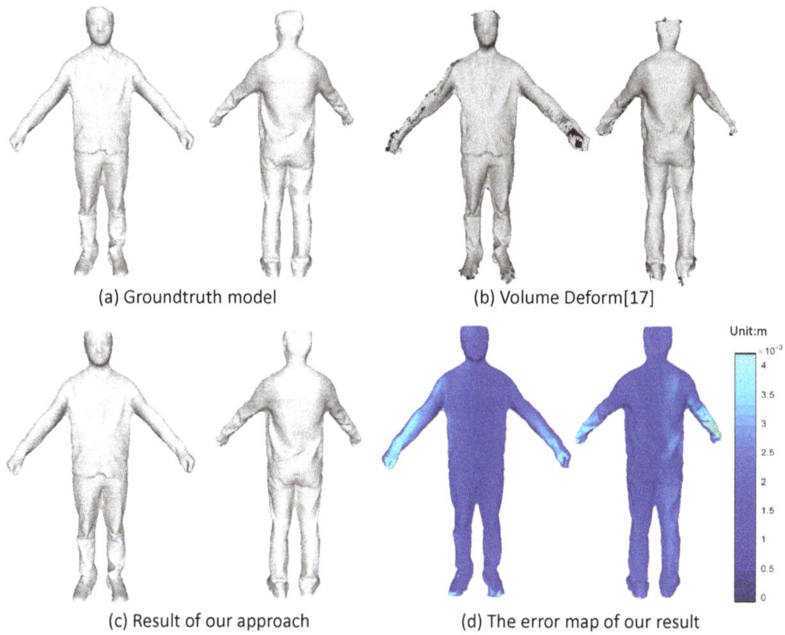

Figure 9. Quantitative evaluation on synthetic dataset.

5.2. Qualitative Evaluation on Captured Subjects

For the qualitative evaluation, we have captured several sequences of human subjects with Microsoft Kinect V2. The human subject is asked to rotate in front of the Kinect sensor.

First, we compare our results with a 3D self-portrait [12], which takes eight partial pieces as input. We run the method on one of our captured sequences for which the non-rigid motion is minimal among all the sequences and the subject has tried to stay at the same pose during rotation. We have manually selected eight frames from the sequence that evenly distributed across a cycle. The comparison results are shown in Figure 10. As the almost inevitable non-rigid motion problem during rotation, the misalignment still exists for the 3D self-portrait method especially around the arms, which can be seen in Figure 10b. On the contrary, we are able to align those partial pieces successfully under our framework, as we have kept tracking the non-rigid motion continuously. The results of our method is displayed in Figure 10c.

Figure 10. Comparison with 3D self-portrait. (**a**) some sampled partial pieces; (**b**) the results from 3D self-portrait [12]; (**c**) the results we get with our approach. We have colored the deformed pieces with different colors to better demonstrate the alignment results.

To compare with previous dynamic fusion methods, we implement a sequential dynamic fusion method [17] that fuse the frames incrementally but without concerning the loop closure. Figure 11 shows the comparison results of the upper body of some human subjects. Figure 12 presents some results on the full body modeling, which is more challenging considering the inevitable occlusion and large deformation for the legs. As compared to the method [17], which shows large gaps in the recovered model, we are able to get a complete and watertight model since we have enforced the loop closure constraints explicitly to solve error accumulation problem. Although we haven't achieved real-time performance, we can get much better results as compared with the dynamic fusion methods. In addition, since we haven't enforced any constraints on the subjects, we are also able to deal with more general cases where the human subject is holding something or carrying a backpack. We can also reconstruct the girl in a shirt, which has experienced free-form deformation as she moves.

As shown in these figures, the recovered color maps of those models are quite clear and edges are sharp. We can see the textures on the surface clearly. This is achieved by our registration method that has incorporated both geometric and appearance constraints. The parts that haven't been observed (e.g., under the chin or inner side of the arm) are colored as black. This could be filled up by color of neighboring vertices, while we haven't put our effort in this.

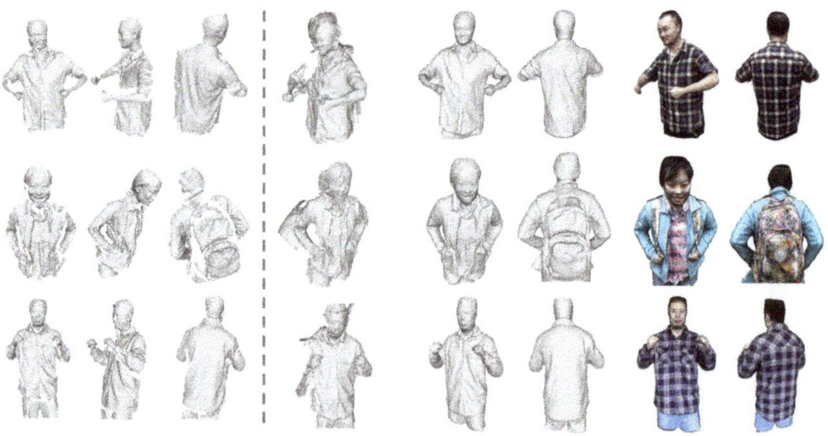

Figure 11. Qualitative evaluation on upper body models. (**a**) some sampled partial pieces; (**b**) the results from VolumeDeform [17]; (**c**) the complete models we get with our approach; (**d**) our recovered colored models.

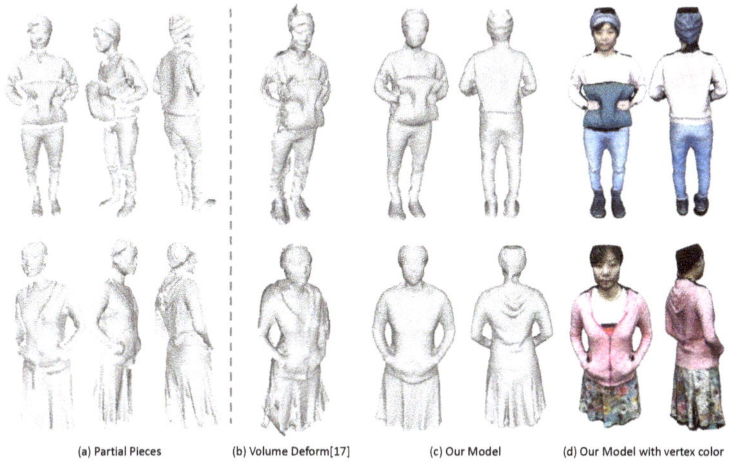

Figure 12. Qualitative evaluation on full body models. (**a**) some sampled partial pieces; (**b**) the results from VolumeDeform [17]; (**c**) the complete models we get with our approach; (**d**) our recovered colored models.

5.3. Applications

Given the complete 3D model that we have recovered from our proposed framework, we are able to drive or deform the model for some model completion applications. That is, given a depth frame as input that has quite limited coverage of the model, we can perform the completion by deforming the 3D model that we have got to get it aligned with the current input. We have employed the registration technique that we have proposed in Section 4.1.2 to accomplish this task. Some completion results are shown in Figure 13.

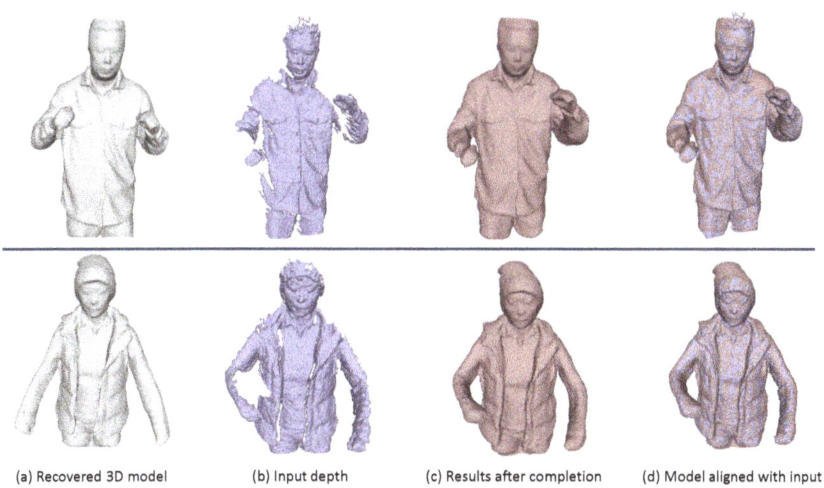

Figure 13. Applications on model completion. (**a**) the recovered 3D models in canonical space from our proposed framework; (**b**) some input depth frames which capture only partial of the model; (**c**) the results after model completion; (**d**) the aligned meshes of our model after completion and the input meshes.

6. Conclusions

In this paper, we have proposed a framework to reconstruct the 3D shape and appearance of the deformable objects under the dynamic scenario. To tackle the drifting problem during the sequential fusion, we have partitioned the entire sequence into several segments, from which we have reconstructed partial scans. A global non-rigid registration approach is applied to align all those pieces together into a consistent canonical space. We achieve this with our loop closure constraints to help eliminate the accumulation error. Afterwards, the recovered model gets updated with our novel model update method to arrive at our final model with accurate geometry and high fidelity of color maps. During the non-rigid alignment and loop closure procedure, we have exploited both geometric and appearance information to resolve the ambiguity of matching. The framework has been validated on both synthetic and real datasets. We are able to recover 3D models with accuracy in millimeters as demonstrated from our quantitative evaluation. Experiments on real datasets demonstrate the capability of our framework to reconstruct complete and watertight deformable objects with high fidelity color maps.

Looking into the future, we would like to further improve our method by replacing the per vertex color representation of the mesh with textures to get even higher quality of mesh appearance. The changing topology could be another direction that we will investigate as for now the topology is restricted to be constant throughout the sequence. In addition, our method relies on the success of building up partial scans, which might fail in case of fast motion. We believe that it could be solved by adopting the learning based approaches to find correspondences instead of using nearest search or projective association. Various applications (e.g., model based view synthesis) could be developed based on our work.

Acknowledgments: This work was supported by the US NSF (IIS-1231545, IIP-1543172), US Army Research grant W911NF-14-1-0437, the National Natural Science Foundation of China (No. 51475373, 61603302, 51375390, 61332017), the Key Industrial Innovation Chain of Shaanxi Province Industrial Area (2016KTZDGY06-01, 2015KTZDGY04-01), the Natural Science Foundation of Shaanxi (No. 2016JQ6009), and the "111 Project" (No.B13044).

Author Contributions: All authors have made substantial contributions to the study including conception, algorithm design and experiments; Sen Wang and Xinxin Zuo wrote the paper.

Conflicts of Interest: The authors declare no conflict of interest.

References

1. Whelan, T.; Salas-Moreno, R.; Glocker, B.; Davison, A.; Leutenegger, S. ElasticFusion: Real-Time Dense SLAM and Light Source Estimation. *Int. J. Robot. Res.* **2016**, *35*, 1697–1716, doi:10.1177/0278364916669237.
2. Mur-Artal, R.; Montiel, J.; Tardós, J. ORB-SLAM: A Versatile and Accurate Monocular SLAM System. *IEEE Trans. Robot.* **2015**, *31*, 1147–1163, doi:10.1109/TRO.2015.2463671.
3. Endres, F.; Hess, J.; Engelhard, N.; Sturm, J.; Cremers, D.; Burgard, W. An evaluation of the RGB-D SLAM system. In Proceedings of the 2012 IEEE International Conference on Robotics and Automation (ICRA), Saint Paul, MN, USA, 14–18 May 2012; pp. 1691–1696.
4. Starck, J.; Hilton, A. Surface Capture for Performance-Based Animation. *IEEE Comput. Graph. Appl.* **2007**, *27*, 21–31, doi:10.1109/MCG.2007.68.
5. Aguiar, E.D.; Stoll, C.; Theobalt, C.; Ahmed, N.; Seidel, H.-P.; Thrun, S. Performance capture from sparse multi-view video. *ACM Trans. Graph.* **2008**, *27*, 98, doi:10.1145/1399504.1360697.
6. Vlasic, D.; Baran, I.; Matusik, W.; Popović, J. Articulated mesh animation from multi-view silhouettes. *ACM Trans. Graph.* **2008**, *27*, 97, doi:10.1145/1399504.1360696.
7. Waschbüsch, M.; Würmlin, S.; Cotting, D.; Sadlo, F.; Gross, M. Scalable 3D video of dynamic scenes. *Vis. Comput.* **2005**, *21*, 629–638, doi:10.1007/s00371-005-0346-7.
8. Dou, M.; Fuchs, H.; Frahm, J.M. Scanning and tracking dynamic objects with commodity depth cameras. In Proceedings of the 2013 IEEE Symposium on Mixed and Augmented Reality (ISMAR), Adelaide, Australia, 1–4 October 2013; pp. 99–106.

9. Tong, J.; Zhou, J.; Liu, L.; Pan, Z.; Yan, H. Scanning 3D full human bodies using Kinects. *IEEE Trans. Vis. Comput. Graph.* **2012**, *18*, 643–650, doi:10.1109/TVCG.2012.56.
10. Alexiadis, D.S.; Zarpalas, D.; Daras, P. Real-Time, Full 3-D Reconstruction of Moving Foreground Objects From Multiple Consumer Depth Cameras. *IEEE Trans. Multimed.* **2013**, *15*, 339–358, doi:10.1109/TMM.2012.2229264.
11. Dou, M.; Khamis, S.; Degtyarev, Y.; Davidson, P.; Fanello, S.R.; Kowdle, A.; Escolano, S.O.; Rhemann, C.; Kim, D.; Taylor, J.; et al. Fusion4D: real-time performance capture of challenging scenes. *ACM Trans. Graph.* **2016**, *35*, 114, doi:10.1145/2897824.2925969.
12. Li, H.; Vouga, E.; Gudym, A.; Luo, L.; Barron, J.T.; Gusev, G. 3D self-portraits. *ACM Trans. Graph.* **2013**, *32*, 187, doi:10.1145/2508363.2508407.
13. Cui, Y.; Chang, W.; Nolly, T.; Stricker, D. Kinectavatar: Fully automatic body capture using a single Kinect. In Proceedings of the Asian Conference on Computer Vision (ACCV), Daejeon, Korea, 5–9 November 2012; pp. 133–147.
14. Anguelov, D.; Srinivasan, P.; Koller, D.; Thrun, S.; Rodgers, J.; Davis, J. SCAPE: shape completion and animation of people. *ACM Trans. Graph.* **2005**, *24*, 408–416, doi: 10.1109/TVCG.2012.56.
15. Gall, J.; Stoll, C.; de Aguiar, E.; Theobalt, C.; Rosenhahn, B.; Seidel, H.-P. Motion capture using joint skeleton tracking and surface estimation. In Proceedings of the 2009 IEEE Conference on Computer Vision and Pattern Recognition (CVPR), Miami, FL, USA, 20–25 June 2009; pp. 1746–1753.
16. Newcombe, R.A.; Fox, D.; Seitz, S.M. DynamicFusion: Reconstruction and tracking of non-rigid scenes in real time. In Proceedings of the 2015 IEEE Conference on Computer Vision and Pattern Recognition (CVPR), Boston, MA, USA, 7–12 June 2015; pp. 343–352.
17. Innmann, M.; Zollhöfer, M.; Nießner, M.; Theobalt, C.; Stamminger, M. VolumeDeform: Real-Time Volumetric Non-rigid Reconstruction. In Proceedings of the 2016 European Conference on Computer Vision (ECCV), Amsterdam, The Netherland, 8–16 October 2016; pp. 362–379.
18. Slavcheva, M.; Baust, M.; Cremers, D.; Ilic, S. KillingFusion: Non-rigid 3D Reconstruction without Correspondences. In Proceedings of the 2017 IEEE Conference on Computer Vision and Pattern Recognition (CVPR), Honolulu, HI, USA, 21–26 July 2017; pp. 343–352.
19. Zollhöfer, M.; Izadi, S.; Rehmann, C.; Zach, C.; Fisher, M.; Wu, C.; Fitzgibbon, A.; Loop, C.; Theobalt, C.; Stamminger, M. Real-time non-rigid reconstruction using an RGB-D camera. *ACM Trans. Graph.* **2014**, *33*, 156, doi:10.1145/2601097.2601165.
20. Guo, K.; Xu, F.; Wang, Y.; Liu, Y.; Dai, Q. Robust Non-rigid Motion Tracking and Surface Reconstruction Using L0 Regularization. In Proceedings of the 2015 IEEE International Conference on Computer Vision (ICCV), Santiago, Chile, 7–13 December 2015; pp. 3083–3091.
21. Bogo, F.; Black, M.J.; Loper, M.; Romero, J. Detailed Full-Body Reconstructions of Moving People from Monocular RGB-D Sequences. In Proceedings of the 2015 IEEE International Conference on Computer Vision (ICCV), Santiago, Chile, 7–13 December 2015; pp. 2300–2308.
22. Zhang, Q.; Fu, B.; Ye, M.; Yang, R. Quality dynamic human body modeling using a single low-cost depth camera. In Proceedings of the 2014 IEEE Conference on Computer Vision and Pattern Recognition (CVPR), Columbus, OH, USA, 23–28 June 2014; pp. 676–683.
23. Zhu, H.Y.; Yu, Y.; Zhou, Y.; Du, S.D. Dynamic human body modeling using a single RGB camera. *Sensors* **2016**, *16*, 402, doi:10.3390/s16030402.
24. Guo, K.W.; Xu, F.; Yu, T.; Liu, X.; Dai, Q.; Liu, Y. Real-time Geometry, Albedo and Motion Reconstruction Using a Single RGBD Camera. *ACM Trans. Graph.* **2017**, *36*, 32, doi:10.1145/3083722.
25. Yu, T.; Guo, K.; Xu, F.; Dong, Y.; Su, Z.; Zhao, J.; Li, J.; Dai, Q.; Liu, Y. BodyFusion: Real-time Capture of Human Motion and Surface Geometry Using a Single Depth Camera. In Proceedings of the 2017 IEEE International Conference on Computer Vision (ICCV), Venice, Italy, 22–29 October 2017; pp. 910–919.
26. Dou, M.; Taylor, J.; Fuchs, H.; Fitzgibbon, A.; Izadi, S. 3D scanning deformable objects with a single RGBD sensor. In Proceedings of the 2015 IEEE Conference on Computer Vision and Pattern Recognition (CVPR), Boston, MA, USA, 7–12 June 2015; pp. 493–501.
27. Sumner, R.W.; Schmid, J.; Pauly, M. Embedded deformation for shape manipulation. *ACM Trans. Graph.* **2007**, *26*, 80, doi:10.1145/1276377.1276478.

28. Li, H.; Sumner, R.W.; Pauly, M. Global correspondence optimization for non-rigid registration of depth scans. In Proceedings of the 2008 Eurographics Association Symposium on Geometry Processing, Copenhagen, Denmark, 2–4 July 2008; pp. 1421–1430.
29. Brox, T.; Malik, J. Large Displacement Optical Flow: Descriptor Matching in Variational Motion Estimation. *IEEE Trans. Pattern Anal. Mach. Intell.* **2011**, *33*, 500–513, doi:10.1109/TPAMI.2010.143.

sensors

MDPI

Article

Video-Based Person Re-Identification by an End-To-End Learning Architecture with Hybrid Deep Appearance-Temporal Feature

Rui Sun, Qiheng Huang *, Miaomiao Xia and Jun Zhang

School of Computer Science and Information Engineering, Hefei University of Technology, Feicui Road 420, Hefei 230000, China; sunrui@hfut.edu.cn (R.S.); 18225514947@163.com (M.X.); zhangjun@hfut.edu.cn (J.Z.)

* Correspondence: jchqh123@163.com; Tel.: +86-151-566-99439

Received: 31 August 2018; Accepted: 26 October 2018; Published: 29 October 2018

Abstract: Video-based person re-identification is an important task with the challenges of lighting variation, low-resolution images, background clutter, occlusion, and human appearance similarity in the multi-camera visual sensor networks. In this paper, we propose a video-based person re-identification method called the end-to-end learning architecture with hybrid deep appearance-temporal feature. It can learn the appearance features of pivotal frames, the temporal features, and the independent distance metric of different features. This architecture consists of two-stream deep feature structure and two Siamese networks. For the first-stream structure, we propose the Two-branch Appearance Feature (TAF) sub-structure to obtain the appearance information of persons, and used one of the two Siamese networks to learn the similarity of appearance features of a pairwise person. To utilize the temporal information, we designed the second-stream structure that consisting of the Optical flow Temporal Feature (OTF) sub-structure and another Siamese network, to learn the person's temporal features and the distances of pairwise features. In addition, we select the pivotal frames of video as inputs to the Inception-V3 network on the Two-branch Appearance Feature sub-structure, and employ the salience-learning fusion layer to fuse the learned global and local appearance features. Extensive experimental results on the PRID2011, iLIDS-VID, and Motion Analysis and Re-identification Set (MARS) datasets showed that the respective proposed architectures reached 79%, 59% and 72% at Rank-1 and had advantages over state-of-the-art algorithms. Meanwhile, it also improved the feature representation ability of persons.

Keywords: person re-identification; end-to-end architecture; appearance-temporal features; Siamese network; pivotal frames

1. Introduction

Person re-identification (person Re-ID) aims at matching a target person across non-overlapping cameras at different times or different locations. It not only has important significance in video surveillance systems and the public security field, but is also a crucial challenge in the field of multi-camera visual sensor networks [1]. In real world situations, because multi-camera visual sensor networks capture the video clip of the target person, research on video-based person re-identification is necessary and inevitable for public safety. Video-based person re-identification is the task of utilizing a sequence of images/tracklets to match the person. At present, an increasing number of exiting research works [2–5] focus on video-based person re-identification.

More specifically, the process of video-based person Re-ID is to give a probe video and search the same person as the probe video in a large gallery of videos. As the probe video and gallery videos are taken from different cameras, they may suffer from inherent challenges such as lighting variations, camera viewpoint changes, background clutter or occlusions, and the person's appearance similarity during person matching. In general, video-based person Re-ID is beneficial to improve the results of person Re-ID under the complex and difficult conditions described above. The reason for this fact is that video-based person Re-ID has the following advantages over still-image-based person Re-ID. Firstly, videos contain more information than a single still image contains. Given the availability of video clips, we can obtain temporal information related to a person's motion. If the person suffers from problems including occlusion, background clutter, and appearance similarity, the person's appearance information, based on a single still-image, is incomplete or missing. However, the use of potential temporal information based on image sequence can effectively alleviate the lack of motion information. What is more, videos provide a large number of the same person's samples, so we can obtain more abundant appearance information to against camera viewpoint changes.

On the other hand, the use of video also brings several challenges for identifying the person. Firstly, some low-resolution image frames may appear in the captured video clips, which lead to inaccurate appearance information. Secondly, when the target pedestrian is obstructed or interfered with by objects or different persons in a video fragment, it becomes difficult to obtain the person's appearance information in the current image sequence. Lastly, although the temporal information in the video is an important clue to identify pedestrians, the movement of different persons may also be similar, which means that purely using temporal information will cause misunderstanding. As shown in Figure 1, in this work, we define the appearance of ambiguity image frames and occluded image frames as interference frames in the video, and others image frames ("good" frames) that contain the full clear persons in the video as pivotal frames. Therefore, the following issues in video-based person Re-ID should be considered. (1) How to establish a stable pedestrian appearance representation model, that enables elimination of the effects of interference frames on individuals' representation in videos? (2) How to effectively harness two types of complementary information including appearance features and temporal features in the video to compare the degree of similarity between different persons, so that the role of pivotal frames is fully realized?

Figure 1. An illustration of the interference frames and pivotal frames definition. The green box indicates the pivotal frame ("good" frame).The red dotted box indicates the interference frame with low-resolution image, occlusion, and background clutter.

To address the first problem, for one thing, previous works have adopted new features [6], appearance feature models, and semantic attribute features [7,8], which extract robust and discriminative information to represent a person. However, we can observe that not all images are informative in a given video, and severe interference frames cause previous methods to obtain erroneous information. For another thing, the current common research idea [5,9] is to adopt a combination of convolutional neural network (CNN) and recurrent neural network (RNN) to extract the space-time features of each image frame, and aggregate them into a single feature vector by the pooling operation. Although these methods have achieved good results, the interference frames in the video will influence the final feature information. Simultaneously, such methods do not make full use of the person's appearance information. To sum up, in this work, we propose a Two-branch Appearance Feature (TAF) sub-structure which consists of the walking cycle analysis model [2], the two-branch Inception-V3 network, and the fusion layer, to select pivotal frames ("good" frames) and discard interference frames, then learn the global and local discriminative appearance feature information.

To deal with the second problem, the current work [10] mainly focuses on the integration of two types of features before learning the distance between persons. Appearance features and temporal features are different modal information. We believe that information maybe lost due to information inequality when these two types of features are combined. In this paper, inspired by a previous literature [11], instead of merging the temporal features and the appearance features of pivotal frames, we learn the independent distances of the two types of features separately. Hence, we designed a hybrid end-to-end deep learning architecture for further learning the feature representation and the independent distance metric. The hybrid end-to-end architecture consists of a two-stream appearance-temporal deep feature structure and two Siamese networks. The integrated architecture separately obtains the person's appearance features and temporal features through the hybrid feature structure, whilst using the two Siamese networks to learn the independent distances of the two types of features.

In summary, the main contributions of this paper are three-fold as follows.

(1) We propose a Two-branch Appearance Feature (TAF) sub-structure consisting of the walking cycle model, the two-branch Inception-V3 network, and the saliency learning fusion layer, which is used to learn the global and local appearance features of persons. This sub-structure is useful for discarding interference frames with occlusion and background clutter in the video, and selecting informative pivotal frames. The features of these pivotal frames can promote the representation learning ability of two-branch Inception-V3 network. Simultaneously, the fusion layer can improve the fusion effect and the learning result of local information.
(2) We design a two-stream hybrid end-to-end deep learning architecture that combines feature learning and metric learning, which uses a hybrid deep feature structure and two Siamese networks to obtain a person's features and separately achieve the independent distance metric of appearance features and temporal features. Note that it can obtain better appearance information and temporal information by having two independent feature sub-structures.
(3) We evaluate our proposed architecture on three public video datasets, including PRID-2011 dataset [12], iLIDS-VID dataset [13], and MARS (Motion Analysis and Re-identification Set) dataset [14]. Extensive comparative experiments show that our proposed video-based person Re-ID architecture achieves comparable results to the existing state-of-the-art methods.

The remainder of this paper is organized as follows. Section 2 reviews the work related to person Re-ID. Section 3 gives a complete explanation of the architecture proposed in this paper and a detailed introduction to each part of the architecture. Section 4 conducts an experimental evaluation of the performance of the proposed algorithm on public datasets. Finally, Section 5 summarizes the work of this paper.

2. Related Work

Person re-identification has attracted the attention of many researchers in recent years. With the development of person re-identification works, we believe that the study of person re-identification can be roughly divided into three groups: image-based person re-identification [15–21], video-based person re-identification [2–5,9,22], and image to video person re-identification [23]. Typically, most existing person Re-ID algorithms focus on three key steps: feature extraction [15–19], distance measure [20,21], and end-to-end learning methods [11,24–27]. To obtain reliable feature representations, the features adopted in the existing person Re-ID work can be divided into hand-designed features [15–19] and deep learning features [28]. Hand-designed features are commonly used for the color and texture features [15], SIFT features [16], and color names features [17], etc. At the same time, there are good representation capabilities in hand-designed features such as GOG [18] and LOMO [19]. In order to learn a robust distance measure, many scholars have proposed effective metric models, including KISSME [20], XQDA [19], FDA [21], etc. To fully understand the relevant algorithms to our proposed architecture in this paper, we will mainly introduce the research development of video-based person Re-ID and the current status of end-to-end deep learning algorithms in person Re-ID.

2.1. Video-Based Person Re-Identification

The research in video-based person re-identification is based on person Re-ID in multi-frame images. At present, more and more video-based person Re-ID methods are emerging. We believe that video-based person Re-ID can be divided into traditional methods and deep learning methods. In terms of traditional algorithms, the work of a past literature [2] uses the discriminative selection and ranking (DVR) method to select discriminative video fragments and extract their HOG3D features for matching. Another previous paper [3] proposes the STFV3D algorithm to extract spatiotemporal features (learn Fisher vectors) with spatial alignment. The top-push distance metric method [4] establishes a top-push constraint metric to improve the intra-distance and inter-distance between persons. In terms of deep learning algorithms, in a past paper [5], a novel recurrent neural network architecture is proposed to obtain space-time features in video. A previous literature [9] proposes an end-to-end learning architecture integrated by Convolutional Neural Networks (CNNs) and Bidirectional Recurrent Neural Networks (BRNNs) to match person in the video. In another past paper [22], a novel joint Spatial and Temporal Attention Pooling Network (ASTPN) is proposed as feature extractor to obtain features for video-based person Re-ID.

2.2. End-To-End Deep Learning on Person Re-Identification

With the wide applications of deep learning, end-to-end deep learning algorithms have appeared in many researches of person re-identification. The essence of an end-to-end learning algorithm is to completely connect the feature representation with the distance metric and jointly identify the same person. The binary input and the ternary input are the common strategy in person Re-ID algorithms of end-to-end learning. Literature [24] proposes a novel Deep Metric Learning (DML) method that jointly learns color features, texture features, and metrics in a unified framework. In a past paper [11], Chen et al. propose a novel deep end-to-end network to automatically learn the spatial-temporal fusion features, and utilize the Siamese to train sample pair. A previous work [25] presents a novel multi-channel parts-based Convolutional Neural Network (CNN) model under the triplet framework for person Re-ID. A different past work [26] also proposes a new end-to-end Comparative Attention Network (CAN) with triplet loss to learn the discriminative features of person images. For quaternary input, a past work [27] designs a quadruplet loss to ensure that model outputs have a larger interclass variation and a smaller intra class variation compared to the triplet loss. In our paper, we borrow the idea from the Siamese network of binary input, and employ two Siamese networks to learn the independent distance metric of different features. This effectively improves the performance of video-based person Re-ID.

3. The Proposed Hybrid End-To-End Deep Learning Architecture

3.1. Architecture Overview

The hybrid end-to-end deep learning architecture of our proposed method is shown in Figure 2. The hybrid end-to-end architecture consists of two-stream deep feature structure and two Siamese networks. The two-stream deep feature structure is composed of the Two-branch Appearance Feature sub-structure and the Optical flow Temporal Feature sub-structure, which can obtain abundant appearance information and stability temporal information of the pairwise person. It then employs two Siamese networks to compare the similarities between different persons of each type of feature.

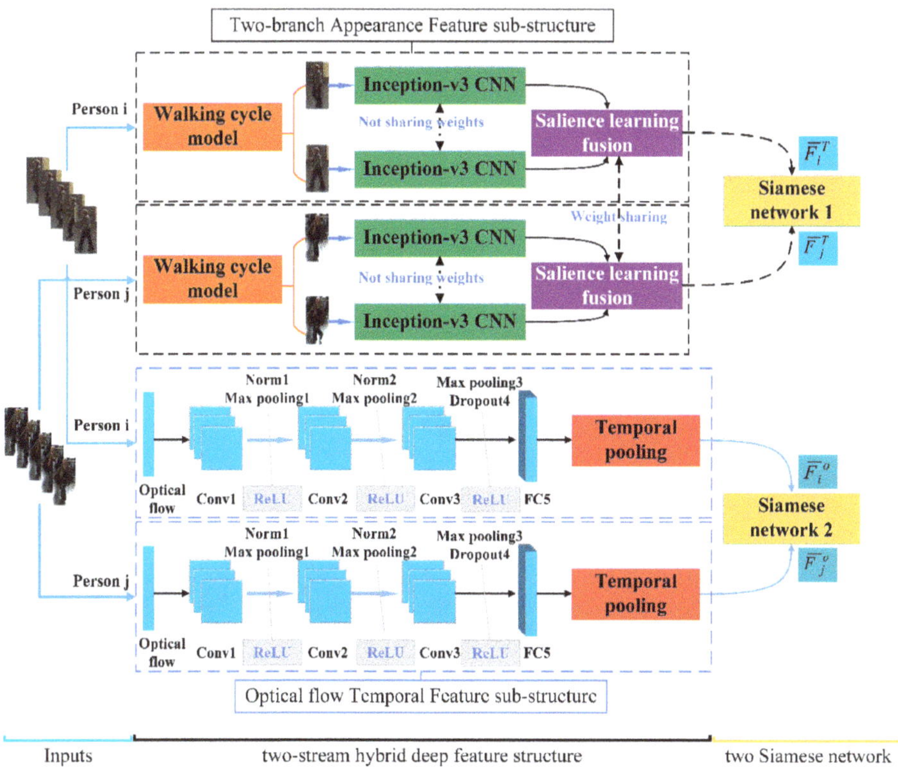

Figure 2. The framework of the proposed hybrid end-to-end deep learning architecture. The architecture consists of inputs, a two-stream hybrid deep feature structure, and two Siamese networks. The two-stream hybrid deep feature structure is composed of the Two-branch Appearance Feature sub-structure and the Optical flow Temporal Feature sub-structure, which can obtain abundant appearance information and stability temporal information of the pairwise person. Then, two Siamese networks are employed to compare the similarities between different persons of each type of feature.

In detail, for the first-stream feature substructure, we take the video of the original RGB image frames with persons i and j as inputs to the Two-branch Appearance Feature (TAF) sub-structure, respectively. A key process of the TAF sub-structure is that the walking cycle analysis model is used to select the pivotal frames N (N represents the number of pivotal frames) in the image sequence, then the pivotal frames are fed into a two-branch Incpetion-V3 network to learn the global appearance feature information. In addition, the fusion layer with weights sharing is applied to learn the salient and local features, whilst also fuse the global and local appearance information in the pivotal frames. Similarly, for the second-stream feature sub-structure, we use the optical flow image of video with persons i and j as inputs to the Optical flow Temporal Feature (OTF) sub-structure, respectively. Then, the CNN architecture and the temporal pooling generate temporal information. Finally, the obtained appearance features $(\overline{F}_i^T, \overline{F}_j^T)$ and temporal features $(\overline{F}_i^o, \overline{F}_j^o)$ are separately trained for similarity between features through two Siamese networks.

3.2. Input Data Acquisition

Multi-camera visual sensor networks are an important source of data acquisition for video-based person re-identification. The three public video datasets used in this paper all capture persons through multiple non-overlapping visual sensing cameras, as shown in Figure 3. Specifically, the PRID-2011 dataset [12] consists of image sequences extracted from multiple person trajectories recorded from two different static surveillance cameras. The iLIDS-VID dataset [13] is created from the persons observed in two non-overlapping camera views from the i-LIDS Multiple-Camera Tracking Scenario (MCTS) dataset which was captured at an airport arrival hall with a multi-camera CCTV network. The MARS dataset [14] was collected from six near-synchronized cameras on the campus of Tsinghua university. There were five 1080 × 1920 HD cameras and one 640 × 480 HD camera.

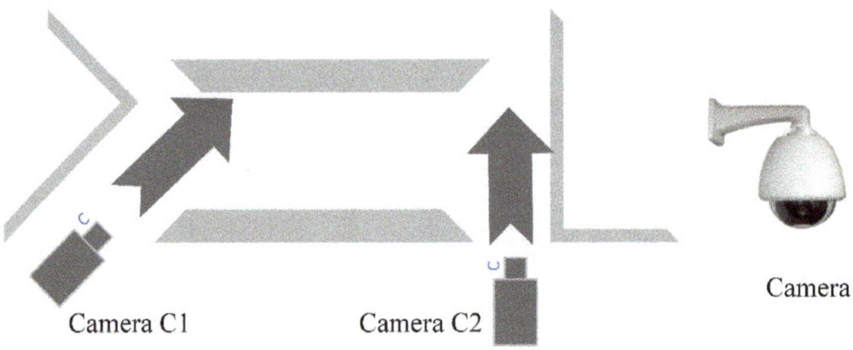

Figure 3. The data acquisition of multi camera visual sensor networks. These scenes are captured under disjoint cameras.

3.3. Two-Branch Appearance Feature Substructure (The First-Stream)

In order to select the pivotal frames in the videos and obtain more distinguishing global and local appearance features of persons, we designed a Two-branch Appearance Feature (TAF) substructure consisting of a walking cycle analysis model, a two-branch Inception-V3 network and a salience-learning fusion layer. The appearance feature sub-structure will be described in detail below.

3.3.1. Walking Cycle Analysis Model

Given a video of the target person, in order to select reliable pivotal frames in the video and discard the interference frames, we consider employing the walking cycle analysis model [2,3,29] to obtain the pivotal frames as the input of the two-branch Inception-V3 network. This is done to prevent the appearance features are not affected by low-resolution image, complex background interference, and occlusion. In this model, we first extract the Flow Energy Profile (FEP) signal [2]. The FEP is a one-dimensional signal which represents the motion energy intensity induced by the activity of human muscles during walking [30], and is approximately estimated by optic flow computation. For each successive and raw RGB image frame, r_{it} of person i in the video $S_i = \{r_{i1}, r_{i2}, \cdots, r_{it}, \cdots, r_{iL}\}$, we calculate its flow energy by the connection between optical flow fields in the horizontal direction, v_x, and the vertical direction, v_y, as shown in Equation (1):

$$E_{r_{it}} = \sum_{(x,y)\in p} \left\| [v_x(x,y), v_y(x,y)] \right\|_2, \tag{1}$$

where $E_{r_{it}}$ represents the FEP value of the r_{it}-th frame, and p is an image of the lower body of a person.

The rough estimated FEP value of the walking cycle is prone to instability due to background noise and occlusion interference. According to this situation, the literature [3] uses the discrete Fourier transform method to convert the original FEP value into the frequency domain, thus obtaining a more accurate walking cycle model. In order to obtain a discriminative pivotal frame, we use this method [3] to convert the video sequence into a walking cycle. During the walking cycle, the image frames corresponding to the maximum and minimum FEP values can improve the result of person representation, so our paper selects them as the pivotal frames N (N represents the number of pivotal frames). As shown in Figure 4, from the graph of FEP value, the local maximum of energy value $E_{r_{it}}$ corresponds to the walking posture when the person's legs overlap. Conversely, the local minimum value represents the person's posture when their legs are farthest away. Through the analysis of the walking cycle model, the pivotal frames $S_{iPF} = \{r_{i1}, r_{i2}\}$ are extracted from the video $S_i = \{r_{i1}, r_{i2}, \cdots, r_{it}, \cdots, r_{iL}\}$ as the input of the two-branch Inception-V3 network.

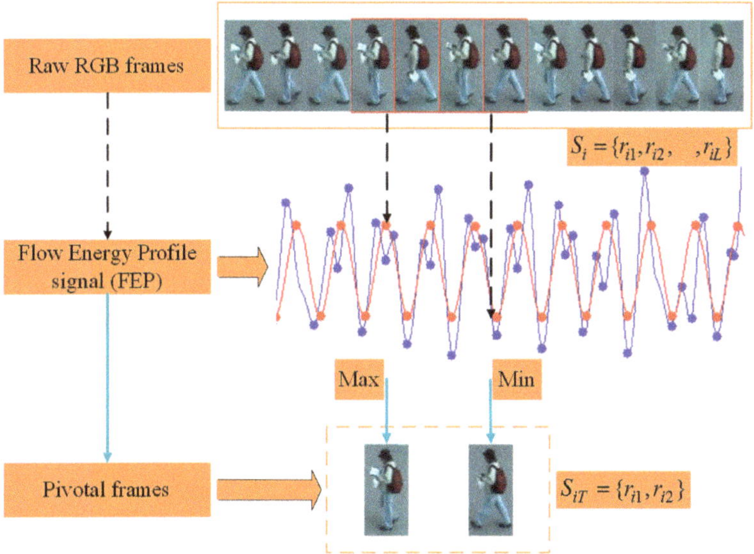

Figure 4. Pivotal frames extraction. The image frames with the golden box in the raw RGB frames is a partially selected video segment. The red curve in the Flow Energy Profile (FEP) signal is the regular FEP value and the blue curve is the rough FEP value.

Remarks. For the selection of pivotal frame's number, our paper adopts the strategy of selecting even frames. Because the strategy of selecting the pivotal frames in our paper is to select the image frames corresponding to the maximum and minimum FEP values in the video, selecting the even pivotal frames not only discards the interference frames, but also preserves the complete appearance posture of the person.

3.3.2. Two-BranchInception-V3 Network

Although the CNN has successfully demonstrated breakthroughs in person re-identification, changing the structure of the CNN from different perspectives enable achieve different performance. A straightforward way to improve CNN performance is to increase the number of layers in the network. Due to the deepening of the number of network layers, the number of network parameters and the computational cost will increase dramatically. At the same time, a deepened network and limited training samples may also cause serious overfitting problems. Hence, we construct the two-branch global feature learning module using the 42-layer Inception-V3 network [31]. Compared to other frameworks, such as VGG-Net [32] or Res-Net [33], we decided that the Inception-V3 network was more suitable for learning global features due to its high computational cost efficiency (higher modeling capacity at a smaller parameter size) and its capability for learning more discriminative appearance features at varying pivotal frames.

For the two-branch Inception-V3 network, although the two branches of model that learns the global features are the same Inception-V3 network, they do not share the weight parameters of the network. Intuitively, the pivotal frames $S_{iPF} = \{r_{i1}, r_{i2}\}$ of person i are input into the two-branch Inception-V3 network, respectively, and the person's global appearance features ($f_{i1}^{T}{}'$ and $f_{i2}^{T}{}'$) are obtained by learning.

Remarks. Since the selected pivotal frames correspond to different postures of the person walking in the video, and the interference frames with the occlusion are discarded, it is thus stable and easy to extract the person's feature information from the pivotal frames with clear appearance. Furthermore, the two-branch Inception-V3 network can learn person's appearance information in two different poses. The above observations are the main reasons for learning more discriminative appearance features at varying pivotal frames.

3.3.3. Salience-Learning Fusion Layer

This fusion layer is designed to learn the local features and fuse the output of two-branch Inception-V3 network. Due to the phenomenon of feature redundancy in the feature map learned from the above multi-layer Inception-V3 network, and because some feature channels may capture interfering information about aperson, we suggest that the salience-learning fusion strategy can automatically discover and emphasize important local information, such as the head, torso, package, etc.

As shown in Figure 5, we input features $f_{i1}^{T}{}'$ and $f_{i2}^{T}{}'$ into the salience-learning fusion layer, which can learn the different weights of each different feature channel. Then, we use the Eltwise operation to sum the feature channels. Finally, the information of each feature map is fused together through the fully connected layer, and we get the final appearance feature $\overline{F}_i^T$. This is the main reason why the layer can extract local appearance information and also reduce the feature dimension.

Remarks. In this work, we use the salience-learning fusion layer to exploit visual saliency. The strategy of salience-learning [26] has been successfully applied to person re-identification. As can be seen from Figure 8, the target person with the package can be accurately identified. Specifically, the layer combines the global and local features of a person in different poses, and the different weights of each feature channel can automatically calculate the positions in different visual saliency features. At the same time, the Eltwise-sum operation links the locations of local visual features.

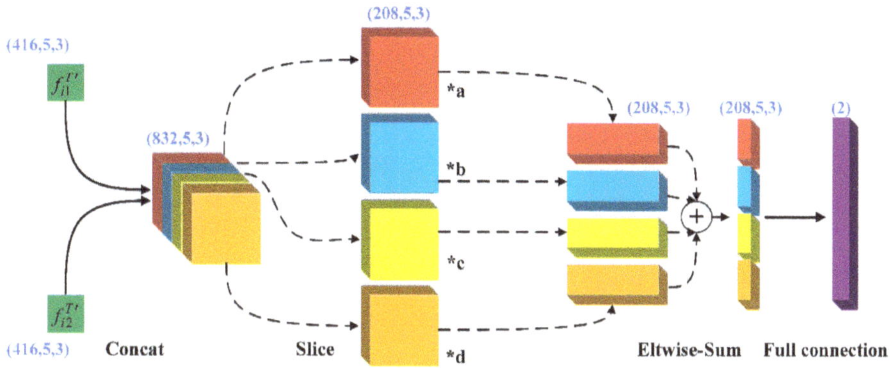

Figure 5. The process and detailed parameters of the salience-learning fusion layer. The blue font in the figure is the size of each layer input. The letters ("a" "b", "c", and "d") are the weights learned.

3.4. Optical Flow Temporal Feature Substructure (The Second-Stream)

Although spatial appearance features are more discriminative than temporal features for person Re-ID [19], the temporal feature information can compensate for the errors caused by persons of similar appearance. The Optical Flow Temporal Feature (OTF) sub-structure combines the optical flow images with the CNN to obtain temporal features. The reason why the RNN is not added because the optical flow images contain temporal information associated with the pedestrian motion, and the temporal information learned in the optical flow images is mapped to the temporal feature map, no longer use the RNN to get the temporal information. Finally, the temporal pooling method is used to aggregate the sequence-level temporal features into a single temporal feature. The OTF sub-structure is described in detail later in this section.

3.4.1. The CNN Architecture

In this paper, the input of the OTF sub-structure (the second-stream) is the image frame of the optical flow information corresponding to the video, which is the same as the literature [11]. Specifically, we define the optical flow images in the video S_{io} as $S_{io} = \{o_{i1}, o_{i2}, \cdots, o_{it}, \cdots, o_{iL}\}$, where o_{iL} represents the optical flow image and L is the video length. The method of obtaining o_{it} is the same as the method described in Section 3.3, computed using the Lucas–Kanade optical flow technique [34].

As shown in Figure 2, we employed a previously proposed CNN architecture [11] to obtain the temporal information. However, in our case, some of the parameters in the architecture were modified. Figure 2 shows the network architecture, and Table 1 demonstrates the parameters of this CNN architecture. The CNN architecture is composed of three convolutional layers, two fully connected layers and a dropout layer. Note that the process steps of each convolutional layer are convolutional, nonlinear activation functions, and pooling. We chose the rectified linear unit (ReLU) as the activation function and set the pooling operation to max-pooling. We input the optical flow image frames $S_{io} = \{o_{i1}, o_{i2}, \cdots, o_{it}, \cdots, o_{iL}\}$ of person i into the CNN architecture and generate the output temporal feature vector $F_i^o = \{f_{i1}^o, f_{i2}^o, \cdots, f_{it}^o, \cdots, f_{iL}^o\}$ after passing the CNN. The process of the above CNN architecture can be expressed by Equations (2) and (3) as follows

$$f_{it}^{o\prime} = Maxpool(relu(Conv(o_{it}))), 1 \leq t \leq L, \quad (2)$$

$$f_{it}^{o} = connect(f_{it}^{o\prime}), 1 \leq t \leq L, \quad (3)$$

where o_{it} denotes the optical flow image at t moment, $f_{it}^{o\prime}$ is the feature vector through three convolutional layers, and f_{it}^{o} is the temporal feature vector after the CNN architecture.

Table 1. Parameters of the CNN architecture.

Layers	Network Parameter/Type
Conv1	Filter 5 × 5/stride 2/pad 4
Max pool1	Filter 2 × 2/stride 2
Conv2	Filter 5 × 5/stride 2/pad 4
Max pool2	Filter 2 × 2/stride 2
Conv3	Filter 5 × 5/stride 2/pad 4
Max pool3	Filter 2 × 2/stride 2
Dropout	dropout ratio 0.5

3.4.2. Temporal Pooling

To aggregate the temporal information of all time steps in the OTF sub-structure, the multi-frame feature vector is aggregated into a single feature vector using the temporal pooling method. The implementation of these functions can be achieved by mean pooling, max pooling, and sum pooling, but it was proven [5] that mean pooling is more suitable for aggregating information in person Re-ID. The median value is tested as well to remove gross errors. In our paper, we adopt the same temporal mean-pooling method to take the temporal feature vector $F_i^o = \{f_{i1}^o, f_{i2}^o, \cdots, f_{it}^o, \cdots, f_{iL}^o\}$ from the CNN architecture as inputs, and then produce a single feature vector $\overline{F}_i^o$ to represent the final temporal feature of person i in the video. This process can be expressed by the following Equation (4).

$$\overline{F}_i^o = \frac{1}{L}\sum_{t=1}^{L} f_{it}^o, \tag{4}$$

where f_{it}^o is the temporal feature at time t, and $t \in [1, L]$. $\overline{F}_i^o$ is the final temporal feature of person i generated by the OTF sub-structure.

3.5. *Two Siamese Networks*

The Siamese network is a measure of the similarity of two objects, which consist of two substructures with shared weights [35]. Each substructure is used as a feature extractor to output the trained feature vectors, and the Siamese network compares these feature vectors using Euclidean distance. The essence of this network comparison idea is to try to reduce the distance between feature vectors of the same class and increase the distance between feature vectors of different classes. Thus, the similarity of a pair of inputs is distinguished by a margin. Fortunately, this property is close to the distance metric learning algorithm in the person Re-ID, so the Siamese network has been applied to person Re-ID work. For video-based person Re-ID, the Siamese network can use the features of a pair of image sequences to train similarity.

Concretely, in our paper, as shown in Figure 2, we constructed two Siamese network [8] to learn the independent distance of the TAF sub-structure and OTF sub-structure, respectively. For the first Siamese network, the final appearance feature vector $\overline{F}_i^T$ and $\overline{F}_j^T$ obtained by the pivotal frames of person i and j through the TAF substructure are taken as inputs. The similarity loss function $Sim(\bullet)$ of the generic first Siamese network is defined as follows

$$Sim(\overline{F}_i^T, \overline{F}_j^T) = \begin{cases} \frac{1}{2}\|\overline{F}_i^T - \overline{F}_j^T\|^2 & i = j \\ \frac{1}{2}[\max(M - \|\overline{F}_i^T - \overline{F}_j^T\|, 0)]^2 & i \neq j \end{cases}, \tag{5}$$

where M represents the margin value in the Siamese network. Similarly, the second-stream Siamese pseudo-network employs the same function with different types of feature vector inputs, as shown in Equation (6):

$$Sim(\overline{F}_i^o, \overline{F}_j^o) = \begin{cases} \frac{1}{2}\|\overline{F}_i^o - \overline{F}_j^o\|^2 i = j \\ \frac{1}{2}[\max(M - \|\overline{F}_i^o - \overline{F}_j^o\|, 0)]^2 & i \neq j \end{cases}, \tag{6}$$

It should be noted that $\overline{F}_i^o$ and $\overline{F}_j^o$ denote the temporal feature vectors of person i and j. To sum up, the joint objective function Sim_{obj} combined with the two-stream Siamese network is shown in Equation (7):

$$Sim_{obj} = \partial_T Sim(\overline{F}_i^T, \overline{F}_j^T) + \partial_o Sim(\overline{F}_i^o, \overline{F}_j^o), \tag{7}$$

where ∂_T, ∂_o represents the loss weight, and $\partial_T > \partial_o$. The reason for setting these weights is the effectiveness of the appearance features compared to the temporal features.

4. Training and Test

4.1. Training (Joint Multiple Loss)

During the training phase, we adopted a joint training method similar to those previously described in the literature [11]. The core of this strategy is to combine the objective function of two Siamese networks and the objective function of the predicted person's identity to train our proposed appearance feature learning substructure. In order to take full advantage of label information, we used the Softmax loss function [36] to predict the person's identify. In our work, as shown in the Figure 6 for feature vector $\overline{F}_i^T$ in the first-stream substructure, the posterior probability of predicting person i is as follows

$$P_{iT}(\widetilde{m}_i = m_i | S_i) = \frac{\exp(W_{m_i} \overline{F}_i^T)}{\sum_{n=1}^{n_{id}} \exp(W_n \overline{F}_i^T)}, \tag{8}$$

where m_i represents the category label of the person i training video sample S_i, $\widetilde{m}_i$ is the predicted label, and W_n refers to the Softmax function parameter of the person's class n. The training loss is computed as

$$Loss_i^{soft} = \sum \log(P_{iT}(\widetilde{m}_i = m_i | S_i)), \tag{9}$$

Note that the use of the Softmax loss function during training is only for the appearance feature sub-structure (the first-stream).

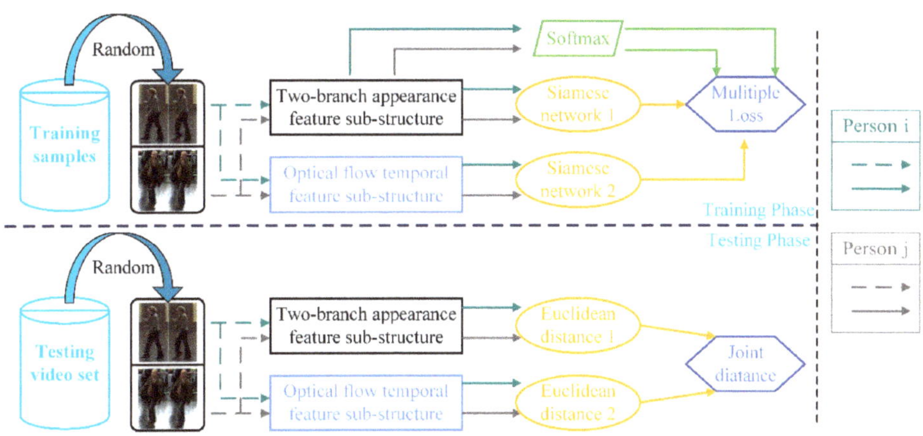

Figure 6. The training steps and test steps of the proposed hybrid end-to-end deep learning architecture. The upper left side of the figure is the training phase, and the lower-left side is the test phase. The right side of the figure is an annotation of the arrows of different colors.

Therefore, the loss function $Loss$ of the entire architecture is as follows

$$Loss = Sim_{obj} + Loss_i^{soft} + Loss_j^{soft}, \tag{10}$$

where $Loss_i^{soft}$ and $Loss_j^{soft}$ are the Softmax functions of persons i and j, respectively.

4.2. Test (Re-Identification)

In the testing phase, given a probe video and candidate video set, the test of the pedestrian re-identification algorithm is compared the distance between the probe video and each the videos in the candidate video set. Therefore, our goal is to calculate and rank the distances between the person's features. In our work, as shown in Figure 6, we use Euclidean distance to express the similarity of persons. For the final appearance feature vectors ($\overline{F}_i^T$ and $\overline{F}_j^T$) and temporal feature vectors ($\overline{F}_i^o$ and $\overline{F}_j^o$), the independent Euclidean distances are calculated as follows

$$d^T = \|\overline{F}_i^T - \overline{F}_j^T\|, \tag{11}$$

$$d^o = \|\overline{F}_i^o - \overline{F}_j^o\|, \tag{12}$$

where d^T and d^o represent the distance of appearance features and the distance of temporal features, respectively. Finally, the weighting merges the above distances and sorts them:

$$d = \partial_T d^T + \partial_o d^o, \tag{13}$$

where d is the joint distance between persons. The data selection principle for training phase and testing phase is specified in Section 5.1.3.

5. Experiments

In this section, we evaluated the proposed architecture of the three video datasets. The first part of our experimental work was mainly to compare experiments with other algorithms, and the other part was to verify the effectiveness of some factors in the proposed method.

5.1. ExperimentalSetup

5.1.1. Datasets

The details of the three datasets are as follows, and Table 2 and Figure 7 show the basic information and some person samples, respectively.

Table 2. Detailed information of the experimental datasets.

Dataset	Persons	Cameras	Videos	Resolution	Evaluation
PRID-2011	385/749	2	400	64 × 128	CMC
iLIDS-VID	300	2	600	64 × 128	CMC
MARS	1261	6	20,478	128 × 256	CMC & mAP

The PRID-2011 dataset [12] is composed of images captured by two cameras (A and B) from outdoor non-overlapping perspectives. There are 385 identities and 749 identities in cameras A and B, respectively, and 200 persons with the same identity under both cameras. Note that there are 400 video sequences for 200 subjects. The video sequence length of each pedestrian is between 5 and 675 frames. The design peculiarity of this dataset is the challenges of persons with simple background interference, less occlusions, and lighting variations.

The iLIDS-VID dataset [13] consists of 600 video sequences of 300 identities, also captured from two non-overlapping cameras view. Each video sequence of the dataset is between 23 and 192 frames in length. The challenges of this dataset include camera-view changes, illumination variations, complex cluttered background, and serious occlusions.

The MARS dataset [14] is a relatively new video person Re-ID dataset. The dataset is derived from an extension of the Market1501 dataset [37] with 1261 pedestrians and 20,478 tracklets. These tracklets were captured by six cameras and collected using a DPM detector [38] and a GMMCP tracker [39]. Furthermore, there are 3278 distracted tracklets in the dataset due to false detection and association.

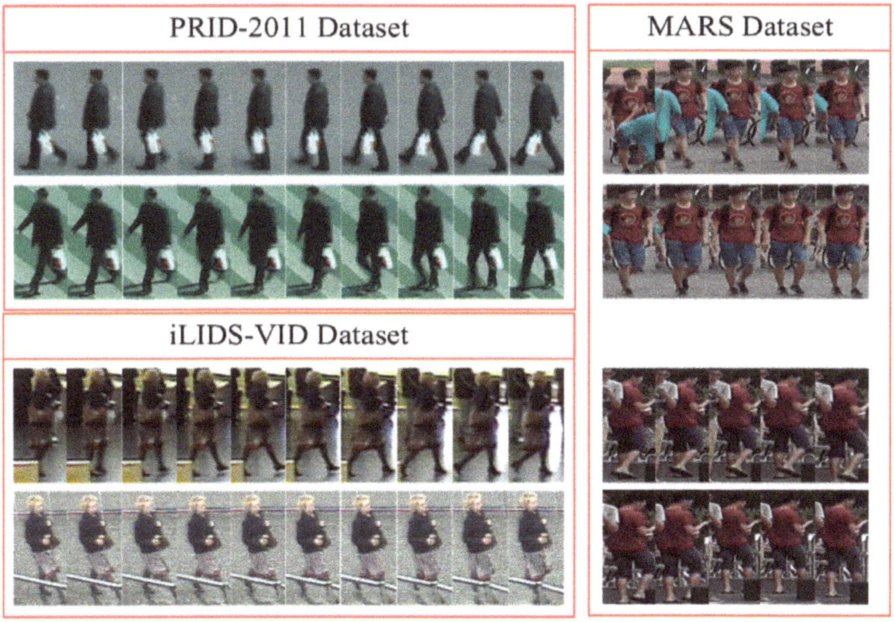

Figure 7. Some samples of persons from different camera-views in three public datasets, including the PRID-2011 dataset, the iLIDS-VID dataset and the Motion Analysis and Re-identification Set(MARS) dataset.

5.1.2. Evaluation Protocol

In order to evaluate the effectiveness of the person Re-ID algorithm, we adopted the cumulative matching characteristic (CMC) curve [40] and the mean average precision (mAP) [14] as evaluation criteria. The CMC value refers to the expectation of a correct match in the rank-k (%) position. The CMC curve refers to the curve of correct match results in the rank-k (%). The mAP considers both the precision and recall of multiple same persons in a gallery. For the PRID-2011 dataset and iLIDS-VID dataset, we used the CMC value to evaluate the performance of the algorithm. For the MARS dataset, both CMC curve and mAP were adopted. The experimental results were the average values after ten random experiments.

5.1.3. Implementation Details

In terms of data preparation, we followed an experimental data selection principle similar to the literature [13] on the PRID-2011 dataset and the iLIDS-VID dataset. Specifically, we used the 177 persons out of 354 videos from the camera A and B on the PRID-2011 dataset. On the iLIDS-VID dataset, we used the 400 videos of 200 persons for the experiment. Similarly, we randomly selected the videos from one camera-view for the training samples, and other videos from the other camera-view for testing. Finally, for the MARS dataset, we followed the experimental data selection principle as described in the literature [14]. The dataset was divided into 625 persons for training, and the rest of the persons for testing. In addition, since we consider the pairwise input of the Siamese network, the person's videos in the training set were randomly combined into positive sample pairs and negative sample pairs. The sequence length on the three datasets was set to 16, as in the literature [11]. In cases where the person sequence was shorter than 16, we use the entire sequence.

In terms of architecture parameter settings, our experiments were conducted under the Caffe [41] deep learning framework. When we trained the deployed network architecture on the deep learning framework, some necessary training parameters needed to be set: initial learning rate was set to 0.0001, the momentum to 0.9, max iterations to 30,000, and the learning rate decline policy was "inv". Then, the M (margin value of the Siamese network) was set to 2. Lastly, the optimization method during training was the stochastic gradient descent method.

5.2. Comparative Experiment

In order to verify the performance of the proposed architecture on the PRID-2011 dataset, iLIDS-VID dataset, and MARS dataset, we established a comparative experiment to compare our video-based person Re-ID architecture with other state-of-the-art algorithms.

5.2.1. Results on PRID-2011 Dataset

For the PRID-2011dataset, we compared the performance of our proposed architecture with eleven state-of-the-art methods, including DVR [2], DVDL [42], STFV3D [3], RMLLC-SLF [43], TDL [4], RFA [44], CNN-RNN [5], CNN-BRNN [9], CRF [10], ASTPN [22], TSSCN [11], and TAM-SRM [45]. The experimental results in the CMC values are shown in Table 3. The black bold in Table 3 indicates the highest correct recognition rate. Note that among these approaches, the first four methods are based on the traditional person Re-ID method, and the remaining are based on deep learning algorithms. As can be seen from Table 3, Rank-1, Rank-5, and Rank-20 of our proposed method reached 79%, 92%, and 99%, respectively. In the comparison methods, in addition to the TAM-SRM algorithm, the Rank-1 recognition rate was improved compared to the rest of the algorithms. Concurrently, our method was also 1% higher on Rank-1 than the similar TSSCN method. These results all show the good performance of our proposed algorithm on the PRID-2011 dataset. Remarks, Figure 8 shows the re-identification sorting results of some persons in the PRID-2011 dataset.

Table 3. Comparison experiment with two types of state-of-the-art algorithms for the PRID-2011 dataset in terms of CMC values.

PRID-2011 Dataset				
Category	**Methods**	**Rank-1**	**Rank-5**	**Rank-20**
Traditional	DVR	40	72	92
	DVDL	40	70	86
	STFV3D	42	72	92
	RMLLC-SLF	50	78	97
	TDL	57	80	94
Deep Learning	RFA	58	86	98
	CNN-RNN	65	90	97
	CNN-BRNN	72	92	98
	CRF	77	93	98
	ASTPN	77	**95**	99
	TSSCN	78	94	99
	TAM-SRM	79	94	99
	Our	**79**	92	**99**

Figure 8. The re-identification results of some people in the proposed architecture in the PRID-2011 dataset. The first column in the figure represents the probe video. The second column is the result of sorting the top ten with the distances, where the green boxes indicate the first and the same person, and the red boxes are the wrong match.

5.2.2. Results on iLIDS-VID Dataset

For the iLIDS-VID dataset, the comparison method we used was consistent with the experiment on the PRID-2011 dataset. The experimental results are shown in Table 4, the black bold indicates the best recognition rate. Rank-1, Rank-5, and Rank-20 of our method on the iLIDS-VID dataset reached 59%, 82%, and 96%, respectively. Compared to the comparison method, our method had a slight gap with the CRF method, the TSSCN method, and the ASTPN method. Our analysis considered that the number of training samples in the dataset was small, and the challenges were complex, including background interference and severe occlusion.

Table 4. Comparison experiment with two types of state-of-the-art algorithms for theiLIDS-VID dataset in terms of CMC values.

iLIDS-VID Dataset				
Category	**Methods**	**Rank-1**	**Rank-5**	**Rank-20**
Traditional	DVR	40	61	82
	DVDL	26	48	69
	STFV3D	37	64	87
	RMLLC-SLF	59	85	96
	TDL	56	88	98
Deep Learning	RFA	49	76	90
	CNN-RNN	65	**90**	97
	CNN-BRNN	55	85	95
	CRF	61	85	97
	ASTPN	**62**	86	**98**
	TSSCN	60	86	97
	TAM-SRM	55	87	97
	Our	59	82	96

5.2.3. Results on MARS Dataset

The MARS dataset is a large-scale dataset for video-based person Re-ID; we compared our method with six state-of-the-art methods, including the ASIPN [22], CAR [29], Zheng et al. [14], CRF [10], TAM-SRM [45], and Li et al. [46]. The experimental results are shown in Table 5 and Figure 9, and the bold black indicates the highest recognition rate. Rank-1, Rank-5, and Rank-20 of our method for the MARS dataset reached 73%, 91%, and 97%, respectively. In the comparative method, the framework proposed in this paper outperformed the TAM-SRM method by 2% and 1% on Rank-1 and Rank-5, respectively. Simultaneously, our method was also superior to the method of Zheng et al. [14] and the TAM-SRM algorithm in terms of the mAP evaluation criterion. The above results indicate that our architecture had good performance on the MARS dataset. In particular, the method of Li et al. [46] obviously outperformed the proposed algorithm by a large margin. The method of Li et al. [46] takes full advantage of the labeled information of the pretrained model on image-based person re-identification datasets to train. The results provide evidence that we can improve the training ability of video-based person re-identification models by using labeled information on image-based datasets.

Table 5. Comparison experiment with other state-of-the-art algorithms on the MARS dataset in terms of CMC values and mean average precision (mAP).

	MARS Dataset			
Methods	**Rank-1**	**Rank-5**	**Rank-20**	**mAP**
ASTPN	44	70	81	-
CAR	56	70	80	-
Zheng et al.	68	83	89	49.3
CRF	71	89	96	-
TAM-SRM	71	90	**98**	50.7
Li et al.	**82**	-	-	**65.8**
Our	73	**91**	97	52.4

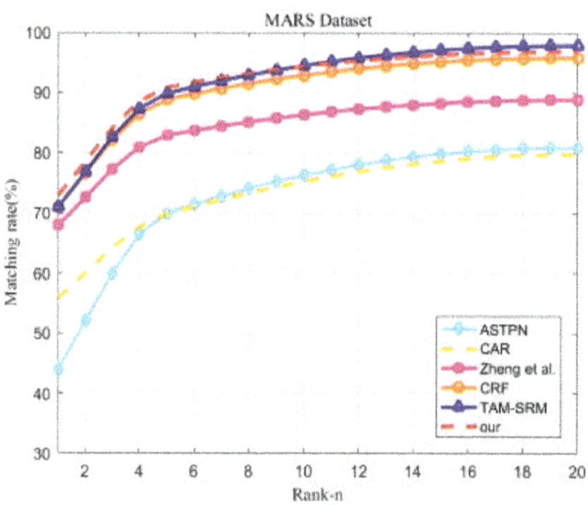

Figure 9. Experimental results of the comparison with other state-of-the-art algorithms for the MARS dataset in term of the CMC curve. Because the work of Li et al. [46] only reported the Rank-1 value, their results cannot be drawn as a curve.

5.3. *Verification Experiment ofKey Components*

In this section, we performed in-depth experiments on the PRID-2011 dataset to verify the effectiveness of four key components, including the pivotal frame's number, the different Inception-V3 structure and network, the different weights with two-stream architecture, and the independent effectiveness of each stream feature's sub-structure. The specific experimental results and analysis are as follows. Note that, when we verified the effectiveness of one component, the other two components were kept unchanged. Therefore, we changed this component to conduct the verification experiment.

5.3.1. Effectiveness of the Pivotal Frame's Number

As shown in the experimental results of the first to third rows in Table 6, the selection of different numbers of pivotal frames yielded different recognition rates. We can observe that when the number of pivotal frames equals 2, $N = 2$, the best performance and recognition rates were achieved. Note that we ensured that the other two components were "Our (Inception-V3-5c)" and "Our ($\partial_T = 0.7, \partial_o = 0.3$)" when we completed the experiment. The experimental results also show that the number of pivotal frames is an important factor in the appearance feature substructure of our proposed. Simultaneously, pivotal frames also help the TAF model get better appearance feature. For the effectiveness of the pivotal frame's number, considering the complete appearance feature sub-structure, the increase in the number of pivotal frames is only the repeated accumulation of the two walking postures of a person. Minor changes in the appearance of the person are likely to cause the fitting of the global feature representation on the deep Inception-V3 network, and the increase in the number of pivotal frames may increase the likelihood of similar poses between different persons.

Table 6. Verification experiment results for the pivotal frame's number for the PRID-2011 dataset in terms of CMC values.

PRID-2011 Dataset			
Methods	**Rank-1**	**Rank-5**	**Rank-20**
Our ($N = 2$)	**79**	**92**	**99**
Our ($N = 4$)	72	90	96
Our ($N = 6$)	73	85	92
Our ($N = 8$)	73	89	93

5.3.2. Effectiveness of the Different Inception-V3 Structures and Different Network

In order to verify that the Inception-V3 network can extract distinguishing features for pivotal frames, we compared different Inception-V3 structures with the Res-Net (50) network [33]. Comparing the results of lines 1–4 in Table 7 we can see that using the Inception-V3 network to perform the extraction of appearance features consistently improves the matching performance. Note that "Inception-V3-3c", "Inception-V3-4e", and "Inception-V3-5c" refer to the outputs of the "3c", "4e", and "5c" modules in the Inception-V3 network, respectively. In particular, the "Inception-V3-5c" structure in the Inception-V3 network performed better than the rest of structure, with improvements of approximately 14% and 8% on Rank-1, respectively. These results verify that the "Inception-V3-5c" structure can learn a rich global appearance feature and effectively improve the person Re-ID recognition rate.

Table 7. Verification experiment results for the different Inception-V3 structures and different networks on the PRID-2011 dataset in terms of CMC values.

PRID-2011 Dataset			
Methods	**Rank-1**	**Rank-5**	**Rank-20**
Res-Net (50)	66	82	90
Our(Inception-V3-3c)	65	85	95
Our(Inception-V3-4e)	71	88	95
Our(Inception-V3-5c)	**79**	**92**	**99**

5.3.3. Effectiveness of the Different Weights with Two-Stream Architecture

In the hybrid end-to-end learning architecture, the appearance features and temporal features of persons can be extracted separately. In order to verify the importance of each stream feature structure, from the 1 to 5 rows in Table 8, we performed a verification experiment of two streams networks with five different weights. It can be seen that when the weight is $\partial_T = 0.7, \partial_o = 0.3$, the optimal result of our architecture was 79% for Rank-1. Note that when there was no temporal feature (OTF) sub-structure, the Rank-1 recognition rate was 70%. After adding the temporal feature (OTF) sub-structure, the recognition rate was significantly improved. The experimental results prove that the temporal features of the OTF model are beneficial to the method proposed in video-based person Re-ID.

Table 8. Verification experiment results for the different weights with the two-stream architecture on the PRID-2011 dataset in terms of CMC values.

PRID-2011 Dataset			
Methods	**Rank-1**	**Rank-5**	**Rank-20**
Our($\partial_T = 0.5, \partial_o = 0.5$)	73	88	95
Our($\partial_T = 0.6, \partial_o = 0.4$)	76	**93**	97
Our($\partial_T = 0.7, \partial_o = 0.3$)	**79**	92	**99**
Our($\partial_T = 0.8, \partial_o = 0.2$)	77	92	97
Our($\partial_T = 1, \partial_o = 0$)	70	86	93

5.3.4. Independent Effectiveness of Each Stream Feature Substructure

In this subsection, we performed a comparison experiment on the PRID-2011 dataset to verify the independent effectiveness of each stream feature's sub-structure. From rows 1 to 5 in Table 9, we chose the independent feature substructure (TAF sub-structure and OTF sub-structure) to be compared with related algorithms, including CNN-RNN [5], CNN-BRNN [9], and CRF [10]. The results showed that the TAF sub-structure reaches 70%, 88%, and 95% on Rank-1, Rank-5, and Rank-20, respectively. Compared with the three other algorithms, the results of the independent TAF sub-structure were better than the CNN-RNN algorithm, and lower than the other two algorithms. For the independent OTF sub-structure, the Rank-1, Rank-5, and Rank-20 reached 57%, 74%, and 89%, respectively. However, the results of the OTF structure were lower than results of the other three algorithms. Among the three algorithms, they all use appearance feature information and temporal feature information to represent the person.

Table 9. Verification experiment results for the independent effectiveness of each stream feature substructure on the PRID-2011 dataset in terms of CMC values.

PRID-2011 Dataset			
Methods	**Rank-1**	**Rank-5**	**Rank-20**
Our ($\partial_T = 1, \partial_o = 0$)	70	86	93
Our ($\partial_T = 0, \partial_o = 1$)	57	74	89
CNN-RNN	65	90	97
CNN-BRNN	72	92	98
CRF	**77**	**93**	**98**

6. Conclusions

In this paper, we proposed a hybrid end-to-end deep learning architecture for video-based person re-identification. The architecture consists of the two-stream hybrid feature structure and two Siamese networks. The two-stream hybrid deep feature structure includes the Two-branch Appearance Feature sub-structure and the Optical flow Temporal Feature sub-structure, which can separately learn appearance and temporal information. For the video-based person re-identification, our method showed, in a large number of experiments on three datasets, that separate feature structures were superior in their ability to learn appearance features and temporal features, as well as the independent distances of different modal features. In future, we will add semantic features to enrich the feature learning model and improve the loss function to optimize the distance metric.

Author Contributions: R.S. and Q.H. conceived the research idea and proposed the architecture. Q.H. and M.X. investigate related work. Q.H. established and completed the experiments. R.S. and Q.H. wrote the paper. R.S., J.Z., and Q.H. gave many suggestions and helped revise the paper.

Funding: The research was funded by the National Natural Science Foundation of China, grant number [61471154] and Anhui Province science and technology project, grant number [1704d0802181].

Conflicts of Interest: The authors declare no conflicts of interest.

References

1. Zheng, C.; Liang, X.; Matsuyama, T. Generic learning-based ensemble framework for small sample size face recognition in multi-camera networks. *Sensors* **2014**, *14*, 23509–23538. [CrossRef] [PubMed]
2. Wang, T.; Gong, S.; Zhu, X.; Wang, S. Person re-identification by discriminative selection in video ranking. *IEEE Trans. Pattern Anal. Mach. Intell.* **2016**, *38*, 2501–2514. [CrossRef] [PubMed]
3. Liu, K.; Ma, B.; Zhang, W.; Huang, R. A spatio-temporal appearance representation for video-based pedestrian re-identification. In Proceedings of the IEEE International Conference on Computer Vision, Santiago, Chile, 7–13 December 2015.
4. You, J.; Wu, A.; Li, X.; Zheng, W.-S. Top-push video-based person re-identification. In Proceedings of the IEEE Conference on Computer Vision and Pattern Recognition, Las Vegas, NV, USA, 27–30 June 2016.
5. McLaughlin, N.; del Rincon, J.M.; Miller, P. Recurrent convolutional network for video-based person re-identification. In Proceedings of the IEEE Conference on Computer Vision and Pattern Recognition, Las Vegas, NV, USA, 27–30 June 2016.
6. Lisanti, G.; Masi, I.; Bagdanov, A.D.; Del Bimbo, A. Person re-identificaiton by iterative re-weighted sparse ranking. *IEEE Trans. Pattern Anal. Mach. Intell.* **2015**, *37*, 1629–1642. [CrossRef] [PubMed]
7. Schumann, A.; Stiefelhagen, R. Person re-identification by deep learning attribute-complementary information. In Proceedings of the IEEE Conference on Computer Vision and Pattern Recognition Workshops, Honolulu, HI, USA, 21–26 July 2017.
8. Matsukawa, T.; Suzuki, E. Person re-identification using cnn features learned from combination of attributes. In Proceedings of the International Conference on Pattern Recognition, Cancun, Mexico, 4–8 December 2016.
9. Zhang, W.; Yu, X.; He, X. Learning bidirectional temporal cues for video-based person re-identification. *IEEE Trans. Circuits Syst. Video Technol.* **2017**, *28*, 2768–2776. [CrossRef]
10. Chen, L.; Yang, H.; Zhu, J.; Zhou, Q.; Wu, S.; Gao, Z. Deep spatial-temporal fusion network for video-based person re-identification. In Proceedings of the IEEE Conference on Computer Vision and Pattern Recognition Workshops, Honolulu, HI, USA, 21–26 July 2017.
11. Chung, D.; Tahboub, K.; Delp, E.J. A two stream siamese convolutional neural network for person re-identification. In Proceedings of the IEEE International Conference on Computer Vision, Venice, Italy, 22–29 October 2017.
12. Hirzer, M.; Beleznai, C.; Roth, P.M.; Bischof, H. Person reidentification by descriptive and discriminative classification. In Proceedings of the Scandinavian Conference on Image Analysis, Ystad, Sweden, 23–25 May 2011.
13. Wang, T.; Gong, S.; Zhu, X.; Wang, S. Person re-identification by video ranking. In Proceedings of the European Conference on Computer Vision, Switzerland, Zurich, 6–12 September 2014; Springer: Cham, Switzerland, 2014.
14. Zheng, L.; Bie, Z.; Sun, Y.; Wang, J.; Su, C.; Wang, S.; Tian, Q. Mars: A video benchmark for large-scale person re-identification. In Proceedings of the European Conference on Computer Vision, Amsterdam, The Netherlands, 11–14 October 2016; Springer: Cham, Switzerland, 2016.
15. Gray, D.; Tao, H. Viewpoint invariant pedestrian recognition with an ensemble of localized features. In Proceedings of the 10th European Conference on Computer Vision, Marseille, France, 12–18 October 2008.

16. Zhao, R.; Ouyang, W.; Wang, X. Unsupervised salience learning for person re-identification. In Proceedings of the IEEE Conference on Computer Vision and Pattern Recognition, Portland, OR, USA, 23–28 June 2013.
17. Yang, Y.; Yang, J.; Yan, J.; Liao, S.; Yi, D.; Li, S.Z. Salient color names for person re-identification. In Proceedings of the European Conference on Computer Vision, Zurich, Switzerland, 6–12 September 2014; Springer: Cham, Switzerland, 2014.
18. Matsukawa, T.; Okabe, T.; Suzuki, E.; Sato, Y. Hierarchical gaussian descriptor for person re-identification. In Proceedings of the IEEE Conference on Computer Vision and Pattern Recognition, Las Vegas, NV, USA, 27–30 June 2016.
19. Liao, S.; Hu, Y.; Zhu, X.; Li, S.Z. Person re-identification by local maximal occurrence representation and metric learning. In Proceedings of the IEEE Conference on Computer Vision and Pattern Recognition, Boston, MA, USA, 7–12 June 2015.
20. Kostinger, M.; Hirzer, M.; Wohlhart, P.; Roth, P.M.; Bischof, H. Large scale metric learning from equivalence constraints. In Proceedings of the IEEE Conference on Computer Vision and Pattern Recognition, Providence, RI, USA, 16–21 June 2012.
21. Li, Y.; Wu, Z.; Karanam, S.; Radke, R. Multi-shot human re-identification using adaptive fisher discriminant analysis. In Proceedings of the British Machine Vision Conference, Swansea, UK, 7–10 September 2015; Springer: Cham, Switzerland, 2015.
22. Xu, S.; Cheng, Y.; Gu, K.; Yang, Y.; Chang, S.; Zhou, P. Jointly attentive spatial-temporal pooling networks for video-based person reidentification. In Proceedings of the IEEE International Conference on Computer Vision, Venice, Italy, 22–29 October 2017.
23. Zhu, X.; Jing, X.-Y.; You, X.; Zuo, W.; Shan, S.; Zheng, W.-S. Image to video person re-identification by learning heterogeneous dictionary pair with feature projection matrix. *IEEE Trans. Inf. Forensics Secur.* **2017**, *13*, 717–732. [CrossRef]
24. Yi, D.; Lei, Z.; Liao, S.; Li, S.Z. Deep metric learning for person re-identification. In Proceedings of the International Conference on Pattern Recognition, Stockholm, Sweden, 24–28 August 2014.
25. Cheng, D.; Gong, Y.; Zhou, S.; Wang, J.; Zheng, N. Person re-identification by multi-channel parts-based cnn with improved triplet loss function. In Proceedings of the IEEE Conference on Computer Vision and Pattern Recognition, Las Vegas, NV, USA, 27–30 June 2016.
26. Liu, H.; Feng, J.; Qi, M.; Jiang, J.; Yan, S. End-to-end comparative attention network for person re-identification. *IEEE Trans. Image Process.* **2017**, *26*, 3492–3506. [CrossRef] [PubMed]
27. Chen, W.; Chen, X.; Zhang, J.; Huang, K. Beyond triplet loss: A deep quadruplet network for person re-identification. In Proceedings of the IEEE Conference on Computer Vision and Pattern Recognition, Honolulu, HI, USA, 21–26 July 2017.
28. Li, W.; Zhao, R.; Xiao, T.; Wang, X. Deepreid: Deep filter pairing neural network for person re-identification. In Proceedings of the IEEE Conference on Computer Vision and Pattern Recognition, Columbus, OH, USA, 23–28 June 2014.
29. Zhang, W.; Hu, S.; Liu, K. Compact appearance learning for video-based person re-identification. *IEEE Trans. Circuits Syst. Video Technol.* **2018**, *99*, 1. [CrossRef]
30. Waters, R.; Morris, J. Electrical activity of muscles of the trunk during walking. *J. Anat.* **1972**, *111*, 191–199. [PubMed]
31. Szegedy, C.; Vanhoucke, V.; Ioffe, S.; Shlens, J.; Wojna, Z. Rethinking the inception architecture for computer vision. In Proceedings of the IEEE Conference on Computer Vision and Pattern Recognition, Las Vegas, NV, USA, 27–30 June 2016.
32. Simonyan, K.; Zisserman, A. Very deep convolutional networks for large-scale image recognition. In Proceedings of the International Conference on Learning Representations, San Diego, CA, USA, 7–9 May 2015.
33. He, K.; Zhang, X.; Ren, S.; Sun, J. Deep residual learning for image recognition. In Proceedings of the IEEE Conference on Computer Vision and Pattern Recognition, Las Vegas, NV, USA, 27–30 June 2016.
34. Lucas, B.D.; Kanade, T. An iterative image registration technique with an application to stereo vision. In Proceedings of the International Joint Conference on Artificial Intelligence, Vancouver, BC, Canada, 24–28 August 1981.
35. Bromley, J.; Bentz, J.W.; Bottou, L.; Guyon, I.; LeCun, Y.; Moore, C.; Sackinger, E.; Shah, R. Signature verification using a "siamese" time delay neural network. *Int. J. Pattern Recognit. Artif. Intell.* **1994**, *7*, 25–44.

36. Sun, Y.; Chen, Y.; Wang, X.; Tang, X. Deep learning face representation by joint identification-verification. In Proceedings of the 27th International Conference on Neural Information Processing Systems, Montreal, QC, Canada, 8–13 December 2014.
37. Zheng, L.; Shen, L.; Tian, L.; Wang, S.; Wang, J.; Tian, Q. Scalable person re-identification: A benchmark. In Proceedings of the IEEE International Conference on Computer Vision, Santiago, Chile, 7–13 December 2015.
38. Felzenszwalb, P.F.; Girshick, R.B.; McAllester, D.A.; Ramanan, D. Object detection with discriminatively trained part-based models. *IEEE Trans. Pattern Anal. Mach. Intell.* **2010**, *32*, 1627–1645. [CrossRef] [PubMed]
39. Dehghan, A.; Assari, S.M.; Shah, M. GMMCP tracker: Globally optimal generalized maximum multi clique problem for multiple object tracking. In Proceedings of the IEEE Conference on Computer Vision and Pattern Recognition, Boston, MA, USA, 7–12 June 2015.
40. Bolle, R.M.; Connell, J.H.; Pankanti, S.; Ratha, N.K.; Senior, A.W. The relation between the roc curve and the cmc. In Proceedings of the Fourth IEEE Workshop on Automatic Identification Advanced Technologies, Buffalo, NY, USA, 17–18 October 2005.
41. Jia, Y.; Shelhamer, E.; Donahue, J.; Karayev, S.; Long, J.; Girshick, R.; Guadarrama, S.; Darrell, T. Caffe: Convolutional architecture for fast feature embedding. In Proceedings of the 22nd ACM international conference on Multimedia, Orlando, FL, USA, 3–7 November 2014.
42. Karanam, S.; Li, Y.; Radke, R.J. Person re-identification with discriminatively trained viewpoint invariant dictionaries. In Proceedings of the IEEE International Conference on Computer Vision, Santiago, Chile, 7–13 December 2015.
43. Chen, J.; Wang, Y.; Tang, Y. Person re-identification by exploiting spatio-temporal cues and multi-view metric learning. *IEEE Signal Process. Lett.* **2016**, *23*, 998–1002. [CrossRef]
44. Yan, Y.; Ni, B.; Song, Z.; Ma, C.; Yan, Y.; Yang, X. Person re-identification via recurrent feature aggregation. In Proceedings of the European Conference on Computer Vision, Amsterdam, The Netherlands, 8–16 October 2016; Springer: Cham, Switzerland, 2016.
45. Zhou, Z.; Huang, Y.; Wang, W.; Wang, L.; Tan, T. See the forest for the trees: Joint spatial and temporal recurrent neural networks for video-based person re-identification. In Proceedings of the IEEE Conference on Computer Vision and Pattern Recognition, Honolulu, HI, USA, 21–26 July 2017.
46. Li, S.; Bak, S.; Carr, P.; Wang, X. Diversity regularized spatiotemporal attention for video-based person re-identification. In Proceedings of the IEEE Conference on Computer Vision and Pattern Recognition, Salt Lake City, UT, USA, 18–22 June 2018.

sensors

MDPI

Article

IrisDenseNet: Robust Iris Segmentation Using Densely Connected Fully Convolutional Networks in the Images by Visible Light and Near-Infrared Light Camera Sensors

Muhammad Arsalan, Rizwan Ali Naqvi, Dong Seop Kim, Phong Ha Nguyen, Muhammad Owais and Kang Ryoung Park *

Division of Electronics and Electrical Engineering, Dongguk University, 30 Pildong-ro 1-gil, Jung-gu, Seoul 100-715, Korea; arsal@dongguk.edu (M.A.); rizwanali@dongguk.edu (R.A.N.); k_ds1028@naver.com (D.S.K.); stormwindvn@dongguk.edu (P.H.N.); malikowais266@gmail.com (M.O.)

* Correspondence: parkgr@dongguk.edu; Tel.: +82-10-3111-7022; Fax: +82-2-2277-8735

Received: 2 April 2018; Accepted: 8 May 2018; Published: 10 May 2018

Abstract: The recent advancements in computer vision have opened new horizons for deploying biometric recognition algorithms in mobile and handheld devices. Similarly, iris recognition is now much needed in unconstraint scenarios with accuracy. These environments make the acquired iris image exhibit occlusion, low resolution, blur, unusual glint, ghost effect, and off-angles. The prevailing segmentation algorithms cannot cope with these constraints. In addition, owing to the unavailability of near-infrared (NIR) light, iris recognition in visible light environment makes the iris segmentation challenging with the noise of visible light. Deep learning with convolutional neural networks (CNN) has brought a considerable breakthrough in various applications. To address the iris segmentation issues in challenging situations by visible light and near-infrared light camera sensors, this paper proposes a densely connected fully convolutional network (IrisDenseNet), which can determine the true iris boundary even with inferior-quality images by using better information gradient flow between the dense blocks. In the experiments conducted, five datasets of visible light and NIR environments were used. For visible light environment, noisy iris challenge evaluation part-II (NICE-II selected from UBIRIS.v2 database) and mobile iris challenge evaluation (MICHE-I) datasets were used. For NIR environment, the institute of automation, Chinese academy of sciences (CASIA) v4.0 interval, CASIA v4.0 distance, and IIT Delhi v1.0 iris datasets were used. Experimental results showed the optimal segmentation of the proposed IrisDenseNet and its excellent performance over existing algorithms for all five datasets.

Keywords: iris recognition; iris segmentation; semantic segmentation; convolutional neural network (CNN); visible light and near-infrared light camera sensors

1. Introduction

In the last two decades, biometrics have been completely incorporated into our daily life. Owing to research efforts, biometrics are now adopted to various applications such as person authentication and identification at airports or in national databases. Biometrics in both physiological and behavioral forms are delivering an efficient platform for security metrics [1]. Physiological biometrics include fingerprint recognition [2], finger vein pattern recognition [3], face recognition [4], iris recognition [5], and palmprint recognition [6]. Iris recognition has been proven as innovative reliable biometric widely used in security, authentication, and identification systems [7].

Iris recognition shares the following advantages of secure biometrics: iris features are unique even in the case of twins, left eye iris features of an individual are different from right eye iris features, iris features are naturally complex to be created artificially, and iris features are permanent and remain same throughout a human's lifespan [8].

Innovative research in iris recognition is related to iris applications in mobile and handheld devices. To provide reasonable security to mobile devices, iris patterns are being used to replace passwords or lock-patterns, which makes the device more user-friendly and convenient. Currently, mobile and handheld devices are mainly secured with fingerprint security, but there have been spoofing cases and it is inconvenient to develop separate fingerprint scan hardware in the smart device and rather easy to implement an iris-based system using a frontal camera [9]. For individuals working in intensive security departments, usage of personal digital assistant (PDA) and notepads is much needed, but this essential hardware lacks a security mechanism, which might lead to fraudulent usage. Therefore, building light algorithms that can be efficiently used with mobile devices is a new-era requirement [10]. The major challenge here is to achieve high performance on a mobile platform because of limitation of space, power, and cost of the system. In conclusion, each stage of iris recognition should be designed such that it reduces process time with improved reliability of recognition with less user cooperation in different light environments [11].

Most iris recognition systems consist of five elementary steps: iris image acquisition, pre-processing, iris boundary segmentation, iris feature extraction, and matching for authentication or identification. The acquired image is pre-processed to eliminate the noise captured in the first step. Then, in the third step, iris boundaries are segmented using various algorithms to extract the iris from the image. In the fourth step, the iris features are extracted from the segmented image, and usually, a code is generated with the template using the encoding scheme. Finally, in the last step, the codes are matched for user authentication or identification [12].

Why Is Iris Segmentation Important?

There are two types of image acquisition environments: ideal and non-ideal. In ideal environments, the iris area is not affected by eyelids and eyelashes, and images are under ideal light conditions. Therefore, recognition rates are high and conventional methods can perform well. On the other hand, in non-ideal environments, the images contain blurs, off-angles, non-uniform light intensities, and obstructions. Therefore, in both ideal and non-ideal environments, a real iris boundary without occlusion is required for better error-free features, so a segmentation algorithm is needed to separate each type of noise from the iris image and provide a real iris boundary even in non-ideal situations [13]. A good segmentation algorithm significantly affects the accuracy of the overall iris recognition system and can handle errors generated by occlusions of eyelashes, motion blurs, off-angle irises, specular reflections, standoff distances, eyeglasses, and poor illuminations [14]. Previous research showed that the error generated in the iris segmentation stage is propagated in all subsequent stages of recognition [15]. Proenca et al. analyzed 5000 images of UBRIS, CASIA, and ICE databases, and concluded that the incorrect segmentation in horizontal and vertical directions affects the recognition errors [16].

2. Related Work

The present schemes for iris segmentation can be categorized into five main implementation approaches: iris circular boundary detection without eyelid and eyelash detection, iris circular boundary detection with eyelid and eyelash detection, active contour-based iris segmentation, region growing and watershed-based iris segmentation, and finally, the most elegant, deep-learning-based iris segmentation.

2.1. Iris Circular Boundary Detection without Eyelid and Eyelash Detection

These methods are usually developed for ideal environments and consider the iris and pupil as a circle and do not deal with occlusions. Hough transform (HT) detects a circular iris boundary in iris images, and determines the circularity of the objects based on edge-map voting considering the given limits of the iris or pupil radii, well known as Wilde's method [17]. Many variants of Daugman's method using iris circular boundary detection have been developed [18]. Khan et al. proposed a gradient-based method in which 1-D profile lines were drawn on the iris boundary, the gradient was computed on each profile line, and the maximum change represented the iris boundary [19]. Ibrahim et al. used a two-stage method for a pupillary boundary circular moving window accompanied by probability, where the iris boundary was detected using the gradient on rows with the pupil [20]. Huang et al. used radial suppression-based edge detection and thresholding to detect the iris circular boundary [21]. Jan et al. used pre-processing for specular reflection and detected the boundaries using HT assisted by gray-level statistics, thresholding, and geometrical transforms [22]. Ibrahim et al. proposed an automatic pupil and iris localization method in which the pupillary boundary was detected by automatic adaptive thresholding and the iris boundary was detected by the first derivative of each row with the pupil [23]. Umer et al. first found the iris inner boundary based on restricted HT. For finding the outer iris boundary, inversion transform, image smoothing, and binarization were performed, and finally, restricted HT was computed for finding the external boundary [24].

2.2. Iris Circular Boundary Detection with Eyelid and Eyelash Detection

In this sub-section, we explain the conventional methods which initially compute the iris as a circular object but try to approximate the real iris boundary by using other methods such as eyelash and eyelid detection. Daugman adopted the integro-differential operator to approximate the iris circular boundaries, and detected the eyelid with eyelash by using an additional algorithm [25]. Jeong et al. proposed an effective approach using two circular edge detector in combination with AdaBoost for inner and outer iris boundary detection. To reduce the error, eyelid and eyelash detection was performed [26]. Parikh et al. first found the iris boundary based on color-clustering. To deal with off-angle iris images, he detected two circular boundaries from both right and left of the iris, where the overlapped area of these two boundaries represented the outer iris boundary [27]. Pundlik et al. used the graph cut-based approach for iris segmentation in non-ideal cases, where the eyelashes were separated from the images by image texture using a Markov random field. A energy minimization scheme based on a graph cut was used to segment the iris, pupil, and eyelashes [28]. Zuo et al. proposed a method of non-ideal iris segmentation where non-ideal factors such as off-angle, blur, contrast, and unbalanced illumination were detected and compensated for each eye separately. Both pupillary and iris boundaries were detected by a rotated ellipse fitting in combination with occlusion detection [29]. Hu et al. proposed a novel method for color iris segmentation based on the fusion strategy using three models and by choosing the best strategy automatically, where limbic boundaries were segmented using Daugman's integrodifferential operator with high-score fitting based on the iris center [30].

2.3. Active Contours for Iris Segmentation

Active contours are a step toward detecting the real boundary. Shah et al. used the geodesic active contour-based approach to extract the iris contour from the surrounding structures. Because the active contour can assume shapes and segment multiple objects, iteratively fashioned boundaries of the iris are found by the guidance of global and local properties of the image [31]. Koh et al. used the combination of active contour and HT to locate the outer and inner iris boundaries for non-ideal situations [32]. Abdullah et al. proposed an accurate and fast method for segmenting the iris boundary by using Chan–Vese active contour and morphology [33]. They proposed an active contour-based fusion technique with shrinking and expanding iterations. A pressure force applied to the active contour model was used to make the method robust. The non-circular iris normalization technique was adopted as a new closed eye detection method [34].

2.4. Region Growing/Watershed-Based Iris Segmentation Methods

These types of method are similar to those used for detecting the true iris boundary. Tan et al. proposed a region growing-based approach. After the rough iris and non-iris boundaries are found, a novel integro-differential constellation is constructed with a clustered region growing algorithm. For accurate detection of inner and outer iris boundaries, eight-neighbor connection and point-to-region distance were evaluated [35]. Patel et al. proposed a region growing of pupil circle and the method based on binary integrated curve of intensity to reduce the difficulties created by non-ideal segmentation conditions. The approach avoided the eyelid portion, and hence, was close to the real boundary [36]. Abate et al. proposed an iris segmentation method for the images captured in visible light on mobile devices. To detect the iris boundary in a noisy environment, he used a watershed transform named watershed-based iris detection (BIRD). The watershed algorithm is a growing process performed generally on the gradient image. To reduce the number of watershed regions, the seed selection process is used [37].

2.5. CNN for Iris Segmentation

To solve the problems of previous methods and lessen the computational burden of pre- and post-processing, convolutional neural network (CNN)-based iris segmentation is proposed. CNN provides a strong platform for segmentation tasks such as brain tumor segmented using several kernels [38]. So far, iris segmentation has been rarely researched using CNN whereas it is mostly used for iris recognition purposes. Ahuja et al. proposed two convolution-based model to verify a pair of periocular images including the iris patterns [39]. Zhao et al. proposed a new semantic-assisted convolutional neural network (SCNNs) to match the periocular images [40]. Al-waisy et al. proposed an efficient real-time multimodal biometric system for iris detection [41]. Gangwar et al. proposed DeepIrisNet for cross-sensor iris recognition [42]. Lee et al. proposed a method for iris and periocular recognition based on three CNNs [43].

Considering the iris segmentation, Liu et al. proposed two modalities using fully convolutional networks (FCNs), where multi-scale FCNs (MFCNs) and hierarchical CNNs (HCNNs) were used to find the iris boundaries in non-cooperative environments without using handcrafted features [44]. Arsalan et al. used a two-stage deep-learning-based method to identify the true iris boundary, and modified HT to detect a rough iris boundary, which is provided to the second stage, which utilizes the deep learning model to identify the 21×21 mask as an iris or non-iris pixel [45]. As CNN-based segmentation schemes require considerable labeled data, Jalilian et al. proposed a new domain adaption method for CNN training with a few training data [46]. These schemes have better accuracies compared to the previous methods, but the accuracy of iris segmentation can be further enhanced.

To address the issues of accurate segmentation without prior pre-processing and to develop a robust scheme for all types of environments, this study presents a densely connected fully convolutional network (IrisDenseNet)-based approach to detect an accurate iris boundary with better information gradient flow due to dense connectivity.

Table 1 shows a comparison between prevailing methods and the proposed IrisDenseNet.

Table 1. Comparisons between the proposed and existing methods for iris segmentation.

Type	Methods	Strength	Weakness
Iris circular boundary detection without eyelid and eyelash detection	Iris localization by circular HT [17,22,24]	These methods show a good estimation of the iris region in ideal cases	These types of methods are not very accurate for non-ideal cases or visible light environments
	Integro-differential operator [18]		
	Iris localization by gradient on iris-sclera boundary points [19]	A new idea to use a gradient to locate iris boundary	The gradient is affected by eyelashes and true iris boundary is not found
	The two-stage method with circular moving window [20]	Pupil based on dark color approximated simply by probability	Calculating gradient in a search way is time-consuming
	Radial suppression-based edge detection and thresholding [21]	Radial suppression makes the case simpler for the iris edges	In non-ideal cases, the edges are not fine to estimate the boundaries
	Adaptive thresholding and first derivative-based iris localization [23]	Simple way to obtain the boundary based on the gray level in ideal cases	One threshold cannot guarantee good results in all cases
Iris circular boundary detection with eyelid and eyelash detection	Two-circular edge detector assisted with AdaBoost eye detector [26]	Closed eye, eyelash and eyelid detection is executed to reduce error	The method is affected by pupil/eyelid detection error
	Curve fitting and color clustering [27]	Upper and lower eyelid detections are performed to reduce the error	The empirical threshold is set for eyelid and eyelash detection, and still true boundary is not found
	Graph-cut-based approach for iris segmentation [28]	Eyelashes are removed using Markov random field to reduce error	A separate method for each eyelash, pupil, iris detection is time-consuming
	Rotated ellipse fitting method combined with occlusion detection [29]	Ellipse fitting gives a good approximation for the iris with reduced error	Still, the iris and other boundaries are considered as circular
	Three model fusion-based method assisted with Daugman's method [30]	Simple integral derivative as a base for iris boundaries is a quite simple way	High-score fitting is sensitive in ideal cases, and can be disturbed by similar RGB pixels in the image
Active contour-based methods	Geodesic active contours, Chan–Vese and new pressure force active contours [31–34]	These methods iteratively approximate the true boundaries in non-ideal situations	In these methods, many iterations are required for accuracy, which takes much processing time
Region growing and watershed methods	Region growing with integro-differential constellation [35]	Both iris and non-iris regions are identified along with reflection removal to reduce error	The rough boundary is found first and then a boundary refinement process is performed separately
	Region growing with binary integrated intensity curve-based method [36]	Eyelash and eyelid detection is performed along with iris segmentation	The region growing process starts with the pupil circle, so the case of visible light images where the pupil is not clear can cause errors
	Watershed BIRD with seed selection [37]	Limbus boundary detection is performed to separate sclera, eyelashes, and eyelid pixels from iris	Watershed transform shares the disadvantage of over-segmentation, so circle fitting is used further

Table 1. *Cont.*

Type	Methods	Strength	Weakness
Deep-learning-based methods	HCNNs and MFCNs [44]	This approach shows the lower error than existing methods for non-ideal cases	The similar parts to iris regions can be incorrectly detected as iris points
	Two-stage iris segmentation method using deep learning and modified HT [45]	Better accuracy due to CNN, which is just applied inside the ROI defined in the first stage	Millions of 21×21 images are needed for CNN training and pre-processing required to improve the image
	IrisDenseNet for iris segmentation (Proposed Method)	Accurately find the iris boundaries without pre-processing with better information gradient flow. With robustness for high-frequency areas such as eyelashes and ghost regions	Due to dense connectivity, the mini-batch size should be kept low owing to more time required for training

3. Contribution

This study focuses on an iris image of low quality in non-cooperative scenarios where segmentation is quite difficult with the existing methods. The proposed IrisDenseNet accurately identifies the iris boundary even in low qualified iris images, such as side views, glasses, off-angle eye images, rotated eyes, non-uniform specular reflection, and partially opened eyes. Following are the five novelties of this study:

- IrisDenseNet is an end-to-end segmentation network that uses the complete image without prior pre-processing or other conventional image processing techniques with the best information gradient flow, which prevents the network from overfitting and vanishing gradient problem.
- This study clarifies the power of dense connectivity with a visual difference between the output feature maps from the convolutional layers for dense connectivity and normal connectivity.
- IrisDenseNet is tested with noisy iris challenge evaluation part-II (NICE-II) and various other datasets, which include both visible light and NIR light environments of both color and greyscale images.
- IrisDenseNet is more robust for accurately segmenting the high-frequency areas such as the eyelashes and ghost region present in the iris area.
- To achieve fair comparisons with other studies, our trained IrisDenseNet models with the algorithms are made publicly available through [47].

4. Proposed Method

4.1. Overview of the Proposed Architecture

Figure 1 shows the overall flowchart of the proposed IrisDenseNet for iris end-to-end segmentation. The input image is given to the IrisDenseNet fully convolutional network without any pre-processing. The network applies the convolutions and up-sampling via pooling indices, and on the basis of learning, provides semantic segmentation mask for a true iris boundary. Figure 2 shows an overview of the proposed IrisDenseNet model for iris segmentation with dense connectivity. The network has two main parts: densely connected encoder and SegNet decoder. In Figure 2, convolutional layers (Conv), batch normalization (BN) and rectified linear unit (ReLU) indicate a convolution layer, a batch normalization layer, and a rectified liner unit layer, respectively. To ensure the information flow between the network layers, dense connectivity is introduced by the direct connections from any layer to all subsequent layers in a dense block. In this study, overall, five dense blocks are used, which are separated by transition layers (a combination of Conv 1 × 1 and max-pool layers).

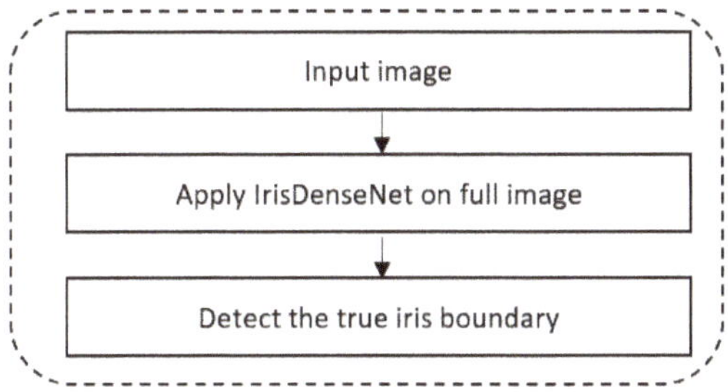

Figure 1. Flowchart of the proposed method.

4.2. *Iris Segmentation Using IrisDenseNet*

In the last decade, CNN has been proved as the most powerful tool for image-related tasks using deep learning. CNN delivers good performances in computer vision applications such as human detection [48], open and close eye detection [49], gender recognition [50], pedestrian detection [51], banknote classification [52], appearance-based gaze estimation [53], and object detection using faster R-CNN [54]; more CNN applications can be found in [55].

To ensure the reliability and accuracy of CNNs, this paper proposes a combination of two foundation methods: (i) densely connected convolutional networks with strengthening feature propagation (DenseNet) [56] and (ii) SegNet [57] deep convolutional encoder–decoder network. SegNet is a practical fully convolutional network for pixel-wise semantic segmentation, which uses the encoder of a 13-layered VGG16 identical core network with fully connected layers removed. The decoder up-samples the low-resolution input feature maps with pooling indices. In our study, we use the SegNet-Basic decoder. SegNet uses a VGG16 identical network owing to the drawback of overfitting and vanishing gradient. Densely connected convolutional network (DenseNet) [56] is proved to be more robust than VGG-net [58] with better information gradient flow due to dense connectivity. In DenseNet, each convolutional layer is connected to all convolutional layers in a feed-forward fashion, which is very useful for strengthening the feature propagation in the subsequent layers in a dense block, as shown in Figure 2.

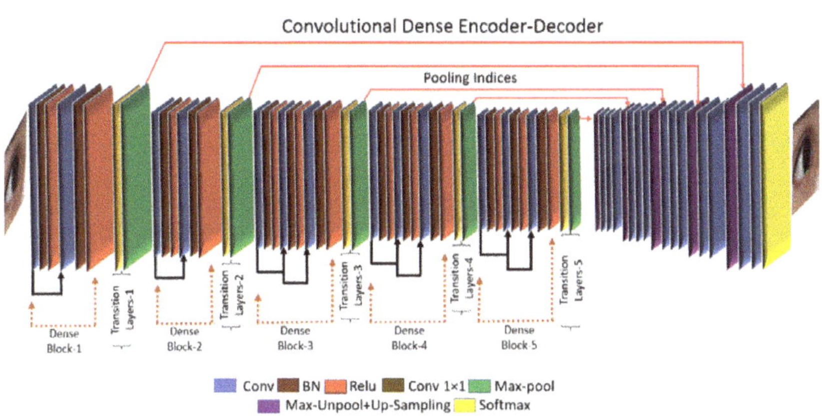

Figure 2. Overview of the proposed method.

Through this connectivity, IrisDenseNet exploits the network potential through the feature reused from the previous layer, which results in enhanced efficiency. Direct connection is basically the feature concatenation achieved by concatenation layers with multiple inputs and one output. Figure 3 shows one dense block separately, and includes the basic convolutional layers (Conv), batch normalization (BN) and rectified linear unit (ReLU), which includes the connections of the layers via concatenation. The pooling indices that are provided after each dense block by max-pooling in the transition layer in the decoder part are shown in Figure 4. These pooling indices are used to un-pool and up-sample for pixel-wise semantic segmentation of iris and non-iris boundaries. The IrisDenseNet encoder and decoder are explained in detail in Sections 4.2.1 and 4.2.2, respectively.

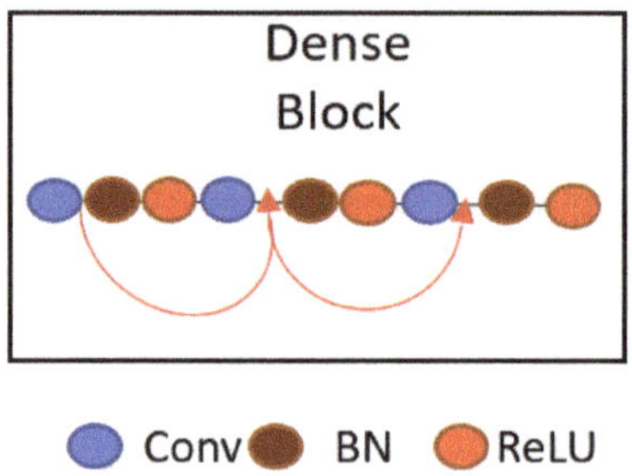

Figure 3. Dense connectivity within the dense block by feature concatenation.

4.2.1. IrisDenseNet Dense Encoder

Owing to the advantages of dense connectivity explained in Section 4.2, this study is based on a densely connected encoder in which five dense blocks are used to improve the performance. Finally, with the SegNet decoder (described in Section 4.2.2), a densely connected convolutional network for iris segmentation (IrisDenseNet) is created. The IrisDenseNet dense encoder consists of 18 convolutional layers, including five 1×1 Conv layers used as the bottleneck layer in transition layers after each dense block, which is useful in reducing the number of input feature maps for computational efficiency. Figure 4 shows the complete dense connectivity with the operation of transition layers. The transition layers are basically a combination of Conv 1×1 and max-pooling, separate two adjacent dense blocks. The IrisDenseNet have the following five key differences compared to DenseNet [56].

- DenseNet is using 3 dense blocks for CIFAR and SVHN datasets and 4 dense blocks for ImageNet classification whereas IrisDenseNet uses 5 dense blocks for each dataset.
- In DenseNet, all dense blocks have four convolutional layers [56], whereas IrisDenseNet has two convolutional layers in the first two dense blocks and 3 convolutional layers for the remaining dense blocks.
- In IrisDenseNet, the pooling indices after each dense block are directly fed to the respective decoder block for the reverse operation of sampling.
- In DenseNet, fully connected layers are used for classification purpose, but in order to make the IrisDenseNet fully convolutional, the fully connected layers are not used.
- In DenseNet, the global average pooling is used in the end of the network whereas in IrisDenseNet, global average pooling is not used to maintain the feature map for decoder operation.

The dense connectivity has following advantages over simple connections:

- It substantially reduces the vanishing gradient problem in CNNs, which increases the network stability.

- Dense connectivity in the encoder strengthens the features flowing through the network.
- It encourages the feature to be reused due to direct connectivity, so the learned features are much stronger than those of normal connectivity.

A detailed layer-wise structure is provided in Table 2 for better understanding, which shows that in a dense block, due to feature map concatenation, the output feature size for all layers in a corresponding dense block always remains the same, which guarantees strong features. As shown in Figure 4, the features through the red and half-circle lines (including arrows) in each dense block are concatenated with the output features of convolution layer, and this concatenation layer is implemented by depthConcatenationLayer() function. In details, there are two concatenation layers in each dense blocks 3, 4, and 5. For example, the first concatenation layer (Cat-3 of Table 2) obtains the output features by concatenating the output features of the 1st convolution layer (Conv-3_1 of Table 2) with the output features of the 2nd convolution layer (Conv-3_2 of Table 2). The second concatenation layer (Cat-4 of Table 2) obtains the output features by concatenating the output features of the first concatenation layer (Cat-3 of Table 2) with the output features of the 3rd convolution layer (Conv-3_3 of Table 2). Same procedure is applied to dense blocks 4 and 5, also.

Table 2. IrisDenseNet connectivity and output feature map size of each dense block (Conv, BN, and ReLU represent convolutional layer, batch normalization layer, and rectified linear unit layer, respectively. Cat, B-Conv, and Pool indicate concatenation layer, bottleneck convolution layer, and pooling layer, respectively) (Here, dense blocks 1 and 2 have the same number of convolution layers, and dense blocks 3, 4, and 5 have the same number of convolution layers) (Convolutional layers with "*" mean that these layers include BN and ReLU. Transition layers are a combination of max-pooling and B-Conv)

Block	Name/Size	No. of Filters	Output Feature Map Size (Width × Height × Number of Channel)
Dense Block-1	Conv-1_1*/3 × 3 × 3	64	300 × 400 × 64
	Conv-1_2*/3 × 3 × 64	64	
	Cat-1	-	300 × 400 × 128
Transition layer-1	B-Conv-1/1 × 1	64	300 × 400 × 64
	Pool-1/2 × 2	-	150 × 200 × 64
Dense Block-2	Conv-2_1*/3 × 3 × 64	128	150 × 200 × 128
	Conv-2_2*/3 × 3 × 128	128	
	Cat-2	-	150 × 200 × 256
Transition layer-2	B-Conv-2/1 × 1	128	150 × 200 × 128
	Pool-2/2 × 2	-	75 × 100 × 128
Dense Block-3	Conv-3_1*/3 × 3 × 128	256	75 × 100 × 256
	Conv-3_2*/3 × 3 × 256	256	
	Cat-3	-	75 × 100 × 512
	Conv-3_3*/3 × 3 × 256	256	75 × 100 × 256
	Cat-4	-	75 × 100 × 768
Transition layer-3	B-Conv-3/1 × 1	256	75 × 100 × 256
	Pool-3/2 × 2	-	37 ×50 × 256
Dense Block-4	Conv-4_1*/3 × 3 × 256	512	37 ×50 × 512
	Conv-4_2*/3 × 3 × 512	512	
	Cat-5	-	37 ×50 × 1024
	Conv-4_3*/3 × 3 × 512	512	37 ×50 × 512
	Cat-6	-	37 ×50 × 1536
Transition layer-4	B-Conv-4/1 × 1	512	37 ×50 × 512
	Pool-4/2 × 2	-	18 × 25 × 512
Dense Block-5	Conv-5_1*/3 × 3 × 512	512	18 × 25 × 512
	Conv-5_2*/3 × 3 × 512	512	
	Cat-7	-	18 × 25 × 1024
	Conv-5_3*/3 × 3 × 512	512	18 × 25 × 512
	Cat-8	-	18 × 25 × 1536
Transition layer-5	B-Conv-5/1 × 1	512	18 × 25 × 512
	Pool-5/2 × 2	-	9 × 12 × 512

4.2.2. IrisDenseNet Decoder

As described in Section 4.2.1, the SegNet-Basic decoder is used to up-sample the dense feature provided by the dense encoder in this study. The decoder basically utilizes the pooling indices along with dense features, and the feature maps are again passed through convolution filters for a reverse process to the encoder to obtain the segmentation mask with same size as that of the input. In the end, each pixel is classified by the soft-max function independently as iris or non-iris via the pixel classification layer.

Figure 4 shows the overall dense encoder–decoder operation. Note that the decoder un-pools and up-samples the pooling in reverse order. The dense features from the last dense block (Dense block 5) relate to the first decoder layers, whereas those from the first dense block relate to the last decoder layers. Table 2 shows that the output feature size is smallest with dense block 5 and largest with dense block 1. Note that the decoder gets the pooling indices from the transition layers, which are a combination of the 1 × 1 convolution layer and a max-pooling layer. The decoder is explained in detail in [57].

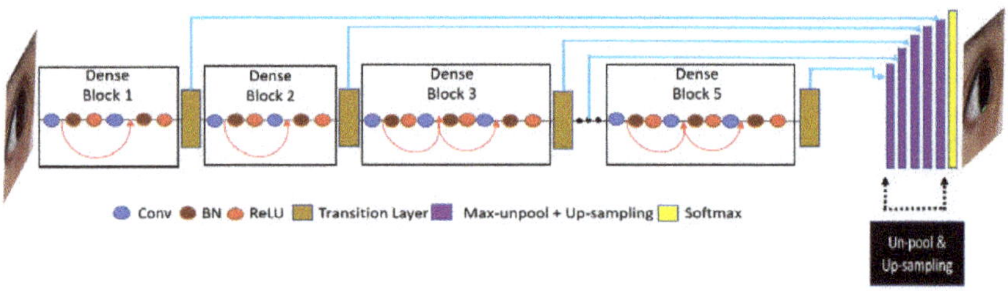

Figure 4. Overall connectivity diagram of IrisDenseNet dense encoder–decoder.

5. Experimental Results

5.1. Experimental Data and Environment

In this study, the NICE-II dataset is used as iris images in visible light environment. The NICE-II database is used for NICE-II competition for iris recognition [59], and consists of 1000 enormously noisy iris images selected from the UBIRIS.v2 database. The database contains 400 × 300 pixel images captured from a Canon EOS 5D camera of 171 classes walking 4–8 m away from the camera. The database includes intruded difficulties such as motion blurs, eyelash and eyelid occlusions, glasses, off-angles, non-uniform light illuminations, and partially captured iris images with irregular rotations. The ground-truth images are publicly available with the database for comparison. In Figure 5, sample images from the NICE-II database are shown with corresponding ground-truth images.

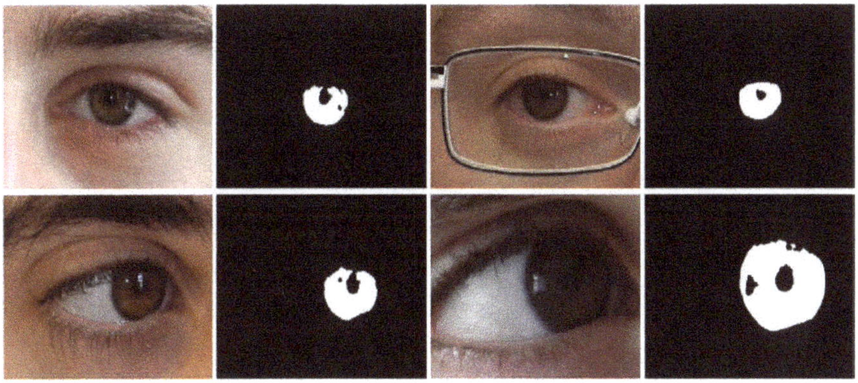

Figure 5. Noisy iris challenge evaluation part-II (NICE-II) sample images with corresponding ground truths.

In this study, half of the total NICE-II images are used for training and the remaining are used for testing purposes based on two-fold cross-validation. To ensure the accuracy of the segmentation, data augmentation is performed as described in Section 5.2. The training and testing of IrisDenseNet are performed on Intel® Core™ i7-3770K CPU @ 3.50 GHz (4 cores) with 28 GB RAM and NVIDIA GeForce GTX 1070 (1,920 Cuda cores) with graphics memory of 8 GB (NVIDIA, Santa Clara, CA, USA) [60] using MATLAB R2017b [61]. We did not use any pretrained models of ResNet-50, Inception-v3, GoogleNet, and DenseNet in our research. Instead, we implemented our IrisDenseNet by using MATLAB functions. In addition, we performed the training our whole network of IrisDenseNet (training from the scratch) with our experimental dataset.

5.2. Data Augmentation

In this study, semantic segmentation of the iris boundary using the full image is proposed, which is much dependent on considerably large amount of image and ground-truth data. Owing to the limited number of images, data augmentation is used to increase the volume of data for training. In this study, each image of training data is augmented 12 times. The augmentation is performed in the following ways:

- Cropping and resizing with interpolation
- Flipping the images only in horizontal direction
- Horizontal translation
- Vertical translation

5.3. IrisDenseNet Training

To segment the iris in a challenging situation, no pre-processing is performed for training. If models that are very deep with the linearity of ReLU are used, the weight initialization with a pre-trained model helps in convergence when training from scratch [62]. Using the same concept, weights are initialized by VGG-16 trained on ImageNet [63] for better training. To train all datasets, a fixed learning rate of 0.001 is used with stochastic gradient descent, which helps to reduce the difference between the desired and calculated outputs by a gradient derivative [64] with a weight decay of 0.0005.

IrisDenseNet is trained using 60 epochs. The mini-batch size is kept 4 with shuffling for the NICE-II dataset. The cross-entropy loss proposed in [65] is used as an unbiased function to train the network, where the loss is calculated over all pixels that are available in a mini-batch acceding to classes (iris and non-iris).

Considering the iris image from the NICE-II database in Figure 6a, the iris size is usually much smaller than the non-iris areas, from which we can deduce that the number of non-iris pixels is much larger than that of the iris pixels. Therefore, during training over a dataset, there can be a large difference of frequency in each class, as shown in Figure 6b.

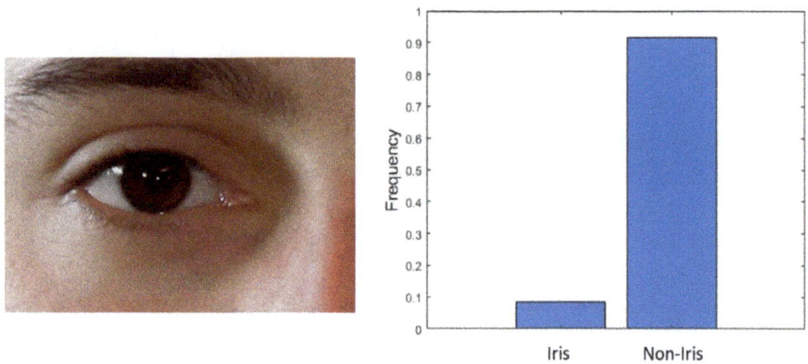

(a) (b)

Figure 6. Difference in frequency between iris and non-iris classes. (**a**) NICE-II original input image. (**b**) Difference in frequency of iris and non-iris pixels in NICE-II training dataset.

The frequencies of the iris and non-iris pixels indicate that the non-iris class dominates much during training, so there is a need to maintain a balance between these two classes during training. This type of frequency balancing is used when some classes are underrepresented in the training data. In this study, median frequency balancing [66] is used, where a weight is assigned to the cross-entropy loss. This weight is calculated from the training data by median class frequency (for both iris and non-iris classes) by the following given formulas:

$$W_{c1} = \frac{Median\_freq}{f_{c1}} \text{ and } W_{c2} = \frac{Median\_freq}{f_{c2}} \quad (1)$$

In Equation (1), W_{c1} and W_{c2} are the class weights for iris and non-iris classes, respectively, f_{c1} indicates the number of iris pixels over the total number of pixels in images, and f_{c2} indicates the number of non-iris pixels over the total number of pixels in images. *Median_freq* is the median of f_{c1} and f_{c2}. With frequency balancing, a weight smaller than 1 is assigned to a larger (non-iris) class and a weight larger than 1 is assigned to a smaller (iris) class in the cross-entropy loss during training.

Figure 7a,b show the accuracy and loss curves for training from the 1st- and 2nd-fold cross-validation, respectively. The x-axis for each curve shows the number of epochs and the y-axis shows the training accuracy and loss. During the training, it is important to achieve accuracy close to maximum and a loss close to the minimum. The loss is dependent on the learning rate, so the learning rate is empirically found by conducting an experiment to achieve the minimum loss. With an increased learning rate, the training loss can decrease dramatically, but it is not certain that the training would converge to the valley point for the loss. In our proposed method, we achieve a training accuracy approaching 100% and a loss approaching 0%. As described in Section 3, our trained models with algorithms are made publicly available through [47] to make fair comparisons with other studies.

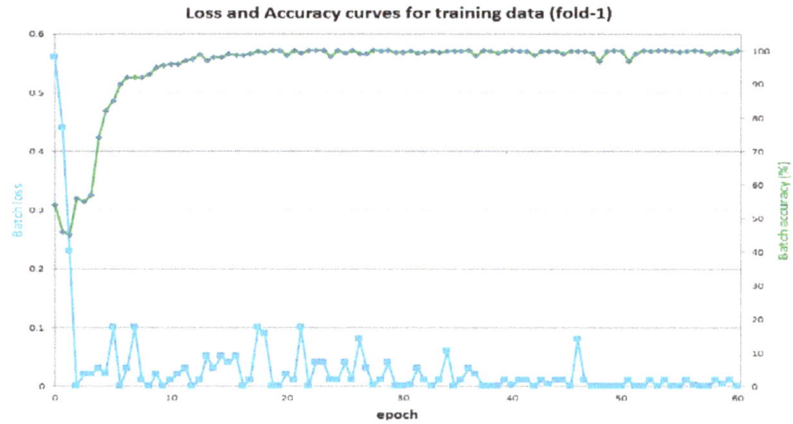

(**a**)

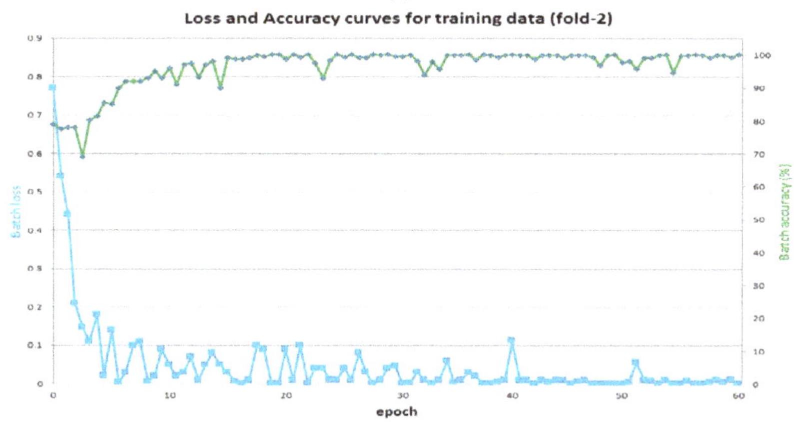

(**b**)

Figure 7. Training accuracy and loss curves from (**a**) 1st-fold cross-validation and (**b**) 2nd-fold cross-validation.

5.4. Testing of IrisDenseNet for Iris Segmentation

To obtain the segmentation results from the proposed IrisDenseNet, the input image is provided to the trained model and there is no pre-processing involved during training and testing. The input image is passed through the dense encoder and decoder in a forward fashion. The output of the trained model is a binary segmentation mask, which is used to generate and evaluate the segmentation results by using our trained model. The performance of the proposed IrisDenseNet is evaluated using the NICE-I evaluation protocol [67], which is being used by many researchers to evaluate the segmentation performance. E_i is computed by exclusive-OR (XOR) between the resultant image ($I_i(m', n')$) and ground-truth image ($G_i(m', n')$) given as

$$E_i = \frac{1}{m \times n} \sum_{(m',n')} I_i(m', n') \otimes G_i(m', n') \quad (2)$$

where $m \times n$ is the image size (by width and height of the image). For each image, E_i is calculated as the pixel classification accuracy. The overall average segmentation error E_a is calculated by averaging the classification error (E_i) over all images in the database:

$$E_a = \frac{1}{t} \sum_i E_i \quad (3)$$

where t represents the total number of images to be evaluated. The value of E_a always lies within [0, 1]. If E_a is close to "0," it shows the minimum error, whereas if E_a is close to "1,"it shows the largest error.

5.4.1. Result of Excessive Data Augmentation

Data augmentation is a method of increasing the training data for better accuracy, but this accuracy is strongly dependent upon the augmentation type. The data augmentation varies with the application, so in case of iris, it is experimentally observed that excessive data augmentation results in reduced accuracy, as shown in Table 3. As explained in Section 5.2, each image of training data was augmented 12 times in our study. In the case of excessive data augmentation, each image of training data was augmented 25 times. Moreover, if we augment the data by changing the contrast and brightness of the iris image, the overall performance degrades in terms of segmentation accuracy as shown in Table 3. The reason why the lower accuracy is obtained by excessive data augmentation is due to the overfitting of training. The reason why the lower accuracy is obtained by data augmentation by changing the contrast and brightness is that the augmented data based on this scheme do not reflect the characteristics of testing data.

Table 3. Comparative accuracies according to various augmentation methods.

Method	E_a
Excessive data augmentation	0.00729
Data augmentation by changing the contrast and brightness of iris image	0.00761
Proposed data augmentation in Section 5.2	0.00695

5.4.2. Iris Segmentation Results Obtained by the Proposed Method

Figure 8 shows the good segmentation results obtained by IrisDenseNet for the NICE-II dataset. As explained in Section 5.4, the performance of the proposed method is measured as E_a, which is the average error for the database. To pictorially represent the result, two error types, false positive and false negative, are defined. The former represents the error of a non-iris pixel being misclassified as an iris pixel, whereas the later represents the error of an iris pixel being misclassified as a non-iris pixel. The false positive and negative errors are represented in green and red, respectively. The true positive case is that the iris pixel is correctly classified as an iris pixel, which is represented in black. As shown in Figures 8 and 9, the false positive error is caused by the eyelash pixel or pixel close to the reflection noise or pupil, whose value is similar to that of the iris pixel, whereas the false negative error is caused by reflection noise caused from glasses or with a dark iris area.

5.4.3. Comparison of the Proposed Method with Previous Methods

In this section, experimental results of the NICE-II dataset are given by IrisDenseNet on the basis of E_a of Equation (3), and SegNet-Basic is tested on the same dataset for comparison. To clarify the power of dense features, a visual comparison of convolutional features by SegNet and IrisDenseNet is also provided in Section 6.1. Table 4 represents the comparative analysis of NICE-II with existing methods based on the NICE-I evaluation protocol. Note that SegNet-Basic is a general semantic segmentation method used for road scene segmentation (road, cars, building, pedestrian, etc.) and the dataset of CamVid road scene segmentation [57,68] with 11 classes. For comparison, the number of outputs for the original SegNet-Basic is changed from 11 to 2 iris and non-iris classes. As shown in Table 4, we can confirm that our IrisDenseNet shows an error lower than those generated in the previous method. The reason why SegNet-Basic shows lower accuracy than our IrisDenseNet is that the scheme of re-using features through dense connections is not adopted in SegNet-Basic.

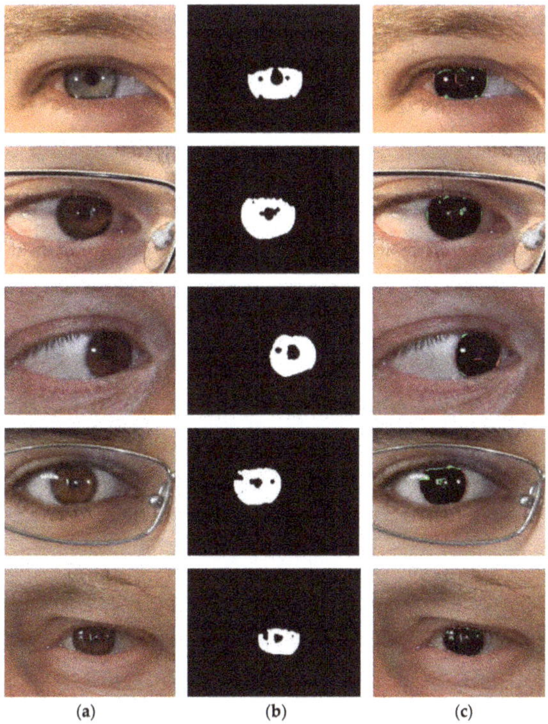

Figure 8. Examples of NICE-II good segmentation results obtained by IrisDenseNet. (**a**) Original image. (**b**) Ground-truth image. (**c**) Segmentation result obtained by IrisDenseNet (The false positive and negative errors are shown in green and red, respectively. The true positive case is shown in black).

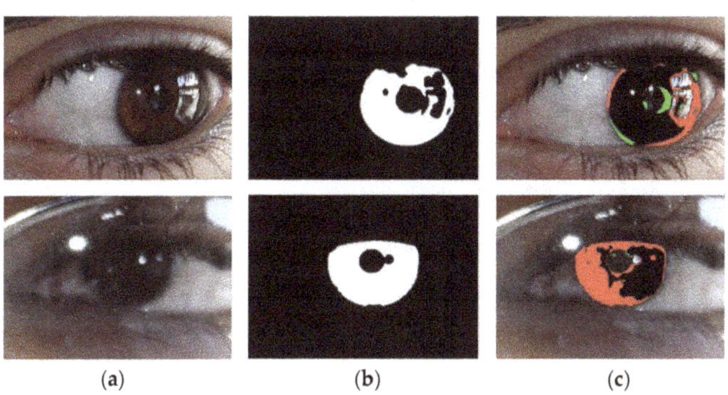

Figure 9. Examples of incorrect iris segmentation by our method. (**a**) Original input images. (**b**) Ground-truth images. (**c**) Segmentation results (The false positive and negative errors are presented as green and red, respectively. The true positive case is shown in black).

Table 4. Comparisons of the proposed method with previous methods using NICE-II dataset.

Method	E_a
Luengo-Oroz et al. [69]	0.0305
Labati et al. [70]	0.0301
Chen et al. [71]	0.029
Jeong et al. [26]	0.028
Li et al. [72]	0.022
Tan et al. [73]	0.019
Proença et al. [74]	0.0187
de Almeida [75]	0.0180
Tan et al. [76]	0.0172
Sankowski et al. [77]	0.016
Tan et al. [35]	0.0131
Haindl et al. [78]	0.0124
Zhao et al. [79]	0.0121
Arsalan et al. [45]	0.0082
SegNet-Basic [57]	0.00784
Proposed IrisDenseNet	0.00695

5.4.4. Iris Segmentation Error with Other Open Databases

This study includes additional experiments with four other open databases: Mobile iris challenge evaluation (MICHE-I) [80], CASIA-Iris-Interval (CASIA v4.0 interval) [81], CASIA-Iris-Distance (CASIA v4.0 distance) [81], and IIT Delhi (IITD) Iris Database (v1.0) [82]. Experiments conducted in different environments such as visible light and NIR light show the robustness of the proposed method.

MICHE-I is a challenging dataset captured with mobile devices to ensure the developing algorithms in non-ideal difficult situations. This database is collected using three smartphones: iPhone5 with 8 MP (72 dpi) back camera and 1.2 MP (72 dpi) frontal camera, Samsung Galaxy S4 with 13 MP (72 dpi) back camera and 2 MP (72 dpi) frontal camera, Samsung Galaxy Tab2 with 0.3 MP frontal camera (with pixel resolutions of 1536 $\times$ 2048, 2322 $\times$ 4128, and 640 $\times$ 480 for iPhone5, Galaxy S4, and Galaxy Tab2, respectively). This database is collected with 195 subjects over two visits with a distance 5–25 feet under non-ideal conditions (indoor and outdoor), including motion blur, occlusions, off-angle, and non-uniform illuminations. Ground-truth data are not provided in MICHE-I database; therefore, we use the images whose ground truth can be correctly identified by their provided algorithm [74] according to the instruction of MICHE. The total numbers of images in sub-datasets from iPhone5, Galaxy S4, and GalaxyTab2 are 611, 674, and 267, respectively. In detail, the numbers of images for training (testing) are 311 (300), 340 (334), and 135 (132) in the sub-datasets from iPhone5, Galaxy S4, and GalaxyTab2, respectively.

Figure 10 shows the sample images for MICHE-I with corresponding ground truths for iPhone5 (left image), Samsung Galaxy S4 (middle image), and Samsung Galaxy Tab2 (right images).

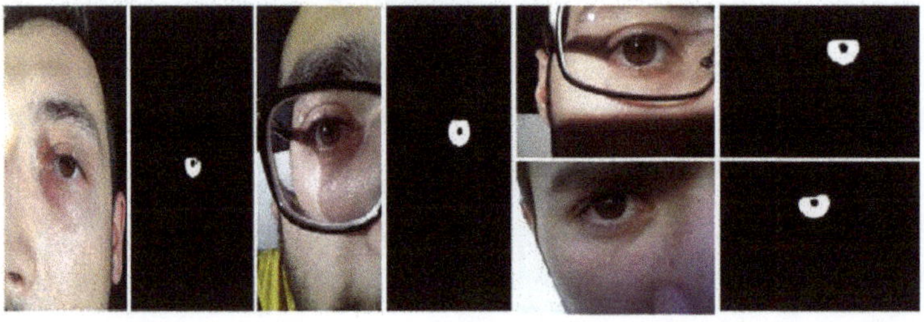

Figure 10. Mobile iris challenge evaluation (MICHE-I) sample images with corresponding ground truths.

The CASIA v4.0 interval dataset is captured with a CASIA self-developed camera with a special circular NIR LED array and appropriate luminous intensity for iris imaging. Owing to this suitable camera setup and novel design, very clear iris images with very detailed features are captured. The images are grayscale and have a pixel resolution of 320 × 280. The ground truth for the CASIA v4.0 interval database is provided by the IRISSEG-EP [83]. Figure 11 shows the sample images and corresponding ground truth.

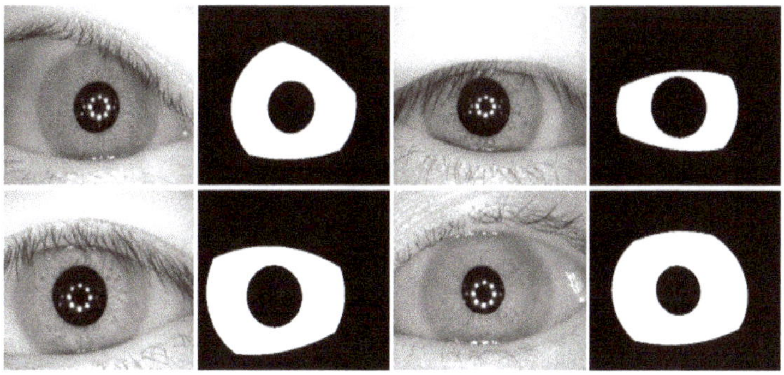

Figure 11. The institute of automation, Chinese academy of sciences (CASIA) v4.0 interval sample images with corresponding ground truths.

The CASIA v4.0 distance database contains iris images from a long-range multimodal biometric image acquisition and recognition system (LMBS). These images are captured using a high-resolution NIR camera. This database contains 2567 images from 142 subjects captured from 3 m distance from the camera. The ground truth for CASIA v4.0 distance is not publicly available. Instead, 400 iris images from 40 subjects are manually labeled and publicly available for research purposes through [47]. Figure 12 shows the sample images for CASIA v4.0 distance with provided corresponding ground truths.

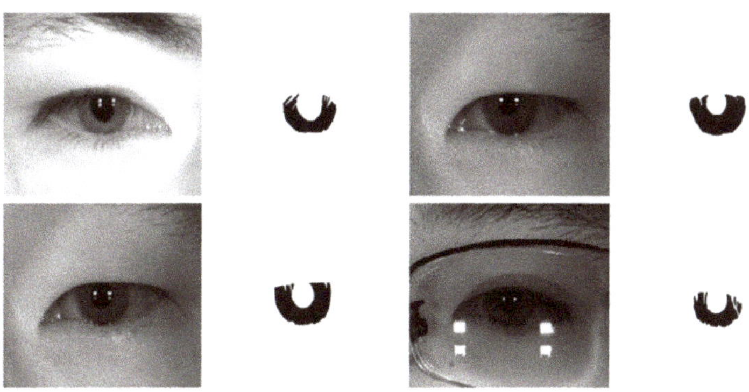

Figure 12. CASIA v4.0 distance sample images with corresponding ground truths.

The IITD Iris database consists of 2240 iris image from 224 subjects with a JPC1000 digital CMOS camera in NIR light environment with a pixel resolution of 320 × 240 [82]. This image database is taken in an indoor controlled environment with user cooperation for a frontal view of iris, so no off-angle iris is involved in this database. The ground truth for IITD is also provided by the IRISSEG-EP [83]. Figure 13 shows the sample images and corresponding ground truth for IITD iris database.

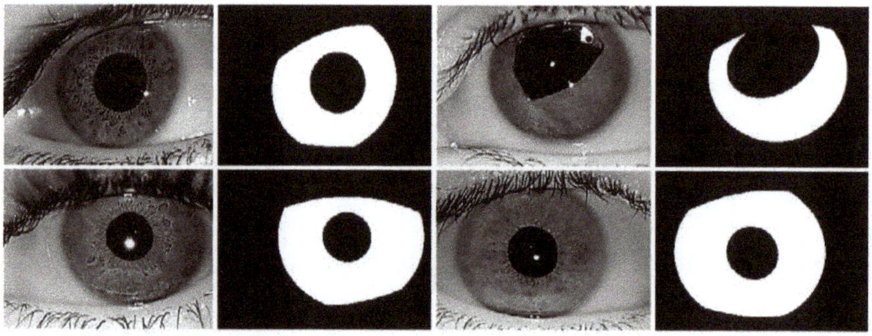

Figure 13. IIT Delhi (IITD) v1.0 sample images with corresponding ground truths.

Figures 14–17 show the good segmentation results for MICHE-I, CASIA v4.0 interval, CASIA 4.0 distance, and IITD databases, respectively. The false positive and negative errors are presented as green and red, respectively. The true positive is presented as black.

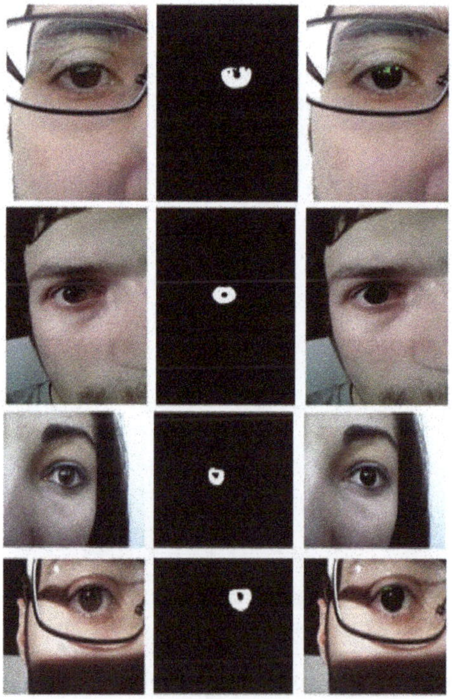

(a) (b) (c)

Figure 14. Examples of correct segmentation results in MICHE-I database by our method. (**a**) Original image. (**b**) Ground-truth image. (**c**) Segmentation result by IrisDenseNet (The false and negative errors are presented as green and red, respectively. The true positive case is presented as black).

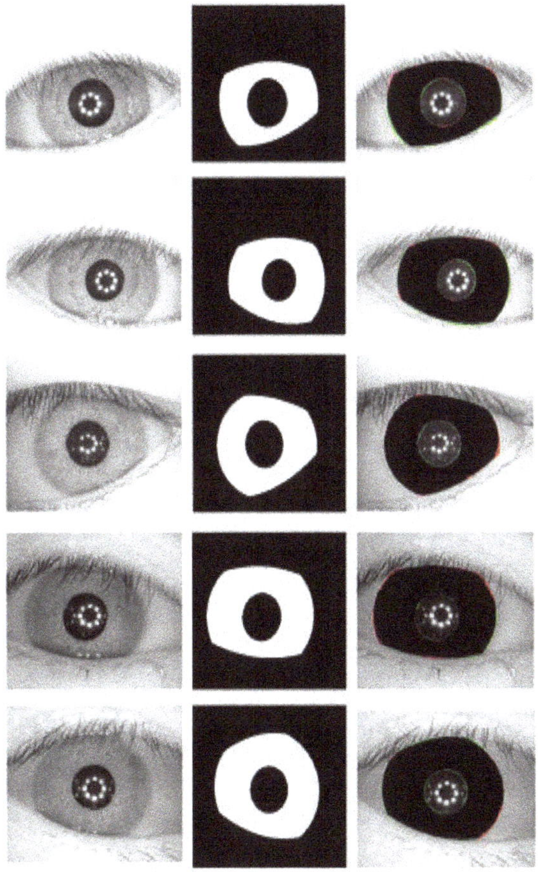

(**a**) (**b**) (**c**)

Figure 15. Examples of correct segmentation results in CASIA v4.0 interval database by the proposed method. (**a**) Original image. (**b**) Ground-truth image. (**c**) Segmentation result by IrisDenseNet (The false positive and negative errors are presented as green and red, respectively. The true positive case is presented as black).

We show the examples of bad segmentation results by IrisDenseNet only for MICHE-I in Figure 18 because there is no considerable error found in other datasets. The false positive errors are caused by reflection noise created by environmental light, whereas the false negative errors are caused by severely dark iris values.

Tables 5 and 6 show the comparisons of accuracies with MICHE-I and CASIA v4.0 distance databases, respectively, based on the NICE-I evaluation protocol. As shown in Tables 5 and 6, our method outperforms previous methods. The reason why SegNet-Basic shows lower accuracy than our IrisDenseNet is that the scheme of re-using features through dense connections is not adopted in SegNet-Basic.

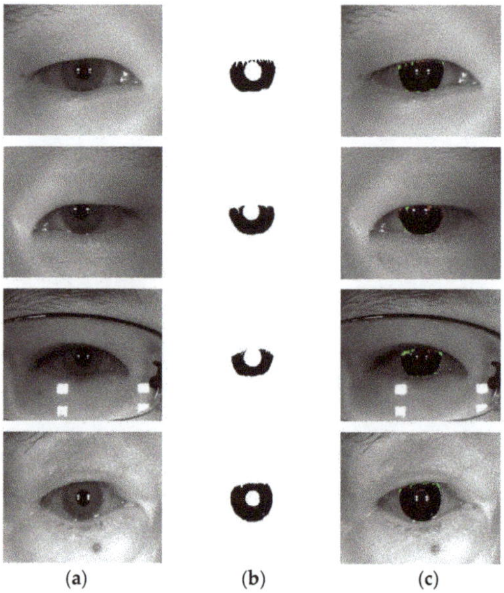

Figure 16. Examples of correct segmentation results in CASIA v4.0 distance database by the proposed method. (**a**) Original image (**b**) Ground-truth image. (**c**) Segmentation result by IrisDenseNet (The false positive and negative errors are presented as green and red, respectively. The true positive case is presented as black).

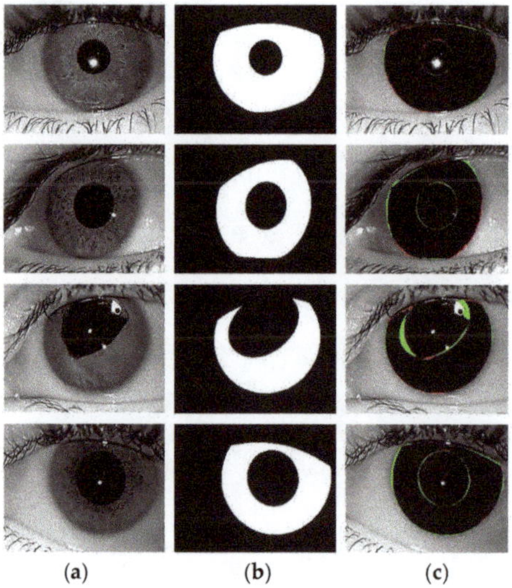

Figure 17. Examples of correct segmentation results in IITD database by the proposed method. (**a**) Original image. (**b**) Ground-truth image. (**c**) Segmentation result by IrisDenseNet (The false positive and negative errors are presented as green and red, respectively. The true positive case is presented as black).

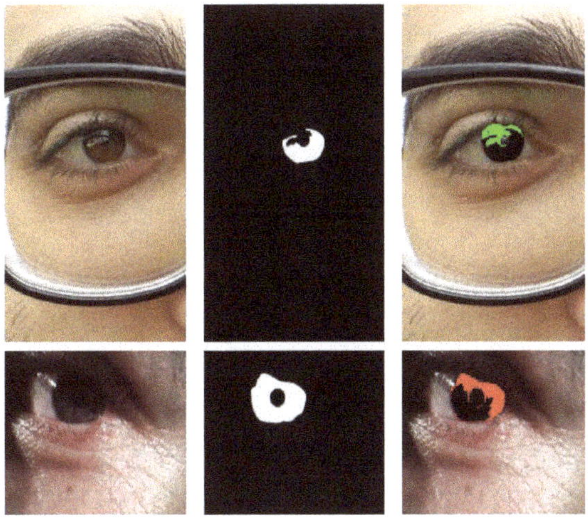

Figure 18. Examples of MICHE-I incorrect segmentation results by our method. (**a**) Original image. (**b**) Ground-truth image. (**c**) Segmentation result by IrisDenseNet (The false positive and negative errors are presented as green and red, respectively. The true positive case is presented as black).

Table 5. Comparison of the proposed method with previous methods using MICHE-I dataset based on NICE-I evaluation protocol.

Method		E_a
Hu et al. [30]	Sub-dataset by iPhone5	0.0193
	Sub-dataset by Galaxy S4	0.0192
Arsalan et al. [45]	Sub-dataset by iPhone5	0.00368
	Sub-dataset by Galaxy S4	0.00297
	Sub-dataset by Galaxy Tab2	0.00352
SegNet-Basic [57]	Sub-dataset by iPhone5	0.0025
	Sub-dataset by Galaxy S4	0.0027
	Sub-dataset by Galaxy Tab2	0.0029
Proposed IrisDenseNet	Sub-dataset by iPhone5	0.0020
	Sub-dataset by Galaxy S4	0.0022
	Sub-dataset by Galaxy Tab2	0.0021

Table 6. Comparison of the proposed method with previous methods using CASIA v4.0 distance database based on NICE-I evaluation protocol.

Method	E_a
Tan et al. [73]	0.0113
Liu et al. [44] (HCNNs)	0.0108
Tan et al. [76]	0.0081
Zhao et al. [79]	0.0068
Liu et al. [44] (MFCNs)	0.0059
SegNet-Basic [57]	0.0044
Proposed IrisDenseNet	0.0034

To evaluate the accuracies with CASIA v4.0 interval and IITD databases by fair comparative analysis with other researchers for the same datasets, another evaluation protocol is used, which is named as RPF-measure protocol, which was used in [82] for evaluating the iris segmentation performance. The RPF measure is basically recall (R), precision (P), and F-measure, which is similarly used to evaluate the performance of an algorithm based on ground-truth images. The RPF measure is a better way to measure the strength and weakness of an algorithm; to measure from this protocol, the essentials, true positive (*TP*), false positive (*FP*), and false negative (*FN*), should be calculated. Based on them, RPF measure can be computed by Equations (4)–(6), where #*TP*, #*FN*, and #*FP* represent the numbers of *TP*, *FN*, and *FP*, respectively.

$$\mathrm{R} = \frac{\#TP}{\#TP + \#FN} \tag{4}$$

$$\mathrm{P} = \frac{\#TP}{\#TP + \#FP} \tag{5}$$

$$\mathrm{F} = \frac{2\mathrm{RP}}{\mathrm{R}+\mathrm{P}} \tag{6}$$

where F-measure is basically the harmonic mean of both R and P, and prevents the evaluation from overfitting or underfitting of iris pixels. The comparisons with CASIA v4.0 interval and IITD databases are conducted based on the RPF-measure protocol. However, very few researchers have addressed these two databases for segmentation purposes due to unavailability of ground-truth images. To compare with other researchers, Gangwar et al. [84] used various algorithms such as GST [85], Osiris [86], WAHET [87], IFFP [88], CAHT [89], Masek [90], and integro-differential operator (IDO) [25] with the same database. Therefore, for comparison, the comparative results with these algorithms are presented in Table 7, and it can be found that the proposed IrisDenseNet outperforms other methods. The reason why SegNet-Basic shows lower accuracy than our IrisDenseNet is that the scheme of re-using features through dense connections is not adopted in SegNet-Basic.

Table 7. Comparison of the proposed method with previous methods using CASIA v4.0 interval and IITD databases based on the RPF-measure evaluation protocol. A smaller value of σ and a higher value of μ show better performance. (unit: %) (The resultant values of GST [85], Osiris [86], WAHET [87], IFFP [88], CAHT [89], Masek [90], IDO [25], and IrisSeg [84] are referred from [84]).

DB	Method	R		P		F	
		μ	σ	μ	σ	μ	σ
CASIA V4.0 Interval	GST [85]	85.19	18	89.91	7.37	86.16	11.53
	Osiris [86]	97.32	7.93	93.03	4.95	89.85	5.47
	WAHET [87]	94.72	9.01	85.44	9.67	89.13	8.39
	IFFP [88]	91.74	14.74	83.5	14.26	86.86	13.27
	CAHT [89]	97.68	4.56	82.89	9.95	89.27	6.67
	Masek [90]	88.46	11.52	89	6.31	88.3	7.99
	IDO [25]	71.34	22.86	61.62	18.71	65.61	19.96
	IrisSeg [84]	94.26	4.18	92.15	3.34	93.1	2.65
	SegNet-Basic [57]	99.60	0.66	91.86	2.65	95.55	1.40
	Proposed Method	97.10	2.12	98.10	1.07	97.58	0.99
IITD	GST [85]	90.06	16.65	85.86	10.46	86.6	11.87
	Osiris [86]	94.06	6.43	91.01	7.61	92.23	5.8
	WAHET [87]	97.43	8.12	79.42	12.41	87.02	9.72
	IFFP [88]	93.92	10.62	79.76	11.42	85.83	9.54
	CAHT [89]	96.8	11.2	78.87	13.25	86.28	11.39
	Masek [90]	82.23	18.74	90.45	11 .85	85.3	15.39
	IDO [25]	51.91	15.32	52.23	14.85	51.17	13.26
	IrisSeg [84]	95.33	4.58	93.70	5.33	94.37	3.88
	SegNet-Basic [57]	99.68	0.51	92.53	2.05	95.96	1.04
	Proposed Method	98.0	1.56	97.16	1.40	97.56	0.84

6. Discussion and Analysis

6.1. Power of Dense Connectivity

In this research, the dense connectivity between the layers for better semantic segmentation is used. The conventional methods eliminate the spatial information during the continuous layer by convolution and this continuous elimination of spatial acuity is not good for minor high frequency features in the image. Therefore, the decent way is to import the features from the previous layers to maintain high frequency component during convolution. This maintenance of features from previous layers is done by direct dense connection as shown in Figure 4.

IrisDenseNet is a densely connected fully convolutional network that proceeds in the feed-forward fashion with dense features for better performance. The above discussion about dense features is related to the theoretical discussion, and in this section, a practical comparison of the simple feature with the dense feature is performed using visual images from the convolutional layers. Note that if we compare SegNet with IrisDenseNet in terms of architecture, we can find one unique similarity that both networks use pooling indices for up-sampling, as shown in Figures 2 and 4. Therefore, a careful analysis of the IrisDenseNet network shows that each dense block is separated with a transition layer that has the same pooling layer for pooling indices. Therefore, the difference between the simple and dense features can be fairly compared with each dense block just before the pooling.

In this study, to explain the power of dense connectivity, the reference convolutional features are compared. These are reference features obtained before the 4th max-pooling (Pool-4 of Table 2) layer for both SegNet and IrisDenseNet, as shown in Figures 19 and 20. Note that the output features from Pool-4 contain 512 channels, and for simplicity, the last 64 channels (449th to 512th channels) are visualized. These features present noticeable visual difference. With a careful analysis of the output (shown in Figures 19 and 20), following important observations can be deduced:

- The real power of dense connectivity is evident from the visual features before Pool-4 for both SegNet (Figure 19) and IrisDenseNet (Figure 20). The Pool-4 features are the 2nd-last pooling index features. The Pool-4 features from SegNet include much more noises than those from IrisDenseNet, which can reduce the error of detecting correct iris pixels.

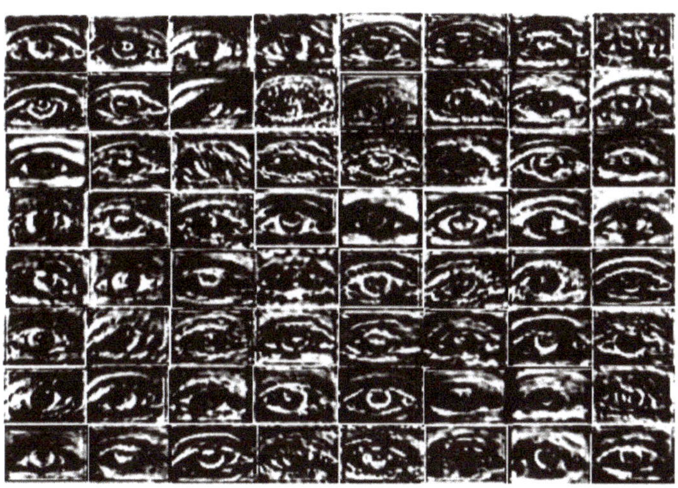

Figure 19. SegNet-Basic last 64 channel (the 449th to 512th channels) features before the 4th max-pooling (Pool-4).

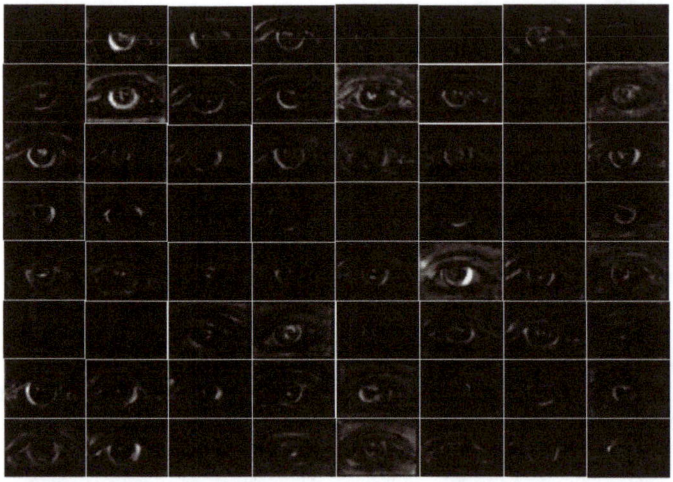

Figure 20. IrisDenseNet last 64 channel (the 449th to 512th channels) features for the 4th dense block before the 4th max-pooling (Pool-4 of Table 2).

6.2. Comparison of Segmentation Results (IrisDenseNet vs. SegNet)

IrisDenseNet uses the concept of feature reuse, which strengthens the features in a dense block for better segmentation. When comparing the SegNet iris segmentation results with IrisDenseNet results, the following differences are found.

- The segmentation results obtained by IrisDenseNet dense features show a thinner and finer iris boundary as compared to SegNet, which substantially reduces the error rate for the proposed method, as shown in Figure 21a,b.
- IrisDenseNet is more robust for the detection of the pupil boundary as compared to SegNet, as shown in Figure 22a,b.
- IrisDenseNet is more robust for the ghost region in the iris area as compared to SegNet, as shown in Figure 23a,b.

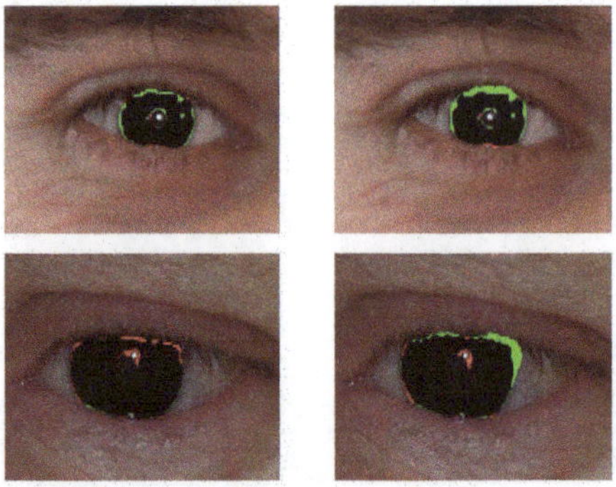

Figure 21. Comparison of iris thin boundary. Segmentation results obtained by (**a**) IrisDenseNet and (**b**) SegNet.

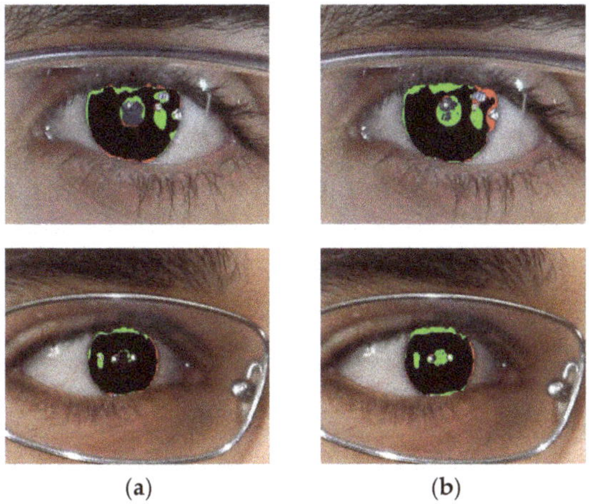

Figure 22. Comparisons of pupil boundary detection. Segmentation results obtained by (a) IrisDenseNet and (b) SegNet.

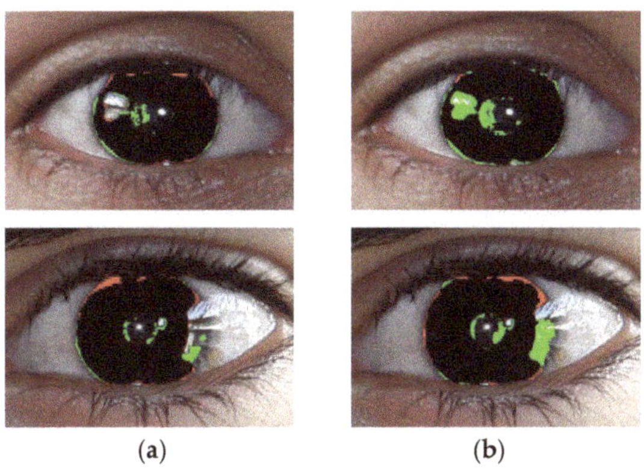

Figure 23. Comparisons of iris detection affected by ghost effect. Segmentation results obtained by (a) IrisDenseNet and (b) SegNet.

7. Conclusions

In this study, we proposed a robust IrisDenseNet with dense concatenated features to segment the true iris boundaries in non-ideal environments even with low-quality images. To achieve better segmentation quality, the encoder was empowered with a dense connection in which layers have direct concatenated connections with all preceding layers in a dense block. This connectivity enhances the capability of the network and enables the feature reuse for better performance. The proposed method provides an end-to-end segmentation without any conventional image processing technique. Experiments conducted with five datasets showed that our method achieved higher accuracies for end-to-end iris segmentation than the state-of-the-art methods for both visible and NIR light environments.

Our method was innovatively powered by dense features but still has a deep network. Therefore, it is difficult to be trained with a GPU with low memory and larger mini-batch size. In the future, we would optimize the network further and reduce the number of layers to make it memory-efficient for mobile and handheld devices with reduced parameters and multiplications. In addition, we will use this method for other applications such as finger vein, human body, or biomedical image segmentations.

Author Contributions: M.A. and K.R.P. designed the overall system for iris segmentation. In addition, they wrote and revised the paper. R.A.N., D.S.K., P.H.N. and M.O. helped to design the comparative CNN architecture and experiments.

Acknowledgments: This research was supported by the Bio & Medical Technology Development Program of the NRF funded by the Korean government, MSIT (NRF-2016M3A9E1915855), by the National Research Foundation of Korea (NRF) grant funded by the Korea government (Ministry of Science and ICT) (NRF-2017R1C1B5074062), and by the Basic Science Research Program through the National Research Foundation of Korea (NRF) funded by the Ministry of Education (NRF-2017R1D1A1B03028417).

Conflicts of Interest: The authors declare no conflict of interest.

References

1. Bowyer, K.W.; Hollingsworth, K.P.; Flynn, P.J. A survey of iris biometrics research: 2008–2010. In *Handbook of Iris Recognition*; Advances in Computer Vision and Pattern Recognition; Springer: London, UK, 2016; pp. 23–61.
2. Jain, A.K.; Arora, S.S.; Cao, K.; Best-Rowden, L.; Bhatnagar, A. Fingerprint recognition of young children. *IEEE Trans. Inf. Forensics Secur.* **2017**, *12*, 1501–1514. [CrossRef]
3. Hong, H.G.; Lee, M.B.; Park, K.R. Convolutional neural network-based finger-vein recognition Using NIR Image Sensors. *Sensors* **2017**, *17*, 1297. [CrossRef] [PubMed]
4. Bonnen, K.; Klare, B.F.; Jain, A.K. Component-based representation in automated face recognition. *IEEE Trans. Inf. Forensics Secur.* **2013**, *8*, 239–253. [CrossRef]
5. Viriri, S.; Tapamo, J.R. Integrating iris and signature traits for personal authentication using user-specific weighting. *Sensors* **2012**, *12*, 4324–4338. [CrossRef] [PubMed]
6. Meraoumia, A.; Chitroub, S.; Bouridane, A. Palmprint and finger-knuckle-print for efficient person recognition based on Log-Gabor filter response. *Analog Integr. Circuits Signal Process.* **2011**, *69*, 17–27. [CrossRef]
7. Alqahtani, A. Evaluation of the reliability of iris recognition biometric authentication systems. In Proceedings of the International Conference on Computational Science and Computational Intelligence, Las Vegas, NV, USA, 15–17 December 2016; pp. 781–785.
8. Bowyer, K.W.; Hollingsworth, K.; Flynn, P.J. Image understanding for iris biometrics: A survey. *Comput. Vis. Image Underst.* **2008**, *110*, 281–307. [CrossRef]
9. Schnabel, B.; Behringer, M. Biometric protection for mobile devices is now more reliable. *Opt. Photonik* **2016**, *11*, 16–19. [CrossRef]
10. Kang, J.-S. Mobile iris recognition systems: An emerging biometric technology. *Procedia Comput. Sci.* **2012**, *1*, 475–484. [CrossRef]
11. Barra, S.; Casanova, A.; Narducci, F.; Ricciardi, S. Ubiquitous iris recognition by means of mobile devices. *Pattern Recognit. Lett.* **2015**, *57*, 66–73. [CrossRef]
12. Albadarneh, A.; Albadarneh, I.; Alqatawna, J. Iris recognition system for secure authentication based on texture and shape features. In Proceedings of the IEEE Jordan Conference on Applied Electrical Engineering and Computing Technologies, The Dead Sea, Jordan, 3–5 November 2015; pp. 1–6.
13. Hajari, K.; Bhoyar, K. A review of issues and challenges in designing iris recognition systems for noisy imaging environment. In Proceedings of the International Conference on Pervasive Computing, Pune, India, 8–10 January 2015; pp. 1–6.
14. Sahmoud, S.A.; Abuhaiba, I.S. Efficient iris segmentation method in unconstrained environments. *Pattern Recognit.* **2013**, *46*, 3174–3185. [CrossRef]
15. Hofbauer, H.; Alonso-Fernandez, F.; Bigun, J.; Uhl, A. Experimental analysis regarding the influence of iris segmentation on the recognition rate. *IET Biom.* **2016**, *5*, 200–211. [CrossRef]
16. Proença, H.; Alexandre, L.A. Iris recognition: Analysis of the error rates regarding the accuracy of the segmentation stage. *Image Vis. Comput.* **2010**, *28*, 202–206. [CrossRef]
17. Wildes, R.P. Iris recognition: An emerging biometric technology. *Proc. IEEE* **1997**, *85*, 1348–1363. [CrossRef]

18. Roy, D.A.; Soni, U.S. IRIS segmentation using Daughman's method. In Proceedings of the International Conference on Electrical, Electronics, and Optimization Techniques, Chennai, India, 3–5 March 2016; pp. 2668–2676.
19. Khan, T.M.; Aurangzeb Khan, M.; Malik, S.A.; Khan, S.A.; Bashir, T.; Dar, A.H. Automatic localization of pupil using eccentricity and iris using gradient based method. *Opt. Lasers Eng.* **2011**, *49*, 177–187. [CrossRef]
20. Ibrahim, M.T.; Khan, T.M.; Khan, S.A.; Aurangzeb Khan, M.; Guan, L. Iris localization using local histogram and other image statistics. *Opt. Lasers Eng.* **2012**, *50*, 645–654. [CrossRef]
21. Huang, J.; You, X.; Tang, Y.Y.; Du, L.; Yuan, Y. A novel iris segmentation using radial-suppression edge detection. *Signal Process.* **2009**, *89*, 2630–2643. [CrossRef]
22. Jan, F.; Usman, I.; Agha, S. Iris localization in frontal eye images for less constrained iris recognition systems. *Digit. Signal Process.* **2012**, *22*, 971–986. [CrossRef]
23. Ibrahim, M.T.; Mehmood, T.; Aurangzeb Khan, M.; Guan, L. A novel and efficient feedback method for pupil and iris localization. In Proceedings of the 8th International Conference on Image Analysis and Recognition, Burnaby, BC, Canada, 22–24 June 2011; pp. 79–88.
24. Umer, S.; Dhara, B.C. A fast iris localization using inversion transform and restricted circular Hough transform. In Proceedings of the 8th International Conference on Advances in Pattern Recognition, Kolkata, India, 4–7 January 2015; pp. 1–6.
25. Daugman, J. How iris recognition works. *IEEE Trans. Circuits Syst. Video Technol.* **2004**, *14*, 21–30. [CrossRef]
26. Jeong, D.S.; Hwang, J.W.; Kang, B.J.; Park, K.R.; Won, C.S.; Park, D.-K.; Kim, J. A new iris segmentation method for non-ideal iris images. *Image Vis. Comput.* **2010**, *28*, 254–260. [CrossRef]
27. Parikh, Y.; Chaskar, U.; Khakole, H. Effective approach for iris localization in nonideal imaging conditions. In Proceedings of the IEEE Students' Technology Symposium, Kharagpur, India, 28 February–2 March 2014; pp. 239–246.
28. Pundlik, S.J.; Woodard, D.L.; Birchfield, S.T. Non-ideal iris segmentation using graph cuts. In Proceedings of the IEEE Computer Society Conference on Computer Vision and Pattern Recognition Workshops, Anchorage, AK, USA, 23–28 June 2008; pp. 1–6.
29. Zuo, J.; Schmid, N.A. On a methodology for robust segmentation of nonideal iris images. *IEEE Trans. Syst. Man Cybern. Part B Cybern.* **2010**, *40*, 703–718.
30. Hu, Y.; Sirlantzis, K.; Howells, G. Improving colour iris segmentation using a model selection technique. *Pattern Recognit. Lett.* **2015**, *57*, 24–32. [CrossRef]
31. Shah, S.; Ross, A. Iris segmentation using geodesic active contours. *IEEE Trans. Inf. Forensics Secur.* **2009**, *4*, 824–836. [CrossRef]
32. Koh, J.; Govindaraju, V.; Chaudhary, V. A robust iris localization method using an active contour model and Hough transform. In Proceedings of the 20th International Conference on Pattern Recognition, Istanbul, Turkey, 23–26 August 2010; pp. 2852–2856.
33. Abdullah, M.A.M.; Dlay, S.S.; Woo, W.L. Fast and accurate method for complete iris segmentation with active contour and morphology. In Proceedings of the IEEE International Conference on Imaging Systems and Techniques, Santorini, Greece, 14–17 October 2014; pp. 123–128.
34. Abdullah, M.A.M.; Dlay, S.S.; Woo, W.L.; Chambers, J.A. Robust iris segmentation method based on a new active contour force with a noncircular normalization. *IEEE Trans. Syst. Man Cybern. Syst.* **2017**, *47*, 3128–3141. [CrossRef]
35. Tan, T.; He, Z.; Sun, Z. Efficient and robust segmentation of noisy iris images for non-cooperative iris recognition. *Image Vis. Comput.* **2010**, *28*, 223–230. [CrossRef]
36. Patel, H.; Modi, C.K.; Paunwala, M.C.; Patnaik, S. Human identification by partial iris segmentation using pupil circle growing based on binary integrated edge intensity curve. In Proceedings of the International Conference on Communication Systems and Network Technologies, Katra, India, 3–5 June 2011; pp. 333–338.
37. Abate, A.F.; Frucci, M.; Galdi, C.; Riccio, D. BIRD: Watershed based iris detection for mobile devices. *Pattern Recognit. Lett.* **2015**, *57*, 41–49. [CrossRef]
38. Pereira, S.; Pinto, A.; Alves, V.; Silva, C.A. Brain tumor segmentation using convolutional neural networks in MRI images. *IEEE Trans. Med. Imaging* **2016**, *35*, 1240–1251. [CrossRef] [PubMed]
39. Ahuja, K.; Islam, R.; Barbhuiya, F.A.; Dey, K. A preliminary study of CNNs for iris and periocular verification in the visible spectrum. In Proceedings of the 23rd International Conference on Pattern Recognition, Cancún, Mexico, 4–8 December 2016; pp. 181–186.

40. Zhao, Z.; Kumar, A. Accurate periocular recognition under less constrained environment using semantics-assisted convolutional neural network. *IEEE Trans. Inf. Forensics Secur.* **2017**, *12*, 1017–1030. [CrossRef]
41. Al-Waisy, A.S.; Qahwaji, R.; Ipson, S.; Al-Fahdawi, S.; Nagem, T.A.M. A multi-biometric iris recognition system based on a deep learning approach. *Pattern Anal. Appl.* **2017**, *20*, 1–20. [CrossRef]
42. Gangwar, A.; Joshi, A. DeepIrisNet: Deep iris representation with applications in iris recognition and cross-sensor iris recognition. In Proceedings of the IEEE International Conference on Image Processing, Phoenix, AZ, USA, 25–28 September 2016; pp. 2301–2305.
43. Lee, M.B.; Hong, H.G.; Park, K.R. Noisy ocular recognition based on three convolutional neural networks. *Sensors* **2017**, *17*, 2933.
44. Liu, N.; Li, H.; Zhang, M.; Liu, J.; Sun, Z.; Tan, T. Accurate iris segmentation in non-cooperative environments using fully convolutional networks. In Proceedings of the IEEE International Conference on Biometrics, Halmstad, Sweden, 13–16 June 2016; pp. 1–8.
45. Arsalan, M.; Hong, H.G.; Naqvi, R.A.; Lee, M.B.; Kim, M.C.; Kim, D.S.; Kim, C.S.; Park, K.R. Deep learning-based iris segmentation for iris recognition in visible light environment. *Symmetry* **2017**, *9*, 263. [CrossRef]
46. Jalilian, E.; Uhl, A.; Kwitt, R. Domain adaptation for CNN based iris segmentation. In Proceedings of the IEEE International Conference on the Biometrics Special Interest Group, Darmstadt, Germany, 20–22 September 2017; pp. 51–60.
47. Dongguk IrisDenseNet CNN Model (DI-CNN) with Algorithm. Available online: http://dm.dgu.edu/link.html (accessed on 18 February 2018).
48. Kim, J.H.; Hong, H.G.; Park, K.R. Convolutional neural network-based human detection in nighttime images using visible light camera sensors. *Sensors* **2017**, *17*, 1065.
49. Kim, K.W.; Hong, H.G.; Nam, G.P.; Park, K.R. A study of deep CNN-based classification of open and closed eyes using a visible light camera sensor. *Sensors* **2017**, *17*, 1534. [CrossRef] [PubMed]
50. Nguyen, D.T.; Kim, K.W.; Hong, H.G.; Koo, J.H.; Kim, M.C.; Park, K.R. Gender recognition from human-body images using visible-light and thermal camera videos based on a convolutional neural network for image feature extraction. *Sensors* **2017**, *17*, 637. [CrossRef] [PubMed]
51. Kang, J.K.; Hong, H.G.; Park, K.R. Pedestrian detection based on adaptive selection of visible light or far-infrared light camera image by fuzzy inference system and convolutional neural network-based verification. *Sensors* **2017**, *17*, 1598. [CrossRef] [PubMed]
52. Pham, T.D.; Lee, D.E.; Park, K.R. Multi-national banknote classification based on visible-light line sensor and convolutional neural network. *Sensors* **2017**, *17*, 1595. [CrossRef] [PubMed]
53. Zhang, X.; Sugano, Y.; Fritz, M.; Bulling, A. Appearance-based gaze estimation in the wild. In Proceedings of the IEEE Conference on Computer Vision and Pattern Recognition, Boston, MA, USA, 7–12 June 2015; pp. 4511–4520.
54. Ren, S.; He, K.; Girshick, R.; Sun, J. Faster R-CNN: Towards real-time object detection with region proposal networks. *IEEE Trans. Pattern Anal. Mach. Intell.* **2017**, *39*, 1137–1149. [CrossRef] [PubMed]
55. Lemley, J.; Bazrafkan, S.; Corcoran, P. Deep learning for consumer devices and services: Pushing the limits for machine learning, artificial intelligence, and computer vision. *IEEE Consum. Electron. Mag.* **2017**, *6*, 48–56. [CrossRef]
56. Huang, G.; Liu, Z.; van der Maaten, L.; Weinberger, K.Q. Densely connected convolutional networks. In Proceedings of the IEEE Conference on Computer Vision and Pattern Recognition, Honolulu, HI, USA, 21–26 July 2017; pp. 2261–2269.
57. Badrinarayanan, V.; Kendall, A.; Cipolla, R. SegNet: A deep convolutional encoder-decoder architecture for image segmentation. *IEEE Trans. Pattern Anal. Mach. Intell.* **2017**, *39*, 2481–2495. [CrossRef] [PubMed]
58. Huang, G.; Liu, S.; van der Maaten, L.; Weinberger, K.Q. CondenseNet: An efficient DenseNet using learned group convolutions. *arXiv*, 2017; arXiv:1711.09224.
59. NICE.II. Noisy Iris Challenge Evaluation-Part II. Available online: http://nice2.di.ubi.pt/index.html (accessed on 28 December 2017).
60. Geforce GTX 1070. Available online: https://www.nvidia.com/en-us/geforce/products/10series/geforce-gtx-1070/ (accessed on 12 January 2018).

61. Matlab R2017b. Available online: https://ch.mathworks.com/help/matlab/release-notes.html (accessed on 12 January 2018).
62. He, K.; Zhang, X.; Ren, S.; Sun, J. Delving deep into rectifiers: Surpassing human-level performance on ImageNet classification. In Proceedings of the IEEE Conference on Computer Vision, Santiago, Chile, 7–13 December 2015; pp. 1026–1034.
63. Simonyan, K.; Zisserman, A. Very deep convolutional networks for large-scale image recognition. In Proceedings of the 3rd International Conference on Learning Representations, San Diego, CA, USA, 7–9 May 2015; pp. 1–14.
64. Zhang, T. Solving large scale linear prediction problems using stochastic gradient descent algorithms. In Proceedings of the 21st International Conference on Machine Learning, Banff, Canada, 4–8 July 2004; pp. 919–926.
65. Long, J.; Shelhamer, E.; Darrell, T. Fully convolutional networks for semantic segmentation. In Proceedings of the IEEE Conference on Computer Vision and Pattern Recognition, Boston, MA, USA, 7–12 June 2015; pp. 3431–3440.
66. Eigen, D.; Fergus, R. Predicting depth, surface normals and semantic labels with a common multi-scale convolutional architecture. In Proceedings of the IEEE Conference on Computer Vision, Santiago, Chile, 7–13 December 2015; pp. 2650–2658.
67. NICE.I. Noisy Iris Challenge Evaluation-Part I. Available online: http://nice1.di.ubi.pt/ (accessed on 4 January 2018).
68. Brostow, G.J.; Fauqueur, J.; Cipolla, R. Semantic object classes in video: A high-definition ground truth database. *Pattern Recognit. Lett.* **2009**, *30*, 88–97. [CrossRef]
69. Luengo-Oroz, M.A.; Faure, E.; Angulo, J. Robust iris segmentation on uncalibrated noisy images using mathematical morphology. *Image Vis. Comput.* **2010**, *28*, 278–284. [CrossRef]
70. Labati, R.D.; Scotti, F. Noisy iris segmentation with boundary regularization and reflections removal. *Image Vis. Comput.* **2010**, *28*, 270–277. [CrossRef]
71. Chen, Y.; Adjouadi, M.; Han, C.; Wang, J.; Barreto, A.; Rishe, N.; Andrian, J. A highly accurate and computationally efficient approach for unconstrained iris segmentation. *Image Vis. Comput.* **2010**, *28*, 261–269. [CrossRef]
72. Li, P.; Liu, X.; Xiao, L.; Song, Q. Robust and accurate iris segmentation in very noisy iris images. *Image Vis. Comput.* **2010**, *28*, 246–253. [CrossRef]
73. Tan, C.-W.; Kumar, A. Unified framework for automated iris segmentation using distantly acquired face images. *IEEE Trans. Image Process.* **2012**, *21*, 4068–4079. [CrossRef] [PubMed]
74. Proenca, H. Iris recognition: On the segmentation of degraded images acquired in the visible wavelength. *IEEE Trans. Pattern Anal. Mach. Intell.* **2010**, *32*, 1502–1516. [CrossRef] [PubMed]
75. De Almeida, P. A knowledge-based approach to the iris segmentation problem. *Image Vis. Comput.* **2010**, *28*, 238–245. [CrossRef]
76. Tan, C.-W.; Kumar, A. Towards online iris and periocular recognition under relaxed imaging constraints. *IEEE Trans. Image Process.* **2013**, *22*, 3751–3765. [PubMed]
77. Sankowski, W.; Grabowski, K.; Napieralska, M.; Zubert, M.; Napieralski, A. Reliable algorithm for iris segmentation in eye image. *Image Vis. Comput.* **2010**, *28*, 231–237. [CrossRef]
78. Haindl, M.; Krupička, M. Unsupervised detection of non-iris occlusions. *Pattern Recognit. Lett.* **2015**, *57*, 60–65. [CrossRef]
79. Zhao, Z.; Kumar, A. An accurate iris segmentation framework under relaxed imaging constraints using total variation model. In Proceedings of the IEEE Conference on Computer Vision, Santiago, Chile, 7–13 December 2015; pp. 3828–3836.
80. De Marsico, M.; Nappi, M.; Riccio, D.; Wechsler, H. Mobile iris challenge evaluation (MICHE)-I, biometric iris dataset and protocols. *Pattern Recognit. Lett.* **2015**, *57*, 17–23. [CrossRef]
81. CASIA-Iris-Interval Database. Available online: http://biometrics.idealtest.org/dbDetailForUser.do?id=4 (accessed on 28 December 2017).
82. IIT Delhi Iris Database. Available online: http://www4.comp.polyu.edu.hk/~csajaykr/IITD/Database_Iris.htm (accessed on 28 December 2017).

83. Hofbauer, H.; Alonso-Fernandez, F.; Wild, P.; Bigun, J.; Uhl, A. A ground truth for iris segmentation. In Proceedings of the 22nd International Conference on Pattern Recognition, Stockholm, Sweden, 24–28 August 2014; pp. 527–532.
84. Gangwar, A.; Joshi, A.; Singh, A.; Alonso-Fernandez, F.; Bigun, J. IrisSeg: A fast and robust iris segmentation framework for non-ideal iris images. In Proceedings of the International Conference on Biometrics, Halmstad, Sweden, 13–16 June 2016; pp. 1–8.
85. Alonso-Fernandez, F.; Bigun, J. Iris boundaries segmentation using the generalized structure tensor. A study on the effects of image degradation. In Proceedings of the 5th International Conference on Biometrics: Theory, Applications and Systems, Arlington, VA, USA, 23–27 September 2012; pp. 426–431.
86. Petrovska, D.; Mayoue, A. Description and documentation of the BioSecure software library. In *Technical Report, Proj. No IST-2002-507634-BioSecure Deliv*; BioSecure: Paris, France, 2007.
87. Uhl, A.; Wild, P. Weighted adaptive hough and ellipsopolar transforms for real-time iris segmentation. In Proceedings of the 5th IEEE International Conference on Biometrics, New Delhi, India, 29 March–1 April 2012; pp. 283–290.
88. Uhl, A.; Wild, P. Multi-stage visible wavelength and near infrared iris segmentation framework. In Proceedings of the 9th International Conference on Image Analysis and Recognition, Aveiro, Portugal, 25–27 June 2012; pp. 1–10.
89. Rathgeb, C.; Uhl, A.; Wild, P. Iris biometrics: From segmentation to template security. In *Advances in Information Security*; Springer: New York, NY, USA, 2013.
90. Masek, L.; Kovesi, P. *MATLAB Source Code for a Biometric Identification System Based on Iris Patterns*; The School of Computer Science and Software Engineering, The University of Western Australia: Perth, Australia, 2003.

sensors

MDPI

Article

Star Image Prediction and Restoration under Dynamic Conditions

Di Liu [1], Xiyuan Chen [1,*], Xiao Liu [1,2] and Chunfeng Shi [1]

1 Key Laboratory of Micro-Inertial Instrument and Advanced Navigation Technology, Ministry of Education, School of Instrument Science and Engineering, Southeast University, Nanjing 210096, China; sdili_liudi@163.com (D.L.); sidescan@126.com (X.L.); shchfeng@163.com (C.S.)

2 School of Electrical Engineering and Automation, Qilu University of Technology, Jinan 250353, China

* Correspondence: chxiyuan@seu.edu.cn; Tel.: +86-25-8379-2010

Received: 24 March 2019; Accepted: 16 April 2019; Published: 20 April 2019

Abstract: The star sensor is widely used in attitude control systems of spacecraft for attitude measurement. However, under high dynamic conditions, frame loss and smearing of the star image may appear and result in decreased accuracy or even failure of the star centroid extraction and attitude determination. To improve the performance of the star sensor under dynamic conditions, a gyroscope-assisted star image prediction method and an improved Richardson-Lucy (RL) algorithm based on the ensemble back-propagation neural network (EBPNN) are proposed. First, for the frame loss problem of the star sensor, considering the distortion of the star sensor lens, a prediction model of the star spot position is obtained by the angular rates of the gyroscope. Second, to restore the smearing star image, the point spread function (PSF) is calculated by the angular velocity of the gyroscope. Then, we use the EBPNN to predict the number of iterations required by the RL algorithm to complete the star image deblurring. Finally, simulation experiments are performed to verify the effectiveness and real-time of the proposed algorithm.

Keywords: star image prediction; star sensor; Richardson-Lucy algorithm; neural network

1. Introduction

Along with the development of navigation technology, the requirement for a spacecraft attitude measurement is becoming higher and higher [1,2]. In general, star sensors and gyroscopes are often used in spacecraft to measure the attitude information. The star sensor is supposed to be the most accurate attitude-measuring device in stable conditions [3]. However, under dynamic conditions, frame loss and blurring of the star image may occur, which leads to decreased accuracy or even failure of the star centroid extraction and attitude determination. Therefore, only by solving the frame loss and blurring problem of the star image, can the star sensor maintain good performance under dynamic conditions. Because gyroscopes have a relatively high measurement accuracy and excellent dynamic performance in a short period, using the gyroscope to assist in improving the dynamic performance of the star sensor has become a hot topic [4–9].

In the process of spacecraft motion, due to the influence of external interference and the limitation of the star sensor, the star sensor is prone to frame loss, which can result in a lack of coherence in the process of moving image tracking and even loss of key motion features. Therefore, how to eliminate the frame loss error has become a research hotspot in the field of image processing. Currently, the primary methods for eliminating frame loss error includes the frame loss error elimination based on the support vector machine (SVM) [10,11], frame loss error elimination based on iterative error compensation [12,13] and frame loss error elimination based on adaptive minimum error threshold segmentation [14]. These methods eliminate the interference noise in the image and compensate the frame loss error, but still cannot avoid the frame loss. To overcome the shortcomings of the above methods, a method for eliminating the frame loss by using a motion image-tracking model is presented in [15], since the frame loss of the star image is mainly affected by the exposure time and readout time of the star sensor [2]. Therefore, in [16–18], parallel processing is used to overlap exposure time and readout time to reduce the frame loss of the star image. In [19,20], the authors used image intensifiers to increase the sensitivity of the image sensor, thereby reducing the occurrence of the frame loss in the star sensors. In [21], Wang et al. proposed using field programmable gate arrays (FPGAs) to improve the processing ability of the star sensor to reduce the readout time. Yu et al. [22] proposed a method to reduce the occurrence of the frame loss by using an intensified star sensor. Although FPGAs and image intensifiers can assist the star sensor in reducing the occurrence of the frame loss, the additional FPGA and image intensifier lead to an increase in the weight and power consumption of the star sensor and limit its application in micro-spacecraft.

The motion blur of the star image is another important reason that affects the dynamic performance of the star sensor. To improve the dynamic performance of the star sensor, many scholars have done a lot of research in the field of image processing, especially on the star image deblurring algorithms [23]. According to whether the point spread function (PSF) is known or not, the deblurring methods can be classified into two typical forms: Blind image deblurring (BID) with unknown PSF, and non-blind image deblurring (NBID) with known PSF [24]. Mathematically, the process of NBID is an inverse problem, and an NBID algorithm has a good real-time performance. Currently, most BID algorithms perform blur kernel estimation and image deblurring simultaneously, and recursively to approach the sharp image [25–30]. Therefore, BID methods have poor real-time performance. Because star sensors are widely used in spacecraft, the real-time requirements are high. Therefore, we intend to study an NBID algorithm for star image deblurring.

Two problems should be solved in the process of restoring the blurred star image. One is how to determine the blur kernel, and the other is to choose which deblurring method to use. The gyroscope can be used to measure the angular rates of the carrier and is easy to integrate, and the blur kernel parameters (blur angle and blur length) can be calculated according to the angular rate information output by the gyroscope. In this paper, a gyroscope is used to assist in the calculation of blur kernels. For the star image deblurring, there are two commonly used NBID algorithms. One is the Wiener filter [31,32]. Quan et al. [31] proposed a Wiener filter based on the optimal window technique for recovering the blurred star image. Ma et al. [32] proposed an improved one-dimensional Wiener filtering method for star image deblurring. Although the two methods are better in real time, they also amplify the noise in the image. The other is the Richardson–Lucy (RL) algorithm, which can effectively suppress the noise in the deblurred star image [33,34]. However, the iterative convergence criterion is not given in the RL algorithm, and the optimal number of iterations needs to be obtained through constant-trying with large time-consumption. If the amount of blurred star image to be processed is enormous, this is a disadvantage that cannot be ignored.

In this paper, to solve the shortcomings of the above methods and further improve the performance of the star sensor under highly dynamic conditions, we propose an improved gyroscope-assisted star image prediction method and RL non-blind deblurring algorithm. In the star image prediction method, considering the second-order distortion of the star sensor lens, a prediction model between the angular rates of the gyroscope and the position of the star spot is established. For the improved RL algorithm, first, we analyze the point spread function (PSF) model of the star sensor under different motion conditions, and then the ensemble back-propagation neural network (EBPNN) prediction model based on the improved bagging method is constructed to predict the number of termination iterations required by the conventional RL algorithm, which is used to overcome the disadvantage of traditional RL algorithm that needs to set the number of iterations manually.

The rest of this paper is organized as follows. In Section 2, we introduce the star image prediction model in the case of the frame loss of the star image. The improved RL algorithm is described in Section 3. In Section 4, simulation results are shown to demonstrate the effectiveness of our method. Finally, we give a conclusion in Section 5.

2. Prediction Model of the Star Image

The star sensor is a vision sensor that can be used to measure the attitude of a spacecraft [35]. To obtain the high-precision attitude information of the spacecraft, we must ensure that the image sensor of the star sensor can output the star image continuously. Due to the highly dynamic motion of the spacecraft, frame loss of the star image often occurs. Therefore, it is especially important to ensure that the star sensor can output the high-precision attitude information under the condition of the frame loss of the star image. In this section, we will show how to predict the position of the star spot based on the angular rates of the gyroscope in the presence of distortion of the star sensor lens. In Figure 1, the star sensor obtains the direction vector of the navigation star in the celestial inertial coordinate system by observing the stars on the celestial sphere. At time t, the attitude matrix of the star sensor in the celestial coordinate system is $A(t)$, the star sensor can detect the direction vector v_i of the navigation star in the celestial coordinate system, and its image vector can be represented as W_i in the star sensor coordinate system. The image coordinate of the principal point of the lens of the star sensor is (x_0, y_0), the coordinates of the navigation star S_i on the image plane is (x_i, y_i). Since the optical lens of the star sensor mainly has a second-order radial distortion, the ideal image coordinate (x_i', y_i') of the navigation star S_i can be expressed as,

$$\begin{cases} x_i' - x_0 = (x_i - x_0)(1 + k_x' \cdot r^2), \\ y_i' - y_0 = (y_i - y_0)(1 + k_y' \cdot r^2), \end{cases} \tag{1}$$

where, $r = \sqrt{(x_i - x_0)^2 + (y_i - y_0)^2}$, k_x' and k_y' represent the second-order radial distortion coefficients in the X and Y directions, respectively.

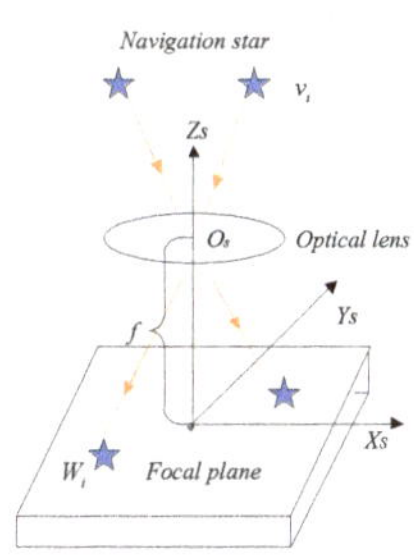

Figure 1. Star image model of the star sensor.

Assuming that the focal length of the star sensor is f, the direction vector W_i can be given by

$$W_i = \frac{1}{\sqrt{[(x_i - x_0)(1 + k_x' \cdot r^2)]^2 + \left[(y_i - y_0)(1 + k_y' \cdot r^2)\right]^2 + f^2}} \begin{bmatrix} (x_i - x_0)(1 + k_x' \cdot r^2) \\ (y_i - y_0)(1 + k_y' \cdot r^2) \\ -f \end{bmatrix}. \tag{2}$$

According to the attitude matrix $A(t)$ of the star sensor, the relationship between the vectors W_i and v_i can be obtained,

$$W_i = A(t) \cdot v_i, \tag{3}$$

where, the attitude matrix $A(t)$ can be solved by the N vector method, Trial method, Quest method, Q-method and Least square method [36]. In this paper, we use the angular velocity information of the gyroscope to calculate the attitude matrix $A(t)$.

In Figure 2, $O_S XYZ$ represents the star sensor coordinate system, $O_C uv$ represents the image plane coordinate system, the projection point of the principal point O_S of the lens of the star sensor on the image plane is O_C, $O_C O_S$ is consistent with the principal optical axis of the star sensor lens and its length is equal to the focal length f. w_x, w_y and w_z represent the three-axis angular rates of the star sensor at instant t, which can be measured by the gyroscope. P denotes the position of the navigation star on the star image at instant t, O_CP denotes the direction vector of the navigation star under the coordinate system of the star sensor, and the star spot P shifts to P' at instant $t + \Delta t$. According to Equation (3), the direction vectors $\overrightarrow{O_SP}$ and $\overrightarrow{O_SP'}$ can be expressed as,

$$\begin{cases} \overrightarrow{O_SP} = W_i(t) = A(t) \cdot v_i, \\ \overrightarrow{O_SP'} = W_i(t+\Delta t) = A(t+\Delta t) \cdot v_i, \end{cases} \tag{4}$$

where, $A(t+\Delta t) = A_t^{t+\Delta t} \cdot A(t)$, $A(t+\Delta t)$ denotes the attitude matrix at instant $t + \Delta t$.

$$\begin{aligned} A_t^{t+\Delta t} &= I - (w(t)\times) \cdot \Delta t = I - \begin{bmatrix} 0 & -w_z(t) & w_y(t) \\ w_z(t) & 0 & -w_x(t) \\ -w_y(t) & w_x(t) & 0 \end{bmatrix} \cdot \Delta t \\ &= \begin{bmatrix} 1 & w_z(t)\cdot\Delta t & -w_y(t)\cdot\Delta t \\ -w_z(t)\cdot\Delta t & 1 & w_x(t)\cdot\Delta t \\ w_y(t)\cdot\Delta t & -w_x(t)\cdot\Delta t & 1 \end{bmatrix}, \end{aligned} \tag{5}$$

where $(w(t)\times)$ represents the cross-product matrix of the star sensor angular rates vector $w(t)$.

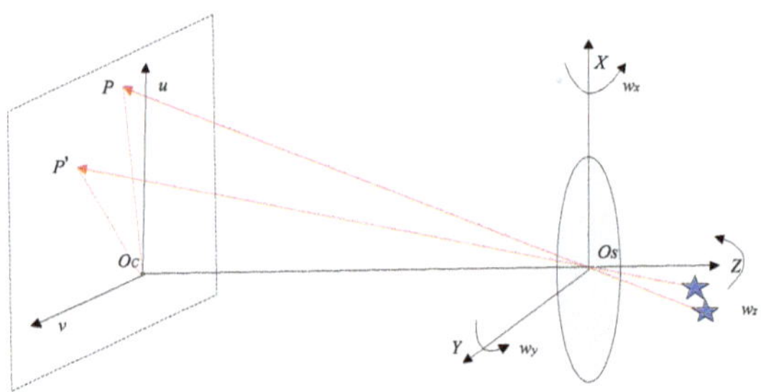

Figure 2. Prediction model of the star spot.

According to Equations (4) and (5), the relationship between $W_i(t)$ and $W_i(t+\Delta t)$ can be obtained,

$$W_i(t+\Delta t) = A_t^{t+\Delta t} \cdot W_i(t), \tag{6}$$

where, we can calculate W_i through the star image. According to Equations (1) and (6), we can obtain the position prediction model as follows,

$$\begin{cases} x_i'(t+\Delta t) = \dfrac{(x_i(t)-x_0)(1+k_x'\cdot r^2)+x_0+((y_i(t)-y_0)(1+k_y'\cdot r^2)+y_0)\cdot w_z(t)\cdot\Delta t+f\cdot w_y(t)\cdot\Delta t}{(-((x_i(t)-x_0)(1+k_x'\cdot r^2)+x_0)\cdot w_y(t)\cdot\Delta t+((y_i(t)-y_0)(1+k_y'\cdot r^2)+y_0)\cdot w_x(t)\cdot\Delta t)/f+1} \\ y_i'(t+\Delta t) = \dfrac{((y_i(t)-y_0)(1+k_y'\cdot r^2)+y_0)-((x_i(t)-x_0)(1+k_x'\cdot r^2)+x_0)\cdot w_z(t)\cdot\Delta t-f\cdot w_x(t)\cdot\Delta t}{(-((x_i(t)-x_0)(1+k_x'\cdot r^2)+x_0)\cdot w_y(t)\cdot\Delta t+((y_i(t)-y_0)(1+k_y'\cdot r^2)+y_0)\cdot w_x(t)\cdot\Delta t)/f+1} \end{cases}. \tag{7}$$

3. Improved Star Image Deblurring Algorithm

Generally, establishing a PSF under a specific motion is the key to star image recovery. In this section, first, we analyze the PSF model of the blurred star image caused by the rotation of the star sensor around the optical axis and non-optical axis and calculate the PSF in the corresponding motion condition through the angular velocity information of the gyroscope. Then, we introduce an improved RL algorithm to recover the blurred star image.

3.1. Motion Blur Model of the Star Image

To better recover the blurred star image, the primary task is to obtain the PSF. Therefore, it is necessary to analyze the mechanism of the star image blurs. The star sensor is a navigation device that acquires the attitude by utilizing star observations. Because the star sensor needs to photograph the sky with a dark background, in order to increase the number of navigation stars in the star image, it needs to increase the exposure time appropriately. If the star sensor has a wide range of motion during the exposure time, the same star will be imaged at different locations on the star image, which will result in blurring of the star image. Mathematically, the model of star image blurring can be written as,

$$g(x,y) = f(x,y) \otimes h(x,y) + n(x,y), \tag{8}$$

where $f(x,y)$, $g(x,y)$, and $h(x,y)$ denote the sharp star image, the blurred star image, and the PSF, respectively; $\otimes$ represents two-dimensional convolution operator, and $n(x,y)$ denotes the image noise.

Due to the different motion types of the star, sensors produce different PSFs, so PSF is important for describing the model of the blurred star image. Since the distance from the navigation star to the earth is much larger than the distance from the star sensor to the earth, the linear motion has less effect on the star image blur, and this effect can be ignored. Therefore, we mainly analyze the model of the blurred star image generated by the angular motion.

In Figure 3a, the star image blur caused by the angular motion is shown. Since the exposure time of the star sensor is short, the angular velocity of the star sensor can be considered to be constant during the exposure time. Moreover, the star sensor coordinate system is coincident with the body-fixed frame. In Figure 3b, the model of the blurred star image generated by the star sensor rotating around the X-axis is shown, the initial angle between the starlight direction and the principal optical axis of the star sensor is α, and the projection of the navigation star is P in the star image. When the star sensor rotates clockwise around the X-axis at an angular velocity w_x, and during the exposure time Δt, the rotational angle is $\Delta\alpha = w_x\Delta t$, and the star spot moves from P to P' in the image plane. The geometric relationship between P and P' is,

$$L_{PP'} = f \cdot [\tan(\alpha + \Delta\alpha) - \tan\alpha] / d_{ccd}, \tag{9}$$

where $L_{PP'}$ represents the distance from P to P' quantized by pixels, d_{ccd} denotes the pixel size, and f is the focal length of the star sensor.

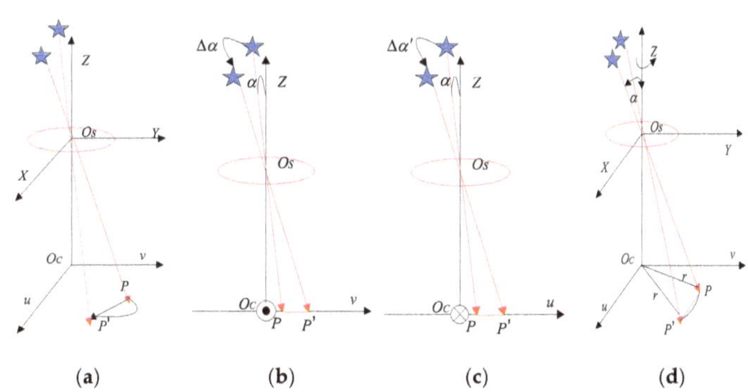

Figure 3. Motion blur star image model. (**a**) Blurred star image generated by the angular motion; (**b**) blurred star image generated by the rotation of the star sensor around the X-axis; (**c**) blurred star image generated by the rotation of the star sensor around the Y-axis; (**d**) blurred star image generated by the rotation of the star sensor around the Z-axis.

As a result of the short exposure time of the star sensor, $\Delta\alpha$ is quite small, the first order Taylor-expansion for $\tan(\alpha+\Delta\alpha)$ can be obtained.

$$\begin{aligned}\tan(\alpha+\Delta\alpha) &\approx \tan\alpha + (\tan\alpha)'\cdot\Delta\alpha \\ &= \tan\alpha + \left(\frac{\sin^2\alpha+\cos^2\alpha}{\cos^2\alpha}\right)\cdot\Delta\alpha \\ &= \tan\alpha + (\tan^2\alpha+1)\cdot\Delta\alpha.\end{aligned} \tag{10}$$

Substituting Equation (10) into (9), we have

$$L_{PP'} = f\cdot(\tan^2\alpha+1)\cdot\Delta\alpha/d_{ccd}. \tag{11}$$

In general, the rotational motion characteristics of the star sensor in the O_SX and O_SY directions are the same. As shown in Figure 3c, during the exposure time Δt, the star sensor rotates clockwise around the Y-axis at an angular velocity w_y, the rotational angle is $\Delta\alpha' = w_y\Delta t$, the star spot shifts along the u axis in the image plane, and its translation vector can be obtained.

$$L_{PP'} = f\cdot(\tan^2\alpha+1)\cdot\Delta\alpha'/d_{ccd}. \tag{12}$$

When the star sensor rotates around the X-axis and Y-axis with angular rates w_x and w_y, respectively, and after the exposure time Δt, the rotation angle of the star sensor is $\Delta\alpha'' = w_{xy}\cdot\Delta t = \sqrt{w_x^2+w_y^2}\Delta t$, and the translation vector of the star spot is

$$L_{PP'} = f\cdot(\tan^2\alpha+1)\cdot\Delta\alpha''/d_{ccd}. \tag{13}$$

In general, when the star sensor rotates around the cross bore-sight direction (O_SX and O_SY directions), the blur kernel angle θ of the star image can be given by

$$\theta = \arctan\left[\frac{\tan(\alpha+\Delta\alpha)-\tan\alpha}{\tan(\alpha+\Delta\alpha')-\tan\alpha}\right]. \tag{14}$$

Then, the PSF of the blurred star image is expressed as [37,38],

$$h_1(x,y) = \begin{cases} 1/L_{PP'},\ if\ y/x = \sin|\theta|/\cos|\theta|, 0\le x\le L_{PP'}\cdot\cos|\theta| \\ 0,\ otherwise \end{cases}. \tag{15}$$

In Figure 3d, the star sensor rotates clockwise around the Z-axis at an angular rate w_z, point $P(u,v)$ does a circular motion with O_C as the center and $r=\sqrt{u^2+v^2}$ as the radius. The rotation angle of the star sensor is $\Delta\alpha''' = w_z\cdot\Delta t$ during the exposure time Δt. Since the exposure time of the star sensor is short, the arc length PP' can be approximated as the chord length $\Delta PP'$. Inspired by reference [39], the motion of the star spot can be regarded as a uniform linear motion on the focal plane. The displacement of the star spot in the direction of the X- and Y-axis can be expressed as,

$$\begin{cases} \Delta PP'_u \approx -v\cdot w_z\cdot\Delta t, \\ \Delta PP'_v \approx u\cdot w_z\cdot\Delta t. \end{cases} \tag{16}$$

The star image blur kernel angle θ and the $\Delta PP'$ are given by

$$\theta = \arctan(\Delta PP'_u/\Delta PP'_v), \tag{17}$$

$$\begin{aligned}\Delta PP' &= \sqrt{\Delta PP'^2_u + \Delta PP'^2_v} \\ &= \sqrt{v^2\cdot w_z^2\cdot\Delta t^2 + u^2\cdot w_z^2\cdot\Delta t^2} \\ &= |w_z|\cdot\Delta t\cdot r.\end{aligned} \tag{18}$$

According to the geometric relation in Figure 3d,

$$\tan\alpha = r \cdot d_{ccd}/f. \tag{19}$$

Substituting Equation (19) into Equation (18), Equation (18) can be rewritten as

$$\Delta PP' = |w_z| \cdot \Delta t \cdot f \cdot \tan\alpha / d_{ccd}. \tag{20}$$

Therefore, when the star sensor rotates around the Z-axis, the PSF of the blurred star image is expressed as,

$$h_2(x,y) = \begin{cases} 1/\Delta PP', if\ y/x = \sin|\theta|/\cos|\theta|, 0 \leq x < \Delta PP' \cdot \cos|\theta| \\ 0, otherwise \end{cases}. \tag{21}$$

In summary, according to Equations (15) and (21), the model of the multiple-blurred star image is given by

$$g(x,y) = f(x,y) \otimes h_1(x,y) \otimes h_2(x,y) + n(x,y), \tag{22}$$

where the $h_1(x,y)$ and $h_2(x,y)$ need to be calculated based on the angular velocity w_x, w_y and w_z of the star sensor. In this paper, we use a gyroscope to provide the angular velocity $[w_{bx}\ w_{by}\ w_{bz}]$ of the spacecraft. Therefore, the angular velocity $[w_x\ w_y\ w_z]$ of the star sensor is expressed as,

$$\left[w_x, w_y, w_z\right]^T = C_b^s\left[w_{bx}, w_{by}, w_{bz}\right]^T, \tag{23}$$

where C_b^s denotes the rotation matrix from the body coordinate system to the star sensor coordinate system. Because the star sensor is fixed on the spacecraft, C_b^s can be calibrated in advance.

After obtaining the PSF, the NBID algorithm is used to recover the blurred star image.

3.2. Richardson-Lucy (RL) Algorithm

The NBID algorithm includes both linear and nonlinear algorithms. The most common linear NBID algorithms include the inverse filtering algorithm, Wiener filtering algorithm, and least squares algorithm [3]. Compared with the linear NBID algorithm, nonlinear NBID algorithm has a better effect in suppressing noise and preserving image edge information. Currently, the RL algorithm [40] is the most widely used non-linear iterative restoration algorithm. The RL algorithm is a blurred image deconvolution algorithm that extends from the maximum a posteriori probability estimate. This method assumes that the noise in the image follows a Poisson distribution, and the likelihood probability of the image is

$$p(g/f) = \prod_{x,y} \frac{((f(x,y) \otimes h(x,y))^{g(x,y)} e^{-(f(x,y) \otimes h(x,y))}}{g(x,y)!}, \tag{24}$$

where, (x,y) denotes the pixel coordinate, $g(x,y)$ represents the blurred image, $h(x,y)$ denotes the PSF, and $\otimes$ denotes the two-dimensional convolution operator.

To get the maximum likelihood solution of the sharp image $f(x,y)$, we minimize the energy function.

$$E(f) = \sum_{x,y} \{(f(x,y) \otimes h(x,y)) - g(x,y) \cdot \log(f(x,y) \otimes h(x,y))(x,y)\}. \tag{25}$$

By deriving the $E(f)$ and normalizing the blur kernel $h(x, y)$, the RL algorithm iteratively updates the image by

$$f^{n+1}(x,y) = [\frac{g(x,y)}{f^n(x,y) \otimes h(x,y)} \otimes h(-x,-y)] f^n(x,y), \quad (26)$$

where n represents the iteration number.

The RL algorithm has two important properties [40]: Non-negativity and energy preserving. It constrains the non-negative of estimated values of the sharp image and preserves the total energy of the image in the iteration so that the RL algorithm has excellent performance in the star image deblurring. However, the iterative convergence criterion is not given in the RL algorithm, and the optimal number of iterations need to be obtained through constant-trying with large time-consumption. This shortcoming of the RL algorithm cannot be ignored if we are dealing with a large number of the blurred star image. Therefore, it is necessary to study an improved RL algorithm which automatically sets the number of iterations.

3.3. Improved RL Algorithm

To overcome the shortcomings of the RL algorithm, we propose an improved RL algorithm, and the flow diagram is shown in Figure 4. First, we set the parameters of the star sensor including the field of view, focal length, star magnitude limit, resolution of the star image, etc. We use these parameters to simulate a large number of sharp star image and the corresponding blurred star image. Second, according to the angular rates of the gyroscope output, we calculate the PSF of each blurred star image and use the RL algorithm to deblur the star image and record the optimal number of iterations used. The optimal number of iterations and the sum of the Magnitude of Fourier Coefficients (SUMFC) of the PSF of the blurred star image are used for the training of the ensemble back-propagation neural network (EBPNN) [41]. After the training is completed, the optimal iteration number prediction model of the RL algorithm can be obtained. Finally, when the navigation system is used, the PSF of the blurred star image is obtained according to the angular velocity of the gyroscope, and the SUMFC of the PSF is used as the input of the prediction model. The star image is deblurred according to the number of iterations required by the RL algorithm of the prediction model output. Especially, when predicting the number of iterations, the ensemble back-propagation neural network (EBPNN) prediction model based on the improved bagging method uses the SUMFC of the PSF of the blurred star image as the input.

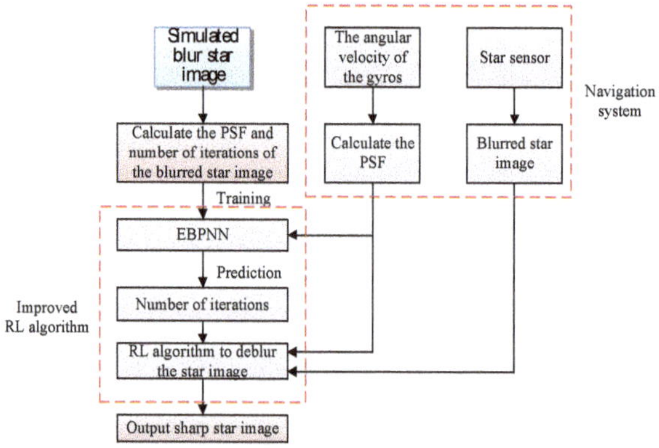

Figure 4. Flow diagram of the improved Richardson-Lucy (RL) algorithm for star image deblurring.

We use different PSFs to blur the sharp star image (Figure 5a). The relationship between the SUMFC of PSFs and the corresponding number of iterations required by the RL algorithm is shown in Figure 6. We can see that there is an obvious non-linear relationship between them, which prompts us to use EBPNN to predict the optimal number of iterations of the RL algorithm.

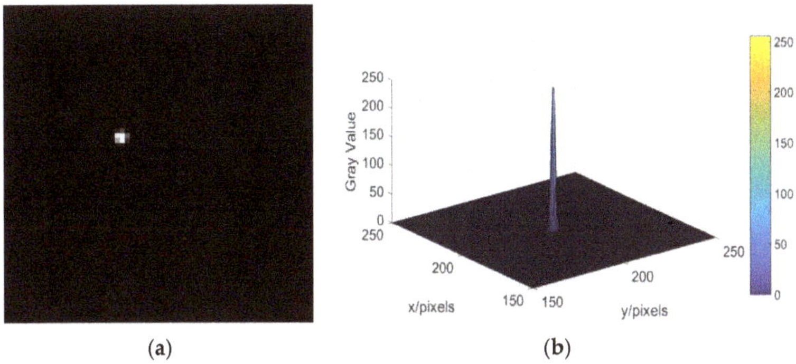

Figure 5. Original sharp star image and its gray distribution. (**a**) Original sharp star image; (**b**) gray distribution of star spot.

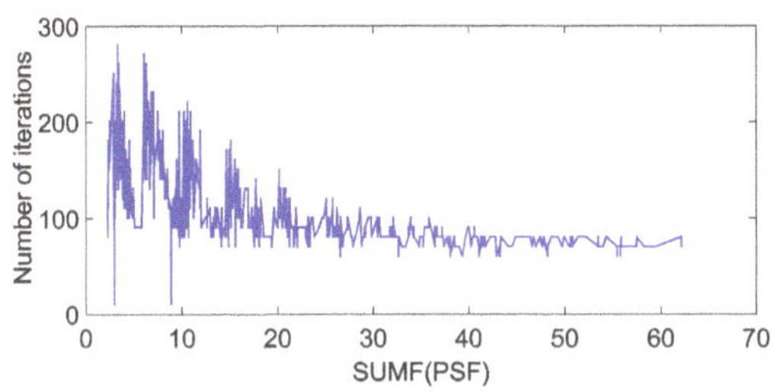

Figure 6. The relationship between the magnitude of Fourier coefficients (SUMFC) of the point spread function (PSF) and the corresponding optimal number of iterations.

The performance of a single back-propagation (BP) neural network is limited. It takes a long time to learn, and its objective function is easy to fall into a local minimum. Therefore, we use the integration strategy based on the improved bagging method to integrate the single neural network. The bagging method [42] is based on the re-sampling and self-help technology. The self-help learning sample set $D_i(i = 1, 2, \ldots)$ is retrieved from the original training set D, the size of each self-learning sample set is equivalent to the original training set, and each self-learning sample trains a single BP neural network. The bagging method increases the diversity of neural network by re-selecting the training set, thereby improving the generalization ability and prediction accuracy of the EBPNN.

In order to further improve the prediction accuracy of the ensemble neural network, we introduce a just-in-time learning algorithm to optimize the sample sets $D_i(i = 1, 2, \ldots)$ obtained by the bagging method. Suppose two input samples x_i and x_q, where x_q is the currently acquired input sample and x_i is a training sample in $D_i(i = 1, 2, \ldots)$. The distance and angle between them can be calculated by the following equation.

$$\begin{cases} d(x_i, x_q) = \sqrt{||x_i - x_q||^2}, \\ \theta(x_i, x_q) = \arccos\frac{x_i^T x_q}{||x_i||^2 ||x_q||^2}. \end{cases} \tag{27}$$

The similarity between x_i and x_q is

$$S(x_i, x_q) = \alpha e^{-d(x_i, x_q)} + (1 - \alpha)\cos(\theta(x_i, x_q)), \tag{28}$$

where, α is the weighting factor, the larger the $S(x_i, x_q)$ value, the higher the similarity between x_i and x_q.

We select the k-group data closest to the currently acquired one sample x_q from the training sample set $D_i(i = 1,2,\ldots)$ and arrange the new sample set in descending order.

$$\begin{cases} D_i' = \left\{(x_{1,i}, y_{1,i}), (x_{2,i}, y_{2,i}), \cdots, (x_{k,i}, y_{k,i})\right\}, i = 1,2,\ldots, \\ S(x_1, x_q) > S(x_2, x_q) > \cdots > S(x_k, x_q), \end{cases} \tag{29}$$

where $y_{k,i}$ denotes the expected output value corresponding to $x_{k,i}$ in training sets $D_i(i = 1,2,\ldots)$.

Therefore, the local modeling problem is transformed into an optimization problem.

$$J(\delta) = \min_{\delta} \sum_{i=1}^{k} (y_i - \hat{y}(\delta, x_i))^2 \cdot S(x_i, x_q). \tag{30}$$

Minimize $J(\delta)$ to obtain the model parameter δ at the current moment, and then obtain its local model:

$$y_q = \hat{y}(\delta, x_q). \tag{31}$$

In particular, we find that the computational complexity of the EBPNN model increases with the increase of the number of BPNN models, but the prediction accuracy of the EBPNN model does not always increase with it, sometimes it even decreases. Therefore, after considering the computational complexity and prediction accuracy of the EBPNN model, we decide to use three sub-BP neural network models to construct the EBPNN model. As shown in Figure 7, three BP neural networks are trained by different sample sets $D_i'(i = 1,2,3)$, and the integrated prediction model is obtained by aggregating the three BP neural networks. When the EBPNN is used for prediction, we use the weighted method to integrate the output of each neural network and take the integrated result as the output of EBPNN. In the process of integrating the output of each BP neural network, first, we calculate the average training errors $e_i(i = 1,2,3)$ of three sub-models on their respective training sample set. Then, we construct a weighting vector w of $1 \times n$ dimensions, the value of n is the same as the number of sub-BP neural network models, so $n = 3$, $w_i = 1/e_i$, $(i = 1,2,3)$. Finally, we calculate the prediction results of three sub-models for the input data x_q by Equation (31) and form a 1×3-dimensional output vector y_q'. The final prediction result of the EBPNN is expressed as,

$$y = \frac{w \cdot {y_q'}^T}{\sum\limits_{i=1}^{3} w_i}. \tag{32}$$

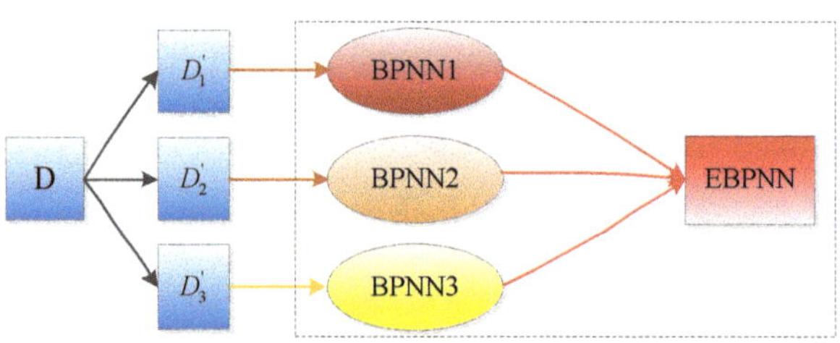

Figure 7. The ensemble back-propagation neural network.

To verify the effectiveness of the EBPNN prediction model, we analyzed the accuracy of the iteration times estimated by the model. In the training stage of the EBPNN model, each BP neural network adopts a three-layer structure. The nodes of the input layer, hidden layer, and output layer are set to 1, 10 and 1, respectively. The sigmoid function is used as the activation function. The original training set D contains 1708 samples. In Figure 8, we show the number of iterations predicted by the

EBPNN model and compare it with the optimal number of iterations. We can see that the number of iterations estimated by the EBPNN almost coincides with the optimal number of iterations, and the error between them is small. Therefore, the performance of the EBPNN prediction model can meet our requirements.

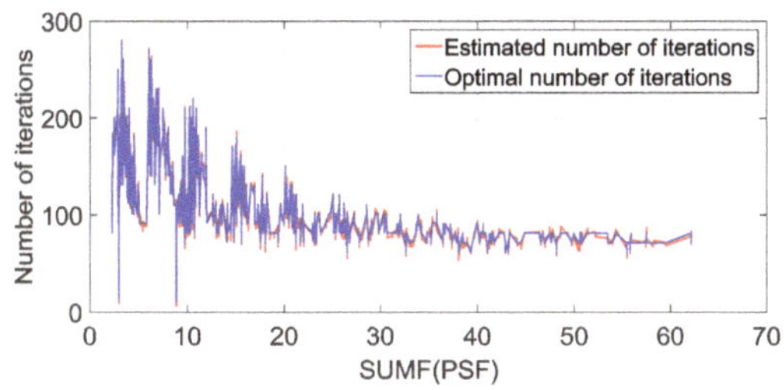

Figure 8. Comparison between the optimal number of iterations and the estimated number of iterations.

After EBPNN predicts the number of iterations, we use the improved RL algorithm to obtain the sharp star image, and then we can accurately estimate the attitude information by the star image segmentation, star extraction, star identification, star matching, and other operations [43].

4. Simulation Results and Analysis

In order to prove the effectiveness of the star image prediction method and the improved RL algorithm in the highly dynamic environment, we compare and analyze the prediction accuracy of the star spot, and the accuracy of the attitude estimation before and after the star image deblurring in the following section.

4.1. Star Image Prediction Experiment

In this section, to validate the star image prediction method, we need to simulate the star image acquired by the star sensor at a different time. In the process of star image simulation, we determine the position of the navigation star in the star image based on the bore-sight direction of the star sensor and the right ascension and declination of the navigation star. Since the star sensor is fixed on the spacecraft, it can obtain different star images as the spacecraft moves. We assume that the exposure time of the star sensor is 0.01 s, the field of view is $20° \times 20°$, the image sensor size is 865 pixels × 865 pixels, the pixel size is 20 μm, the focal length is 49 mm, and select the stars brighter than 3m in Yale Bright Star Catalogue as the guide star catalog. We use these parameters and the spacecraft trajectory to simulate the images at different times and use them as the ground truth of the star image. According to the above parameters, the resolution of the star image we simulated is 865×865. To speed up the processing of the star image, we intercept the 512×512 size as the star image to be processed. The trajectory of the spacecraft we simulated is shown in Figure 9. And 1500 frames of consecutive star images are simulated, the first and the 1500th frame star image are shown in Figure 10.

To validate the star image prediction method, we predict the star image based on the first frame star image and the angular velocity of the gyroscope, and compare it with the ground truth of the star image. Figure 11a,b show the ground truth of the 1500th frame star image and the 1500th frame star image predicted by the proposed algorithm. To more intuitively demonstrate the accuracy of the prediction algorithm, in Table 1, the centroid coordinates of the star spot in the real and predicted 1500th frame star image is shown, where (x, y) represents the centroid coordinate of the star spot in the real star image, (x', y') is the centroid coordinate of the predicted star spot. Δx and Δy represent the difference of the horizontal and vertical coordinates between the true star spot and the predicted star spot, respectively. As seen from Table 1, the maximum error of the horizontal and vertical coordinates of the star spot predicted by our method within 15 s is 0.89 and 0.50 pixels, respectively.

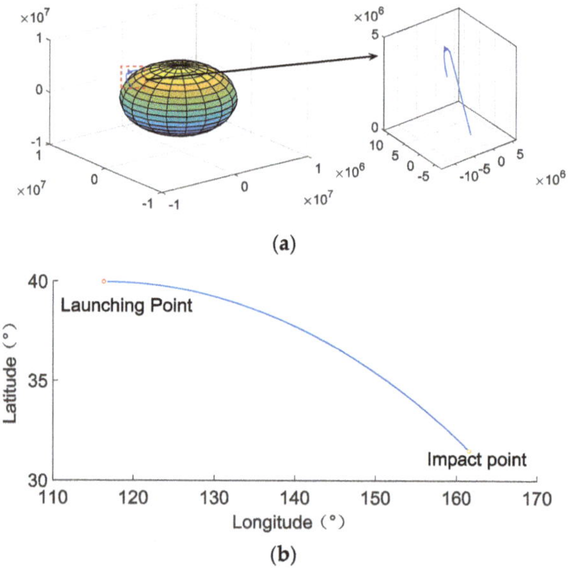

Figure 9. The spacecraft trajectory. (**a**) Three-dimensional trajectory of spacecraft; (**b**) projection of the spacecraft trajectory on the surface of the Earth.

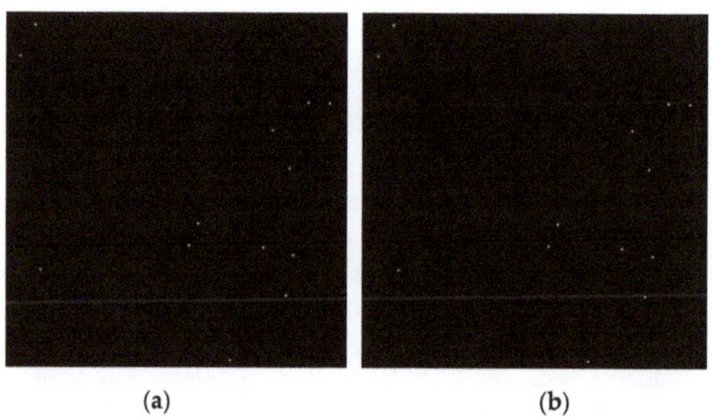

Figure 10. Star image simulation result. (**a**) The first frame star image; (**b**) the 1500th frame star image.

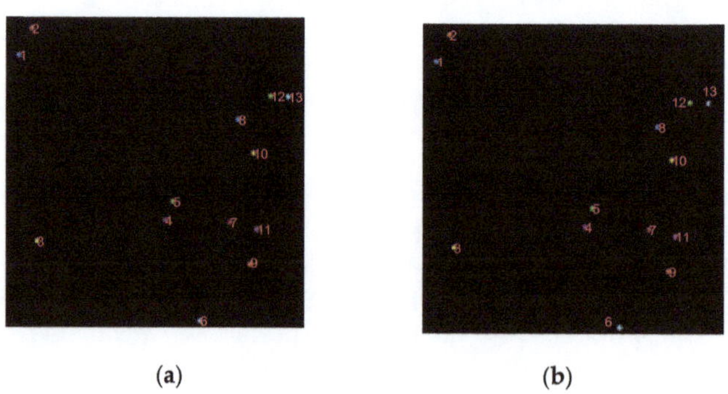

Figure 11. True star image versus predicted star image. (**a**) The true value of the 1500th frame star image; (**b**) the 1500th frame star image predicted by the proposed algorithm.

Table 1. Comparison of the coordinates between the ideal and the predicted star spots in the 1500th star image.

Star Number	Ideal Star Spot Coordinate		Predicted Star Spot Coordinate		Predicted Star Spot Coordinate Error	
	x	y	x'	y'	Δx	Δy
1	24	63.50	23.25	63.62	0.75	−0.12
2	46.5	19	45.62	19.25	0.87	−0.25
3	54.50	371.50	54.23	371.76	0.26	−0.26
4	277	336.50	276.60	336.60	0.39	−0.10
5	290.50	305	290.11	305.27	0.38	−0.27
6	336.50	502.71	336.50	502.71	0	0
7	386.50	340	386	339.64	0.50	0.35
8	400.50	170	400	169.78	0.50	0.21
9	420.71	409.50	420.50	409.00	0.21	0.50
10	425.72	225.88	425.40	225.60	0.32	0.28
11	431.64	351	431.50	350.71	0.14	0.28
12	455	130.50	454.23	130.23	0.76	0.26
13	486.40	131.60	485.50	131.50	0.89	0.09

To further analyze the prediction algorithm, according to the first frame star image shown in Figure 10a, we successively predicted the position of stars in 1500 star images, and analyze the mean value of the estimation error of the star spot position in each predicted star image. As shown in Figure 12, the mean value of the coordinate errors of the predicted star spot increases with the increase of the estimated number of frames, but the mean errors could stay in a small range. Therefore, in the case of the short-term frame loss, the proposed method can achieve an accurate prediction of the star spot.

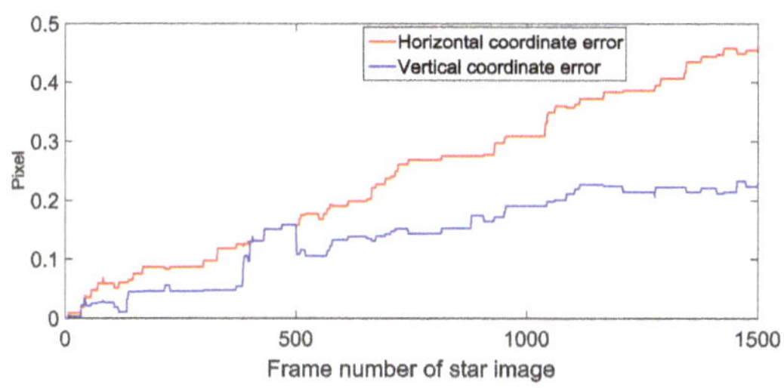

Figure 12. Predicted star position error.

4.2. Experiments on Star Image Deblurring

In this section, we present some examples to validate the proposed gyro-assisted improved RL algorithm. First, we analyze the blurring of the star image when the star sensor rotates around the X-axis, the Y-axis, the Z-axis, the X- and Y-axis, and the three axes simultaneously. Then, we add the Gaussian white noise with zero mean and variance 0.01 to the blurred star images. Finally, the blurred star image is deblurred by our proposed algorithm, and we compare the deblurred star image with the original sharp star image. Figure 13 shows the magnified original star image, blurred star images caused by the star sensor rotate around the X- and Y-axis, ($w_x = 10°/s$, $w_y = 10°/s$), deblurred star image, and the gray distribution of the star spot in them. As can be seen from Figure 13, the gray value of the star spot in the blurred star image decreases significantly, and after deblurring the star image, the smearing phenomenon is obviously suppressed, the gray value and the gray distribution of star spot are closer to those in original star image.

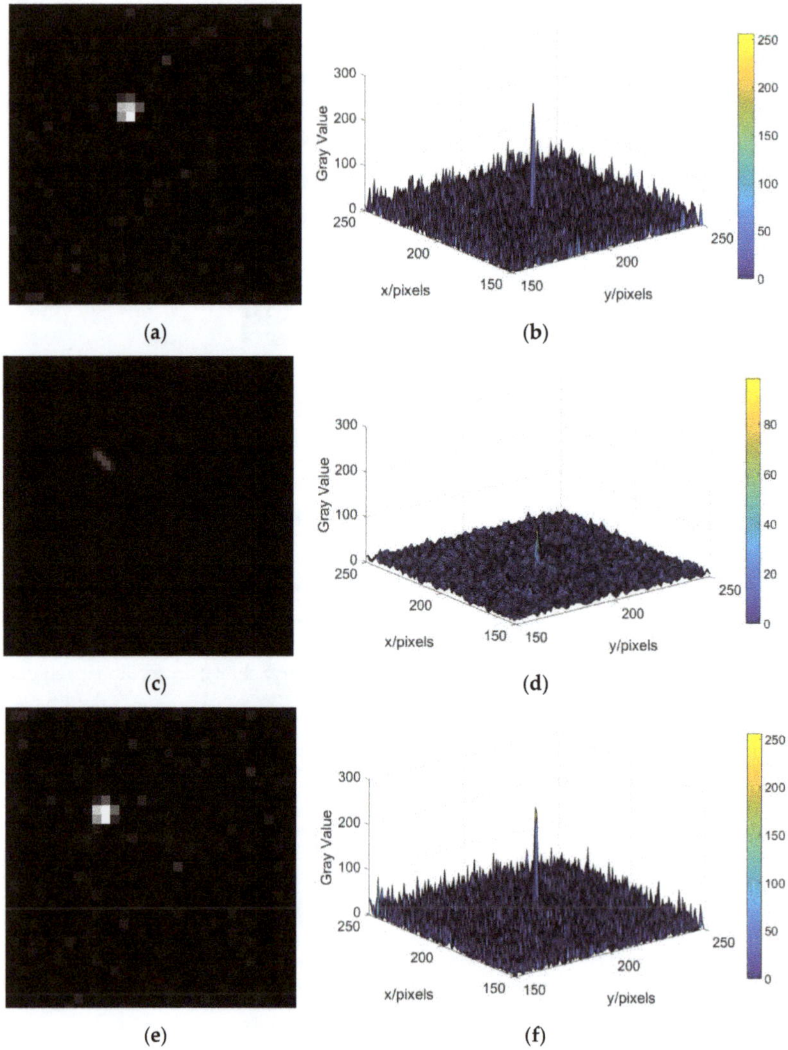

Figure 13. The magnified star image and the gray distribution of the star spot in the case of Gaussian white noise. (**a**) The magnified original star image; (**b**) gray value distribution of star spot in the original star image; (**c**) the magnified blurred star image ($w_x = w_y = 10^\circ/s$); (**d**) gray value distribution of star spot in the blurred star image ($w_x = w_y = 10^\circ/s$); (**e**) the magnified deblurred star image ($w_x = w_y = 10^\circ/s$); (**f**) gray value distribution of star spot in the deblurred star image ($w_x = w_y = 10^\circ/s$).

The star sensor is an attitude measurement device. To more intuitively reflect the deblurring performance of the proposed algorithm, we compare the attitude information of the spacecraft estimated by the star image before and after deblurring. The star image observed by the star sensor at a certain time is shown in Figure 14. First, we perform an angular motion blurring on the observed star image, then we use the proposed algorithm and the automatic iterative RL algorithm to deblur the star image, and compare the attitude information estimated by the deblurred images. The automatic iterative RL algorithm calculates the mean square error (MSE) of the currently restored image by automatically increasing the number of iterations, and compares it with the MSE of the image restored by the last iteration. If the MSE of the currently restored image is higher than the last iteration recovery result, the last iteration number is considered to be the optimal number of iterations, and the restored image is the optimal restoration result. The attitude estimation results are shown in Tables 2–6, and the "Fail" indicates that the attitude information of the spacecraft cannot be estimated by the star image because the degree of blurring of the star image is too high.

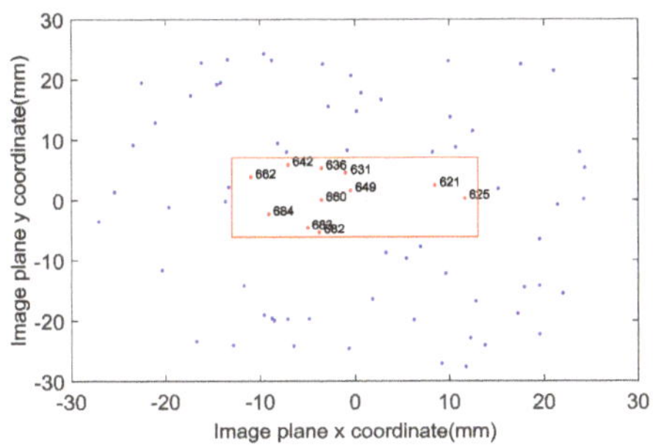

Figure 14. Star spots observed by a star sensor.

From Tables 2–6, it can be seen that the attitude estimation failed when the angular velocity of the star sensor rotating around the X-axis, the Y-axis, the Z-axis, the X-and the Y-axis, and the three axes exceeds $w_x = 25°/s$, $w_y = 30°/s$, $w_z = 25°/s$, $w = [20, 20, 0]°/s$ and $w = [15, 15, 15]°/s$, respectively. After the blurred star image is restored by the proposed algorithm and the automatic iterative RL algorithm, the maximum angular velocity of the attitude can be estimated to be expanded to $w_x = 75°/s$, $w_y = 75°/s$, $w_z = 80°/s$, $w = [75, 75, 0]°/s$, and $w = [55, 55, 55]°/s$, respectively, these two methods have a similar performance, and the attitude errors are kept in a small range. This is because with the increase of the angular velocity of the star sensor, the blur extent of the star image gets bigger, and the gray value of the star spot decreases significantly. When the gray value of a blurred star is lower than the threshold for star image segmentation and the blurred star can hardly be detected. However, after the restoration of the blurred star image, the gray value of the star spot is improved, and the gray distribution of the star spot is closer to the true distribution so that the star spot can also be extracted under high dynamic conditions. Finally, the attitude of the spacecraft can be estimated by these star spots.

Table 2. Comparison of attitude estimation in the case of Gaussian white noise (Vary w_x).

w_x (deg/s)	Attitude Errors (Blurred Star Image) (arc-second)			Attitude Errors (Restored Star Image by our Method) (arc-second)			Attitude Errors (Restored Star Image by Iterative RL Algorithm) (arc-second)		
	Pitch	Yaw	Roll	Pitch	Yaw	Roll	Pitch	Yaw	Roll
1	14.61	14.59	10.28	14.61	14.59	10.28	14.61	14.59	10.28
5	14.63	13.02	2.33	14.61	14.59	10.28	14.61	14.59	10.28
10	141.49	159.97	17.98	32.05	17.39	5.26	58.82	22.07	4.94
15	26.76	4.71	7.07	24.77	11.80	5.09	14.61	14.59	10.28
20	181.60	136.94	5.69	24.77	11.80	5.09	24.04	19.74	0.04
25	Fail	Fail	Fail	31.08	67.23	5.44	48.33	69.82	0.37
35	Fail	Fail	Fail	43.03	6.38	5.03	14.61	14.59	10.28
45	Fail	Fail	Fail	14.61	14.59	10.28	14.61	14.59	10.28
55	Fail	Fail	Fail	14.61	14.59	10.28	31.08	67.23	5.44
65	Fail	Fail	Fail	64.17	34.57	10.08	75.65	45.95	4.80
75	Fail	Fail	Fail	Fail	Fail	Fail	Fail	Fail	Fail

Table 3. Comparison of attitude estimation in the case of Gaussian white noise (Vary w_y).

w_y (deg/s)	Attitude Errors (Blurred Star Image) (arc-second)			Attitude Errors (Restored Star Image by Our Method) (arc-second)			Attitude errors (Restored Star Image by Iterative RL Algorithm) (arc-second)		
	Pitch	Yaw	Roll	Pitch	Yaw	Roll	Pitch	Yaw	Roll
1	31.08	67.23	5.44	31.08	67.23	5.44	31.08	67.23	5.44
5	115.63	61.76	52.20	43.03	6.38	5.03	43.03	6.38	5.03
10	11.50	76.89	2.08	14.61	14.59	10.28	14.61	14.59	10.28
15	11.67	61.47	6.26	14.61	14.59	10.28	24.77	11.80	5.09
20	154.24	149.17	10.11	136.04	113.32	0.70	82.09	103.43	0.50
25	104.74	147.65	0.63	14.61	14.59	10.28	43.60	79.70	20.88
30	Fail	Fail	Fail	220.53	211.73	8.91	56.75	92.69	9.77
40	Fail	Fail	Fail	80.44	65.31	4.72	81.32	67.23	5.44
50	Fail	Fail	Fail	49.10	34.30	4.87	53.03	33.80	5.03
60	Fail	Fail	Fail	131.35	108.60	30.05	134.43	105.09	30.27
70	Fail	Fail	Fail	58.07	57.61	0.29	80.44	65.31	4.72
75	Fail	Fail	Fail	Fail	Fail	Fail	Fail	Fail	Fail

Table 4. Comparison of attitude estimation in the case of Gaussian white noise (Vary w_z).

w_z (deg/s)	Attitude Errors (Blurred Star Image) (arc-second)			Attitude Errors (Restored Star Image by Our Method) (arc-second)			Attitude Errors (Restored Star Image by Iterative RL Algorithm) (arc-second)		
	Pitch	Yaw	Roll	Pitch	Yaw	Roll	Pitch	Yaw	Roll
1	14.61	14.59	10.28	14.61	14.59	10.28	14.61	14.59	10.28
5	103.24	73.51	12.20	4.38	40.73	5.28	14.61	14.59	10.28
10	60.43	55.49	25.06	14.61	14.59	10.28	14.61	14.59	10.28
15	157.89	162.40	15.28	14.61	14.59	10.28	14.61	14.59	10.28
20	84.80	136.06	3.85	14.61	14.59	10.28	14.61	14.59	10.28
25	Fail	Fail	Fail	14.61	14.59	10.28	14.61	14.59	10.28
35	Fail	Fail	Fail	14.61	14.59	10.28	34.94	19.01	19.09
45	Fail	Fail	Fail	30.03	66.27	20.87	14.61	14.59	10.28
55	Fail	Fail	Fail	14.61	14.59	10.28	91.86	76.68	4.65
65	Fail	Fail	Fail	75.67	96.89	0.44	112.50	111.72	16.04
75	Fail	Fail	Fail	92.57	106.52	5.68	170.85	140.59	4.22
80	Fail	Fail	Fail	Fail	Fail	Fail	Fail	Fail	Fail

Table 5. Comparison of attitude estimation in the case of Gaussian white noise (Vary w_x and w_y).

Angular Velocity (deg/s)		Attitude Errors (Blurred Star Image) (arc-second)			Attitude Errors (Restored Star Image by Our Method) (arc-second)			Attitude Errors (Restored Star Image by Iterative RL Algorithm) (arc-second)		
w_x	w_y	Pitch	Yaw	Roll	Pitch	Yaw	Roll	Pitch	Yaw	Roll
1	1	31.30	28.17	16.23	14.61	14.59	10.28	14.61	14.59	10.28
5	5	203.95	191.12	22.83	13.14	30.54	0.11	14.61	14.59	10.28
10	10	29.66	18.84	4.72	14.61	14.59	10.28	14.61	14.59	10.28
15	15	335.46	369.20	31.02	14.61	14.59	10.28	14.61	14.59	10.28
20	20	Fail	Fail	Fail	14.61	14.59	10.28	14.61	14.59	10.28
30	30	Fail	Fail	Fail	58.82	22.07	4.94	56.75	92.69	9.77
40	40	Fail	Fail	Fail	14.61	14.59	10.28	4.68	39.04	15.47
50	50	Fail	Fail	Fail	49.10	34.30	4.87	41.74	9.47	10.16
60	60	Fail	Fail	Fail	14.61	14.59	10.28	31.08	67.23	5.44
70	70	Fail	Fail	Fail	126.33	125.57	0.77	120.24	97.62	0.61
75	75	Fail	Fail	Fail	Fail	Fail	Fail	Fail	Fail	Fail

Table 6. Comparison of attitude estimation in the case of Gaussian white noise (Vary w_x, w_y and w_z).

Angular Velocity (deg/s)			Attitude Errors (Blurred Star Image) (arc-second)			Attitude Errors (Restored Star Image by Our Method) (arc-second)			Attitude Errors (Restored Star Image by Iterative RL Algorithm) (arc-second)		
w_x	w_y	w_z	Pitch	Yaw	Roll	Pitch	Yaw	Roll	Pitch	Yaw	Roll
1	1	1	66.08	27.77	16.17	14.61	14.59	10.28	14.61	14.59	10.28
5	5	5	56.02	46.97	37.18	14.61	14.59	10.28	14..61	14.59	10.28
10	10	10	101.42	84.97	12.21	58.82	22.07	4.94	31.08	67.23	5.44
15	15	15	Fail	Fail	Fail	5.46	41.82	20.65	14.61	14.59	10.28
25	25	25	Fail	Fail	Fail	83.89	119.70	9.66	42.49	49.42	25.86
35	35	35	Fail	Fail	Fail	80.51	43.55	20.27	176.13	2014.12	29.58
45	45	45	Fail	Fail	Fail	148.94	104.19	29.98	132.07	131.26	30.66
55	55	55	Fail	Fail	Fail	Fail	Fail	Fail	Fail	Fail	Fail

To verify the real-time performance of the proposed algorithm in the case of Gaussian noise, we use the proposed algorithm and the automatic iterative RL algorithm to restore the blurred star image caused by the star sensor rotating around the Z-axis and compare the time consumed by the two methods. As shown in Figure 15, the real-time performance of the proposed algorithm is significantly better than the iterative RL algorithm. This is mainly because the proposed algorithm can use the ensemble neural network based on the improved bagging method to quickly predict the number of iterations required by the RL algorithm, while the iterative RL algorithm requires a step-by-step iteration to optimize the number of iteration steps required.

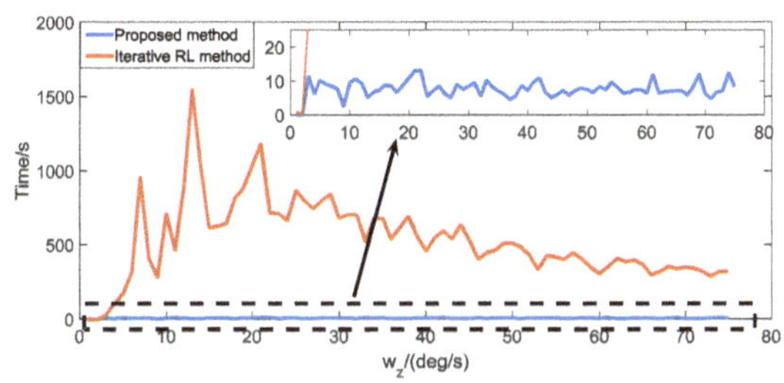

Figure 15. Comparison of running time between the proposed method and the iterative RL method in the case of Gaussian noise.

Second, in the case where the blurred star image is contaminated by Poisson noise, we present the deblurring performance of the proposed method and compare it with the automatic iterative RL algorithm. Figure 16 shows the magnified original star image, blurred star images caused by star sensor rotate around the X- and Y-axis, ($w_x = w_y = 10°/s$), deblurred star image, and the gray distribution of the star spot in the case of Poisson noise. Combined with Tables 7–11, we can see that in the case of Poisson noise, the attitude estimation failed when the angular velocity of the star sensor rotating around the X-axis, the Y-axis, the Z-axis, the X-and the Y-axis, and the three axes exceeds $w_x = 40°/s$, $w_y = 35°/s$, $w_z = 35°/s$, $w = [30, 30, 0]°/s$ and $w = [15, 15, 15]°/s$, respectively. After the blurred star image is restored by the proposed algorithm and the automatic iterative RL algorithm, the maximum angular velocity of the attitude can be estimated to be expanded to $w_x = 160°/s$, $w_y = 160°/s$, $w_z = 170°/s$, $w = [120, 120, 0]°/s$, and $w = [80, 80, 80]°/s$, respectively, these two methods have a similar performance, and the attitude errors are kept in a small range. Figure 17 shows the real-time performance of the proposed algorithm and the iterative RL algorithm when dealing with the blurred star image caused by the star sensor rotating around the Z-axis, and the result shows that the real-time performance of our algorithm is better than the iterative RL algorithm when the degree of the blurred star image is large.

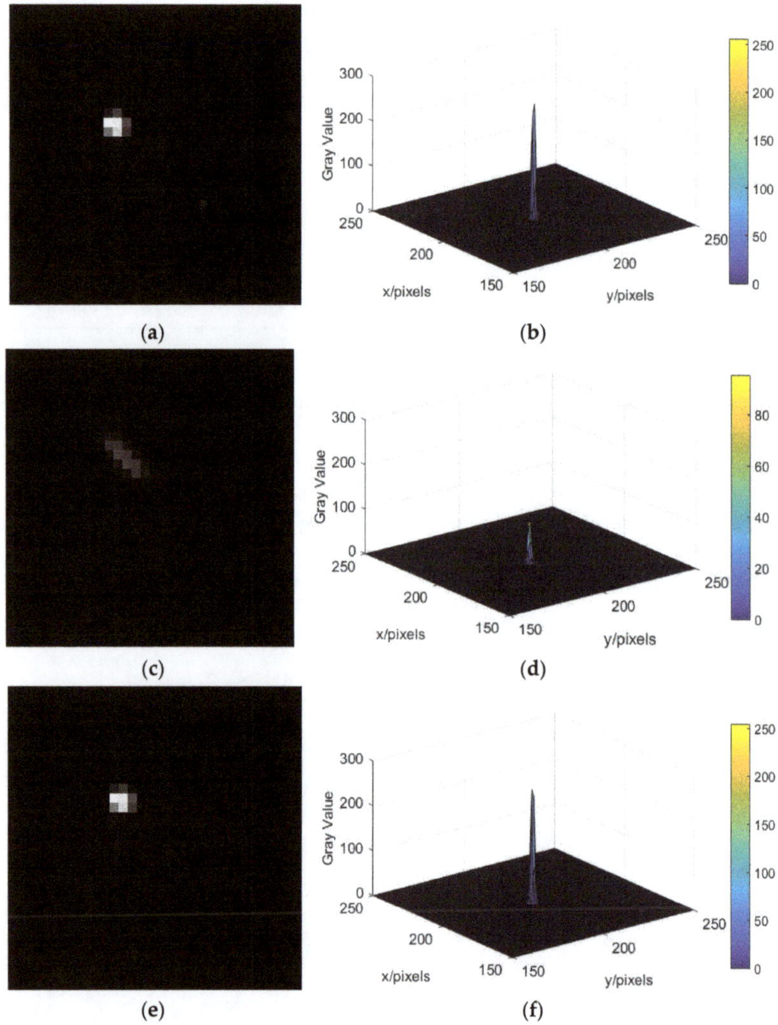

Figure 16. The magnified star image and the gray level distribution of the star spot in the case of Poisson noise. (**a**) The magnified original star image; (**b**) gray level distribution of star spot in the original star image; (**c**) the magnified blurred star image ($w_x = w_y = 10°/s$); (**d**) gray level distribution of star spot in the blurred star image ($w_x = w_y = 10°/s$); (**e**) the magnified deblurred star image ($w_x = w_y = 10°/s$); (**f**) gray level distribution of star spot in the deblurred star image ($w_x = w_y = 10°/s$).

In summary, the proposed method and the iterative RL algorithm significantly improve the dynamic performance of the star sensor and have similar performance. However, the real-time performance of our algorithm is better than the iterative RL algorithm, especially in the case of Gaussian white noise.

Table 7. Comparison of attitude estimation in the case of Poisson noise (Vary w_x).

w_x (deg/s)	Attitude Errors (Blurred Star Image) (arc-second)			Attitude Errors (Restored Star Image by Our Method) (arc-second)			Attitude Errors (Restored Star Image by Iterative RL Algorithm) (arc-second)		
	Pitch	Yaw	Roll	Pitch	Yaw	Roll	Pitch	Yaw	Roll
1	14.61	14.59	10.28	14.61	14.59	10.28	14.61	14.59	10.28
20	32.73	32.90	15.31	14.61	14.59	10.28	14.61	14.59	10.28
35	47.34	86.91	27.38	14.61	14.59	10.28	15.05	16.10	14.61
40	Fail	Fail	Fail	14.61	14.59	10.28	14.61	14.59	10.28
55	Fail	Fail	Fail	108.82	137.30	5.82	65.28	94.19	26.05
70	Fail	Fail	Fail	81.34	73.45	5.61	2.64	12.00	20.58
85	Fail	Fail	Fail	5.77	37.94	0.10	33.45	76.93	15.70
100	Fail	Fail	Fail	14.61	14.59	10.28	24.77	11.80	5.09
115	Fail	Fail	Fail	32.61	11.32	30.77	58.32	14.25	15.18
130	Fail	Fail	Fail	77.30	40.39	20.17	155.73	162.04	11.18
155	Fail	Fail	Fail	43.03	6.38	5.03	115.90	122.36	10.87
160	Fail	Fail	Fail	Fail	Fail	Fail	Fail	Fail	Fail

Table 8. Comparison of attitude estimation in the case of Poisson noise (Vary w_y).

w_y (deg/s)	Attitude Errors (Blurred Star Image) (arc-second)			Attitude Errors (Restored Star Image by Our Method) (arc-second)			Attitude Errors (Restored Star Image by Iterative RL Algorithm) (arc-second)		
	Pitch	Yaw	Roll	Pitch	Yaw	Roll	Pitch	Yaw	Roll
1	14.61	14.59	10.28	14.61	14.49	10.28	14.61	14.59	10.28
15	79.44	29.91	24.85	14.61	14.49	10.28	48.33	69.82	0.37
30	142.86	135.25	5.44	31.08	67.23	4.87	14.61	14.59	10.28
35	Fail	Fail	Fail	14.61	14.59	10.28	14.61	14.59	10.28
60	Fail	Fail	Fail	29.23	21.90	20.51	89.29	132.34	16.03
75	Fail	Fail	Fail	31.08	67.23	5.44	35.21	85.90	26.01
90	Fail	Fail	Fail	14.61	14.59	10.28	58.07	57.61	0.29
105	Fail	Fail	Fail	134.41	104.39	24.90	154.70	146.49	6.03
120	Fail	Fail	Fail	136.04	113.32	0.70	82.09	103.43	0.50
135	Fail	Fail	Fail	102.04	79.48	0.54	127.91	105.30	14.66
155	Fail	Fail	Fail	119.10	74.69	0.48	137.62	93.05	14.73
160	Fail	Fail	Fail	Fail	Fail	Fail	Fail	Fail	Fail

Table 9. Comparison of attitude estimation in the case of Poisson noise (Vary w_z).

w_z (deg/s)	Attitude Errors (Blurred Star Image) (arc-second)			Attitude Errors (Restored Star Image by Our Method) (arc-second)			Attitude Errors (Restored Star Image by Iterative RL Algorithm) (arc-second)		
	Pitch	Yaw	Roll	Pitch	Yaw	Roll	Pitch	Yaw	Roll
1	14.61	14.59	10.28	14.61	14.59	10.28	14.61	14.59	10.28
15	124.07	82.00	14.15	14.61	14.59	10.28	15.01	14.94	0.09
30	171.49	129.07	26.77	14.61	14.59	10.28	14.61	14.59	10.28
35	Fail	Fail	Fail	14.61	14.59	10.28	14.61	14.59	10.28
45	Fail	Fail	Fail	14.61	14.59	10.28	4.38	40.73	5.28
60	Fail	Fail	Fail	14.61	14.59	10.28	49.10	34.30	4.87
75	Fail	Fail	Fail	14.52	58.08	20.74	28.99	7.05	15.25
90	Fail	Fail	Fail	101.47	129.93	4.50	115.88	64.16	9.72
105	Fail	Fail	Fail	22.90	22.86	25.87	88.02	109.29	15.91
120	Fail	Fail	Fail	31.08	67.23	5.44	33.47	54.99	0.21
150	Fail	Fail	Fail	15.34	15.26	0.09	14.61	14.59	10.28
165	Fail	Fail	Fail	87.48	43.19	14.98	75.17	74.57	9.85
170	Fail	Fail	Fail	Fail	Fail	Fail	Fail	Fail	Fail

Table 10. Comparison of attitude estimation in the case of Poisson noise (Vary w_x and w_y).

Angular Velocity (deg/s)		Attitude Errors (Blurred Star Image) (arc-second)			Attitude Errors (Restored Star Image by Our Method) (arc-second)			Attitude Errors (Restored Star Image by Iterative RL Algorithm) (arc-second)		
w_x	w_y	Pitch	Yaw	Roll	Pitch	Yaw	Roll	Pitch	Yaw	Roll
1	1	40.61	18.59	22.13	14.61	14.59	10.28	14.61	14.59	10.28
15	15	51.70	33.10	21.44	14.61	14.59	10.28	14.61	14.59	10.28
25	25	372.32	260.78	13.32	14.61	14.59	10.28	49.10	34.30	4.87
30	30	Fail	Fail	Fail	14.61	14.59	10.28	43.08	6.38	5.03
45	45	Fail	Fail	Fail	14.71	14.46	10.27	14.71	14.46	10.27
60	60	Fail	Fail	Fail	43.60	79.70	20.88	43.03	6.38	5.03
75	75	Fail	Fail	Fail	80.44	65.31	4.72	75.75	118.93	16.02
90	90	Fail	Fail	Fail	52.75	59.83	25.98	58.82	22.07	4.94
105	105	Fail	Fail	Fail	58.82	22.07	4.94	60.44	25.31	4.72
115	115	Fail	Fail	Fail	138.30	159.19	0.82	138.30	159.19	0.82
120	120	Fail	Fail	Fail	Fail	Fail	Fail	Fail	Fail	Fail

Table 11. Comparison of attitude estimation in the case of Poisson noise (Vary w_x, w_y and w_z).

Angular Velocity (deg/s)			Attitude Errors (Blurred Star Image) (arc-second)			Attitude Errors (Restored Star Image by Our Method) (arc-second)			Attitude Errors (Restored Star Image by Iterative RL Algorithm) (arc-second)		
w_x	w_y	w_z	Pitch	Yaw	Roll	Pitch	Yaw	Roll	Pitch	Yaw	Roll.
1	1	1	59.74	47.73	20.42	14.61	14.59	10.28	14.61	14.59	10.28
5	5	5	67.70	46.17	16.68	14.61	14.59	10.28	14..61	14.59	10.28
10	10	10	88.61	125.45	16.57	14.61	14.59	10.28	14.61	14.59	10.28
15	15	15	Fail	Fail	Fail	14.61	14.59	10.28	80.44	65.31	4.72
25	25	25	Fail	Fail	Fail	17.43	10.02	5.16	40.75	11.52	15.49
35	35	35	Fail	Fail	Fail	53.52	24.12	25.38	53.52	24.12	25.38
45	45	45	Fail	Fail	Fail	93.01	128.85	36.48	150.77	171.60	0.99
55	55	55	Fail	Fail	Fail	87.19	79.21	0.45	97.60	67.91	30.26
65	65	65	Fail	Fail	Fail	42.59	42.40	0.16	47.73	83.74	10.54
75	75	75	Fail	Fail	Fail	42.34	71.39	41.44	42.34	71.39	41.44
80	80	80	Fail	Fail	Fail	Fail	Fail	Fail	Fail	Fail	Fail

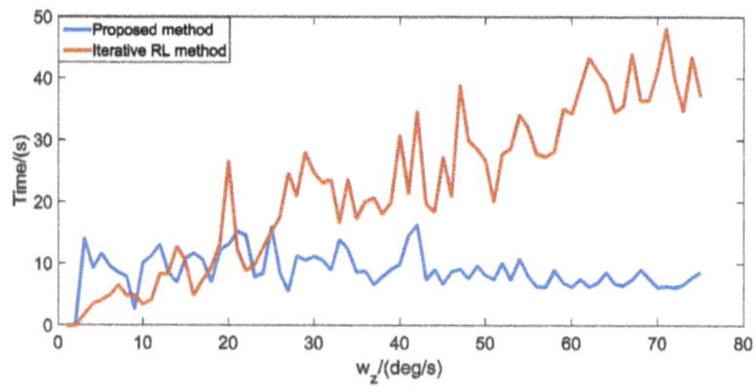

Figure 17. Comparison of running time between the proposed method and the iterative RL method in the case of Poisson noise.

5. Conclusions

In this paper, we improve the dynamic performance of the star sensor by using the star image prediction method and the star image deblurring method. Taking into account the distortion of the star sensor lens, we use the information provided by the star sensor and the gyroscope to establish a star spot prediction model. Also, for the blurred star image problem, we proposed an improved Richardson-Lucy (RL) algorithm based on the EBPNN.

Experimental results demonstrate that the proposed methods are effective in improving the dynamic performance of the star sensor. The maximum error of the star image prediction algorithm is 0.89 pixels in 15 s and the attitude errors calculated from the star image restored by the improved RL algorithm can be kept in a small range. Compared with the iterative RL algorithm, the improved RL algorithm proposed in this paper has better real-time performance.

Author Contributions: X.C. designed and conceived this study; D.L. and C.S. performed the experiments and wrote the paper; D.L. and X.L. developed the program used in the experiment; X.C. reviewed and edited the manuscript. All authors read and approved this manuscript.

Funding: This work was supported by the National Natural Science Foundation of China (No. 61873064, 51375087), Transformation Program of Science and Technology Achievements of Jiangsu Province (No. BA2016139), Postgraduate Research & Practice Innovation Program of Jiangsu Province (No. KYCX18_0073) and Scientific Research Foundation of Graduate School of Southeast University (No. YBPY1931).

Conflicts of Interest: The authors declare no conflict of interest.

References

1. Inamori, T.; Hosonuma, T.; Ikari, S.; Saisutjarit, P.; Sako, N.; Nakasuka, S. Precise attitude rate estimation using star images obtained by mission telescope for satellite missions. *Adv. Space Res.* **2015**, *55*, 1199–1210. [CrossRef]
2. Zhang, S.; Xing, F.; Sun, T.; You, Z.; Wei, M. Novel approach to improve the attitude update rate of a star tracker. *Opt. Express* **2018**, *26*, 5164–5181. [CrossRef]
3. Sun, T.; Xing, F.; You, Z.; Wang, X.; Li, B. Deep coupling of star tracker and MEMS-gyro data under highly dynamic and long exposure conditions. *Meas. Sci. Technol.* **2014**, *25*, 085003. [CrossRef]
4. Lu, J.; Lei, C.; Yang, Y. A dynamic precision evaluation method for the star sensor in the stellar-inertial navigation system. *Sci. Rep.* **2017**, *7*, 4356. [CrossRef]
5. Tan, W.; Dai, D.; Wu, W.; Wang, X.; Qin, S. A comprehensive calibration method for a star tracker and gyroscope units integrated System. *Sensors* **2018**, *18*, 3106. [CrossRef] [PubMed]
6. Ma, L.; Zhan, D.; Jiang, G.; Fu, S.; Jia, H.; Wang, X.; Huang, Z.; Zheng, J.; Hu, F.; Wu, W.; et al. Attitude-correlated frames approach for a star sensor to improve attitude accuracy under highly dynamic conditions. *Appl. Opt.* **2015**, *54*, 7559–7566. [CrossRef]
7. Yan, J.; Jiang, J.; Zhang, G. Dynamic imaging model and parameter optimization for a star tracker. *Opt. Express* **2016**, *24*, 5961–5983. [CrossRef]
8. Gao, Y.; Qin, S.; Jiang, G.; Zhou, J. Dynamic smearing compensation method for star centroding of star sensors. In Proceedings of the IEEE Conference on Metrology for Aerospace, Florence, Italy, 21–23 June 2016; pp. 221–226.
9. Jiang, J.; Yu, W.; Zhang, G. High-accuracy decoupling estimation of the systematic coordinate errors of an INS and intensified high dynamic star tracker based on the constrained least squares method. *Sensors* **2017**, *17*, 2285. [CrossRef]
10. Sharma, V.K.; Mahapatra, K.K. Visual object tracking based on sequential learning of SVM parameter. *Digit. Signal Process.* **2018**, *79*, 102–115. [CrossRef]
11. Shi, R.; Wu, G.; Kang, W.; Wang, Z.; Feng, D. Visual tracking utilizing robust complementary learner and adaptive refiner. *Neurocomputing* **2017**, *260*, 367–377. [CrossRef]
12. Danelljan, M.; Hager, G.; Khan, F.S.; Felsberg, M. Learning spatially regularized correlation filters for visual tracking. In Proceedings of the IEEE Conference on Computer Vision, Santiago, Chile, 11–18 December 2015; pp. 4310–4318.
13. Fan, Z.; Ji, H.; Zhang, Y. Iterative particle filter for visual tracking. *Signal Process. Image Commun.* **2015**, *36*, 140–153. [CrossRef]
14. Jin, J.; Yan, L. Adaptive image tracking algorithm based on improved particle filter and sparse representation. *J. Comput. Appl. Softw.* **2014**, *31*, 152–155.
15. Wu, L. Research on method of eliminating frame error in moving image tracking. *Comput. Simul.* **2016**, 198–201.
16. Zhong, H.; Yang, M.; Lu, X. Increasing update rate for star sensor by pipelining parallel processing method. *Opt. Precis. Eng.* **2009**, *17*, 2230–2235.

17. Mao, X.; Liang, W.; Zheng, X. A parallel computing architecture based image processing algorithm for star sensor. *J. Astronaut.* **2011**, *32*, 613–619.
18. Zhou, Q.; Mao, X.; Zhang, Q. A image processing algorithm with marker for high-speed and multi-channel star sensor. *J. Harbin Inst. Technol.* **2016**, *48*, 119–124.
19. Katake, A.B. Modeling, Image Processing and Attitude Estimation of High Speed Star Sensors. Ph.D. Thesis, Texas A&M University, College Station, TX, USA, August 2006.
20. Katake, A. StarCam SG100: A high-update rate, high-sensitivity stellar gyroscope for spacecraft. In Proceedings of the Conference on Sensors, Cameras, and Systems for Industrial/Scientific Applications XI, San Jose, CA, USA, 19–21 January 2010; pp. 1–10.
21. Wang, X.; Wei, X.; Fan, Q.; Li, J.; Wang, G. Hardware implementation of fast and robust star centroid extraction with low resource cost. *IEEE Sens. J.* **2015**, *15*, 4857–4865. [CrossRef]
22. Yu, W.; Jiang, J.; Zhang, G. Star tracking method based on multiexposure imaging for intensified star trackers. *Appl. Opt.* **2017**, *56*, 5961–5971. [CrossRef]
23. Wang, S.; Zhang, S.; Ning, M.; Zhou, B. Motion blurred star image restoration based on MEMS gyroscope aid and blur kernel correction. *Sensors* **2018**, *18*, 2662. [CrossRef]
24. Zhu, H.; Deng, L.; Bai, X.; Li, M.; Cheng, Z. Deconvolution methods based on φ HL regularization for spectral recovery. *Appl. Opt.* **2015**, *54*, 4337–4344. [CrossRef] [PubMed]
25. Liu, G.; Chang, S.; Ma, Y. Blind image deblurring using spectral properties of convolution operators. *IEEE Trans. Image Process.* **2014**, *23*, 5047–5056. [CrossRef]
26. Ren, W.; Cao, X.; Pan, J.; Guo, X.; Zuo, W.; Yang, M. Image deblurring via enhanced low-rank prior. *IEEE Trans. Image Process.* **2016**, *25*, 3426–3437. [CrossRef] [PubMed]
27. Ma, L.; Zeng, T. Image deblurring via total variation based structured sparse model selection. *J. Sci. Comput.* **2016**, *67*, 1–19. [CrossRef]
28. Lu, Q.; Zhou, W.; Fang, L.; Li, H. Robust blur kernel estimation for license plate images from fast moving vehicles. *IEEE Trans. Image Process.* **2016**, *25*, 2311–2323. [CrossRef] [PubMed]
29. Chen, S.; Shen, H. Multispectral image out-of-focus deblurring using interchannel correlation. *IEEE Trans. Image Process.* **2015**, *24*, 4433–4445. [CrossRef] [PubMed]
30. Xue, F.; Blu, T. A novel SURE-based criterion for parametric PSF estimation. *IEEE Trans. Image Process.* **2015**, *24*, 595–607. [CrossRef] [PubMed]
31. Quan, W.; Zhang, W. Restoration of motion-blurred star image based on Wiener filter. In Proceedings of the IEEE Conference on Intelligent Computation Technology and Automation, Shenzhen, China, 28–29 March 2011; pp. 691–694.
32. Ma, X.; Xia, X.; Zhang, Z.; Wang, G.; Qian, H. Star image processing of SINS/CNS integrated navigation system based on 1DWF under high dynamic conditions. In Proceedings of the IEEE Conference on Position, Location and Navigation Symposium, Savannah, GA, USA, 28–29 March 2011; pp. 514–518.
33. Jiang, J.; Huang, J.; Zhang, G. An accelerated motion blurred star restoration based on single image. *IEEE Sens. J.* **2017**, *17*, 1306–1315. [CrossRef]
34. Ma, L.; Bernelli-Zazzera, F.; Jiang, G.; Wang, X.; Huang, Z.; Qin, S. Region-confined restoration method for motion-blurred star image of the star sensor under dynamic conditions. *Appl. Opt.* **2016**, *55*, 4621–4631. [CrossRef]
35. Anderson, E.H.; Fumo, J.P.; Erwin, R.S. Satellite ultraquiet isolation technology experiment (SUITE). In Proceedings of the IEEE Conference on Aerospace, Big Sky, MT, USA, 25 March 2000; pp. 299–313.
36. Rad, A.M.; Nobari, J.H.; Nikkhah, A.A. Optimal attitude and position determination by integration of INS, star tracker, and horizon sensor. *IEEE Aerosp. Electron. Syst. Mag.* **2014**, *29*, 20–33. [CrossRef]
37. Moghaddam, M.E.; Jamzad, M. Blur identification in noisy images using radon transform and power spectrum modeling. In Proceedings of the 12th IEEE International Workshop on Systems, Signals and Image Processing, Chalkida, Greece, 22–24 September 2005; pp. 347–352.
38. Aizenberg, I.; Paliy, D.V.; Zurada, J.M. Blur identification by multilayer neural network based on multivalued neurons. *IEEE Trans. Neural Netw.* **2008**, *19*, 883–898. [CrossRef]
39. Liu, C.; Hu, L.; Liu, G.; Yang, B.; Li, A. Kinematic model for the space-variant image motion of star sensors under dynamical conditions. *Opt. Eng.* **2015**, *54*, 063104. [CrossRef]

40. Yang, H.; Huang, H.; Lai, S. A novel gradient attenuation Richardson–Lucy algorithm for image motion deblurring. *Signal Process.* **2014**, *103*, 399–414. [CrossRef]
41. Tao, H.; Lu, X. Smoky vehicle detection based on multi-feature fusion and ensemble neural networks. *Multimed. Tools Appl.* **2018**, *77*, 32153–32177. [CrossRef]
42. Breiman, L. Bagging predictors. *Mach. Learn.* **1996**, *24*, 123–140. [CrossRef]
43. Wu, X.; Wang, X. Multiple blur of star image and the restoration under dynamic conditions. *Acta Astronaut.* **2011**, *68*, 1903–1913.

sensors

MDPI

Article

An Optimized Tightly-Coupled VIO Design on the Basis of the Fused Point and Line Features for Patrol Robot Navigation

Linlin Xia [1,*], Qingyu Meng [1], Deru Chi [1], Bo Meng [2] and Hanrui Yang [1]

1 School of Automation Engineering, Northeast Electric Power University, Jilin 132012, China; 13846877678@163.com (Q.M.); ANATKH237@163.com (D.C.); yanghanrui1208@163.com (H.Y.)
2 School of Computer Science, Northeast Electric Power University, Jilin 132012, China; mengbo_nannan@163.com
* Correspondence: xiall521@neepu.edu.cn; Tel.: +86-432-6480-6483

Received: 29 January 2019; Accepted: 28 April 2019; Published: 29 April 2019

Abstract: The development and maturation of simultaneous localization and mapping (SLAM) in robotics opens the door to the application of a visual inertial odometry (VIO) to the robot navigation system. For a patrol robot with no available Global Positioning System (GPS) support, the embedded VIO components, which are generally composed of an Inertial Measurement Unit (IMU) and a camera, fuse the inertial recursion with SLAM calculation tasks, and enable the robot to estimate its location within a map. The highlights of the optimized VIO design lie in the simplified VIO initialization strategy as well as the fused point and line feature-matching based method for efficient pose estimates in the front-end. With a tightly-coupled VIO anatomy, the system state is explicitly expressed in a vector and further estimated by the state estimator. The consequent problems associated with the data association, state optimization, sliding window and timestamp alignment in the back-end are discussed in detail. The dataset tests and real substation scene tests are conducted, and the experimental results indicate that the proposed VIO can realize the accurate pose estimation with a favorable initializing efficiency and eminent map representations as expected in concerned environments. The proposed VIO design can therefore be recognized as a preferred tool reference for a class of visual and inertial SLAM application domains preceded by no external location reference support hypothesis.

Keywords: tightly-coupled VIO; SLAM; fused point and line feature matching; pose estimates; simplified initialization strategy; patrol robot; map representation

1. Introduction

When robots operate under an unknown environment, an absolute external location reference such as a Global Positioning System (GPS) may be not available, and the no-prior-knowledge based navigating technology will be highly required. Thus, the individual intelligent robot should have the ability to estimate its own location using the carried sensors, such as Inertial Measurement Units (IMUs), laser radars, cameras, et al. [1–3]. For the navigation and perception problems of patrol robots working in the substations, the electromagnetic interferences will influence the signal transmissions, which therefore does not allow for the GPS receiver to assist the patrol robots with continuous and steady signal supports. In contrast to the existing navigation modes performed by dedicated external sensors, the robust solutions mainly lie mainly in utilizing the essential visual functions of cameras to build an environment map in real-time and estimate the position of the robot within the map simultaneously. This problem is called simultaneous localization and mapping (SLAM). It is noteworthy that SLAM may not only contribute to the acquisition and identification of the scene knowledge by some appropriate mode, but that it may also improve navigation performances with steady pose estimates [4]. One of the most significant SLAM results is proposed by Davison A.J., who pioneered the updating of the states of cameras and landmark points by an extended Kalman filter (EKF) and addressed the real-time SLAM problems for practical applications [5]. Klein G. extended the above model using a nonlinear optimization. He explicitly structured the SLAM system in terms of the front-end and the back-end, and improved the matched back-end framework by having the fused global constraints of the state variables be optimal rather than the pure iterations of EKF [6].

The above methods form the basis of feature-based methods for an efficient pose estimation [7–9]. Under the simple circumstances where the illumination changes slowly, or the cameras equipped are at a low speed movement, the direct methods are generally simpler to apply in practice, directly recovering the camera motion by minimizing a pixel-level intensities-based measurement error with no need to detect feature points [10–12]. Lately, there has been more research in the area of SLAM-based robot localization. In cases where the accurate pose estimates and large-scale scene reconstructions for mapping tasks are desired, the feature-based methods are more suitable for robotic applications.

Some research focuses on eliminating the accumulative positioning errors mainly caused by the incorrect feature points matching among images [13,14]. Actually, considering the fact that the cameras in motion find it difficult to present the expected brilliant images continuously, and in view of the fact that in some cases the cameras are working under the scenes with poor visibility or the 'understanding' of scenes can not be achieved in terms of textures, a visual inertial odometry (VIO) scheme is generally preferred, by fusing the inertial recursion (IMUs present) and SLAM calculation (cameras present) in robotics, to satisfy a long-term positioning accuracy and a matched favorable navigation stability in a short-time rapid maneuver.

By a method in which the state of the camera and the state of IMU are either directly incorporated in one state estimator or not, the typical VIO may be classified into a loosely-coupled mode and tightly-coupled mode. A loosely-coupled VIO separately estimates the relative motion by two state estimators, viz., the state of the camera and the state of IMU are separately estimated, and the VIO makes a fusion of these two results. A tightly-coupled VIO fuses raw measurements from the camera and IMU, explicitly estimating the relative motion by one state estimator, and this is generally fulfilled by constructing the joint nonlinear loss functions associated with the state variables. By contrast, the tightly-coupled mode presents a better accuracy and robustness.

For the state estimation, a filter-based method and optimization-based method are both possible [15–18]. The tightly-coupled mode fully takes into account the coupling between the used sensors. The optimization-based method explicitly incorporates the raw measurements of sensors and globally optimizes the sensor states by one estimator. As a mainstream framework, the tightly-coupled optimization-based VIO has been greatly extended theoretically. In principle, the system state of a VIO is expressed by typically integrating the pose (such as a rotation and translation by IMUs/cameras), velocity and zero bias (such as an inherent gyro bias and accelerometer bias by IMUs). The system state estimation of a VIO can converge to the desired state by optimizing the previously-constructed loss functions with respect to the state. It should also be noted that the initial values of the state variables for the global optimization are given by a system initialization module. To guarantee the long-term and steady availability in cases where limited numbers of feature points or textures are present, some research has been developed to improve the feature extraction pattern by fusing the line features or plane features in the VIO front-end, enabling the cameras to efficiently keep tracking. These solutions are equivalent to exerting some additional constraints to the entire pose estimation tasks [19,20].

The maturation and development of the above techniques underpin a successful robot application in the power patrol inspection. Accordingly, the efficiently initialized VIO permits the robot to perform accurate localization and navigation tasks [21,22]. Based on the above discussion, an optimized VIO system is presented to take into account the problems associated with the initialization efficiency and feature matching results.

The main contributions to this paper are shown in the following aspects.

1. First, during the course of a VIO initialization, the constant-velocity constraints are applied to the robots in motion. The consuming time for calculating the camera rotation between frames, is, in consequence, much less than that under the non-restriction conditions, accelerating the acquisition process of the initial state variables (including the pose, velocity, zero bias, etc.) dynamically.
2. Second, as a consequence of explicitly taking into account the textures of the electrical equipment in the work volume, the improved VIO characterized by the feature matching in terms of point features and line features enables the camera movement estimation (such as the rotation or translation) to be more accurate and smooth.
3. Third, the sparse maps represented by the point features and line features are constructed as expected under the sliding window optimization model. The introduction of this practical optimization model improves the efficiencies of the state estimation and mapping. Additionally, both dataset tests and substation scene tests for the robot routing inspection applications have been conducted, and the detailed evaluation results are given.

The outline of the remainder of the paper is as follows. The following section mainly discusses the VIO anatomy, besides the detailed description of the VIO front-end, including the reprojection errors associated with the points features and line features; additionally, the IMU pre-integration model is given, and the superiority of the fused-point and line feature-matching based method in accurate pose estimates over the direct method and simple point feature-matching based method is numerically proven by multiple sets of simulations. In Section 3, a simplified VIO initialization strategy is proposed and discussed, which subsequently includes a gyro bias estimation, accelerometer bias and gravity estimation, and scale factor and velocity estimation; furthermore, the laboratory test on the comparative time consumption by three typical feature-based visual odometries (VO) is highlighted. The matched state variable optimization tasks in the VIO back-end are emphasized in Section 4; specifically, the sliding window model for the accumulated error reduction and the visual measurement model for the two Jacobian matrix calculations with respect to the reprojection errors defined in Section 2, are respectively established. Section 5 carries out the experiments on dataset tests and real substation scene tests, and presents the main conclusions of this investigation.

2. Overall Description of Tightly-Coupled VIO

The physical structure of VIO can be divided into two parts: an IMU and a monocular camera. The embedded IMU provides the VIO system with an orthogonal 3-axial acceleration and angular rate in the body (robot) coordinate frame. The camera is mounted on the stationary base of the robot, providing the VIO system with sequential image information, by which it estimates the robot pose in the world coordinate frame and which can be further applied to represent and address the structure from motion (SFM) problem [23,24]. The essential part of integrating these two components consists in updating the state variables of the tightly-coupled VIO system as time evolves, so as to efficiently obtain the global optimum solutions of the state variables.

2.1. VIO Anatomy

Denote the world coordinate frame of the VIO system by W, which is referred to as the absolute reference used to denote the position and orientation of the objects in the concerned scenes. Denote the IMU coordinate frame (body coordinate frame) and the camera coordinate frame by B and C, respectively. A transformation between W and B is represented by a homogeneous transform matrix $T_{WB} = (R_{WB}|_W p_B)$, where R_{WB} represents the rotation and $_W p_B$ represents the displacement. Let $_W v_B$ denote the robot velocity expressed in the world coordinate frame. Denote the gyro bias and accelerometer bias by b_g and b_a, respectively. Figure 1 presents the diagrammatic representation of a VIO state estimator algorithm.

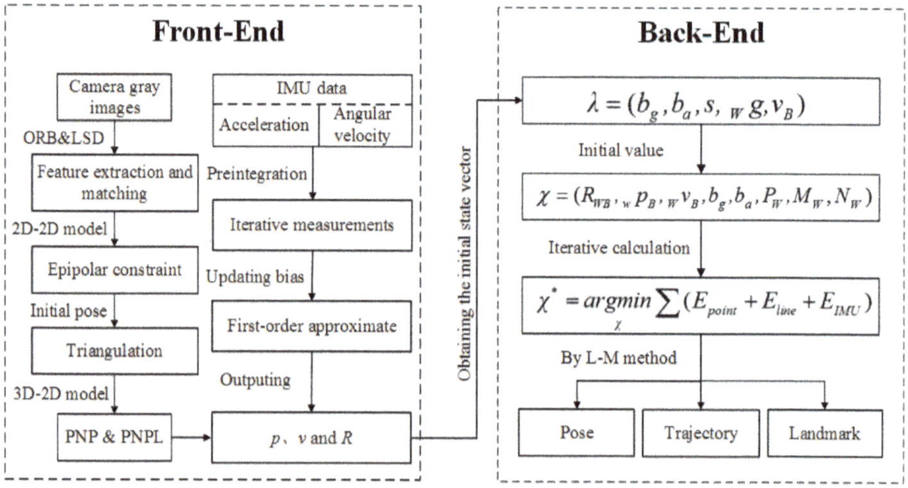

Figure 1. Flow chart of a VIO state estimator algorithm.

As illustrated, Figure 1 shows how information flows forward from the front-end to the back-end of the process. The VIO front-end collects the manipulated inputs from the IMU and the camera, and after obtaining the raw pose estimates of the robot in motion it turns to the VIO back-end to calculate the initial state vector λ. As mentioned above, the fused point and line feature-matching based method is conducted for the ideal pose estimates, on basis of the gray images.

The VIO back-end is used to optimize the state vector χ from λ. Let:

$$\begin{gathered} \lambda = \left(b_g, b_a, s, {}_W g, {}_W v_B\right) \\ \chi = \left(R_{WB}, {}_w p_B, {}_W v_B, b_g, b_a, P_W, M_W, N_W\right) \\ \chi^* = \arg\min_{\chi} \sum_{k} \left(E_{point} + E_{line} + E_{IMU}\right) \end{gathered} \tag{1}$$

where s represents the scale factor of the monocular camera, and ${}_W g$ represents the gravity vector expressed in the world coordinate frame. χ represents the VIO state vector and χ^* represents the loss function with respect to χ. P_W and (M_W, N_W) respectively represent the point features and line features of the images in the world coordinate frame. E_{point} and E_{line} are, respectively, the constructed quadratic form functions of the point feature reprojection error and line feature reprojection error. E_{IMU} is also a quadratic form function of the IMU error, which in nature denotes the constraints between the current frame and the previous keyframe in terms of a series of variable errors, like the rotation, position, velocity and bias [25]. Minimize the loss function χ^* by means of a typical Levenberg-Marquardt iterative calculation to assure the global optimization results, viz., the VIO can put out the globally optimal pose, trajectory, and landmark position in the world coordinate frame.

Note that the relative position and orientation between the camera and the IMU are fixed once the installation is done. Analogously, the transformation relationship between C and B can be represented by a homogeneous transform matrix $T_{CB} = (R_{CB}|{}_C p_B)$, where R_{CB} represents the rotation and ${}_C p_B$ represents the displacement. More specifically, T_{CB} essentially has a major impact on the precision and stability of the VIO system, which should therefore be calibrated with some mathematical means beforehand. Referring to the existing well-developed ways [26], the typical hand-eye calibration method is adopted in this paper.

2.2. Reprojection Error of the Camera

As described above, the VIO system fuses the point features and line features derived from the camera images. For the point features, the reprojection error denotes the distance (on the imaging plane) of the projection position of 3-D points from the detected position, minimizing this error by means of identifying the matched transform matrix, which then indicates that the pose optimization process is fully implemented. Suppose $P_i = (X_i, Y_i, Z_i)$ is the position of the ith feature point in 3-D space and u_i is the detected projection position of P_i on the imaging plane, the constructed reprojection error in terms of the point features can be defined as [27]:

$$r_{point} = u_i - \frac{1}{z_i} Kexp(\xi^{\wedge})P_i \tag{2}$$

where, z_i is the depth of P_i, and K is the intrinsic matrix of the camera. ξ is the Lie algebraic representation of the pose, and it follows that:

$$\xi^{\wedge} = \begin{bmatrix} 0 & -\xi_3 & \xi_2 \\ \xi_3 & 0 & -\xi_1 \\ -\xi_2 & \xi_1 & 0 \end{bmatrix} \tag{3}$$

For a line segment with the ends $M, N \in R^3$, the line reprojection error denotes a sum of point-to-line distances between the projected line segment l ends (m, n) and the detected line segment l' ends (M', N') on the imaging plane; it follows that [28]:

$$r_{line}(M', N', l, \xi, K) = r_{pl}^2(M', l, \xi, K) + r_{pl}^2(N', l, \xi, K) \tag{4}$$

where, $r_{pl}^2(M', l, \xi, K)$ represents the distance between the detected position of M' and line l, similarly, $r_{pl}^2(N', l, \xi, K)$ represents the distance between the detected position of N' and line l. The normalized form l may be defined as:

$$l = (l_1, l_2, l_3) = \frac{m_d^h \times n_d^h}{\left|m_d^h \times n_d^h\right|} \tag{5}$$

where m_d^h and n_d^h respectively indicate the corresponding homogeneous coordinates of the two ends of l. The graphic interpretation of the point/line feature reprojection error is illustrated by the points and line segments in Figure 2.

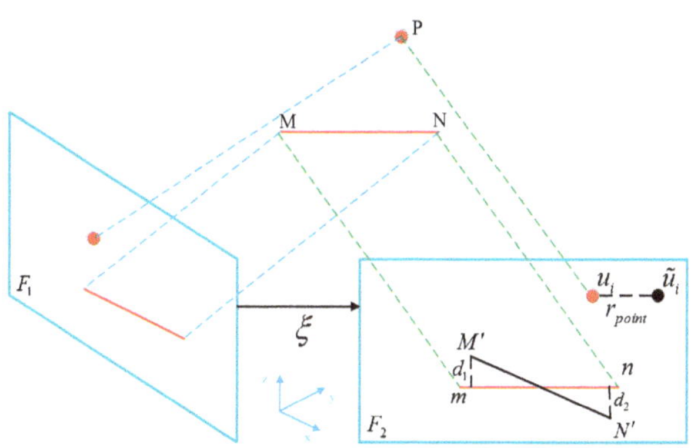

Figure 2. The graphic interpretation of the point/line feature reprojection error.

2.3. IMU Pre-Integration

The output frequency of the IMUs is generally dozens of times that of the cameras, which then indicates during the course of the data fusion that the VIO collects multiple sets of IMU measurement data in a single sampling interval $[i, i+1]$ (between two keyframes).

Let $_B\widetilde{a}(t)$ and $_B\widetilde{\omega}(t)$ respectively denote the measured angular rate and acceleration. We have:

$$_B\widetilde{a}(t) = R_{BW}(_Wa(t) - {_W}g) + b^a(t) + \eta^a(t) \tag{6}$$

$$_B\widetilde{\omega}(t) = {_B}\omega(t) + b^g(t) + \eta^g(t) \tag{7}$$

where $_Wa(t)$ and $_W\omega(t)$ are the angular rate and acceleration to be estimated. $\eta^a(t)$ and $\eta^g(t)$ are white noise. The accelerometer bias $b^a(t)$ and the gyro bias $b^g(t)$ are subject to random walk noise.

The $(i+1)$th updated R_{WB}^{i+1}, $_Wv_B^{i+1}$ and $_Wp_B^{i+1}$ can be given by [29]:

$$R_{WB}^{i+1} = R_{WB}^{i} Exp((\widetilde{\omega}_i - b_i^g - \eta_i^g)\Delta t_{i,i+1}) \tag{8}$$

$$_Wv_B^{i+1} = {_W}v_B^i + {_W}g\Delta t_{i,i+1} + R_{WB}^i(\widetilde{a}_i - b_i^a - \eta_i^a)\Delta t_{i,i+1} \tag{9}$$

$$_Wp_B^{i+1} = {_W}p_B^i + v_i\Delta t_{i,i+1} + \frac{1}{2}{_W}g\Delta t_{i,i+1}^2 + \frac{1}{2}R_{WB}^i(\widetilde{a}_i - b_i^a - \eta_i^a)\Delta t_{i,i+1}^2 \tag{10}$$

where $\Delta t_{i,i+1}$ is the time interval between two keyframes. The relative motion between two keyframes can be defined in terms of the pre-integrated $\Delta R_{i,i+1}$, $\Delta v_{i,i+1}$ and $\Delta p_{i,i+1}$, shown as follows:

$$\Delta R_{i,i+1} \doteq R_i^T R_{i+1} = Exp((\widetilde{\omega}_i - b_i^g - \eta_i^g)\Delta t_{i,i+1}) \tag{11}$$

$$\Delta v_{i,i+1} \doteq R_i^T(v_{i+1} - v_i - {_W}g\Delta t_{i,i+1}) = \Delta R_{i,i+1}(\widetilde{a}_i - b_i^a - \eta_i^a)\Delta t_{i,i+1} \tag{12}$$

$$\begin{aligned} \Delta p_{i,i+1} &\doteq R_i^T(p_{i+1} - p_i - v_i\Delta t_{i,i+1} - \frac{1}{2}{_W}g\Delta t_{i,i+1}^2) \\ &= \Delta v_{i,i+1}\Delta t_{i,i+1} + \frac{1}{2}\Delta R_{i,i+1}(\widetilde{a}_i - b_i^a - \eta_i^a)\Delta t_{i,i+1}^2 \end{aligned} \tag{13}$$

Note that it is supposed that bias b^a and bias b^g are constant during the time interval from t to $t + \Delta t_{i,i+1}$, as indicated in Equations (11)–(13), and for this to be the case they should be initially calibrated in practice. Define the change of b^a (and b^g) as the disturbance δb and linearize it with first-order approximation; consequently, we obtain the $(i+1)$th state estimates in terms of the ith state estimates and the residual error:

$$R_{WB}^{i+1} = R_{WB}^i \Delta R_{i,i+1} Exp(J_{\Delta R}^g b_i^g) \tag{14}$$

$$_Wv_B^{i+1} = {_W}v_B^i + g_W\Delta t_{i,i+1} + R_{WB}^i(\Delta v_{i,i+1} + J_{\Delta v}^g b_i^g + J_{\Delta v}^a b_i^a) \tag{15}$$

$$_Wp_B^{i+1} = {_W}p_B^i + {_W}v_B^i\Delta t_{i,i+1} + \frac{1}{2}g_W\Delta t_{i,i+1} + R_{WB}^i(\Delta p_{i,i+1} + J_{\Delta p}^g b_i^g + J_{\Delta p}^a b_i^a) \tag{16}$$

where $J_{(\cdot)}^g$ and $J_{(\cdot)}^a$ are the Jacobian matrices of the pre-integrated measurements with respect to δb at the sampling point i.

The pose estimation and IMU pre-integration form the front-end tasks of the designed VIO. To evaluate the performances of the VIO, we carry out a set of numerical simulations. Two images (F_1, F_2) derived from fr1/desk of the TUM RGB-D datasets [30] are arbitrarily designated as the testing samples, the fused point and line feature-matching based method and the simple point feature-matching based method, together with the direct method. are conducted under different optimization strategies, including non-optimization, typical Gauss-Newton (G-N) optimization and Levenberg-Marquardt (L-M) optimization for the first round and convergence achieved respectively. The comparative results are shown in Table 1, in terms of the transform matrix $T_{F_1F_2}$ and RMSE (root mean squared error) values.

Table 1. The comparative pose measurement results.

		Simple Point Feature-Matching Based Method	Fused Point and Line Feature-Matching Based Method	Direct Method
Non optimization	$T_{F_1F_2}$	$\begin{bmatrix} 0.9973 & -0.033 & -0.0647 & -0.0808 \\ 0.0343 & 0.9992 & 0.0199 & -0.0858 \\ 0.0639 & -0.0221 & 0.9977 & 0.993 \\ 0 & 0 & 0 & 1 \end{bmatrix}$	$\begin{bmatrix} 0.9979 & -0.0379 & -0.0514 & -0.1126 \\ 0.0397 & 0.9986 & 0.0355 & -0.1137 \\ 0.0499 & -0.0374 & 0.998 & 0.2248 \\ 0 & 0 & 0 & 1 \end{bmatrix}$	×
	RMSE	0.7329	0.2128	-
G-N for 1 round	$T_{F_1F_2}$	×	$\begin{bmatrix} 0.998 & -0.0379 & -0.0514 & -0.1125 \\ 0.0397 & 0.9986 & 0.03533 & -0.1127 \\ 0.0499 & -0.0373 & 0.998 & 0.2248 \\ 0 & 0 & 0 & 1 \end{bmatrix}$	×
	RMSE	-	0.1020	-
G-N for convergence achieved	$T_{F_1F_2}$	$\begin{bmatrix} 0.998 & -0.0373 & -0.0516 & -0.1045 \\ 0.04 & 0.9985 & 0.037 & -0.1198 \\ 0.05 & -0.039 & 0.998 & 0.2334 \\ 0 & 0 & 0 & 1 \end{bmatrix}$	$\begin{bmatrix} 0.9979 & -0.0373 & -0.0516 & -0.1045 \\ 0.0392 & 0.9985 & 0.0371 & -0.1198 \\ 0.0502 & -0.039 & 0.9979 & 0.2334 \\ 0 & 0 & 0 & 1 \end{bmatrix}$	$\begin{bmatrix} 0.9999 & -0.0037 & -0.0005 & 0.0035 \\ 0.0034 & 0.9999 & -0.0005 & 0.002 \\ 0.0004 & 0.0005 & 1 & -0.0005 \\ 0 & 0 & 0 & 1 \end{bmatrix}$
	RMSE	0.1010	0.1009	0.2926
L-M for 1 round	$T_{F_1F_2}$	$\begin{bmatrix} 0.9973 & -0.0368 & 0.0623 & -0.0325 \\ 0.0383 & 0.999 & 0.0223 & -0.0303 \\ 0.0614 & -0.0247 & 0.9978 & ,0.234 \\ 0 & 0 & 0 & 1 \end{bmatrix}$	$\begin{bmatrix} 0.998 & -0.0373 & -0.0516 & -0.1045 \\ 0.03919 & 0.999 & 0.0371 & -0.1198 \\ 0.0502 & -0.0391 & 0.9979 & 0.2334 \\ 0 & 0 & 0 & 1 \end{bmatrix}$	$\begin{bmatrix} 0.9999 & -0.0037 & -0.0022 & 0.0282 \\ 0.0037 & 0.9999 & -0.0001 & 0.003 \\ 0.0022 & 0.001 & 0.9999 & -0.044 \\ 0 & 0 & 0 & 1 \end{bmatrix}$
	RMSE	0.1320	0.1011	0.3376
L-M for convergence achieved	$T_{F_1F_2}$	$\begin{bmatrix} 0.9999 & -0.0372 & -0.0516 & -0.1045 \\ 0.0392 & 0.9985 & 0.037 & -0.1198 \\ 0.0502 & -0.039 & 0.9989 & 0.2334 \\ 0 & 0 & 0 & 1 \end{bmatrix}$	$\begin{bmatrix} 0.9978 & -0.0368 & -0.0623 & -0.0325 \\ 0.0383 & 0.999 & 0.0224 & -0.0304 \\ 0.0614 & -0.0247 & 0.9987 & 0.2341 \\ 0 & 0 & 0 & 1 \end{bmatrix}$	$\begin{bmatrix} 0.9999 & -0.0034 & -0.0023 & 0.0286 \\ 0.0038 & 0.9999 & -0.0008 & 0.0016 \\ 0.0023 & 0.0008 & 0.9999 & -0.0448 \\ 0 & 0 & 0 & 1 \end{bmatrix}$
	RMSE	0.0601	0.0486	0.2917

As in Table 1, since the direct method estimates the robot pose directly by minimizing a pixel-level intensities-based measurement error, which in nature belongs to the optimization problem, when non optimization is adopted the direct method itself is not available at all. For the first-round G-N optimization, the direct method and the simple point feature-matching based method both fail to result in valid estimates, which is mainly because the trust region problem is not fully taken into account during the optimization process, and consequently an oversized step is employed by mistake. By contrast, the fused point and line feature-matching based method presents a better robustness under a wider range of optimization strategies without any load in complexity; specifically, with the L-M optimization conditions its pose estimation precision is generally best (a lower RMSE between the estimated $T_{F_1F_2}$ and the true transform matrix given in fr1/desk TUM). The following section concentrates on fulfilling the VIO initialization design for a better state initializing efficiency.

3. VIO Initialization Design

The behavior of the VIO highly depends on the initial values of the system states. A proposed method of initializing the VIO states consists of previously setting a constant velocity for a patrol robot in operation. Moreover, it assumes that the rotation is steadily unchangeable. The simplified solution, therefore, is expected to improve the initializing efficiency of an actual VIO without any decrease in the precision. Quite simply, the accuracy of the estimated gravity is evaluated by reference to its true value (since the magnitude of the true gravity is known), so that the effectiveness of the simplified VIO initializing strategy can be verified. The detailed procedures are shown below.

3.1. Gyro Bias Estimation

Assume that the relative rotation defined in the pre-integration module is constant, and that the velocity difference is zero during the given time interval $[i, i+1]$, $[i+1, i+2]$, … ; we have:

$$\Delta R_{i,i+1} = \Delta R_{i+1,i+2}, \Delta v_{i,i+1} = \Delta v_{i+1,i+2} = 0 \quad (17)$$

Define the residual error $r_{\Delta R_{i,i+1}}$ by integrating the terms from the camera calculation and gyro pre-integration. It follows that [31]:

$$r_{\Delta R_{i,i+1}} = \sum_{i=1}^{N-1} Log((\Delta R_{i,i+1} Exp(J_{\Delta R}^{g} b_i^{g}))^T R_{BW}^{i+1} R_{WB}^{i}) \quad (18)$$

where $R_{WB} = R_{WC}R_{CB}$ (R_{WC} is derived from the monocular camera). N is the number of keyframes.

The gyro bias b_i^g is estimated by minimizing $r_{\Delta R_{i,i+1}}$ with the L-M calculation. Among some typical feature point methods such as ORB (Oriented Brief) feature, SURF (Speeded Up Robust Features) feature and SIFT (Scale Invariant Feature Transform) feature, the process of feature extraction and matching cost more execution time. To quantitatively illustrate the time taken for each step of the VIO pose estimation, Table 2 presents the comparative time consumption results through three typical feature-based visual odometries (VO) with a computer Lenovo Y510 (Inteli5-4200MQ, 2.5GHz CPU, 8GB RAM, Lenovo Grope, Beijing, China,) under an Ubuntu 16.04 environment. The images that are used are coming from the fr1_xyz of TUM dataset.

Table 2. The comparative time consumption results (s).

	Feature Extraction	Descriptor Calculation	Feature Matching	Pose Estimation	Total
ORB	0.0101	0.0087	0.0118	0.0009	0.0315
SURF	0.0435	0.0095	0.0274	0.0014	0.0818
SIFT	0.9228	0.0125	0.0285	0.0012	0.9650

As described, the main idea of the VIO initialization lies in calculating the rotation matrix of each frame according to the results from the first two frames on the basis of keeping the rotation constant, rather than repetitively performing a routine feature extraction and feature matching. This is illustrated by the comparative time consumed for the bias estimation in Figure 3; we arbitrarily designate different numbers of the images for testing, and compare the corresponding consumption time by the method in this paper and the typical methods in [22,31]. Clearly, continuously estimating the rotation between the frames reveals its poor efficiency when a larger number of frames are concerned; therefore, the proposed method shows its superiority in dealing with the bias estimation in large-scale scene information.

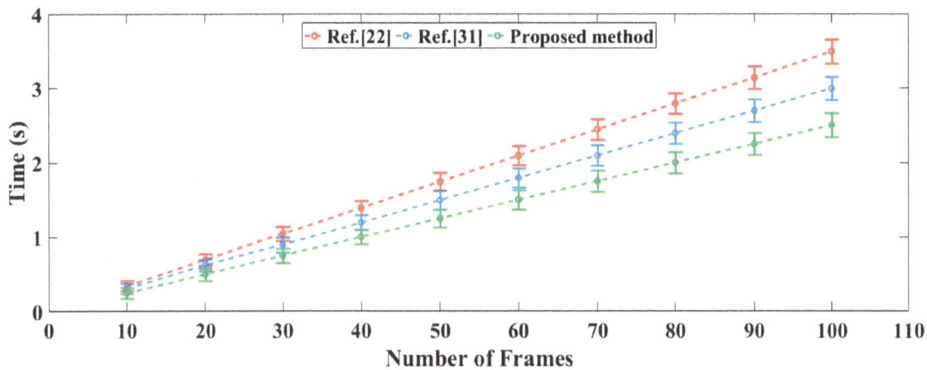

Figure 3. The time consumed for the bias estimation.

3.2. *Accelerometer Bias and Gravity Estimation*

The residual error of relative velocity $r_{\Delta v_{i,i+1}}$ may be directly defined on the basis of the constant velocity hypothesis with the known b_i^g, viz., the accelerometer bias is fully taken into account in this case, which is quite different from that adopted in [31]. We define:

$$r_{\Delta v_{i,i+1}} = \sum_{i=1}^{N-1} (\underbrace{{}_W v_B^{i+1} - {}_W v_B^{i}}_{0} - g_W \Delta t_{i,i+1} - R_{WB}^{i}(\Delta v_{i,i+1} + J_{\Delta v}^{g} b_i^{g} + J_{\Delta v}^{a} b_i^{a})) \tag{19}$$

Analogously, the estimates of the accelerometer bias b_i^a and the gravity g_W are solved by forming a least-square problem with manipulated VIO inputs. It is noted that, in view of the VIO computational load, only three keyframes with a strong parallax excitation are used to establish the fewer simultaneous equations, and this simplified scheme is sufficiently accurate to deal with a wider range of accelerometer bias phenomena.

We further optimize the gravity g_W and parameterize it as:

$$\hat{g}_W = g \cdot \overline{g}_W + \omega_1 b_1 + \omega_2 b_2 \tag{20}$$

where g is the magnitude of the gravity, and $\overline{g}_W$ is the direction vector of the current gravity $\hat{g}_W$. b_1 and b_2 are two orthogonal bases on the tangent plane and can be easily determined by the Gram-Schmidt process. ω_1 and ω_2 are the corresponding 2D components to be estimated. Substitute Equation (20) into Equation (19) and solve it by Singular Value Decomposition (SVD) [32]. This process is iterated several times until $\hat{g}_W$ converges.

3.3. Scale Factor and Velocity Estimation

The scale uncertainty of the monocular cameras may lead to an ambiguous estimate trajectory. The scale factor s is therefore introduced to represent the position transformation between the camera and IMU, and it follows that [33]:

$$_W p_B = s_W p_C + R_{WC\,C} p_B \tag{21}$$

Substitute Equation (21) into Equation (16) and ignore the accelerometer bias. We have:

$$\begin{aligned}&\left[R^{i\ T}_{WB}(R^i_{WC} - R^{i+1}_{WC})_C p_B + \frac{1}{2}R^{i\ T}_{WB} g_W \Delta t_{i,i+1} + \Delta p_{i,i+1}\right] \\ &= \left[R^{i\ T}_{WB}({}_W p^{i+1}_C - {}_W p^i_C) - R^{i\ T}_{WB}\Delta t_{i,i+1}\right]\begin{bmatrix} s \\ {}_W v^i_B \end{bmatrix}\end{aligned} \tag{22}$$

Substitute the relative velocity of the pre-integration measurements (expressed in Equation (12)) into Equation (22), and let $\Delta t_{i,i+1}$ and $\Delta t_{i+1,i+2}$ respectively denote the time interval between Keyframe 1 to Keyframe 2 and Keyframe 2 to Keyframe 3. Eliminate the unknown, and we can get $\hat{z}_{i,i+1,i+2}$, similar to [31]. Thus, s can be calculated from the residual error equation below:

$$s^* = \underset{s}{argmin}\left(\begin{array}{c} \hat{z}_{i,i+1,i+2} - [s({}_W p^{i+1}_C - {}_W p^i_C)\Delta t_{i+1,i+2} - s({}_W p^{i+2}_C - {}_W p^{i+1}_C)\Delta t_{i,i+1} \\ +\frac{1}{2} g_W(\Delta t^2_{i,i+1}\Delta t_{i+1,i+2} + \Delta t^2_{i+1,i+2}\Delta t_{i,i+1})] \end{array}\right) \tag{23}$$

In Equation (22), so far, the unknown ${}_W v^i_B$ is solvable. For the first (K−1) keyframes, the corresponding velocity can be explicitly calculated. Conversely, the current (the Kth) keyframe should be given by Equation (15).

4. Tightly-Coupled Information Fusion Based on Sliding Window

The VIO system may proceed, in this phase, by realizing the initialization of the variables illustrated above. The core points consist in continuously optimizing the joint loss functions of each error term (including E_{point}, E_{line} and E_{IMU}). However, since the front-end of the VIO collects a large amount of input information from the camera and IMU, a heavy emphasis should be placed upon the real-time state estimation of the VIO that has to cope with the potential tracking failures. Considering the computational load in the back-end of the VIO, a practical sliding window scheme is developed to perform the efficient state optimization [34].

4.1. Sliding Window Model

The sliding window in the VIO mainly marginalizes out certain states of the system by a Schur complement, and the reinsertion of these as prior information (the prior term E_{prior}) would allow the loss functions to be formed and optimized. That is, E_{prior} further supplies the system state with observable constraints. Suppose that the ith system state vector (in terms of discrete moment) is $\chi_i = (R^i_{WB}, {}_w p^i_B, {}_W v^i_B, b^i_g, b^i_a, P^i_W, M^i_W, N^i_W)$, the matched error terms, can therefore be expressed as:

$$E_{point} = \sum_{k \in K_V}\sum_{i \in \beta} \rho({r^{i,k}_{point}}^T \Sigma^{-1}_{r^{i,k}} r^{i,k}_{point}) \tag{24}$$

$$E_{line} = \sum_{k \in K_V}\sum_{j \in \eta} \rho({r^{j,k}_{line}}^T \Sigma^{-1}_{r^{j,k}} r^{j,k}_{line}) \tag{25}$$

$$E_{IMU} = \sum_{i,j \in K_I}\left[\rho(r^T_{\Delta R} r^T_{\Delta v} r^T_{\Delta p})\Sigma_I (r^T_{\Delta R} r^T_{\Delta v} r^T_{\Delta p})^T + \rho(r^T_{\Delta b}\Sigma_R r_{\Delta b})\right] \tag{26}$$

where K_V and K_I respectively represent the sets of visual and inertial measurements in the current sliding window, and P_W and (M_W, N_W) respectively represent the point features and line features which are observed at least twice in the current sliding window. $\Sigma_{r_{i,k}}^{-1}$ and $\Sigma_{r_{j,k}}^{-1}$ respectively represent the information matrix of the point feature reprojection error and line feature reprojection error. Σ_I and Σ_R are also information matrices, respectively representing the pre-integration information matrix and bias random walk information matrix. ρ is the robust kernel, piece-wisely expressed as:

$$\rho(s) = \begin{cases} \frac{1}{2}s^2 & |s| \leq \delta \\ \delta(|s| - \frac{1}{2}\delta) & Others \end{cases} \tag{27}$$

where $\rho(\cdot)$ is in the Huber norm (δ being a pre-set threshold). $r_{\Delta R}$ and $r_{\Delta v}$ are defined in Equations (18) and (19). Analogously, the definitions of $r_{\Delta p}$ and $r_{\Delta b}$ are also derived from the pre-integration measurements, and we have:

$$r_{\Delta p} = {}_W p_B^j - {}_W p_B^i - {}_W v_b^i \Delta t_{ij} - \frac{1}{2} g_W \Delta t_{ij}^2 - R_{WB}^i (\Delta p_{i,i+1} + J_{\Delta p}^g b_i^g + J_{\Delta p}^a b_i^a) \tag{28}$$

$$r_b = r_b^j - r_b^i \tag{29}$$

The marginalization result can be denoted as the prior term E_{prior}, and it follows that:

$$E_{prior} = ||r_{prior} - H_{prior}\chi||^2 \tag{30}$$

where r_{prior} represents the prior information after marginalization, and H_{prior} represents the Hessian matrix constrained by the pose, landmark position and IMU measurements.

The modified loss function in a linear combination form can therefore be further written as:

$$F_{loss} = \sum_i \left(E_{point} + E_{line} + E_{IMU} + E_{prior}\right) \tag{31}$$

The typical optimization strategy of F_{loss} is similar to Visual-Inertial System (VINS) [35]. Given the frames in the optimization window, the decision-making pattern of the end-back of the VIO is diagrammatically represented in Figure 4. In the figure, the green circle in the figure indicates the pose of the keyframes, the gray circle indicates the pose of the non-keyframes, the yellow square indicates the measurements of the features, the red square indicates the inertial constraints of the IMU, and the purple square and the arrow indicate the information that is marginalized. The red cross indicates the measurements that was discarded. Two cases are discussed: ① if the current inserted frame is not a keyframe, the visual measurement, together with the current pose estimate, would be explicitly neglected, viz., the IMU constraints would only be marginalized out; ② if the current frame is a keyframe, the visual measurement and the pose estimate of the oldest keyframe in the sliding window would be marginalized out and the current keyframe would be kept accordingly.

Owing to the specific forms of the variables to be optimized in the sliding window model, the following work will turn to the definition of the vertices/edges in the graph optimization model by means of a G^{2}o optimization framework and to the estimation of the state variables by means of an L-M iterating calculation [36].

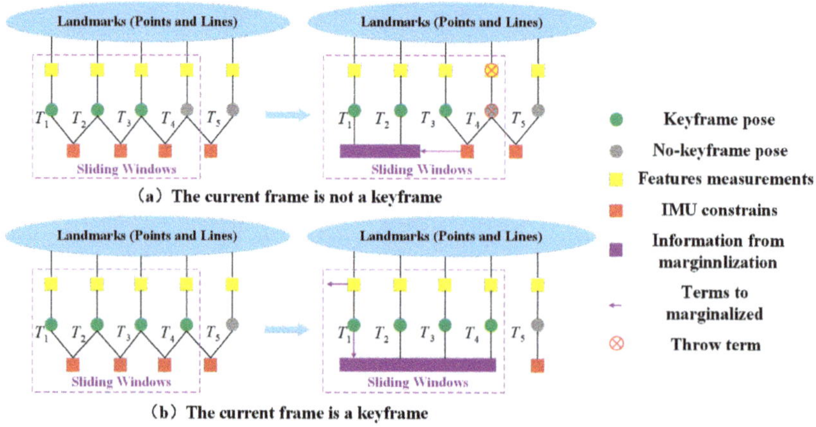

Figure 4. Decision-making pattern of the sliding window model, (**a**) the inserted frame is not a keyframe and (**b**) the inserted frame is a keyframe.

4.2. Visual Measurement Model

For the loss function represented by Equation (31), the optimization means recurrently performing the linear expansion of Equation (31) around the current estimated value, which therefore implies its principal of calculating the Jacobian matrices of the residual functions with respect to the state variables. Specifically, the method chosen to solve the Jacobian matrix of the point reprojection error with respect to the pose should be the typical chain rule [37], which yields:

$$\frac{\partial r_{point}}{\partial \delta\xi} = -\frac{\partial r_{point}}{\partial P_C}\frac{\partial P_C}{\partial \delta\xi} \tag{32}$$

with

$$\frac{\partial r_{point}}{\partial P_C} = \begin{bmatrix} \frac{f_x}{Z} & 0 & -\frac{f_x X}{Z^2} \\ 0 & \frac{f_y}{Z} & -\frac{f_y Y}{Z^2} \end{bmatrix} \tag{33}$$

$$\frac{\partial P_C}{\partial \delta\xi} = [-P_C{}^{\wedge}, I_{3\times3}] \tag{34}$$

where $\delta\xi$ is the disturbance of the pose, $P_C = [X, Y, Z]^T$ is the coordinate of the landmark in the camera coordinate frame, and f_x and f_y are the focal length parameters in K. $I_{3\times3}$ is an identity matrix.

For the Jacobian matrix of line reprojection error with respect to the pose, let $\ell = [n, v]^T$ be the Plücker coordinate of the line feature [38], and let the homogeneous coordinates of M' and N' be $M' = (u_1, v_1, 1)^T$ *and* $N' = (u_2, v_2, 1)^T$ respectively. We have:

$$\frac{\partial r_{line}}{\partial \delta\xi} = -\frac{\partial r_{line}}{\partial l}\frac{\partial l}{\partial \ell}\frac{\partial \ell}{\partial \delta\xi} \tag{35}$$

with

$$\frac{\partial r_{line}}{\partial l} = \begin{bmatrix} \frac{u_1 l_2^2 - l_1 l_2 v_1 - l_1 l_3}{(l_1^2 + l_2^2)^{\frac{3}{2}}} & \frac{v_1 l_1^2 - l_1 l_2 u_1 - l_2 l_3}{(l_1^2 + l_2^2)^{\frac{3}{2}}} & \frac{1}{(l_1^2 + l_2^2)^{\frac{1}{2}}} \\ \frac{u_2 l_2^2 - l_1 l_2 v_2 - l_1 l_3}{(l_1^2 + l_2^2)^{\frac{3}{2}}} & \frac{v_2 l_1^2 - l_1 l_2 v_2 - l_2 l_3}{(l_1^2 + l_2^2)^{\frac{3}{2}}} & \frac{1}{(l_1^2 + l_2^2)^{\frac{1}{2}}} \end{bmatrix} \tag{36}$$

$$\frac{\partial l}{\partial \ell} = \begin{bmatrix} f_y & 0 & 0 & 0 & 0 & 0 \\ 0 & f_x & 0 & 0 & 0 & 0 \\ -f_y c_x & -f_x c_y & f_x f_y & 0 & 0 & 0 \end{bmatrix} \tag{37}$$

$$\frac{\partial \ell}{\partial \delta \xi} = \begin{bmatrix} -[R_{CW} n_W]^\wedge - [t_{CW}^\wedge R_{CW} v_W]^\wedge & -[R_{CW} v_W]^\wedge \\ -[R_{CW} v_W]^\wedge & 0 \end{bmatrix} \tag{38}$$

where v is the direction vector of the line, and n is the normal vector of the plane formed by the line and origin point; they are both in the Plücker coordinate frame. In addition to the Jacobian matrices of the point/line reprojection error with respect to the pose, analogously, the Jacobian matrices of the point/line position in space could be formulized as the similar forms to those in Equations (32) and (35), due to the limits of the space. Please see [39] for details.

5. Experimental Section

The experimental observations consist of dataset tests and substation scene tests. The behaviors of the VIO on the datasets largely reflect its actual performances, so the process of evaluating the performances of the designed VIO consists of first testing it in the public datasets.

5.1. Dataset Tests and Analyses

The public dataset European Robotics Challenge (EUROC) [40] provides a series of information (such as images, accelerations and angular rates, etc.) invoking a micro aerial vehicle (MAV) equipped with a stereo camera and an IMU in either ① a cluttered workspace scene or ② an industrial machine hall scene. Moreover, the derived information (11 sequences in total) is classified into three grades: "easy", "medium" and "difficult", depending for example on the velocity of the aerial vehicle, the texture status of the scene, or the lighting conditions nearby. Also, EUROC presents the standard trajectories captured by the VICON motion capture system with reliable navigating parameters (so-called 'Ground Truth') available to users, including the position, attitude, velocity of the MAV in 3D space and some other inertial data, such as the gyro bias and the accelerometer bias obtained by the IMU. Specifically, the V1_01_easy sequence and the MH_04_difficult sequence are designated as the testing samples, and are therefore more appropriate to reflect the strong information domain coverage. In contrast, the state estimates are compared with those extracted by the existing eminent VIOs, such as OKVIS, VIORB, VINS, etc. One thing that should be noted is that, since EUROC doesn't explicitly provide the Ground Truth scale, we therefore extract it by collecting the translation results from ORB-SLAM2 and translation references provided by Ground Truth. Once we obtained the translation transformation between the first two keyframes in ORB-SLAM2, the truth scale would be a calculation of the translation transformation to the references. Note also that the EUROC dataset presents the stereo images at 20 Hz with IMU measurements at 200 Hz and a trajectory Ground Truth with a higher updating frequency. Hence, the efficient state estimate comparison can only depend upon the accurate alignment of the timestamps. Among these, the VIO trajectory comparison is fulfilled by means of the evo tool [41], and the position error comparison is conducted by the script that TUM provides.

5.1.1. VIO Initialization Results

The initialization results are illustrated by the convergence procedures of the initialization state with respect to two typical sequences (V1_01_easy and MH_04_difficult) in Figure 5, and the initialization state is constructed of ① the accelerometer bias, ② the gyro bias in orthogonal tri-axes, ③ the condition number (referring to the data adaptation), ④ the scale factor of the monocular camera, and ⑤ the orthogonal tri-axial component of the gravity vector. Quite clearly, all of these five sets of variables converge for $t > 8$ s. Specifically, the accelerometer bias and gravity vector appear convergent after 2 s, and the accelerometer bias converges to almost zero even under the MH_04_difficult sequence circumstances, while in contrast to this the gyro bias appears larger yet, with more stable characteristics; the reasons for this consist in the fact that we merely calculated and corrected the gyro bias by means of the pose transformation directly derived from the camera, whereas the estimations for the accelerometer bias were implicitly performed by the precise least-square iterations. By comparison, the initialization performances for the MH_04_difficult sequence are slightly inferior, because the condition number illustrated in Figure 6c approximately converges until $t = 8$ s; by then, the observabilities for the initialization state variables are satisfied. Meanwhile, the estimated scale factor, as shown, may be considered to be a true value for $t > 8$ s; the camera trajectory can therefore be recognized as being precisely recovered as expected.

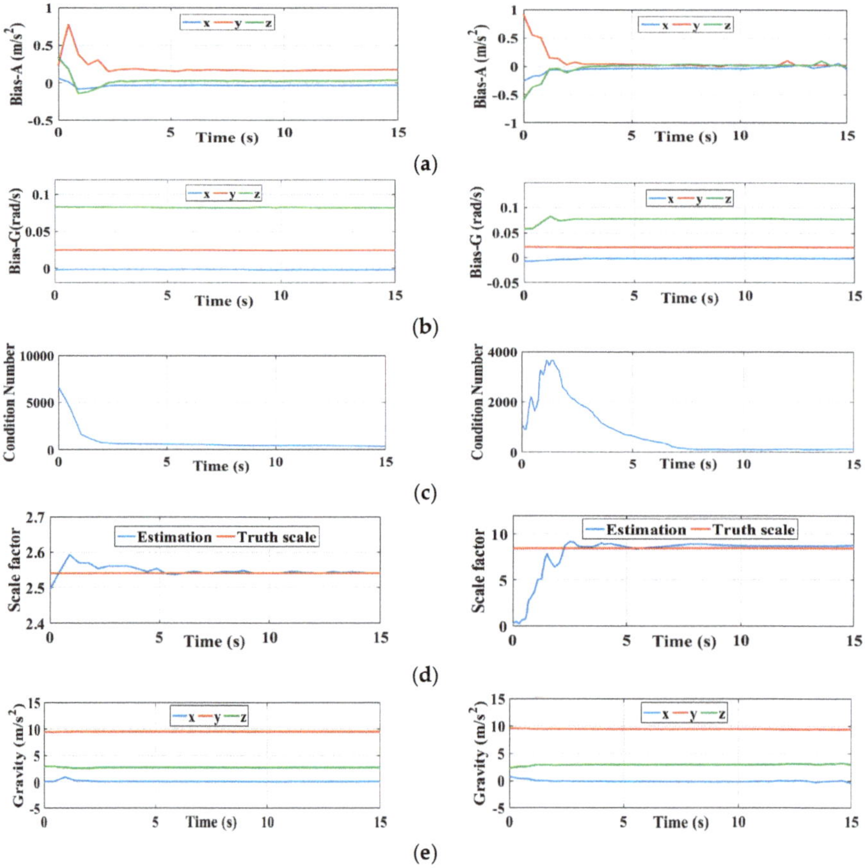

Figure 5. The convergence procedures of the initialization states for the V1_01_easy & MH_04_difficult sequences, (**a**) Initialization results of accelerometer bias; (**b**) Initialization results of gyro bias; (**c**) Calculation of the condition number; (**d**) Initialization results of scale factor; (**e**) Initialization results of gravity vector.

5.1.2. Navigation Performance Evaluations

The feature extraction results are diagrammatically illustrated by Figure 6. As shown, in cases where the scene textures appear clear with an ideal illumination, a large amount of point features and line features are captured as expected (see Figure 6a). Additionally, even though the MH_04_difficult sequence supplies the system with an unstable illumination for representing the MAV in motion circumstances (see Figure 6b), the VIO front-end can still extract enough features and consequently stabilize the dynamic VIO. Here, four representative pictures are selected to describe the scenes that are considered.

(**a**)

(**b**)

Figure 6. Feature extraction performances of the VIO front-end: (**a**) V1_01_easy sequence; (**b**) MH_04_difficult sequence.

The performances of the VIO designed above are diagrammatically given in 3D space, being characterized by absolute positioning errors (APEs). APE is often used as the absolute trajectory error, and the corresponding poses are directly compared between the estimate and reference, and given a pose relation.

Figure 7a–k corresponds to 11 sequences at different difficulty levels. Furthermore, more detailed analyses related to the two typical sequences (V1_01_easy and MH_04_difficult) are illustrated by planar trajectories, as shown in Figure 8. In Figure 7, the dotted lines represent the Ground Truth trajectories (reference), the color lines represent the estimated trajectories by the designed VIO; the closer the color of the lines approaches to red, the greater the APE, and vice versa. As we can see, the designed VIO presents stable tracking performances for all difficulty levels, even for a fast camera movement or un-ideal illumination circumstances (as V2_03_ difficult and MH_05_difficult denote); no 'tracking lost' appears.

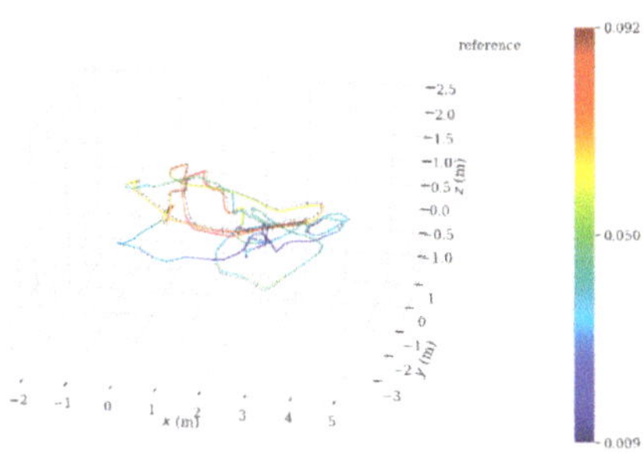

(**a**)

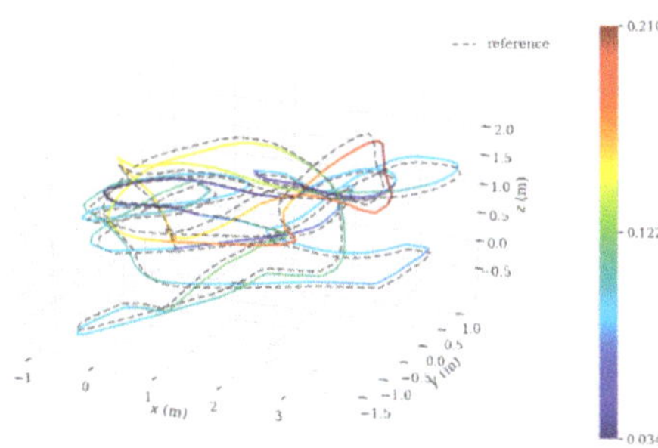

(**b**)

Figure 7. *Cont.*

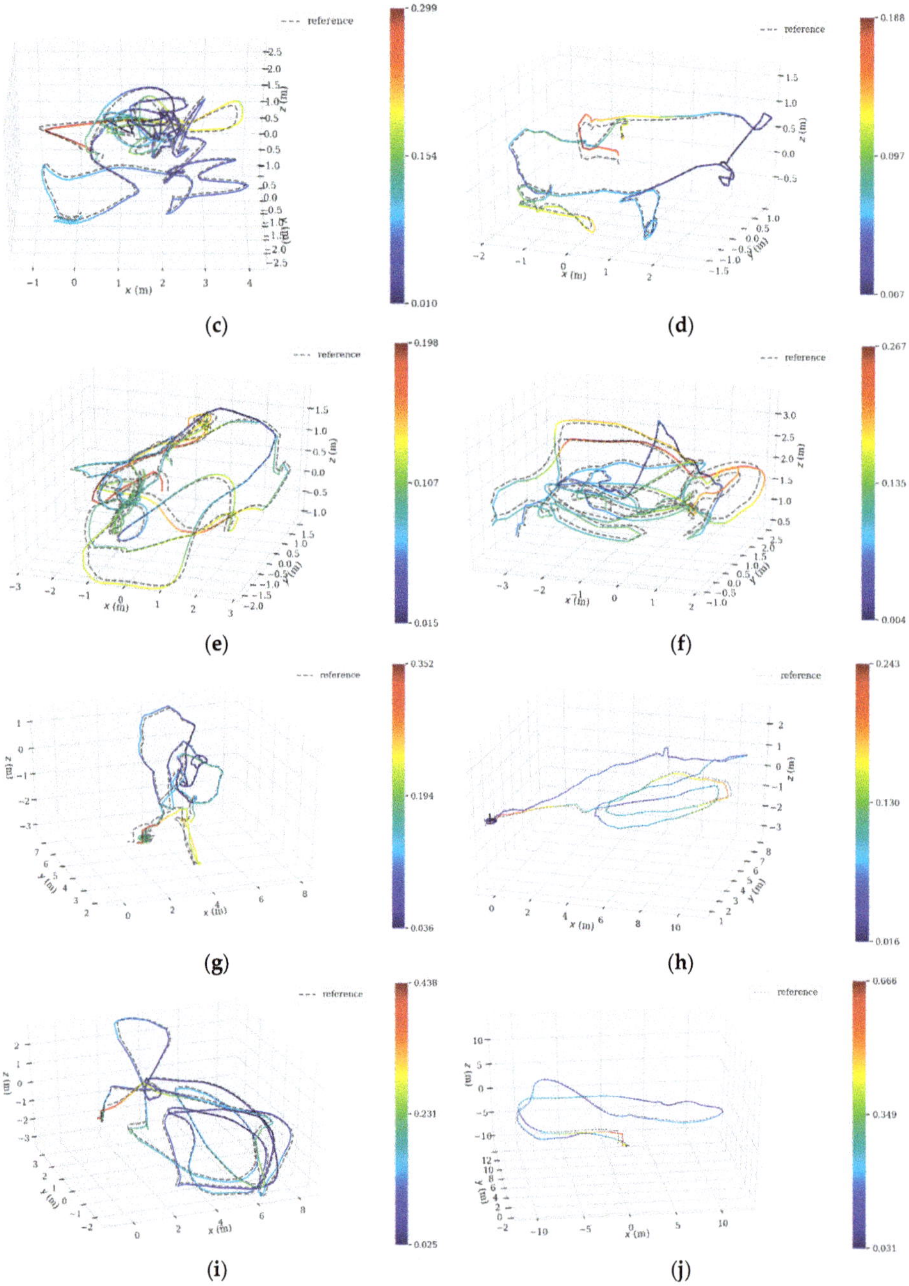

Figure 7. *Cont.*

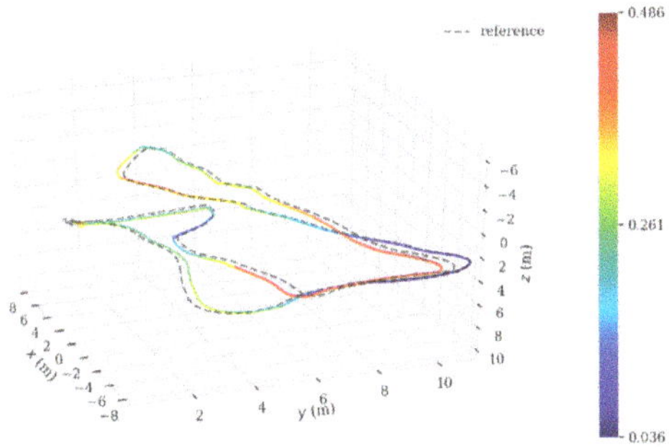

(**k**)

Figure 7. VIO Performances when dealing with sequences at different difficulty levels. (**a**) V1_01_ easy sequence; (**b**) V1_02_ medium sequence; (**c**) V1_03_ difficult sequence; (**d**) V2_01_ easy sequence; (**e**) V2_02_ medium sequence; (**f**) V2_03_ difficult sequence; (**g**) MH_01_easy sequence; (**h**) MH_02_easy sequence; (**i**) MH_03_medium sequence; (**j**) MH_04_difficult sequence; (**k**) MH_05_difficult sequence.

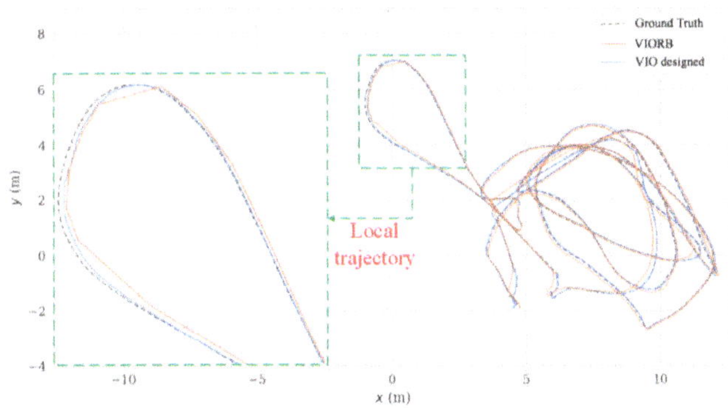

(**a**)

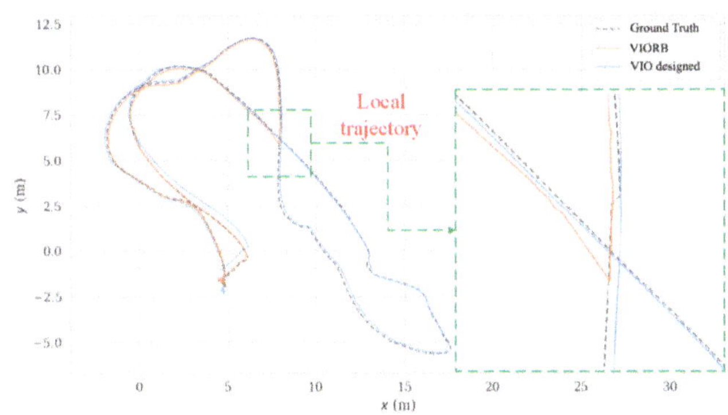

(**b**)

Figure 8. VIO planar trajectory comparisons, (**a**) V1_01_easy sequence; (**b**) MH_04_difficult sequence.

The corresponding trajectory comparisons by VIORB (merely with point-based SLAM) and the designed VIO (with fused point and line based SLAM) are given in Figure 8 with a more detailed APE (see Table 3). Considering the fact that the dynamics of the MAV in space are irregular, the 3D trajectory comparisons, would therefore be insufficiently visible; we are, accordingly, mainly concerned with the projected planar trajectory for further analyses (take typical sequence V1_01_easy and sequence

MH_04_difficult, for example). In Figure 8, the dotted lines represent the projected Ground Truth trajectories, and the orange full lines and blue full lines respectively denote the trajectories by VIORB and the designed VIO. Figure 8b shows that the VIORB scheme failed to dynamically track the desired Ground Truth trajectory stably. Quite clearly, the orange full line shows its interruption in tracking, which is mainly caused by a lack of environmental textures. Even though the loop closure detection part could help VIORB by restarting the positioning tracing thread according to the previous scene information, the short-term tracking failures could be never acceptable for the actual robot inspection applications. Compared with VIORB, the generated trajectories by the designed VIO kept close to the Ground Truth trajectories (being collected by Vicon). The amplified local trajectories clearly show its superior performances in precision.

Table 3. The comparative absolute positioning errors in the European Robotics Challenge (EUROC) datasets.

	Ref. [42]	Ref. [25]	Ref. [18]	Ref. [35]	Ref. [39]	VIO Designed
V1_01_easy	0.1167	0.0958	0.0716	0.0544	0.0591	0.0524
V1_02_medium	0.1392	0.0964	0.0912	0.0849	0.0766	0.0724
V1_03_diffcult	0.1934	×	0.1742	0.1597	0.1302	0.1102
V2_01_easy	0.1267	0.0858	0.1017	0.0712	0.0502	0.0413
V2_02_medium	0.2049	0.1525	0.1876	0.1638	0.0945	0.0815
V2_03_diffcult	×	0.2588	0.2719	0.2347	0.2609	0.2176
MH_01_easy	0.2557	0.1537	0.1647	0.1221	0.0731	0.0513
MH_02_easy	0.1861	0.1595	0.1573	0.1287	0.2327	0.0407
MH_03_medium	0.2176	0.1719	0.2077	0.1365	0.1122	0.1065
MH_04_diffcult	0.3037	0.3165	0.3921	0.1894	0.1394	0.1377
MH_05_diffcult	0.3509	×	×	0.2173	0.2569	0.1546

This high precision can also be indicated by the tri-axial APE in the world coordinate frame in Figure 9, and the VIO designed in this paper supplies the combined system with less APE along the X & Y directions in statistics. Two essential enhancements actually facilitate this good result: one is the fused line feature constraints, which further improved the pose transformation precision between the images; the other is the introduced sliding window, which efficiently reduced the data dimension for the back-end optimization. These enhancements are encouragingly achieved with no sacrifices in the VIO operating efficiency.

The corresponding visualized APE distributions are shown in Figure 10a,b, which also statistically shows the max values (red lines), the median values (yellow lines), the min values (green lines) and the concentrated error distributions, being termed 'mean value domain' (blue and orange blocks). Here, the remaining points represent the outliers with less weight. As we see, the positioning accuracy by the designed VIO over that by VIORB approaches 4 cm for the V1_01_easy sequence, whose value would be impressively over 16 cm for the MH_04_difficult sequence. Table 3 also gives the detailed APE for the total 11 sequences in terms of the comparison between 5 typical VIOs and the VIO designed in this paper. It can be concluded that the proposed VIO steadily presents its superiorities when dealing with the datasets with different difficulty levels.

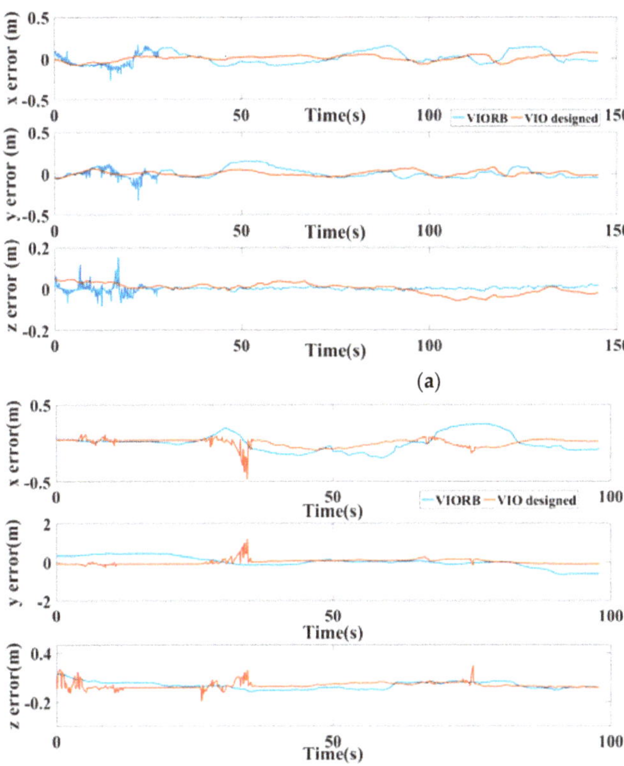

		VIORB	VIO designed
X axis	Max	0.2573	0.1663
	Min	0	0
	Mean	0.0961	0.0249
	Median	0.1051	0.0240
	Rmse	0.1117	0.0942
Y axis	Max	0.3960	0.3231
	Min	0.0009	0
	Mean	0.0905	0.0241
	Median	0.0778	0.0186
	Rmse	0.1241	0.0919
Z axis	Max	0.0759	0.0196
	Min	0.0010	0
	Mean	0.0365	0.0137
	Median	0.0388	0.0096
	Rmse	0.0409	0.0562

(a)

		VIORB	VIO designed
X axis	Max	0.2441	0.4770
	Min	0	0
	Mean	0.0910	0.0431
	Median	0.0802	0.0334
	Rmse	0.1151	0.0556
Y axis	Max	0.6622	1.1919
	Min	0	0
	Mean	0.2165	0.0983
	Median	0.1460	0.0977
	Rmse	0.2918	0.1308
Z axis	Max	0.1467	0.2488
	Min	0	0
	Mean	0.0394	0.0343
	Median	0.0380	0.0314
	Rmse	0.0400	0.0466

(b)

Figure 9. Tri-axial absolute positioning error, (**a**) V1_01_easy sequence; (**b**) MH_04_difficult sequence.

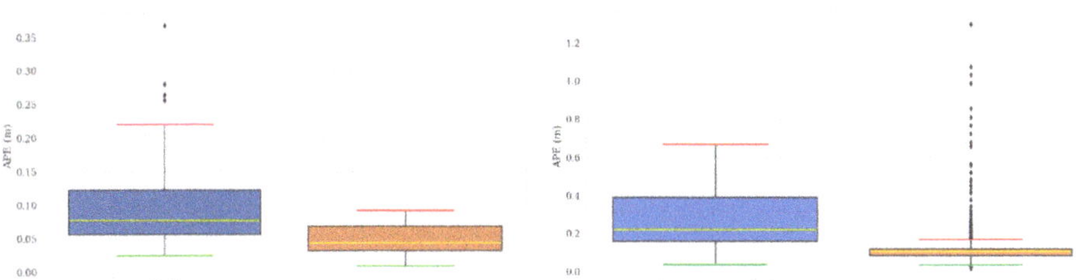

(**a**) (**b**)

Figure 10. Absolute positioning error distribution, (**a**) V1_01_easy sequence; (**b**) MH_04_difficult sequence.

5.1.3. Mapping Results

As an illustration of how the point and line features can be fused to support the operations of the VIO front-end, the sparse maps in terms of the fused point and line features for the V1_01_easy sequence and MH_04_difficult sequence are respectively shown in Figure 11. The green lines represent the trajectories of the keyframes, the blue lines represent the selected keyframes for the sliding window optimization, the black points or lines represent the fixed features in 3D space which have been marginalized out, and the red or pink points and lines represent the features which are still in their early optimizing phase. The results indicate that the designed VIO powerfully provides additional structured supports for the typical sparse maps, and this efficient mapping therefore means that it can be recognized as an eminent tool for the solution of scene reconstructions under complex human interaction situations, being preferred for assisting the practical location, navigation and obstacle avoidance tasks.

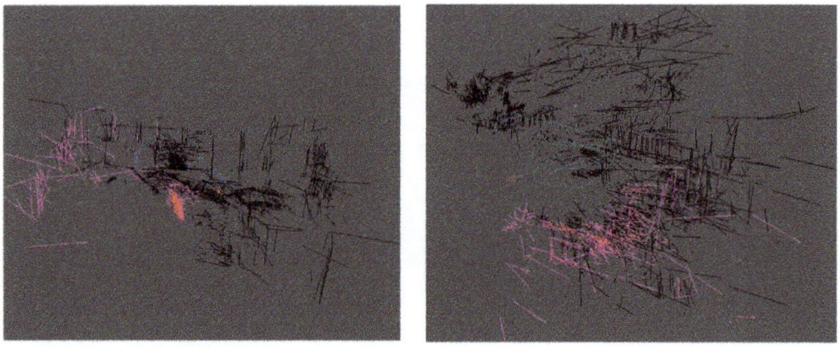

(a) **(b)**

Figure 11. Sparse maps in terms of fused point and line features, (**a**) V1_01_easy sequence; (**b**) MH_04_difficult sequence.

5.2. *Substation Scene Tests and Evaluations*

The positioning performances are experimentally assessed to evaluate the universal applicability of the VIO designed in practice. The substation scene tests are conducted based upon campus substation (100 m × 40 m rectangle) observations and subsequent laboratory analyses. Table 4 presents the calibration parameters of the camera and IMU we use.

Table 4. Calibration parameters of camera and IMU.

Camera Intrinsic	Focal length: $f_x = 363.034$ pixel, $f_y = 364.019$ pixel Principal point of photograph: [366.871, 243.308] Radial distortion: [−3.08252, 8.42513, −1.50093, 2.01707]
Camera/IMU Extrinsic	$T_{CB} = \begin{bmatrix} -0.00647 & -0.99995 & -0.00764 & 0.00534 \\ 0.99998 & -0.00647 & -0.00009 & -0.04303 \\ 0.00005 & -0.00764 & 0.99997 & 0.02303 \\ 0 & 0 & 0 & 1 \end{bmatrix}$
Image parameters	Image resolution: 752 × 480 pixel

Let the robot move around the rectangle with a lower constant velocity; the monocular camera embedded simultaneously entered the working state and was set to initialize the state variables λ by the initialization strategy described in Section 3, once the user workstation obtained the moderate convergent behaviors of the initial state variables. This, then, permitted the robot to perform higher-speed moving tasks (keep walking around the substation). Given the collected information by the user workstation, as shown in Figure 12, the state variables converge for $t > 6.4$ s, as we expected. With a controllable constant velocity, it is relatively efficient to initialize a VIO system. Figure 12e also presents an increase in speed for $t > 9$ s.

The feature extraction results of the VIO front-end in the substation scene is shown in Figure 13; obviously, the VIO front-end is capable of acquiring abundant point and line features even in cases where the illumination changes frequently (the snow diffuse reflection happens). As in Figure 14, the trajectory drawn according to the camera motion is rectangle distributed, which favorably conforms to the planar geometric appearance of the substation. The fused line features is therefore proven to improve the VIO accuracy both for translation and rotation, and to further improve the VIO robustness under the un-ideal illumination environments.

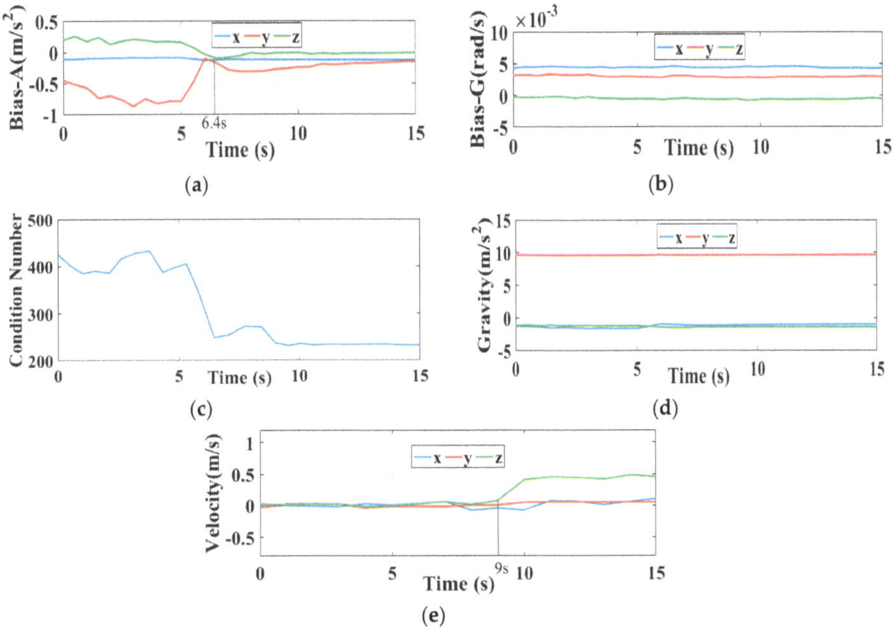

Figure 12. The convergence procedures of the initialization states in the substation scene tests, (**a**) Initialization results of accelerometer bias; (**b**) Initialization results of gyro bias; (**c**) Calculation of the condition number; (**d**) The initialization results of the gravity vector; (**e**) The initialization results of velocity.

Figure 13. Feature extractions of the VIO front-end in the substation scene.

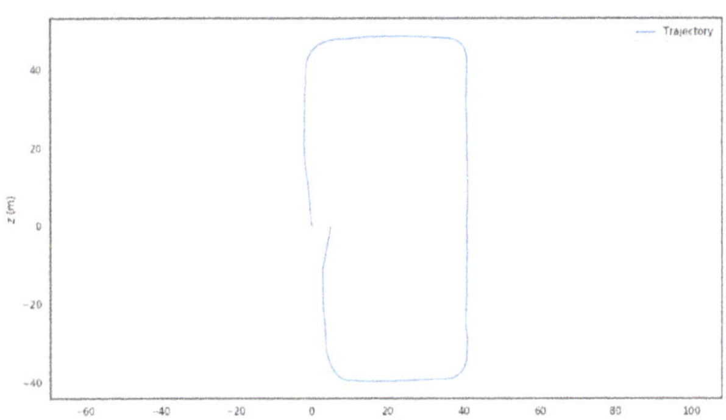

Figure 14. Rectangular trajectory drawn according to the camera motion.

6. Conclusions

An optimized tightly-coupled VIO model which combines an efficient initializing strategy and fused point and line feature matching ideas was employed for navigating and mapping tasks of patrol robots in substations. After exhibiting favorable performances in initializing efficiency, pose estimation and trajectory tracking in a public dataset, this was further experimentally assessed by a campus substation application. It illustrated that, for the feature extraction and matching tasks in the VIO front-end, the fused point and line based method is generally preferred with an L-M optimization strategy; the optimized VIO presents its superiorities even though it is dealing with datasets with different difficulty levels. With respect to the point features and line features, the sparse maps are constructed under the sliding window optimization model, providing the VIO with a necessary location, navigation and obstacle avoidance references. The experimental results showed that a shortened initialization time was derived in practice and that the designed VIO could still accurately fulfill the point and line feature extractions and recover the motion trajectory under un-ideal illumination circumstances. The proposed VIO model therefore fairly meets the SLAM requirements with no external absolute location reference supports.

Author Contributions: L.X., Q.M. and D.C. devised the research and wrote the paper; L.X. and B.M. polished the English expression; Q.M. and H.Y. designed the experiments. All authors have read and approved the final manuscript.

Funding: This research was funded by National Nature Science Foundation of China under Grant 61503073, 61703090, and Natural Research Fund of Science and Technology Department, Jilin Province under Grant 20170101125JC.

Acknowledgments: Special thanks go to Xun Xu, Senior Research Fellow of University of Wollongong, Australia for his cordial help to the successful accomplishment of this paper.

Conflicts of Interest: The authors declare no conflict of interest.

References

1. Liang, X.; Chen, H.; Li, Y. Visual Laser-SLAM in Large-Scale Indoor Environments. In Proceedings of the IEEE International Conference on Robotics & Biomimetics, Qingdao, China, 3–6 December 2016; pp. 19–24.
2. Zhang, Z.; Liu, S.; Tsai, G. PIRVS: An Advanced Visual-Inertial SLAM System with Flexible Sensor Fusion and Hardware Co-Design. In Proceedings of the IEEE International Conference on Robotics and Automation (ICRA), Brisbane, Australia, 21–25 May 2018; pp. 3826–3832.
3. Teng, Z.J.; Qu, Z.Q.; Zhang, L.Y. Research on Vehicle Navigation BD/DR/MM Integrated Navigation Positioning. *J. Northeast Electr. Power Univ.* **2017**, *37*, 98–101. (In Chinese)
4. Guo, X.L.; Yang, T.T.; Zhang, Y.C. Gesture Recognition Based on Kinect Depth Information. *J. Northeast Dianli Univ.* **2016**, *36*, 90–94 (In Chinese).
5. Davison, A.J.; Reid, I.D.; Molton, N.D. MonoSLAM: Real-Time Single Camera SLAM. *IEEE Trans. Pattern Anal. Mach. Intell.* **2007**, *6*, 1052–1067. [CrossRef]
6. Klein, G.; Murray, D. Parallel Tracking and Mapping for Small AR Workspaces. In Proceedings of the IEEE and ACM International Symposium on Mixed and Augmented Reality, Nara, Japan, 13–16 November 2007; pp. 1–10.
7. Mur-Artal, R.; Montiel, J.M.M.; Tardos, J.D. ORB-SLAM: A Versatile and Accurate Monocular SLAM System. *IEEE Trans. Robot.* **2015**, *31*, 1147–1163. [CrossRef]
8. Zhou, H.; Zou, D.; Pei, L. StructSLAM: Visual SLAM with Building Structure Lines. *IEEE Trans. Veh. Technol.* **2015**, *64*, 1364–1375. [CrossRef]
9. Benedettelli, D.; Garulli, A.; Giannitrapani, A. Cooperative SLAM Using M-Space Representation of Linear Features. *Robot. Auton. Syst.* **2012**, *60*, 1267–1278. [CrossRef]
10. Engel, J.; Schöps, T.; Cremers, D. LSD-SLAM: Large-Scale Direct Monocular SLAM. In Proceedings of the European Conference on Computer Vision (Computer Vision—ECCV 2014), Zurich, Switzerland, 6–12 September 2014; pp. 834–849.

11. Forster, C.; Pizzoli, M.; Scaramuzza, D. SVO: Fast Semi-Direct Monocular Visual Odometry. In Proceedings of the IEEE International Conference on Robotics and Automation (ICRA), Hong Kong, China, 31 May–7 June 2014; pp. 1–8.
12. Engel, J.; Koltun, V.; Cremers, D. Direct Sparse Odometry. *IEEE Trans. Pattern Anal. Mach. Intell.* **2018**, *40*, 611–625. [CrossRef]
13. Tian, Y.Y.; Tan, Q.C. Filter Noise Analysis Based on Sub-Pixel Edge Orientation Algorithm. *J. Northeast Dianli Univ.* **2016**, *36*, 43–47. (In Chinese)
14. Hu, J.P.; Li, L.; Xie, Q.; Zhang, D.C. A Novel Segmentation Approach for Glass Insulators in Aerial Images. *J. Northeast Electr. Power Univ.* **2018**, *38*, 87–92. (In Chinese)
15. Weiss, S.; Siegwart, R. Real-Time Metric State Estimation for Modular Vision-Inertial Systems. In Proceedings of the 2011 IEEE International Conference on Robotics and Automation (ICRA), Shanghai, China, 9–13 May 2011; pp. 4531–4537.
16. Ethzasl_sensor_fusion. Available online: https://github.com/ethz-asl/ethzasl_sensor_fusion (accessed on 3 October 2018).
17. Falquez, J.M.; Kasper, M.; Sibley, G. Inertial Aided Dense & Semi-Dense Methods for Robust Direct Visual Odometry. In Proceedings of the IEEE/RSJ International Conference on Intelligent Robots & Systems, Daejeon, Korea, 9–14 October 2016; pp. 3601–3607.
18. Leutenegger, S.; Lynen, S.; Bosse, M. Keyframe-Based Visual-Inertial Odometry Using Nonlinear Optimization. *Int. J. Robot. Res.* **2014**, *34*, 314–334. [CrossRef]
19. Gomez-Ojeda, R.; Zuñiga-Noël, D.; Moreno, F.A. PL-SLAM: A Stereo SLAM System through the Combination of Points and Line Segments. *arXiv* **2017**, arXiv:1705.09479, 1–12. [CrossRef]
20. Hsiao, M.; Westman, E.; Kaess, M. Dense planar-inertial slam with structural constraints. In Proceedings of the 2018 IEEE International Conference on Robotics and Automation (ICRA), Brisbane, Australia, 21–25 May 2018.
21. Huang, W.; Liu, H. Online Initialization and Automatic Camera-IMU Extrinsic Calibration for Monocular Visual-Inertial SLAM. In Proceedings of the 2018 IEEE International Conference on Robotics and Automation (ICRA), Brisbane, Australia, 21–25 May 2018; pp. 5182–5189.
22. Qin, T.; Shen, S. Robust Initialization of Monocular Visual-Inertial Estimation on Aerial Robots. In Proceedings of the 2017 IEEE/RSJ International Conference on Intelligent Robots and Systems (IROS), Vancouver, BC, Canada, 24–28 September 2017; pp. 24–28.
23. Locher, A.; Havlena, M.; Van Gool, L. Progressive Structure from Motion. In Proceedings of the European Conference on Computer Vision (ECCV), Munich, Germany, 8–14 September 2018; pp. 22–38.
24. Saputra, M.R.U.; Markham, A.; Trigoni, N. Visual SLAM and structure from motion in dynamic environments: A survey. *ACM Comput. Surv. (CSUR)* **2018**, *51*, 1–36. [CrossRef]
25. Mur-Artal, R.; Tardos, J.D. Visual-Inertial Monocular SLAM with Map Reuse. *IEEE Robot. Autom. Lett.* **2017**, *2*, 796–803. [CrossRef]
26. Sun, J.; Wang, P.; Qin, Z. Effective Self-Calibration for Camera Parameters and Hand-Eye Geometry Based on Two Feature Points Motions. *IEEE/CAA J. Autom. Sin.* **2017**, *4*, 370–380. [CrossRef]
27. Liu, Y.; Chen, Z.; Zheng, W. Monocular Visual-Inertial SLAM: Continuous Preintegration and Reliable Initialization. *Sensors* **2017**, *17*, 2613. [CrossRef]
28. Zuo, X.; Xie, X.; Liu, Y. Robust Visual SLAM with Point and Line Features. In Proceedings of the IEEE/RSJ International Conference on Intelligent Robots and Systems, Vancouver, BC, Canada, 24–28 September 2017; pp. 1–8.
29. Forster, C.; Carlone, L.; Dellaert, F. On-Manifold Preintegration for Real-Time Visual-Inertial Odometry. *IEEE Trans. Robot.* **2016**, *33*, 99–120. [CrossRef]
30. RGB-D SLAM Dataset and Benchmark. Available online: https://vision.in.tum.de/data/datasets/rgbd-dataset (accessed on 11 June 2018).
31. Mu, X.; Chen, J.; Zhou, Z. Accurate Initial State Estimation in a Monocular Visual-Inertial SLAM System. *Sensors* **2018**, *18*, 506.
32. Zhou, S.; Yang, F. Inverse Quadratic Eigenvalues Problem for Mixed Matrix and Its Optimal Approximation. *J. Northeast Electr. Power Univ.* **2018**, *38*, 85–89. (In Chinese)
33. Ruotsalainen, L.; Kirkko-Jaakkola, M.; Rantanen, J.; Mäkelä, M. Error Modelling for Multi-Sensor Measurements in Infrastructure-Free Indoor Navigation. *Sensors* **2018**, *18*, 590. [CrossRef]

34. Liu, Y.; Yang, D.; Li, J. Stereo Visual-Inertial SLAM with Points and Lines. *IEEE Access* **2018**, *6*, 69381–69392. [CrossRef]
35. Qin, T.; Li, P.; Shen, S. VINS-Mono: A Robust and Versatile Monocular Visual-Inertial State Estimator. *IEEE Trans. Robot.* **2018**, *34*, 1004–1020. [CrossRef]
36. Kümmerle, R.; Grisetti, G.; Strasdat, H. G^2o: A General Framework for Graph Optimization. In Proceedings of the IEEE International Conference on Robotics and Automation, Shanghai, China, 9–13 May 2011; pp. 3607–3613.
37. Qin, T.; Li, P.; Shen, S. Relocalization, Global Optimization and Map Merging for Monocular Visual-Inertial SLAM. In Proceedings of the IEEE International Conference on Robotics and Automation (ICRA), Brisbane, Australia, 21–25 May 2018; pp. 1197–1204.
38. Pumarola, A.; Vakhitov, A.; Agudo, A. PL-SLAM: Real-time Monocular Visual SLAM with Points and Lines. In Proceedings of the IEEE International Conference on Robotics and Automation (ICRA), Singapore, Singapore, 29 May–3 June 2017; pp. 1–6.
39. He, Y.; Zhao, J.; Guo, Y. PL-VIO: Tightly-Coupled Monocular Visual-Inertial Odometry Using Point and Line Features. *Sensors* **2018**, *18*, 1159. [CrossRef] [PubMed]
40. Burri, M.; Nikolic, J.; Gohl, P.; Schneider, T.; Rehder, J.; Omari, S.; Achtelik, M.W.; Siegwart, R. The EuRoC Micro Aerial Vehicle Datasets. *Int. J. Robot. Res.* **2016**, *35*, 1157–1163. [CrossRef]
41. Available online: https://github.com/MichaelGrupp/evo (accessed on 6 December 2018).
42. Kasyanov, A.; Engelmann, F.; Stückler, J. Keyframe-Based Visual-Inertial Online SLAM with Relocalization. *arXiv* **2017**, arXiv:1702.02175, 1–8.

sensors

MDPI

Article

High-Accuracy Globally Consistent Surface Reconstruction Using Fringe Projection Profilometry

Xu Cheng [1], Xingjian Liu [1], Zhongwei Li [1,*], Kai Zhong [1,*], Liya Han [1], Wantao He [2], Wanbing Gan [3], Guoqing Xi [3], Congjun Wang [1] and Yusheng Shi [1]

1 State Key Laboratory of Material Processing and Die & Mould Technology, Huazhong University of Science and Technology, Wuhan 430074, China; xu_cheng@hust.edu.cn (X.C.); xingjianliu@hust.edu.cn (X.L.); hly1993@hust.edu.cn (L.H.); walden@263.net (C.W.); shiyusheng@hust.edu.cn (Y.S.)

2 School of Mechanical Engineering, Heilongjiang University of Science and Technology, Harbin 150022, China; wantaohe@hust.edu.cn

3 Hubei Tri-Ring Forging Co., Ltd, Gucheng 441700, China; wbgan1@163.com (W.G.); gqxi11@163.com (G.X.)

* Correspondence: zwli@hust.edu.cn (Z.L.); kaizhong@hust.edu.cn (K.Z.); Tel.: +86-027-8755-9545 (Z.L.)

Received: 16 January 2019; Accepted: 4 February 2019; Published: 6 February 2019

Abstract: This paper presents a high-accuracy method for globally consistent surface reconstruction using a single fringe projection profilometry (FPP) sensor. To solve the accumulated sensor pose estimation error problem encountered in a long scanning trajectory, we first present a novel 3D registration method which fuses both dense geometric and curvature consistency constraints to improve the accuracy of relative sensor pose estimation. Then we perform global sensor pose optimization by modeling the surface consistency information as a pre-computed covariance matrix and formulating the multi-view point cloud registration problem in a pose graph optimization framework. Experiments on reconstructing a 1300 mm × 400 mm workpiece with a FPP sensor is performed, verifying that our method can substantially reduce the accumulated error and achieve industrial-level surface model reconstruction without any external positional assistance but only using a single FPP sensor.

Keywords: quality control; fringe projection profilometry; depth image registration; 3D reconstruction

1. Introduction

Fringe projection profilometry provides a convenient way to measure dense and accurate three dimensional (3D) surface point cloud of target objects. It plays an increasingly important role in various fields such as industrial quality inspection, prototyping, culture heritage preservation and movie industry [1–5]. Owing to the limited field of view (FOV) and object self-occlusion, 3D point cloud obtained from a single viewpoint only contains partial surface shape data. To reconstruct complete surface models, 3D measurements from multiple viewpoints are deserved to cover the whole object, and their sensor poses need to be precisely tracked to further transform these partial surface point clouds into a global coordinate system [6–9].

Existing sensor pose tracking solutions are mostly based on external assistance methods, such as attaching artificial markers or using external positional equipment such as a laser tracker or optical coordinate measuring machines (CMMs) [10], their usage flexibility is inherently limited. Alternatively, sensor poses can also be directly estimated by using 3D registration techniques [11–13] to compute the relative pose between sequential two measurements. However, sensor pose estimation drifts inevitably exist due to 3D registration inaccuracy. Small sensor pose estimation error which may seem negligible on a local scale, can drastically accumulate along a long scanning trajectory [12,14]. The accumulated error directly leads to surface point clouds inconsistency between the first and last scans and finally breaks the reconstruction result.

Different optimization methods have been adopted to solve the accumulated error problem. Among them, bundle adjustment (BA) is one of the most well-known approaches that performs global optimization by minimizing the reprojection error across different frames. Specifically, BA is conducted by firstly identifying the same visual feature points appearing in multiple frames, and then adjusting the estimated 3D locations of feature points together with the camera poses [7,9]. Nevertheless, BA only optimizes sparse 3D feature points and camera poses, thus it does not guarantee local shape consistency of the reconstructed 3D models [14]. Besides, visual feature detection is the prerequisite for BA optimization, it cannot be fulfilled when the color image is not valid or the target object surface is textureless (e.g., industrial parts).

Instead of optimizing the accumulated error to solve surface inconsistency, Zhou et al. [15] and Whelan el al. [16] chose to deform inconsistency local point clouds together using non-rigid 3D registration techniques, consumer RGB-D sensors are taken as the depth input in their works. Shape deformation provides a simple yet useful approach to obtain globally consistent models, especially in some applications such as indoor reconstruction [12] where surface consistency instead of the accuracy is of the most importance. However, shape deformation is not desired in our problem, because it directly ruins the surface measurement accuracy. Furthermore, since FPP sensor provides high-accuracy surface point cloud measurements, theoretically when sufficient accurate sensor poses are recovered, the individual local 3D point clouds should be able to integrate into a globally consistent model using only rigid transformations.

Differently, Cao et al. [17] and Yue et al. [18] optimized the accumulated error by first identifying the loop closures formed through successful 3D registration between each current frame and other earlier frames, and then performing a pose graph optimization [19] to reduce the sensor poses drifts. However, in their works the loop closures are identifying either by manually checking the 3D point cloud overlapping ratio [17], or by using the measurement system setup information [18], which prevents their further usage in a practical 3D scanning system. Moreover, the pose graph optimization in [17,18] only optimized the inconsistency between two associated sensor poses and their relative pose constraint; it ignores important surface consistency information in the 3D registration process [6].

According to the above analysis, the key to accurate surface reconstruction lies in the reduction of accumulated sensor pose estimation error. In this paper, we present a flexible and accurate method for high-accuracy globally consistent surface reconstruction using a single FPP sensor. The accumulated error problem is addressed from two aspects: (1) observing the underlying principle that surface curvature remains invariant against measurement viewpoint changes, a novel 3D registration method is proposed which fuses both dense geometric and curvature consistency constraints to joint optimize the relative sensor pose estimation. The introduction of curvature consistency constraint implicitly pays attention to high-curvature surfaces, which helps to generate more accurate 3D registration results [20]. (2) We utilize 6-DOF pose distances for adaptive keyframe determination, and use a two-step checking scheme for automatic loop closure detection. By modelling the surface inconsistency information as a pre-computed covariance matrix and formulating the multi-view point cloud registration problem in a pose graph optimization framework, the accumulated error can be effectively reduced to obtain the final accurate sensor pose estimations.

The effectiveness of our proposed method is demonstrated by reconstructing a 1300 mm $\times$ 400 mm workpiece with a FPP sensor. Results show that the proposed method substantially reduced the accumulated error, making the sensor pose estimation accuracy match the measurement accuracy well. Our method shows the ability to accomplish industrial-level surface model reconstruction without any external positional assistance but only using a single FPP sensor.

2. Measurement Principle

In our FPP sensor, a series of sinusoidal fringes along the horizontal axes of projector image frame with constant phase shifting are projected onto a target object, and two cameras capture the distorted fringe images synchronously. The captured images can be expressed as:

$$I_i(x,y) = A_i(x,y) + B_i(x,y)\cos(\phi(x,y) + \delta_i), \quad i = 1,2,3,\ldots,n \tag{1}$$

where (x,y) is the pixel coordinates and is omitted in the following expression, I_i denotes the recorded intensity, A_i indicates the average intensity, B_i represents the modulation intensity, δ_i is the constant phase-shift, n is the phase shift number, and ϕ is the desired phase information. By solving Equation (1), the phase value ϕ can be obtained according to:

$$\phi = -\arctan\left(\sum_{i=1}^{n} I_i \sin(\delta_i) / \sum_{i=1}^{n} I_i \cos(\delta_i)\right). \tag{2}$$

The arctangent function in Equation (2) will result in a phase value within the range of $[-\pi,\pi]$ with 2π discontinuities. In our sensor, multi-frequency heterodyne technology is adopted to construct the continuous phase map [21], so that the correspondence between two camera views can be established unambiguously. Finally, the 3D result can be obtained according to the pre-calibrated camera intrinsic and external parameters. The measurement principle of the FPP sensor is shown in the Figure 1 below.

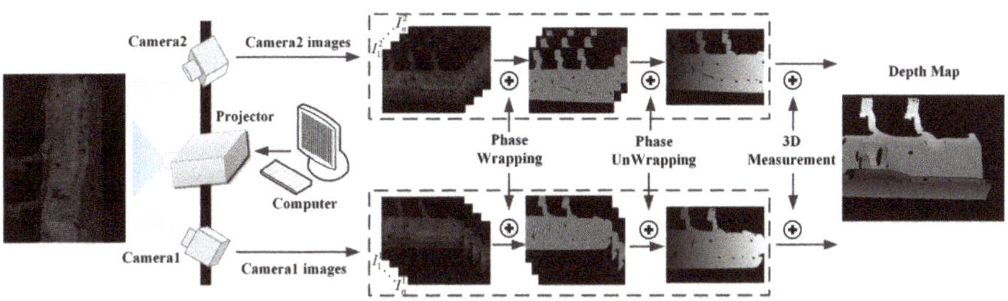

Figure 1. The 3D measurement principle of fringe projection profilometry (FPP sensor).

3. Relative Sensor Pose Estimation

The relative sensor pose estimation between sequential two measurements (also called as frames in the following) is the basis to obtain the initial global sensor pose estimation of each measurement. In this section, we will introduce the proposed method which estimates the relative sensor pose (a rigid transformation) by 3D registering two depth maps to jointly optimize the dense geometric and curvature inconsistency errors. The whole process is conducted by first computing the curvature map of each depth map, and then iteratively performing data association and error minimization steps.

3.1. Curvature Map Estimation

Similarly to depth map (also called as depth image), curvature map is a 2D image in which the value of each pixel is the surface curvature value instead of the depth value. Specifically, for each pixel $\mathbf{x} = (u,v)^\mathsf{T}$ in the depth map with valid depth $z(\mathbf{x})$, its corresponding 3D point coordinate $\mathbf{p}(\mathbf{x})$ can be computed using the inverse of projection function $\Pi(\cdot)$ as:

$$\mathbf{p}(\mathbf{x}) = \Pi^{-1}(\mathbf{x}, z(\mathbf{x}))$$
$$= z(\mathbf{x})(\frac{u - c_x}{f_x}, \frac{v - c_y}{f_y}, 1)^{\mathsf{T}}, \tag{3}$$

where f_x, f_y are the focal lengths and c_x, c_y are the principle point, respectively. The mean curvature of each point on the surface is represented using a surface variation notion in [22]. Hence, the surface curvature value $\kappa(\mathbf{x})$ at pixel $\mathbf{x}$ is estimated by taking the eigen-analysis of the covariance matrix of the local neighbor points of point $\mathbf{p}(\mathbf{x})$. The covariance matrix is defined as:

$$\mathbf{C}(\mathbf{x}) = \sum_i^k (\mathbf{p}_i - \bar{\mathbf{p}})(\mathbf{p}_i - \bar{\mathbf{p}})^{\mathsf{T}}, \quad \bar{\mathbf{p}} = \frac{1}{k}\sum_i^k \mathbf{p}_i, \tag{4}$$

where $\mathbf{p}_i$ is one of the nearest neighbor points of $\mathbf{p}(\mathbf{x})$. Then $\kappa(\mathbf{x})$ can be computed as:

$$\kappa(\mathbf{x}) = \frac{\lambda_0}{\lambda_0 + \lambda_1 + \lambda_2}, \tag{5}$$

where $\lambda_0 \leq \lambda_1 \leq \lambda_2$ are the eigenvalues of the covariance matrix $\mathbf{C}(\mathbf{x})$.

To speed up the nearest neighbor search, we take advantage of the organized point cloud structure embedded in the depth map, only taking adjacent pixels as candidate neighbors. Meanwhile, the geometric continuity constraints are also considered to filter the potential depth gaps by specifying a maximum allowed distance. Pixel $\mathbf{x}_i$ is the nearest neighbor of pixel $\mathbf{x}$, only when it satisfies $\|\mathbf{x} - \mathbf{x}_i\| \leq \sigma_1$, and $\|\mathbf{p}(\mathbf{x}) - \mathbf{p}(\mathbf{x}_i)\| \leq \sigma_2$, where σ_1 and σ_2 represent the pixel and point nearest neighbor distance threshold, respectively. In this paper, we set $\sigma_1 = 3$ and $\sigma_2 = 1.1$ mm (with average point cloud density as 0.275 mm) to allow approximate 30 nearest neighbor points for curvature value estimation.

Figure 2a shows a depth map measured with the FPP sensor, Figure 2b shows the estimated curvature map using our method. Figure 2c is the corresponding 3D point cloud whose color is mapped from the curvature map, and the local detail is displayed in Figure 2d. It can be seen that the estimated curvature map exhibits high consistency with the point cloud surface variation. Furthermore, by carefully handling the discontinuous boundary case, the curvature values at boundary points can also be robustly estimated, as shown in Figure 2d.

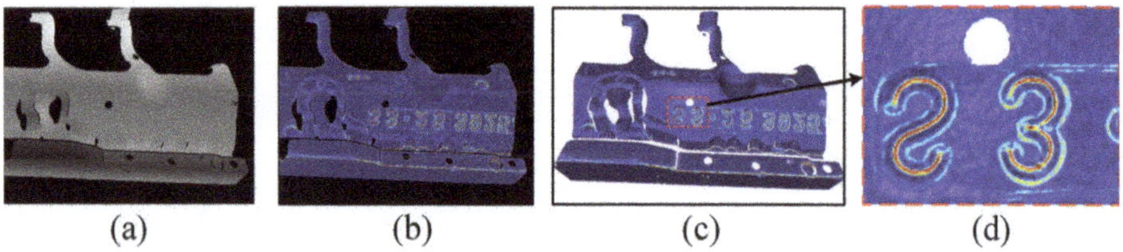

Figure 2. (**a**) A depth map acquired with the FPP sensor, (**b**) Its corresponding curvature map estimated using our method, (**c**) The rendered 3D point cloud with its color mapped from the curvature map, (**d**) Local details of curvature information at local point cloud surface.

3.2. *Data Association*

Data association is to identify the corresponding points between two sequential frames, the correspondence set is then fed to the optimization process to find the optimal relative sensor pose estimation. Assuming small camera motion between sequential frames, the projective data association algorithm [12] is conducted to produce the point correspondences set. Given the relative sensor pose estimation $\mathbf{T}_{i-1,i}$ between current frame f_i and its previous frame f_{i-1}, then for each pixel $\mathbf{x}$ with valid depth in f_i, we first transform its corresponding 3D point $\mathbf{p}(\mathbf{x})$ into the local coordinate system of previous frame f_{i-1} as $\mathbf{T}_{i-1,i}\mathbf{p}(\mathbf{x}) = (x, y, z)^\intercal$. Then the corresponding pixel of $\mathbf{x}$ in frame f_{i-1} can be computed with perspective projection:

$$\begin{aligned}\bar{\mathbf{x}} &= \Pi(\mathbf{T}_{i-1,i}\mathbf{p}(\mathbf{x})) \\ &= \frac{1}{z}\mathbf{K}\mathbf{T}_{i-1,i}\mathbf{p}(\mathbf{x}) \\ &= (f_x\frac{x}{z} + c_x, f_y\frac{y}{z} + c_y)^\intercal,\end{aligned} \tag{6}$$

where $\mathbf{K}$ is the camera intrinsic matrix. Note that for simplicity of notation, we omit the conversions between vectors and its homogeneous vectors throughout this paper.

With the projective data association, multiple pixels in source depth image f_i may correspond to a common pixel in target depth image f_{i-1}. To solve the many-to-one problem, the z-buffer technique is adopted, for each pixel in target depth map f_{i-1} we only keep the corresponding pixel in source depth map f_i with minimum depth. All corresponding points pairs together construct the corresponding set $\mathcal{K}_{i-1,i} = \{(\mathbf{x}, \bar{\mathbf{x}})\}$ between frame f_i and f_{i-1}.

3.3. Minimization

The relative sensor pose optimization function E_{reg} is defined as:

$$E_{reg} = E_{geo} + \lambda E_{cur}, \tag{7}$$

where E_{geo} denotes the geometric inconsistency error, E_{cur} denotes the curvature inconsistency error, λ is the weight of the curvature inconsistency error.

The geometric error is defined as the point-to-plane error [11] between current and previous frames:

$$E_{geo} = \sum_{(\mathbf{x},\bar{\mathbf{x}})\in\mathcal{K}_{i-1,i}} ||(\exp(\hat{\xi})\mathbf{T}_{i-1,i}\mathbf{p}_i(\mathbf{x}) - \mathbf{p}_{i-1}(\bar{\mathbf{x}})) \cdot \mathbf{n}_{i-1}(\bar{\mathbf{x}})||^2, \tag{8}$$

in which $(\mathbf{x}, \bar{\mathbf{x}})$ is one corresponding pixels pair in the corresponding set $\mathcal{K}_{i-1,i}$, $\mathbf{p}_i(\mathbf{x})$ is the local 3D point in the current frame f_i, $\mathbf{p}_{i-1}(\bar{\mathbf{x}})$ and $\mathbf{n}_{i-1}(\bar{\mathbf{x}})$ are the corresponding 3D point and normal, respectively. $\mathbf{T}_{i-1,i}$ is the current estimation of the relative sensor pose between the two frames. $\exp(\hat{\xi}) \in \mathbb{SE}(3)$ is the incremental transformation to be estimated in each iteration, in which $\xi = (\omega, \mathbf{t})^\intercal = (\alpha, \beta, \gamma, t_x, t_y, t_z)^\intercal \in \mathbb{R}^6$.

The curvature inconsistency error E_{cur} is defined as the curvature value inconsistency between the warped curvature map of current frame f_i and the curvature map of previous frame f_{i-1}:

$$\begin{aligned}E_{cur} &= \sum_{(\mathbf{x},\bar{\mathbf{x}})\in\mathcal{K}_{i-1,i}} ||\kappa_i(\mathbf{x}) - \kappa_{i-1}(\bar{\mathbf{x}})||^2 \\ &= \sum_{(\mathbf{x},\bar{\mathbf{x}})\in\mathcal{K}_{i-1,i}} ||\kappa_i(\mathbf{x}) - \kappa_{i-1}(\Pi(\exp(\hat{\xi})\mathbf{T}_{i-1,i}\mathbf{p}_i(\mathbf{x}))||^2,\end{aligned} \tag{9}$$

where $\kappa_i(\mathbf{x})$ is the curvature value at pixel $\mathbf{x}$ of the current frame, $\kappa_{i-1}(\bar{\mathbf{x}})$ is the curvature value at pixel $\bar{\mathbf{x}}$ of the previous frame.

Assuming the incremental pose transformation $\exp(\hat{\xi})$ to optimize at each iteration is small, it can be linearized as $\exp(\hat{\xi}) \approx \mathbf{I} + \hat{\xi}$, where $\hat{\xi} \in \mathfrak{se}(3)$ is the corresponding Lie algebra element:

$$\hat{\xi} = \begin{bmatrix}[\omega]_\times & \mathbf{t} \\ \mathbf{0}^\intercal & 0\end{bmatrix} = \begin{bmatrix}0 & -\gamma & \beta & t_x \\ \gamma & 0 & -\alpha & t_y \\ -\beta & \alpha & 0 & t_z \\ 0 & 0 & 0 & 0\end{bmatrix}, \tag{10}$$

the $[\cdot]_\times : \mathbb{R}^3 \to \mathfrak{so}(3)$ is a linear skew-symmetric operator (see [23] for details).

With this linearization and simple notation $\dot{\mathbf{p}}_{i-1}(\mathbf{x}) = \mathbf{T}_{i-1,i}\mathbf{p}_i(\mathbf{x})$, the error term E_{geo} becomes:

$$\begin{aligned} E_{geo} &\approx \sum_{(\mathbf{x},\bar{\mathbf{x}})\in\mathcal{K}_{i-1,i}} \|((\mathbf{I}+\hat{\xi})\dot{\mathbf{p}}_{i-1}(\mathbf{x}) - \mathbf{p}_{i-1}(\bar{\mathbf{x}}))\cdot \mathbf{n}_{i-1}(\bar{\mathbf{x}})\|^2 \\ &= \sum_{(\mathbf{x},\bar{\mathbf{x}})\in\mathcal{K}_{i-1,i}} \| \begin{bmatrix} \mathbf{p}_{i-1}(\mathbf{x}) \times \mathbf{n}_{i-1}(\bar{\mathbf{x}}) \\ \mathbf{n}_{i-1}(\bar{\mathbf{x}}) \end{bmatrix}^{\mathsf{T}} \xi + (\dot{\mathbf{p}}_{i-1}(\mathbf{x}) - \mathbf{p}_{i-1}(\bar{\mathbf{x}}))\cdot \mathbf{n}_{i-1}(\bar{\mathbf{x}})\|^2 \\ &= \|\mathbf{J}_{geo}\xi + \mathbf{r}_{geo}\|^2, \end{aligned} \tag{11}$$

where $\mathbf{J}_{geo}$ is the Jacobian matrix and $\mathbf{r}_{geo}$ is the residual vector. Similarly, the error term E_{cur} becomes:

$$\begin{aligned} E_{cur} &\approx \sum_{(\mathbf{x},\bar{\mathbf{x}})\in\mathcal{K}_{i-1,i}} \|\kappa_i(\mathbf{x}) - \kappa_{i-1}(\Pi((\mathbf{I}+\hat{\xi})\dot{\mathbf{p}}_{i-1}(\mathbf{x})))\|^2 \\ &= \sum_{(\mathbf{x},\bar{\mathbf{x}})\in\mathcal{K}_{i-1,i}} \|\kappa_i(\mathbf{x}) - \kappa_{i-1}(\frac{1}{z}\mathbf{K}(\mathbf{I}+\hat{\xi})\dot{\mathbf{p}}_{i-1}(\mathbf{x}))\|^2 \\ &\approx \sum_{(\mathbf{x},\bar{\mathbf{x}})\in\mathcal{K}_{i-1,i}} \| -\frac{\partial \kappa_{i-1}(\bar{\mathbf{x}})}{\partial \bar{\mathbf{x}}} \frac{\partial \bar{\mathbf{x}}}{\partial \hat{\xi}\dot{\mathbf{p}}_{i-1}(\mathbf{x})} \frac{\partial \hat{\xi}\dot{\mathbf{p}}_{i-1}(\mathbf{x})}{\partial \xi}\xi + \kappa_i(\mathbf{x}) - \kappa_{i-1}(\frac{1}{z}\mathbf{K}\dot{\mathbf{p}}_{i-1}(\mathbf{x}))\|^2 \\ &= \|\mathbf{J}_{cur}\xi + \mathbf{r}_{cur}\|^2. \end{aligned} \tag{12}$$

With the above linearization, minimization of Equation (7) allows to solve the following linear system:

$$(\mathbf{J}_{geo}^{\mathsf{T}}\mathbf{J}_{geo} + \lambda \mathbf{J}_{cur}^{\mathsf{T}}\mathbf{J}_{cur})\xi = -(\mathbf{J}_{geo}^{\mathsf{T}}\mathbf{r}_{geo} + \lambda \mathbf{J}_{cur}^{\mathsf{T}}\mathbf{r}_{cur}). \tag{13}$$

In each iteration, we compute Jacobian $\mathbf{J}_{geo}$, $\mathbf{J}_{cur}$ and residual $\mathbf{r}_{geo}$, $\mathbf{r}_{cur}$ at current relative sensor pose estimation $\mathbf{T}_{i-1,i}$, and solve the linear system in Equation (13) to find the ξ that best satisfies the geometric and curvature consistency constraint. Then the relative pose $\mathbf{T}_{i-1,i}$ is updated to $\exp(\hat{\xi})\mathbf{T}_{i-1,i}$, and taken as the initialization for the next iteration.

When the optimization converges, the $\mathbf{T}_{i-1,i}$ is taken as the final relative sensor pose estimation between two frames. We fix the sensor pose of the first frame f_1 as $\mathbf{T}_1 = \mathbf{I}$ and regard it as the world coordinate system. Then the initial global sensor pose of frame f_i is computed as $\mathbf{T}_i = \mathbf{T}_{i-1}\mathbf{T}_{i-1,i}$.

Figure 3 shows the 3D registration results comparison between the proposed method and two other methods. The sensor pose estimation accuracy is directly reflected in the surface shape consistency of two registered point clouds. When independently visual inspecting each registration result, each method seems to converge to a correct result. However, when comparing the registration results between Figure 3b–d, it is not hard to see that the relative sensor pose estimation accuracy of our method outperforms the other two methods.

Figure 3. (**a**) Initial relative pose between source (green) and target (yellow) point cloud, (**b**) Registration result by only minimizing geometric error in Equation (8), (**c**) Point-to-plane ICP performed on 3D point cloud with a max distance threshold to eliminate outliers, (**d**) Minimizing both of the geometric error and curvature error as proposed in this paper.

Figure 4a,b represents the curvature value difference map between source and target point cloud before and after the 3D registration, respectively. The curvature difference map is built on the target frame f_{i-1}, correspondences are built using the above data association method. Gray pixels indicate that no correspondence is built for these pixels. It can be seen that the curvature value difference from Figure 4a,b decreases dramatically over the whole map, which demonstrates the significance of introducing curvature map consistency into the 3D registration constraints.

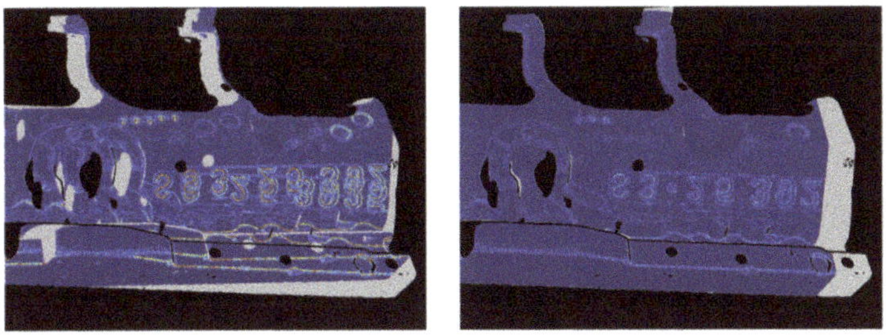

(a) (b)

Figure 4. Curvature difference map (**a**) Before registration, (**b**) After registration.

4. Global Sensor Pose Optimization

Though fusing curvature consistency information improves the accuracy of the estimated relative sensor poses, the global sensor pose drift will inevitably accumulate during a long scanning process. To reduce the accumulated error and obtain globally consistent 3D models, successful relative pose estimation to much earlier frames (also called as building *loop closure*) is deserved. In this section, we will first introduce how to automatically build a series of loop closures with the proposed adaptive keyframe selection and the two-step checking method. We will then introduce our method which performs multi-view point cloud registration in a pose graph optimization framework [19].

4.1. Keyframe Selection

Detecting loop closure for every new-income measurement is not optimal; it will greatly increase the computation cost after a long time scanning. Therefore, we only detect loop closure for selected keyframes. We utilize 6-DOF (degree of freedom) pose distance metrics to determine when to add a new keyframe for further loop closure detection. For each new input frame f_j, we evaluate the relative pose distances between it and the last added keyframe f_{i-1}^k. In which, the rotation distance is measured as the rotation angle using the Rodrigues' formula:

$$d(\mathbf{R}_j, \mathbf{R}_{i-1}^k) = |\arccos(\frac{trace(\mathbf{R}_j^\top \mathbf{R}_{i-1}^k) - 1}{2})|. \tag{14}$$

The translation distance is computed as:

$$d(\mathbf{t}_j, \mathbf{t}_{i-1}^k) = \|\mathbf{t}_j - \mathbf{t}_{i-1}^k\|, \tag{15}$$

If either the rotation or translation distance exceeds its corresponding threshold $\sigma_\mathbf{R}$ or $\sigma_\mathbf{t}$, the current frame f_j is marked as a new keyframe f_i^k. We set $\sigma_\mathbf{R} = 20°$, $\sigma_\mathbf{t} = 130$ mm in our paper. Figure 5 shows the keyframe selection results using the total 146 depth maps acquired with our FPP sensor (see Section 5). Gray points identify the 34 selected keyframes out of a total 146 depth maps.

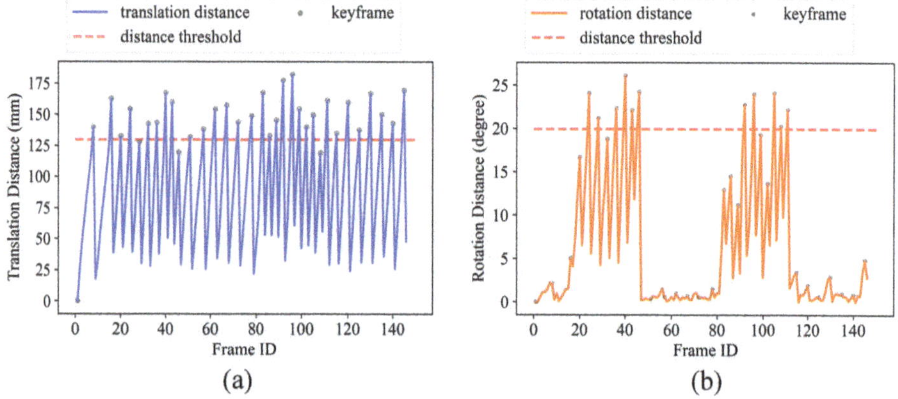

Figure 5. (**a**,**b**) show the translation and rotation distance between each frame with its previous keyframe, respectively.

4.2. Loop Closure Detection

For each new added keyframe f_i^k, we use a two-step checking scheme to detect whether it forms correct loop closures with previous keyframes. If two keyframes construct a loop closure, then they must fulfil: (1) the overlapping area between two point clouds is enough, (2) the mean absolute error (MAE) between them is small.

The overlapping area ratio is crucial for arbitrary two frames with loop closures, as small overlapping area ratios are prone to correspond to non-loop-closure connection. In this paper, we propose to use the projective association algorithm to efficiently compute the overlapping area ratio between two keyframes. When a new keyframe f_i^k arrives, we compute its depth valid map V_i^k for each pixel. $V_i^k(\mathbf{x}) = 1$ for each pixel where its depth is valid, and $V_i(\mathbf{x}) = 0$ when depth is not valid. Then for a pair of keyframes f_i^k and f_j^k, we obtain the correspondence set $\mathcal{K}_{i,j}^k = \{(\mathbf{x}, \bar{\mathbf{x}})\}$ using the data association method in Section 3.2. Note that, the relative sensor pose between f_i^k and f_j^k is computed as $\mathbf{T}_{i,j}^k = \mathbf{T}_i^{k^{-1}} \mathbf{T}_j^k$ here. A correspondence pair $(\mathbf{x}, \bar{\mathbf{x}})$ is identified as overlapped when $V_j^k(\bar{\mathbf{x}}) = 1$. We collect all overlapped point pairs, the overlapping ratio is computed as $\tau_o = N/M$, where N is the overlapped points number, M is the total number of points with valid depth. If the overlapping ratio τ_o is larger than the threshold σ_o, we mark keyframe f_i^k and f_j^k as a candidate loop closure. Figure 6a shows the overlapping ratios between the 34th keyframe (frame 145) with all its previous keyframes, we set the overlapping ratio threshold $\sigma_o = 0.65$ in this paper. We select frame 36, 96, 120 and 140 to visualize the correctness of our proposed method as shown in Figure 6b, dotted line sketches the scanning path.

We then check the dense geometric consistency to further validate the correctness of these candidate loop closures. A candidate loop closure (f_i^k, f_j^k) is considered as reliable only if the MAE of the correspondence points between two frames is below a threshold σ_r:

$$\frac{1}{|\mathcal{K}_{i,j}|} \sum_{(\mathbf{x},\bar{\mathbf{x}}) \in \mathcal{K}_{i,j}} \|\mathbf{T}_{i,j}^k \mathbf{p}(\mathbf{x}) - \mathbf{p}(\bar{\mathbf{x}})\| < \sigma_r. \tag{16}$$

If the two-step checks all passed, the two frames are further registered together to construct a loop closure.

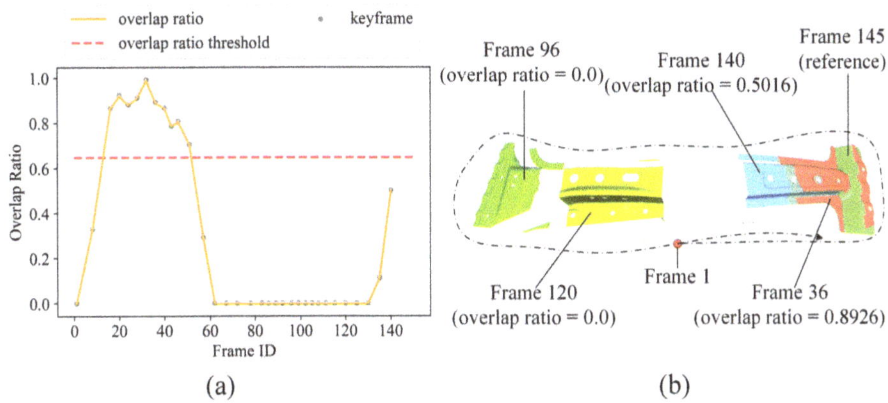

Figure 6. (**a**) The computed overlapping ratios between frame 145 with all its previous keyframes and (**b**) Frame 36, 96, 120, 140 and the reference frame 145.

4.3. Graph Based Sensor Pose Optimization

Removing the accumulated error to get globally consistent model needs to eliminate the surface inconsistencies across all associated point clouds. We define the surface inconsistency as a error term $F_{i,j}$ in terms of the dense geometric registration error between frame f_i and f_j, as:

$$\begin{aligned} F_{i,j} &= \sum_{\mathbf{p}_i,\mathbf{p}_j} \|(\mathbf{T}_j\mathbf{p}_j - \mathbf{T}_i\mathbf{p}_i)\|^2 \\ &= \sum_{\mathbf{p}_i,\mathbf{p}_j} \|\mathbf{T}_i^{-1}\mathbf{T}_j\mathbf{p}_j - \mathbf{p}_i\|^2 \\ &\approx \sum_{\mathbf{p}_i,\mathbf{p}_j} \|\mathbf{T}_i^{-1}\mathbf{T}_j\mathbf{T}_{j,i}\mathbf{p}_i - \mathbf{p}_i\|^2. \end{aligned} \tag{17}$$

Note that $\mathbf{T}_i$, $\mathbf{T}_j$ is obtained through the relative sensor pose estimation in Section 3.3, $\mathbf{T}_{j,i}$ is obtained through the loop closure detection in Section 4.2. Inconsistency exists between $\mathbf{T}_{j,i}$ and $\mathbf{T}_i$, $\mathbf{T}_j$ due to the accumulated error. Line (17) holds by restricting the corresponding points $(\mathbf{p}_i, \mathbf{p}_j)$ must fulfil $\|\mathbf{T}_{j,i}\mathbf{p}_i - \mathbf{p}_j\| < \epsilon$, we set $\epsilon = 1.0$ mm in this paper.

Then by approximating $\mathbf{T}_i^{-1}\mathbf{T}_j\mathbf{T}_{j,i} = \mathbf{I} + \hat{\xi}_{i,j}$, Equation (17) can be written as:

$$\begin{aligned} F_{i,j} &\approx \sum_{\mathbf{p}_i,\mathbf{p}_j} \|\hat{\xi}_{i,j}\mathbf{p}_i\|^2 \\ &= \sum_{\mathbf{p}_i,\mathbf{p}_j} \| \begin{bmatrix} -[\mathbf{p}_i]_\times & \mathbf{I} \end{bmatrix} \xi_{i,j}\|^2, \end{aligned} \tag{18}$$

in which $\xi_{i,j}$ actually measures the inconsistency between sensor pose $\mathbf{T}_i$, $\mathbf{T}_j$ and their relative pose constraint $\mathbf{T}_{i,j}$. Define $\mathbf{G}_i = \begin{bmatrix} -[\mathbf{p}_i]_\times & \mathbf{I} \end{bmatrix}$, we obtain:

$$\begin{aligned} F_{i,j} &\approx \sum_{\mathbf{p}_i,\mathbf{p}_j} \|\mathbf{G}_i\xi_{i,j}\|^2 \\ &= \xi_{i,j}^\top \sum_{\mathbf{p}_i,\mathbf{p}_j} \mathbf{G}_i^\top\mathbf{G}_i\xi_{i,j} \\ &= \xi_{i,j}^\top\Omega_{i,j}\xi_{i,j}. \end{aligned} \tag{19}$$

Equation (19) shows the surface inconsistency term $F_{i,j}$ can be represented with the sensor pose inconsistency term $\xi_{i,j}$ and a covariance matrix $\Omega_{i,j} = \sum_{\mathbf{p}_i,\mathbf{p}_j} \mathbf{G}_i^\top \mathbf{G}_i$, it is constant and can be pre-computed for each term during the 3D registration process.

Let $\mathcal{C}$ be the set of indices for which a connection between two sensor poses exists, then the multi-view point cloud registration problem can be formulated as:

$$\begin{aligned} F &= \sum_{(i,j)\in\mathcal{C}} F_{i,j} \\ &= \sum_{(i,j)\in\mathcal{C}} \xi_{i,j}^\top \Omega_{i,j} \xi_{i,j}. \end{aligned} \tag{20}$$

This exactly defines a pose graph optimization, which can be directly solved using the *g2o* library [19]. Figure 7 shows the pose graph constructed with our method. Vertices represent the 6-DoF sensor poses, edges represent the constraints between sensor poses. The pose graph is visualized with the *g2o viewer* software.

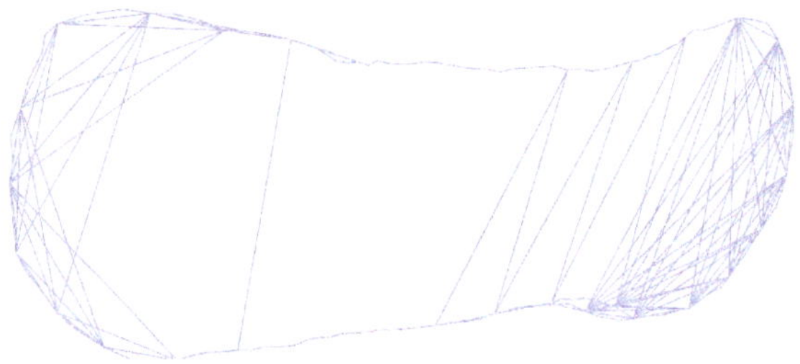

Figure 7. It shows a pose graph consists of 146 pose vertices, 229 edges (84 loop closure edges inside).

5. Experiment

In the experiment, a FPP sensor is constructed using (1) a Texas Instruments LighterCrafter4500 board (Texas Instruments, Dallas, TX, USA) for fringe patterns projection, (2) two Basler acA1300-30gm cameras (Basler AG, Ahrensburg, Germany) simultaneously capturing the modulated images with pixel resolution of 1296 × 966. The proposed method is validated by scanning a 1300 mm × 400 mm sheet metal using the FPP sensor as shown in Figure 8, the 3D measurement and model reconstruction are conducted on a desktop PC with a 3.3 GHz Intel Xeon CPU and 16 GB RAM. By moving the FPP sensor around, complete scan of the sheet metal with totally 146 frames (depth maps) acquired is accomplished.

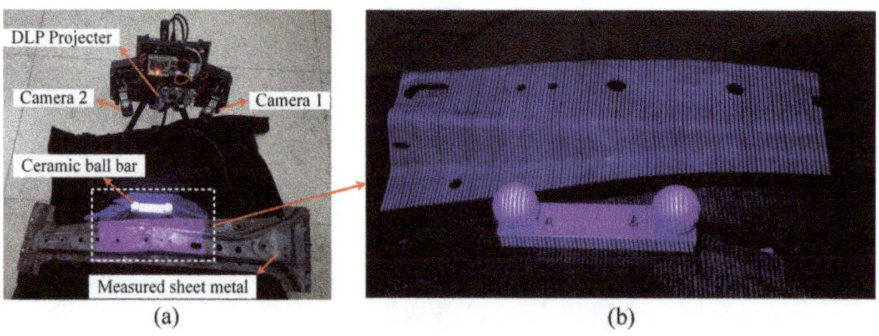

Figure 8. (**a**) The measurement scene, (**b**) Sinusoidal fringe pattern projected onto the measured object.

To test and verify the accuracy and effectiveness of the proposed relative sensor pose estimation method and the global optimization method, a ceramic ball bar is placed beside the measured sheet metal. The reconstruction accuracy can then be well examined by qualitatively observing the surface consistency and quantitatively analyzing the size fitting results of the reconstructed ceramic ball bar.

5.1. Relative Sensor Pose Estimation Accuracy

The accuracy of our proposed relative sensor pose estimation method is tested first. The sensor pose of each frame relative to the world coordinate system (frame 1) is separately estimated by (1) jointly optimizing the geometric and curvature consistency constraints (our method), (2) only optimizing the geometric consistency constraint for comparison. With the estimated sensor poses, 3D point cloud of each frame is transformed to the world coordinate system and further voxel downsampled to a unified 3D point cloud. Figure 9a shows the reconstructed surface of sheet metal with our method, it shows that the overall shape of our reconstruction result matches the actual sheet metal shape well. The point clouds are rendered with *Open3D* library [24].

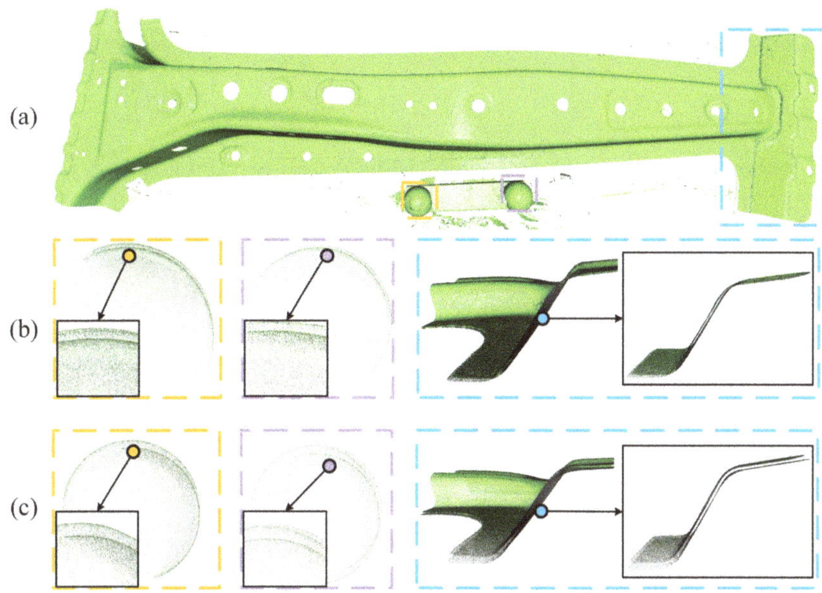

Figure 9. (**a**) The reconstructed surface and (**b**) Its local details using both geometric and curvature consistency constraints, (**c**) The corresponding local details using only geometric consistency constraints (its complete surface model not displayed here).

On the other side, sensor pose estimation error inevitably accumulated in the reconstruction process, which leads to obvious surface shape artifacts, as shown in Figure 9b,c. In which, Figure 9b shows the local surface inconsistency at 3 difference places using our method, Figure 9c shows the corresponding results using only geometric consistency constraints. With this comparison, it is not hard to see that introducing the curvature consistency constraint effectively improves the sensor pose estimation accuracy, which provides a good foundation for further global optimization.

5.2. Global Sensor Pose Optimization Accuracy

Based on the sensor pose estimation results above, the global optimization is performed by (1) keyframe selection, (2) loop closure detection and (3) pose graph optimization. Then the globally optimized reconstruction result is obtained with the optimized sensor poses. Figure 10a,b show the optimized surface model and its local details, respectively. With the global model optimization, we obtained globally consistent surface model, surface inconsistencies due to the accumulated error are well optimized as shown in Figure 10b.

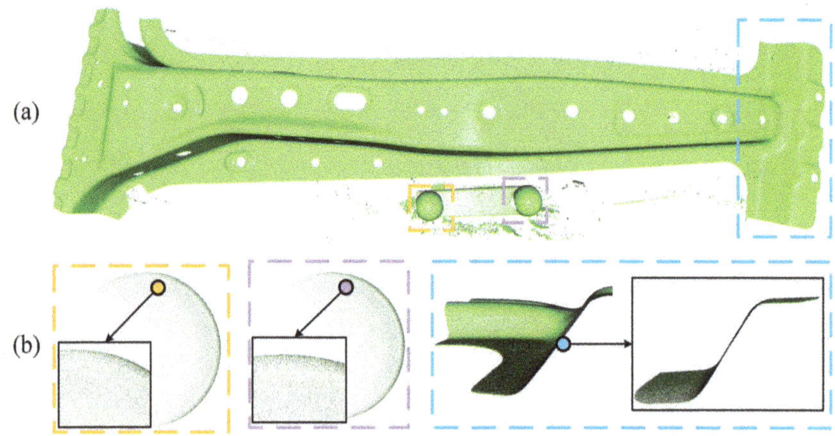

Figure 10. (**a**) The reconstructed surface after global optimization, (**b**) Its local details.

To further quantitatively analyze the accuracy improvement with the global optimization, we computed the relative translation and rotation changes of each keyframe pose before and after global optimization, as shown in Figure 11, the optimized poses are taken as the reference values here. It demonstrates that even very small translation estimation inaccuracy (less than 2.0 mm) and rotation estimation inaccuracy (less than 0.10°) in the reconstruction range of 1300 mm × 400 mm, are enough to cause obvious surface inconsistency (as shown in Figure 9b), and lead to reconstruction results that are unusable for high-accuracy dimensional inspection.

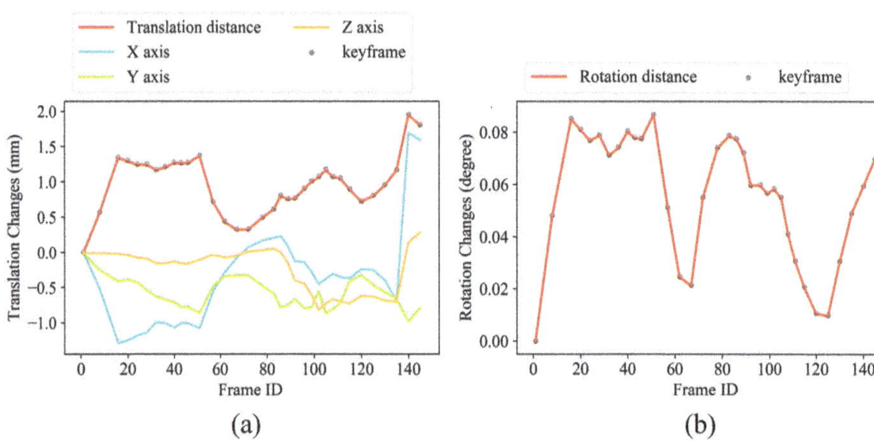

Figure 11. (**a**) Relative translation and (**b**) Rotation changes of each keyframe pose.

Meanwhile, the absolute accuracy of the reconstructed surface model can be directly and precisely tested by comparing (1) diameter fitting values of two spheres, (2) standard deviation values of Euclidean distances between sphere surface 3D points and the fitted sphere surface, (3) Euclidean distance between two sphere centers. The comparison is made between the not-optimized model, globally-optimized model and the ground truth. The ground-truth is obtained with the fitting values of frame 130, because two spheres are both measured in this frame, the fitting values are only related to the measurement accuracy of our FPP sensor, and are not affected by any sensor pose estimation error. Specifically, for each kind of data source, we manually cropped the corresponding points that belong to the two sphere surfaces, and fitted the diameter and standard deviation values using the *Geomagic* software.

Table 1 shows the comparison results of diameter and standard deviation fitting values of two spheres. The standard deviation values directly reflect the surface consistency of our reconstruction model. After the global optimization, it decreases from 0.1971 mm to 0.0282 mm for sphere 1, and decreases from 0.2534 mm to 0.0301 mm for sphere 2. Furthermore, the standard deviation value of globally-optimized model is very close to the value of a single measurement (frame 130), which demonstrates that our reconstructed surface exhibits very good shape consistency.

Table 1. Comparison of the diameter and standard deviation fitting results between not-optimized and globally-optimized model.

	Data Source	Diameter (mm)	Standard Deviation (mm)
Sphere 1	not-optimized model	44.0074	0.1971
	globally-optimized model	**43.9713**	**0.0282**
	Frame 130	44.1121	0.0164
Sphere 2	not-optimized model	43.8685	0.2534
	globally-optimized model	**44.0624**	**0.0301**
	Frame 130	44.0881	0.0258

We also compared the difference of the sphere center distances between not-optimized and globally-optimized models, as shown in Table 2. The absolute error of sphere center distance relative to the ground truth decreases from 0.2080 mm to 0.0205 mm, the relative error relative to the ground truth decreases from 0.1387% to 0.0137%.

Table 2. Sphere center distance fitting results with the absolute and relative errors relative to the ground truth.

Data Source	Sphere Center Distance (mm)	Absolute Error (mm)	Relative Error (%)
not-optimized model	149.7950	0.2080	0.1387
globally-optimized model	**149.9825**	**0.0205**	**0.0137**
Frame 130	150.0030	/	/

Both of the above two comparison results explain the surface shape inconsistency refinement from Figure 9a,b to Figure 10a,b, and illustrate that with the global optimization (1) the accumulated error is substantially reduced to less than 1/10 of the not-optimized reconstruction result, (2) the final sensor pose estimation accuracy can well match the measurement accuracy of our FPP sensor.

6. Conclusions

In this paper, we present a high-accuracy globally consistent surface reconstruction method using fringe projection profilometry. The accumulated sensor pose estimation error problem is solved with a first relative sensor pose estimation step and a following global sensor pose optimization step. The former step tries to reduce the accumulated error by maximizing the relative sensor pose estimation accuracy; it helps to ensure the initial sensor poses lie in the convergence basin of the following global optimization method. The latter step globally optimizes the sensor poses through a multi-view point cloud registration formulated in the pose graph optimization framework. Besides, adaptive keyframe selection and loop closure detection method are proposed to efficiently and automatically build point cloud connections and their relative pose constraints, which are the prerequisites of global sensor pose optimization. By qualitatively observing and quantitatively analyzing the reconstruction results of a 1300 mm $\times$ 400 mm workpiece, we validated the effectiveness and accuracy of our method.

Our method demonstrates the ability to accomplish industrial-level surface model reconstruction without any external positional assistance but only using a single FPP sensor.

Since our reconstruction method is based on 3D registration, it also shares some limitations similar to most 3D registration based surface reconstruction methods [7,12,16]. For example, when the target object is near a plane, 3D registration may not converge to a correct result due to insufficient geometric constraint [11], which will stop the sensor poses from being robustly tracked. A possible solution is to further exploit the usage of surface textures constraint to help the robust tracking of sensor poses.

Author Contributions: X.C. conceived the main idea, designed the main algorithm and wrote the original draft. X.C. and L.H. wrote the algorithm, X.C. designed the main experiments under the supervision of Z.L., K.Z., C.W. and Y.S. The experimental results were analyzed by X.C., X.L. L.H. and K.Z. And W.G. and G.X gave suggestions on the experiments and provided the measured workpiece. X.L., K.Z. and W.H. reviewed and edited the original draft.

Funding: This research was funded by National Key Research and Development Program of China (No. 2018YFB1105800, 2017YFB1103200, 2018YFB110170), National Natural Science Foundation of China (No. 51505169, 51675165).

Conflicts of Interest: The authors declare no conflict of interest.

References

1. Zhang, S. Recent progresses on real-time 3D shape measurement using digital fringe projection techniques. *Opt. Lasers Eng.* **2010**, *48*, 149–158, doi:10.1016/j.optlaseng.2009.03.008. [CrossRef]
2. Zhong, K.; Li, Z.; Li, R.; Shi, Y.; Wang, C. Pre-calibration-free 3D shape measurement method based on fringe projection. *Opt. Express* **2016**, *24*, 14196–14207, doi:10.1364/OE.24.014196. [CrossRef] [PubMed]
3. Li, B.; Zhang, S. Superfast high-resolution absolute 3D recovery of a stabilized flapping flight process. *Opt. Express* **2017**, *25*, 27270–27282. [CrossRef] [PubMed]
4. Zuo, C.; Feng, S.; Huang, L.; Tao, T.; Yin, W.; Chen, Q. Phase shifting algorithms for fringe projection profilometry: A review. *Opt. Lasers Eng.* **2018**, *109*, 23–59, doi:10.1016/j.optlaseng.2018.04.019. [CrossRef]
5. Han, L.; Cheng, X.; Li, Z.; Zhong, K.; Shi, Y.; Jiang, H. A Robot-Driven 3D Shape Measurement System for Automatic Quality Inspection of Thermal Objects on a Forging Production Line. *Sensors* **2018**, *18*, 4368. [CrossRef] [PubMed]
6. Choi, S.; Zhou, Q.; Koltun, V. Robust reconstruction of indoor scenes. In Proceedings of the 2015 IEEE Conference on Computer Vision and Pattern Recognition (CVPR), Boston, MA, USA, 7–12 June 2015; pp. 5556–5565; doi:10.1109/CVPR.2015.7299195. [CrossRef]
7. Dai, A.; Nießner, M.; Zollhöfer, M.; Izadi, S.; Theobalt, C. BundleFusion: Real-Time Globally Consistent 3D Reconstruction Using On-the-Fly Surface Reintegration. *ACM Trans. Graph.* **2017**, *36*, doi:10.1145/3072959.3054739. [CrossRef]
8. Endres, F.; Hess, J.; Sturm, J.; Cremers, D.; Burgard, W. 3-D Mapping With an RGB-D Camera. *IEEE Trans. Robot.* **2014**, *30*, 177–187, doi:10.1109/TRO.2013.2279412. [CrossRef]
9. Henry, P.; Krainin, M.; Herbst, E.; Ren, X.; Fox, D. RGB-D mapping: Using Kinect-style depth cameras for dense 3D modeling of indoor environments. *Int. J. Rob. Res.* **2012**, *31*, 647–663, doi:10.1177/0278364911434148. [CrossRef]
10. Cuypers, W.; Gestel, N.V.; Voet, A.; Kruth, J.P.; Mingneau, J.; Bleys, P. Optical measurement techniques for mobile and large-scale dimensional metrology. *Opt. Lasers Eng.* **2009**, *47*, 292–300, doi:10.1016/j.optlaseng.2008.03.013. [CrossRef]
11. Rusinkiewicz, S.; Levoy, M. Efficient variants of the ICP algorithm. In Proceedings of the Third International Conference on 3-D Digital Imaging and Modeling, Quebec City, QC, Canada, 28 May–1 June 2001; pp. 145–152; doi:10.1109/IM.2001.924423. [CrossRef]
12. Newcombe, R.A.; Izadi, S.; Hilliges, O.; Molyneaux, D.; Kim, D.; Davison, A.J.; Kohi, P.; Shotton, J.; Hodges, S.; Fitzgibbon, A. KinectFusion: Real-time dense surface mapping and tracking. In Proceedings of the 2011 10th IEEE International Symposium on Mixed and Augmented Reality, Basel, Switzerland, 26–29 October 2011; pp. 127–136; doi:10.1109/ISMAR.2011.6092378. [CrossRef]

13. Kerl, C.; Sturm, J.; Cremers, D. Dense visual SLAM for RGB-D cameras. In Proceedings of the 2013 IEEE/RSJ International Conference on Intelligent Robots and Systems, Tokyo, Japan, 3–7 November 2013; pp. 2100–2106; doi:10.1109/IROS.2013.6696650. [CrossRef]
14. Cao, Y.P.; Kobbelt, L.; Hu, S.M. Real-time High-accuracy Three-Dimensional Reconstruction with Consumer RGB-D Cameras. *ACM Trans. Graph.* **2018**, *37*, 171:1–171:16, doi:10.1145/3182157. [CrossRef]
15. Zhou, Q.; Miller, S.; Koltun, V. Elastic Fragments for Dense Scene Reconstruction. In Proceedings of the 2013 IEEE International Conference on Computer Vision (ICCV), Sydney, Australia, 1–8 December 2013; pp. 473–480; doi:10.1109/ICCV.2013.65. [CrossRef]
16. Whelan, T.; Kaess, M.; Johannsson, H.; Fallon, M.; Leonard, J.J.; McDonald, J. Real-time large-scale dense RGB-D SLAM with volumetric fusion. *Int. J. Robot. Res.* **2015**, *34*, 598–626, doi:10.1177/0278364914551008. [CrossRef]
17. Cao, Y.; Xu, B.; Ye, Z.; Yang, J.; Cao, Y.; Tisse, C.L.; Li, X. Depth and thermal sensor fusion to enhance 3D thermographic reconstruction. *Opt. Express* **2018**, *26*, 8179–8193, doi:10.1364/OE.26.008179. [CrossRef] [PubMed]
18. Yue, H.; Yu, Y.; Chen, W.; Wu, X. Accurate three dimensional body scanning system based on structured light. *Opt. Express* **2018**, *26*, 28544–28559. [CrossRef] [PubMed]
19. Kümmerle, R.; Grisetti, G.; Strasdat, H.; Konolige, K.; Burgard, W. G2o: A general framework for graph optimization. In Proceedings of the 2011 IEEE International Conference on Robotics and Automation, Shanghai, China, 9–13 May 2011; pp. 3607–3613.
20. Lefloch, D.; Kluge, M.; Sarbolandi, H.; Weyrich, T.; Kolb, A. Comprehensive Use of Curvature for Robust and Accurate Online Surface Reconstruction. *IEEE Trans. Pattern Anal. Mach. Intell.* **2017**, *39*, 2349–2365, doi:10.1109/TPAMI.2017.2648803. [CrossRef] [PubMed]
21. Towers, C.; Towers, D.; Jones, J. Absolute fringe order calculation using optimised multi-frequency selection in full-field profilometry. *Opt. Lasers Eng.* **2005**, *43*, 788–800, doi:10.1016/j.optlaseng.2004.08.005. [CrossRef]
22. Pauly, M.; Gross, M.; Kobbelt, L.P. Efficient simplification of point-sampled surfaces. In Proceedings of the conference on Visualization '02, Boston, MA, USA, 27 October–1 November 2002; pp. 163–170; doi:10.1109/VISUAL.2002.1183771. [CrossRef]
23. Barfoot, T.D. *State Estimation for Robotics*, 1st ed.; Cambridge University Press: New York, NY, USA, 2017.
24. Zhou, Q.Y.; Park, J.; Koltun, V. Open3D: A Modern Library for 3D Data Processing. *arXiv* **2018**, arXiv:1801.09847.

sensors

MDPI

Article

Lane Marking Detection and Reconstruction with Line-Scan Imaging Data

Lin Li [1,*], Wenting Luo [1,*] and Kelvin C. P. Wang [2]

1 School of Transportation and Civil Engineering, Fujian Agriculture and Forestry University, Fuzhou 350002, China

2 School of Civil and Environmental Engineering, Oklahoma State University, Stillwater, OK 74078, USA; kelvin.wang@okstate.edu

* Correspondence: lilin@fafu.edu.cn (L.L.); luowenting531@gmail.com (W.L.); Tel.: +86-591-8384-5921 (W.L.)

Received: 15 April 2018; Accepted: 17 May 2018; Published: 20 May 2018

Abstract: Lane marking detection and localization are crucial for autonomous driving and lane-based pavement surveys. Numerous studies have been done to detect and locate lane markings with the purpose of advanced driver assistance systems, in which image data are usually captured by vision-based cameras. However, a limited number of studies have been done to identify lane markings using high-resolution laser images for road condition evaluation. In this study, the laser images are acquired with a digital highway data vehicle (DHDV). Subsequently, a novel methodology is presented for the automated lane marking identification and reconstruction, and is implemented in four phases: (1) binarization of the laser images with a new threshold method (multi-box segmentation based threshold method); (2) determination of candidate lane markings with closing operations and a marching square algorithm; (3) identification of true lane marking by eliminating false positives (FPs) using a linear support vector machine method; and (4) reconstruction of the damaged and dash lane marking segments to form a continuous lane marking based on the geometry features such as adjacent lane marking location and lane width. Finally, a case study is given to validate effects of the novel methodology. The findings indicate the new strategy is robust in image binarization and lane marking localization. This study would be beneficial in road lane-based pavement condition evaluation such as lane-based rutting measurement and crack classification.

Keywords: laser sensor; line scan camera; lane marking detection; support vector machine (SVM); image binarization; lane marking reconstruction

---

1. Introduction

Road lane markings deteriorate from routine use, which can lead to unexpected traffic accidents for road users [1]. Usually, lane marking data can be acquired by various approaches, such as visual cameras, GPS sensors, radar sensors, and laser sensors [2–4]. Each acquisition method has its own advantages and limitations in different application fields. Previous studies indicate that lane marking data captured by visual cameras are widely used for autonomous driving navigation and traffic surveillance [2,5,6], based on which numerous efforts have been made to detect, locate, and track lane markings in the spatial domain. However, the study of lane marking detection and location for use in road condition evaluation is neglected.

Generally the detection and localization of lane markings can be roughly implemented in a three-step process: (1) extraction of the lane marking features though pre-processing operations (i.e., exposure correction and shadow removal ...) [7–9]; (2) obtaining the location of true lane marking through a series of related process (i.e., thresholding, particle filtering, model fitting ...) [10,11]; and (3) tracking the detected lane marking with different techniques (i.e., temporal consistency, position consistency, Hough transform...) [12–14]. However, unexpected challenges always appear in lane marking detection and localization due to various interferences such as illumination conditions (occlusion, night time ...), camera location and orientation, environmental factors (i.e., foggy days, cloudy and rainy days ...), the appearance of the lane markings, the type of road, and so on [2]. To deal with the abovementioned problems, numerous vision-based lane marking detection and localization algorithms have been proposed, which for structured roads can be roughly grouped into two categories: feature-based methods and model-based techniques [6,15–18].

Feature-based methods identify road lane markings with low-level features such as line edges and colors [19]. Traditional edge-based segmentation methods such as the watershed transformation [20], the OTSU segmentation method [21], and Canny edge detectors [22] are used to identify lane markings. However, these traditional methods are susceptible to the effects of occlusions and intensity noise, and thus produce unsatisfactory identification results. Color representation is a widely used technique in image processing, which captures the feature information of lane markings in several color spaces (i.e., RGB, HSI and XYZ) [23–27]. The authors in [28] compared the effectiveness of color representation in HSI and RGB space, and then developed an adaptive method for lane marking identification in HSI color space. Although HSI-based color representation can alleviate the influence of brightness changes, it tends to confuse true targets with noises when the color information is similar. Moreover, color representation cannot comprehensively disclose lane marking features so that its use should be in combination with other non-color features such as lane edges or corners, painted lines, etc. [29–31]. The authors in [32] analyzed low-level features by using an adaptive segmentation method, and then an efficient line segment detector was proposed for lane marking detection. However, one explicit limitation exists for feature-based methods, that is, it requires the well-painted road or strong lane edges, therefore, it may suffer from background noises.

Model-based methods use a few parameters or templates to represent the lines by assuming straight lines or parabolic curves [6,33]. These techniques are more robust in noise removal, probably due to their high-level processing instead of pixel-based processing. Deformable template models that describe road edges in terms of their curvature, orientation, and offset are proposed to locate the lane boundaries [34,35]. These models are deformable so that they can best fit or match the underlying intensity variation [36], which enables them to detect lane markings in situations with shadows and broken segments since thresholding of the intensity information is ignored. A lane detection and tracking algorithm was initiated based on B-snakes [11]. This method can describe a lane through a wide range of lane structures since this model can form an arbitrary shape by a set of control points. Linear-parabolic lane models are proposed for lane departure warning systems, in which the linear function and quadratic function are used to model the lane markings in the near field and far field, respectively [33,37]. Hough Transform (HT) and its variants (e.g., improved HT, randomized HT, hierarchical HT) are widely used for straight or curved lane marking detection [2,38–41]. However, one primary limitation of this method is how to model arbitrary road shape. Furthermore, model parameters' setting and computation are an iterative trial-and-error process, which requires both human expertise and labor.

Note that the abovementioned approaches may perform well for the color images captured by an on-board camera of a vehicle and fulfill their application in driving assistance systems. However, studies on lane-based infrastructure performance assessment using 2D laser images are neglected.

Although lots of efforts have been made on pavement distress identification and rutting measurement in the past several decades [42], road lane boundaries cannot be accurately positioned, thus resulting in the inaccuracy of lane-based distress classification and performance assessments. To implement lane-based distress evaluation (i.e., pavement cracks, rutting measurement) using 2D laser images, a robust lane detection and localization approach is presented in this study. Firstly, 2D laser image data are collected by the Digital Highway Data Vehicle (DHDV) which is a real-time multi-functional system for roadway data acquisition, and then sigmoid correction method is used for background noise removal and contrast enhancement. Subsequently a new thresholding strategy is proposed to binarize laser images, based on which the pixel-based contour traversal method is developed to produce the contour boxes used as basic elements for lane marking identification. Thirdly, a Linear Support Vector Machine (LSVM) is introduced to determine proper vector weights and bias to discriminate true lane markings from noises based on contour box attributes. Finally, true lane markings along the traveling direction can be continuously reconstructed using the geometry information of the previous and current frames or images. To validate effects of the new methodology on lane marking detection and localization, a 2.286 km-long pavement section (including 1000 laser images) is chosen as a test bed. The performance of the new methodology is evaluated using Precision-Curve (PR) analysis. Results indicate the new methodology is robust and reliable in lane marking detection and localization for laser images. This study would be beneficial in continuous measurement and evaluation of lane-based pavement distress for project- and network-level pavement survey.

2. Data Acquisition System

The DHDV is a real-time multi-functional system for roadway data acquisition and analysis, particularly for pavement surface distress survey, roughness- and safety-related pavement performance evaluation [42]. The PaveVision3D Ultra (3D Ultra for short) system is the latest imaging sensor technology that enables one to acquire both 2D and 3D laser imaging data from pavement surfaces through two separate left and right sensors. The system is made up of eight high resolution cameras and two sets of lasers and is capable of constructing 4096 × 2048 images of full-lane width pavement surface with complete and continuous coverage. The subsystems of the DHDV vehicle include one 8-core computer, a Waylink Power Chassis (WPC), a WayLink Control Chassis (WCC), a differential GPS receiver or Inertial Measuring Unit (IMU), a Distance Measuring Instrument (DMI), and laser imaging sensors, as illustrated in Figure 1.

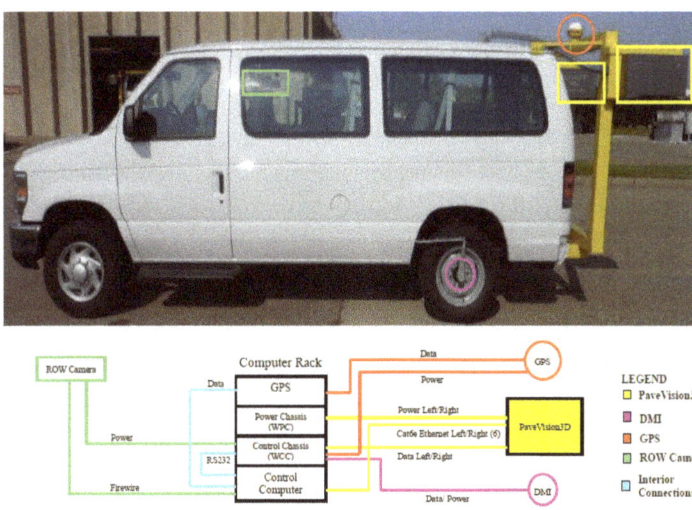

Figure 1. Generic DHDV subsystem overview.

With the high-power line laser projection system and custom optic filters, the DHDV can work at highway speeds during daytime and nighttime and maintain image quality and consistency. That means the images are shadow-free at any time of the day. Figure 2 demonstrates the wiring of the cameras and lasers to the computer rack inside the vehicle. The cameras and lasers are powered by WPC and triggered by the WCC. The WCC connects to the Control Computer. The cameras are mounted on an aluminum alignment frame spaced equidistant from previously calibrated readings. The cameras and lasers reside inside two water-tight, aluminum containers, which are mounted on the external DHDV frame. The calibrated spacing of the cameras ensures that captured laser images can cover four-meter-wide pavements. The height of the sensors has been specifically designed for cameras to accurate capture data within the laser illumination ranges.

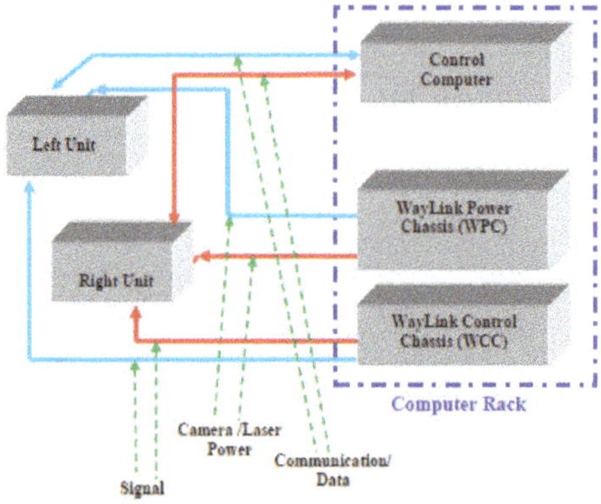

Figure 2. Line scan camera wiring diagram.

Figure 3a shows the interior appearance. Figure 3b shows rear view of the working DHDV equipped with the 3D Ultra technology. The camera and laser working principle are depicted in Figure 3c,d. By illuminating a surface using a line laser and shooting both 2D and 3D images using the corresponding cameras, the surface intensity and height variation information can be captured, in which surface height information is calculated from the distance from the camera to pavement based on the laser points (termed as the triangulation principle).

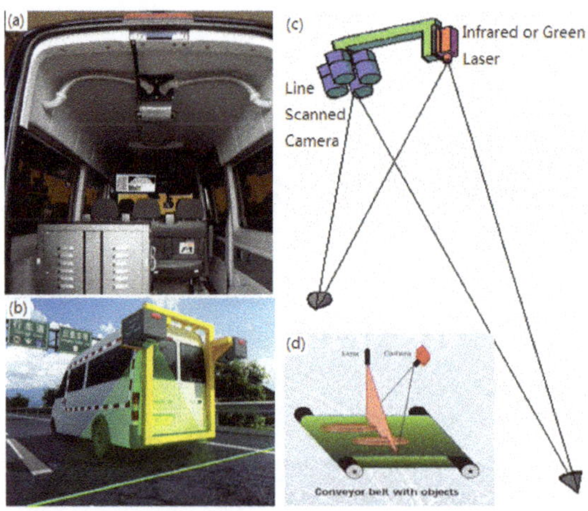

Figure 3. Photos of (**a**) DHDV interior appearance (**b**) DHDV rear view with PaveVision3D sensors; and (**c**,**d**) Pavevision3D working principle.

From Figure 3b, it can be observed that the width of laser images acquired from DHDV is more than the width of highway lanes (e.g., 3.66 m in United States) [43]. Accordingly, the exact detection and location of road lane marking are significant for lane-based pavement distress measurement and evaluation.

3. Methodologies

To achieve this objective, a series of image processing techniques are presented in this paper, which can be classified into four phases, as illustrated in Figure 4. The first phase is to binarize 2D the laser images with sigmoid correction and a new threshold method; the second phase is to delineate all contour boxes or candidate lane markings based on closing operation and marching square algorithm; the third phase is to separate out true lane marking from candidate lane marking using LSVM based on contour box attributes; and the last phase is to reconstruct broken and inconsecutive segments and form the continuous lane marking along traveling direction. As a consequence, the exact location of lane marking of the entire pavement section can be obtained, and the lane-based pavement distress survey can be performed.

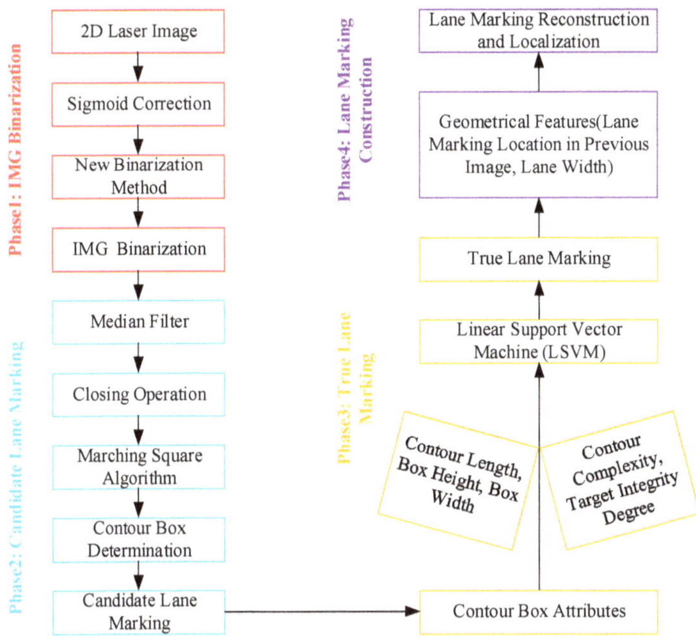

Figure 4. Schematic of the new methodology for automated identification and localization of lane marking.

3.1. *Image Binarization*

During laser image data collection, some unexpected errors or intensity noises (i.e., whitening strips in travel direction) might be produced due to the presence of non-uniformity of laser intensity, lens distortion, physical installation locations of cameras. Therefore, maximally suppressing effects of noises on target detection is critical for the laser image binarization.

3.1.1. Data Preprocessing

To maximally suppress background noises and enhance the contrast between targets (lane marking) and background noises, histogram equalization and sigmoid correction are introduced, in which the method that produces better pre-processing results would be used in this paper.

Histogram equalization is a widely used method in image contrast enhancement [44]. The basic idea behind this method is to redistribute all pixel values to be as close as possible to a specified desired histogram. Its mathematical description can be given in (1) and (2):

$$P_r(r^k) = n^k/n \tag{1}$$

$$T(r^i) = r \times \sum_{i=0}^{k-1} P_r(r^i) \tag{2}$$

where r represents the grayscale range of 2D image data (in this case r = 255), $P_r(r^k)$ stands for the frequency of grayscale value of r^k; n^k is the number of grayscale value of r^k; n is the total of all pixels; $T(r^i)$ represents the new grayscale value for the grayscale of r^i.

Sigmoid correction method uses a continuous non-linear function to transform the normalized pixel values of input images to the pixel values of output images [45], and its mathematical equation can be described in (3):

$$\mathrm{I}_{out} = \frac{1}{1 + e^{gain \times (cutoff - I_{in})}} \tag{3}$$

where I_{in} and I_{out} respectively represent the normalized pixel values of input and output images; *gain* is the multiplier in exponential's power of sigmoid function; *cutoff* is the shift value of the characteristic curve in horizontal direction. Note that both *gain* and *cutoff* should be properly initialized before use.

Note that sigmoid function is 'S' shaped, as shown in Figure 5. Figure 5a shows the transform trend ranged at [−0.5, 0.5] decreases sharply with the decrease of gain, and it becomes approximately linear when the gain variable equals to 2. The cutoff variable shifts the curve characteristics in the horizontal direction, as shown in Figure 5b. In this study, the gain of 10 and the cutoff of 0.5 are chosen after several rounds of trial and error.

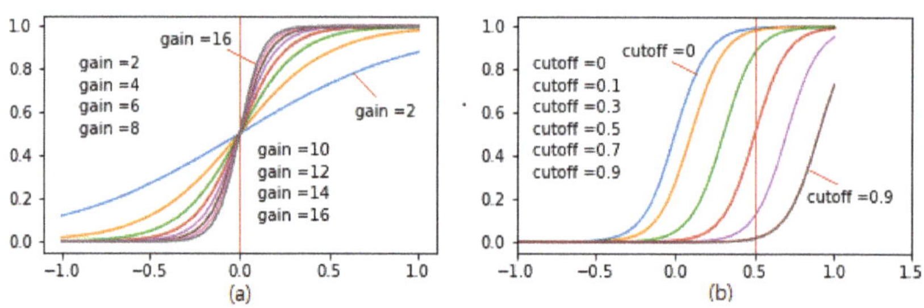

Figure 5. Sigmoid functions of (**a**) with different gains (cutoff = 0); (**b**) with different cutoffs (gain = 10).

To examine the effects of the two techniques on background noise removal and contrast enhancement, two laser images (Original_IMG1 and Original_IMG2) are chosen as test specimens, as shown in Figure 6a,d. It can be observed that both images contain whitening strips or noises, as red rectangle marks. Subsequently, the two methods are respectively applied on the two images for noise removal. Figure 6b,c,e,f represent the pre-processing results of Original_IMG1 and Original_IMG2 with the two different techniques. Note that the sigmoid correction method has better performance in separating background from foreground (lane marking) than histogram equalization. For the sigmoid correction method, the background pixels become much darker than that in the original images, that is, the influences of background noises on laser image binarization are greatly suppressed. Meanwhile, intensities of foreground pixels are increased, that is, lane marking would be easier to be identified out in the process of image binarization. Therefore, the sigmoid correction is chosen and used for background noise removal and contrast enhancement.

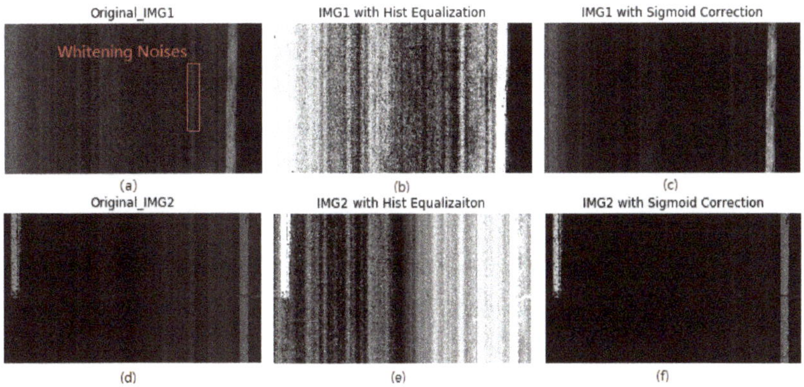

Figure 6. Photographs of noise removal: (**a**) raw image1 with whitening noises; (**b**) IMG1 with histogram equalization; (**c**) IMG1 with sigmoid correction; (**d**) Original_IMG2; (**e**) IMG2 with histogram equalization; (**f**) IMG2 with sigmoid correction.

3.1.2. New Binarization Method

Once noise removal and contrast enhancement are accomplished, the following task is image binarization. In this study, two methods, namely OTSU method and minimum threshold method are examined for this purpose. The OTSU method is a clustering-based image thresholding method [46]. The algorithm assumes that the image contains two classes of pixels following bi-modal histogram (foreground pixels and background pixels), and then it calculates the optimum threshold separating the two classes so that their combined spread (intra-class variance) is minimal. The mathematical description is given in (4)–(6):

$$\sigma^2_{\text{intra}} = \omega_0(t)\omega_1(t)[\mu_0(t) - \mu_1(t)]^2 \tag{4}$$

$$\omega_0(t) + \omega_1(t) = 1 \tag{5}$$

$$\omega_0(t)\mu_0(t) + \omega_1(t)\mu_1(t) = \mu(t) \tag{6}$$

where weights ω_0 and ω_1 are the probabilities of the two classes separated by a threshold t; ${\sigma^2}_{\text{intra}}$ are variances of these two classes, μ_0 and μ_1 respectively represent the means of these two classes.

The minimum threshold method [47,48] is suitable for binarizing images with two spikes or maxima so that the algorithm requires keep calculating and smoothing the histogram of the input image until there are only two maxima. Subsequently the threshold can be determined by the minimum value between the two maxima. However, in fields the laser image may not have the two maxima, and thus the threshold method would fail in image processing. To deal with this problem, the minimum thresholding method is modified to adapt the binarization of the image with one spike, and its mathematical expression is given in (7):

$$T = \begin{cases} f(h_1 + T_m)/2 \\ f(\min(h_i)) \end{cases}, \quad h_i \in (h_1, h_2) \tag{7}$$

where T is the minimum threshold; h_1 and h_2 represents the two maxima of the histograms of the input image; T_{m} is the maxima intensity of input image; f is used to calculate the threshold.

Figure 7a,d show two 2D laser images and their histogram distribution, respectively. Note that IMG2 has the two spikes, and both methods produce excellent binarization results for IMG2 since the histogram distribution of IMG2 has two maxima. It can be found that the two methods perform well in binarization for laser images that have two spikes in their histogram distribution, based on which the optimal threshold can be determined, as shown in Figure 7e,f. For IMG1, however, both methods produce the poor binarization results since it only has one single maximum. In this case, the OTSU method produces a false positive (FP) result, while the modified minimum threshold method produces a false negative (FN) result, as the red circles show in Figure 7b,c, respectively. It can be concluded that both methods fail to binarize the laser image that has one single spike in its histogram distribution.

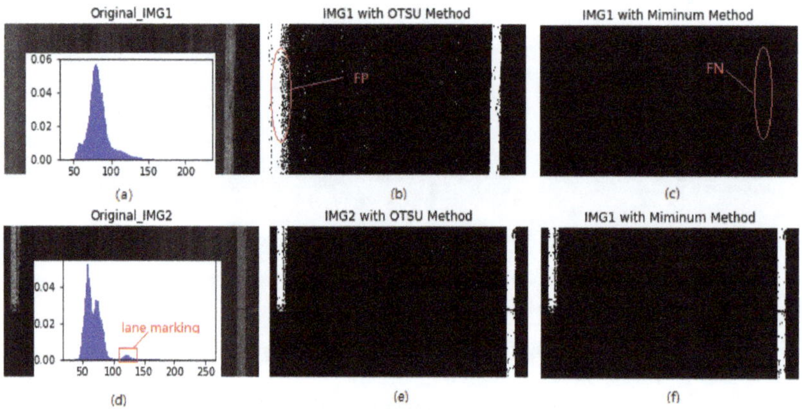

Figure 7. Binarization results: (**a**) IMG1 and its histogram; (**b**) binarized IMG1 with OTSU method; (**c**) binarized IMG1 with minimum method; (**d**) IMG2 and its histogram; (**e**) binarized IMG2 with OTSU method; (**f**) binarized IMG2 with minimum method.

To investigate the cause why the two methods fail in Original_IMG1 binarization, the sum of pixel intensity in the vertical direction is projected onto the x-axis for IMG1 and IMG2, as plotted in Figure 8a,b, respectively. In this study one laser image is obtained by merging pixel data derived from the left and right cameras. Note that IMG2 has a strong contrast between background and foreground pixels for both sides of the laser image, that is, the foreground and background are apparent and easily distinguished, as shown in Figure 8b. For the left-sided lane marking of IMG1 in Figure 8a, however, a low contrast is observed, indicating the background and foreground are indistinct and thus are cumbersome to separate out. To deal with the issue that may be caused by the non-uniformity of laser intensity, the multi-box segmentation-based threshold method is proposed.

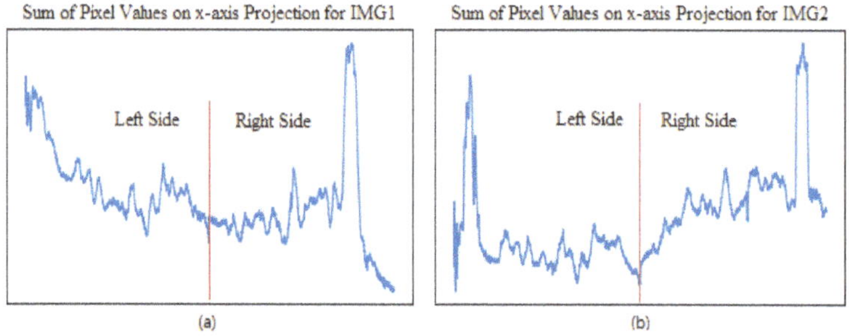

Figure 8. Pixel intensity sum's projection on x-axis for: (**a**) IMG1; and (**b**) IMG2.

The basic idea behind the new binarization method is to divide one laser image into multiple small segmentation regions, and subsequently the threshold operation is performed on each individual segmentation region. Its implementation can be elaborated below: (1) partition 2D laser image into the left and right sides (i.e., IMG_L and IMG_R) since each 2D laser image is made of two components derived from two different cameras mounted on DHDV, and thus the better binarization result might be obtained once the left and right sides are separated out; (2) divide both left and right sides of images into multiple small regions (i.e., IMG_L_1, … , IMG_L_N, N is the number of small segmentation regions for left side) along traveling direction, and the corresponding threshold can be obtained; (3) recalculate the new threshold for each small region based on minimum square error method; and (4) reconstruct the binarized images by merging all small segmentation boxes in sequence. The new threshold for each segmentation box can be calculated using (8)–(10):

$$\hat{\beta}_0 = \frac{\sum_{i=1}^{n} X_i^2 \sum_{i=1}^{n} Y_i - \sum_{i=1}^{n} X_i \sum_{i=1}^{n} X_i Y_i}{n \sum_{i=1}^{n} X_i^2 - \left(\sum_{i=1}^{n} X_i\right)^2} \tag{8}$$

$$\hat{\beta}_1 = \frac{\sum_{i=1}^{n} X_i Y_i - \sum_{i=1}^{n} X_i \sum_{i=1}^{n} Y_i}{n \sum_{i=1}^{n} X_i^2 - \left(\sum_{i=1}^{n} X_i\right)^2} \tag{9}$$

$$Y_{Newi} = \hat{\beta}_0 + \hat{\beta}_1 X_i \tag{10}$$

where X_i, Y_i represent the i-th small segmentation region in sequence and its corresponding threshold, respectively; n is the number of small segmentation regions for each side of image; $\hat{\beta}_0$, $\hat{\beta}_1$ refer to the regression coefficients of the ordinary least square errors. Y_{New_i} is the new threshold for the segmentation region i.

Figure 9 shows the working principle of the new binarization method. Firstly, the left side of IMG1 is partitioned into 16 small segmentation regions (X_i), and the modified minimum threshold method is used on each small region to calculate thresholds (Y_i). The calculated threshold for each small region are shown on Figure 9a. Note that the different segmentation regions have different thresholds, and the two adjacent regions may even have a sharp variation in threshold (i.e., region IDs 2 and 3). The large variation in threshold may be caused by two underlying reasons: (1) the inconsistency or ununiform of pixel intensity of images, and (2) the drawback or limitation of the threshold method.

To deal with this issue, the minimum square error method is used to recalculate thresholds for each segmentation region based on the pre-calculated thresholds (Y_i) from 16 segmentation regions. Once the coefficients of linear regression model are obtained, the new threshold (Y_{New_i}) for each segmentation region can be recalculated, as shown in Figure 9b. Note that the new thresholds between the adjacent segmentation regions display smooth changes, with a threshold value of approximately 137. Finally, the left side of IMG1 can be reconstructed by merging all small regions that have been binarized with the new threshold, as shown in Figure 9c.

Figure 10a–h show the effects of the new binarization method, OSTU method, and the modified minimum threshold method on laser images. It can be found that the new threshold method produces the best binarization results. For IMG2, all three methods can produce decent binarization results for lane markings, except for several whitened spots. For IMG1, the OSTU threshold method produces a false positive binarization result, and the modified minimum threshold method produces a false negative binarization results. The new threshold method produces an excellent binarization result for IMG1, and the true positive and true negative binarization results are produced. Therefore, in this paper, the new method, namely the multi-box segmentation-based traversal method, is used for 2D laser image binarization.

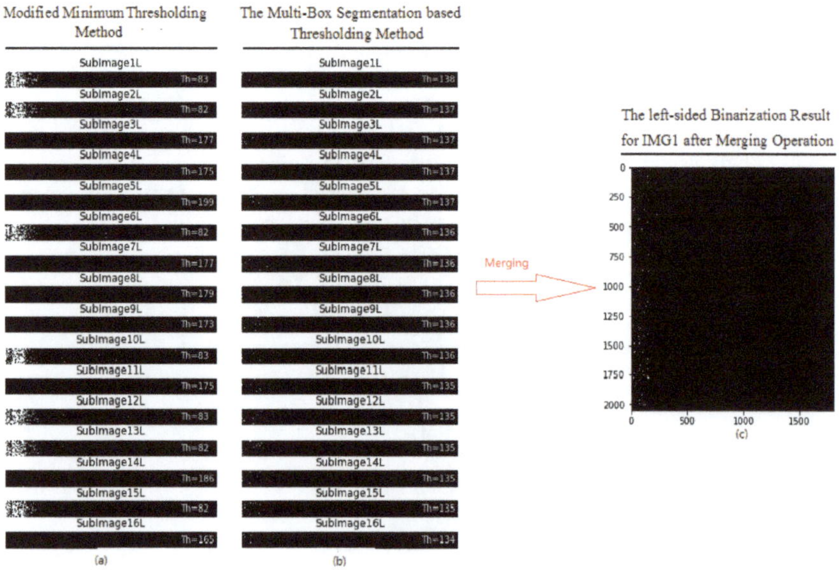

Figure 9. Binarization with the new method: (**a**) segmentation regions with their thresholds; (**b**) segmentation regions with new thresholds; and (**c**) image binarization result with new threshold.

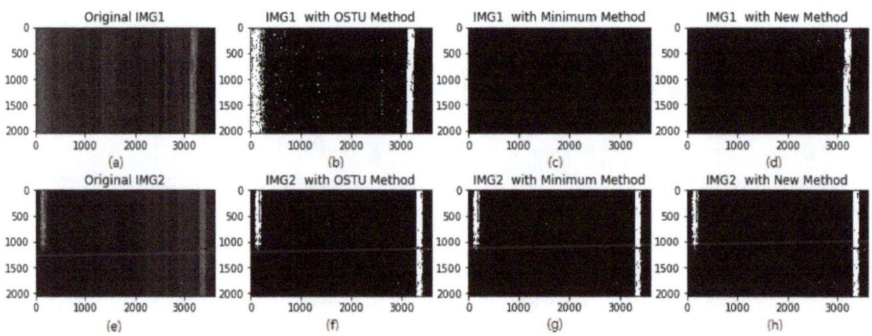

Figure 10. Comparison of binarization results: (**a–d**) IMG1 and its corresponding threshold methods; and (**e–h**) IMG2 and its corresponding threshold methods.

3.2. Candiate Lane Marking

Once 2D laser images are binarized with the new threshold method, the following task is to determine whether any whitened strips in binary images belong to lane markings or not. Firstly, a median filter is employed to eliminate the discrete spots or small blobs that are produced in binarization. Usually the discrete spots or small blobs can be assumed as fake targets and should be eliminated. Secondly, morphological closing operation and marching square algorithm are used to obtain the contour of each whitening strip or blob, and then contour box-based method is proposed to frame each whitening strip or blob. In this study, each contour box is considered as a candidate lane marking, and is taken as a basic element for the true lane marking identification.

3.2.1. Closing Operation

Due to the existence of noises such as the whitening aggregates and others, the binarized images may contain some discrete pixels or spots. To eliminate the influence of discrete spots on true lane marking identification, a median filter is employed to remove the discrete none-zero pixels.

Pavement distress such as cracking or potholes will appear during pavement aging. As a result, one entire lane marking or whitening strips may be broken into several segments by cracks, which results in extra difficulties in true lane marking identification. To deal with this issue, the morphological closing operation is used to stitch the separated whitening strips with gaps in between and produce one well-connected strip, and simultaneously the discrete white pixels are eliminated. The morphological closing operation is defined as a dilation followed by an erosion [49]. The closing operation can remove small bright spots and patch small dark cracks in lane markings. Erosion removes the non-zero pixels from object boundaries to shrink the boundaries, while the dilation operation adds binary pixels with non-zero values to the boundaries of objects in an image to fill the gaps and enlarge boundaries [42]. The number of pixels added or removed from the objects in an image depends on the size and shape of the structuring element used to process the image. The structuring element defines the neighborhood of the pixel of interest. In this study, the structuring element with a size of 15 $\times$ 15-pixel matrix is used after several trials and errors.

In Figure 11a,d, the discrete spots and lane marking gaps are marked using red circles and rectangles, respectively. Firstly, median filtering is used to remove the discrete spots, as shown in Figure 11b,e. It can be observed that the discrete spots inside circles are totally removed. Subsequently, closing operations is employed to stitch lane marking with gaps in between and produces one independent and well-connected strip. From Figure 11c,f, it can be observed that the gap or crack inside rectangles are fully filled up. Accordingly, both median filter and closing operation are robust in eliminating discrete spots and patching up lane marking gaps, which are crucial for removing fake targets and determining candidate lane markings.

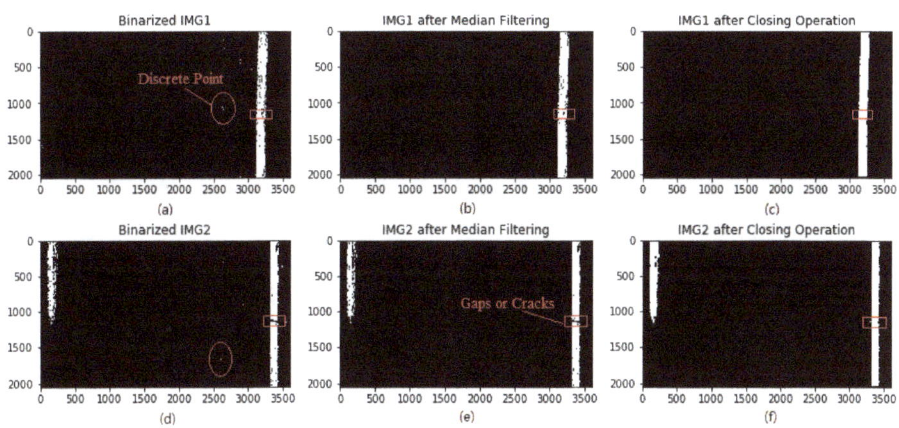

Figure 11. Photographs of morphological operation: (**a**,**d**) images after binarization; (**b**,**e**) binarized images after median filtering; (**c**,**f**) binarized images after closing operation.

3.2.2. Marching Square Algorithm

All candidate lane markings should be found before true lane marking identification. To achieve this goal, a marching square algorithm is introduced to generate the contour of the segmentation region for a two-dimensional image [50]. For one binary image, every 2 $\times$ 2 block of pixels (see Figure 12) forms a contouring box or cell, so the entire image can be represented by numerous contouring boxes. The important thing in marching square algorithm is the "sense of direction". The moving direction you head are with respect to your current positioning, which depends on the way you entered the pixel you are standing on. Therefore, it's important to keep track of your current orientation.

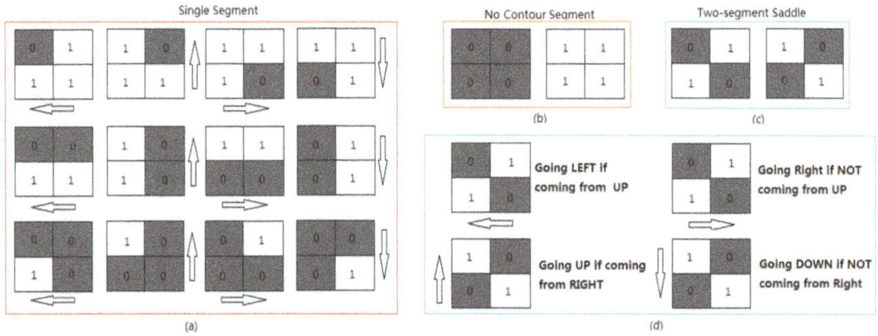

Figure 12. Photographs of (**a**) 12-moving direction for single segment; (**b**) no contour segment; (**c**) two-segment saddle; and (**d**) the 4-moving direction for the two-segment saddle.

The algorithm can be described as follows: (1) assume that you stand on the start pixel of one image binary; (2) observe the up, left, and up left pixel values, and then pick next moving direction based on Figure 12. For 'single segment' case, it easy to determine the next moving direction by matching the right contouring box, as shown in Figure 12a. For two-segment saddle (see Figure 12c), each contouring box can be divided into two states and their moving direction, as given in Figure 12d; and (3) keep moving until you get back the start position, and pixels you walked over would be the contour of the pattern.

The marching square algorithm is used on binary images (i.e., IMG1 and IMG2) that have been pre-processed with median and closing operations, and then the contours of candidate lane marking can be obtained, as shown in Figure 13, which shows that IMG1 only has one contour box, indicating only one candidate lane marking needs to be judged whether it belongs to true lane marking or not. Figure 13b shows there are eight contour boxes for IMG2, indicating there are eight candidate lane markings that need to be validated which one or two belong to true lane marking or not.

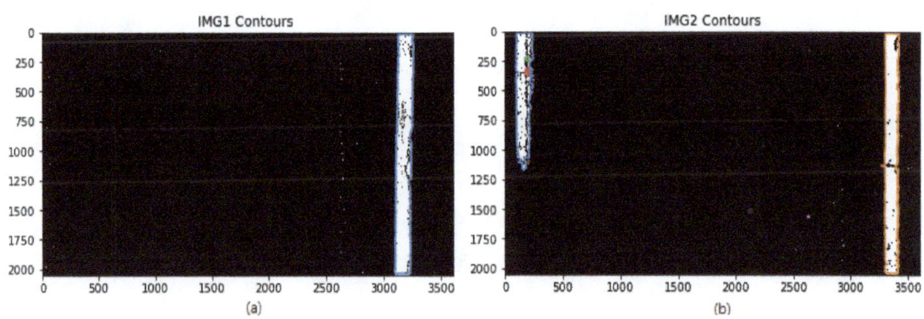

Figure 13. Photographs after the use of marching square algorithm: (**a**) one contour box for IMG1; and (**b**) eight contour boxes for IMG2 (as different colors show).

3.3. *True Lane Marking*

Contour box attributes (i.e., box width, box height, contour complexity, contour length, and target integrity degree) for each candidate lane marking are calculated along with contour box determination. They are stored into arrays and used for separating true lane marking from noises. In this study contour box attributes are defined below:

3.3.1. Contour Box Attributes

Contour box width and height are pixel differences between the minimum and maximum coordinates of contouring box in x-axis and y-axis, respectively. Contour length is the number of pixels that comprise object contours. Contour complexity is calculated by the contour length divided by the perimeter of boundary box. Contour complexity should approximate to 1 if the candidate lane marking belongs to true lane marking. Target integrity degree I_t equals to one minus the root of square sum of gradients regions ∇x and ∇y, which is used to help judge whether candidate lane marking belongs to true lane marking or not. The target integrity degree is close to one if the candidate lane marking is true lane marking. The mathematical description of target integrity degree is given in (11):

$$I_t = (1 - \sqrt{(\partial z/\partial x)^2 + (\partial z/\partial y)^2}) \times 100\% \tag{11}$$

where I_t represents the target (lane marking) integrity degree; z represents the binary values at point (x, y); $\partial z/\partial x$ denotes the first-derivative of binary image in the x direction; $\partial z/\partial y$ denotes the first-derivative of binary image in the y direction.

In general, each candidate lane marking belongs to either a true lane marking or noises, which depends on four contour box attributes: contour box width, contour box height, contour complexity, and target integrity degree. Table 1 shows contour box attributes of each candidate lane marking. In addition, the sum of pixel intensity for each contour box is projected onto the X-axis, as shown in Figure 14. IMG1 has one single contour box namely BoxID1, and its binary projection on X-axis is plotted in Figure 14a. IMG2 has eight contour boxes namely from BoxID1 to BoxID8, and their binary projections on X-axis are plotted in Figure 14b–i, respectively. It is apparent that IMG1 has one true lane marking based on its pixel projection on X-axis. IMG2 has a pair of lane marking, based on its pixel projection on X-axis in Figure 14b,c. For other contour boxes, their binary projections on X-axis are not apparent and can be negligible, and thus these contour boxes or candidate lane markings do not belong to true lane marking.

Table 1. Summary of Contour Box Attributes.

IMG ID	Box ID	Box Width	Box Height	Contour Length	Contour Complexity	∇x	∇y	$I_t(\%)$
IMG1	ID1	170	2019	4310	1.0	0.004	0.012	98.8
IMG2	ID1	152	1149	2630	1.0	0.008	0.015	98.4
	ID2	151	2019	4314	1.0	0.004	0.013	98.6
	ID3	34	34	96	0.7	0.057	0.054	92.2
	ID4	36	61	161	0.8	0.049	0.054	92.6
	ID5	5	3	13	0.8	0.533	0.267	40.4
	ID6	9	10	27	0.7	0.178	0.222	71.5
	ID7	5	6	16	0.7	0.267	0.333	57.3
	ID8	3	5	12	0.8	0.267	0.533	40.4

In summary, it can be preliminarily concluded that IMG1_ID1, IMG2_ID1 and IMG_ID2 belong to true lane markings based on their contour box attributes and binary projections on the X-axis. To efficiently separate out true lane marking from fake targets, linear support vector machine is presented in this study.

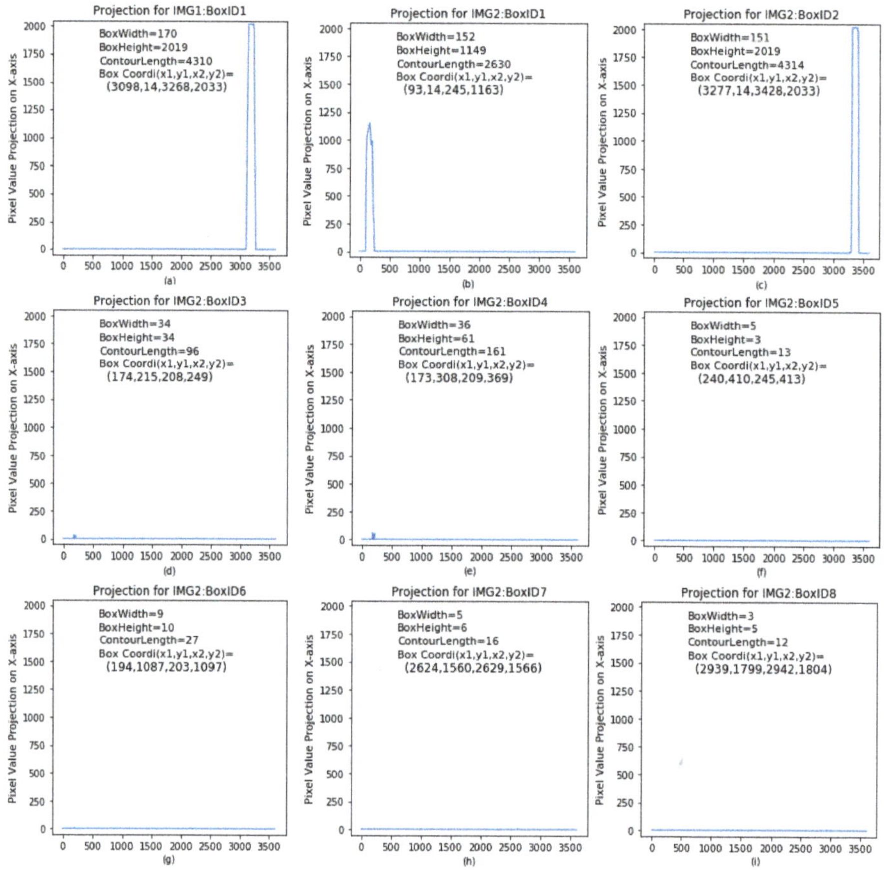

Figure 14. Photographs of the binary pixel's projection on X-axis for each contour box: (**a**) BOXID1 for IMG1; and (**b**–**i**) BOXID1 to 8 for IMG2.

3.3.2. Linear Support Vector Machine (LSVM)

A Linear Support Vector Machine (LSVM) is used to separate out true lane markings from candidate lane markings based on three variables since contour box height may be very low in laser images due to the presence of dash lane markings. SVM model is a representation of the samples as points in space and is mapped so that the samples of the separate categories are divided by a clear gap that is as wide as possible [51,52]. Typically, this clear gap is defined as the hyper plane, and the distance between hyper plane and the corresponding support vectors equals to $1/||w||$.

Once the hyper plane is located, the new sample is then mapped into that same space and predicted to belong to a classification based on which side of the hyperplane they fall. The key of the LSVM is to determine the vector weights W and the bias b of the hyperplane $g(X)$. The hyperplane can be mathematically expressed using (12):

$$g(X) = W^T X + b \tag{12}$$

where $X = [x_w, x_c, x_t]$ is a 3-dimentional vector (inputs), x_w, x_c, x_t represent the contour box width, contour complexity and target integrity degree, respectively; $W = [w_w, w_c, w_t]$ are three vector weights or the normal vector to hyper plane; b is the bias of the hyperplane.

To use the vector weight W and the bias b to separate out true lane marking from candidate lane marking, they should be computed first based on the labeled training data $[X^p, \delta^p]]$. p represents the training sample number. Y is either 1 or -1, denoting the class to which the input vector X belongs, if the predicted $g(X)$ is larger than zero, the input vector belongs to true lane marking, otherwise it belongs to noise box, which can be described using (13):

$$Y^p(W^T X^p) + b \geq 1 \tag{13}$$

To calculate the maximum-margin hyper plane, the cost function $\Phi(W) = \frac{1}{2} W^T W$ is introduced and minimized. Equation (13) is one equality constraint of cost function. It is well known that the Lagrange function is widely used to deal with the optimization problem that finds the local minima or maxima of a function. In this study it is introduced to find the optimal solutions of W_0 and b_0, and its mathematical expression is (14):

$$L(W, b, \alpha) = \frac{1}{2} W^T W - \sum_{p=1}^{P} \alpha_p [Y^p (W^T X^p + b) - 1] \tag{14}$$

where $L(W, b, \alpha)$ is the Lagrange function or expression; α_p is the Lagrange multiplier, and its value is no less than 0.

To minimize Lagrange function, the calculation of partial derivatives of $L(W,b,\alpha)$ with respect to vector weights and bias can be mathematically expressed in (15) and (16). Subsequently, the calculated vector weights are given in (17), and one equality constraint is obtained and given in (18):

$$\frac{\partial L(W, b, \alpha)}{\partial w} = 0 \tag{15}$$

$$\frac{\partial L(W, b, \alpha)}{\partial b} = 0 \tag{16}$$

$$W = \sum_{p=1}^{P} \alpha_p Y^p X^p \tag{17}$$

$$\sum_{p=1}^{P} \alpha_p Y^p = 0 \tag{18}$$

Using (17) to replace W in (14), the Lagrange function can be rewritten as (19). According to the Kuhn Tucker theory [53], the optimal solution for (19) can be deduced and rewritten as (20):

$$L(W, b, \alpha) = \sum_{p=1}^{P} \alpha_p - \frac{1}{2} \sum_{p=1}^{P} \sum_{j=1}^{P} \alpha_p \alpha_j Y^p Y^j (X^p)^T X^j \tag{19}$$

$$\alpha_p [Y^p (W^T X^p + b) - 1] = 0, \alpha_p > 0 \tag{20}$$

Assume the optimal Lagrange multiplier is $\{\alpha_{0p}, \alpha_{1p}, \Lambda, \alpha_{0p}\}$, the optimal weight vector can be calculated and rewritten as (21), and the optimal bias can be calculated using (22). Once W_0 and b_0 are calculated, the hyperplane coefficients can be determined accordingly:

$$W_0 = \sum_{p=1}^{P} \alpha_{0p} Y^p X^p = \sum_{ASV} \alpha_{0s} Y^s X^s \tag{21}$$

$$b_0 = 1 - {W_0}^T X^s \tag{22}$$

where X^s is the support vector sample; ASV is defined as all support vectors; α_{0s} is the Lagrange multiplier of the support vector sample X^s; Y^s is the classification label for the support vector sample X^s.

Eight continuous 2D laser images are chosen to illustrate how LSVM works. 38 contour boxes (p = 38) and their corresponding contour box attributes are obtained via a series of image processing operations. Subsequently the LSVM model is employed to fit sample features X with classification labels Y. The weight vector $W_0 = [w_{0w}, w_{0c}, w_{0t}] = [2.38092890 \times 10^{-2}, 7.31285305 \times 10^{-5}, -1.41721958 \times 10^{-5}]$ and the bias $b_0 = -1.92861422$ are trained. Finally, the hyperplane or decision boundary can be plotted as seen in Figure 15.

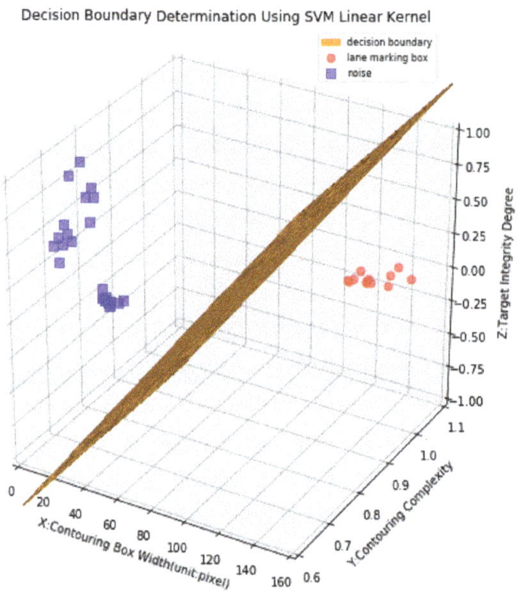

Figure 15. Illustration of hyperplane to separate true lane marking box from noise box.

As a result, the category that the contouring box belongs to can be determined based on (23). If the sign of the function $f(X)$ is positive, the contouring box is a true lane marking box, otherwise it is a noise box:

$$f(X) = \text{sgn}(W_0^T + b_0) \tag{23}$$

3.4. Lane Marking Reconstruction

In this study the 2D laser image contains either one or a pair of lane markings, as shown in Figure 16a,d,g. For images having a pair of lane markings, it is easy to reconstruct the continuous lane markings based on the identified lane markings, as shown in Figure 16b,c,e,f. However, for images having only one lane marking, it is a challenge to determine the exact location of the other one lane marking, and two variables, namely lane marking location in previous image and lane width are proposed to solve this problem. Finally, a pair of lane markings for each laser image can be reconstructed, as shown in the right Figure 16h,i.

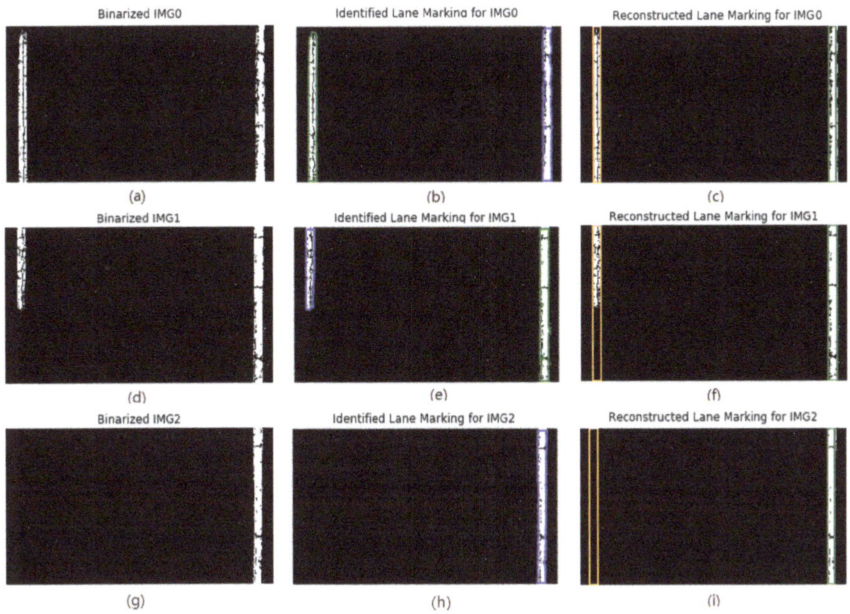

Figure 16. Binarization, Identification, and Reconstruction for solid and dash lane markings: (**a**–**c**) for IMG0; (**d**–**f**) for IMG1; and (**g**–**i**) for IMG2.

The lane width depends on the distance between the coordinates of the left and right lane markings. The coordinates of the left and right lane markings for current and previous images are stored in the vectors $X_l = \left[x_l^c, x_l^p\right]^T$, $X_r = \left[x_r^c, x_r^p\right]^T$, respectively. Eventually, a pair of lane marking along traveling direction can be continuously reconstructed with (24) and (25):

$$D_{cp} = \begin{cases} \left|x_l^c - x_l^p\right| \\ \left|x_r^c - x_r^p\right| \end{cases}, \quad D_{cp} \leq T_{os} \tag{24}$$

$$D_{lr} = \left|x_l^c - x_r^c\right|, \quad D_{lr} \leq T_w \tag{25}$$

where D_{cp} is the offset of left or right lane marking locations between previous and current images; T_{os} refers to the tolerable range of lane marking offsets; D_{lr} is the actual lane width; T_w represents the tolerable range of lane widths.

4. Case Study

To validate the effectiveness of the new methodology in lane marking identification and localization, a 7500 ft-long asphalt pavement section is chosen as a test bed in this study. Data collection starts at GPS coordinate of 34.8681, −92.401996, and ends at the GPS coordinate of 34.881418, −92.39309, located at 17468 to 16420 Maumelle Blvd. in Maumelle, AR, USA. The test section consists of 1000 laser images, and each image may either contain or not contain lane marking. In this study the binarization, identification, and localization of lane markings are validated.

4.1. Binarization Result Analysis

To quantitatively describe binarization results of lane marking, three evaluation metrics namely precision, recall, and F-score are introduced. For each lane marking, it can be regarded as "True Positive (TP)" if the automatic binarization result exactly matches with the manual survey result (ground truth); otherwise, it would be considered as the "False Negative (FN)". For non-lane marking, it can be considered as "True Negative (TN)" if the binarized non-lane marking still is non-lane marking; otherwise, it would be considered as the "False Positive (FP)". In this study TP and TN are regarded as the acceptable binarization results, while FP and FN are considered as the unacceptable binarization results.

Once the TP, TN, FP, and FN are determined, three evaluation metrics can be calculated, as described in Equations (26)–(28). Generally, the larger the evaluation metrics is, the better the performance of the test algorithm is [54]. An ideal or robust algorithm would have values of all evaluation metric approximating to one:

$$Precision = TP/(TP + FP) \tag{26}$$

$$Recall = TP/(TP + FN) \tag{27}$$

$$F = \frac{2 \times Precision \times Recall}{Precision + Recall} \tag{28}$$

Several methods, the namely OTSU threshold method [46], minimum threshold method [47], Yen's method [55], Li's cross entropy method [56], ISODATA method [57], and the new method are used to verify the binarization effects, as summarized in Table 2.

Table 2. Comparison of Binarization Results with Various Methods.

Binarization Methods	# of IMGs	Precision	Recall	F-Measure
OTSU	1000	0.873	0.898	0.885
Minimum	1000	0.866	0.842	0.854
Yen's Method	1000	0.553	0.900	0.685
Li's Method	1000	0.644	0.732	0.685
ISODATA	1000	0.813	0.843	0.828
New Method	1000	0.967	0.961	0.964

Note that the new method produces the best binarization results when compared with the other five binarization methods, with a precision of 0.97, recall of 0.96, and F-measure of 0.96, followed by is the OTSU threshold method, minimum threshold method, ISODATA method, Yen's method, and Li's cross entropy method. Therefore, it can be concluded that the new binarization method is robust for 2D laser image binarization in this calculation example.

4.2. Identification and Reconstruction Result Analysis

To validate the effects of the new method on road lane marking detection, the detection result from the new method is compared with that from two widely used methods, namely the Hough linear transform and linear-parabolic lane method. The laser image has a size of 2048 × 3604 pixels. Two laser images are chosen to demonstrate the implementation of lane marking detection and reconstruction. The colorful lines and solid rectangles of IMG1 in Figure 17a–c show the lane marking detection results based on the three methods. For the lane marking reconstruction, both Hough linear transform and linear-parabolic method cannot successfully reconstruct the dash lane marking in IMG2, as shown in Figure 17d,e, however, the new method can efficiently reconstruct the dash lane marking, as shown in Figure 17f.

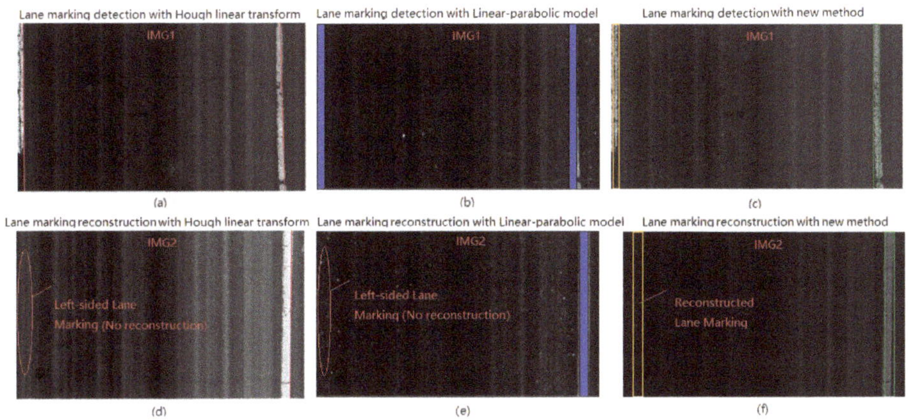

Figure 17. Comparison of detection and reconstruction results of lane marking with: (**a**,**d**) Hough linear transform; (**c**,**e**) linear-parabolic method; (**d**,**f**) the new method.

In this study the precision, recall, and F-measure are used to evaluate the effects of three methods on lane marking detection. The lane marking detection accuracy with the three methods are given in Table 3. It can be observed that the new method produces the best detection result among them, with a precision of 0.95, recall of 0.93, and F-measure of 0.94, based on 1000 test laser images, followed by the linear-parabolic lane method which produces a detection result with a precision of 0.91, recall of 0.89, and F-measure of 0.90. The Hough linear transform produces a result with a F-measure of 0.88. The corresponding results based on the three methods are given in Table 3.

Table 3. Comparison of Lane Marking Identification Results with Various Methods.

Detection Methods	# of IMGs	Computing Time(s)/Frame	Precision	Recall	F-Measure
Hough Linear Transform	1000	1.135	0.91	0.86	0.88
Linear-parabolic Lane Method	1000	1.124	0.91	0.89	0.90
Newly Proposed Method	1000	1.423	0.95	0.93	0.94

The three methods are implemented using Python & OpenCV running on an Intel(R) Core(TM) i7-7700K @4.2 GHz computer. The processing times for the three methods are given in Table 3. With the new method, the processing times for image binarization, candidate lane marking determination, and true road lane marking detection and reconstruction are 1.263, 0.156 s, and 0.004 s, respectively. The total processing time is about 1.423 s per frame, which is slightly longer than that of the other two methods. Therefore, the new method is not suitable for real-time processing of lane marking detection and is recommended to be used for image post-processing with the purpose of pavement performance evaluation.

In addition, a precision of 0.95, recall of 0.91, and F-measure of 0.94 are obtained for the lane marking reconstruction results based on 1000 test laser images. It can be concluded that the new method is robust for lane marking detection and reconstruction. The exact identification and localization of lane marking are crucial for pavement lane-based study, such as crack detection and classification, rutting measurement and evaluation, etc.

5. Conclusions and Recommendations

In this paper a new methodology is proposed to detect and locate road lane markings with 2D laser images collected from a DHDV. Firstly, the multi-box segmentation-based traversal method to binarize 2D laser images is presented, and excellent binarization results are produced when compared with other methods such as the OTSU method, minimum method, ISODATA method, Yen's method, and Li's cross entropy method, with a precision of 0.97, and recall of 0.96. Subsequently the morphological closing method and marching square method are employed to determine the contours of the potential lane markings, where generally one contouring box represents one candidate lane marking. Thirdly, a linear support vector machine is used to distinguish true lane markings from candidate lane markings based on contour box attributes, with a precision of 0.95, recall of 0.93, and F-measure of 0.94. The new method produces the better detection results when compared with the Hough linear transform and linear-parabolic lane methods. Finally, the continuous true lane markings along the traveling direction are reconstructed with the location of adjacent lane markings and road lane width. The findings indicate that the proposed methodology is robust for the detection and location of road lane markings in 2D laser images, which would benefit in road lane-based pavement distress measurement and evaluation, such as pavement cracking detection and classification, rutting measurement and so on.

Although LSVM based on contour box attributes can efficiently separate out true lane markings from fake targets, the effects of pedestrian crosswalks and lane direction arrows on lane marking identification cannot be avoided. As a future improvement, a new strategy could be developed to solve this issue, and simultaneously examine lane-based crack detection and classification.

Author Contributions: Lin Li and Wenting Luo conceived and designed the method and performed experiments; Kelvin C.P. Wang revised the paper and guided the overall process of this paper.

Funding: This work was supported by "Digital Fujian" Key Laboratory of Internet Things for Intelligent Transportation Technology and funded by Chinese National Natural Fund for Young Scholars under grant No. 51608123, Fujian Natural Science Funds under grant No. 2017J01475 and 2017J01682.

Acknowledgments: The authors would like to thank Lexing Xie, Lingfeng Huang and Dihua Chen who helped process the laser imaging data.

Conflicts of Interest: The authors declare no conflict of interest.

References

1. Cha, Y.J.; Choi, W.; Büyüköztürk, O. Deep Learning-Based Crack Damage Detection Using Convolutional Neural Networks. *Comput. Aided Civ. Infrastruct. Eng.* **2017**, *32*, 361–378. [CrossRef]
2. Gopalan, R.; Hong, T.; Shneier, M. A learning approach towards detection and tracking of lane markings. *IEEE Trans. Intell. Transp. Syst.* **2012**, *13*, 1088–1098. [CrossRef]
3. Wang, K.C.; Hou, Z.; Gong, W. Automated road sign inventory system based on stereo vision and tracking. *Comput. Aided Civ. Infrastruct. Eng.* **2010**, *25*, 468–477. [CrossRef]
4. Yan, L.; Liu, H.; Tan, J.; Li, Z.; Xie, H.; Chen, C. Scan line based road marking extraction from mobile LiDAR point clouds. *Sensors* **2016**, *16*, 903. [CrossRef] [PubMed]
5. Cáceres Hernández, D.; Kurnianggoro, L.; Filonenko, A.; Jo, K.H. Real-time lane region detection using a combination of geometrical and image features. *Sensors* **2016**, *16*, 1935. [CrossRef] [PubMed]
6. Giralt, J.; Rodriguez-Benitez, L.; Moreno-Garcia, J.; Solana-Cipres, C.; Jimenez, L. Lane mark segmentation and identification using statistical criteria on compressed video. *Integr. Comput. Aided Eng.* **2013**, *20*, 143–155.
7. Otsuka, Y.; Muramatsu, S.; Takenaga, H. Multitype lane markers recognition using local edge direction. In Proceedings of the IEEE 2002 Intelligent Vehicle Symposium, Versailles, France, 17–21 June 2002.
8. Rasmussen, C. Combining laser range, color, and texture cues for autonomous road following. In Proceedings of the ICRA'02, IEEE International Conference on Robotics and Automation, Washington, DC, USA, 11–15 May 2002.
9. Tapia-Espinoza, R.; Torres-Torriti, M. A comparison of gradient versus color and texture analysis for lane detection and tracking. In Proceedings of the IEEE 2009 6th Latin American Robotics Symposium (LARS), Valparaiso, Chile, 29–30 October 2009.
10. Li, Q.; Zheng, N.; Cheng, H. Springrobot: A prototype autonomous vehicle and its algorithms for lane detection. *IEEE Trans. Intell. Transp. Syst.* **2004**, *5*, 300–308. [CrossRef]
11. Wang, Y.; Teoh, E.K.; Shen, D. Lane detection and tracking using B-Snake. *Image Vis. Comput.* **2004**, *22*, 269–280. [CrossRef]
12. Apostoloff, N.; Zelinsky, A. Robust vision based lane tracking using multiple cues and particle filtering. In Proceedings of the 2003 IEEE Intelligent Vehicles Symposium, Columbus, OH, USA, 9–11 June 2003.

13. Dickmanns, E.D.; Mysliwetz, B.D. Recursive 3-D road and relative ego-state recognition. *IEEE Trans. Pattern Anal. Mach. Intell.* **1992**, *14*, 199–213. [CrossRef]
14. Kim, Z. Robust lane detection and tracking in challenging scenarios. *IEEE Trans. Intell. Transp. Syst.* **2008**, *9*, 16–26. [CrossRef]
15. Hillel, A.B.; Lerner, R.; Levi, D.; Raz, G. Recent progress in road and lane detection: A survey. *Mach. Vis. Appl.* **2014**, *25*, 727–745. [CrossRef]
16. Hoang, T.M.; Baek, N.R.; Cho, S.W.; Kim, K.W.; Park, K.R. Road lane detection robust to shadows based on a fuzzy system using a visible light camera sensor. *Sensors* **2017**, *17*, 2475. [CrossRef] [PubMed]
17. Kaur, G.; Kumar, D. Lane detection techniques: A review. *Int. J. Comput. Appl.* **2015**, *112*, 569–602.
18. Mandlik, P.T.; Deshmukh, A. A Review on Lane Detection and Tracking Techniques. Available online: https://pdfs.semanticscholar.org/8ff2/852ceae1b44b873243de8e6c2dd1192f574b.pdf (accessed on 25 October 2017).
19. Guo, J.; Tsai, M.-J.; Han, J.-Y. Automatic reconstruction of road surface features by using terrestrial mobile lidar. *Autom. Constr.* **2015**, *58*, 165–175. [CrossRef]
20. Beucher, S.; Bilodeau, M. Road segmentation and obstacle detection by a fast watershed transformation. In Proceedings of the 1994 IEEE Intelligent Vehicles' 94 Symposium, Paris, France, 24–26 October 1994.
21. Mu, C.; Ma, X. Lane detection based on object segmentation and piecewise fitting. *Indones. J. Electr. Eng. Comput. Sci.* **2014**, *12*, 3491–3500. [CrossRef]
22. Li, Y.; Iqbal, A.; Gans, N.R. Multiple lane boundary detection using a combination of low-level image features. In Proceedings of the 2014 IEEE 17th International Conference on Intelligent Transportation Systems (ITSC), Qingdao, China, 8–11 October 2014.
23. Abdel-Hakim, A.E.; Farag, A.A. Color segmentation using an eigen color representation. In Proceedings of the IEEE 2005 8th International Conference on Information Fusion, Philadelphia, PA, USA, 25–28 July 2005.
24. Chiu, K.-Y.; Lin, S.-F. Lane detection using color-based segmentation. In Proceedings of the 2005 IEEE Intelligent Vehicles Symposium, Las Vegas, NV, USA, 6–8 June 2005.
25. Parajuli, A.; Celenk, M.; Riley, H.B. Robust lane detection in shadows and low illumination conditions using local gradient features. *Open J. Appl. Sci.* **2013**, *3*, 68–74. [CrossRef]
26. Sun, T.-Y.; Tsai, S.-J.; Chan, V. HSI color model based lane-marking detection. In Proceedings of the 2006. ITSC'06, IEEE Intelligent Transportation Systems Conference, Toronto, ON, Canada, 17–20 September 2006.
27. Werman, M.; Omer, I. Image specific color representation. In Proceedings of the IEEE Conference on Computer vision and pattern recognition (CVPR), Washington, DC, USA, 27 June–2 July 2004.
28. Tran, T.-T.; Bae, C.S.; Kim, Y.-N.; Cho, H.-M.; Cho, S.-B. An adaptive method for lane marking detection based on HSI color model. In Proceedings of the International Conference on Intelligent Computing, Changsha, China, 18–21 August 2010; Springer: Berlin/Heidelberg, Germany, 2010.
29. Meuter, M.; Muller-Schneiders, S.; Mika, A. A novel approach to lane detection and tracking. In Proceedings of the 12th International IEEE Conference on 2009 ITSC'09 Intelligent Transportation Systems, St. Louis, MO, USA, 4–7 October 2009.
30. Saudi, A.; Teo, J.; Ahmad Hijazi, M.H. Fast lane detection with randomized hough transform. In Proceedings of the ITSim IEEE 2008 International Symposium on, Information Technology, Kuala Lumpur, Malaysia, 26–28 August 2008.
31. Truong, Q.-B.; Lee, B.-R. New lane detection algorithm for autonomous vehicles using computer vision. In Proceedings of the IEEE ICCAS 2008 International Conference on Control, Automation and Systems, Seoul, Korea, 14–17 October 2008.
32. Liu, W.; Li, S. An effective lane detection algorithm for structured road in urban. In Proceedings of the International Conference on Intelligent Science and Intelligent Data Engineering, Natal, Brazil, 29–31 August 2012; Springer: Berlin/Heidelberg, Germany, 2012.
33. Jung, C.R.; Kelber, C.R. A robust linear-parabolic model for lane following. In Proceedings of the 17th IEEE Brazilian Symposium on Computer Graphics and Image Processing, Curitiba, Brazil, 20 October 2004.
34. Kluge, K.; Lakshmanan, S. A deformable-template approach to lane detection. In Proceedings of the IEEE Intelligent Vehicles' 95 Symposium, Detroit, MI, USA, 25–26 September 1995.
35. Lakshmanan, S.; Grimmer, D. A deformable template approach to detecting straight edges in radar images. *IEEE Trans. Pattern Anal.Mach. Intell.* **1996**, *18*, 438–443. [CrossRef]
36. Kaliyaperumal, K.; Lakshmanan, S.; Kluge, K. An algorithm for detecting roads and obstacles in radar images. *IEEE Trans. Veh. Technol.* **2001**, *50*, 170–182. [CrossRef]

37. Jung, C.R.; Kelber, C.R. Lane following and lane departure using a linear-parabolic model. *Image Vis. Comput.* **2005**, *23*, 1192–1202. [CrossRef]
38. Ghazali, K.; Xiao, R.; Ma, J. Road lane detection using H-maxima and improved hough transform. In Proceedings of the 2012 IEEE Fourth International Conference on Computational Intelligence, Modelling and Simulation (CIMSiM), Kuantan, Malaysia, 25–27 September 2012.
39. Măriut, F.; Foşalău, C.; Petrisor, D. Lane mark detection using Hough Transform. In Proceedings of the 2012 International Conference and Exposition on IEEE Electrical and Power Engineering (EPE), Iasi, Romania, 25–27 October 2012.
40. Satzoda, R.K.; Sathyanarayana, S.; Srikanthan, T. Hierarchical additive Hough transform for lane detection. *IEEE Embed. Syst. Lett.* **2010**, *2*, 23–26. [CrossRef]
41. Voisin, V.; Avila, M.; Emile, B.; Begot, S.; Bardet, J.-C. Road markings detection and tracking using hough transform and kalman filter. In Proceedings of the 2005 International Conference on Advanced Concepts for Intelligent Vision Systems, Antwerp, Belgium, 20–23 September 2005; Springer: Berlin/Heidelberg, Germany, 2005.
42. Li, L.; Wang, K.C. Bounding Box–Based Technique for Pavement Crack Classification and Measurement Using 1 mm 3D Laser Data. *J. Comput. Civ. Eng.* **2016**, *30*, 04016011. [CrossRef]
43. U.S. Department of Transportation Federal Highway Administration. *Flexibility in Highway Design*; Federal Highway Administration: Washington, DC, USA, 1997.
44. Pizer, S.M.; Austin, J.D.; Cromartie, R.; Geselowitz, A.; Romeny, B.T.H.; Zimmerman, J.B.; Zuiderveld, K. Algorithms for adaptive histogram equalization. In Proceedings of the Physics and Engineering of Computerized Multidimensional Imaging and Processing, Irvine, CA, USA, 2–4 April 1986.
45. Braun, G.J.; Fairchild, M.D. Image lightness rescaling using sigmoidal contrast enhancement functions. *J. Electron. Imaging* **1999**, *8*, 380–394. [CrossRef]
46. Otsu, N. A threshold selection method from gray-level histograms. *IEEE Trans. Syst. Man Cybern.* **1979**, *9*, 62–66. [CrossRef]
47. Glasbey, C.A. An analysis of histogram-based thresholding algorithms. *CVGIP Graph. Models Image Process.* **1993**, *55*, 532–537. [CrossRef]
48. Prewitt, J.; Mendelsohn, M.L. The analysis of cell images. *Ann. N. Y. Acad. Sci.* **1966**, *128*, 1035–1053. [CrossRef] [PubMed]
49. Cord, A.; Chambon, S. Automatic road defect detection by textural pattern recognition based on AdaBoost. *Comput. Aided Civ. Infrastruct. Eng.* **2012**, *27*, 244–259. [CrossRef]
50. Mantz, H.; Jacobs, K.; Mecke, K. Utilizing Minkowski functionals for image analysis: A marching square algorithm. *J. Stat. Mech. Theory Exp.* **2008**, *2008*, P12015. [CrossRef]
51. Catanzaro, B.; Sundaram, N.; Keutzer, K. Fast support vector machine training and classification on graphics processors. In Proceedings of the 25th International Conference on Machine Learning, Helsinki, Finland, 5–9 July 2008.
52. Chou, J.S.; Pham, A.D. Smart artificial firefly colony algorithm-based support vector regression for enhanced forecasting in civil engineering. *Comput. Aided Civ. Infrastruct. Eng.* **2015**, *30*, 715–732. [CrossRef]
53. Gale, D.; Kuhn, H.W.; Tucker, A.W. Linear programming and the theory of games. *Act. Anal. Prod. Alloc.* **1951**, *13*, 317–335.
54. Boyd, K.; Eng, K.H.; Page, C.D. Area under the precision-recall curve: Point estimates and confidence intervals. In Proceedings of the Joint European Conference on Machine Learning and Knowledge Discovery in Databases, Prague, Czech Republic, 23–27 September 2013; Springer: Berlin/Heidelberg, Germany, 2013.
55. Yen, J.-C.; Chang, F.-J.; Chang, S. A new criterion for automatic multilevel thresholding. *IEEE Trans. Image Process.* **1995**, *4*, 370–378. [PubMed]
56. Li, C.H.; Lee, C. Minimum cross Entropy thresholding. *Pattern Recognit.* **1993**, *26*, 617–625. [CrossRef]
57. Ridler, T.; Calvard, S. Picture thresholding using an iterative selection method. *IEEE Trans. Syst. Man Cybern.* **1978**, *8*, 630–632.

sensors

MDPI

Article

Improved Seam-Line Searching Algorithm for UAV Image Mosaic with Optical Flow

Weilong Zhang [1], **Bingxuan Guo** [1,2,*], **Ming Li** [1,2,3,*], **Xuan Liao** [1] **and Wenzhuo Li** [1]

1 State Key Laboratory of Information Engineering in Surveying Mapping and Remote Sensing, Wuhan University, Wuhan 430079, China; zhangweilong@whu.edu.cn (W.Z.); liaoxuan@whu.edu.cn (X.L.); alvinlee@whu.edu.cn (W.L.)
2 Collaborative Innovation Center of Geospatial Technology, Wuhan University, Wuhan 430079, China
3 School of Resource and Environmental Science, Wuhan University, Wuhan 430079, China
* Correspondence: 00201550@whu.edu.cn (B.G.); lisouming@whu.edu.cn (M.L.); Tel.: +86-138-7115-8036 (B.G.); +86-138-7102-9525 (M.L.)

Received: 6 March 2018; Accepted: 12 April 2018; Published: 16 April 2018

Abstract: Ghosting and seams are two major challenges in creating unmanned aerial vehicle (UAV) image mosaic. In response to these problems, this paper proposes an improved method for UAV image seam-line searching. First, an image matching algorithm is used to extract and match the features of adjacent images, so that they can be transformed into the same coordinate system. Then, the gray scale difference, the gradient minimum, and the optical flow value of pixels in adjacent image overlapped area in a neighborhood are calculated, which can be applied to creating an energy function for seam-line searching. Based on that, an improved dynamic programming algorithm is proposed to search the optimal seam-lines to complete the UAV image mosaic. This algorithm adopts a more adaptive energy aggregation and traversal strategy, which can find a more ideal splicing path for adjacent UAV images and avoid the ground objects better. The experimental results show that the proposed method can effectively solve the problems of ghosting and seams in the panoramic UAV images.

Keywords: UAV image; dynamic programming; seam-line; optical flow; image mosaic

1. Introduction

Image mosaics have a long history starting in the early days of computer vision and photogrammetry. With the rise of UAV remote sensing technologies, this research has become paramount to many applications based on UAV survey including 3D reconstruction, ecological farming, disaster emergency management, and photography activity. These are due to UAV remote sensing technology's strengths of low-cost, high–Speed, and easy accessibility [1–5]. However, there are three disadvantages, its low flight altitude, the camera perspective constraints, and the small coverage area of a single UAV image. In many applications mentioned above, in order to expand the image coverage area to capture more information from the target area, multiple UAV images are collected, leading to the need of mosaic multiple images to form a panoramic image. Furthermore, high-altitude wind has a significant impact on the UAV platform due to its light-weight, problems such as irregular image overlaps and uneven image exposure are introduced into the adjacent images [6]. Therefore, images captured from an unstable UAV platform will lead to a vulnerable stitched image with ghosting, blur, dislocation, and color inconsistency. Overall, there are many challenges about the state-of-the-art image mosaic methods.

In response to these difficulties, this paper proposes a new UAV image mosaic method. The method solves the dislocation and ghosting problem cause by selecting the optimal seam-line in the building-intensive areas. In this new method, we first introduce the optical flow to construct the energy function for seam-line searching, it can factor the image structure information into the seam-line optimization better. Secondly, a new seam-line search strategy is presented. In this method, its basic idea is to determine the geometric errors introduced by perspective errors, camera distortions, and radiation errors by analyzing the mapping relationships between the left and right images, then using these errors to aid in the seam-line search process.

2. Related Work

There are various methods for seamless mosaic of UAV remote sensing images have been investigated [7–17]. Among them, seam-line based methods are intended to reduce grayscale and geometric differences. They look for the least-cost seam-line in the overlapping region of adjacent images by constructing an energy function. This paper will focus on the seam-line search methods based on dynamic programming and optical flow.

2.1. Methods Based on Dynamic Programming

This is a kind of mainstream image mosaic method. The methods in [11–13] focus on the energy difference between the images and their effects are superior, but they still present some problems. For example, dynamic programming-based methods in [11–13] adopt Dijkstra's shortest path algorithm to search for the optimal seam-lines, which address the ghosting and dislocation problems because of the movements of the objects and registration errors, but they suffer from low search efficiency and complex weight determination. The ant colony method in [14] is also based on dynamic programming, which can evade the areas where the color contrast is larger between images, while it will easily lead the search processing to the local optimum due to its sensitivity to the number of ants. Furthermore, there are some other methods based on the snake model [15], and some based on a morphological model [16,17]. Although these methods can almost ensure the consistence of the geometric structure and evade the phenomenon of ghosting in the overlapping regions under some conditions, they are still unable to ensure that ghosting and seams can be overcame at the same time—especially when there is a significant brightness difference between adjacent images. Meanwhile, these methods are unable to achieve satisfactory results when there are rich texture structures, registration errors, and radiation brightness differences between images. Furthermore, most of the current seam-line search methods based on dynamic programming theory rely strongly on image direction that leads to a low robustness with their energy functions.

2.2. Methods Based on Optical Flow

Optical flow is the pattern of apparent motion of objects, surfaces, and edges in a visual scene caused by the relative motion between an observer and a scene [18,19]. The American psychologist Gibson introduced the concept of optical flow in the 1940s [20]. Sequences of ordered images allow the estimation of motion as either instantaneous image velocities or discrete image displacements [21]. David and Weiss introduce gradient-based optical flow [22]. John, David, and Steven provide a performance analysis of a number of optical flow techniques. It emphasizes the accuracy and density of measurements [23]. So far, there are many methods to calculate the optical flow, and these methods have great differences. There is still no systematic classification of the existing methods. Here, we divide the optical flow calculation methods into the following four categories: methods based on gradient, methods based on matching, methods based on frequency, and methods based on Bayesian. Among them, gradient-based methods are simple computation and effective, so they have been widely studied. Lucas–Kanade method and Horn–Schunk method are representative methods, they are used to calculate the motion of a partial pixels of images (called sparse optical flow) and the motion of all pixels of images (called dense optical flow), respectively [24,25]. The energy function constructed in this paper needs the optical flow value of each pixel in the overlapping area of adjacent images, so we use the dense optical flow method to calculate them. In the image mosaic, using the methods based on optical flow to estimate the camera's motion parameters has the following advantages over the methods based on feature matching. It is unnecessary to extract image features. They are not sensitive to noise. Moreover, they can be applied to complex scenes and do dense optical flow calculation on the entire image without extrapolation of interpolation. Nevertheless, methods based on optical flow also have some weaknesses. Specifically, feature matching–based methods can be applied to the adjacent image with large difference and correct mark the corresponding points of adjacent images. However, methods based on optical flow assume that the change between images is continuous and the difference between adjacent images is very small, which greatly limits the application of these methods. For UAV image mosaic, the difference between adjacent images may be very large due to the fast flight and illumination changes of UAV. Therefore, it is difficult to create UAV image mosaics only by the methods based on optical flow. Nonetheless, the optical flow information of pixels in the overlap area of adjacent images can well provide the structural information of the images, which is conducive to searching for the optimal seam-line [26].

In this paper, the optical flow information of the pixels in the overlapped area of the adjacent images is used to construct the improved dynamic programming energy function, trying to find the best seam-line between the adjacent images and realizing the seamless mosaic of UAV images. The reminder of this paper is organized as follows. In Section 2, we explain the methodology of our proposed new method based on Duplaquet's method in detail. Experiments and results are described and analyzed in Section 3. Discussion and conclusions are drawn in Section 4.

3. Methodology

It is well–known that image registration is a key technology in the research of image mosaic method. The same is true of the research in this paper. Before the seam-line search, this paper uses the classic SIFT(scale-invariant feature transform)-based image feature extraction and matching algorithm for the registration of experimental images, in which the false matching points are removed by the RANSAC (random sample consensus) algorithm. Then, the experimental image pairs in this paper are transformed into the same coordinate system. Finally, these registered images are used for subsequent experiments.

3.1. Classic Duplaquet's Method

In 1958, Bellman proposed the optimization theroy for multi-stage problems. He transformed the multi-stage process into a series of single-stage solution problems, and created a dynamic programming method [27]. Based on Bellman's theory, Duplaquet proposed an improved method to search for image seam-lines based on dynamic programming idea [28]. Formula (1) is the energy criterion defined in the classic Duplaquet's method

$$C(x, y) = C_{dif}(x, y) - \lambda C_{edge}(x, y) \tag{1}$$

where $C_{dif}(x, y)$ is the mean value of the gray scale difference of the pixel in the overlapping regions between adjacent images, $C_{edge}(x, y)$ is the gradient minimum of the pixel in the overlapping areas, x, y are the pixel coordinates, and λ is a weighting factor, which can be used to adjust the proportion of gray scale difference and structure difference in the energy function, the value of λ is -0.15 in classic Duplaquet's method.

3.2. Problems Analysis of Duplaquet's Method

The energy criterion in the Duplaquet's method only considers the horizontal and vertical gradients, and compares the pixels in three adjacent directions near the current pixel, as shown in Figure 1. P is current pixel, m and n respectively present the pixel width and height of the overlapping region. When the overlapped region has dense tall ground objects (e.g., buildings or trees), the seam-lines output from the Duplaquet's method are likely through the edges of the buildings due to the inconsistent deformation from the image point to the roof point (as in Figure 2). Thus, it is easy to produce ghosting and seams in the stitched images.

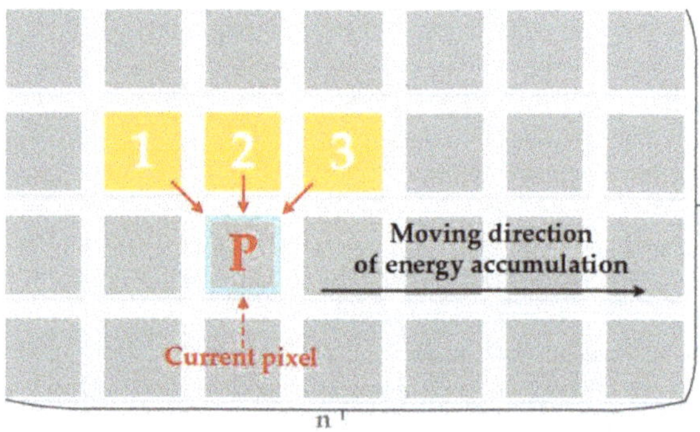

Figure 1. The schematic diagram of Duplaquet's energy criterion.

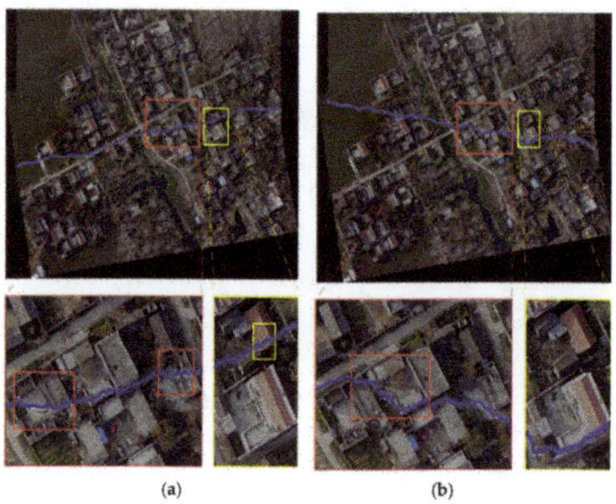

Figure 2. The mosaic results using the existing methods for two image pairs. (**a**) The Duplaquet's method; (**b**) the method introduces the fourth horizontal direction based on the Duplaquet's method.

As shown in Figure 2. There are two experimental results based on the existing methods. The seam-lines across the houses can be easy to see in stitched images. Among them, the Figure 2a is the result of the Duplaquet's method, the Figure 2b is the result of another existing method which introduces the fourth horizontal direction based on the Duplaquet's method, its seam-line across the houses less than the Duplaquet's method, but the seam-line still deviates from the ideal seam-line. Nowadays, some researchers believe that the energy function is poorly fitted, making it difficult to find the optimal seam-line. For this reason, these researchers attempt to modify the energy function based on dynamic programming. However, they overlook the optimality of the corresponding model. This also happens in the methods included in the OpenCV library. One of reasons for these problems is that the Duplaquet's method cannot ensure the best seam-line by using the classical Sobel operator to calculate the approximate gradient of the pixels based on the horizontal and vertical templates

(Formula (2)) without considering diagonal directions in the calculation process [29]. In Formula (2), D_x and D_y are the gradients of the pixel (x, y) in the vertical and horizontal directions, respectively.

$$D_x = \begin{bmatrix} -1 & 0 & 1 \\ -2 & 0 & 2 \\ -1 & 0 & 1 \end{bmatrix} D_y = \begin{bmatrix} -1 & -2 & -1 \\ 0 & 0 & 0 \\ 1 & 2 & 1 \end{bmatrix} \tag{2}$$

Specifically, the Duplaquet's method has the following three problems: (1) The gradient guidance direction of the energy function does not support omnidirectional searching. (2) The energy function is direction-dependent, and the energy aggregation considers only three directions, and the direction of energy traversal are limited from left to right, as well as from top to bottom. (3) The energy function getting local optimal solution is easy due to the impact of the two factors mentioned above. These will directly lead to the optimal seam-line susceptible to dense ground objects.

3.3. Improved Seam-Line Search Method

This paper introduces the optical flow value of the pixels in the overlapped regions for seam-line searching, and proposes a new method for finding optimal seam-lines by improving gradient guidance direction, energy accumulation directions that include energy aggregation directions, and energy traversal direction.

3.3.1. Gradient Calculation

The Duplaquet's method only considers the horizontal and vertical gradients in the energy criterion; it often fails to obtain the optimal seam-line. To solve this problem, this paper uses a new gradient operator based on the classical Sobel operator, which considers eight-directional neighborhood information of current pixel and the similarity of its surrounding structure [30]. The new approach of gradient calculation is

$$\begin{aligned} D_{0^\circ} &= \begin{bmatrix} 1 & 2 & 1 \\ 0 & 0 & 0 \\ -1 & -2 & -1 \end{bmatrix} D_{45^\circ} = \begin{bmatrix} 2 & 1 & 0 \\ 1 & 0 & -1 \\ 0 & -1 & -2 \end{bmatrix} D_{90^\circ} = \begin{bmatrix} 1 & 0 & -1 \\ 2 & 0 & -2 \\ 1 & 0 & -1 \end{bmatrix} D_{135^\circ} = \begin{bmatrix} 0 & -1 & -2 \\ 1 & 0 & -1 \\ 2 & 1 & 0 \end{bmatrix} \\ D_{180^\circ} &= \begin{bmatrix} -1 & -2 & -1 \\ 0 & 0 & 0 \\ 1 & 2 & 1 \end{bmatrix} D_{225^\circ} = \begin{bmatrix} -2 & -1 & 0 \\ -1 & 0 & 1 \\ 0 & 1 & 2 \end{bmatrix} D_{270^\circ} = \begin{bmatrix} -1 & 0 & 1 \\ -2 & 0 & 2 \\ -1 & 0 & 1 \end{bmatrix} D_{315^\circ} = \begin{bmatrix} 0 & 1 & 2 \\ -1 & 0 & 1 \\ -2 & -1 & 0 \end{bmatrix} \end{aligned} \tag{3}$$

In Formula (3), $D_{0^\circ}, D_{45^\circ}, D_{90^\circ}, D_{135^\circ}, D_{180^\circ}, D_{225^\circ}, D_{270^\circ}, D_{315^\circ}$ are the gradients of pixel (x, y) in eight directions, respectively.

3.3.2. Directionality of Energy Accumulation

In order to solve the direction-dependent problem in energy accumulation, this paper introduces a fourth horizontal direction in energy accumulation, as is shown in Figure 3. This change can get better seam-line which can be seen easily in Figure 2b. It is closer to the ideal seam-line by using this method than the Dulapquet's method, but it is obviously insufficient.

Since the optimal seam-line searching is not only affected by the directions of energy aggregation, but also affected by the directions of energy traversal. Therefore, this paper redefines the new energy criterion and adds the new aggregate directions to our dynamic programming algorithm with a stereo dual-channel energy accumulation strategy. It improves the searching scheme of optimal seam-line. As shown in Figure 4, there is a schematic diagram of our optimal seam-line search strategy, which optimizes the seam-line search criteria by detecting the eight pixels (contain the horizontal direction) surrounding the current pixel.

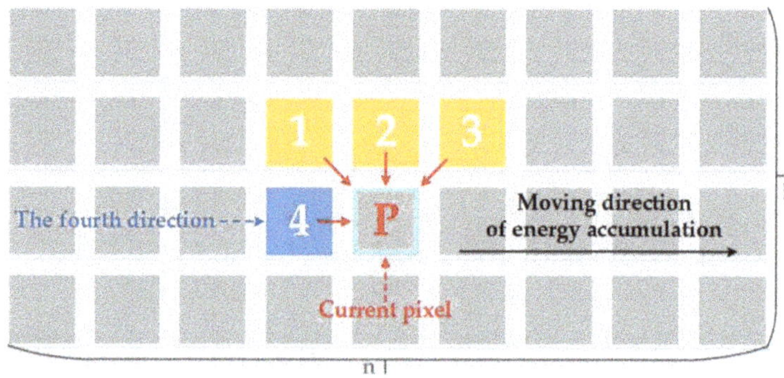

Figure 3. A schematic diagram of energy accumulation by improving energy guidelines.

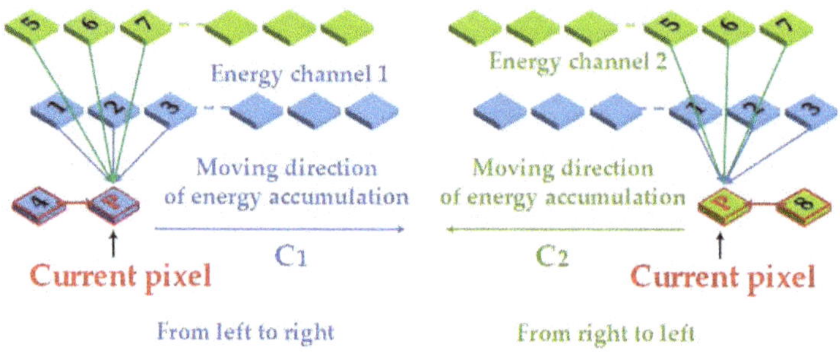

Figure 4. A schematic diagram of our search strategy.

In Figure 4, *P* is the current pixel. This paper redefines the nine related directions surrounding *P* as follows: 0 (initial invalid direction), 1 (top-left of *P* for energy channel 1), 2 (top of *P* for energy channel 1), 3 (top-right of *P* for energy channel 1), 4 (left of *P* for energy channel 1), 5 (top-left of *P* for energy channel 2), 6 (top of *P* for energy channel 2), 7 (top-right of *P* for energy channel 2), 8 (right of *P* for energy channel 2). Seam-line searching is an aggregation process of minimum energy. Each seam-line consists of neighborhood pixels with smallest energy value. In our method, the longest seam-line is the optimal seam-line.

3.3.3. Calculation of Optical Flow Value of Pixels in the Overlapped Region

In this paper, in order to take into account the image structure information better in the overlapped region of the adjacent images, we use the optical flow value of pixels in the overlapped region as a constraint condition for the construction of seam-line energy function. According to Section 2, this paper uses a dense optical flow method for optical flow calculation. The H–S method proposed by Horn and Schunck is a very popular dense optical flow method, which is easy to calculate [25]. The pixel displacement in the overlapping area can be calculated by H–S method, it can obtain the optical flow value of each pixel in the overlapped region. L–K is a sparse optical flow method, which can only calculate the optical flow value of part pixels. Others need to be obtained by interpilation. This is also the main reason for using the H–S method in this paper. Formula (4) is H–S method's objective function.

$$\min_{u,v} E_{flow}(u,v) = \iint [((T(x,y) - I(x+u,y+v))^2 + a \cdot (u_x^2 + u_y^2 + v_x^2 + v_y^2))]dxdy \tag{4}$$

In Formula (4), E_{flow} is the value of optical flow, u and v is the displacement in the x and y axis directions. T is the reference image, I is the current image, a is a weight factor, u_x, v_x, u_y, v_y are the first derivative of u and v in the x and y directions, respectively.

3.3.4. Energy Function

Based on the analysis of the theoretical model, we constructed a mathematical abstract expression of the theoretical model. Assuming that image f_1 and image f_2 are an original image pair to be stitched, the energy function is defined as

$$E = \sum_{(x,y)\in O} B(x,y)\sigma(|f_1(x,y) - f_2(x,y)|) + \sum_{(x,y)\in O}\left(\max\left(\underset{0\le k\le 7}{d_k}(f_1(x,y)) - \underset{0\le k\le 7}{d_k}(f_2(x,y))\right)\right) + \sum_{(x,y)\in O} E_{flow}(x,y) + \sum_{(x,y)\notin O} N(x,y) \quad (5)$$

In Formula (5), E is the energy value of the current pixel, $B(x, y)$ determines whether the current pixel $P(x, y)$ is in the boundary of the overlapped region, when $B(x, y) = 1$, it means that it is not in the boundary region, and when $B(x, y) = 10$, it is in the boundary region. $\sigma(*)$ is the Gaussian smoothing term, which uses the information in the local window to enhance the local influence of the current pixel. $f_1(x, y), f_2(x, y)$ are pending images to be stitched. O is the overlapped area. $d(*)$ represents the gradient function of one of the eight directions. $N(x, y)$ is the energy value of the invalid area, which is the constant term, and the value is 100 times larger than the maximum value of O. $E_{flow}(x, y)$ is the optical flow constraint item. In the actual processing, each data item and smoothing item are to be normalized, and the boundary effect is considered. That is, a large weight is given at the boundary area and the invalid area.

3.3.5. Computation Procedure

Because the overlapped region of adjacent UAV images is often irregular, it needs to be handled properly to facilitate the calculation. As Figure 5 shows, the irregular overlapped area is in Figure 5b, it can be extended to a regular area by using the minimum exterior rectangle of the overlapped region. Let us say that the overlapped region is $m \times n$.

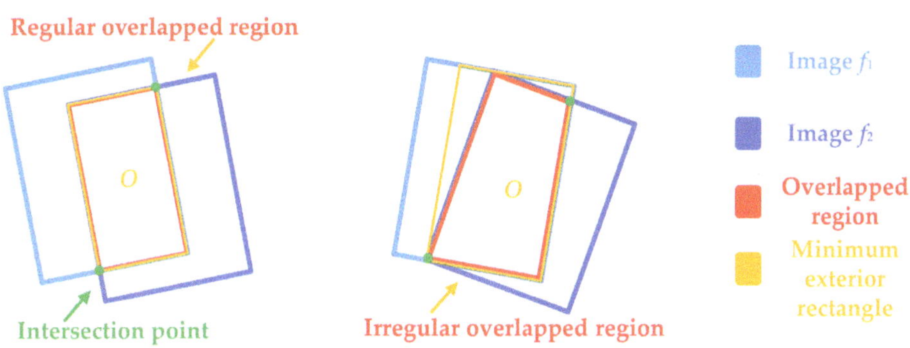

Figure 5. Processing principle of irregular overlapped region: (**a**) Regular overlapped region; (**b**) Irregular overlapped region.

The image energy function E can be calculated by Formulas (3)–(5). A method flow chart of this paper is shown in Figure 6.

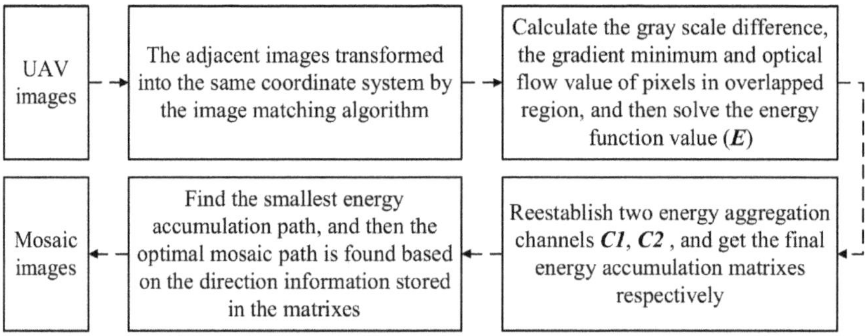

Figure 6. A flow chart of our method.

In the actual calculation process, each data item and smoothing item should be normalized. Furthermore, the boundary effect of overlapped regions needs to be taken into account, i.e., assigning a greater weight to invalid regions and boundaries. Therefore, the specific steps of our method are as following: Define the overlapped region of the adjacent images to be O, the buffer area of O boundary is W (set its width as 20 pixels, and W is an empirical value, the invalid area is N (extend area), and the boundary intersection J. Set $W \in [1,10]$, the closer to the boundary, the larger the value is, and set $N = 100 \times \max(O)$, $J = -1000 \times \max(O)$). At the same time, energy aggregation channels $C1$ and $C2$ have the same size as the minimum exterior rectangle; each pair of corresponding elements in these two matrices hold two scalar numbers representing the current aggregation value and the current path direction of the seam-line. For the first row of the matrices $C1$ and $C2$ assigned with the first row of E as the initial value, and set them corresponding direction as zero. The energy aggregation channel matrixes start to make a difference from the second row, which are divided into two aggregation processes from left to right and from right to left (see in Figure 4). For the energy aggregation channel $C1$, its aggregation process is from the left to the right; the current pixel only considers the directions of 1, 2, 3, 5, 6, 7, and 4. For the energy aggregation channel $C2$, its aggregation process is from the right to the left, and the current pixel only considers the directions of 1, 2, 3, 5, 6, 7, and 8. When the aggregation is finished, the minimum energy values are found from the last row in $C1$ and $C2$ respectively, and then an optimal mosaic path is found based on the direction information stored in the matrixes. In addition, in order to ensure that the seam-lines start and end at the intersection points, this paper selects two special intersection points (see that in Figure 5) that have the smallest energy value above, so that the seam-lines can be guided and adsorbed.

4. Experimental Results Analysis

4.1. Experimental Environment and Data

In order to verify the effectiveness of our method (Our-flow-DP), this paper not only utilized the UAV images from different regions with different flight altitudes and cameras, but also compared the experimental results with the classic Duplaquet's dynamic programming method using three search directions (Duplaquet3-DP), dynamic programming method based on Duplaquet using four search directions (Duplaquet4-DP), and the dynamic programming methods from OpenCV (OpenCV-DP). In this paper, we used Visual C++ based on OpenCV open source library to program the proposed improvement method. The experimental images are divided into four data sets; among them, the data in Figure 7a were acquired by Canon IXUS 220HS (Canon, Oita, Japan) in Paris, the height of the UAV is approximately 250 m. The data in Figure 7b were acquired by DJ-Phantom4 (DJ, Shenzhen, China) at Wuhan University square, the height of the UAV is approximately 115 m. The data in Figure 7c were acquired by DJ-Phantom4 (DJ, Shenzhen, China) at Hongshan district of Wuhan city, the height of the

UAV is approximately 116 m. The data in Figure 7d were acquired by ILCE-QX1 (Sony, Chonburi, Thailand) in Jiashan County, China, the height of the UAV is approximately 150 m. The experimental computer environment is Windows 7 operating system, CPU: Intel (R) Core (TM) i7-4790, RAM: 32 GB.

Figure 7. Four experimental UAV image pairs from four data sets: (**a**) The first image pair; (**b**) the second image pair; (**c**) the third image pair; (**d**) the fourth image pair.

4.2. Experimental Results Analysis

4.2.1. Comparison of Mosaic Results with Four Different Image Pairs

So as to verify the effect of this paper proposed method, Duplaquet3-DP, Duplaquet4-DP, OpenCV-DP, and Our-Flow-DP were used to search the optimal seam-lines of image pairs in Figure 7 with irregular overlapped regions. Figures 8–11 are the respective results. It can be seen from Figures 8–11 that the optimal seam-lines are obviously different with the four test methods. From the local zoom view of Figures 8–11, we can find that the optimal seam-lines searched by Our-Flow-DP are basically following along the road direction, which avoid the ground buildings, this will greatly reduce the probability of dislocation and ghosting because of image geometric errors. The other three methods place the seam-lines across the edges of houses, and present a ghosting and seam phenomenon. Especially in Figure 8, the other three methods have poor mosaic effects due to the dense distribution of buildings and the large changes in height. Furthermore, there was a problem of house information loss around seam-line edge in stitched images.

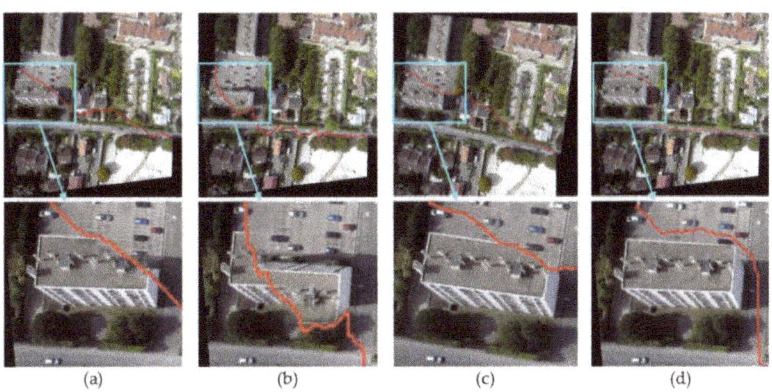

Figure 8. The seam-lines of different search methods for Figure 7a: (**a**) Dulapquet3-DP; (**b**) Dulapquet4-DP; (**c**) OpenCV-DP; (**d**) Our-Flow-DP.

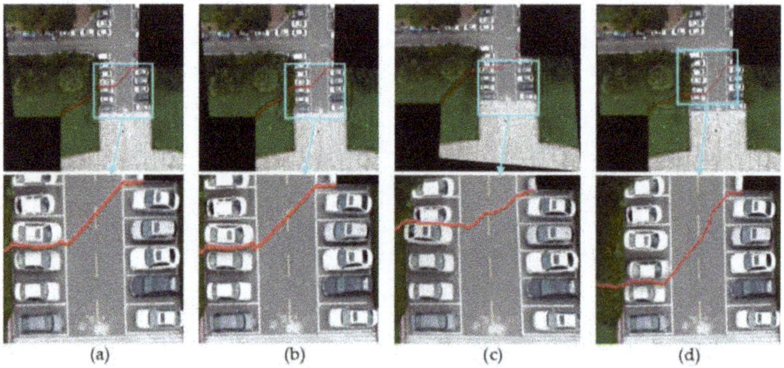

Figure 9. The seam-lines of different search methods for Figure 7b: (**a**) Dulapquet3-DP; (**b**) Dulapquet4-DP; (**c**) OpenCV-DP; (**d**) Our-Flow-DP.

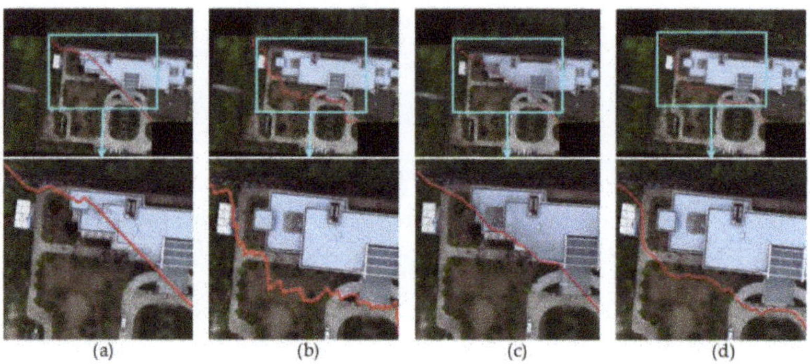

Figure 10. The seam-lines of different search methods for Figure 7c: (**a**) Dulapquet3-DP; (**b**) Dulapquet4-DP; (**c**) OpenCV-DP; (**d**) Our-Flow-DP.

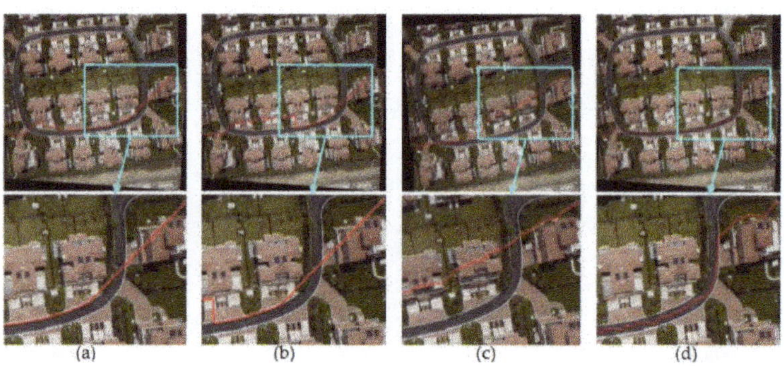

Figure 11. The seam-lines of different search methods for Figure 7d: (**a**) Dulapquet3-DP; (**b**) Dulapquet4-DP; (**c**) OpenCV-DP; (**d**) Our-Flow-DP.

4.2.2. Comparison of Four Methods under the Condition of Image Rotation

Images in Figure 7 were rotated from the horizontal to the vertical firstly. Then, we used the four methods mentioned above to search the optimal seam-lines for vertical and horizontal images respectively. Figures 12–15 show the results of them. In Figures 12, 14, and 15, the partially enlarged pictures illustrated that the optimal seam-lines searched by Our-Flow-DP basically no change before and after rotation, they always were good at avoiding the ground buildings and tall trees in the overlapped regions of adjacent images. In Figure 13, our seam-lines changed slightly, but they were less affected by the cars and tall trees in the overlapped regions than others, and the directions and movements of the seam-lines basically avoided the cars. In contrast, the seam-lines of the other three methods all crossed the edges of the buildings in different places before and after rotation, and the directions and movements of the seam-lines have an obvious change in Figures 12, 14, and 15. In Figure 13, the seam-lines of Duplaquet4-DP is more susceptible to tall trees than Duplaquet3-DP and OpenCV-DP. From the above results analysis, Our-Flow-DP is more independent than the other three methods in direction, and it can best avoid houses and tall trees for the best seam-lines searching when there are a large number of buildings and tall tress distribution in images, this is crucial for finding the most suitable seam-lines for adjacent images. Therefore, due to the specific improvements to the above issues of dynamic programming mentioned in Section 3.3, our method has advantages in adaptability and robustness for different UAV images. The minimum value of our energy function is almost no relationship to the direction of energy aggregation and traversal, and it can better take into account the structural information of the adjacent images.

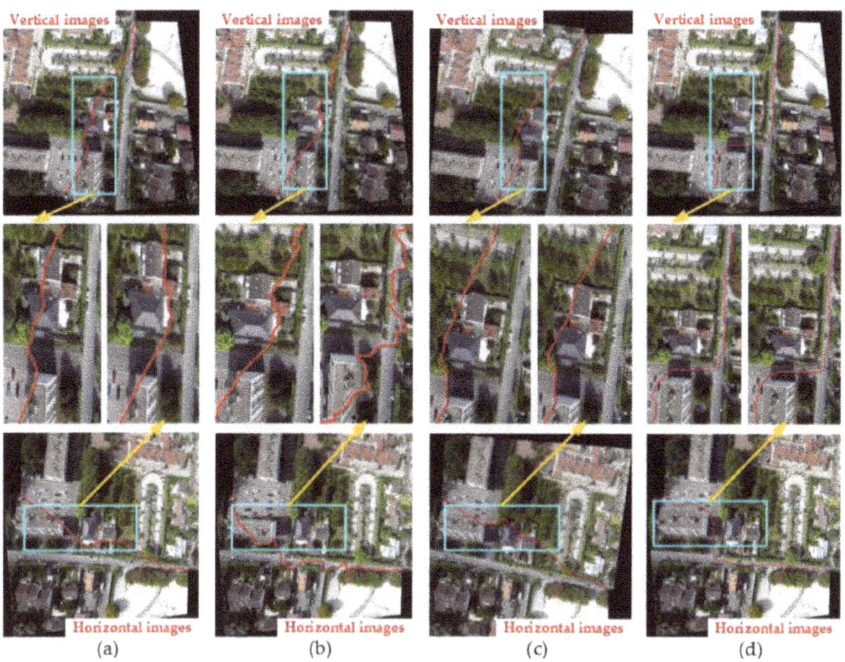

Figure 12. The seam-lines of four search methods for Figure 7a: (**a**) Dulapquet3-DP; (**b**) Dulapquet4-DP; (**c**) OpenCV-DP; (**d**) Our-Flow-DP.

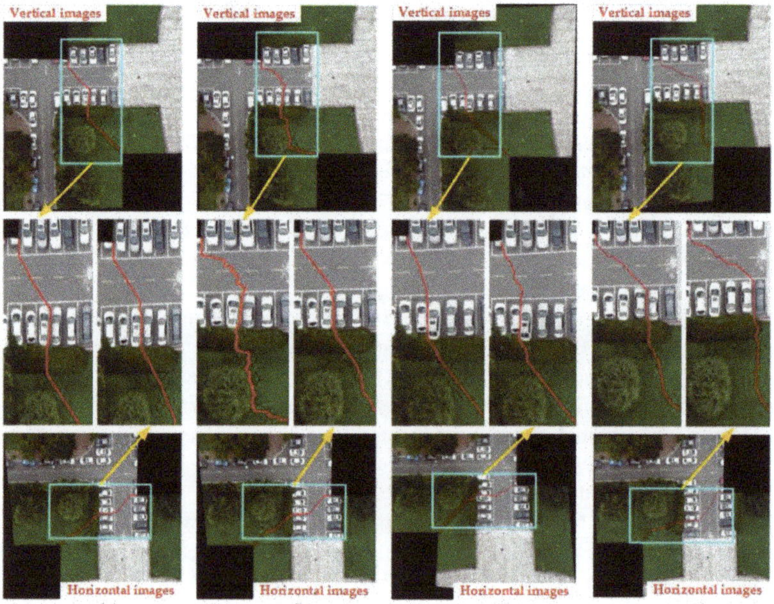

Figure 13. The seam-lines of four search methods for Figure 7b: (**a**) Dulapquet3-DP; (**b**) Dulapquet4-DP; (**c**) OpenCV-DP; (**d**) Our-Flow-DP.

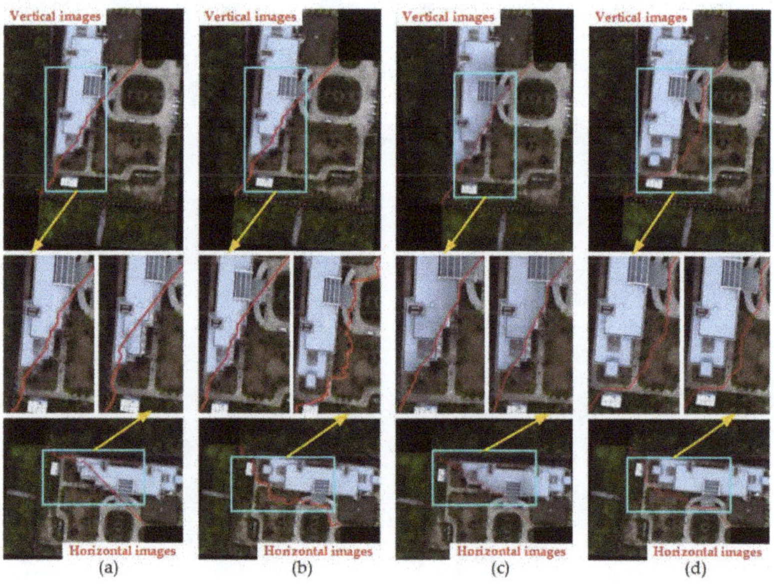

Figure 14. The seam-lines of four search methods for Figure 7c: (**a**) Dulapquet3-DP; (**b**) Dulapquet4-DP; (**c**) OpenCV-DP; (**d**) Our-Flow-DP.

Figure 15. The seam-lines of four search methods under different situations for Figure 7d: (**a**) Dulapquet3-DP; (**b**) Dulapquet4-DP; (**c**) OpenCV-DP; (**d**) Our-Flow-DP.

4.2.3. Efficiency Comparison of Our Energy Accumulation Processing

The improved method proposed in this paper found the almost best seam-lines in the previous experiments. Since Our-Flow-DP is based on the classical Duplaquet method, this section will compare the energy accumulation time efficiency of Duplaquet3-DP, Duplaquet4-DP, and Our-Flow-DP. Firstly, we assumed that Our-Flow-DP, Duplaquet3-DP, and Duplaquet4-DP could find the same optimal seam-lines. Their time efficiency difference can be quantitatively analyzed from the method's complexity. In this paper, the direction of energy aggregation form three aggregation directions increased to eight is mainly an improvement. Setting that the time complexity of the Dulapquet3-DP is $O(z^3)$, the time complexity of the Dulapquet4-DP is $O(z^4)$, and Our-Flow-DP is $O(z^8)$, where z is the total number of pixels within the minimum exterior rectangle of the overlapped region, z can be expressed as the product of m and n, m is the width of the minimum exterior rectangle and n is the length of the minimum exterior rectangle. Both m and n are measured by unit pixel. However, because the local energy minima exists in the energy function of the Duplaquet3-DP and Duplaquet4-DP, they result in a lot of time consumption. Therefore, the above assumption is invalid, that is to say, they cannot get the same optimal seam-lines.

Four experimental image pairs were selected in Figure 7 to verify the above conclusions. In order to speed up the calculation, it is generally necessary to zoom the image at a certain scale. Therefore, the size of the overlapped region is not same to the size of the original image overlapped region. The experimental results can be seen in Table 1. The efficiency of our method is more than 37–148 times that of the other three methods. It proved the convergence speed of our energy function was faster than others. In addition, it further pointed out that the theory and the results of the proposed method were obviously different with the classic Duplaquet's method. The theoretical improvement and experimental comparison have proven that this paper proposed a global and non-direction optimization method, which not only has the best seam-line, but also has better time efficiency.

Table 1. Time efficiency comparison of energy accumulation processing with different methods

Image Pair	Figure 7a		Figure 7b		Figure 7c		Figure 7d	
Rotation	Horizontal	Vertical	Horizontal	Vertical	Horizontal	Vertical	Horizontal	Vertical
z = m × n	956 × 522		635 × 362		745 × 560		1473 × 642	
Location	Paris		Square		Hongshan		Jiashan	
Duplaquet3-DP	5088 ms	4911 ms	1625 ms	1538 ms	3373 ms	3244 ms	14,501 ms	14,503 ms
Duplaquet4-DP	5075 ms	4862 ms	1580 ms	1516 ms	3321 ms	3238 ms	14,547 ms	14,229 ms
Our-Flow-DP	90 ms	55 ms	43 ms	24 ms	76 ms	44 ms	201 ms	98 ms
Multiple	56	89	38	64	44	74	72	148
	56	88	37	63	43	73	72	145

5. Conclusions

This paper selected the essential problems of dynamic programming algorithms for image seam-line optimization, and introduced the optical flow value of the pixels in the overlapped regions for seam-line searching. At last, on the basis of classic dynamic programming algorithm, this paper proposes a new improved dynamic programming algorithm to search the optimal seam lines. Meanwhile, this paper carried out a detailed theoretical study and a lot of UAV image mosaic experiments. The superiority and efficiency of the method proposed in this paper are verified by the credible experiments of different image pairs with irregular overlapped regions. It can be seen from the experimental results in this paper, the UAV image mosaic results are better than the comparison methods. Furthermore, the proposed method is proven invariant to image rotation, and the improved dynamic programming algorithm works more efficiently. It is worth mentioning that the improved method in this paper is even better than the OpenCV method, which is an open source method and has been constantly updated. In the future, we will continue to improve our existing deficiencies to achieve a more perfect and robust method of rapid UAV image mosaic. In addition, a real-time UAV image mosaic will be our main research direction.

Acknowledgments: This study is supported by the national Key Research and Development Program of China (2016YFB0502202), the Fundamental Research Funds for the Central Universities (2042018kf0013), the China Postdoctoral Science Foundation (2017M622520), the Postdoctoral Science Foundation of Hubei (2017Z1), and the LIESMARS Special Research Expenses.

Author Contributions: W.Z. and M.L. proposed the methodology and wrote the paper; X.B., X.L., and W.L. conceived, designed, and performed the experiments.

Conflicts of Interest: The authors declare no conflicts of interest.

References

1. Chen, S.; Laefer, D.; Mangina, E.; Mangina, E. State of Technology Review of Civilian UAVs. *Recent Pat. Eng.* **2016**, *10*, 160–174. [CrossRef]
2. Byrne, J.; Keeffe, E.; Lenon, D.; Laefer, D. 3D reconstructions using unstabilized video footage from an unmanned aerial vehicle. *J. Imaging* **2017**, *3*, 15. [CrossRef]
3. Sun, H.; Li, L.; Ding, X. The precise multimode GNSS positioning for UAV and its application in large scale photogrammetry. *Geo-Spat. Inf. Sci.* **2016**, *19*, 188–194. [CrossRef]
4. LI, D.; Li, M. Research advance and application prospect of unmanned aerial vehicle remote sensing system. *Geomat. Inf. Sci. Wuhan Univ.* **2014**, *39*, 505–513.
5. Zhang, W.; Li, M.; Guo, B.; Li, D.; Guo, G. Rapid texture optimization of three-dimensional urban model based on oblique images. *Sensors* **2017**, *17*, 911. [CrossRef] [PubMed]
6. Li, M.; Li, D.; Fan, D. A study on automatic UAV image method for paroxysmal disaster. *Int. Arch. Photogramm. Remote Sens. Spat. Inf. Sci.* **2012**, *B6*, 123–128. [CrossRef]
7. Wang, W.; Michale, K. A variational approach for image mosaic. *SIAM J. Imaging Sci.* **2013**, *6*, 1318–1344. [CrossRef]
8. Tao, M.; Johnson, M.; Paris, S. Error tolerant image compositing. *Int. J. Comput. Vis.* **2013**, *103*, 178–189. [CrossRef]

9. Ghosh, D.; Kaabouch, N. A survey on image mosaicing techniques. *J. Vis. Commun. Image Represent.* **2016**, *34*, 1–11. [CrossRef]
10. Bu, S.; Yong, Z.; Gang, W.; Liu, Z. Map2DFusion: Real-time incremental UAV image mosaicing based on monocular slam. In Proceedings of the 2016 IEEE/RSJ International Conference on Intelligent Robots and Systems (IROS), Daejeon, Korea, 9–14 October 2016; pp. 4564–4571.
11. Chon, J.; Kim, H.; Lin, C. Seam-line Determination for Image Mosaicking: A Technique Minimizing the Maximum Local Mismatch and the Global Cost. *ISPRS J. Photogramm. Remote Sens.* **2010**, *65*, 86–92. [CrossRef]
12. Zhao, Y.; Han, T.; Feng, S.; Miao, C. Remote Sensing Image Mosaic by Incorporating Segmentation and the Shortest Path. *Geo-Inform. Resour. Manag. Sustain. Ecosyst.* **2013**, *398*, 684–691.
13. Pan, J.; Zhou, Q.; Wang, M. Seamline Determination Based on Segmentation for Urban Image Mosaicking. *IEEE Geosci. Remote Sens. Lett.* **2014**, *11*, 1335–1339. [CrossRef]
14. Amrita; Neeru, N. A Novel Approach for Medical Image Stitching Using Ant Colony Optimization. *J. Eng. Res. Appl.* **2014**, *4*, 21–28.
15. Shashank, K.; Sivachaitanya, N.; Mainkanta, G.; Balaji, C.; Murthy, V. A Survey and Review over Image Alignment and Stitching Methods. *J. Electron. Commun. Technol.* **2014**, *5*, 50–52.
16. Aghamohamadnia, M.; Abedini, A. A Morphology-stitching Method to Improve Landsat SLC-off Images with Stripes. *Geodesy Geodyn.* **2014**, *5*, 27–33. [CrossRef]
17. Jufriadif, N. Edge Detection on Objects of Medical image with Enhancement Multiple Morphological Gradient method. In Proceedings of the International Conference on Electrical Engineering, Boumerdes, Algeria, 29–31 October 2017; pp. 1–7.
18. Chen, J.; Cai, Z.; Lai, J.; Xie, X. Fast Optical Flow Estimation Based on the Split Bregman Method. *IEEE Trans. Circuits Syst. Video Technol.* **2018**, *28*, 664–678. [CrossRef]
19. Liu, B.; Wei, W.; Pan, Z.; Wang, S. Fast Algorithms for large Displacement Variation Optical Flow Computation. *J. Image Graph.* **2017**, *22*, 66–74.
20. Clerc, M.; Mallat, S. The Texture Gradient Equation for Recovering Shape from Texture. *IEEE Trans. Pattren Anal. Mach. Intell.* **2013**, *24*, 536–549. [CrossRef]
21. Zhou, X.; Yang, C.; Yu, W. Moving Object Detecting Contiguous Outliers in the Low-Rank Representation. *IEEE Trans. Pattren Anal. Mach. Intell.* **2013**, *35*, 597–610. [CrossRef] [PubMed]
22. Xue, T.; Rubinstein, M.; Wadhwa, N.; Levin, A.; Durand, F. Refraction Wiggles for Measuring Fluid Depth and Velocity from Video. In Proceedings of the European Conference on Computer Vision, Zurich, Switzerland, 6–12 September 2014; Volume 8691, pp. 767–782.
23. Butler, D.; Wulff, J.; Stanley, G.; Black, M. A Naturalistic Open Source Movie for Optical Flow Evaluation. In Proceedings of the European Conference on Computer Vision, Florence, Italy, 7–13 October 2012; Volume 7577, pp. 611–625.
24. Liu, C.; Yuen, J.; Torralba, A. SIFT Flow: Dense Correspondence across Scenes and Its Applications. *Dense Image Corresp. Comput. Vis.* **2016**, *33*, 15–49.
25. Hildreth, E.; Koch, C. The Analysis of Visual Motion: From Computational Theory to Neuronal Mechanisms. *Annu. Rev. Neurosci.* **2014**, *10*, 477–533. [CrossRef] [PubMed]
26. Xu, J.; Yuan, F.; Kou, Y. An image method based on the seamline and dynamic datum. *Bull. Surv. Mapp.* **2014**, *9*, 42.
27. Li, W.; Zhao, S.; Zhang, W.; Liu, X.; Yu, G. A Mosaic Method for UAV Images Based on Filtering. *Geomat. Inf. Sci. Wuhan Univ.* **2017**, *42*, 1–8.
28. Duplaquet, L. Building Large Images Mosaics with Invisible Seam-lines. *Proceed SPIE* **1998**, *3387*, 369–377.
29. Li, Z.; Zheng, J.; Zhu, Z.; Yao, W.; Wu, S. Weighted Guided Image Filtering. *IEEE Trans. Image Process.* **2014**, *24*, 120–129. [PubMed]
30. Zheng, Y.; Zhang, Y.; Wang, Z.; Zhang, J.; Fan, S. Edge Detection Algorithm Based on the Eight Directions Sobel Operator. *Comput. Sci.* **2013**, *40*, 354–356.

sensors

MDPI

Article

2D Rotation-Angle Measurement Utilizing Least Iterative Region Segmentation

Chenguang Cao and Qi Ouyang *

School of Automation, Chongqing University, Chongqing 400044, China; guangcc@foxmail.com
* Correspondence: yangqi@cqu.edu.cn; Tel.: +86-139-8365-1722

Received: 3 March 2019; Accepted: 3 April 2019; Published: 5 April 2019

Abstract: When geometric moments are used to measure the rotation-angle of plane workpieces, the same rotation angle would be obtained with dissimilar poses. Such a case would be shown as an error in an automatic sorting system. Here, we present an improved rotation-angle measurement method based on geometric moments, which is suitable for automatic sorting systems. The method can overcome this limitation to obtain accurate results. The accuracy, speed, and generality of this method are analyzed in detail. In addition, a rotation-angle measurement error model is established to study the effect of camera pose on the rotation-angle measurement accuracy. We find that a rotation-angle measurement error will occur with a non-ideal camera pose. Thus, a correction method is proposed to increase accuracy and reduce the measurement error caused by camera pose. Finally, an automatic sorting system is developed, and experiments are conducted to verify the effectiveness of our methods. The experimental results show that the rotation angles are accurately obtained and workpieces could be correctly placed by this system.

Keywords: geometric moments; camera pose; rotation-angle; measurement error

1. Introduction

An automatic sorting system has the advantages of high efficiency, low error rate, and low labor cost. It is widely used in several fields, such as vegetable classification [1,2], the postal industry [3], waste recycling [4,5], and medicine [6]. To meet the requirements of intelligent sorting in industrial environments, a vision system is often used to sense, observe, and control the sorting process. In this system, the pose and position of the workpiece are obtained using a camera with image processing, and the actuator is driven according to these parameters. Consequently, the adaptive ability of the automatic sorting system will be improved. Generally, pose is described by angles in space [7]. For a plane workpiece, only one angle is needed. To place a plane workpiece correctly, the rotation angle of each workpiece must be calculated. Because workpieces are placed arbitrarily in the sorting area, they should be placed in the storage area in one pose. Therefore, the rotation-angle is an important parameter that is used to plan a path for the actuator. An incorrect rotation-angle will lead to an error in path planning, causing the workpieces to be placed incorrectly.

Rotation-angle measurement is an important component of visual measurement and has been substantially studied. As a result, various visual rotation-angle measurement methods have emerged, and they are used in different fields. Existing rotation-angle measurement methods are mainly classified into four categories. The first one is template matching, in which the rotation angle is calculated through a similarity measurement. This method is simple, but it has a high computational cost and is slow. The main challenges for template matching are the reduction of its computational cost and improvement of its efficiency [8]. The second category is polar transform. The advantage of this method is that any rotation and scale in Cartesian coordinates are represented as shifts in the angular and the log-radius directions in log-polar coordinates, respectively [9]. Then, the rotation angle is obtained between two images. The third category of methods takes advantage of the feature line and feature point in images. The feature line is obtained by Hough transformation or from the image moment [10]. The angle between the matching lines in two images is calculated, and it could be regarded as the rotation angle. Such methods are simple and suitable for fast detection. Feature points are some local features that are extracted from the image. They remain constant for rotation, scaling, and various types of illumination. Scale-invariant feature transform (SIFT) is always used to obtain feature points [11], and the rotation angle is calculated by matching points in different images. In [12], a isosceles triangle is established and then the rotation-angle between two points could be obtained by solving the triangle formed by origin coordinates and position of these two points. The fourth category of methods requires auxiliary equipment, which mainly include a calibration board and projector. In [13], a calibration pattern with a spot array was installed at the rotor. The rotation angle of the spot array is detected with the equation of coordinate rotation measurement. The standard deviation of rotation-angle measurement is smaller than 3 arcsec. In [14], a practical method to measure single-axis rotation angles with conveniently acquirable equipment was presented. Experiments achieved a measurement accuracy of less than 0.1° with a camera and a printed checkboard. The Moire fringe is an optical phenomenon used in rotation-angle measurement. The principle of measurement is that the width of the Moire fringe varies as the angle between the grating lines varies [15]. Lensless digital holographic microscopy is used to accurately measure ultrasmall rotation angles [16]. Furthermore, white-light interferometry was used in a previous study to measure one-dimensional rotation angles [17]. In that study, the rotation angle was measured with an optical plane-parallel plate with a standard refractive index. The phase change of the interference spectrum of the interferometer was output during the rotation of the plane workpiece.

It should be noted that although many methods have been developed, these methods are not suitable for automatic sorting system. This is because the method used in the automatic sorting system has three requirements. Firstly, the rotation angle needs to be calculated correctly in a short time. Secondly, auxiliary equipment should not be used, because of the continuous movement of the workpieces. Thirdly, the method has generality and can calculate the rotation-angle of different workpieces. Therefore, a new rotation-angle measurement method which satisfies the above conditions is needed.

In the present paper, an improved rotation-angle measurement method based on geometric moments is proposed. The improved method is suitable for workpieces of all shapes and could overcome a limitation of geometric moments when calculating the rotation-angle. The analysis of speed and accuracy of the proposed method shows that it can meet the requirements of automatic sorting systems. In addition, a rotation-angle measurement model is established, and the relationship between camera pose and rotation-angle measurement error is investigated. Subsequently, a correction method is presented to reduce the measurement error caused by camera pose. Experimental results show that this method is accurate and suitable for rotation-angle measurement. The remainder of this paper is organized as follows. Section 2 reviews the concept of image moment and clarifies the limitation of rotation-angle measurement based on geometric moments. Section 3 describes the rotation-angle measurement method in detail. Section 4 establishes a rotation-angle measurement model and illustrates that a measurement error can be caused by camera pose. Subsequently, a correction method for rotation-angle measurement error is presented. In Section 5, an automatic sorting system is set up, and experimental results are discussed. Section 6 draws conclusions.

2. Basic Theory

2.1. Image Moment

The concept of moments was initially proposed in classical mechanics and statistics. At present, it is widely used in image recognition [18,19], image segmentation [20], and digital compression [21].

The geometric moment of an image is the simplest, and lower moments have a clear physical meaning in an image.

Area is expressed by the zeroth moment:

$$M_{00} = \sum_{i=1}^{n}\sum_{j=1}^{m} I(i,j). \tag{1}$$

Center of mass is expressed by the first moment:

$$\begin{cases} M_{10} = \sum_{i=1}^{n}\sum_{j=1}^{m} i \cdot I(i,j) \\ M_{01} = \sum_{i=1}^{n}\sum_{j=1}^{m} j \cdot I(i,j) \end{cases} \tag{2}$$

$$x_c = \frac{M_{10}}{M_{00}}, y_c = \frac{M_{01}}{M_{00}}. \tag{3}$$

The second moments are defined as:

$$\begin{cases} M_{20} = \sum_{i=1}^{n}\sum_{j=1}^{m} i^2 \cdot I(i,j) \\ M_{02} = \sum_{i=1}^{n}\sum_{j=1}^{m} i \cdot j \cdot I(i,j) \\ M_{11} = \sum_{i=1}^{n}\sum_{j=1}^{m} j^2 \cdot I(i,j) \end{cases} \tag{4}$$

Orientation is used to describe how the object lies in the field of view and it could be expressed by this three second moments:

$$\tan 2\theta = \frac{b}{a-c} \tag{5}$$

where

$$\begin{cases} a = \frac{M_{20}}{M_{00}} - x_c^2 \\ b = \frac{M_{11}}{M_{00}} - x_c y_c \\ c = \frac{M_{02}}{M_{00}} - y_c^2 \end{cases} \tag{6}$$

θ is an angle which is defined by the direction of the axis of least inertia [22]. It is worth noting that the summation is used in Equations (1), (2) and (4) because we are dealing with discrete images rather than continuous images.

Higher moments contain details of the image that are relatively more sensitive to noise. Redundancies will be shown in the operation of a higher moment owing to its nonorthogonality. Therefore, many new moments [18,21,23] have been proposed.

2.2. Deficiency of Rotation-Angle Measurement Based on Geometric Moments

The rotation-angle measurement method based on geometric moments is advantageous because of its high accuracy and speed. However, there is a limitation when this method is used, as illustrated in Figure 1. The figure shows two workpieces S_1 and S_2 with different poses. Their centers of mass A_1 and A_2 are represented by the yellow $*$ symbols, and their axes are represented by green dotted lines. The angles θ_1 and θ_2 of the two axes are equal. The object position and pose are expressed by P_{ob}, and the angle of its axis is θ_{ob}.

Assuming that the rotation direction is counterclockwise, when S_1 and S_2 need to be placed into P_{ob}, the minimum rotating angle around the center of mass is obtained as follows:

$$r_n = \begin{cases} \theta_{ob} - \theta_n + 180°, n = 1. \\ \theta_{ob} - \theta_n, \quad\quad n = 2. \end{cases} \tag{7}$$

where r is the rotation angle.

The difference between r_1 and r_2 is 180° because θ_1 is equal to θ_2. S_1 and S_2 will be rotated by the same angle if the angle of the axis is regarded as the rotation angle. Therefore, the same rotation angle is obtained with dissimilar poses, which is an error of the measurement. The reason is that S_1 and S_2 are non-centrosymmetric about points A_1 and A_2, respectively. Thus, for non-centrosymmetric workpieces, the same rotation-angle would be obtained with dissimilar poses when the geometric moments are used for rotation-angle measurement. An automatic sorting system using this method is only suitable for center-symmetrical workpieces. This limitation significantly decreases the generality of the system.

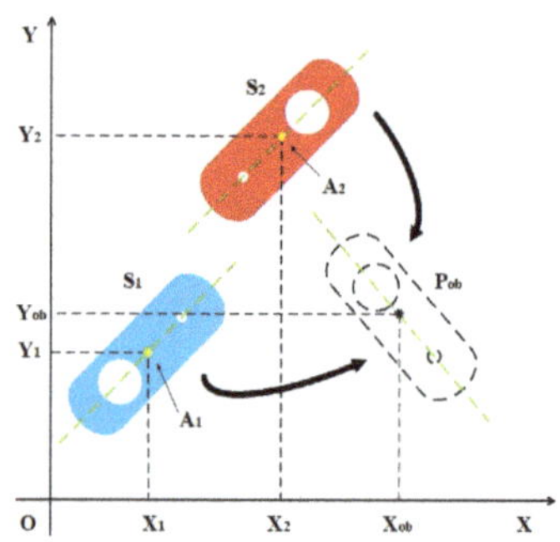

Figure 1. Case where the same rotation angle is obtained with dissimilar poses when using image geometric moments.

3. Method for Rotation-Angle Measurement

An improved method is presented here to overcome the limitation described in the previous section. The method is called the least iterative region segmentation (LIRS) method which consists of three steps and geometric information is used to overcome the limitation caused by the shape of the workpiece. The following two points are made before the LIRS method is introduced:

(1) We assume that the plane workpiece is uniform and the center of mass is located on the workpiece.
(2) We assume that the optical axis of the camera is perpendicular to the work plane.

The LIRS method is illustrated in detail below.

3.1. Image Preprocessing

An image point will be deviated from its ideal position in the presence of lens distortion [24], resulting in distorted images. Therefore, the calibration is used to improve the accuracy of rotation-angle measurement [25]. Moreover, the complicated background and the surface texture of a workpiece will appear as noise in rotation-angle measurement. Therefore, image processing is required to acquire a superior binary image. Common methods for image processing include denoising, grayscale, image morphology, and binarization. The image needs to be segmented into several pieces when more than one workpiece exists because only one workpiece can be handled at a time.

3.2. Least Iterative Region Segmentation Method

The coordinates system is established as shown in Figure 2. The red region is a workpiece. I_{sw} is a regions which is the minimum enclosing rectangle of the workpiece. I'_{sw} is also a regions which boundary is violet dotted line. The axis is shown as blue dotted line and the angle of the axis θ is obtained from Equation (5). The point A is the center of mass which coordinate is $(\overline{x}, \overline{y})$.

3.2.1. Judgment of Centrosymmetry

Two steps are required to judge whether the workpiece is centrosymmetric. Firstly, the center of I_{sw} should be calculated and the region I_{sw} needs to be extended to I'_{sw}, if A is not the center of I_{sd}. After extention, the point A is the center of I'_{sd}. Secondly, the region I'_{sd} rotated by 180° is convolved with the original region. The workpiece is centrosymmetric about the center of mass if the result is greater than a threshold. The angle in the counterclockwise direction between the two axes can be regarded as the rotation angle. Otherwise, the next step should be carried out.

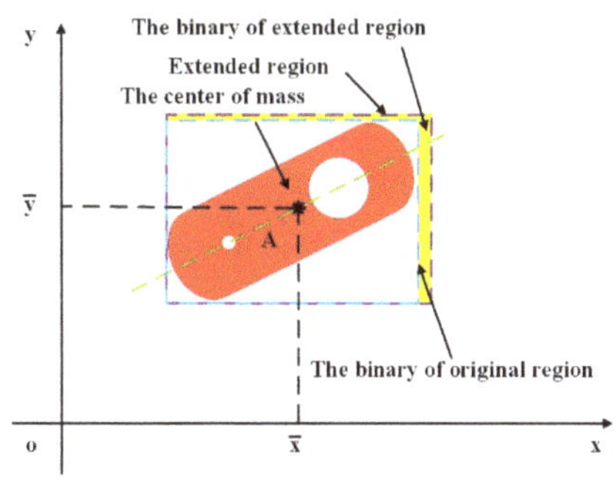

Figure 2. Schematic of image extension.

Template matching is always used for recognition. Therefore, this step can be changed to evaluate whether the template is centrosymmetric. The next step will be performed when the workpiece matches the asymmetric template. In this manner, the judgment of centrosymmetry will be completed before the region segmentation, and the efficiency of LIRS will be improved.

3.2.2. Region Segmentation and Identification

The purpose of this subsection is to find a separation line which divide the workpiece into two parts with different areas. A new rectangular coordinate system is established with center of mass as its origin, as shown in Figure 3. Then, a separation line through the origin is drawn as follows:

$$y - kx = 0, \tag{8}$$

where $k = \tan(\theta + n\alpha)$, α is the deviation angle which range is $[0°, 360°)$ and n is the iteration number which initial value is 1.

After θ and n are assigned, the equation of the separation line would be obtained. Then the workpiece could be divided into two parts D_1 and D_2 according to the relationship between the point and the line. The areas of D_1 and D_2 are $\Gamma(D_1)$ and $\Gamma(D_2)$. When the $\Gamma(D_1)$ is equal $\Gamma(D_2)$, we need to add 1 to n and divide the workpiece with the new separation line. The iteration will be stopped until the condition $\Gamma(D_1) \neq \Gamma(D_2)$ is met. The larger between the two parts is marked as D_l while the other is marked as D_s. The workpiece must be divided into two regions with different areas by the separation line because it is non-centrosymmetric about the center of mass.

To improve the efficiency of division, the threshold method is used. Firstly, the threshold function $B(P) = y - kx$ is established and $P(x, y)$ is a point in the workpiece. The segmentation function is set up as expressed by Equation (9), and $P(x, y)$ can be assigned to a region according to the polarity of the thresh function. Therefore, the workpiece is divided into two parts according to the relationship between the point and the separation line.

$$\begin{cases} B(P) > 0, P \in D_1 \\ B(P) > 0, P \in D_2 \end{cases} \tag{9}$$

There are two point which need attention:

(1) The deviation angle needs to be selected reasonably. We should avoid choosing the symmetry axis or its perpendicular axis as the separation line because these axes divide a symmetric workpiece into two parts with the same area.
(2) The area of the workpiece will not be exactly equal after the workpiece is rotated at different angles because the images captured by the industrial camera have been already discretized by a charge-coupled device and a discretization error will always exist. To eliminate the effect of discretization on the measurement, a threshold is employed. The areas of D_l and D_s are considered equal when the absolute area difference is less than the threshold.

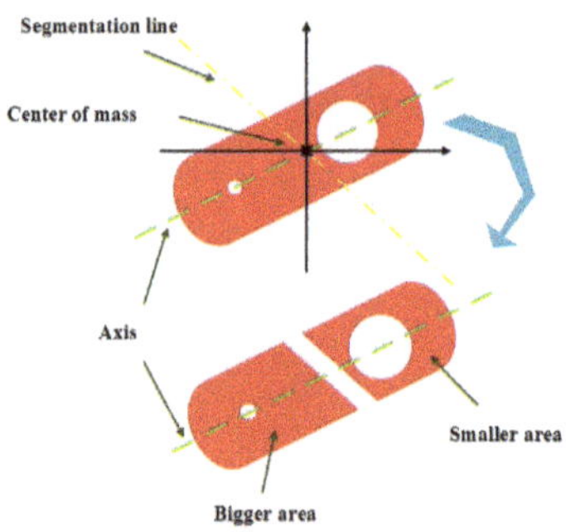

Figure 3. Image segmentation with a separation line.

3.2.3. Rotation-Angle Calculation

After segmentation, a direction vector $\vec{p}$ can be established from $(\overline{x_l}, \overline{y_l})$ to $(\overline{x_s}, \overline{y_s})$, where $(\overline{x_l}, \overline{y_l})$ is the center of mass of D_l and $(\overline{x_s}, \overline{y_s})$ is the center of mass of D_s. The two coordinates are calculated by Equation (3). The direction vector can be used to calculate the rotation angle because of rotation invariance. Assuming that a pose is represented by vector $\vec{p} = (x_o, y_o)$, the rotation angle is obtained by employing:

$$\Theta = \frac{\vec{p} \times \vec{q}}{|\vec{p}||\vec{q}|} = \frac{\Delta x x_o + \Delta y y_0}{\sqrt{(\Delta x)^2 + (\Delta y)^2}\sqrt{x_0^2 + y_0^2}} \tag{10}$$
$$\Delta x = x_s - x_l, \Delta y = y_s - y_l$$

$$\Lambda = \vec{p} \times \vec{q} \tag{11}$$

$$\theta = f(\Theta, \Lambda) \tag{12}$$

where, Θ is a cosine value and the Λ is symbol which polarity is decided by the relationship between $\vec{q}$ and $\vec{p}$. f is a function which calculate the rotation-angle based on Θ and the Λ. The value range of θ is $[0^\circ, 360^\circ)$.

The result of the LIRS method is shown in Figure 4. The green dotted lines are the separation lines and the blue dotted lines are axes. The red arrows are the direction vectors and the purple arrows are the object vectors. Although the slopes of the two axes are the same, the direction vectors are different.

The angle in the counterclockwise direction between two direction vectors could be regarded as the rotation-angle. The result shows that the LIRS method can effectively measure the rotation-angle of the workpiece, and it overcomes the limitation of the conventional rotation-angle measurement method based on geometric moments.

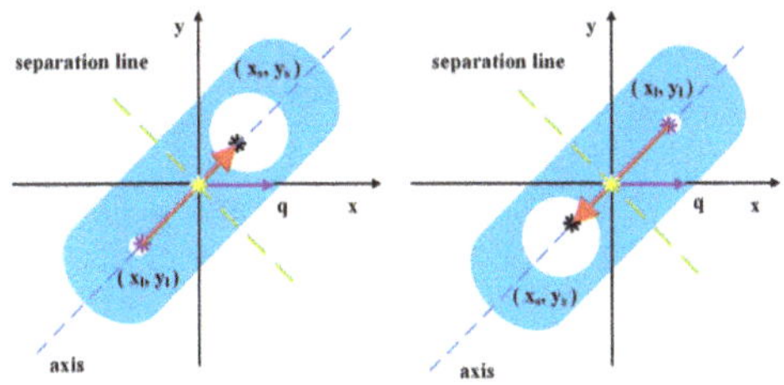

Figure 4. Result of the LIRS method.

3.3. *Evaluation of LIRS Method*

Efficiency, accuracy, and application range are the three most important indexes of automatic sorting systems. They are affected by the performance of the vision algorithm. Therefore, the applicability of the LIRS method in industrial environments needs to be evaluated. In this section, the accuracy, speed, and generality of LIRS as well as the image size are analyzed in detail.

A schematic of the rotation-angle measurement assessment system is shown in Figure 5. The experimental set up consists of a CCD, a computer, a support, and rotary equipment, which includes a pedestal and a rotor. A dial is fixed on the surface of the rotor, and the workpiece is placed on the dial. The workpiece is rotated by the rotor.

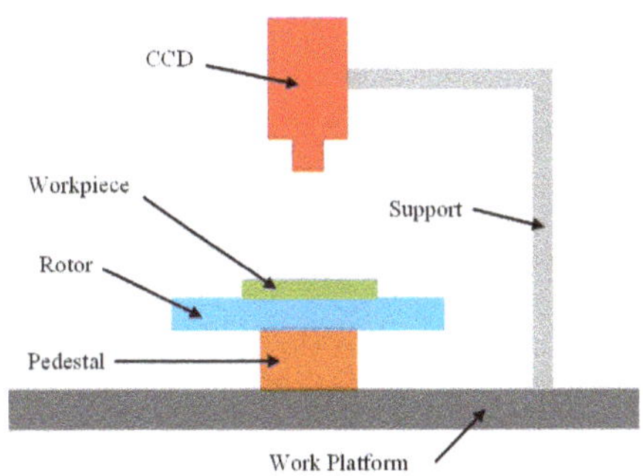

Figure 5. Schematic of the rotation-angle measurement assessment system.

The CCD model is MER-200-14GC. It has a 12-mm lens, and its image resolution is 1628 × 1236. A support with three degrees of freedom is used to adjust the camera pose. For convenience, the camera optical axis is made perpendicular to the work plane by adjusting the support. A photograph of the experimental set up is shown in Figure 6.

The LIRS method is coded in C++ and compiled for 64 bits under Windows 10, and OpenCV 3.2 is used to process images. The program is executed using an Intel(R) Core(TM)i5-6300HQ CPU running at 2.30 GHz with 16 GB RAM. The workpiece is rotated from 1° to 360° in steps of 1°. One image is obtained per rotation, and the first image is regarded as the reference. Subsequently, rotation angles are calculated between the reference and the other images.

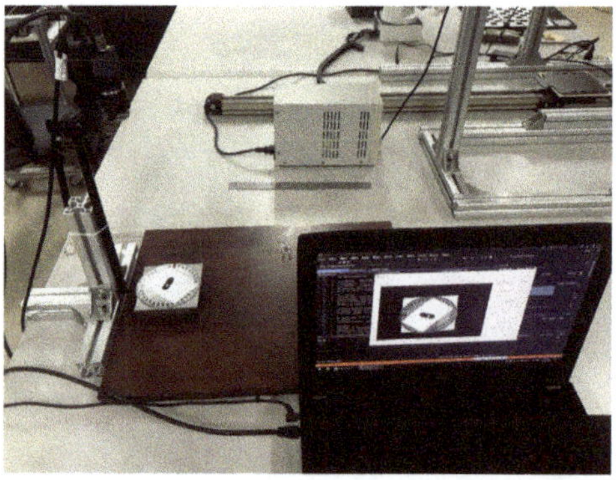

Figure 6. Experimental set up of the LIRS assessment system.

3.3.1. Accuracy

The measurement error is shown in Figure 7. The maximum measurement error is less than 0.1°, which indicates that the LIRS method has a high accuracy. In other words, the LIRS method can be used to realize rotation-angle measurement with the whole angle range of 1°–360° in an automatic sorting system.

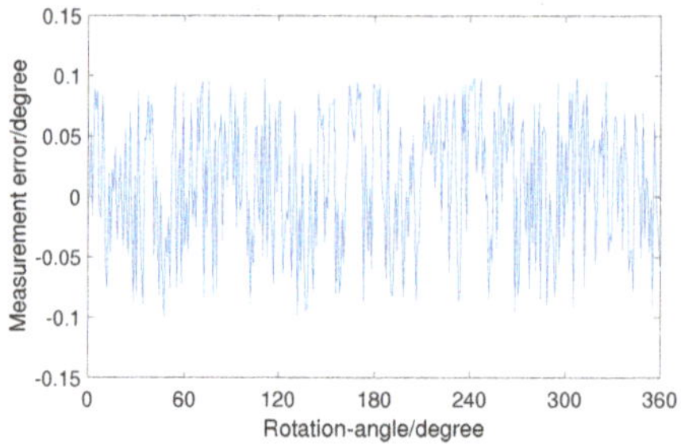

Figure 7. Measurement error of rotation angle in the experiment when the rotation angle is 1°–360°.

3.3.2. Time Consumption

The time consumption of LIRS method is shown in Figure 8. The average time to calculate a rotation-angle is 62.1136 ms. The time-consumption curve shows large fluctuations because the images have different sizes. Each image shows the region of interest (ROI), which is determined based on the minimum external rectangle. The size of the ROI differs after rotation, as shown in Figure 9. Therefore, the time consumption shows large fluctuations when the whole angle range is measured.

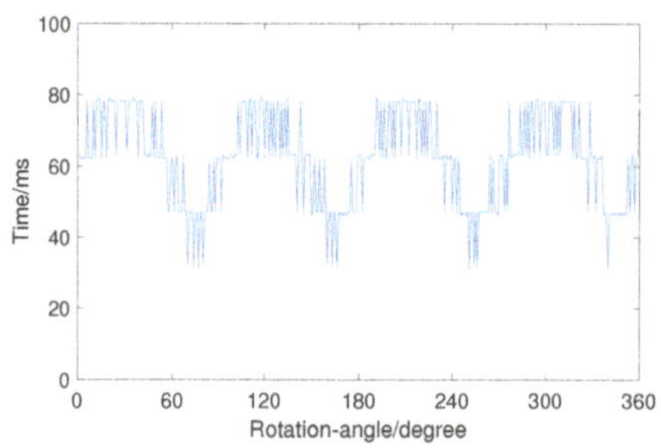

Figure 8. Time consumption of rotation-angle measurement when the rotation angle is 1°–360°.

There may be several workpieces in an image, and the execution of this program is sequential. Therefore, the time consumption is high. If the program is run in a field-programmable gate array (FPGA) device, the parallel-computing features of the FPGA device can be used to reduce the operating time substantially, further improving the efficiency of the LIRS method.

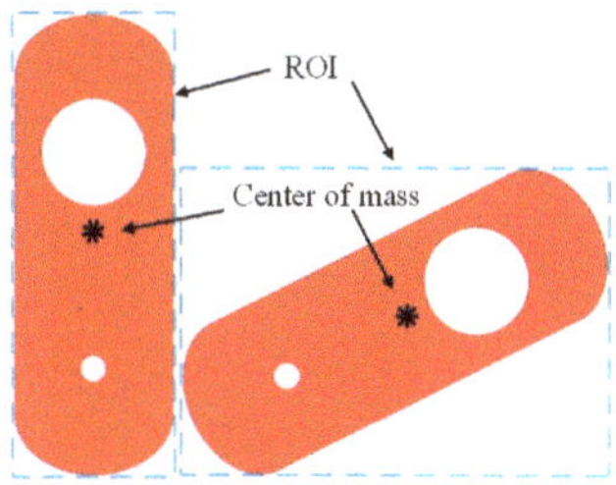

Figure 9. Schematic of ROI selection with the minimum external rectangle.

3.3.3. Generality

The LIRS method is designed to overcome the limitation of rotation-angle measurement methods based on geometric moments. The LIRS method has an iteration number n and a deviation angle α, which can adjust the orientation of the separation line. The LIRS method can find a separation line for all non-centrosymmetric workpieces. This separation line will be determined uniquely after the deviation angle is selected. Therefore, the LIRS method has a higher flexibility and a better generality compared to the conventional method because it is suitable for workpieces of all shapes.

3.3.4. Image Size

The relationship between the length of the direction vector and the measurement error should be consider since discretization error exist. Assume that the length of the direction vector is l. Take the starting point of the vector as the center and draw an one-pixel circle. The maximum directions which the vector could represent is equal to A_n, which is also the number of pixels on the circle. Therefore, with more pixels on the circle, the direction vector can represent more directions. As Figure 10 shows, $L = 70, 130, 180$ are selected, and the maximum numbers of angles are $A_n = 636, 792, 1312$, respectively.

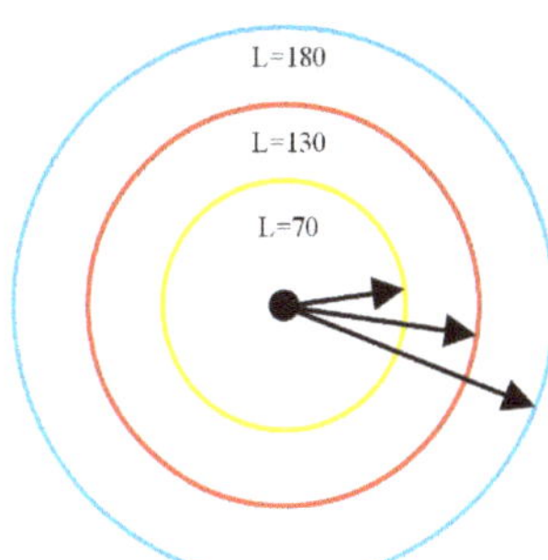

Figure 10. Schematic of the image size.

The discretization error between the measured value and the actual value decreases when l is larger. As the number of direction increases, the discrete values are closer to being continuous values. Consequently, the accuracy of the LIRS method is increased. For the same workpiece, the length of the direction vector can be increased by selecting a suitable lens and reducing the distance between the camera and the workpiece. However, this will increase the size of the ROI and time consumption. It is necessary to obtain the optimal solution between time consumption and accuracy.

4. Rotation-Angle Measurement Model

4.1. Modeling

When the optical axis is non-perpendicular to the work plane, a dimensional measurement error will occur. In other words, the accuracy of dimensional measurement is affected by camera pose. However, the relationship between the accuracy of rotation-angle measurement and camera pose has not been studied. Therefore, a rotation-angle measurement model needs to be established. Figure 11 shows the basic geometry of the ideal camera model. Three steps are necessary because only an ideal camera model is addressed [26]. For convenience, the pose in which the optical axis is perpendicular to the work plane is called the ideal pose. All other camera poses are non-ideal.

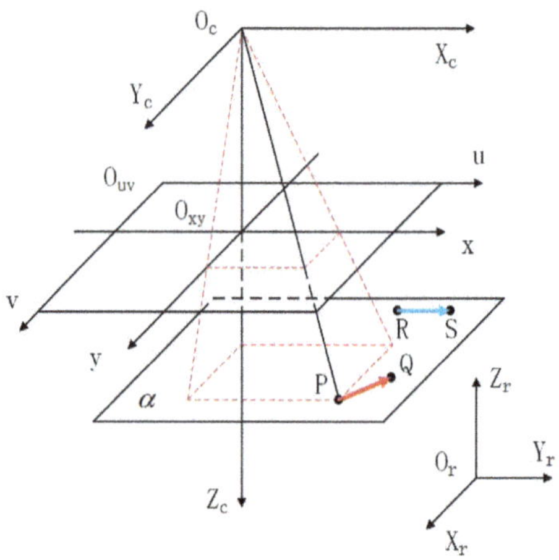

Figure 11. Geometry of the ideal camera model.

There are four coordinate systems in this model. The camera coordinate system is composed of X_c, Y_c, and Z_c axes and the point O_c. The robot coordinate system is treated as the world coordinate system, which is composed of X_r, Y_r, and Z_r axes and the point O_r. The pixel coordinate system is composed of u and v axes and the point O_{uv}. The image coordinate system is composed of x and y axes and the point O_{xy}. The work plane is represented by α. For convenience, we assume that the Z_c axis is perpendicular to the work plane, and the height of the workpiece is neglected. For any point P in α, its image coordinates can be expressed as Equations (13)–(17).

$$Z_c \begin{bmatrix} u \\ v \\ 1 \end{bmatrix} = \begin{bmatrix} f/dx & 0 & u_0 \\ 0 & f/dy & v_0 \\ 0 & 0 & 1 \end{bmatrix} \begin{bmatrix} X_c \\ Y_c \\ Z_c \end{bmatrix} \tag{13}$$

$$\begin{bmatrix} X_c \\ Y_c \\ Z_c \end{bmatrix} = R_z R_y R_x \left(\begin{bmatrix} X_w \\ Y_w \\ Z_w \end{bmatrix} - \begin{bmatrix} t_x \\ t_y \\ t_z \end{bmatrix} \right), \tag{14}$$

$$R_x = \begin{bmatrix} \cos\alpha & -\sin\alpha & 0 \\ \sin\alpha & \cos\alpha & 0 \\ 0 & 0 & 1 \end{bmatrix}, \tag{15}$$

$$R_y = \begin{bmatrix} 0 & 0 & 1 \\ 0 & \cos\beta & -\sin\beta \\ 0 & \sin\beta & \cos\beta \end{bmatrix}, \tag{16}$$

$$R_z = \begin{bmatrix} \cos\gamma & 0 & -\sin\gamma \\ 0 & 1 & 0 \\ -\sin\gamma & 0 & \cos\gamma \end{bmatrix}, \tag{17}$$

where f is focal length, dx and dy are the distances between adjacent pixels in the u and v axes, respectively. u_0 and v_0 are row and column numbers of the center. $[t_x, t_y, t_z]'$ is a translation vector from the robot coordinate to the camera coordinate system. R_x, R_y, and R_z are three rotation matrixes, which are multiplied in the order of Equation (14). α, β, and γ are three angles. Equation (13) describes the relationship between camera coordinates system and pixel coordinates system. Equation (14) describes the relationship between robot coordinate system and camera coordinate system. The equation of coordinate transformation between pixel coordinate system and the robot coordinate system is established by using this two equations.

For convenience, vector $\overrightarrow{RS} = (1,0,0)$ is considered as the object pose, and the workpiece is abstracted as a vector $\overrightarrow{PQ} = (\Delta x, \Delta y, 0)$. $\overrightarrow{RS}$ and $\overrightarrow{PQ}$ are represented by blue and red arrow in Figure 11, respectly. The work plane in robot coordinates is $Zr = Z$. The center of mass is treated as the starting point P, and the center of mass of region D_s is regarded as the ending point Q. Therefore, the angle in the counterclockwise direction between $\overrightarrow{PQ}$ and $\overrightarrow{RS}$ can be regarded as the rotation angle.

The ideal value of the rotation angle is

$$\theta_i' = \arccos \frac{\overrightarrow{PQ} \times \overrightarrow{RS}}{\left|\overrightarrow{PQ}\right|\left|\overrightarrow{RS}\right|}, \tag{18}$$

$$\theta_i = f\left(\theta_i'\right), \tag{19}$$

where f is an adjusting function that makes the value range of the rotation angle $[0°, 360°)$.

The measured value is obtained by substituting Equation (13) into Equation (18) and simplifying, as expressed by Equation (20)

$$\theta_r' = \arccos \frac{f_{n1} + t_y f_{n2}\left(1 - c^2\right) + C\left(A f_{n2} - t_y k \sin y\right)}{0.5\sqrt{f_{d1}^2 + f_{d2}^2}\sqrt{f_{d3} t_y^2 + f_{d4}}}, \tag{20}$$

$$\theta_r = f\left(\theta_r'\right), \tag{21}$$

where

$$\begin{cases} f_{n1} = kA\sin\beta - d^2 - \cos\alpha \\ f_{n2} = -t_y + kt_x + \frac{b}{t_z - Z} \\ f_{n3} = -2t_y + kt_x + \frac{b}{t_z - Z} \\ f_{d1} = k\cos\beta - Bf_{n2} - D \\ f_{d2} = \cos\alpha - f_{n2}\sin\alpha \\ f_{d3} = 3 - 2\left(C^2 - B^2\right) - \cos 2\alpha \\ f_{d4} = 3 - 2\left(A^2 - D^2\right) + \cos 2\alpha + 8t_y AC \\ A = \sin\alpha\cos\beta \\ B = \cos\alpha\sin\beta \\ C = \cos\alpha\cos\beta \\ D = \sin\alpha\sin\beta \end{cases} \tag{22}$$

$y = kx + b$ is a line corresponding to $\overrightarrow{PQ}$ in robot coordinates.

Thus, the rotation-angle measurement model has been established, and the difference between θ_i and θ_r is the rotation-angle measurement error.

4.2. Simulation and Discussion

It can be seen that the measured value is affected by several parameters, which can be divided into two categories. The first includes α, β, t_x, t_y, and the difference between the work plane and optical center $t_z - Z$. These six parameters will be confirmed after the camera is installed. There are only two angles in the model, and γ is not included. It can be seen that γ is uncorrelated with the measured value, and camera rotation around the its optic axis can be neglected in installation. Thus, camera-installation flexibility is improved in the automatic sorting system. The second category includes k and b. k is the tangent value of the rotation angle, and b is the position of the vector with the angle α. When the vector moves along the line, the measured value remains invariant. Otherwise, it will be changed. This means that different measured values would be obtained for some vectors that have the same rotation angle but dissimilar positions. This case would result in measurement error.

Figure 12 shows curves of rotation-angle measurement error when four vectors move along the line $y = 50$. An approximately linear relationship exists between displacement and measurement error. The polarity and the rate of error are related to the vectors. This means that different measured values would be obtained when the workpiece is located at different positions with the same rotation angle. Figure 13 shows the rotation-angle measurement error in simulation with four values of α and β. The vector rotates around its starting point in steps of 1°. For vectors with different values of α and β, the measurement-error curves are different. The measured values are different, when the same vector is selected with different values of α and β. That is, when the same workpiece is measured with different camera poses, the measured values are different. The polarity and value of the error is related to the camera pose.

To reduce the rotation-angle measurement error to zero, the following condition should be met:

$$\theta_i - \theta_r = 0. \tag{23}$$

Then, Equation (24) will obtained:

$$\alpha = 0, \beta = 0. \tag{24}$$

It can be seen that the measurement error is always present only if the camera is in a non-ideal pose.

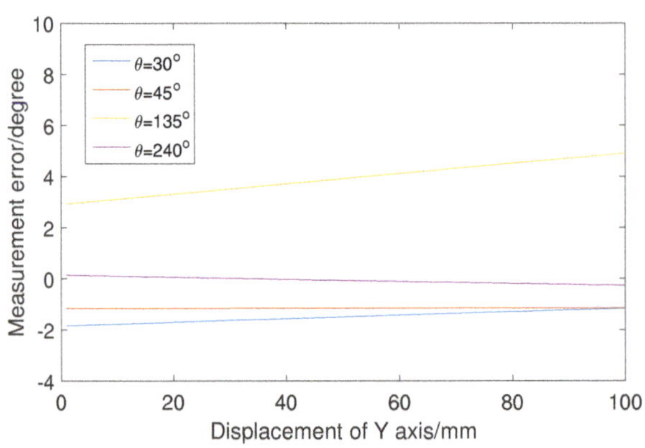

Figure 12. Measurement error in the simulation experiment when vectors move along the line $y = 50$.

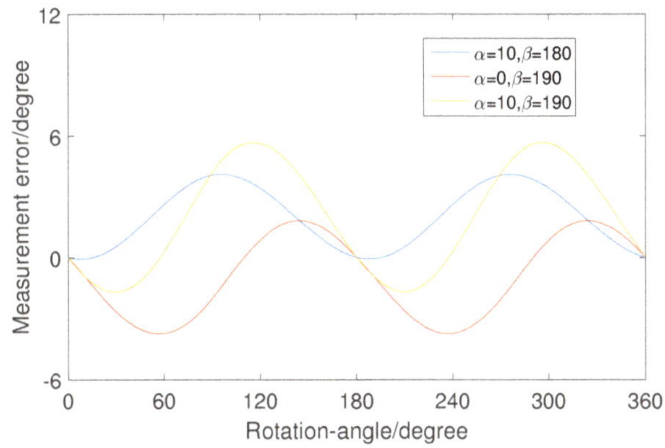

Figure 13. Measurement error in the simulation experiment when the vector rotates around its starting point.

4.3. Method for Correction of Rotation-Angle Measurement Error

To meet the condition of perpendicularity, camera should be adjusted by the support before the measurement. The rotation-angle measurement error will always exist when the camera is in a non-ideal pose, reducing the accuracy of rotation-angle measurement. To make the measured value accurate, it is necessary to keep the camera in the ideal pose. In other words, the optical axis needs to be adjusted to be perpendicular to the work plane. However, this condition cannot be met easily in industrial environments, because of camera-installation errors or position limitations. The actual pose could not be coinciding with the ideal pose completely. Therefore, the rotation-angle measurement error needs to be corrected.

When the camera is in a non-ideal pose, the Z_c coordinate of a point on the work plane will be changed form a constant to a variable. The relationship between the image coordinates and camera coordinates can be expressed as follows:

$$\Delta u = f\frac{X_1Z_2 - X_2Z_1}{dxZ_1Z_2}, \Delta v = f\frac{Y_1Z_2 - Y_2Z_1}{dyZ_1Z_2}, \tag{25}$$

where (X_1, Y_1, Z_1) and (X_1, Y_1, Z_1) are two camera coordinates in the work plane. dv and du are the differences of image coordinates. There is no linear relationship between $\sqrt{(\Delta u)^2 + (\Delta v)^2}$ and $\sqrt{(X_1 - X_2)^2 + (Y_1 - Y_2)^2}$. Therefore, the image will be distorted. This is the primary cause of the rotation-angle measurement error.

A rotation-angle error measurement correction (REMC) method with an error-correction matrix is presented to reduce the rotation-angle measurement error. A binary function ω is employed to multiply with Z_c and keep the result constant. A linear relationship will be kept between the image coordinates and camera coordinates after mapping. The REMC method is illustrated in detail below.

A correction matrix A is introduced as follows:

$$\begin{bmatrix} u' \\ v' \\ 1 \end{bmatrix} = \frac{1}{\omega}\begin{bmatrix} a_{11} & a_{12} & a_{13} \\ a_{21} & a_{22} & a_{23} \\ a_{31} & a_{32} & a_{33} \end{bmatrix}\begin{bmatrix} u \\ v \\ 1 \end{bmatrix}, \tag{26}$$

$$\omega = ua_{31} + va_{32} + a_{33}. \tag{27}$$

The relationship between the image coordinate system (u, v) and camera coordinate system (X_c, Y_c, Z_c) can be expressed as follows:

$$u = \frac{fX_c}{Z_c dx} + u_0, u = \frac{fY_c}{Z_c dy} + v_0. \tag{28}$$

Then, Equation (13) can be rewritten as follows:

$$Z_c\omega \begin{bmatrix} u' \\ v' \\ 1 \end{bmatrix} = \begin{bmatrix} F'_{11} & F'_{12} & F'_{13} \\ F'_{21} & F'_{22} & F'_{23} \\ F'_{31} & F'_{32} & F'_{33} \end{bmatrix} \left(\begin{bmatrix} X_w \\ Y_w \\ Z_w \end{bmatrix} - \begin{bmatrix} t_x \\ t_y \\ t_z \end{bmatrix} \right), \tag{29}$$

where F is a coefficient matrix and $Z_c\omega$ can be expressed as

$$Z_c\omega = a_{31} f_x X_c + a_{32} f_y Y_c + s(a_{31} u_0 + a_{32} v_0 + a_{33}). \tag{30}$$

The work plane in the camera coordinate system can be expressed as follows:

$$aX_c + bY_c + cZ_c - A = 0, \tag{31}$$

where a, b, c, and A are constant parameters. Three parameters a_{31}, a_{32}, and a_{33} must exist to ensure the equation holds:

$$Z_c w = A. \tag{32}$$

Then, $(u', v', 1)$ can be obtained as follows:

$$\begin{bmatrix} u' \\ v' \\ 1 \end{bmatrix} = \frac{1}{A} \begin{bmatrix} F'_{11} & F'_{12} & F'_{13} \\ F'_{21} & F'_{22} & F'_{23} \\ F'_{31} & F'_{32} & F'_{33} \end{bmatrix} \left(\begin{bmatrix} X_w \\ Y_w \\ Z_w \end{bmatrix} - \begin{bmatrix} t_x \\ t_y \\ t_z \end{bmatrix} \right). \tag{33}$$

A is the Z_c value of the work plane in the camera coordinate system. It can be seen that the Z_c value can remain invariant during the mapping process. Therefore, the measurement error caused by camera pose will be reduced when (u', v') is used to calculate the rotation angle.

The experimental system is shown in Figure 6. The optical axis is adjusted using the support to be non-perpendicular to the work plane, and the obtained results are listed in Table 1. It can be seen that the REMC method can reduce the rotation-angle measurement error caused by a non-ideal camera pose, and the error is less than 0.1°. Therefore, the proposed method is effective and meets the requirements.

The correction matrix is selected as follows:

$$A = \begin{bmatrix} 4.5985 & 0.0779219 & -1904.15 \\ 0.0572827 & 4.58486 & -2660.6 \\ 2.07701 \times 10^{-5} & 4.75542 \times 10^{-5} & 1 \end{bmatrix}. \tag{34}$$

Table 1. Experimental results obtained when the REMC method is employed in the experiment under a non-ideal camera pose.

Ideal Value	Measured Value	Correction Value	Error
30°	31.15°	30.11°	0.11°
60°	62.95°	60.06°	0.06°
120°	121.97°	119.04°	0.04°
150°	150.31.32°	150.03°	0.03°
210°	210.21°	209.09°	0.09°
240°	242.61°	240.05°	0.05°
310°	311.73°	319.03°	0.03°
330°	330.82°	330.06°	0.06°

5. Experiment

An automatic sorting system with machine vision is established, as shown in Figure 14. A robot (Dobot Magician) with a four degrees of freedom robot is used in this system. Four stepping motors are used to drive a manipulator, which moves with a re-orientation accuracy of 0.2 mm. The software is coded by MFC with OpenCV 3.2 and consisted of three parts: (1) a camera and a robot control system including initialization, start and stop functions, and parameter setting; (2) a real-time display system consisting of an image display and information display; and (3) an information storage system designed to save important data during program operation. The correction matrix A is selected as follows:

$$A = \begin{bmatrix} 1.45457 & 0.383242 & -902.439 \\ -0.220291 & 1.86919 & -318.915 \\ 1.57966 \times 10^{-6} & 0.000271224 & 1 \end{bmatrix}. \tag{35}$$

The workpiece is a uniform-thickness thin sheet with two holes of different diameters. The experimental result is shown in Figure 15. The image with the blue external rectangle and yellow point is shown on the main interface. Key information is shown in the message region. The results show that the rotation angles are obtained accurately, and the workpieces could be placed correctly by this system. Therefore, the LIRS and REMC method could be used in automatic sorting systems in industrial environments.

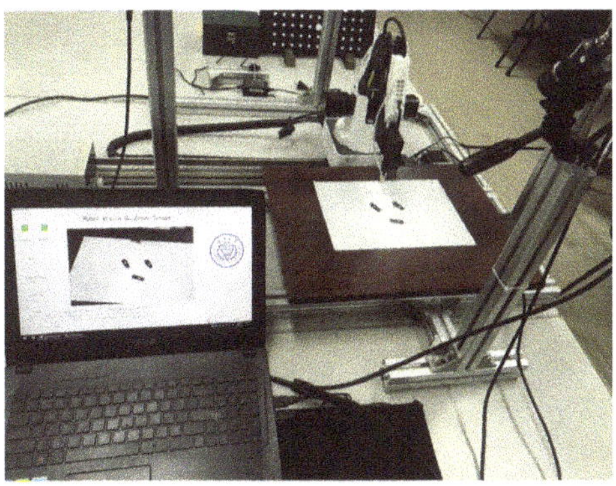

Figure 14. Automatic sorting system with machine vision.

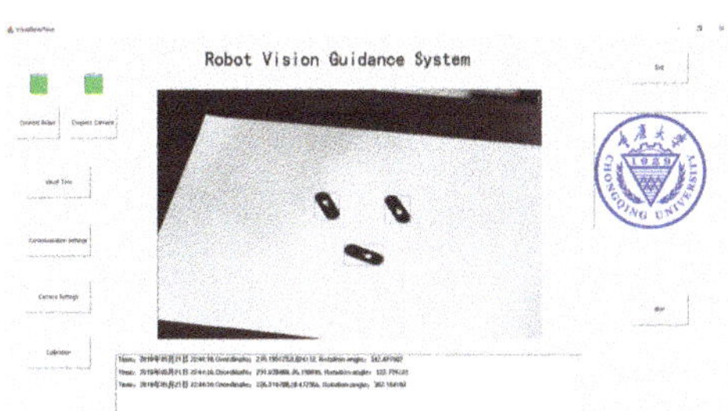

Figure 15. Result of the experiment.

6. Conclusions

The rotation angle is an important parameter in an automatic sorting system. To accurately measure the rotation angles of plane workpieces for an automatic sorting system, the LIRS method was proposed. This method overcomes limitation of the conventional method based on geometric moments, and it is suitable for workpieces of all shapes. Experimental results show that the measurement error of the LIRS method is less than 0.1°, and the measurement range is between 0° and 360°. Therefore, the LIRS method meets the requirements of automatic sorting in industrial environments. However, the average measurement time is approximately 62.1136 ms, which leaves much room for improvement.

A model was established for studying the relationship between camera pose and rotation-angle measurement error. Then, a formula for calculating the error was derived. The simulation results show that the measurement error will always exist when the camera is in a non-ideal pose. The value and polarity of the measurement error are related to the camera pose and location of the workpiece. Subsequently, the REMC method was designed to correct the rotation-angle measurement error. The experimental results show that the REMC method is effective, and the measurement error with the REMC method is less than 0.12°.

Finally, an automatic sorting system with the LIRS and REMC method was established, and sorting experiments were conducted. The two proposed methods yielded accurate rotation angles, and plane workpieces could be placed correctly by this system.

Author Contributions: Conceptualization, C.C.; methodology, Q.O.; formal analysis, C.C.; software, C.C.; investigation: C.C.; data curation, C.C.; validation, Q.O.; writing—original draft preparation, C.C.; writing—review and editing, Q.O.; supervision, Q.O.; project administration, Q.O.

Funding: This work is supported by the National Natural Science Foundation of China (No. 51374264) and Overseas Returnees Innovation and Entrepreneurship Support Program of Chongqing (No. CX2017004).

Conflicts of Interest: The authors declare no conflict of interest.

References

1. Cho, N.H.; Chang, D.I.; Lee, S.H.; Hwang, H.; Lee, Y.H.; Park, J.R. Development of automatic sorting system for green pepper using machine vision. *J. Biosyst. Eng.* **2007**, *30*, 110–113.
2. Zheng, H.; Lu, H.F.; Zheng, Y.P.; Lou, H.Q.; Chen, C.Q. Automatic sorting of Chinese jujube (*Zizyphus jujuba*, Mill. cv. 'hongxing') using chlorophyll fluorescence and support vector machine. *J. Food Eng.* **2010**, *101*, 402–408. [CrossRef]
3. Basu, S.; Das, N.; Sarkar, R.; Kundu, M.; Nasipuri, M.; Basu, D.K. A novel framework for automatic sorting of postal documents with multi-script address blocks. *Pattern Recognit.* **2010**, *43*, 3507–3521. [CrossRef]
4. Mesina, M.B.; de Jong, T.P.R.; Dalmijn, W.L. Automatic sorting of scrap metals with a combined electromagnetic and dual energy X-ray transmission sensor. *Int. J. Miner. Process.* **2007**, *82*, 222–232. [CrossRef]
5. Jiu, H.; Thomas, P.; Bian, Z.F. Automatic Sorting of Solid Black Polymer Wastes Based on Visual and Acoustic Sensors. *Energy Procedia* **2011**, *11*, 3141–3150.
6. Wilson, J.R.; Lee, N.Y.; Saechao, A.; Tickle-Degnen, L.; Scheutz, M. Supporting Human Autonomy in a Robot-Assisted Medication Sorting Task. *Int. J. Soc. Robot.* **2018**, *10*, 621–641. [CrossRef]
7. Urizar, M.; Petuya, V.; Amezua, E.; Hernandez, A. Characterizing the configuration space of the 3-SPS-S spatial orientation parallel manipulator. *Meccanica* **2014**, *49*, 1101–1114. [CrossRef]
8. De Saxe, C.; Cebon, D. A Visual Template-Matching Method for Articulation Angle Measurement. In Proceedings of the IEEE International Conference on Intelligent Transportation Systems, Las Palmas, Spain, 15–18 September 2015; pp. 626–631.
9. Matungka, R.; Zheng, Y.F.; Ewing, R.L. Image registration using adaptive polar transform. In Proceedings of the 15th IEEE International Conference on Image Processing, San Diego, CA, USA, 12–15 October 2008; pp. 2416–2419.
10. Revaud, J.; Lavoue, G.; Baskurt, A. Improving Zernike Moments Comparison for Optimal Similarity and Rotation Angle Retrieval. *IEEE Trans. Pattern Anal. Mach. Intell.* **2008**, *31*, 627–636. [CrossRef]
11. Delponte, E.; Isgro, F.; Odone, F.; Verri, A. SVD-matching using SIFT features. *Graph. Models* **2006**, *68*, 415–431. [CrossRef]
12. Munoz-Rodriguez, J.A.; Asundi, A.; Rodriguez-Vera, R. Recognition of a light line pattern by Hu moments for 3-D reconstruction of a rotated object. *Opt. Laser Technol.* **2004**, *37*, 131–138. [CrossRef]
13. Li, W.M.; Jin, J.; Li, X.F.; Li, B. Method of rotation angle measurement in machine vision based on calibration pattern with spot array. *Appl. Opt.* **2010**, *49*, 1001–1006. [CrossRef] [PubMed]

14. Dong, H.X.; Fu, Q.; Zhao, X.; Quan, Q.; Zhang, R.F. Practical rotation angle measurement method by monocular vision. *Appl. Opt.* **2015**, *54*, 425–435. [CrossRef]
15. Fang, J.Y.; Qin, S.Q.; Wang, X.S.; Huang, Z.S.; Zheng, J.X. Frequency Domain Analysis of Small Angle Measurement with Moire Fringe. *Acta Photonica Sin.* **2010**, *39*, 709–713. [CrossRef]
16. Wu, Y.M.; Cheng, H.B.; Wen, Y.F. High-precision rotation angle measurement method based on a lensless digital holographic microscope. *Appl. Opt.* **2018**, *57*, 112–118. [CrossRef] [PubMed]
17. Yun, H.G.; Kim, S.H.; Jeong, H.S.; Kim, K.H. Rotation angle measurement based on white-light interferometry with a standard optical flat. *Appl. Opt.* **2012**, *51*, 720–725. [CrossRef] [PubMed]
18. Hu, M.K. Visual pattern recognition by moment invariants. *IRE Trans. Inf. Theory* **1962**, *8*, 179–187.
19. Khotanzad, A.; Hong, Y.H. Invariant Image Recognition by Zernike Moments. *IEEE Trans. Pattern Anal. Mach. Intell.* **1990**, *12*, 489–497. [CrossRef]
20. Marin, D.; Aquino, A.; Gegundez-Arias, M.E.; Bravo, J.M. A new supervised method for blood vessel segmentation in retinal images by using gray-level and moment invariants-based features. *IEEE Trans. Med. Imaging* **2011**, *30*, 146–158. [CrossRef] [PubMed]
21. Mandal, M.K.; Aboulnasr, T.; Panchanathan, S. Image indexing using moments and wavelets. *IEEE Trans. Consum. Electron.* **1996**, *42*, 557–565. [CrossRef]
22. Berthold, K.P.H. *Robot Vision*; MIT: Boston, MA, USA, 1987; p. 50.
23. Liu, Z.J.; Li, Q.; Xia, Z.W.; Wang, Q. Target recognition of ladar range images using even-order Zernike moments. *Appl. Opt.* **2012**, *51*, 7529–7536. [CrossRef]
24. Ouyang, Q.; Wen, C.; Song, Y.D.; Dong, X.C.; Zhang, X.L. Approach for designing and developing high-precision integrative systems for strip flatness detection. *Appl. Opt.* **2015**, *54*, 8429–8438. [CrossRef] [PubMed]
25. Munoz-Rodriguez, J.A. Online self-camera orientation based on laser metrology and computer algorithms. *Opt. Commun.* **2011**, *284*, 5601–5612. [CrossRef]
26. Tian, J.D.; Peng, X. Three-dimensional digital imaging based on shifted point-array encoding. *Appl. Opt.* **2005**, *44*, 5491–5496. [CrossRef] [PubMed]

sensors

MDPI

Article

A Stereo-Vision System for Measuring the Ram Speed of Steam Hammers in an Environment with a Large Field of View and Strong Vibrations

Ran Chen, Zhongwei Li *, Kai Zhong *, Xingjian Liu, Yonghui Wu, Congjun Wang and Yusheng Shi

State Key Laboratory of Material Processing and Die & Mould Technology, Huazhong University of Science and Technology, Wuhan 430074, China; chenran@hust.edu.cn (R.C.); xingjianliu@hust.edu.cn (X.L.); wuyonghui@hust.edu.cn (Y.W.); walden@hust.edu.cn (C.W.); shiyusheng@hust.edu.cn (Y.S.)

* Correspondence: zwli@hust.edu.cn (Z.L.); kaizhong@hust.edu.cn (K.Z.)

Received: 24 January 2019; Accepted: 22 February 2019; Published: 26 February 2019

Abstract: The ram speed of a steam hammer is an important parameter that directly affects the forming performance of forgers. This parameter must be monitored regularly in practical applications in industry. Because of the complex and dangerous industrial environment of forging equipment, non-contact measurement methods, such as stereo vision, might be optimal. However, in actual application, the field of view (FOV) required to measure the steam hammer is extremely large, with a value of 2–3 m, and heavy steam hammer, at high-speed, usually causes a strong vibration. These two factors combine to sacrifice the accuracy of measurements, and can even cause the failure of measurements. To solve these issues, a bundle-adjustment-principle-based system calibration method is proposed to realize high-accuracy calibration for a large FOV, which can obtain accurate calibration results when the calibration target is not precisely manufactured. To decrease the influence of strong vibration, a stationary world coordinate system was built, and the external parameters were recalibrated during the entire measurement process. The accuracy and effectiveness of the proposed technique were verified by an experiment to measure the ram speed of a counterblow steam hammer in a die forging device.

Keywords: speed measurement; stereo-vision; large field of view; vibration; calibration

1. Introduction

The ram speed of a steam hammer reflects the energy of a forging's deformation and directly affects the forming performance of forging equipment. Thus, it is an important parameter in the forging process. Due to their advantages of long life, high efficiency, and great energy, hammers [1] driven by steam are used in forging, especially for large workpieces. However, the ram speed of the steam hammer varies as the number of its use cycles increases, which may deteriorate its performance in forming workpieces. Therefore, it must be monitored regularly in practical applications. Currently, contact sensors, such as acceleration sensors [2] and inertial sensors [3,4], are widely used for speed measurement. Although these sensors are relatively accurate, they need to be affixed to the object being measured. Considering that the temperature of the steam hammer is very high, reaching around 300 °C, it is difficult to obtain stable and accurate measurement data when using contact speed sensors, which greatly limits their application.

To overcome the impact of high temperatures, non-contact methods are a good choice. There are various non-contact speed measurement systems, such as the Global Positioning System (GPS) [5,6], laser Doppler velocimetry [7], radar velocimetry [8], and stereo-vision techniques [9–13]. However, when GPS is used indoors, the signal can be disturbed by the building, which ultimately affects its measurement accuracy. Devices for laser Doppler velocimetry and radar velocimetry should be installed on the moving line of a steam hammer and facing the measurement surface. This installation requirement is hard to achieve owing to the complex and dangerous environment in and near forging equipment. In addition, non-contact methods require the sensors to remain stationary during the measurement process, but the high speed and weight of a steam hammer usually cause a strong vibration. Thus, it is difficult to keep a non-contact system stationary in the vibrating environment, which often leads to unreliable measurement results for non-contact measurement methods. Despite the diversity of non-contact speed measurement options, stereo-vision techniques have received increasing attention due to their outstanding advantages, such as easy-to-use setup, multipoint measurement and visualization, and wide range of resolution and applicability. However, in actual application, the size of the steam hammer and its travel distance are large, with a value up to 2 m. To measure the ram speed of steam hammer, the measurement FOV of the stereo-vision system should be larger than that size, which can lead to inaccurate measurement results. The main reason for this is that it is difficult to obtain accurate calibration results in a large FOV. Traditional calibration methods [14,15] require a high-precision calibration target with a size similar to that of the range of the measurement FOV. This creates challenges for a large FOV stereo-vision system because fabricating a high-precision large calibration target is difficult and expensive. In addition, strong vibration can cause the speed measurement to fail. First, strong vibration may change the relative pose between the two cameras. Second, the measurement coordinate system moves in a strongly vibrating environment.

To solve these problems, a bundle-adjustment-based system calibration method is proposed to obtain accurate calibration results when the calibration target is not precisely manufactured. The calibration method is divided into calibrating internal parameters and external parameters. To eliminate the influence of strong vibration, the external parameters are recalibrated during the whole measurement process, and the world coordinate system is based on the steam hammer bracket, which is relatively stationary in respect to the steam hammer. Compared with traditional contact measurement techniques, the stereo-vision method has the advantages of compact configuration, ease of use, and capability for non-contact and multipoint deflection measurement. Additionally, in contrast with non-contact methods, the proposed technique offers an outstanding advantage of flexible system configuration and insensitivity to strong vibration due to its stationary measurement coordinate system. The accuracy and effectiveness of the proposed method were verified by experiments that measured the ram speed of a counterblow steam hammer in a die forging device. In addition, this technique can also be used in displacement measurement in a vibrating environment with a large FOV.

This paper is organized as follows: Section 2 introduces the system configuration, the procedures and principles involved in this system, an overview of the stereo-vision method, the system calibration method for a large FOV, the external parameters calibration method of the stereo-vision system, and speed solutions. Section 3 shows the experimental validations, and Section 4 summarizes the study.

2. Stereo-Vision System for Measuring Steam Hammer Speed

2.1. System Configuration

The stereo-vision system we developed for measuring the ram speed of a steam hammer is shown in Figure 1. The system consists of two high-speed digital monochrome video cameras (Photron FASTCAM Mini UX100 type 200K-M-16GB, resolution of 1280 × 1024 pixels with 12-bit quantization, maximum frame rate of 4000 frames per second at full resolution, Photron, Tokyo, Japan), two Schneider Xenoplan 1.9/35 lenses with a fixed focal length of 35 mm (Schneider, Bad Kreuznach, Germany), two white LED light sources with a power of 500 W each, an aluminum beam 2 m long, two tripods, and a laptop computer (Dell Precision 4800, Intel® Core i7-4800MQ, 2.7 GHz, 8 GB RAM, Dell, Texas, USA). The video cameras were tightly clamped to the aluminum beam and could be adjusted readily. The two video cameras were synchronized with an internal trigger signal. Images captured by the cameras were stored in the camera's memory and transmitted to the computer through a gigabit network cable after image acquisition was complete.

It is important to note that the steam hammer has a high ram speed, about 3 m/s, and the duration of impact between the hammer and forging is very short. Thus, to accurately measure the ram speed of the steam hammer, the stereo-vision system uses two high-speed cameras. To ensure the image quality of high-speed cameras with a large FOV, their FOV is illuminated by two high-power LED light sources. To increase the image contrast and signal-to-noise ratio, retro-reflection targets were used which made by mixing glass, silver powder, and resin according to a specific mixing ratio.

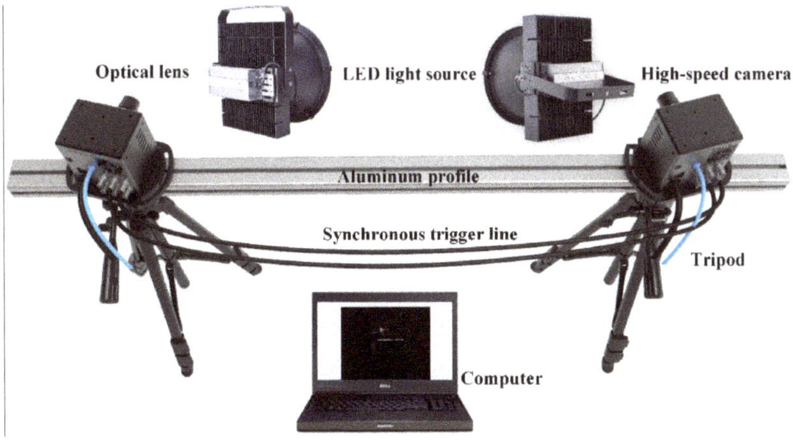

Figure 1. Stereo-vision system for measuring the ram speed of the steam hammer.

2.2. Measurement Principles

2.2.1. Stereo-Vision Theory

In the stereo-vision system, the camera can be modeled by a standard pinhole model. An arbitrary 3D point **P** in the world coordinate system is denoted as $\mathbf{P}_w$; the ray departing from **P** and passing through the *ith* camera lens is captured on the camera sensor formed the image point $\mathbf{p}_i$. In practice, although the lens aberrations distort the shape of the images, the imaging process can be described with a nonlinear camera model [16]:

$$\begin{cases} s_i\widetilde{\mathbf{p}}'_i = \mathbf{A}_i[\mathbf{R}_i|\mathbf{t}_i]\widetilde{\mathbf{P}}_w \\ \mathbf{p}_i = \mathbf{p}'_i + \theta(\mathbf{k}_i;\mathbf{p}'_i) \end{cases}, \text{with} \mathbf{A}_i = \begin{bmatrix} ax_i & 0 & u_i \\ 0 & ay_i & v_i \\ 0 & 0 & 1 \end{bmatrix}, \quad (1)$$

where s_i is a scaling factor, $\widetilde{\bullet}$ is the homogenous coordinate, $\bullet'$ is the ideal image point without distortion; $\mathbf{A}_i$ is the intrinsic camera matrix which contains the normalized horizontal and vertical direction focal lengths (ax_i, ay_i) and the principal point coordinates (u_i, v_i) in the image coordinate system; $\theta(\mathbf{k}_i;\bullet)$ is the lens distortion that parameterized by the distortion coefficients $\mathbf{k}_i$; and $\mathbf{R}_i$ and $\mathbf{t}_i$ are the rotation matrix and translation vector, respectively, which are referred to as extrinsic parameters.

In the binocular vision system, if the corresponding points $\mathbf{p}_1$ and $\mathbf{p}_2$ are determined, the 3D coordinates of the homogenous point $\mathbf{P}_w$ can be reconstructed by a least-squares solution according to Equation (2):

$$\begin{cases} s_1\widetilde{\mathbf{p}}_1 = \mathbf{A}_1[\mathbf{R}_1|\mathbf{t}_1]\widetilde{\mathbf{P}}_w \\ s_2\widetilde{\mathbf{p}}_2 = \mathbf{A}_2[\mathbf{R}_2|\mathbf{t}_2]\widetilde{\mathbf{P}}_w \end{cases}, \quad (2)$$

where $s_1, s_2, \mathbf{A}_1, \mathbf{A}_2, \mathbf{R}_1, \mathbf{t}_1, \mathbf{R}_2, \mathbf{t}_2$ can be accurately calibrated before measurement, which will be described in the next Sections 2.2.2 and 2.2.3.

In the stereo-vision system, the cameras are fixed with respect to each other, the rotation matrix **R** and translation vector **t** can be introduced to represent the relative motion between two cameras, which is defined by:

$$\begin{cases} \mathbf{R} = \mathbf{R}_2\mathbf{R}_1^{-1} \\ \mathbf{t} = \mathbf{t}_2 - \mathbf{R}_2\mathbf{R}_1^{-1}\mathbf{t}_1 \end{cases} \tag{3}$$

2.2.2. System Calibration for Large FOV

The essence of stereo-vision system calibration is to solve the internal and external parameters using the known coordinates of multiple 3D points and the corresponding 2D image points. The calibration accuracy directly determines the measurement accuracy of the stereo-vision system. In this paper, a bundle-adjustment-principle-based system calibration method in a large FOV is proposed, and the calibration process is divided into calibrating internal parameters and external parameters.

As shown in Figure 2, a cross-shaped calibration target with ring-coded points and circular points is used. Because the calibration target is portable and easy to manufacture, it is convenient for in-situ calibration, especially for large FOV calibration. The ring-coded points on the calibration target can be identified from an image regardless of their rotation or scale. It is used to correlate the reference points in different images to make the calibration process completely automated. The ring-coded points can be identified by an accurate gray-gradient-based method [17] previously proposed by the authors. In addition, each circular point's identification number can be easily determined according to the positional relationship between the circular points and ring-coded points.

Figure 2. Cross-shaped calibration target.

After identifying the reference points in different images, the corresponding points in these images can be determined. The corresponding points can be also related by epipolar geometry, which can be described by the fundamental matrix **F**; it contains all the geometric information between two cameras, including the intrinsic parameters and relative rigid motion:

$$\widetilde{\mathbf{p}}_2'^{T}\mathbf{F}\widetilde{\mathbf{p}}_1' = 0, \text{with } \mathbf{F} = A_2^{-T}[\mathbf{t}]_\times \mathbf{R} A_1^{-1}, \tag{4}$$

where $[\mathbf{t}]_\times$ is the anti-symmetric matrix by **t**, and for any vector **y**, $[\mathbf{t}]_\times\mathbf{y} = \mathbf{t}\times\mathbf{y}$.

To obtain the relative rigid motion, we assume that the cameras are an ideal perspective imaging system that the distortion coefficients $\mathbf{k}_i$ are 0; thus, we have $\widetilde{\mathbf{p}}' = \widetilde{\mathbf{p}}$, and the fundamental matrix **F** can be estimated by the classic eight-point algorithm and iterative optimization based on minimizing the distances between reference points and epipolar lines. Furthermore, the initial value of intrinsic parameters in $\mathbf{A}_i$ can be deduced from the nominal parameters of camera and lens. For example, the principal point coordinates (u_i, v_i) are assumed to be at the image center. The initial value of the normalized horizontal and vertical focal lengths (ax_i, ay_i) can be computed by $ax_i = f_i/dx_i$ and $ay_i = f_i/dy_i$, where f_i is the nominal focal length, dx_i and dy_i are the horizontal and vertical pixel size of the camera sensor, respectively. Although the initial value of intrinsic matrix is not very accurate, it is sufficient to gain the essential matrix $\mathbf{E} = [\mathbf{t}]_{\mathrm{x}}\mathbf{R}$ according to Equation (3). Furthermore, the relative rigid motion of the cameras can be calculated at a low precision level by singular-value-deposition of $\mathbf{E}$ [18]. After retrieving the relative rigid motion between cameras, the 3D coordinate of the homogenous control points could be reconstructed by a least-squares solution according to Equation (2).

When calibrating the camera internal parameters, the first image is taken as a reference. Then, the initial 3D coordinates of reference points on the calibration target and the initial external parameters can be solved by the above method. To obtain accurate camera internal parameters and accurate 3D coordinates of reference points, the non-linear least-squares optimization method is used to refine the optical geometry together with the 3D coordinates of reference points. The corresponding cost function is built according to discrepancies between reference points and the expected re-projection data:

$$Cst = \sum_{i=1}^{M}\sum_{j=1}^{N}\left\|\mathbf{p}^{ij} - \hat{\mathbf{p}}^{ij}(\mathbf{A}, \mathbf{K}, \mathbf{R}^{i}, \mathbf{t}^{i}, \mathbf{P}_{w}^{j})\right\|^{2}, \tag{5}$$

where superscripts i, j denote the sequence number of images and the reference points respectively; M is the total number of images; N is the total number of reference points; $\mathbf{P}_w^j$ are the jth reference points in the world coordinate system; $\mathbf{p}^{ij}$ are the image points of jth reference points in the ith image plane; and $\hat{\mathbf{p}}^{ij}$ are the re-projection of $\mathbf{P}_w^j$ in the image plane. This problem can be solved using the Levenberg-Marquardt algorithm [19].

2.2.3. External Parameters Self-Calibration

In a typical stereo-vision system, the cameras are fixed with respect to each other, and one of the cameras is chosen as the world coordinate system. If the stereo-vision system is vibrating during the measurement, it will move two camera coordinate systems and change the relative pose between the two cameras. To obtain accurate measurement results in a vibrating environment, the external parameters of two cameras need to be corrected in real time, and the world coordinate system has to be stationary during the measurement.

To gain an accurate solution of the relatively rigid motion $\mathbf{R}, \mathbf{t}$ of the two cameras, the non-linear least-squares optimization method is used. According to Equation (5), the optimization cost function can be described by Equation (6):

$$Cst = \sum_{i=1}^{M}\sum_{j=1}^{N}\left\|\mathbf{p}_1^{ij} - \hat{\mathbf{p}}_1^{ij}(\mathbf{A}_1, \mathbf{K}_1, \mathbf{R}_1^{i}, \mathbf{t}_1^{i}, \mathbf{P}_{w}^{j})\right\|^{2} + \sum_{i=1}^{M}\sum_{j=1}^{N}\left\|\mathbf{p}_2^{ij} - \hat{\mathbf{p}}_2^{ij}(\mathbf{A}_2, \mathbf{K}_2, \mathbf{R}_2^{i}, t_2^{i}, \mathbf{R}, \mathbf{t}, \mathbf{P}_{w}^{j})\right\|^{2}. \tag{6}$$

During iterative computation, the intrinsic camera matrixes, lens distortion parameters, and 3D coordinates of the reference points are fixed; only the external parameters of two cameras and the relative rigid motion are refined.

After the optimization, the translation vector is given in a metric reconstruction coordinate system, and it needs to be transformed into the Euclidean coordinate system using a scaling factor. If the distance between any two control points in the Euclidean coordinate system is known, it can be used to calculate the scaling factor. After calibrating the external parameters of both cameras, the 3D coordinates of target points can be calculated according to Equation (2). Note that the initial value of external parameters is important for the iterative computation, which may cause the iteration to not converge or fall into a local minimum. To solve this problem, an accurate non-iterative method [20] is used to calculate the initial value of external parameters.

If the world coordinate system is fixed on one of the cameras in the stereo-vision system, it will move in the vibration environment and the measurement error will increase. To solve this problem, the world coordinate system is defined by reference points that are stationary during the measuring process. During the forging process, the bracket of the forging machine can be considered to be stationary. Therefore, the reference points are placed on the bracket, which will be described in Section 3.2.

2.2.4. Speed Solution

To compute the ram speed of a steam hammer, the measured points should be stable with respect to the steam hammer. Then, those measured points can move with the same speed and in the same direction as the steam hammer. Then, successive frame images of the stereo-vision system are used to compute the speed as follows:

$$v = \frac{\Delta s}{\Delta t}, \quad (7)$$

where v is the average speed of the measured point in time interval Δt, and Δs is the displacement of the measured point in time interval Δt.

It should be noted that, according to Equation (7), the speed measurement error is inversely proportional to the time interval. In actual application, the time interval is very short, only 0.25 ms, which will result in a large error in the speed measurement. In addition, if the displacement measurement error is smaller than the displacement Δs in time interval Δt, the measurement speed will be wrong. To solve this problem, the speed is calculated by calculating the derivative of the displacement curve, and the displacement curve is fitted by the least-squares method. If sufficient displacement data are measured, the speed can be accurately calculated.

3. Experiments

To verify the accuracy and effectiveness of the proposed method, we programmed it using Microsoft Visual Studio 2015 with C++ (Microsoft, Redmond, USA). The experimental arrangement is shown in Figure 3. The steam hammer in this experiment is a counterblow steam hammer in which each hammer has a weight of about 100 tons. In this section, the experiments are as follows: (1) Large FOV calibration experiment. A scale bar with known coded feature distance was measured 10 times while it was positioned in the measurement volume. This allowed the measured distance and reference values to be compared to evaluate the measurement accuracy. (2) External parameter calibration experiments. The 3D coordinates of reference points on the forging machine's bracket were measured when the forging machine was operated, and the distances between them were analyzed to evaluate the dynamic measurement accuracy. (3) Ram speed measurement. The ram speed of the steam hammer was measured to verify the capability and effectiveness of the stereo-vision system.

Figure 3. Experimental setup.

3.1. Large FOV Calibration Experiment

To test the measurement accuracy of the developed stereo-vision system, a scale bar with two ring-coded circle features was selected as the measured object according to the VDI-2634 Optical 3D measurement accuracy test standard. The distance between the centers of these two circles was accurately known. In the experiment, the internal parameters of the stereo-vision system were calibrated by placing the cross-shaped calibration target at 12 different orientations and positions within the measurement volume, then the external parameters could be calibrated using images of the calibration target. The scale bar was placed at 10 different orientations and positions within the measurement volume. The measurement range was approximately 3000 mm wide, 2400 mm tall, and 1000 mm deep, and the measurement distance was approximately 5400 mm. The measurement error is defined as the difference between the measured length of the ring-coded circle centers and the known value of 1001.762 mm. The results are shown in Figure 4. It can be seen that the measurement error of the stereo vision system was less than 0.15 mm, which verified that the stereo vision system can obtain accurate 3D results.

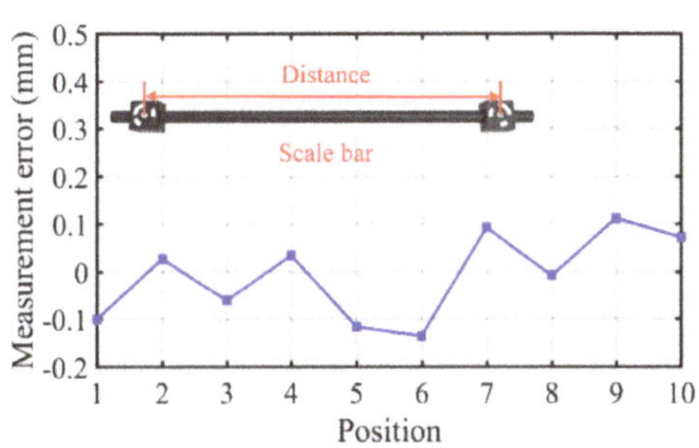

Figure 4. Measurement accuracy test results.

3.2. External Parameter Self-Calibration Experiments

The stereo-vision system was vibrating while it measured the ram speed of the steam hammer. Therefore, it is necessary to evaluate the dynamic measurement accuracy of the stereo-vision system, which can be regarded as the difference of the measured distance between the measured points during the measurement. In other words, it is actually an analysis of the stability of measuring the distance between two measured points during the measurement. As shown in Figure 5, a number of retro-reflection targets on the forging machine's bracket were chosen as reference points for synchronously calibrating the external parameters during the measurement. Before measurement, the 3D coordinates of all the reference points were accurately calibrated by the optical coordinate measuring system Creaform MaxSHOT 3D (Creaform, Lévis, Canada) [21], so the world coordinate system of the stereo-vision system is the measurement coordinate system of the optical coordinate measuring system. The retro-reflection targets on the steam hammer are the measured points.

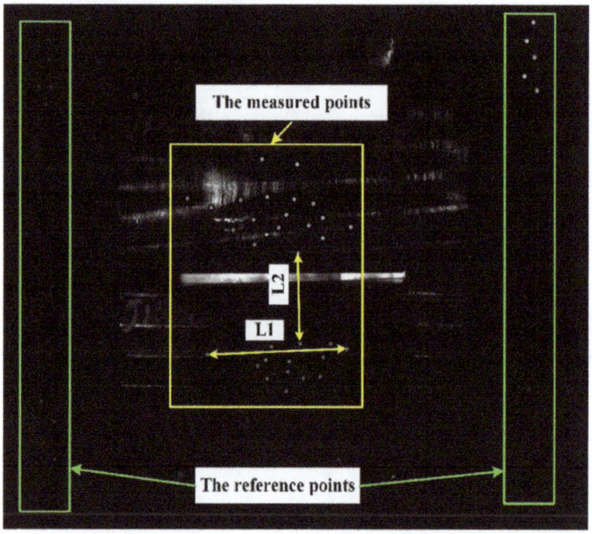

Figure 5. Reference points and measured points.

During the experiment, the upper and lower hammers moved downwards and upwards, respectively; 1000 images were captured by each camera with an acquisition speed of 4000 frames per second and exposure time of 0.000125 s. The measurement range was approximately 3000 mm wide, 2400 mm tall, and 1000 mm deep, and the measurement distance was approximately 5400 mm. As shown in Figure 5, two distances were measured within the measurement area. As the hammer moved, L2 changed, but L1 did not change. The error is defined as the difference between the measured distance of L1 and its known value. The measurement results are shown in Figure 6. Figure 6a shows that by using the self-calibration method, the measurement results were more accurate than without using the self-calibration method, and the measurement accuracy was less than 0.2 mm by using the self-calibration method. After the upper and lower steam hammers were in contact (where L2 reaches its minimum value), the measurement accuracy was significantly reduced when the self-calibration method was not used, as shown in Figure 6a,b. The reason for this is that when the upper and lower hammers were in contact, the vibration was at a maximum. This is shown in Figure 6c, where the distance curve is smoother when using the self-calibration than without using it. This also proves that the self-calibration method can effectively improve the measurement accuracy.

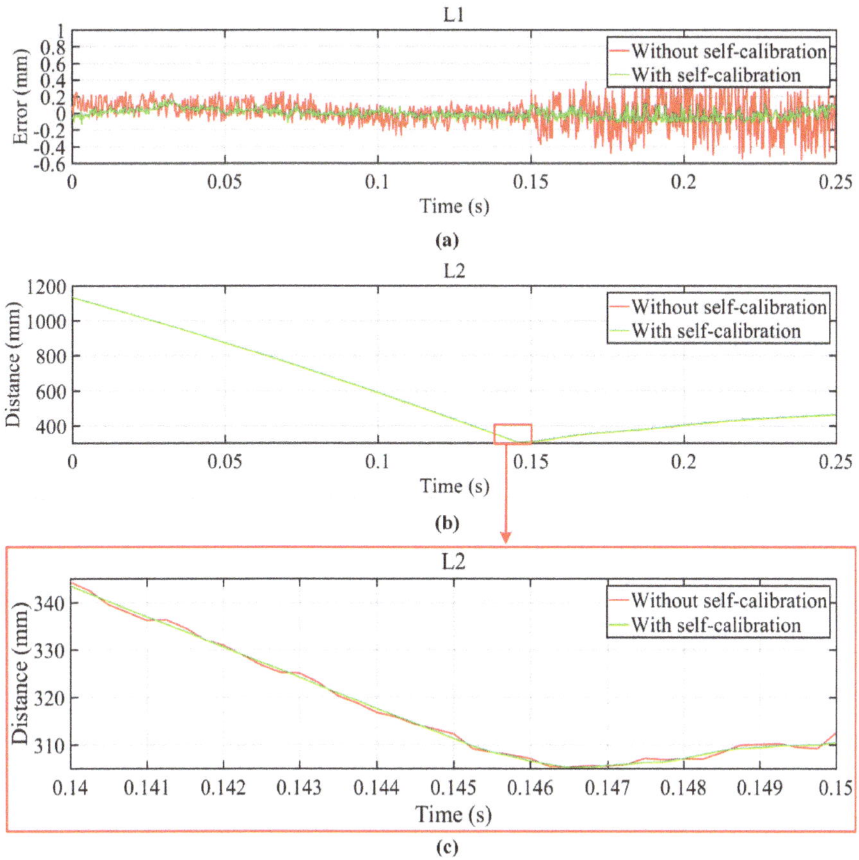

Figure 6. Dynamic measurement results for (**a**) distance error of L1, (**b**) distance of L2, and (**c**) distance of L2 at 0.14–0.15 s.

3.3. Ram Speed Measurement

In this experiment, a number of retro-reflection targets on the upper and lower hammers were measured to analyze their ram speeds, as shown in Figure 7a. The exposure time, measurement range, and measurement distance were the same as those described in Section 3.2. During the experiment, 4000 images were captured by each camera, but some scattered debris blocked the measurement points. Only five points on the upper hammer (shown in Figure 7a) could be measured during the whole process. The displacements of these points are shown in Figure 7b. It can be seen that the five displacement curves were almost the same because the upper hammer is a rigid body and the displacement of all points on the upper hammer should be the same at the same time. The displacement curves can also show the movement stage of the steam hammer as the displacement of points became larger during the striking stage. After the displacement reached the maximum, the steam hammer began to rebound and then rise under mechanical force, during which the displacement of these points became smaller.

Using this information, the speed can be calculated according to the speed solution described in Section 2.2. The speed of the upper and lower hammers is the average speed of the measured points on the upper and lower hammers separately, as shown in Figure 8. In the comparison of the speed of the steam hammer, the direction of the upper steam hammer's velocity is downward and the direction of the lower steam hammer is upward. As shown in Figure 8a, the maximum speed of the upper and lower hammers is 3.149 m/s, and the rebounding speed is −0.7855 m/s. Figure 8b shows the speed history for 0.50–0.55 s, where the speeds of the upper and lower hammers were almost the same; this is consistent with the structure of the counterblow steam hammer. This proves that this method can achieve high accuracy in speed measurement.

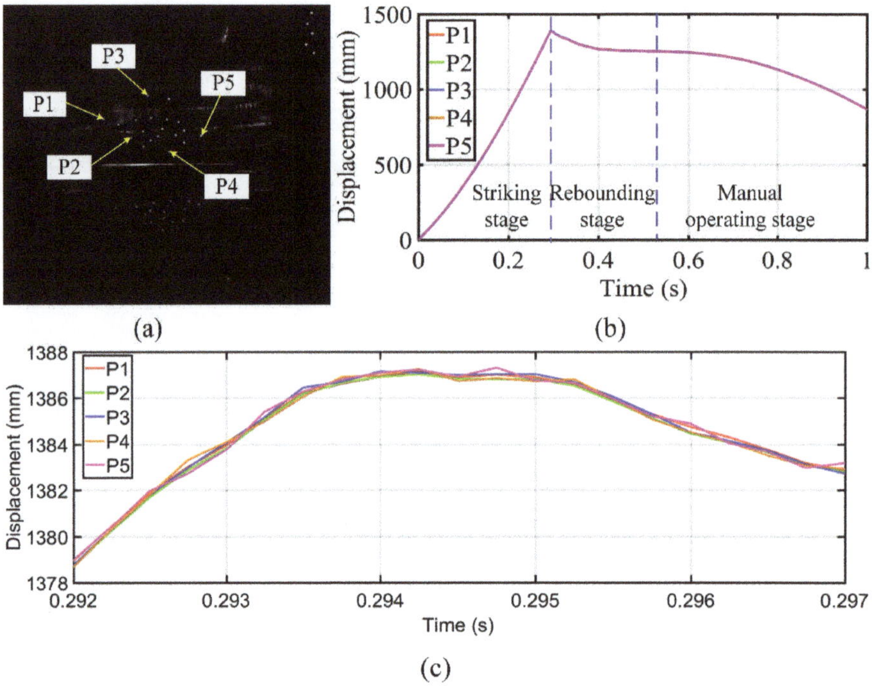

Figure 7. Displacement measurement for upper steam hammer: (**a**) measurement points and displacement of measured points at (**b**) 0–1 s and (**c**) 0.292–0.297 s.

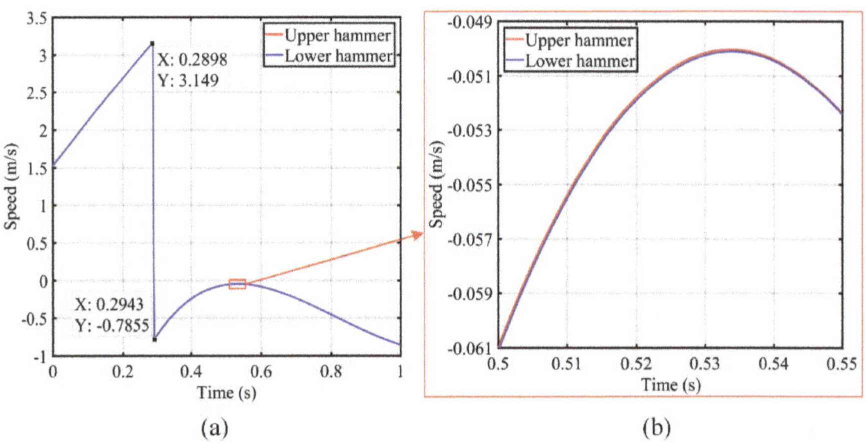

Figure 8. Ram speed of steam hammer at (**a**) 0–1 s and (**b**) 0.50–0.55 s.

4. Conclusions

This paper presents a stereo-vision system for measuring the ram speed of a steam hammer. To solve the problem of calibrating a stereo-vision system with a large FOV, a bundle-adjustment-principle-based method is proposed to realize high-accuracy calibration. This method can obtain accurate calibration results when the calibration target is not precisely manufactured. To decrease the influence of strong vibrations, a stationary world coordinate system was devised, and the external parameters were recalibrated during the entire measurement process. The accuracy and effectiveness of the proposed technique were verified by an experiment to measure the ram speed of counterblow steam hammers in a die forging device. In addition, this technique can be also used for displacement measurement in an environment with strong vibrations and a large FOV.

Author Contributions: R.C., Z.L., K.Z. and X.L. conceived the study idea and designed the experiments; R.C. and Y.W. performed the experiments and analyzed the data; Z.L. and Y.S. contributed the experimental platform; R.C., K.Z., X.L. and C.W. wrote the paper. All authors read and approved the manuscript.

Acknowledgments: This work was supported by the National Key R&D Program of China (SQ2018YFB110170, 2017YFB1103200), the National Natural Science Foundation of China (No. 51505169, No. 51675165).

Conflicts of Interest: The authors declare no conflict of interest.

References

1. Steam Hammer. Available online: https://en.wikipedia.org/w/index.php?title=Steam_hammer&oldid=867390576 (accessed on 21 December 2018).
2. Jeon, S.; Tomizuka, M. Benefits of Acceleration Measurement in Velocity Estimation and Motion Control. *IFAC Proc. Vol.* **2004**, *37*, 217–222. [CrossRef]
3. Park, S.K.; Suh, Y.S. A Zero Velocity Detection Algorithm Using Inertial Sensors for Pedestrian Navigation Systems. *Sensors* **2010**, *10*, 9163–9178. [CrossRef] [PubMed]
4. Dadashi, F.; Crettenand, F.; Millet, G.P.; Aminian, K. Front-Crawl Instantaneous Velocity Estimation Using a Wearable Inertial Measurement Unit. *Sensors* **2012**, *12*, 12927–12939. [CrossRef] [PubMed]
5. Rampinini, E.; Alberti, G.; Fiorenza, M.; Riggio, M.; Sassi, R.; Borges, T.O.; Coutts, A.J. Accuracy of GPS Devices for Measuring High-intensity Running in Field-based Team Sports. *Int. J. Sports Med.* **2015**, *36*, 49–53. [CrossRef] [PubMed]
6. Witte, T.H.; Wilson, A.M. Accuracy of non-differential GPS for the determination of speed over ground. *J. Biomech.* **2004**, *37*, 1891–1898. [CrossRef] [PubMed]
7. Truax, B.E.; Demarest, F.C.; Sommargren, G.E. Laser Doppler velocimeter for velocity and length measurements of moving surfaces. *Appl. Opt.* **1984**, *23*, 67–73. [CrossRef] [PubMed]
8. Ludloff, A.; Minker, M. Reliability of Velocity Measurement by MTD Radar. *IEEE Trans. Aerosp. Electron. Syst.* **1985**, *AES-21*, 522–528. [CrossRef]
9. Dahmouche, R.; Ait-Aider, O.; Andreff, N.; Mezouar, Y. High-speed pose and velocity measurement from vision. In Proceedings of the 2008 IEEE International Conference on Robotics and Automation, Pasadena, CA, USA, 19–23 May 2008; pp. 107–112.
10. Karayel, D.; Wiesehoff, M.; Özmerzi, A.; Müller, J. Laboratory measurement of seed drill seed spacing and velocity of fall of seeds using high-speed camera system. *Comput. Electron. Agric.* **2006**, *50*, 89–96. [CrossRef]
11. Ait-Aider, O.; Andreff, N.; Martinet, P.; Lavest, J. Simultaneous pose and velocity measurement by vision for high-speed robots. In Proceedings of the 2006 IEEE International Conference on Robotics and Automation, ICRA 2006, Orlando, FL, USA, 15–19 May 2006; pp. 3742–3747.
12. Pumrin, S.; Dailey, D.J. Roadside camera motion detection for automated speed measurement. In Proceedings of the Proceedings. In Proceedings of the IEEE 5th International Conference on Intelligent Transportation Systems, Singapore, 6 September 2002; pp. 147–151.
13. Malki, S.; Deepak, G.; Mohanna, V.; Ringhofer, M.; Spaanenburg, L. Velocity Measurement by a Vision Sensor. In Proceedings of the 2006 IEEE International Conference on Computational Intelligence for Measurement Systems and Applications, La Coruna, Spain, 12–14 July 2006; pp. 135–140.
14. Zhang, Z. A flexible new technique for camera calibration. *IEEE Trans. Pattern Anal. Mach. Intell.* **2000**, *22*, 1330–1334. [CrossRef]

15. Tsai, R. A versatile camera calibration technique for high-accuracy 3D machine vision metrology using off-the-shelf TV cameras and lenses. *IEEE J. Robot. Autom.* **1987**, *3*, 323–344. [CrossRef]
16. Heikkila, J.; Silven, O. A four-step camera calibration procedure with implicit image correction. In Proceedings of the Proceedings of IEEE Computer Society Conference on Computer Vision and Pattern Recognition, San Juan, PR, USA, 17–19 June 1997; pp. 1106–1112.
17. Chen, R.; Zhong, K.; Li, Z.; Liu, M.; Zhan, G. An accurate and reliable circular coded target detection algorithm for vision measurement. In Proceedings of the Optical Metrology and Inspection for Industrial Applications IV, International Society for Optics and Photonics, Beijing, China, 24 November 2016; Volume 10023, p. 1002319.
18. Hartley, R.; Zisserman, A. *Multiple View Geometry in Computer Vision*, 2nd ed.; Cambridge University Press: Cambridge, UK, 2004.
19. Marquardt, D.W. An Algorithm for Least-Squares Estimation of Nonlinear Parameters. *SIAM J. Appl. Math.* **1963**, *11*, 431–441. [CrossRef]
20. Lepetit, V.; Moreno-Noguer, F.; Fua, P. EPnP: An Accurate O(n) Solution to the PnP Problem. *Int. J. Comput. Vis.* **2008**, *81*, 155. [CrossRef]
21. MaxSHOT 3D Handheld Optical Coordinate Measuring System | Creaform. Available online: https://www.creaform3d.com/en/metrology-solutions/optical-measuring-systems-maxshot-3d (accessed on 21 December 2018).

sensors

MDPI

Article

Pose Estimation of Sweet Pepper through Symmetry Axis Detection

Hao Li [1], **Qibing Zhu** [1,*], **Min Huang** [1], **Ya Guo** [1] **and Jianwei Qin** [2]

1 Key Laboratory of Advanced Process Control for Light Industry, Ministry of Education, Jiangnan University, Wuxi 214122, China; 6161905027@vip.jiangnan.edu.cn (H.L.); huangmin2004@jiangnan.edu.cn (M.H.); Guoy@jiangnan.edu.cn (Y.G.)

2 USDA/ARS Environmental Microbial and Food Safety Laboratory, Beltsville Agricultural Research Center, Bldg., 303, BARC-East, 10300 Baltimore Ave., Beltsville, MD 20705-2350, USA; jianwei.qin@ars.usda.gov

* Correspondence: zhuqibing@jiangnan.edu.cn; Tel.: +86-159-6175-3162

Received: 20 August 2018; Accepted: 7 September 2018; Published: 13 September 2018

Abstract: The space pose of fruits is necessary for accurate detachment in automatic harvesting. This study presents a novel pose estimation method for sweet pepper detachment. In this method, the normal to the local plane at each point in the sweet-pepper point cloud was first calculated. The point cloud was separated by a number of candidate planes, and the scores of each plane were then separately calculated using the scoring strategy. The plane with the lowest score was selected as the symmetry plane of the point cloud. The symmetry axis could be finally calculated from the selected symmetry plane, and the pose of sweet pepper in the space was obtained using the symmetry axis. The performance of the proposed method was evaluated by simulated and sweet-pepper cloud dataset tests. In the simulated test, the average angle error between the calculated symmetry and real axes was approximately 6.5°. In the sweet-pepper cloud dataset test, the average error was approximately 7.4° when the peduncle was removed. When the peduncle of sweet pepper was complete, the average error was approximately 6.9°. These results suggested that the proposed method was suitable for pose estimation of sweet peppers and could be adjusted for use with other fruits and vegetables.

Keywords: pose estimation; symmetry axis; point cloud; sweet pepper

1. Introduction

Fruit harvesting is an important part of the entire production process of fruit farming. In the production process, prevalent harvesting methods are still based on high-cost, -intensity, and -risk manual harvesting, in which the labor force employed accounts for 33% to 50% of the total labor force [1]. In the Washington State alone, the harvesting cost for handpicking apple is approximately $1150 to $1700 per acre per year [2]. A total of $21 million was used for personal injury compensation related to manual harvesting between 1996 and 2001 in the USA [3]. More importantly, the agricultural labor force worldwide exhibits a declining trend with aging population, which aggravates the problem of labor shortage and labor cost increase, and eventually affects sustainable fruit cultivation development [4]. Thus, automated harvesting systems must be developed to meet the increasing labor demand, to decrease human risks of injuries in orchards, and to decrease the harvesting cost by saving time, money, and energy to benefit producers and consumers [5].

Since Schertz and Brown [6] first introduced the concept of automatic harvesting as an alternative to mechanical harvesting, the development of an automated crop harvesting system has been a crucial topic in the field of agricultural engineering. After approximately 50 years of development, many automatic harvesting machines based on various working principles and structural forms were studied and proposed [7–9]. A harvesting robot based on machine vision technology is the most important fruit harvesting machine because it can automatically obtain a variety of information (such as plant structure, fruit location, and surrounding environment), and thus, facilitate the appropriate picking action to reduce the damage to the fruit and plant.

Fruit detection in plants provides fundamental information for developing harvesting robots. The accuracy in detection of a fruit is easily influenced by uncertain and variable lighting conditions in the field; variable and complex canopy structures; and varying colors, shapes, and sizes of fruits [10]. Numerous methods were established to improve the accuracy of fruit detection in similar outdoor environments. These methods mainly focus on the following: (1) acquiring an image using different types of visual sensors (i.e., black/white (B/W), colored, spectral, and thermal cameras); (2) extracting color, geometric, and texture features from the acquired images using different imaging analysis techniques; (3) and identifying fruit object from the whole images using different machine learning methods with different extracted features. Edan et al. [11] used a B/W camera to detect melons using the intensity levels of reflectance and texture and analyzing shape; 82–88% of the fruits were detected under real field conditions. Cohen [12] identified 85% of the apples using combined color and texture analyses with a standard color charge-coupled device (CCD) camera. Safren et al. [13] used principle component analysis to reduce high-dimensional data from hyperspectral camera. Homogeneous objects were extracted and classified, and morphological operations, watershed analysis, and blob analysis were then conducted. The integration of these methods led to a fruit segmentation accuracy of 88.1%.

Fruit localization in trees is another important ability of harvesting robots, which locate the fruit in a three-dimensional (3D) coordinate system, and are used to guide the manipulator and end-effector move to the desired position for picking action. Mehta et al. [14] calculated the citrus position detected by a single camera to the robot base; they reported that the accuracy of estimating the position was approximately 15 mm. Bulanon et al. [15] used a color camera and a laser sensor, which were mounted together in a cylindrical manipulator controlled by a visual servo method. The target fruit center was aligned at the center of image by visual servoing, and a laser range sensor measured the distance to the fruit. The accuracy of the system for estimating the distance to the fruit was ± 3 mm. Plebe and Grasso [16] used stereo cameras to locate oranges in a 3D coordinate system by stereo matching based on artificial neural networks (ANNs). Gongal et al. [17] used a time-of-flight (TOF) camera to determine the 3D coordinates of the fruit in apple trees.

After the detection and localization of the fruit, automatic harvesting robots implement the detachment action. Studies suggested that the efficiency of fruit detachment can be improved if the fruit is rotated or twisted in a particular manner relative to the orientation of the fruit and peduncle [18,19]. Researchers attempted to detach the fruit and peduncle through peduncle detection. Sa et al. [20] utilized color and geometry information acquired from a red/green/blue-depth (RGB-D) sensor coupled on a support vector machine for peduncle detection. The performance of the proposed method was evaluated using qualitative and quantitative results (the area under the curve (AUC) of the detection precision–recall curve). The method achieved an AUC of 0.71 for peduncle detection of field-grown sweet peppers. Eizentals and Oka [21] presented a peduncle estimation algorithm for automatic harvesting of Japanese green pepper. In the laboratory test, the mean total error for the affine transformation was less than 25 mm in 42 of 49 positions, less than 20 mm in 28 of 49 positions, and less than 15 mm in 19 of 49 positions. However, the direct detection of peduncle with a machine vision system remains challenging because it is small and often occluded by the fruit. Future research in machine vision systems should focus on estimating the orientation of the fruit and the orientation and location of the peduncle. Determination of fruit geometric parameters, such as symmetry of the shape, can provide means to estimate the direction of the fruit [5]. However, few studies investigate pose estimation based on the symmetry axis of the fruit.

Sweet pepper naturally has a regular shape and a symmetry axis (Figure 1). Hence, the method proposed to estimate the pose of sweet pepper by detecting the symmetry axis is feasible. In contrast to previous methods, the proposed method estimates sweet-pepper posture based on the symmetry axis, regardless of whether the peduncle is small or occluded.

Figure 1. Example of the symmetry axis passing through the peduncle (the axis is manually annotated).

Overall, this study aimed to estimate the pose of sweet peppers and to provide effective guidance for the end-effector of an automatic harvesting robot. The specific goal was to calculate the symmetry axis of sweet pepper.

2. Dataset and Methods

2.1. Dataset

The proposed pose estimation algorithm was tested on a set of manually annotated 3D images of sweet pepper and peduncle to evaluate its performance. The dataset obtained from an Intel RealSense SR300 camera was reported by Sa et al. [18]. In their work, the preregistered 3D models of the scene containing the peduncle and sweet pepper were obtained. They also used a statistical outlier remover and a voxel grid down sampler supported from the point cloud library [22] to filter the dataset. The dataset included 27 point clouds with peduncle (Figure 2a) and point clouds with removed peduncle (Figure 2b).

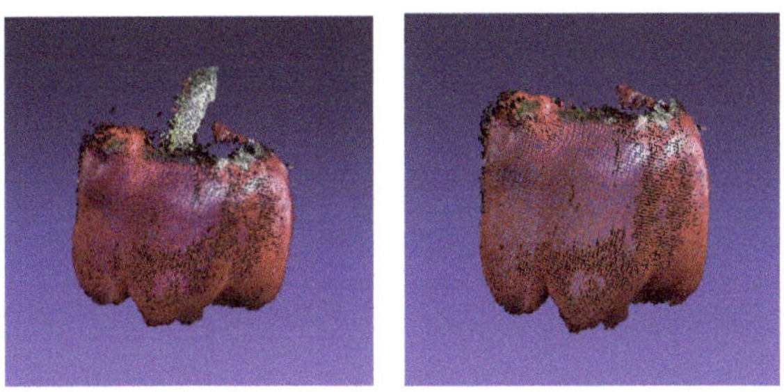

(**a**) (**b**)

Figure 2. Three-dimensional (3D) pepper point cloud in MeshLab. (**a**) Full pepper; (**b**) sweet pepper with removed peduncle.

The real symmetry axis could not be obtained because the sweet-pepper point cloud provided by Sa et al. was not manually calibrated. To verify the accuracy of the algorithm, we simulated a dataset to compare errors between the calculated and real axes. The sweet-pepper point cloud was simulated by the fitted ellipsoid, whose symmetry axis could be manually determined. An ellipsoid was produced as follows:

$$\frac{(x-x_0)^2}{{x_1}^2}+\frac{(y-y_0)^2}{{y_1}^2}+\frac{(z-z_0)^2}{{z_1}^2}=1, \tag{1}$$

where x_0, y_0, and z_0 represent the coordinates of the center of the point cloud; and x_1, y_1, and z_1 represent the half-axis of the point cloud on the x-, y-, and z-axes, respectively. The Intel RealSense

SR300 depth camera has a working range of 20–150 cm. As the distance increases, the accuracy of the camera decreases. In the simulation, the depth between the ellipsoid and the viewpoint z_0 increased from 25 cm to 70 cm with a step of 1 cm; thus, 45 point clouds were tested in each experiment using the proposed algorithm. The values of x_0 and y_0 were set to 3 and 5 cm, respectively, to simulate the offset of the point cloud from the camera. To simulate the size of the sweet pepper in space, we set the size of the point cloud to approximately 5 cm × 5 cm × 4 cm (x_1 = 2.5 cm, y_1 = 2.5 cm, z_1 = 2 cm). The depth camera uses KinFu technology, which made an incomplete 3D reconstruction of the sweet-pepper cloud point (the part located opposite the camera was invisible). Therefore, we artificially removed this part from the viewpoint of the ellipsoid (Figure 3).

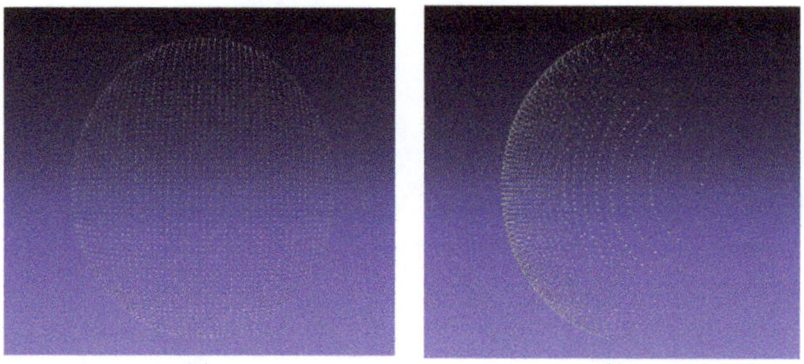

(**a**) (**b**)

Figure 3. Simulated point cloud 20 cm away from the camera viewed from different perspectives. (**a**) Front view of the simulated point cloud; (**b**) side view of the simulated point cloud.

2.2. Method

The symmetry axis, rather than peduncle, should be the focus in developing a sweet-pepper pose estimation scheme. The geometric information in the 3D fruit point cloud was used. In the main part of the algorithm, the normal to the local plane at each point in the fruit point cloud was calculated first. The point cloud was then separated by u^2 ($u > 4$, where u is an integer) candidate planes in the spherical coordinate system established with the centroid of the crop (not the centroid of the point cloud) as the origin. A scoring strategy was employed to calculate the scores for each plane separately, and the plane with the lowest score was selected as the symmetry plane of the point cloud. The symmetry axis could finally be calculated using the selected symmetry plane. An overview of our method is illustrated in Figure 4. The different parts of the proposed algorithm are described in the sections below.

2.3. Calculation of Normal

The normal is an essential property of point clouds. An estimation of the normal plays an important role in point cloud processing. However, point clouds are prone to containing noise, outliers, and holes because of unavoidable noise, physical errors, and occlusions during acquisition. The three main methods of point-cloud normal-vector estimation are partial surface fitting, the Delaunay/Voronoi method, and robust statistical methods [23–25]. Assuming that the sampling plane of the point cloud is smooth everywhere, a local neighborhood of any point can be fitted by the plane. Therefore, a method based on partial surface fitting was used to calculate the normal vector for each point p in the sweet pepper cloud. The normal n at a point p was calculated by fitting a plane to all the points in the neighborhood of that point. The plane fit was found with the eigenvectors of the covariance matrix M. The normal n of the plane was the eigenvector that corresponded to the smallest eigenvalue of the covariance matrix, as follows:

$$M = \frac{1}{k}\sum_{i=1}^{k}(p_i - \overline{p})(p_i - \overline{p})^T, \tag{2}$$

where k is the number of neighbors of p for fitting a plane; $i = 1, \ldots, k$; and $\overline{p}$ is the centroid of k neighbor points.

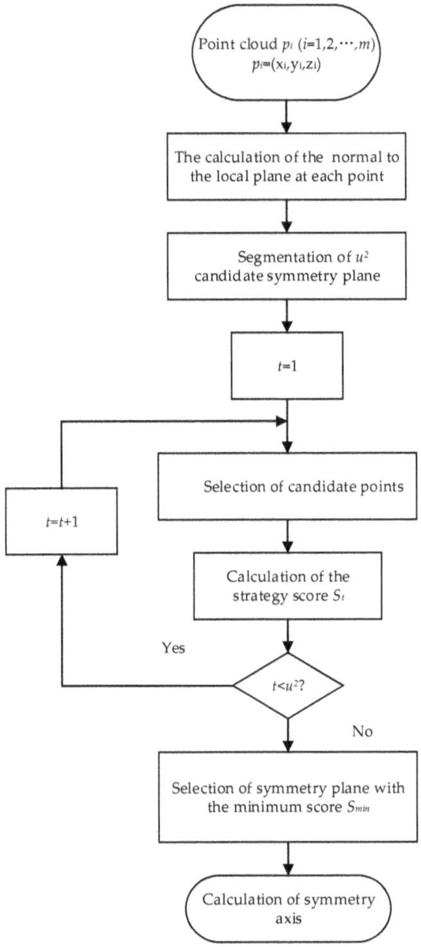

Figure 4. Overview of sweet-pepper pose estimation method.

2.4. *Candidate Symmetry Plane*

Before the candidate plane is calculated, the centroid of the fruit p_c should be obtained. The weighted average of the coordinate value of each point cannot be used as the centroid of the fruit because the part that is far away from the camera cannot be obtained. The sweet-pepper point cloud was similar to a part of a spheroid. Therefore, the centroid of the sweet pepper could be calculated indirectly. A predetermined number of random points generated based on normal distribution within the vicinity of the original point cloud data were randomly generated. For each random point, the root-mean-square error between each point in the random point cloud and the original cloud data was calculated, as follows:

$$d_r = \sqrt{\frac{1}{m}\sum_{i=1}^{m}(p_r - p_i)^2}, \tag{3}$$

where d_r is the root-mean-square error of the distances between each point p_r in the random point cloud and the original point cloud data, and m is the number of points in the point cloud. Finally, the stochastic point with the smallest root-mean-square error was regarded as the centroid p_c of the original point cloud data.

In centering the centroid of point cloud p_c, a spherical coordinate system was established. The u^2 ($u > 4$, and u is an integer) candidate symmetry planes were then selected using the spherical coordinate system division method, which includes the following steps:

Step 1: The ranges of the horizontal and vertical segmentation angles were defined as φ and θ, respectively. φ had the entire range ($\varphi\epsilon(0,\pi)$), and θ was only for half a sphere ($\theta\epsilon\left(-\frac{\pi}{2},\frac{\pi}{2}\right)$).

Step 2: φ and θ were divided as follows:

$$\varphi_{size} = \frac{(\varphi(2) - \varphi(1))}{u}, \tag{4}$$

$$\theta_{size} = \frac{(\theta(2) - \theta(1))}{u}, \tag{5}$$

where φ_{size} is the φ variation range, and θ_{size} is the θ variation range. $\varphi(2)$ and $\varphi(1)$ are the maximum and minimum values of φ, respectively. Similarly, $\theta(2)$ and $\theta(1)$ are the maximum and minimum values of θ, respectively.

Step 3: For each φ, θ was changed u times to obtain the u^2 unit-plane normal vector n_p, which was calculated as follows:

$$\alpha = (g-1) \cdot \varphi_{size} + \frac{\varphi_{size}}{2} + \varphi(1), \tag{6}$$

$$\beta = (h-1) \cdot \theta_{size} + \frac{\theta_{size}}{2} + \theta(1), \tag{7}$$

$$x = r \cdot \sin(\alpha) \cdot \cos(\beta), \tag{8}$$

$$y = r \cdot \sin(\alpha) \cdot \sin(\beta), \tag{9}$$

$$z = r \cdot \cos(\alpha), \tag{10}$$

where α is the azimuth in the spherical coordinate system, β is the elevation angle in the spherical coordinate system, g is the number of changes in the horizontal segmentation angle, and h is the number of changes in the vertical segmentation angle; x, y, and z are the coordinate values of the normal vectors calculated, and r is the radius of the unit circle. The point cloud could finally be divided by u^2 candidate planes.

2.5. Calculation of Symmetry Axis

The score of each candidate symmetry plane was calculated according to the preset scoring strategy [26]. Each candidate plane was traversed to determine the one with the lowest score and to select it as the symmetry plane of the target object. The calculation of the score is described below.

Assuming that points are generated from a fluoroscopic imaging device from a single viewpoint, the symmetric partners (the symmetry points of the original points with respect to the candidate symmetry plane) of many points are invisible, and the points without visible symmetry partners are useless for score calculation of the candidate symmetry planes. Therefore, before calculating the score of the candidate symmetry plane, the points without the symmetric partner should be processed.

When viewing an object from one side of a symmetry plane, most of the visible points observed will be those that share the same side with the camera. For the same reason, the points observed from the other side of the camera should have visible symmetry points (Figure 5).

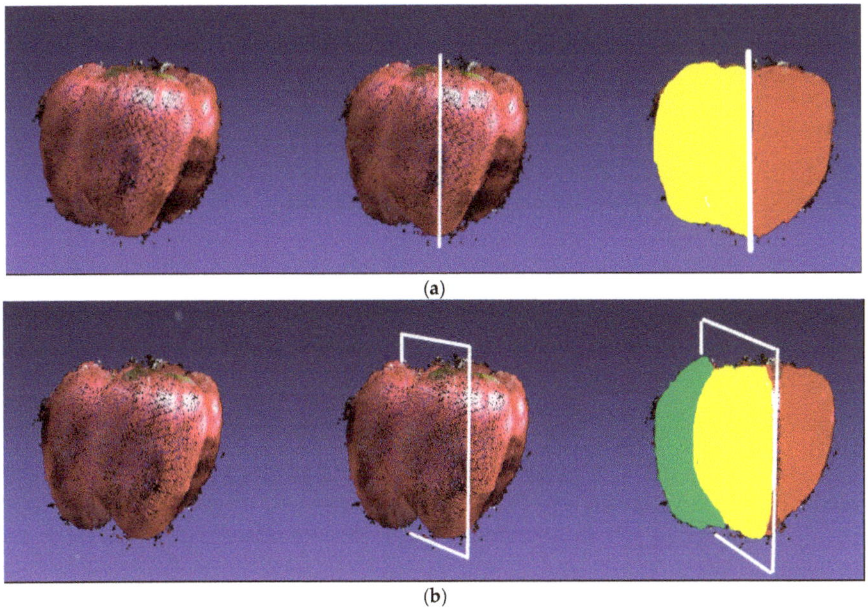

Figure 5. Symmetry planes and visible symmetric partners. For each of the two cases, we show a sweet pepper imaged from a particular point of view (left), estimated symmetry plane (middle, depicted by the intersection of the plane with the sweet pepper), and a color map of points with and without symmetric partners. (**a**) A sweet pepper imaged frontally has virtually all of its points (yellow) in the range data possessing symmetric partners (red). Hence, all points on the left of the symmetry plane are marked yellow; (**b**) a sweet pepper imaged obliquely has the same symmetry plane (now shown rotated). Only the points marked yellow have visible symmetric partners (red), and the green part has invisible symmetry partners.

These points were found on the closer side of the candidate plane and the points without partners on the farther side were excluded from the score calculation using surface normals. Point $\dot{p}$ on the surface with an estimated surface normal n_x was visible from viewpoint p_v if

$$\left(n_x, p_v - \dot{p}\right) > 0, \tag{11}$$

where p_v is the coordinate of the viewpoint, and the default is (0,0,0). Therefore, points with a symmetric partner and a surface normal that points away from the camera were determined. Point $\dot{p}$ with estimated normal n_x was reflected over the candidate symmetry plane with center point p_c and normal n_p by

$$\ddot{p} = \dot{p} - 2 \cdot n_p \cdot d_x, \tag{12}$$

where d_x is the signed distance between the point $\dot{p}$ and the plane. Correspondingly, $\dot{p}$'s normal was reflected by

$$\widetilde{n}_x = n_x - 2 \cdot n_p \cdot d_n, \tag{13}$$

where $\widetilde{n}_x$ is the normal of $\ddot{p}$, and d_n is the signed distance between the normal's head and the candidate plane centered at the camera's axis origin with normal n_p. Thus, $\dot{p}$ had no symmetric partner if:

$$\left(\widetilde{n}_x, p_v - \ddot{p}\right) \leq 0. \tag{14}$$

The score of each candidate point was calculated according to a preset scoring strategy. Specifically, the candidate point $\dot{p}$ was symmetrically transformed with respect to the candidate symmetry plane to obtain symmetrical point $\ddot{p}$. Searching the point with the smallest Euclidean distance from point $\ddot{p}$, point $\dddot{p}$, which is the nearest neighbor to $\ddot{p}$, could be obtained, (Figure 6a). The distance d between $\ddot{p}$ and $\dddot{p}$ and the angle ξ between the normal vector of points $\ddot{p}$ and $\dddot{p}$ were also calculated (Figure 6b).

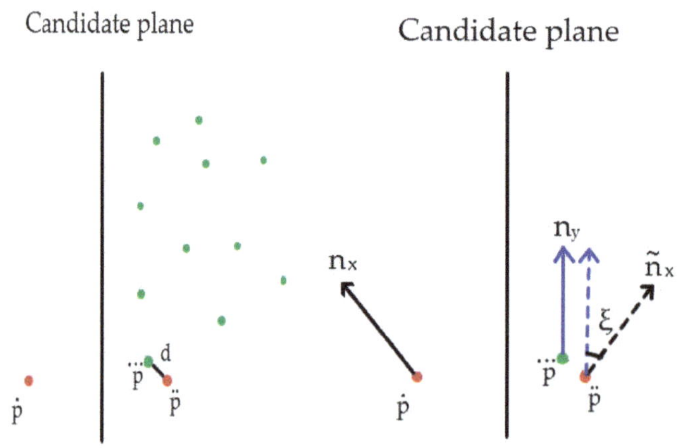

Figure 6. Schematic of reflection score calculation for point $\dot{p}$. (**a**) Determining the closest point $\dddot{p}$ using the reflection point $\ddot{p}$ of $\dot{p}$ and calculating the distance between $\dddot{p}$ and $\ddot{p}$; (**b**) calculating the normal difference ξ between $\dddot{p}$ and $\ddot{p}$.

The score of each candidate point was calculated using Equation (14), and the scores were added as the score of candidate planes; the plane with the lowest score was the symmetry plane.

$$S = d + \tau \cdot \xi, \quad (15)$$

where S is the score of the candidate point, and τ is the weight of the normal differences relative to the point distances. Parameter τ was determined simply by testing the different values and choosing the best one, which eventually was chosen to be $\frac{3}{\pi}$. The symmetry axis of pepper could be calculated based on the symmetry plane of the point cloud by

$$\gamma = n_p \times \frac{p_v - p_c}{||p_v - p_c||}, \quad (16)$$

where γ is the symmetric axis vector, n_p is the normal vector of the symmetry plane, p_v is the coordinate of the viewpoint with a default value of (0,0,0), and p_c is the centroid of the fruit. The vector γ starting from p_c was considered the symmetry axis of sweet pepper.

3. Results and Discussion

3.1. Results for Simulation Dataset

The symmetry axes of the fitting point cloud were calculated using the proposed algorithm. The results of the point cloud that was 20 cm away from the camera and its true symmetry axes are shown in Figure 7.

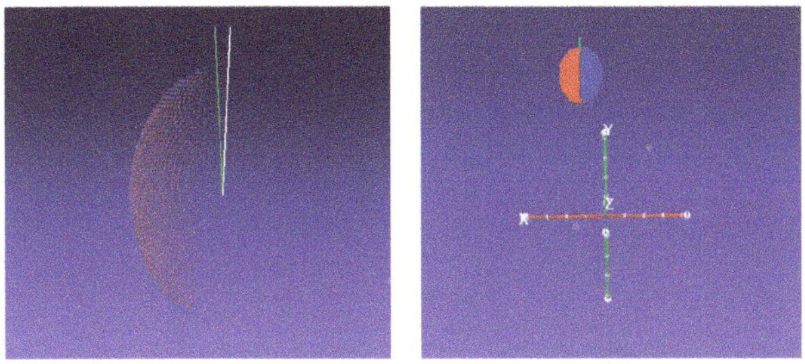

(**a**) (**b**)

Figure 7. Errors between true and calculated symmetry axes, where the point cloud is rendered in two colors to show the symmetry plane. The green line indicates the calculated symmetry axis, and the white line indicates the real symmetry axis. (**a**) Close-range view; (**b**) camera view.

The presented method was tested based on 45 point clouds 50 times. The mean errors between the two axes in the simulated experiments are shown in Figure 8, where the average error was 6.4710°.

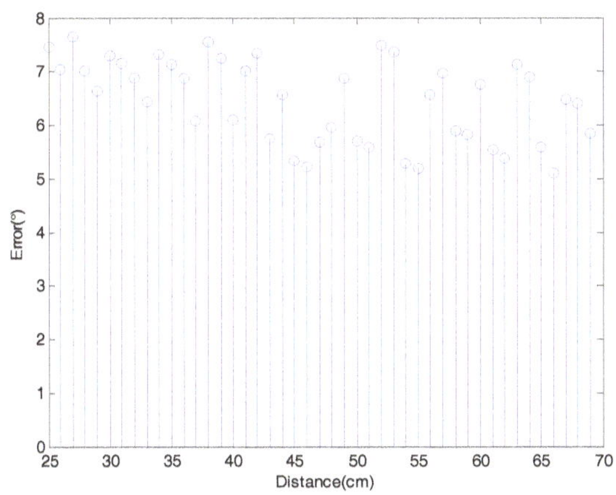

Figure 8. Mean error between real and calculated axes for each distance of the simulated point cloud from the camera.

Figure 8 shows that the error for each distance between the real and the calculated symmetry axis varied between 5° and 8° as the distance between the point cloud and the viewpoint increased. Given that the distance of the fruit point cloud from the camera was not more than 70 cm when using the Intel RealSense SR300, the upper limit distance of the simulation experiment was set to 70 cm. From a distance change of 25 cm to 70 cm, the average deviation of each distance does not vary by more than 3° because the angle is invariant for the perspective projection and should not change with increasing distance from the camera. Therefore, although the change in distance causes reduced accuracy of the point cloud obtained from the camera, it has no effect on the proposed algorithm for pose estimation.

3.2. Results for Sweet-Pepper Dataset Analysis

The normal vector of each point in the point cloud could be directly calculated because the data were prefiltered. The sweet pepper cloud was separated from (u = 6) planes using the methods mentioned in the previous section (Figure 9). The score of each plane was calculated.

(**a**)

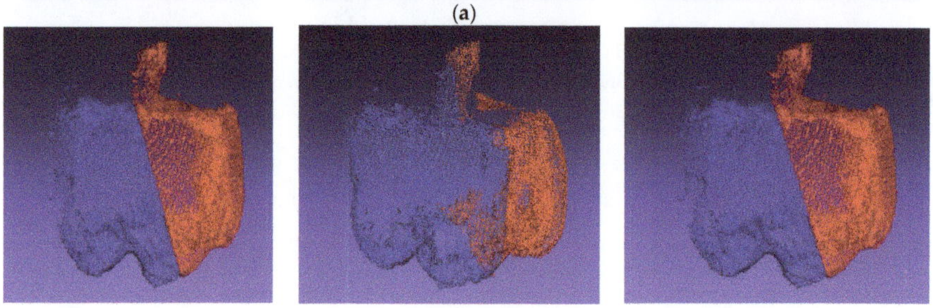

(**b**)

Figure 9. Six effects of u^2 segmentation. The sides of the candidate plane are shown in red and blue. (**a**) The left picture shows the symmetry plane of the point cloud without peduncle; (**b**) the left picture shows the symmetry plane of the point cloud with peduncle.

The best plane was selected as the symmetry plane of the point cloud. The axis was calculated based on the obtained symmetry plane (Figure 10).

(**a**)

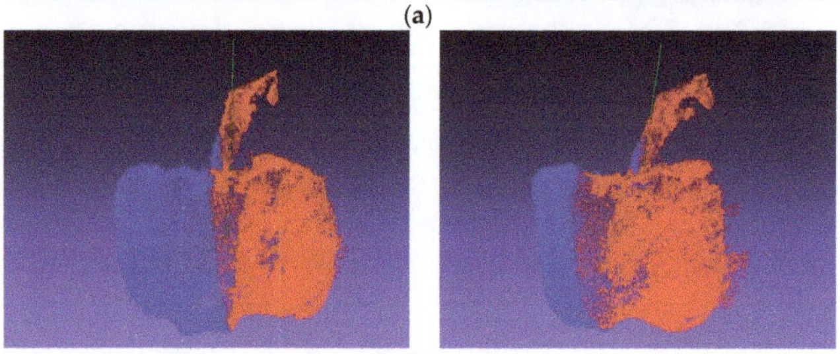

(**b**)

Figure 10. Sweet-pepper cloud with calculated symmetry axes in different perspectives, where the green dotted lines indicate symmetry axes. (**a**) Sweet pepper without peduncle; (**b**) sweet pepper with peduncle.

The real symmetry axis could not be obtained because the sweet-pepper point cloud provided by Sa et al. was not manually calibrated. The fruit peduncle was used to calculate the symmetry axis of the sweet pepper in our experiments. The weighted average of the point cloud coordinates of the peduncle cloud was first considered as the center of the peduncle. The extension line of the center of the peduncle and the center of the fruit in the sweet-pepper point cloud experiment was considered to be the true axis of symmetry. The qualitative results are shown in Figure 11.

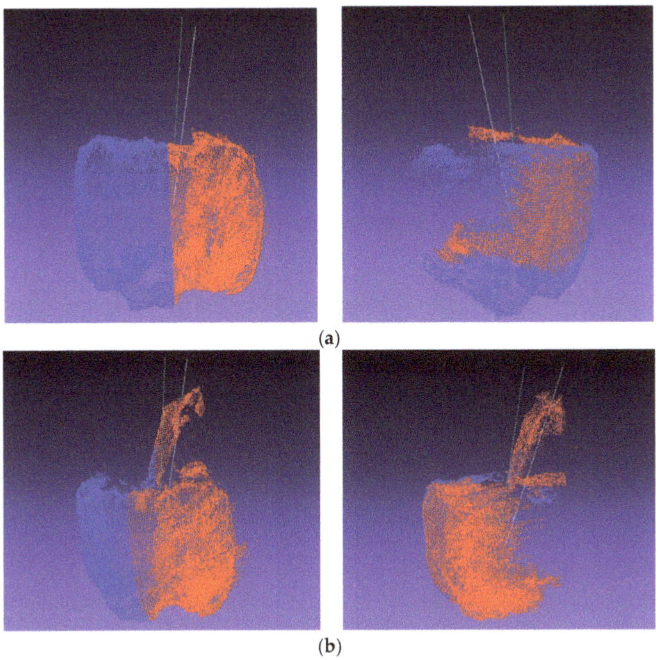

Figure 11. Calculated symmetry axes and the peduncle with standard symmetry axes from different views, where the green lines indicate the calculated symmetry axes, and the white lines indicate the real symmetry axes. (**a**) Sweet pepper without peduncle; (**b**) sweet pepper with peduncle.

The error between the two axes of experiments on sweet pepper without peduncle are shown in Figure 12. The error ranged from 3.6957° to 9.3433°, and the average error was 7.3729°.

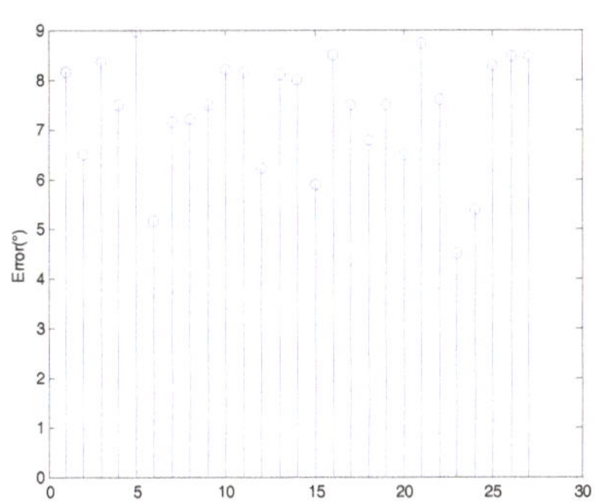

Figure 12. Errors between calculated and real symmetry axes when the peduncle was removed.

When the peduncle was complete, the quantitative results shown in Figure 13 were obtained. The error ranged from 4.4879° to 8.9578°, and the average error was 6.9343°.

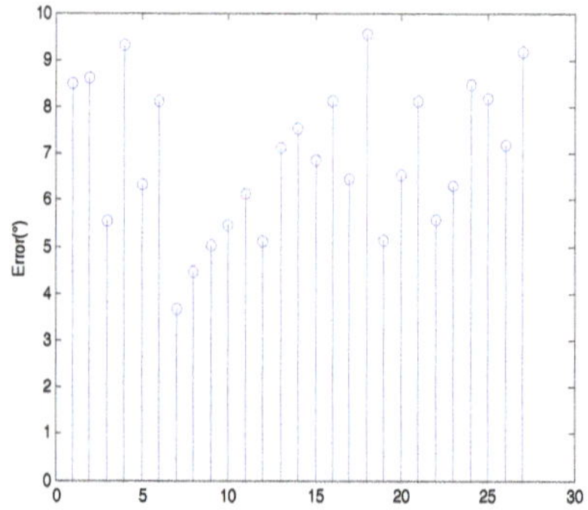

Figure 13. Errors between calculated and real symmetry axes when the peduncle is complete.

Figures 12 and 13 show the slightly different results of the experiments using the sweet-pepper point cloud with peduncle and without peduncle. When the peduncle was complete, the average error of the experiment was 6.9343°, which was lower than the average error (7.3729°) of the experiment on sweet pepper without peduncle. The sweet pepper with peduncle allowed obtaining more geometric information than the sweet pepper without the peduncle.

3.3. Discussion

The results of sweet-pepper point cloud analysis and experiments of the simulated point cloud were compared. The error based on the sweet-pepper point cloud was larger than the average based on the simulated data. On one hand, the standard axis of symmetry in the experiment of simulating a point cloud was determined by the ellipsoid equation, and the extension line of the center of the peduncle and the center of the fruit in the sweet-pepper point cloud experiment was artificially regarded as the standard symmetry axis. This finding may affect the experiment error calculation on the sweet-pepper point cloud. On the other hand, the simulated point cloud was more regular than the shape of the sweet-pepper point cloud. Despite the error, the proposed method for detection of the symmetry axis could still estimate the pose of sweet pepper when the peduncle was invisible and visible.

Fruit pose estimation through the direct detection of peduncles using machines remains challenging because peduncles are small and often occluded by the fruit. In contrast to other methods, the proposed pose estimation method does not depend on the peduncle of the sweet pepper. Given that the shape of the sweet-pepper point cloud is more complex than those of other fruits, such as apples, pose estimation, applied here to sweet pepper, will be easier when the algorithm is applied to fruits with regular shapes.

In this study, the pose of non-occluded sweet pepper was estimated. To improve the accuracy when the fruit is occluded, future research should focus on two aspects: firstly, the point cloud without serious occlusion (the occlusion area accounts for less than one-half of the cross-sectional area of the fruit) can be fitted by least squares [27], and the approach can be improved based on the fitted point cloud; secondly, seriously occluded sweet pepper (the occlusion area accounts for more than one-half of the cross-sectional area of the fruit) can be detected and labeled, such that the robot would perform localization and movement planning, and thus, prioritize normal sweet peppers for harvesting.

4. Conclusions

This study proposed a machine vision approach to estimate the pose of sweet pepper based on its symmetry axis. The method can also overcome the case where the peduncle is too thin to detect and when self-occlusion of the peduncle occurs. In the simulated test, the mean angle error between the calculated symmetry axis and the real axis was approximately 6.5°. In a particularly challenging sweet-pepper cloud dataset test, the average error was approximately 7.4° when the peduncle was removed. When the peduncle of sweet pepper was complete, the average error was approximately 6.9°. The proposed method can not only estimate the pose of fruit to provide effective guidance for the end-effector in the detachment process, but can also be used to detect the symmetry axis of regular objects. However, the improvement required for the presented approach is remarkable, especially for the occluded fruit. For point clouds without serious occlusion, the improvements can be made based on the fitted point cloud by least squares. For seriously occluded sweet pepper, the robot would perform localization and movement planning to prioritize normal sweet peppers for harvesting.

Author Contributions: M.H., H.L., and Q.Z. conceived and designed the experiments; M.H. and Y.G. performed the experiments; Q.Z. and J.Q. analyzed the data; M.H. and H.L. wrote the paper. All authors collaborated on the interpretation of the results and on the preparation of the manuscript.

Funding: This research is supported by the National Natural Science Foundation of China (Grant No. 61772240, 61775086), the Fundamental Research Funds for the Central Universities (JUSRP51730A), and the Prospective Joint Research Foundation of Jiangsu Province of China (BY2016022-32), as well as sponsorship by the 111 Project (B12018).

Acknowledgments: Min Huang, Qibing Zhu, and Hao Li would like to thank Sa for providing the sweet-pepper point cloud dataset.

Conflicts of Interest: The authors declare no conflicts of interest.

References

1. Lu, J.; Wang, X.F.; Hou, D.J. Development of machine vision system for fruit harvesting robot. *Hubei Agric. Sci.* **2012**, *51*, 4705–4708.
2. Gallardo, R.K.; Taylor, M.R.; Hinman, H. *Cost Estimates of Establishing and Producing Gala Apples in Washington*; Extension Fact Sheet FS005E; WSU Extension: Wenatchee, WA, USA, 2010.
3. Hofmann, J.; Snyder, K.; Keifer, M. A descriptive study of workers' compensation claims in Washington State orchards. *Occup. Med.* **2012**, *56*, 251–257. [CrossRef] [PubMed]
4. Fennimore, S.A.; Doohan, D.J. The challenges of specialty crop weed control, future directions. *Weed Technol.* **2008**, *22*, 364–372. [CrossRef]
5. Gongal, A.; Amatya, S.; Karkee, M.; Zhang, Q.; Lewis, K. Sensors and systems for fruit detection and localization. *Comput. Electron. Agric.* **2015**, *116*, 8–19. [CrossRef]
6. Schertz, C.E.; Brown, G.K. Basic considerations in mechanizing citrus harvest. *Trans. ASABE* **1968**, *11*, 343–346.
7. Sarig, Y. Robotics of fruit harvesting: A state-of-the-art review. *J. Agric. Eng. Res.* **1993**, *54*, 265–280. [CrossRef]
8. Grift, T.; Zhang, Q.; Kondo, N.; Ting, K. A review of automation and robotics for the bioindustry. *J. Biomech. Eng.* **2008**, *1*, 37–54.
9. Li, B.; Vigneault, C.; Wang, N. Research development of fruit and vegetable harvesting robots in China. *Stewart Postharvest Rev.* **2010**, *6*, 1–8.
10. Li, P.; Lee, S.H.; Hsu, H.Y. Review on fruit harvesting method for potential use of automatic fruit harvesting systems. *Procedia Eng.* **2011**, *23*, 351–366. [CrossRef]
11. Edan, Y.; Rogozin, D.; Flash, T.; Miles, G.E. Robotic melon harvesting. *IEEE Trans. Robot. Autom.* **2000**, *16*, 831–835. [CrossRef]
12. Cohen, O.; Linker, R.; Naor, A. Estimation of the Number of Apples in Color Images Recorded in Orchards. *Comput. Comput. Technol. Agric. IV* **2011**, *344*, 630–642.
13. Safren, O.; Alchanatis, V.; Ostrovsky, V.; Levi, O. Detection of Green Apples in Hyperspectral Images of Apple-Tree Foliage Using Machine Vision. *Trans. ASABE* **2007**, *50*, 2303–2313. [CrossRef]
14. Mehta, S.S.; Burks, T.F. Vision-based control of robotic manipulator for citrus harvesting. *Comput. Electron. Agric.* **2014**, *102*, 146–158. [CrossRef]
15. Bulanon, D.M.; Burks, T.F.; Alchanatis, V.; Noguchi, N. A multispectral imaging analysis for enhancing citrus fruit detection. *Environ. Control. Boil.* **2010**, *48*, 81–91. [CrossRef]

16. Plebe, A.; Grasso, G. Localization of spherical fruits for robotic harvesting. *Mach. Vis. Appl.* **2001**, *13*, 70–79. [CrossRef]
17. Gongal, A.; Silwal, A.; Amatya, S.; Karkee, M.; Zhang, Q.; Lewis, K. Apple crop-load estimation with over-the-row machine vision system. *Comput. Electron. Agric.* **2015**, *120*, 26–35. [CrossRef]
18. Bulanon, D.M.; Kataoka, T. Fruit detection system and an end effector for robotic harvesting of Fuji apples. *Agric. Eng. Int. CIGR E J.* **2010**, *12*, 203–210.
19. Tong, J.; Zhang, Q.; Karkee, M.; Jiang, H.; Zhou, J. Understanding the dynamics of hand picking patterns of fresh market apples. In Proceedings of the Annual International Meeting of the American Society of Agricultural and Biological Engineers, Montreal, QC, Canada, 13–16 July 2014; p. 141898024.
20. Sa, I.; Lehnert, C.; English, A.; Mccool, C.; Dayoub, F.; Upcroft, B.; Perez, T. Peduncle detection of sweet pepper for autonomous crop harvesting—Combined color and 3-D information. *IEEE Robot. Autom. Lett.* **2017**, *2*, 765–772. [CrossRef]
21. Eizentals, P.; Oka, K. 3D pose estimation of green pepper fruit for automated harvesting. *Comput. Electron. Agric.* **2016**, *128*, 127–140. [CrossRef]
22. Nguyen, T.T.; Vandevoorde, K.; Wouters, N.; Kayacan, E.; Baerdemaeker, J.G.D.; Rusu, R.B.; Cousins, S. 3D is here: Point cloud library (PCL). In Proceedings of the IEEE International Conference on Robotics and Automation, Shanghai, China, 9–13 May 2011; Volume 47, pp. 1–4.
23. Lee, K.M.; Meer, P.; Park, R.H. Robust adaptive segmentation of range images. *IEEE Trans. Pattern Anal. Mach. Intell.* **1998**, *20*, 200–205.
24. Lange, C.; Polthier, K. Anisotropic smoothing of point sets. *Comput. Aided Geom. Des.* **2005**, *22*, 680–692. [CrossRef]
25. Ouyang, D.; Feng, H.Y. On the normal vector estimation for point cloud data from smooth surfaces. *Comput. Aided Des.* **2005**, *37*, 1071–1079. [CrossRef]
26. Barnea, E.; Mairon, R.; Ben-Shahar, O. Colour-agnostic shape-based 3D fruit detection for crop harvesting robots. *Biosyst. Eng.* **2016**, *146*, 57–70. [CrossRef]
27. Duncan, K.; Sarkar, S.; Alqasemi, R.; Dubey, R. Multi-scale superquadric fitting for efficient shape and pose recovery of unknown objects. In Proceedings of the IEEE International Conference on Robotics and Automation, Karlsruhe, Germany, 6–10 May 2013; pp. 4238–4243.

sensors

MDPI

Article

Comparative Analysis of Warp Function for Digital Image Correlation-Based Accurate Single-Shot 3D Shape Measurement

Xiao Yang [1,2], Xiaobo Chen [1,2] and Juntong Xi [1,2,3,*]

1 School of Mechanical Engineering, Shanghai Jiao Tong University, Shanghai 200240, China; yangxiao1992@sjtu.edu.cn (X.Y.); xiaoboc@sjtu.edu.cn (X.C.)
2 Shanghai Key Laboratory of Advanced Manufacturing Environment, Shanghai 200030, China
3 State Key Laboratory of Mechanical System and Vibration, Shanghai 200240, China
* Correspondence: jtxi@sjtu.edu.cn; Tel.: +86-21-3420-5693

Received: 11 February 2018; Accepted: 12 April 2018; Published: 16 April 2018

Abstract: Digital image correlation (DIC)-based stereo 3D shape measurement is a kind of single-shot method, which can achieve high precision and is robust to vibration as well as environment noise. The efficiency of DIC has been greatly improved with the proposal of inverse compositional Gauss-Newton (IC-GN) operators for both first-order and second-order warp functions. Without the algorithm itself, both the registration accuracy and efficiency of DIC-based stereo matching for shapes with different complexities are closely related to the selection of warp function, subset size, and convergence criteria. Understanding the similarity and difference of the impacts of prescribed subset size and convergence criteria on first-order and second-order warp functions, and how to choose a proper warp function and set optimal subset size as well as convergence criteria for different shapes are fundamental problems in realizing efficient and accurate 3D shape measurement. In this work, we present a comparative analysis of first-order and second-order warp functions for DIC-based 3D shape measurement using IC-GN algorithm. The effects of subset size and convergence criteria of first-order and second-order warp functions on the accuracy and efficiency of DIC are comparatively examined with both simulation tests and real experiments. Reference standards for the selection of warp function for different kinds of 3D shape measurement and the setting of proper convergence criteria are recommended. The effects of subset size on the measuring precision using different warp functions are also concluded.

Keywords: single-shot 3D shape measurement; digital image correlation; warp function; inverse compositional Gauss-Newton algorithm

1. Introduction

Optical 3D shape measurement has become one of the research hotspots in the field of measurement due to the advantages of high precision, non-contact, and high speed, etc. Laser scanning [1–3], structured light [4,5], and digital image correlation (DIC) [6–8] are commonly used for accurate 3D shape measurement. According to previous researches [6,9], all of the three methods can achieve the same level of precision. The principle of laser scanning can be briefly summarized as: a laser line stripe plane is projected onto a measuring surface, then a laser stripe is formed and modulated by the depth of the surface. By calibrating the line stripe plane previously and recording the laser stripe by a well-calibrated camera, the 3D information along the stripe line on the surface can be characterized. For structured light measurement, coded fringe patterns are projected onto a measuring surface, the captured images are processed by relative decoding method, whereby an exact phase is computed for each pixel. The phase value is used as a measure for getting depth information of the pixel during

3D reconstruction. Laser scanning is robust to severe environment, but it needs several scans to obtain a complete shape. Structured light measurement is fast at obtaining full-field shape, which can be classified into single-shot and multiple-shot methods according to the number of projected fringe patterns. Multiple-shot structured light measurement can achieve high precision but is sensitive to vibration. Single-shot structured light measurement does not have synchronization problem between projector and camera(s) but is inaccurate at large slope or discontinuities [10]. DIC-based shape measurement is an accurate single-shot method, which is usually accompanied with speckle projection to enhance the surface characteristic, but the calculation amount is much larger than laser scanning and structured light measurement.

The principle basis of DIC-based 3D shape measurement is binocular stereovision. The key component of 3D reconstruction by the way of stereovision is stereo matching. DIC is adopted as a region-matching algorithm to get the disparity of the same characteristic in the left (reference) image and right (target) image: it is assumed that a warp function can be used to describe the mapping relation of two local regions around the same characteristic with proper warp parameters. The warp parameters are optimized by sub-pixel registration algorithm, which is usually the most time-consuming step. The forward additive Newton-Raphson (FA-NR) algorithm is a typical iterative updating method, which is widely used with first-order [11,12] and second-order [13,14] warp functions in last decade. However, the limitation of FA-NR is that the Hessian matrix must be re-computed and inverted in each iteration, which leads to a heavy calculation burden. A more efficient algorithm called inverse compositional Gauss-Newton (IC-GN) [15] was proven to have the same accuracy as classical forward additive image alignment algorithm, but the Hessian matrix remains the same in each iteration of IC-GN [16]. Pan et al. first combined IC-GN and zero-mean normalized sum of square difference (ZNSSD) criterion in DIC with first-order warp function [17]. Since then, almost all the researches related to DIC adopted IC-GN algorithm for sub-pixel registration, which can be summed up as first-order and second-order IC-GN. First-order IC-GN is extensively used for real-time human pulse monitoring [18], real-time dynamic strain measurement [19], and 3D shape measurement [7]. It is worth noting that first-order warp function is a linear transformation, which can only characterize local translation, rotation, and uniform mapping. Therefore, Gao et al. [20] and Bai et al. [21] proposed operators for second-order IC-GN, which is effective to handle non-uniform complex mapping. Additionally, some researches have been done to study the factors that may influence the efficiency or accuracy of DIC, such as subset size [22], convergence criteria [23]. However, only first-order warp function is used in the studies. As far as we know, there is no comparative analysis about the measurement effectiveness and different characteristics of first-order and second-order warp functions until now. Therefore, it is hard to select a proper warp function and set optimal parameters according to the characteristics of different measurements.

In this work, we present a comparative analysis of first-order and second-order warp functions for DIC-based 3D shape measurement using IC-GN algorithm. The influences of convergence criteria, subset size on the convergence efficiency and accuracy are comparatively studied by simulations and real tests. The remainder of this paper is organized as follows: The principle of DIC-based single-shot 3D shape measurement is introduced in Section 2. Experimental results and discussions are reported in Section 3. Finally, conclusions are drawn in Section 4.

2. Principle of DIC-Based Single-Shot 3D Shape Measurement

A single-shot stereo system, composed of two Charge Coupled Device (CCD) cameras and a digital projector, was introduced in our previous work [7]. A speckle pattern is projected onto the measuring object to enhance surface characteristics. With accurate stereo calibration and rectification [24], the same points locates on the same row of left image and right image due to the epipolar constraint. DIC can be used as a local stereo matching method to measure the disparity of the two same points that locate on the left and right images. Speckle projection-based DIC has been proven to have good performances in single-shot 3D measurement and the principle is introduced in this section [25,26].

2.1. Warp Function of DIC

The captured images of a same local region from two different angles of view have obvious difference due to rotation and deformation. By setting a reference subset in the reference image, the position and shape of the relative target subset in the target image can be described by a warp function with proper parameters.

The first-order and second-order warp functions can be represented as:

$$\mathbf{W}_1(x,y;\mathbf{p}_1) = \begin{bmatrix} x' \\ y' \\ 1 \end{bmatrix} = \begin{bmatrix} 1+u_x & u_y & u \\ v_x & 1+v_y & v \\ 0 & 0 & 1 \end{bmatrix} \begin{bmatrix} x \\ y \\ 1 \end{bmatrix} \tag{1}$$

$$\mathbf{p}_1 = \left(u, u_x, u_y, v, v_x, v_y\right)^T \tag{2}$$

$$\Delta\mathbf{p}_1 = \left(\Delta u, \Delta u_x, \Delta u_y, \Delta v, \Delta v_x, \Delta v_y\right)^T \tag{3}$$

$$\mathbf{W}_2(x,y;\mathbf{p}_2) = \begin{bmatrix} x' \\ y' \\ 1 \end{bmatrix} = \begin{bmatrix} \frac{1}{2}u_{xx} & u_{xy} & \frac{1}{2}u_{yy} & 1+u_x & u_y & u \\ \frac{1}{2}v_{xx} & v_{xy} & \frac{1}{2}v_{yy} & v_x & 1+v_y & v \\ 0 & 0 & 0 & 0 & 0 & 1 \end{bmatrix} \begin{bmatrix} x^2 \\ xy \\ y^2 \\ x \\ y \\ 1 \end{bmatrix} \tag{4}$$

$$\mathbf{p}_2 = \left(u, u_x, u_y, u_{xx}, u_{xy}, u_{yy}, v, v_x, v_y, v_{xx}, v_{xy}, v_{yy}\right)^T \tag{5}$$

$$\Delta\mathbf{p}_2 = \left(\Delta u, \Delta u_x, \Delta u_y, \Delta u_{xx}, \Delta u_{xy}, \Delta u_{yy}, \Delta v, \Delta v_x, \Delta v_y, \Delta v_{xx}, \Delta v_{xy}, \Delta v_{yy}\right)^T \tag{6}$$

where (x, y) denotes the local coordinate of the pixel in reference subset, (x', y') is the mapped coordinate of (x, y). $\mathbf{W}_1(x, y; \mathbf{p}_1)$ and $\mathbf{W}_2(x, y; \mathbf{p}_2)$ are the first-order and second-order warp functions with parameter vector $\mathbf{p}_1$ and $\mathbf{p}_2$, respectively. $\Delta\mathbf{p}_1$ and $\Delta\mathbf{p}_2$ denote the incremental parameter vectors. u and v denote the displacement components of center pixel of the reference subset in x direction and y direction, respectively. The other parameters are the first-order gradient components (i.e., u_x, u_y, v_x, v_y) and second-order gradient components (i.e., u_{xx}, u_{xy}, u_{yy}, v_{xx}, v_{xy}, v_{yy}).

2.2. Principle of DIC-Based Stereo Matching Using IC-GN Algorithm

Figure 1 [7] shows the schematic principle of DIC-based stereo matching using IC-GN algorithm. The first-order and second-order warp functions are adopted in Figure 1a,b, respectively. Subscript 1 and 2 are used hereinafter to distinguish the first-order and second-order IC-GN algorithms: IC-GN_1 and IC-GN_2. f and g denote the gray level intensities of reference subset and target subset, respectively. A whole DIC process using IC-GN algorithm can be concluded as three steps. Firstly, compute the optimal parameter incremental vector $\Delta\mathbf{p}$ according to current $\mathbf{p}$, which need to be estimated before the first iteration. The most commonly used ZNSSD criterion is employed in this step [20].

$$C_{ZNSSD}(\Delta\mathbf{p}) = \sum_{y=-M}^{y=M} \sum_{x=-M}^{x=M} \left[\frac{f(\mathbf{W}(x,y;\Delta\mathbf{p})) - \overline{f}}{\Delta f} - \frac{g(\mathbf{W}(x,y;\mathbf{p})) - \overline{g}}{\Delta g} \right]^2 \tag{7}$$

$$\Delta f = \sqrt{\sum_{y=-M}^{y=M} \sum_{x=-M}^{x=M} \left(f(x,y) - \overline{f}\right)^2}, \ \Delta g = \sqrt{\sum_{y=-M}^{y=M} \sum_{x=-M}^{x=M} \left(g(x',y') - \overline{g}\right)^2} \tag{8}$$

where x' and y' in Equation (8) are usually sub-pixel values, $g(x', y')$ is calculated by B-spline interpolation [18]. $\overline{f}$ and $\overline{g}$ are the mean values of gray level intensities of the reference subset and target subset. $\overline{f}$ is constant during the iterations, while $\overline{g}$ need to be calculated in each iteration. Equation (7) can be simplified by first-order Taylor expansion with respect to $\Delta\mathbf{p}$:

$$C_{ZNSSD}(\Delta\mathbf{p}) = \sum_{y=-M}^{y=M}\sum_{x=-M}^{x=M}\left[\frac{f\left(\mathbf{W}(x,y;0)+\nabla f\left(\frac{\partial\mathbf{W}}{\partial\mathbf{p}}\right)\Delta\mathbf{p}-\overline{f}\right)}{\Delta f}-\frac{g(\mathbf{W}(x,y;\mathbf{p}))-\overline{g}}{\Delta g}\right]^2 \tag{9}$$

where $\nabla f = (\partial f/\partial x, \partial f/\partial y)$ is the gray level intensity gradient in x and y directions of the reference subset. $\frac{\partial\mathbf{W}}{\partial\mathbf{p}}$ is the Jacobian of the warp function. For first-order and second-order warp functions, the Jacobians can be expressed respectively as:

$$\frac{\partial\mathbf{W}_1}{\partial\mathbf{p}_1} = \begin{bmatrix} 1 & x & y & 0 & 0 & 0 \\ 0 & 0 & 0 & 1 & x & y \end{bmatrix} \tag{10}$$

$$\frac{\partial\mathbf{W}_2}{\partial\mathbf{p}_2} = \begin{bmatrix} 1 & x & y & \frac{1}{2}x^2 & xy & \frac{1}{2}y^2 & 0 & 0 & 0 & 0 & 0 & 0 \\ 0 & 0 & 0 & 0 & 0 & 0 & 1 & x & y & \frac{1}{2}x^2 & xy & \frac{1}{2}y^2 \end{bmatrix} \tag{11}$$

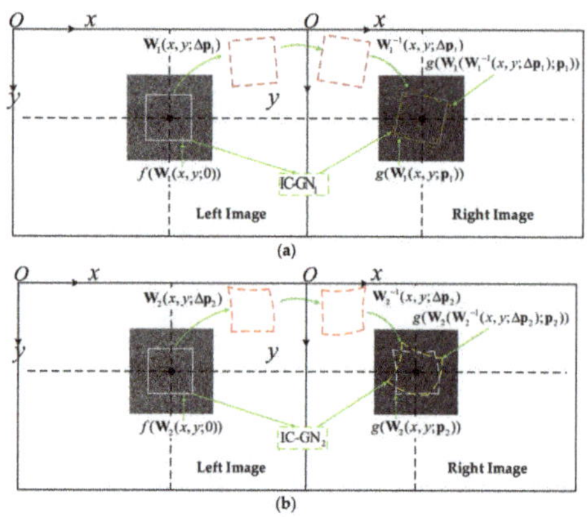

Figure 1. Schematic principle of DIC-based stereo matching using IC-GN algorithm: (**a**) First-order warp function; and (**b**) Second-order warp function.

From Equation (9) $\Delta\mathbf{p}$ can be solved by least-squares method:

$$\Delta\mathbf{p} = -\mathbf{H}^{-1}\sum_{y=-M}^{y=M}\sum_{x=-M}^{x=M}\left\{\left[\nabla f\left(\frac{\partial\mathbf{W}}{\partial\mathbf{p}}\right)\right]^T\left[f(\mathbf{W}(x,y;0))-\overline{f}-\frac{\Delta f}{\Delta g}g(\mathbf{W}(x,y;\mathbf{p}))+\frac{\Delta f}{\Delta g}\overline{g}\right]\right\} \tag{12}$$

$$\mathbf{H} = \sum_{y=-M}^{y=M}\sum_{x=-M}^{x=M}\left\{\left[\nabla f\left(\frac{\partial\mathbf{W}}{\partial\mathbf{p}}\right)\right]^T\left[\nabla f\left(\frac{\partial\mathbf{W}}{\partial\mathbf{p}}\right)\right]\right\} \tag{13}$$

where $\mathbf{H}$ is the Hessian matrix in the IC-GN algorithm, which is constant during the iterations because ∇f and $\frac{\partial\mathbf{W}}{\partial\mathbf{p}}$ are independent of the target subset.

Secondly, exert $\Delta\mathbf{p}$ on the reference subset to get the incremental warp $\mathbf{W}(x,y;\Delta\mathbf{p})$. Subsequently, compose current warp $\mathbf{W}(x,y;\mathbf{p})$ with the inverse incremental warp $\mathbf{W}^{-1}(x,y;\Delta\mathbf{p})$ to obtain an updated warp:

$$\mathbf{W}(x,y;\mathbf{p}) = \mathbf{W}(x,y;\mathbf{p})\cdot\mathbf{W}^{-1}(x,y;\Delta\mathbf{p}) \tag{14}$$

Thirdly, repeat the above two steps with the updated **p** obtained by Equation (14) until preset convergence conditions have been met. In Equation (14), the warp function must be invertible. The first-order warp function can be inverted directly, while the second-order warp function need to be expanded to make it invertible [20].

There are usually two steps to get dense disparity map in DIC-based stereo matching, namely seed point generation and seed point propagation. Scale-invariant feature transform (SIFT) [27] is a classical feature detection method, features extracted by which is invariant to affine transformation, rotation, and scale. In this paper, SIFT-based feature detection, feature matching [28], and affine transformation are adopted to estimate initial values for **p** to generate seed points. The detailed procedure can be found in our previous work [7]. The initial values for $\mathbf{p}_1$ can be estimated directly. For IC-GN_2, the initial values for the second-order components (i.e., u_{xx}, u_{xy}, u_{yy}, v_{xx}, v_{xy}, v_{yy}) of $\mathbf{p}_2$ are set to zeros. To improve the calculation efficiency, a fast recursive scheme [29] and reliability-guided seed point propagation [14] are utilized. Based on the disparity map, 3D reconstruction can be finished via triangulation.

3. Experiments and Discussions

To conduct the comparative analysis quantitatively, two groups of experiments are investigated. In the first group, numerical simulations with two speckle images generated by computer are conducted to compare performances of first-order and second-order warp functions. In the second group, a set of experiments with different real objects are performed to evaluate the applicability and efficiency of first-order and second-order warp functions for the measurement of surfaces with different complexities. All the experiments are executed on a normal Intel(R) Core(TM) i7-4710MQ CPU 2.50 GHz laptop by C++ language with the additional library of Open Source Computer Vision (OpenCV).

In the following experiments, the modulus of the incremental displacement components Δu and Δv, $\|\Delta \mathbf{P}_{main}\| = \sqrt{\Delta u^2 + \Delta v^2}$, is used to examine the convergence. Also the optimized ZNSSD correlation coefficient is converted to zero-mean normalized cross-correlation (ZNCC) coefficient, which is equivalent to ZNSSD but more straightforward [30]. The judging conditions for the success of IC-GN are that $\|\Delta \mathbf{P}_{main}\|$ is less than the preset convergence threshold and the optimized ZNCC coefficient is larger than 0.8, as well as the number of iterations is less than 30.

3.1. Comparative Analysis by Numerical Simulations

In the following tests, a simulated image pair is equalized as a rectified stereo image pair: the displacements between the two images only occur along the along the x-axis. Therefore, the measurement of the displacements between the reference image and target image is equivalent to the process of stereo matching (getting dense disparity map) in DIC-based 3D shape measurement. As shown in Figure 2, the reference image and target images are generated by the well-known simulation algorithm proposed by Zhou [31] and widely used in previous researches [32–34]:

$$I(x,y) = \sum_{k=1}^{S} I_0 \exp\left(-\frac{(x-x_k)^2 + (y-y_k)^2}{r^2}\right) \tag{15}$$

where I is the generated intensity of the simulated speckle image. S is the total number of speckles, r is speckle size. (x_k, y_k) is a randomly generated speckle position. I_0 is the peak intensity of each speckle, which is usually set to be 255.

In Figure 2a, there are totally 150,000 randomly generated speckles in the reference image with a resolution of 1280 × 960 pixels, and the speckle radius is 1.2 pixels. Figure 2b is the corresponding target image generated by exerting specific displacements on the reference image:

$$U(x,y) = \begin{cases} \sin(2\pi E(x)/t)\sin(2\pi E(y)/t) & x < 640 \\ E(x)E(y) & x \geq 640 \end{cases} \tag{16}$$

$$E(x) = \begin{cases} e^{-(x-u_1)^2/(2\sigma_1{}^2)} & x < 640 \\ e^{-(x-u_2)^2/(2\sigma_2{}^2)} & x \geq 640 \end{cases} \tag{17}$$

$$E(y) = \begin{cases} e^{-(y-v_1)^2/(2\sigma_1{}^2)} & x < 640 \\ e^{-(y-v_2)^2/(2\sigma_2{}^2)} & x \geq 640 \end{cases} \tag{18}$$

Two different forms of displacements along the x-axis are exerted on the reference image according to the displacement function $U(x,y)$. The displacements for the left part and right part are generated by an analogous sinusoidal-Gaussian function and an analogous Gaussian function, respectively. In Figure 2a, two preset regions of interest (ROI) are marked by yellow rectangles in the left part (ROI_1) and right part (ROI_2). (u_1, v_1) and (u_2, v_2) are the coordinates of the center pixels of ROI_1 and ROI_2, which are set to be $(320, 480)$ and $(960, 480)$, respectively. σ denotes the Gaussian Root-Mean-Square (RMS) width, where σ_1 and σ_2 are set to be 50 and 200, respectively. t is the period of sinusoidal function, which is set to be 1. The displacement fields of ROI_1 and ROI_2 are shown in Figure 2c,d, it is obvious that the displacement field of ROI_1 is much more complex than that of ROI_2.

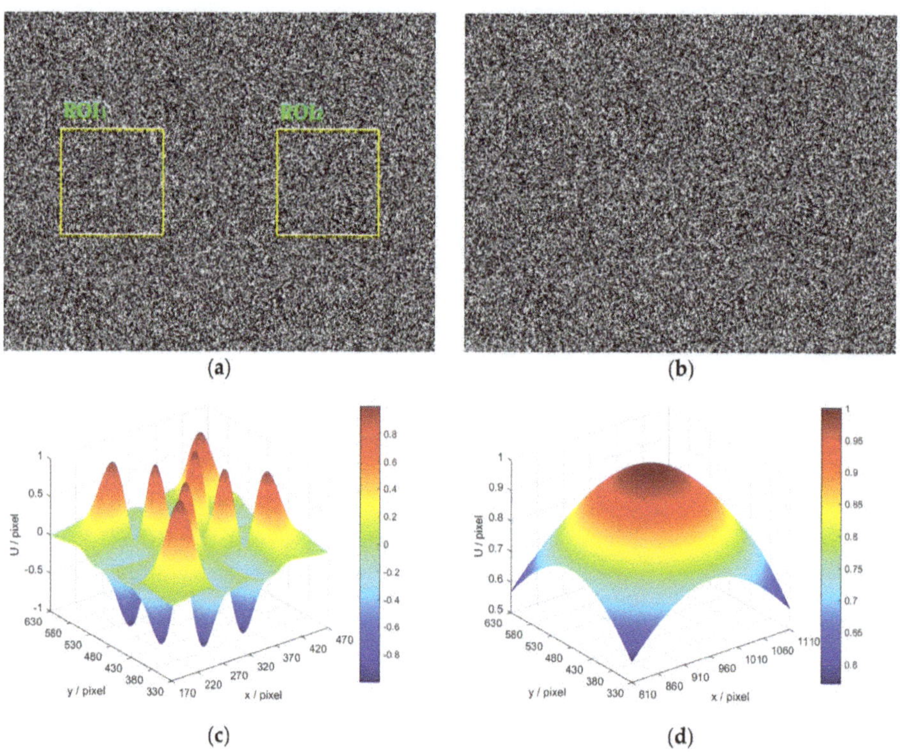

Figure 2. Synthetic speckle images: (**a**) Simulated reference image; (**b**) Simulated target image; (**c**) Theoretical displacements along x-axis of ROI_1; and (**d**) Theoretical displacements along x-axis of ROI_2.

The displacements of all the pixels in ROI_1 and ROI_2 are measured by $IC\text{-}GN_1$ and $IC\text{-}GN_2$. The measured data are analyzed statistically:

$$s_U = \sqrt{\sum_{i=1}^{i=N}(|U_{mei} - U_{thi}| - \bar{e}_U)^2/(N-1)}, \bar{e}_U = \frac{1}{N}\sum_{i=1}^{i=N}|U_{mei} - U_{thi}| \tag{19}$$

$$RMSE_U = \sqrt{\sum_{i=1}^{i=N}(U_{mei} - U_{thi})^2/N} \tag{20}$$

where $\bar{e}_U$ is the mean bias error, s_U is the standard deviation, and $RMSE_U$ is the root-mean-square error (RMSE). U_{mei} and U_{thi} denote the measured and theoretical displacements along the x-axis of the sampling pixel with index i. N is the number of sampling pixels. It is necessary to state here that the influence of subset size and convergence criterion on the accuracy of $IC\text{-}GN_1$ and $IC\text{-}GN_2$ in different displacement fields are compared.

3.1.1. Comparative Analysis with Different Subset Sizes

Three groups of data (namely, measured data of ROI_1, ROI_2, both ROI_1 and ROI_2) are analyzed with subset size changed from 15 × 15 to 35 × 35 pixels, where the three groups of data are denoted as ROI_1, ROI_2, and $ROI_{1\&2}$ hereinafter. Figure 3 shows the s_U and $RMSE_U$ as a function of subset size, where the convergence threshold for $\|\Delta\mathbf{P}_{main}\|$ is set to be 0.001. The corresponding data are listed in Table 1. To compare the characteristics of the errors measured by $IC\text{-}GN_1$ and $IC\text{-}GN_2$ of the two displacement fields, the error distribution maps with a specific subset size as 27 × 27 pixels are shown in Figure 4.

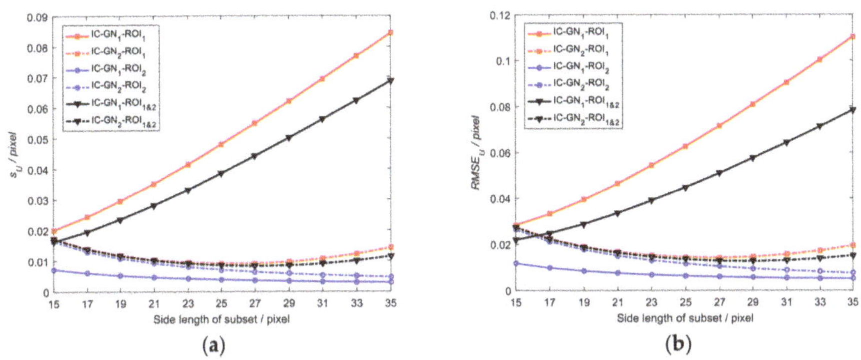

Figure 3. Measured displacement errors (s_U, $RMSE_U$): (**a**) Standard deviation as a function of side length of subset; and (**b**) RMSE as a function of side length of subset.

It can be easily seen in Figure 3 that for $IC\text{-}GN_1$, s_U and $RMSE_U$ of ROI_1 both increase as the subset size becomes larger. However, s_U and $RMSE_U$ of ROI_2 decrease as the subset size becomes larger. For $IC\text{-}GN_2$, s_U and $RMSE_U$ of ROI_1 get the minimums with the subset size of 27 × 27 pixels. s_U and $RMSE_U$ of ROI_2 decrease as the subset size becomes larger. For both $IC\text{-}GN_1$ and $IC\text{-}GN_2$, s_U and $RMSE_U$ of ROI_2 are always smaller than that of ROI_1, which indicates that the precision of IC-GN can be reduced by complex displacement field. It should be noted that $IC\text{-}GN_2$ is more accurate than $IC\text{-}GN_1$ for ROI_1. However, $IC\text{-}GN_1$ is more accurate for ROI_2 with all tested subset sizes. The errors of $ROI_{1\&2}$ are the tradeoff of errors of ROI_1 and ROI_2.

Table 1. Comparison of measured displacement errors with different subset sizes (SS) by IC-GN_1 (1st) and IC-GN_2 (2nd) of ROI_1, ROI_2, and $ROI_{1\&2}$ (unit: pixel).

SS	s_U-ROI_1		s_U-ROI_2		s_U-$ROI_{1\&2}$		$RMSE_U$-ROI_1		$RMSE_U$-ROI_2		$RMSE_U$-$ROI_{1\&2}$	
	1st	2nd	1st	2nd	1st	2nd	1st	2nd	1st	2nd	1st	2nd
15	0.02029	0.01755	0.00749	0.01661	0.01621	0.01709	0.02869	0.02800	0.01211	0.02673	0.02202	0.02738
17	0.02467	0.01422	0.00632	0.01324	0.01949	0.01375	0.03365	0.02284	0.01018	0.02149	0.02486	0.02218
19	0.02981	0.01207	0.00552	0.01108	0.02354	0.01159	0.03982	0.01942	0.00886	0.01800	0.02884	0.01872
21	0.03553	0.01065	0.00496	0.00952	0.02815	0.01012	0.04688	0.01713	0.00793	0.01540	0.03362	0.01629
23	0.04173	0.00977	0.00450	0.00826	0.03319	0.00908	0.05468	0.01563	0.00719	0.01342	0.03900	0.01457
25	0.04830	0.00941	0.00416	0.00736	0.03858	0.00851	0.06307	0.01484	0.00665	0.01192	0.04485	0.01346
27	0.05517	0.00941	0.00389	0.00667	0.04422	0.00827	0.07194	0.01457	0.00625	0.01073	0.05106	0.01280
29	0.06228	0.00994	0.00369	0.00612	0.05009	0.00844	0.08120	0.01497	0.00592	0.00978	0.05757	0.01264
31	0.06960	0.01099	0.00350	0.00567	0.05615	0.00904	0.09079	0.01601	0.00569	0.00903	0.06433	0.01299
33	0.07707	0.01254	0.00335	0.00530	0.06233	0.01006	0.10065	0.01763	0.00555	0.00841	0.07128	0.01381
35	0.08466	0.01454	0.00321	0.00498	0.06862	0.01148	0.11072	0.01985	0.00548	0.00786	0.07839	0.01510

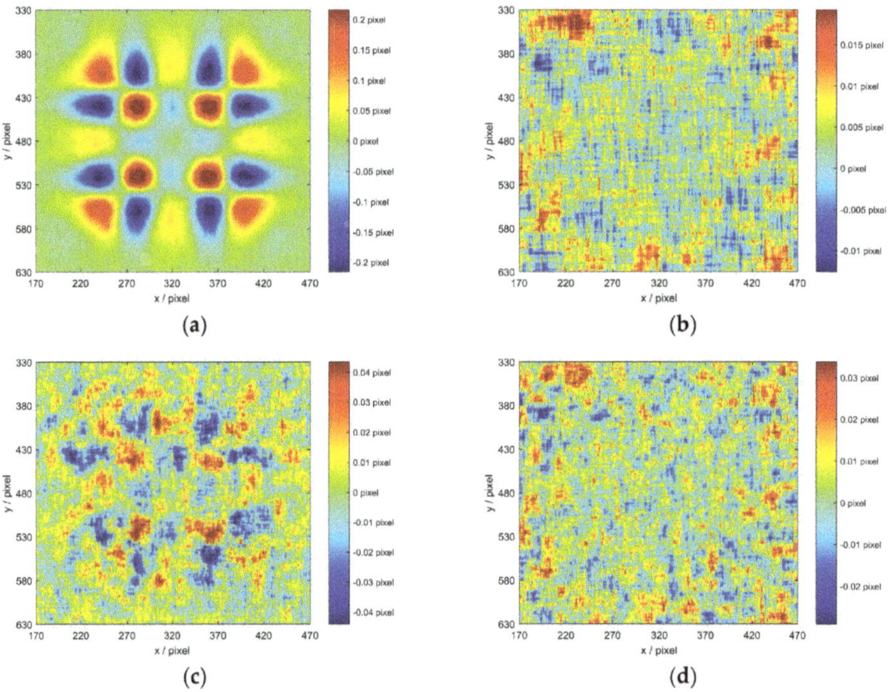

Figure 4. Error distribution maps with a subset size of 27 × 27 pixels: (**a**) Error distribution map of ROI_1 measured by IC-GN_1; (**b**) Error distribution map of ROI_2 measured by IC-GN_1; (**c**) Error distribution map of ROI_1 measured by IC-GN_2; and (**d**) Error distribution map of ROI_2 measured by IC-GN_2.

The error distribution maps of ROI_1 and ROI_2 measured by IC-GN_1 and IC-GN_2 are shown in Figure 4a–d, respectively. By horizontal comparison, it is obvious that the errors of ROI_1 measured by IC-GN_1 and IC-GN_2 are both mainly concentrate on the peak areas of the shape of displacement field, while the error distribution of ROI_2 likes a random distribution. By vertical comparison, we can see that the concentrated errors in the peak areas measured by IC-GN_1 can be suppressed by IC-GN_2, while the errors of ROI_2 measured by IC-GN_2 are about double of that measured by IC-GN_1. Therefore, it can be concluded that IC-GN_2 is more accurate for complex displacement (disparity) field measurement, while IC-GN_1 is more accurate for general uniform displacement (disparity) field measurement.

3.1.2. Comparative Analysis with Different Convergence Criteria

In Figure 3b, the curves of $RMSE_U$ of $ROI_{1\&2}$ measured by $IC\text{-}GN_1$ and $IC\text{-}GN_2$ have an intersection around the side length of subset of 17 pixels. Therefore, the subset size is set to be 17 × 17 pixels to compare the performances of $IC\text{-}GN_1$ and $IC\text{-}GN_2$ under different convergence criteria. As shown in Figure 5, the convergence threshold for $\|\Delta \mathbf{P}_{main}\|$ is set to be 0.1, 0.01, 0.001, and 0.0001, respectively. To compare the convergence efficiency of $IC\text{-}GN_1$ and $IC\text{-}GN_2$ under different convergence thresholds, the average numbers of iterations (denoted as $\overline{n}_{itor}$) of ROI_1, ROI_2, and $ROI_{1\&2}$ are listed in Table 2.

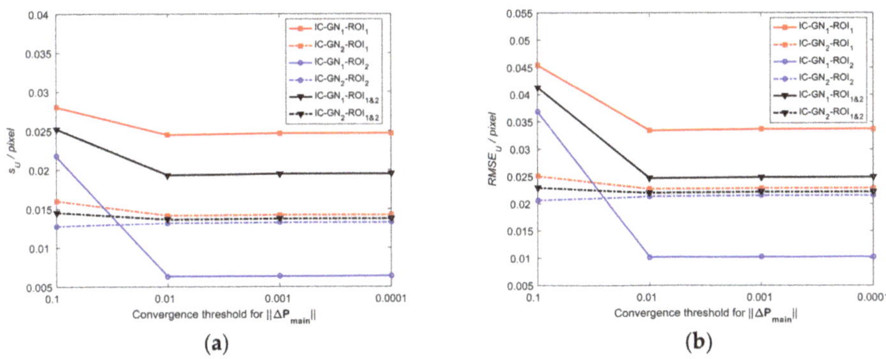

Figure 5. Measured displacement errors ($s_U, RMSE_U$) under different convergence criteria: (**a**) Standard deviation as a function of convergence threshold for $\|\Delta \mathbf{P}_{main}\|$; and (**b**) RMSE as a function of convergence threshold for $\|\Delta \mathbf{P}_{main}\|$.

Table 2. Comparison of average number of iterations of matched pixels in ROI_1, ROI_2, and $ROI_{1\&2}$ with different convergence thresholds for $IC\text{-}GN_1$ and $IC\text{-}GN_2$.

Threshold for $\|\Delta P_{main}\|$	$\overline{n}_{itor}$-ROI_1		$\overline{n}_{itor}$-ROI_2		$\overline{n}_{itor}$-$ROI_{1\&2}$	
	$IC\text{-}GN_1$	$IC\text{-}GN_2$	$IC\text{-}GN_1$	$IC\text{-}GN_2$	$IC\text{-}GN_1$	$IC\text{-}GN_2$
0.1	1.0110	1.4293	1.0024	1.3989	1.0063	1.4141
0.01	1.4927	2.4875	1.3874	2.4457	1.4401	2.4666
0.001	2.4787	3.8182	2.3830	3.7693	2.4308	3.7937
0.0001	3.6110	5.1762	3.5212	5.1098	3.5661	5.1430

It can be concluded from Figure 5 that the same characteristic of $IC\text{-}GN_1$ and $IC\text{-}GN_2$ for the three groups is that the errors under the convergence thresholds of 0.01, 0.001, and 0.0001 are almost the same from each other. The difference is that the errors of $IC\text{-}GN_1$ under the convergence threshold of 0.1 are significantly larger than that under the other thresholds, while the errors of $IC\text{-}GN_2$ under the convergence threshold of 0.1 are slightly larger or smaller than under the other thresholds. Furthermore, s_U and $RMSE_U$ of $ROI_{1\&2}$ measured by $IC\text{-}GN_2$ under the convergence threshold of 0.1 are smaller than that measured by $IC\text{-}GN_1$ under any one of the tested convergence thresholds. Also, it is evident in Table 2 that the preset convergence threshold directly affects the convergence efficiency. For all ROI_1, ROI_2, and $ROI_{1\&2}$, the average numbers of iterations of $IC\text{-}GN_2$ under the convergence threshold of 0.1 are about the same as those of $IC\text{-}GN_1$ under the convergence threshold of 0.01. If only $ROI_{1\&2}$ is considered, $IC\text{-}GN_2$ under the convergence threshold of 0.1 is more accurate than $IC\text{-}GN_1$ under the convergence threshold of 0.01. Considering both the efficiency and accuracy, conclusions can be drawn that the convergence threshold of 0.01 is the best choice for $IC\text{-}GN_1$, while 0.1 is more suitable for $IC\text{-}GN_2$.

3.2. Comparative Anslysis by Real Tests

As Shown in Figure 6a [7], a single-shot stereo system is used to perform real experiments, which is composed of two CCD cameras with a resolution of 1280 × 960 pixels (Basler acA1300-30 gm. Manufactured by Basler AG, Ahrensburg, Germany. Supplied by Shanghai Vision-Light Tech Co., Ltd. Pudong New Area, Shanghai, China), two camera lenses (Computar 8 mm 1:1.4 2/3. Manufactured by Computar®, Tokyo, Japan. Supplied by Shanghai Vision-Light Tech Co., Ltd. Pudong New Area, Shanghai, China), and a projector with a resolution of 1140 × 912 pixels (TI DLP LightCrafter4500. Manufactured by TEXAS INSTRUMENTS, Dallas, Texas, America. Supplied by Texas Instruments Semico . . . es (Shanghai) Co. Ltd. Pudong New Area, Shanghai, China). Figure 6b shows the projected speckle pattern with the same resolution as the projector: there are totally 120,000 speckles with a fixed radius of 1.2 pixels.

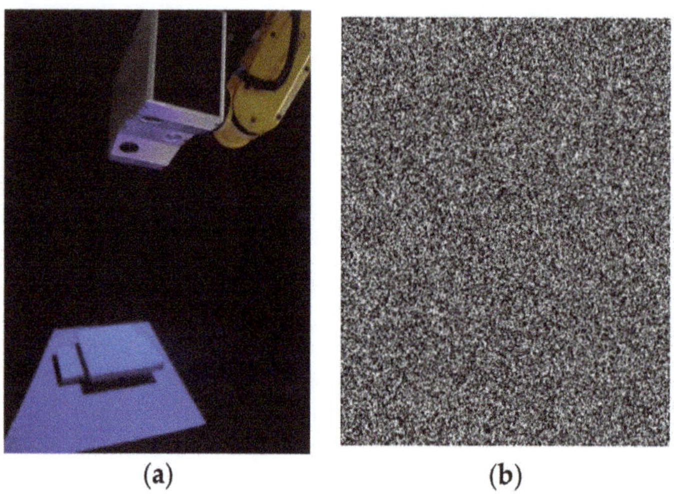

Figure 6. Experimental setup for real tests: (**a**) A single-shot stereo system with speckle projection; and (**b**) Projected speckle pattern.

Three objects are employed to compare the real measurement performances of IC-GN_1 and IC-GN_2. The measured surfaces are shown in Figure 7a–c, namely, plane surface, cylinder surface, and back surface of a plaster head (named as head hereinafter for short). The plane surface and cylinder surface are used as standard surfaces, which are measured by a Coordinate Measuring Machine (CMM (2 + (L/350) μm. Manufactured by Thome Präzision GmbH, Messel, Germany. Supplied by THOME China, Minhang District, Shanghai, China)). The 3D coordinates of the measured points are fitted into plane and cylinder surface by least square method, respectively, and the fitted results are listed in Table 3. In the calculations of real tests, a ROI is set in the left image of each rectified stereo image pair. The shape of the ROI in the head is much more complex compared to that of the plane or cylinder. The disparity maps and 3D shapes of the three ROIs are shown in Figure 8 to enable a visual comparison. In the following comparative analysis, both IC-GN_1 and IC-GN_2 are used for all the ROIs except for that in Figure 8, which refers to different warp function for different ROI: the ROIs in the plane and cylinder are measured by IC-GN_1 under a convergence threshold of 0.01, and the ROI in the head is measured by IC-GN_2 under a convergence threshold of 0.1. In addition, the subset size is set to be 27 × 27 pixels in Figure 8.

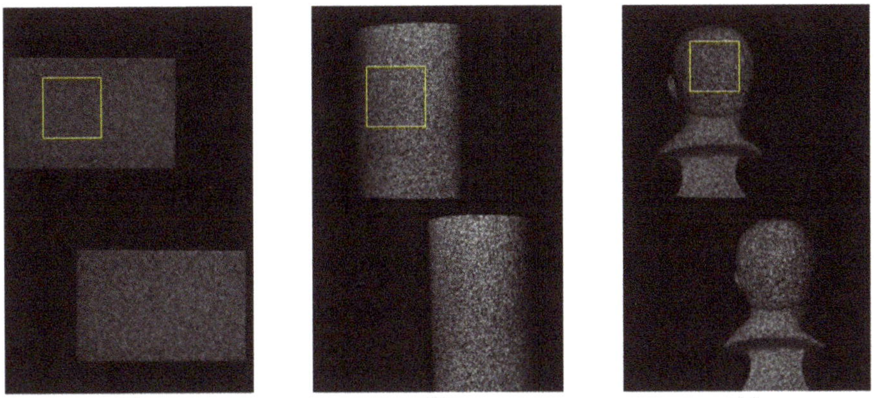

(a) (b) (c)

Figure 7. Rectified stereo image pairs for real tests, the left images are listed on the up row and the corresponding right images are listed in the bottom row: (**a**) Plane surface; (**b**) Cylinder surface; and (**c**) Back surface of a plaster head.

Table 3. Plane and cylinder surface fitting results of 3D coordinates measured by CMM.

	Point Number	Standard Deviation	Positive Maximum	Negative Maximum
Plane	15	0.001 mm	0.003 mm	−0.004 mm
Cylinder	44	0.004 mm	0.011 mm	−0.008 mm

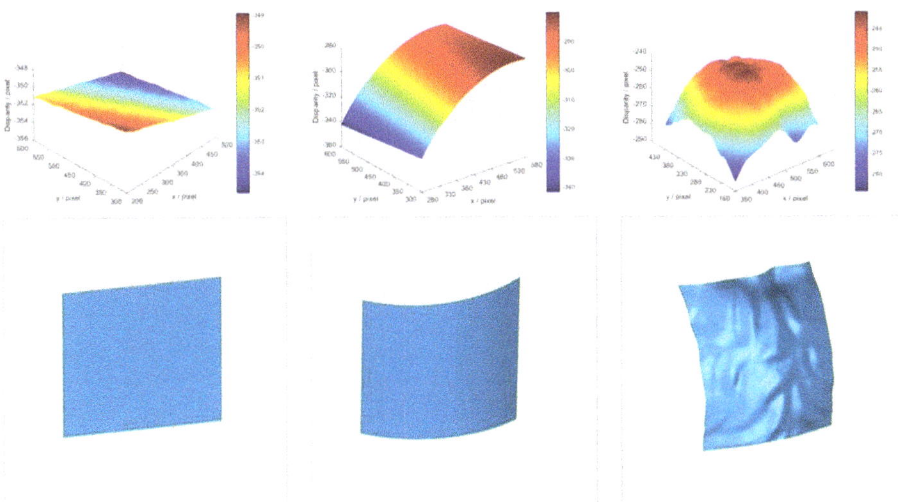

(a) (b) (c)

Figure 8. Measured disparity maps and corresponding 3D shapes, the disparity maps are listed on the up row and the corresponding 3D shapes are listed in the bottom row: (**a**) The ROI of plane; (**b**) The ROI of cylinder; and (**c**) The ROI of head.

To verify the conclusions drawn by simulation tests. The pixels in each ROI are matched by IC-GN_1 and IC-GN_2 with the convergence threshold of 0.001, and the subset size ranges from 15 × 15 to 35 × 35 pixels. For the plane surface and cylinder surface, the standard deviation (denoted as s) of plane or cylinder surface fitting in each measurement is plotted in Figure 9a.

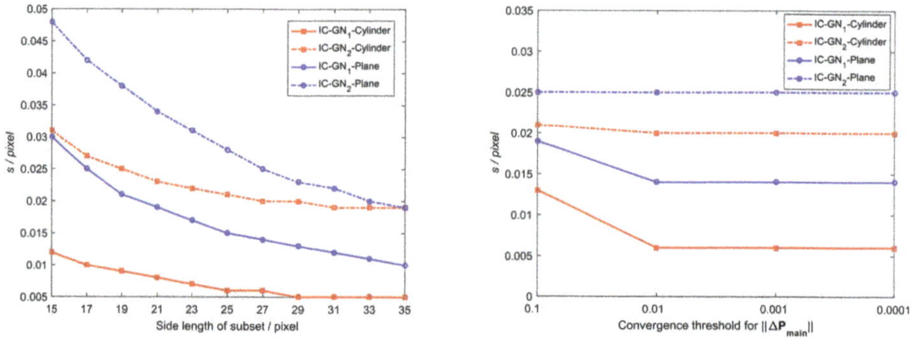

(a) (b)

Figure 9. Comparisons of IC-GN$_1$ and IC-GN$_2$ by the standard deviations of plane fitting and cylinder surface fitting: (**a**) Comparison with the change of subset size; and (**b**) Comparison with the change of convergence threshold.

It can be seen that IC-GN$_1$ is always more accurate than IC-GN$_2$ with all tested subset sizes for both surfaces. To compare the measuring abilities of IC-GN$_1$ and IC-GN$_2$ for different surfaces, the statistics of matching rates with different subset size of each ROI are listed in Table 4. The matching rate is denoted as r_m, which is the ratio of number of matched pixels to the number of total pixels (denoted as N_{pix}) in the ROI. The matching rates of IC-GN$_1$ and IC-GN$_2$ are all equal or very close to 100% for the plane and cylinder surfaces. However, the matching rates of IC-GN$_1$ for the ROI of head are all below 70%, while the matching rates of IC-GN$_2$ are all very close to 100%. Therefore, the measurement ability of IC-GN$_1$ for complex shape measurement is limited, which is almost unrelated to the change of subset size.

Table 4. Statistics of matching rates with different subset sizes measured by IC-GN$_1$ and IC-GN$_2$.

SS	Plane			Cylinder			Back of Head		
	N_{pix}	r_m (%)		N_{pix}	r_m (%)		N_{pix}	r_m (%)	
		IC-GN$_1$	IC-GN$_2$		IC-GN$_1$	IC-GN$_2$		IC-GN$_1$	IC-GN$_2$
15	90,000	99.93	99.84	90,000	100	99.01	62,500	63.95	98.77
17	90,000	99.99	99.97	90,000	100	99.29	62,500	64.52	98.93
19	90,000	100	99.97	90,000	100	99.31	62,500	65.09	99.00
21	90,000	100	99.98	90,000	100	99.35	62,500	65.55	98.99
23	90,000	100	99.98	90,000	100	99.38	62,500	65.62	98.99
25	90,000	100	99.98	90,000	100	99.38	62,500	66.27	98.98
27	90,000	100	99.98	90,000	100	99.38	62,500	67.01	99.01
29	90,000	100	99.98	90,000	100	99.35	62,500	67.71	98.99
31	90,000	100	99.98	90,000	100	99.40	62,500	68.15	98.97
33	90,000	100	99.98	90,000	100	99.41	62,500	68.60	98.89
35	90,000	100	99.98	90,000	100	99.38	62,500	68.80	98.88

The accuracy of IC-GN$_1$ and IC-GN$_2$ are also compared under different convergence thresholds with a specific subset size of 27 × 27 pixels. The standard deviations of plane fitting and cylinder surface fitting versus convergence threshold are plotted in Figure 9b. It can be seen that for IC-GN$_1$, only the differences of standard deviations under convergence thresholds of 0.1 and 0.01 are relevant.

For IC-GN$_2$, the standard deviations are almost the same under different convergence thresholds. As shown in Figure 10, the 3D data of the ROI of head measured by IC-GN$_2$ under two different convergence thresholds are compared. That means for every pixel in the ROI that has been matched, the spatial distance of the corresponding two 3D points reconstructed under the two convergence thresholds are calculated. The distance distribution maps by comparison of convergence threshold of 0.1 to 0.01 and 0.001 are shown in Figure 10a,b, respectively. Furthermore, comparison of shapes measured by IC-GN$_2$ and structured light of the head are shown in Figure 11.

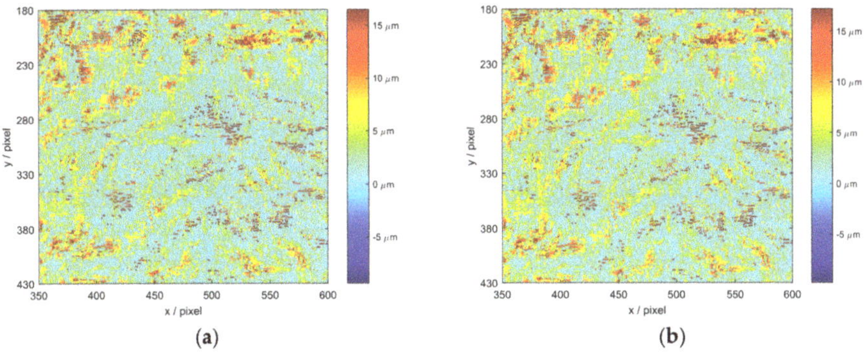

Figure 10. Distribution maps of spatial distance of the ROI in the head, which are measured by IC-GN$_2$ under two different convergence thresholds: (**a**) Under convergence thresholds of 0.1 and 0.001; and (**b**) Under convergence thresholds of 0.01 and 0.001.

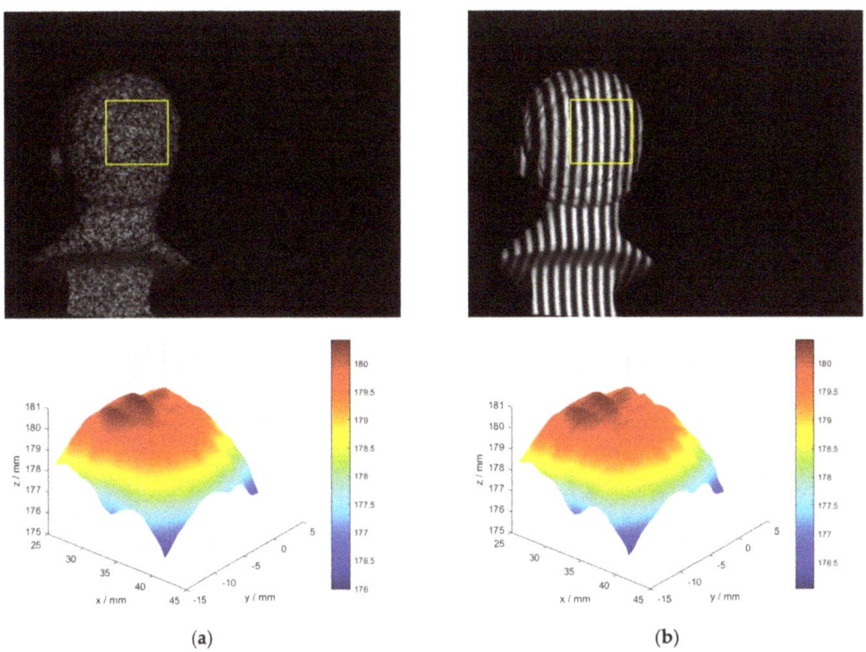

Figure 11. Rectified left image of head and reconstructed 3D data of the ROI marked by yellow rectangle: (**a**) Measured by IC-GN$_2$ under the convergence threshold of 0.1; and (**b**) Measured by three-frequency three-step structured light.

It needs to declare that the distance values for unmatched pixels are set to be zeros in Figure 10. There is no significant difference between Figure 10a,b; the corresponding standard deviations are 4.318 μm and 4.496 μm. To further verify the measurement effectiveness of IC-GN_2 under the threshold of 0.1, the head is measured at the same position by both IC-GN_2 and three-frequency three-step structured light using the same system. The same ROI is set in the rectified left images of DIC measurement and structured light measurement, and the shapes of the ROI measured by IC-GN_2 and structured light are shown in Figure 11a,b, respectively. For every pixel in the ROI, the spatial distance of the two 3D coordinates measured by IC-GN_2 and structured light is calculated. The standard deviation of all the calculated distance values is 0.023 mm, which is in the same level of precision of the above plane fitting and cylinder surface fitting. Therefore, conclusions can be drawn from the above comparisons that the convergence threshold of 0.01 is suitable for IC-GN_1, while 0.1 is recommended for IC-GN_2. The conclusions are consistent with that drawn in the simulation tests.

4. Conclusions

In this paper, a comparative analysis of first-order and second-order warp functions for DIC-based stereo 3D shape measurement is presented. Both simulation tests and real experiments with different objects are performed to compare the impacts of subset size and convergence criteria on the measuring ability, efficiency, and precision by IC-GN using first-order and second-order warp functions. Conclusions are summarized as follows:

(1) The first-order warp function is more suitable for surfaces with a shape of flat or small curvature, such as plane, cylinder, and flat Gaussian surface, etc. Under the same convergence criteria, IC-GN_1 is always more efficient and accurate than IC-GN_2 with all tested subset sizes.

(2) The second-order warp function is more suitable for surfaces with a complex shape or large curvature, such as the tested back surface of head and analogous sinusoidal-Gaussian surface, etc. IC-GN_1 is not capable or accurate enough for such kind of 3D shape measurement; the matching rate of tested ROI of head is under 70% with any of the tested subset size.

(3) The convergence thresholds for IC-GN_1 and IC-GN_2 are recommended to be that the variation of the modulus of incremental displacement vector is less than 0.01 pixel, and 0.1 pixel, respectively. Both the recommended convergence thresholds can achieve considerable measurement precision compared to smaller thresholds according to the simulation tests and real experiments.

Acknowledgments: This work is supported by the National Natural Science Foundation of China (51575354), the National Key Technology Research and Development of the Ministry of Science and Technology of China (973 Program 2014CB046604), the Ministry of Industry and Information Technology of China (17GFB-ZB02-194), and the Shanghai Municipal Science and Technology project (16111106102).

Author Contributions: Xiao Yang and Juntong Xi designed the experiments; Xiao Yang and Xiaobo Chen performed the experiments and analyzed the data; and Xiao Yang wrote the paper.

Conflicts of Interest: The authors declare no conflict of interest.

References

1. Chi, S.; Xie, Z.; Chen, W. A laser line auto-scanning system for underwater 3D reconstruction. *Sensors* **2016**, *16*, 1534. [CrossRef] [PubMed]
2. Yu, C.; Chen, X.; Xi, J. Modeling and calibration of a novel one-mirror galvanometric laser scanner. *Sensors* **2017**, *17*, 164. [CrossRef] [PubMed]
3. Jung, J.; Yoon, S.; Ju, S.; Heo, J. Development of kinematic 3D laser scanning system for indoor mapping and as-built bim using constrained slam. *Sensors* **2015**, *15*, 26430–26456. [CrossRef] [PubMed]
4. Chen, X.; Xi, J.T.; Jiang, T.; Jin, Y. Research and development of an accurate 3D shape measurement system based on fringe projection: Model analysis and performance evaluation. *Precis. Eng.* **2008**, *32*, 215–221.
5. Nguyen, T.T.; Slaughter, D.C.; Max, N.; Maloof, J.N.; Sinha, N. Structured light-based 3D reconstruction system for plants. *Sensors* **2015**, *15*, 18587–18612. [CrossRef] [PubMed]

6. Kieu, H.; Pan, T.; Wang, Z.; Le, M.; Nguyen, H.; Vo, M. Accurate 3D shape measurement of multiple separate objects with stereo vision. *Meas. Sci. Technol.* **2014**, *25*, 1–7. [CrossRef]
7. Yang, X.; Chen, X.; Xi, J. Efficient background segmentation and seed point generation for a single-shot stereo system. *Sensors* **2017**, *17*, 2782. [CrossRef] [PubMed]
8. Yan, T.H.; Yong, S.; Zhang, Q.C. Precise 3D shape measurement of three-dimensional digital image correlation for complex surfaces. *Sci. China Technol. Sci.* **2017**, *61*, 68–73. [CrossRef]
9. Nguyen, H.; Wang, Z.; Quisberth, J. Accuracy Comparison of Fringe Projection Technique and 3D Digital Image Correlation Technique. In *Advancement of Optical Methods in Experimental Mechanics*; Springer: Cham, Switzerland, 2016; pp. 195–201.
10. Zhang, Z.H. Review of single-shot 3D shape measurement by phase calculation-based fringe projection techniques. *Opt. Las. Eng.* **2012**, *50*, 1097–1106. [CrossRef]
11. Xie, H. Full-field strain measurement using a two-dimensional savitzky-golay digital differentiator in digital image correlation. *Opt. Eng.* **2007**, *46*, 033601.
12. Huang, J.; Pan, X.; Peng, X.; Yuan, Y.; Xiong, C.; Fang, J.; Yuan, F. Digital image correlation with self-adaptive gaussian windows. *Exp. Mech.* **2013**, *53*, 505–512. [CrossRef]
13. Lu, H.; Cary, P.D. Deformation measurements by digital image correlation: Implementation of a second-order displacement gradient. *Exp. Mech.* **2000**, *40*, 393–400. [CrossRef]
14. Pan, B. Reliability-guided digital image correlation for image deformation measurement. *Appl. Opt.* **2009**, *48*, 1535–1542. [CrossRef] [PubMed]
15. Baker, S.; Dellaert, F.; Matthews, I. Aligning Images Incrementally Backwards. 2001. Available online: http://pdfs.semanticscholar.org/11e4/f603e2cacf4533a919ba3fbdf79939423c74.pdf (accessed on 2 February 2018).
16. Baker, S.; Matthews, I. Lucas-kanade 20 years on: A unifying framework. *Int. J. Comput. Vis.* **2004**, *56*, 221–255. [CrossRef]
17. Pan, B.; Li, K.; Tong, W. Fast, robust and accurate digital image correlation calculation without redundant computations. *Exp. Mech.* **2013**, *53*, 1277–1289. [CrossRef]
18. Dai, X.; He, X.; Shao, X.; Chen, Z. Real-time 3D digital image correlation method and its application in human pulse monitoring. *Appl. Opt.* **2016**, *55*, 696.
19. Wu, R.; Kong, C.; Li, K.; Zhang, D. Real-time digital image correlation for dynamic strain measurement. *Exp. Mech.* **2016**, *56*, 1–11. [CrossRef]
20. Gao, Y.; Cheng, T.; Su, Y.; Xu, X.; Zhang, Y.; Zhang, Q. High-efficiency and high-accuracy digital image correlation for three-dimensional measurement. *Opt. Lasers Eng.* **2015**, *65*, 73–80. [CrossRef]
21. Bai, R.; Jiang, H.; Lei, Z.; Li, W. A novel 2nd-order shape function based digital image correlation method for large deformation measurements. *Opt. Las. Eng.* **2017**, *90*, 48–58. [CrossRef]
22. Pan, B.; Xie, H.; Wang, Z.; Qian, K.; Wang, Z. Study on subset size selection in digital image correlation for speckle patterns. *Opt. Express* **2008**, *16*, 7037. [CrossRef] [PubMed]
23. Pan, B. An evaluation of convergence criteria for digital image correlation using inverse compositional gauss–newton algorithm. *Strain* **2014**, *50*, 48–56. [CrossRef]
24. Silva, L.C.; Petraglia, M.R.; Petraglia, A. A robust method for camera calibration and 3-D reconstruction for stereo vision systems. In Proceedings of the 2004 12th European Signal Processing Conference, Vienna, Austria, 6–10 September 2004; pp. 1151–1154.
25. Asundi, A.; Pan, B.; Xie, H.; Gao, J. Improved speckle projection profilometry for out-of-plane shape measurement. *Appl. Opt.* **2008**, *47*, 5527–5533.
26. Barone, S.; Neri, P.; Paoli, A.; Razionale, A. Digital image correlation based on projected pattern for high frequency vibration measurements. *Procedia Manuf.* **2017**, *11*, 1592–1599. [CrossRef]
27. Lowe, D.G. Distinctive image features from scale-invariant keypoints. *Int. J. Comput. Vis.* **2004**, *60*, 91–110. [CrossRef]
28. Muja, M. Flann-Fast Library for Approximate Nearest Neighbors User Manual. 2009. Available online: https://www.cs.ubc.ca/research/flann/uploads/FLANN/flann_manual-1.6.11.pdf (accessed on 2 February 2018).
29. Huang, J.; Zhu, T.; Pan, X.; Qin, L.; Peng, X.; Xiong, C.; Fang, J. A high-efficiency digital image correlation method based on a fast recursive scheme. *Meas. Sci. Technol.* **2010**, *21*, 35101–35112. [CrossRef]
30. Pan, B.; Xie, H.; Wang, Z. Equivalence of digital image correlation criteria for pattern matching. *Appl. Opt.* **2010**, *49*, 5501–5509. [CrossRef] [PubMed]

31. Zhou, P.; Goodson, K.E. Subpixel displacement and deformation gradient measurement using digital image/speckle correlation (disc). *Opt. Eng.* **2001**, *40*, 1613–1620. [CrossRef]
32. Huang, J.; Pan, X.; Shanshan, L.I.; Peng, X.; Xiong, C.; Fang, J. A digital volume correlation technique for 3-D deformation measurements of soft gels. *Int. J. Appl. Mech.* **2011**, *3*, 335–354. [CrossRef]
33. Yuan, Y.; Huang, J.; Peng, X.; Xiong, C.; Fang, J.; Yuan, F. Accurate displacement measurement via a self-adaptive digital image correlation method based on a weighted znssd criterion. *Opt. Lasers Eng.* **2014**, *52*, 75–85. [CrossRef]
34. Yuan, Y.; Zhan, Q.; Xiong, C.; Huang, J. Digital image correlation based on a fast convolution strategy. *Opt. Lasers Eng.* **2017**, *97*, 52–61. [CrossRef]

MDPI
St. Alban-Anlage 66
4052 Basel
Switzerland
Tel. +41 61 683 77 34
Fax +41 61 302 89 18
www.mdpi.com

Sensors Editorial Office
E-mail: sensors@mdpi.com
www.mdpi.com/journal/sensors

www.ingramcontent.com/pod-product-compliance
Lightning Source LLC
Chambersburg PA
CBHW041749220926
51847CB00014B/149
9783039283385